HUMAN EVOLUTIONARY BIOLOGY

Wide-ranging and inclusive, this text provides an invaluable review of an expansive selection of topics in human evolution, variation, and adaptability for professionals and students in biological anthropology, evolutionary biology, medical sciences, and psychology. The chapters are organized around four broad themes, with sections devoted to phenotypic and genetic variation within and between human populations, reproductive physiology and behavior, growth and development, and human health from evolutionary and ecological perspectives. An introductory section provides readers with the historical, theoretical, and methodological foundations needed to understand the more complex ideas presented later. Two hundred discussion questions provide starting points for class debate and assignments to test student understanding.

Michael P. Muehlenbein is an Assistant Professor of anthropology at Indiana University, Bloomington. He holds an MsPH in both tropical medicine and biostatistics from Tulane University, as well as an MPhil and PhD in biological anthropology from Yale University. His research interests are focused most recently on (1) evaluating hormone-mediated immune functions in reference to evolutionary and life history theories, and (2) investigating potential zoonotic and anthropozoonotic pathogen transmission associated with primate-based ecotourism. He has received teaching awards for his graduate and undergraduate courses on human biological variation, behavioral endocrinology, evolutionary medicine, and global health. In addition to running an endocrinology and infectious disease laboratory in Indiana, he presently conducts fieldwork in the United States, Malaysia, Dominica, and the Dominican Republic.

HUMAN EVOLUTIONARY BIOLOGY

Edited by

MICHAEL P. MUEHLENBEIN

Indiana University, Bloomington

CAMBRIDGE
UNIVERSITY PRESS

University Printing House, Cambridge CB2 8BS, United Kingdom

One Liberty Plaza, 20th Floor, New York, NY 10006, USA

477 Williamstown Road, Port Melbourne, VIC 3207, Australia

314-321, 3rd Floor, Plot 3, Splendor Forum, Jasola District Centre, New Delhi - 110025, India

79 Anson Road, #06-04/06, Singapore 079906

Cambridge University Press is part of the University of Cambridge.

It furthers the University's mission by disseminating knowledge in the pursuit of education, learning and research at the highest international levels of excellence.

www.cambridge.org
Information on this title: www.cambridge.org/9780521705103

First published 2010

A catalogue record for this publication is available from the British Library

Library of Congress Cataloging in Publication data
Human evolutionary biology / edited by Michael P. Muehlenbein.
    p.   cm.
  ISBN 978-0-521-87948-4 (hardback) – ISBN 978-0-521-70510-3 (pbk.)
  1. Human evolution.   2. Human biology.   I. Muehlenbein, Michael P., 1976–
II. Title.
  GN281.H8477 2010
  599.93′8–dc22

                                                                                2010016390

ISBN 978-0-521-87948-4 Hardback
ISBN 978-0-521-70510-3 Paperback

Contents

Contributors

Barry Bogin
Department of Human Sciences
Loughborough University
Loughborough, UK

Richard G. Bribiescas
Department of Anthropology
Yale University,
New Haven, CT, USA

Tom D. Brutsaert
Department of Exercise Science
Syracuse University
Syracuse, NY, USA

Noël Cameron
Department of Human Sciences
Centre for Human Development and Ageing
Loughborough University
Loughborough, UK

Benjamin C. Campbell
Department of Anthropology
University of Wisconsin
Milwaukee, WI, USA

Douglas E. Crews
Department of Anthropology
Ohio State University
Columbus, OH, USA

Peter S. W. Davies
Children's Nutrition Research Centre
Discipline of Paediatrics and Child Health
The University of Queensland
Royal Children's Hospital
Herston, Australia

S. Boyd Eaton
Department of Anthropology
Emory University
Atlanta, GA, USA

Guillaume Evanno
Department of Ecology and Evolution
University of Lausanne
Lausanne, Switzerland

Paul W. Ewald
Department of Biology
University of Louisville
Louisville, KY, USA

Mark V. Flinn
Department of Anthropology
University of Missouri
Columbia, MO, USA

A. Roberto Frisancho
Department of Anthropology
University of Michigan
Ann Arbor, MI, USA

Douglas J. Futuyma
Department of Ecology and Evolution
State University of New York, Stony Brook
Stony Brook, NY, USA

Peter B. Gray
Department of Anthropology and Ethnic
Studies
University of Nevada, Las Vegas
Las Vegas, NV, USA

Nina G. Jablonski
Department of Anthropology
Pennsylvania State University
University Park, PA, USA

Grazyna Jasienska
Department of Epidemiology and Population
Studies
Institute of Public Health
Jagiellonian University, Collegium Medicum
Krakow, Poland

James Holland Jones
Department of Anthropological Sciences
Stanford University
Stanford, CA, USA

Laura L. Jones
Division of Epidemiology and Public Health
Nottingham City Hospital
Nottingham, UK

Hillard S. Kaplan
Department of Anthropology
University of New Mexico
Albuquerque, NM, USA

Peter T. Katzmarzyk
Pennington Biomedical Research Center
Baton Rouge, LA, USA

Christopher W. Kuzawa
Department of Anthropology
Northwestern University
Evanston, IL, USA

Jane B. Lancaster
Department of Anthropology
University of New Mexico
Albuquerque, NM, USA

William R. Leonard
Department of Anthropology
Northwestern University
Evanston, IL, USA

Michael A. Little
Department of Anthropology
State University of New York, Binghamton
Binghamton, NY, USA

Jonathan Marks
Department of Anthropology
University of North Carolina, Charlotte
Charlotte, NC, USA

Thomas W. McDade
Department of Anthropology
Northwestern University
Evanston, IL, USA

Robert J. Meier
Department of Anthropology
Indiana University
Bloomington, IN, USA

Lauren A. Milligan
Department of Anthropology
University of California, Santa Cruz
Santa Cruz, CA, USA

Michael P. Muehlenbein
Department of Anthropology
Indiana University
Bloomington, IN, USA

Alexia J. Murphy
Children's Nutrition Research Centre
Discipline of Paediatrics and Child Health
The University of Queensland

Royal Children's Hospital
Herston, Australia

Colleen H. Nyberg
Department of Anthropology
University of Massachusetts Boston
Boston, MA, USA

Ivy L. Pike
Department of Anthropology
University of Arizona
Tucson, AZ, USA

Jennifer A. Raff
Department of Anthropology
University of Utah
Salt Lake City, UT, USA

Lawrence M. Schell
Department of Anthropology
State University of New York, Albany
Albany, NY, USA

David P. Schmitt
Department of Psychology
Bradley University
Peoria, IL, USA

Michael E. Steiper
Department of Anthropology
Hunter College
New York, NY, USA

James A. Stewart
Department of Social and Behavioral Sciences
Columbus State Community College
Columbus, OH, USA

Trudy R. Turner
Department of Anthropology
University of Wisconsin-Milwaukee
Milwaukee, WI, USA

Stanley J. Ulijaszek
Institute of Social and Cultural Anthropology
University of Oxford
Oxford, UK

Claus Wedekind
Department of Ecology and Evolution
University of Lausanne
Lausanne, Switzerland

Brant Wenegrat
Department of Psychiatry and Behavioral
Sciences
Stanford University
Palo Alto, CA, USA

Preface

In review of a different text, Moses Hadas (1900–1966) remarked that "this book fills a much-needed gap." Unlike the text he was referring to, the topic of human evolutionary biology deserves no such gap in our understanding. To identify one's place in nature and to appreciate how human evolution has been guided by the same evolutionary principles that guide other organisms is humbling and necessary. We are products of evolution, and this is reflected throughout our biology and behaviors.

The purpose of our text is to provide thorough and modern reviews of a wide range of pertinent aspects of human evolutionary biology and contemporary human biological variation. The history of research on human biological variation is a long one, and includes studies on general human adaptability, variations in growth patterns, body sizes and shapes, genetic diversity, and race concepts. More recently, the study of human biology has included analyses of reproductive physiology and behavior within evolutionary and ecological frameworks. Other advancements include the science of evolutionary medicine, in which evolutionary research on health and disease is used to elucidate medical research and practices.

The text before you is different from most others. Unlike traditional texts on evolutionary biology, our book focuses specifically on humans and the application of evolutionary theories on understanding modern human variation and adaptability. Unlike traditional texts on human biology, our book does not include detailed descriptions of all human physiological and anatomical systems. You will also not find detailed accounts of the history of human evolution, where we came from and how we are related to other species. What you will find are historical perspectives on the study of human evolutionary biology, detailed reviews of modern methods for studying human evolutionary biology, descriptions of fundamental research on genotypic and phenotypic variation within and between contemporary human populations, comprehensive discussions on human reproductive physiology and behavior as well as evolutionary medicine.

It is not possible to produce one single inclusive text on such a diverse topic. Supplemental materials to our book would include, among others, detailed reviews of basic evolutionary biology, molecular biology, biological anthropology, behavioral ecology, and statistics. We have, however, tried to produce a text readable to a wide audience and organized in an intuitive fashion. Part I of the book begins where it should, with basic and detailed reviews of theory, history, and methods in human evolutionary biology. This includes introductions to evolutionary theory, human adaptability, genetics, demography, evolutionary endocrinology, anthropometry, and nutrition/energetics, as well as the history of the study of human biology. We even introduce readers to some of the ethical considerations for human biology research. Clearly the purpose of Part I is to provide readers with a basic foundation in theory and methodology that can be used as a basis for understanding some of the more complex problems presented by authors throughout the remainder of the volume.

Part II of the book focuses on phenotypic and genotypic variation. This has been the bread and butter of research on contemporary human biological variation. Body size and shape, skin color, and adaptations to high altitude are all revisited. Chapters 13 and 14 provide detailed accounts of classic (e.g., serological, etc.) and DNA markers of human variation, and Chapter 15 importantly addresses the "race concept," a long-held discussion in physical and cultural anthropology. A chapter on human behavioral endocrinology logically bridges our section on phenotypic/genotypic variation with the following Part III: reproduction. This section begins with more discussion on human behaviors, specifically mate choice. Female reproductive ecology/physiology is addressed in Chapters 19 and 20, and male reproductive ecology/physiology in Chapter 21. Unfortunately, we were unable to provide at this time a chapter on female reproductive senescence (i.e., menopause).

As Part IV focuses on growth and development, it contains discussions on the evolution of, and variation in, rates and patterns of somatic growth. This includes classic and modern ideas on the sensitivity of human development to environmental factors like nutrition, disease, and even social challenges. Chapter 26 shows us how human life histories, cognition, and body/brain development are intimately intertwined. Its discussion on human longevity also serves as a logical transition to our final section of the book. Part V focuses on various aspects of human health from evolutionary and ecological perspectives. This includes basic discussions of immunity and infectious diseases, the evolution of chronic diseases (including the metabolic syndrome and mental disorders), the infectious causes of some chronic diseases, and human senescence. We conclude the section and book with discussions on some of the cultural determinants of health that have and will continue to be influential on human evolution.

We have intended this text to be used as a general reference for professional scholars as well as graduate/ postgraduate students and advanced undergraduates in evolutionary biology, biological anthropology, and other academic programs. The ultimate goal is to forward research on human evolutionary biology by identifying gaps in our understandings of this inclusive discipline. Admittedly, it may be difficult to cover all chapters of this book in a single semester in a classroom setting. In this case, instructors may choose to focus only on certain sections. Alternatively, instructors may choose to have different students present the contents of different chapters to their classmates. It is never too early to learn how to give an effective presentation. Readers are also encouraged to utilize the questions listed at the end of each chapter to facilitate discussion. Identifying and stimulating future directions of research should be the primary goal.

Lastly, credit must be given where credit is due. Acknowledgments belong to Dr. Dominic Lewis and others at Cambridge University Press for being brave enough to tackle this project. The friends, families, and colleagues who assisted each contributing author are too numerous to list, but all deserve recognition. We do not do this by ourselves or for ourselves.

Michael P. Muehlenbein
Indiana University

". . . And we must acknowledge, as it seems to me, that man with all his noble qualities, with sympathy which feels for the most debased, with benevolence which extends not only to other men but to the humblest living creature, with his god-like intellect which has penetrated into the movements and constitution of the solar system – with all these exalted powers – Man still bears in his bodily frame the indelible stamp of his lowly origin."

Charles Darwin (1809–1882), *The Descent of Man and Selection in Relation to Sex* (1871)

Part I

Theory and Methods

"The natural phenomena of the evolutionary history of man claim an entirely peculiar place in the wide range of the scientific study of nature. There is surely no subject of scientific investigation touching man more closely, or in the knowledge of which he is more deeply concerned, than the human organism itself; and of all the various branches of the science of man, or anthropology, the history of his natural evolution should excite his highest interest."

Ernst Haeckel (1834–1919), *The Evolution of Man* (1892)

1 Evolutionary Theory

Douglas J. Futuyma

Our contemporary understanding of evolutionary processes builds on theory developed during the "Evolutionary Synthesis" of the 1930s and 1940s, when Darwin's ideas, especially on natural selection, were joined with Mendelian genetics. Since, then, of course, our understanding of evolution has been greatly advanced by the discoveries in molecular genetics, as well as by continuing elaboration of the "neo-Darwinian" theory that issued from the Evolutionary Synthesis (Futuyma, 1998, 2009).

A capsule summary of contemporary theory, to be followed by more detailed explication, is as follows. Elementary evolutionary change consists of changes in the genetic constitution of a population of organisms, or in an ensemble of populations of a species. These genetic changes may be reflected in change of the population mean or variance of phenotypic characteristics. Any change requires that genetic variation originate by mutation of DNA sequences, and/or by recombination. The minimal evolutionary process is an increase in the frequency of a mutation, or a set of mutations, within a population, and the corresponding decrease in frequency of previously common alleles. Such frequency changes are the consequence of random genetic drift (leading to occasional fixation of nearly neutral genetic variants) or of diverse forms of natural selection. Successive such changes in one or more characteristics cumulate over time, yielding potentially indefinite divergence of a lineage from the ancestral state. Different populations of a species retain similarity due to gene flow and perhaps uniform selection, but can diverge due to differences in mutation, drift, and/or selection. Some of the consequent genetic differences can generate biological barriers to gene exchange between populations, resulting in the formation of different biological species.

THE ORIGIN OF VARIATION

Mutational changes in DNA sequences are of many kinds, ranging from single base-pair alterations to insertions, deletions, and rearrangements of genetic material, and even changes in ploidy. Many mutations have no effect on phenotype or fitness (are *selectively neutral*), such as synonymous mutations in protein-coding regions, which do not alter amino acid sequence, and mutations that occur in pseudogenes and other apparently nonfunctional regions. There exists greater potential for fitness effects of nonsynonymous mutations in coding regions, or of mutations in regulatory sequences. The rate of mutation (usually on the order of 10^{-9} per base pair per gamete) is usually too low to be a significant factor in driving allele frequency change within a population, but it can determine the rate of DNA sequence divergence in the long term, and can influence the equilibrium level of standing genetic variation. Considerable contemporary research concerns whether or not rates and directions of phenotypic evolution are often constrained by the supply of suitable mutations (Houle, 1998; Blows and Hoffmann, 2005). Mutation is a random process, in the sense that the probability of occurrence of a particular mutation is not affected by environmental circumstances which would make it advantageous. That is, there is no known mechanism by which the mutational process can be directed by the environment in advantageous directions.

VARIATION WITHIN POPULATIONS

Based on studies of many species, it appears that most populations carry substantial sequence variation in many gene loci, and that there exists some heritable variation in many or most "quantitative" phenotypic traits (continuous traits such as size, as well as the number of highly repeated unit characters, such as hairs or scales). Presence of two or more fairly common alleles or genotypes within a population is referred to as *polymorphism*. The level of variation is enhanced by mutation, recombination (often but not always), gene flow from other, genetically differentiated, populations, and some forms of natural selection (e.g., frequency-dependent selection, below). It is eroded by genetic drift and by most forms of natural selection (including

Human Evolutionary Biology, ed. Michael P. Muehlenbein. Published by Cambridge University Press. © Cambridge University Press 2010.

directional and stabilizing selection on quantitative traits). The analysis of genetic variation is based on the frequencies of the alleles and genotypes at individual genetic loci (for an introduction to population genetics, see Hartl and Clark, 1997) For sexually reproducing populations, the Hardy–Weinberg (H-W) theorem states that the frequency of each allele (p_i for allele i) will remain constant from generation to generation unless perturbed by mutation, gene flow, sampling error (genetic drift), or selection, and that the frequencies of the several genotypes will likewise remain constant, at values given by the binomial theorem (p_i^2 for homozygote A_iA_i, and $2p_ip_j$ for heterozygote A_iA_j), if mating occurs at random. A single generation of random mating establishes H-W genotype frequencies at any autosomal locus. Furthermore, alleles at two or more polymorphic loci will become randomized with respect to each other (a state of *linkage equilibrium*) due to recombination. These principles have important consequences; for example, at H-W equilibrium, a rare allele exists mostly in heterozygous state, and so is concealed if it is recessive. In fact, rare, deleterious recessive alleles exist at a great many loci in populations of most outcrossing species, including humans. The frequency of heterozygotes ("heterozygosity") at a locus in H-W equilibrium ($2p_ip_j$) is often used as a measure of genetic variation at that locus, since variation is maximized when allele frequencies are equal.

Phenotypic variation in most quantitative traits is continuous or almost so, because it is *polygenic*, based on segregating alleles at several or many loci, and also includes environmental effects on the development or expression of a character (Falconer and Mackay, 1996; Barton and Keightley, 2002). At many of the segregating loci, the individual effects of alleles on the character typically are small, relative to the range of variation, but substantially larger effects are commonly contributed by segregating alleles at a few loci. The variance in phenotype (V_P) includes a genetic component (genetic variance, V_G) and an environmental component (V_E), and often an interaction effect ($V_{G.E}$) as well. An important component of V_G is the "additive genetic variance" (V_A), which is described by the correlation between the phenotype of parents and their offspring; it is this component of variation that is most important for evolution by natural selection. This component consists of the "additive" effects of alleles, that is, the phenotypic effect of each allelic substitution, averaged over all the genetic backgrounds in which it occurs. V_A depends on the number of loci contributing to the character, on the evenness of allele frequencies at each locus, and on the average magnitude of the phenotypic effect of different alleles. ($V_A = 2\Sigma p_ip_ja^2$ in the simplest case, where a is the average phenotypic effect and Σ indicates summation over loci.) The ratio V_A/V_P is

termed the *heritability* (in the narrow sense), defined more narrowly than the "broad sense heritability" V_G/V_P. Because V_A is a function of allele frequencies, and V_P includes the environmental variance V_E, an estimate of the heritability of a trait is valid only for the particular population and the particular environment in which it was estimated, since other populations might differ in both these respects. Although many or most characters are genetically variable, we do not know what fraction of this variation can contribute to evolution by natural selection, since it is possible that a considerable portion of the variation may be deleterious under most circumstances.

The "mapping," or relationship, between a phenotypic character state (e.g., body mass) and the environment (e.g., caloric intake) is a genotype' *norm of reaction*. Genotypes may differ in their norms of reaction; for example, some people may gain more weight on a given diet than others. Such differences give rise to a genotype X environment interaction, expressed at the population level by the variance component V_{GXE}. The "mapping" between genotypes and phenotypes, even within a constant environment, often depends on developmental processes. For example, a trait may simply increase or decrease additively and gradually as + or − alleles (those that increase or decrease the character) are substituted in the genotype; or there may be nonlinear effects, so that the character suddenly changes from one to another discretely different state when the number of + alleles crosses a threshold.

A gene commonly affects two or more characters (*pleiotropy*), and so can contribute to a *genetic correlation* (r_G) between them. Another possible cause of genetic correlation is *linkage disequilibrium*, nonrandom association of certain alleles at two or more loci within a population (e.g., an excess of AB and ab combinations and a deficiency of Ab and aB). A genetic correlation caused by pleiotropy may be the net effect of both positive and negative components, since alleles at some loci may affect both characters in the same direction, and at other loci, in opposite directions. The value of r_G depends on the frequencies and phenotypic effects of all contributing loci. It is estimated by the covariance between characters over a set of families, just as the genetic variance is estimated for a single character. Genetic correlations are important because if the population mean of one character is altered, perhaps by natural selection, the other character will also be changed.

GENETIC DRIFT

Random genetic drift is simply random change in the frequency of alleles (and consequently, of genotypes). The genes carried by a generation of newly formed zygotes in a population are a sample of the genes carried by the previous generation, to which the parents belong. Because of random *sampling error*, the frequency (p) of an allele, say A_i, among the zygotes is unlikely to be exactly the same as in the previous generation, since there is likely to have been random mortality and random variation in female reproduction (fecundity) and male reproduction (number of mates) among individuals in the previous generation (here we are considering random, not selective, variation in survival and reproduction). So although the allele frequency in a new generation of N zygotes (carrying $2N$ genes if the species is diploid) is p on average (the same as in the previous generation), the frequency distribution of possible allele frequencies has a *variance*, given by the binomial expression Var $(p) = p(1-p)/(2N)$. This may be conceptualized as the variation among a large number of possible samples of $2N$ genes. The greater Var (p) is, the greater the random change in allele frequency is likely to be, from generation to generation, and thus the faster the process of evolutionary change by genetic drift. The expression for Var (p) tells us that this happens faster, the smaller the population size N. N in this theory refers to the *effective size* of the population, which is smaller than the "census size" if individuals vary in reproductive rate, if the sex ratio among breeding individuals departs from 1:1, or if the population fluctuates in size.

Since this variance holds in each generation, p fluctuates at random from generation to generation with no corrective tendency to return to its starting point, in a "random walk" to a boundary from which no return is possible: either loss of the allele A_i from the population or *fixation* of the allele A_i, i.e., attainment of $p = 1$. (Movement away from this boundary is possible, however, if new variation enters the population by mutation or by gene flow from other populations.) Hence, genetic drift results in loss of genetic variation within a population.

If a number of separate populations of the species all began with the same initial p, different populations would have different random paths, and in principle A_i would become fixed in some and lost in others; thus, genetic drift results in variation (divergence) among populations. An allele is more likely to be lost than to be fixed if its frequency is near zero, and conversely if its frequency is near 1.0; in fact, the probability, at any time, t, that an allele will eventually become fixed is p_t, its frequency at that time. A new mutation often exists, at first, as a single gene copy among the $2N$ genes carried by the N individual organisms in a population, so its initial frequency is $1/(2N)$, and this is its probability of fixation (if it is selectively neutral).

Since DNA sequence data have become available, another theoretical approach to studying the dynamics of genetic variation, *coalescent theory*, has become prominent (Hein et al., 2005). Looking back in time from the present, the gene copies (at a particular gene locus) in the population today are descended from only some of the genes carried by the previous generation's zygotes, due to sampling error; those zygotes in turn carried genes descended from only some of those in *their* parents' generation; and so on. Pursuing this logic, it is inevitable that all the gene copies in the population today are descended from one single ancestral gene copy (one DNA molecule) at some time in the past. The descendants of that gene form lineages of genes, replicating down through the generations to the present time, the set of lineages forming a gene tree which, like a phylogenetic tree of species, portrays their ancestry back ("coalesces to") the common ancestral (CA) gene, which existed t_{CA} generations ago. That ancestor was one of some number (say, $2N$) of genes in the population at that time, but the descendants of those other genes have not persisted to the present time, due to random genetic drift. (When this history was first described for human mitochondrial DNA, the catchy phrase "mitochondrial Eve" was applied to the female that carried the ancestor of all human mitochondrial genomes. Some people wrongly supposed that this meant the ancestral human population consisted of only one woman [and presumably one man].) The speed of genetic drift depends on population size, so it will not be surprising to learn that for a population of constant effective population size N ($2N$ genes at a diploid locus), the average time back to the common ancestor of all contemporary genes, t_{CA}, is $4N$ generations (e.g., four million if the effective population size is one million individuals).

A gene tree, representing the history of common ancestry of a sample of gene copies from one or more populations of a species, can be estimated by phylogenetic methods, using as characters the mutational differences (e.g., nucleotide substitutions) that have accrued among the lineages during their descent from their common ancestor.

THE NEUTRAL THEORY OF MOLECULAR EVOLUTION

Building on these principles, Motoo Kimura pioneered the development of a neutral theory of molecular evolution that is the basis for analyzing DNA sequence variation within and among species, and is often considered the "null hypothesis" against which alternative hypotheses, such as natural selection, must be compared (Kimura, 1983; Nei and Kumar, 2000). Mutational changes occur at many sites in a DNA sequence, at a total rate of, say, u_T per gene per generation. Of these, suppose some fraction f is selectively neutral, so the neutral mutation rate is $u = fu_T$. (The fraction f may depend on the functional role of a DNA sequence or the effect of a nucleotide change; for instance, a synonymous mutation in a functional gene or any mutation in a nonfunctional sequence such as a pseudogene is more likely to be selectively neutral than a nonsynonymous mutation in a gene with a critical function.) Since $2N$ genes are carried by (diploid) zygotes in each generation, the total number of new neutral mutations in the population each generation is $2Nu$, on average. We know from genetic drift theory that the probability of fixation of a new neutral mutation is $1/(2N)$ in a diploid population of constant size N, so $2Nu \times 1/(2N) = u$ new mutations occur each generation that will eventually be fixed. The time to fixation, we have just learned, is $4N$ generations, on average. Since this is the case each generation, u mutations should be fixed in a population every generation on average. In other words, population-wide substitutions of nucleotides in a DNA sequence occur at a roughly constant rate, so DNA sequence evolution theoretically acts as a *molecular clock*, accumulating ut substitutions over the course of t generations. If two populations (or species) are derived from a common ancestor and do not exchange genes for t generations, and if mutations at different sites in the DNA sequence are fixed in each population, the difference D between sequences taken from the two populations will be $D = 2ut$. If u (the neutral mutation rate, which can vary among genes because of functional differences or DNA repair processes) can be calibrated, then the time since the two populations separated can be estimated from the observed difference D, as $t = D/2u$. (Calibration is usually based on geologically dated events, such as fossils of the studied lineage or related lineages, or separation of two land masses on which related taxa reside.)

Eventually, D increases at a lower rate and levels off, because mutational substitutions occur repeatedly at the same sites within the sequence, erasing evidence of previous substitutions. This happens sooner for rapidly than slowly evolving sequences. According to the neutral theory, evolutionary change is more rapid if mutations do not affect organismal function, since mutations that affect protein function are more likely to be deleterious and eliminated by natural selection. Consequently, evolution is predicted, and found, to be more rapid at third-base than second-base positions in codons, because third-base mutations are more often synonymous. Sequence evolution is also more rapid in nonfunctional sequences, such as pseudogenes, than in functional sequences. (Indeed, the rate of sequence evolution between species is now used by molecular biologists to target functionally important versus less important sequences. This and other lines of evidence

suggest that some supposedly nonfunctional, "junk," DNA sequences may have unknown functions, perhaps in gene regulation.) The ratio $\omega = k_A/k_S$, where k_A and k_S are the rates of nonsynonymous and synonymous nucleotide substitution, respectively, is often used as an index of the degree to which a protein-coding DNA sequence has been evolving neutrally, relatively free of functional constraint. If all mutations have been selectively neutral, ω should equal 1.

Genetic variation is lost from a population by genetic drift, as we have seen. However, it is regenerated by mutations at many sites in a DNA sequence, and at equilibrium there exists variation in nucleotide sequence within a population, when the rate of input by neutral mutation balances the rate of loss by genetic drift. A measure of polymorphism is the expected proportion of base pairs that differ between two gene copies taken at random (π) from a population. At equilibrium this equals $4Nu$, i.e., it is proportional to the population size and the mutation rate. Consequently, effective population size can be estimated from $\pi/4u$.

Because of polymorphism, the history of population separation may not be the same as the history of any one gene locus. Suppose an ancestral population divides into two populations (or species) A and B at time t_1, and B later separates into populations B_1 and B_2 at time t_2. Populations B_1 and B_2 are more closely related to each other, by definition, than they are to population A. If population B became fixed for a new mutation, and thus for a different sequence than population A, the mutation would be inherited by populations B_1 and B_2 and provide evidence of their sister-group relationship. Suppose, however, that populations A and B, and their common ancestor, have effective size N, and that the time between successive splits (between t_1 and t_2) is less than the $4N$ generations required for one or another sequence variant to be fixed in each population by genetic drift. If the common ancestor is polymorphic for sequences x and y (perhaps differing by a new mutation in sequence x), fixation may not occur until after the three populations have become separate. Then one sequence (say, x) may be fixed in both A and one of the derived B-populations (say B_1), and the other sequence (y) may become fixed in B_2. The phylogeny of genes may be accurate (the gene copies in B_1 are most closely related to those in A), but it would differ from the phylogeny of the populations. Therefore, it is important to use information from several or many independently inherited genes when analyzing the historical relationships among populations or species that have become separated during a short time span.

Summarizing this section, note that for selectively neutral mutations, whose fate is unaffected by natural selection, the theory of genetic drift and the related neutral theory of molecular evolution provide a basis for many important inferences: e.g., inferring effective population size, time since separation of populations (or since speciation), historical relationships among populations, and whether or not natural selection has affected DNA sequence divergence and polymorphism.

NATURAL SELECTION

There are so many nuances to the concept of natural selection that a simple, comprehensive definition is difficult to devise, but it may suffice, for present purposes, to define it as *consistent (nonrandom) differences in the rate of survival or reproduction among classes of entities that differ in inheritable characteristics*. The term "reproductive success" is often used for "survival and reproduction," since survival to reproductive age is prerequisite for reproduction. "Entities" is deliberately vague, because selection can (in principle) act among various kinds of biological "individuals," such as genes or larger sections of genetic material, individual organisms, groups of conspecific organisms, species, or clades (Williams, 1992). We speak of "classes" of genes, individuals, etc., because we cannot tell if a difference in reproductive success is nonrandom from information about a single individual of each kind; we require samples of similar genes or individuals in order to see if there is a consistent difference between different types of alleles or phenotypically different organisms. Natural selection, in distinction from genetic drift, is marked by a consistent difference in mean reproductive success within a given environment, not a random, unpredictable difference; thus natural selection is the antithesis of chance.

MODES OF SELECTION

Most analyses of evolution by natural selection are concerned with *individual selection*: differences in fitness, owing to a genetically variable phenotypic character, among individual organisms within a population. In the simplest models, the character is affected by variation at a single locus, and we suppose that the fitness of each genotype can be estimated. In practice, this can be difficult, because fitness, defined as a genotype's relative rate of increase, i.e., growth in numbers from one generation to the next, depends on several life-history parameters. The rate of increase is a complex function of the probability of survival at each age from birth to the oldest reproductive age class, and on the age-specific values of female reproduction (fecundity) and male reproduction (affected by mating success and sometimes by sperm competition). (In some cases, it may be affected also by other complicating factors, such as genetic compatibility among uniting gametes.)

Let us consider selection among individual organisms in a sexually reproducing population that differ in genotype at a single locus with two alleles, A and a. In the simplest case, the fittest of the three genotypes AA, Aa, and aa is a homozygote. If aa is rare, because the environment previously favored AA and has only recently changed so that aa is now the fittest genotype, we speak of *directional selection* for aa. Once aa becomes the prevalent genotype, allele A, as well as any other disadvantageous alleles that may arise by mutation, are reduced in frequency, and selection is often termed *purifying*. These are two faces of the same coin, selection that fixes the allele that, in homozygous state, maximizes fitness. The frequency q of the advantageous allele a attains the deterministic equilibrium $q = 1$ if only selection is operating, but if other alleles repeatedly arise by mutation, the equilibrium frequency will be set by the mutation rate relative to the strength of purifying selection ("mutation/selection balance"). Similarly, if a locally disadvantageous allele (perhaps A) that is advantageous in a different geographic population enters the population by gene flow, the genetic equilibrium is determined by the relative strength of gene flow and purifying selection. Gene flow from other populations can sometimes severely diminish the degree of adaptation of populations to their local environment.

Suppose the advantageous allele a is very rare, either because it has recently originated by mutation or because it has formerly been disadvantageous but nevertheless persisted in the population due to mutation/selection balance. If the frequencies of A and a are p and q respectively, the Hardy–Weinberg frequencies of the two genotypes that contain the a allele, Aa and aa, are $2pq$ and q^2, and the vast majority of the a genes are carried by heterozygotes. (For example, if $q = 0.01$, $2pq = 0.0198$, $q^2 = 0.0001$, and the ratio of heterozygotes to homozygotes is 198:1.) Whether or not the a allele can increase (or "invade" the population) depends almost entirely on the fitness of Aa relative to the prevalent homozygous genotype (AA); at this stage the fitness of aa is almost irrelevant because it is so rare. This means that even if aa is the fittest genotype, the a allele will not increase if it reduces the fitness of the heterozygote. This illustrates that natural selection acts only in the present, and cannot look forward toward the best possible outcome. It also shows the value of the Hardy–Weinberg principle.

Directional (or purifying) selection eliminates genetic variation, but several other modes of selection (*balancing selection*) may maintain genetic polymorphism. The simplest model is *heterozygous advantage*, in which the fitness of Aa is greater than that of either AA or aa, and all three genotypes segregate each generation due to random mating. Several hemoglobin polymorphisms in human populations, including sickle cell hemoglobin, are the best-known of the few well-documented examples of this mode of selection. Unquestionably more important is *frequency-dependent selection*, in which the fitness of each genotype is a decreasing function of its own frequency in the population, relative to other genotypes; that is, each genotype is more and more advantageous, the rarer it is. Many biological phenomena, including competition for resources, social interactions, and resistance to different genotypes of parasites, can give rise to such frequency-dependent effects. Mathematically, this is a powerful way of maintaining multiple alleles in a population, and cases are known in which 100 or more alleles appear to be maintained this way. Variable selection, in which different homozygotes are advantageous at different times or in different microhabitats within the area occupied by a breeding population, can also maintain polymorphism, although this is by no means guaranteed: mathematical models show that even if both homozygotes (*AA* and *aa*) are advantageous in different environmental states, only a rather narrow range of combinations of selection intensities and environmental frequencies will maintain all the genotypes indefinitely. (Note that persistence of both homozygotes because of their variable fitnesses also implies persistence of heterozygotes, due to random mating.)

The phenotypic implications of these genetic models depend on the relation between genotype and phenotype. In simple cases, in which there is either complete dominance of one allele or additive inheritance (in which the heterozygote's phenotype is intermediate), persistent genetic polymorphism implies persistence of two or three phenotypic classes, respectively. Most of the consequences of the single-locus models carry over into thinking about the effects of selection on a polygenic phenotypic trait, in which each variable locus contributes a small amount to overall variation. We consider the simplest case, an additive character, measured in, say, millimeters, for which "+" and "−" alleles at each of k loci add or subtract the same amount. The mean and variance of the character are determined by the frequency of the alleles at all of the loci; the mean will clearly be higher (and the variance lower) if most of the + alleles have high frequency. However, an intermediate mean could result from many possible allele frequency arrays, ranging from a highly variable population with $p = 0.5$ (i.e., + and − equally frequent) at each locus, to fixation of a single genotype that is homozygous for + at half of the loci and for − at the other half.

Directional selection on the character occurs when there is a monotonic relationship (at least over part of the range of possible phenotypes) between phenotype and fitness. For example, selection may favor larger phenotypes, namely those with more + alleles in their genetic make-up, so + alleles rise in frequency. If the fitness/phenotype relationship is "open-ended" (e.g., the unlikely circumstance that bigger is always better), selection will ultimately favor the genotype with + alleles only (and subsequently, any mutations with still greater effects), so + alleles become fixed at all loci, genetic variation is eliminated, and evolution ceases except insofar as variation continues to arise by mutation. Thus the magnitude of the "mutational variance," the per-generation increment in the variance of the character due to new mutations, will then limit the rate of subsequent response to selection.

What if the relationship between fitness and phenotype is not monotonic, but instead has an intermediate maximum ("optimum") that lies above the current mean phenotype? Directional selection will increase the frequency of + alleles and bring the mean to the new optimum. The character then becomes subject to *stabilizing selection*: deviations in either direction from the mean are disadvantageous. Many different combinations of + and − alleles can add up to give the same optimal intermediate phenotypic value; some of them are highly heterozygous, and others are homozygous for + alleles at some loci and for − alleles at other loci. Mathematical theory has shown that one or another of the homozygous genotypes will eventually replace all the other genotypes, so that genetic variation will be eliminated by stabilizing selection.

Studies of natural populations have shown that the most common forms of selection on quantitative phenotypic characteristics are stabilizing selection and *disruptive* (also called *diversifying*) selection, in which two or more phenotypes have higher fitness than do the intermediates between them (Endler, 1986). Disruptive selection at a single locus generally implies that the heterozygote for two alleles *A* and *a* has lower fitness than both homozygotes. Such a polymorphism is unstable, however, and the population will become fixed for the initially more common allele. In models of disruptive selection on an additive polygenic character, variation is not maintained indefinitely; instead, the population mean evolves to one or the other of the superior phenotypes, and stabilizing selection then takes over and reduces variation. In both the single-locus and polygenic models, variation is maintained only if disruptive selection is frequency-dependent, such that the fitness of the superior genotypes declines as they become more abundant. The simplest example would be if the genotypes are each adapted for a different food or other limiting resource, so that competition becomes more intense, and fitness declines, as a particular genotype becomes more abundant and depletes its resource.

I have introduced frequency-dependent selection as a negative feedback loop that can maintain stable coexistence of different genotypes in a single breeding population. It is possible, however, to imagine that the fitness of individuals of a particular genotype increases as the genotype's frequency increases. This would be a form of positive feedback that hastens fixation of the genotype, eliminating variation. Such selection is easily envisioned for many social behavioral traits, in which conformity to a predominant behavior pattern might be advantageous.

COMPONENTS OF FITNESS

Genotypes may differ in fitness due to one or more components, most of which are generally considered life history features (Stearns, 1992). These components contribute to the rate of increase (numbers/time) of a genotype, relative to others. One may think of a population of organisms as consisting of subpopulations of different genotypes (or of alleles) that are all growing like a bank account, with compound interest. All else being equal, a difference (in, say, survival probability or fecundity) expressed at an earlier age generally has a bigger impact on growth in numbers (fitness) than a similar difference expressed at a later age. Suppose individuals reproduce from age three until ten, and then die. A mutation that increases the chance of survival from age eight to nine has a smaller selective advantage than one that provides a similar survival advantage from age two to three, because survival enhancement in the older age classes will have much less effect on the number of offspring they might yet produce (and the number of genes passed on). Similarly, a mutation that increases reproductive output at age three has a greater impact on the increase of the mutation's frequency than if it affects reproduction at age ten, because (a) fewer individuals survive to age ten, so they don't get the benefit of the mutation; and (b) the mutation expressed at the younger age effectively shortens the generation time, so more descendants (grandchildren, great-grandchildren ...) are produced per time unit than are produced by the genotype whose reproductive capacity is enhanced only at an older age.

Consequently, mutations that enhance survival or the number of offspring (e.g., number of eggs or young) are expected to increase fitness, but the magnitude of increase depends on the age at which these effects come into play. Moreover, there may exist trade-offs between different fitness components, or between a given component expressed at different ages, partly because an organism must partition energy or nutrients (e.g., protein) among different functions (the *principle of allocation*). For example, if reproduction reduces growth, it may be advantageous to delay reproduction until the individual is larger, which may ensure a longer life and higher fecundity

that together more than make up for the reproduction foregone at an earlier age. On the other hand, if abundant inescapable predators make death at an early age virtually inevitable, selection will favor early reproduction, and mutations that defer senescence or enhance fecundity at advanced ages may well be disadvantageous (if these effects reduce early fecundity). The evolution of *intrinsic* senescence and mortality may therefore be affected by the age distribution of *extrinsic* mortality factors. Many potential adaptations have both benefits and costs, which may be environment-dependent.

A fitness component of particular interest is reproductive success achieved through success in mating, which Darwin termed *sexual selection* (Andersson, 1994). In many species of animals, the variation in reproductive success, and therefore the potential intensity of sexual selection, is greater in males than in females. This difference is generally ascribed to the smaller and far more abundant gametes of males than females, but sexual selection acts more strongly on females of some species (e.g., phalaropes and some pipefish and seahorses), in which investment in paternal care of offspring limits the number of a male's potential mates. (Thus the "choosier" sex, that exerts stronger sexual selection on the opposite sex, usually expends greater parental effort.) The two most commonly discussed modes of sexual selection are conflict between males, with winners gaining access to more females, and female "choice" of some males over others, based on one or more characteristics that usually are actively displayed to females. (In many cases, the same trait seems to play a role in both male–male and male–female interactions.) There is considerable evidence that conflict between males selects for larger size, greater weaponry, and many other kinds of traits that are used to establish dominance. The equilibrium mean value of such a trait will be set by balance between the reproductive advantage it provides and disadvantages such as its energy costs or effects on susceptibility to predation. Indeed, male investment in features that enhance mating success, such as mating activity, weaponry, or display features, may reduce investment in maintenance (e.g., immune system) and survival.

There is considerable evidence from birds, insects, and other animals that female "choice" imposes sexual selection, but there is considerable uncertainty about why females choose particular male phenotypes, such as males with more vigorous displays or more highly elaborated ornaments or vocalizations. According to one hypothesis, exaggerated male features indicate high physiological vigor that may stem from superior genetic constitution (the "*good genes*" hypothesis), and females that choose such males will have fitter offspring. There is some support both for this hypothesis and for several contenders. In models of *runaway sexual selection*, a nonrandom association (linkage disequilibrium) develops between genes that affect a male ornament and genes that affect the degree of female preference for this character. Females that prefer more highly ornamented males have daughters that inherit this preference (as well as unexpressed genes for large male ornamentation) and sons that inherit larger ornaments (as well as unexpressed genes for heightened female preference). (Note that most features expressed by a single sex are encoded in the genome of both sexes.) Therefore, any increase in the average male ornament in the population will cause a correlated increase in the average female preference, and vice versa, ratcheting both toward more extreme values until the process is halted either by counteracting selection or by running out of genetic variation.

In a twist on sexual selection theory, females and males are engaged in sexual conflict: males reduce females' fitness in various ways (e.g., incessantly attempting to mate), females are selected to resist, and selection favors males with ever more stimulating characteristics that can overcome female resistance (Arnqvist and Rowe, 2005). The scope for such interactions appears greater than was formerly supposed, because it is clear that females of many species mate with multiple males, even in species that form a supposedly monogamous pair bond. Thus males have the potential of siring offspring by mating not only with unmated, but also with previously mated, females. The consequences include competition between sperm from different males. Probably because of the strong, long-continued selection exerted by sperm competition and sexual selection, reproductive characteristics, including male display features, genitalic morphology, proteins from accessory reproductive glands, sperm morphology, and cell-surface proteins of gametes, are rapidly evolving characteristics that often are the major differences among closely related species.

MODELING ADAPTATION

In considering components of fitness, we have moved from the very general theories of population and quantitative genetics, which apply to unspecified genes and characters, to models of the evolution of specific classes of characters, such as life history features. Evolutionary analyses of adaptive evolution of specific classes of characters employ several approaches to modeling (Bulmer, 1994). The evolution of some features is best analyzed by *genetic models*. This is true of models of sexual selection by female choice, for example, because linkage disequilibrium is an essential component and it requires an explicit genetic approach. The major alternative is *optimization*, an approach that

attempts to specify what the optimal character state ought to be, given some assumptions about benefits, costs, and constraints. This approach assumes that there has been enough time and enough genetic variation for natural selection to bring the mean character state in a population nearly to its optimum value, and that the genetic details do not matter very much. Whether or not these are reasonable assumptions may depend on empirical information about such things as genetic variation and evolutionary history (e.g., inferences about how long a species has probably been subject to consistent environmental selection).

Optimization is a common approach in the fields of functional morphology and physiology, in which it is assumed that fitness is enhanced by maximizing some function, subject to constraints such as costs in energy or materials, or compromises with other functions. For example, aerodynamic models have been used to model flight and optimal wing morphology in birds, in which compromises among speed, maneuverability, and energy expenditure are taken into account. Among nonsocial aspects of behavior, models of optimal foraging describe when a foraging animal should give up searching in one patch and move to another.

Social interactions entail complexities that make genetic modeling difficult, and have been analyzed almost entirely by optimal models. The complexity arises from the frequency-dependent fitness of different trait values: the optimal behavior of an individual often depends on the behavior of other members of the population. Among the most widely used approaches is game theory (Maynard Smith, 1982). Suppose, for example, that the problem is whether or not parental care, by either or both mated partners, will evolve by individual selection. One might postulate two "strategies," "Stay and provide care to offspring" and "Defect and attempt to reproduce again." For each possible pair (S♀/S♂, S♀/D♂, D♀/S♂, D♀/D♂), one postulates for each partner the expected reproductive "payoff," which depends on both the benefit to each partner (in terms of surviving offspring from this mating) and the costs to each (in terms of the likely reproductive success sacrificed). The average fitness of each strategy, for each sex, is then its payoff averaged over the possible pairings, and weighted by their frequency in a random-pairing population. The best strategy, within the set of strategies considered (here, S and D), is the one that, if fixed in the population, will remain fixed even if individuals with alternative strategies attempt to invade. This is the *evolutionarily stable strategy*, or *ESS*.

LEVELS OF SELECTION

Natural selection was defined above as "consistent (nonrandom) differences in the rate of survival or reproduction among classes of entities that differ in inheritable characteristics." These "entities" may be at different, nested levels, and the effects of selection at different levels may be opposite (Okasha, 2006). Consider, for example, the level "individual organism" and the level "somatic cell lineage" within a multicellular organism. If a cell lineage experiences a mutation that causes rapid, unrestricted cell division, that lineage has a "selective advantage" relative to other cells, and will constitute an increasing proportion of cells within the domain of the single organism (Nowak, 2006). This proliferation – cancer – is clearly disadvantageous to the higher-level entity (the organism), if it occurs before or during the organism's reproductive ages. Selection among genetically variable individual organisms will favor genotypes that have the ability to suppress cancerous tumors.

We may likewise distinguish selection among individual organisms with different genotypes (the level of selection assumed so far in this chapter) from selection at the level of the individual gene (locus). In asexual organisms, there is little conflict between selection at these levels, because the fate of a gene (survival, passage to subsequent generations) depends on that of the rest of the genotype to which it is bound. But in sexually reproducing organisms, conflicts can arise. A famous example is the "t locus" in house mice. More than 90% of the sperm of males heterozygous for a normal allele (T) and one of several recessive alleles (t) carry the t allele (an example of *meiotic drive*). Some of the recessive alleles cause embryonic death, and others male sterility, in homozygous condition. The differential transmission of T and t alleles constitutes differential "reproduction" at the gametic level (*genic selection*), opposing differential survival of individual mice (*individual selection*). Genic selection accounts for many phenomena, such as the proliferation of transposable elements ("selfish genetic elements"): DNA sequences that replicate more frequently than most of the genome.

Genic selection provides one way of viewing the evolution of co-operation, which stands in contrast to the selfish individualism that generally characterizes individual selection (Dawkins, 1982; Sober and Wilson, 1998). Cells in multicellular organisms co-operate because they are (generally) genetically identical: a gene in a liver cell is replicated by virtue of the replication of identical copies in the germ cell line – and the fate of the germ cell line depends on the gene copies functioning in the liver. Likewise, the rate of increase of a parent's gene over generations depends on the survival and replication of copies of that gene in the parent's offspring – and so alleles that program parental care may increase in frequency. This is an example of *kin selection*: selection in which alleles differ in fitness by influencing the effect of their bearers on the reproductive success of individuals (kin) who carry the same allele due to common descent. (In this case, the "bearers" are parents, and the "kin" are their offspring.) In the same way, genes that enhance their bearers' propensity to help more distant relatives may increase in frequency – but the consequent increase in the relatives' fitness must be greater, since their probability of sharing the "helping allele" is lower. William Hamilton formalized this relationship in what has become known as Hamilton's rule, which states that "altruism" spreads if $rb > c$: an altruistic trait can increase in frequency if the benefit (b) received by the donor's relatives, weighted by their relationship (r) to the donor, exceeds the cost (c) of the trait to the donor's fitness. The relationship, r, between donor and recipient is the fraction of the donor's genes that, on average, are identical by descent with any of the recipient's genes. For example, $r = \frac{1}{2}$ between parent and offspring, so an allele for parental care should spread, even if it costs the parent her life and subsequent reproduction, as long as her care results in survival of more than two extra offspring (compared to a parent that does not provide care). (Kin selection is only one of several explanations of the evolution of co-operation among genes, cells, or conspecific organisms. For example, reciprocity ["reciprocal altruism"] may evolve if individuals recognize one another and can benefit others or not, depending on their history of behavior.)

Because of kin selection, the family (mated pair and associated offspring) is an obvious context in which co-operation may evolve. Nevertheless, intrafamilial interactions are riddled with conflict. *Sexual conflict* inevitably arises from the sex difference in gamete size (and some other features in certain species): male fitness can be increased by mating with many females, whereas all of a female's eggs can usually be fertilized by a single male. Female fitness is more likely to be enhanced by her offspring's survival, which may be increased by parental care or by "genetic quality." Parental care increases the fitness of both parents, but it entails costs, including lost reproductive opportunities. If offspring were as likely to survive with uniparental care as with biparental care, selection would favor defection by one sex – the one for which parental care is more costly (Clutton-Brock, 1991). A further complication is that if a caregiver were not actually the parent of some or all of the offspring, he (or she) would have less of a genetic interest in their survival. "Extrapair copulation," common in seemingly monogamous species of birds, therefore alters the costs and benefits of parental care. In some species of primates and other mammals, a male that replaces another male kills his new mate's offspring, since he has no genetic interest in them, and killing them enables him to father his own offspring faster. (Killing some offspring can also be advantageous to parents if, by reducing competition among the remaining offspring, it maximizes the number of healthy survivors.)

Trivers (1974) first pointed out that because a parent maximizes her (his) fitness by allocating care among a number of offspring, present and future, the optimal investment of care in any one offspring is lower from the parent's perspective than from the offspring's. The consequent *parent–offspring conflict* has many consequences for behavior, and even for pregnancy. Haig (1993, 1997) has described how genes expressed in human and mouse fetuses that enhance uptake of sugar and other nutrients from the mother are counteracted by maternally expressed genes that prevent the fetus from extracting too much. For example, insulin levels are increased in pregnant women, decreasing blood glucose (just when one might expect it to be higher), in order to counteract a fetal hormone that has the opposite effect.

If an individual with an "altruism allele" dispenses benefits indiscriminately to both related and unrelated individuals, the survival of both that allele and its nonaltruistic ("selfish") alternative allele are equally enhanced, but the donor suffers a cost (*c*), so the selfish allele increases. As a rule, individual selfishness increases within populations, even if the population as a whole suffers. In principle, extinction of whole populations of selfish individuals, and survival of populations of co-operative individuals, could cause evolution of co-operation in the species as a whole. This would be *group selection* (also called population selection and interdemic selection), in opposition to individual selection. Some authors hold that group selection can play a role in evolution, especially if the groups are very small and temporary (Wilson, 1983). (For example, if some groups of nestling birds include co-operators and others do not, the overall productivity of those nests with co-operators may be higher, under some circumstances.) However, because populations of co-operators are likely to be invaded by immigrant selfish genotypes ("cheaters"), which rapidly increase within these populations, co-operation is unstable. Most evolutionary theorists therefore conclude that group selection, evolution by differential extinction, or reproductive productivity of whole populations, is unlikely to play a major role in evolution. The important consequence is that we generally do not expect the evolution of characteristics that benefit the population or species, but which are disadvantageous to the individual or its kin.

Selection among groups, as long as it does not oppose individual selection within groups, may however have evolutionary consequences. These may be most evident when the groups are species. As a consequence of characteristics of their constituent individuals, or of properties of the species as a whole (e.g., size of geographic range), species may differ in their probability of extinction or of speciation per unit time. Stephen Jay Gould and others, pointing out that these probabilities are the species-level analogs of differential mortality and reproductive rates of individual organisms, have proposed that *species selection* has shaped the frequency distribution of traits in large clades and biotas (Gould, 2002). The fraction of plant species with flowers, for example, greatly increased during the Mesozoic, and the world's mammal fauna became increasingly dominated by placental eutherians in the late Mesozoic and Tertiary. Identifying the trait(s) that have been the "target" of species selection can be difficult, but is sometimes possible by comparing the diversification rate of multiple clades that have independently evolved a postulated diversity-enhancing feature, compared to the diversification rate of sister clades that lack the feature. Herbivory in insects, sexual dichromatism in passerine birds, and low body mass in several orders of mammals seem to be associated with increased diversification (Coyne and Orr, 2004).

SPECIATION

There are several contending concepts of "species" because this word serves several purposes. As a term in classification, it may simply label phenotypically distinguishable populations. For instance, morphologically different sections of a single lineage in the fossil record are sometimes given different names, and may be termed "chronospecies." Many systematists use a *phylogenetic species concept* (PSC), according to which a species is a diagnosably different cluster of organisms, within which there is a parental pattern of ancestry and descent. This would include asexual forms. Most evolutionary biologists concerned with evolutionary processes use one or another version of the *biological species concept* (BSC), articulated by Ernst Mayr in 1942. A biological species is a group of actually or potentially interbreeding populations that are reproductively isolated (by biological differences) from other such groups. The reproductive isolating barriers (RIBs), which are usually genetically based, include *prezygotic* barriers that reduce gene exchange before zygote formation (e.g., differences in habitat association, timing of reproduction, behavior, pollination [for plants], and failure of gametes to unite) and *postzygotic* barriers that act after zygote formation, and are expressed as diminished survival or reproduction of hybrid genotypes. The BSC applies only to sexual organisms and may be more difficult to apply in practice than the PSC, since the potential ability of spatially separated (allopatric) populations to interbreed may be difficult to evaluate. However, reproductive isolation

(RI) plays a critical role in long-term evolution, since it enables populations, even if they eventually meet and overlap, to retain their distinct characteristics and to generate clades of species that subsequently elaborate those differences. It is even possible that speciation facilitates sustained evolution of morphological and other characteristics, by preventing the "slippage" that interbreeding with other populations would cause (Futuyma, 1987).

There is abundant evidence that many species of animals and plants form by genetic divergence of spatially disjunct (allopatric) or neighboring (parapatric) populations of an ancestral species, since this reduces gene flow enough to allow divergent changes in allele frequencies by natural selection or genetic drift. Whether or not species are often formed sympatrically, i.e., by evolution of an intrinsic separation of a single randomly mating population into two reproductively isolated forms, is controversial. Except for speciation by chromosome doubling (polyploidy), which is common in plants but rare in animals, the evolution of RI between populations, like the evolution of most phenotypic differences, is gradual: arrays of populations can be found that display all degrees of pre- or postzygotic isolation from none to complete. Genetic analyses, correspondingly, show that RI is polygenic, based on contributions from several or many genes. Thus it is common to find partially reproductively isolated populations (semispecies) that are not readily classified as the same or as different species. Moreover and very importantly, hybridization between such forms in nature can result in some parts of the genome readily "introgressing," or penetrating from one semi-species into the other, while other parts of the genome remain much more differentiated, because stronger divergent natural selection on these regions counter-acts gene flow.

Because RIBs are usually polygenic, full RI requires that two distinct clusters of different alleles be formed at loci that affect a RIB ($AABBCC...$ and $aabbcc...$), whereas recombination generates allelic mixtures ($AaBBCc...$, etc.) that are only partially reproductively isolated and so form a "bridge" for gene exchange between diverging populations. An extrinsic barrier (e.g., topography or unsuitable habitat) between diverging populations, if it is seldom surmounted by dispersing individuals, reduces or eliminates this problem, which is why allopatric speciation is easy. After allopatric populations have diverged sufficiently, they may expand their ranges and overlap without interbreeding. Conversely, sympatric speciation is generally considered difficult because divergent selection, unaided by an extrinsic barrier, must overcome recombination. Several models have been proposed by which this may occur, but the conditions and assumptions required for sympatric speciation are fairly stringent (Gavrilets,

2004). Only a few well-documented cases of sympatric speciation (completed or in process) satisfy skeptics (Coyne and Orr, 2004).

What causes the evolution of RI between spatially separated populations? As long as the populations are allopatric, there cannot be selection to avoid hybridization, so RI must evolve as a by-product of genetic divergence that transpires for other reasons. Genetic drift, ecological selection, and sexual selection are the postulated processes, but it is only recently that this problem has been explicitly and rigorously studied.

Divergent sexual selection, resulting in different female preferences and male display traits, can result in behavioral isolation. There is considerable indirect evidence for this hypothesis, in that the diversification rate of clades of birds and insects with features indicative of strong sexual selection has been higher than that of sister clades lacking those features. The high rate of evolutionary divergence of genitalia, sperm, and other features associated with reproduction is consistent with this hypothesis. Diverse lines of evidence have also recently pointed to divergent ecological selection as a cause of RI. In some cases the effect of selection is fairly direct: beak size differences among Darwin's finches, for example, are adaptations for feeding, but also are used by females to discriminate among males. In other cases, genes underlying ecological divergence may contribute to RI pleiotropically; examples include a correlation between copper tolerance and partial hybrid sterility among monkeyflower populations, and between adaptation to different host plants and behavioral isolation in some herbivorous insects.

Genetic drift plays a role in Ernst Mayr's (1954) influential hypothesis of *founder-effect speciation* (also called *peripatric speciation*). Mayr believed that selective advantage of one allele over another depends strongly on which alleles it interacts with at other loci (epistasis for fitness). He postulated that a population founded by few individuals will undergo genetic drift at some loci, so that some previously rare alleles become common by chance. This change in the "genetic environment" alters selection at interacting loci, so that previously disadvantageous alleles become advantageous. Consequently, natural selection, initiated by genetic drift, alters the genetic constitution of the newly founded population so that it may become reproductively isolated from the parent population. There is little evidence for Mayr's hypothesis, which is considered implausible by some theoreticians, but new theoretical developments suggest that it still warrants consideration, albeit in modified form (Gavrilets, 2004).

If some genetic incompatibility has evolved between allopatric populations, and if they subsequently expand their range, mate, and form hybrids with low fitness (i.e., low survival or reproduction), individuals that mate conspecifically will have more successful offspring than those that hybridize, and alleles that enhance (reinforce) mating discrimination may be transmitted and increase in frequency. Such *reinforcement* of prezygotic isolation is the major context in which there is direct selection for RI. It appears to be a fairly common component of speciation in some taxa.

The divergence of populations into distinct species often includes ecological differentiation, often initiated in allopatry due to environmental differences among regions, and sometimes enhanced via *character displacement*, the evolution of accentuated ecological difference between sympatric populations of two species, to reduce competition. Bursts of speciation, accompanied by ecological divergence, are referred to as adaptive radiations, macroevolutionary episodes that account for much of the extraordinary adaptive variety of organisms (Schluter, 2000).

FROM MICROEVOLUTION TO MACROEVOLUTION

Macroevolution, or evolution above the species level (Levinton, 2001), includes the evolution of "higher taxa," which, most systematists today insist, should be monophyletic groups of species that are recognized by shared derived characters (synapomorphies), and which usually are phenotypically quite different from related higher taxa. At least among extant animals, for example, mammals differ substantially in skeletal and other features from other amniotes. During the 1930s and 1940s, the major authors of the Evolutionary Synthesis provided both theory and evidence that the evolution of the distinctive features of higher taxa consist simply of incremental changes in each of the differentiating characters, attributable to the processes, outlined above, that operate within and during the formation of species (Simpson, 1953). The many distinctive characters of mammals, for example, are a product of *mosaic evolution*: largely independent evolution, often at different rates, of many distinct features. Each such feature, it was postulated, evolves by successive small steps rather than by large discrete jumps. Such *gradual evolution* has been paleontologically documented for many skeletal features of mammals, and for many other lineages. Often, evolution of a character is accelerated, and takes surprising directions, because a feature may serve a new and different function.

This neo-Darwinian view affirmed Darwin's hypothesis that evolution is a gradual process – although, to be sure, it can sometimes be quite rapid, which makes finding intermediate stages in an already very incomplete fossil record even less likely. Gradualism has been

repeatedly challenged, and it is certainly not possible to affirm that mutations of large effect, that radically alter developmental pathways, have never contributed to evolution. Some evolutionary changes have been discontinuous, such as neoteny in salamanders that retain larval morphology into reproductive adulthood, and which is based on one or a few gene substitutions; but this is an abbreviation of a developmental pathway that requires the action of many genes for its continuation. Mutations in key regulatory genes, such as the *Hox* genes that are important in establishing the fundamental body plans of animals, can have drastic effects, sometimes switching development of one body region into another; but whether or not comparable mutations in these genes were the origin of major evolutionary changes in body plan is not known. The complex developmental pathways that these genes now control may have evolved incrementally.

Classical evolutionary and embryological studies of morphological differences among taxa identified several common patterns, such as (a) allometric differences: evolved differences in the rate at which one structure or dimension grows compared to other structures (compare limb proportions in humans and apes); (b) heterochrony: differences in the timing of developmental events, including initiation or cessation of a structure's growth (as in neotenic salamanders); (c) heterotopy: differences in the location on the body where a feature develops (e.g., bone in the skin of armadillos); (d) individualization of repeated elements (e.g, heterodont dentition of mammals compared to homodont dentition of most "reptiles"); (e) "standardization," or restriction of variation (e.g., digit number was higher and more variable in the earliest tetrapods than in their pentadactylous descendants). A major challenge is to understand how differences in genes are translated into such phenotypic changes, via their effects on the processes by which such characters develop. Increasingly, this challenge is being met in the field of evolutionary developmental biology (EDB, or "evo-devo"), which is based on increased understanding of how certain genes regulate the time and place of expression of other genes, often in hierarchical sequences of control (Carroll, 2007). For example, changes in the expression pattern of certain *Hox* genes along the embryonic anterior-posterior axis in turn control activation of other genes that govern the form of vertebrae. In snakes, most precaudal vertebrae have characteristic thoracic form, rather than sacral and other vertebral forms, due to changes in the domains of expression of certain *Hox* genes. Moreover, new regulatory connections between genes have often evolved. For instance, although the canonical role of *Hox* genes is specification of anterior-posterior domains of differentiation, a *Hox* locus also controls the development of bristles on the legs of *Drosophila*.

Major challenges lie ahead in understanding how changes in gene regulatory pathways evolve, and the extent to which understanding the genetic basis of development can predict possible constraints on the kind of phenotypic variation that can arise and be available to natural selection. Moreover, development is often responsive to environmental signals, so a genotype often expresses phenotypic plasticity, the ability to develop a variety of adaptive phenotypic states (its norm of reaction) under different environmental conditions (Pigliucci, 2001). Conversely, canalization, or developmental buffering, can evolve: the capacity of a genotype to produce a consistent phenotype despite potentially destabilization by environmental variation or gene mutations. Adaptive phenotypic plasticity and canalization are under genetic control, but how they evolve is little understood.

EVOLUTIONARY THEORY TODAY

When Darwin referred to "my theory," he was speaking of a little-tested hypothesis. Since then, all the major elements of his theory – the common ancestry of all organisms, the bifurcation of lineages in the origin of species, the evolution of adaptations by the action of natural selection on hereditary variation, the diversification of species by adaptation to different ecological niches, or places in the "economy of nature" – have been abundantly supported, elaborated, and extended. When biologists speak today of "evolutionary theory," they refer not to a speculation or hypothesis, but to a mature scientific theory, an accepted complex of general principles that explain a wide variety of natural phenomena, as quantum theory and atomic theory do in the physical sciences.

No biologist today would think of publishing "new evidence for evolution" – it hasn't been a scientific issue for a century. However, although the major principles of evolutionary theory have been increasingly supported despite, and indeed because of, vast changes in biological knowledge, many questions remain unanswered and substantial controversy persists, as it does in all active sciences, about some important questions such as gradual versus discontinuous major changes in phenotype. Today, research in evolutionary biology is flourishing as never before. Because of increasingly powerful and affordable molecular and computing technologies, documentation of newfound phenomena that call for evolutionary explanation (e.g., in genomics), and the development of new theory addressing both new and old questions, evolutionary biology is now occupied with the phylogeny of all major groups of organisms, discerning and explaining patterns of evolution of genes and genomes, explaining puzzling behaviors and other phenotypic traits, using advances in developmental biology to understand the evolution of phenotypes, and many other concerns. New approaches are being developed to answer old but difficult questions, such as the causes of speciation, the evolution of functionally integrated characteristics, and the conditions under which populations adapt to environmental change or become extinct – one of the most important questions now, when humans are changing the Earth at a frightening pace.

DISCUSSION POINTS

1. Although evolution consists of genetic change, we often do not have direct evidence that differences among individuals, populations, or related species in a feature of interest have a genetic foundation. Are there ways in which we can have more or less confidence that interesting variation represents evolved differences rather than direct environmental effects on the phenotype?
2. What kinds of evidence might we use to judge whether a character of interest, or a difference between populations or species, is an adaptation rather than the consequence of genetic drift?
3. In what ways is evolution a matter of chance (random) versus a nonrandom process?
4. How do functionally complex characters, such as the vertebrate eye, evolve? Cast your answer both in historical, phenotypic terms and in terms of genetics and selection.
5. In what ways may DNA sequence data inform us about the history of differentiation among human populations and the causes of the differences among them?
6. Taking into account principles of evolutionary genetics and possible scenarios for environmental change, do you expect adaptive evolutionary changes to occur in human populations in the next few hundred years? If so, discuss what changes may occur, and how. If not, why?

REFERENCES

Andersson, M. B. (1994). *Sexual Selection*. Princeton: Princeton University Press.

Arnqvist, G. and Rowe, L. (2005). *Sexual Conflict*. Princeton: Princeton University Press.

Barton, N. H. and Keightley, P. D. (2002). Understanding quantitative genetic variation. *Nature Reviews Genetics*, **3**, 11–21.

Blows, M. W. and Hoffmann, A. A. (2005). A reassessment of genetic limits to evolutionary change. *Ecology*, **86**, 1371–1384.

Bulmer, M. G. (1994). *Theoretical Evolutionary Ecology*. Sunderland, MA: Sinauer.

Carroll, S. B. (2007). *The Making of the Fittest: DNA and the Ultimate Forensic Record of Evolution*. New York: W. W. Norton.

Clutton-Brock, T. H. (1991). *The Evolution of Parental Care*. Princeton: Princeton University Press.

Coyne, J. A. and Orr, H. A. (2004). *Speciation*. Sunderland, MA: Sinauer.

Dawkins, R. (1982). *The Extended Phenotype*. Oxford: Oxford University Press.

Endler, J. A. (1986). *Natural Selection in the Wild*. Princeton: Princeton University Press.

Falconer, D. S. and Mackay, T. F. C. (1996). *Introduction to Quantitative Genetics*. Harlow, Essex: Longmans.

Futuyma, D. J. (1987). On the role of species in anagenesis. *American Naturalist*, **130**, 465–473.

Futuyma, D. J. (1998). *Evolutionary Biology*, 3rd edn. Sunderland, MA: Sinauer.

Futuyma, D. J. (2009). *Evolution*. Sunderland, MA: Sinauer 2nd edn.

Gavrilets, S. (2004). *Fitness Landscapes and the Origin of Species*. Princeton: Princeton University Press.

Gould, S. J. (2002). *The Structure of Evolutionary Theory*. Cambridge, MA: Harvard University Press.

Haig, D. (1993). Genetic conflicts in human pregnancy. *Quarterly Review of Biology*, **68**, 495–532.

Haig, D. (1997). Parental antagonism, relatedness asymmetries, and genomic imprinting. *Proceedings of the Royal Society of London. Series B*, **264**, 1657–1662.

Hartl, D. L. and Clark, A. G. (1997). *Principles of Population Genetics*. Sunderland, MA: Sinauer.

Hein, J., Schierup, M. H. and Wulff, C. (2005). *Gene Genealogies, Variation and Evolution. A Primer in Coalescent Theory*. Oxford: Oxford University Press.

Houle, D. (1998). How should we explain variation in the genetic variance of traits? *Genetica*, **102/103**, 241–253.

Kimura, M. (1983). *The Neutral Theory of Molecular Evolution*. Cambridge: Cambridge University Press.

Levinton, J. S. (2001). *Genetics, Paleontology, and Macroevolution*, 2nd edn. Cambridge: Cambridge University Press.

Maynard Smith, J. (1982). *Evolution and the Theory of Games*. Cambridge: Cambridge University Press.

Mayr, E. (1954). Change of genetic environment and evolution. In *Evolution as a Process*, J. Huxley, A. C. Hardy, and E. B. Ford (eds). London: Allen and Unwin, pp. 157–180.

Nei, M. and Kumar, S. (2000). *Molecular Evolution and Phylogenetics*. New York: Oxford University Press.

Nowak, M. A. (2006). *Evolutionary Dynamics: Exploring the Equations of Life*. Cambridge, MA: Harvard University Press.

Okasha, S. (2006). *Evolution and the Levels of Selection*. Oxford: Oxford University Press.

Pigliucci, M. (2001). *Phenotypic Plasticity: Beyond Nature and Nurture*. Baltimore: Johns Hopkins University Press.

Schluter, D. (2000). *The Ecology of Adaptive Radiation*. Oxford: Oxford University Press.

Simpson, G. G. (1953). *The Major Features of Evolution*. New York: Columbia University Press.

Sober, E. and Wilson, D. S. (1998). *Unto Others: the Evolution and Psychology of Unselfish Behavior*. Cambridge, MA: Harvard University Press.

Stearns, S. C. (1992). *The Evolution of Life Histories*. Oxford: Oxford University Press.

Trivers, R. (1974). Parent–offspring conflict. *American Zoologist*, **11**, 249–264.

Williams, G. C. (1992). *Natural Selection: Domains, Levels, and Challenges*. New York: Oxford University Press.

Wilson, D. S. (1983). The group selection controversy: history and current status. *Annual Review of Ecology and Systematics*, **14**, 159–187.

2 The Study of Human Adaptation

A. Roberto Frisancho

INTRODUCTION

Ever since hominids left Africa, they have expanded throughout the world and have adapted to diverse environments, and acquired specific biological and cultural traits that have enabled them to survive in a given area. The conceptual framework of research in biological anthropology is that evolutionary selection processes have produced the human species and that these processes have produced a set of genetic characteristics, which adapted our evolving species to their environment. Current investigations have demonstrated that the phenotype measured morphologically, physiologically, or biochemically is the product of genetic plasticity operating during development. Within this framework, it is assumed that some of the biological adjustments or adaptations people made to their natural and social environments have also modified how they adjusted to subsequent environments. The adjustments we have made to improve our adaptations to a given environment have produced a new environment to which we, in turn, adapt in an ongoing process of new stress and new adaptation.

HOMEOSTASIS AND ENVIRONMENTAL STRESS

Central to the study of adaptation is the concept of homeostasis and environmental stress. Environmental stress is defined as any condition that disturbs the normal functioning of the organism. Such interference eventually causes a disturbance of internal homeostasis. Homeostasis means the ability of the organism to maintain a stable internal environment despite diverse, disruptive, external environmental influence (Proser, 1964). On a functional level, all adaptive responses of the organism or the individual are made to restore internal homeostasis. These controls operate in a hierarchy at all levels of biological organization, from a single biochemical pathway, to the mitochondria of a cell, to cells organized into tissues, tissues into organs and systems of organs, to entire organisms. For example, the lungs provide oxygen to the extracellular fluid to continually replenish the oxygen that is being used by the cells, the kidneys maintain constant ion concentrations, and the gastrointestinal system provides nutrients.

Humans living in hot or cold climates must undergo additional functional adjustments to maintain thermal balance; these may comprise adjustments to the rate of metabolism, avenues of heat loss, heat conservation, respiration, blood circulation, fluid and electrolyte transport, and exchange. In the same manner, persons exposed to high altitudes must adjust through physiological, chemical, and morphological mechanisms, such as increase in ventilation, increase in the oxygen-carrying capacity of the blood resulting from an increased concentration of red blood cells, and increased ability of tissues to utilize oxygen at low pressures. Failure to activate the functional adaptive processes may result in failure to restore homeostasis; which in turn results in the maladaptation of the organism and eventual incapacitation of the individual. Therefore homeostasis is a part and function of survival. The continued existence of a biological system implies that the system possesses mechanisms that enable it to maintain its identity, despite the endless pressures of environmental stresses (Proser, 1964). The complementary concepts of homeostasis and adaptation are valid at all levels of biological organization; they apply to social groups as well as to unicellular or multicellular organisms (Proser, 1964).

Homeostasis is a function of a dynamic interaction of feedback mechanisms whereby a given stimulus elicits a response aimed at restoring the original equilibrium. Several mathematical models of homeostasis have been proposed. In general, they show (as schematized in Figure 2.1) that when a primary stress disturbs the homeostasis that exists between the organism and the environment the organism must resort either to biological or cultural–technological responses in order to function normally. For example, when faced with heat stress, the organism may simply reduce its metabolic activity so all heat-producing processes are slowed down, or may increase the activity of the

Human Evolutionary Biology, ed. Michael P. Muehlenbein. Published by Cambridge University Press. © Cambridge University Press 2010.

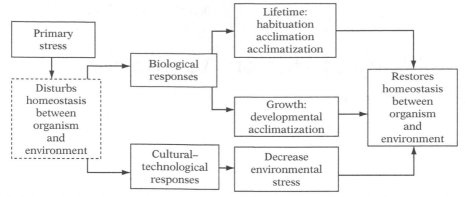

2.1. Schematization of adaptation process and mechanisms that enable individual or population to maintain homeostasis in the face of primary environmental disturbing stress. From Frisancho (1993).

heat-loss mechanisms. In either case, the organism may maintain homeostasis, but the physiological processes will occur at a different set point. The attainment of full homeostasis or full functional adaptation, depending on the nature of the stress, may require short-term responses, such as those acquired during acclimation or acclimatization, or may require exposure during the period of growth and development as in developmental acclimatization.

In theory, the respective contributions of genetic and environmental factors vary with the developmental stage of the organism – the earlier the stage, the greater the influence of the environment and the greater the plasticity of the organism (Proser, 1964; Timiras, 1972; Frisancho, 1975, 1993). However, as will be shown, the principle does not apply to all biological parameters; it depends on the nature of the stress, the developmental stage of the organism, the type of organism, and the particular functional process that is affected. For example, an adult individual exposed to high-altitude hypoxia through prolonged residence may attain a level of adaptation that permits normal functioning in all daily activities and as such we may consider him adapted. However, when exposed to stress that requires increased energy, such as strenuous exercise, this individual may not prove to be fully adapted. On the other hand, through cultural and technological adaptation, humans may actually modify and thus decrease the nature of the environmental stresses so that a new microenvironment is created to which the organism does not need to make any physiological responses. For example, cultural and technological responses permit humans to live under extreme conditions of cold stress with the result that some of the physiological processes are not altered. However, on rare occasions, humans have been able to completely avoid an environmental stress. Witness the fact that the Eskimos, despite their advanced technological adaptation to cold in their everyday hunting activities, are exposed to periods of cold stress and in response have developed biological

processes that enable them to function and to be adapted to their environment.

Not all responses made by the organism can be considered adaptive. Although a given response might not be adaptive per se, through its effect on another structure or function it may prove beneficial to the organism's function. Conversely, a given adaptive response may aid the organism in one function, but actually have negative effects on other functions or structures. Thus, within all areas of human endeavor a given trait is considered adaptive when its beneficial effects outweigh the negative ones. In theory, this is a valid assumption, but in practice, due to the relative nature of adaptation, it is quite difficult to determine the true adaptive value of a given response. Every response must be considered in the context of the environmental conditions in which the response was measured and within the perspective of the length of time of the study and the subject population.

ADAPTIVE PROCESSES

The term adaptation is used in the broad generic sense of functional adaptation, and it is applied to all levels of biological organization from individuals to populations. A basic premise of this approach is that adaptation is a process whereby the organism has attained a beneficial adjustment to the environment (Lewontin, 1957; Mayr, 1963; Proser, 1964; Dubos, 1965; Baker, 1966; Lasker, 1969; Mazess, 1973; Frisancho, 1975, 1993). This adjustment can be either temporary or permanent, acquired either through short-term or lifetime processes, and may involve physiological, structural, behavioral, or cultural changes aimed at improving the organism's functional performance in the face of environmental stresses. If environmental stresses are conducive to differential mortality and fertility, then adaptive changes may become established in the population through changes in genetic composition and thus attain a level of genetic adaptation. In this context, functional adaptation, along with cultural and genetic adaptation, is viewed as part of a continuum in an adaptive process that enables individuals and populations to maintain both internal and external environmental homeostasis. Therefore the concept of adaptation is applicable to all levels of biological organization from unicellular organisms to the largest mammals and from individuals to populations. This broad use of the concept of adaptation is justified not only in theory but also because it is currently applied to all areas of human endeavor so that no discipline can claim priority or exclusivity in the use of the term (Dubos, 1965). Functional adaptation involves changes in organ system function, histology, morphology, biochemical composition, anatomical relationships, and body composition; either independently or integrated in the organism as a whole. These changes can occur through acclimation, habituation, acclimatization or genetic adaptation.

ACCLIMATION

Acclimation refers to the adaptive biological changes that occur in response to a single experimentally induced stress (Eagan, 1963; Folk, 1974) rather than to multiple stresses as occurring in acclimatization. As with acclimatization, changes occurring during the process of growth may also be referred to as developmental acclimation (Timiras, 1972; Frisancho, 1975, 1993).

HABITUATION

Habituation implies a gradual reduction of responses to, or perception of, repeated stimulation (Eagan, 1963; Folk, 1974). By extension, habituation refers to the diminution of normal neural responses, for example, the decrease of sensations such as pain. Such changes can be generalized for the whole organism (general habituation) or can be specific for a given part of the organism (specific habituation). Habituation necessarily depends on learning and conditioning; which enable the organism to transfer an existing response to a new stimulus. A common confusion is that habituation can lead to adaptation. However, the extent to which these physiological responses are important in maintaining homeostasis depends on the severity of environmental stress. For example, with severe cold stress or low oxygen availability, failure to respond physiologically may endanger the well-being and survival of the organism. Likewise, getting used to tolerating high levels of noise implies ignoring the stress, which eventually can lead to deafness. In other words, habituation is a process that in the long run produces negative side effects.

ACCLIMATIZATION

Acclimatization refers to changes occurring within the lifetime of an organism that reduce the strain caused by stressful changes in the natural climate or by complex environmental stresses (Eagan, 1963; Bligh and Johnson, 1973; Folk, 1974). If the adaptive traits are acquired during the growth period of the organism, the process is referred to as either developmental adaptation or developmental acclimatization (Timiras, 1972; Frisancho, 1975, 1993). Studies on acclimatization are done with reference to both major environmental stresses and several related secondary stresses. For example, any difference in the physiological and structural characteristics of subjects prior to and after residence in a tropical environment is interpreted as a result of acclimatization to heat stress. In addition, because tropical climates are also associated with nutritional and disease stresses, individual or population differences in function and structure may also be related to these factors. On the other hand, in studies of acclimation any possible differences are easily attributed to the major stress to which the experimental subject has been exposed in the laboratory. For understanding the basic physiological processes of adaptation, studies on acclimation are certainly better than those of acclimatization. However, since all organisms are never exposed to a single stress, but instead to multiple stresses, a more realistic approach is that of studying acclimatization responses. Thus, studies on both acclimation and acclimatization are essential for understanding the processes whereby the organism adapts to a given environmental condition. This rationale becomes even more important when the aim is to understand the mechanisms whereby humans adapt to a given climatic area, since humans in a given area are not only exposed to diverse stresses but have also modified the nature and intensity of these stresses, as well as created new stresses for themselves and for generations to come.

DEVELOPMENTAL ACCLIMATIZATION

The concept of developmental acclimatization (also referred to as developmental adaptation) is based upon the fact that the organism's plasticity and susceptibility to environmental influence is inversely related to developmental states of the organism, so that the younger the individual the greater is the influence of the environment and the greater the organism's plasticity (Frisancho, 1975, 1993; Frisancho and Schechter, 1997). Hence, variability in physiological traits can be traced to the developmental history of the individual.

ACCOMMODATION AND ADAPTATION

The term accommodation is used to describe responses to environmental stresses that are not wholly successful because, even though they favor survival of the individual, they also result in significant losses in some important functions (Waterlow, 1990). For example, subjects when exposed to a low intake of leucine for three weeks can achieve body leucine balance at the expense of reducing protein synthesis and protein turnover (Young and Marchini, 1990). Since low protein synthesis and protein turnover diminishes the individual's capacity to successfully withstand major stresses, such as infectious diseases (Frenk, 1986), under conditions of low-dietary protein intake achieving body leucine balance represents only a temporary accommodation, which in the long run is not adaptive. In other words, accommodation is a stopgap that ultimately produces negative side effects.

INDIVIDUALS VERSUS POPULATIONS

Whatever the method employed, geographical or experimental research in human adaptation is concerned with populations, not with individuals; although the research itself is based on individuals. There are two related reasons for this.

The first is a practical consideration. Studying all members of a given population, unless its size is small enough, is too difficult to be attempted by any research team. Therefore, according to the objectives of the investigation, the research centers on a sample that is considered representative of the entire population. Based on these studies, the researchers present a picture of the population as a whole, with respect to the problem being investigated.

The second reason is a theoretical one. In the study of adaptation, we usually focus on populations rather than on individuals because it is the population that survives and perpetuates itself. In the investigation of biological evolution, the relevant population is the breeding population because it is a vehicle for the gene pool, which is the means for change and hence evolution. The study of an individual phenomenon is only a means to understand the process. The adaptation of any individual or individuals merely reflects the adaptation that has been achieved by the population of which he is a member.

CULTURAL AND TECHNOLOGICAL ADAPTATION

Cultural adaptation refers to the nonbiological responses of the individual or population to modify or ameliorate an environmental stress. As such, cultural adaptation is an important mechanism that facilitates human biological adaptation (Thomas, 1975; Rappaport, 1976; Moran, 1979). It may be said that cultural adaptation during both contemporary times and in an evolutionary perspective, represents humanity's most important tool. It is through cultural adaptation that humans have been able to survive and colonize far into the zones of extreme environmental conditions. Humans have adapted to cold environments by inventing fire and clothing, building houses, and harnessing new sources of energy. The construction of houses, use of clothing in diverse climates, certain behavioral patterns, and work habits, represent biological and cultural adaptations to climatic stress. The development of medicine, from its primitive manifestations to its high levels in the present era, and the increase of energy production associated with agricultural and industrial revolutions are representative of human cultural adaptation to the physical environment.

Culture and technology have facilitated biological adaptation, yet they have also created, and continue to create, new stressful conditions that require new adaptive responses. A modification of one environmental condition may result in the change of another, and such a change may eventually result in the creation of a new stressful condition. Advances in the medical sciences have successfully reduced infant and adult mortality to the extent that the world population is growing at an explosive rate, and unless world food resources are increased, the twenty-first century will witness a world famine. Western technology, although upgrading living standards, has also created a polluted environment that may become unfit for good health and life. If this process continues unchecked, environmental pollution will eventually become another selective force to which humans must adapt through biological or cultural processes, or else face extinction. Likewise, cultural and technological adaptation has resulted in the rapid increase of energy availability and has decreased energy expenditure; causing a disproportionate increase in the development of degenerative diseases associated with metabolic syndrome. This mismatch between biology and lifestyle threatens our survival as human species (Eaton et al., 2002). Therefore, adaptation to the world of today may be incompatible with survival in the world of tomorrow unless humans learn to adjust their cultural and biological capacities.

GENETIC ADAPTATION AND ADAPTABILITY

Genetic adaptation refers to specific heritable characteristics that favor tolerance and survival of an individual or a population in a particular total environment. A given biological trait is considered genetic when it is unique to the individual or population and when it can be shown that it is acquired through biological inheritance. A genetic adaptation becomes established through the action of natural selection (Livingstone, 1958; Neel, 1962; Mayr, 1997; Neel et al., 1998). Natural selection refers to the mechanisms whereby the genotypes of those individuals showing the greatest adaptation or "fitness" (leaving the most descendents through reduced mortality and increased fertility) will be perpetuated, and those less adapted to the environment will contribute fewer genes to the population gene pool. Natural selection favors the features of an organism that bring it into a more efficient relationship with its environment. Those gene combinations fostering the best-adapted phenotypes will be "selected for," and inferior genotypes will be eliminated.

The selective forces for humans, as for other mammals, include the sum total of factors in the natural environment. All the natural conditions, such as hot and cold climates and oxygen-poor environments, are potential selective forces. Food is a selective force by its own abundance, eliminating those susceptible to obesity and cardiac failures, or by its very scarcity, favoring smaller size and slower growth. In the same manner, disease is a powerful selective agent, favoring in each generation those with better immunity. The natural world is full of forces that make some individuals, and by inference some populations, better adapted than others because no two individuals or populations have the same capacity of adaptation. The maladapted population will tend to have lower fertility and/or higher mortality than that of the adapted population.

The capacity for adaptation (adaptability) to environmental stress varies between populations and even between individuals. The fitness of an individual or population is determined by its total adaptation to the environment – genetic, physiological, and behavioral (or cultural). Fitness, in genetic terms, includes more than just the ability to survive and reproduce in a given environment; it must include the capacity for future survival in future environments. The long-range fitness of a population depends on its genetic stability and variability. The greater the adaptation, the longer the individual or population will survive, and the greater the advantage in leaving progeny resembling the parents. In a fixed environment, all characteristics could be under rigid genetic control with maximum adaptation to the environment. On the other hand, in a changing environment a certain amount of variability is necessary to ensure that the population will survive environmental change. This requirement for variability can be fulfilled either genetically or phenotypically or both. In most populations a compromise exists between the production of a variety of genotypes

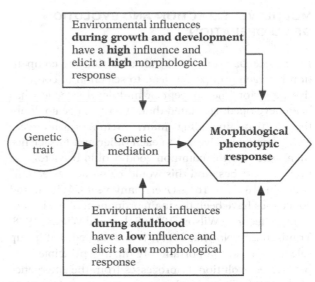

2.2. Schematization of interaction of genetic trait and environment and the phenotypic morphological outcome. Morphological and physiological diversity reflects the responses and adaptations that the organism makes to a particular environment during and after development. From Frisancho (1993).

and individual flexibility. Extinct populations are those which were unable to meet the challenges of new conditions. Thus, contemporary fitness requires both genetic uniformity and genetic variability.

Contemporary adaptation of human beings is both the result of their past and their present adaptability (Lasker, 1969; Frisancho, 1993). It is this capacity to adapt that enables them to be in a dynamic equilibrium in their biological niche. It is the nature of the living organism to be part of an ecosystem whereby it modifies the environment and, in turn, is also affected by such modification. The maintenance of this dynamic equilibrium represents homeostasis; which, in essence, reflects the ability to survive in varying environments (Dubos, 1965; Proser, 1964). The ecosystem is the fundamental biological entity – the living individual satisfying its needs in a dynamic relation to its habitat. In Darwinian terms, the ecosystem is the setting for the struggle for existence, efficiency and survival are the measures of fitness, and natural selection is the process underlying all products (Proser, 1964).

In general, the morphological and functional features reflect the adaptability or capacity of the organism to respond and adapt to a particular environment (Figure 2.2). The effect and responses to a given environmental condition are directly related to the developmental stage of the organism; so that the younger the age, the greater the effect and the greater the flexibility to respond and adapt. Conversely, the later the age and, especially during adulthood, the effect of the environment is less likely to be permanent and the capacity to respond and adapt is also diminished when compared to a developing organism.

MULTILEVEL SELECTION AND EVOLUTION OF CO-OPERATION

Ever since Darwin (1871) indicated that the competition between groups can lead to selection of cooperative behavior, the concept of multilevel selection has been developed. He stated that, "there can be no doubt that a tribe including many members who were always ready to give aid to each other and to sacrifice themselves for the common good, would be victorious over other tribes; and this would be natural selection" (Darwin, 1871, p. 166). Over many years, Wilson and colleagues have been the main proponent of the idea of group selection (Wilson, 1975; Sober and Wilson, 1998; Traulsen and Nowak, 2006). It is assumed that group selection is an important organizing principle that permeates evolutionary processes from the emergence of the first cells to development of nations. According to multilevel selection, groups consist of genetically unrelated individuals, and successful groups attract new individuals, which learn the strategies of others in the same group. A population can be subdivided into groups, and the individuals interact with other members of the group, and depending on their reproductive fitness, individuals can lead to larger groups that split more often. In other words, higher-level or group selection emerges as a by-product of individual reproduction.

A fundamental condition for the success of the group, therefore, must be co-operation among individuals, and thus group selection favors co-operative altruistic behavior and opposes defectors. The fitness of an individual, and the group at large, also depends on the altruistic behavior of nonrelatives. When an altruist gives an alarm call, it benefits not only his or her relatives, but also other unrelated members of the group because a primate troop does not only include relatives. Thus, altruism can be selected if these nonrelated individuals can be counted on to reciprocate the favor when the need arises. A recipient of an altruistic behavior who fails to reciprocate is a cheater. Studies of nonhuman primates indicate that a cheater may gain in the short run by receiving aid without any costs to their own fitness (Strier, 2000). However, reciprocity is necessary for future support in the long run because the cheater's fitness is lower when compared to the individual who reciprocated. In view of the fact that primates constantly need to protect themselves from neighboring communities and predators, one can assume that reciprocal altruism must have been selected for because it enhanced their fitness, not because the animals are conscious of their motives or the reproductive consequences of their behavior. Mathematical models indicate that a single co-operator has a greater fixation probability than a single defector (Traulsen and Nowawk, 2006). Hence, this simple condition ensures that selection favors co-operators and opposes defectors.

The concept of group selection has been a major tenet of behavioral ecology. The major premise of behavioral evolutionary ecology is that genetic and behavioral traits are two distinct expressions of a single evolutionary process (Trivers, 1971; Cronk, 1991; Strier, 2000; Silk, 2001). In behavioral ecology, behaviors are treated like any other biological trait and are potentially subject to natural selection. The processes involved in behavioral evolution are equivalent to those in genetic evolution: natural selection influences the frequency of a trait transmitted from parent to offspring through differential fertility and mortality. In the evolutionary perspective, biological structures have been custom tailored to motivate behaviors that are likely to enhance individual fitness. Therefore, behavioral variants with a high fitness have been favored and these perpetuate the evolutionary origin of fitness-enhancing biological traits. It follows then, that the behavioral traits that enhance fitness also accentuate biological fitness. In other words, a change occurring in one system entails a change in the fitness governing evolution in the other system. Therefore, both genetic and behavioral selection tend to favor those existing variants whose net effect is to increase the average fitness of the individual and population to the prevailing conditions. Studies of primates indicate that they use a diversity of behaviors that increase the likelihood of gaining access to mates and guarantee the survival of their offspring; which, in turn, insures the passing of their traits to the next generation. In this context, behavioral actions that lead to a higher reproductive success will become adaptive and the genes associated with such behavior will be transferred to the next generation faster than those that are less adaptive. Therefore, the differences in fitness between individuals and populations will determine the behavioral pattern of a given primate group. In other words, a specific behavioral strategy that contributes to the survival and reproductive fitness of the individual – and eventually the population – becomes part of the genetic milieu of the species.

In summary, co-operation and altruism evolve by group selection or multilevel selection. Human behavioral ecology rests upon a foundation of evolutionary theory, which include sexual selection, whereby individuals within one sex secure mates and produce offspring at the expense of other individuals within the same sex, which can cause changes in gene frequency across generations that are driven at least in part by interactions between related individuals referred to as kin selection, and be expressed as the sum of an individual's own reproductive success.

CURRENT DIRECTIONS IN ADAPTATION RESEARCH

In the 1970s I postulated the hypothesis of developmental adaptation to explain the enlarged lung volume and enhanced aerobic capacity that characterize the Andean high altitude natives. According to the developmental adaptation hypothesis, "adult biological traits are the result of the effects of the environment and the physiological responses that the organism makes during the developmental state" (Frisancho, 1970, 1975, 1977). This concept is based upon the fact that the organism's plasticity and susceptibility to environmental influence is inversely related to developmental states of the organism, so that the younger the individual the greater is the influence of the environment and the greater the organism's plasticity (Frisancho, 1975, 1977, 1993). Hence, variability in physiological traits can be traced to the developmental history of the individual (Figure 2.2). Currently this concept has been applied to explain the variability in adult behavioral traits such as in learning and crime and delinquency (Yueh-Au Chien, 1994; Sroufe et al., 2005; Kruger et al., 2008), in sensory inputs and auditory spatial processing (Martin and Martin, 2002), in tolerance to surgical intervention (Faury et al., 2003), in variability in oxygen consumption and mitochondrial membrane potential in energy metabolism of rat cortical neurons (Schuchmann et al., 2005), and in variability in increased risk of adult obesity and cardiovascular problems associated with the metabolic syndrome (Barker, 1994). A common denominator of all these studies is that humans and many other organisms are conditioned by experiences during development and a developmental experiences is an important contributor to variability in adult phenotypic behavioral and biological traits.

In this section I will summarize the evidence supporting the applicability of the concept of developmental adaptation to account for the origins of the high risk of the adult metabolic syndrome incorporating information derived from a thrifty gene, thrifty phenotype, and epigenetics. The evolution of the metabolic syndrome is also discussed at length in Chapter 30 of this volume.

DEVELOPMENTAL ADAPTATION AND THE THRIFTY GENOTYPE

Neel and colleagues (Neel, 1962; Neel et al., 1998) attempted to explain the epidemic proportions of diabetes in Native American populations, such as the Pima Native Americans, by postulating the existence of a "thrifty gene" that increased the risk of type II diabetes. According to this hypothesis, the basic defect in diabetes mellitus was a quick insulin trigger. Insulin's main function is to assist in the homeostasis of glucose in the blood. Specifically, when blood glucose levels are too high, the pancreas releases insulin to increase tissue uptake of glucose, thus reducing blood glucose levels. Conversely, when blood glucose levels are low, the organism secretes glucagon and growth hormone, which in turn, induce the release of stored glucose and fatty acids into the blood stream raising serum glucose levels. The insulin response is to activate an uptake of glucose into the muscle cells for storage, and in liver cells it influences the conversion of glucose to fatty acids for storage in fat (adipose) tissue. This response was an asset during times of abundance because it would allow an individual to build-up energy reserves more quickly and thus better survive times of food scarcity. Under these conditions, the thrifty gene was selected to regulate efficient intake and utilization of fuel stores. In other words, during periods of food shortage and famine, those with the thrifty genotype would have a selective advantage because they relied on larger, previously stored energy to maintain homeostasis; whereas those without "thrifty" genotypes would be at a disadvantage and less likely to survive and reproduce. However, under modern conditions of abundant food and sedentary lifestyle, this genotype becomes perversely disadvantageous. With a constant abundance of food, insulin levels remain high, resulting in tissues becoming less sensitive to the effects of insulin. This reduced sensitivity to the effects of insulin results in chronically elevated blood glucose levels type II diabetes and related chronic health problems (e.g., obesity).

A test of the genetic predisposition to type II diabetes involved a comparative study of the Pima Indians of southern Arizona and the Pima Indians of the Sierra Madre mountains of northern Mexico (Knowler et al., 1990; Price et al., 1992). These two groups, which were separated 700 to 1000 years ago, differ in their lifestyle. The Arizona Pima live under conditions of access to a high fat, highly refined diet and low energy expenditure. In contrast, the Mexican Pima still pursue a much more traditional lifestyle and have a diet based on the occasional intake of lamb and poultry, but mainly on beans, corn, and potatoes, grown by traditional, and physically very energy-demanding, techniques. These two groups differ significantly in the frequency of obesity and diabetes. The Arizona Pima adults have a body mass index (BMI) of 33.4 kg/m^2; compared to a BMI of 24.9 kg/m^2 in the Mexican Pima (Ravussin et al., 1994). Likewise, in the Arizona Pima 37% of men and 54% of women were diabetic, while in the Mexican Pima only 2 of 19 women and 1 of 16 men were diabetic (Knowler et al., 1990; Price et al., 1992). In other words, although the Mexican Pima share the "thrifty gene" with Arizona Pima, their increased frequency of obesity and diabetes is more evidence that an abundance of fatty foods and modern sedentary lifestyles are the real culprits. Thus, it is not the presence of a "thrifty gene" alone that results in increased rates of diabetes, but rather the interaction with modern dietary and lifestyle conditions that results in increased rates of the chronic health problems.

In summary, the thrifty genotype hypothesis has been used to explain the epidemic levels of obesity and diabetes among non-Western populations, such as South Pacific Islanders, sub-Saharan Africans, Native Americans in the southwestern United States, Inuit, Australian aborigines, etc. (Eaton et al., 1988; O'Dea, 1991); all of whom were newly introduced to industrialized diets and environments. The fact that the frequency of type II diabetes has recently increased among Europeans that were not subjected to periodic famines cannot be attributed to the action of a so-called "thrifty" gene.

DEVELOPMENTAL ADAPTATION AND THE THRIFTY PHENOTYPE

Recently, Barker and colleagues (Barker, 2007; Hales and Barker, 1992) have reported an inverse relationship between birthweight and the risk of hypertension, cardiovascular disease, and type II diabetes in adulthood when the individual is well nourished postnatally. To account for these observations, Barker and colleagues proposed that adverse effects in utero induce cellular, physiological, and metabolic compensatory responses, such as insulin resistance, high blood plasma levels of fatty acids, which result in energy conservation and reduced somatic growth that enable the fetus to survive undernutrition. This response is referred to as the thrifty phenotype hypothesis (Armitage et al., 2005). These responses that were adaptive under poor prenatal conditions become a problem if food becomes abundant. In this view, thrifty physiological mechanisms are adaptive in nutritionally poor environments, but in rich environments are maladaptive. That is, what was positive under reduced availability of nutrients, particularly during periods of rapid development, becomes negative in rich environments because it facilitates nutrient absorption and hence increases the risk of adult obesity and the suite of risk factors for cardiovascular disease known as the metabolic syndrome (Figure 2.3).

In summary, it appears that nutrition and other environmental factors during prenatal and early postnatal development influence cellular plasticity; thereby altering susceptibility to adult cardiovascular disease, type II diabetes, obesity, and other chronic diseases referred as the adult metabolic syndrome. This hypothesis is supported by the finding that the offspring of women who were starved and became pregnant during

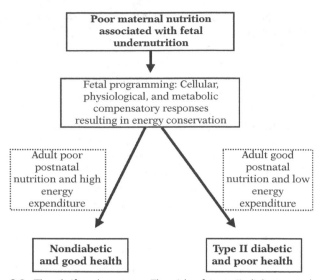

2.3. The thrifty phenotype. The risk of type II diabetes and metabolic syndrome in adulthood is associated with prenatal undernutrition resulting in efficient physiological adaptation that becomes detrimental when food is abundant and energy expenditure is low.

the Dutch famine of World War II were found to have impaired glucose tolerance and increased adiposity in adulthood (Stein et al., 1975, 2007).

DEVELOPMENTAL ADAPTATION AND EPIGENETICS

Epigenetics refers to the transmission of phenotypic traits from one generation to the next that do not depend on differences in DNA sequence (Waddington, 1952; Jablonka, 2004; Holliday, 2006). During the last two decades, there has been an accumulation of observations indicating that the expression of DNA traits can be affected by environmental factors acting during development. Specifically, experimental studies showed that identical twin mice differ in the color of fur; one has brown fur and will grow up to be lean and healthy, while the other has yellow fur and becomes obese and prone to cardiovascular disease. The different phenotypes are due to the addition of a methyl group (-CH$_3$); which is referred to as methylation.

Methylation

Methylation refers to the altering of the genetic environment through the addition of a methyl group (-CH$_3$) to the fifth position of cytosine, which is largely confined to CpG dinucleotides. This addition, by modifying the CpG islands, prevents signaling molecules from reaching the promoter site to turn the gene on and prevent the expression of the dark coat color. In other words, the additional methyl group attaches to and shuts off the gene that controls dark fur color and allows the yellow color to be expressed. Thus, the process of methylation works as a kind epigenome that dictates which genes in the genome are turned on and which are not. This process can differ even between identical twins.

Recently, experimental studies indicate that bisphenol A (BPA) can alter gene expression and affect adult phenotype by modifying CpG methylation at critical epigenetically labile genomic regions (Waterland and Jirtle, 2004). Bisphenol A is used in the production of polycarbonate and plastic containers and in the organism acts like the body's own hormones. Thus, there is concern that long-term exposure to BPA may induce chronic toxicity in humans (vom Saal and Hughes, 2005). Fortunately, the effects of methylation are not permanent but reversible, as shown by the fact that the yellow agouti (Avy), whose diet was supplemented with folic acid, vitamin B_{12}, choline, betaine, and zinc, counteracted the DNA methylation and changed coat color from yellow to dark brown (Dolinoy and Jirtle, 2008), which is associated with a low risk of cardiovascular disease.

Transgenerational epigenetic effects

It has been suggested that the epigenetic modifications brought about by parental conditions may be expressed even in grandchildren. Extensive records of a population in Overkalix cohorts, northern Sweden, found an association between grandparental prepubertal slow growth periods (SGP) or rapid growth periods (RGP), and parental periods of low or high food availability, with grandchildren's mortality and disease risk (Kaati et al., 2007). If the SGP of the grandparent was during a period of high food availability, then the male grandchild had reduced longevity but an increased mortality. The extent to which these associations represent multigenerational epigenetic effects is unwarranted, in part because ruling out genetic and societal confounders, and in the absence of molecular evidence, is extremely difficult. Hence, future research must be focused on long-term transgenerational studies whereby many birth cohorts are studied using intensive prenatal and perinatal genotyping across generations. Only then can variability in the expression of phenotypic traits can be attributed to epigenetic changes.

In summary, epigenetic effects exist that are not necessarily adaptive, and in many of these cases, the inherited phenotype is actually detrimental to the organism. Environmental exposure to nutritional, chemical, and physical factors can alter gene expression and affect adult phenotype: a process known as epigenetics. In all of these studies, the extent of DNA methylation depends on and is inversely related to the developmental state of the organism; so that the younger the individual, the greater the epigenetic marks, including CpG methylation. Despite the great interest in molecular genetics, there is scant incontrovertible evidence indicating epigenetic effects in humans. Considering society's increased concern about environmental pollutants, this area of research should be a good direction for human biologists.

OVERVIEW

The term *adaptation* encompasses the physiological, cultural, and genetic adaptations that permit individuals and populations to adjust to the environment in which they live. These adjustments are complex, and the concept of adaptation cannot be reduced to a simple rigid definition without oversimplification. The functional approach in using the adaptation concept permits its application to all levels of biological organization from unicellular to multicellular organisms, from early embryonic to adult stages, and from individuals to populations. In this context, human biological responses to environmental stress can be considered as part of a continuous process whereby past adaptations are modified and developed to permit the organism to function and maintain equilibrium within the environment to which it is daily exposed.

The mechanisms for attaining full functional adaptation include acclimation, acclimatization, habituation, and accommodation. The role played by each of these processes depends on the nature of the stress or stresses, the organ system involved, and the developmental stage of the organism. It is emphasized that the goal of the organism's responses to a given stress is to maintain homeostasis within an acceptable normal range with itself and with respect to other organisms and the environment (as schematized in Figure 2.1). Such adaptations are usually reversible, but the reversibility depends on the developmental stage of the organism at which the adaptive response occurs and the nature of the environmental stress. This characteristic allows organisms to adapt to a wide range of environmental conditions. Furthermore, an adaptation is always a compromise between positive and negative effects. Every adaptation involves a cost. The process of adaptation is always positively beneficial; without which the organism would be worse off, however the organism has to pay a price for the benefit. The benefit derived from a given response depends on the circumstances and the conditions where it occurs. As recently pointed out (Young and Marchini, 1990), every adaptation involves a choice. For example, a man has 6 hours in which to walk 11 km. If he walks slowly, he saves energy expenditure, and therefore it may be adaptive if the energy resources are limited; however he has no time left to do anything else. On the other hand, if he walks fast, he saves time at the cost of using more energy. Thus, the adaptive importance of a given type of response depends on the conditions.

The concept of developmental adaptation has become a major focus for studying the origins of human diversity (Figure 2.2). The applicability of this research strategy is based upon the premise that human biological responses to environmental stress represent a continuous process whereby past adaptations are modified and developed to permit the organism to function and maintain equilibrium within the environment to which it is daily exposed. From the studies of the thrifty genotype and thrifty phenotype, and their relationship to the etiology of metabolic syndrome, it is evident that what was positive under reduced availability of nutrients, particularly during periods of rapid prenatal development, becomes negative in rich environments. Research in epigenetics may provide the bridge between the thrifty genotype and thrifty phenotype to unravel the interrelations of how the impact of early diet helps how the organism adapts to a given environmental condition that differs in nutritional resources; resulting in the diverse phenotypic expression of physiology, body size, and health risk of contemporary and past populations. The study of individuals exposed to stressful conditions in natural and laboratory environments is one of the most important approaches for understanding the mechanisms whereby human populations adapt to a given environment. Knowledge of human adaptation is basic in our endeavors to understand past and present human variation in morphology and physiological performance. The insights we have gained during the last decade have stimulated new approaches to study human adaptation, not only for understanding behavior, but of how ecology shapes function both in the present and in the past, and for understanding variability in the immune system, thermo-regulation, coping with limited and excessive amounts of foods, low oxygen pressure, and high and low solar radiation.

DISCUSSION POINTS

1. What are the four processes of functional adaptation?
2. Why does the organism have to respond to environmental stimuli? Give examples.
3. When does a given functional adaptation become a genetic adaptation?
4. Discuss the role of multiple-level selection in the expression of biological and behavioral traits.
5. Compare accommodation and adaptation. Give specific examples.
6. Discuss how technological and cultural adaptations have created new environmental stresses. Give examples.
7. Compare developmental and adult adaptation. Which is more likely to be reversible and why?
8. Discuss the applicability of the concept of developmental adaptation to the hypothesis of thrifty genotype and thrifty phenotype that account for the increased frequency of the adult metabolic syndrome among native and nonnative populations.
9. Discuss the relationship of the concept of developmental adaptation to the field of epigenetics.

REFERENCES

Armitage, J. A., Taylor, P. D. and Poston, L. (2005). Experimental models of developmental programming: consequences of exposure to an energy rich diet during development. *Journal of Physiology*, **565**, 3–8.

Baker, P. T. (1966). Human biological variation as an adaptive response to the environment. *Eugenics Quarterly*, **13**, 81–91.

Barker, D. J. P. (1994). *Mothers, Babies, and Disease in Later Life*. London: BMJ Publishing.

Barker, D. J. P. (2007). The origins of the Developmental Origins Theory. *Journal of Internal Medicine*, **261**, 412–417.

Bligh, J. and Johnson, K. G. (1973). Glossary of terms for thermal physiology. *Journal of Applied Physiology*, **35**, 941–961.

Cronk, L. (1991). Human behavioral ecology. *Annual Review of Anthropology*, **20**, 25–53.

Darwin, C. (1871). *The Descent of Man and Selection in Relation to Sex*, 2nd edn. London: Murray.

Dolinoy, D. C. and Jirtle, R. L. (2008). Environmental epigenomics in human health and disease. *Environmental and Molecular Mutagenesis*, **49**, 4–8.

Dubos, R. (1965). *Man Adapting*. New Haven, CT: Yale University Press.

Eagan, C. J. (1963). Introduction and terminology. *Federation Proceedings*, **22**, 930–932.

Eaton, S. B., Konner, M. and Shostak, M. (1988). Stone agers in the fast lane: chronic degenerative diseases in evolutionary perspective. *American Journal of Medicine*, **84**, 739–749.

Eaton, S. B., Strassman, B. I., Nesse, R. M., et al. (2002). Evolutionary health promotion. *Preventative Medicine*, **34**, 109–118.

Faury, G., Pezet, M., Knutsen, R. H., et al. (2003). Developmental adaptation of the mouse cardiovascular system to elastin haplo insufficiency. *Journal of Clinical Investigation*, **112**, 1419–1428.

Folk, G. E. Jr (1974). *Textbook of Environmental Physiology*. Philadelphia: Lea and Febiger.

Frenk, S. (1986). Metabolic adaptation in protein-energy malnutrition. *Journal of the American College of Nutrition*, **5**, 371–381.

Frisancho, A. R. (1970). Developmental responses to high altitude hypoxia. *American Journal of Physical Anthropology*, **32**, 401–407.

Frisancho, A. R. (1975). Functional adaptation to high altitude hypoxia. *Science*, **187**, 313–319.

Frisancho, A. R. (1977). Developmental adaptation to high altitude hypoxia. *International Journal of Biometeorology*, **21**, 135–146.

Frisancho, A. R. (1993). *Human Adaptation and Accommodation*. Ann Arbor, MI: University of Michigan Press.

Frisancho, A. R. and Schechter, D. E. (1997). Adaptation. In *History of Physical Anthropology: an Encyclopedia*, vol. 1. Garland, TX: Garland Press, pp. 6–12.

Hales, C. N. and Barker, D. J. (1992). Type 2 (non-insulin-dependent) diabetes mellitus: the thrifty phenotype hypothesis. *Diabetologia*, **35**, 595–601.

Holliday, R. (2006). Epigenetics: a historical overview. *Epigenetics*, **1**, 76–80.

Jablonka, E. (2004). Epigenetic epidemiology. *International Journal of Epidemiology*, **33**, 929–935.

Kaati, G., Bygren, L. O., Pembrey, M., et al. (2007). Transgenerational response to nutrition, early life circumstances and longevity. *European Journal of Human Genetics*, **15**, 784–790.

Knowler, W. C., Pettitt, D. J., Saad, M. F., et al. (1990). Diabetes mellitus in the Pima Indians: Incident, risk factors and pathogenesis. *Diabetes/Metabolism Research and Reviews*, **6**, 1–27.

Kruger, D. J., Reisch, T. and Zimmerman, M. A. (2008). Time perspective as a mechanism for functional developmental adaptation. *Journal of Social, Evolutionary, and Cultural Psychology*, **2**, 1–22.

Lasker, G. W. (1969). Human biological adaptability. *Science*, **166**, 1480–1486.

Lewontin, R. C. (1957). The adaptations of populations to varying environments. In *Cold Spring Harbor Symposia on Quantitative Biology*, vol. 22. Boston, MA: Cold Spring Harbor Laboratory of Quantitative Biology.

Livingstone, F. B. (1958). Anthropological implications of sickle-cell gene distribution in West Africa. *American Anthropologist*, **60**, 533–562.

Martin, P. and Martin, M. (2002). Proximal and distal influences on development: The model of developmental adaptation. *Developmental Review*, **22**, 78–96.

Mayr, E. (1963). *Animal Species and Evolution*. Cambridge, MA: The Belknap Press of Harvard University Press.

Mayr, E. (1997). The objects of selection. *Proceedings of the National Academy of Sciences of the United States of America*, **94**, 2091–2094.

Mazess, R. B. (1973). Biological adaptation: aptitudes and acclimatization. In *Biosocial Interrelations in Population Adaptation*, E. S. Watts, F. E. Johnson and G. W. Lasker (eds). The Hague: Mouton, pp. 9–18.

Moran, E. F. (1979). *Human Adaptability: an Introduction to Ecological Anthropology*. North Scituate, MA: Duxbury Press.

Neel, J. V. (1962). Diabetes mellitus: A "thrifty" genotype rendered detrimental by "progress"? *American Journal of Human Genetics*, **14**, 353–362.

Neel, J. V., Weder, A. B. and Julius, S. (1998). Type II diabetes, essential hypertension, and obesity as "syndromes of impaired genetic homeostasis": the "thrifty genotype" hypothesis enters the twenty-first century. *Perspectives in Biology and Medicine*, **42**, 44–74.

O'Dea, K. (1991). Traditional diet and food preferences of Australian aboriginal hunter-gatherers. *Philosophical Transactions of the Royal Society of London. Series B*, **334**, 233–240.

Price, R. A., Lunetta, K., Ness, R., et al. (1992). Obesity in Pima Indians. Distribution characteristics and possible thresholds for genetic studies. *International Journal of Obesity and Related Metabolic Disorders*, **16**, 851–857.

Proser, C. L. (1964). Perspectives of adaptation: theoretical aspects. In *Handbook of Physiology*, vol. 4, D. B. Dill, E. F. Adolph and C. G. Wilber (eds). Washington, DC: American Physiological Society, pp. 11–26.

Rappaport, R. A. (1976). Maladaptation in social systems. In *Evolution in Social Systems*, J. Friedman and M. Rowlands (eds). London: Gerald Duckworth, pp. 49–71.

Ravussin, E., Valencia, M. E., Esparza, J., et al. (1994). Effects of a traditional lifestyle on obesity in Pima Indians. *Diabetes Care*, **17**, 1067–1074.

Schuchmann, S., Buchheim, K., Heinemann, U., et al. (2005). Oxygen consumption and mitochondrial membrane potential indicate developmental adaptation in energy metabolism of rat cortical neurons. *European Journal of Neuroscience*, **21**, 2721–2732.

Silk, J. (2001). Ties that bind: the role of kinship in primate societies. In *New Directions in Anthropological Kinship*, L. Stone (ed.). Lanham, MD: Rowman and Littlefield Publishers, pp. 71–92.

Sober, E. and Wilson, D. S. (1998). *Unto Others: the Evolution and Psychology of Unselfish Behavior*. Cambridge, MA: Harvard University Press.

Sroufe, L. A., Egeland, B., Carlson, E., et al. (2005). *The Development of the Person: the Minnesota Study of Risk and Adaptation from Birth to Adulthood*. New York: Guilford Press.

Stein, Z., Susser, M., Saenger, G., et al. (1975). *Famine and Human Development. The Dutch Hunger Winter of 1944–45*. New York: Oxford University Press.

Stein, A. D., Kahn, H. S., Rundle, A., et al. (2007). Anthropometric measures in middle age after exposure to famine during gestation: Evidence from the Dutch famine. *American Journal of Clinical Nutrition*, **85**, 869–876.

Strier, K. B. (2000). *Primate Behavioral Ecology*. Boston: Allyn and Bacon.

Thomas, R. B. (1975). The ecology of work. In *Physiological Anthropology*, A. Damon (ed.). New York: Oxford University Press, pp. 59–79.

Timiras, P. S. (1972). *Developmental Physiology and Aging*. New York: Macmillan.

Traulsen, A. and Nowak, M. A. (2006). Evolution of cooperation by multilevel selection. *Proceedings of the National Academy of Sciences of the United States of America*, **103**, 10952–10955.

Trivers, R. L. (1971). The evolution of reciprocal altruism. *Quarterly Review of Biology*, **46**, 35–57.

Vom Saal, F. S. and Hughes, C. (2005). An extensive new literature concerning low-dose effects of bisphenol A shows the need for a new risk assessment. *Environmental Health Perspectives*, **113**, 926–933.

Waddington, C. H. (1952). *Epigenetics of Birds*. Cambridge: Cambridge University Press.

Waterland, R. and Jirtle, R. (2004). Early nutrition, epigenetic changes at transposons and imprinted genes, and enhanced susceptibility to adult chronic diseases. *Nutrition*, **20**, 63–68.

Waterlow, J. C. (1990). Nutritional adaptation in man: General introduction and concepts. *American Journal of Clinical Nutrition*, **51**, 259–263.

Wilson, D. S. (1975). A theory of group selection. *Proceedings of the National Academy of Sciences of the United States of America*, **72**, 143–146.

Young, V. R. and Marchini, J. S. (1990). Mechanisms and nutritional significance of metabolic responses to altered intakes of protein and amino acids, with reference to nutritional adaptation in humans. *American Journal of Clinical Nutrition*, **51**, 270–289.

Yueh-Au Chien, C. (1994). *Developmental Adaptation of Social Learning Theory: the Etiology of Crime and Delinquency*. Denver, CO: University of Colorado Press.

3 History of the Study of Human Biology

Michael A. Little

INTRODUCTION

The field of human biology is broad based but with its principal origins in studies of variation in living populations within physical or biological anthropology. Today, human biology incorporates a majority of scientists trained in anthropology, but also counts among its members scientists trained in other specialties of the human sciences. During the late nineteenth and early twentieth centuries, our stem science – physical anthropology – was oriented toward skeletal studies, gross anatomy, and human variation as represented by race. Skeletal studies included those of both living and prehistoric populations. Anatomically oriented studies of the skeleton and of the living were focused on structure and origins, but with less interest in function and evolutionary causality. In his *Manual of Physical Anthropology*, Juan Comas (1960) presented an excellent overview of the history of physical anthropology from its earliest origins, including the origins of and connections with natural history, racial classification, craniometry, prehistory and paleoanthropology, and evolution. In later chapters, he reviewed the histories of growth studies, somatology, constitutional typology, craniology, osteology, and racial classification.

The beginning of an integrated human biology in the United States dates back to the late 1920s when Darwin's ideas of population variation, adaptation, selection, and evolution began to be reconceptualized by a number of leading scientists. At that time, evolutionary theory, and concepts from behavioral sciences, demography, genetics, and child growth began to be consolidated into a science of human biology. In later decades, body composition, physiology, nutrition, and ecology were added to the mix leading to an even greater understanding of the variations in, the evolution of, and the characteristics of humans in all of their complex biobehavioral states.

A history of anything is a reflection of *traditions*, *connections*, and *innovations*. These can be found as well in a history of human biology. In the narrative that follows, the development of the science of human biology will be traced through the contributions of key authors, their ideas, and the role of key institutions in its history.

RACE AND TYPOLOGY

The tangled roots of interests in human biology and human variation lie deep in antiquity. In the early modern era, Europeans became acutely aware of human population variation during the Age of Exploration and the opening of new regions of the world in the fifteenth, sixteenth, and seventeenth centuries (Eiseley, 1958). Newly discovered populations from the Americas, Africa, the Pacific, and Asia, were classified, along with Europeans, into broad racial groups according to physical characteristics. Such racially based physical characteristics were erroneously thought to be tightly bound to mental, emotional, intellectual, and cultural attributes, as well. Some races were identified as clearly inferior to others on a typological scale – an extension of *The Scale of Being* – from primitive to advanced. Ideas of fixed, unchanging racial categories were tightly bound to Judeo-Christian religious beliefs associated with, among other things, the creation or origin of humans. Some believed that humans originated from a single creation, *monogenism*, where the more primitive races had then degenerated from the original people living in the Garden of Eden. Another belief was in *polygenism*, where God had created several independent peoples, some who were elevated to civilization and others doomed at the outset to perpetual savagery. In both of these beliefs of human creation, Western, civilized populations were held, with a few exceptions, to be superior to non-Western races from Asia, Africa, the New World, and the Pacific.

Within Western societies, the two sexes were considered unequal – men were believed to be superior to women, by virtue of their very nature, and women were often viewed as incapable of rational thought (Stocking, 1987). Women were more emotional than men and tied more closely to nature through their role in reproduction

Human Evolutionary Biology, ed. Michael P. Muehlenbein. Published by Cambridge University Press. © Cambridge University Press 2010.

and childbirth. The upper socioeconomic classes were believed to be innately superior to the lower classes, because class differences were thought to have a hereditary basis. Hence, aristocratic males ("good breeding") were alleged to be at the pinnacle of God's creations. There *were* some who questioned the dominance of nature over nurture, however. For example, in the last chapter of the popularly known *Voyage of the Beagle* (Darwin, 1839), a thoughtful Charles Darwin noted, in the context of slavery: "... if the misery of the poor be caused not by the laws of nature, but by our institutions, great is our sin ..." (Darwin, 1839, p. 433). Questions about the contributions of nature (heredity) and nurture (environment) to human diversity persist and are debated up to the present.

Eighteenth-century racial classifications were established by Carolus Linnaeus (1707–1778), Georges Louis Leclerc, the Comte de Buffon (1707–1788), Georges Cuvier (1769–1832), and, of course, Johann Friedrich Blumenbach (1752–1840) (Gould, 1996; Molnar, 2002). Classifications ranged from Linnaeus's Asian, European, African, and American groups to Blumenbach's Caucasian, Mongolian, Ethiopian, American, and Malayan.

Twentieth-century classifications of human populations have been devised in numerous combinations and have been based on a variety of criteria. Earnest A. Hooton (1931), the Harvard professor who trained most of the modern generation of physical anthropologists before World War II, identified "composite races," in addition to the "primary races and subraces." The composite races were formed by admixture of primary races. Carleton Coon was one of Hooton's early students who had written a book classifying *The Races of Europe* (Coon, 1939). Several years later, Coon et al. (1950), in a mid-twentieth century approach, presented a six-fold geographical classification and then divided these major races into thirty subpopulations, whose characteristics were based on external physical and skull attributes. In that same year, Boyd (1950) identified six races according to their blood group genetics. About a decade later, Garn (1961), who was a coauthor of the Coon et al. (1950) book, refined these classifications and identified geographical, local, and micro-races, in a hierarchy of populations and subpopulations. These three works were different from previous efforts at classification in that they attempted to apply contemporary evolutionary, genetic, and ecological principles to the identification of racial (population) variation around the world. They were transitional in the sense that they applied modern theory to an outdated typological system in which boundaries between populations were fixed. There has been considerable controversy in scientific writings over the intervening years about the utility and social impact of racial classification and the validity of the concept itself. Works by Shipman (1994), Marks (1995), Brace (2005)

and the first chapter by Mielke et al. (2006) provide histories of these controversies about race. Fundamentally, all classifications of this nature are *essentialist* in character, that is, they focus on fixed *types* that obscure the variation that contributes to human population diversity around the globe. It was the rejection of typological thinking (fixed species and races) that enabled Darwin to conceptualize ideas of variation, adaptation, and natural selection (Mayr, 1972), and it was this same transition in thinking about human populations that moved human biologists toward a truly scientific approach in this field.

FRANZ BOAS AND HIS CONTRIBUTIONS

Although the late nineteenth and early twentieth centuries were characterized by beliefs in fixed racial types, accompanied by beliefs in the superiority of some racial groups over others (racism or racialism), there was a contrary opinion in the form of Franz Boas (1858–1942). Boas, often referred to as the founder of American anthropology and the four-field approach, made remarkable contributions to dispelling the myth of fixed or pure races and the importance of the environment in structuring the character of human populations (Figure 3.1). These interests in human plasticity in the context of race were almost certainly based on his research on child and adolescent growth and development (Tanner, 1959, 1981). Hence, he can be identified also as one of the founders of human biology in the United States because of his contributions to: (1) debunking the idea of fixed races; (2) establishing a migration research design that

3.1. Franz Boas (1858–1942) in 1906. Photograph with permission from the American Philosophical Society.

continues to be used up to the present; (3) incorporating the social and material environment as influencing human biology (plasticity); and (4) making numerous discoveries about the patterns of growth in children and adolescents.

The migrant study, designed to test the idea of races as static types, was initiated in 1908 with modest funding from the US Immigration Commission. The pilot study began in June 1908 with studies of height and cephalic index of Eastern European Jewish boys in New York City Schools (Stocking, 1974). The results of the pilot study and the more extensive study of Bohemians, Sicilians, Neapolitans, Polish, Hungarians, and Scots established differences between those born to parents before and those born to parents after they migrated to the United States (Tanner, 1959). Head form or cephalic index differed between the two groups, refuting the idea of fixed races and demonstrating the influence of the environment on human variation in physical characteristics. The completed study, published by Boas (1912), was accompanied by the publication several years later of the raw data of more than 18 000 subjects who were measured in the migration study (Boas, 1928). The reanalysis of these data stimulated a new controversy (Sparks and Jantz, 2002; Gravlee et al., 2003) over Boas's analyses and interpretations and whether he really demonstrated plasticity in these migrant populations.

Relethford (2004), in exploring this controversy, identified three ways that craniometric variation can change over time: (1) developmental plasticity through environmental change; (2) long-term changes through natural selection; and (3) within-group and among-groups variation by gene flow. Each of these has been shown to operate, but what Relethford (2004) noted was that the debate about Boas's study centered on the relative importance of these three causes of craniometric change. Based on the two major studies, the question that Relethford raised, "... is whether developmental plasticity has a significant effect on craniometric variation" (Relethford, 2004, p. 380). Relethford approached this question by inspecting the changes in craniometric variation among the groups Boas studied. Three of the seven European ethnic groups showed no statistical difference between US-born and European-born migrants (Hungarians, Polish, and Scots). The remaining four groups did show statistically significant differences, but these were relatively slight differences. Based on this, Relethford (2004) suggested that these data *do* demonstrate developmental plasticity, but this plasticity does not obscure the underlying genetic differences that separate the ethnic groups. In other words, both genetic characters and developmental plasticity contribute to the variation, but the genetic contribution to the total variation, in this case, is the stronger of the two. Relethford's (2004) conclusion was anticipated by Boas in a paper published years after the original, where Boas

acknowledged the plasticity in the immigration study when he stated: "These changes do not obliterate the differences between genetic types [characters] but they show that the type as we see it contains elements that are not genetic but an expression of the influence of the environment" (Boas, 1936, p. 523).

Boas's growth studies date back to 1888 when he took an academic position at Clark University in Worcester, Massachusetts. His first important discovery was based on data gathered by the growth studies pioneer and Harvard professor, Henry Bowditch (Tanner, 1959). Boas discovered that the asymmetrical distributions in Bowditch's data on height during adolescence could be explained by the individual variations in growth rates, which he called "tempo of growth." This is the first indication of Boas's sensitivity to the importance of longitudinal growth data in uncovering subtle changes in growth during childhood and adolescence. In 1891, he initiated a longitudinal survey of Worcester school children that confirmed his sense of the value of longitudinal growth data. Between 1892 and 1941, Boas published numerous papers on growth in *Science*, *Human Biology*, and other journals in which he made other significant discoveries, each a reflection of his remarkable knowledge of statistics and his genius. Over the years, he: (1) produced the first National Growth Standards or norms; (2) introduced the concept of *physiological or developmental age* and observations that males were behind females as early as five years of age; (3) observed that working class and poor children from large sibships tend to be smaller on average than those from small sibships; (4) established relationships between "age of peak velocity" during adolescence and other measures of size during adolescence and adulthood; and (5) observed that children from the Horace Mann School of Columbia University had become larger between 1909 and 1935 (now known as the *secular trend in growth*) (Tanner, 1959, 1981).

PRE-WORLD WAR II AND HUMAN POPULATION BIOLOGY

In addition to Boas's contributions to a developing field of human biology through his growth studies, there are several other lines of continuity from pre-World War II to the present. In 1929, Raymond Pearl (1879–1940) founded the journal *Human Biology* and served as editor until his death in 1940 (Crawford, 2004). Pearl was an accomplished population biologist who was Professor of Biometry and Vital Statistics in the School of Hygiene and Public at Health at Johns Hopkins University and who had broad interests in genetics, fertility, evolution, nutrition, disease, duration of life, senescence, and physical anthropology (Figure 3.2). According to Kingsland (1984, p. 8), "Pearl considered himself to be first

3.2. Raymond Pearl (1879–1940) sometime in the late 1920s or early 1930s. Photograph with permission from the American Philosophical Society.

and foremost a human biologist"; that is, he was totally committed to a human population biology that was quite akin to some of the holistic and integrated science practiced today.

The journal, *Human Biology*, published a variety of papers in, among other topics, genetics, growth, health, human population studies, mathematical and statistical modeling, and human evolution. Goldstein (1940), in a survey of the first decade of *Human Biology* (*HB*) and two decades (1920s and 1930s) of the *American Journal of Physical Anthropology* (*AJPA*), suggested that the journal *HB* published articles more in the realm of "biological anthropology" whereas the *AJPA* was more prone to publish in "anatomical anthropology." These patterns of topical publication imply that the earliest development of a professional human biology "identity" began about this time. As Goldstein (1940) reported, there were strong connections between Pearl and his journal and physical anthropology – connections with the journal that have continued to the present. Many early papers in *Human Biology* were published on topics related to physical anthropology (Lasker, 1989). Raymond Pearl and Franz Boas were acquainted at that time because Pearl was active in a number of professional societies in which Boas was a member, including the American Association of Physical Anthropologists (AAPA). In fact Pearl was well known among physical anthropologists of this time as evidenced by his election as the third president of the AAPA from 1934 to 1936. Moreover, both Pearl and Boas were members of the National Academy of Sciences, both were sophisticated biostatisticians, and Boas published several papers on growth in the journal *HB* while Pearl

was editor in the 1930s. It is probably the case that Aleš Hrdlička, one of the founders and first president of the AAPA, and Pearl did not share the same scientific philosophy because of Hrdlička's strong dislike and avoidance of statistics and Pearl's commitment to their use (Lasker, 1989). This added an additional dimension to the differences between the *AJPA* and *HB*, since Hrdlička was editor of the former and Pearl was editor of the latter.

Perhaps the most extensive pre-war research in human biology was that of human growth conducted at several centers in the United States. Some of these longitudinal growth studies (sequential measurements over time of the same individuals) were the Fels Longitudinal Study (Yellow Springs Ohio), the Bolton-Brush Study at Western Reserve University (Cleveland, Ohio), the Berkeley Growth Study at the Institute of Human Development, University of California (Berkeley), the Child Research Council Study at the University of Colorado (Denver), and the Harvard School of Public Health Growth Study (Boston) (Roche, 1992, pp. 1–2). Each of these was founded from the 1920s to the early 1930s principally to determine the effects of the Great Depression and to assess the means to help impoverished children. As Tanner noted: these longitudinal studies were part of "... a powerful child welfare movement [that] arose in the 1920s and provided the soil for a crop of longitudinal studies, whose harvesting shaped the whole pattern of human auxology in the years 1935–1955" (Tanner, 1981, p. 299). Here he uses the term *auxology*, which refers to the study of human physical growth, which he (James Tanner) pioneered in the 1950s and 1960s in the United Kingdom (Figure 3.3). It is not clear to what extent Boas's work

3.3. James M. Tanner (1920–) at a conference in 1982. Photograph courtesy of Barbara Garn.

stimulated these longitudinal growth studies, but principle investigators must have known the value of longitudinal series from Boas's work before the turn of the twentieth century and later papers in the 1930s. For example, when Frank Shuttleworth (1889–1958) analyzed the Harvard Growth Study data, he employed some of the same procedures that Boas employed in analyzing data of adolescent girls (Tanner, 1981, p. 310).

Boas's influence was also seen in students from Harvard and elsewhere who used his migration model in their own research during the 1930s and 1940s (Little and Leslie, 1993). For example, Harry Shapiro and Frederick Hulse (1939) studied Japanese migrants to Hawai'i, Marcus Goldstein (1943) measured Mexican-Americans, and Gabriel Lasker (1946) studied immigrant and American-born Chinese. The migration-research design was also used in a great deal of human biology research up to the end of the twentieth century.

THE WORLD WAR II YEARS

The enormity of World War II brought a halt to a great deal of academic research in the United States and elsewhere, but stimulated research that could be directly applied to the war effort. Some of this research carried over to the post-war years and not only led to additional academic research, but even contributed to changes in thinking about the causes of human variation. Much of the wartime research centered on the maintenance and health of US troops in stressful environments and ameliorating the stress of warfare and hunger in the civilian populations in Europe and Asia. During the war, US military personnel were stationed and fighting in Europe, East Asia, Southeast Asia, North Africa, the Pacific, and Alaska and the Aleutian Islands. Hence, they were expected to be able to work strenuously at high levels while at the same time being exposed to climatic variation that ranged from Arctic and temperate zone winter cold to tropical and temperate summer heat. In addition to stresses from dietary restriction and starvation and from climatic extremes, there was interest in the interaction between diet and climate, particularly in the context of providing enough food calories for troops fighting in cold or hot geographic zones (Mitchell and Edman, 1951). Two of the most prominent wartime studies will be surveyed briefly: the semi-starvation study conducted at the University of Minnesota and the desert physiology study conducted by the Rochester Desert Unit whose research was conducted at several places in the United States.

The Minnesota semi-starvation study was initiated by Ancel Keys (1904–2004), a stress physiologist best known for having developed K-rations during the early part of the war, and several colleagues including Josef Brožek (1913–2004), a psychologist who later distinguished himself in studies of body composition. The study began in 1944 with 36 conscientious objectors who volunteered for the full-year study that consisted of an equilibration period on a normal diet, 3 months of semi-starvation (1600 kcal/day), and a period of refeeding. In addition, the volunteers were expected to walk 22 miles per week to increase their energy expenditure. The study was intensive, with almost daily biochemical, physiological, psychological, and body composition measurements taken. The results of the study were published in two massive volumes that stand as state-of-the-art research even today, particularly since it probably would not be permitted to conduct such a high-risk project during present times (Keys et al., 1950).

The Rochester Desert Unit was charged with the research task of determining water and food requirements, sweat rates, energy balance, heat tolerance, work capacity, and the probability of survival while living under the hot-dry conditions of desert environments. Much of the desert research was conducted on military personnel on maneuvers in southern California. Other studies were conducted of men on life rafts without water, and some limited tests were conducted under hot-wet conditions in Florida to simulate troops in tropical forests. In the preface to the volume that reported this research, the authors identify this work as a part of the "... recently developing field of environmental physiology" (Adolph and Associates, 1947, p. vii). What was not known then was that this work was to stand as the first major work in climatic stress physiology, and one that would serve as the basis for heat-stress studies both in environmental physiology and human biology.

In addition to the pioneering research in nutritional and environmental physiology represented by these two wartime studies, a relatively new area of human variation was being developed that would be explored by anthropologists and human biologists interested in growth, environmental stress, exercise physiology, and nutrition – *body composition*. These early studies introduced hard-tissue anthropologists (bone and teeth) to the importance of soft tissue (muscle and adipose tissue) in human function, human growth, and in studies of human adaptation to the environment.

POST-WAR UNITED STATES AND UNITED KINGDOM

There were parallel transformations of physical anthropology in the years directly following World War II in both the United States and in the United Kingdom. In the United States, Sherwood L. Washburn (1911–2000) successfully promoted a scientific agenda of problem solving, application of evolutionary theory, and understanding of human variation rather than racial

3.4. Sherwood L. Washburn (1911–2000) at the 1952 Wenner-Gren Foundation International Symposium of Anthropology. Photograph courtesy of the Wenner-Gren Foundation for Anthropological Research.

typology and simply descriptive anthropometric survey (Washburn, 1951). Washburn, who had been trained at Harvard, among many others, under Earnest A. Hooton (1887–1954), rejected many of the traditional ideas of his mentor and began an active program to bring his new ideas to other anthropologists, especially younger ones (Figure 3.4). With support from a private anthropological foundation in 1946, he organized the Viking Fund Summer Seminars in Physical Anthropology. These Summer Seminars were dynamic "state-of-the-art" meetings in which many new ideas in physical anthropology were explored for the first time. Washburn, despite having received his PhD degree only six years earlier (1940), already was a dynamic force in the field of physical anthropology. He was Secretary-Treasurer of the AAPA (1943–1946), had been teaching anatomy at Columbia University since 1939, and persuaded Paul Fejos (1897–1963), Director of the Viking Fund (later the Wenner-Gren Foundation), to sponsor these Summer Seminars held in New York City. In addition, Washburn enlisted a new Harvard PhD, Gabriel W. Lasker (1912–2002) to edit a new *Yearbook of Physical Anthropology* to report on the Summer Seminars and other news of the profession as well as to reprint important papers not easily found in American anthropology journals. The *Yearbook of Physical Anthropology* has continued to the present from the year of its first publication in 1946.

In 1950, the fifth year of the Summer Seminars, Washburn organized a major conference that served to bring together physical anthropologists and human geneticists with hopes to produce a fresh synthesis in human variation and evolution. The conference was the 15th Cold Spring Harbor Symposium on Quantitative Biology (Warren, 1951) and was initiated by the Cold Spring Harbor Institute Director, Milislav Demerec (1895–1966). Washburn's co-organizer of the Symposium, called "Origin and Evolution of Man," was the distinguished geneticist, Theodosius Dobzhansky (1900–1975), whom he had met at Columbia University. Although the conference was successful in bringing together geneticists, evolutionists, and anthropologists, there were some differences among the more traditional physical anthropologists, particularly the typological "constitutionalists" such as William Sheldon and Earnest Hooton on the one hand, and the more forward-looking physical anthropologists and geneticists on the other concerning topics of race and population variation. "Constitutional somatology" is an outmoded and typological area of study of body form and associated behavior that has a long history (see Comas, 1960, 319 ff.) and was promoted by Sheldon (Sheldon et al., 1940) and Hooton (1939) at the Cold Spring Harbor meeting. Many years later, the daughter of the Institute Director, Rada Dyson-Hudson, who had attended the Symposium as a young Swarthmore College student, remarked that after the afternoon sessions, the geneticists would gather to discuss the talks given that day, while the anthropologists would head for the nearest local tavern! Despite these differences, new ground had been broken, and Washburn's "scientific and evolutionary" physical anthropology was gaining momentum.

In the United Kingdom, the transformation also involved experimental scientific approaches to investigating human variation and the application of evolutionary theory as a basis for explaining human variation. The United Kingdom's transformation moved in two directions: first, toward an adaptive and ecological view of human biological function, and second toward an evolutionary view of human genetic population structure. The leaders in this movement were Joseph S. Weiner (1915–1982) and his colleague Nigel A. Barnicot (1914–1975) (Harrison, 1982). Weiner, who was trained in physiology, anatomy, and anthropology in South Africa, came to England in 1937. With postwar academic posts at Oxford with the distinguished anatomist Le Gros Clark, and later at the London School of Hygiene and Tropical Medicine, he contributed to the training of a whole generation of human biologists, promoted the transformation of physical anthropology, and was instrumental in organizing and managing human adaptability research during the International Biological Programme (IBP) from 1962 to 1974 (Harrison and Collins, 1982). Barnicot was trained in zoology and physiology, and was in the Anthropology Department at University College, London for most of his professional life. He worked on blood genetics, and skin and hair color in West African populations, on studies of nonhuman primates and, in later years, on the human biology of Tanzanian Hadza hunter-gatherers (Sunderland, 1975).

Joseph Weiner was also a strong supporter of biocultural approaches in human biology, in addition to his commitment to solid experimental studies in human physiology. In a 1958 paper on "... training in physical anthropology and human biology," he noted, "... it should be emphasized that in their [students'] research interests an important field of overlap exists between biological and cultural anthropologists" (Weiner, 1958, p. 47). Shortly before his death and nearly 25 years after the earlier statement, Weiner wrote:

The simple fact that his unit of study is a defined community has made it imperative for the human biologist to take full account of the sociocultural properties of his community. Population structural analysis ... cannot be pursued without close attention to social factors whatever parameter is under examination. Genetic analysis is inseparable from demographic and mating patterns; nutrition and energetics are inseparable from food production and food distribution, and land holding; climatic adaptation must encompass the technology of housing and clothing; biomedical fitness is related to population size, sanitary systems, health services and life-style (Weiner, 1982, p. 19).

Several other leading human biologists in the United Kingdom who were junior to Weiner and influenced by his ideas maintained this biocultural perspective. They are Geoffrey A. Harrison, Derek F. Roberts, and James M. Tanner. Roberts conducted pioneering demographic and genetic research on Southern Sudan peoples (Roberts, 1956). Harrison and Tanner, along with Barnicot and Weiner, wrote the first modern textbook in *Human Biology* in 1964, that continued the British tradition of anthropological and biocultural perspectives in human biology (Harrison et al., 1964). In each section of the book, culture figured prominently in examples that were presented. Overseas field experience and a commitment to comparative studies were attributes that exemplified most of this generation of biomedically trained human biologists. Each of these attributes led to an appreciation of the need to study human biology in the context of human behavior and culture.

In the United States, with the stimulus of Washburn's ideas on science and evolution, Earnest Hooton's former students at Harvard were moving the human biology of living populations forward. Washburn contributed very little to the growth of human biology because his interests were largely in skeletal biology, functional morphology, and primatology. Hooton's students, who completed their PhDs between 1940 and 1956, were to become the leaders in human biology during the next generation. These included, with PhD year in parenthesis: Alice Brues (1940) on the genetics of eye, skin, and hair; Marshall T. Newman (1941) on the peopling of the American Southeast; Joseph B. Birdsell (1942) on Australian Aborigine populations; Gabriel W. Lasker

3.5. Gabriel W. Lasker (1912–2002) at his desk. Photograph courtesy of Bernice A. Kaplan.

(1945) on Chinese migrants; James N. Spuhler (1946) on human genetics; Stanley M. Garn (1948) on human hair composition and distribution; William S. Laughlin (1949) on Aleut populations; Edward E. Hunt, Jr (1951) on Micronesian populations; and Paul T. Baker (1956) on desert-heat stress. With the exception of Alice Brues and Paul Baker, all of these young anthropologists attended some or most of the six Summer Seminars organized by Washburn that were held in New York City, and many contributed directly to these sessions, as well. Gabriel Lasker's (1999) memoir describes some of his associations with Hooton's students at this time.

Again, in parallel with the United Kingdom, Americans Gabriel W. Lasker and Paul T. Baker provided inspiration and guidance to junior colleagues and students by promoting a biocultural framework. As noted, Lasker (Figure 3.5) had built on Boas's migration model in studies of Chinese (Lasker, 1946) and Mexican (Lasker, 1952) migrants. In addition to his research, which he pursued actively for more than 50 years, a major and long-term contribution to students and junior colleagues was by way of support of their papers submitted to the journal *Human Biology*, which he edited for 35 years. Baker (Figure 3.6), published widely with his own strong biocultural commitment, and also socialized his students about the value of biocultural research. In a paper published four and a half decades ago on "Climate, culture, and evolution," he stated: "To completely reject the concept of climatic selection without cultural involvement would be premature, but from anthropology's present theoretical framework it is more accurate always to consider the role of culture when trying to formulate an evolutionary process related to climate" (Baker, 1960, p. 5). Twenty years later

3.6. Paul T. Baker (1927–2007) (left) and Stanley M. Garn (1922–2007) (right) some time during the late 1950s or early 1960s. Photograph courtesy of Barbara Garn.

Baker (1982) continued to emphasize this theme about the importance of the linkage between human biology and culture in a Huxley Memorial Lecture. In his 1996 Pearl Memorial Lecture, he summarized his ideas and underlined the "... need for all human biologists to have some training in relevant aspects of cultural anthropology" (Baker, 1997). Such a perspective seemed not only intuitively correct and logical to Baker, but it also led to more productive lines of research and testable hypotheses. Weiss and Chakraborty (1982, p. 383) observed that the complex influence of culture on evolution "... is a point still largely misunderstood or ignored by many researchers without anthropological training." Harrison (1982, p. 471) suggested that it is much easier to give "equal attention ... to cultural as to biological ..." processes in the United States than in Europe, because the subfields of anthropology can often be found in the same department in the United States.

Human biology research and student training during the immediate post-war years was supported enormously by the private Wenner-Gren Foundation for Anthropological Research (founded as The Viking Fund in 1941). Partly because of the close personal relationship between its Director, Paul Fejos, and Sherwood Washburn, the Foundation supported the Summer Seminars, the *Yearbook of Physical Anthropology*, and other activities associated with the development of physical anthropology during that time (Szathmáry, 1991). During these early years, and up to the present, it is hard to imagine how human biology, within the larger field of physical anthropology, could have developed into a mature science without the vision of Paul Fejos and later directors of the Wenner-Gren and the financial support of the Foundation (Baker and Eveleth, 1982).

AFTER THE MIDDLE OF THE TWENTIETH CENTURY

After 1950, new directions were taken in climatic morphology and physiology, genetics, disease, and adaptation to the environment. The book by Coon et al. (1950) outlined applications to climatic rules in humans including Bergmann's rule (body size) and Allen's rule (size and shape of extremities), and may have stimulated several biogeographic studies of human form. For example, Derek F. Roberts (1952, 1953) from the United Kingdom and Eugene Schreider (1950, 1951) from France explored relationships between body size and basal metabolism and climate in human populations around the world. They found that humans, as with other warm-blooded animals, conformed to Allen's and Bergmann's rules. Other anthropologists from both sides of the Atlantic conducted similar studies of the climatic environment and adaptation of humans to both heat and cold stress (Newman, 1953, 1956; Weiner, 1954; Newman and Munro, 1955; Baker and Daniels, 1956; Baker, 1958a, 1958b). Some of these studies were experimental (either laboratory or field) and were influenced by the physiological studies conducted of heat and cold stress of South Africans (Wyndham et al., 1952), Australian Aborigines (Scholander et al., 1958), Arctic Indians (Irving et al., 1960), Inuit/Eskimos (Brown and Page, 1952; Meehan, 1955), Alakaluf Indians (Hammel, 1960), and Laplanders (Scholander et al., 1957). These and other studies demonstrated quite clearly that human populations distributed around the world had adapted *morphologically*, *physiologically*, and *behaviorally* to the climatic conditions of their environments.

The difference between the physiologists' studies and the anthropologists' studies was in the anthropologists' understanding of the role of culture in these patterns of adaptation to temperature extremes (Baker, 1960). The anthropologists were more concerned with the whole culture/population and variations in responses by age and sex, whereas the physiologists' contributions were more focused on physiological mechanisms in small samples of men tested under controlled laboratory conditions. Both kinds of studies enriched what we learned and were complementary.

One of the post-war military institutions that stimulated work in anthropology and human biology was the Natick, Massachusetts Quartermaster Corps Climatic Research Laboratory (Francesconi et al., 1986). Being close to Harvard University it could draw on talented individuals from this university and also influence research directions there and elsewhere, as well. Many of the young scientists at the Climate Research Laboratory were environmental physiologists who became well-known scientists in later years, while a few of the scientists were trained in physical anthropology (e.g., Paul Baker and Russell Newman). Since the

Natick laboratories were also engaged in clothing and nutrition studies, there were also strong interests in body size (anthropometry) and body composition variations in military personnel. These government studies contributed to the mix of interests in nutrition, disease, body composition, and climate in the context of human adaptation to variable environments.

By the early 1960s, studies of body composition in human biology were on the rise (Brožek and Henschel, 1961; Brožek, 1963; Garn, 1963). Interests among anthropologists were derived, in part, from knowledge of skeletal biology and were linked closely to studies of nutritional adaptation (Brožek, 1956; Newman, 1960, 1962), and child growth (Garn and Shamir, 1958). Anthropometric (Garn, 1961) and physique (Brožek, 1956) surveys were carried out to evaluate nutritional status, and associations between growth and nutrition in stressful environments were studied (Roberts, 1960; Schraer and Newman, 1958). Josef Brožek (1999) provided a personal history of body composition studies in human biology.

During this period, human population genetics focused almost exclusively on blood polymorphisms. A good example was Alice Brues's (1954) important paper demonstrating that selection must have operated on the ABO blood group system because of the non-random distribution of ABO allele frequencies around the world. Electrophoresis was first used to separate proteins, and it was at this time that "... the staggering magnitude of [human] genetic variation ..." became apparent (Cavalli-Sforza et al., 1994, p. 3). One of the great discoveries of the 1950s was the relationship between the sickle cell gene and malaria and the protection afforded against malaria by the sickle cell heterozygote. James Neel (1915–2000) had worked out the genetics of sickle cell disease (Neel, 1949), and then described the anemias of sickle cell disease and thalassemia in his 15th Cold Spring Harbor Symposium paper (Neel, 1951). Neel proposed mutation or selection as hypotheses to explain the high prevalence of these dangerous anemias in the Mediterranean and in Africa, but it remained for Anthony Allison to show that the major force was natural selection. Allison (1954) demonstrated that the sickle cell gene was a balanced polymorphism maintained by the selective pressure of malaria. He described in some detail the history of his early work and the conditions surrounding the discovery in an autobiographical paper (Allison, 2002).

James Neel established the first human genetics department at the University of Michigan in 1956. At that time, this program was one of the strongest in the country, partly because of its connection with anthropology. William J. Schull worked with Neel in human genetics and had a joint appointment in anthropology, while James N. Spuhler (1917–1992) and Frank B. Livingstone (1928–2005) were conducting genetic research in the anthropology department. Another major contribution in human genetics in the 1950s was Livingstone's (1958) paper on the anthropology of sickle cell in West Africa. Based on fieldwork in Liberia and extensive literature research on West Africa, he demonstrated the relationships among malaria, sickle cell distribution, mosquito ecology, agriculture (forest transformation), and language and human migration. Only four years after Allison's (1954) publication, Livingstone had shown the importance of culture in structuring evolutionary change through its influence on malaria prevalence and gene-frequency distributions. Culture change had produced evolutionary change!

Other important genetics work in the 1950s and 1960s by anthropologists outside the United States and the United Kingdom included research by Jean Hiernaux (1966), from Paris, who studied blood-group genetics in African populations. Another major figure is Francisco Salzano (Salzano and Callegari-Jacques, 1988), from Brazil, who first did post-doctoral research with James Neel at the University of Michigan in the late 1950s, and who went on to become the major human geneticist to study tropical forest Native Americans in South America.

HUMAN ADAPTABILITY, ECOLOGY, AND INTERNATIONAL PROGRAMS

Among human biologists, the 1960s ushered in a mature sense of scientific problem solving, an increasing commitment to understanding evolutionary process and adaptation to the environment, and a growing interest in integrated, collaborative studies. In anthropology, broadly, there was a movement toward empiricism with interests in ecological approaches and cultural materialism. At the same time, human biologists were exploring ecological models in the context of adaptation to the environment (Baker, 1962; Little, 1982). In 1964, four distinguished British human biologists published the first edition of an important introductory textbook (Harrison et al., 1964). In this book, Geoffrey A. Harrison and Nigel A. Barnicot dealt with genetics and phenotypic variation, James M. Tanner covered human growth, and Joseph S. Weiner wrote the sections on human adaptation and human ecology. This was an important work because it defined the field of human biology, was synthetic, and it was state of the science for the 1960s. Weiner's final section on "Human Ecology" was the first definition of this important perspective in human biology: topics included nutrition, disease, climate, and population (demography), and a central emphasis of the work was on "ecological adaptive processes."

Two other important events took place in 1964. Firstly, the International Council of Scientific Unions (ICSU) (now the International Council for Science) in Paris established the International Biological

Programme (IBP) with a planning phase from 1964 to 1967, a research phase from 1967 to 1972, and a synthesis phase to follow from 1972 to 1974. The orientation of this worldwide program was ecological, and its theme was "The Biological Basis of Productivity and Human Welfare." A number of sections of the IBP were designed to cover various components of ecology and ecosystems studies. A separate section called Human Adaptability (HA) was to cover "the ecology of mankind" from a variety of perspectives including health and welfare, environmental physiology, population genetics, child growth, anthropology, and demography (Weiner, 1965). Secondly, although some preliminary preparation had begun in 1962, an important HA planning conference was held in Austria at the Wenner-Gren Foundation Burg Wartenstein Conference Center in the July of that year. It was at that meeting that Weiner (1966) outlined the major categories of planned research, and the other, more than 20 internationally represented, participants reported on current knowledge over a wide variety of topics in human biology for North and South American, European, Asian, African, and Australian populations (Baker and Weiner, 1966). Weiner, who was International Convener (director) of the Human Adaptability Section of the IBP prepared a handbook (Weiner and Lourie, 1969) of standardized methods, and at the end of the IBP published a compendium of the completed research (Collins and Weiner, 1977). The planning and research that followed resulted in the participation of 40 nations, the completion of more than 230 projects, and several thousand publications under the HA banner. In a retrospective review of the HA research, there has been some criticism based on topical omissions, but the contributions far outweigh the shortcomings (Ulijaszek and Huss-Ashmore, 1997). Harrison (1997, p. 25) noted, in a contribution to the same review: "Notwithstanding its limitations, it played a major part in converting the old defunct physical anthropology into the vibrant and exciting component of biological anthropology as it is today." In addition, it is quite clear that the associations made between the British and American human biologists during the IBP were of remarkable value by serving to cross-fertilize ideas and to reinforce the biocultural and environmental perspectives shared by most of the participants (Baker, 1988).

One of the major conceptual contributions of the IBP was in the organization of multidisciplinary research (Little et al., 1997). Because complex ecosystems required broad expertise, many individuals from different fields of science were required to collaborate. Ecologists from the IBP were organized according to major ecosystems (e.g., tropical forests, arctic tundra, deserts, and grasslands), while some of the HA projects focused on single human populations in special environments. Two early projects were the Kalahari Research Group to study the !Kung Bushmen initiated in 1963 (not IBP affiliated; Lee and DeVore, 1976) and the Andean Biocultural Studies initiated in 1962 (IBP affiliated, Baker and Little, 1976). Two other important IBP projects included the International Studies of Circumpolar Peoples, a four-nation investigation of the health of Inuit/Eskimos in Alaska, Canada, and Greenland (Milan, 1980), Population Genetics of the American Indian, a project that focused on the Yanomama horticulturalists from Brazil and Venezuela (Neel et al., 1977), and the Solomon Islands Project, organized by Albert Damon and others at Harvard (Damon, 1974; Howells 1987). These and other HA projects were discussed in detail by Hanna et al. (1972) and from theoretical perspectives by Lasker (1969).

Other major contributions from the IBP were the surveys of human growth (Eveleth and Tanner, 1976) and human gene frequencies (Collins and Weiner, 1977) that were conducted on populations around the globe that contributed primary data. In many of these smaller Third World projects, research design and "problem orientation" were subordinated to data collection, but these important data sets, then, became available for comparative, historical, and in-depth analyses for future investigators (Weiner, 1977).

Following the close of the IBP in the 1970s, a new international program arose from UNESCO (United Nations Educational, Scientific, and Cultural Organization) called the Man and the Biosphere Program (MaB). Additional multidisciplinary projects were affiliated with the MaB program including the Multinational Andean Genetic and Health Project (Schull and Rothhammer, 1990) and the Samoan Migrant Project (Baker et al., 1986). Other projects, such as the South Turkana Ecosystem Project (Little and Leslie, 1999), the Ituri Forest Project (Bailey, 1991), and the Siberian Evenki Project (Crawford et al., 1992) followed somewhat later. These multidisciplinary projects promoted international collaboration, gathered fundamental data on populations that were on the brink of extinction, were instrumental in developing new field and experimental methods, and provided research training for a whole new generation of human biologists.

DNA ANALYSIS AND MOLECULAR GENETICS

Weiss and Chakraborty (1982) provided a detailed review of the history of human genetics up to the late 1970s. The late 1970s and 1980s, however, marked a dramatic transition in human population genetics from a reliance on phenotypic proteins and inference about genes (phenotype-based approach) to the direct measures of genes via DNA (the new molecular genetics). This transition was based on the new laboratory methods that enabled DNA to be extracted and purified from Blood and other tissues (Crawford, 2000). These methods included: use of restriction enzymes to clip nucleotide segments of DNA (restriction fragment length polymorphisms), DNA hybridization, and the polymerase chain reaction for amplification of DNA. Some of the new DNA methods were used in human biology for phylogenic reconstruction of our ancestors, to reconstruct human population movements over space and time, and for forensic and ancient DNA analysis. New methods of DNA extraction and analysis also led to the Human Genome Project and the controversial Human Genome Diversity Project.

Mitochondrial DNA (mtDNA), which is only transmitted through the maternal line and does not undergo recombination, has been very useful in a variety of approaches to human evolution (Cann, 1986). One of the most remarkable studies of mtDNA was the so-called "Mitochondrial Eve" observation that startled scientists interested in modern human origins. Cann et al. (1987) found in a sample of 147 people from around the world that "All these mitochondrial DNAs stem from one woman who is postulated to have lived about 200 000 years ago, probably in Africa" (Cann et al., 1987, p. 31). This and other DNA work led to the "Out of Africa" hypothesis on modern human origins (Stoneking and Cann, 1989; Vigilant et al., 1991). More recent research incorporated the Y chromosome, which is inherited through paternal lineages (Hammer, 1995).

A variety of studies have been conducted to trace population distributions and migrations in the historic and prehistoric past. Cavalli-Sforza et al. (1988) and Sokal (1988) combined genetic, archaeological, and linguistic information to reconstruct population expansion in Europe. This geographic genetic research continues up to the present (Barbujani, 2000). Another area of interest that had been dominated by archaeological evidence soon saw the application of genetic data to the question of the peopling of the New World (O'Rourke, 2000). Based on mtDNA and Y-chromosome markers, Merriwether (2002) suggested a single-wave migration, possibly originating from ancient Mongolia through Siberia and into the New World between 20 000 and 30 000 BP. All of these dates fall within the late Pleistocene when the Wisconsian glaciation covered parts of North America and the cold Beringia land bridge between Siberia and Alaska was exposed.

By the late 1980s, it was feasible to sequence the whole human genome of nucleotides and DNA on the human chromosomes. This culminated in the Human Genome Project (HGP), a massive DNA-sequencing effort that was estimated to cost several billion dollars. In the United States, the Department of Energy (DOE) and the National Institutes of Health (NIH) supported the HGP (Crawford, 2000). The work conducted by an international consortium began in 1990 and was completed in 2003, several years ahead of schedule. The principal concern of human population biologists about the HGP was that it would provide a *generic* genome; that is, little or no information would be provided on human genetic variation in individuals and populations around the world. Such variation was of value to studies of human evolution and population history, as well as applied research in the genetic bases for disease (resistance, susceptibility, environmental interactions, etc.) and genetic epidemiology.

In response to these concerns, the Human Genome Diversity Project (HGDP) was founded in 1991 under the leadership of L. L. Cavalli-Sforza, the distinguished Stanford University geneticist. Also, the project was identified as urgent because of the disappearance of many small Mendelian populations. The proposed HGDP sparked a controversy for a variety of factors: first, there was distrust by anthropologists of the geneticists in their conceptualization of race (linked to typological race and eugenics); second, there were concerns by Native American groups about their being exploited again, about race stereotyping, and about the potential use of DNA for commercial gain (patenting genes); third, there was an element of political naïveté of the organizing geneticists about conflicting beliefs about race and, initially, there were few anthropologists included in the program and its design (Reardon, 2005). Because of conflicts and protests the original plan to establish cell lines of DNA samples from 25 individuals from each of 400 populations was never satisfied, and funding was not forthcoming. Later the HGDP was revived under the title of HGDP-CEPH (Centre d'Etude du Polymophisme Humain), and now DNA is available for 1064 individuals from 52 populations. These DNA samples have begun to be used for research (Ramachandran, 2005).

During the past half century, there has been increasing scientific activity in what Derek Roberts (1965) first referred to as "anthropological genetics." Crawford (2007) outlined the differences between "anthropological genetics" and "human genetics," where the former incorporates a biocultural perspective, focuses on the population and, often, non-Western populations, and centers on questions of interest to anthropologists, particularly evolutionary questions. Research in molecular anthropology, genetic epidemiology, forensic anthropology via DNA analysis, human origins, and the history of human migration and dispersal are all areas of exploration that are being pursued by anthropological geneticists.

REPRODUCTION AND CHILD GROWTH

Interests in reproduction in anthropology date back to the early part of the twentieth century and are linked closely to two primary emphases in human biology:

evolution/natural selection and human health. In evolution, differential fertility is one of the main driving forces of selection, and fertility (measured as number of live-born offspring) is also a prime indicator of adult health and well-being. Studies of growth are an extension of reproduction and, of course, date back in human biology to Franz Boas.

Pioneering studies by Sophie Aberle, who was trained in genetics at Stanford and medicine at Yale, were conducted of sex and reproduction in San Juan Pueblo Indians in New Mexico (Aberle, 1928, 1931). She was a member of the National Research Council Committee for Research in Problems of Sex (Aberle, 1953) and worked for the Bureau of Indian Affairs. From her research she found that the poor reproductive pattern of Pueblo Indians was not based on reduced fertility, which in fact was very high, but rather was based on extraordinarily high infant mortality rates (about 25%). Ashley Montagu (1905–1999) conducted early work on female sexual maturation by discovering post-menarcheal sterility in adolescent girls (Ashley-Montagu, 1939a, 1939b, 1946). He found that, for a year or more following menarche in girls, there was a markedly reduced fecundity or even sterility, and "... that puberty and the power to procreate are not synchronous events ..." (Ashley-Montagu, 1939b, p. 213). The sociologists, Kingsley Davis and Judith Blake (Davis and Blake 1956) formulated a groundbreaking model of factors affecting fertility in the context of social behavior. This provided the basis for systematic hypothesis testing and refinement of the model, which was done by Bongaarts (1978). Campbell and Wood (1988) further refined the model for "proximate determinants" of fertility in non-Western (noncontraceptive) as well as Western populations. One of the first studies to document an important variable, not considered by Davis and Blake (1956), was the central importance of breast-feeding in fertility control by Konner and Worthman (1980) in studies of !Kung Bushmen. This and other research work stimulated a number of anthropological studies of nursing, energetics, and fecundity in traditional populations (Bentley, 1985; Gray, 1994; Vitzthum, 1994).

An important series of studies was conducted in the 1960s and 1970s that explored relationships among energy balance (diet and activity), body composition (fat and muscle), and reproduction (age of menarche) in adolescent girls. The work by Rose Frisch (Frisch and Revelle, 1970; Frisch and McArthur, 1974) led to a hypothesis that there was a critical weight, and, then later, a critical level of body fat that both triggered menarche and maintained normal menstrual cycles. Frisch's ideas were severely criticized, partly because of her unyielding adherence to this basic hypothesis. Subsequent research has demonstrated that reproductive controls on the ovarian cycle are more complex than just being a function of the energy of body stores,

although energy balance does play an important role. However, Frisch's research was truly pioneering in that it contributed to the development of a new *ecology of reproduction* that began to explore the evolution and ecology of reproductive function in a variety of Western and non-Western peoples (Howell, 1979; Ellison, 1990, 2001; Leslie et al., 1994). These and other investigations explored relationships among nutritional status and availability of food resources, disease, physical activity, endocrine function, body composition, behavior, growth, and reproduction – truly an integrated, behavioral and environmental science of human reproduction.

At the same time as this new research direction in fertility and reproduction was being taken, so were new discoveries being made in infant, child, and adolescent growth studies. Surveys of the worldwide variation in human growth, which were originally compiled from the IBP HA studies, synthesized the available knowledge on population variations in human growth up to the late 1980s (Eveleth and Tanner, 1976, 1990). Somewhat later, Michelle Lampl conducted new longitudinal research of individual growth patterns (Lampl et al., 1992) in an ingenious study of infant length, where some infants were measured every day for more than a year. She and her colleagues found that rather than being a continuous process as believed, growth proceeded in "incremental bursts" (saltatory growth) followed by periods of stasis. Some measurements of infants showed daily increases in length of up to 1 cm! This work was extended to adolescents, who also showed saltatory growth (Lampl and Johnson, 1993), and has led to new lines of research in bone and soft tissue growth and in the endocrine control of growth.

In the 1980s and 1990s, increasing research in growth has been directed toward health and the total life span from conception to senescence and death. This research has had both basic and applied components related to health across what has been called the "life span approach" to the study of human biology (Leidy, 1996). A general principle of this approach is that "... no single stage of a person's life (childhood, middle age, old age) can be understood apart from its antecedents and consequences" (Riley, 1979, p. 4). Activity levels, diet, and exposure to adverse substances during childhood all influence health during adulthood and senescence. It is also the case that longitudinal study of the same individuals through time, in the tradition of Franz Boas, is one of the best methods to demonstrate these life span correlations. David J. P. Barker (Barker et al., 1989; Barker, 1992) developed an extension of this approach to understanding the life span. He discovered from early twentieth-century archival birthweight data that the lower the birthweights of infants, the greater the risk of coronary heart disease in adulthood. He also found that lower birthweight increases the risk of hypertension, stroke, and adult diabetes. These findings led to what is now known as the "fetal origins hypothesis" of adult disease that is currently being explored in a variety of research lines today, particularly in the context of juvenile and adult obesity.

Another important developing research area in human growth focuses on the evolution of the unique properties of human growth (Bogin, 1997, 1999). Humans have a unique growth pattern that is similar to nonhuman primates, but differs to the extent that we have a long period of nursing and post-weaning dependence (birth to seven years of age), a period of rapid growth that defines adolescence, a delayed onset of reproduction, and menopause in females. These, and other unique attributes of human development, facilitate development of the brain, manual dexterity, and the social and sexual skills needed for life in a complex society. Evolution of growth has been studied in both living primate and paleoanthropology studies. Also within the evolutionary framework, "life history theory" centers on the allocation of energy to somatic functions, such as growth and maintenance of the body, and to reproductive functions, such as gestation, lactation, and child rearing (Hill, 1993). This research paradigm reintegrates areas of reproduction, demography, and growth in new and interesting ways.

BIOMEDICAL ANTHROPOLOGY, HEALTH, AND EPIDEMIOLOGY

There are deep traditions of medical studies in anthropology and human biology. As noted above, physical anthropology was linked to anatomy during the nineteenth and early twentieth centuries, and many physical anthropologists were employed in medical schools because of their expertise in gross anatomy. Some were also trained as physicians. These biomedical interests in physical anthropology expanded to include epidemiology, public health, child growth, and nutrition, all in the context of disease or other conditions of ill health. Johnston and Low (1984) identify biomedical anthropology as being (1) based on "... the application of anthropological theory to problems of health and disease" (1984, p. 215) (bioculturally centered); and (2) focused on a "biological outcome" (disease centered).

Many biomedical contributions have been made with anthropological knowledge in epidemiology, where population, the target disease, culture, and the environment intersect in complex ways. Frank Livingstone's (1958) research on sickle cell in West Africa was conducted from the anthropological side to the biomedical side, whereas Carleton Gajdusek's (1977) Nobel Prize winning research on Kuru in New Guinea was conducted from the biomedical side to be informed by anthropology. Correspondingly, Baruch Blumberg's

(2006) Nobel Prize winning discovery of the Australia antigen and hepatitis B virus was carried out while investigators searched for serum protein polymorphisms in non-Western (anthropological) populations.

Human biologists have also taken advantage of unique opportunities to explore diseases or health threats in traditional, modernizing, and modern populations (Garruto et al., 1999). The stresses of high altitude hypoxia were explored in the Andes (Baker and Little, 1976; Schull and Rothhammer, 1990), and the high prevalence of amyotrophic lateral sclerosis and Parkinsonian dementia in the Chamorros population were studied in Guam (Garruto et al., 1989). Of increasing interest in human biology are the effects of modernization in non-Western populations, migration from traditional to modern societies, and life-style transitions in Western populations. For example, the nutritional and life-style transition in the West, particularly the United States, has led to high caloric intakes, reduced physical activity, and high rates of obesity, cardiovascular disease, hypertension, and diabetes. At the same time, the epidemiological transition has led to declines in infectious diseases and increases in chronic diseases of aging and the diseases just noted that are associated with life style. An exception to declines in infectious diseases is AIDS and other emerging diseases, some of which are a function of human population and social disruption (see Chapter 27 of this volume). All of these old and newly emerging diseases offer opportunities for human biologists to explore them in biological, cultural, and environmental frameworks.

SOCIETIES AND JOURNALS IN HUMAN BIOLOGY

There are several professional societies devoted to human biology. The oldest is the Society for the Study of Human Biology (SSHB), which was founded in Britain in 1958. About a decade later, the International Association of Human Biologists (IAHB) was founded at a Wenner-Gren Conference in Austria. The Human Biology Association (HBA) was established in 1974 (originally named the Human Biology Council), and most recently the American Association of Anthropological Genetics (AAAG) was founded in 1993. The affiliation of these societies with the three main journals in human biology is listed in Table 3.1. *Human Biology* is the oldest journal, dating back to 1929; the British *Annals of Human Biology* was founded in 1974; and the *American Journal of Human Biology* is the most recent publication, having begun in 1989. Among the three journals in human biology, there are about 2500 pages of articles and reviews published each year. Currently, *Human Biology* is the official publication of AAAG and publishes principally in population biology and

TABLE 3.1. A chronology of physical anthropology and human biology journals, societies, and major editors.

1918	*American Journal of Physical Anthropology* founded and edited by Aleš Hrdlička until 1942
1929	*Human Biology* founded and edited by Raymond Pearl until his death in 1940
1930	American Association of Physical Anthropologists established
1946	*Yearbook of Physical Anthropology* founded by Sherwood L. Washburn, edited by Gabriel W. Lasker
1953	*Human Biology* edited by Gabriel W. Lasker until 1987
1958	Society for the Study of Human Biology (SSHB) established in the United Kingdom (UK)
1963	SSHB and journal *Human Biology* become affiliated with James M. Tanner co-editor in the UK
1967	International Association of Human Biologists established at Wenner-Gren Conference in Austria
1972	SSHB and *Human Biology* become disaffiliated because of disagreements with the publisher, Wayne State University Press
1974	*Annals of Human Biology* founded in the UK by the SSHB
1974	Human Biology Council (HBC) established and affiliated with *Human Biology*
1988	HBC and *Human Biology* become disaffiliated because of disagreements with the publisher, Wayne State University Press
1988	*Human Biology* edited by Michael H. Crawford
1989	*American Journal of Human Biology* founded and affiliated with the HBC
1990	*American Journal of Human Biology* edited by Robert M. Malina
1994	Human Biology Council name changed to Human Biology Association
1995	American Association of Anthropological Genetics becomes affiliated with *Human Biology*
2002	*American Journal of Human Biology* edited by Peter T. Ellison

genetics; *Annals of Human Biology* is the official publication of SSHB; and the *American Journal of Human Biology* is the official publication of HBA. Histories of each of these publications and their associations can be found in Tanner (1999), Crawford (2004), and Little and James (2005).

Three other societies and their journals are linked to human biology: the American Association of Physical Anthropologists (*American Journal of Physical Anthropology*, AJPA), the Society for the Study of Social Biology (*Social Biology*, SB), and The Galton Society in the United Kingdom (*Journal of Biosocial Science*, JBS) (Johnston and Little, 2000; Alfonso and Little, 2005). The *AJPA* publishes about 15–20% of its articles in human biology, and both the *SB* and *JBS* publish a majority of the papers on topics in public health and health-related demography. Over the years, the vast majority of editors of these six journals have been physical/biological anthropologists and human biologists.

WHAT IS HUMAN BIOLOGY?

From its earliest beginnings, human biology has made unique contributions to scientific inquiry for a number of reasons. Firstly, a principal pursuit has been to understand and explain the basis for human variation in all its dimensions, at the individual and the population levels. Secondly, because of ties with anthropology, explanations have drawn on knowledge of both the biology and the culture of humans. This has been a consistent practice dating back more than a hundred years to Franz Boas, the founder of American anthropology, and one of the principal founders of human biology. Thirdly, understanding of these biocultural relations has been couched in evolutionary and adaptational theory. That is, evolutionary theory contributes to or explains how humans interact with their environments. Fourthly, although genetics structures human variation, humans are flexible in their adaptation to the environment; hence, there is an acute awareness of human plasticity in response to the physical and the social environment. Finally, there has been a willingness among human biologists to incorporate knowledge and expertise from other fields within the biological, medical, and social sciences in order to enrich our understanding of this most complex of species – *Homo sapiens*.

DISCUSSION POINTS

1. Why is "race" an inappropriate concept to apply to an understanding of human variation?
2. Did Franz Boas's championing of plasticity and his studies of human migrants demonstrate that genetics played only a minor role in human variation?
3. Was all research in human biology halted during World War II? If not, then what kind of research was done?
4. Who were the principal figures in the United States who contributed to the modernization of human biology? And what were their major

contributions? Who were the principal figures in the United Kingdom?

5. The 1960s were dominated by the International Biological Programme and its Human Adaptability Component that included a number of human biologists from around the world. What were some of the basic research themes and why was this an important vehicle for the promotion of human biology research?

6. An important transformation in human genetics studies occurred in the late 1970s. What was this transformation and how did it influence our interpretation of human variation?

7. Why is reproduction of interest to human biologists? What are some new areas of research?

8. Why is health of interest to human biologists? What can knowledge of human biology contribute to our understanding of HIV?

9. What are the journals that are devoted to human biology research? Do they differ in the kinds of articles that they publish?

REFERENCES

Aberle, S. B. (1953). *Twenty-Five Years of Sex Research: History of the National Research Council Committee for Research in Problems of Sex, 1922–1947*. Philadelphia: Saunders.

Aberle, S. B. D. (1928). Frequency of childbirth among Pueblo Indians. *Anatomical Record*, **38**, 1–2.

Aberle, S. B. D. (1931). Frequency of pregnancies and birth intervals among Pueblo Indians. *American Journal of Physical Anthropology*, **16**, 63–80.

Adolph, E. F. and Associates. (1947). *Physiology of Man in the Desert*. New York: Wiley.

Alfonso, M. P. and Little, M. A. (transl. and ed.) (2005). Juan Comas's summary history of the American Association of Physical Anthropologists (1928–1968). *Yearbook of Physical Anthropology*, **48**, 163–195.

Allison, A. C. (1954). Protection afforded by sickle-cell trait against subtertian malaria infection. *British Medical Journal*, **1**, 290–294.

Allison, A. C. (2002). The discovery of resistance to malaria of sickle-cell heterozygotes. *Biochemistry and Molecular Biology Education*, **30**, 279–287.

Ashley-Montagu, M. F. (1939a). Adolescent sterility. *Quarterly Review of Biology*, **14**(1), 13–34.

Ashley-Montagu, M. F. (1939b). Adolescent sterility (concluded). *Quarterly Review of Biology*, **14**(2), 192–219.

Ashley-Montagu, M. F. (1946). *Adolescent Sterility. A Study in the Comparative Physiology of the Infecundity of the Adolescent Organism in Mammals and Man*. Springfield, IL: C. C. Thomas.

Bailey, R. C. (1991). *The Behavioral Ecology of Efe Pygmy Men in the Ituri Forest, Zaïre. Anthropological Papers of the Museum of Anthropology*, no. **86**. Ann Arbor: University of Michigan.

Baker, P. T. (1958a). The biological adaptation of man to hot deserts. *American Naturalist*, **92**, 337–357.

Baker, P. T. (1958b). Racial differences in heat tolerance. *American Journal of Physical Anthropology*, **16**, 287–305.

Baker, P. T. (1960). Climate, culture, and evolution. *Human Biology*, **32**, 3–16.

Baker, P. T. (1962). The application of ecological theory to anthropology. *American Anthropologist*, **64**, 15–22.

Baker, P. T. (1982). The adaptive limits of human populations. *Man (New Series)*, **19**, 1–14.

Baker, P. T. (1988). Human population biology: A developing paradigm for biological anthropology. *International Social Science Journal*, **116**, 255–263.

Baker, P. T. (1997). The Raymond Pearl Memorial Lecture, 1996: The eternal triangle – genes, phenotype, and environment. *American Journal of Human Biology*, **9**, 93–101.

Baker, P. T. and Daniels, F. Jr (1956). Relationship between skinfold thickness and body cooling for 2 hours at 15°C. *Journal of Applied Physiology*, **8**, 409–416.

Baker, P. T. and Little, M. A. (eds) (1976). *Man in the Andes: a Multidisciplinary Study of High-Altitude Quechua*. Stroudsburg, PA: Dowden, Hutchinson and Ross.

Baker, P. T. and Weiner, J. S. (eds) (1966). *The Biology of Human Adaptability*. London: Clarendon Press.

Baker, P. T., Hanna, J. M. and Baker, T. S. (eds) (1986). *The Changing Samoans: Health and Behavior in Transition*. New York: Oxford University Press.

Baker, T. S. and Eveleth, P. B. (1982). The effects of funding patterns on the development of physical anthropology. In *A History of American Physical Anthropology: 1930–1980*, F. Spencer (ed.). New York: Academic Press, pp. 31–48.

Barbujani, G. (2000). Geographic patterns: how to identify them and why. *Human Biology*, **72**, 133–153.

Barker, D. J. P. (1992). *The Fetal and Infant Origins of Adult Disease*. London: BMJ Books.

Barker, D. J. P., Osmond, C., Winter, P. D., et al. (1989). Weight in infancy and death from ischaemic heart disease. *Lancet*, **2**, 577–580.

Bentley, G. R. (1985). Hunter-Gatherer energetics and fertility: a reassessment of the !Kung San. *Human Ecology*, **13**, 79–109.

Blumberg, B. S. (2006). The curiosities of hepatitis B virus: prevention, sex ratio, and demography. *Proceedings of the American Thoracic Society*, **3**, 14–20.

Boas, F. (1912). *Changes in Bodily Form of Descendants of Immigrants*. New York: Columbia University Press.

Boas, F. (1928). *Materials for the Study of Inheritance in Man*. New York: Columbia University Press.

Boas, F. (1936). Effects of American environment on immigrants and their descendants. *Science*, **84**, 522–525.

Bogin, B. (1997). Evolutionary hypotheses for human childhood. *Yearbook of Physical Anthropology*, **40**, 63–89.

Bogin, B. (1999). Evolutionary perspectives on human growth. *Annual Review of Anthropology*, **28**, 109–153.

Bongaarts, J. (1978). A framework for analyzing the proximate determinants of fertility. *Population and Development Review*, **4**, 105–132.

Boyd, W. C. (1950). *Genetics and the Races of Man*. Boston: Little, Brown.

Brace, C. L. (2005). *"Race" is a Four-Letter Word*. New York: Oxford University Press.

Brown, G. M. and Page, J. (1952). The effect of chronic exposure to cold on temperature and blood flow to the hand. *Journal of Applied Physiology*, **5**, 221–227.

Brožek, J. (1956). Physique and nutritional status of adult men. *Human Biology*, **28**, 124–140.

Brožek, J. (1963). Quantitative description of body composition: physical anthropology's fourth dimension. *Current Anthropology*, **4**, 3–39.

Brožek, J. (1999). Human biology: from a love to profession and back again. *American Journal of Human Biology*, **11**, 143–155.

Brožek, J. and Henschel, A. (eds) (1961). *Techniques for Measuring Body Composition*. Proceedings of a Conference, Quartermaster Research and Engineering Center, Natick, MA. Washington, DC: National Academy of Sciences – National Research Council.

Brues, A. M. (1954). Selection and polymorphism in the ABO blood groups. *American Journal of Physical Anthropology*, **12**, 559–597.

Campbell, K. L. and Wood, J. W. (1988). Fertility in traditional societies. In *Natural Human Fertility: Social and Biological Determinants*, P. Diggory, M. Potts and S. Teper (eds). Basingstoke: Macmillan Press, pp. 39–69.

Cann, R. L. (1986). Nucleotide sequences, restriction maps, and human mitochondrial DNA diversity. In *Genetic Variation and Its Maintenance: with Particular Reference to Tropical Populations*, D. F. Roberts and G .F. De Stefano (eds). Society for the Study of Human Biology Symposium Series, no. 27. Cambridge: Cambridge University Press, pp. 77–86.

Cann, R. L., Stoneking, M. and Wilson, A. C. (1987). Mitochondrial DNA and human evolution. *Nature*, **325**, 31–36.

Cavalli-Sforza, L. L., Piazza, A., Menozzi, P., et al. (1988). Reconstruction of human evolution: bringing together genetic, archaeological, and linguistic data. *Proceedings of the National Academy of Sciences of the United States of America*, **85**, 6002–6006.

Cavalli-Sforza, L. L., Menozzi, P. and Piazza, A. (1994). *The History and Geography of Human Genes*. Princeton: Princeton University Press.

Collins, K. J. and Weiner, J. S. (1977). *Human Adaptability: a History and Compendium of Research*. London: Taylor and Francis.

Comas, J. (1960). *Manual of Physical Anthropology*. English edition. Springfield, IL: C. C. Thomas.

Coon, C. S. (1939). *The Races of Europe*. New York: Macmillan.

Coon, C. S., Garn, S. M. and Birdsell, J. B. (1950). *Races: a Study of the Problems of Race Formation in Man*. Springfield, IL: C. C. Thomas.

Crawford, M. H. (2000). Anthropological genetics in the twenty-first century: Introduction. *Human Biology*, **72**, 3–13.

Crawford, M. H. (2004). History of human biology (1929–2004). *Human Biology*, **76**, 805–815.

Crawford, M. H. (2007). Foundations of anthropological genetics. In *Anthropological Genetics: Theory, Methods and Applications*, M. H. Crawford (ed.). Cambridge: Cambridge University Press, pp. 1–16.

Crawford, M. H., Leonard, W. R. and Sukernik, R. I. (1992). Biological diversity and ecology in the Evenkis of Siberia. *Man and the Biosphere Northern Sciences Network Newsletter*, **1**, 13–14.

Damon, A. (1974). Human ecology in the Solomon Islands: biomedical observations among four tribal societies. *Human Ecology*, **2**, 191–215.

Darwin, C. R. (1839). *Journal of Researches into the Geology and Natural History of the Various Countries Visited by H.M.S. Beagle, Under the Command of Captain FitzRoy, R.N. from 1832 to 1836*. London: Henry Colburn.

Davis, K. and Blake, J. (1956). Social structure and fertility: an analytic framework. *Economic Development and Culture Change*, **4**, 211–235.

Eiseley, L. (1958). *Darwin's Century: Evolution and the Men Who Discovered It*. Garden City, NY: Doubleday.

Ellison, P. T. (1990). Human ovarian function and reproductive ecology: new hypotheses. *American Anthropologist*, **92**, 933–952.

Ellison, P. T. (ed.). (2001). *Reproductive Ecology and Human Evolution*. New York: Aldine de Gruyter.

Eveleth, P. B. and Tanner, J. M. (eds) (1976). *Worldwide Variation in Human Growth*. International Biological Programme 8. Cambridge: Cambridge University Press.

Eveleth, P. B. and Tanner, J. M. (eds) (1990). *Worldwide Variation in Human Growth*, 2nd edn. Cambridge: Cambridge University Press.

Francesconi, R., Byrom, R. and Mager, M. (1986). The United States Army Research Institute of Environmental Medicine: First Quarter Century. *The Physiologist*, **29**(5) Suppl., 58–62.

Frisch, R. E. and McArthur, J. W. (1974). Menstrual cycles: fatness as a determinant of minimum weight for height necessary for their maintenance or onset. *Science*, **185**, 949–951.

Frisch, R. E. and Revelle, R. (1970). Height and weight at menarche and a hypothesis of critical body weights and adolescent events. *Science*, **169**, 397–399.

Gajdusek, D. C. (1977). Unconventional viruses and the origin and disappearance of Kuru. *Science*, **197**, 943–960.

Garn, S. M. (1961). *Human Races*. Springfield, IL: C. C. Thomas.

Garn, S. M. (1963). Human biology and research in body composition. *Annals of the New York Academy of Sciences*, **110**, 429–446.

Garn, S. M. and Shamir, Z. (1958). *Methods for Research in Human Growth*. Springfield, IL: C. C. Thomas.

Garruto, R. M., Way, A. B., Zansky, S., et al. (1989). Natural experimental models in human biology, epidemiology, and clinical medicine. In *Human Population Biology: a Transdisciplinary Science*, M. A. Little and J. D. Haas (eds). New York: Oxford University Press, pp. 82–109.

Garruto, R. M., Little, M. A., James, G. D., et al. (1999). Natural experimental models: the global search for biomedical paradigms among traditional, modernizing, and modern populations. *Proceedings of the National Academy of Sciences of the United States of America*, **96**, 10536–10543.

Goldstein, M. S. (1940). Recent trends in physical anthropology. *American Journal of Physical Anthropology*, **26**, 191–209.

Goldstein, M.S. (1943). *Demographic and Bodily Changes in Descendants of Mexican Immigrants*. Austin: Institute of Latin American Studies, University of Texas.

Gould, S.J. (1996). *The Mismeasure of Man, Revised and Expanded*. New York: W. W. Norton.

Gravlee C.C., Bernard H.R. and Leonard, W.R. (2003). Heredity, environment, and cranial form: A reanalysis of Boas's immigrant data. *American Anthropologist*, **105**, 125–138.

Gray, S.J. (1994). Comparison of effects of breast-feeding practices on birth-spacing in three societies: nomadic Turkana, Gainj, and Quechua. *Journal of Biosocial Science*, **26**, 69–90.

Hammel, H.T. (1960). *Thermal and Metabolic Responses of the Alakaluf Indians to Moderate Cold Exposure*. US Air Force Systems Command Research and Technology Division, Technical Report WAAD-TR-60-633. Ohio: Wright-Patterson Air Force Base.

Hammer, M.F. (1995). A recent common ancestry for human Y-chromosomes. *Nature*, **378**, 376–378.

Hanna, J.M., Friedman, S.M. and Baker, P.T. (1972). The status and future of US Human Adaptability research in the International Biological Program. *Human Biology*, **44**, 381–398.

Harrison, G.A. (1982). The past 50 years of human population biology in North America: An outsider's view. In *A History of American Physical Anthropology 1930–1980*, F. Spencer (ed.). New York: Academic Press, pp. 467–472.

Harrison, G.A. (1997). The role of the Human Adaptability International Biological Programme in the development of human population biology. In *Human Adaptability: Past, Present, and Future*, S.J. Ulijaszek and R. Huss-Ashmore (eds). Oxford: Oxford University Press, pp. 17–25.

Harrison, G.A. and Collins, K.J. (1982). In memoriam: Joseph Sydney Weiner (1915–1982). *Annals of Human Biology*, **9**, 583–592.

Harrison, G.A., Weiner, J.S., Tanner, J.M., et al. (1964). *Human Biology: an Introduction to Human Evolution, Variation, and Growth*. Oxford: Oxford University Press.

Hiernaux, J. (1966). Peoples of Africa from 22° to the equator: Current knowledge and suggestions for future research. In *The Biology of Human Adaptability*, P.T. Baker and J.S. Weiner (eds). London: Clarendon Press, pp. 91–110.

Hill, K. (1993). Life history theory and evolutionary anthropology. *Evolutionary Anthropology*, **2**(3), 78–88.

Hooton, E.A. (1931). *Up from the Ape*. New York: Macmillan.

Hooton, E.A. (1939). *The American Criminal: an Anthropological Study*. Cambridge, MA: Harvard University Press.

Howell, N. (1979). *Demography of the Dobe !Kung*. New York: Academic Press.

Howells, W.W. (1987). Introduction. In *The Solomon Islands Project: a Long-Term Study of Health, Human Biology, and Culture Change*, J.S. Friedlaender (ed.). Oxford: Clarendon Press, pp. 3–13.

Irving, L., Andersen, K.L., Bolstad, A., et al. (1960). Metabolism and temperature of Arctic Indian men during a cold night. *Journal of Applied Physiology*, **15**, 635–644.

Johnston, F.E. and Little, M.A. (2000). History of human biology in the United States of America. In *Human Biology: an Evolutionary and Biocultural Perspective*, S. Stinson, B. Bogin, R. Huss-Ashmore and D. O'Rourke (eds). New York: Wiley-Liss, pp. 27–46.

Johnston, F.E. and Low, S.M. (1984). Biomedical anthropology: an emerging synthesis in anthropology. *Yearbook of Physical Anthropology*, **27**, 215–227.

Keys, A., Brožek, J., Henschel, A., et al. (1950). *The Biology of Human Starvation*. Minneapolis: University of Minnesota Press.

Kingsland, S. (1984). Raymond Pearl: On the frontier in the 1920s. *Human Biology*, **56**, 1–18.

Konner, M. and Worthman, C. (1980). Nursing frequency, gonadal function, and birth spacing among !Kung hunter-gatherers. *Science*, **207**, 788–791.

Lampl, M. and Johnson, M.L. (1993). A case study of daily growth during adolescence: a single spurt or changes in the dynamics of saltatory growth? *Annals of Human Biology*, **20**, 595–603.

Lampl, M., Veldhuis, J.D. and Johnson, M.L. (1992). Saltation and status: a model of human growth. *Science*, **258**, 801–803.

Lasker, G.W. (1946). Migration and physical differentiation. A comparison of immigrant and American-born Chinese. *American Journal of Physical Anthropology*, **4**, 273–300.

Lasker, G.W. (1952). Environmental growth factors and selective migration. *Human Biology*, **24**, 262–289.

Lasker, G.W. (1969). Human biological adaptability. *Science*, **166**, 1480–1486.

Lasker, G.W. (1989). Genetics in the journal *Human Biology*. *Human Biology*, **61**, 615–627.

Lasker, G.W. (1999). *Happenings and Hearsay: Experiences of a Biological Anthropologist*. Detroit: Savoyard Books.

Lee, R.B. and DeVore, I. (1976). *Kalahari Hunter-Gatherers: Studies of the !Kung San and their Neighbors*. Cambridge, MA: Harvard University Press.

Leidy, L.E. (1996). Lifespan approach to the study of human biology: an introductory overview. *American Journal of Human Biology*, **8**, 699–702.

Leslie, P.W., Campbell, K.L. and Little, M.A. (1994). Reproductive function in nomadic and settled women of Turkana, Kenya. In *Human Reproductive Ecology: Interactions of Environment, Fertility, and Behavior*, K.L. Campbell and J.W. Wood (eds). *Annals of the New York Academy of Sciences*, **709**, 218–220.

Little, M.A. (1982). The development of ideas on human ecology and adaptation. In *A History of American Physical Anthropology: 1930–1980*, F. Spencer (ed.). New York: Academic Press, pp. 405–433.

Little, M.A. and James, G.D. (2005). A brief history of the Human Biology Association: 1974–2004. *American Journal of Human Biology*, **17**, 141–154.

Little, M.A. and Leslie, P.W. (1993). Migration. In *Research Strategies in Human Biology: Field and Survey Studies*, G.W. Lasker and C.G.N. Mascie-Taylor (eds). Cambridge Studies in Biological Anthropology. Cambridge: Cambridge University Press, pp. 62–91.

Little, M.A. and Leslie, P.W. (eds) (1999). *Turkana Herders of the Dry Savanna: Ecology and Biobehavioral Response of Nomads to an Uncertain Environment*. Oxford: Oxford University Press.

Little, M. A., Leslie, P. W. and Baker, P. T. (1997). Multidisciplinary research of human biology and behavior. In *History of Physical Anthropology: an Encyclopedia*, F. Spencer (ed.). New York: Garland Publishing, pp. 695–701.

Livingstone, F. B. (1958). Anthropological implications of sickle cell gene distribution in West Africa. *American Anthropologist*, **60**, 533–562.

Marks, J. (1995). *Human Biodiversity: Genes, Race, and History*. New York: Aldine de Gruyter.

Mayr, E. (1972). The nature of the Darwinian revolution. *Science*, **176**, 981–989.

Meehan, J. P. (1955). Individual and racial variations in vascular response to cold stimulus. *Military Medicine*, **116**, 330–334.

Merriwether, D. A. (2002). A mitochondrial perspective on the peopling of the New World. In *The First Americans: the Pleistocene Colonization of the New World*, N. G. Jablonski (ed.). *Memoirs of the California Academy of Sciences*, no. **27**, pp. 295–310.

Mielke, J. H., Konigsberg, L. W. and Relethford, J. H. (2006). *Human Biological Variation*. New York: Oxford University Press.

Milan, F. A. (1980). *The Human Biology of Circumpolar Populations*. Cambridge: Cambridge University Press.

Mitchell, H. H. and Edman, M. (1951). *Nutrition and Climatic Stress: with Particular Reference to Man*. Springfield, IL: C. C. Thomas.

Molnar, S. (2002). *Human Variation: Races, Types, and Ethnic Groups*, 5th edn. Upper Saddle River, NJ: Prentice Hall.

Neel, J. V. (1949). The inheritance of sickle cell anemia. *Science*, **110**, 64–66.

Neel, J. V. (1951). The population genetics of two inherited blood dyscrasias in man. In *Origin and Evolution of Man*, K. B. Warren (ed.). *Cold Spring Harbor Symposia on Quantitative Biology*, vol. **15**. Cold Spring Harbor, NY: The Biological Laboratory, pp. 141–158.

Neel, J. V., Layrisse, M. and Salzano, F. M. (1977). Man in the tropics: The Yanomama Indians. In *Population Structure and Human Variation*, G. A. Harrison (ed.). Cambridge: Cambridge University Press, pp. 109–142.

Newman, M. T. (1953). The application of ecological rules to the racial anthropology of the aboriginal New World. *American Anthropologist*, **55**, 311–327.

Newman, M. T. (1956). Adaptation of man to cold climates. *Evolution*, **10**, 101–105.

Newman, M. T. (1960). Adaptations in the physique of American Aborigines to nutritional factors. *Human Biology*, **32**, 288–313.

Newman, M. T. (1962). Ecology and nutritional stress in man. *American Anthropologist*, **64**, 22–33.

Newman, R. W. and Munro, E. H. (1955). The relation of climate and body size in US males. *American Journal of Physical Anthropology*, **13**, 1–17.

O'Rourke, D. H. (2000). Genetics, geography, and human variation. In *Human Biology: an Evolutionary and Biocultural Perspective*, S. Stinson, B. Bogin, R. Huss-Ashmore et al. (eds). New York: Wiley-Liss, pp. 87–133.

Ramachandran, S. (2005). Support from the relationship of genetic and geographic distance in human populations for a serial founder effect originating in Africa. *Proceedings of the National Academy of Sciences of the United States of America*, **102**, 15942–15947.

Reardon, J. (2005). *Race to the Finish: Identity and Governance in an Age of Genomics*. Princeton: Princeton University Press.

Relethford, J. H. (2004). Boas and beyond: Migration and craniometric variation. *American Journal of Human Biology*, **16**, 379–386.

Riley, M. W. (1979). Introduction: life-course perspectives. In *Aging from Birth to Death: Interdisciplinary Perspectives*, M. W. Riley (ed.). Boulder, CO: Westview Press, pp. 3–13.

Roberts, D. F. (1952). Basal metabolism, race and climate. *Journal of the Royal Anthropological Institute*, **82**, 169–183.

Roberts, D. F. (1953). Body weight, race and climate. *American Journal of Physical Anthropology*, **11**, 533–558.

Roberts, D. F. (1956). A demographic study of a Dinka village. *Human Biology*, **28**, 323–349.

Roberts, D. F. (1960). Effects of race and climate on human growth as exemplified by studies on African children. In *Human Growth*, J. M. Tanner (ed.). Oxford: Pergamon Press, pp. 59–72.

Roberts, D. F. (1965). Assumption and fact in anthropological genetics. *Journal of the Royal Anthropological Institute*, **95**, 87–103.

Roche, A. F. (1992). *Growth, Maturation and Body Composition: the Fels Longitudinal Study 1929–1991*. Cambridge: Cambridge University Press.

Salzano, F. M. and Callegari-Jacques, S. M. (1988). *South American Indians: a Case Study in Evolution*. Oxford: Clarendon Press.

Scholander, P. F., Andersen, K. L., Krog, J., et al. (1957). Critical temperature in Lapps. *Journal of Applied Physiology*, **10**, 231–234.

Scholander, P. F., Hammel, H. T., Hart, S. J., et al. (1958). Cold adaptation in Australian Aborigines. *Journal of Applied Physiology*, **13**, 211–218.

Schraer, H. and Newman, M. T. (1958). Quantitative roentgenography of skeletal mineralization in malnourished Quechua Indian boys. *Science*, **128**, 476–477.

Schreider, E. (1950). Geographic distribution of the body weight/body surface ratio. *Nature*, **165**, 286.

Schreider, E. (1951). Anatomic factors in body heat regulation. *Nature*, **167**, 823–824.

Schull, W. J. and Rothhammer, F. (eds) (1990). *The Aymara: Strategies in Human Adaptation to a Rigorous Environment*. Dordrecht, the Netherlands: Klewer.

Shapiro, H. L. and Hulse, F. S. (1939). *Migration and Environment: a Study of the Physical Characteristics of the Japanese Immigrants to Hawaii and the Effects of Environment on Their Descendants*. London: Oxford University Press.

Sheldon, W. H., Stevens, S. S. and Tucker, W. B. (1940). *The Varieties of Human Physique: an Introduction to Constitutional Psychology*. New York: Harper.

Shipman, P. (1994). *The Evolution of Racism: Human Differences and the Use and Abuse of Science*. New York: Simon and Schuster.

Sokal, R. R. (1988). Genetic, geographic, and linguistic distances in Europe. *Proceedings of the National Academy of Sciences of the United States of America*, **85**, 1722–1726.

Sparks C.S. and Jantz, R.L. (2002). A reassessment of human cranial plasticity: Boas revisited. *Proceedings of the National Academy of Sciences of the United States of America*, **99**, 14636–14639.

Stocking, G.W. Jr (ed.) (1974). *The Shaping of American Anthropology, 1883–1911: a Franz Boas Reader*. New York: Basic Books.

Stocking, G.W. Jr (1987). *Victorian Anthropology*. New York: The Free Press.

Stoneking, M. and Cann, R.L. (1989). African origin of human mitochondrial DNA. In *The Human Revolution: Behavioral and Biological Perspectives on the Origin of Modern Humans*, P.A. Mellars and C.B. Stringer (eds). Princeton: Princeton University Press, pp. 17–30.

Sunderland, E. (1975). In Memoriam: Nigel Ashworth Barnicot (1914–1975). *Annals of Human Biology*, **2**, 399–403.

Szathmáry, E.J.E. (1991). *Reflections on Fifty Years of Anthropology and the Role of the Wenner-Gren Foundation: Biological Anthropology*. In Report for 1990 and 1991: Fiftieth Anniversary Issue. New York: Wenner-Gren Foundation for Anthropological Research, pp. 18–30.

Tanner, J.M. (1959). Boas' contributions to knowledge of human growth and form. In *The Anthropology of Franz Boas: Essays on the Centennial of His Birth*, W. Goldschmidt (ed.). Memoir no. 89 of the American Anthropological Association. San Francisco: Howard Chandler, pp. 76–111.

Tanner, J.M. (1981). *A History of the Study of Human Growth*. Cambridge: Cambridge University Press.

Tanner, J.M. (1999). The growth and development of the *Annals of Human Biology*: A 25-year retrospective. *Annals of Human Biology*, **26**, 3–18.

Ulijaszek, S.J. and Huss-Ashmore, R. (eds) (1997). *Human Adaptability: Past, Present, and Future*. Oxford: Oxford University Press.

Vigilant, L., Stoneking, M., Harpending, H., et al. (1991). African populations and the evolution of human mitochondrial DNA. *Science*, **253**, 1503–1507.

Vitzthum, V.J. (1994). Comparative study of breast-feeding structure and its relation to human reproductive ecology. *Yearbook of Physical Anthropology*, **37**, 307–349.

Warren, K.B. (ed.) (1951). *Origin and Evolution of Man*. Cold Spring Harbor Symposia on Quantitative Biology, vol. **15**. Cold Spring Harbor, NY: The Biological Laboratory.

Washburn, S.L. (1951). The new physical anthropology. *Transactions of the New York Academy of Sciences*, Series II, **13**, 298–304.

Weiner, J.S. (1954). Nose shape and climate. *American Journal of Physical Anthropology*, **12**, 1–4.

Weiner, J.S. (1958). Courses and training in physical anthropology and human biology. In *The Scope of Physical Anthropology and Its Place in Academic Studies*, D.F. Roberts and J.S. Weiner (eds). Oxford: Society for the Study of Human Biology and the Wenner-Gren Foundation for Anthropological Research, pp. 43–50.

Weiner, J.S. (1965). *International Biological Programme Guide to the Human Adaptability Proposals*. London: International Council of Scientific Unions, Special Committee for the IBP.

Weiner, J.S. (1966). Major problems in human population biology. In *The Biology of Human Adaptability*, P.T. Baker and J.S. Weiner (eds). Oxford: Clarendon Press, pp. 1–24.

Weiner, J.S. (1977). History of the Human Adaptability Section. In *Human Adaptability: a History and Compendium of Research*, K.J. Collins and J.S. Weiner (eds). London: Taylor and Francis, pp. 1–31.

Weiner, J.S. (1982). History of physical anthropology in Great Britain. *International Association of Human Biologists, Occasional Papers*, **1**(1), 1–33.

Weiner, J.S. and Lourie, J.A. (1969). *Human Biology: a Guide to Field Methods*. IBP Handbook No. 9. Philadelphia: Davis.

Weiss, K.M. and Chakraborty, R. (1982). Genes, populations, and disease (1930–1980): A problem-oriented review. In *A History of American Physical Anthropology, 1930–1980*, F. Spencer (ed.). New York: Academic Press, pp. 371–404.

Wyndham, C.H., Bouwer, W.M., Devine, M.G., et al. (1952). Physiological responses of African laborers at various saturated air temperatures, wind velocities, and rates of energy expenditure. *Journal of Applied Physiology*, **5**, 290–298.

4 Genetics in Human Biology

Robert J. Meier and Jennifer A. Raff

By now you most likely have discerned that human biologists focus much of their research on variation. Their studies have investigated humans at multiple levels of organization and interaction, from within cells to between large populations. The primary subject of this chapter will be genes, and our aims are to define what genes are and what they do, how they become variable, how they are transmitted between generations, and how they undergo evolutionary processing. More formally, the areas to be addressed are Mendelian genetics, human genetics, molecular genetics, and population genetics. In addition, there will be some discussion of newly developing research areas of interest to human biologists, for instance, epigenetics. Along the way we will point out where certain topics covered here are addressed, and often more fully presented, in other chapters of the volume. We will begin with a brief look back into history when earlier notions of hereditary transmission began to be transformed into increasingly more accurate foundations that eventually led to our current understanding of the nature of genes.

PARTICULATE THEORY OF INHERITANCE

A prevailing notion up through the nineteenth century was that parents passed on to their offspring equal portions of their traits, such as stature or skin color, that blended into an inseparable mixture. Thus, for example, a mating between a tall and short parent would result in children of intermediate height, who themselves would then go on to produce children of intermediate height. Hereditary transmission of traits according to this scheme was known as the "blending theory" of inheritance. This scheme appears to have some plausibility in selected observations, as, in many cases, children do in fact look intermediate to their parents.

However, a major problem with the "blending theory" was that, over time, it should cause variation to diminish in each generation, and ultimately to disappear. (This conundrum was a significant issue for Darwin, who, while recognizing the importance

of variation for the process of evolution, could not develop a theory of evolution that did not involve inheritance blending; his critics were quick to point this problem out to him.) In reality, it was apparent that variation was not only maintained but could even be enhanced over time (through selective breeding – domesticated animals, plants). Hence, somehow parental contributions were temporarily combined in children but then were subject to redistribution in subsequent generations. And the discovery of this notion of heredity, called the "particulate theory" of inheritance, is largely attributed to Gregor Mendel.

Gregor Mendel (1822–1884) was a clergyman, a monk, and later an abbot at a monastery in Brno (now in the Czech Republic), where he taught high science and mathematics, while also carrying out extensive research in plant hybridization (on the common pea – *Pisum sativum*). One of the principal experimental findings, reported in this translated and reprinted 1865 paper (Mendel, 1962), was that certain paired "differentiating characters," such as pea-pod color, present in parents as either yellow or green, did not "blend" in the F_1 hybrid generation. Instead, there was almost exclusively one color form. However, both yellow and green pod colors once again appeared in the F_2 generation produced from the cross of F_1 hybrids. Hence, his "characters" seemed to be transmitted unchanged from parents to subsequent generations. This finding is now known as the Principle of Segregation, which will be defined more precisely below.

After completing numerous hybrid and descendent crosses involving a total of seven paired "differentiating characters" of the pea plants, and after compiling a truly remarkably close statistical conformity to expected probabilities, Mendel concluded that each of these "characters" was inherited without change and without affecting the occurrence of each other. We now refer to this interpretation as the Principle of Independent Assortment, which also will be covered below.

Mendel's research and his Principles of Inheritance apparently were not known to Charles Darwin (1809–1882), who incidentally had devised an ingenious

Human Evolutionary Biology, ed. Michael P. Muehlenbein. Published by Cambridge University Press. © Cambridge University Press 2010.

notion of inheritance known as Pangenesis, that involved units of heredity he called gemmules. Since Darwin proposed that gemmules were subject to modification during an organism's lifetime, Pangenesis is often likened to a prevailing theory of the day, that of "inheritance of acquired characteristics," usually closely identified with the writings of an eighteenth/ nineteenth century French naturalist, Jean-Baptiste Lamarck.

In actuality, it was not until around the turn of the twentieth century that Mendel's work was "rediscovered" by two European researchers, Carl Correns and Hugo de Vries, and possibly a third, Erich von Tschermak. De Vries is generally credited as the first to acknowledge Mendel's prior groundbreaking research. Around this time, in 1903, other historically relevant events occurred, including the relabeling by Johannsen of Mendel's "characters" as "genes" and the formation of the Sutton–Boveri hypothesis that stated that "characters" were carried on chromosomes (see Peters, 1962).

MENDELIAN GENETICS

With this abbreviated background into the history of genetics, we can now proceed to introduce the basic concepts and terminology associated with Mendelian genetics; that is, a description of how gene transmission and inheritance operates. Illustrations and examples will apply to humans, but the discussion will build on the essentials of Mendel's and other early relevant scientific discoveries in genetics that may have dealt with nonhuman organisms for which an understanding of basic principles applies equally.

Mendel's Principles of Inheritance can be illustrated through two widely known human blood groups, the ABO and Rhesus (Rh) systems. (See Chapter 13 of this volume for an in-depth coverage of these and other classical genetic markers.) Of clinical significance (in terms of adverse blood transfusion reactions due to mismatching blood types), the ABO system is inherited according to three gene variants, or more formally termed "alleles," A, B, and O. Parents will produce gametes that contain only one allele of the gene-pair found in their genotype. Then at fertilization, any two of these alleles combine in forming a genotypic makeup of AA, AO, BB, BO, OO, or AB. Genotypes are designated as homozygous if they have the same alleles, as in AA, BB, and OO, or heterozygous if they have different alleles, as in AO, BO, and AB.

Phenotypic expression in the ABO system, which can be observed in a serological test, is controlled by inheritance rules in which the A and B alleles are dominant over O, and are codominant to each other. In the terminology that is consistent with Mendel's use, the O allele is recessive to both A and B.

Parents	AO	×	AO
Gametes	A, O		A, O
			_A_ _O_
F_1 offspring	A |	AA	AO
	O |	AO	OO

4.1. Principle of Segregation.

Considering advancements following Mendel's work, codominance means that A and B alleles are expressed equally in the phenotype. In sum, the observed blood types are A, B, AB, and O. Of note here, both the A and B blood types could be concealing the presence of an O allele that would reappear for example, if parental genotypes AO and BO were to mate and produce offspring having the OO genotype. In this illustration of the Principle of Segregation, alleles making up the parental genotypes AO and AO separate cleanly during the formation of gametes, and this can result in novel recombined paired alleles or genotypes, as shown in Figure 4.1.

This mating results in a phenotypic segregation ratio as 3(A):1(O), and genotypic ratio of 1(AA):2(AO):1(OO). An exception to the segregation principle occurs when homologous chromosomes fail to separate during meiosis, a condition termed nondisjunction. Nondisjunction of chromosome 21 is a major cause of Down syndrome, and sex chromosome aneuploidy (abnormal chromosome complement) has resulted in such conditions as XO (Turner female), XXY (Kleinfelter male), XXX females, and XYY males.

In a simplified version of the Rh system, which, once again is serologically important for transfusion purposes, two alleles are present, D and d. These combine to form three genotypes, DD, Dd, and dd, yet produce only two phenotypes, D-positive and D-negative, since the d allele is recessive to the D allele. Mendel's Principle of Independent Assortment can be demonstrated by the fact that the inheritance of a person's ABO blood type is not conditioned by that person's Rh blood type. Gametes contain all possible combinations of the two alleles at the two loci. This is illustrated by the Punnett square presented in Figure 4.2. The mating pair in Figure 4.2 yielded a phenotypic segregation ratio of 9:3:3:1, or:

- nine persons who are blood type AD-positive
- three persons who are blood type AD-negative
- three persons who are blood type OD-positive
- one person who is blood type OD-negative.

The ABO and Rh blood groups are inherited independent of each other because they happen to be located on separate chromosomes and thus are not

Parents	AO, Dd		×	AO, Dd		
Gametes	AD, Ad			AD, Ad		
	OD, Od			OD, Od		
F_1 offspring		_AD_	_Ad_	_OD_	_Od_
	AD |	AA DD	AA Dd	AO DD	AO Dd
	Ad |	AA Dd	AA dd	AO Dd	AO dd
	OD |	AO DD	AO Dd	OO DD	OO Dd
	Od |	AO Dd	AO dd	OO Dd	OO dd

4.2. Principle of Independent Assortment.

linked. Of experimental importance and historical significance, Mendel's seven paired sets of "differentiating characters" of garden peas were each located on separate chromosomes and hence were not linked, and all assorted independently from one another. It is not known for sure if Mendel's choice of traits was fortuitous or by design, or perhaps some of each. Interestingly, later experiments between 1905 and 1908 on inheritance patterns in sweet peas (a close relative of garden peas) done by Bateson and Punnett did demonstrate linkage, yet they seemed uncertain of how to interpret their results (Bateson and Punnett, 1962).

Looking ahead to our discussion of evolutionary genetics, it is important to point out here that Mendel's two principles of inheritance are significant ways for secondarily producing genetic variation that ultimately arises through mutation. Segregation of alleles and independent assortment, and also the reshuffling of maternal and paternal chromosomes, produces offspring that possess multiple combinations of their parental genotypes. Indeed, sexual reproduction offers continuing replenishment each generation of more variation which evolution can operate upon from new variants supplied by mutation. As if there might even be a call for more variants, chromosomes regularly break up linked alleles and recombine them through a crossing-over process. When crossing-over takes place between nearby loci, previously linked alleles are recombined. However, the probability of a cross-over event occurring depends on how closely the linked loci are positioned. Tightly linked loci may greatly reduce the likelihood of a crossing-over involving them. Conversely, widely separated loci may have a 50% chance of a cross-over between them, and that amounts to independent assortment. This process perhaps is most important for evolution if it takes place between homologous chromosomes, and is equal and reciprocal. Overall, the rate of crossing-over is relatively low and

thus does not contribute to genetic variation nearly as much as segregation within loci and independent assortment of chromosomes.

Additional sources of genetic variation are in the investigative stage at this time, and not yet fully understood in terms of frequency nor in evolutionary significance. Included here are transposons ("jumping genes") that are readily found throughout the genome as repeated DNA sequences. These sometimes are referred to as "junk" DNA, but their functions, possibly regulatory in nature, are beginning to be understood (Ahnert et al., 2008). Finally, there is some speculation that horizontal gene transfer, which is known to have played an important role in bacterial evolution, might also be found in higher organisms.

We leave this section on Mendelian genetics with a note of appreciation for the work that built a foundation upon which Darwin's evolutionary thinking could thrive, but only after some needed reconciliation between geneticists and proponents of evolution. Human biology researchers were obviously beneficiaries of this important historical development. This matter will be covered under the topic of "a Modern Synthesis" later in the chapter.

HUMAN GENETICS AND MODES OF INHERITANCE

As noted above, alleles at a given locus interact with each other in expressing a phenotype. For all the traits studied by Mendel, this was one of dominance and recessivity. Of some historical interest, it was initially thought that the dominance referred to traits that would become predominant or that their frequency would increase at the expense of recessive traits. That notion was dispelled by G. H. Hardy (1962) in his 1908 paper that essentially established the Hardy (later joined with Weinberg) Principle of Equilibrium. This development will be taken up later.

Here, we generalize on the concept of allelic interaction. To do so, first requires that a distinction be made among chromosomes found in a karyotype. Karyotype refers to the total complement of chromosomes present in the cell nucleus, at one time thought in humans to number 16, later 48, and then shown for certain to be 46 (Tjio and Levan, 1956). Chromosomes occur in 23 homologous pairs, each half of which had descended from the maternal and paternal lines. Twenty-two pairs are designated as autosomes, while the twenty-third pair constitutes the sex chromosomes, either XX for females or XY for males. Figure 4.3 shows a karyotype for a human male, denoting chromosomes by number and by groupings.

With this background on chromosomes we can now distinguish the kinds of interactions that alleles

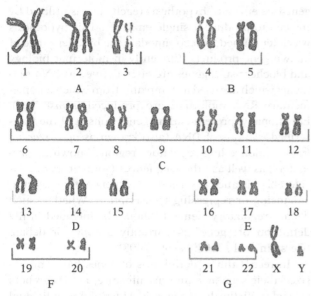

4.3. Karyotype of a human male. From http://homepages.uel.uk/V.K.Sieber/human.htm.

can express. This is simply stated in terms of phenotypic traits, or modes of inheritance categorized as autosomal dominants or recessives, and sex-linked dominants or recessives. Just as in the experiments performed by Mendel, human researchers examined the outcomes of many matings across successive generations of many pedigrees to establish these modes of inheritance. Twin studies were also effectively employed to elucidate inheritance patterns, which is now more precisely aided through methods of genetic analysis and DNA testing. These issues will be discussed later.

Most basic phenotypic expressions were observed during the discovery of human blood groups during the first half of the twentieth century. We have already cited the ABO blood group that was discovered in 1900 within which the *A* and *B* alleles were found to be dominant to the recessive *O* allele. To this can be added a host of phenotypic traits, with dubious functional purpose, that are shown in human pedigrees to be autosomally inherited, and for the most part, to be either dominant or recessive in their expression. This list includes such traits as earlobe attachment, hitchhiker's thumb, absence of a widow's peak, and no freckles, all of which are deemed to be recessively inherited. At one time, it was thought that eye (iris) color was determined by a simple Mendelian inheritance pattern whereby brown was dominant and blue recessive, but that model has since been replaced by one that posits several genes in control of the trait (Duffy et al., 2007). As would be expected for autosomal traits, there should be an equal frequency expressed in both sexes.

On the other hand, X-linked traits, if they are recessive occur mostly in hemizygous (haploid X) males.

However, if dominant they will appear more often in females, who can be either homozygous or heterozygous and express the trait. Obviously, there is no father-to-son inheritance of X-linked traits. Normally, Y-linked traits only are found in males, except for very rare instances of crossing-over to the X chromosome. Examples of X-linked recessives are hemophilia and color-blindness (the latter does have other loci that affect its expression). An example of X-linked dominant trait is a human blood group, Xga. Sex-linkage on the Y-chromosome of course relates to loci that control the development of maleness, such as the testis-determining locus.

As might be anticipated, pedigree analysis and genetic family histories did not always yield unambiguous results. We have already mentioned one of these in the case of codominance in the ABO and MN blood groups, where two alleles can be equally expressed. In general, one can observe a range of expression due to incomplete dominance at a locus. This has been found for the phenylthiocarbamide (PTC) taster trait, whose strength of expression is due its interaction with another locus. The PTC trait is discussed further in Chapter 13 of this volume.

Another exception to a single heredity transmission model is that of genetic heterogeneity. For example, retinitis pigmentosa, an eye pathology, may be inherited as an autosomal dominant, autosomal recessive, or be X-linked.

Two additional departures from common modes of inheritance are incomplete penetrance, i.e., whether the possessed genotype is expressed at all, and variable expressivity, where persons with the same genotype express a trait to varying degrees in their phenotype. For medically important traits, this variability presents as clinical severity. Reduced penetrance and variable expressivity in the phenotype probably result from extenuating circumstances, implicating aging and environmental effects, or possibly interaction with other genes. Late-onset diseases, such as Huntington's disease (under autosomal-dominant control), do not manifest themselves in 100% of affected persons, but this may be due to these individuals not surviving long enough for the diseases to become clinically diagnosed.

In completing this section on modes of inheritance it is appropriate to include a brief mention of mitochondria. In humans, these organelles are exclusively inherited from the mother, although there has been a reported case of paternal inheritance (Schwartz and Vissing, 2002). Each mitochondrion contains many copies of mitochondrial DNA (mtDNA), and since each copy forms an intact circular structure composed of 16 569 base pairs, they can be considered equivalent to a single gene or locus. Due to a high mutation rate that provides an abundance of genetic variation, mtDNA has been of tremendous value in tracing ancient human matrilineal ancestry and evolution, in seeking out more recent genetic relationships, and in determining personal identification. Some of this research can be complicated due to mtDNA-related diseases (such as Leber hereditary optic neuropathy – LHON) in which cells can have a mixture of normal and mutant mitochondria, a condition known as heteroplasmy (Jacobi et al., 2001).

MOLECULAR GENETICS

Molecular genetics examines the mechanisms by which genetic material and environmental conditions act in concert to influence the phenotype of an organism throughout its development and life, as well as the mechanisms that allow the transmission of an organism's genetic material to its offspring. In this section, we briefly survey these important phenomena, with a particular focus on current findings in human genetics.

Genes

The usual starting point for a discussion of molecular genetics is the definition of the gene itself. The concept of a gene is somewhat nebulous, having been shaped by findings in molecular biology, developmental biology, evolutionary biology, and (most recently) large-scale analyses of the genome, transcriptome, and proteome. In the classical definition, genes are conceptualized as operational units of inheritance with discrete physical locations on chromosomes (hence "locus" is often used near-synonymously with "gene"). A functional definition of the gene was provided by Beadle and Tatum (1941); their elegant mutational analysis on *Neurospora* strains supported the conceptualization of the "one

gene-one enzyme" hypothesis (each gene is utilized by the cell to produce a single enzyme). This hypothesis was later refined to accommodate the broader range of known gene products. Currently in molecular biology and biochemistry, genes are often defined as DNA segments which are used as templates from which complementary RNA molecules are produced (*transcribed*). Functionally, protein-coding genes consist of the transcribed sections of DNA (also known as *open reading frames*), noncoding regulatory regions known as promoters, as well as other sequences (sometimes at considerable distance from coding sequence) that assist in enhancing or repressing transcription. Whether such distant regulatory elements should be included in the definition of "gene" is currently a topic of debate (Gerstein et al., 2007; Pesole, 2008).

In recent times, definitions of genes have emphasized their sequence and architecture, as researchers struggle with the daunting task of identifying individual genes within huge stretches of genomic DNA. Identifying genes on the basis of their DNA sequence is considerably less straightforward for eukaryotic genomes than prokaryotic genomes. Within an open reading frame (*ORF*), a typical eukaryotic gene will have DNA sequence that encodes amino acids (*exons*), often separated by noncoding DNA sequences (*introns*) which are removed (*spliced*) from the RNA prior to translation. The number of introns present within genes varies considerably, but the average vertebrate gene contains 5–8, and introns seem to be surprisingly well conserved (Koonin, 2009). By contrast, prokaryotic genomes do not have introns; instead, there is a direct correspondence between the sequence of prokaryotic DNA and the RNA product of that gene (Figure 4.4).

It is no exaggeration to state that insights from the sequencing of the genomes of humans and major model organisms have revolutionized our understanding

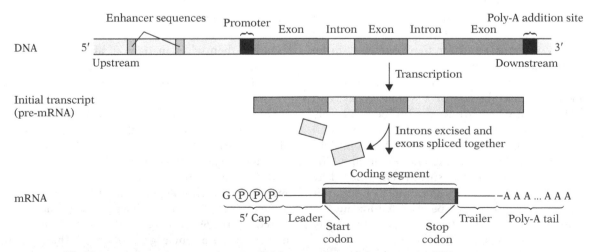

4.4. Diagram of a eukaryotic gene, its initial transcript, and the mature mRNA transcript. From Futuyma (1998), p. 44.

of molecular genetics. For example, emerging research on the transcriptome (the totality of transcripts within a cell) has revealed that our understanding of genes as discrete loci on chromosomes is too simplistic; transcribed regions overlap each other, introns can be used for coding functional products, and a single gene can produce several types of protein. Therefore, although a gene can still be conceptualized as a stretch of DNA sequence that is used to produce RNA or protein, there is still considerable discussion regarding the limitations of this definition (Beurton et al., 2000; Gerstein et al., 2007; Pesole, 2008).

DNA and RNA

The complex molecule that comprises the genetic material – DNA – has been well characterized biochemically since its structure was published by Watson and Crick (1953). DNA is composed of a backbone of covalently bonded phosphates and 2′ deoxyribose (five-carbon) sugars to which nitrogenous bases (adenine, guanine, thymine, and cytosine) are attached, forming a long asymmetric polymer. The terminal phosphate end of the DNA strand is designated as the 5′ end, and the terminal hydroxyl end is designated as the 3′ end. Each of the four bases is able to form noncovalent hydrogen bonds with another base: adenine (A) forms two hydrogen bonds with thymine (T), and guanine (G) forms three hydrogen bonds with cytosine (C). Because of this specific pairing, a strand of DNA is able to form a stable association with another strand having the same sequence of bases in reverse polarity. That is, the 5′ end of one strand sits opposite to the 3′ end of the complementary strand. Although the hydrogen bonds between bases are individually weak, in aggregate they produce a stable (though not unbreakable) double helical structure (Alberts et al., 2002) (Figure 4.5).

The structure of RNA is quite similar to DNA, except that ribose is used as the sugar instead of deoxyribose, and the base uracil (U) is used in place of thymine (T). RNA is often single stranded, and is less stable than DNA. As we discuss later, mRNA undergoes extensive editing in order to generate a molecule stable enough to serve as a template for protein synthesis. Although DNA constitutes the genetic material of all known living organisms, virus genomes are composed of RNA, and it has been hypothesized that early forms of life on earth also utilized RNA as their genetic material (for a review of the "RNA World" hypothesis, see Bartel and Unrau, 1999).

Chromosome architecture

Eukaryotic chromosomes are composed of long stretches of double helical DNA, packaged with

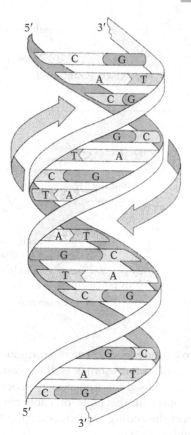

4.5. The DNA double helix. From Futuyma (1998), p. 44.

associated proteins (most notably histones) and RNA molecules. Except during mitosis, this DNA-protein-RNA complex exists in a fairly uncondensed form known as chromatin. Chromatin can be chemically altered and further condensed with the help of histones and RNA to prevent or facilitate access by transcriptional machinery in a process known as *chromatin remodeling*. Portions of the chromosome thus rendered inactive are known as *heterochromatin*, while genes within the less condensed *euchromatin* can be expressed (Wallrath, 1998).

The entire complement of human chromosomes contains an estimated three-billion nucleotide pairs (ENCODE Project Consortium, 2007). As described previously in this chapter, chromosomes are the vehicles for transmitting genetic information from parents to offspring. As such, in addition to being used as the template for the synthesis of proteins and RNA needed for cellular functions, each chromosome also must be duplicated and apportioned correctly into the daughter cells during cell division.

Certain structural features of eukaryotic chromosomes play key roles in these processes. Centromeres are heterochromatic regions found in the center (*metacentric*) or towards the end (*acrocentric*) of chromosomes. Centromeres function during cell division as attachment points for microtubules through structures known as kinetochores, which position chromosomes

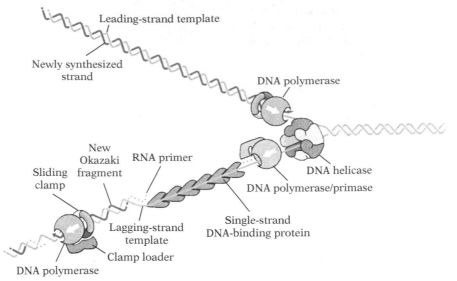

4.6. The mammalian replication fork. From Alberts et al. (2002), p. 254.

during mitosis and meiosis. The other important structural feature of chromosomes is the telomere. Telomeres are composed of noncoding, repeating DNA sequences found at the ends of each chromosome, which protect the coding sequences during DNA replication (Greider, 1998).

Replication

At the end of their description of the structure of DNA, Watson and Crick noted in perhaps the most famous understatement in biology: "It has not escaped our notice that the specific pairing we have postulated immediately suggests a possible copying mechanism for the genetic material" (1953, p. 737). The generation of new cells and the transmission of genetic information from parent to offspring requires the precise duplication of an organism's genome. Understanding the chemical properties of the DNA strands was the key to understanding the mechanisms of this process.

Because the nucleotide bases pair only with their proper complementary base (A with T, G with C), DNA replication is *semiconservative*. That is, each strand of the original chromosome is used as a template to generate a complementary daughter strand. In human chromosomes, replication begins simultaneously at several points per chromosome known as *origins of replication*, located between 5 and 300 kilobases apart from each other. Numerous proteins are required to initiate DNA synthesis. Topoisomerases, helicases, and other associated enzymes work to unwind the two strands of DNA ahead of the replication machinery, stabilize the separated strands and relieve upstream torsional stress caused by the unwinding of DNA. The enzyme that synthesizes the daughter strand of DNA,

known as DNA polymerase, cannot initiate DNA synthesis de novo; it can only add bases to an already existing nucleotide chain. Therefore, short regions of RNA sequence (called *primers*) are created by enzymes on both template strands in order to allow the polymerase to bind and begin synthesis.

Once the DNA polymerase has bound to the origin (assisted by additional proteins), DNA synthesis begins along both strands of the replication fork, terminating when double-stranded DNA is encountered at the next replication origin. DNA polymerase can only add deoxyribonucleotides to the 3′ end of a growing DNA strand. Therefore, synthesis proceeds continuously opposite to the 3′–5′ template strand (called the *leading strand*) but can only synthesize short fragments opposite the 5′–3′ (*lagging*) strand, as their synthesis moves the polymerases away from the replication fork. Numerous short DNA segments, known as *Okazaki fragments*, are thus produced repeatedly by the polymerases, with the intervening gaps filled by an additional enzyme (Meyers, 2007).

Given the polymerase's directional constraints, the replication of the ends of lagging strand eukaryotic chromosomes is problematic. After the RNA primers are removed from the newly synthesized complement, and the Okazaki fragments are ligated together, there still remains an unsynthesized segment at the very end of the DNA strand. Although telomeres protect the ends of coding region sequence from degradation, over successive rounds of replication, chromosomes become progressively shorter without the action of an enzyme called telomerase, which adds DNA repeats to the end of the 3′ end. Telomere shortening appears to be associated with cell senescence (aging) (for a review, see Greider, 1998).

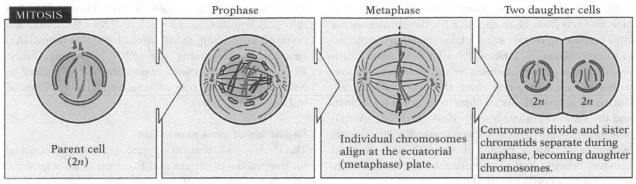

4.7. Mitosis. From Futuyma (1998), p. 32.

An extremely important property of DNA polymerase is its 3'–5' exonuclease ability, which serves as a "proofreader" for the genome. As DNA bases are added to the daughter strands, an occasional erroneous base is incorporated. DNA polymerase detects such mismatches and excises the inappropriate base such that the error rate for DNA replication is estimated to be only one in a billion bases. Although proofreading likely slows the progress of the polymerase, its processivity is enhanced by the presence of clamp proteins, which help to maintain contact between the polymerase and the DNA template (Figure 4.6).

The somatic cell cycle and mitosis

The majority of cells in the human body (*somatic* cells) do not contribute genetic information to the next generation. Instead, they perform a variety of roles in growth and development which constantly require the generation of new cells. The process by which a diploid somatic cell divides to produce two identical diploid daughter cells is known as *mitosis*. The cell cycle consists of four discrete phases: G_1 (during which the cell assesses, via checkpoints, that it is ready for proliferation and DNA replication); S (during which chromosomes are duplicated as described above); G_2 (during which the cell pauses to verify via additional checkpoints that the chromosomes are ready for mitosis); and M (during which chromatin condenses and mitosis occurs).

Mitosis itself consists of several steps. During prophase, chromosomes condense within the nucleus; it is here that the classical "X" shape of sister chromatids (joined at the centromere) can be first observed. The cell undergoes a number of changes to prepare for mitosis. Cellular structures which organize microtubules – *centrosomes* and *centrioles* – duplicate and migrate to opposite poles in the cell. From these positions, they generate long microtubule polymers which make contact with kinetochores on the centromeres of both sister chromatids. The nuclear envelope breaks down during prometaphase, and the chromosomes are "captured" (at the kinetochores of both sister chromatids) by microtubules, which serve to position chromosomes correctly during mitotic events. During metaphase, the chromosomes line up at the approximate center of the cell – an extremely important step for ensuring the correct distribution of each chromosome into the daughter cells. During anaphase the cohesion between the sister chromatids is dissolved and they are pulled apart towards the poles. The nuclear envelope is reformed around the chromosomes during telophase, and following cell division they decondense into transcriptionally active chromatin. Thus, at the end of mitosis each daughter cell has received one copy (in the form of one of the sister chromatids) of each chromosome (for a detailed description of mitosis, see Alberts et al., 2002) (Figure 4.7).

Meiosis

Meiosis is a process by which a diploid germline cell is divided twice to produce four haploid daughter cells called *gametes*. Gametes from each parent are combined during fertilization to produce a diploid zygote, with both halves of its genome contributed equally by the two parents. The initial stages of meiosis are similar to mitosis. DNA replication produces a copy of each chromosome, and microtubules are produced at the spindle poles. However, unlike mitosis, the microtubules attach to only one kinetochore on each pair of joined sister chromatids. During prophase I, homologous chromosomes (one inherited from each parent) pair with each other and recombination takes place between them. As described previously in this chapter, recombination is a major generator of genetic diversity. During metaphase I, the chromosomes form two rows in the center of the cell, with each chromosome positioned across from its homolog. This alignment is critical to ensure the proper ploidy of the gametes. During anaphase I, the homologous chromosomes are pulled apart, one to each pole; both sister chromatids from each chromosome remain attached to each other. Cell division takes place during telophase I, producing two daughter cells, each with only one copy of each duplicated chromosome. During the next phases of meiosis (prophase II–metaphase II), the chromosomes (still consisting of two sister chromatids) are attached to microtubules via kinetochores on both sister chromatids. In a process similar to mitosis, the chromosomes form a single line along the center of the cell. During anaphase II, sister chromatid cohesion is lost, and the sister chromatids are pulled apart, one to each pole of the cell, which subsequently divides during telophase II. As a result of this process, haploid gametes bearing half of the number of chromosomes seen in diploid cells are produced.

Problems with the carefully orchestrated alignments of the chromosomes can lead to nondisjunction – a condition in which the daughter cells receive an improper number of chromosomes. This is responsible for trisomy diseases as have been discussed above.

From DNA to phenotype

Information contained within the genome is utilized by the cell to make proteins and RNA critical for the development and function of all organisms. This process has been characterized as the "Central Dogma" of biology by Francis Crick: DNA is transcribed into RNA, and RNA is translated (ultimately) into protein in a unidirectional process. Subsequent research, however, continues to provide numerous exceptions to this model. Here, we will briefly review the processes of DNA transcription, post-transcription processing, and translation of RNA into protein in eukaryotes, with a special emphasis on lessons learned thus far from the human genome and transcriptome.

Transcription

Transcription is the means by which DNA is used as a template for the production of a complementary RNA molecule. These gene products have diverse roles within the cell. A subset of RNA transcripts is translated into chains of amino acids by cellular machinery to be used as components of proteins. RNA used to convey the genetic instructions to the translation machinery for protein production is known as "messenger" RNA (mRNA). As will be discussed below, the process of creating and using mRNA for protein synthesis in eukaryotic cells is highly complex.

Less than 2% of the human genome contains DNA that codes for protein. Yet, it has recently been reported that the vast majority of the human genome analyzed thus far is transcriptionally active (Birney et al., 2004; ENCODE Project Consortium, 2007). Far from being "junk DNA" as previously labeled, noncoding DNA appears to play an important role in diverse cellular processes. It is known that this (noncoding) RNA is used in many different cellular functions. For example, transfer RNA (tRNA) and ribosomal RNA

(rRNA) serve essential roles in translation of mRNA into polypeptides. Small nuclear RNAs (snRNAs) are involved in splicing, small nucleolar RNAs (snoRNAs) function in processing the rRNA transcript (Eddy, 1999), and long intronic noncoding RNAs appear to have an important role in gene regulation (reviewed in Louro et al., 2009).

Regulation of gene expression

The timing and location of gene expression is critical for normal development and function of an organism, and the regulation of gene activity is an extremely complex process. Although a subset of genes (known as *housekeeping* genes) is constitutively active, environmental conditions can require precise changes in the expression of many genes. Although gene expression occurs during post-transcriptional processing of mRNAs, during translation, and during post-translational modification of polypeptides, its regulation mainly occurs at the transcriptional level.

A major mechanism of transcriptional control involves the structural or chemical alteration of DNA itself. For example, acetylation of histones can cause them to "loosen" DNA, making it more accessible for transcriptional machinery. DNA methylation is a common mechanism for gene silencing. In mammals, this usually consists of methylation (addition of CH_3 groups) of cytosine bases when they are immediately 5′ to guanine bases (designated as "CpG" motifs, which are commonly found in promoters).

One well-studied example of long-term gene regulation is dosage compensation in mammals. Because sex determination in mammals is achieved by the presence of a Y chromosome, male mammals are haploid for the genes on the X chromosome. Dosage compensation is a phenomenon in which all of the genes on one randomly chosen X chromosome in each cell in a female mammal are suppressed so that the levels of gene product will be comparable between the sexes. This appears to be achieved by two mechanisms; chromatin remodeling (induced by the coating of the X chromosome with RNA) initially silences X chromosomal genes during development, followed by DNA methylation to achieve permanent suppression (Cedar and Bergman, 2009).

Transcription is also controlled at the level of individual genes by *cis* regulatory elements (sequences of DNA on the same strand as the transcribed gene). Located within and just upstream of the promoter, these sites bind a family of proteins, called *transcription factors* that are required for transcription initiation; by increasing or decreasing the levels of transcription factors the cell is able to control the rate of transcription of particular genes. Additional control over gene expression is brought about by proteins that bind to genetic elements more distant from the gene, and are able to enhance or repress transcription.

Mechanisms of transcription

Transcription begins with the binding of a transcription factor to the promoter region of a gene. In many eukaryotic genes, the region of the promoter that binds the transcription factor is located 20–30 bases upstream of the transcription start site, and is named the "TATA" box (after a commonly found motif TATAAAA). The transcription factor subsequently recruits to the promoter numerous functionally related proteins that collectively are known as the *preinitiation complex*. Among these proteins is one of several DNA-dependent RNA polymerase enzymes, each of which produces a specific type of RNA. Pol I produces the large rRNA subunit, pol III produces the small rRNA subunit, tRNA, and the small nuclear RNAs, and pol II is used to make mRNA. Unlike DNA polymerase, RNA polymerase is able to bind to and use a single-stranded template. Proteins within the preinitiation complex unwind the DNA helix in the region to be transcribed; only one strand (sometimes referred to as the *Watson* strand) is used as the template for RNA production. Assisted by proteins known as elongation factors, the RNA polymerase generates an RNA polymer complementary to the template sequence – with the exception that uracil is utilized instead of thymine – and terminates transcription at a specific sequence (Meyers, 2007).

Post-transcriptional processing in eukaryotes

Noncoding RNAs undergo various structural modifications, depending on their cellular role. For mRNA, the transcript (*pre-mRNA*) is not immediately translated into a polypeptide sequence, but instead undergoes significant post-transcriptional processing. Specific sequences must be added to the beginning (*cap*) and the end (*tail*) of the pre-mRNA, which specify its fate and protect it from premature degradation.

Additionally, introns must be spliced from the transcript. The completion of the sequencing of the human genome revealed an estimated 20 000–25 000 protein coding genes – a considerably lower number than had previously been predicted based on the number of proteins known (International Human Genome Sequencing Consortium, 2001). Alternative splicing is one mechanism which helps account for the vast diversity of proteins that are produced by a relatively few number of human genes. Variation in transcript splicing frequently results in multiple alternative RNA molecules produced from the same gene.

Translation

After being exported out of the nucleus into the cytoplasm, mRNA is used as a template to generate a chain of amino acids, which are specified by three nucleotide "words" known as *codons*. For a given sequence, there are three possible reading frames; the correct reading frame is specified by an initiation codon (AUG). The template is "read" by protein/RNA complexes known as ribosomes. Ribosomes recruit tRNAs, which are adaptor molecules whose structure contains a three letter "anti-codon" complementary to an mRNA codon, and can bind the appropriate amino acid that codon specifies. Once the mRNA/ribosome/initiator tRNA complex has assembled at the first codon (assisted by numerous associated proteins), the ribosome moves along the mRNA in three base increments, recruiting the appropriate tRNA for each codon. The tRNA transfers its attached amino acid to the growing polypeptide chain, then disassociates from the complex. When the translation machinery reaches the "stop" codon in the mRNA, translation is completed and the complex dissociates.

Proteins

Proteins are three-dimensional macromolecules, composed of one or more linear amino acid polymers (*polypeptides*). Twenty amino acids are utilized by cells to create proteins. These amino acids differ considerably in chemical properties such as size, shape, charge, and affinity for water, and allow for considerable diversity in the form (and therefore function) of proteins.

Proteins have myriad roles in the cell, ranging from structural (such as forming the collagen in skin) to participation in the complex signaling pathways of nearly all biological processes. Thus, the mechanisms behind genetic phenomena such as dominance and recessivity can often be understood biochemically. A loss of function (LOF) mutation, for example, frequently generates a premature stop codon within an exon of the affected gene. The translation of the mRNA product results in a truncated, malfunctioning protein whose inactivity causes the LOF phenotype.

One illustration of this phenomenon can be seen in spinal muscular atrophy (SMA), a neuromuscular disease caused by the progressive degeneration of motor neurons. It has several phenotypes, ranging from mild (muscle weakness) to severe (lethality in infancy). Genetically, SMA is associated with an autosomal recessive LOF mutation caused by deletion of exons 7–8 of the survival motor neuron I (*SMN1*) gene; individuals with one functional copy of this gene do not develop SMA. Individuals also maintain variable numbers of a second version of the *SMN* gene (*SMN2*); copy numbers of *SMN2* are inversely proportional to disease severity. *SMN2* has a near-identical gene sequence to *SMN1*. However, a single mutation in *SMN2* results in a truncated protein product (due to improper splicing) in all but 10% of mRNA transcripts. Multiple copies of *SMN2*, therefore, result in higher levels of functional SMN protein, accounting for the variability of the disease phenotype (Lunke and El-Osta, 2009).

SOME TOOLS OF MOLECULAR GENETICS

DNA isolation

The first step in the extraction of DNA is the breakdown of cellular membranes and other protein components of the cell. After mechanical homogenization of the sample tissue (if applicable), protein is digested with a protease, usually proteinase K. Sodium dodecyl sulfate (SDS) and ethylenediaminetetraacetate (EDTA) are added to this reaction to denature DNAses. DNA is then isolated from the digested protein and other cellular constituents by one of several methods, including phenol-chloroform extraction (which isolates the DNA in the aqueous phase), column chromatography, or binding to silica. DNA can be further concentrated and purified by ethanol precipitation (Sambrook et al., 1989).

Polymerase chain reaction

The development of the polymerase chain reaction (PCR) technique by Kary Mullis in 1984 was a breakthrough for molecular genetics. Polymerase chain reaction enables the generation of millions of copies of any DNA sequence by essentially adopting the cell's own mechanisms for DNA replication. Utilizing a machine known as a thermocycler, double-stranded DNA is "denatured" into single-stranded molecules at high temperatures (95°C) by breaking the hydrogen bonds between nucleotide pairs on the complementary strands. Because the DNA polymerase enzyme only binds to double-stranded DNA, short DNA primers specific for the target region are annealed to both the template strands at a slightly lower temperature (50–60°C). The temperature of primer annealing depends upon the length of the primer sequence and its nucleotide composition. The temperature is increased for the extension reaction (72°C), during which the DNA polymerase adds dNTPs to the growing synthetic strand. This process is repeated for multiple cycles (30–45), generating exponentially increasing numbers of copies with each cycle (Mullis and Faloona, 1987; Innis et al., 1990; Erlich et al., 1991; Meyers, 2007).

Sequencing: Sanger method

Determining the sequence of a particular stretch of DNA begins with a pool of copies of the template of interest. In a process similar to the first steps of PCR, the double-stranded template is melted at high temperatures, and the temperature is then lowered for primer annealing. An extension reaction using DNA polymerase is performed in the presence of a mixture of deoxynucleoside triphosphates containing equal quantities of all four bases (abbreviated dNTPs), as well as dideoxynucleoside triphosphates (ddNTPs) which lack a 3' hydroxyl. When ddNTPs are randomly incorporated to the growing synthetic strand instead of dNTPs, the reaction is terminated. This termination step generates a pool of different-sized fragments complementary to the parent sequence, each terminating at a progressively later nucleotide position. Older sequencing methods incorporated radioactively labeled dATP into the synthetic strands, which were separated by size on a polyacrylamide gel and visualized on an X-ray film. More recently, fluorescently labeled ddNTPs are utilized for this purpose, and automated capillary sequencers have replaced the slab polyacrylamide gels as the means of visualizing the sequences. This has increased the speed and ease of sequencing by many orders of magnitude (Green et al., 1999; Meyers, 2007).

Sequencing: shotgun

The first drafts of the human genome sequence were published in 2001 (and completed in 2003), thus ushering in the genomic (or alternatively called the *postgenomic*) era of molecular genetics (International Human Geome Sequencing Consortium, 2001; Venter et al., 2001). Genome sequencing makes use of the shotgun sequencing method, which circumvents the size limitation on sequencing. Briefly, long DNA molecules are broken into random fragments for library generation. These DNA fragments are cloned into vectors, most often either the bacteriophage M13 or plasmids. Clones are sequenced in enough quantity to provide approximately six- to eight-fold coverage of the genome. The sequences are assembled into contigs (overlapping fragments used to infer contiguous sequences), and assessed for quality; gaps or poor quality regions are then manually resequenced. Now that the molecular nature of the gene and the essentials of DNA sequencing have been discussed, we can turn our attention to the combined expression of multiple genes as they play a role in the adaptive process.

POLYGENIC INHERITANCE AND ADAPTATION STUDIES

Several chapters in this volume deal with phenotypic traits whose inheritance patterns are more complex than those we have been discussing under Mendelian and molecular genetics. The formal distinction is that of monogenic versus polygenic inheritance, although it is unlikely that, even for monogenic or single-gene traits, there are absolutely no other genes involved in their expression. For polygenic inheritance, it is taken for granted that two or more loci participate in gene–gene interactions and also in gene–environment interactions, which will be discussed later. Expression of these traits in animals and plants is what the early evolutionists, including Darwin, observed and studied across biogeographic localities and through geologic time periods. In animal species, researchers noted variation in tooth and body size, body colorization, and novel development of anatomical structures, and acknowledged the adaptive significance of such features as eyes and wings.

This section will cover the basic genetics underlying polygenic inheritance, variously referred to as quantitative or multifactorial inheritance. For illustration, we will employ human eye color, or more precisely, iris color variation among individuals. With but a minimal observation study, iris color can be shown to differ much more than presumed under an earlier monogenic model whereby two phenotypes were expressed, dominant brown over recessive blue. A single Mendelian locus segregating two alleles would yield three genotypes, meaning there was the potential for this model to only account for three phenotypes of dark brown, light brown, and blue, if incomplete dominance was invoked. However, these predictions did not match up with observed variation. Early geneticists who pondered the matter of inheritance of complex traits, such as stature, that showed nearly continuous variation among individuals in a large sample, from taller to shorter, correctly reasoned that there must be more than one set of genes involved. Consequently, they extended Mendelian genetics to the level of polygenic inheritance. Following their lead, if we assume that iris color is controlled by two independent and interacting loci, each segregating two alleles, then the number of genotypes expands to five, which, in turn, corresponds better to categories observed for iris color variation. If yet one more locus is added, then the model would contain three loci, two alleles each, for a total of seven genotypes.

In general, for two allelic models, the number of genotypes can be calculated as $2^n + 1$, where n specifies the number of loci. As the number of loci becomes large, the distribution of genotype classes begins to approach a normal- or bell-shaped curve. Continuously varying traits in humans, such as stature, approximate normal distributions in large samples, in part due to their being under polygenic control, and another major part accounted for by environmental effect. The basic concept of a normal distribution in polygenic inheritance can be traced back to the early part of the twentieth century in such renowned figures as Sir Francis Galton and Sir Ronald A. Fisher (Strachan and Read, 2004).

The model described above would permit the classification of iris colors through a broad range of expression from the darkest of brown, down through hazel and on to the lightest of blue. This model attempts to represent a genetic basis of iris color that is under the control of multiple genes that determine the concentration of the brown pigment melanin, from heaviest to lightest. Melanin concentration as related to skin color is taken up in depth in Chapter 12 of this volume. This model of iris color inheritance further assumes that all of the loci contribute equally to melanin production, and that there are no other pigments involved and no environmental effects on iris color. Another feature of this model is that different genotypes can specify the same phenotype, a phenomenon sometimes referred to as polytypy. It further explains how blue-eyed parents can produce brown-eyed children as melanin concentration alleles at different loci are inherited in a set that represents one of the possible combinations.

We will deal with the role of the environment in the next section, and this effect could mean that a single genotype can result in different phenotypic expression of particular traits or in affecting several aspects of the phenotype (called pleiotropy). An example of pleiotropy is found in sickle cell anemia with multiple cytological, physiological, neurological, and other effects. It is also important to point out that for some traits or conditions hereditary transmission and observed variation remains unclear but are suspected to result from genes having a major effect along with additive influences of other interacting genes. These rather complex situations are being investigated through quantitative trait loci (QTL) analysis, which attempts to map a trait to multiple chromosomes. Positive QTL results for adult height in humans fit a model of a major recessive gene combined with other significantly contributing loci on chromosomes 6, 9, and 12 (Xu et al., 2002).

We can conclude this section by pointing out that many of the adaptive traits covered in Part II of this volume are polygenic expressions whose more precise mode of inheritance forms the basis of ongoing human biological research.

HEREDITY, ENVIRONMENT, AND HERITABILITY

In this section, we will concentrate on an earlier methodology that has some historical significance in human biology. Indeed, it had attempted to shed light on the vexing question of nature versus nurture, of what is more important, genes or environment. However, as we will see, this turns out to be a scientifically invalid question to pose, since it cannot be directly tested. As described above, polygenic traits are under the control of multiple genes that are subject to environmental effects. What was needed in this and related cases was a way to sort out the respective contributions of heredity and environment, that is, to provide an estimate of heritability. To some degree this could be done by using a basic developmental difference found in human twin types, namely, whether they were monozygotic (single-egg or identical) or dizygotic (two-egg or fraternal). The basic methodology of twin study is described in the following section.

Monozygotic (MZ) twin pairs were presumed to differ only with respect to changes brought about by environmental factors, since they were deemed to be genetically identical. A little later, we will have an opportunity to modify that claim. In contrast, differences between dizygotic (DZ) twin pairs would be due to both heredity and environmental differences. Therefore, if the differences found in MZ pairs were subtracted from those found in DZ pairs, what remain should be differences only due to heredity. Based on this simple calculation, a population derived statistic called heritability could be estimated. Heritability (h^2) is defined as the percentage of the total variance (having both genetic and environmental components) that is due solely to genetic differences, in this case, among twin pairs. This statement is expressed in the following formula:

$$h^2 = DZ_{variance} - MZ_{variance}/DZ_{variance}.$$

Several other formulas for estimating heritability have been derived, but we will employ this one here for illustrative purposes. In this application, h^2 is considered to be a "narrow sense" version that recognizes that total phenotypic variance is equal to only the additive genetic variance (and not the variance from all genetic effects) plus environmental variance. It must be emphasized that heritability does not in any way inform us about how polygenic traits are inherited or the relative importance of either heredity or environment in the individual phenotypic expression of these traits. Obviously, both heredity and environment are necessary. However, heritability estimates, which can range from 0–1, have proven especially useful for plant and animal breeders who were looking for ways to improve yields, either by experimentally altering the parental stock and/or by changing the environmental conditions that could affect growth and development outcomes.

An early and nonexperimental application for investigating human variation through the twin study outlined above was conducted by Osborne and DeGeorge (1959). They took a series of anthropometric measurements by twin type (MZ vs. DZ), and assessed the resulting data for pattern of inheritance. Their analysis did not directly calculate heritabilities, but did provide variance data from which heritability estimates could be made. These estimates are presented in Table 4.1 for adult height and weight.

A reasonable interpretation of these results would be that height variation is more strongly attributed to genetic differences while weight variation is more subject to environmental factors. This conclusion does

TABLE 4.1. Heritability (h^2) estimates for height and weight.*

Variable	Male	Female
Height	0.79	0.92
Weight	0.05	0.42

Note: *Heritability estimates are based on variance data found in Osborne and DeGeorge (1959).

conform to an expected higher amount of plasticity in body weight as compared with stature, as adult body weight is highly modifiable through diet and exercise. The authors of this study were well aware that this kind of comparison was made possible by the twin study, but they were rather pointed in recognizing limitations of the method when remarking:

Preoccupation with the problem of establishing the relative importance of heredity or environment rather than with that of understanding their interactions has resulted in failure to explore adequately the possibilities for extending the use of twin analysis (Osborne and DeGeorge, 1959, p. 24)

Their plea for concentrating on gene–environment interactions resonates very well with current thinking. Indeed, over the decades following this study, heritabilities from twin studies came under strong criticism on a number of grounds, particularly in the area of behavioral genetics, and especially with respect to traits such as personality and IQ.

Twins are not representative of the whole population, and it does not appear appropriate to consider the social environment within which they developed as comparable to that of singletons. One consequence of this fact is that MZ environmental variance could be an underestimate and thereby inflate the value of h^2. In general, heritability estimates are just that, they are specific to the sample from which they were calculated and therefore cannot be uncritically generalized to other populations. And they are never applicable toward explaining phenotypic expression at the level of individuals, as is often done in the popular media. Heritability estimates have been shown to differ widely when made on what were presumed to be comparable samples and they are subject to change over the course of childhood development (and individual life history, as with intelligence quotient estimates). These are all good reasons for heeding the advice of Osborne and DeGeorge and paying more attention to the interaction of genes and their environment.

Gene by environment interaction

To counter many of the objections and limitations found in the simple method of estimating heritability illustrated above, newer approaches have been devised most of which take into account what is now understood to be an inevitable gene and environment interaction. The untestable question of what is more important, genes or environment, has given way to an investigation into the environmental factors that interact with the expression of genotypes. A straightforward example here can assist in defining this direction of research.

For certain persons, consumption of foods containing the amino acid phenylalanine during early childhood can result in detrimental brain development and a disorder called phenylketonuria (PKU). This inherited condition is due to a mutation in an enzyme responsible for converting phenylalanine to tyrosine. It is an autosomal recessive disorder, and, because this means that a homozygous genotype is present, the condition should be expressed. Yet, its expression is dependent upon the nutritional environment of the child. If the child's diet is entirely free of phenylalanine, then PKU is not manifested. In this instance, a given genotype may predispose a person to getting or make him or her susceptible to or resistant to a particular unhealthy phenotype, but that outcome will not occur unless the appropriate environmental conditions are present.

Many examples of gene by environment interaction are now known, some having medical relevance as PKU, and they have been incorporated in newborn screening programs in order to lessen or prevent harmful phenotypic expressions early in life. Thus, the old saying "biology is destiny" takes on new meaning in these cases. Other examples of gene by environment interactions can lead to enhanced outcomes at a behavioral level, as in the case of a child born with musical talents who is then raised by parents who themselves are musicians, thereby receiving more than ample opportunities for those talents to flourish.

Continuing discoveries in gene by environment research are enlightening in and of themselves but on the horizon are still more revealing lines of inquiry that undoubtedly will expand our understanding of the mutually interacting roles of heredity and environment. These roles are considered next.

NON-MENDELIAN INHERITANCE AND EPIGENETICS

Non-Mendelian inheritance simply refers to heritable tendencies that are not in accordance with Mendelian principles of segregation and independent assortment. Not always so simple or clearly understood, they amount to changes in phenotypes sustained over generations without alteration of genotypic combinations.

The phenomenon of non-Mendelian inheritance has been known for some time when applied to viruses and bacteria, but more recently it has been observed in higher life forms, including mammals and then also in humans. The discussion here will focus on humans, particularly in relation to the rapidly growing research effort in epigenetics. There is a long history of the use of the term "epigenetics" but currently it pertains to variable heritable traits that are not based on changes in DNA sequence.

A clear-cut case of non-Mendelian inheritance, representing one of the many kinds of epigenetic effects, was discovered in persons who carried a particular mutation, a partial deletion of the long arm of chromosome 15. It turned out that this chromosomal variant manifested itself differently depending upon whether it was inherited through the mother or the father. If transmitted from the mother, children were affected with Angelman's syndrome, and if transmitted through the father, they inherited Prader–Willi syndrome. These differences occur because of a process known as imprinting. More specifically, maternal and paternal chromosomes were differentially imprinted such that they produced clinically distinct abnormalities during embryonic development.

While, in the above example, the actual mechanism(s) of imprinting is not entirely understood, research has shown that imprinting can occur through methylation at specific junctions along a DNA sequence which can be maintained over successive generations. Methylation is the attachment of a CH_3 group preferentially at cytosine nucleotide positions, which effectively silences or turns off this portion of DNA. Its principal effect is one of epigenetic regulation of gene expression during both embryonic and postnatal development. Especially interesting have been the results of studies showing how MZ twins become somewhat phenotypically differentiated due to imprinting (Fraga et al., 2005; Kaminsky et al., 2009). This finding, of course, implicates subsequent transmission of environmental influences (mediated through DNA) acquired during an individual's lifetime, but imprinting probably should not be construed as a form of Lamarkian inheritance.

Imprinting will very likely take on increasing significance in human biology studies. (The reader can tap into this active research area through a website – geneimprint.com). Likewise, epigenetics has entered into discussions of evolutionary theory (Jablonka and Lamb, 2005), indicating an ongoing rethinking of the Modern Synthesis, to be discussed next. Lastly, we would remind the reader that Chapter 2 of this volume had a section devoted to developmental adaptation and epigenetics, and some of the points discussed there have been further elaborated here.

A MODERN SYNTHESIS AND EVOLUTIONARY GENETICS

Mendelian genetics as espoused by early twentieth-century laboratory researchers seemed to be at odds with post-Darwin's practitioners of natural selection theory, whereas today we fully embrace and necessarily incorporate genetic principles as vital to a more complete understanding of evolutionary processes. There have been several plausible explanations as to why genetics and evolution were not initially synthesized into a new paradigm, not the least of which is the fact that the two camps, laboratory experimentalists versus field naturalists, were engaged in separate research pursuits that in turn did not offer abundant opportunities for fruitful interactions of ideas. However, a meeting of the minds did take place around the 1920s when at first there was the recognition of neo-Darwinism, which added mutation and genetic variation into its formulation of natural selection theory, later followed by what came to be labeled the "Modern Synthesis." Fundamentally, the Modern Synthesis proposes that evolution occurs as a result of selection acting upon genetic variation generated within natural populations through mutation and recombination, as well as from gene flow between groups, and is subject to random drift, particularly in small, widely scattered populations. Over extended periods of time, multiple generations and the possible subdivision of original populations, evolution leads to genetic divergence and speciation. This section briefly describes the tenets that will be discussed next under the topic of population genetics.

With the application of more recent terminology, the Modern Synthesis might well be viewed as conjoining what are now described as microevolution and macroevolution, the comfortable coupling of which continues to be questioned by some theoreticians. In fact, it is important to note that the Modern Synthesis is continually under revision as important questions (and the accumulation of new data) about the mode and tempo of evolution are investigated and debated. For example, what determines the rate of evolutionary change? Does evolution proceed gradually or in relatively rapid bursts, as hypothesized in the model of punctuated equilibrium? At what level does selection operate, on the gene, the individual or kin group, or all of the above? (Refer back to Chapter 1 of this volume for an extended discussion of this matter.) Future developments in genetics and biology can be expected to enliven these debates and also enlighten our understanding of evolution.

HUMAN POPULATION GENETICS AND THE HARDY–WEINBERG EQUILIBRIUM

The Modern Synthesis united genetics with evolutionary biology, and, in particular, highlighted the importance of population genetics, which essentially defined micro-evolutionary studies. This section will describe the fundamentals of human population genetics and also provide examples from human biology that apply to microevolution.

In an abbreviated form, population genetics is the study of the behavior of genes in populations. This means that there is an emphasis upon gene frequency analysis, noting whether or not there are changes in frequency over generations, and, if so, identifying the underlying causes for those changes. Conversely, there are also reasons why gene frequencies are maintained or remain stable over successive generations. To illustrate these points, let us begin with a simple gene, or actually, allele frequency calculation.

The example of the MN blood group system will be used. This system has two codominant alleles, M and N, which yield three distinct genotypes, MM, MN, and NN. If we were to have studied a sample of 100 persons and found their genotypic proportions to be 25 of MM, 50 of MN, and 25 of NN, then we can directly count the number of M and N alleles in this sample.

Let p = frequency of M and q = frequency of N.

Then p = all of $MM(25) + \frac{1}{2}$ of $MN(25) = 50/100 = 0.5$.

And q = all of $NN(25) + \frac{1}{2}$ of $MN(25) = 50/100 = 0.5$.

Note that: $p + q = 1.0$.

Through this example, we have broached the most important concept in population genetics, the Hardy–Weinberg Theorem. Hardy, a British mathematician, and Weinberg, a German physician, independently discovered that, under certain specifiable conditions, genotypic proportions within a population will remain in equilibrium and will not change over successive generations. Genotypic proportions will in fact correspond to an expansion of a binomial expression containing p and q allele frequencies. In algebraic form, this would read:

$$(p + q)^2 = p^2 + 2pq + q^2.$$

For the Hardy–Weinberg Equilibrium to be present, the following conditions are expected to apply:

1. the population is infinitely large in numbers of individuals;
2. there is random mating; and
3. there is no evolution.

We will discuss each of these variables in turn, bringing out some major features that will help to illustrate the power and profundity that comes from testing populations with such a simple mathematic formulation, the Hardy–Weinberg Equilibrium test, hereafter designated H-W. Applying the H-W test to the MN blood group discussed above yields equilibrium expected genotype frequencies of $p^2 + 2pq + q^2$, or numerically as $25(MM) + 50(MN) + 25(NN)$, exactly

TABLE 4.2. Hypothetical frequencies for applying the Hardy–Weinberg Equilibrium test.

Genotype	*MM*	*MN*	*NN*	Total
Frequency	9	42	49	100 persons

the same genotype proportions observed. What would be the result if, instead, the sample had yielded genotypic frequencies shown in Table 4.2? Are these proportions in H-W equilibrium?

Testing for H-W equilibrium follows these basic steps:

Step 1 Calculate allele frequencies from the Observed genotypic frequencies:

$$p(M) = 9 + 21/100 = 0.3 \text{ and } q(N) = 21 + 49/100 = 0.7.$$

Step 2 Calculate Expected genotypic frequencies from allele frequencies:

$$(0.3 + 0.7)^2 = 0.9 + 0.42 + 0.49 \text{ or } MM = 9, MN = 42,$$
$$NN = 49 \text{ persons.}$$

Step 3 Compare Expected with Observed frequencies.

In Table 4.2, Observed genotypic frequencies do match the Expected frequencies. However, if they did not, then we would employ a statistical test to ascertain the significance of the difference between them.

Step 4 Do a Chi-square test of proportionality between Observed and Expected genotype frequencies.

Step 5 Interpret test results in light of the conditions that are needed for H-W equilibrium to be present; if H-W equilibrium is not found, then consider which of these conditions appears most likely to account for the disruption of the equilibrium.

When we have discussed all of the conditions that are necessary for H-W equilibrium to exist, we will work a second problem later where it is not present, and thus will require further interpretation that incorporates information from the following discussion.

FACTORS THAT AFFECT THE HARDY–WEINBERG EQUILIBRIUM

Effective population size

To be sure, no natural population, human or otherwise, is infinitely large in size, yet H-W has often been found when various human groups were tested. Even so, caution is certainly in order whenever H-W results are interpreted for populations with less than hundreds of individuals. This caution is especially warranted because evolution expressed as differential fertility does not act on a population as a whole but on what has

been called the breeding (effective) population. The number of breeding adults has been approximated to be about one-third of the total population census, with prereproductive (children) and postreproductive (elderly) segments comprising the remainder. In more formal terms, breeding population is defined in population genetics as "effective population size." The effective population number is more accurately estimated through derived formulas that generally reach lower values than the number of breeding adults due to such variables as relatedness among individuals, sibship size variation, sex ratio imbalance, and the relative stability of population size. A critical consideration here is how all of these variables can affect the amount of genetic variation that is available for evolutionary processes to act upon. This matter has particular relevance to our upcoming discussion of random genetic drift, a process that is greatly enhanced by small population size. An additional note here is that gene pool will be used when designating the complete genetic makeup of a population. Furthermore, it should be realized that field research studies most often will be based upon samples of populations and, therefore, may not faithfully reflect the actual variation contained in the total gene pool.

Random mating

What appears to be another stipulation that cannot be met for H-W equilibrium to be present in natural populations is that of random mating. Random mating in a formal sense means that there is an equal probability of selecting any potential mate from an available pool. In spite of some difficulties in precisely defining what a pool is, no human groups anywhere on the globe have ever been found to choose mates solely on a random basis. We will discuss ways human mating systems are, in fact, socially and culturally influenced below. (Also see Chapters 17 and 18 of this volume for extended discussions of mate choice.) For purposes of conforming to H-W equilibrium, it is required that the gene or more precisely the locus, under question be distributed through randomly mating. The MN blood group can be used here to illustrate this point. Since very few people know what their MN blood type is, it is highly unlikely that it has been used to decide among potential mates. Thus the MN locus can be said to mate randomly. By extension, all other such loci undergo random mating if they are not involved with mate choice.

Departures from random mating of individuals

Before proceeding, we will provide some background on mating systems that do not conform to random mating, namely, inbreeding and assortative mating. Inbreeding refers to matings between persons who are genetically related to one another more closely than chance, i.e., siblings, first cousins, and so on. Some societies have engaged in systematic forms of close-relative inbreeding, such as the royal families of Egypt and Britain. Ethnographers have reported on preferred mating systems between cross-cousins, that is, cousins who are related from a son through to his father's sister's daughter, or a son through to his mother's brother's daughter in many societies in forming kinship alliances (Levi-Strauss, 1969). In this instance of cousin marriages, a certain proportion of the genotypes found in offspring of these unions can be expected to be not only homozygous, but also autozygous. Autozygous genotypes contain alleles that are Identical by Descent or different copies of the same allele, by virtue of being inherited from a common ancestor through pedigree pathways. These pathways are used to calculate an inbreeding coefficient which then can be used for estimating the proportion of loci that are expected to be autozygous due to inbreeding.

We will use cousin matings to illustrate these calculations. The inbreeding coefficient formula for general application is:

$$F_x = \sum [(1/2)^n (1 + F_A)],$$

where n is the number of individuals in a common ancestor pathway summed over all paths, and F_A is the inbreeding coefficient for the final ancestor in each path. Figure 4.8 shows a pedigree representing first-cousin matings; applying the above formula yields:

$$F_x = (1/2)^5 + (1/2)^5 = 0.0625 \text{ or } 1/16.$$

Referring once again to the MN blood group, the proportion of loci that are expected to be of autozygous

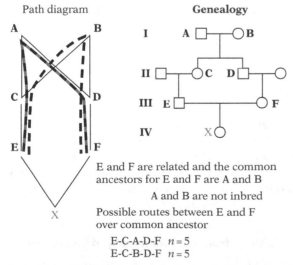

E and F are related and the common ancestors for E and F are A and B

A and B are not inbred

Possible routes between E and F over common ancestor

E-C-A-D-F $n = 5$
E-C-B-D-F $n = 5$

4.8. Pedigree and path diagram of offspring from a first-cousin mating. From http://www.husdyr.kvl.dk/htm/kc/popgen/genetics/4/2.htm.

MM genotype based on first-cousin matings can be determined by the following formula:

$$P_{MM} = p_M F + p_M^2 (1 - F).$$

In this first-cousin example, 1/16 of the loci are predicted to be autozygous. Stated in an equivalent manner, the chances of randomly selecting an autozygous locus in offspring of first cousins is 1/16. Very important to our discussion is that there is an increased proportion of homozygosity under inbreeding in comparison with random mating. This, of course, disrupts H-W equilibrium and leads to a loss of genetic variation. In addition, increased homozygosity may have an evolutionary consequence whenever selection has an opportunity to act against harmful recessive genotypes. However, inbreeding per se does not change allele frequencies, it only redistributes alleles into homozygote classes and away from heterozygotes.

It is presumed that inbreeding is harmful due to an increased likelihood of exposing relatively rare recessive alleles that can cause birth defects and generally be detrimental to normal growth and health. This situation has been termed inbreeding depression. In population genetics, the consequences of inbreeding that lead to a loss of average fitness make up part of what is known as the genetic load. The theory of genetic load was developed by Crow and Kumira (1963) and has been studied in a number of human populations. For illustration purposes, we will discuss a report by McKusick et al. (1971) here. In this study, they investigated several subgroups of Old Order Amish, a religious sect composed of farming communities located from Pennsylvania to Indiana. The communities are rural and somewhat isolated, and have small memberships and practice endogamy, which makes them closed to outsiders. These social conditions raised the degree of relatedness among individuals which, in turn, meant higher inbreeding levels. The study reported the presence of several rare genetic disorders, five of them inherited as autosomal recessives, and two as autosomal dominants. Accompanying pedigrees documented the transmission of the disorders through successive generations.

The adverse genetic consequences noted above form a major rationale that prompted societal attempts to disallow close-relative matings through various inbreeding avoidance practices that might be informal or more institutionalized through marriage rules. For instance, many US states prohibit marriages between persons who are cousins, or require the cousins to be beyond a certain age before marrying. Conversely, there has also been a recent discussion on whether these marriage restrictions regarding related persons need to be revised or even repealed (Paul and Spencer, 2008).

Assortative mating is another major departure from random mating of individuals. It appears in two forms, positive assortative mating (or homogamy), when phenotypes of mated pairs are more similar than would be expected under random mating, and negative assortative mating (or heterogamy), in which the mated pairs are more dissimilar than they would be under random mating. As is probably obvious to even the most casual observer, humans have a strong tendency to choose mates like themselves in several respects, such as sharing cultural traits of language and religion or biological traits of age and stature. The last-cited trait is of further interest in that, for many societies, there is a "male taller norm" that characterizes unions (Gillis and Avis, 1980).

Documentation of negative assortative mating has been scarce and contradictory. One such study dealt with the Hutterites, a religious community residing in North America. In investigating their human leukocyte antigen (HLA) genotypes, researchers found evidence indicating that there was an avoidance of marriages between persons possessing the same genotypes (Ober et al., 1997). However, a similar study done on South American Indian tribes failed to find this form of non-random mating pattern for HLA genotypes (Hedrick and Black, 1997). An accompanying invited editorial comment on these two conflicting studies (Beauchamp and Yamazaki, 1997) recommended that continued research be done to resolve what they determined to be a clearly plausible explanation for why great variation in HLA genotypes is being maintained in human populations. (See Chapter 13 of this volume for more discussion of the HLA system.)

Assortative mating pertains to specific phenotypic traits, usually of multifactorial inheritance, which, unlike inbreeding, can affect any of the entire genome simultaneously in terms of similar genotypes. Furthermore, similar to inbreeding, positive assortative mating likely leads to an increase in proportion of homozygotes; however, in contrast to inbreeding, this results in increasing the variance within a population due to the formation of more distinct phenotypic grouping of individuals. For instance, tall and short stature tends to form a bimodal distribution composed of the assortatively mated pairs. The correlation coefficient is used to measure the strength of the association for traits or variables undergoing assortative mating. This coefficient is denoted by an r-value which can range from 1 (complete positive correlation) to -1 (complete negative correlation). In comparing statures for husbands and wives, r-values tend to be in the range of 0.2–0.3, although, for spouses of female MZ twins, it was found to be at nearly 0.5 (Meier and Jamison, 1990). Lastly, it should be noted that, like inbreeding, assortative mating by itself only changes genotypic frequencies, thus upsetting H-W equilibrium, but does not directly alter gene frequencies. It is now time to take up this last point, and examine the forces of evolution.

MICROEVOLUTIONARY PROCESSES OF ALLELE FREQUENCY CHANGE

The crux for testing for H-W equilibrium is that it assists us in knowing whether or not evolution is occurring, and can further point to the agent of gene frequency change should H-W equilibrium not be present. The four such evolutionary processes or agents of change are natural selection, mutation, gene drift, and gene flow, to which can be added recombination and population structure as factors or variables that can facilitate the action of all of these processes. (The reader is reminded that Chapter 1 of this volume covered this material in depth, so the purpose here is to fill out some of that discussion with examples taken from human biology studies.)

For the sake of simplicity, two fundamental statements of evolution can be written, first as depicting the essence of Darwinian thinking and then according to a Modern Synthesis viewpoint, each shown in Figure 4.9.

These formulations demonstrate the nature of scientific advance in which the profound new discovery made by Darwin has been added to and refined through continued research. We can now proceed to discuss important aspects of microevolution that underlie the formulations.

EVOLUTIONARY SOURCES OF GENETIC VARIATION

Mutation

A primary source of new genetic material is mutation. Of potential evolutionary significance, point mutations of nucleotides (substitutions, insertions, deletions of base-pairs) have been shown to be highly relevant. An oft-cited example of the evolutionary importance of mutations is human hemoglobin, whereby a point mutation on one of the molecular chains (beta) making up the protein offered a selective advantage to persons carrying the sickle cell trait when they were exposed to malarial infection, as will be discussed shortly. However, most point mutations are found to be neutral to the action of selection, while, on the opposite side, major mutational changes and chromosomal aberrations very likely lead to dire outcomes in terms of reduced chances for survival and lowering of fertility. It is sometimes stated that mutations are generally harmful. This could be clarified to mean that mutations usually occur without respect to the adaptive needs of organisms, that is to say they occur randomly. Therefore, there is little likelihood that newly arising mutations will be beneficial to enhancing either the survival or reproductive outcome of individuals. On the other hand, should the mutational change turn out to be beneficial, it is then possible for its frequency to increase at a rate that is affected by its relative

Darwinian Evolution:

Time + Variation + Natural Selection → Evolutionary adaptation

Modern Synthetic Theory:

Time	+	Variation	+	Population structure	+	Selection	→	Evolution
		↓		↓		↓		↓
		(includes mutation, gene flow and recombination)		*(includes random drift related to population size and composition)*		*(leads to adaptation and extinction)*		*(as genetic change in gene frequencies and DNA sequences)*

4.9. Fundamental statements of evolutionary theory.

benefit and whether it is expressed as a dominant or recessive allele. Unfortunately, in the case of bacteria, certain mutational change that becomes adaptively useful may cause drug resistance because of the overuse of antibiotics. It might be noted that there has been a discussion in the literature regarding directed or adaptive mutation, where nucleotide changes may not always occur randomly (Hastings et al., 2004; Galhardo et al., 2007). While this idea is highly intriguing, additional research certainly is needed to verify that some kind of nonrandom mutation actually occurs.

Mutations alter allele frequencies as an existing gene is chemically changed. For example, if, in the MN blood group discussed above, some M alleles mutated to N, then the frequency of $p(M)$ would decrease and $q(N)$ would correspondingly increase. Obviously, if the mutation of the N to M allele occurred at the same rate (signified as μ), there would be a zero net change in allele frequencies, although evolution can still be said to have taken place.

Gene flow

A secondary source of genetic variation arrives through the process of gene flow, otherwise known as admixture, or migration (signified as m) of people who interbreed to some degree with individuals from other groups they encounter. There are many variables and obstacles that make gene flow selective to certain groups and individuals within groups, all of which can affect gene frequencies. One of these is that persons who migrate tend to be in a better state of health than those who are not so able. Then, if mobility was in part attributable to genetic variants unique to the migrants, a consequence will be the spread of these

novel genes. While this complicated interplay between natural selection and gene flow will be difficult to document, especially for recent human groups (cf. Gross, 2006), it probably does underlie the successful migration and adaptation that had occurred in much earlier Pleistocene populations (Leonard, 2003).

There are also alternate forms of migration, for instance, one-way gene flow in which one population admixes exclusively with another, or two-way exchange of genes among two or more groups. Added to these forms are the variables of how much difference there is between the gene pools involved before admixing occurred, and how many individuals participated in the process adjusted by the relative sizes of the groups involved. Considering all of these factors, it becomes apparent that accurately estimating gene flow becomes complex and altogether not very meaningful.

There has been an indeterminately long history of human expansions, conquests, and voluntary and forced movements that represent a complex array of admixing of once separated peoples. While formulas have been derived to estimate the amount of admixture within a human group, major uncertainties regarding gene frequencies of original parental groups usually raises questions as to the accuracy of these calculations. An enlightening realization of the long history of human population movements and intermingling is that any attempts at racial classification become highly arbitrary (See Chapter 15 of this volume for a full discussion of this and other matters pertinent to studying the nature of human variation.)

Once again, just using the MN blood group, gene flow alters allele frequencies. If, for example, a population high in genotype MM admixes with another high in NN, then this process produces an elevated proportion of *MN* genotypes, which, in turn, disrupts H-W equilibrium. Overall, the gene flow process becomes highly relevant in the context of investigating population history and structure, which has been a focused interest within the field known as anthropological genetics. Results of some of this research are summarized in Chapter 13 of this volume.

EVOLUTIONARY PROCESSES ACTING UPON VARIATION

The remaining two processes, random genetic drift and natural selection, are active in changing allele frequencies and can lead, in the case of genetic drift, to a loss of variation, while selection can result in a loss, an increase or even maintain a stable level of genetic variation. We will look at genetic drift first.

Random genetic drift

Random genetic drift is basically sampling error due to small numbers. Hence, its effect is strongest in populations having small effective size. Drift is the opposite of random sampling that is expected to be representative of the variation that exists in sampled populations. Instead, random here refers to the unpredictability of the direction of gene frequency change in the short term, say over each successive generation. A graph of genetic drift shows fluctuating frequencies of a single allele (Figure 4.10).

Ultimately, over many generations, there is a tendency for allele frequencies to drift toward the loss of one allele and the corresponding fixation of the other allele. This, of course, results in a reduction in genetic variation. The time to fixation will be dependent upon the size of the population and how significant the other evolutionary processes are with regard to the locus under study. As a summarizing statement within

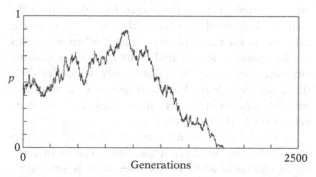

4.10. Fluctuating allele frequency simulating random genetic drift in a population of 1000 individuals. The allele frequency, designated as *p* along the *y*-axis, begins at about 0.5 before reaching fixation over the course of more than 1750 generations, as plotted on the *x*-axis. From http://darwin.eeb.uconn.edu/simulations/drift.html.

population genetics theory, drift is considered most effective whenever the mutation rate (μ), net migration/gene flow rate (m) and the selection coefficient (s) $< 1/2N$, where N is the population size. In particular, for those loci that are selectively neutral, it is expected that random drift will be the primary agent of change in frequency. As will be seen in Chapter 14 of this volume, allelic frequencies at neutral loci are decidedly important in studying human population genetic relationships and in constructing phylogenetic trees.

Genetic drift has been described and, at times, documented in gene frequency analysis for extant human populations that had experienced sampling error episodes in their histories. One of these circumstances involves a small group of migrants founding a new community ("founder effect") and another reflects a markedly reduced number of survivors following a catastrophe ("survivorship effect"). In either case, the newly formed group does not accurately represent the genetic variation that had been present in the parent population from which it was derived. Crucial for assigning genetic drift as a cause, this change has to be attributable to random processes.

The founder effect or principle probably has been cited most often, as in the case of reconstructing the effective population size and genetic makeup of the earliest migrant groups to the New World (Hey, 2005). Founder effect has also been employed to explain unusual gene frequencies observed in religious isolates, which will be discussed below to illustrate the essential features of genetic drift.

A religious isolate refers to a community that maintains strict within-group marriage rules, that is, the practice of endogamy is required. This, of course, means that gene flow from outside would not influence gene frequency distributions. Beginning in the late 1940s, a number of these religious communities have been studied with regard to genetic drift, and from that time period there appeared a classic example. Glass et al. (1952) conducted their research among the Old Order "Dunker" (Old Order German Baptist Brethren) living in Franklin County, Pennsylvania. This community has its origins in Rhineland, Germany, and descended from a small number of founding migrants who came to the United States about 200 years ago. The researchers investigated several genetic traits, including the major blood groups. Since we have been referring to the MN blood group, that system can serve as our reference point. The pertinent findings are that the frequency of the *M* allele in the Dunker Isolate was $p(M) = 0.655$, while, in the comparative groups, it was $p(M) = 0.548$ in West Germany, and $p(M) = 0.540$ for the United States. Of course, $q(N)$ would just be $1-p$. The genotypic frequency differences between the Dunker Isolate and the other two groups were highly significant. There is naturally some concern whether these comparative samples are truly representative of especially the US population, but, at face value, they clearly suggest that the founder Dunker Isolate had diverged genetically from its source population in West Germany, and had not converged upon the US sample. The other genetic traits under study generally confirmed these findings. As for interpretation, after ruling out any real likelihood of selection acting at the MN locus, the authors concluded that these results are most likely attributable to genetic drift. As another example of founder effect, it has been reported in a study done in Quebec, Canada, that this kind of demographic event accounted for 85% of cases of the rare mitochondrial disease called Leber hereditary optic neuropathy (LHON), a disorder that was mentioned earlier (Laberge et al., 2005).

Survivorship effect probably has been repeated untold times throughout human history, and indeed population "bottlenecks," due to sharply reduced numbers caused by loss of life, or due to small group migration of founders as described above, form a major point of reference for tracing our species' Pleistocene origins (for a review see Hawks et al., 2000). As an example from a recent time period, Roberts (1971) reported on an isolated island population in the south Atlantic, Tristan da Cunha, originally settled by a small number of founders in the nineteenth century, subsequently experienced two episodes of population bottlenecks over approximately a century. Through some careful tracing of pedigrees, and scouring records pertaining to migration to and from the island, he was able to ascertain the relative genetic contributions (differential reproduction) of island members up to the time of the study in 1961. Some were much more prolific than others. Of considerable interest here is that, while at least one of the bottleneck reductions was deemed to be due to sheer accident, and likely therefore to be relegated as contributing toward genetic drift, the other earlier bottleneck could well have led to loss of genetic variation in the remaining gene pool by selective migration of certain families. In both cases, marked population reduction likely led to increased levels of inbreeding in this small, highly isolated island community, and as a consequence elevated the risk of exposing genetic diseases. There had been outbreaks of respiratory diseases on Tristan da Cunha, which led a research team to search for a genetic basis of asthma in a setting (founder and survivorship effects, and inbreeding) thought to be conducive for localizing genes (Zamel et al., 1996). The results from this study did support a major gene involvement in the expression of asthma.

Selection

The last of the evolutionary processes to cover is the one that undoubtedly ranks at the forefront of any discussion concerning evolution, that is, natural selection. Darwin's theory is that the underlying force that explains adaptive change within evolutionary lines, and diversification and speciation between lines, is selection. Natural selection has been thoroughly tested and overwhelmingly confirmed over the past century and a half. Darwin's prediction that "[m]uch light would be thrown ..." [on human evolution] (Darwin, 1892, p. 304) has been fully realized. While this chapter deals only with some highlights of evolutionary genetics applied to humans, the reader will find that this volume as a whole is replete in its coverage of natural selection outcomes within our species.

Of the many forms that selection can take (see Chapter 1 of this volume for a complete discussion of these, and other chapters for topical applications of selection theory), this section will focus on only two, directional and balancing selection pertaining to single locus, and two segregating allele systems. Again, there will be examples drawn from human biology research.

Directional selection

Initially, we will use the MN blood group to illustrate selection and related concepts in a hypothetical case. Say that for whatever reason, persons possessing the MM genotype had a selective advantage over the other two genotypes, MN and NN. By this we mean that MM individuals had either a greater likelihood of surviving or having a higher reproductive success, or both. What would be expected over time is that the frequency of the M allele would increase relative to the N allele, and, hence, the direction of systematic change would shift toward fixing the M allele and losing the N allele. However, that tendency is mediated by the extent of the differential between the three genotypes that can be measured in terms of their respective fitnesses.

For this calculation, fitness refers to genotypic reproductive success, and of course, to have survived to sexual maturity. Fitness can be denoted as w, while, on the opposite side of the ledger, the selection coefficient is designated as s, such that $w + s = 1$. At the extremes in the range of these values, when $w = 1$, this would designate an optimal genotype for the existing environmental conditions. Conversely, whenever $s = 1$, this genotype is lethal both in the sense of mortality or sterility, and, in either instance, does not contribute genetically to the next generation.

Table 4.3 presents an example of selection acting against the dominant genotypes which results in directional change in allele frequencies. This is sometimes referred to as negative selection since it acts to remove deleterious alleles.

Achondroplasia is a form of dwarfism characterized by reduced limb growth but normal proportions of the torso. It results from an autosomal dominant

TABLE 4.3. Selection against the dominant using achondroplasia as an example.

Genotype	AA	Aa	aa
Relative fitness (w)	0	0.20*	1.00
Frequency	$p^2(1-s)$	$2pq(1-t)$	q^2

Note: Selection coefficients are $s = 1.00$ and $t = 0.80$.
*This fitness value was taken from Bodmer and Cavalli-Sforza, 1976.

TABLE 4.4. Heterozygote advantage and balancing selection in a malarial environment.

Genotype	AA	AS	SS
Advantage	–	Resistance to malaria	Resistance to malaria
Disadvantage	Susceptibility to malaria	–	Sickle cell anemia
Relative Fitness	1−s	1	1−t
Frequency	$p^2(1-s)$	$2pq(1)$	$q^2(1-t)$

Note: Selection coefficients are $s = 0.12$ and $t = 0.86$ based on estimates taken from Bodmer and Cavalli-Sforza (1976).

point mutation (substitution of arginine for glycine) in the fibroblast growth factor receptor gene (*FGFR3*, located on the short arm of chromosome 4) that leads to shortening of the long bones of arms and legs. About 80% of cases are the result of sporadic new mutations with some evidence for paternal age effect.

From Table 4.3 it can be seen that *AA* and *Aa* have reduced fitnesses in comparison with *aa*. Unfortunately, the homozygous *AA* genotype is either stillborn or lives for only a short period following birth. On the other hand, the heterozygote *Aa* generally has a normal life span. Their reduced fitness is not due to fertility problems but to social or cultural conditions that implicate constraints on finding available marriage partners.

From a microevolutionary perspective, fitness values and selection coefficients in Table 4.3 indicate that, over generations, there will be a reduction in the dominant allele *A*, with a corresponding increase in *a*. There are equations used in population genetics to estimate the rate of gene frequency change due to directional selection, and also models for predicting an equilibrium, say, between selection and mutation. We will cover that model a little later on.

Modeling of directional evolutionary change for polygenic traits having adaptive value is not so precise, simply because the loci involved generally are not known, and as we discussed earlier, these traits are subject to multiple environmental interactions. For instance, the major trend of increasing brain size throughout early human evolution is clearly demonstrable under a positive direction selection model, and presumed to be the result of expanding brains for evolutionary success. Even though it is not possible to be very specific with regard to the genetic basis of this change, a possible breakthrough has occurred recently (Evans et al., 2005; Mekel-Bobrov et al., 2005), yet this discovery also has been questioned (Timpson et al., 2007).

Balancing selection

We now shift to another selection model that has received considerable attention in human biology, balancing selection, in which the heterozygote genotype maintains an advantage over homozygote forms. This, of course, includes the sickle cell locus that has been

shown to be implicated in human groups living in endemic malarial regions. A more complete story of this connection can be found in Chapters 13, 14, and 27 of this volume. Here, we focus upon the model of selection for the heterozygote to illustrate some basic points regarding population genetics. Table 4.4 lays out a framework of this mode of selection, involving a single locus, with two alleles, *A* and *S*.

It can be seen that genotypes *AA* and *SS* are at a disadvantage for separate reasons, while *AS* heterozygotes enjoy an advantage on the grounds that they have a functional level of normal hemoglobin and are also resistant to malaria. Historical empirical data are available to demonstrate just how much this advantage meant in terms of survivorship to adulthood in an African group, the Yorubas, from Nigeria, in terms of selection coefficient estimates found in Table 4.4. Converting these estimates into fitness values means that, when the study was conducted some 40 years ago, for every 100 surviving *AS* heterozygotes, 88 *AA* individuals lived to adulthood, while only 14 were *SS* survivors. Given this differential selection process, the *S* allele is expected to rapidly decrease, but of course, it won't entirely disappear because of the counter selection against the *A* allele of the other homozygote class. These opposing selective forces could lead to an equilibrium whereby allele frequencies no longer change but are maintained as a balanced polymorphism, so long as environmental conditions remain fairly stable and no other evolutionary forces are operating. Under these assumptions, the frequency of the *S* allele will attain equilibrium according to the following formula: $\hat{q} = s/s + t$, or numerically as: $\hat{q} = 0.12/0.12 + 0.86 = 0.122$. The *S* allele would then be maintained in the gene pool at about the level of 12%, if conditions remained fairly constant. This means, for example, that the level of malarial stress did not change, and there were no interactions with other evolution processes, such as gene flow from outside the region. Of historical significance, the *S* allele has increased in

TABLE 4.5. Test of the Hardy–Weinberg Equilibrium.

Genotype	MM	MN	NN	Total
Observed frequency	10	80	10	100 persons
Allele frequency	$p(M) = 0.5$ and $q(N) = 0.5$			
Expected frequency	25	50	25	100 persons
Obs – Exp	–15	30	–15	
$(\text{Obs} - \text{Exp})^2/\text{Obs}$	9	18	9	
Chi-square $(x^2) = \sum [(\text{Obs} - \text{Exp})^2/\text{Obs}] = 36$				

frequency in generations of African-Americans descended from ancestors who were slaves, due in part to relaxation of selection constraints and reduced malarial disease stress in the New World (Cavalli-Sforza and Bodmer, 1999).

The Hardy–Weinberg test

Now that we have described all of the factors that can alter genotypic frequencies, and also microevolutionary processes that change gene frequencies, we can do a test for the H-W equilibrium, following the steps outlined earlier (Table 4.5).

With this x^2 value, a probability table in a standard statistics textbook can be consulted, and it would be found that this is a highly significant finding. Hence, it would lead to a justifiable conclusion that H-W equilibrium was not present in this sample. Therefore, it is appropriate to consider which of the equilibrium conditions were not met. Note that in this sample there is a marked deficiency of homozygotes and a corresponding high gain in heterozygotes. What could account for this pattern?

We can start with the sample size, which is relatively small and thus might be implicated. Then again, the research sampling might well be an accurate determination of the makeup of the gene pool, and the disproportionality of genotypes is due to a random genetic drift event such as seen in the founder effect. Finally, and perhaps most obviously, heterozygote advantage jumps out as a possibility. In this case, additional research is required to demonstrate how selection has been operating to bring about an excess survival/reproduction of heterozygotes at the detriment of homozygote classes. You will find another study problem that deals with a test of H-W equilibrium at the end of the chapter.

INTERACTION OF MICROEVOLUTIONARY PROCESSES

The above discussion of the four evolutionary processes as separate entities is not entirely realistic in that natural populations are very likely subjected to interacting forces.

Selection and mutation

The most obvious of these might be selection acting as a filtering agent to remove deleterious mutations, at a rate dependent upon whether the mutation is expressed as a dominant or recessive, and then depending how harmful the mutation is, that is, how severe the selection coefficient is. Considering these variables, an equilibrium of allele frequencies between selection and mutation can be reached. This equilibrium can be expressed in the formula, $\hat{p} = \mu/s$. For an extreme example, newly arising mutations that are lethal autosomal dominants would be restricted to a frequency of that of their mutation rate, since they would be eliminated as soon as they appeared in each generation. In the case of achondroplasia presented above, an equilibrium value would be: $\hat{p} = 0.00001/0.8 = 0.0000125$, or only slighter higher than the mutation rate because of the marked selection coefficient. Of course, this calculation only considers the interaction of mutation and selection, while other factors may also be involved and alter the estimate.

Selection and genetic drift

A second kind of interaction takes place between selection and genetic drift. To a certain degree, it can be argued that selection and drift are mutually exclusive forces, one being systematic and the other random. In more formal population genetic terminology, this distinction is sometimes expressed as the difference between deterministic and stochastic processes. As a consequence, drift is most likely to act upon neutral loci and not on those genes that are vitally important for an organism's survival and under heavy selection pressure. However, when effective population sizes become drastically reduced, there are opportunities for drift to occur even on genes that are not neutral, although they probably also have relatively low selection coefficients. An example of genetic drift/selection interaction was discussed earlier for the island population on Tristan da Cunha.

Genetic drift and gene flow

Yet a third kind of interaction often occurs between genetic drift and gene flow. Human populations, particularly prehistoric groups that resided in small groups and were dispersed over large areas, were highly prone to the action of genetic drift and its consequence of loss of variation. Yet, there were also regular contacts among these dispersed groups, and the resultant gene flow would have introduced some genetic variation back into their gene pools. In this situation, the dual action of gene flow with respect to variation is apparent in that it increases variation into populations that are receiving the migrants, but, as a consequence, the amount of genetic difference between the populations involved with the gene flow process is lessened. As a further consideration, genetic drift may not have been so prominent since population size was effectively extended beyond each of the local groups.

SHIFTING BALANCE THEORY

This discussion of microevolution and interacting forces naturally leads to a discussion of the shifting balance theory wherein all four processes enter into a coherent, but not universally accepted, model of evolution. For some background, two prominent founders of population genetics engaged in a heated battle in the mid-twentieth century as they offered opposing viewpoints concerning the big picture of evolution. One of them, Sir Ronald A. Fisher (1890–1962), proposed that a species, or at least a large panmictic (random-mating) population, adapted as a whole through the selection of combinations of alleles that increased overall fitness. In contrast, Sewall Wright (1889–1988) constructed the shifting balance theory that required a species, or large populations, to be composed of many small, partially isolated breeding units (demes), some of which would evolve to increase average fitness of the total population, while the others would be replaced by these more successful demes. Random genetic drift, and its attendant nature of producing, by chance, novel genotypic combinations of alleles and genotypes, played an essential role in the shifting balance theory. The theory takes its name by explaining how a one-time balance of genotypic combinations can be shifted to a new and more fit combination as likely demanded in a varied landscape of adaptive requirements, and if evolution of the population is to continue.

An alternative of course is extinction. Reduced to its essentials, this shifting balance originates with mutations taking place in an array of demes, randomly resulting in new combinations only a few of which become selectively advantageous. Thus, reproductive success that leads to population growth which then promotes migration into areas occupied by other less successful demes will replace them in part via gene flow. This discussion very briefly describes the process of inter-demic selection as proposed by Sewall Wright (1969).

A succinct review of the shifting balance theory can be found in Crow (1986), a colleague of Wright's at the University of Wisconsin-Madison. Crow considered that both Wright and Fisher may have been correct on some points while, in more recent discussions of the matter, sides are still being drawn, either in

TABLE 4.6. Microevolutionary processes and genetic variation.

	Genetic variation	
Process	Within a population	Between populations
Mutation	Increases	Increases
Gene flow	Increases	Decreases
Genetic drift	Decreases	Increases
Selection	Increases/decreases and maintains variation	Increases/decreases

critique of (Coyne et al., 2000) or in support of (Wade and Goodenough, 2000) the theory.

As a summing up of this section, Table 4.6 organizes how evolutionary processes affect genetic variation. Please note that selection can yield multiple outcomes with respect to variation.

CONCLUSION

The major aim of this chapter has been to show the many ways in which genetics has been incorporated into human biological research. We have discussed four areas of genetics, namely, Mendelian, non-Mendelian, molecular, and population or evolutionary genetics. Also included in this discussion were examples of studies that were mainly conducted with an eye toward understanding genetic variation, most notably in the context of microevolutionary processes of mutation, gene flow, random drift, and selection. Along the way, we indicated topics that have been addressed in other chapters of the book. The reader is encouraged to seek out these topics in order to gain an enhanced appreciation for the central role that genetics has played in human biology.

DISCUSSION POINTS

1. What exactly is meant by the term "recombination" in generating genetic variation?
2. What are the major similarities and differences between monogenic and polygenic inheritance? Discuss what you think would be the evolutionary impact on traits that are Mendelian versus polygenic.
3. Define the following in terms of function, give a brief description of biochemistry and/or structural components, and briefly describe the relationships between them: *chromosome, gene, gene product, amino acid, protein.*

TABLE 4.7. Sample genotypic distribution.

MM	MN	NN	Total
20	25	75	120 persons

4. What are the major mechanisms by which cells regulate gene expression? Describe possible outcomes of the failure of such mechanisms.

5. Suppose that the genotypic distribution in a sample was as shown in Table 4.7. Conduct a Hardy–Weinberg test on this sample, and if the results show that it is not in equilibrium, offer possible explanations for why it is not.

6. What do you think are the most important questions human biologists should address with regard to better understanding genetic variation in populations?

USEFUL ONLINE RESOURCES

The Genome Database: http://www.gdb.org/

Online Mendelian Inheritance in Man: http://www3.ncbi.nlm.nih.gov/Omim/

National Center for Biotechnology Information: http://www.ncbi.nlm.nih.gov/

REFERENCES

Ahnert, S. E., Fink, T. M. A. and Zinovyev, A. (2008). How much non-coding DNA do eukaryotes require? *Journal of Theoretical Biology*, **252**, 587–592.

Alberts, B., Johnson, A., Lewis, J., et al. (2002). *Molecular Biology of the Cell*, 4th edn. New York: Garland Science.

Bartel, D. P. and Unrau, P. J. (1999). Constructing an RNA world. *Trends in Cell Biology*, **9**, M9–M13.

Bateson, W. and Punnett, R. C. (1962). Experimental studies in the physiology of heredity. Reprinted in *Classic Papers in Genetics*, J. A. Peters, (ed.). Englewood Cliffs, NJ: Prentice-Hall, pp. 42–60.

Beadle, G. W. and Tatum, E. L. (1941). The genetic control of biochemical reactions in *Neurospora*. *Proceedings of the National Academy of Sciences of the United States of America*, **27**, 499–506.

Beauchamp, G. K. and Yamazaki, K. (1997). HLA and mate selection in humans: commentary. *American Journal of Human Genetics*, **61**, 494–496.

Beurton, P. J., Falk, R. and Rheinberger, H.-J. (eds) (2000). *The Concept of the Gene in Development and Evolution: Historical and Epistemological Perspectives*. Cambridge: Cambridge University Press.

Birney, E., Andrews, T. D., Bevan, P., et al. (2004). An overview of Ensembl. *Genome Research*, **14**, 925–928.

Bodmer, W. F. and Cavalli-Sforza, L. L. (1976). *Genetics, Evolution, and Man*. San Francisco, CA: W. H. Freeman and Company.

Cavalli-Sforza, L. L. and Bodmer, W. F. (1999). *The Genetics of Human Populations*. Mineola, NY: Dover Publications.

Cedar, H. and Bergman, Y. (2009). Linking DNA methylation and histone modification: patterns and paradigms. *Nature Reviews Genetics*, **10**(5), 295–304.

Coyne, J., Barton, N. and Turelli, M. (2000). Is Wright's shifting balance process important in evolution? *Evolution*, **54**, 306–317.

Crow, J. F. (1986). *Basic Concepts in Population, Quantitative, and Evolutionary Genetics*. New York: W. H. Freeman and Company.

Crow, J. F. and Kumira, M. (1963). The theory of genetic load. In *Genetics Today, Proceedings of the XI International Congress of Genetics, The Hague*. Oxford: Pergamon Press, pp. 495–506.

Darwin, C. (1892). *The Origin of Species*. New York: D. Appleton and Co.

Duffy, D. L., Montgomery, G. W., Zhen Zhen, Z., et al. (2007). A three-single-nucleotide polymorphism haplotype in Intron 1 of *OCA2* explains most human eye-color variation. *American Journal of Human Genetics*, **80**, 241–252.

Eddy, S. (1999). Noncoding RNA genes. *Current Opinion in Genetics and Development*, **9**, 695–699.

ENCODE Project Consortium (2007). Identification and analysis of functional elements in 1% of the human genome by the ENCODE pilot project. *Nature*, **447**, 799–816.

Erlich, H. A., Gelfand, D. and Sninsky, J. J. (1991). Recent advances in the polymerase chain reaction. *Science*, **252**, 1643–1650.

Evans, P. D., Gilbert, S. L., Mekel-Bobrov, N., et al. (2005) *Microcephalin*, a gene regulating brain size, continues to evolve adaptively in humans. *Science*, **309**, 1717–1720.

Fraga, M. F., Balletar, E., Paz, M. F., et al. (2005). Epigenetic differences arise during the lifetime of monozygotic twins. *Proceedings of the National Academy of Sciences of the United States of America*, **102**, 10604–10609.

Futuyma, D. (1998). *Evolutionary Biology*, 3rd edn. Sunderland, MA: Sinauer Associates.

Galhardo, R. S., Hastings, P. J. and Rosenberg, S. M. (2007). Mutation as a stress response and the regulation of evolvability. *Critical Review of Biochemical and Molecular Biology*, **42**, 399–435.

Gerstein, M. B., Bruce, C., Rozowsky, J. S., et al. (2007). What is a gene, post-ENCODE? History and updated definition. *Genome Research*, **17**, 669–681.

Gillis, J. S. and Avis, W. E. (1980). The male taller norm in mate selection. *Perspectives in Social Psychology Bulletin*, **6**, 396–401.

Glass, B., Sacks, M. S., Johns, E. F., et al. (1952). Genetic drift in a religious isolate; an analysis of the causes of variation in blood group and other gene frequencies in a small population. *American Naturalist*, **86**, 145–159.

Green E. D., Birren B., Klapholz S., et al. (eds) (1999). *Genome Analysis: a Laboratory Manual*. Cold Spring Harbor, NY: Cold Spring Harbor Laboratory Press.

Greider, C. W. (1998). Telomeres and senescence: the history, the experiment, the future. *Current Biology*, **8**, R178–R181.

Gross, L. (2006). Clues to our past: mining the human genome for signs of recent selection. *PLoS Biology*, **4**(3), e94.

Hardy, G. H. (1962). Mendelian proportions in a mixed population. Reprinted in *Classic Papers in Genetics*, J. A. Peters, (ed.). Englewood Cliffs, NJ: Prentice-Hall, pp. 60–62.

Hastings, P. J., Slack, A., Petrosino, J. F., et al. (2004). Adaptive amplification and point mutation are independent mechanisms: evidence for various stress-inducible mutation mechanisms. *PLoS Biology*, **2**(12), e399.

Hawks, J., Hunley, K., Lee, S.-H., et al. (2000). Population bottlenecks and Pleistocene evolution. *Molecular Biology and Evolution*, **17**, 2–22.

Hedrick, P. W. and Black, F. L. (1997). HLA and mate selection: no evidence in South Amerindians. *American Journal of Human Genetics*, **61**, 505–511.

Hey, J. (2005). On the number of New World founders: a population genetic portrait of the peopling of the Americas. *PLoS Biology*, **3**(6), e193.

Innis, M. A., Gelfand, D. H., Sninsky, J. J., et al. (eds) (1990). *PCR Protocols: a Guide to Methods and Applications*. San Diego, CA: Academic Press.

International Human Genome Sequencing Consortium (2001). Initial sequencing and analysis of the human genome. *Nature*, **409**, 860–921.

Jablonka, E. and Lamb, M. J. (2005). *Evolution in Four Dimensions*. Boston: MIT Press.

Jacobi, F. K., Leo-Kottler, B., Mittelviefhaus, K., et al. (2001). Segregation patterns and heteroplasmy prevalence in Leber's hereditary optic neuropathy. *Investigative Ophthalmology and Visual Science*, **42**, 1208–1214.

Kaminsky, Z. A., Tang, T., Wang, S.-C., et al. (2009). DNA methylation profiles of monozygotic and dizygotic twins. *Nature Genetics*, **41**, 240–245.

Koonin, E. V. (2009). Evolution of genome architecture. *International Journal of Biochemistry and Cell Biology*, **41**, 298–306.

Laberge, A.-M., Jomphe, M., Houde, L., et al. (2005). A *fille du Roy* introduced the *T14484C* Leber hereditary optic neuropathy mutation in French Canadians. *American Journal of Human Genetics*, **77**, 313–317.

Leonard, W. R. (2003). Food for thought. *Scientific American*, **13**, 64–74.

Levi-Strauss, C. (1969). *The Elementary Structures of Kinship*. Boston: Beacon Press.

Louro, R., Smirnova, A. S. and Verjovski-Almeida, S. (2009). Long intronic noncoding RNA transcription: expression noise or expression choice? *Genomics*, **93**, 291–298.

Lunke, S. and El-Osta, A. (2009). The emerging role of epigenetic modifications and chromatin remodeling in spinal muscular atrophy. *Journal of Neurochemistry*, **109**, 1557–1569.

McKusick, V. A., Hostetler, J. A., Egeland, J. A., et al. (1971). The distribution of certain genes in the old order Amish. Reprinted in *Human Populations, Genetic Variation, and Evolution*, L. Newall Morris, (ed.). San Francisco: Chandler Publishing Company, pp. 358–380.

Meier, R. J. and Jamison, P. L. (1990). Assortative mating in monzygotic twins. *Social Biology*, **37**, 128–136.

Mekel-Bobrov, N., Gilbert, S. L., Evans, P. D., et al. (2005). Ongoing adaptive evolution of *ASPM*, a brain size determinant in *Homo sapiens*. *Science*, **309**, 1720–1722.

Mendel, G. (1962). Experiments in plant-hybridization. Reprinted in *Classic Papers in Genetics*, J. A. Peters, (ed.). Englewood Cliffs, NJ: Prentice-Hall, pp. 1–20.

Meyers, R. A. (ed.) (2007). *Genomics and Genetics: from Molecular Details to Analysis and Techniques*. Weinheim, Germany: Wiley-VCH.

Mullis, K. B. and Faloona, F. A. (1987). Specific synthesis of DNA in vitro via a polymerase-catalysed chain reaction. *Methods in Enzymology*, **155**, 335–351.

Ober, C., Weitkamp, L. R., Cox, N., et al. (1997). HLA and mate choice in humans. *American Journal of Human Genetics*, **61**, 497–504.

Osborne, R. H. and DeGeorge, F. V. (1959). *Genetic Basis of Morphological Variation: an Evaluation and Application of the Twin Study Method*. Cambridge, MA: Harvard University Press.

Paul, D. B. and Spencer, H. G. (2008). "It's OK, we're not cousins by blood": the cousin marriage controversy in historical perspective. *PLoS Biology*, **6**(12), e320.

Pesole, G. (2008). What is a gene? An updated operational definition. *Gene*, **417**, 1–4.

Peters, J. A. (ed.) (1962). *Classic Papers in Genetics*. Englewood Cliffs, NJ: Prentice-Hall.

Roberts, D. F. (1971). The demography of Tristan da Cunha. *Population Studies*, **25**, 465–479.

Sambrook, J., Fritsch, E. F. and Maniatis, T. (1989). *Molecular Cloning: a Laboratory Manual*. Cold Spring Harbor, NY: Cold Spring Harbor Laboratory.

Schwartz, M. and Vissing, J. (2002). Paternal inheritance of mitochondrial DNA. *New England Journal of Medicine*, **347**, 576–580.

Strachan, T. and Read, A. P. (2004). *Human Molecular Genetics*, 3rd edn. New York: Garland Science, Taylor and Francis.

Timpson, N., Heron, J., Smith, G. D., et al. (2007). Comment on paper by Evans et al. and Mekel-Bebrov et al. on evidence for positive selection of *MCPH1* and *ASPM*. *Science*, **317**, 1036.

Tjio, J. H. and Levan, A. (1956). The chromosome number in man. *Hereditas*, **42**, 1–6.

Venter, J. C., Adams, M. D., Myers E. W., et al. (2001). The sequence of the human genome. *Science*, **291**, 1304–1351.

Wade, M. and Goodenough, C. (2000). The ongoing synthesis: a reply to Coyne, Barton, and Turelli. *Evolution*, **54**, 317–324.

Wallrath, L. (1998). Unfolding the mysteries of heterochromatin. *Current Opinion in Genetics and Development*, **8**(2), 147–153.

Watson, J. D. and Crick, F. H. C. (1953). Molecular structure of nucleic acids: a structure for deoxyribose nucleic acid. *Nature*, **171**, 737–738.

Wright, S. (1969). *Evolution and the Genetics of Populations. Volume 2: The Theory of Gene Frequencies*. Chicago: University of Chicago Press.

Xu, J., Bleecker, E. R., Jongepier, H., et al. (2002). Major recessive gene(s) with considerable residual polygenic effect regulating adult height: confirmation of genome-wide scan results for chromosomes 6, 9, and 12. *American Journal of Human Genetics*, **71**(3), 646–650.

Zamel, N., McClean, P. A., Sandell, P. R., et al. (1996). Asthma on Tristan da Cunha: looking for the genetic link. The University of Toronto Genetics of Asthma Research Group. *American Journal of Respiratory Critical Care Medicine*, **153**, 1902–1906.

5 Demography

James Holland Jones

INTRODUCTION

Demography lies at the heart of every statement about selection. Fitness is generally understood to be the relative proportion that an individual unit contributes to a population of such units. These populations could easily be of individuals, genes, or even groups. The way that a unit comes to make a larger contribution to a population is to have a growth rate greater than that of competitors, leading to a more technical definition of fitness as the instantaneous rate of increase. Since fitness is a growth rate, considerations of the size and composition of population – i.e., demographic considerations – are central to any evolutionary story. Life history theory is the evolutionary study of the major events of the life cycle. It is the body of theory that explains why characteristics such as life span, age at maturity, and the tempo and duration of reproduction vary between species. Life history theory is intimately related to demography as explaining the age pattern of reproductive investments – what Schaffer (1983) refers to as "the general life history problem" – is a fundamentally demographic question.

In this chapter, I will focus the discussion on formal demography, the collection of mathematical and statistical tools for enumerating populations, measuring their vital rates (i.e., rates of birth, death, and marriage) and related quantities, and projecting how they will change in structure, size, and composition. These same tools, which were largely developed by social scientists, appear repeatedly in the ecological and evolutionary literature. For example, the formalism linking birth and death rates to population structure and the population rate of increase also happens to define fitness in an age-structured population (Charlesworth, 1994). This formalism will be discussed extensively in "The Euler–Lotka characteristic equation" section below.

Humans are dioecious organisms requiring the successful union of the gametes of both male and female to reproduce. As such, a full accounting of human reproduction and population renewal requires that we simultaneously monitor male and female segments of the population. Human male and female life cycles are, indeed, quite different. For example, while more boys are born on average than girls, males experience higher mortality rates throughout life. Men typically begin reproduction later than women, but can continue reproducing after women cease reproduction. While such observations suggest that modeling both sexes would be important, a common demographic conceit is to focus attention on the female segment of the population and assume what is known as "female demographic dominance." The idea behind female demographic dominance is that women are, in essence, the rate-limiting factor for reproduction. One-sex models are mathematically simpler and manage to capture a remarkable amount of the variation that exists in human populations.

The environment, whether natural, social, or otherwise human-constructed, is not constant. When birth, death, and migration rates depend on environmental inputs (e.g., mortality rates may be high during a drought or hard winter), these rates will vary over time. Furthermore, populations of anthropological interest are frequently small and so we should expect a considerable amount of variation (i.e., "error") in our demographic measurements that results simply from sampling and finite-population effects. These observations suggest that stochastic models – models that account for the varying nature of demographic rates – are necessary for understanding human populations (Jones, 2005). However, stochastic population models are much more mathematically complicated than deterministic models that assume constant rates, and stochastic models ultimately take as their foundation deterministic models. In this chapter, I will therefore focus on one-sex deterministic population models.

A number of excellent demography textbooks currently exist for the reader interested in pursuing this material further. Preston et al. (2001) is a fairly complete introductory text in formal demography, with a focus on the analysis of large state-level populations. Hinde (1998) is a somewhat more basic

Human Evolutionary Biology, ed. Michael P. Muehlenbein. Published by Cambridge University Press. © Cambridge University Press 2010.

introduction, while Siegel and Swanson (2004) provide a quite complete reference on demographic methods. Keyfitz and Caswell (2005) is the third edition of Keyfitz's classic text on applied mathematical demography and is written at a more advanced level. The comprehensive reference for matrix population models is Caswell (2001).

I begin in the "Exponential growth" section below by reviewing the basic features of populations that grow at a constant rate in continuous time. Exponential population increase is fundamental to much of the formalism of demography. In the following "The life table" section, I introduce the life table, a scheme for representing the mortality experience of an age-structured population. Life table analysis lies at the heart of demography and probably represents the bulk of anthropological applications of demographic techniques. One important subset of life table analysis is the use of model schedules of mortality, which is also reviewed. In the "Fertility" section below, I discuss methods for the analysis of fertility. In the following "The Euler–Lotka characteristic equation" section, I introduce the stable population model. This model links the major features of demography – age-specific schedules of mortality and reproduction, age-structure, and the growth rate – into a single formalism. In addition to being an elegant representation of the demography of populations, the stable model can be used to help model populations lacking vital-event registration or where demographic data are incomplete or missing. Any set of tools that help with missing data must be useful for anthropology. In the "Population projection: matrix models" section below, I introduce population projection methods and evolutionary demography. The same formalism that allows us to project age-structured populations forward in time also helps us understand how selection shapes the life cycle.

EXPONENTIAL GROWTH

A population closed to immigration and emigration, in which births and deaths can happen at any time and the rates of births and deaths are constant, will grow at a rate $r = b - d$, where b is the per capita number of births in a unit of time and d is the per capita number of deaths. This growth rate is the per capita, *instantaneous rate of increase* of the population:

$$r = \frac{1}{N} \cdot \frac{dN}{dt},\qquad(5.1)$$

where N is the size of the population. It is worth noting here that, by definition, this also means that $r = d \log N(t)/dt$. That is, the rate of increase of a population is equal to the rate of change in the natural logarithm of the population size.

Rearranging Equation 5.1 and integrating to solve for $N(t)$, the size of the population at time t, we see that

$$N(t) = N(0)e^{rt},\qquad(5.2)$$

where $N(0)$ is the initial population size.

Dividing both sides by $N(0)$, we see that the ratio of population sizes t years apart is e^{rt} and the ratio of population sizes one year apart will be e^r. To solve for the doubling time of a population, substitute $N(t) = 2$ and $N(0) = 1$ and we see that $t = \log(2)/r$, where all the logarithms are assumed to the base e (i.e., natural logarithms). Since $\log(2) \approx 0.69$, this yields the heuristic that the doubling time of a population growing r percent annually is approximately $70/r$. A population growing 2% annually has a doubling time of approximately 35 years (and a tripling time of 55 years).

We can also use the exponential growth equation to estimate the growth rate of a population if we have available censuses at two separate times. Without loss of generality, call the population size at the first census N_0 and the population size at the second census N_t. The estimated growth rate is simply

$$\hat{r} = \frac{\log(N_t) - \log(N_0)}{t}.\qquad(5.3)$$

For example, Goodall (1986) tabulates population counts of the Kasekela community of chimpanzees for the Gombe National Park, Tanzania. In 1965, she counted a total of 51 individuals summed across all stage/sex classes, while there were 36 total individuals in the Kasekela community in 1983. This leads to an estimated growth rate of $\hat{r} = -0.019$, a decline of nearly 2% annually. This summary measure hides a great deal of complexity, including a serious epidemic and the fissioning of the community, but this is the nature of simple summary measures of complex phenomena.

THE LIFE TABLE

The life table is a means of representing the mortality experience of a population. It takes as its object of study a hypothetical cohort (i.e., a group of people born at the same moment) that is closed to migration. This cohort is followed through time until every one of its members die. In reality, we are rarely faced with analyzing the mortality experience of a true cohort. The data that we collect typically come from a particular period and represent a cross-section of the population in a particular moment in time. Many of the mathematical and statistical techniques surrounding the life table involve translating the observed period data into a synthetic cohort that can be analyzed. The distinction between period and cohort is fundamental in demography. In general, we can only analyze cohort measures retrospectively. The majority of anthropological applications will be based on period measures. This arises, as much as anything, because of the typically small samples available in anthropological populations.

A graphical device that helps to keep the concepts of period and cohort – really between calendar time and age – straight is the Lexis diagram. In a Lexis diagram, the horizontal axis represents calendar time, while the vertical axis represents age. A person who is born at some time t is plotted with a *lifeline* that intercepts the horizontal axis at t and then increases with a slope of unity until death. Consider the Lexis diagram in Figure 5.1. It spans the period between 1960 and 1970, in which 5 births, labeled 1–5, occur. Individuals 2, 3, and 5 survive to the end of the observed period, whereas individuals 1 and 4 die in their 6th and 3rd years respectively (i.e., having reached the exact ages of 5 and 2).

Mortality, like many other features of the human life cycle, changes systematically with age and an adequate description of the mortality experience of a human population should attempt to reduce the units being described to be as homogeneous as possible. For the rest of this chapter, I will assume that we are dealing with an *age-structured* population. That is, we keep track of the size and changes that happen to the population classified by an exhaustive set of age categories. Begin by defining a series of nonoverlapping age classes, x of width n. In the standard notation of demography a measure of interest is subscripted by its age, x and presubscripted by the width of that interval n. For a life cycle with K age-classes, n could conceivably be a vector of length k specifying different age-class widths. Let the mid-interval population size for age-class x be $_nK_x$. For example, in Madagascar in 1966, there were $_5K_{20} = 225\,887$ women alive aged

20–24. In human demography, it is usual to set $n = 5$ for all ages greater than $x = 5$. As mortality changes very rapidly in the first five years of life, most analyses of human mortality break the first five years into two age classes, the first lasting one year and the second lasting four years. Thus, it is standard in human demography for x to include the classes 0, 1–4, 5–9, 10–14, . . . The five-year age classes are known as *quinquennia*. All deaths beyond a certain age are typically lumped together and that last age class is open (i.e., $n = \infty$). This last open age class has historically been 85 years for state level societies with vital event registration systems. However, the drastic reduction in old-age mortality has led to vital-event tabulations being extended to ages as old as 110. Life tables constructed for age class where $n > 1$ are known as *abridged life tables*.

The data underlying the life table are the number of observed deaths by age and the population at risk for death in those ages. If we assume that the population is closed to net migration, then the mid-interval population size represents an adequate measure of exposure to risk. The estimate of the mortality rate for any age x is therefore $_nM_x = {_nd_x}/{_nK_x}$. Continuing with the example of 20-year-olds in Madagascar, the number of deaths in 1966 to women ages 20 to 24 was 3162, for a central mortality rate of 0.014. The quantity $_nM_x$ is an empirical estimator for what is known as the *period central death rate* of the population. It is important to note that this is a rate and not a probability of death in an interval. The central death rate is directly analogous to the growth rate of a population, r, discussed in the above "Exponential growth" section. In particular, it is the per capita rate of decrease (whereas r is usually the per capita rate of increase) and it applies only to a cohort. In order to proceed with the construction of a life table, we need to convert the mortality rates to probabilities. The key ingredient for this conversion is the $_na_x$ schedule, where $_na_x$ denotes the number of person-years lived by individuals dying between ages x and $x + n$. Let $_nq_x$ denote the probability of dying in the interval $[x, x + n)$.[1] The Greville equation states that, given an estimate of the central death rate, $_nm_x$ (typically the observed death rate $_nM_x$), the conversion from $_nM_x$ and $_nq_x$ is given by:

$$_nq_x = \frac{n \cdot {_nM_x}}{1 + {_nM_x}(n - {_na_x})}. \tag{5.4}$$

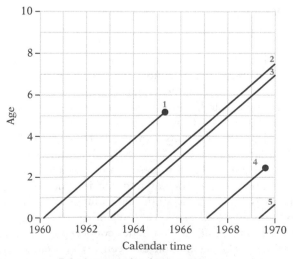

5.1. An example of a Lexis diagram.

---

[1] The notation $[x, x + n)$ indicates that for some age y in the interval $x \leq y < x + n$. That is, it includes exact age x but not exact age $x + n$, which is in the next interval. Thus the 5-year age interval beginning at the exact age of 20 includes individuals who were aged 20–24 on their last birthday.

Assuming that deaths are distributed evenly across the age interval, we can convert the value of $_5M_{20}$ to $_5q_{20}$ for Malagasy women in 1966, to get:

$$_5q_{20} = \frac{n \cdot _nM_x}{1 + _nM_x(n - _na_x)} = \frac{5 \cdot 0.014}{1 + 0.014(5 - 2.5)} = 0.068. \quad (5.5)$$

Thus a force of mortality of 1.4% acting constantly over a 5-year interval yields a 6.8% probability of death in the interval.

The schedule of $_na_x$ can be specified in a variety of ways. The most common means of specifying the $_na_x$ schedule is by using coefficients from a regression of person-years lived on the mortality rate. This is the technique employed by Keyfitz and Flieger (1990) and Coale et al. (1983). In general, the only troublesome parts of $_na_x$ schedule are in the beginning and the end of life. For five-year age-classes greater than age five, it is reasonable to assume that deaths are distributed randomly across the interval. If this is true then a value of $_na_x = 2.5$ is appropriate (as used above). Details can be found in Keyfitz (1977) and Preston et al. (2001). When mortality is high, people tend to die earlier in the interval, making the values of $_na_x$ less than half the interval width.

Once the interval death probabilities, $_nq_x$ have been calculated, the rest of the life table follows in a straightforward manner. Denote survivorship by l_x. Since the life table represents a synthetic cohort, the value of l_0 is somewhat arbitrary. This first value of the survivorship column is known as the *radix*. It is common in the human demographic literature for the radix to be set to 100 000. Under these circumstances, the l_x column is interpreted as the total number of the initial cohort still living at exact age x. Howell (1979, p. 81, Table 4.1), presents a life table for ever-born children from 165 !Kung women. Using a radix of 100 000, Howell calculates that $l_{20} = 56 228$, meaning that 56.23% of children ever born survive to their 20th birthday. In biological applications, the radix of the life table is usually set to $l_0 = 1$, and l_x is interpreted as the probability of surviving to exact age x. This is the approach I adopt here. Given a radix of $l_0 = 1$, all other survivorship values follow as:

$$l_x = \prod_{a=0}^{a=x-n} (1 - _nq_a), \quad (5.6)$$

$l_0 = 1$ by definition since everyone is alive when they are born, even if only briefly.

The other elements of the life table include $_nd_x = l_x - l_{x+n}$, which is the fraction of deaths in the interval when the radix is unity and the number of deaths when the radix represents some cohort size, and $_nL_x = \int_x^{x+n} l(x)dx$, which is the number of person-years lived in the interval. These person-years include those lived by both those who live through the interval and those who die during the interval. So we can write $_nL_x = nl_x + _na_x\,_nd_x$, since there are $_nd_x$ deaths in the

TABLE 5.1. Columns of the life table.

Element	Definition
x	Exact age of the start of the interval
$_na_x$	Average number of years lived by individuals who die in the interval
$_nM_x$	Death rate in the interval $[x, x + n)$
$_nq_x$	Probability of dying in the interval $[x, x + n)$
l_x	Cumulative probability of survival to exact age x
$_nd_x$	Number of deaths in the interval $[x, x + n)$
$_nL_x$	Person-years lived in the interval $[x, x + n)$
T_x	Person-years of life remaining at exact age x
e_x	Expectation of remaining life at exact age x

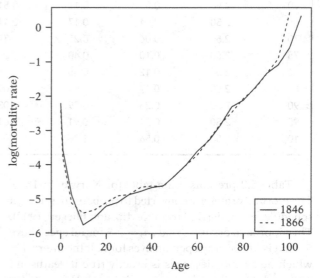

5.2. The natural logarithm of the central mortality rate for historical Norway, 1846 and 1866.

interval and the average amount of life lived by those who die is $_na_x$. T_x is the number of person-years of remaining life, which is not especially interesting except that it is required to calculate life expectancy, $e_x = \int_x^\infty l(x)dx$. Life expectancy is calculated in the life table as $e_x = T_x/l_x$. The elements of the life table are summarized in Table 5.1.

Survivorship integrates all previous mortality experience. Indeed, for a probability distribution for age at death $F(x)$, survivorship is equivalent to the complement cumulative distribution of age at death: $l(x) = 1 - F(x)$. Survivorship is frequently plotted to provide a graphical summary of the mortality experience of a population. However, the logarithm of the central mortality rate is a much better plot as it gives far more information about the changing nature of age-specific mortality throughout the life cycle. An example of such a plot is given in Figure 5.2, in which I compare the mortality rate for Norway in 1846 and 20 years later in 1866.

TABLE 5.2. Life table for Norway (1846). Data from the Human Mortality Database (Wilmoth, 2007).

x	$_na_x$	$_nM_x$	$_nq_x$	L_x	$_nd_x$	$_nL_x$	T_x	e_x
0	0.26	0.11	0.10	1.00	0.10	0.92	49.82	49.82
1	1.50	0.03	0.10	0.90	0.09	3.36	48.89	54.40
5	2.50	0.01	0.04	0.80	0.03	3.95	45.53	56.58
10	2.50	0.00	0.02	0.78	0.01	3.85	41.58	53.63
15	2.50	0.00	0.02	0.76	0.01	3.78	37.73	49.44
20	2.50	0.01	0.03	0.75	0.02	3.69	33.96	45.37
25	2.50	0.01	0.03	0.73	0.02	3.59	30.26	41.56
30	2.50	0.01	0.04	0.71	0.03	3.47	26.68	37.76
35	2.50	0.01	0.04	0.68	0.03	3.33	23.21	34.13
40	2.50	0.01	0.05	0.65	0.03	3.18	19.88	30.49
45	2.50	0.01	0.05	0.62	0.03	3.03	16.70	26.85
50	2.50	0.01	0.06	0.59	0.04	2.87	13.66	23.08
55	2.50	0.02	0.09	0.55	0.05	2.65	10.80	19.48
60	2.50	0.03	0.13	0.51	0.06	2.37	8.15	16.11
65	2.50	0.04	0.17	0.44	0.08	2.02	5.78	13.09
70	2.50	0.06	0.25	0.37	0.09	1.60	3.76	10.27
75	2.50	0.10	0.40	0.27	0.11	1.10	2.16	7.89
80	2.50	0.12	0.46	0.16	0.08	0.63	1.06	6.47
85	2.50	0.17	0.61	0.09	0.05	0.31	0.43	4.91
90	2.50	0.26	0.79	0.03	0.03	0.10	0.13	3.62
95	2.50	0.34	0.91	0.01	0.01	0.02	0.02	2.89
100	2.50	0.56	1.16	0.00	0.00	0.00	0.00	2.70

Table 5.2 presents a life table for Norway in 1846. The mortality rates are converted to probabilities using a schedule for $_na_x$ derived from Keyfitz and Flieger (1990). This approach assumes that $_na_x = 2.5$ for all ages above 5. That is, the average person lives for half the interval in which he or she dies. This is exactly true if deaths are uniformly distributed across the interval. However, for the youngest ages, $_na_x$ is usually much less than half the interval. If a baby is going to die – particularly in a low-mortality regime – it is most likely to die shortly after birth. Under the Keyfitz system, $_1a_0$ is estimated as a linear function of the mortality rate in the first year, $_1a_0 = 0.07 + 1.7_1M_0$. The coefficients of this model were derived from the empirical examination of many high-quality mortality data sets. The life table in Table 5.2 was calculated from data on age-specific deaths and mid-interval populations available on the Human Mortality Database (Wilmoth, 2007) using software that I have developed in the R statistical programming language (Jones, 2007a; R Development Core Team, 2008). The software allows users to set different radix values and to choose between different $_na_x$ schedules.

Model life tables

For many anthropological applications, vital-event registration is nonexistent. This also happens to be true for much of historical demography and the demography of the developing world more generally. Furthermore, many populations of anthropological interest are small and the estimation of age-patterns of vital events can be complicated by the very high sampling variance. In effect, the demographic signal is frequently swamped by the noise attributable to chance events acting on small populations. For these and other applications, *model life tables* have been developed to assist with: (1) smoothing mortality data from small samples; and (2) providing full age-specific mortality schedules when mortality estimates are available for only a few ages (or not at all!).

Model life tables capitalize on the universality of the human life cycle. Populations that live in similar ecologies, broadly speaking, respond with similar patterns of mortality. The most commonly used model life tables were produced by Coale and Demeny, originally in 1966, with a second edition approximately 20 years later (Coale et al., 1983). Coale and Demeny present four "regional" model life table families. All the families are derived from historical mortality data, largely of European origin. The family determines the overall shape of the mortality schedule (e.g., the relative contribution of infant or old-age mortality). Within each family, Coale and Demeny present 25 levels of mortality indexed by e_x. Level 1 gives $e_0 = 20$, while level 25 gives $e_0 = 80$. For a more complete discussion of the details of the Coale–Demeny model life tables, see Jones (2007b).

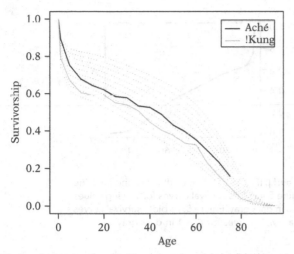

5.3. Survival curves for the !Kung (gray) and Aché (black) superimposed on Coale–Demeny West model life table survivorship schedules.

For the present discussion, it suffices to say that the "West" model life table family is the most general pattern and the one most likely to be relevant to anthropological applications. Howell (1979) used Coale–Demeny West (CDW) model life tables to smooth the age-structure of the !Kung in order to calculate a life table. In contrast, Hill and Hurtado (1996) eschewed the use of model life tables in their construction of a life table for the Aché. In Figure 5.3, I plot the survival curves for the two hunter-gatherer populations along with the CDW l_x schedules for levels 5–15 ($e_0 = 30 - 55$).

The two curves are very similar, though there is the hint that the Aché (in which the CDW model life tables were not used in the calculation of l_x) may show a different pattern of early life mortality, a point emphasized by Hill and Hurtado (1996) and discussed in Jones (2007b). The apparent departure of the Aché mortality schedule from the CDW family raises important questions for the general application of these model life tables. To be included in Coale and Demeny's collection of life tables, a population could not be at war or have recently experienced any major demographic shocks. Given the ubiquity of conflict in human societies, these conditions for inclusion in the Coale–Demeny system suggest that the model life table families may not provide a completely representative picture of the general human mortality experience. It is difficult to overstate how commonplace the usage of these model life tables is. As such, it is nearly impossible to know the degree to which more anthropological populations deviate in biologically significant ways from the pattern of the Coale–Demeny model life table families. These observations suggest caution in the uncritical use of Coale–Demeny (or any other) model life tables. However, model life tables nonetheless remain an important demographic tool that can

help augment incomplete or missing age-specific mortality data. Future work incorporating demographic shocks into model life tables may help advance the utility of this demographic tool. One example of such an approach is the INDEPTH system of AIDs-decremented model life tables for African populations (INDEPTH Network, 2002). The Coale–Demeny model life tables can be generated using software recently developed and discussed in Jones (2007a).

Primatologists have developed model life tables for a variety of nonhuman primate species based on the mortality records of large captive populations (Gage and Dyke, 1988; Dyke et al., 1993, 1995).

A useful adjunct to model life tables are the so-called relational life table methods (Brass, 1971, 1975). Anthropological and historical demographers, and those working in areas lacking vital-event registration, are frequently faced with the situation where only the partial mortality schedule for a population is known. Brass's basic idea is to take the observed values and regress them on some known standard. The parameters estimated from this regression allow the construction of a full mortality schedule. Ordinary least-squares (OLS) regression assumes that both dependent and independent variables are normally distributed, yet mortality probabilities, survivorships, etc., are bounded on the interval [0,1] and are not normally distributed (nor are the errors associated with them). Brass suggested using a logit transform on the mortality probabilities. For some variable $0 \leq x \leq 1$, the logit transformation of x is

$$\hat{Y} = \text{logit}(x) = \log\left(\frac{x}{1-x}\right). \tag{5.6}$$

This transforms a variable that ranges over [0, 1] to one that ranges over $[-\infty, \infty]$. The original variable x is recovered through the inverse logit transform, $x = e^{\hat{Y}}/(1 + e^{\hat{Y}})$.

Let $q_x = 1 - l_x$ (note the difference between this cumulative q_x, which is the probability of dying before exact age x and the life table $_nq_x$, the probability of dying between ages x and $x + n$). The Brass relational model is thus:

$$\text{logit}(q_x) = \alpha + \beta \, \text{logit}(q_x^{(S)}). \tag{5.7}$$

The two parameters of the relational model are the level α and the shape β. Figure 5.4 plots a CDW six-model survivorship schedule with varying values of α (Figure 5.4a) and β (Figure 5.4b). In Figure 5.4a, values of $\alpha < 0$ are drawn in light gray, while values of $\alpha > 0$ are drawn in dark gray. Values of $\alpha < 0$ mean that the level of cumulative mortality at any given age is lowered; thus, the survivorship is shifted up. Similarly, values of $\alpha > 0$ shift the survivorship curve down. As these are simply shifts, varying the level of α is shape-preserving. The effects of varying β are more complex as β controls

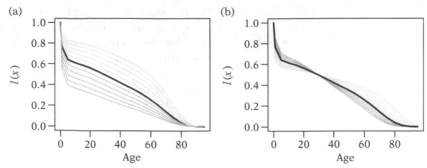

5.4. Plots showing the effect of varying α (Figure 5.4a) and β (Figure 5.4b). In both plots, the baseline l_x value is a Coale–Demeny West 6 (thick black line). Figure 5.4a plots survival curves for varying values of α: curves for $h\alpha < 0$ in light gray, while curves for $\alpha > 0$ in dark gray. Figure 5.4b plots survival curves for varying values of β: curves for $0 < \beta < 1$ in light gray while curves for $\beta > 1$ in dark gray.

the shape of the relational mortality schedule. Note that β is only defined when it is greater than zero. Otherwise, there would be negative deaths. When $\beta < 1$ early mortality is accentuated. Survivorship is lower prior to an inflection point in midlife and is greater after this point. When $\beta > 1$ mortality in the first half of life is lower while it is greater following the inflection point.

FERTILITY

Define the age-specific fertility rate (ASFR) as the number of births that occur to women whose last birthday was x, divided by the number of woman-years lived in that age class:

$$_nF_x = \frac{B_x}{_nL_x},\qquad(5.8)$$

where $_nB_x$ is the total number of births and $_nL_x$ is the woman-years lived between ages x and $x + n$.

The total fertility rate (TFR) is the sum of the ASFRs from the earliest age of reproduction in the population to the latest:

$$TFR = n \cdot \sum_{x=\alpha}^{\beta-n} {_nF_x}.\qquad(5.9)$$

The TFR is the number of offspring a woman would have if she survived the duration of the reproductive span. The TFR is probably the most important measure of fertility and it is frequently used as an index for the overall pace of reproduction and development. The TFR can have either a cohort or a period interpretation, though the period interpretation is more common. As a period measure, TFR describes the number of children a hypothetical woman would bear if she survived the entirety of her reproductive span and bore children at the rates characteristic of the period.

While the period interpretation of TFR is more common (because the data are more easily collected),

period TFRs, as a measure of fertility behavior, can show substantial distortion. The classical references for this point are Hajnal (1947), Henry (1953), and Ryder (1965, 1986). The distortion results from the fact that TFR contains two contributing forces. First is the intensity of reproduction. Populations in which the intensity of reproduction is higher will, all things being equal, have high TFRs. The second contributor to TFR is timing: when do women begin reproduction, when do they stop, are there interruptions? Consider a case where some extreme event (e.g., a war or natural disaster) caused all women in a particular population to stop reproducing for the duration of the event. If these women ceased reproduction at the same age as typical in the population, they would have lower cohort TFRs to go with the lower period TFRs measured for a period overlapping the crisis. However, if these women extended their reproductive span beyond the normal period of fertility to make up for the years of lost reproduction, they could have the same cohort TFR as previous cohorts, but the period TFR calculated for the period overlapping the crisis will still be lower. Furthermore, the period TFR for periods subsequent to the crisis, in which women extend their reproductive span, will be inflated since women in these periods are reproducing later in life than they otherwise would. Thus period TFR can decrease, increase, and decrease again without a fundamental change in the intensity of reproduction.

While TFR is probably the most commonly employed measure of fertility, it is a very poor measure of population growth because it fails to account for mortality. We define the *net reproduction ratio* (NRR) as the sum of the product of ASFRs and person-years lived in the age-class

$$NRR = \sum_{x=\alpha}^{\beta-n} {_nF_x}\,{_nL_x}.\qquad(5.10)$$

The NRR is a discrete-time analog to the measure R_0 discussed below in the continuous time framework (see Equation 5.24). If we consider only female births (and work with a female life table), then in the case of $NRR = 1$, each woman replaces herself with exactly one daughter and the population will neither grow nor decline. This is known as replacement level fertility.

The Princeton indices

Given the weakness of TFR as a measure of overall fertility, and the fact that the NRR confounds fertility with mortality, there was great interest in developing alternative measures of the overall fertility of populations, particularly as a means of measuring the impact on fertility of the use of contraceptives. A number of indices have been developed, but the most widely used of these indices are due to Coale (1969). These are frequently collectively known as the "Princeton indices." Ideally, a measure of the overall fertility of a population would use age-standardization common in the epidemiology and the demography of contemporary state-level societies. The general form of such an age-standardized measure would be:

$$I = \frac{\sum_\alpha^\beta F_x K_x}{\sum_\alpha^\beta F_x^{(S)} K_x}, \tag{5.11}$$

where α is age at first reproduction, β is the age at last reproduction, F_x is the ASFR for age x, K_x is the total fraction of the (female) population in age class x, and $F_x^{(S)}$ is some standardized value of an ASFR.

However, Coale noted that we frequently lack detailed data on the age of mothers, particularly when dealing with historical data as he was. It was therefore necessary to develop measures of overall fertility that used total births rather than births broken down by age class. First, we note that the numerator of this expression is simply the total number of births, $B = \sum_\alpha^\beta F_x K_x$. The question then is what fertility schedule with which to standardize? Coale used the Hutterites, an Anabaptist population living in the upper Midwestern United States and Canada whose fertility was extensively studied (Eaton and Mayer, 1953). The Hutterites represent one of the highest natural fertility populations ever studied and therefore make a natural point for comparison. Letting $H_x \equiv F_x^{(S)}$ be the ASFR of Hutterite women age x, Coale's standardized index of general fertility is

$$I_f = \frac{B}{\sum_\alpha^\beta H_x K_x}. \tag{5.12}$$

There is a long tradition in demography of restricting attention to marital fertility. In most populations, this is a reasonable restriction since the vast majority of fertility happens within marriage. The legal framework surrounding marriage means that births are more likely to be recorded if they happened within a marital union, making data on marital fertility more reliable than overall fertility data. Denote the number of married women in age class x as $K_x^{(M)}$ and the fertility rate of married women age x as $F_x^{(M)}$. The total number of births can thus be written as:

$$B = \sum_\alpha^\beta F_x K_x = B_M = \sum_\alpha^\beta K_x^{(M)} F_x^{(M)}. \tag{5.13}$$

The *index of general fertility*, I_f can now be decomposed into two components, I_g and I_m, known as the *index of marital fertility* and the *index of proportion married* respectively. Putting this all together, we have:

$$I_f = \overbrace{\frac{\sum_\alpha^\beta F_x^{(M)} K_x^{(M)}}{\sum_\alpha^\beta H_x K_x^{(M)}}}^{I_g} \times \overbrace{\frac{\sum_\alpha^\beta H_x K_x^{(M)}}{\sum_\alpha^\beta H_x K_x}}^{I_m}, \tag{5.14}$$

bearing in mind that we still only observe B and have assumed that:

$$B = \sum_\alpha^\beta K_x^{(M)} F_x^{(M)}. \tag{5.15}$$

Howell (1979) calculated Coale's fertility indices for the !Kung hunter-gatherers of Botswana. She showed that while the !Kung I_m value was as high or higher than that for a range of other natural-fertility populations, their values of I_g and therefore I_f were surprisingly low. Howell attributes this pattern of early and nearly universal marriage and low marital fertility to energetics, citing the critical fat hypothesis (Frisch, 1978). More recent research suggests that the critical fat hypothesis is probably incorrect (Ellison, 1981; Ellison et al., 1993). Harpending (1994) suggests that the low fertility seen among the !Kung actually reflects the high frequency of secondary sterility in Southern African populations (Bushmen included) due to high prevalence of bacterial sexually transmitted infections.

Henry (1961) pioneered the study of natural fertility. The term "natural fertility" can cause considerable confusion among the uninitiated. A population is considered to be characterized by natural fertility if there is no parity-specific fertility control. Thus, a population that contracepted to increase birth-spacing, as long as this was done in a parity-independent manner, is considered to be natural fertility.

Model fertility schedules

As with mortality, we frequently do not have complete fertility schedules for populations of anthropological interest. A number of authors have developed model fertility schemes, but here I focus on the model of Coale and Trussell (1978). Coale and Trussell employ parametric model fertility schedules derived from earlier work (Coale and Trussell, 1974). The pattern of age-specific fertility can be manipulated to achieve a wide range of fertility schedules. The Coale–Trussell model expresses the realized ASFRs as a function of a synthetic natural fertility schedule and two parameters M and m. These parameters characterize the overall level of marital fertility and the degree of departure from natural fertility respectively. The basis for the schedule is a synthetic natural fertility schedule derived from Henry's (1961) classic work on natural fertility.

The realized age-specific fertility for each age a between ages 20 and 44 is

$$r_a = n_a M e^{mv_a}, \qquad (5.16)$$

where v_a is a set of empirically derived deviations presented in Coale and Trussell (1974) and updated in Coale and Trussell (1978), and n_a is the synthetic natural fertility schedule derived by Coale and Trussell from sources in Henry (1961). The age-specific fertilities for ages 15–19 and 45–49 are interpolated linearly between zero and the values for age classes 20–24 and 40–44 respectively. M and m are parameters that measure the level of overall fertility and the deviation from the Henry synthetic natural fertility schedule respectively.

Dividing both sides of Equation 5.16 by n_a and taking logarithms provides a simple means for estimating the parameters m and M from the Coale–Trussell model. The equation,

$$\log(r_a/n_a) = \log(M) + m\, v_a, \qquad (5.17)$$

suggests that we can regress observed age-specific (standardized by Henry's synthetic fertility schedule) fertilities onto the deviations using OLS to estimate M and m. An interesting feature of Aché fertility is how fertile women were in their late 30s. We can use the Coale–Trussell model to demonstrate the extent of the late-age fertility. A regression of observed Aché ASFRs on the Coale–Trussell deviations yields an estimate of $M = 0.52$. This means that the overall level of Aché fertility in the forest period was approximately half

that of the most fecund synthetic natural fertility population. Surprisingly, $m = -0.796$, indicating that Aché women are far more fecund at later ages than the synthetic natural-fertility schedule would predict. The remarkable departure of the Aché fertility pattern from the Coale–Trussell–Henry standard is shown in Figure 5.5. Compare this plot to Figure 5.6b in which all the values of the parameter m are positive (which is the expectation based on the interpretation of m as a measure of parity-specific control).

The natural fertility populations in Henry's sample all had very high early fertility, in contrast to the Aché. A hypothesis to explain the deviation of the Aché from the expected pattern can be derived from life history theory. Assume that early and late fertility trade-off. In natural fertility agrarian populations with high energy flux, more rapid growth rates, and therefore earlier ages of menarche (Ellison, 1981), we might expect high

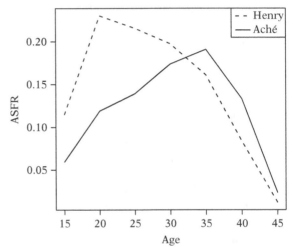

5.5. Synthetic natural fertility schedule from Henry (1961) and used by Coale and Trussell (1978) as the basis for their model fertility schedule along with the fit to the Coale–Trussell model for the Aché hunter-gatherers of Paraguay (Hill and Hurtado 1996).

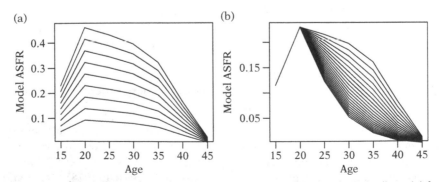

5.6. Illustration of the effects of varying the two parameters of the Coale–Trussell model fertility schedules. Figure 5.6a shows the effect of varying the level parameter from $M = 1$ to $M = 0.2$. Figure 5.6b shows the effect of varying the deviation parameter from $m = 0.1$ to $m = 2$. Values of $m > 1$ cause more severe deviations from the natural fertility shape (i.e., indicate more parity-specific control).

fertility in the early reproductive years. The trade-off between early and late fertility means that these women will have lower fertility late in their reproductive careers. The Aché, who were energy stressed and had relatively late ages at menarche (Bribiescas, 1996; Hill and Hurtado, 1996), did not have high early fertility and so did not suffer the later-life cost in fertility. This hypothesis suggests that our understanding of human fertility, which derives almost exclusively from populations with agriculture, may not well represent the experience of hunter-gatherers throughout human history.

A more sophisticated means for estimating the parameters of the Coale–Trussell model by a Poisson regression with offsets was suggested by Bronström (1985).

A major weakness of the Coale–Trussell model is that, while m models the departure from natural fertility (i.e., the application of parity-specific control) there is no obvious behavioral implication associated with a particular value of m.

Proximate determinants of fertility

A great number of ecological, social, behavioral, and economic factors can contribute to variation in fertility. Davis and Blake (1956) noted, and Bongaarts (1978) formalized the idea that all the myriad potential ultimate sources of variation in fertility must pass through a small set of proximate determinants of a biosocial nature. These are the so-called proximate determinants of fertility. Bongaarts (1978) suggests that there are three general types of proximate fertility determinant: exposure factors, deliberate marital control factors, and natural marital control factors. His list of proximate determinants is provided in Table 5.3. The proximate determinants framework has found much support in the literature on human reproductive ecology (Wood, 1994). Variations on the Bongaarts classification scheme exist, notably that used by Wood and colleagues (e.g., Wood, 1994).

TABLE 5.3. Bongaarts's proximate determinants of fertility.

1. Exposure factors
   (a) Proportion married
2. Deliberate marital fertility control factors
   (a) Contraception
   (b) Induced abortion
3. Natural marital fertility factors
   (a) Lactational infecundability
   (b) Frequency of intercourse
   (c) Sterility
   (d) Spontaneous intrauterine mortality
   (e) Duration of the fertile period

THE EULER–LOTKA CHARACTERISTIC EQUATION

Populations renew themselves. A fraction of the babies who are born at some time t grow up to eventually have babies themselves. Our goal is to write an expression for the number of births at time t, $B(t)$, as a function of the number of births that happened prior to t. The number of births that occur at time t is composed of two components: (1) births to women already alive at time $t = 0$; and (2) births to women born since $t = 0$.

$$B(t) = \int_0^t N(a,t)m(a)da + G(t), \qquad (5.18)$$

where $N(a,t)$ is the number of women age a alive at time t, $m(a)$ is the ASFR of women age a, and $G(t)$ are births from women alive at $t = 0$.

Equation 5.18 is the *renewal equation*. It shows how the present births were generated by previous births – that is, how the population renews itself. It is also very general. We typically want to posit something more specific in which the number of women alive is itself a function of schedules of vital events. Invoking a number of assumptions, Lotka derived the closed-form solution for the renewal equation and explored its implications in a series of papers in the early twentieth century (Lotka, 1907; Sharpe and Lotka, 1911; Lotka, 1922):

$$1 = \int_\alpha^\beta e^{-ra}l(a)m(a)da, \qquad (5.19)$$

where α is age at first reproduction, β is the age at last reproduction, r is the instantaneous rate of increase, and $m(a)$ is the fertility rate of age-a women.

Equation 5.19 is known as the *characteristic equation* of Euler and Lotka. The unique value of r that equates the two sides of four is known as the *intrinsic rate of increase* and it is conceptually identical to the r discussed in the above "Exponential growth" section. There is no analytical solution to Equation 5.19. Rather, it is typically solved for iteratively. Alternatively, there are approximations. The interested reader is referred to the classic work of Coale (1972) for a lucid discussion of the characteristic equation.

There are a number of assumptions that are used to derive the characteristic equation. For most practical purposes the most important of these is that the rates have been operating in a constant manner for a long time. In particular, the rates have to be constant for a period of time that exceeds the age of last reproduction. When this is so, then the $G(t)$ term in Equation 5.18 is zero. A population in which this assumption holds is said to be "stable." This terminology unfortunately frequently causes confusion. A stable population can grow or decline. A population in which $r = 0$ is said to be *stationary*. Stationary populations clearly do not grow.

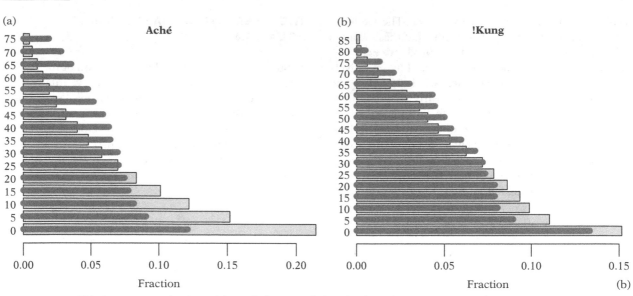

5.7. Age structure for the stable equivalent populations for the Aché (Figure 5.7a) and !Kung (Figure 5.7b). Dark bars represent the stable age distribution for a stationary population with the same $l(x)$ schedule as the respective populations.

The characteristic equation relates four fundamental quantities of demography in a single equation. Specifically, the characteristic equation ties together the schedule of age-specific mortality, the schedule of age-specific fertility, the age-structure of the population, and the rate of increase of the population.

For any given set of $l(x)$ and $m(x)$ schedules, there is a stable population that would exist if those rates applied for a long period of time. The stable population model that corresponds to a given set of vital rates is known as the *stable equivalent population*.

Two other equations besides the characteristic equation define the body of theory known as stable population theory (Coale, 1972). The first specifies the age structure in the stable population:

$$c(x) = bl(x)e^{-rx}, \qquad (5.20)$$

where $c(x)$ is the number of people in age class x in the stable population and b is the crude birth rate of the population.

We can use Equation 5.20 to compare the stable equivalent age structure to the stationary age structure of two hunter-gatherer populations. Figure 5.7 compares the two theoretical age structures of the Aché (Figure 5.7a) and !Kung (Figure 5.7b). Both populations had positive growth rates and, as a result, both plots show that the stable-equivalent population is younger than the stationary population corresponding to the $l(x)$ schedule.

It is the formula for the crude birth rate that rounds out the three primary equations that define the stable model:

$$b = \frac{1}{\int_\alpha^\beta l(x)e^{-rx}dx}. \qquad (5.21)$$

In a stationary population, the proportionate age structure of the population is given simply by the survivorship. This reflects the well-known relationship between age-structure and standing crop in wildlife management. Similarly, let life expectancy be the sum of the survivorship function:

$$e_0 = \int_0^\infty l(x)dx. \qquad (5.22)$$

In a stationary population, the following statement is then clearly true:

$$be_0 = 1. \qquad (5.23)$$

That is, the birth rate is simply the reciprocal of the average life span in the population. When $r > 0$, each value of $l(x)$ is discounted multiplicatively by a term e^{-rx}. As r gets very large, this discount factor approaches 0. This means that as r gets large, b will also get large. But since the discount term does not appear in the expression for life expectancy, the product of life expectancy and the gross birth rate will now exceed unity.

There are a variety of other derived quantities of interest. Define $\varphi(x) = l(x)m(x)$, the age-specific *net maternity rate*. The sum of the net maternity rate across all ages yields the *NRR*, R_0:

$$R_0 = \int_\alpha^\beta l(x)m(x)dx = \int_\alpha^\beta \varphi(x)dx. \qquad (5.24)$$

The NRR measures the relative size of the population from one generation to the next. That is, for a population growing at rate r with generation time T,

$$R_0 = e^{rT}. \tag{5.25}$$

While this relationship defines a generation, we typically want to relate a generation to the birth and death rates observed in the population. The generation time is essentially the average age of childbearing in the population. However, there is more than one way to measure the average age of childbearing. One is the average age of childbearing in a stable population, A_B:

$$A_B = \int_\alpha^\beta a e^{-ra} l(a) m(a) da. \tag{5.26}$$

The second is the cohort average age of childbearing,

$$\mu = \frac{\int_\alpha^\beta a l(a) m(a) da}{\int_\alpha^\beta l(a) m(a) da}. \tag{5.27}$$

In a stationary population, $r = 0$. Note that Equation 5.24 for R_0 and the characteristic equation (Equation 5.19) are identical except for the discount factor e^{-ra}. When $r = 0$, $e^{-ra} = 1$ for any a. We see that the expression for R_0 is in fact a special case of the characteristic equation. That is, it is the special case where the sum of net maternity is exactly unity. In the stationary population, each woman will replace herself on average, with exactly one daughter.

Reproductive value

In a stable population, the expected number of daughters born to a woman age x is $l(x)m(x)$ and the expected number of offspring born in her lifetime is simply the sum of these, R_0. However, when the size of a population is changing, offspring produced later in life will constitute a different proportion of the total population (i.e., fitness) than offspring born early in life. For concreteness' sake, say a population is growing at a rate of 1% annually. In the 25-year reproductive span of an individual woman, the population will have increased by over 28% ($=e^{0.01 \cdot 25}$). Reproductive value accounts for such changes in the size of the population and the impacts on individual fitness.

Reproductive value measures the net present value of offspring produced at a particular age (Fisher, 1958). The probability of a woman surviving to some age x given that she has already survived to age a is $l(x)/l(a)$ and she will have $l(x)m(x)$ offspring. When offspring are produced at age x the population has grown by a factor of e^{x-a}. We assemble these facts into a formulation for reproductive value:

$$v(a) = \int_a^\infty e^{-r(x-a)} \frac{l(x)}{l(a)} m(x) dx. \tag{5.28}$$

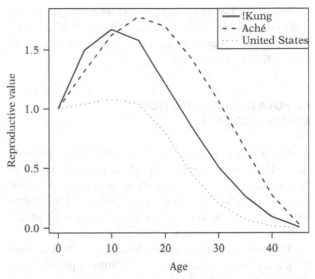

5.8. Age-specific reproductive value curves for three populations: !Kung (Howell, 1979), Aché (Hill and Hurtado, 1996), and the United States in 1967 (Keyfitz and Flieger, 1990).

A little algebra yields the form of the reproductive value equation originally presented by Fisher (1958) and that appears in most textbooks:

$$v(a) = \frac{e^{ra}}{l(a)} \int_a^\infty e^{-rx} l(x) m(x) dx. \tag{5.29}$$

For any $a > 0$, the cohort into which the woman is born will have declined according to the age-schedule of mortality. The term outside the integration can thus be seen as a premium for having survived to age a. The term inside the integration is the discounted net maternity at each age $x \geq a$.

Reproductive value curves for age-structured populations have a characteristic shape. Figure 5.8 plots the age-specific reproductive value for three populations: !Kung (Howell, 1979), Aché (Hill and Hurtado, 1996), and the United States in 1967 (Keyfitz and Flieger, 1990). Reproductive value at birth (i.e., $a = 0$) is conventionally set to $v(a) = 1$. As age increases, so does $v(a)$ until reaching a maximum around the age of first reproduction, α.[2] From this point reproductive value declines, reaching zero at the age of last reproduction, ta. In Figure 5.8, both the Aché and the !Kung show the premium received by women who survive to the first reproductive age class (15). Because mortality is so high, $v(a)$ increases steeply with age before declining. Only 59% of !Kung girls and 64% of Aché girls survived

---

[2] Age at first reproduction is slightly complicated for Leslie matrix models with five-year age classes, since the fertility values are averaged across adjacent ages for a birth-flow population. This means that there is typically nonzero fertility for ages before actual reproduction is observed. These are the types of modeling compromises that are necessary when trying to represent a fundamentally continuous process in discrete time.

to their 15th birthday. In contrast 97.5% of American girls survived to their 15th birthday in 1967; consequently, their reproductive value rose only nominally before first reproduction.

POPULATION PROJECTION: MATRIX MODELS

Stable population theory is constructed in continuous time. However, demographic data typically come in discrete age classes. Population projection, it turns out, is much simpler to do in a discrete-time framework than in the continuous time framework of the stable population model.

Consider a population divided into k nonoverlapping age classes, n years wide. For human populations, it is common to divide the reproductive life span into 9–11 5-year age-classes (quinquennia), depending upon the age of last reproduction. Ten is probably the most common number of age classes and that is what I will discuss. The ages are thus: 0–4,5–9, 10–14,. . .,45–49. Note that unlike the case of the life table, these age classes must all be the same width. In order to project the population from time t to time $t + 5$, we need 10 equations. For example, to project the age-class zero individuals to the next age class, we have $n_5 = P_0 n_0$, where P_0 is the probability of surviving from age $n = 0$ to age $n = 5$. Each equation that moves the cohort forward in the life cycle is similarly sparse. For the human life cycle divided into quinquennia and age of last reproduction in the 45–49 age class, there are 9 such equations. The last of the 10 equations accounts for the production of age-class zero individuals (i.e., reproduction) and takes the form $n_0 = F_0 n_0 + F_5 n_5 +, . . ., +F_k n_k$, where the F_j are the ASFRs (some of which may be zero).

Using so many equations to perform a population projection is cumbersome to say the least. Leslie (1945) noted that matrix algebra can greatly simplify working with such systems of linear equations. All the equations can be compactly represented in a $k \times k$ matrix, known as a Leslie matrix. The Leslie matrix is square and sparse, with the age-specific survival probabilities along the subdiagonal, age-specific fertilities along the first row and zeros everywhere else. The 10×10 Leslie matrix for the human life cycle takes the form:

$$\mathbf{A} = \begin{bmatrix} 0 & 0 & F_{10} & F_{15} & F_{20} & F_{25} & F_{30} & F_{35} & F_{40} & F_{45} \\ P_0 & 0 & 0 & 0 & 0 & 0 & 0 & 0 & 0 & 0 \\ 0 & P_5 & 0 & 0 & 0 & 0 & 0 & 0 & 0 & 0 \\ 0 & 0 & P_{10} & 0 & 0 & 0 & 0 & 0 & 0 & 0 \\ 0 & 0 & 0 & P_{15} & 0 & 0 & 0 & 0 & 0 & 0 \\ 0 & 0 & 0 & 0 & P_{20} & 0 & 0 & 0 & 0 & 0 \\ 0 & 0 & 0 & 0 & 0 & P_{25} & 0 & 0 & 0 & 0 \\ 0 & 0 & 0 & 0 & 0 & 0 & P_{30} & 0 & 0 & 0 \\ 0 & 0 & 0 & 0 & 0 & 0 & 0 & P_{35} & 0 & 0 \\ 0 & 0 & 0 & 0 & 0 & 0 & 0 & 0 & P_{40} & 0 \end{bmatrix}. \quad (5.30)$$

Matrix $\mathbf{A}$ can be represented as a directed graph (Harary, 1969) in which the nodes represent the age classes and the directed edges represent the transitions (i.e., the fertilities and survival probabilities). Figure 5.9 presents the life cycle graph corresponding to matrix $\mathbf{A}$ in Equation 5.30. Survival transitions are represented by the horizontally aligned edges directed to the right, while fertility is represented by the left-directed arcs back to node 1(age class 0–4). In addition to providing an appealing graphical analog to the matrix formalism, the life cycle graph provides important information about the dynamical properties of the system (discussed below).

An initial 10×1 population vector n_t can be projected from time t to time $t + 5$; we simply premultiply $n_{(0)}$ by $\mathbf{A}$:

$$\mathbf{n}_{t+5} = \mathbf{A}\mathbf{n}_t. \quad (5.31)$$

The matrix algebra formalism introduced by Leslie has more to recommend it than simply a compact mechanism for projecting a population. A question that arises naturally in the analysis of systems of linear equations is whether there exists a scalar value (i.e., a single number), λ, that can substitute for the matrix $\mathbf{A}$ in projecting the population:

$$\mathbf{A}\mathbf{u} = \lambda\mathbf{u}, \quad (5.32)$$

for some vector $\mathbf{u}$. Equation 5.32 generally has a solution. Details of the solution to this equation is beyond the scope of this chapter.[3] When it does, the scalar λ is known as an *eigenvalue* and the vector $\mathbf{u}$ is its corresponding *eigenvector*. In fact, there are k distinct eigenvalues and eigenvectors for the $k \times k$ matrix $\mathbf{A}$. However, when $\mathbf{A}$ fulfills two conditions, it is guaranteed that $\mathbf{A}$ will have one eigenvalue which is positive, real, and strictly greater than the other $k - 1$ eigenvalues. This is known as the *dominant eigenvalue* of matrix $\mathbf{A}$ and it is the asymptotic rate of increase of the age-structured population. The eigenvector that corresponds to the dominant eigenvalue is known as the dominant right eigenvector and it corresponds to the stable age distribution of the population between ages 0 and k. The conditions that guarantee the existence of a single dominant eigenvalue are known as: (1) irreducibility; and (2) primitivity. A non-negative matrix is irreducible if all of its states can communicate with each other – that is, if there is a path from between all the nodes of the life cycle graph. Such a graph is said to be strongly connected (Harary, 1969; Caswell, 2001). An irreducible matrix is primitive if all loops of the life cycle graph are relatively prime to each other. This ensures that the growth of the population will be (asymptotically) aperiodic and that the population eventually converges to its stable age distribution.

---

[3] The interested reader can consult Caswell (2001), or any textbook in linear algebra.

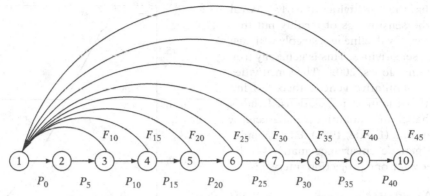

5.9. Life cycle diagram corresponding to the Leslie matrix of Equation 5.30.

Assume that a population is in the stable age distribution. Then let the initial population vector be $\mathbf{n}_0$, the population at some time t is then

$$\mathbf{n}_t = \lambda^t \mathbf{n}_0. \tag{5.33}$$

Using the spectral decomposition of the projection matrix, we can say what the population vector will be at time t starting from any arbitrary initial population vector. The spectral decomposition of $\mathbf{A}$, the details of which are beyond this chapter, uses all k eigenvalues and eigenvectors and for a deterministic population with constant vital rates, will give an exact projection. See, for example, Caswell (2001) for details.

The stable age distribution is given by the dominant right eigenvector, $\mathbf{u}$. There also exists a left eigenvector for each eigenvalue (matrix multiplication is, in general, not commutative). Denote this eigenvector $\mathbf{v}$ and it corresponds to reproductive values of the ages 0 through k.

Fitness sensitivities and elasticities

A population with projection matrix $\mathbf{A}$ will grow asymptotically at rate λ. This growth rate is the fitness measure of the age-structured population (Charlesworth, 1994; Caswell, 2001). Clearly, if we increase the values of the survival probabilities and fertilities in the projection matrix, λ will increase. But which matrix entries will have the greatest impact on the growth rate? Knowing the rates to which λ is most sensitive provides important evolutionary information.

We want to change the values of $\mathbf{A}$ by a small amount (i.e., perturb them) and calculate the change in λ that results. Perturbations of the characteristic equation were used by Hamilton (1966) in his foundational study of senescence. Caswell (1978) devised a simple formula for the sensitivity of the growth rate λ using the left and right eigenvectors of the projection matrix.

Let a_{ij} be the ijth element of the projection matrix $\mathbf{A}$. The sensitivity of λ to a small change in a_{ij} is

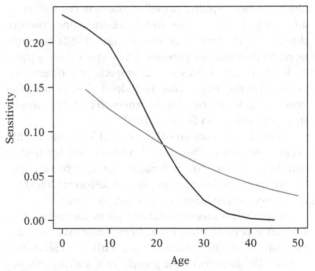

5.10. Fitness sensitivities for Madagascar (1966). Sensitivities with respect to survival are in black, and sensitivities with respect to fertility are in gray.

$$\frac{\partial \lambda}{\partial a_{ij}} = v_i u_j. \tag{5.34}$$

Equation 5.34 assumes that the eigenvectors have been scaled in such a way that $<\mathbf{vw}> = 1$. The element a_{ij} in the Leslie matrix represents the transition rate from stage j to stage i.[4] The sensitivity of λ to a perturbation in a_{ij} is thus the product of the reproductive value of the receiving age and the stable age fraction of the sending age.

Since λ is the fitness measure for an age-structured population, the s_{ij} measure the force of selection on each life cycle transition, a_{ij}. Hamilton (1966) used sensitivities of survival to measure the decline in the force of selection with age. Figure 5.10 shows this age-related decline in the force of selection. In addition to

---

[4] Note that the ecologists' convention of Leslie matrices of the transitions going from column to row is backwards from most social science applications of transition matrices.

this decline, note that the sensitivities of early survival start higher than the sensitivities of fertility but that the sensitivities of fertility decline less steeply with age than do the survival sensitivities. This is generally true for human populations (Jones, 2009). The sensitivities also appear in the quantitative genetic theory of life history evolution in structured populations. Lande's equation for the change in a quantitative character **z** uses the sensitivities of λ (Lande, 1982). For a population with additive genetic covariance matrix **G**, the change in character **z** for one projection interval is

$$\Delta \mathbf{z} = \lambda^{-1} \mathbf{G} \, \mathbf{s}, \qquad (5.35)$$

where **s** is a vector of all the sensitivities in the life cycle.

Thus, transitions to which λ is highly sensitive will change more rapidly, provided there is (a) sufficient additive genetic variance, and (b) a covariance structure that allows the trait to change (Lande, 1982). Strongly negative correlations between traits – particularly traits with high sensitivities – will impede the directional change of the trait. This is indeed the quantitative genetic basis of the classic trade-offs of life history theory (reviewed in Stearns, 1992).

Sensitivities measure the force of selection on life cycle transitions. They are also important for understanding the population dynamics of structured populations in variable environments. See Jones (2005) for discussion in the human evolutionary context.

Sensitivities measure the change in the growth rate λ given a perturbation on a linear scale. We can also measure the perturbation on a logarithmic scale. Let e_{ij} denote the *elasticity* of the growth rate λ to a perturbation of element a_{ij},

$$e_{ij} = \frac{\partial \lambda}{\partial a_{ij}} \cdot \frac{a_{ij}}{\lambda} = \frac{\partial \log \lambda}{\partial \log a_{ij}}. \qquad (5.36)$$

This logarithmic scale measures proportional sensitivities of λ to perturbations. That is, if we perturb vital rate a_{ij} by 1%, by what percentage will λ increase? For an elasticity of $e_{ij} = 0.1$, λ would increase by 0.1%.

Elasticities have a number of desirable properties. First, the sum of all the elasticities in a life cycle is unity. Elasticities can therefore be seen, albeit in a rather restricted way, as the fraction of total selection that a particular life cycle transition experiences. From Figure 5.11, we can see that survival to age 5 has an elasticity of 0.184. Thus infant survival accounts for more than 18% of the total selection on the human life cycle, at least for Madagascar in 1966. The fraction of total selection that prereproductive survival accounts for in the human life cycle is remarkably constant across populations with very different vital rates (Jones, 2009). The degree to which elasticities are limited as a measure of total selection derives from the fact that both sensitivities and elasticities are

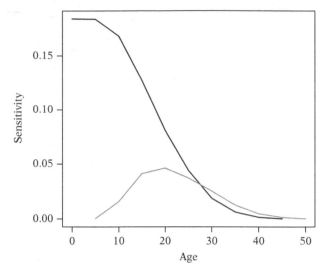

5.11. Fitness elasticities for Madagascar (1966). Elasticities with respect to survival are in black, and elasticities with respect to fertility are in gray.

partial derivatives: they measure the local slope of a perturbation holding every other transition constant. An actual environmental perturbation rarely changes only one transition.

A second interesting feature of elasticities is that the sum of elasticities of transitions entering a node in the life cycle graph must equal the elasticities of transitions leaving that node. For age-structured models, this means that elasticity of survival in the reproductive ages is necessarily less than in the prereproductive ages since there are two transitions out of every reproductive age class and only one transition out of each prereproductive age class.

CONCLUSIONS

In this chapter, I have focused primarily on the classic methods of formal demography as they apply to questions of human evolutionary biology. Given the strongly methodological nature of this chapter, I have not attempted to provide a comprehensive review of anthropological or human evolutionary demography. The review of anthropological demography by Howell (1986) still provides pointers to much of the significant demographic work in anthropology. Hill (1993), Hill and Kaplan (1999), and Mace (2000) provide reviews of life history theory and its applications to evolutionary anthropology. Since Howell's (1986) review, there have been a number of laudable studies of the demography of small-scale populations, including Hill and Hurtado's (1996) monograph on the Aché, and Early and Peters's (1990) demographic study of Yanomama. Outstanding evolutionary demographic work continues to be carried out by human behavioral ecologists working in populations around the world (Pennington and Harpending, 1991; Borgerhoff Mulder, 1992; Roth, 1993; Leslie and Winterhalder, 2002; Sear et al., 2002; Sear et al., 2003; Gurven et al., 2007). Roth (2004) presents a novel integration of both evolutionary biology and culture in anthropological demography.

The tools of formal demography have a direct bearing on evolutionary questions. Natural selection is fundamentally a demographic process. It results from the differential survival and reproduction of heritable variants of phenotypes. Humans are long-lived and our vital rates change dramatically across the life cycle. Such age-structure complicates simplistic arguments about selection and makes the use of demographic models that incorporate age-structure absolutely paramount if we are to understand the process of selection (Charlesworth, 1994).

While the mathematical formalism of demography may be unfamiliar to many anthropologists, the calculation of all the quantities discussed in this chapter is completely straightforward given appropriate software. Matlab is one excellent option. Several population biology texts have been written that contain extensive example code for many of the calculations I have discussed in this chapter (Caswell, 2001; Morris and Doak, 2002). Furthermore, there are a number of very good books on general scientific computing with Matlab (Davis and Sigmon, 2004). Another increasingly attractive alternative is the R statistical programming language (R Development Core Team, 2008). R is a state-of-the-art statistical and numerical programming environment, is available on a variety of computing platforms, and is freely available on the Internet (http://www.r-project.org). All the calculations and all the figures except Figure 5.9 that I have produced for this chapter were carried out in a recently developed open-source software library for R. This package is freely available and contains a great deal of documentation and worked examples (Jones, 2007a).

TABLE 5.4. Keyfitz (1977) values for the US female Leslie matrix.

Age class	Survival probability	Age-specific fertility	Population size ($\times 10^6$)
0	0.99661	0	10.136
5	0.99834	0.00103	10.006
10	0.99791	0.08779	9.065
15	0.99682	0.34873	8.045
20	0.99605	0.47607	6.546
25	0.99472	0.33769	5.614
30	0.99229	0.18333	5.632
35	0.98866	0.07605	6.193
40	0.98304	0.01744	6.345
45	–	0.00096	5.796

DISCUSSION POINTS

1. Lotka's characteristic equation connects four important demographic phenomena. What are they?

2. Why does reproductive value increase until the age at first reproduction and then decline thereafter?

3. Keyfitz (1977) outlines the method of population projection using Leslie matrices and presents data on the female population of the United States in 1964 to illustrate. His values for the US female Leslie matrix are reproduced in Table 5.4.

   (a) Construct a 10 × 10 Leslie matrix for the female population of reproductive age using these data. Project it forward to the year 2004.

   (b) How do you think your projection compares to the actual population structure of the United States in the year 2004?

   (c) What is the growth rate of the population? What is the stable age distribution? How different is the initial population from the stable population? Show this graphically.

   (d) Imagine you had the ability to change the vital rates. Which element of the projection matrix, if perturbed, would increase the growth rate of the population the most? Explain based both on the mechanics of the calculation and the biology.

4. Table 5.5 presents life table survivorship (l_x) and age-specific fertility values for the Hutterites, an anabaptist sect living in the Dakotas in the mid 1950s. Plot the age-specific survival, and net maternity functions. What is the value of R_0 for the Hutterites in 1953? What about the total fertility rate (TFR)? What does this tell us about population growth?

5. Table 5.6 presents Leslie matrix entries for the Hutterites. Construct a Leslie matrix (don't forget to ensure that it is irreducible). What is the annual rate of increase? What are the elasticities of the growth rate with respect to perturbations of the projection matrix? What perturbation would have the greatest impact on the rate of increase? Is it feasible for this value to change much? Do you think that fertility can increase much in this population?

6. Why do we care about irreducibility and primitivity of Leslie matrices? What happens when we repeatedly multiply a matrix that is reducible? What happens when we repeatedly multiply a matrix that is primitive? Can you think of cases where a demographic projection matrix might be reducible? Primitive?

7. Total fertility rate (TFR) is probably the most widely used measure of fertility. However, it is frequently criticized for not representing the fertility experience

TABLE 5.5. Survival and age-specific fertility values for the Hutterites, 1953.

Age	l_x	m_x
0	1.00	0.00
1	0.96	0.00
5	0.96	0.00
10	0.95	0.00
15	0.95	0.02
20	0.95	0.29
25	0.95	0.42
30	0.95	0.35
35	0.95	0.22
40	0.94	0.10
45	0.94	0.02
50	0.94	0.00
55	0.93	0.00
60	0.93	0.00
65	0.92	0.00
70	0.90	0.00
75	0.88	0.00
80	0.83	0.00
85	0.77	0.00

TABLE 5.6. Leslie matrix entries for the Hutterites demographic data.

Age	P_i	F_i
0	0.96	0.00
1	1.00	0.00
5	0.99	0.00
10	1.00	0.02
15	1.00	0.37
20	1.00	0.85
25	1.00	0.92
30	1.00	0.68
35	0.99	0.38
40	1.00	0.14
45	1.00	0.02
50	0.99	0.00
55	1.00	0.00
60	0.99	0.00
65	0.98	0.00
70	0.98	0.00
75	0.94	0.00
80	0.93	0.00

of real women. In what way is TFR an artificial measure? How might the way it is constructed distort our understanding of actual women's fertility?

8. Using Lotka's characteristic equation as a guide, explain what happens to the population structure when survivorship increases while fertility is held constant. What happens when fertility decreases when survivorship is held constant? What are the demographic circumstances in which each of these scenarios might apply?

REFERENCES

Bongaarts, J. (1978). A framework for analyzing the proximate determinants of fertility. *Population and Development Review*, **4**, 105–132.

Borgerhoff Mulder, M. (1992). Demography of pastoralists: preliminary data on the Datoga of Tanzania. *Human Ecology*, **20**, 383–405.

Brass, W. (1971). On the scale of mortality. In *Biological Aspects of Demography*, W. Brass (ed.). London: Taylor and Francis, pp. 69–110.

Brass, W. (1975). *Methods for Estimating Fertility and Mortality from Limited and Defective Data*. Chapel Hill: University of North Carolina International Program of Laboratories for Population Statistics.

Bribiescas, R. G. (1996). Testerone levels among Aché hunter-gatherer men: a functional interpretation of population variation among adult males. *Human Nature*, **7**, 163–188.

Bronström, G. (1985). Practical aspects on the estimation of parameters in Coale's model for marital fertility. *Demography*, **22**, 625–631.

Caswell, H. (1978). A general formula for the sensitivity of population growth rate to changes in life history parameters. *Theoretical Population Biology*, **14**, 215–230.

Caswell, H. (2001). *Matrix Population Models: Construction, Analysis and Interpretation*. Sunderland, MA: Sinauer.

Charlesworth, B. (1994). *Evolution in Age-Structured Populations*. Cambridge: Cambridge University Press.

Coale, A. J. (1969). The decline of fertility in Europe from the French Revolution to World War II. In *Fertility and Family Planning: a World View*, S. B. Behrman, L. Corsa and R. Freedman (eds). Ann Arbor: University of Michigan Press.

Coale, A. J. (1972). *The Growth and Structure of Human Populations: a Mathematical Investigation*. Princeton: Princeton University Press.

Coale, A. J. and Trussell, J. (1974). Model fertility schedules: variations in the age-structure of childbearing. *Population Index*, **40**, 185–258.

Coale, A. J. and Trussell, J. (1978). Technical note: finding two parameters that specify a model schedule of marital fertility. *Population Index*, **44**, 202–213.

Coale, A. J., Demeny, P. and Vaughn, B. (1983). *Regional Model Life Tables and Stable Populations*. New York: Academic Press.

Davis, K. and Blake, J. (1956). Social structure and fertility: an analytic framework. *Economic Development and Cultural Change*, **4**, 211–235.

Davis, T. and Sigmon, K. (2004). *MATLAB Primer*. New York: Chapman and Hall.

Dyke, B., Gage, T. B., Ballou, J. D., et al. (1993). Model life tables for the smaller New World monkeys. *American Journal of Primatology*, **29**, 268–285.

Dyke, B., Gage, T. B., Alford, P. L., et al. (1995). Model life table for captive chimpanzees. *American Journal of Primatology*, **37**, 25–37.

Early, J. D. and Peters, J. F. (1990). *The Population Dynamics of the Mucajai Yanomama*. San Diego, CA: Academic Press.

Eaton, J. W. and Mayer, A. J. (1953). The social biology of very high fertility among the Hutterites: the demography of a unique population. *Human Biology*, **25**, 206–264.

Ellison, P. T. (1981). Threshold hypotheses, developmental age, and menstrual function. *American Journal of Physical Anthropology*, **54**, 337–340.

Ellison, P. T., Panter-Brick, C., Lipson, S. F., et al. (1993). The ecological context of human ovarian function. *Human Reproduction*, **8**, 2248–2258.

Fisher, R. A. (1958). *The Genetical Theory of Natural Selection*. New York: Dover.

Frisch, R. E. (1978). Population, food intake, and fertility. *Science*, **199**, 22–30.

Gage, T. B. and Dyke, B. (1988). Model life tables for the larger Old World monkeys. *American Journal of Primatology*, **16**, 305–320.

Goodall, J. (1986). *The Chimpanzees of Gombe: Patterns of Behavior*. Cambridge: Harvard University Press.

Gurven, M., Kaplan, H. et al. (2007). Mortality experience of Tsimane Amerindians of Bolivia: regional variation and temporal trends. *American Journal of Human Biology*, **19**, 376–398.

Hajnal, J. (1947). The analysis of birth statistics in the light of the recent international recovery of the birth-rate. *Population Studies*, **1**, 137–164.

Hamilton, W. D. (1966). The moulding of senescence by natural selection. *Journal of Theoretical Biology*, **12**, 12–45.

Harary, F. (1969). *Graph Theory*. Reading, MA: Addison-Wesley.

Harpending, H. (1994). Infertility and forager demography. *American Journal of Physical Anthropology*, **93**, 385–390.

Henry, L. (1953). *Fécondité des mariages: nouvelle méthode de mesure*. Paris: Presses Universitaires de France.

Henry, L. (1961). Some data on natural fertility. *Eugenics Quarterly*, **8**, 81–91.

Hill, K. (1993). Life history theory and evolutionary anthropology. *Evolutionary Anthropology*, **2**, 78–88.

Hill, K. and Hurtado, A. M. (1996). *Aché Life History*. New York: Aldine de Gruyter.

Hill, K. and Kaplan, H. (1999). Life history traits in humans: theory and empirical studies. *Annual Review of Anthropology*, **28**, 397–430.

Hinde, A. (1998). *Demographic Methods*. London: Arnold.

Howell, N. (1979). *The Demography of the Dobe !Kung*. New York: Academic Press.

Howell, N. (1986). Demographic anthropology. *Annual Review of Anthropology*, **15**, 219–246.

INDEPTH Network (2002). *Population, Health, and Survival at INDEPTH sites*. Ottawa: IDRC.

Jones, J. H. (2005). Fetal programming: adaptive life-history tactic or making the best of a bad start? *American Journal of Human Biology*, **17**, 22–33.

Jones, J. H. (2007). demog R: a package for evolutionary demographic analysis in R. *Journal of Statistical Software*, **22**, 1–28.

Jones, J. H. (2009). The force of selection on the human life cycle. *Evolution and Human Behavior*, **30**(5), 305–314.

Keyfitz, N. (1977). *Introduction to the Mathematics of Populations*. Menlo Park, CA: Addison-Wesley.

Keyfitz, N. and Caswell, H. (2005). *Applied Mathematical Demography*. New York: Springer.

Keyfitz, N. and Flieger, W. (1990). *World Population Growth and Aging: Demographic Trends in the Late Twentieth Century*. Chicago: University of Chicago Press.

Lande, R. A. (1982). A quantitative genetic theory of life history evolution. *Ecology*, **63**, 607–615.

Leslie, P. and Winterhalder, B. (2002). Demographic consequences of unpredictability in fertility outcomes. *American Journal of Human Biology*, **14**, 168–183.

Leslie, P. H. (1945). On the use of matrices in certain population mathematics. *Biometrics*, **33**, 213–245.

Lotka, A. J. (1907). Relation between birth rates and death rates. *Science*, **26**, 21–22.

Lotka, A. J. (1922). The stability of the normal age distribution. *Proceedings of the National Academy of Sciences of the United States of America*, **8**, 339–345.

Mace, R. (2000). Evolutionary ecology of human life history. *Animal Behaviour*, **59**, 1–10.

Morris, W. F. and Doak, D. F. (2002). *Quantitative Conservation Biology*. Sunderland, MA: Sinauer.

Pennington, R. and Harpending, H. (1991). Effect of infertility on the population structure of the Herero and Mbanderu of Southern Africa. *Social Biology*, **38**, 127–139.

Preston, S. H., Heuveline, P. and Guillot, F. (2001). *Demography: Measuring and Modelling Population Processes*. Oxford: Blackwell Publishers.

R Development Core Team (2008). *R: A language and Environment for Statistical Computing*. Vienna, Austria: R Foundation for Statistical Computing.

Roth, E. A. (1993). A reexamination of Rendille population regulation. *American Anthropologist*, **95**, 597–611.

Roth, E. A. (2004). *Culture, Biology, and Anthropological Demography*. Cambridge: Cambridge University Press.

Ryder, N. (1965). The cohort as a concept in the study of social change. *American Sociological Review*, **30**, 843–861.

Ryder, N. (1986). Observations on the history of cohort fertility in the United States. *Population and Development Review*, **12**, 617–643.

Schaffer, W. M. (1983). The application of optimal control theory to the general life history problem. *American Naturalist*, **121**, 418–431.

Sear, R., Steele, F., Mcgregor, A. A., et al. (2002). The effects of kin on child mortality in rural Gambia. *Demography*, **39**, 43–63.

Sear, R., Mace, R. and Mcgregor, I. A. (2003). The effects of kin on female fertility in rural Gambia. *Evolution and Human Behavior*, **24**, 25–42.

Sharpe, F. and Lotka, A. J. (1911). A problem in age-distribution. *Philosophical Magazine*, **21**, 435–438.

Siegel, J. S. and Swanson, D. A. (2004). *The Methods and Materials of Demography*. Amsterdam: Elsevier.

Stearns, S. C. (1992). *The Evolution of Life Histories*. Oxford: Oxford University Press.

Wilmoth, J. (2007). *Human Mortality Database*. Berkeley, CA: University of California Press and Max Planck Institute for Demographic Research.

Wood, J. W. (1994). *Dynamics of Human Reproduction: Biology, Biometry, Demography*. New York: Aldine de Gruyter.

6 History, Methods, and General Applications of Anthropometry in Human Biology

Noël Cameron and Laura L. Jones

INTRODUCTION

If it is accepted that a core element of human evolutionary biology is about understanding the nature and meaning of morphological variation within and between the various species of primates that form our phylogenetic ancestors, then anthropometry is the essential tool used in describing morphological variation. Without a standardized form of measurement comparisons between the morphological characteristics of any two or more individuals is impossible. In the acceptance of that simple principle, the complexity of anthropometry is glimpsed. For anthropometry to be useful in describing morphology it must involve standardized instruments being applied to defined landmarks. The instruments must measure in the same units, to the same degree of precision, and the ability of the observer to repeat the measurement and obtain the same result must be within a range of error that does not significantly alter the outcome of the measurement.

So, modern anthropometry requires a universally understood terminology applied to morphological landmarks, universally applied units of measurement (or at worst units of measurement that have a constant relationship), and instrumentation that is appropriately designed to measure to a degree of precision that will be useful in describing similarities and differences in size. In addition, the anthropometrist must be able to use anthropometric instruments with an acceptable degree of reliability. It has taken over 350 years since a German physician, Johann Sigismund Elsholtz (1623–1688), submitted his graduate thesis entitled "Anthropometria" to the University of Padua in 1654 (Tanner, 1981) for us to be able to be reasonably certain that most of these requirements have been met.

TERMINOLOGY

Sakai (2007) maintains that the historical development of anatomical terminology from the ancient to the modern can be divided into five stages. The initial stage is represented by Galen's (129 – *ca*. 200 or 216) use of colloquial Greek words for a limited number of anatomical terms. Galen is thought to have published over 600 texts but few survived following the destruction of the library at Alexandria sometime prior to 800 AD. Yet his influence lasted for the better part of 1500 years until the second stage when Vesalius (1514–1564) worked on a translation of Galen's Greek works into Latin. Through dissection Vesalius was able to demonstrate many of the errors in Galen's work and his resulting magnus opus, *De Humani Corporis Fabrica* in 1543, is viewed as the foundation document of modern human anatomical description and a good deal of its terminology. The third stage was expansion of anatomical terminology during the sixteenth and seventeenth centuries through the work of the botanist and anatomist Caspar Bauhin (1560–1624) in Basel and in Paris through Franciscus Sylvius (1614–1672) on the brain and vascular system. The fourth stage is characterized by the many anatomical textbooks published in Latin in the seventeenth century, and in modern languages in the eighteenth and nineteenth centuries although there was no universal agreement as to terminology, thus anatomical terms for the same structure were differently expressed by different authors. By the late nineteenth century about 50 000 terms were in use for various body parts. The terminology depended in part on the anatomist's school and national tradition. Vernacular translations of Latin and Greek, as well as various eponymous terms, were barriers to effective international communication. There was thus disagreement and confusion among anatomists regarding anatomical terminology. The final stage was initiated between 1887 and 1895 at the Ninth Congress of the *Anatomische Gesellschaft* in Basel when the 50 000 anatomical terms were reduced to 5528 and the resultant *Basle Nomina Anatomica* became adopted by a significant number of countries. Some, however, maintained their own regional variations and it was not until 1956, when the first edition of *Nomina Anatomica* was produced by the International

Federation of Associations of Anatomists, that some wider standardization was achieved. Finally, in 1998, the *Terminologia Anatomica* became accepted by all major anatomy associations as the only international standard for anatomical terminology. So, at the start of this millennium, we finally have the ability to unequivocally define fixed anatomical landmarks to which measuring instruments can be applied.

MEASUREMENT UNITS

The adoption of an internationally agreed system of measurements seems a self-evident requirement for a science that espouses internationalism. It is therefore surprising that anthropology and human biology still discuss their results using two measurement systems; Imperial and SI units (SI is the abbreviation of *"Système International d'Unités"*) or foot–pound–second versus meter–kilogram–second. As of 2007, the United States is partnered with Liberia and Myanmar as the only three countries not to have adopted SI units as their primary system of measurement. The consequences of this lack of universal conversation was amply (and expensively) demonstrated on September 23, 1999 when NASA's $125 million Mars Climate Orbiter was lost on entry to the Mars atmosphere 10 months and 416 million miles after launch because the Lockheed constructors of the spacecraft used Imperial units of the Poundal and the Jet Propulsion Laboratory controlling the craft used SI units of Newtons (1 Newton = 7.23 Poundals). No such expense is likely to occur in anthropology by the application of two measuring systems, but clearly standardization is an absolutely fundamental requirement for anthropometric measurement. Standardization of measurement became a cause of concern during the enlightenment in eighteenth-century France in which the *Ancien Régime* employed 250 000 units of measurement (Alder, 2002) that differed mainly depending on whether one was buying or selling the commodity to be measured; clearly a short foot of cloth was advantageous to the weaver and disadvantageous to the tailor.

The desire for *"Liberté, égalité et fraternité"* that led to the French Revolution at the end of the eighteenth century, and with it the change of mindset from dogmatism and imperial decree to evidence-based knowledge, was the perfect historical moment to espouse a universal system of measurement. In 1791, the French Academy of Sciences suggested that a meter should be defined as one ten-millionth of the earth's meridian along a quadrant (i.e., one ten-millionth of the distance between the equator and the North Pole). The meridian system was both simple and scientific as the unit of length (meter) could be used as the basis for measures of both mass and volume. In addition, the metric system was decimal (base 10) and so larger and smaller multiples of each unit could be created through the multiplication or division of the unit by the number 10 and its powers. The French Government officially adopted the metric system in 1795 and in 1799 a scientific conference involving representatives from Denmark, Italy, the Netherlands, Spain, and Switzerland was held to validate the system and standard prototypes were commissioned.

The original platinum one-meter standard was made and placed in the Archives de la République in 1799. However, it was too small, by 0.2 mm, as the French astronomers Delambre and Méchain, who had spent seven years determining the length of the arc from Dunkirk through Paris to Barcelona, had failed to compensate adequately for the earth's tendency to flatten out due to its rotation. In 1899, the London firm Johnson, Matthey and Co. made a new prototype meter in the shape of a modified X-shaped cross-section. It was a bar made of platinum iridium alloy with lines inscribed at each end and the meter was defined as the distance between the two graduation lines at 0°C. The metric system was officially named the *"Système International d'Unités"* or SI in 1960 and the need for a more precise way to define a meter was highlighted. After a number of revisions, the most recent definition of a meter, based on an unchanging value, came in 1983. The General Conference on Weights and Measures defined a meter as "the length of the path travelled by light in a vacuum during a time interval of 1/299 792 458 of a second" and this remains the definition today.

While the actual length of the meter has changed very little since it was established, the precision by which it is measured has been noticeably enhanced. This point is perhaps the most important for contemporary anthropometry. Anthropometry is used in a range of scientific disciplines including medicine, evolutionary biology, anthropology, human biology, ergonomics, sports science, and forensic medicine to name but a few. The extent to which differences between values are deemed to be significant, either statistically or, more importantly, biologically depends on the precision with which we measure. Differences of 200 g in the weight of adult humans are of little importance but such differences in weight between babies at birth are associated with a range of constraints upon intrauterine growth such as maternal smoking and malnutrition. Thus precision becomes an appropriate cause for concern in anthropometry.

INSTRUMENTS AND TECHNIQUES

The development of anthropometric instrumentation, unlike the turbulent history of the acceptance of anatomical terminology and measurement units, is marked by the agreement with which instrument design is characterized. The fact that an instrument to measure length must necessarily incorporate a rigid rod and two projecting arms, one fixed and the other movable, appears to have been accepted from the very first anthropometric instruments of which we have descriptions. Johann Sigismund Elsholtz's thesis in 1654 on "Anthropometria" contains an illustration of what must have been one of the first "anthropometers." This instrument has a transverse rod (the regula) moving up and down a vertical rod, which Elsholtz describes as "similar to the draughtsman's rule except that at one end it has a special wider portion composed of two projecting angular pieces made so that when held against any object having a regular section it will be kept in a straight line ..." (Tanner, 1981, p. 45).

Whilst the development of instrumentation took place across a broad series of scientific disciplines aimed at measuring the human body it was in the field of Auxology – the study of human growth – that development of appropriate instrumentation was of singular importance. This is because the measurement of human growth deals with a constantly changing object. Whilst we now appreciate that children grow according to an aperiodic series of saltatory bursts punctuating periods of stasis (Lampl et al., 1992), the majority of our knowledge of the process of growth is based on measurements taken too infrequently to be able to record that pattern. When measured on a monthly or, more commonly, three-monthly basis the pattern of human growth appears as a smooth curve. The first longitudinal study of human growth, and with it the demonstration that accurate and repeatable measurement techniques were being used, is usually credited to Count Philibert Guéneau de Montbeillard, whose study on the growth of his son was published in a supplement to Buffon's *Histoire naturelle* (Tanner, 1962). Between 1759 and 1777 he measured his son at six-monthly intervals. Scammon (1927) translated these raw data into a height distance graph and D'Arcy Thompson (1942) derived a velocity graph which demonstrate by their smooth nature that de Montbeillard must have used a standard measurement technique and appropriate instrumentation. To illustrate the constancy of technique and instrumentation over the last two and a half centuries, Tanner (1981) records two examples. The first records the growth of children from the Carlschule in Stuttgart between 1772 and 1794 (see Hartmann, 1970 in Tanner, 1981) and the second, in the nineteenth century, Quetelet's (1870) measurements on his own son and daughter from age 5 to maturity as well as on the two daughters of a friend, one between the ages of 10 and 17 and the other between the ages of 12 and 17. Quetelet in particular was known from his cross-sectional studies of neonates

to have used a measuring board known as a "mecometre" developed by a French physiologist, François Chaussier, and described by Murat (1816; cited by Tanner, 1981) as consisting of a square wooden rod, marked in decimeters, centimeters, and millimeters. There were two ends made of copper, one fixed, and the other movable. As Quetelet had used such a purpose-built instrument to measure neonates it seems reasonable that he also used purpose-built apparatus to measure his children. At about the same time in England, Galton (1874) was commenting to the Anthropological Institute on the report by Fergus and Rodwell (1874) of the growth of Marlborough College schoolboys. Fergus and Rodwell had measured the height, weight, chest, upper-arm, and head circumferences of 550 boys aged 10–19 years. Although the data were cross-sectional, it is of interest that they used a stadiometer, described by them as a "vertical board provided with a sliding square at right angles to it." Galton described it more fully in a footnote to his comments as: "A bracket sliding between vertical guides, and balanced by a counter-weight, acting over two pulleys (which would) be found easy, quick and sure in its action. The vertical board and foot-piece may be dispensed with, if the guides can be nailed to the wall" (Galton, 1874, p. 126). The development of such technology for measuring height means that researchers were well aware of the difficulty of obtaining accurate measurements and were not prepared simply to use the anthropological instrumentation which existed to measure skeletal remains.

Anthropologists were also well aware that the standardization of anthropometric techniques and procedures was a necessity. Until 1870 the Broca techniques of anthropometry were universal. Growing individualism and the isolationism of Germany resulted in three conferences dealing with German anthropological anthropometry during the nineteenth century: the Congress of the German Anthropological Society in 1874, the Craniometric Conference in Munich in 1877, and the Berlin Conference in 1880. The result of these conferences was the "Frankfurt Agreement" adopted at the 13th General Congress of the German Anthropological Society held at Frankfurt-on-Main. Thus the French and German schools of anthropometry were established. Further attempts at the unification of anthropological anthropometry occurred during the 1890s in Paris under the individual initiative of R. Collignon (1892) and at the 12th International Congress of Prehistoric Anthropology and Archaeology of 1892 in Moscow. Hrdlička (1947) thought that neither of these attempts accomplished anything substantial, but standardizations became at least a recognized necessity. An International Agreement for the Unification of Anthropometric Measurements to be made on the "Living Subject" was drawn up in Geneva at the International Congress of 1912. Although this agreement did not deal with children, it gave general principles of measurement and detailed definitions of 49 measurements.

The anthropometry peculiar to somatic growth that we use today had its recent development in the American longitudinal studies of the first half of the 20th century. From 1904 to 1948, 17 such studies were started and 11 completed. Their complexity varied from the relatively simple elucidation of the development of height and weight to data yielding correlations between behavior, personality, social background, and physical development. One factor common to all was the desire to maintain accuracy of measurement. Administrative problems and staff changes meant that this was not always possible but runs of 10 years with the same measurement team are to be found in the Berkeley Growth Study of 1927 (Bayley, 1940), and the Yale Study of the same year (Gesell and Thompson, 1938).

Research workers were aware of the need for comparability of measurements and published precise accounts of their methods and techniques, with suitable adaptations for the measurement of growth. The three most important and informative accounts from this period are those of Frank Shuttleworth (1937) for the Harvard Growth Study of 1922 (made in the School of Education), Harold Stuart (1939) for the Center for Research in Child Health and Development Study of 1930, and Katherine Simmons in her reports of the Brush Foundation's Studies of 1931 (Simmons, 1944). Stuart (1939) provides the most complete account of measurements used for auxology using the techniques that resulted from the International Congress of 1912, "with diversions from these techniques at appropriate times." H. V. Meredith (1936) is unique in his perception of the problems involved in the measurement of human growth at the time of the early US studies. The multitude of papers he published on the growth of children from Iowa City (Iowa), Massachusetts, Alabama, Toronto (Canada), and Minnesota contain excellent examples of reliability control. He was convinced that long-term studies of physical growth would only be valid if preliminary detailed investigations were made into the accuracy with which body dimensions could be measured during growth. If the growth increment from one age to the next was less than the 90th centile of the differences between repeated measurements of a chosen dimension, he thought it unwise to take the measurement. Detailed descriptions of the measurements his team used are included in his reports. He and his colleague Virginia Knott (1941) criticized the sparsity of modern techniques and the lack of information pertaining to their reliability. These American studies almost exclusively used accepted anthropological instrumentation such as Martin anthropometers, which would be modified

when they did not precisely meet the needs of the auxological situation. An anthropometer may, for instance, have been fixed to a board to facilitate the measurement of recumbent length. In addition, American studies promoted the development of instrumentation such as skinfold calipers. They did much to influence more recent researchers in the field of growth and development on both sides of the Atlantic.

In Britain, few longitudinal growth studies were undertaken before 1949. Fleming's study of stature and head measurements (Fleming, 1933) was the only one to be published for use as standards of reference by 1950 (Tanner, 1952), although other longitudinal studies had been undertaken which would later produce worthwhile results. Alexander Low had measured 900 newborn babies between 1923 and 1927 and followed 65 of the boys and 59 of the girls with annual measurements until the age of five years (Low, 1952). These data were later to provide Tanner and his colleagues with the opportunity to revisit some of these subjects as adults and repeat the measurements and so investigate the relationship between growth in the preschool years and eventual adult size (Tanner et al., 1956). At the Institute of Social Medicine in Oxford Professor John Ryle set up the Oxford Child Health Survey in 1944. This was originally designed to investigate illness experience in the first five years of life and the effect on growth. In the end it provided considerable data to allow the original development of bone-specific-scoring systems for skeletal maturity assessments (Acheson, 1966) and longitudinal data up to the age of 18 years.

The Harpenden Longitudinal Growth Study set up by J. M. Tanner and R. H. Whitehouse in 1949, became the strongest influence in British studies of human growth, and did much to advance auxological anthropometry. Whitehouse measured all the subjects from the beginning of the study and was to stay with the study for its duration into the 1970s. He created a virtual record in taking every measurement on every child on every occasion for 25 years. The team of Tanner and Whitehouse radically altered the approach to auxological anthropometry. Whitehouse was dissatisfied with the instrumentation available and developed the Harpenden range of instruments that are recognized today as being among the best in the world. They are accepted internationally for their accuracy, consistency, and ease of use. Their principal advance was to eliminate graduated rules for measuring linear distances and instead to use counter mechanisms. These counters were turned by a simple ratchet system and displayed the measurement in millimeters. Reading errors were thus minimized.

The International Children's Centre Coordinated Longitudinal Growth Studies had a major effect on standardizing anthropometric measurements and growth study design. Growth studies were set up in France, Sweden, Britain, Switzerland, Belgium, Dakar (Senegal), Uganda, and Louisville (Kentucky), and coordinated by meetings of the growth study teams every two years. Initially these teams discussed their methods and eventually their results. The papers and articles that resulted directly or indirectly from these studies form a bibliography of 948 references (International Children's Centre, 1977). The International Biological Programme (IBP) that brought together scientists from all over the world under the umbrella of research in "Human Biology" during the years 1962–1972 gave rise to one of the standard texts for research into the human biological sciences. Weiner and Lourie's (1969) IBP Handbook, now revised and in its second edition (Weiner and Lourie, 1981), forms the source for many scientists who wish to use standard techniques. Its value is in its acceptance, by many, as the source, not only for techniques of measurement but also for many other techniques that are applied in research on human adaptability. Weiner and Lourie were not attempting to illustrate new concepts of technique but rather to illustrate the most appropriate techniques available which would allow the greatest degree of comparability. In parallel to the work of the IBP, a conference, chaired by Hetzberg, was held in 1967 on the standardization of anthropometric techniques and terminology. This meeting was attended by anthropologists, engineers, dental and medical researchers, physical educationists, and statisticians and helped to improve the comparability of data across different fields by providing a range of standardized anthropometric techniques and terminology (Hertzberg, 1968).

Noël Cameron's *The Measurement of Human Growth* in 1984 (Cameron, 1984) and Tim Lohman's *Anthropometric Standardization Reference Manual* (Lohman et al., 1988) have now become the standard reference texts for anthropometry. The former was the result of Cameron's liaison with Tanner and Whitehouse in London and the latter resulted from a National Institutes of Health (NIH) conference held to help standardize anthropometric measurements in 1988.

RELIABILITY

It is important to define the terms accuracy, precision, reliability, and validity because an understanding of these concepts is fundamental to obtaining anthropometric measurements. According to the *Oxford English Dictionary* accuracy is described as "... careful, precise, in exact conformity with truth." In most measurement situations we do not know what "truth" is. For example, when we measure a child's height we only have the estimation of that height, we do not know

what the actual true height is. We can improve the accuracy (and decrease error) by ensuring that we use an appropriate, specific, purpose-made, valid, calibrated instrument to measure the child and by using a properly trained observer. Accuracy may be determined from a test–retest experiment in which the standard error of measurement (S_{meas}) or technical error of measurement (TEM) is calculated (Cameron, 1978, 1983, 1984; Lohman et al., 1988). Precision is either the proximity with which an instrument can measure a particular dimension or the smallest unit of measurement chosen by the observer. Note that this is different from accuracy in that it relates to the smallest unit of measurement possible (or chosen) with a particular instrument. So a stadiometer that measures height to the nearest centimeter is not as precise as one that measures to the nearest millimeter. Clearly accuracy and precision are related in that reliability can only be measured within the limits imposed by the precision of the instrument or observer. Thus two observers may appear to have equal reliability when they use a stadiometer to measure height to the last completed centimeter but a more precise instrument, that measures in millimeters, might demonstrate that one of the observers is more accurate than the other. Reliability is the extent to which an observer or instrument consistently and accurately measures whatever is being measured on a particular subject. Note that reliability involves three sources of error; the observer, the instrument, and the subject. Also, reliability involves two characteristics; consistency and accuracy. Reliability can also be assessed in a test–retest experimental design in order to calculate the standard deviation of differences (SD) (Cameron, 1978, 1983, 1984). Finally, validity is the extent to which a measurement procedure measures or assesses the variable of interest. For instance, the measurement of total body fat may be measured by dual energy X-ray absorptiometry (DXA) or by densitometry. The former is more direct than the latter and is thus more valid. Similarly, growth in stature is most validly assessed by measuring height rather than by measuring leg length. These concepts are discussed in detail in Cameron (1978, 1983, 1984). Within the measurement of human growth it is important to know one's accuracy and reliability both for the absolute (cross-sectional) determination of a child's height and for the (longitudinal) determination of growth velocity. Obviously when taking two measurements of height to calculate growth velocity the estimation is affected by two sources of error, one for each height estimation. The interval between measurement occasions will be determined, to some extent by the reliability of the observer. If, for instance, the observer has a reliability, or SD, of 0.3 cm and the child is growing at a rate of 4 cm/year then growth will not be certain to have occurred (with 95% confidence limits) until the difference in heights between two measurement occasions is greater than 1.96×0.3 cm, or 0.59 cm. It will take this child 54 days to grow 0.6 cm and thus measurements taken on a monthly or even two-monthly basis will be subject to observer error. The minimal time between measurement occasions for this child should be at least three months to ensure that false growth due to observer error is minimal.

Generally linear dimensions, e.g., height, are more accurate than those involving soft tissue, e.g., skinfolds. Linear dimensions are most often taken between bony landmarks with little or no compressible tissues to interfere with accurate measurement. Soft tissue measurements, such as arm circumference, involve compressible subcutaneous fat and muscle that will reduce accuracy and reliability. It is thus of some importance for the observer to be aware of exactly what is being measured within a particular dimension. For instance an observer measuring triceps skinfold without an understanding of the anatomical relationship between subcutaneous fat and muscle is unlikely to appreciate the importance of the need to separate the subcutaneous fat layer from the muscle layer during measurement. Similarly, an observer ignorant of skeletal anatomy will not appreciate the need for straightening the vertebral column or of applying pressure to the mastoids prior to the measurement of stature.

DATA EDITING

Once anthropometric data have been collected there is a need to ensure that these data are free from errors. Data editing involves the processes required to convert recorded data into an analysis data set. This process involves cleaning the captured data set ("capturing" in this context refers to the process of inputting data from the raw paper files into a computerized database) to identify and deal with missing and erroneous data. It involves a number of systematic stages that need to be completed before analysis is initiated. The stages of cleaning are dependent on whether the data have been collected in a cross-sectional (XS) or longitudinal research design, but initially the stages are the same for both types of data.

STAGES OF DATA CLEANING

The first action on the captured database is to generate descriptive statistics including the sample size, measures of central tendency, SD, variance, skewness, and range. Each of these statistics provides a specific piece of information that guides the process of editing and cleaning. Sample size (N) for each variable identifies missing data – a complete data set should have the same N for each variable unless the research design excludes certain cases. Measures of central tendency including the mean, median, and mode provide clues as to the Gaussian (normal) nature of the distribution and the potential significance of departures from normality. The means three moments (SD, variance, and skewness) provide information on variability and importantly whether the shape of the distribution departs significantly from normal. The range provides the largest and smallest values of a variable and thus is an immediate check on whether all values fall within the expected limits. This is particularly important for the recognition of erroneous values in categorical variables that have predetermined upper and lower bounds.

1. Identification of missing data. Having identified the absence of data through the sample size for each variable, each missing value in the database should be cross checked with the original hard copy to ensure that missing data are actually missing rather than missed in the capturing process. Non-missing data should be captured. Prior to considerations regarding the type of analysis to be undertaken it is not necessary to make decisions regarding the imputation of missing values. The purpose of editing is to present a clean (error-free) database for statistical analysis.

2. Identification of potentially erroneous data through the detection of outliers (known as flagging). Outliers will be values that are at the extremes of the expected distribution. Within categorical variables these will commonly be outside the predetermined range. Outliers within continuous variables, however, present a different paradigm in that they may actually be extremes of normal and not erroneous. Outliers may be illustrated through frequency distributions, standard deviation checks, and bivariate scatter plots.

   a. Frequency distributions are used with discrete categorical variables to highlight cases that fall outside the expected range.

   b. Ninety-five percent of the normal (Gaussian) distribution is within ±2 SD, thus values of continuous variables that lie outside this range may be either erroneous or extremes of normal. This "standard deviation check" can be employed for all continuous variables but the extent of the range (e.g., ±2, 3, or 4 SD) will depend on the importance of correctly identifying true extremes. Where doubt exists as to the authenticity of an outlier then when possible "witness variables" should be employed to help make cleaning decisions. For example if a participant's weight lies outside of ± 2 SD, then one can examine skinfold or girth data (if these were collected) to determine if it is plausible that the child is either over or underweight.

c. Bivariate scatter plots provide a pictorial representation of the relationship between two variables, e.g., height and weight, and thus also the presence of outliers who do not conform to the expected distribution of the scatter.

Although it should only be done with extreme caution, erroneous data points can either be excluded from analysis or the analysis should be executed both with and without the erroneous data points and significance of the difference between both outcomes reviewed in the light of the hypothesis being tested.

3. The third stage involves checking the captured data of a 5% random sample with the raw data file. This is because not all errors result in extreme outliers but some may be contained within the normal distribution of the data. A 1% error limit should be imposed, meaning that if there are errors in more than 1% of the selected cases then a further 5% random should be selected and the process repeated. This process should be repeated to a maximum of three iterations. If greater than 1% of these data are found to be erroneous in each of the random sample checks then a repeat of cleaning stages 1 to 3 should be undertaken.

If stage 3 is completed successfully then cleaning is complete for XS data and analysis may begin. If longitudinal data have been collected then stages 4–5 should be employed.

4. Flagging variables need to be created between longitudinal data collection points.

a. Longitudinal research designs investigate the biology of change and commonly have an expected directionality, either positive or negative. For example if the height of a child was measured at 9 and 10 years of age, height is expected to increase or show positive change. Flag variables where the value at the first time point is subtracted from the value at the second time point should be created and the expected direction of change should be checked for all of the derived values.

b. If a case has, for example, shown an inappropriate direction of change then the raw hard data files at each time point should be cross-checked prior to a decision regarding exclusion.

c. With nonlinear dimensions such as weight, longitudinal cleaning can be more difficult as it is feasible that a case particularly in constrained environments may have lost weight between time points. Where flags are raised with weight,

one should study each case individually and decide with reference to the context of the study, and the change in witness variables, if a case may have lost weight or if the case needs to be excluded from further analysis.

5. Once stages 1 to 3 have been completed on a cross-sectional basis and stage 4 has been completed on a longitudinal basis, stage 3 (5% random sample checks) should be repeated using the complete longitudinal data set. If no 5% random sample errors are found then the longitudinal data can be used for analysis.

A comprehensive record of every cleaning change and the reason why changes were made should be maintained (such as in the form of a data cleaning report) so that each correction can be subsequently traced. Once cleaning has been completed and comprehensively documented, one may then begin to derive variables from the cleaned data such as body mass index (BMI) and standardized scores.

TECHNIQUES IN ANTHROPOMETRIC MEASUREMENT

The organization of this chapter is such that a measurement technique is described with the instrumentation recommended for the most accurate and reliable results. The measurements chosen for description are those that are the most relevant for human and evolutionary biologists; linear dimensions of hard and soft tissue and in particular lengths, diameters, circumferences, and skinfolds. The following measurements will be described: weight; linear dimensions (e.g., stature and supine length); circumference/girths (e.g., waist and hip); skinfolds (e.g., triceps and subscapular). A detailed glossary of anatomical surface landmarks can be found towards the end of this chapter.

MEASURING PROCEDURES

The accuracy with which the following measurements may be obtained can be maintained at a high level by following a few simple rules of procedure:

1. Ensure that the subject is in the minimum of clothing or at least in clothing that in no way interferes with the identification of surface landmarks.

2. Familiarize the subject with the instrumentation, which may appear frightening to the very young subject, and ensure that he or she is relaxed and happy. If necessary involve the parents to help in this procedure by conversing with the child.

3. Organize the laboratory so that the minimum of movement is necessary and so that the ambient temperature is comfortable and the room well lit.

4. Place the recorder in such a position that he or she can clearly hear the measurements and is seated comfortably at a desk with enough room to hold recording forms, charts, and so on.

5. Measure the left-hand side of the body unless the particular research project dictates that the right-hand side should be used or unless the comparative projects have used the right-hand side.

6. Mark the surface landmarks with a water soluble felt-tip pen prior to starting measurements.

7. Apply the instruments gently but firmly. The subject will tend to pull away from the tentative approach but will respond well to a confident approach.

8. Call out the results in whole numbers; for example, a height of 112.1 cm should be called out as "one, one, two, one," not as "one hundred and twelve point one," nor as "eleven, twenty-one." Inclusion of the decimal point may lead to recording errors and combinations of numbers may sound similar; for example, "eleven" may sound similar to "seven."

9. If possible, measure the subject twice for all dimensions but particularly for skinfolds. The recorder should check that the retest value is within the known reliability of the observer. If it is not then a third measurement is indicated. For a final value average the two that fall within the limits.

10. Do not try to measure too many subjects in any one session. Fatigue will detract from accurate measurement for which concentration is vital.

11. Greater co-operation from the subject will result if the appearance of confidence and efficiency is given by being clean and smartly dressed.

TRAINING

The training of anthropometrists is extremely important and should be planned carefully. Practical instruction from a trained anthropometrist may take only a matter of days but the new observer will need to practice his or her techniques over a number of weeks to acquire the required accuracy. It is a good idea for the learner to pursue regular reliability checks on his or her accuracy and to refer back to the expert for checks on technique. As with all practical techniques, the objective view is usually more critical and therefore more helpful. Familiarity with the instruments is vital to the precise collection of anthropometric data. The observer should be able to service and calibrate the instruments to the required degree of precision.

WEIGHT

The measurement of weight should be the simplest and most accurate of the anthropometric measurements.

Assuming that the scales are regularly calibrated, the observer ensures that the subject is either dressed in the minimum of clothing or a garment of known weight that is supplied by the observer. The subject stands straight, but not rigid or in a "military position," and is instructed to "stand still." If the instrument is a beam balance then the observer moves the greater of the two counter-weights until the nearest 10 kg point below the subject's weight is determined. The smaller counter-weight is then moved down the scale until the nearest 100 g mark below the point of over balance is reached and this is recorded as the true weight. This procedure is necessary to determine weight to the last completed unit. If the weight is taken as the nearest 100 g above true weight then that 100 g is greater than actual weight and the last unit has not been completed. Determining the weight of neonates can be a noisy and tearful procedure but need not be if the help of the mother is solicited. The observer simply weighs mother and child together and then transfers the baby to his or her assistant's arms and weighs the mother by herself. The baby's weight can thus be determined by difference, (weight of mother + baby) − (weight of mother), and the child is left relatively undisturbed.

WEIGHING SCALES

The measurement of weight should be one of the easiest of anthropometric measurements and yet results are often inadequate due to inappropriate instrumentation. Precision to 0.1 kg (100 g) is acceptable as long as regular calibration ensures the minimum chance of error and appropriate steps are taken to tare for any clothing the subjects are required to wear. The mechanical instrument best suited to repeatedly accurate weight measurements is that designed on the balance-arm principle with two balance arms. One major balance arm measures to greater than 110 kg in steps of 10 kg and the other, minor arm, to 10 kg in steps of 100 g. Such an instrument is capable of measuring individuals from birth through to adulthood with appropriate modification to include a baby-pan and seat for subjects unable to stand and/or too large for the baby-pan. Electronic balances are also available at less than the cost of mechanical machines and appear well suited to growth clinic use as long as their power source (usually batteries) is regularly checked.

STATURE (FIGURE 6.1)

The subject presents for the measurement of stature dressed in the minimum of clothing, preferably just underclothes, but if social custom or environmental conditions do not permit this then at the very least

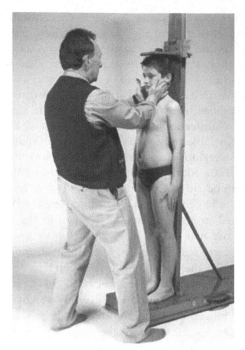

6.1. The measurement of stature.

without shoes and socks. The wearing of socks will not, of course, greatly affect height, but socks may conceal a slight raising of the heels that the observer from his or her upright position may not notice. The subject is instructed to stand upright against the stadiometer such that his or her heels, buttocks, and scapulae are in contact with the backboard, and the heels are together. As positioning is of the greatest importance the observer should always check that the subject is in the correct position by starting with the feet and checking each point of contact with the backboard as he or she moves up the body. Having got to the shoulders he or she then checks that they are relaxed, by running his or her hands over them and feeling the relaxed trapezius muscle. The observer then checks that the arms are relaxed and hanging loosely at the sides. The head should be positioned in the "Frankfurt plane," and the headboard of the instrument then moved down to make contact with the vertex of the skull. To ensure that the Frankfurt plane has been achieved the observer may find it helpful to use either of the following techniques. The observer bends down so that his or her eyes are level with the plane and notes that the lower orbits of the eyes and the external auditory meatii are in a horizontal line. Alternatively, when using instruments with a counter, he or she may grip the head with open hands and pivot it backwards and forwards, in a nodding motion, and at the same time observe the counter. The counter should register the greatest height when the head is tilted not too far forward or backwards. It is thus a relatively easy matter to ensure correct positioning.

It is advisable to place a weight, of about 0.5 kg, on the headboard. This weight presses down on the hair thus flattening any hairstyle and overcomes the natural friction of the machine so that any upward or downward movement during the measurement is recorded on the counter. With the subject in the correct position he or she is instructed: "Take a deep breath and stand tall." This is done to straighten out any kyphosis or lordosis and produce the greatest unaided height. It is at this point that the observer applies pressure to the mastoid processes – not to physically raise the head but to hold it in the position that the subject has lifted it to by breathing deeply. The subject is then instructed to "Relax" or to "Let the air out" and "Drop the shoulders." The shoulders are naturally raised when the subject takes a deep breath and thus tension is increased in the spinal muscles and prevents total elongation of the spine. Relaxing or breathing out releases this tension and commonly produces an increase of about 0.5 cm in absolute height. The effect of this pressure or traction technique is to counteract the effect of diurnal variation that works to reduce stature during the normal course of a day. This reduction may be as much as 20 mm (Strickland and Shearin, 1972) but the pressure technique reduces that to less than 4.6 mm over the whole day (Whitehouse et al., 1974). Stature is read to the last completed unit whether from a counter or graduated scale. Height is not rounded up to the nearest unit as this will produce statistical bias and almost certainly invalidate estimates of height velocity.

STADIOMETERS

The stadiometer is composed of a horizontal headboard and vertical backboard. The backboard must be so designed that it maintains a vertical position whether free-standing or wall-mounted and the headboard must always move freely over the surface of the backboard. The method of recording the measurement must be accurate and reliable and, if necessary, capable of easy calibration. Usually a counter mechanism or graduated rule is used for this purpose. It has been found that counter mechanisms lead to fewer reading errors but, in an uncontrolled environment, they may be broken by abuse. Reading from graduated rules requires the observer to be at the same height as the cursor to avoid parallax errors. It is very important to appreciate that in the absence of a purpose-built stadiometer of the type described below, it is not acceptable to use less suitable equipment. For instance, the rule attached to the traditional balance, to be found in doctor's surgeries throughout the world, is completely unacceptable. It is usually extremely unstable and imprecise. It is better in situations when purpose-built machines are not available to use a vertical wall and a book held at right-angles on the subject's head. With the subject properly positioned the book can be placed on the head so that one side touches the wall and the other the top of the head. A mark is then made on the wall at the inferior surface of the book. The distance from the floor to the book can then be measured with a tape measure, after the child has moved away.

HARPENDEN STADIOMETER

This stadiometer is a counter recording instrument in which the counter gives a reading in millimeters over a range of 600–2100 mm. It is wall-mounted and made of light alloy with a wooden headboard fixed to a metal carriage that moves freely on ball-bearing rollers. The face of the stadiometer is finished in plastic for easy cleaning. The complete instrument is 232 cm tall and weighs 12.7 kg.

These stadiometers have been in use in many clinics, hospital outpatient departments, and growth centers for a number of years. When treated properly they give consistently accurate results, but the counter will break if the headboard is "raced" up or down the backboard. For this reason it is recommended that the headboard is always locked or moved to its topmost position when not in use to prevent children or inexperienced adults from breaking the counter. Calibration of this instrument is straightforward and takes very little time. A metal rod of known length is placed between the headboard and the floor so that it stands vertically. If the counter does not record the correct length of the rod then it may be loosened by undoing the two metal retaining screws, and pulled away from the main fiber cog of the carriage. In this position the small metal cog of the counter may be turned until the counter records the true length of the metal rod. The counter is then pressed against the back-plate so that the teeth of the counter cog and carriage cog engage and the retaining screws are tightened. The headboard is then moved up and down the backboard a number of times to ensure that the counter continues to give an accurate reading. If not, the counter must be replaced. It is recommended that the instrument be calibrated prior to every measuring session, particularly if the stadiometer is left in a situation that allows public access.

SUPINE OR RECUMBENT LENGTH

The measurement of supine length requires two observers, one to hold the head and the other to hold the feet and move the cursor. The subject (usually an infant) lies on the supine-length table or in the neonatometer such that the head is positioned in a supinated Frankfurt plane and the vertex of the head touches the fixed end of the apparatus. The head is held in this position throughout the measurement by an observer who constantly checks that the correct position is being maintained and that contact between head and headboard is constant. The second observer checks that the rest of the body is relaxed and that the subject is not arching the spine or bending the knees. This observer holds the feet such that the ankles are at right angles and the toes not bending over to interfere with the cursor. The cursor is then moved into contact with the feet and slight pressure is applied to the ankles to straighten the legs. This normally causes the head to be moved away from the headboard so that the other observer must gently pull the head back into contact with it. This dual pulling of the subject has the same effect as deep breathing in the measurement of stature – to overcome diurnal variation in posture.

Depending on the age of the subject, various problems arise during this measurement. Very young children will automatically bend the knees and the observer must apply downward pressure on them with his or her forearm or elbow. The shoulders will also be lifted off the board and the observer holding the head must use the index fingers to press them gently back into contact. It is sometimes necessary to release one of the feet if the child fights so strongly that accurate measurement is compromised and indeed in the very young this is often easier than trying to struggle with both feet. It should be emphasized that these problems will arise more often if steps are not always taken to relax the subject and familiarize it with the apparatus. Cuddly toys suspended above the table or pictures on the ceiling are good methods of attracting the attention of slightly older subjects but on the very young it is a great help to allow the mother to lean over the child and talk to it to reassure it.

HARPENDEN NEONATOMETER

The neonatometer is constructed as a rectangular light-alloy frame with a curved metal headrest at one end and a cursor carriage at the other. In common with the other Harpenden instruments, the recording is by a counter mechanism. The important addition to this instrument is the locking mechanism attached to the cursor that locks the footboard when a pressure of 0.5 kg is exerted against it. Such a mechanism prevents the observer from having to fight with the unruly baby to maintain the leg in a straight position for longer than a few seconds. The highly portable nature of the instrument allows it to be placed over the recumbent baby rather than disturbing the baby by placing the subject inside the instrument. A short version is made to fit inside most incubators. This version measures over the range 180–600 mm compared with 188–750 mm for the standard model. This instrument is necessary for any neonatal clinic but the general growth clinic, dealing with all ages of subjects, can perfectly well measure supine length accurately with a longer, all age, instrument.

HARPENDEN INFANTOMETER

This instrument was designed to fill the instrument gap between neonates and school-age children, measuring over a range of 300–940 mm. Bearing in mind the fact that many studies on this age range of subjects are set in the home rather than the growth center, it is designed as a portable instrument weighing about 6.75 kg. It is constructed of light-alloy with a flat headboard and footboard fixed to a movable cursor and counter recording mechanism. As with the neonatometer, a locking device is fitted to aid measurement when the subjects are active.

HARPENDEN INFANT MEASURING TABLE

This instrument is the non-portable version of the infantometer. The base is constructed of light-alloy with a fixed wooden headboard and footboard fixed to a carriage and counter mechanism as with the other instruments. There is no locking mechanism. The measuring range is 230–1200 mm so it is suitable for post-neonates up to pre-school children. The lack of a locking mechanism reflects the fact that the subject should be more co-operative and that the measuring position can be maintained for longer.

HARPENDEN SUPINE MEASURING TABLE

This is the full-length supine table similar in construction to the stadiometer. It is recommended that this instrument is mounted on permanent wall-brackets, but adjustable legs may be supplied at an additional cost. As in the infant measuring table, the head and footboards are made of wood and the latter is fixed to a cursor and counter mechanism. The measuring range is 300–2100 mm and so accommodates all age ranges of children and adults.

CALIBRATION

Calibration of all these instruments is very easy, and is similar to that employed for calibrating the stadiometers. A metal rod of known length is used to ensure that the counter is reading correctly. If not, then suitable adjustment can be made by loosening the retaining screws of the counter and turning its metal cog to the correct measurement. The footboard is moved forwards and backwards a few times to check the reliability, and if this is suspect a new counter is fitted.

ABDOMINAL CIRCUMFERENCE; WAIST CIRCUMFERENCE

Some confusion exists in the literature with regard to the level of measurement. However, the minimum circumference between the iliac crests and lower ribs would appear to be the most reliable to determine. The general technique is for the subject to stand erect facing the observer with the arms away from the body. The tape is passed around the body and tightened at the required level ensuring that it is horizontal and not compressing the soft tissue.

BUTTOCK CIRCUMFERENCE; HIP CIRCUMFERENCE; HIP GIRTH

Hip or buttock circumferences should be measured at the level of the greatest protrusion of the buttocks when the subject is standing erect with the feet together. The subject stands sideways to the observer with the feet together and arms folded. The observer passes the tape around the body at the level of the most prominent protrusion of the buttocks so that it lightly touches but does not compress the skin. In most cases the subject will be dressed in underclothes which will obviously affect the measurement, and so the observer should either provide standard thin undergarments or insist that the subject be naked.

TAPE MEASURES

Many tape measures are available that are suitable for anthropometric use. Suitability depends on fulfilling five criteria:

1. Flat cross-section. Tapes with a curved cross-section are difficult to bend maintaining a smooth outline.
2. Millimeter graduations. The graduations must be in centimeters and millimeters and preferably marked on both edges of the tape. Thus at the crossover position it makes little difference whether the lead of the tape is above or below the reading.
3. Blank leader strip. The tapes should contain a blank leader strip prior to the graduations commencing. This enables the observer to hold the leading part without obscuring the zero value.

4. Metal or fiberglass construction. It has always been recommended that only steel tapes should be used so that they did not stretch or deteriorate with use. Fiberglass tapes are now being manufactured that are guaranteed not to stretch.
5. Minimum length of 1 m.

We have not described any single tape in detail because various types are available that fulfill most or all of these criteria.

SKINFOLDS

The technique of picking up the fold of subcutaneous tissue measured by the skinfold caliper is often referred to as a "pinch" (Cameron, 1978, 1983, 1984), but the action to obtain the fold is to sweep the index or middle finger and thumb together over the surface of the skin from about 6 to 8 cm apart. This action may be simulated by taking a piece of paper and drawing a, say, 10 cm line on its surface. If the middle finger and thumb are placed at either end of this line and moved together such that they do not slide over the surface of the paper but form a fold of paper between them then that is the action required to pick up a skinfold. To "pinch" suggests a small and painful pincer movement of the fingers and this is not the movement made. The measurement of skinfolds should not cause undue pain to the subject, who may be apprehensive from the appearance of the calipers and will tend to pull away from the observer, and, in addition, a pinching action will not collect the quantity of subcutaneous tissue required for the measurement. In addition, the observer must be careful to open the caliper prior to removing it from the fold of skin as failure to do this can result in a painful scratch for the participant.

The measurement of skinfolds is prone to many sources of error. Location of the correct site is critical (Ruiz et al., 1971), but greater errors may arise from the consistency of subcutaneous tissue and the individual way in which each observer collects the fold of tissue. The novice should practice skinfold measurements more than other anthropometric techniques. The observer will thus obtain an awareness of how a correct skinfold should "feel" and thus be aware of those occasions when a true skinfold is not being obtained.

TRICEPS SKINFOLD

The level for the triceps skinfold is the same as that for the arm circumference – mid-way between the acromion and the olecranon when the arm is bent at a right angle. It is important that the skinfold is picked up both at a midpoint on the vertical axis of the upper-arm and a midpoint between the lateral and medial surfaces of the arm. If the subject stands with his back to the observer and bends the left arm the observer can palpate the medial and lateral epicondyles of the humerus. This is most easily done with the middle finger and thumb of the left hand, which will eventually grip the skinfold. The thumb and middle finger are then moved upwards, in contact with the skin, along the vertical axis of the upper-arm until they are at a level about 1.0 cm above the marked midpoint. The skinfold is then lifted away from the underlying muscle fascia with a sweeping motion of the fingers to the point at which the observer is gripping the "neck" of the fold between middle finger and thumb. The skinfold caliper, which is held in the right hand with the dial upwards, is then applied to the neck of the skinfold just below the middle finger and thumb at the same level as the marked midpoint of the upper arm. The observer maintains his or her grip with the left hand and releases the trigger of the skinfold caliper with his right to allow the caliper to exert its full pressure on the skinfold. In almost every case the dial of the caliper will continue to move but should come to a halt within a few seconds at which time the reading is taken to the last completed 0.1 mm. In larger skinfolds the caliper may take longer to reach a steady state but it is unusual for this to be longer than seven seconds. Indeed, if the caliper is still moving rapidly it is doubtful that a true skinfold has been obtained and the observer must either try again or admit defeat. This situation is only likely to occur in the more obese subject with skinfolds greater than 20–25 mm – that is, above the 97th centile of British charts. Within the 97th and 3rd centiles skinfolds are relatively easy to obtain but they do require a great deal of practice.

SUBSCAPULAR SKINFOLD (FIGURE 6.2)

The point of measurement is located immediately below the inferior angle of the scapula. The subject stands with his or her back to the observer and his or her shoulders relaxed and arms hanging loosely at the sides of the body. This posture is most important to prevent movement of the scapulae; if the subject folded his or her arms, for instance, the inferior angle of the scapula would move laterally and upwards and therefore no longer be in the same position relative to the layer of fat. The skinfold is picked up, as for triceps skinfold, by a sweeping motion of the middle finger and thumb, and the caliper applied to the neck of the fold immediately below the fingers. The fold will naturally be at an angle laterally and downwards and will not be vertical. Once again, the dial of the caliper will show some movement that should soon cease.

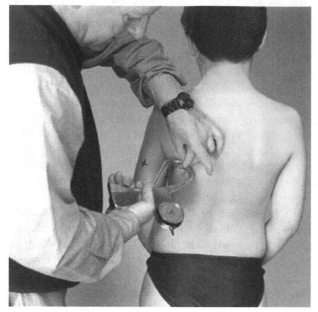

6.2. The measurement of subscapular skinfold.

dial 6 cm in diameter, with an almost linear scale and divisions 150% of natural size.

HARPENDEN SKINFOLD CALIPERS

The Harpenden caliper (Tanner and Whitehouse, 1955) resulted from the investigation of Edwards et al. (1955) which recommended a design that included jaw faces of size 6×15 mm, with well-rounded edges and corners, pressure at the faces of 10 g/mm^2 that does not vary by more than 2.0 g/mm^2 over the range of openings 2–40 mm, and a scale such that readings can be taken to the nearest 0.1 mm.

These calipers may be calibrated by fixing them in a bench clamp so that the jaws are parallel to the floor. When a weight equivalent to 10 times jaw face area is applied (90 g if the face is 6×15 mm as recommended) the jaws should stay closed but if another gram is added they should open to their fullest extent. There is always some leeway around this figure but it has been demonstrated that pressures between 9 g/mm^2 and 13 g/mm^2 make little difference to the skinfold reading.

HOLTAIN (TANNER–WHITEHOUSE) SKINFOLD CALIPERS

The Holtain/Tanner–Whitehouse skinfold caliper is the improved version of the Harpenden caliper. The design principle regarding jaw pressure, jaw face area, and readability are maintained but the Holtain caliper is lighter than the Harpenden model and easier to hold, thus making repeated measurements less tiring and perhaps creating greater accuracy.

Calibration may be checked in the same way as for the Harpenden model. Both models require little or no servicing beyond cleaning and care. There has been some criticism of these calipers due to inconsistency of spring pressure but if they are calibrated when first received and checked regularly thereafter, as with all other instruments, no problems should arise.

SKINFOLD CALIPERS

Skinfold calipers are designed to measure the thickness of a fold of subcutaneous fat that has been picked up at a specific landmark on the body by the anthropometrist. The compressible nature of these skinfolds means that the calipers have had to be designed to exert a constant pressure at all settings of the caliper jaws. Constant-pressure calipers have been developed in the United States and the United Kingdom that exert a pressure of 10 g/mm^2 of the jaw face area at all settings. We describe below the three clinic/research calipers that have been most commonly used and tested for comparability although other calipers are available (e.g., the Lafayette skinfold caliper), but they have not, as yet, been extensively tested in clinical and research situations.

LANGE CALIPER

The Lange caliper was introduced by Lange and Brozek (1961) to provide for "persistent demands for a light compact skinfold caliper." It is composed of a slender handle opposed by a thumb lever. Pressure on this lever opens the jaws uniformly to a maximum of 6 cm with a reading accuracy to +1 mm. The lever is released to clamp these jaws on to the skinfold. The jaws have an area of approximately 30 mm^2 and a constant pressure of 10 g/mm^2 irrespective of the size of the skinfold, and they are pivoted to adjust automatically for parallel measurement of the skinfold. The reading is displayed by a fine pointer on a semicircular

DIAMETERS

Bi-acromial diameter (Figure 6.3)

Bi-acromial diameter is the distance between the tips of the acromial processes. It is measured from the rear of the subject with the anthropometer. The position of the lateral tips of the acromials is slightly different in each subject and it is therefore necessary for the observer to carefully palpate their exact position in each subject before applying the instrument. This is most easily done with the subject standing with their back to the observer such that the observer can run his or her hands over the shoulders of the subject. This tactile

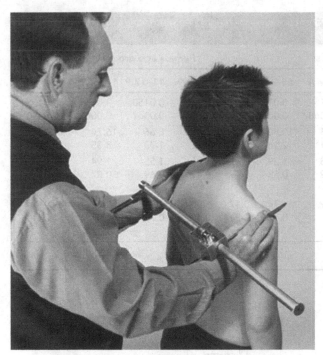

6.3. The measurement of bi-acromial diameter.

awareness of the positioning of the acromials is an important part of the measurement procedure because it allows the observer to be confident of the measurement points when he or she applies the instrument. Having felt the position of the acromia and that the subject's shoulders are relaxed, the observer applies the anthropometer blades to the lateral tips of the processes. The anthropometer is held so that the blades rest medially to the index fingers and over the angle formed by the thumb and index finger. The index fingers rest on top of the blades, to counteract the weight of the bar and counter mechanism, and the middle fingers of each hand are free to palpate the measurement points immediately prior to measurement. In this position the observer can quite easily move the blades of the anthropometer so long as it is of the counter type. Other anthropometers have too great a frictional force opposing such easy movement and must be held by the main bar so that the blades are remotely applied to the marked acromial processes. The blades must be pressed firmly against these protuberances so that the layer of tissues which covers them is minimized. To ensure that the correct measurement is being made it is a simple matter to roll the blades up and over the acromia and then outwards and downwards so that the observer feels the blades drop over the ends of the acromia.

Bi-iliac diameter

The subject stands with their back to the observer, feet together and hands away from his or her sides to ensure a clear view of the iliac crests. As for bi-acromial

diameter, it is a good measurement procedure for the observer to feel the position and shape of the crests prior to applying the instrument, especially when the subject has considerable fat deposits in that region. The anthropometer is held as for bi-acromial diameter and applied to the most lateral points of the iliac crests. This will be more easily accomplished if the anthropometer is slightly angled downwards and the blades applied to the crests at a point about 1 inch (2–3 cm) from the tips. To ensure that the most lateral points have been obtained it is a useful point of technique to "roll" the blades over the crests. It will be seen on the counter of the instrument that at a particular point the distance between the crests is greatest; this is the point of measurement.

ANTHROPOMETERS

Whilst anthropometers are not required for the measurements described above the anthropometer is one of the most versatile anthropometric instruments and can be used for measuring any linear dimension such as height, sitting-height, or arm span in addition to limb segment lengths. Because of this versatility and usefulness anthropometers should be chosen with care. Two basic designs of anthropometer are available; that known as the Martin anthropometer, and the Harpenden anthropometer.

MARTIN ANTHROPOMETER

The Martin anthropometer is used universally in physical anthropology. It is manufactured by GPM (Switzerland) and is thus sometimes referred to as the GPM anthropometer. The original version is composed of four metal rods with graduations in millimeters and centimeters engraved on them. A sliding cursor runs the length of the rods when they are joined together to give a maximum reading of 2 m. Curved or straight blades may be inserted into the cursor housing and fixed end so that the distance between them may be read from the graduated main beam. The major disadvantage becomes apparent when this anthropometer is used to measure children. In these situations it is important to be able to feel the landmarks of the body as the blades of the anthropometer are applied to the marked positions. It must be possible, therefore, to move the cursor housing whilst holding the tips of the blades. The frictional forces involved in the Martin anthropometer make this operation virtually impossible. A more recent version of the Martin instrument is now available from GPM. The main difference is that the beam has a square rather than round cross-section but the problem with frictional forces still remains.

TABLE 6.1. Total technical error of measurement (TEM) reference values for height, weight, triceps, and subscapular skinfold measurements split by gender and age group.

Measurement	Coefficient of reliability	Male age groups (Years)					Female age groups (Years)				
		1–4.9	5–10.9	11–17.9	18–64.9	65+	1–4.9	5–10.9	11–17.9	18–64.9	65+
Height (cm)	0.095	0.0103	0.0130	0.0169	0.0152	0.0152	0.0104	0.0138	0.0150	0.0139	0.0135
	0.099	0.0046	0.0058	0.0076	0.0068	0.0068	0.0047	0.0062	0.0067	0.0062	0.0060
Weight (kg)	0.095	0.21	1.20	5.94	13.06	10.80	0.22	1.61	8.66	16.74	11.70
	0.099	0.04	0.24	1.19	2.61	2.16	0.04	0.32	1.73	3.35	2.34
Triceps skinfold (mm)	0.095	0.61	0.97	1.45	1.38	1.29	0.65	1.05	1.55	1.94	1.86
	0.099	0.28	0.43	0.65	0.62	0.58	0.29	0.47	0.69	0.87	0.83
Subscapular skinfold (mm)	0.095	0.43	0.87	1.55	1.79	1.74	0.47	1.08	1.74	2.39	2.27
	0.099	0.19	0.39	0.69	0.8	0.78	0.21	0.48	0.78	1.07	1.02

Source: Adapted from Ulijaszek and Kerr (1999).

HARPENDEN ANTHROPOMETER

The Harpenden instrument was designed to overcome this problem of frictional forces when holding the instrument by the tips of the blades. The cursor runs on miniature ball-bearing rollers allowing a free movement that is without crossplay. As with the other Harpenden instruments this is a counter display instrument giving readings over the range 50–570 mm. As with the Martin anthropometer, beam extensions may be added to the main bar but because of the counter display, constants, equivalent to the length of the beams, must be added on to the counter reading.

ACCEPTABLE ERROR LIMITS

Following on from the methods section, Table 6.1 provides a summary of the accepted upper limits for total TEM for two levels of reliability (95% and 99%) for a number of anthropometric measurements split by gender and chronological age. These TEM values provide a reference for anthropometrists on which to base decisions about the acceptability of measurement error within their study (Ulijaszek and Kerr, 1999).

SURFACE LANDMARKS

Acromion process (lateral border of the acromion) (Figures 6.4 and 6.5)

The acromion projects forwards from the lateral end of the spine of the scapula with which it is continuous. The lower border of the crest of the spine and the lateral border of the acromion meet at the acromial angle which may be the most lateral point of the acromion. Great diversity in the shape of the acromion

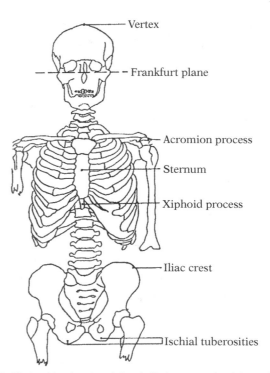

6.4. Skeletal landmarks of the skull, thorax, and pelvic girdle.

between individuals means that sometimes the acromial angle is not the most lateral point. Palpation of the most lateral part may best be performed by running the anthropometer blades laterally along the shoulders until they drop below the acromia. If the blades are then pushed medially the most lateral part of the acromia must be closest to the blades and may be felt below the surface marks left by the blades. There is the possibility of the inexperienced anthropometrists confusing the acromio-clavicular joint with the lateral end of the acromion. Great care must be taken to distinguish between these two landmarks prior to measurement.

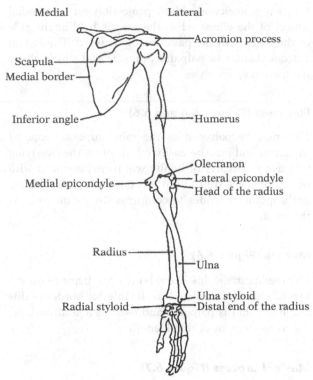

6.5. Skeletal landmarks from posterior view of scapula and arm.

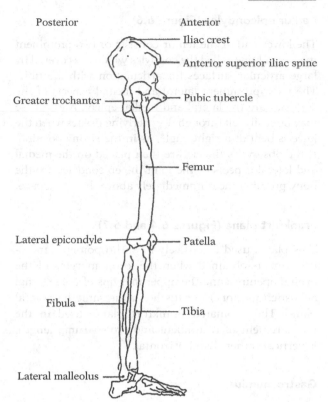

6.6. Skeletal landmarks from lateral view of pelvic girdle and leg.

Anterior superior iliac spine (Figure 6.6)

This is the anterior extremity of the ilium, which projects beyond the main portion of the bone and may be palpated at the lateral end of the fold of the groin. It is important to distinguish the iliac crest from the anterior spine when measuring bi-iliac diameter.

Biceps brachii

The biceps brachii is the muscle of the anterior aspect of the upper arm. Its two heads, the short and the long, arise from the coracoid process and the supraglenoid tubercle of the scapula respectively and are succeeded by the muscle bellies before they end in a flattened tendon that is attached to the posterior part of the radial tuberosity. When relaxed the muscle belly has its greatest bulge towards the radius, but when contracted with the arm flexed the belly rises to a point nearer the shoulder. Thus relaxed and contracted arm circumferences, taken at the maximum bulge of the muscle, are not at exactly the same level.

Distal end of the radius (Figure 6.5)

This is the border of the radius proximal to the distal-superior borders of the lunate and scaphoid and medial to the radial styloid. It may be palpated by moving the fingers medially and proximally from the radial styloid (see Radial styloid, below).

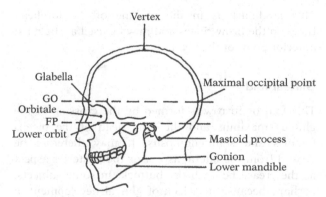

6.7. Skeletal and surface landmarks of the head. FP, Frankfurt plane; GO, Glabella-occipital plane.

External auditory meatus (Figure 6.7)

This landmark, used to obtain the Frankfurt plane, is also called the external acoustic meatus and leads to the middle ear from the external auricle. In terms of a surface landmark it is therefore simply present as a hole in the external ear and may therefore be easily seen. The tragus, the small curved flap that extends posteriorly from the front of the external ear, overlaps the orifice of the meatus and may be used to gauge the level of the orifice.

Femur epicondyles (Figure 6.6)

The lower end of the femur consists of two prominent masses of bone called the condyles which are covered by large articular surfaces for articulation with the tibia. The most prominent lateral and medial aspects of the condyles are the lateral and medial epicondyles. These may be easily felt through the overlying tissues when the knee is bent at a right angle, as in the sitting position. If the observer's fingers are then placed on the medial and lateral aspects of the joint the epicondyles are the bony protuberances immediately above the joint space.

Frankfurt plane (Figures 6.4 and 6.7)

This plane, used extensively in anthropometric measurement, is obtained when the lower margins of the orbital openings and the upper margins of the external acoustic (auditory) meatus lie in the same horizontal plane. The supinated Frankfurt plane, used in the measurement of recumbent and crown–rump length, is vertical rather than horizontal.

Gastrocnemius

This is the most superficial of the group of muscles at the rear of the lower leg and forms the belly of the calf.

Glabella (Figure 6.7)

This landmark is in the midline of the forehead between the brow ridges and may be used as the most anterior point of the head.

Gluteal fold

This fold or furrow is formed by the crossing of the gluteus maximus and the long head of the biceps femoris and semitendinosus. It may therefore be viewed from the lateral aspect or the posterior aspect as the crease beneath the buttock. In some subjects, perhaps because of a lack of gluteal development, a crease may not be present. In this case the level of the gluteal fold is judged from the lateral profile of the buttocks and posterior thigh.

Head of the radius (Figure 6.5)

This may be palpated as the inverted, U-shaped bony protuberance immediately distal to the lateral epicondyle of the humerus when the arm is relaxed with the palm of the hand facing forwards.

Humeral epicondyles (Figure 6.5)

These are the nonarticular aspects of the condyles on the lower surface of the humerus. The medial epicondyle forms a conspicuous blunt projection on the medial aspect of the elbow when the arm is held at the side of the body with the palm facing forward. The lateral epicondyle may be palpated opposite and a little above the medial epicondyle.

Iliac crest (Figures 6.4 and 6.6)

This may be palpated as the most superior edge of the ilium and may be easily felt through the overlying soft tissue. Greater difficulty will be experienced with the more obese subject but it is quite possible with the anthropometer blades to compress the tissue and feel the crest.

Malleoli (Figure 6.6)

The medial malleolus is the bony protuberance on the medial side of the ankle. It is the inferior border of this malleolus that is palpated and used as a landmark for the measurement of tibial length.

Mastoid process (Figure 6.7)

This is the conical projection below the mastoid portion of the temporal bone. It may be palpated immediately behind the lobule of the ear and is larger in the male than in the female.

Mid-axillary line

The axilla is the pyramidal region situated between the upper parts of the chest wall and the medial side of the upper arm. The mid-axillary line is normally taken as the line running vertically from the middle of this region to the iliac crest.

Mid-inguinal point (inguinal crease)

The inguinal ligament runs from the anterior superior iliac spine to the pubic tubercle at an angle of 35–40 degrees and is easily observed in all individuals. The midpoint between the anterior spine and the pubic tubercle on the line of the inguinal ligament is taken as the mid-inguinal point.

Midpoint of the arm

The midpoint of the arm, used for arm circumference, is taken as the point on the lateral side of the arm midway between the lateral border of the acromion and the olecranon when the arm is flexed at 90 degrees. This may be most easily determined by marking the lateral border of the acromion and applying a tape measure to this point. If the tape is allowed to lie over the surface of the arm, the midpoint may easily be calculated and marked. Alternatively, tape measures do exist with a zero midpoint specifically designed to determine this landmark. It has been common to refer to this point, and the circumference or girth at this level, as the "mid upper-arm" landmark/circumference. This terminology is specifically not used here because it is anatomically incorrect to describe this area the brachium or arm as the "upper-arm". Anatomically the brachium (arm) and the ante-brachium (forearm) form the upper limb.

Occiput (Figure 6.7)

The occipital bone is situated at the back part and base of the cranium. The occiput is the most posterior part of this bone and may be clearly seen from the side view of the subject.

Olecranon (Figure 6.5)

The olecranon is the most proximal process of the ulna and may be easily observed when the arm is bent as the point of the elbow.

Patella (Figure 6.6)

The patella is the sesamoid bone in front of the knee joint embedded in the tendon of the quadriceps muscle. It is flat, triangular below and curved above. When the subject is standing erect its lower limit lies above the line of the knee joint and its upper border may be palpated at the distal end of the quadriceps muscle.

Pinna of the ear

The pinna of the ear is more correctly called the lobule and is the soft part of the auricle that forms the ear-lobe.

Radial styloid (Figure 6.5)

The radial styloid is the distal projection of the lateral surface of the radius. It extends towards the first metacarpal and may be palpated as a bony projection on the lateral surface of the wrist when the hand is relaxed.

Scapula (Figure 6.5)

The scapula is the large, triangular flattened bone on the posterolateral aspect of the chest, and is commonly known as the shoulder blade. Its medial border slopes downwards and laterally to the inferior angle that may be easily palpated, and lies over the seventh rib or seventh intercostal space when the arm is relaxed.

Sternum (Figure 6.4)

The sternum or breastbone is the plate of bone inclined downwards and a little forwards at the front of the chest. It is composed of three parts; the manubrium at the top, the body or mesosternum at the center, and the xiphoid process at the lower end. The mesosternum and xiphoid process are important landmarks in anthropometry. The mesosternum is marked by three transverse ridges or sternabrae and the junction between the third and fourth sternabra form a landmark in chest measurement. The fourth sternabra may not be easily palpated but the junction lies below the more easily palpated third sternabrae. The xiphoid process may be palpated by following the line of the sternum to its end. The sternum is considerably larger in males than in females.

Trapezius

The trapezius is a flat, triangular muscle extending over the back of the neck and the upper thorax.

Triceps

The triceps muscle is the large muscle on the posterior side of the upper arm. When the arm is actively extended two of the three triceps heads may be seen as medial and lateral bulges.

Trochanters (Figure 6.6)

The greater and lesser trochanters are projections at the proximal end of the femur. The lesser trochanter cannot be palpated on the living subject because it lies on the posterior surface of the femur and is covered by the large gluteal muscles. The greater trochanter, however, is palpable as the bony projection on the lateral surface of the upper thigh approximately a hand's breadth below the iliac crest.

Ulna styloid (Figure 6.5)

The styloid process of the ulna is present as a short, rounded projection at the distal end of the bone. It may be easily palpated on the posterior-medial aspect of the wrist opposite and about 1 cm above the styloid process of the radius.

Umbilicus

The umbilicus, or naval, is clearly observable in the center of the abdomen. It is variable in position, lying lower in the young child due to lack of abdominal development.

Vertex of the skull (Figure 6.7)

This is the top-most point of the skull and theoretically comes into contact with the stadiometer headboard when height is being properly measured. With the head in the Frankfurt plane the vertex is slightly posterior to the vertical plane through the external auditory meatus and may be easily palpated.

TECHNOLOGICAL ADVANCES IN ANTHROPOMETRIC ASSESSMENT

Whilst traditional anthropometry has many benefits, it also has limitations in that it is subject to a number of potential sources of error, it is time consuming and expensive to measure large numbers of participants (Azouz et al., 2006), and it provides limited information about body shape (Robinette et al., 1997). The use of semi- and fully automated data acquisition equipment is becoming increasingly commonplace. There has been rapid advancement of automated anthropometric data collection instruments in recent decades, in particular, the advancement of digital human models which are generated from three-dimensional (3D) whole body scans. From these digital models, one-dimensional (1D) measurements of body dimensions such as lengths (circumferences are more problematic) can be derived (Robinette and Daanen, 2006) quickly and accurately for large numbers of participants. In addition, new or different measurements can be derived from stored digital images post clinic or field appointment. In the Civilian American and European surface anthropometry resource (CAESAR) study, some 4400 participants (aged 18–65 years) from the United States, the Netherlands, and Italy were scanned and digital body images generated (Robinette et al., 2002). From the scan data, 60 1D linear dimensions were derived and a further 40 measures were collected using standard anthropometric techniques (i.e., with tape measures and calipers). In a study comparing the precision of the CAESAR scan-derived 1D measurements with the allowable errors reported in the US Army ANSUR traditional anthropometric survey (Gordon et al., 1989); Robinette and Daanen (2006) showed that the CAESAR measures were more reliable than the traditional measures reported in the ANSUR survey, in that mean absolute differences (MAD), indicating the error between repeated measures of the same participant, were on average less than 5 mm for the CAESAR measurements compared to an average allowable error of 6.2 mm for the ANSUR survey. Robinette and Daanen (2006) concluded that these types of scan-extracted 1D linear measurements were as good if not better than traditionally collected anthropometric measurements. One must however acknowledge that the CAESAR study did not use scan-derived measures for circumferences as Perkins et al. (2000) have shown that scan-extracted circumferences are more error prone and less reliable compared to traditional anthropometric assessments of body circumference.

OTHER APPLICATIONS

Of course, the use of anthropometry for the assessment of child growth and development is only one application and has been used here for illustration purposes. Anthropometry is regularly used in a large number of other contexts such as the assessment of body composition and nutritional status (see the following chapter for a detailed summary), directional and fluctuating asymmetry, digit ratios, and for diagnosing diseases such as body dysmorphia. Asymmetry in bilateral anatomical structures such as facial features, hands, and feet is commonplace, for example one hand my be slightly bigger than the other, and is termed directional asymmetry if the trait is consistently larger on one side of the body within a population. In contrast, fluctuating asymmetry refers to traits that are not unidirectional within a population. It is thought that genetic and/or environmental insults influence an individual's ability to maintain bilateral symmetry in morphological traits during growth and development. Directional asymmetry is a measure of laterality and is thought to be driven by exposure to sex steroids whereas fluctuating asymmetry is thought to reflect developmental instability (Martin et al., 2008). Differences in bilateral anatomic traits can easily be assessed using standard anthropometric techniques. A common bilateral trait examined using anthropometry is the 2D:4D ratio (a ratio of the length of the index and ring fingers on one hand), which is a sexually dimorphic trait established in utero and has been associated with sexuality, behavior, and health (Manning, 2002). Finally, anthropometry can be used in clinical settings to help physicians diagnose certain disorders, for example body dysmorphic disorder (BDD). A person with BDD has a preoccupation with an imagined/non-existent or slight defect in their appearance (most commonly skin, hair, and/or facial features) which usually begins during adolescence (Albetini and Phillips, 1999). Anthropometry can be used by clinicians to help diagnose BDD by establishing if the complaints of the patient are medical in nature (i.e., fall outside of expected ranges for a population) or psychological as they cannot be explained by physical disease, substance abuse, or other mental disorders.

DISCUSSION POINTS

1. How do you decide to exclude potentially erroneous data points within an analysis data set?
2. Why is it important to have quality control checks in studies using anthropometric measurements?
3. What different considerations apply to the cleaning of cross-sectional as opposed to longitudinal data?

4. What principles should be employed to decide on the level of precision required in any particular research design?
5. What are the advantages and disadvantages of using automated anthropometric data collection tools?

REFERENCES

Acheson, R. M. (1966). Maturation of the skeleton. In *Human Development*, F. Faulkner (ed.). Philadelphia: Saunders.

Albetini, R. S. and Phillips, K. A. (1999). Thirty-three cases of body dysmorphic disorder in children and adolescents. *Journal of the American Academy of Child and Adolescent Psychiatry*, **38**(4), 453–459.

Alder, K. (2002). *The Measure of All Things*. New York: Free Press.

Azouz, Z. B., Rioux, M., Shu, C., et al. (2006). Characterizing human shape variation using 3D anthropometric data. *Visual Computer*, **22**, 302–314.

Bayley, N. (1940). *Studies in the Development of Young Children*. Berkeley, CA: University of California Press.

Cameron, N. (1978). The methods of auxological anthropometry. In *Human Growth*, F. Faulkner and J. M. Tanner (eds). New York: Plenum.

Cameron, N. (1983). The methods of auxological anthropometry. In *Human Growth*, 2nd edn, F. Faulkner and J. M. Tanner (eds). New York: Plenum.

Cameron, N. (1984). *The Measurement of Human Growth*. London: Croom-Helm.

Collignon, R. (1892). Projet d'entente internationale au sujet des researches anthropometrique dans les conseils de revision. *Bulletins et mémoires de la société d'anthropologie de Paris*, **4**, 186.

Edwards, D. A. W., Hammond, W. H., Healy, M. J. R., et al. (1955). Design and accuracy of calipers for measuring subcutaneous tissue thickness. *British Journal of Nutrition*, **9**, 133–143.

Fergus, W. and Rodwell, G. F. (1874). On a series of measurements for statistical purposes recently made at Marlborough College. *Journal of the Anthropological Institute*, **4**, 126–130.

Fleming, R. M. (1933). Special report senior Medical Research Council, no. 190. In J. M. Tanner (1952), The assessment of growth and development in children. *Archives of Disease in Childhood*, **27**, 10.

Galton, F. (1874). Notes on the Marlborough School Statistics. *Journal of the Anthropological Institute*, **4**, 130–135.

Gesell, A. and Thompson, H. (1938). *The Psychology of Early Growth*. New York: Macmillan.

Gordon, C. C., Bradtmiller, B., Clausen, C. E., et al. (1989). *1987–1988 Anthropometric Survey of US Army Personnel. Methods and Summary Statistics. Natick/TR-89-044*. Natick, MA: US Army Natick Research Development and Engineering Center.

Hartmann, W. (1970). Beobachtungen zur acceleration des langenwachstums in der zweiten hälfte des 18 jahrhunderts. Thesis, University of Frankfurt-on-Main, Germany.

Hertzberg, T. (1968). The conference on standardization of anthropometric techniques and terminology. *American Journal of Physical Anthropology*, **28**, 1–16.

Hrdlička, A. (1947). *Hrdlicka's Practical Anthropometry*, 3rd edn. Philadelphia: Wistar Institute.

International Children's Centre (1977). *Growth and Development of the Child: International Children's Centre Coordinated Studies Publications Index 1951–1976*. Paris: International Children's Centre.

Knott, V. B. (1941). Physical measurements of young children. *University of Iowa Studies in Child Welfare*, **18**, 3.

Lampl, M., Veldhuis, J. D. and Johnson, M. L. (1992). Saltation and stasis: A model of human growth. *Science*, **258** (5038), 801–803.

Lange, K. O. and Brozek, J. (1961). A new model of skinfold caliper. *American Journal of Physical Anthropology*, **19**, 98–99.

Lohman, T. G., Roche, A. F. and Martorell, R. (1988). *Anthropometric Standardization Reference Manual*. Champaign, IL: Human Kinetics Books.

Low, A. (1952). *Growth of Children*. Aberdeen: Aberdeen University Press.

Manning, J. T. (2002). *Digit Ratio: a Pointer to Fertility, Behavior, and Health*. Piscataway, NJ: Rutgers University Press.

Martin, J. T., Puts, D. A. and Breedlove, S. M. (2008). Hand asymmetry in heterosexual and homosexual men and women: relationship to 2D:4D digit ratios and other sexually dimorphic anatomic traits. *Archives of Sexual Behaviour*, **37**, 119–132.

Meredith, H. V. (1936). The reliability of anthropometric measurements taken on eight- and nine-year old white males. *Child Development*, **7**, 262.

Murat, J. A. (1816). Foetus. *Dictionnaire des sciences médicales*, **16**, 49–80.

Perkins, T., Burnsides, S., Robinette, K. M., et al. (2000). Comparative consistency of univariate measures from traditional and 3-D scan anthropometry. In *Proceedings of the SAE International Digital Human Modelling for Design and Engineering International Conference and Exposition*. Dearborn, MI: CD-ROM, PDF file 2145.

Quetelet, A. (1870). *Anthropometrie, ou mesure des differentes facultés de l'homme*. Brussels: Marquardt.

Robinette, K. M. and Daanen, H. A. (2006). Precision of the CAESAR scan-extracted measurements. *Applied Ergonomics*, **37**, 259–265.

Robinette, K. M., Vannier, M., Rioux, M., et al. (1997). *3-D Surface Anthropometry: Review of Technologies*. Neuilly-sur-Seine, France: North Atlantic Treaty Organization Advisory Group for Aerospace Research and Development, Aerospace Medical Panel.

Robinette, K. M., Blackwell, S., Daanen, H. A., et al. (2002). *Civilian American and European Surface Anthropometry Resource (CAESAR). Final Report, Summary*, vol. 1. AFRL-HE-WP-TR-2002-0169. Wright-Patterson AFB, OH: Human Effectiveness Directorate, Crew System Interface Division; and Warrendale, PA: SAE International.

Ruiz, L., Colley, J. R. T. and Hamilton, P. J. S. (1971). Measurement of triceps skinfold thickness. *British Journal of Preventative and Social Medicine*, **25**, 165–167.

Sakai, T. (2007). Historical evolution of anatomical terminology from ancient to modern. *Anatomical Science International*, **82**(2), 65–81.

Scammon, R. E. (1927). The first seriatim study of human growth. *American Journal of Physical Anthropology*, **10**, 329.

Shuttleworth, F. K. (1937). Sexual maturation and the physical growth of girls aged 6 to 19. *Monographs of the Society for Research in Child Development*, **2**(5), serial no. 12.

Simmons, K. (1944). The Brush Foundation study of child growth and development II. Physical growth and development. *Monographs of the Society for Research in Child Development*, **9**(1), serial no. 37.

Strickland, A. L. and Shearin, R. B. (1972). Diurnal height variation in children. *Pediatrics*, **80**, 1023.

Stuart, H. C. (1939). Studies from the Center for Research in Child Health and Development, School of Public Health, Harvard University: I. The center, the group under observation and studies in progress. *Monographs of the Society for Research in Child Development*, **4**(1).

Tanner, J. M. (1952). The assessment of growth and development in children. *Archives of Disease in Childhood*, **27**, 10–33.

Tanner, J. M. (1962). *Growth at Adolescence*. Oxford: Blackwell Scientific Publications.

Tanner, J. M. (1981). *A History of the Study of Human Growth*. Cambridge: Cambridge University Press.

Tanner, J. M. and Whitehouse, R. H. (1955). The Harpenden skinfold caliper. *American Journal of Physical Anthropology*, **13**, 743–746.

Tanner, J. M., Healy, M. J. R., Lockhart, R. D., et al. (1956). Aberdeen Growth Study 1. The prediction of adult body measurements from measurements taken each year from birth to five years. *Archives of Disease in Childhood*, **31**, 372–381.

Thompson, D. W. (1942). *On Growth and Form*. Cambridge: Cambridge University Press.

Ulijaszek, S. J. and Kerr, D. A. (1999). Anthropometric measurement error and the assessment of nutritional status. *British Journal of Nutrition*, **82**, 165–177.

Weiner, J. S. and Lourie, J. A. (1969). *Human Biology: a Guide to Field Methods*, IBP Handbook no. 9. Oxford: Blackwell Scientific Publications.

Weiner, J. S. and Lourie, J. A. (1981). *Practical Human Biology*. London: Academic Press.

Whitehouse, R. H., Tanner, J. M. and Healy, M. J. R. (1974). Diurnal variation in stature and sitting-height in 12–14 year old boys. *Annals of Human Biology*, **1**, 103–106.

7 Energy Expenditure and Body Composition: History, Methods, and Inter-relationships

Peter S. W. Davies and Alexia J. Murphy

ENERGY EXPENDITURE AND BODY COMPOSITION

Energy expenditure and body composition are closely related. The energy expenditure of virtually any animal can be measured but of course this energy is not being expended equally throughout the body. Organs such as the brain, liver, heart, and kidneys have, relative to their weight, a high-energy output, whilst, for example, muscle mass, although being a substantial component of body weight, has on a per kilogram basis a lower energy output. When these individual organs or organ systems are combined, essentially the fat free mass (FFM) is the major contributor to energy expenditure, with the remaining fat mass (FM) being more energetically inert. Thus, methods of studying both energy metabolism and body composition have often developed in parallel, with the need being to adjust one for the other. This chapter aims to provide a history of some of those methods as well as some theoretical and practical information regarding their use before finally considering how they relate and how the relationships influence our understanding of both areas of biology.

INTRODUCTION TO THE STUDY OF ENERGY METABOLISM

There are a number of fundamental maxims that underpin large areas of modern science. It is significant that many of these laws and principles were described in a concise form in the eighteenth and nineteenth centuries when some of the great men and women of science were laying down the foundations and laws that govern much of modern physical and biological studies. There are two such laws that form the basis of studies pertaining to energy metabolism and hence energetics in the human. Firstly, Hermann Helmholtz described in 1847 the law of the conservation of energy thus; "in all processes occurring in an isolated system the energy of the system remains constant." This classical description was based upon the earlier work of a compatriot of Helmholtz, Robert Mayer. Secondly, Germain Hess, around the same time, described the law of constant heat summation, i.e., the heat released by a number of reactions is independent of the chemical pathways involved and is only, and totally, dependent upon the end products. This is often simply referred to as Hess's law. These two basic laws enable us to study energy metabolism and hence energetics in the human by calorimetry, the measurement of heat production.

The pioneers in this field used animals to attempt to understand the relationships between heat production, metabolism, and life. One of the first important questions to be asked at this time was what is the source of animal heat? This question posed by the French *Académie des Sciences* in 1822 as a subject for a prize, led two scientists, Despretz and Dulong to independently devise, construct and describe the first true calorimeters. These calorimeters were, however, based upon Lavoisier's apparatus that had been designed and then neglected by science some 40 years earlier to measure the heat production of small mammals. The equipment devised by both Despretz and Dulong consisted of a small chamber surrounded by a water jacket. When placed in the chamber heat produced by the animal was transferred to the water and the temperature change recorded. Any gases produced by the animal were also collected for analysis. The techniques available for respiratory gas analysis were enhanced by the work of Regnault and Reiset in order to study the effect of the consumption of differing foods on expired gas composition. This area of study led to the design of the first closed-circuit indirect calorimeter, that is, equipment that would allow the measurement, or at least calculation, of energy expenditure, by the determination of carbon dioxide production, oxygen consumption, and nitrogen balance. This magnificently designed and constructed piece of apparatus used

Human Evolutionary Biology, ed. Michael P. Muehlenbein. Published by Cambridge University Press. © Cambridge University Press 2010.

three large volumetric flasks as the supply of air, and thus was a "closed" system. It was used to measure the effect of differing foodstuffs on expired gases in dogs, pigeons, and other animals. Initial scientific observation by these workers and others then began to give way to experimentation. By 1860, the effect of starvation on energetics, or metabolism as it was then termed, was being studied in Munich by Bischof and Voit. These experiments led to the development of further apparatus. Thus, the group in Germany built the first open-circuit indirect calorimeter, which in fact was large enough to accommodate a man, if not in great comfort. In this case atmospheric air was allowed into the chamber and hence the term "open" calorimeter.

The measurement of energetics by calorimetry was now beginning to bifurcate. While Rubner spent much time and effort perfecting a direct calorimeter to measure heat production in dogs, Pettenkofer endeavored to perfect indirect calorimetry. Once it was recognized that the two approaches could be complementary rather than antagonistic it became important to know the extent to which they gave the same answer! This question was tackled by Rubner in 1894 using dogs as the experimental animal. There was agreement of better than 1% between the two methodologies. The equivalent experiment was, however, not carried out in humans for almost a further 10 years. However, when completed in 1903 the classic work of Atwater and Benedict (1903) showed an equal measure of agreement.

The value of being able to assess energy expenditure in order to estimate energy requirements was soon appreciated by the meat and livestock industry, and much of the incentive for the creation of improved direct and indirect calorimeters was shown by the food industry. In human studies the necessity for confinement in a closed chamber, regardless of size, in order to assess energy expenditure was causing frustration for investigators. By the beginning of the twentieth century portable apparatus was being designed in order to study energy expenditure and energy balance during physical activity. The first truly successful method of assessing energy output during physical activity was developed by Douglas (1911). His system, basically still in widespread use today, used a large rubber-lined bag, usually carried on the subject's back, into which all expired air was collected. The volume of the expired gases collected as well as the composition of those gases allowed a calculation of both carbon dioxide production rate and oxygen consumption. The major limitation then, as today, was the size of the bag. The logical development of this system occurred many years later. The expired gases were not collected but metered whilst in situ with a small sample being kept for gas analysis. This was then the basis of the Kofranyi–Michaelis instrument (Kofranyi and Michaelis, 1940). Again, this type of apparatus is

still used today. A significant publication around this time appeared in the *Journal of Physiology (London)* in which Weir described an equation for the mathematical conversion of data relating to oxygen consumption and carbon dioxide production to a measure of energy expenditure (Weir, 1949). This equation continues to be widely used today.

In the first decade of the twentieth century questions were posed relating to energetics and energy metabolism in health and disease, and the possible clinical applications of such knowledge was being investigated notably in North America. Large calorimeters were built for this purpose by Williams, Lusk, and Dubois in order to study what were termed "metabolic disorders." About this time, as well, the first work involving infants and children was reported, notably the studies of Benedict and Talbot (Benedict, 1914; Benedict and Talbot, 1914; Benedict, 1919). These two pioneers of energy metabolism studies in infancy and childhood developed an indirect calorimeter for the measurement of basal metabolic rate in early infancy. Small infants were kept in this chamber for a number of hours until, in the words of Benedict (1919), "the child became accustomed to the conditions and fell asleep." Larger chambers were later built to accommodate older children. The largest of these chambers was installed at the New England Home for Little Wanderers, where children up to the age of 14 could be measured. The chamber was designed to have a minimal dead space, with only a small window allowing illumination into the chamber and visibility out of it.

Larger chambers that allow the analysis of respiratory gases are now available, which are large enough to allow a certain amount of activity or "free living." Nevertheless such cambers, with volumes in the order of 30 000 L, require periods of up to 8 hours before equilibrium is reached and measurements can be taken. Fast-response algorithms have been developed to enable results to be achieved more quickly, but such chambers cannot still be used easily in many populations such as infants, children, and the sick.

A significant advancement in the ability to study energetics in humans in what has been termed the "free living" situation occurred in the 1980s with the refinement of the so-called doubly labeled water technique for calculating the carbon-dioxide production rate using two stable isotopes in the form of water (Schoeller and Van Santen, 1982; Coward et al., 1988). This method is described in further detail later in this chapter.

MEASUREMENTS OF ENERGY EXPENDITURE

Direct calorimetry

There are few if any large direct calorimeters still in use. They were technically demanding and expensive to maintain and whilst in some ways they are the true calorimeters, in that they actually measured heat production by a variety of different methods, their disadvantages outweighed their advantages. Whilst sometimes referred to as the gold standard for the assessment of energy expenditure the method has been superseded by other less difficult and sometimes less expensive approaches.

Indirect calorimetry

Indirect calorimetry in various forms continues to be used in many research centers and clinical situations. Indirect calorimetry, as its name implies, does not measure heat production but is based on the measurement of oxygen consumption and carbon dioxide production that occurs with the oxidization of protein, carbohydrate, fat, and alcohol. The amount of oxygen consumed and carbon dioxide produced is used in the Weir equation to calculate the amount of energy expended (Weir, 1949).

There are still a number of large indirect calorimeters in operation that allow measurements of energy expenditure to be made in humans for periods of typically between 24 and 72 hours. Such indirect calorimeters can be found, for example, at the University of Maastricht in the Netherlands and the University of Wollongong in Australia. These chambers are much improved in many ways in comparison to the early chambers described previously, and now contain amenities such as televisions, telephones, computers, etc. These apparatus require careful calibration and maintenance.

Measurements of energy expenditure, usually at rest, over significantly shorter periods can be achieved with any one of an array of commercially available apparatus. Facemasks, mouthpieces, or ventilated hoods are the most popular and prominent methods of achieving gas collection. Such technology can be used in infants, children, and adults both in health and in disease (Singhal et al., 1993; Wells and Davies, 1995). Other apparatus are designed for measuring the energy cost of activity and thus energy expenditure in a range of situations can be assessed using this methodology (Littlewood et al., 2002). In young children this can be challenging but is achievable. Many of the newer pieces of apparatus have been validated against existing indirect methods (Duffield et al., 2004; Perret and Mueller, 2006).

The traditional and sometimes termed gold standard method for calibration of indirect calorimeters is the alcohol burn. This method, first designed many years ago, by Carpenter and Fox (1923), is based on the fact that 1 g of ethanol consumes 1.70 L of oxygen and produces 0.97 L of carbon dioxide when burnt

(value at standard temperature and pressure [STP]). However, it cannot be guaranteed that all the alcohol has been burnt and that none has been lost to evaporation when using this approach.

Doubly labeled water

This method is, in many ways an example of indirect calorimetry. The doubly labeled water technique is the first noninvasive method available to measure daily energy expenditure accurately (Schoeller and Van Santen, 1982). This method involves enriching the body water with isotopes of oxygen (oxygen-18) and hydrogen (deuterium) and then measuring the difference in turnover rates of these isotopes in body fluid samples. The difference in the rate of turnover of the two isotopes can be used to calculate the carbon dioxide production rate. The mean respiratory quotient (carbon dioxide production rate/oxygen production rate) is assumed and therefore the oxygen production rate and consequently energy expenditure can be calculated.

Doubly labeled water is a preferred method of energy expenditure measurement as it requires limited subject effort, is noninvasive, and measurements are performed in real life conditions. The disadvantages of the method are that it is expensive, requires complex analysis, and is not suitable for large population studies. The sources of error of the method include analytical errors with mass spectrometry, biological variation in the isotope enrichment and isotope fractionations, and the assumptions of the method. The method is in use in a relatively small number of centers but has been validated against indirect calorimetry in a number of differing populations from premature infants to adults (Schoeller and Van Santen, 1982; Roberts et al., 1986; Schoeller et al., 1986; Coward, 1988).

Heart rate and activity monitoring

The prediction of total energy expenditure from heart rate monitoring and activity monitors are basic field measures. Heart rate monitoring is a useful field method where actual energy expenditure is derived from the regression of oxygen production versus heart rate. Each individual needs to be "calibrated," that is, an individual relationship between heart rate and oxygen consumption needs to be determined. Following this the heart rate monitor is then worn, sometimes for many days, after which the individual heart beats are related to oxygen consumption and hence energy expenditure. The advantage of this method is that it is inexpensive, in comparison with doubly labeled water for example, but the disadvantages of this method are that factors other than oxygen production affect heart rate. Also it is inaccurate at low levels of activity, is time consuming, and is sometimes seen as having a significant participant burden. Nevertheless, the method has been validated against the doubly labeled water technique and is an accepted field method (Livingstone et al., 1990, 1992).

Activity monitors vary considerably from pedometers to significantly more complicated devices that aim to predict energy expenditure. The pedometer simply measures steps or movement in one direction, whereas accelerometers can measure body acceleration in several planes. In accelerometry, an equipment-specific algorithm is often used to convert the activity counts in each vector to energy expenditure. Although this method is simple and can be used in population studies, there are limitations with this method when attempting to quantify total energy expenditure because of the need to relate physical activity "counts" to the energy cost of the various activities undertaken.

INTRODUCTION TO BODY COMPOSITION METHODS

Body composition concepts have always been a vital component of biological studies. The earliest record of scientists investigating body composition was in the 1850s when European chemists were developing chemical analytical techniques in animal tissues. By the early 1900s fetuses and newborns were being chemically analyzed for Na, K, Cl, Ca, P, N, water, and fat (Moulton, 1923; Givens and Macy, 1933; Iob and Swanson, 1934). In 1906 the German physiologist, Adolf Magnus-Levy, reported the importance of expressing body tissue composition in a fat free basis and the concept of fat free body mass was formed (Magnus-Levy, 1906). This was an important new concept that continues to influence body composition studies to this time. Another important and long-lasting concept, that of relating height and weight to adjust one for the other, was also first described around this time when, in 1871, Quetelet reported that weight increased in proportion to height squared, which is the basis of the body mass index (Quetelet, 1871).

In the 1930s, researchers collected anthropometric measurements providing the basis of today's reference data and worked towards developing indirect methods of determining human body composition. Albert R. Behnke was a pioneer in the area of body composition research, with densitometry being one of the first indirect body composition methods. In 1933, Behnke and colleagues proposed that fat could be estimated from a measurement of body density. In 1942 they reported the assumptions inherent to densitometry, that the chemical composition of FFM is different to FM and is assumed to remain constant and is known (Behnke et al., 1933, 1942). In 1942, Behnke and colleagues, again, were the first to develop an underwater weighing system that included adjustment for residual trapped air in the lungs (Behnke et al., 1942). Another researcher who contributed to the early era of body composition research was Francis Moore. His research in the 1940s was the first to focus on the study of biochemical phases and ignited the surge of research into this area of body composition methods (Moore, 1946).

The late 1950s saw investigation of whole body counters to measure total body potassium (TBK) and the link between the body's potassium-40 content and FFM was reported (Anderson and Langham, 1959; Allen et al., 1960; Forbes et al., 1961). In 1961 Forbes and colleagues estimated fat and lean contents using whole body counting (Forbes et al., 1961). Also in the early 1960s, Thomasset (1962, 1963) introduced the bioelectrical impedance analysis (BIA) method, evolving from Pace and Rathburn's findings in 1945 that water is not present in stored fat and that water occupies a fixed fraction of the FFM (Pace and Rathburn, 1945). By the early 1970s many new medical methods were being introduced for body composition assessment – in vivo neutron activation (Anderson et al., 1964), computerized tomographic (CT) imaging (Hounsfield, 1973), and total body electrical conductivity.

From 1979, Steve Heymsfield led the reintroduction of anatomy to the research of body composition by identifying the value of CT scans to provide organ-specific tissue volumes (Heymsfield et al., 1979). In 1981, Peppler and Mazess introduced the concept of dual photon absorptiometry (Peppler and Mazess, 1981), and in 1984 Foster and colleagues reported the use of magnetic resonance imaging (MRI) as a body composition measurement (Foster et al., 1984). With the development of these new techniques, by the 1990s researchers were able to measure both the anatomical and chemical content of the body, thus gaining the best possible insight into human body composition.

Body composition research in the last 20 years has been strongly focused on improving techniques and extending their validity to clinical and specific populations. The most recent development of a new body composition device is based on one of the oldest principles of body composition research, densitometry. Underwater weighing was the only available measurement of densitometry until recently, when in 1995 an instrument based upon air displacement, plethysmography, became available (McCrory et al., 1995). The Bod Pod® (Life Measurements Instruments, Inc., Concord, CA) is suitable for use in adults and children, and in 2003 the Pea Pod® (Life Measurements Instruments, Inc., Concord, CA) was introduced for infants up to six months of age (Urlando et al., 2003). Another method recently developed uses resonant cavity perturbation techniques to measure total body water (TBW) (Stone and Robinson, 2003).

MEASUREMENTS OF BODY COMPOSITION

Compartmental models of body composition

Two compartment

The two-compartment (2C) model is the basic division of the body compartments into fat, FM, and FFM. Fat free mass is defined as all the tissues of the body minus the extractable fat, which is termed the FM. The 2C model has been used in body composition research since the 1940s and continues to play a vital role (Behnke et al., 1942). The most commonly used 2C models are based on the measurements of body density, TBK, and TBW. The 2C model assumes that the chemical composition of FFM body tissue stores remain constant; however, it is known that the composition of the FFM is variable in children and disease states. Therefore the 2C model is inadequate for individuals who deviate from the healthy adult constants on which they are based.

Three compartment

The three-compartment (3C) model expands on the 2C model and controls for the variability of one of the FFM components, with the 3C model dividing the FFM into water and remaining solids (Siri, 1961). Water impacts significantly on the variability of the FFM density as it has the lowest density, but comprises the largest mass and volume of the FFM components (Siri, 1961). The 3C model requires that a measurement of densitometry is combined with a measurement of TBW. As the 3C model is still assuming the protein and mineral content of the FFM is constant, the estimated values for the solids compartment would be incorrect in clinical patients with depleted body protein mass and bone mineral content (BMC).

Four compartment

The four-compartment (4C) model breaks the body into FM, water, mineral, and protein. The 4C model takes into account the individual variability in the composition of the FFM and is considered the best commonly available method for measuring body composition as the more components of FFM that can be measured, the better the accuracy. In the 4C model, the mineral component can be measured by BMC and protein mass by neutron activation. However, neutron activation is not widely available, so the common 4C chemical model measures BMC with dual energy X-ray absorptiometry (DXA) and this model assumes that protein mass is proportional to the BMC. This model requires the measurement of body weight, body volume by densitometry, TBW by deuterium dilution, and BMC by DXA.

Another, less commonly applied 4C model, divides the body into three cellular components: body cell mass (BCM), extracellular fluid (ECF), and extra-cellular solids (ECS). For this model the BCM is measured by TBK, ECF by bromide or sulfate dilution, and ECS by total body calcium or BMC. The disadvantages of the 4C model are that the techniques are technically demanding, expensive, and not widely available in all setting.

METHODS OF MEASURING BODY COMPOSITION

Anthropometry

As described in the previous chapter, anthropometric measures are amongst the oldest body composition methods still applied and allow the evaluation of body composition outside the laboratory. Anthropometry includes measurements of weight, height, circumferences, and skinfolds. These methods are often used as they are simple, inexpensive, safe, and portable, however they are not recommended for individual clinical subject evaluations or for examining short-term changes in body fat (Burden et al., 2005). Anthropometric techniques usually demonstrate the largest standard error and lowest correlation coefficients when compared against other techniques for estimating total body fat such as DXA, BIA, or in vivo neutron activation analysis (IVAA) (Eisenmann et al., 2004; Daniel et al., 2005). These latter techniques are sometimes referred to as criterion methods or gold standards, but care should be taken with this approach as certainly all of these methods have error that cannot be discounted.

Indices of weight and height have been developed to provide a simple method of measuring body composition. The majority of studies that define obesity in disease states rely on the body mass index (BMI), which, as stated earlier, is weight divided by height squared. Although this method is widely used and recommended by the World Health Organization, it is limited as a measure of body composition, as it is not able to quantify FM or distinguish between fat and lean tissue, and subjects of the same BMI may differ widely in fatness (Ellis, 2001). The normal relationship between height and weight is altered in disease states and with the limited sensitivity of BMI z-scores to predict increased body fat in both a clinical and a healthy population, BMI is considered a poor measure of body fat in clinical situations (Warner et al., 1997; Wells et al., 2002; Eto et al., 2004).

Circumferences, commonly the mid-arm and waist, are also quick and simple methods that are taken to represent body composition, particularly in low socio-economic countries. Waist-circumference measurements can provide a validated measure of visceral adipose tissue (Janssen et al., 2002; Zhu et al., 2002;

Bosy-Westphal et al., 2006) and mid-arm circumference has been shown to represent malnutrition (Kumar et al., 1996; Powell-Tuck and Hennessy, 2003).

Skinfold-measurements are commonly used because of their low cost, portability, and simplicity. The measurement of skinfold thickness is taken by grasping the skin between thumb and forefinger and measuring this thickness with calipers. Duplicate measurements are recommended to improve accuracy and reproducibility. A number of measurements at different sites can be used in age and gender-specific equations to determine body composition (Durnin and Womersley, 1974). Skinfold-measurements are based upon two assumptions; that the thickness of subcutaneous fat represents a constant proportion of the total body fat, and that the measurement sites represent the thickness of the total body fat. Neither of these assumptions has been proven. The high variability of skinfold-measurements may be due to the calipers used, the technique applied, the increased error with high fat content, and the inappropriate application of prediction equations.

Whole body potassium counting

Whole body counting is the method used to determine TBK. Total body potassium is represented by potassium-40, a naturally occurring radioactive isotope that emits a gamma ray. Potassium-40 is primarily found intracellular and not in stored fat. As the subject is measured by the counter, the gamma rays emitted by the potassium at 1.46 MeV are detected by sodium iodide crystals in either single or multidetector configurations (Figure 7.1). Depending on the scanning system, measurements are taken over a few minutes to an hour.

As 98% of TBK is located in the BCM, whole body counting of TBK is considered the best body composition index for identifying the BCM (Pierson and Wang, 1988). Body cell mass is defined as the cellular component of the body that contains the energy metabolizing, work performing tissue; for example, the muscles and organs (Moore et al., 1963). Body cell mass measurements by TBK can be used in health and disease, because potassium concentrations in BCM are constant and kept within strict limits by homeostatic mechanisms (Edmonds et al., 1975). Unlike other methods, TBK measurements will not be affected by limitations such as cellular fluid shifts, so changes in TBK will be identifying true changes in BCM and not just reflecting the changes in weight (Trocki et al., 1998). The other advantages of this method are that it is noninvasive and requires limited effort by the subject. The disadvantages of this method are that it is not widely available, that equipment is expensive, and that it can be time consuming.

Densitometry

Air displacement plethysmography (ADP) is a relatively new method available to measure body composition and represents a preferred alternative to underwater weighing. The only available system for ADP is the Bod Pod® (Life Measurements Instruments, Inc., Concord, CA) (Dempster and Aitkens, 1995) (Figure 7.2). The Bod Pod® system is divided into two chambers, a

7.1. Whole body counter.

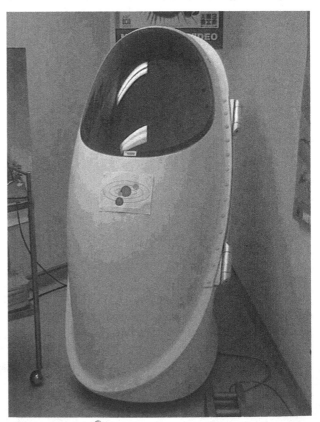

7.2. The Bod Pod® (Life Measurements Instruments, Inc., Concord, CA) body composition system.

measurement chamber and a reference chamber, with a computer-operated oscillating diaphragm between the chambers. The sinusoidal volume perturbations produced by the diaphragm result in small pressure changes and the ratio of the pressures indicate the volume of the measurement chamber in adiabatic condition, using the principle of Poisson's law:

$$(P_1/P_2) = (V_1/V_2)$$

The Bod Pod® measurement of body volume firstly requires a calibration of the chamber at 0 L, to establish baseline, and at 50 L. The subject, dressed in minimal clothing and a hair cap, then sits in the chamber for two 50-second volume measurements. The two measurements must be within 150 ml or 0.2% of each other, whichever is the smallest, with a maximum of three attempts. Thoracic gas volume is then measured within the Bod Pod® or can be estimated. The raw body volume given by the Bod Pod® requires adjustment for the volume of isothermal air found in the lungs and near the body surface, as it compresses 40% or more under pressure changes. To correct for the isothermal air, body volume is adjusted for thoracic gas volume and surface area artefact. Once corrected body volume is known, the principles of densitometry are applied and body density is calculated from body mass and volume. Body density can be used to estimate body fat with a 2C body composition model, or be used in combination with other methods for a 4C body composition model.

The advantages of the Bod Pod® are that it is quick, simple, and noninvasive. Studies have found that it is a reliable and valid measurement in adults and children (McCrory et al., 1995; Nunez et al., 1999). The limitations of the Bod Pod® lie with the limitations of densitometry when used as a 2C technique. Assuming the density of the FFM is similar in all ages and disease states will lead to overestimation of fatness in conditions where fluid retention and under mineralization decreases the density of the FFM.

Dual energy X-ray absorptiometry

Dual energy X-ray absorptiometry (DXA) was originally designed for measuring the amount of bone mineral in the body; however, it can also measure FM and FFM, and is becoming a favorable body composition assessment technique. A DXA scan requires an X-ray source to produce a broad photon beam that is filtered, yielding two different energy peaks. The photons pass through the body's tissues and the resulting attenuation between the two energy peaks is characteristic for each tissue. The concept of DXA technology is that photon attenuation is a function of tissue composition, with bone, lean tissue, and fat being distinguishable by

their X-ray attenuation properties. Bone is composed of calcium and phosphorus so it has high attenuation, lean tissues are composed of oxygen and electrolytes with a medium attenuation, and FM is predominately hydrogen and carbon with low attenuation properties.

Using a series of assumptions and algorithms, the attenuation for fat, lean, and bone allows the development of pixel-by-pixel estimation of body composition. For a whole body measurement approximately 40% of pixels are classified as containing bone, the remaining pixels are used to estimate the body's lean-to-fat ratio. It is assumed that the lean tissue over bone has the same fat-to-lean ratio as that for nonbone pixels in the same scan region, so this estimated value is applied to the lean tissue component in the adjacent bone pixels. Therefore, the lean-to-fat composition of the total lean tissue mass is based on sampling only one-half of the whole body.

Although DXA studies are increasing in popularity in nutritional studies, such studies should be interpreted with caution because, as previously stated, the technique does not represent a reference technique. Studies have shown that the bias of DXA is unreliable for monitoring body composition longitudinally or in case-controlled studies as results vary with gender, size, fatness, and disease state (Williams et al., 2005). Issues such as hydration and tissue thickness have also been investigated for their effect on DXA measurements (Jebb et al., 1993; Kohrt, 1998). As DXA assumes that the lean tissue is normally hydrated, the addition of fluid results in an underestimation of FM changes.

There are also several instrumental factors that may affect DXA measurements of body composition. Results have been shown to differ with manufacturer, software version, and beam mode (Kistorp et al., 2000). When DXA measurements are performed in the same subjects but on different brands of equipment, there has been shown to be significant differences in the body composition estimates (Tothill et al., 1993). The difference may be due to the differing algorithms used to divide the soft tissue mass between lean and fat compartments or the number of pixels assigned as containing bone. Despite the limitations of DXA, the technology provides a body composition technique that with some further research will be useful in many settings. The measurements are simple, quick and painless to perform, give immediate results, require minimal radiation dose, and are available in many clinical settings.

Hydrometry

Total body water can be measured using the dilution principle, which states the volume of the body is equal to the amount of tracer added to the body divided by

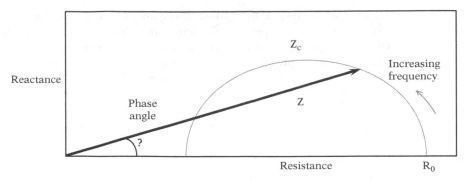

7.3. Cole–Cole plot.

the concentration of the tracer in the compartment (Edelman, 1952). The most commonly used tracer in this method is deuterium, but may also include tritiated water or oxygen-18 labeled water. The assumptions of the technique are that when the tracer is ingested, the tracer is distributed equally only in the exchangeable pool, the rate of equilibration is rapid, and neither the tracer nor body water is metabolized in the equilibration time.

After providing a predose sample of blood, urine, or saliva, the subject drinks a dose of the labeled water. Once the equilibrium time of 4–5 hours has passed, a postdose sample of bodily fluid is analyzed by mass spectrometry. The sample is corrected for excretion, exchange with nonaqueous hydrogen or oxygen, and isotope fractionation. From the measurement of TBW, FFM can be estimated, which requires an assumed value for FFM hydration. The assumed constant is not consistent over age or health status (Fomon et al., 1982; Lohman, 1986).

With this method, TBW can be measured with an accuracy of approximately 1–2%. The advantages of isotope dilution are that it is accurate and requires minimal co-operation so it can be used in a range of ages. However, a disadvantage of this method is that the analysis is time consuming and the entire approach not suitable in abnormal hydration states if body composition is the final outcome required.

Bioelectrical impedance analysis

Bioelectrical impedance analysis (BIA) measures the impedance of the body tissues to the flow of a low level alternating current. The principle of BIA is that when a current is passed through the body it will only pass through the water and electrolyte containing tissues that have low impedance, not the body fat or bone which have poor conduction properties (Nyboer, 1959; Thomasset, 1962). Therefore as impedance is proportional to body water, we can determine the volume of TBW.

For whole-body BIA measurements, four electrodes are placed on the wrists and ankles and a tetra polar arrangement is utilized (Lukaski et al., 1985). Segmental measurements are also possible by altering electrode placement to the specific segment required. A current at 50 kHz is passed through one set of electrodes, while the voltage drop is measured, and impedance derived, by the other set of electrodes. There are two assumptions involved in this measurement to determine body volume; the body is a collection of cylinders with their length proportional to their height, and the reactance contributing to impedance is small, with the resistance considered equivalent to impedance.

The standard BIA method uses just one current at 50 kHz, but multifrequency analysis is also possible. By passing currents between 10 kHz and 1 MHz through the body it is possible to measure intra and extracellular fluid. At zero frequency the current can not pass through the cell membrane, so ECF can be determined and at infinite frequency both intra and extracellular fluid are penetrated so both can be examined. Measurements are not possible at zero (R_0) and infinite frequency, so the value at characteristic frequency (Z_c) is mathematically derived by fitting the shape of the reactance versus resistance curve (Cole and Cole, 1941) (Figure 7.3).

Considerations for all BIA measurements include room temperature, body position, lead placement, prior activity, and food intake. The benefits of this method are that it is portable, inexpensive, and simple. The problem of this method is that the assumptions are not entirely true, measurement error is likely in subjects with altered hydration levels or clinical conditions, and conversion to FFM is population specific.

Additional methods

Several methods which are less commonly used in body composition research because of the cost and availability are computed tomography (CT), magnetic resonance imaging (MRI), in vivo neutron activation analysis (IVAA), and total body electrical conductivity (TOBEC). Computed tomography and MRI are considered the most accurate methods available for quantification of total and regional adipose and skeletal muscle tissue. Computed tomography uses the relationship between differences in X-ray attenuation and differences in the physical density of tissues to construct a two-dimensional image and determine cross-sectional area of the fat, bone, muscle, and organs. Magnetic resonance imaging estimates the volume of fat tissue by analyzing the absorption and emission of energy in the radio-frequency range of the electromagnetic spectrum.

In vivo neutron activation analysis is used in body composition to quantify elements in the body including hydrogen, carbon, nitrogen, oxygen, calcium, and phosphorous. This method uses a neutron field to induce a nuclear reaction in the body's atoms dependent on the energy of the neutrons. Total body electrical conductivity can be used to measure TBW. This technique uses coils to generate an electromagnetic field, with an electrical current produced in the conductive tissues of the body and the difference between the coil impedance when empty and containing a body measured.

Another new method is the three-dimensional body scanner which can be used to measure body volume. The use of a digitized optical method and computer to generate a three-dimensional photonic image of an object was developed in the 1950s and was used as a technology for whole-body surface anthropometric measurements in humans (Hertzberg et al., 1957). The newly developed 3DPS system (Hamamatsu Photonics KK, Hamamatsu, Japan) collects a maximum of 2 048 000 data points in 10 s and generates values for total and regional body volumes and dimensions. There is much potential in this new method; however, complete validation studies are still required.

RELATIONSHIP BETWEEN ENERGY METABOLISM, BODY SIZE, AND COMPOSITION

These two biological parameters are intimately related and it is of extreme importance that the relationship is understood when undertaking studies in the field. Whilst it is almost self evident that there will be relationships between FFM and resting metabolic rate, for example, unless understood and accounted for comparisons of energy expenditure between species or within species when there are major discrepancies in body size or body composition will be confounded.

This relationship and potential problem was apparent to some of the earliest students of energy metabolism. An initial attempt to remove the confounding affect of size, shape, and body composition was proposed by Sarrus and Rameaux (1839), as the surface law, which was thought to enable comparisons of energy expenditure between different animals.

This law states that, when expressed relative to body surface area, energy expenditure (or metabolic rate as it was termed), was constant in adult homeotherms. The surface law became very popular and rapidly became entrenched in physiological doctrine. The theoretical bases of the law expounded at the time were summarized much later by Kleiber (1947). He stated that the metabolic rate of animals must be proportional to their body surface area. This statement was based upon the following observations:

1. The rate of heat transfer between animal and environment is proportional to the body surface area.
2. The intensity of flow of nutrients, in particular oxidizable material and oxygen, is a function of the sum of internal surfaces which in turn is proportional to the body surface.
3. The rate of supply of oxidizable material and oxygen to the tissues is a function of the mean intensity of the blood current, which is proportional to the area of the blood vessels, which in turn is proportional to the area of the body.
4. The composition of the animal is a function of their body size. The composition may be meant either anatomically; the larger the animal the lower is the ratio of the mass of metabolically active organs to the mass of metabolically inert organs; or the composition may be meant chemically; the larger the animal the lower its percentage of "active protoplasm."
5. The cells of the body have an inherent requirement of oxygen consumption per unit weight, which is smaller the larger the animal.

An example of the popularity of this law, at this time, was that when Mitchell et al. (1940) produced data from the rat that did not fit the law it was suggested that they had made fundamental errors in their calculation of body surface area. The possibility that the law itself was flawed did not seem an option!

Mathematically, bodies of similar shape have surface areas proportional to the squares of their linear dimensions. Similarly, their volumes are proportional to the cubes of their linear dimensions. So if density is constant then surface area is proportional to two-thirds power of body weight. In this way the surface area law came to be interpreted as metabolic rate expressed relative to body weight raised to the two-thirds power. This expression of energy expenditure relative to body weight was accepted, almost exclusively, and used throughout human and animal studies for 70 years.

The first three decades of the twentieth century saw the realization that while the surface law might be the most appropriate method of standardizing metabolic rate, surface area itself could not be defined sufficiently well. It was being suggested that another power function of weight be sought that might relate to metabolic size and in 1932 Kleiber suggested the three-quarters power as the best function of body weight to standardize metabolic rate in adult homeotherms (Kleiber, 1932).

Almost immediately Brody and colleagues (Brody and Proctor, 1932; Brody et al., 1932) put forward a more defined, and somewhat specific, power of 0.734 based on the analysis of data from a wide range of mammals. An official "seal of approval" was given to this power function by the US National Research Council in 1935, although the committee stated that whether the change from 0.75 to 0.734 was either biologically or statistically valid was uncertain.

Later Kleiber (1932) showed that metabolic rate was best expressed to body weight$^{0.75}$ in a group of mammals ranging from a mouse to cow, differing in weight by a factor of almost 30 000. Importantly, in this particular study, the best power was derived statistically without recourse to a physiological model.

Interestingly, it has been suggested that elastic criteria impose limits on biological proportions, and consequently on metabolic rates (McMahon, 1973). This paper shows elegantly that when one considers fundamental aspects of size and shape it can be shown that maximal power output in animals, at least, is proportional to (weight$^{3/8}$)^2 which is equivalent, of course, to weight$^{0.75}$.

Appropriate methods of expressing energy expenditure relative to body weight were still actively being sought more than 20 years later (Sinclair, 1971). The surface law was still a consideration although attitudes towards this model were now less intransigent, as shown by the fact that when the data produced by Sinclair for energy expenditure in neonates did not fit the model, the model was questioned and not, as previously, the data.

ADJUSTMENT RELATIVE TO BODY COMPOSITION

At the same time that body weight was being adjusted by using a power function, the concept that energy expenditure was best expressed relative to "active tissue mass" was being put forward. This concept was first voiced by two scientists previously mentioned, Benedict and Talbot in 1914 (Benedict and Talbot, 1914). The major problem of this method of expression, acknowledged by these workers, was that the calculation of active tissue mass has been defined in three different ways – as lean body mass, FFM, and cell mass.

Owen et al. (1987), determined that resting metabolic rate was best predicted in adult men when FFM was included in the regression equations, and others (Ravussin and Bogardus, 1989) have addressed in detail the expression of energy expenditure relative to FFM. These authors suggest that FFM should be used as the denominator in comparisons of energy expenditure between individuals. Indeed, they take the concept a step further. They suggest that because of a mathematical bias it is incorrect to express metabolic rate data per kilogram FFM. Also, that regression analysis should be used to take into account the effect of FFM upon total energy expenditure. This is a slightly different approach to using a power function or weight of FFM but nevertheless the same effect is achieved.

Here is the fundamental issue. Whilst it might seem intuitive that dividing energy expenditure by body weight or FFM "adjusts" for body weight or FFM, this is not necessarily the case.

Log–log regression

Expressing resting metabolic rate (RMR), in this example as kcal/kg FFM, is equivalent to the expression kcal/kg FMM[1]. The power here of 1, is obviously not the power function to effect an appropriate adjustment. It might be tempting to try an adjustment such as kcal/kg FFM$^{0.75}$ as suggested by Kleiber (1947) for body weight. However, it is not necessary to guess or assume an appropriate power function as it can be simply calculated using log–log regression. The simplest approach, which has an obvious connection with linear regression, is to consider the correlation between the logarithm of the index RMR/FFM^P and the logarithm of FFM. The log index can be rearranged as follows:

$$\log (\text{RMR}/\text{FFM}^P) = \log (\text{RMR}) - p/\log (\text{FFM}),$$

which shows why logarithms are useful in this case. They allow the index to be expressed as a linear function of log (RMR) and log (FFM), which is then suitable for analysis by linear regression. If the log index is to be uncorrelated with log (FFM) then p must be chosen to remove all the log (FFM) information from log (RMR). This is equivalent to saying that p should be the slope of the regression line relating log (RMR) to log (FFM). Natural logarithms, to the base e, have some advantages over base 10 logarithms.

In reality, the value of p is unlikely to be a simple, round number such as 0.75 say, and the value, like any other regression coefficient will have a standard error. Thus, often in these circumstances a value of p is chosen that is numerically convenient and statistically within the confidence interval for p.

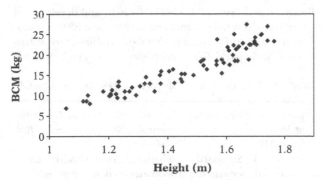

7.4. Relationship between body cell mass (BCM) and height.

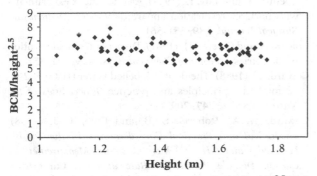

7.5. Relationship between body cell mass (BCM)/height$^{2.5}$ and height.

An example of this approach is shown in Figures 7.4 and 7.5. The first figure shows the relationship between BCM and height in a group of 73 healthy females between the ages of 6 and 17 years. Clearly the two are related with a correlation of 0.94. Following log–log regression of these data, the value p was determined to be 2.35 and a statistically valid and more convenient power of 2.50 was chosen. In Figure 7.5, the relationship shown is that between BCM divided by height raised to the power 2.50. Again clearly there is now no relationship and so expressing BCM in this way removes the influence of height.

A number of other studies have also found that a square root function normalizes or adjusts a number of different physiological variables. Total energy expenditure, as measured using the doubly labeled water technique, was best expressed as kcal/kg body weight$^{0.5}$ in a cohort of infants aged between 6 and 26 weeks (Davies et al., 1989). This adjustment was also appropriate for measurements of sleeping metabolic rate in infants at 12 weeks of age. Interestingly the value of 0.5 was statistically appropriate when sleeping metabolic rate was expressed as kcal/kg body weight$^{0.5}$ and as kcal/kg FFM$^{0.5}$.

Further support for using kg$^{0.5}$ as an adjustment for body weight has been provided by a study on adult Gambian women (Lawrence, 1988). In this study there was a significant negative correlation between energy expenditure per kg body weight and body weight during many activities including sitting and standing.

This correlation was estimated in nearly every case when energy expenditure was expressed per kg$^{0.5}$ body weight. Lawrence (1988) explained this by the fact that the percentage differences in body weight when raised to the power 0.5 are virtually the same as the percentage differences in energy expended during a particular task if half the total energy is expended in carrying out a fixed amount of external work. Consequently the energy expenditure when expressed per kg$^{0.5}$ body weight would be similar between subjects. This explanation was offered to account for observations of adult Gambian women. Basal metabolic rate has also been shown to be relatively constant when expressed on a per kg$^{0.5}$ body weight basis.

One should now pose the question why the weight power is close to 0.5 in humans, appreciably less than Kleiber's power of 0.75 in animals. Put statistically the correlation between body size and energy expenditure across species is very close to 1, whereas in humans it is much lower, about 0.7. For prediction purposes the slope of the regression line is equal to the ratio of the standard deviations of the y and x variables (i.e., logged energy expenditure and body weight) multiplied by the correlation, r, between them. So if the ratio of the standard deviations is 0.75 (i.e., that proposed by Kleiber) and the correlation is about 0.70, the best slope for prediction purposes will be close to 0.5.

Evidence to support the expression of energy expenditure relative to body weight$^{0.5}$ can also be found in the literature relating to animal husbandry. Millward and Garlick (1976) suggested that heat production (energy expenditure) raised to the power 0.56 might be a general physiological relationship. This speculation was based upon the findings that total energy expenditure in growing pigs was best expressed relative to body weight$^{0.56}$ (Kielanowski, 1969; Thorbeck, 1969) and also in rats (Walker and Garret, 1970). The same relationship had also been reported in pigs some 30 years previously (Brierem, 1939).

Adjusting metabolic variables for differences in body composition and body size is often necessary. But, inappropriate adjustment can distort the picture and make interpretation difficult, or worse, lead to inaccurate conclusions. Whilst the simplicity of dividing metabolic variables by body weight or FFM, is appealing, it is clear that in most cases it is inappropriate.

DISCUSSION POINTS

1. Consider the relative advantages and disadvantages of indirect and direct calorimetry to assess energy metabolism, in infants and in children.
2. How would you design a study to evaluate the ability of a new body composition method to assess percentage body fat?

3. Describe three circumstances where an inappropriate expression of energy expenditure relative to body composition could lead to inappropriate conclusions.

4. Write a paragraph that explains the doubly labeled water method that would be suitable for inclusion in an information package for parents of children to be studied.

5. What are the factors that may cause errors in determining fat free mass (FFM) from bioelectrical impedance analysis (BIA)?

6. List the disadvantages of a method based on the two-compartment (2C) model of body composition?

7. What adjustments need to be made to body volume when measuring a child in the Bod Pod®?

8. Discuss the use of magnetic resonance imaging (MRI) as a method for determining body composition.

REFERENCES

Allen, T. H., Anderson, E. C. and Langham, W. H. (1960). Total body potassium and gross body composition in relation to age. *Journal of Gerontology*, **15**, 348–357.

Anderson, E. C. and Langham, W. H. (1959). Average potassium concentration of the human body as a function of age. *Science*, **130**, 713–714.

Anderson, J., Osborn, S. B., Tomlinson, R. W. S., et al. (1964). Neutron-activation analysis in man in vivo: a new technique in medical investigation. *Lancet*, **ii**, 1201–1205.

Atwater, W. O. and Benedict, F. G. (1903). *Experiments on the Metabolism of Matter and Energy in the Human Body, Agricultural Bulletin Number 136*. Washington, DC: US Department of Agriculture.

Behnke, A. R., Osserman, E. F. and Welham, W. C. (1933). Lean body mass. *Archives of Internal Medicine*, **91**, 585.

Behnke, A. R., Feen, B. G. and Welham, W. C. (1942). The specific gravity of healthy men. *JAMA: Journal of the American Medical Association*, **118**, 495–498.

Benedict, F. G. (1914). The basal metabolism of boys from 1 to 13 years of age. *Proceedings of the National Academy of Sciences of the United States of America*, 7–10.

Benedict, F. G. (1919). Energy requirements of children from birth to puberty. *Boston Medical and Surgical Journal*, **181**, 107–139.

Benedict, F. G. and Talbot, F. B. (1914). *The Gaseous Metabolism of Infants with Special Reference to Its Relation to Pulse Rate and Muscular Activity*. Washington, DC: Carnegie Institute, p. 168.

Bosy-Westphal, A., Geisler, C., Onur, S., et al. (2006). Value of body fat mass versus anthropometric obesity indices in the assessment of metabolic risk factors. *International Journal of Obesity*, **30**, 475–483.

Brierem, K. (1939). Der Energieumsatz bei den Schweinen. *Tierernährung*, **11**, 487–528.

Brody, S. and Proctor, T. (1932). Growth and development with special reference to domestic animals: further investigations of surface area in metabolism. *University of Missouri Agricultural Experiment Station, Research Bulletin*, **116**.

Brody, S., Proctor, T. and Ashworth, J. (1932). Basal metabolism, endogenous nitrogen, creatinine and neutral sulphur excretions as functions of body weight. *University of Missouri Agricultural Experiment Station, Research Bulletin*, **220**.

Burden, S. T., Stoppard, E., Shaffer, J., et al. (2005). Can we use mid upper arm anthropometry to detect malnutrition in medical inpatients? A validation study. *Journal of Human Nutrition and Dietetics*, **18**, 287–294.

Carpenter, T. and Fox, E. (1923). Alcohol check experiments with portable respiration apparatus. *Boston Medical and Surgical Journal*, **189**, 551–561.

Cole, K. S. and Cole, R. H. (1941). Dispersion and absorption in dielectrics. *Journal of Chemical Physics*, **9**, 341–351.

Coward, W. (1988). The doubly-labelled water (H_2O)-H-2-O-18 method – principles and practice. *Proceedings of the Nutrition Society*, **47**, 209–218.

Coward, W. A., Roberts, S. B. and Cole, T. J. (1988). *Theoretical and Practical Considerations in the Doubly Labelled Water ($^2H_2{}^{18}O$) Method for the Measurement of Carbon Dioxide Production Rate in Man*. Cambridge: Medical Research Council, Dunn Nutrition Laboratory, pp. 1–12.

Daniel, J. A., Sizer, P. S. J. and Latman, N. S. (2005). Evaluation of body composition methods for accuracy. *Biomedical Instrumentation and Technology*, **39**, 397–405.

Davies, P. S., Cole, T. J. and Lucas, A. (1989). Adjusting energy expenditure for body weight in early infancy. *European Journal of Clinical Nutrition*, **43**, 641–645.

Dempster, P. and Aitkens, S. (1995). A new air displacement method for the determination of human body composition. *Medicine and Science in Sports and Exercise*, **27**, 1692–1697.

Douglas, C. G. (1911). A method for determining the total respiratory exchange in man. *Journal of Physiology*, **42**, xvii–xviii.

Duffield, R., Dawson, B., Pinnington, H., et al. (2004). Accuracy and reliability of a Cosmed K4b(2) portable gas analysis system. *Journal of Science and Medicine in Sport*, **7**, 11–22.

Durnin, J. V. and Womersley, J. (1974). Body fat assessed from total body density and its estimation from skinfold thickness: measurements on 481 men and women aged from 16 to 72 years. *British Journal of Nutrition*, **32**, 77–97.

Edelman, I. S. (1952). Body composition: studies in the human being by the dilution principle. *Science*, **115**, 447–454.

Edmonds, C. J., Jasani, B. M. and Smith, T. (1975). Total body potassium and body fat estimation in relationship to height, sex, age, malnutrition and obesity. *Clinical Science and Molecular Medicine*, **48**, 431–440.

Eisenmann, J. C., Heelan, K. A. and Welk, G. J. (2004). Assessing body composition among 3- to 8-year-old children: anthropometry, BIA, and DXA. *Obesity Research*, **12**, 1633–1640.

Ellis, K. J. (2001). Selected body composition methods can be used in field studies. *Journal of Nutrition*, **131**, S1589–S1595.

Eto, C., Komiya, S., Nakao, T., et al. (2004). Validity of the body mass index and fat mass index as an indicator of obesity in children aged 3–5 years. *Journal of Physiological Anthropology and Applied Human Science*, **23**, 25–30.

Fomon, S. J., Haschke, F., Ziegler, E. E., et al. (1982). Body composition of reference children from birth to age 10 years. *American Journal of Clinical Nutrition*, **35**, 1169–1175.

Forbes, G. B., Gallup, J. and Hursh, J. B. (1961). Estimation of total body fat from potassium-40 content. *Science*, **133**, 101–102.

Foster, M. A., Hutchison, J. M. S., Mallard, J. R., et al. (1984). Nuclear magnetic resonance pulse sequence and discrimination of high- and low-fat tissue. *Magnetic Resonance Imaging*, **2**, 187–192.

Givens, M. H. and Macy, I. G. (1933). The chemical composition of the human fetus. *Journal of Biological Chemistry*, **102**, 7–17.

Hertzberg, H., Dupertuis, C. and Emanual, I. (1957). Sterophotogrammetry as an anthropometric tool. *Photogrammetric Engineering and Remote Sensing*, **23**, 942–947.

Heymsfield, S. B., Olafson, R. P., Kutner, M. N., et al. (1979). A radiographic method of quantifying protein-calorie undernourishment. *American Journal of Clinical Nutrition*, **32**, 693–702.

Hounsfield, G. N. (1973). Computerized transverse axial scanning (tomography). 1. Description of system. *British Journal of Radiology*, **46**, 1016–1022.

Iob, V. and Swanson, W. W. (1934). Mineral growth of the human fetus. *Archives of Disease in Childhood*, **47**, 302–306.

Janssen, I., Heymsfield, S. B., Allison, D. B., et al. (2002). Body mass index and waist circumference independently contribute to the prediction of nonabdominal, abdominal subcutaneous, and visceral fat. *American Journal of Clinical Nutrition*, **75**, 683–688.

Jebb, S. A., Goldberg, G. R. and Elia, M. (1993). DXA measurements of fat and bone mineral density in relation to depth and adiposity. *Basic Life Sciences*, **60**, 115–119.

Kielanowski, J. (1969). Variation in heat production in growing pigs: some observations on the relationship between feed intake and heat production in pigs fed barley and skim-milk. *Publications, European Association of Animal Production*, **12**, 297–289.

Kistorp, C. N., Toubro, S., Astrup, A., et al. (2000). Measurements of body composition by dual-energy X-ray absorptiometry improve prediction of energy expenditure. *Annals of the New York Academy of Sciences*, **904**, 79–84.

Kleiber, M. (1932). Body size and metabolism. *Hilgardia*, **6**, 315–353.

Kleiber, M. (1947). Body size and metabolic rate. *Physical Reviews*, **27**, 511–541.

Kofranyi, E. and Michaelis, H. F. (1940). Ein tragbarer Appart zur bestimmung des Gasstoff wechsels. *Arbeitphysiolologie*, **11**, 148–215.

Kohrt, W. M. (1998). Preliminary evidence that DEXA provides an accurate assessment of body composition. *American Physiological Society*, 372–377.

Kumar, R., Aggarwal, A. K. and Iyengar, S. D. (1996). Nutritional status of children: Validity of mid-upper arm circumference for screening undernutrition. *Indian Pediatrics*, **33**, 189–196.

Lawrence, M. (1988). Predicting energy requirements: is energy expenditure proportional to BMR or to body weight? *European Journal of Clinical Nutrition*, **42**, 919–927.

Littlewood, R., White, M. S., Bell, K. L., et al. (2002). Comparison of the Cosmed K4 b2 and the Deltatrac II™ Metabolic Chart in measuring resting energy expenditure in adults. *Clinical Nutrition*, **21**, 491–497.

Livingstone, M., Prentice, A., Coward, W., et al. (1990). Simultaneous measurement of free-living energy expenditure by the doubly labeled water method and heart-rate monitoring. *American Journal of Clinical Nutrition*, **52**, 59–65.

Livingstone, M. B., Coward, W. A., Prentice, A. M., et al. (1992). Daily energy expenditure in free-living children: comparison of heart-rate monitoring with the doubly labeled water (2H2(18)O) method. *American Journal of Clinical Nutrition*, **56**, 343–352.

Lohman, F. G. (1986). Applicability of body composition techniques and constants for children and youths. *Exercise and Sport Sciences Reviews*, **14**, 325–357.

Lukaski, H. C., Johnson, P. E., Bolonchuk, W. W., et al. (1985). Assessment of fat-free mass using bioelectrical impedance measurements of the human body. *American Journal of Clinical Nutrition*, **41**, 810–817.

Magnus-Levy, A. (1906). *Physiologie des Stoffwechesels. Handbuch der Pathologie des Stoffwechesels*. Berlin: C. von Noorden, p. 446.

McCrory, M. A., Gomez, T. D., Bernauer, E. M., et al. (1995). Evaluation of a new air displacement plethysmograph for measuring human body composition. *Medicine and Science in Sports and Exercise*, **27**, 1686–1691.

McMahon, T. (1973). Size and shape in biology. *Science*, **179**, 1201–1204.

Millward, D. J. and Garlick, P. J. (1976). The energy cost of growth. *Proceedings of the Nutrition Society*, **35**, 339–349.

Mitchell, H. H., Hamilton, T. S. and Haines, W. T. (1940). The utilization by calves of energy in rations containing different percentages of proteins and glucose supplements. *Journal of Agricultural Research*, **61**, 847–864.

Moore, F. D. (1946). Determination of total body water and solids with isotopes. *Science*, **104**, 157–160.

Moore, F. D., Olesen, K. H., McMurray, J. D., et al. (1963). *The Body Cell Mass and its Suppporting Environment*. Philadelphia: W. B. Saunders and Company.

Moulton, C. R. (1923). Age and chemical development in mammals. *Journal of Biological Chemistry*, **57**, 79–97.

Nunez, C., Kovera, A. J., Pietrobelli, A., et al. (1999). Body composition in children and adults by air displacement plethysmography. *European Journal of Clinical Nutrition*, **53**, 382–387.

Nyboer, J. (1959). *Electrical Impedance Plethysmography*. Springfield, IL: Charles C. Thomas.

Owen, O. E., Holup, J. L., D'Alessio, D. A., et al. (1987). A reappraisal of the calorific requirements of man. *American Journal of Clinical Nutrition*, **46**, 875–888.

Pace, N. and Rathburn, E. N. (1945). Studies on body composition. III. The body water and chemically combined nitrogen content in relation to fat content. *Journal of Biological Chemistry*, **158**, 685–691.

Peppler, W. W. and Mazess, R. B. (1981). Total body bone mineral and lean body mass by dual-photon absorptiometry. I. Theory and measurement procedure. *Calcified Tissue International*, **33**, 353–359.

Perret, C. and Mueller, G. (2006). Validation of a new portable ergospirometric device (Oxycon Mobile®) during exercise. *International Journal of Sports Medicine*, **27**, 363–367.

Pierson, R. Jr and Wang, J. (1988). Body composition denominators for measurement of metabolism: What measurements can be believed? *Mayo Clinic Proceedings*, **63**, 947–949.

Powell-Tuck, J. and Hennessy, E. M. (2003). A comparison of mid upper arm circumference, body mass index and weight loss as indices of undernutrition in acutely hospitalized patients. *Clinical Nutrition*, **22**, 307–312.

Quetelet, L. A. J. (1871). *Anthropometric ou mesure des differentes facultés de l'homme*. Brussels: C. Marquardt, p. 479.

Ravussin, E. and Bogardus, C. (1989). Relationship of genetics, age, and physical fitness to daily energy expenditure and fuel utilization. *American Journal of Clinical Nutrition*, **49**, 968–975.

Roberts, S. B., Coward, W. A., Schlingenseipen, K. H., et al. (1986). Comparison of the doubly labelled water method with indirect calorimetry and a nutrient balance study for simultaneous determination of energy expenditure, water intake and metabolizable energy intake in preterm infants. *American Journal of Clinical Nutrition*, **44**, 315–322.

Sarrus and Rameaux (1839). Application des sciences accessories et principalement des mathématiques à la physiologie géneralé. *Bulletin de l'academie royale de medicine*, **3**, 1094–1110.

Schoeller, D. A. and Van Santen, E. (1982). Measurement of energy expenditure in humans by doubly labelled water method. *Journal of Applied Physiology*, **53**, 955–959.

Schoeller, D., Ravussin, E., Schutz, Y., et al. (1986). Energy expenditure by doubly labeled water: validation in humans and proposed calculation. *American Journal of Physiology*, **250**, R823–R830.

Sinclair, J. C. (1971). Metabolic rate and body size of the newborn. *Clinical Obstetrics and Gynecology*, **14**, 840–854.

Singhal, A., Davies, P., Sahota, A., et al. (1993). Resting metabolic rate in sickle cell disease. *American Journal of Clinical Nutrition*, **57**, 32–35.

Siri, W. E. (1961). Body composition from fluid spaces and density: analysis of methods. In *Techniques for Measuring Body Composition*, J. Brozek and A. Henschel (eds). Washington, DC: National Academy of Sciences, National Research Council, pp. 223–243.

Stone, D. A. and Robinson, M. P. (2003). Total body water measurements using resonant cavity perturbation techniques. *Physics in Medicine and Biology*, **49**, 1773–1788.

Thomasset, A. (1962). Bioelectric properties of tissue. Impedance measurement in clinical medicine. Significance of curves obtained. *Lyon Médical*, **94**, 107–118.

Thomasset, A. (1963). Bio-electric properties of tissues. *Lyon Médical*, **209**, 1325–1350.

Thorbeck, G. (1969). Studies on the energy metabolism of growing pigs. *Publications, European Association of Animal Production*, **12**, 281–328.

Tothill, P., Reid, D. M., Avenell, A., et al. (1993). Comparisons between Hologic, Norland and Lunar dual-energy X-ray bone absorptiometers. *Basic Life Sciences*, **60**, 385–388.

Trocki, O., Theodoros, M. T. and Shepherd, R. W. (1998). Lack of sensitivity of weight targets compared with body cell mass for determining recovery from malnutrition in adolescents with anorexia nervosa. *International Journal of Eating Disorders*, **23**, 169–176.

Urlando, A., Dempster, P. and Aitkens, S. (2003). A new air displacement plethysmograph for the measurement of body composition in infants. *Pediatric Research*, **53**, 486–492.

Walker, J. J. and Garret, W. N. (1970). Shifts in the energy metabolism of male rats during their adaption to prolonged undernutrition and during their subsequent realimentation. *Publications, European Association of Animal Production*, **19**, 193–196.

Warner, J. T., Cowan, F. J., Dunstan, F. D., et al. (1997). The validity of body mass index for the assessment of adiposity in children with disease states. *Annals of Human Biology*, **24**, 209–215.

Weir, J. B. (1949). New methods for calculating metabolic rate with special reference to protein metabolism. *Journal of Physiology*, **109**, 1–9.

Wells, J. and Davies, P. (1995). The effect of diet and sex on sleeping metabolic rate in 12-week infants. *European Journal of Clinical Nutrition*, **49**, 329–335.

Wells, J. C. K., Coward, W., Cole, T., et al. (2002). The contribution of fat and fat-free tissue to body mass index in contemporary children and the reference child. *International Journal of Obesity*, **26**, 1323–1328.

Williams, J. E., Wells, J. C. K., Wilson, C. M., et al. (2005). Evaluation of Lunar prodigy dual energy X-ray absorptiometry for assessing body composition in healthy individuals and patients by comparison with the four-component model. *International Journal of Body Composition Research*, **3**, 83.

Zhu, S., Wang, Z., Heshka, S., et al. (2002). Waist circumference and obesity-associated risk factors among whites in the third National Health and Nutrition Examination Survey: Clinical action thresholds. *American Journal of Clinical Nutrition*, **76**, 743.

8 Evolutionary Endocrinology

Richard G. Bribiescas and Michael P. Muehlenbein

INTRODUCTION

Hormones do not fossilize. Yet, arguably, they are as important to understanding the evolution of *Homo sapiens* and other primates as any fossil specimen. The role of hormones in understanding human life history evolution emerges from how genes translate into phenotypes with considerable input from environmental cues. Most hormones are evolutionarily quite conservative, with very similar if not identical chemical structures between species. Many hormones that flow through the veins of humans are identical to those that flow through the most exotic vertebrate. Other hormones and receptors, however, can differ in subtle but important ways between species and even individuals. Hormonal variation, as reflected by circulating levels as well as chemical structure, are of central importance to the evolution of human life histories, both from a macro- and microevolutionary perspective.

The evolutionary significance of hormones is clearly evident in the multitude of functions that are served, including growth, reproduction, metabolism, and senescence, all of which are central to the evolution of human life histories. Hormones are inextricably involved in the optimal allocation of time and energy. Insulin, leptin, and cortisol, for example, initiate and manage the flow and assessment of energetic assets such as glucose and fat. Indeed, hormones are involved in life history trade-offs that influence many aspects of human health (Bribiescas and Ellison, 2008). Testosterone, estradiol, and oxytocin affect behavioral patterns that result in differences in how individuals allocate their time, such as in the trade-off between mate seeking and parenting. In essence, hormones are a common biological currency that humans and other primates share with other organisms. This allows biological anthropologists to assess the evolution of life history patterns in reference to a common physiological aspect, endocrinology (Bribiescas and Ellison, 2008).

Also important to the evolution of life histories is the contribution of hormones to the onset and timing of key life history events (Finch and Rose, 1995). Childhood growth, reproductive maturation, and reproductive senescence all result from changes in hormone production. The significance of some changes, such as the decline in estrogens during menopause, remains to be fully understood from an adaptive perspective; however, the impact of these changes on reproductive investment is unequivocal. In this chapter, we present an overview of how hormones contribute to important life history trade-offs, events, and characteristics in humans. In doing so, we introduce and describe various hormones that are illustrative of human life history evolution. The hormones discussed are not meant to represent an exhaustive list. Only a few representative hormones are discussed to illustrate the evolutionary significance of endocrine function in human life histories.

HOW AND WHAT IS MEASURED MAKES A DIFFERENCE

The amount of hormone that is produced is the most common mode of assessment in contemporary clinical and biological studies, and for good reason. Hormone levels provide useful insights into the physiology of an organism, such as the presence of illness. For example, Graves's disease is the overproduction of thyroid hormone, resulting in greater than expected metabolic rates and unpleasant symptoms such as bulging eyes. A low level of insulin is indicative of type II diabetes. Yet, absolute levels only provide a partial picture. Variation in hormone levels can result from three basic sources; production, clearance, and bioavailability, usually resulting from carrier protein binding. Production is the amount of hormone that is manufactured and secreted into the circulatory system. This is the most common form of variation. However, hormone levels can also be affected by clearance rates, or how fast the hormone is flushed from the body by the liver and kidneys. Finally, bioavailability, or whether the hormone is capable of activating receptors, can be

Human Evolutionary Biology, ed. Michael P. Muehlenbein. Published by Cambridge University Press. © Cambridge University Press 2010.

affected by what proportion of the hormone is bound to a carrier protein. For example, the vast proportion of sex steroid such as estradiol and testosterone in circulation is bound to a carrier protein such as sex hormone binding globulin (SHBG) (Griffin and Ojeda, 2004). Sex hormone binding globulin packages the steroid and allows it to be carried freely by the circulatory system until it is needed. The mechanism that frees the steroid from its carrier protein is not completely understood, but the amount of carrier protein can affect the availability, activity, and influence of a hormone in the body. For example, increases in SHBG probably account from some of the declines in testosterone in some older men (Gray et al., 1991).

Sample collection protocols, time of day, and fluid medium also influence subsequent measurements. Moreover, the rate and pattern of production can provide important information into endocrine physiology. For example, many steroid hormones are under the control of other hormones that are released in a pulsatile manner. The frequency and amplitude of secretion can also serve as a window into endocrine function. Older men for example exhibit changes in pulsatility patterns of luteinizing hormone (LH) and follicle-stimulating hormone (FSH) secretion with age (Takahashi et al., 2007). Luteinizing hormone stimulates the production of testosterone in men and estradiol in women while FSH is involved with gamete production in both men and women. Hormones in all these ways vary between human populations, sexes, and age classes. Through this variation, adaptive plasticity is revealed that indicates that environmental and lifestyle factors can alter hormone and receptor structure and function.

While hormones are commonly segregated into distinct categories, such as those related to growth and reproduction, it is important to note that most, if not all hormones, exhibit complex interactions and cross-talk that span across most physical functions. Organizational classifications used in this chapter are meant to serve as guideposts for discussion and may not necessarily reflect a biological reality in the strict sense. For example, testosterone and estradiol are commonly referred to as "male" and "female" hormones. While it is certainly true that testosterone is often found in greater quantities in male circulation, testosterone also serves important functions within females despite much lower levels. The same applies to estradiol in males.

This chapter serves to not only provide a brief overview of hormone physiology, but to also provide contextual background on how hormones aid in regulating the flow of somatic resources such as fat and glucose and therefore act as proximate mechanisms for adjusting life history trade-offs. Hormones will also be shown to be central to the evolution and maintenance of phenotypic plasticity.

HORMONE FORM, FUNCTION, AND ASSESSMENT

The standard definition of hormones states that they are chemical substances secreted by glands into the circulatory system, ultimately stimulating receptors on distant target tissues. This is mostly true although hormones can also enact local or paracrine actions close to the site of production as well as autocrine actions on the originating cells (Griffin and Ojeda, 2004). Assessment of hormones can be done with a variety of biological substrates and fluids, including blood, urine, feces, hair, saliva, cerebral spinal fluid (CSF), and tissue. Each medium provides unique challenges, limitations, and advantages, depending on the hormone, the research question, and the species. For example, blood is the most common diagnostic fluid in clinical settings while urine, saliva, and feces are most often used in field biology conditions. While blood provides access to most hormones, sample collection is invasive and optimally performed in clinical settings. Saliva is much less invasive but is most appropriate for steroid assessment although progress has been made with some protein hormones (Groschl et al., 2001, 2005). Urine and feces are often used under conditions in which sample collection is opportunistic and subject manipulation is not possible, as in the case of wild nonhuman primate studies (Muehlenbein, 2006). When assessing the range of variability and bioactivity of a specific hormone, it is important to be aware of the limitations and advantages of all of these sources of hormone data.

Blood is often the most direct method of assessing a hormone. However, blood measurements provide a snapshot of hormonal status and unless multiple draws are made, pulsatile and diurnal variation cannot be quantified. Salivary measurements allow for multiple collections over a relatively brief time period but are only practical in human populations and occasionally in nonhuman primates (Tiefenbacher et al., 2003). Moreover, salivary measurements can only observe steroids, and to a much more limited extent, some peptides. Urine assessments dampen pulsatile variability, which can be used to the researcher's advantage. However, assessments of urinary hormones still necessitate an awareness of diurnal variability (Anestis and Bribiescas, 2004).

Steroid hormones

Steroids are small, lipid-soluble molecules derived from cholesterol (Figure 8.1). Consequently, in their unbound form, they pass freely through cell membranes and affect genetic expression and transcription of various agents. Steroids are ancient molecules that are shared in all vertebrates (Norris, 2007). Indeed,

8.1. Steroid hormone synthesis pathways. From GNU Free Documentation License: http://en.wikipedia. org/wiki/File:Steroidogenesis.svg

steroids and their associated ligands may have been important for the evolution of vertebrate phenotypic complexity and share close similarities to steroids in plants and invertebrates, such as phytoestrogens (Thornton, 2001). The role of steroids encompasses reproductive function, metabolism, and behavior. Indeed the brain is rich with steroid receptors. For example, mineralocorticoid and glucocorticoid hormones are among the most ancient steroids with a deep evolutionary history (Baker et al., 2007).

Protein hormones

The second class of hormones consists of large, water-soluble molecules encoded and transcribed from specific genes that can exhibit a significant range of variation in their genetic structure and action (Nilsson et al., 1997; Timossi et al., 2000). Much of the structural variation of

a specific protein hormone both between individuals and populations is not completely understood and often rare, depending on the type of hormone. Protein structure variation may reflect microevolutionary processes that may have favored a particular protein hormone phenotype or limited its range of variation due to strong selection pressure. For example, FSH is highly conserved and is not known to exhibit any variation that affects hormone levels (Lamminen et al., 2005). However variation in upstream regulatory regions of the beta subunit of FSH (SNP, rs10835638; G/T) does affect serum levels (Grigorova et al., 2008). Luteinizing hormone on the other hand exhibits a "wild" and "variant" type that is found in many populations. The "variant" type is less common and is most often exhibited among Australian Aboriginal groups and may be associated with subfertility (Nilsson et al., 1997; Lamminen and Huhtaniemi, 2001).

Also worth mentioning are enzymes that regulate synthesis pathways of steroid hormones. While not hormones themselves, these enzymes control the conversion of cholesterol to a specific steroid, often being constrained to produce one specific steroid hormone before continuing on to its final end product (Figure 8.1). One such enzyme is aromatase which converts testosterone into estradiol. Extreme disruptions of enzyme structure results in disorders such as congenital adrenal hyperplasia, a condition in which the enzymatic pathway to the production of cortisol is disrupted resulting in the erroneous production of an androgen with testosterone-like properties of phenotypic masculinization. Nonpathological variation in enzymatic structure is evident between human populations. However, further research is necessary to determine the fitness implications of population and individual variation (Miller, 2002; Jasienska et al., 2006).

Another set of proteins act as carrier agents for hormones, allowing for efficient dispersal away from the site of production. Structural variations in transporter or binding proteins have become more evident. Variation in thyroid transporting proteins as well as SHBG, for example, have illustrated the potential importance of these agents in hormone activity (van der Deure et al., 2007; Riancho et al., 2008). As an example, the Asp327Asn polymorphism contributes to higher SHBG and testosterone levels among young, middle-aged, and older men (Vanbillemont et al., 2009). Similarly, SHBG polymorphisms were associated with serum levels in women with the AA genotype at the rs1799941 locus exhibiting the highest SHBG levels (Riancho et al., 2008). It is unclear how significant these polymorphisms are to contributing to between individual variation in binding protein levels; however, their potential effects have yet to be fully explored.

Hormone receptors

Both steroid and protein hormones enact their influence by binding to specific proteins on target cells. Indeed the evolution of receptors appears to have been a crucial aspect in the emergence of multicellular organisms (Whitfield et al., 1999). As with protein hormones, receptors are genetically encoded and subject to structural variation. Receptors are found on the surface and within target cells. The density, specificity, and binding capacity of receptors vary depending on the type of tissue, the hormone, genetic variation, and hormonal milleau. For example, exposure to high levels of leptin, a hormone that is secreted by fat cells, results in an increase in receptor resistance (Sahu, 2003). Such variation may reflect an important aspect of molecular phenotypic plasticity in which hormone influence is regulated within the context of environmental conditions.

Genetic variation is evident in hormone receptors and is related to detrimental effects on fertility. For example, point mutations of FSH receptors found in Scandinavian populations are associated with subfertility and sometimes infertility in women, although the effects on male fertility appear to be less severe (Tapanainen et al., 1997, 1998). Luteinizing hormone mutations can also detrimentally affect fertility or cause early or precocious puberty (Latronico and Segaloff, 1999).

ENERGY MANAGEMENT

Energy is often limited in many organisms and must therefore be allocated efficiently between competing needs such as growth, maintenance, and reproduction (Stearns, 1992; Ellison, 2003). Hormones are intricately involved in the regulation of energetic resources. They both sense and reflect the availability of energy substrates such as glucose and fat, regulating the flow of energetic assets between the needs of growth, maintenance, and reproduction. The following descriptions are not meant to be an all encompassing list, but rather a brief overview of the major hormone functions that have received significant attention from human evolutionary biologists.

Growth and organization

Growth is the embodiment of harvested energy from the environment. The amount of energy necessary to create tissue is a function of the amount of mass and the rate at which that mass is created. From these two processes, we can define the size of an organism and the pattern in which that tissue is created, otherwise known as growth rate. Other important factors include the rate of energetic usage by tissue or basal metabolic rate and the type of tissue being formed and supported. For example, muscle and brain tissue are much more metabolically taxing than adipose tissue. Hormones are instrumental for these processes as they influence cellular replication, differentiation, and enlargement as well as the amount of energy to be allocated. For more complete discussions of variability in human growth and the evolution of rates and patterns of human growth, see Chapters 22 and 23 of this volume. Excellent reviews of the endocrinology of growth and development are provided by Cohen and Rosenfeld (2004), Grumbach and Styne (1998), and Reiter and Rosenfeld (1998).

Within the human life cycle, the impacts of hormones on growth and development begin in utero. Glucose and subsequent increases in insulin, growth hormone (GH), and insulin-like growth factor (IGF) are crucial for overall skeletal growth and fat deposition. Müllerian-inhibiting factor, testosterone, and dihydrotestosterone promote sex-specific defeminization and masculinization of the genitalia in male and female fetuses (Grumbach and Conte, 1998). Circulating gonadotropin (LH and FSH) levels in the fetus peak in the second trimester of development, corresponding with maximal follicle development in females (Faiman et al., 1976). In males, testosterone levels rise during mid-gestation and then fall prior to birth (Siiteri and Wilson, 1974). Testosterone levels in males are also likely responsible for differences in energetic demands on the mother as well as muscle and fat deposition compared to female fetuses (Tamimi et al., 2003). Interestingly, it is becoming increasingly evident that energetic status can alter hormonal milieu in utero and potentially influence energetic management or even disease during adulthood (Kuzawa and Adair, 2003; Kuzawa, 2005; Barker et al., 2008).

Parturition is accompanied by decreased estrogen and progesterone and increased gonadotropin levels in infants that persist for the first few months after birth. Fetal thyroid hormone levels also surge at birth, which may facilitate new thermoregulatory requirements. In male infants, there is a second rise in testosterone level that falls again prior to the first year of age (Forest et al., 1974). Like the neonatal surge, the functions of temporary elevations in androgen levels (beyond masculinization of genitalia) are incompletely understood, although they may play important roles in sexual differentiation of the central nervous system (Wilson, 1982) as well as priming of androgen target tissues (De Moor et al., 1973; Davies and Norman, 2002). Nutritional factors in addition to injury and illness (i.e., immune activation) during development may also play important roles in "programming" baseline testosterone secretion for later adulthood (Bribiescas, 2001; Muehlenbein, 2008). Hormonal priming is discussed in more detail in Chapter 21 of this volume.

Childhood growth is marked by a steady increase in body mass, particularly from bone growth. The hypothalamic-pituitary axis is very sensitive to low levels of steroids and thus keeps gonadotropin levels low throughout childhood (Kaplan et al., 1976; Grumbach and Styne, 1998). Adrenarche, around six to eight years of age, marks the onset of adrenal androgen secretion, specifically androstenedione, dehydroepiandrosterone, and dehydroepiandrosterone sulfate (DHEAS) following stimulation by adrenocorticotropin hormone (Odell and Parker, 1985; Parker and Rainey, 2004). Although the roles of adrenal androgens in the onset of puberty are unknown (Parker, 1991), it has been suggested that DHEAS produced during adrenarche may play an important role in human brain maturation, and thus cognitive development (Campbell, 2006). These androgens may also be involved with decreasing hypothalamic sensitivity to the negative feedback effect of circulating sex steroids, thus contributing to the subsequent pubertal growth spurt (Havelock et al., 2004; Campbell, 2006).

Comparative studies show that chimpanzees also exhibit adrenarche while other primates do not (Nadler et al., 1984; Muehlenbein et al., 2001). Genetic sequence comparisons between humans, chimpanzees, rhesus macaques, and baboons of the enzyme P450c17, which is responsible for the conversion of pregnenolone to DHEA, revealed different patterns of DHEA production with very little genetic variation in the P450c17 gene, illustrating the potentially conservative evolutionary nature of adrenarche coupled with complex endocrine regulatory mechanisms that await further description (Arlt et al., 2002).

During childhood, energy is devoted to increasing tissue investment that will increase survivorship. During human evolution, this was made possible only through parental or allocare since child foraging returns tend to be quite low (Hill and Hurtado, 1996; Hewlett and Lamb, 2005). Growth rates decline rapidly during infancy but remain positive and steady throughout childhood until the early stages of puberty when rates of growth of both bone and sexually dimorphic tissue rise in dramatic fashion.

At puberty, the hypothalamus becomes much less sensitive to circulating steroid levels (Plant et al., 1989; Ojeda, 2004b) and begins producing more gonadotropin-releasing hormone (GnRH) in short pulsatile bursts (usually sleep-related) from the arcuate nucleus of the medial basal hypothalamus (King et al., 1985; Wu et al., 1996). Luteinizing hormone and FSH are then released in a pulsatile manner from the anterior pituitary. Leptin, a lipostatic hormone produced by adipose tissue, in conjunction with other growth factors, could also contribute to hypothalamic maturation (Yu et al., 1997).

The frequency and amplitude of gonadotropin pulses increase throughout sexual maturation, causing enhanced steroid secretion from the gonads (see Grumbach and Styne, 1998 for review). Androgens (particularly the conversion of testosterone to dihydrotestosterone) control hair, vocal cord and genitalia development, fat catabolism, and skeletal muscle anabolism in boys. Estrogens from the ovaries control bi-iliac growth, breast development, and fat redistribution in girls, and androgens from the adrenal cortex and ovaries control growth of pubic and axillary hair (Grumbach and Styne, 1998). Interestingly, hormones associated with adiposity, such as leptin, exhibit inverse responses to puberty in males and females, with leptin increasing in females and declining in males (Garcia-Mayor et al., 1997). Within the context of other mammals, this may indicate differential investment in reproductive effort, with adiposity being important to childbearing and survivorship while muscle growth provides males with competitive advantages (Bribiescas, 2001, 2006a).

As is evident, a number of hormones are responsible for growth and differentiation. Growth hormone (GH) is released in a pulsatile pattern from the anterior pituitary gland following stimulation by growth hormone-releasing hormone from the hypothalamus as well as thyroid hormones (Reiter and Rosenfeld, 1998). Somatostatin (somatotropin release-inhibiting factor) inhibits the release of GH. Growth hormone stimulates tissue and skeletal growth primarily by increasing insulin-like growth factors (IGFs) I and II and their variants (IGFs, somatomedins), particularly in the liver and bone (Reiter and Rosenfeld, 1998; Cohen and Rosenfeld, 2004). Gonadal steroids also trigger GH and IGF synthesis and secretion (Attie et al., 1990; Rogol, 1994). The IGF-binding proteins play a number of important roles, including inhibition of IGF actions (Reiter and Rosenfeld, 1998).

Androgens are largely responsible for muscle growth (Herbst and Bhasin, 2004) and development of the hematopoietic system (Jepson et al., 1973). Glucocorticoids increase circulating glucose, fatty acids, and amino acid levels (Parker and Rainey, 2004). Cortisol is also important for lung and intestinal maturation (Ballard, 1979). Other hormones that regulate metabolism, like insulin, glucagon, leptin, and ghrelin (Miers and Barrett, 1998; Dobbins et al., 2004; Klok et al., 2007), will also play important indirect roles in growth and development.

Estrogens (particularly the aromatization of testosterone to estradiol in boys) are important for skeletal development, including epiphyseal fusion (Matkovic, 1996; Juul, 2001). Thyroid hormones (T_3, tri-iodothyronine; T_4, thyroxine) stimulate protein synthesis and lipolysis and are necessary for tissue development (Steinacker et al., 2005). Thyroid hormones are also crucial for epiphyseal growth (Shao et al., 2006), and skeletal tissue modeling and remodeling are largely under the control of parathyroid hormone which alters calcium homeostasis (Hruska et al., 1991; Griffin, 2004b). Vitamin D increases calcium absorption, with estradiol improving calcium retention and preventing bone resorption (Kenny and Raisz, 2002; Heller, 2004). Additional growth-regulating peptides include, among others, the fibroblast and epidermal growth factors (Reiter and Rosenfeld, 1998; Cohen and Rosenfeld, 2004).

Female reproductive endocrinology

Female reproductive functions are also under complex control by the endocrine system. For more detailed discussion of ovarian function, pregnancy, lactation, and menopause, see Chapters 19 and 20 of this volume. Excellent reviews of the endocrinology of female reproductive function are provided by Baird (1984), Carr (1998), Carr and Rehman (2004), Casey and MacDonald (1998), Knobil et al. (1988), Ojeda (2004b), and Wood (1994).

Similar to the neuroendocrine control found in human males, GnRH is released in short pulsatile bursts from the arcuate nucleus ("pulse generator") of the medial basal hypothalamus (Reichlin, 1998). Gonadotropin-releasing hormone stimulates LH and FSH release from the gonadotrophs of the anterior pituitary gland (adenohypophysis) (Ojeda, 2004a). Thecal interstitial cells of a woman's follicles secrete androgens in response to LH (McNatty et al., 1979). Granulosa cells of the ovaries support follicular development in response to FSH, as well as convert androgens to estrogens (McNatty et al., 1979). Some androgens are produced from the adrenal glands, and most of these androgens are aromatized into estrogens in adipose tissue (Carr, 1998). The major estrogens include estradiol-17β, estrone, and estriol. Estrogens are largely responsible for the development of female secondary sexual characteristics, endometrial growth, and ductal development in the breast (Wood, 1994; Carr and Rehman, 2004; Ojeda, 2004b).

The ovulatory (menstrual) cycle of a woman is approximately 28 days long, and divided into four distinct phases: menstruation, follicular phase, ovulation, and luteal phase. In the follicular phase, a dominant follicle develops and inhibits the development of adjacent follicles (Zeleznik, 2004). Luteinizing hormone, estradiol, and progesterone levels rise throughout this proliferation phase of the endometrium (Wood, 1994; Carr, 1998). Just prior to ovulation, estradiol, progesterone, prostaglandin, LH, and FSH levels surge followed by release of the ovum from the follicle. Following ovulation (which takes place usually around day 14), the luteal phase begins with formation of the corpus luteum from follicular cells, and is accompanied by a drop in estradiol and gonadotropin levels (Wood, 1994; Carr, 1998). The corpus luteum produces large amounts of progesterone in order to support zygote implantation and maintenance of the endometrium and myometrium, mucosal development, and glandular development in the breast (Carr, 1998; Casey and MacDonald, 1998). Estradiol reaches a secondary peak in the mid-luteal phase, corresponding with a rise in basal body temperature.

Secondary (nondominant follicles) undergo atresia and apoptosis due primarily to activation of proapoptotic factors and reduced estrogen, FSH, and progesterone levels and increased androgen and prolactin levels in the follicular fluid (Rolaki et al., 2005; Craig et al., 2007). Luteolysis, or degeneration of the corpus luteum, takes place in the absence of fertilization with subsequent declines in progesterone and estradiol and increase in prostaglandin levels (Niswender et al.,

2000). Vascular changes and menstruation ensue. A decrease in inhibin B (which normally inhibits FSH release from the anterior pituitary) and a rise in FSH levels initiate follicular development for the next cycle (Groome et al., 1996).

In the event of fertilization, human chorionic gonadotropin (hCG) is released by the invading embryo (and later by the placenta) in order to maintain the corpus luteum, ensuring continued progesterone secretion as well as fetal gonadal development (Licht et al., 2001). The placenta also releases estriol in large quantities that stimulate development of the myometrium (Conley and Mason, 1990). Progesterone is released from the placenta, inhibiting smooth muscle contraction of the uterine myometrium by inhibiting prostaglandin formation (Sfakianaki and Norwitz, 2006). Progesterone also likely inhibits maternal immune reactions against the fetus (Thongngarm et al., 2003). The placenta, ovaries, and corpus luteum all produce relaxin which induces cervical remodeling to accommodate the pregnancy and later parturition (Sherwood, 2004).

Prolactin (PRL) from the anterior pituitary gland and human placental lactogen (hPL, chorionic somatomammotropin) from the placenta further develop the duct system and tissue of the mammary gland and stimulate milk synthesis (Neville et al., 2002). Human placental lactogen is also involved in maternal metabolic changes such as elevated glucose levels and increased insulin resistance which may lead to gestational diabetes (Grumbach et al., 1968). Other factors of fetal/placental origin (particularly inhibin A) may be responsible for maternal vascular changes, including pre-eclampsia (Rodgers et al., 1988; Bersinger et al., 2002). Such fetal manipulation of maternal resources may be viewed as the outcome of parent–offspring conflict in which the genetic interests of offspring and mothers are not identical (Trivers, 1974; Haig, 1993).

At parturition, maternal corticotrophin-releasing hormone (CRH) levels increase dramatically, possibly under direct fetal control (McLean and Smith, 2001; Snegovskikh et al., 2006). Corticotrophin-releasing hormone stimulates prostaglandin production which initiates labor. Androgens produced by the fetal adrenals are converted into estrogens, and a high ratio of estrogens to progesterone likely contributes to the onset of labor and delivery (Challis et al., 2000). Oxytocin rises both before and after parturition, with the former stimulating muscle contractions associated with labor, and the later stimulating uterine blood vessel coagulation following expulsion of the placenta (Wood, 1994; Blanks and Thornton, 2003).

Prolactin is necessary for milk production, and dopamine acts antagonistically to this purpose (Buhimschi, 2004). Oxytocin, released from the posterior pituitary gland, is responsible for milk ejection and delivery to the infant. Suckling stimuli from the infant maintain elevated levels of prolactin and oxytocin as well as trigger release of β-endorphin. Prolactin acts directly on the ovaries to produce a contraceptive effect (McNeilly et al., 1982), and β-endorphin suppresses pulsatile GnRH release from the hypothalamus (Franceschini et al., 1989). The combined effects contribute to lactational infecundability, although maternal energetic status can attenuate lactational amenorrhea. For example, Toba women of Argentina resume menstrual cycling in response to rising C-peptide (insulin) levels despite intense nursing (Ellison and Valeggia, 2003).

Cyclic ovarian function and menstruation cease in menopause due to a loss of follicles (for review, see Sievert, 2006). Responsiveness of the ovaries to gonadotropins decreases with elevated levels of LH and FSH accompanied by low levels of estradiol, androgens, inhibin B, and progesterone (Sherman et al., 1976; Metcalf et al., 1982). Dysregulation of endocrine activity can further produce vascular dilation which may lead to "hot flashes" (Meldrum, 1983).

Male reproductive endocrinology

More specific aspects of male reproductive endocrinology are covered in greater detail in Chapter 21 of this volume. However, a brief summary is presented within the context of how human male reproductive endocrinology has been shaped by natural selection and energetic constraints. Male hormones such as testosterone have both organizational and activational effects on males (Griffin, 2004a). In utero, testosterone and other hormones are responsible for internal and external genital development and organization as well as differences in somatic composition and brain development (Grumbach and Conte, 1998; Tamimi et al., 2003; Knickmeyer and Baron-Cohen, 2006). After a period of childhood quiescence, hypothalamic sensitivity to circulating testosterone dampens, allowing circulating levels to rise and promote the onset of puberty. During adulthood, downstream effects of GnRH and gonadotropins stimulate the production of sperm and sex hormones, specifically testosterone and, to a lesser extent, estradiol. Follicle-stimulating hormone promotes spermatogenesis while inhibin exerts negative feedback on FSH. Luteinizing hormone induces the production of testosterone and, to a lesser extent, estradiol. This system is common among most vertebrates and thus reflects common comparative selective pressures shared by humans (Norris, 2007).

As with other mammals, the primary factors affecting the evolution of male fitness and reproductive endocrinology is access to females as well as paternity uncertainty (Bribiescas, 2001). It is therefore not surprising that with the low metabolic costs associated with spermatogenesis, FSH in males is relatively insensitive to energetic stresses (Klibanski et al., 1981; Bergendahl and Huhtaniemi, 1993). Moreover, variation in FSH levels within the common range of variation is not associated with differences in spermatogenesis. Indeed, spermatogenesis is tolerant of a broad range of FSH exposure (Kumar et al., 1997; Tapanainen et al., 1997). Coupled with the modest association of spermatogenesis with variation in male fertility (Guzick et al., 2001), it is therefore not surprising that hormonal responses to energetic stressors are modest.

As with most other organisms with internal fertilization, high investment in mate access (libido) and tissue that augments competition and female attractiveness is supported by male hormones. Absence or severe suppression of testosterone in particular, can dampen libido and somatic investment (Sinha-Hikim et al., 2002; Gray et al., 2005). However, human males exhibit a broad range of variation between individuals and populations that is poorly understood. Between-population variation may involve an adaptive response to minimize the metabolic costs of testosterone-induced anabolism in the face of chronic caloric deficiencies (Bribiescas, 1996, 2001) and/or pathogen stress (Muehlenbein, 2008). Maintaining low testosterone levels in resource-limited and/or high pathogen-risk environments may avoid some immunosuppression and suspend energetically expensive anabolic functions. Augmenting testosterone levels in the presence of fertile and receptive mates, areas of high food resource availability, and low disease risk habitats will function to maximize lifetime reproductive success (Muehlenbein, 2008). Between-individual variation in testosterone level is also sensitive to a variety of factors, including marital status, fatherhood, and age. Married and pair-bonded men as well as fathers exhibit lower testosterone, perhaps as an indication of greater offspring and mate investment, at the expense of mate seeking (see Chapter 16 of this volume).

Insulin and energy sequestration

Insulin is a protein hormone that is secreted tonically from the pancreas. It is a member of a class of hormones that stimulate growth and regulate cellular glucose uptake (Nussey and Whitehead, 2001). In essence, insulin is an energy sequestering hormone, mopping up glucose in circulation and making it available for cellular needs. A lack of insulin results in type II diabetes, while insensitivity to insulin reflects the type I form. Insulin is most often measured in blood; however, its metabolite, C-peptide, is readily measured in blood and urine, making it a useful proxy for insulin assessment under remote field conditions in humans and nonhuman primates (Meistas et al., 1981; Sherry and Ellison, 2007).

Variation in insulin sensitivity between human populations is widely reported, especially among communities that exhibit unusually high rates of diabetes and obesity, such as Pima Amerindians and Samoans (Zimmet et al., 1996; Hanson et al., 2001). Insulin resistance was initially suggested to underlie high rates of diabetes in these populations, perhaps as the result of selection for greater efficiency for fat deposition, otherwise known as the "thrifty gene hypothesis" (TGH) (Neel, 1962). Although recent refinements of the TGH more readily support the notion that high rates of obesity and diabetes are the result of contemporary changes in diet in high-risk populations, specifically significant increases in carbohydrate consumption (Neel, 1999). More recently, evidence has accumulated for transgenerational effects in which maternal condition exerts downstream effects on offspring diabetes risk (Gluckman and Hanson, 2004).

Insulin also acts as an important ergostat to the hypothalamus and reproductive system. The hypothalamus maintains a significant number of insulin receptors, with potent downstream effects on reproductive function (Bruning et al., 2000). For example, increases in urinary C-peptide is associated with the resumption of ovarian activity and the cessation of postpartum lactational amenorrhea, suggesting that insulin is an active agent in shifting energetic investment between present and future reproductive effort in women (Ellison and Valeggia, 2003).

Thyroid hormones and metabolic regulation

Thyroid hormones, thyroxine (T_4) and tri-iodothyronine (T_3), are produced and secreted by the thyroid gland which is situated around the trachea. Synthesized in association with iodine and the amino acid tyrosine, thyroid hormones are potent regulators of basal metabolic rate (Kronenberg and Williams, 2008). Although most thyroid hormone consists of T_4, T_3 has a greater affinity for target receptors. Thyroxine is commonly converted to T_3 within target cells. The production of thyroid hormones are controlled by thyroid-stimulating hormone (TSH) which is secreted by the pituitary gland, which in turn is stimulated by thyroid-releasing hormone (TRH) from the hypothalamus. Thyroid-binding globulin (TBG) binds to T_3 and T_4 and acts as a carrier protein in circulation.

The lack of thyroid hormones, or hypothyroidism, results in weight gain and sluggishness. Hypothyroidism during infancy can cause cretinism, leading to stunted physical and mental development. Hyperthyroidism or excess thyroid hormone, also known as Graves's disease, results in accelerated basal metabolic rate, weight loss, hyperactivity, and other symptoms such as bulging eyes (Kronenberg and Williams, 2008). Because thyroid hormone synthesis relies on the availability of iodine in circulation, iodine deficiencies can result in goiter, an enlargement of the thyroid gland which is common in many developing countries (Andersson et al., 2005). This enlargement, while potentially disfiguring, results from an adaptive response to increase the iodine absorption ability of the thyroid gland.

Some variation in thyroid hormone physiology within and between human populations is evident, although in most cases, the adaptive significance remains unclear (Aoki et al., 2007). Exceptions are indigenous circumpolar groups who tend to have higher basal metabolic rates, higher annual levels of T_4, and augmented winter increases in T_4, presumably as an adaptive response to cold (Tkachev et al., 1991; Leonard et al., 1999, 2002).

Leptin and fat

Per unit mass, fat is the most efficient mode of somatic energy storage. Indeed, adiposity is crucial for surviving food deficiencies and other sources of energetic depletion such as disease or infection. It would therefore be vitally important to evolve a chemical signal that would alert the brain, particularly the hypothalamus, to fat availability and storage status. Leptin exhibits qualities for such a signal. Leptin is a polypeptide hormone that is secreted primarily by fat cells (adipocytes) (Casanueva and Dieguez, 1999), although other secondary sources have been identified. Leptin is most commonly measured in blood although salivary and urinary assessments have met with limited success (Groschl et al., 2001; Zaman et al., 2003). In essence, leptin often serves as a lipostat, signaling fat availability to receptors within the hypothalamus and other regions. The lack of leptin or receptor insensitivity usually causes hyperphagia and extreme weight gain, most likely due to the brain's perception that the body is experiencing starvation and the lack of adiposity. Because leptin seems to be a mechanism of energetic accounting, the life history implications of the discovery of leptin are potentially profound (Niewiarowski et al., 2000), although the functional complexity of this hormone has only recently begun to be appreciated.

Other proposed functions include influences on immune function, growth, and reproduction. For example, leptin modulates T-cell mediated immunity and reverses starvation-induced immunosuppression in mice (Lord et al., 1998). Leptin administration also accelerates sexual maturation in mice although results in other mammals such as humans have been equivocal (Himms-Hagen, 1999). Here again, as a modulator of immunocompetence, leptin appears to be an important mechanism in energy allocation towards infectious challenges.

Variation in leptin structure and function between species is considerable. Comparative investigations have suggested interesting differences between human and nonhuman leptin despite its relatively conservative chemical structure (Muehlenbein et al., 2003b, 2005). Among chimpanzees, very little is known although preliminary investigations have shown that leptin levels are higher in females, perhaps reflecting the greater metabolic costs of reproduction (Bribiescas and Anestis, in press). However no associations between leptin and body mass in male captive chimpanzees are evident, perhaps illustrating the marginal role of adipose tissue modulation in energy maintenance in this species (Bribiescas and Anestis, in press).

In addition to interspecies variation, between population contrasts in leptin function and associations with body composition are significant. Leptin is lower in males among the Aché of Paraguay even after controlling for adiposity. However, as a population, leptin levels are extremely low despite relatively high fat percentages. Aché women with 33% body fat on average, exhibit leptin levels that are indistinguishable from American anorectic women (7% body fat) (Bribiescas, 2005). Similarly, leptin is a poor reflector of adiposity in Aché men, in contrast to the tight association between leptin and adiposity on lean American men (Bribiescas and Hickey, 2006). While polymorphisms in the leptin molecule and receptor that may influence sensitivity are possible, it is also possible that lifetime energetic conditions can influence adult leptin independent of adiposity although additional data are needed to test this hypothesis.

Ghrelin and hunger

A recently discovered polypeptide hormone that is secreted primarily within the stomach, ghrelin levels are positively associated with hunger and are a potent stimulant of GH secretion. Ghrelin is found in two forms, total and active. The active or acylated form, differs from total ghrelin in that active maintains an N-octanoyl group at the Ser3 position that is believed to be necessary for bioactivity (Kojima and Kangawa, 2005). Assessment of ghrelin is limited to blood samples although salivary measurements have been reported (Groschl et al., 2005).

Ghrelin is a potentially significant mechanistic agent of regulating energy intake through its effect on hunger and satiation. It also plays a significant role in stimulating the production of GH (Kojima and Kangawa, 2005). The seemingly contradictory relationship between hunger-stimulated increases in ghrelin and its GH stimulating effects remain to be elucidated. Very little is known about population variation within humans but available evidence suggests modest functional variation (Chanoine et al., 2003; Shukla et al., 2005; Bribiescas et al., 2008).

Cortisol and stress

One of the most well known and researched hormones, cortisol belongs to a class of steroids known as glucocorticoids. Cortisol is secreted by the adrenal gland in response to the pituitary hormone adrenocorticotropin hormone (ACTH), which in turn results from the hypothalamic hormone, corticotropin releasing factor (CRF) (Kronenberg and Williams, 2008). Cortisol is commonly referred to as the "stress" hormone due to elevations that occur in response to physical or mental discomfort, or even the anticipation of potential discomfort. However, the central function of cortisol is to mobilize energy resources such as glucose and amino acids through muscle breakdown or "catabolism." Cortisol also acts as a potent anti-inflammatory agent. Related steroids such as cortisone are commonly used for the treatment of inflammation (Kronenberg and Williams, 2008).

The evolutionary function of cortisol appears to be to make glucose available for immediate use during times of acute need. As an anti-inflammatory agent, it also acts to postpone attention towards injury and insult in the face of more immediate needs. In the absence of a stressor, cortisol commonly exhibits a strong diurnal signal with levels being higher in the morning and declining into the evening (Rose et al., 1972; Knutsson et al., 1997). Chronically elevated cortisol levels can result in long-term damage to many tissues, including the brain. For example, high glucocorticoid levels can induce severe damage to the hippocampus, an important brain region for memory consolidation (Sapolsky et al., 1990). A much more detailed discussion of cortisol and stress physiology can be found in Chapter 24 of this volume.

TIME MANAGEMENT

In contrast to energy, organisms cannot harvest time. The amount of time an organism has to conduct its daily tasks of feeding, resting, and reproducing is limited by the number hours in a day. One cannot be in two places at once. On a grander scale, organisms have a finite lifetime to grow and reproduce. Even if an individual is able to diminish extrinsic mortality, it is limited by the species-specific rate of senescence (Hill et al., 2001).

Daily activity budgets during human evolution were dictated largely by foraging strategies and efficiency. The physiological mechanisms that influence these daily time budgeting decisions involve intricate neural activity that is obviously extremely complex and far beyond our current realm of understanding, although the underlying strategic behavioral ecology aspects have been discussed widely (Cronk, 1991). However, it is evident that neuroendocrine aspects of some time-budgeting decisions have intricate relationships with daily activity. For example, among wild chimpanzees, individuals tend to engage in hunting when fruit availability is high (Watts and Mitani, 2002). The underlying hormonal process likely involves greater glucose and insulin levels that allow the high energetic output necessary for a successful hunt (Sherry and Ellison, 2007). Low energy status stimulates increases in ghrelin-induced hunger, thereby compelling individuals to pursue food acquisition. Low energy status can also induce hypoinsulinemia, lethargy, and the desire to spend time resting (Elia et al., 1984; Jenike, 1996).

The decision to budget time between present and future reproductive effort, as in the case of parenting versus mate seeking, is a particularly important facet of daily life. Such decisions and the effects of hormonal variation are clearly evident in nonhuman seasonally breeding organisms and, in more subtle ways, humans. Because of the high variance in potential fitness in association with mate availability among males, it is not surprising that shifts in testosterone are evident in males investing in reproductive effort compared to parenting. Human males who are pair bonded or are fathers exhibit lower testosterone levels compared to single men as well as those without children (Gray et al., 2002; Burnham et al., 2003; Gray, 2003). The physiological effects of testosterone variation in association with parenting and pair bonding are unclear, although adaptive alterations of behavior, metabolism, or immunocompetence are possible (Mazur and Michalek, 1998; Bribiescas, 2001; Muehlenbein and Bribiescas, 2005).

In females, the amount of invested time spent in breast-feeding is contingent on the energetic status of the mother. Postpartum amenorrhea is a function of nursing frequency and intensity as well as maternal circulating glucose levels and body mass. Among Toba women of Argentina, postpartum amenorrhea is quite short despite heavy nursing investment. With increases in body mass and insulin (as reflected by urinary C-peptide levels), ovarian function resumes and allows women to begin investment in future reproduction (Ellison and Valeggia, 2003; Valeggia and Ellison, 2004).

Senescence

There are very few hormonal markers of overall senescence. No single hormone is responsible for somatic degeneration. The most salient hormonal markers of senescence are DHEA and DHEAS. Both rise during adolescence and drop steadily with age (Perrini et al.,

2005). This decline seems to be shared with other primates and is therefore evolutionarily conservative (Muehlenbein et al., 2003a; Perret and Aujard, 2005). While the association with aging is well established, the physiological significance of DHEA and DHEAS changes with age remains elusive (Johnson et al., 2002).

Senescence involves intrinsic physiological constraints on life span that are inherent to a particular species. While it is not uncommon for humans (even hunter-gatherers) to live well into their 60s and beyond, the doubling rate of mortality begins to shorten dramatically around the age of 60 (Hill and Hurtado, 1996). In chimpanzees, the same applies at around the age of 30 (Hill et al., 2001). As humans senesce, changes in hormone levels map the process of aging and somatic degeneration. This process is most prominent in regards to female reproductive senescence, otherwise known as menopause. Ova depletion leads to decreases in estrogen levels and greater gonadotropin production. While the timing of menopause and the extraordinary length of postreproductive life is unique to humans, the endocrine signals of menopause seem to be common among mammals and other great apes (Austad, 1994; Videan et al., 2006).

Senescence among males is not characterized by an abrupt cessation of reproductive function. However, significant hormone changes occur such as declines in testosterone and increases in FSH and LH (Harman et al., 2001), which may result in compromised fertility (de La Rochebrochard et al., 2006). Changes in hormonal milieu also results in changes in somatic composition, including a decline in muscle mass, increased adiposity, and lower metabolic rates (Fukagawa et al., 1990). The adaptive significance of these changes are unclear although it has been suggested that age-associated somatic composition changes may indicate a shift from investment in reproductive effort to survivorship or perhaps offspring care (Bribiescas, 2006b).

HORMONES AND PHENOTYPIC PLASTICITY

A cornerstone of life history and evolutionary theory is the importance of phenotypic plasticity or the ability of organisms to modulate a phenotype in response to an environmental challenge. Since environments and selection pressures can change rapidly, it is seldom adaptive for an organism to maintain a rigid set of phenotypes (Schlichting and Pigliucci, 1998). Some phenotypic plasticity responses rely on distinct periods of sensitivity while others are malleable throughout the organism's lifetime. It can therefore be postulated that along with phenotypes themselves, the range of phenotypic plasticity of important traits related to growth, maintenance, and reproduction are themselves adaptive features. But what are the regulatory mechanisms that adjust the range and sensitivity of plastic traits? Hormones are central mechanisms of phenotypic plasticity and modulate gene and phenotypic expression in response to environmental cues (Ketterson and Nolan, 1992; Zera and Harshman, 2001).

If phenotypic plasticity is so advantageous, why have not organisms evolved the ability to maintain total malleability? In addition to the constraints of physics (Pennycuick, 1992), there are costs associated with phenotypic plasticity such the energetic costs of maintaining the physiological capacity to alter phenotypes, as well as the potential of misinterpreting an environmental cue (Relyea, 2002). Phenotypic plasticity also involves trade-offs between competing physical needs. For example, prolactin increases in association with breast-feeding and investment in present offspring tends to suppress ovarian function and investment in future reproduction. The amount of prolactin secreted is directly related to the amount of nursing intensity. However the ovarian suppressive effects of prolactin and this trade-off can be tempered by greater insulin levels caused by enhanced energy availability and somatic condition (Ellison and Valeggia, 2003).

CONCLUSION

In order to gain a more complete picture of the adaptive function of endocrine factors, including the role of hormones in life history trade-offs and events, future investigations need to move beyond simple snapshot measurements of hormone levels. Potential fruitful areas of research include interactions between various hormones and greater awareness that many hormones exhibit actions beyond their standard definitions. For example, testosterone is tagged as a sex hormone whereas many of its effects clearly have important metabolic implications. In a similar fashion, the effects of hormones on immune function merit considerable attention, since it is in this association that trade-offs with maintenance are likely to be evident. Additionally, patterns and amplitude of pulsatility have only barely been appreciated. Such investigations will rely upon more sophisticated multiple-sampling regimens and an understanding of potential hormone pulsatility signals. A greater appreciation of adaptive, nonpathological hormone variation is also needed.

DISCUSSION POINTS

1. What are the common trade-offs associated with hormone variation?
2. How do hormones regulate energetic allocation decisions?

3. How do hormones affect time allocation decisions?
4. Many hormones are involved in growth and reproduction. Are these actions unique to human growth and reproduction, or are they shared with other species? Explain any major differences and why/how natural and sexual selections could have produced such differences.

REFERENCES

Andersson M., Takkouche B., Egli I., et al. (2005). Current global iodine status and progress over the last decade towards the elimination of iodine deficiency. *Bulletin of the World Health Organization*, **83**(7), 518–525.

Anestis, S. F. and Bribiescas, R. G. (2004). Rapid changes in chimpanzee (*Pan troglodytes*) urinary cortisol excretion. *Hormones and Behavior*, **45**(3), 209–213.

Aoki, Y., Belin. R. M., Clickner, R., et al. (2007). Serum TSH and total T$_4$ in the United States population and their association with participant characteristics: National Health and Nutrition Examination Survey (NHANES 1999–2002). *Thyroid*, **17**(12), 1211–1223.

Arlt, W., Martens. J. W., Song, M., et al. (2002). Molecular evolution of adrenarche: structural and functional analysis of p450c17 from four primate species. *Endocrinology*, **143**(12), 4665–4672.

Attie, K. M., Ramirez, N. R., Conte, F. A., et al. (1990). The pubertal growth spurt in eight patients with true precocious puberty and growth hormone deficiency: evidence for a direct role of sex steroids. *Journal of Clinical Endocrinology and Metabolism*, **71**, 975–983.

Austad, S. N. (1994). Menopause: an evolutionary perspective. *Experimental Gerontology*, **29**(3–4), 255–263.

Baird, D. T. (1984). The ovary. In *Reproduction in Mammals*, vol. 3. *Hormonal Control of Reproduction*, C. R. Austin and R. V. Short (eds), 2nd edn. New York: Cambridge University Press, pp. 91–114.

Baker, M. E., Chandsawangbhuwana, C. and Ollikainen, N. (2007). Structural analysis of the evolution of steroid specificity in the mineralocorticoid and glucocorticoid receptors. *BMC Evolutionary Biology*, **7**, 24.

Ballard, P. L. (1979). Glucocorticoids and differentiation. In *Glucocorticoid Hormone Action*, J. D. Baxter and G. G. Rousseau (eds). New York: Springer-Verlag, pp. 493–515.

Barker, D. J., Osmond, C., Thornburg, K. L., et al. (2008). A possible link between the pubertal growth of girls and breast cancer in their daughters. *American Journal of Human Biology*, **20**(2), 127–131.

Bergendahl, M. and Huhtaniemi, I. (1993). Acute fasting is ineffective in suppressing pituitary-gonadal function of pubertal male rats. *American Journal of Physiology*, **264**(5 Pt 1), E717–E722.

Bersinger, N. A., Groome, N. and Muttukrishna, S. (2002). Pregnancy-associated and placental proteins in the placental tissue of normal pregnant women and patients with pre-eclampsia at term. *European Journal of Endocrinology*, **147**, 785–793.

Blanks, A. M. and Thornton, S. (2003). The role of oxytocin in parturition. *British Journal of Obstetrics and Gynaecology*, **110**, 46–51.

Bribiescas, R. G. (1996). Testosterone levels among Aché hunter/gatherer men: a functional interpretation of population variation among adult males. *Human Nature*, **7**(2), 163–188.

Bribiescas, R. G. (2001). Reproductive ecology and life history of the human male. *Yearbook of Physical Anthropology*, **33**, 148–176.

Bribiescas, R. G. (2005). Serum leptin levels in Aché Amerindian females with normal adiposity are not significantly different from American anorexia nervosa patients. *American Journal of Human Biology*, **17**(2), 207–210.

Bribiescas, R. G. (2006a). *Men: Evolutionary and Life History*. Cambridge, MA: Harvard University Press.

Bribiescas, R. G. (2006b). On the evolution of human male reproductive senescence: proximate mechanisms and life history strategies. *Evolutionary Anthropology*, **15**(4), 132–141.

Bribiescas, R. G. and Anestis, S. F. (in press). Leptin associations with age, weight and sex among chimpanzees (*Pan troglodytes*). *Journal of Medical Primatology*.

Bribiescas, R. G. and Ellison, P. T. (2008). How hormones mediate tradeoffs in human health and disease. In *Evolution in Health and Disease*, S. C. Stearns and J. C. Koella (eds), 2nd edn. Oxford: Oxford University Press, pp. 77–94.

Bribiescas, R. G. and Hickey, M. S. (2006). Population variation and differences in serum leptin independent of adiposity: a comparison of Aché Amerindian men of Paraguay and lean American male distance runners. *Nutrition and Metabolism (London)*, **3**, 34.

Bribiescas, R. G., Betancourt, J., Torres, A. M., et al. (2008). Active ghrelin levels across time and associations with leptin and anthropometrics in healthy Aché Amerindian women of Paraguay. *American Journal of Human Biology*, **20**(3), 352–354.

Bruning, J. C., Gautam, D., Burks, D. J., et al. (2000). Role of brain insulin receptor in control of body weight and reproduction. *Science*, **289**(5487), 2122–2125.

Buhimschi, C. S. (2004). Endocrinology of lactation. *Obstetrics and Gynecology Clinics of North America*, **31**, 963–979.

Burnham, T. C., Chapman, J. F., Gray, P. B., et al. (2003). Men in committed, romantic relationships have lower testosterone. *Hormones and Behavior*, **44**(2), 119–122.

Campbell, B. (2006). Adrenarche and the evolution of human life history. *American Journal of Human Biology*, **18**(5), 569–589.

Carr, B. R. (1998). Disorders of the ovaries and female reproductive tract. In *Williams Textbook of Endocrinology*, P. R. Larsen, H. M. Kronenberg, S. Melmed, et al. (eds), 9th edn. Philadelphia: Saunders, pp. 751–818.

Carr, B. R. and Rehman, K. S. (2004). Fertilization, implantation, and endocrinology of pregnancy. In *Textbook of Endocrine Physiology*, J. E. Griffin and S. R. Ojeda (eds), 5th edn. New York: Oxford University Press, pp. 249–273.

Casanueva, F. F. and Dieguez, C. (1999). Neuroendocrine regulation and actions of leptin. *Frontiers in Neuroendocrinology*, **20**, 317–363.

Casey, M. L. and MacDonald, P. C. (1998). Endocrine changes of pregnancy. In *Williams Textbook of*

Endocrinology, P. R. Larsen, H. M. Kronenberg, S. Melmed, et al. (eds), 9th edn. Philadelphia: Saunders, pp. 1259–1271.

Challis, J. R. G., Matthews, S. G., Gibb, W., et al. (2000). Endocrine and paracrine regulation of birth at term and preterm. *Endocrine Reviews*, **21**, 514–550.

Chanoine, J. P., Yeung, L. P. and Wong, A. C. (2003). Umbilical cord ghrelin concentrations in Asian and Caucasian neonates. *Hormone Research*, **60**(3), 116–120.

Cohen, P. and Rosenfeld, R. G. (2004). Growth regulation. In *Textbook of Endocrine Physiology*, J. E. Griffin and S. R. Ojeda (eds), 5th edn. New York: Oxford University Press, pp. 274–293.

Conley, A. J. and Mason, J. I. (1990). Placental steroid hormones. *Ballière's Clinical Endocrinology and Metabolism*, **4**, 249–272.

Craig, J., Orisaka, M., Wang, H., et al. (2007). Gonadotropin and intra-ovarian signals regulating follicle development and atresia: the delicate balance between life and death. *Frontiers in Bioscience*, **12**, 3628–3639.

Cronk, L. (1991). Human behavioral ecology. *Annual Review of Anthropology*, **20**, 25–53.

Davies, M. J. and Norman, R. J. (2002). Programming and reproductive functioning. *Trends in Endocrinology and Metabolism*, **13**, 386–392.

de La Rochebrochard, E., de Mouzon, J., Thepot, F., et al. (2006). Fathers over 40 and increased failure to conceive: the lessons of in vitro fertilization in France. *Fertility and Sterility*, **85**(5), 1420–1424.

De Moor, P., Verhoeven, G. and Heyns, W. (1973). Permanent effects of fetal and neonatal testosterone secretion on steroid metabolism and binding. *Differentiation*, **1**, 241–253.

Dobbins, R. L., Cowley, M. A. and Foster, D. W. (2004). Glucose, lipid, and protein metabolism. In *Textbook of Endocrine Physiology*, J. E. Griffin and S. R. Ojeda (eds), 5th edn. New York: Oxford University Press, pp. 377–406.

Elia, M., Lammert, O., Zed, C., et al. (1984). Energy metabolism during exercise in normal subjects undergoing total starvation. *Human Nutrition. Clinical Nutrition*, **38**(5), 355–362.

Ellison, P. T. (2003). Energetics and reproductive effort. *American Journal of Human Biology*, **15**(3), 342–351.

Ellison, P. T. and Valeggia, C. R. (2003). C-peptide levels and the duration of lactational amenorrhea. *Fertility and Sterility*, **80**(5), 1279–1280.

Faiman, C., Winter, J. S. D. and Reyers, F. I. (1976). Patterns of gonadotropins and gonadal steroids throughout life. *Clinical Obstetrics and Gynecology*, **3**, 467–483.

Finch, C. E. and Rose, M. R. (1995). Hormones and the physiological architecture of life history evolution. *Quarterly Review of Biology*, **70**(1), 1–52.

Forest, M. G., Sizonenko, P. C., Cathiard, A. M., et al. (1974). Hypophyso-gonadal function in humans during the first year of life. I: Evidence for testicular activity in early infancy. *Journal of Clinical Investigation*, **53**, 819–828.

Franceschini, R., Venturini, P. L., Cataldi, A., et al. (1989). Plasma beta-endorphin concentrations during suckling in lactating women. *British Journal of Obstetrics and Gynaecology*, **96**, 711–713.

Fukagawa, N. K., Bandini, L. G. and Young, J. B. (1990). Effect of age on body composition and resting metabolic rate. *American Journal of Physiology*, **259**(2 Pt 1), E233–E238.

Garcia-Mayor, R. V., Andrade, M. A., Rios, M., et al. (1997). Serum leptin levels in normal children: relationship to age, gender, body mass index, pituitary-gonadal hormones, and pubertal stage. *Journal of Clinical Endocrinology and Metabolism*, **82**(9), 2849–2855.

Gluckman, P. D. and Hanson, M. A. (2004). Living with the past: evolution, development, and patterns of disease. *Science*, **305**(5691), 1733–1736.

Gray, A., Feldman, H. A., McKinlay, J. B., et al. (1991). Age, disease, and changing sex hormone levels in middle-aged men: results of the Massachusetts Male Aging Study. *Journal of Clinical Endocrinology and Metabolism*, **73**(5), 1016–1025.

Gray, P. B. (2003). Marriage, parenting, and testosterone variation among Kenyan Swahili men. *American Journal of Physical Anthropology*, **122**(3), 279–286.

Gray, P. B., Kahlenberg, S. M., Barrett, E. S., et al. (2002). Marriage and fatherhood are associated with lower testosterone in males. *Evolution and Human Behavior*, **23**, 193–201.

Gray, P. B., Singh, A. B., Woodhouse, L. J., et al. (2005). Dose-dependent effects of testosterone on sexual function, mood, and visuospatial cognition in older men. *Journal of Clinical Endocrinology and Metabolism*, **90**(7), 3838–3846.

Griffin, J. E. (2004a). Male reproductive function. In *Textbook of Endocrine Physiology*, J. E. Griffin and S. R. Ojeda (eds), 5th edn. New York: Oxford University Press, pp. 226–248.

Griffin, J. E. (2004b). The thyroid. In *Textbook of Endocrine Physiology*, J. E. Griffin and S. R. Ojeda (eds), 5th edn. New York: Oxford University Press, pp. 294–318.

Griffin, J. E. and Ojeda, S. R. (eds) (2004). *Textbook of Endocrine Physiology*, 5th edn. New York: Oxford University Press.

Grigorova, M., Punab, M., Ausmees, K., et al. (2008). FSHB promoter polymorphism within evolutionary conserved element is associated with serum FSH level in men. *Human Reproduction*, **23**(9), 2160–2166.

Groome, N. P., Illingworth, P. J., O'Brien, M., et al. (1996). Measurement of dimeric inhibin B throughout the human menstrual cycle. *Journal of Clinical Endocrinology and Metabolism*, **81**, 1401–1405.

Groschl, M., Rauh, M., Wagner, R., et al. (2001). Identification of leptin in human saliva. *Journal of Clinical Endocrinology and Metabolism*, **86**(11), 5234–5239.

Groschl, M., Topf, H. G., Bohlender, J., et al. (2005). Identification of ghrelin in human saliva: production by the salivary glands and potential role in proliferation of oral keratinocytes. *Clinical Chemistry*, **51**(6), 997–1006.

Grumbach, M. M. and Conte, F. A. (1998). Disorders of sex differentiation. In *Williams Textbook of Endocrinology*, P. R. Larsen, H. M. Kronenberg, S. Melmed, et al. (eds), 9th edn. Philadelphia: Saunders, pp. 1303–1425.

Grumbach, M. M. and Styne, D. M. (1998). Puberty: ontogeny, neuroendocrinology, physiology, and disorders. In *Williams Textbook of Endocrinology*, P. R. Larsen,

H. M. Kronenberg, S. Melmed, et al. (eds), 9th edn. Philadelphia: Saunders, pp. 1509–1625.

Grumbach, M. M., Kaplan, S. L., Sciarra, J. J., et al. (1968). Chorionic growth hormone-prolactin (CPG): secretion, disposition, biological activity in man, and postulated function as the "growth hormone" of the second half of pregnancy. *Annals of the New York Academy of Sciences*, **148**, 501–531.

Guzick, D. S., Overstreet, J. W., Factor-Litvak, P., et al. (2001). Sperm morphology, motility, and concentration in fertile and infertile men. *New England Journal of Medicine*, **345**(19), 1388–1393.

Haig, D. (1993). Genetic conflicts in human pregnancy. *Quarterly Review of Biology*, **68**, 495–532.

Hanson, R. L., Imperatore, G., Narayan, K. M., et al. (2001). Family and genetic studies of indices of insulin sensitivity and insulin secretion in Pima Indians. *Diabetes/Metabolism Research and Reviews*, **17**(4), 296–303.

Harman, S. M., Metter, E. J., Tobin, J. D., et al. (2001). Longitudinal effects of aging on serum total and free testosterone levels in healthy men. Baltimore Longitudinal Study of Aging. *Journal of Clinical Endocrinology and Metabolism*, **86**(2), 724–731.

Havelock, J. C., Auchus, R. J. and Rainey, W. E. (2004). The rise in adrenal androgen biosynthesis: adrenarche. *Seminars in Reproductive Medicine*, **22**(4), 337–347.

Heller, H. J. (2004). Calcium homeostasis. In *Textbook of Endocrine Physiology*, J. E. Griffin and S. R. Ojeda (eds), 5th edn. New York: Oxford University Press, pp. 349–376.

Herbst, K. L. and Bhasin, S. (2004). Testosterone action on skeletal muscle. *Current Opinion on Clinical Nutrition and Metabolic Care*, **7**, 271–277.

Hewlett, B. S. and Lamb, M. E. (2005). *Hunter-Gatherer Childhoods: Evolutionary, Developmental, and Cultural Perspectives*. New Brunswick, NJ: Aldine de Gruyter Transaction.

Hill, K. and Hurtado, A. M. (1996). *Aché Life History: the Ecology and Demography of a Foraging People*. New York: Aldine de Gruyter.

Hill, K., Boesch, C., Goodall, J., et al. (2001). Mortality rates among wild chimpanzees. *Journal of Human Evolution*, **40**, 437–450.

Himms-Hagen, J. (1999). Physiological roles of the leptin endocrine system: differences between mice and humans. *Critical Reviews in Clinical Laboratory Sciences*, **36**(6), 575–655.

Hruska, K. A., Civitelli, R., Duncan, R., et al. (1991). Regulation of skeletal remodeling by parathyroid hormone. *Contributions to Nephrology*, **91**, 38–42.

Jasienska, G., Kapiszewska, M., Ellison, P. T., et al. (2006). CYP17 genotypes differ in salivary 17-beta estradiol levels: a study based on hormonal profiles from entire menstrual cycles. *Cancer Epidemiology, Biomarkers and Prevention*, **15**(11), 2131–2135.

Jenike, M. R. (1996). Activity reduction as an adaptive response to seasonal hunger. *American Journal of Human Biology*, **8**, 517–534.

Jepson, J. H., Gardner, F. H., Gorshein, D., et al. (1973). Current concepts of the action of androgenic steroids on erythropoiesis. *Journal of Pediatrics*, **83**, 703–708.

Johnson, M. D., Bebb, R. A. and Sirrs, S. M. (2002). Uses of DHEA in aging and other disease states. *Ageing Research Reviews*, **1**(1), 29–41.

Juul, A. (2001). The effects of oestrogens on linear bone growth. *Human Reproduction Update*, **7**, 303–313.

Kaplan, S. L., Grumbach, M. M. and Aubert, M. L. (1976). The ontogenesis of pituitary hormones and hypothalamic factors in the human fetus: maturation of the central nervous system regulation of anterior pituitary function. *Recent Progress in Hormone Research*, **32**, 161–243.

Kenny, A. M. and Raisz, L. G. (2002). Mechanisms of bone remodeling: implications for clinical practice. *Journal of Reproductive Medicine*, **47**, 63–70.

Ketterson, E. D. and Nolan, V. Jr (1992). Hormones and life histories: an integrative approach. *American Naturalist*, **140**(suppl. 5), S33–S62.

King, J. C., Anthony, E. L. P. and Fitzgerald, D. M. (1985). Luteinizing hormone-releasing hormone neurons in human preoptic/hypothalamus: differential intraneuronal localization of immunoreactive forms. *Journal of Clinical Endocrinology and Metabolism*, **60**, 88–97.

Klibanski, A., Beitins, I. Z., Badger, T., et al. (1981). Reproductive function during fasting in men. *Journal of Clinical Endocrinology and Metabolism*, **53**(2), 258–263.

Klok, M. D., Jakobsdottir, S. and Drent, M. L. (2007). The role of leptin and ghrelin in the regulation of food intake and body weight in humans: a review. *Obesity Reviews*, **8**, 21–34.

Knickmeyer, R. C. and Baron-Cohen, S. (2006). Fetal testosterone and sex differences. *Early Human Development*, **82**(12), 755–760.

Knobil, E., Neill, J. D., Ewing, L. L., et al. (1988). *The Physiology of Reproduction*. New York: Raven Press.

Knutsson, U., Dahlgren, J., Marcus, C., et al. (1997). Circadian cortisol rhythms in healthy boys and girls: relationship with age, growth, body composition, and pubertal development. *Journal of Clinical Endocrinology and Metabolism*, **82**(2), 536–540.

Kojima, M. and Kangawa, K. (2005). Ghrelin: structure and function. *Physiological Reviews*, **85**(2), 495–522.

Kronenberg, H. and Williams, R. H. (2008). *Williams Textbook of Endocrinology*, 11th edn. Philadelphia: Saunders/Elsevier.

Kumar, T. R., Wang, Y., Lu, N., et al. (1997). Follicle stimulating hormone is required for ovarian follicle maturation but not male fertility. *Nature Genetics*, **15**(2), 201–204.

Kuzawa, C. W. (2005). Fetal origins of developmental plasticity: are fetal cues reliable predictors of future nutritional environments? *American Journal of Human Biology*, **17**(1), 5–21.

Kuzawa, C. W. and Adair, L. S. (2003). Lipid profiles in adolescent Filipinos: relation to birth weight and maternal energy status during pregnancy. *American Journal of Clinical Nutrition*, **77**(4), 960–966.

Lamminen, T. and Huhtaniemi, I. (2001). A common genetic variant of luteinizing hormone; relation to normal and aberrant pituitary-gonadal function. *European Journal of Pharmacology*, **414**(1), 1–7.

Lamminen, T., Jokinen, P., Jiang, M., et al. (2005). Human FSH beta subunit gene is highly conserved. *Molecular Human Reproduction*, **11**(8), 601–605.

Latronico, A. C. and Segaloff, D. L. (1999). Naturally occurring mutations of the luteinizing-hormone receptor: lessons learned about reproductive physiology and G protein-coupled receptors. *American Journal of Human Genetics*, **65**(4), 949–958.

Leidy Sievert, L. (2006). *Menopause: a Biocultural Perspective*. Piscataway, NJ: Rutgers University Press.

Leonard, W. R., Galloway, V. A., Ivakine, E., et al. (1999). Nutrition, thyroid function and basal metabolism of the Evenki of central Siberia. *International Journal of Circumpolar Health*, **58**(4), 281–295.

Leonard, W. R., Sorensen, M. V., Galloway, V. A., et al. (2002). Climatic influences on basal metabolic rates among circumpolar populations. *American Journal of Human Biology*, **14**(5), 609–620.

Licht, P., Russu, V. and Wildt, L. (2001). On the role of human chorionic gonadotropin (hCG) in the embryo-endometrial microenvironment: implications for differentiation and implantation. *Seminars in Reproductive Medicine*, **19**, 37–47.

Lord, G. M., Matarese, G., Howard, J. K., et al. (1998). Leptin modulates the T-cell immune response and reverses starvation-induced immunosuppression. *Nature*, **394**(6696), 897–901.

Matkovic, V. (1996). Skeletal development and bone turnover revisited. *Journal of Clinical Endocrinology and Metabolism*, **81**, 2013–2016.

Mazur, A. and Michalek, J. (1998). Marriage, divorce, and male testosterone. *Social Forces*, **77**(1), 315–330.

McLean, M. and Smith, R. (2001). Corticotrophin-releasing hormone and human parturition. *Reproduction*, **121**, 493–501.

McNatty, K. P., Makris, A., DeGrazia, C., et al. (1979). The production of progesterone, androgens and estrogens by granulose cells, thecal tissue and stromal tissue from human ovaries in vitro. *Journal of Clinical Endocrinology and Metabolism*, **49**, 687–699.

McNeilly, A. S., Glasier, A., Jonassen, J., et al. (1982). Evidence for direct inhibition of ovarian function by prolactin. *Journal of Reproduction and Fertility*, **65**, 559–569.

Meistas, M. T., Zadik, Z., Margolis, S., et al. (1981). Correlation of urinary excretion of C-peptide with the integrated concentration and secretion rate of insulin. *Diabetes*, **30**(8), 639–643.

Meldrum, D. R. (1983). The pathophysiology of postmenopausal symptoms. *Seminars in Reproductive Endocrinology*, **1**, 11–17.

Metcalf, M. G., Donald, R. A. and Livesey, J. H. (1982). Pituitary-ovarian function before, during and after the menopausal transition: a longitudinal study. *Clinical Endocrinology*, **17**, 484–489.

Miers, W. R. and Barrett, E. J. (1998). The role of insulin and other hormones in the regulation of amino acid and protein metabolism in humans. *Journal of Basic Clinical Physiology and Pharmacology*, **9**, 235–253.

Miller, W. L. (2002). Androgen biosynthesis from cholesterol to DHEA. *Molecular and Cellular Endocrinology*, **198**(1–2), 7–14.

Muehlenbein, M. P. (2006). Intestinal parasite infections and fecal steroid levels in wild chimpanzees. *American Journal of Physical Anthropology*, **130**, 546–550.

Muehlenbein, M. P. (2008). Adaptive variation in testosterone levels in response to immune activation: empirical and theoretical perspectives. *Social Biology*, **53**, 13–23.

Muehlenbein, M. P. and Bribiescas, R. G. (2005). Testosterone-mediated immune functions and male life histories. *American Journal of Human Biology*, **17**(5), 527–558.

Muehlenbein, M. P., Campbell, B. C., Phillippi, K. M., et al. (2001). Reproductive maturation in a sample of captive male baboons. *Journal of Medical Primatology*, **30**, 273–282.

Muehlenbein, M. P., Campbell, B. C., Richards, R. J., et al. (2003a). Dehydroepiandrosterone-sulfate as a biomarker of senescence in male non-human primates. *Experimental Gerontology*, **38**(10), 1077–1085.

Muehlenbein, M. P., Campbell, B. C., Richards, R. J., et al. (2003b). Leptin, body composition, adrenal and gonadal hormones among captive male baboons. *Journal of Medical Primatology*, **32**(6), 320–324.

Muehlenbein, M. P., Campbell, B. C., Watts, D. P., et al. (2005). Leptin, adiposity, and testosterone levels in captive male macaques. *American Journal of Physical Anthropology*, **127**, 335–341.

Nadler, R. D., Roth-Meyer, C., Wallis, J., et al. (1984). Hormonal and compartmental correlates during adrenarche in the chimpanzee. *Comptes Rendus de l'Académie des Sciences. Série III, Sciences de la Vie*, **298**(14), 409–413.

Neel, J. V. (1962). Diabetes mellitus: a "thrifty" genotype rendered detrimental by "progress?" *American Journal of Human Genetics*, **14**, 353–362.

Neel, J. V. (1999). The "Thrifty Genotype" in 1998. *Nutrition Reviews*, **57**(5 pt 2), S2–S9.

Neville, M. C., McFadden, T. B. and Forsyth, I. (2002). Hormonal regulation of mammary differentiation and milk secretion. *Journal of Mammary Gland Biology and Neoplasia*, **7**, 49–66.

Niewiarowski, P. H., Balk, M. L. and Londraville, R. L. (2000). Phenotypic effects of leptin in an ectotherm: a new tool to study the evolution of life histories and endothermy? *Journal of Experimental Biology*, **203**(pt 2), 295–300.

Nilsson, C., Pettersson, K., Millar, R. P., et al. (1997). Worldwide frequency of a common genetic variant of luteinizing hormone: an international collaborative research. International Collaborative Research Group. *Fertility and Sterility*, **67**(6), 998–1004.

Niswender, G. D., Juengel, J. L., Silva, P. J., et al. (2000). Mechanisms controlling the function and life span of the corpus luteum. *Physiological Reviews*, **80**, 1–29.

Norris, D. O. (2007). *Vertebrate Endocrinology*. Boston: Elsevier Academic Press.

Nussey, S. and Whitehead, S. A. (2001). *Endocrinology: an Integrated Approach*. Oxford: Bios.

Odell, W. D. and Parker L. N. (1985). Control of adrenal androgen production. *Endocrine Research*, **10**, 617–630.

Ojeda, S. R. (2004a). The anterior pituitary and hypothalamus. In *Textbook of Endocrine Physiology*, J. E. Griffin and S. R. Ojeda (eds), 5th edn. New York: Oxford University Press, pp. 120–146.

Ojeda, S. R. (2004b). Female reproductive function. In *Textbook of Endocrine Physiology*, J. E. Griffin and S. R. Ojeda (eds), 5th edn. New York: Oxford University Press, pp. 186–225.

Parker, K. L. and Rainey, W. E. (2004). The adrenal glands. In *Textbook of Endocrine Physiology*, J. E. Griffin and S. R. Ojeda (eds), 5th edn. New York: Oxford University Press, pp. 319–348.

Parker, L. N. (1991). Adrenarche. *Endocrinology and Metabolism Clinics of North America*, **20**, 71–83.

Pennycuick, C. J. (1992). *Newton Rules Biology: a Physical Approach to Biological Problems*. Oxford: Oxford University Press.

Perret, M. and Aujard, F. (2005). Aging and season affect plasma dehydroepiandrosterone sulfate (DHEA-S) levels in a primate. *Experimental Gerontology*, **40**(7), 582–587.

Perrini, S., Laviola, L., Natalicchio, A., et al. (2005). Associated hormonal declines in aging: DHEAS. *Journal of Endocrinological Investigation*, **28**(3 suppl.), 85–93.

Plant, T. M., Gay, V. L., Marshall, G. R., et al. (1989). Puberty in monkeys is triggered by chemical stimulation of the hypothalamus. *Proceedings of the National Academy of Sciences of the United States of America*, **86**, 2506–2510.

Reichlin, S. (1998). Neuroendocrinology. In *Williams Textbook of Endocrinology*, P. R. Larsen, H. M. Kronenberg, S. Melmed, et al. (eds), 9th edn. Philadelphia: Saunders, pp. 165–248.

Reiter, E. O. and Rosenfeld, R. G. (1998). Normal and aberrant growth. In *Williams Textbook of Endocrinology*, P. R. Larsen, H. M. Kronenberg, S. Melmed, et al. (eds), 9th edn. Philadelphia: Saunders, pp. 1427–1507.

Relyea, R. A. (2002). Costs of phenotypic plasticity. *The American Naturalist*, **159**(3), 272–282.

Riancho, J. A., Valero, C., Zarrabeitia, M. T., et al. (2008). Genetic polymorphisms are associated with serum levels of sex hormone binding globulin in postmenopausal women. *BMC Medical Genetics*, **9**, 112.

Rodgers, G. M., Taylor, R. N. and Roberts, J. M. (1988). Preeclampsia is associated with a serum factor cytotoxic to human endothelial cells. *American Journal of Obstetrics and Gynecology*, **159**, 908–914.

Rogol, A. D. (1994). Growth at puberty: interaction of androgens and growth hormone. *Medicine and Science in Sports and Exercise*, **26**, 767–770.

Rolaki, A., Drakakis, P., Millingos, S., et al. (2005). Novel trends in follicular development, atresia and corpus luteul regression: a role for apoptosis. *Reproductive Biomedicine Online*, **11**, 93–103.

Rose, R. M., Kreuz, L. E., Holaday, J. W., et al. (1972). Diurnal variation of plasma testosterone and cortisol. *Journal of Endocrinology*, **54**(1), 177–178.

Sahu, A. (2003). Leptin signaling in the hypothalamus: emphasis on energy homeostasis and leptin resistance. *Frontiers in Neuroendocrinology*, **24**(4), 225–253.

Sapolsky, R. M., Uno, H., Rebert, C. S., et al. (1990). Hippocampal damage associated with prolonged glucocorticoid exposure in primates. *Journal of Neuroscience*, **10**(9), 2897–2902.

Schlichting, C. and Pigliucci, M. (1998). *Phenotypic Evolution: a Reaction Norm Perspective*. Sunderland, MA: Sinaur.

Sfakianaki, A. K. and Norwitz, E. R. (2006). Mechanisms of progesterone action in inhibiting prematurity. *Journal of Maternal-Fetal and Neonatal Medicine*, **19**, 763–772.

Shao, Y. Y., Wang, L. and Ballock, R. T. (2006). Thyroid hormone and the growth plate. *Reviews of Endocrine and Metabolic Disorders*, **7**, 265–271.

Sherman, B. M., West, J. H. and Korenman, S. G. (1976). The menopausal tradition: analysis of LH, FSH, estradiol, and progesterone concentrations during menstrual cycles of older women. *Journal of Clinical Endocrinology and Metabolism*, **42**, 629–636.

Sherry, D. S. and Ellison, P. T. (2007). Potential applications of urinary C-peptide of insulin for comparative energetics research. *American Journal of Physical Anthropology*, **133**(1), 771–778.

Sherwood, O. D. (2004). Relaxin's physiological roles and other diverse actions. *Endocrine Reviews*, **25**, 205–234.

Shukla, V., Singh, S. N., Vats, P., et al. (2005). Ghrelin and leptin levels of sojourners and acclimatized lowlanders at high altitude. *Nutritional Neuroscience*, **8**(3), 161–165.

Siiteri, P. K. and Wilson, J. D. (1974). Testosterone formation and metabolism during male sexual differentiation in the human embryo. *Journal of Clinical Endocrinology and Metabolism*, **38**, 113–125.

Sinha-Hikim, I., Artaza, J., Woodhouse, L., et al. (2002). Testosterone-induced increase in muscle size in healthy young men is associated with muscle fiber hypertrophy. *American Journal of Physiology. Endocrinology and Metabolism*, **283**(1), E154–E164.

Snegovskikh, V., Park, J. S. and Norwitz, E. R. (2006). Endocrinology of parturition. *Endocrinology and Metabolism Clinics of North America*, **35**, 173–191.

Stearns, S. C. (1992). *The Evolution of Life Histories*. Oxford: Oxford University Press.

Steinacker, J. M., Brkic, M., Simsch, C., et al. (2005). Thyroid hormones, cytokines, physical training and metabolic control. *Hormone and Metabolic Research*, **37**, 538–544.

Takahashi, P. Y., Votruba, P., Abu-Rub, M., et al. (2007). Age attenuates testosterone secretion driven by amplitude-varying pulses of recombinant human luteinizing hormone during acute gonadotrope inhibition in healthy men. *Journal of Clinical Endocrinology and Metabolism*, **92**(9), 3626–3632.

Tamimi, R. M., Lagiou, P., Mucci, L. A., et al. (2003). Average energy intake among pregnant women carrying a boy compared with a girl. *British Medical Journal*, **326**(7401), 1245–1246.

Tapanainen, J. S., Aittomaki, K., Min, J., et al. (1997). Men homozygous for an inactivating mutation of the follicle-stimulating hormone (FSH) receptor gene present variable suppression of spermatogenesis and fertility. *Nature Genetics*, **15**(2), 205–206.

Tapanainen, J. S., Vaskivuo, T., Aittomaki, K., et al. (1998). Inactivating FSH receptor mutations and gonadal dysfunction. *Molecular and Cellular Endocrinology*, **145**(1–2), 129–135.

Thongngarm, T., Jenkins, J. K., Ndebele, K., et al. (2003). Estrogen and progesterone modulate monocyte cell cycle progression and apoptosis. *American Journal of Reproductive Immunology*, **49**, 129–138.

Thornton, J. W. (2001). Evolution of vertebrate steroid receptors from an ancestral estrogen receptor by ligand exploitation and serial genome expansions. *Proceedings of the National Academy of Sciences of the United States of America*, **98**(10), 5671–5676.

Tiefenbacher, S., Lee, B., Meyer, J. S., et al. (2003). Non-invasive technique for the repeated sampling of salivary free cortisol in awake, unrestrained squirrel monkeys. *American Journal of Primatology*, **60**(2), 69–75.

Timossi, C. M., Barrios-de-Tomasi, J., Gonzalez-Suarez, R., et al. (2000). Differential effects of the charge variants of human follicle-stimulating hormone. *Journal of Endocrinology*, **165**(2), 193–205.

Tkachev, A. V., Ramenskaya, E. B. and Bojko J. R. (1991). Dynamics of hormone and metabolic state in polar inhabitants depend on daylight duration. *Arctic Medical Research*, **50**(suppl. 6), 152–155.

Trivers, R. L. (1974). Parent-offspring conflict. *American Zoologist*, **14**, 249–264.

Valeggia, C. and Ellison, P. T. (2004). Lactational amenorrhoea in well-nourished Toba women of Formosa, Argentina. *Journal of Biosocial Science*, **36**(5), 573–595.

van der Deure, W. M., Peeters, R. P. and Visser, T. J. (2007). Genetic variation in thyroid hormone transporters. *Best Practice and Research. Clinical Endocrinology and Metabolism*, **21**(2), 339–350.

Vanbillemont, G., Bogaert, V., De Bacquer, D., et al. (2009). Polymorphisms of the *SHBG* gene contribute to the inter-individual variation of sex steroid hormone blood levels in young, middle-aged and elderly men. *Clinical Endocrinology (Oxford)*, **70**(2), 303–310.

Videan, E. N., Fritz, J., Heward, C. B., et al. (2006). The effects of aging on hormone and reproductive cycles in female chimpanzees (*Pan troglodytes*). *Comparative Medicine*, **56**(4), 291–299.

Watts, D. P. and Mitani, J. C. (2002). Hunting behavior of chimpanzees at Ngogo, Kibale National Park, Uganda. *International Journal of Primatology*, **23**(1), 1–27.

Whitfield, G. K., Jurutka, P. W., Haussler, C. A., et al. (1999). Steroid hormone receptors: evolution, ligands, and molecular basis of biologic function. *Journal of Cellular Biochemistry Supplement*, **32–33**, 110–122.

Wilson, J. D. (1982). Gonadal hormones and sexual behavior. In *Clinical Neuroendocrinology*, G. M. Besser and L. Martini (eds). New York: Academic Press, pp. 1–29.

Wood, J. W. (1994). *Dynamics of Human Reproduction: Biology, Biometry, Demography*. New York: Aldine de Gruyter.

Wu, F. C., Butler, G. E., Kelnar, C. J., et al. (1996). Ontogeny of pulsatile gonadotropin-releasing hormone secretion from midchildhood, through puberty, to adulthood in the human male: a study using deconvolution analysis and an ultrasensitive immunofluorometric assay. *Journal of Clinical Endocrinology and Metabolism*, **81**, 1798–1805.

Yu, W. H., Kimura, M., Walczewska, A., et al. (1997). Role of leptin in hypothalamic-pituitary function. *Proceedings of the National Academy of Sciences of the United States of America*, **94**, 1023–1028.

Zaman, N., Hall, C. M., Gill, M. S., et al. (2003). Leptin measurement in urine in children and its relationship to other growth peptides in serum and urine. *Clinical Endocrinology (Oxford)*, **58**(1), 78–85.

Zeleznik, A. J. (2004). The physiology of follicle selection. *Reproductive Biology and Endocrinology*, **2**, 31.

Zera, A. J. and Harshman, L. G. (2001). The physiology of life history trade-offs in animals. *Annual Review of Ecology and Systematics*, **32**, 95–126.

Zimmet, P., Hodge, A., Nicolson, M., et al. (1996). Serum leptin concentration, obesity, and insulin resistance in Western Samoans: cross sectional study. *British Medical Journal*, **313**(7063), 965–969.

9 Ethical Considerations for Human Biology Research

Trudy R. Turner

The past 25 years have seen an ever increasing emphasis on and discussion of ethics in professional life. The Center for the Study of Ethics in the Professions at the Illinois Institute of Technology currently has a library of over 850 Codes of Ethics for various professions. Professional societies often have ethics modules online. Courses on ethics or ethics training are recommended parts of graduate curricula. Medicine, law, engineering, and business all have ethical standards and codes. The scientific community as a whole also shares a set of guiding principles that have been codified into a code of ethics for research and practice. In addition, each academic discipline has its own set of standards and principles, since each discipline has its own history and its own ethical dilemmas. Here I will briefly review ethical principles common to the scientific community as well as some of the ethical dilemmas faced by human biologists. This is not a comprehensive account. I direct the reader to the volume *Biological Anthropology and Ethics: from Repatriation to Genetic Identity* (Turner, 2005a) for a fuller discussion of the issues presented here.

Codes of ethics exist because every individual faces choices. These codes provide a framework for making informed choices in situations where there are conflicting obligations and responsibilities. The codes provide a framework of general principles for discussion and choice. No code can anticipate each unique situation. Discussion and reflection are vital to anticipate situations that may require quick decisions. Anthropologists (as evidenced in the American Anthropological Association [AAA] Code of Ethics, the American Association of Physical Anthropologists [AAPA] Code of Ethics) recognize a series of responsibilities – to the people with whom they work and whose lives they study, to scholarship, to science, to the public, to students and trainees, to employers, and employees. With these multiple levels of responsibility it can be difficult to determine which takes precedence in a given situation. Linda Wolfe (2005) has reviewed the responsibilities that anthropologists face that are common to all scientists in the practice of their science. These

include: avoid falsifying data or plagiarizing; avoid carelessness when collecting data; avoid falsifying grant records; avoid mistreating or discriminating against others, specifically students, coworkers, and employees; avoid giving professional advice on topics you are not qualified to discuss; avoid falsely representing a professional organization; report conflicts of interest; avoid clandestine research that cannot be published; follow rules of multiple authorship and be an objective peer reviewer. These are well established and agreed upon. However, the most difficult responsibilities a physical anthropologist or human biologist faces are often to the people we study. Discussion of these responsibilities can be subsumed under a general discussion of bioethics.

Bioethics, a special branch of applied ethics, is concerned with human health and human subjects research. Bioethics sets forth standards and principles that have become the model for work in medicine and research. Formal bioethics began after World War II, in the wake of Nazi experimentation, with the Nuremberg Code. This Code sets forth explicitly the principle of voluntary consent and lists criteria that must be met before any experimentation can be done on human subjects (Turner, 2005b). In the decades following Nuremberg, several ethical codes were enacted by the US government, the National Institutes of Health, the World Medical Association, and the Department of Health, Education and Welfare. In 1974, Congress enacted the National Research Act, which mandated an Institutional Review Board (IRB) review for all Public Health Service-funded research, and authorized the establishment of the National Commission for the Protection of Human Subjects of Biomedical and Behavioral Research. The Commission produced a document, known as the Belmont Report. The Belmont Report articulated three ethical principles: autonomy or respect for persons, beneficence, and justice. These principles are usually understood as do no harm, apply the rules of justice and fair distribution, do not deprive persons of freedom and help others (for a fuller discussion, see Stinson, 2005). The

Human Evolutionary Biology, ed. Michael P. Muehlenbein. Published by Cambridge University Press. © Cambridge University Press 2010.

Belmont Report has been codified into federal regulations and is used by Institutional Review Boards (IRBs) in their analysis of research protocols. These IRBs are local and found at institutions conducting or supporting human subjects' research. Institutional Review Boards are responsible for the review and approval of research activities involving human subjects. Their primary mandate is to protect the rights and safeguard the welfare of human research subjects. In 1981, final Department of Health, Education, and Welfare (DHEW) approval was given in 45 CFR 46, Subparts A, B and C (Title 45 Public Welfare, Code of Federal Regulations, Part 46 Protection of Human Subjects, 1991). On March 18, 1983, Subpart D was added to the regulations, providing additional protections for children who are subjects in research. Initially the Department of Health and Human Services (DHHS, the agency that replaced DHEW) regulations applied only to research conducted or supported by DHHS. But, in June 1991, the United States published a common policy for federal agencies conducting or supporting research with human subjects. That policy, which is known as "the Common Rule," extended the provisions of 45 CFR Part 46, to 14 other federal agencies; it now governs most federally supported research. The composition and operation of each university or institution IRB must conform to the terms and conditions of 45 CFR Part 46. (NIH Human Research Protection Program, http://www1.od.nih. gov/oma/manualchapters/intramural/3014/). Since the establishment of the IRB system, other federal commissions, including the National Research Council, the National Bioethics Advisory Commission, and the President's Council on Bioethics, have continued to examine issues concerning human subjects and to prepare updated guidelines. Human subjects research must be overseen by local IRBs. Funding by federal agencies will not be approved without IRB oversight and approval. In multi-institution or multi-national projects more than one IRB may be involved. Since every institution in this country has its own IRB and every country may have its own regulations, approval to do research can be cumbersome. But as Long (2005, p. 278) states, "As a general rule, investigators should simultaneously meet the highest standards of both our own culture and those of the research subjects' culture." According to Long (2005, p. 279) the current guidelines now "mandate that among other things, the researcher is responsible for proper scientific design, monitoring participant rights and welfare in the course of research and ensuring that all personnel on the research team are qualified and trained in human subjects protections."

Until 45 CFR 46 was implemented, oversight of research projects was not well codified. It is certainly possible to look back at research conducted in the years between the Nuremberg Code and 45 CFR 46 and identify practices that would no longer be considered wholly acceptable. However, these practices may have been fully acceptable and even far-sighted at the time research was conducted. Recently a controversy occurred surrounding James Neel and his research among the Yanomami. The controversy erupted a short time before a book by Patrick Tierney, *Darkness in El Dorado* (2000), was published. In proofs of the book, Tierney had accused Neel of starting a measles epidemic by injecting local villagers with a virulent measles vaccine. These charges were withdrawn before the book was published, due to a huge outcry by the scientific community about the validity of these claims. But controversy continued, with some researchers claiming Neel and his team did not do all they could to alleviate the measles epidemic among the Yanomami. Several professional organizations, including the AAA, set up task forces to review all materials. Within the anthropological community the controversy quickly came to concern the seeming conflict between obligations to science and humanitarian efforts. Those members of the task force charged with reviewing the Neel material (Turner and Nelson, 2005) found that Neel worked very hard to alleviate the measles epidemic he found in Venezuela.

There were other issues in the *Darkness in El Dorado* controversy that continue to have resonance for researchers today. These include the nature of informed consent, reciprocation for samples, and disposition of samples.

INFORMED CONSENT

There are several excellent reviews of the history of informed consent. Philosophers, ethicists, historians of science, and attorneys have all written about this (see for example, Beauchamp and Childress, 1989; Gert et al., 1997). Before the Nuremberg Report, most clinical medicine researchers were guided by the principle of beneficence and dealt little with the principle of autonomy. This principle of autonomy or respect for persons, articulated as voluntary or informed consent, was of primary importance in the shaping of the Nuremberg Code, which resulted as a response to Nazi medical experimentation. The Nuremberg Code and others that followed presented an ideal for dealing with human subjects. However, the particulars of application of this ideal to real-life situations was not as well articulated. There were certainly also situations and research projects conducted before the current regulations when principles of informed consent were missing (one of the most egregious examples being the Tuskegee Syphilis Study, which ended in 1972). The real question in many studies conducted before the implementation of the Belmont Report is how informed was informed consent. How well articulated were the goals, methods, and consequences of the research? While these questions are important when dealing with relatively informed Western, English-speaking individuals, how were they handled with non-Western, non-English-speaking, indigenous populations?

Recognizing this as a special case, the World Health Organization (WHO) convened a working group which met in 1962 and 1968 to discuss studies of "long-standing, but now rapidly changing, human indigenous populations" (Neel, 1964). Two reports were produced, both authored by James Neel (1964, 1968), which detailed the relationship and ethical obligations of researcher to study population. Neel particularly emphasized six factors of special importance: (1) The privacy and dignity of an individual must be respected and anonymity of subjects must be maintained. (2) Satisfactory, but carefully considered, recompense should be given for participation in a study. (3) The local population should benefit from the study by medical, dental, and related services. (4) Attempts should be made to maintain congenial social relationships with participants. (5) Learned individuals should be consulted. (6) There should be the utmost regard for cultural integrity of the group.

It is clear these principles were in place during the heyday of studies conducted under the Human Adaptability Section of the International Biological Programme (Collins and Weiner, 1977). Informed consent was sought, but not in the ways it is now sought. Turner and Nelson (2002, 2005) conducted a survey of 14 researchers working in the field during this time. They asked specifically how information was conveyed to individuals involved in studies. Every survey respondent stressed that there were individuals in the populations they worked with who did not participate. Voluntary consent was therefore assumed. Researchers either had government or local permission to conduct their studies. In every case, researchers gave some explanation of the motivation for the study. But some of these explanations were not necessarily complete. Researchers felt that perhaps local populations might not understand precisely the questions they were pursuing. Scientists who were part of the Yanomami expedition in the late 1960s have stated that the Yanomami were told that the researchers were going to look for diseases of the blood. This was true, but there were other things that were researched as well. Some Yanomami that have spoken to outsiders after the publication of the Tierney book have stated that there was an expectation of greater medical benefit from the work.

GROUP CONSENT

Physical anthropologists and human biologists are in a unique position – they are interested in the range of human variation and they frequently study traits that require them to return again and again to local, identified communities. Some researchers have had multi-decade relationships with their study populations. Over the decades, standards of what is included in informed consent have changed. Friedlaender (2005) gives a detailed account of his 35-year relationship with groups in the Solomon Islands and the changing standards of informed consent that he has implemented in his work. The current standard is, of course, full disclosure of the research project and the risks and benefits. This includes returning to the population for additional consent if samples might be used for a related, but not identical project. The best way to describe the current paradigm is in terms of an ongoing relationship between subjects and researchers, with subjects as active participants in research design and implementation.

Researchers are conditioned to think about the impact of research on an individual – on his or her health or psychological well being. It is important now that the researcher think about the impact of the research on the study population. The Belmont Principles protect individual participants in research projects. But many anthropological studies are population based and the findings of these studies can impact and affect whole populations. Consultation and group consent is now sought from populations. But group consent leads to a new suite of questions (enumerated by Juengst, 1999; Turner, 2005b): "Who speaks for the group? If the group is nested within a larger group, who represents the original group? What are the limits of the group? What is the relationship between expatriate groups and the community of origin? Does permission from a national government to conduct research have meaning for the community being studied? How does one obtain informed consent from an individual or a group whose members have little understanding of the project or the risks involved? How can the culture of the population be taken into account in the design or implementation of the project? What are the implications concerning the disclosure of the identity of the group? Can consent be withdrawn sometime in the future? How? Can samples be withdrawn sometime in the future? How? Are there appropriate benefits for the population under study?" This series of questions must be asked by every researcher engaged in research with human populations. In fact, these same questions are asked by cultural anthropologists, archaeologists, skeletal biologists, and any other researcher working with identified human populations.

An example of group consent and consultation can be found in the work of O'Rourke et al. (2005) who have been engaged in ancient DNA research with several populations. Each of the populations O'Rourke has worked with necessitated an individualized approach for access to samples. Some communities requested in-person meetings; others did not. Different communities had different restrictions on the size of samples. Working with Paleo-Indian remains for ancient DNA (aDNA) or skeletal biology studies in this country requires adherence to Native American Graves and Repatriation Act (NAGPRA) regulations. This may mean that some of these studies cannot take place. On the other hand, some scientists (Larsen and Walker, 2005) have been able to open discussions with native peoples on the study and disposition of human remains.

WEIGHING RISKS AND BENEFITS

Distinctions are made in the Belmont Report between biological and behavioral research. In biological or medical research, risks can often be more clearly identified than in behavioral research. But, behavioral research can cause emotional, psychological, or social harm (Stinson, 2005). Embarrassment or social stigma can be real consequences of participation in a research project. An individual may find questions embarrassing or might face social consequences if his or her answers to questions were known. One of the most important risks to an individual is disclosure of identity. Institutional Review Boards are very aware of the risks of this disclosure and will look closely at the ways in which identity can be safeguarded.

Winston and Kittles (2005) describe the challenges to perceived identity that were sometimes generated by the African Ancestry Project. Williams (2005) also discusses some disclosure issues faced by descendents of Thomas Jefferson after a study of DNA from descendents of Sally Hemings and the Jefferson family. Stigmatization can also occur at the group level and this may be especially true for marginalized or identified populations. Members of a group might be stigmatized by having their circumstances discussed. How can one avoid this situation? Researchers feel that a frank and full discussion of this risk can lead to a negotiation between subject and researcher on the presentation of the results and the naming of the group as participant. Williams (2005) also suggests that constant vigilance during the planning and execution of the project be paramount.

COMPENSATION

There is often a huge differential between the researcher and the participant in studies in education, socio-economic status and access to resources. Researchers can and do compensate participants in research studies for their time and effort. But the compensation must not be so great as to compel participation in a study. In addition, this differential may also influence rapport and trust between the investigator and the participant. In medical studies in this country, compensation may take the form of some level of medical care. However, what are appropriate compensations for research studies conducted with non-Western, identified populations? If a study includes medical personnel, some level of medical care may be given to the participants. But this is not necessarily the type of care individuals need. Certainly there are some conditions where antibiotics or analgesics can be useful and even life saving. What if a person is identified as diabetic? A single visit from a medical professional will not be sufficient to help this person. Referrals to more long-term care facilities may be in order. In the past, researchers have given many items as compensation. Researchers usually select these items in consultation with those familiar with the culture. Food items, photos, tools, machetes, and cash have all been given as compensation. Other items have been given to the group or community. One of the more recent examples of compensation involved technology transfer and training of individuals to use this technology (Bamshad, 1999; Jorde, 1999).

DATA SHARING

Circular A-110 of the US Office of Management and Budget stipulates that data collected through grants awarded by federal agencies such as National Institutes of Health (NIH) and National Science Foundation (NSF) are public. Federal agencies encourage the broad and rapid dissemination of information throughout the scientific community reflecting the scientific ideal of an open community of scholars pursuing novel ideas and avenues of research. Major complex electronic databanks already exist. Genetic information is shared via the International Nucleotide Sequence Database, which includes GenBank, the DNA DataBank of Japan, and the European Molecular Biology Laboratory. The NSF has a database initiative. Data used by physical anthropologists, however, is often unique and difficult to obtain. It may not be possible to obtain second sets of blood or saliva samples or measurements, or interviews from members of identified communities. Individual and group consent and confidentiality become major issues if samples are shared. Specific questions about data sharing range from the definition of data to fair use for the individual collecting the data (Turner, 2005c). The physical anthropology program of the NSF has a data-sharing requirement. However, the design implementation of this requirement is up to the individual researcher, but must go beyond publication of results in a scientific journal. Questions related to the ethics and the requirements of data sharing are really just beginning in our community.

IMPLEMENTING ETHICAL STANDARDS

Discussions of ethics usually generate more questions than answers. Field-work situations are unique. Issues change from place to place and from time to time. Ethical standards have also changed over time. There are now laws and standards concerning research and research subjects. In the vast majority of cases, researchers make their best possible efforts to behave according to the letter and the spirit of the law. But there are judgment calls and the IRB system is in place for oversight – so that someone else can add perspective. It is important, though, that ethics not be confined merely to IRB oversight. Continual discussion, especially during training, is important. Many training programs offer specialized courses in ethics. Others include discussion of ethics in every class. In either case, students and faculty need to continually engage each other in an assessment of ethical standards in professional life.

DISCUSSION POINTS

1. In the digital age, the treatment, storage, and transportation of sensitive data are particularly important. Storage of sensitive personal information on a laptop may not be secured. What precautions must be taken to secure this type of information?
2. How can one overcome the logistical issues associated with obtaining additional informed consent from remote populations?
3. What if a previously unknown condition is revealed during genetic screening? In many areas of the world, referring a subject to a physician is not feasible. What do you do?
4. The disparity in wealth between a researcher and a participant may influence rapport between the two. What are some methods to deal with this?
5. There may be situations when a research agenda and humanitarian concerns come into conflict. Discuss the expectations and limits of a researcher's responsibility to the participant community.

ACKNOWLEDGMENTS

I am grateful to Michael Muehlenbein for asking me to participate in this volume. I am also indebted to all the authors of the volume on Biological Anthropology and Ethics for their insights. I am also grateful to two anonymous reviewers who helped greatly with the discussion questions.

REFERENCES

Bamshad, M. (1999). Session one: issues relating to population identification, anthropology, genetic diversity and ethics. Symposium held at the University of Wisconsin-Milwaukee. http://www.uwm.edu/Dept/21st//projects/GeneticDiversity/session1.html (last accessed, November, 2007).

Beauchamp, T. L. and Childress, J. F. (1989). *Principles of Biomedical Ethics*, 3rd edn. New York: Oxford.

Collins, K. J. and Weiner, J. S. (1977). *Human Adaptability: a History and Compendium of Research in the International Biological Programme*. London: Taylor and Francis.

Friedlaender, J. S. (2005). Commentary: changing standards of informed consent: raising the bar. In *Biological Anthropology and Ethics: from Repatriation to Genetic Identity*, T. R. Turner (ed.). Albany, NY: SUNY Press, pp. 263–274.

Gert, B., Culver, C. M. and Clouser, K. D. (1997). *Bioethics: a Return to Fundamentals*. Oxford: Oxford University Press.

Jorde, L. (1999). Session four: successful research collaborations. Anthropology, genetic diversity and ethics. Symposium held at the University of Wisconsin-Milwaukee. http://www.uwm.edu/Dept/21st//projects/GeneticDiversity/jorde.html (last accessed, November, 2007).

Juengst, E. (1999). Session one: issues relating to identifying populations. Anthropology, genetic diversity and ethics. Symposium held at the University of Wisconsin-Milwaukee. http://www.uwm.edu/Dept/21st//projects/GeneticDiversity/juengst.html (last accessed, November, 2007).

Larsen, C. S. and Walker, P. L. (2005). The ethics of bioarchaeology. In *Biological Anthropology and Ethics: from Repatriation to Genetic Identity*, T. R. Turner (ed.). Albany, NY: SUNY Press, pp. 111–119.

Long, J. C. (2005). Commentary: an overview of human subjects research in biological anthropology. In *Biological Anthropology and Ethics: from Repatriation to Genetic Identity*, T. R. Turner (ed.). Albany, NY: SUNY Press, pp. 275–279.

Neel, J. V. (1964). *Research in Human Population Genetics of Primitive Groups*. WHO Technical Report Series No. 279. Geneva: World Health Organization.

Neel, J. V. (1968). *Research on Human Population Genetics*. WHO Technical Report Series No. 387. Geneva: World Health Organization.

O'Rourke, D. H., Hayes, M. G. and Carlyle, S. W. (2005). The consent process and DNA research: contrasting approaches in North America In *Biological Anthropology and Ethics: from Repatriation to Genetic Identity*, T. R. Turner (ed.). Albany, NY: SUNY Press, pp. 231–240.

Stinson, S. (2005). Ethical issues in human biology behavioral research and research with children. In *Biological Anthropology and Ethics: from Repatriation to Genetic Identity*, T. R. Turner (ed.). Albany, NY: SUNY Press, pp. 139–148.

Tierney, P. (2000). *Darkness in El Dorado*. New York: W. W. Norton.

Turner, T. R. (2005a). *Biological Anthropology and Ethics: from Repatriation to Genetic Identity*. Albany, NY: SUNY Press.

Turner, T. R. (2005b). Introduction: ethical concerns in biological anthropology. In *Biological Anthropology and*

Ethics: from Repatriation to Genetic Identity, T. R. Turner (ed.). Albany, NY: SUNY Press, pp. 1–13.

Turner, T. R. (2005c). Commentary: data sharing and access to information. In *Biological Anthropology and Ethics: from Repatriation to Genetic Identity*, T. R. Turner (ed.). Albany, NY: SUNY Press, pp. 281–287.

Turner, T. R. and Nelson, J. D. (2002). *Turner Point by Point. El Dorado Task Force Papers. Final Report*, vol. 1, part 6. Washington, DC: American Anthropological Association.

Turner, T. R. and Nelson, J. D. (2005). Darkness in El Dorado: claims, counter claims and the obligations of researchers. In *Biological Anthropology and Ethics: from Repatriation to Genetic Identity*, T. R. Turner (ed.). Albany, NY: SUNY Press, pp. 165–183.

Williams, S. R. (2005). A case study of ethical issues in genetic research: the Sally Hemings–Thomas Jefferson story. In *Biological Anthropology and Ethics: from Repatriation to Genetic Identity*, T. R. Turner (ed.). Albany, NY: SUNY Press, pp. 185–208.

Winston, C. E. and Kittles, R. A. (2005). Psychological and ethical issues related to identity and inferring ancestry of African Americans. In *Biological Anthropology and Ethics: from Repatriation to Genetic Identity*, T. R. Turner (ed.). Albany, NY: SUNY Press, pp. 209–229.

Wolfe, L. D. (2005). Field primatologists: duties, rights and obligations. In *Biological Anthropology and Ethics: from Repatriation to Genetic Identity*, T. R. Turner (ed.). Albany, NY: SUNY Press, pp. 15–26.

Commentary: a Primer on Human Subjects Applications and Informed Consents

Michael P. Muehlenbein

Investigators utilizing human subjects for any reason in their research are charged with the vital responsibility of ensuring ethical standards of conduct. These standards are outlined by a number of governing bodies, particularly the Department of Health and Human Services, Office for Human Research Protections (http://www.hhs.gov/ohrp/) and Office for Civil Rights (http://www.hhs.gov/ocr/) in the United States. Human subjects committees, empowered by these governing bodies, are then designated at the institutional level to provide oversight of human use in biomedical and behavioral research projects. Principle investigators and their teams are then responsible for ensuring the safety and welfare of subjects and compliance with research protocols. This is initially accomplished by working with institutional review boards (human subjects committees in particular) in a detailed application process prior to the initiation of research.

Below is a brief introduction for newcomers to the human subjects committee application, including the informed consent. This general information has been compiled from a number of online sources (last accessed April, 2010). Application format is certainly specific to individual institutions. However, the general guidelines are usually the same, drawn from basic requirements laid forward by, among others, the US Department of Health and Human Services Code of Federal Regulations, Title 45 (Public Welfare), Part 46 (Protection of Human Subjects) (http://www.hhs.gov/ohrp/humansubjects/guidance/45cfr46.htm) and the Health Insurance Portability and Accountability Act (HIPAA) (http://www.hhs.gov/ocr/hipaa/ and http://privacyruleandresearch.nih.gov/). Specific questions should be directed towards your institution's review board office, its human protections administrator and staff, as well as its HIPAA privacy board. Other useful resources include the following: Public Responsibility in Medicine and Research (http://www.primr.org/); the US National Institutes of Health Office of Extramural Research, certificates of confidentiality (http://grants.nih.gov/grants/policy/coc/index.htm); the US Food and Drug Administration, Guidance or Institutional Review Boards, Clinical Investigators, and Sponsors (http://www.fda.gov/oc/ohrt/irbs/default.htm); the US National Institutes of Health Office of Extramural Research, grants policy on research involving human subjects (http://grants.nih.gov/grants/policy/hs/index.htm); the International Association of Bioethics (http://www.bioethics-international.org/); and the Association for Practical and Professional Ethics (http://www.indiana.edu/~appe/). Some of the basic information on human subjects applications is summarized in Table 9.1.

Any project consisting of greater than minimal risk to subjects requires full committee review. However, US regulations allow certain types of research to be exempt from or expedited through reviews. Those that may be exempt include studies that constitute very minimal risk of harm or discomfort, such as research involving normal educational practices with adults, anonymous surveys and interviews, and analyses of existing data or biological specimens (with no identifiers attached). Other applications can be expedited through the review process if they constitute minimal risk from invasive or noninvasive sample collection (e.g., image, voice, body composition, biological fluid, DNA, etc.), or certain behavioral observations. In all cases, applications are still read by at least an administrator or decentralized reviewer.

Certain research subjects require special attention, and research projects utilizing these subjects must be evaluated via full panel review. This includes projects utilizing students, employees, and incarcerated individuals who are susceptible to coercion by teachers, employers, and parole boards. If students are given course credit or extra credit for participating in research projects, they should be allowed alternative means to obtaining this credit if they do not wish to volunteer as research subjects. Permission from a legally authorized representative is necessary for minors and decisionally impaired individuals,

TABLE 9.1. Basic list of important considerations for preparation of human subjects applications

1. Is your project exempt from review, or can it be expedited? This will depend on the type of data collected, methodology, and the type of population used (i.e., vulnerable populations like minors, prisoners, pregnant women, etc.). See http://www.hhs.gov/ohrp/humansubjects/guidance/45cfr46.htm for categories.

2. For projects being conducted at multiple institutions or with multiple investigators, additional review board submissions may be required. Biomedical and behavioral research conducted outside of the United States is expected to be conducted with the same ethical and regulatory standards as research conducted within the United States.

3. Are all risks to participants minimized, including physical, emotional, monetary and other risks?

4. Include basic and detailed information in your application, but make sure that it is readable to a wide audience unfamiliar with your type of work. Include a brief description of the purpose of the study, the anticipated length and location of your study, the subject pool you are utilizing, including sex, ethnicity and age ranges, your ultimate sample size as well as detailed inclusion and exclusion criteria for your study.

5. Explain how you will contact and enroll subjects. Include any advertisements in your application.

6. The informed consent document should be written in a nontechnical language that is understandable to all participants or their representatives. Explain the purpose and procedures of the research, risks and benefits to participation, how confidentiality will be maintained, how participation may be terminated, important contact information and, most importantly, how participation is completely voluntary. Include the informed consent document in your application.

7. Include a detailed protocol for your project. Include any survey instruments with your application.

8. Thoroughly consider any risks of participation to your subjects. What alternatives to your procedures have been considered and why are they not feasible? How will risks be minimized?

9. How are you going to protect participant confidentiality during data collection, storage, analyses, and presentation?

10. Carefully consider whether participation with monetary or course credit incentives are coercive.

11. Include in your application any documentation of investigator training in the protection of human research participants. Do any of the investigators have significant financial interests in the subject of this research?

12. After approval by the human subjects committee, any changes to your project, including small ones, must be reapproved by the committee before implementation into the protocol.

13. Continuing review applications, following expiration of the initial approvals, require a status report on the project, particularly unanticipated problems involving risks to subjects.

including those with cognitive disorders, those under the influence of drugs or alcohol, or someone under extreme emotional distress. For those participants aged 7–17 in the United States, the participant's assent must be obtained in addition to consent from a parent, advocate, or guardian. Other special subjects include pregnant women, fetuses, and newborns, all of which are particularly sensitive to adverse health outcomes.

Some research may require that a researcher provide participants with false information about the research or withhold information about the real purpose of the research. Such deception may require debriefing after study completion as well as a statement in the informed consent form explaining that the full intent of the study may not be disclosed until completion.

For projects being conducted at multiple institutions or with multiple investigators, additional review board submissions may be required. This is particularly the case with international projects. Biomedical and behavioral research conducted outside of the United States is expected to be conducted with the same ethical and regulatory standards as research conducted within the United States. Research must comply with laws of the host country, usually enforced through collaboration with a local research or educational institution. Some useful international resources include: the International Covenant on Civil and Political Rights (1996) from the

Office of the United National High Commissioner for Human Rights (http://www.ohchr.org/english/law/ccpr.htm); the International Ethical Guidelines for Biomedical Research Involving Human Subjects (2002) from the Council for International Organization of Medical Sciences (http://www.cioms.ch/frame_guidelines_nov_2002.htm); the Medical Ethics Manual (2005) from the World Medical Association (http://www.wma.net/e/ethicsunit/resources.htm); the Universal Declaration on Bioethics and Human Rights (2005) and the Universal Declaration on the Human Genome and Human Rights (1997) from the United Nations Educational, Scientific, and Cultural Organization (www.unesco.org/shs/bioethics). Other Country-specific human subjects research legislation can be found at http://www.hhs.gov/ohrp/international/HSPCompilation.pdf.

Most human subjects committee applications and informed consent documents require similar basic information, with variations depending on level of detail. You are usually first asked to provide a brief description of the purpose of the study, possibly including specific hypotheses or problem statements that are easily interpretable to nonexperts. You should specify the anticipated length of your study, possibly including a general scheduled timeline. You must specify the subject pool you are utilizing, including sex, ethnicity, and age ranges. Does your research focus only on one sex or ethnicity? If so, why? Are your subjects affiliated with a specific group, such as university students or members of a religious sect? Specify your ultimate sample size as well as detailed inclusion and exclusion criteria for your study. If there is a possibility that the investigators can or will withdraw subjects from participation, what are the conditions for withdrawal?

An individual must be free from coercion to make the decision to participate in your study, and thus it is necessary to specify how you will be contacting and enrolling subjects. Will you be using media advertisements, classroom announcements, direct greetings, etc.? Will the way you advertise the study inadvertently coerce someone into participation? Where will recruitment take place? Where will samples be procured or data collected?

An inadequate informed consent can easily delay the human subjects application process. It is a truism that consent must be voluntarily given by each research subject after he or she has been completely informed. The consent document should be addressed to the subject, inviting him or her to participate in the study. It should be written in a nontechnical language that is understandable to all participants or their representatives (a 12-year-old reading level is a good benchmark), and it must be in the subject's native language, in which case a translator/interpreter may have to be employed. In some circumstances, the informed consent document can be read to the participant. Studies involving anonymous participation with only minimal risks to the subject may not require any informed consent, in which case an information sheet is still supplied to participants. Anonymous surveys should still include a statement to the effect that completing the survey indicates that they agree to participate and are of appropriate age to participate. For web-based studies, information about the study should be provided, and participants should be required to click an "agreement" button before proceeding to the study task.

A detailed list of procedures used to gather information must be included in your application and informed consent. Describe the order of events during a typical session with each subject. How much time will it take? Include information regarding follow-up visits, and include copies of all advertisements and survey instruments with your application.

The risks of research with your human subjects must be thoroughly thought out and clearly defined, including all foreseeable immediate and long-term risks of their participation. These may include physical discomfort, injury or illness, anxiety, embarrassment, lost of respect, loss of time and wages, altered behavior, loss of confidentiality, and so on. Are you investigating illegal behaviors or gathering information that could make a subject uninsurable or unemployable?

What is the likelihood of infection if you are collecting blood samples? How will you minimize these risks from occurring? How do the anticipated benefits of this project outweigh the risks? What is the rationale for the necessity of such risk? What alternatives to your procedures have been considered and why are they not feasible? How will adverse events be reported?

Unique identifiers should be used to link subjects with their data and samples in order to maintain confidentiality. The code list linking this information must be kept in a secure place, such as a locked file cabinet or a password-protected computer. How secure is your data if you are conducting the project over the internet? Who will have access to the individual research records? Will access be limited to only the principle investigator, or will access also be granted to research assistants, collaborators, and representatives of funding agencies? To further safeguard subject identity, outcomes of the research are usually only released and published in aggregate form. Information that personally identifies individuals should not be released without prior written permission or as required by Federal or State laws (e.g., disclosure of reportable diseases, abuse, intent to harm oneself or others, etc.). Finally, you must consider how participant confidentiality will be maintained when the study is over. What will happen to the data/samples once you are done with them? Will information be deposited for future use or will it be destroyed? If there is a possibility that a subject's data or samples may be used for another related research project in the future, this must be stated in the original informed consent. Otherwise consent will have to be reobtained in the future.

Participant benefits must be carefully considered. Often times the only benefit to subjects may be their contribution to the body of knowledge. High monetary incentives are coercive, but some payment is frequently provided, at a minimum to cover transportation costs, meals, and lost wages. If the subject withdraws from the study before completion, it will likely be necessary to distribute partial payment for time rendered. Keep in mind that your home institution's accounting office may also require that receipt of each participant's compensation be signed for. Sometimes payee addresses and social security or other identification numbers must be recorded.

Subjects sometimes request to know their individual results, particularly in the cases of genetic, hormonal, and infection diagnostics. This may be problematic, however, because most analyses are not conducted in clinically certified laboratories and thus results are not technically available for nonresearch purposes. If subjects test positive for an infection, will you be providing medical treatment or only informing them to see a physician? Most researchers in human evolutionary biology are not qualified physicians and thus cannot legally provide diagnoses and treatment. Other things to consider may include: if injury results from participation in your study, will subjects be provided with medical care, and will participants be further compensated if your research results in eventual technological development?

Most human subjects applications require contact information from all personnel directly interacting with subjects. Conflicts of interest need to be identified: do any of the investigators have significant financial interests in the outcomes of the study? Furthermore, prior to the release of funds, most granting agencies require documentation of investigator training in the protection of human research participants. This may include simple online presentations and exams or even workshop attendance. Several excellent web-based tutorials can be found at the following: http://ohrp-ed.od.nih.gov/CBTs/Assurance/login.asp; http://cme.cancer.gov/clinicaltrials/learning/humanparticipant-protections.asp; http://irb-prod.cadm.harvard.edu:8153/hirbert/hethr/HethrLogin.jsp; http://www.research.umn.edu/consent/; http://www.yale.edu/training/; http://www.indiana.edu/~rcr/index.php.

It is obvious that contact information of the principle investigator must be supplied on the informed consent documents. Additional information should include contact information for the human subjects committee of the principle investigator's institution as well as any relevant foreign contact information. A statement should highlight the completely voluntary nature of the subject's participation and how refusal to participate or withdrawal at any time will not result in penalty or loss of benefits to which the subject would otherwise be entitled, or affect the subject's medical care if the research is being conducted at a medical institution. Above the signature line, a statement should be provided that the subject has read the consent form, has been given the opportunity to have all of their questions answered to their satisfaction, and agrees to participate. The informed consent should be signed (or otherwise recorded) and witnessed, and a copy supplied to the participant.

Review board approval is usually granted for up to one year, after which a continuation or termination must be applied for. All changes in protocols, even minor ones, require reapproval from the institutional review board before the changed protocols can be implemented. This may all seem like a daunting task, but one that is absolutely necessary for legal, moral, and ethical reasons. There is great satisfaction in doing things right the first time, especially for something as critical as the advancement of science with the assurance of human welfare.

Part II

Phenotypic and Genotypic Variation

"Man is the only creature who refuses to be what he is."

Albert Camus (1913–1960), *The Rebel: an Essay on Man in Revolt* (1951)

10 Body Size and Shape: Climatic and Nutritional Influences on Human Body Morphology

William R. Leonard and Peter T. Katzmarzyk

INTRODUCTION

Since the initial spread of *Homo erectus* from Africa some 1.8 million years ago, the human lineage has colonized every major ecosystem on the planet, adapting to a wide range of environmental stressors (Antón et al., 2002). As with other mammalian species, human variation in both body size and morphology appears to be strongly shaped by climatic factors. The most widely studied relationships between body morphology and climate in mammalian species are those described by "Bergmann's" and "Allen's" ecological rules. Bergmann's rule addresses the relationship between body weight (mass) and environmental temperature, noting that within a widely distributed species, body mass increases with decreasing average temperature (Bergmann, 1847). In contrast, Allen's rule considers the relationship between body proportionality and temperature (Allen, 1877). It finds that individuals of a species that are living in warmer climes have relatively longer limbs, whereas those residing in colder environments have relatively shorter extremities.

The physical basis of both of these ecological rules stems from the differences in the relationship between surface area (cm^2) and volume (cm^3 proportional to mass [kg]) for organisms of different size (Schmidt-Nielson, 1984). Because volumetic measurements increase as the cube of linear dimensions, whereas surface area increases as the square, the ratio of surface area (SA) to volume (or mass) decreases as organisms increase in overall size. In addition, metabolic heat production in all animals is most strongly related to body mass (e.g., Kleiber, 1975; FAO/WHO/UNU, 1985). Thus, in terms of thermoregulation, larger organisms are better suited to colder environments because they produce more heat and have relatively less SA through which to lose that heat. Conversely, small body size is better in warmer conditions, because these organisms will both produce less heat and have relatively greater surface for dissipating that heat.

These morphological and physiological differences between organisms of different size are at the heart of the relationship described by Bergmann's rule.

Allen's rule, on the other hand, considers how changes in shape can alter SA:mass ratios. For organisms of the same size, the more elongated or linear the shape, the greater the SA:mass ratio. Thus, for organisms residing in tropical environs, a linear body plan – with less mass in the trunk and greater mass in the extremities – will best help to facilitate heat dissipation. In contrast, for arctic-adapted organisms, a body build characterized by larger trunk size and shorter limbs will reduce metabolic heat loss by minimizing SA/mass.

Over the last 60 years, numerous studies have demonstrated that contemporary human populations generally conform to the expectations of Bergmann's and Allen's rules, such that populations residing in colder climes are heavier and have shorter relative limb lengths, resulting in a decreased ratio of SA to body mass (Schreider, 1950, 1957, 1964, 1975; Newman, 1953; Roberts, 1953, 1973, 1978; Barnicot, 1959; Baker, 1966; Walter, 1971; Stinson, 1990; Ruff, 1994; Katzmarzyk and Leonard, 1998). The most widely cited research on this topic is the work by D. F. Roberts (1953, 1978). In his 1953 paper, "Body weight, race and climate," Roberts demonstrated a significant negative correlation between body mass and mean annual temperature, indicating that humans appear to conform to Bergmann's rule. In subsequent work, Roberts (1973, 1978) showed that humans also conform to Allen's rule, such that populations living in colder regions have relatively shorter legs and larger relative sitting heights (RSH = [sitting height]/[stature]) than those groups inhabiting warmer regions.

In 1998, we re-examined the influence of temperature on body mass and proportions, drawing on anthropometric data collected after Roberts's pioneering work in 1953 (Katzmarzyk and Leonard, 1998). Our analyses confirmed many of Roberts's original findings. Specifically, we found that the inverse relationship between

Human Evolutionary Biology, ed. Michael P. Muehlenbein. Published by Cambridge University Press. © Cambridge University Press 2010.

mass and temperature continues to persist for both men and women; however, the slope of the regression was significantly shallower than that reported by Roberts. Likewise, the relationship that we found between RSH and temperature was more modest than that of Roberts's sample. These differences partly reflect secular changes in growth and body size over the last half-century, and development of improved technology that moderates extreme temperature exposure during development. These findings underscore the importance of both nutritional and temperature stresses in shaping human variation in body size and shape.

Unlike most chapters in this book, the present chapter will not only provide a review, but also a detailed reanalysis of the influence of climatic and nutritional factors on worldwide human variation in body mass and proportions. We hope that this may serve as an example of the application of modern theory and methods in human evolutionary biology, as outlined above as well as in Chapters 6 and 7 of this volume.

First we examine the relationships between selected anthropometric dimensions and mean annual temperature in the Roberts (1953) sample, and our own worldwide sample (from Katzmarzyk and Leonard, 1998). In exploring the influence of climate on body morphology in the two samples, we utilize indices (the body mass index [BMI] and SA/mass ratios) not considered by Roberts and others in previous work. Next, we consider the magnitude of patterns of change in relations between climate and body size and proportions over the last 50+ years, and consider the reasons for those changes. Finally, we examine the implications that this work has for the use of anthropometric methods of nutritional health assessment in biomedical and public health contexts. Increasingly, weight-for-height measures such as the BMI are being used as a screening tool for assessing the risks of obesity around the world. Our findings suggest that simple indices such as the BMI must be applied with caution when they are used to compare the relative risk of obesity and associated chronic diseases in populations from different ecological/environmental contexts.

SAMPLE AND METHODS

This study draws on the published data on mean stature (cm) and body weight (kg) from 116 adult male samples compiled by Roberts (1953) in his original study on climatic influences on body mass. In addition, the primary sample that we compiled includes mean stature and body weights for 223 male and 198 female samples from studies published between 1953 and 1996. For a subsample of these studies, data on mean

TABLE 10.1. Geographic distribution of studies compiled for the current sample of males and females and Roberts (1953) sample (males only).

| | Current sample | | Roberts sample |
Region	Males	Females	Males
African	44	42	28
Australian	11	8	2
Melanesian	47	41	4
American	55	47	16
European	17	18	20
Central Asian	6	7	2
East Asian	11	8	29
Polynesian	13	12	2
South Asian	4	1	6
Indian	15	14	7
Total	223	198	116

sitting height (cm; $n = 168$ samples; 94 males; 74 females) and mean triceps skinfold thickness (mm; $n = 102$ samples; 56 males, 46 females) were also available. Table 10.1 shows the geographic distribution of the studies compiled in the current sample and in the Roberts (1953) sample. The groupings follow those used by Roberts (1953). Additional information about the composition of the sample and its geographic distribution is presented in Katzmarzyk and Leonard (1998). Mean annual temperatures (°C) for the locales from which the anthropometric data were collected were obtained from climatic tables and atlases (Steinhauser, 1970, 1979; Gentilli, 1977; Schwerdtfeger, 1976; Lydolf, 1971; Willmott et al., 1981).

From the sample means of height and weight compiled from the literature, several derived indices were also calculated. These included: (1) the body mass index (BMI; kg/m^2); (2) body surface area (cm^2); (3) surface area/mass ratio (SA/mass; cm^2/kg); and (4) relative sitting height (RSH). The BMI was calculated as (weight [kg])/(stature [m^2]). Surface area was estimated using the equation of Gehan and George (1970) as recommended by Bailey and Briars (1996):

$$\ln SA = -3.751 + 0.422\ln(\text{stature}) + 0.515\ln(\text{mass})$$

where stature is in cm, mass is in kg, and SA is in m^2. Surface area/mass was then determined as body SA (cm^2)/weight(kg). Relative sitting height was calculated as: $100 \times$ (sitting height[cm])/(stature[cm]).

CLIMATIC INFLUENCES ON BODY WEIGHT AND PROPORTIONS

Table 10.2 presents the descriptive statistics for the anthropometric sample of Roberts (1953), and our

TABLE 10.2. Descriptive statistics of anthropometric dimensions for the D. F. Roberts (1953) sample (males only), and the current sample.

Measure	Roberts (males)		Current (males)		Current (females)	
	n	Mean±SD	n	Mean±SD	n	Mean±SD
Stature (cm)	116	163.7 ± 6.6^a	222	165.4 ± 6.5^d	197	154.3 ± 6.0
Body weight (kg)	116	56.5 ± 7.9^b	223	61.6 ± 9.5^d	198	54.1 ± 9.3
BMI (kg/m^2)	116	21.0 ± 2.0^b	222	22.4 ± 2.5	197	22.6 ± 2.9
Surface area (m^2)	116	1.61 ± 0.14^b	222	1.69 ± 0.15^d	197	1.53 ± 0.15
SA/mass (cm^2/kg)	116	287.1 ± 16.5^b	222	276.8 ± 16.4^d	197	286.8 ± 18.9
Sitting height (cm)	–	–	94	85.1 ± 3.7^d	74	80.7 ± 3.4
RSH (%)	–	–	94	51.7 ± 1.7^c	74	52.3 ± 1.7
Triceps (mm)	–	–	56	7.7 ± 3.4^d	46	14.1 ± 7.0

Notes: Differences between the Roberts sample males and the current sample males are statistically significant at: [a]$P < 0.05$; [b]$P < 0.001$. Differences between current sample males and females are statistically significant at: [c]$P < 0.05$; [d]$P < 0.001$.
BMI, body mass index; RSH, relative sitting height; SA, surface area.

TABLE 10.3. Regression parameters for the relationships of body weight (kg), BMI (kg/m^2), SA/mass (cm^2/kg), and RSH (%) versus mean annual temperature (°C) for the Roberts sample, and the males and females of the current sample.

Sample/measure	n	Regression parameters		
		Y-intercept	b+SE	R
Roberts (males):				
Body weight (kg)	116	65.80	-0.55 ± 0.70	-0.59^{***}
BMI (kg/m^2)	116	23.41	-0.14 ± 0.02	-0.59^{***}
SA/mass (cm^2/kg)	116	267.55	1.15 ± 0.15	0.59^{***}
Current males:				
Body weight (kg)	223	66.86	-0.26 ± 0.06	-0.27^{***}
BMI (kg/m^2)	222	23.62	-0.06 ± 0.02	-0.22^{***}
SA/mass (cm^2/kg)	222	267.00	0.49 ± 0.11	0.29^{***}
RSH (%)	94	52.97	-0.06 ± 0.02	-0.37^{***}
Current females:				
Body weight (kg)	198	59.33	-0.26 ± 0.06	-0.28^{***}
BMI (kg/m^2)	197	24.41	-0.09 ± 0.02	-0.30^{***}
SA/mass (cm^2/kg)	197	273.73	0.66 ± 0.13	0.34^{***}
RSH (%)	74	53.66	-0.07 ± 0.02	-0.45^{***}

Note: Correlations are statistically significant at: [***]$P < 0.001$.
BMI, body mass index; RSH, relative sitting height; SA, surface area.

subsequent sample (Katzmarzyk and Leonard, 1998). Our male sample is significantly taller, heavier (higher body weights and BMIs) and has a significantly lower average SA/mass ratio than the Roberts sample. Note that the relative differences in body weight between the two samples are larger than the changes in stature (+9% vs. +1%). Thus, we find that the prevalence of overweight and obesity (BMI $\geq$ 25 kg/m^2) in the current sample is more than three times that of the Roberts sample (12.2% vs. 3.4%; $P < 0.001$).

The Roberts (1953) paper did not provide data for females. The females of our sample are significantly shorter and lighter than their male counterparts, with shorter sitting heights and lower estimated body surface areas. Conversely, SA/mass ratios and triceps skinfold thicknesses are higher among the women. Body mass indices of the female sample are comparable to those of the males, with similar prevalence of overweight and obesity (12.2% in males; 15.6% in females; n.s.).

The influence of mean annual temperature on selected anthropometric dimensions in both the original Roberts (1953) sample and our more recent sample is explored in Table 10.3 and Figures 10.1–10.4. In each

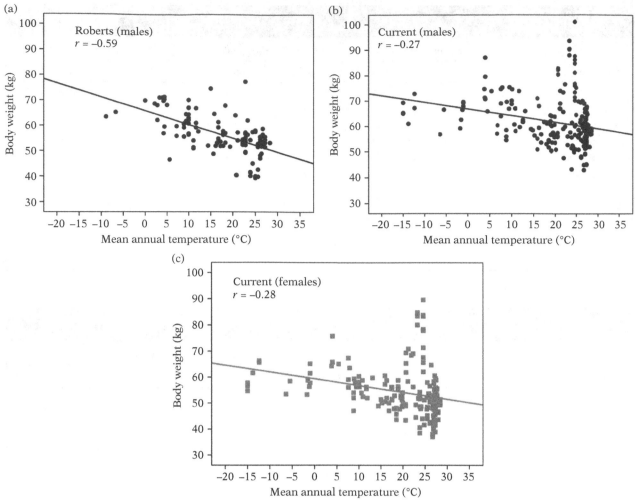

10.1. Relationship between body weight (kg) and mean annual temperature (°C) in **(a)** males of the Roberts (1953) sample, **(b)** males of the current sample, and **(c)** females of the current sample. In all three samples, body weight is inversely related to temperature. The correlation between body weight and temperature is stronger in the Roberts sample compared to males and females of the current sample.

group, weight and the BMI are negatively correlated with mean annual temperature (Figures 10.1 and 10.2). In males of the present sample, the correlations of mean annual temperature with weight and BMI are −0.27, and −0.22, respectively, whereas the correlations for females are −0.28 for body mass and −0.30 for the BMI. In the Roberts (1953) sample, the correlations are stronger ($r = -0.59$ for both weight and BMI), and the slopes of the best fit regressions are significantly steeper than in the current male and female samples (weight: $b = -0.55$ [Roberts] vs. −0.26 [males], −0.26 [females]; $P < 0.001$; BMI: $b = -0.14$ [Roberts] vs. −0.06 [males], −0.09 [females]; $P < 0.001$).

The SA/mass ratios are positively associated with mean annual temperature in all three of the samples (Figure 10.3). This indicates that populations of colder climes have body plans that minimize surface area to mass to reduce metabolic heat loss, whereas those

inhabiting hotter regions have physiques that maximize surface area/mass to promote heat dissipation. However, as noted above, the correlations for the Roberts (1953) sample ($r = 0.59$) are much stronger than in the current male and female samples ($r = 0.29$ for males; $r = 0.34$ for females). Additionally, the regression slopes of the Roberts sample are twice as large as those seen for in the current samples ($b = 1.15$ [Roberts] vs. 0.49 [males] and 0.66 [females]; $P < 0.001$).

Finally, we see that in the current sample RSH is negatively correlated with temperature in both sexes ($r = -0.37$ in males; $r = -0.45$ in females; see Table 10.3 and Figure 10.4). This relationship is consistent with the expectations of Allen's rules, indicating that tropically adapted populations have a more linear body build, with relatively longer limbs and shorter trunks. In contrast, populations of high latitude environments

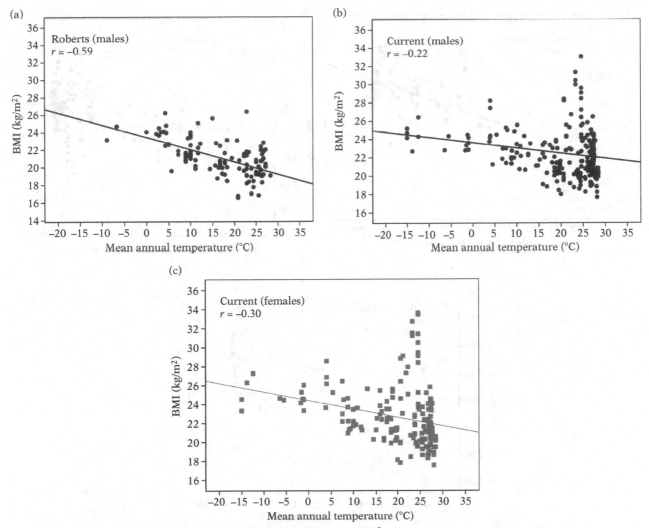

10.2. Relationship between the body mass index (BMI; kg/m²) and mean annual temperature (°C) in **(a)** males of the Roberts (1953) sample, **(b)** males of the current sample, and **(c)** females of the current sample. In all three samples, the BMI is inversely related to temperature. The correlation between BMI and temperature is stronger in the Roberts sample compared to males and females of the current sample.

are characterized by a more stout body plan with shorter extremities and a relatively larger trunk. As with the previously discussed analyses, the correlations between RSH and temperatures reported by Roberts (1978, pp. 21–22) were stronger than those observed in the current sample (−0.62 for males; −0.65 for females). In addition, the regression coefficients for the Roberts (1978) sample were twice those of the current samples ($b = -0.12$ vs. -0.06 for males; -0.13 vs. -0.07 for females).

SECULAR TRENDS IN BODY WEIGHT AND PROPORTIONS

Obesity and associated metabolic disorders have now emerged as major global health problems. According to recent estimates by the World Health Organization

(WHO) (2006), at least 400 million adults worldwide are obese. If current trends continue, this number is projected to increase to 1.1 billion by 2030 (Kelly et al., 2008). These dramatic increases in body size over the last two generations have altered the relationship between climate and body morphology across human populations.

The results presented in Table 10.4 further investigate the temporal changes in the relationship between climate and body morphology within our sample. The table presents results of multiple regression analyses in which both mean annual temperature and year of study publication were entered as independent variables. We see that even after adjusting for the influence of climate, both body weight and BMI have increased over time, whereas SA/mass ratios have significantly declined. In contrast, RSH does not appear to show a consistent pattern of change over time. The temporal

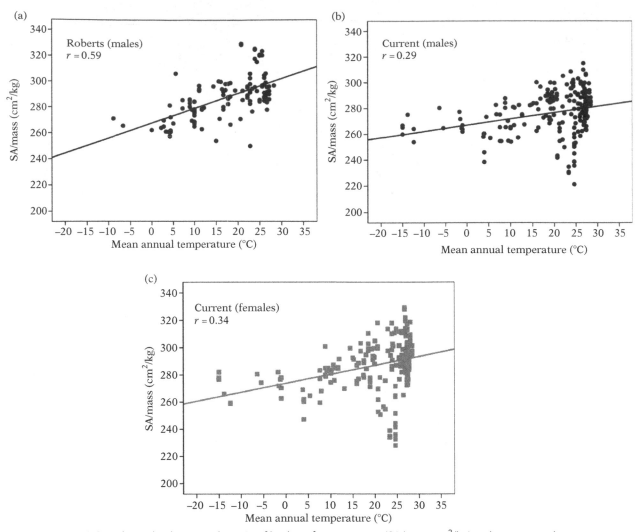

10.3. Relationship between the ratio of body surface area:mass (SA/mass; cm²/kg) and mean annual temperature (°C) in **(a)** males of the Roberts (1953) sample, **(b)** males of the current sample, and **(c)** females of the current sample. In all three samples, the SA/mass is positively related to temperature. The correlation between SA/mass and temperature is stronger in the Roberts sample compared to males and females of the current sample.

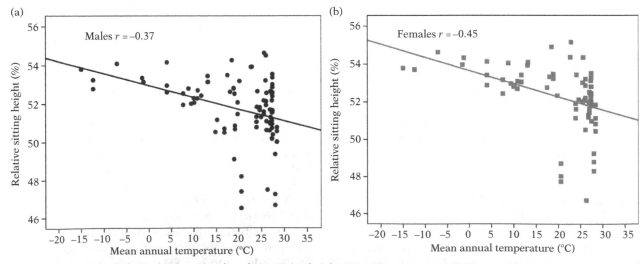

10.4. Relationship between the relative sitting height (RSH; %) and mean annual temperature (°C) in **(a)** males and **(b)** females of the current sample. In both males and females, RSH is inversely related to temperature.

TABLE 10.4. Multiple regression analyses of the influence of mean annual temperature and "year of study publication" on body weight, BMI, SA/mass, and RSH.

Dependent variables	n	Constant	Independent variables		
			Temperature (b + SE)	Year (b + SE)	Model R^2
Males:					
Body weight (kg)	223	−165.05	−0.27 ± 0.06***	0.12 ± 0.06	0.09***
BMI (kg/m^2)	222	−80.75	−0.06 ± 0.02***	0.05 ± 0.02**	0.09***
SA/mass (cm^2/kg)	222	690.90	0.50 ± 0.11***	−0.22 ± 0.11	0.10***
RSH (%)	94	108.67	−0.06 ± 0.02***	−0.03 ± 0.02	0.16***
Females:					
Body weight (kg)	198	−229.70	−0.28 ± 0.06***	0.15 ± 0.06*	0.11***
BMI (kg/m^2)	197	−98.24	−0.10 ± 0.02***	0.06 ± 0.02**	0.14***
SA/mass (cm^2/kg)	197	905.77	0.68 ± 0.13***	−0.32 ± 0.12**	0.15***
RSH (%)	74	33.39	−0.07 ± 0.02***	0.01 ± 0.02	0.21***

Note: *$P < 0.05$; **$P < 0.01$; ***$P < 0.001$.
BMI, body mass index; RSH, relative sitting height; SA, surface area.

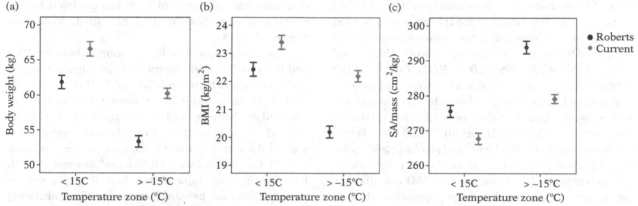

(a)   (b)   (c)   ● Roberts   ◆ Current

10.5. Mean (±SEM) values of **(a)** body weight (kg), **(b)** body mass index (BMI) (kg/m^2), and **(c)** SA/mass (cm^2/kg) for males of the Roberts (1953) and the current samples residing in "colder" (<15°C) and "warmer" (≥15°C) climates. For all three measures, the differences between the Roberts and current samples are much larger in the warmer climate samples.

changes are statistically significant for body weight, BMI, and SA/mass in the female sample. For men, the secular trend is statistically significant for BMI ($b = 0.05$; $P < 0.01$), and approaches statistical significance for the weight ($b = 0.12$; $P = 0.07$) and SA/mass ($b = -0.22$; $P = 0.056$).

The temporal increases in body mass are greater in women than men. The regression coefficients indicate that after controlling for temperature, average gains in body weight were 0.12 kg/year in men and 0.15 kg/year in women, whereas the increases in BMI were 0.05 and 0.06 kg/m^2 a year for men and women, respectively.

Relative sitting heights have shown modest changes over time. Among men, there has been a decline in RSH over the last 50 years, a trend that approaches statistical significance ($b = -0.03$; $P = 0.16$).

Among women, the temporal trend for RSH has been essentially flat ($b = 0.01$; $P = 0.50$).

The marked changes in body size over time are not the same across all regions of the world. Rather, the increases in body mass have been disproportionately larger among populations of the tropics. This point is evident in Figure 10.5, which shows the differences in: (a) weight; (b) BMI; and (c) SA/mass among men of the Roberts (1953) and current samples for groups living in areas with mean annual temperatures of <15°C ("colder") and ≥15°C ("warmer"). Differences in body weight between the two samples are about 50% greater among populations of the warmer climates, as compared to those of colder environments (differences of +6.9 kg [+13%] vs. +4.7 kg [+8%]). Similarly the differences in BMI in the warmer climes are twice those of the colder regions (differences of +2.0 vs. +1.0 kg/m^2).

Conversely, declines in SA/mass ratios are twice as large in colder environs (differences of -0.15 vs. $-0.08\,cm^2/kg$). Thus while there has been a general, worldwide increase in body mass over the last 50 years, the increases appear to be disproportionately larger in tropical regions.

CLIMATE, BODY PROPORTIONS AND THE BMI: IMPLICATIONS FOR THE ASSESSMENT OF NUTRITIONAL STATUS

The results presented thus far clearly indicate that climate and dietary/nutritional factors both play important roles in shaping patterns of interpopulational variation in adult body size and proportions around the globe. This finding has important implications for the use of anthropometric indices for assessing physical nutrition status (Frisancho, 1990, 2008; Gibson, 2005). The implications are particularly critical for considering the use of the BMI for assessing risks of both under- and over-nutrition. Over the last 15 years, the BMI has become the most widely used standard for assessing nutritional status of adults in both the United States and throughout the world (Shetty and James, 1994; WHO, 1995, 2000, 2004). Current WHO (1995) recommendations advocate using the following BMI ranges for discerning different levels of nutritional well-being in adults over the age of 18 years: (1) undernutrition: $<18.5\,kg/m^2$; (2) "healthy": $18.5-24.9\,kg/m^2$; (3) overweight: $25.0-29.9\,kg/m^2$; and (4) obese: $\geq30\,kg/m^2$. Yet, accumulating research in human nutritional science suggests that a single set of BMI cut-offs may not be appropriate for all human populations. In particular, recent work by Deurenberg and colleagues (Deurenberg et al., 1998; Deurenberg-Yap et al., 2000; Deurenberg-Yap and Deurenberg, 2003) indicates that the WHO cut-off points for overweight and obesity do not effectively apply to Asian populations who appear to have a different BMI versus body fat relationship than European populations. These conclusions are consistent with anthropometric studies of indigenous populations living in different environments throughout the world (e.g., Norgan, 1994a, 1994b; Shephard and Rode, 1996; Leonard et al., 2002; Snodgrass et al., 2006).

Much of the debate surrounding the appropriateness of a single set of BMI cut-offs for assessing nutritional status has overlooked the central question of what factors may be responsible for producing different relationships between BMI and body fatness across different populations. It is clear that variation in body proportions and body morphology play a strong role shaping variation in the BMI. Garn et al. (1986a, 1986b), for example, found that BMI was strongly correlated with measures such as RSH and chest breadth

in two large US samples (NHANES I and the Tecumseh Community Health Study). Norgan (1994b) also found RSH to be strongly correlated with the BMI, and argued that differences in body proportions shape the relationship between BMI and body composition across populations.

In light of the strong influence that RSH has on the BMI, it is reasonable to expect that the relationship between BMI and body fatness may vary with climate. The extreme examples of climatic differences in the BMI versus fatness relationship can be seen when we compare data from the Australian Aborigines and Inuit of the Canadian Arctic. Norgan (1994a) has shown that traditionally living Australian Aborigines studied before the 1970s had very low BMIs, suggestive of chronic undernutrition, yet had skinfold thicknesses that indicated adequate nutritional status. Conversely, early work among Inuit men and women has shown that despite having BMIs that were at or above the threshold for "overweight," they were relatively lean, as reflected in both skinfold measures and estimates of body fatness from hydrostatic weighing (Shephard et al., 1973; Rode and Shephard, 1994).

Figure 10.6 examines the relationship between BMI and RSH for men and women of the current sample. The correlations between RSH and BMI in the sample are 0.48 for men and 0.58 for women, similar to those reported by Norgan (1994b) for analyses of 95 male and 63 female samples from human populations around the world. Among men, each percent increase in RSH is associated with a $0.58\,kg/m^2$ increase in BMI. For women, the slope of the best fit regression is steeper, with each percent increase in RSH translating into an increase of $0.84\,kg/m^2$ in BMI. These relationships imply that at the extremes of human RSHs, from 46 to 55%, the corresponding average BMI differences are about $5\,kg/m^2$ in men (mean BMIs of 19.1 vs. 24.3 kg/m^2) and $7\,kg/m^2$ in women (mean BMIs of 17.0 vs. 24.5 kg/m^2). Thus, for adults with extremely linear builds (i.e., low RSHs), average BMIs tend to cluster in the underweight to low healthy range, whereas those with very high RSHs have mean BMIs that approach the overweight range.

It is also clear from Figure 10.6 that populations from colder climates tend to cluster in the upper end of the relationship, having high BMIs and RSHs, despite the recent trend for tropical populations to show more rapid increases in body mass. Body mass indexes of colder climate populations ($<15°C$) are significantly higher than their warmer climate peers in both men (24.3 vs. $21.9\,kg/m^2$; $P < 0.001$) and women (25.0 vs. $21.4\,kg/m^2$; $P < 0.001$). Similarly RSH among populations from cooler climes average 1.4% greater in men (52.8% vs. 51.4%; $P < 0.001$) and 1.6% greater in women (53.4% vs. 51.8%; $P < 0.001$).

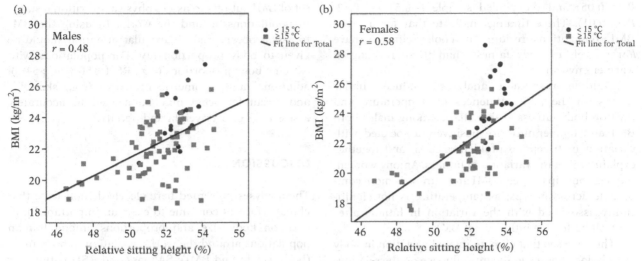

10.6. Relationship between the body mass index (BMI) (kg/m²) and relative sitting height (RSH) (%) in **(a)** males and **(b)** females of the current sample. The BMI is positively correlated with RSH in both sexes, suggesting that body proportion exerts a strong influence on the BMI. Additionally, note that populations from colder climes cluster to the upper right corner of the graph, having high BMIs and RSHs.

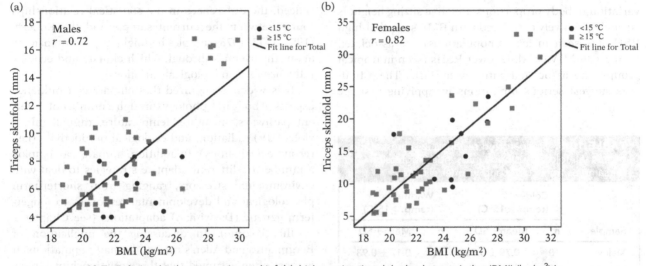

10.7. Relationship between triceps skinfold thickness (mm) and the body mass index (BMI) (kg/m²) in **(a)** males and **(b)** females of the current sample. The triceps skinfold measures are positively correlated with BMI for both sexes. Note, however, that populations from colder climates generally fall below the regression line (particularly among males), suggesting that they are leaner than expected for their BMIs.

To directly test the influence of climate on the relationship between the BMI and body fatness, we draw on the subsample of 102 groups for which we have measurements of triceps skinfold thickness, BMI, and mean annual temperature. Figure 10.7 shows the plot of triceps versus BMI for males and females. As expected, the two measures are strongly positively correlated, $r = 0.72$ in males and 0.82 in females ($P < 0.001$ in both sexes). The populations clustering in the upper right-hand corners of both the male and female graphs (Figures 10.7 a and b) include Samoans and Mexican-Americans, both groups that have shown marked

increases in obesity rates over the last 50 years (McGarvey, 1991; McGarvey et al., 1993; Flegal et al., 2002; Ogden et al., 2006). We also find that in both sexes, the populations from colder regions generally fall below the regression line, suggesting lower than expected levels of body fatness for a given BMI.

Table 10.5 compares the standardized residuals (z-scores) from the sex-specific triceps versus BMI regressions for populations from warmer and colder regions. Populations of colder climes have lower values than those of warmer climates, with the differences being statistically significant for males (-0.79 vs. 0.17;

$P < 0.05$) and the pooled sample (-0.51 vs. 0.12; $P < 0.01$). These findings indicate that for a given BMI, populations residing in cooler climates have *lower* levels of body fatness than those residing in warmer environs.

Multiple regression analyses produce similar results on the joint influences of temperature and BMI on body fatness (Table 10.6). Among males both BMI and temperature are positively associated with variation in triceps skinfold measures, and together explain 57% of the variation in fatness. Among women, the relationship between BMI and fatness is more comparable across groups, as temperature is not significantly associated with the variation in fatness after controlling for the influence of BMI.

Thus, we see that climate-related variation in body morphology has a significant influence on the relationship between the BMI and body fatness. At the same BMI, individuals from colder climates are, paradoxically, *leaner* than expected based on international references, whereas those of warmer regions are *fatter*. These climatic differences appear to be associated with variation in body proportions. Relative sitting height is strongly positively correlated with BMI, such that high RSHs (as seen in arctic populations) are associated with greater BMIs, while lower RSHs (seen in tropical groups) are associated with lower BMIs. These findings suggest serious limitations in applying a single set of BMI cut-offs to assess physical nutritional status in populations around the world. In using the BMI to assess obesity risks, particular attention should be given to body proportionality. For populations with extreme body proportions (e.g., RSH < 50% or >54%), additional anthropometric measures (e.g., skinfolds and circumferences) may be needed to accurately assess risks of overweight and obesity.

DISCUSSION

The analyses presented here clearly demonstrate that climatic factors continue to exert an important influence on body size and proportions among human populations around the world. Variation in body mass (both weight and BMI), SA/mass and RSH within our current sample broadly conforms to the expectations of Bergmann's and Allen's rules. However, it is equally clear that these relationships have not remained static over time, but rather, have changed in response to shifting socioeconomic and ecological circumstances. Indeed, the differences in the statistical relationships found between the current sample and the Roberts (1953, 1973, 1978) samples highlight important insights about the avenues through which climate and ecology influence human biological variation.

It is widely recognized that climate can influence aspects of body morphology through a number of different pathways, including temperature, rainfall, ultraviolet (UV) radiation, and ecological productivity (i.e., resource availability). In addition, humans can employ a number of different adaptive strategies to deal with environmental stressors, ranging from shorter-term physiological and developmental responses to longer-term genetic (Darwinian) adaptations (see Chapter 2 of this volume). In discussing the application of Bergmann's and Allen's rules to human populations, it is most often assumed that these relationships largely (or exclusively) reflect genetic adaptations to thermal stress (e.g., Schreider, 1975; Ruff, 1994). Yet, there is strong evidence to show these patterns are also the product of other environmental factors (such as nutrition) operating through nongenetic adaptive mechanisms.

TABLE 10.5. Standardized residuals (*z*-scores) of the triceps skinfold versus body mass index regression for samples from colder and warmer climates.

Sample	Colder (temp. <15°C)		Warmer (temp. ≥15°C)	
	n	Mean ± SD	*n*	Mean ± SD
Males	10	−0.79 ± 0.91	46	0.17 ± 0.93[*]
Females	9	−0.20 ± 1.25	37	0.05 ± 0.93
Total	19	−0.51 ± 1.09	83	0.12 ± 0.93[**]

Note: Differences between the colder and warmer samples are different at: [*]$P < 0.05$; [**]$P < 0.01$.

TABLE 10.6. Multiple regression analyses of the influence of BMI and mean annual temperature on triceps skinfold thickness.

Dependent variables	*n*	Constant	Independent variables		
			BMI (*b* + SE)	Temperature (*b* + SE)	Model R[2]
Males	56	−15.52	0.96 ± 0.12[***]	0.77 ± 0.03[**]	0.57[***]
Females	46	−28.65	1.81 ± 0.19[***]	0.03 ± 0.05	0.68[***]

Note: [**]$P < 0.01$; [***]$P < 0.001$.
BMI, body mass index.

Improvements in both nutrition and public health over the last half century have contributed to secular trends in growth of stature and body mass in populations throughout the world (e.g., Roche, 1979). Declining rates of childhood malnutrition in the developing world (de Onis et al., 2000) have contributed to the relatively larger increases in body mass among tropical populations documented here (Figure 10.5). These disproportionate increases in mass among populations of the tropics help to explain the marked reductions in the strength of the associations of body weight and BMI with mean annual temperature in the current sample relative to the Roberts sample. These findings further suggest that the very strong inverse relationship between weight and temperature initially reported by Roberts was partly attributable to differences in diet and nutrition, as well as differences in thermal stress. Recent analyses by Kelly et al. (2008) suggest that these trends are continuing, such that the developing regions of the world are projected to have the greatest proportional increases in the number of overweight and obese adults over the next 20 years.

Nutritional and developmental factors also appear to play a role in shaping variation in body proportions. Work by Frisancho et al. (1975, 1980; Stinson and Frisancho, 1978) among Quechua children of the Lamas region of lowland Peru provides important insights into the role of ecology in shaping the developmental changes in body proportions. This research compared the growth of lowland Quechua children from Lamas to Quechua children of the same genetic background living in the highland region of Junin. Stinson and Frisancho (1978) found that the immigrant Quechua children to the warm and humid lowlands had more linear body builds than their peers from the cold, high altitude regions. The authors attributed the differences in body proportions between the two groups to the influence of temperature and altitude stressors. These results also demonstrate the role of developmental acclimatization in promoting significant differences in body proportions among two populations with similar genetic backgrounds living in radically different environments.

A growing body of research is also documenting the influence of nutrition on the development of body proportions during childhood growth. It is now recognized that nutritional stress and poor growth during early childhood disproportionately affects the growth of long bones (Tanner, 1978). Hence, improvements in nutrition during growth and development are associated with not only taller overall stature, but longer relative leg lengths and thus, lower RSHs (Frisancho, 2007).

Such developmental changes in body proportions may help to explain the observed differences in the relationship between RSH and temperature between the current sample and that of Roberts (1978).

Although the multiple regression analyses failed to document a secular trend in body proportions in the current sample (see Table 10.4), is it clear that the relationship between RSH and temperature in the current sample is different from that reported by Roberts (1978). Unfortunately, because we do not have Roberts's raw data on body proportions, we cannot explore the nature of the differences in the same detail that we did with the body weight and BMI differences. Nonetheless, it appears that changes in food availability and Westernization of dietary habits may be responsible for reducing global variation in RSHs, thus explaining the lower correlations between RSH and temperature observed in the current sample.

Finally, this work also has practical applications for the development of anthropometric standards for assessing nutritional status. Our findings highlight some of the limitations in using a single set of BMI norms for assessing risks of under- and over-nutrition around the world (Shetty and James, 1994; WHO, 1995). It appears that climatic influences on body proportionality play a strong role in shaping the relationship between the BMI and body fatness across diverse human populations. Paradoxically, populations of colder climates tend to have *lower* levels of body fatness for a given BMI, a pattern typified by arctic populations such as the Inuit of North America (Shephard and Rode, 1996). In contrast, tropically adapted populations such as the Australian Aborigines have relatively higher levels of fatness than suggested by their BMI values (Norgan, 1994a). These differences emphasize the need for caution when interpreting BMI values among populations with extremely high or low RSHs. For these groups in particular, it will be important to include a broader range of anthropometric measures (e.g., skinfolds, circumferences) to provide an accurate picture of body composition and potential risks of chronic diseases.

CONCLUSIONS

Human biologists have long recognized the important role that climate plays in shaping body size and shape. The analyses presented in this chapter have reexamined the application of Bergmann's and Allen's rules for understanding climatic variation in human body size and proportions. The classic work of D. F. Roberts (1953, 1973, 1978) demonstrated that humans broadly conform to these classic ecological rules. He found that across a diverse sample of human populations, body mass, and RSH were inversely correlated with mean annual temperature, consistent with the predictions of Bergmann's and Allen's rules, respectively.

Our current analyses, drawing on anthropometric studies published after Roberts (1953) initial pioneering work, confirm some, but not all of his conclusions.

The inverse relationships between body mass (as both body weight and BMI) and temperature continue to persist for both men and women; however, the correlations are lower and the slopes of the regressions are shallower than those reported by Roberts. Similarly, the relationships between RSH and temperature in men and women of the current sample are also more modest than those reported by Roberts (1978). These changes in the relationship between body morphology and climate over the last 50 years, in part, reflect secular change in the growth of body size and proportions. Improvements in nutritional health, particularly among impoverished tropical populations, have produced notable changes in body mass and proportions. These results underscore the importance of both nutritional and temperature stresses in shaping developmental changes in human variation in body size and shape. Moreover, they have important implications for the use of anthropometric indexes such as the BMI as tools for assessing nutritional status and chronic disease risks.

DISCUSSION POINTS

1. Discuss the physical principles that are thought to underlie Bergmann's and Allen's "ecological rules."
2. Discuss how ongoing trends in global climate change may influence interpopulational variation in body mass, BMI, and body proportions (e.g., relative sitting height).
3. Discuss the utility and the limitations of the use of the BMI as the preferred measure for assessing risks of overweight and obesity in the United States and around the world.
4. Discuss the pros and cons of having a single set of international BMI norms for quantifying the prevalence rates of obesity and overweight in adult populations.

ACKNOWLEDGMENTS

We are grateful to Professor Michael Muehlenbein for the opportunity to contribute to this volume. Additionally, we thank Dr. Marcia Robertson and two anonymous reviewers for their comments and suggestions on this chapter. This work was supported in part by a grant from the Natural Sciences and Engineering Research Council of Canada (OGP-0116785).

REFERENCES

Allen, J. A. (1877). The influence of physical conditions on the genesis of species. *Radical Review*, **1**, 108–140.

Antón, S. C., Leonard, W. R. and Robertson, M. L. (2002). An ecomorphological model of the initial hominid dispersal from Africa. *Journal of Human Evolution*, **43**, 773–785.

Bailey, B. J. R. and Briars, G. L. (1996). Estimating the surface area of the human body. *Statistics in Medicine*, **15**, 1325–1332.

Baker, P. T. (1966). Human biological variation as an adaptive response to the environment. *Eugenics Quarterly*, **13**, 81–91.

Barnicot, N. A. (1959). Climatic factors in the evolution of human populations. *Cold Spring Harbor Symposia on Quantitative Biology*, **24**, 115–129.

Bergmann, C. (1847). Uber die verhaltniesse der warmeokononomie der thiere zu ihrer grosse. *Gottingen Studien*, **1**, 595–708.

de Onis, M., Frongillo, E. A. and Blossner, M. (2000). Is child malnutrition declining? An analysis of changes in levels of child malnutrition since 1980. *Bulletin of the World Health Organization*, **78**, 1222–1233.

Deurenberg, P., Yap, M. and Van Staveren, W. A. (1998). Body mass index and percent body fat: a meta analysis among different ethnic groups. *International Journal of Obesity*, **22**, 1164–1171.

Deurenberg-Yap, M. and Deurenberg, P. (2003). Is a re-evaluation of WHO body mass index cut-off values needed? The case of Asians in Singapore. *Nutrition Reviews*, **61**, S80–S87.

Deurenberg-Yap, M., Schmidt, G., Van Staveren, W. A., et al. (2000). The paradox of low body mass index and high body fat percentage among Chinese, Malays and Indians in Singapore. *International Journal of Obesity*, **24**, 1011–1017.

Flegal, K. M., Carroll, M. D., Ogden, C. L., et al. (2002). Prevalence and trends in obesity among US adults, 1999–2000. *Journal of the American Medical Association*, **288**, 1723–1727.

Food and Agriculture Organization, World Health Organization, and United Nations University (FAO/WHO/UNU) (1985). *Energy and Protein Requirements. Report of Joint FAO/WHO/UNU Expert Consultation*. WHO Technical Report Series, no. 724. Geneva: World Health Organization.

Frisancho, A. R. (1990). *Anthropometric Standards for the Assessment of Growth and Nutritional Status*. Ann Arbor, MI: University of Michigan Press.

Frisancho, A. R. (2007). Relative leg length as a biological marker to trace the developmental history of individuals and populations: growth delay and increased body fat. *American Journal of Human Biology*, **19**, 500–508.

Frisancho, A. R. (2008). *Anthropometric Standards: an Interactive Nutritional Reference of Body Size and Body Composition for Children and Adults*. Ann Arbor, MI: University of Michigan Press.

Frisancho, A. R., Borkan, G. A. and Klayman, J. E. (1975). Pattern of growth of lowland and highland Peruvian Quechua of similar genetic composition. *Human Biology*, **47**, 233–243.

Frisancho, A. R., Guire, K., Babler, W., et al. (1980). Nutritional influence on childhood development and genetic control of adolescent growth of Quechuas and Mestizos from the Peruvian lowlands. *American Journal of Physical Anthropology*, **52**, 367–375.

Garn, S. M., Leonard, W. R. and Hawthorne, V. M. (1986a). Three limitations of the body mass index. *American Journal of Clinical Nutrition*, **44**, 996–997.

Garn, S. M., Leonard, W. R. and Rosenberg, K. R. (1986b). Body build dependence, stature dependence and influence of lean tissue on the body mass index. *Ecology of Food and Nutrition*, **19**, 163–165.

Gehan, E. A. and George, S. L. (1970). Estimation of human body surface area from height and weight. *Cancer Chemotherapy Reports*, **54**, 225–235.

Gentilli, J. (1977). *Climates of Australia and New Zealand. World Survey of Climatology*, vol. 13. New York: Elsevier Scientific.

Gibson, R. S. (2005). *Principles of Nutritional Assessment*, 2nd edn. Oxford: Oxford University Press.

Katzmarzyk, P. T. and Leonard, W. R. (1998). Climatic influences on human body size and proportions: ecological adaptations and secular trends. *American Journal of Physical Anthropology*, **106**, 483–503.

Kelly, T., Yang, W., Chen, C.-S., et al. (2008). Global burden of obesity in 2005 and projections to 2030. *International Journal of Obesity*, **32**, 1431–1437.

Kleiber, M. (1975). *The Fire of Life: an Introduction to Animal Energetics*, 2nd edn. Huntington, NY: Krieger.

Leonard, W. R., Galloway, V. A., Ivakine, E., et al. (2002). Ecology, health and lifestyle change among the Evenki herders of Siberia. In *The Human Biology of Pastoral Populations*, W. R. Leonard and M. H. Crawford (eds). Cambridge: Cambridge University Press, pp. 206–235.

Lydolf, P. E. (1971). *Climate of the Soviet Union. World Survey of Climatology*, vol. 7. New York: Elsevier Scientific.

McGarvey, S. T. (1991). Obesity in Samoans and a perspective on its etiology in Polynesians. *American Journal of Clinical Nutrition*, **53**, 1586S–1594S.

McGarvey, S. T, Levinson, P. D., Bausserman, L., et al. (1993). Population change in adult obesity and blood lipids in American Samoa from 1976/1978 to 1990. *American Journal of Human Biology*, **5**, 17–30.

Newman, M. T. (1953). The application of ecological rules to the racial anthropology of the aboriginal new world. *American Anthropologist*, **55**, 311–327.

Norgan, N. G. (1994a). Interpretation of low body mass indices: Australian Aborigines. *American Journal of Physical Anthropology*, **94**, 229–237.

Norgan, N. G. (1994b). Relative sitting height and the interpretation of the body mass index. *Annals of Human Biology*, **21**, 79–82.

Ogden, C. L., Carroll, M. D., Curtin, L. R., et al. (2006). Prevalence of overweight and obesity in the United States, 1999–2004. *Journal of the American Medical Association*, **295**, 1549–1555.

Roberts, D. F. (1953). Body weight, race and climate. *American Journal of Physical Anthropology*, **11**, 533–558.

Roberts, D. F. (1973). *Climate and Human Variability. An Addison-Wesley Module in Anthropology*, Number **34**. Reading, MA: Addison-Wesley.

Roberts, D. F. (1978). *Climate and Human Variability*, 2nd edn. Menlo Park, CA: Cummings.

Roche, A. F. (1979). Secular trends in human growth, maturation, and development. *Monographs of the Society for Research in Child Development*, **44**, 1–120.

Rode, A. and Shephard, R. J. (1994). Prediction of body fat content in an Inuit population. *American Journal of Human Biology*, **6**, 249–254.

Ruff, C. B. (1994). Morphological adaptation to climate in modern and fossil hominids. *Yearbook of Physical Anthropology*, **37**, 65–107.

Schmidt-Nielson, K. (1984). *Scaling: Why Animal Size is so Important*. Cambridge: Cambridge University Press.

Schreider, E. (1950). Geographical distribution of the body-weight/body-surface ratio. *Nature*, **165**, 286.

Schreider, E. (1957). Ecological rules and body-heat regulation in man. *Nature*, **179**, 915–916.

Schreider, E. (1964). Ecological rules, body-heat regulation, and human evolution. *Evolution*, **18**, 1–9.

Schreider, E. (1975). Morphological variations and climatic differences. *Journal of Human Evolution*, **4**, 529–539.

Schwerdtfeger, W. (1976). *Climates of Central and South America. World Survey of Climatology*, vol. 12. New York: Elsevier Scientific.

Shephard, R. J. and Rode, A. (1996). *Health Consequences of "Modernization": Evidence from Circumpolar Peoples*. Cambridge: Cambridge University Press.

Shephard R. J., Hatcher, J. and Rode, A. (1973). On the body composition of the Eskimo. *European Journal of Applied Physiology*, **32**, 3–15.

Shetty, P. S. and James, W. P. T. (1994). *Body Mass Index: a Measure of Chronic Energy Deficiency*. FAO Food and Nutrition Paper, no. 50. Rome: FAO.

Snodgrass, J. J., Leonard, W. R., Sorensen, M. V., et al. (2006). The emergence of obesity among indigenous Siberians. *Journal of Physiological Anthropology*, **25**, 75–84.

Steinhauser, F. (1970). *Climatic Atlas of Europe*. Hungary: WMO, BMO, UNESCO.

Steinhauser, F. (1979). *Climatic Atlas of North and Central America*. Hungary: WMO, BMO, UNESCO.

Stinson, S. (1990). Variation in body size and shape among South American Indians. *American Journal of Human Biology*, **2**, 37–51.

Stinson, S. and Frisancho, A. R. (1978). Body proportions of highland and lowland Peruvian Quechua children. *Human Biology*, **50**, 57–68.

Tanner, J. M. (1978). *Fetus into Man: Physical Growth from Conception to Maturity*. Cambridge, MA: Harvard University Press.

Walter, H. (1971). Remarks on the environmental adaptation of man. *Humangenetik*, **13**, 85–97.

Willmott, C. J., Mather, J. R. and Rowe, C. M. (1981). *Average Monthly and Annual Surface Air Temperature and Precipitation Data for the World Parts 1 and 2*, Publications in Climatology, vol. 34. Elmer, NJ: C. W. Thornewaite Associates.

World Health Organization (WHO) (1995). *Physical Status: the Use and Interpretation of Anthropometry*. Report of a WHO Expert Consultation. WHO Technical Report Series, no. 854. Geneva: World Health Organization.

World Health Organization (WHO) (2000). *Obesity: Preventing and Managing the Global Epidemic*. Report of a WHO Consultation on Obesity, Geneva, 3–5 June, 1997. WHO/NUT/NCD/98.1, Technical Report Series, no. 894. Geneva: World Health Organization.

World Health Organization (WHO) (2004). Appropriate body-mass index for Asian populations and its implications for policy and intervention strategies. *Lancet*, **363**, 157–163.

World Health Organization (WHO) (2006). Obesity and overweight. Fact sheet, no. 311, http://www.who.int/mediacentre/factsheets/fs311/en/print.html.

11 Human Adaptation to High Altitude

Tom D. Brutsaert

INTRODUCTION

As of 1995, over 140 million people worldwide lived at altitudes exceeding 2500 m (Niermeyer et al., 2001). The effects of hypobaric hypoxia – defined as a low environmental oxygen partial pressure – on cellular metabolic function, growth and development, physical activity, reproduction, and human health have made high altitude a unique setting in which to investigate human adaptation. This is especially true for traits that are directly or indirectly related to oxygen (O_2) transport. Following Niermeyer et al. (2001), the term *adaptation* will be used to refer to a feature of structure, function, or behavior that is beneficial and enables survival in a specific environment. Such features may be genetic in origin, although features that arise via developmental and/or physiological processes may also be termed adaptations if they enable survival (see Chapter 2 of this volume). The term *genetic adaptation* will refer to a heritable feature that was produced by natural selection (or other force of evolution) altering allele frequencies over time. A *developmental adaptation* will refer to an irreversible feature that confers survival benefit and is acquired through lifelong exposure to an environmental stress or stressors. Developmental features that arise without clear adaptive benefit will be termed *developmental responses*. Finally, the term *acclimatization* will refer to a time-dependent physiological response to high altitude, which may or may not be adaptive.

Around the world, various regional populations show a wide diversity of time-depth exposure to altitude providing a "natural experiment" to test hypotheses of developmental and/or genetic adaptation (Moore, 1990). The term *native* (as in high-altitude or highland native) will be used generally to identify an individual born and raised at high altitude, independent of ancestry. For example in North America, some high-altitude natives of the Rocky Mountains with lowland European origins may trace altitude ancestry back ~150 years. This is not long enough for genetic adaptation, but such altitude natives may show unique features acquired through lifelong exposure to hypobaric hypoxia. The term *indigenous* (as in indigenous altitude native) will specifically refer to an individual with deep altitude ancestry who belongs to a population that may have experienced natural selection in the past. In South America, archaeological evidence suggests the presence of Native Americans as early as 12 000 years ago (Lynch, 1990), and current indigenous high-altitude groups (the Aymara and Quechua) can trace altitude ancestry back to at least 4000–6000 years ago, which marks the domestication of many highland plant and animal species (llama, alpaca, potatoes, etc. ...) (MacNeish and Berger, 1970; Lynch, 1990). On the Himalayan plateau, there is evidence of paleolithic human habitation as early as 25 000–50 000 years ago (Zhimin, 1982). These paleolithic populations may not have been ancestral to current populations, but genetic and linguistic studies of Tibetans suggest a time frame of habitation that goes back at least many thousands of years (Torroni et al., 1994). Thus, in both the Andes and the Himalayas, there has been sufficient time for the operation of natural selection. However, it is worth noting that there is little direct support for the hypothesis of genetic adaptation, i.e., to date there are no known gene (allele) or haplotype frequencies that are unique to an indigenous high-altitude native population with demonstrable effects on mortality/fertility under the conditions of hypobaric hypoxia. This situation may change rapidly as several research groups are now actively working on the genetics of indigenous high altitude populations.

This chapter is organized into eight sections as follows: Section I defines hypoxia and provides an introduction to the well-known biological effects of low environmental O_2. Section II reviews the physiology of O_2 transport as a basic understanding is critical for subsequent discussions of human adaptation. Section III provides several examples of the process of acclimatization to altitude, mainly to distinguish acclimatization from other modes of adaptation that are more central to the interests of human evolutionary physiology, i.e., genetic and developmental adaptation.

Human Evolutionary Biology, ed. Michael P. Muehlenbein. Published by Cambridge University Press. © Cambridge University Press 2010.

Sections IV and V, respectively, focus on the physical work capacity (exercise) and reproductive performance of highland natives in hypoxia. These two areas have been widely studied at high altitude and are considered as important "adaptive domains," following a conceptual framework that was originally described by Mazess (1978). An adaptive domain is an area of life where the relative benefit (or significance) of a specific adaptive response can be evaluated or defined. For example, if large lungs are adaptive in hypoxia, then benefit must be defined with reference to an adaptive domain like reproductive performance, i.e., do women with larger lungs produce babies with lower mortality risk? In most cases, adaptive significance is difficult to ascertain. Thus, Section VI evaluates specific structural and/or functional traits of the oxygen transport system that differ between highland and lowland populations, without a priori reference to an adaptive domain. Where possible, the significance of a specific trait is considered relative to health and disease, reproductive performance, exercise performance, growth and development, and/or nutrition and energy use. Section VII focuses on genetic studies of altitude adaptation, and Section VIII offers areas of future research.

SECTION I: WHAT IS ENVIRONMENTAL HYPOXIA?

Environmental hypoxia is defined with reference to "normoxia," which is simply the O_2 availability at sea-level with 760 mmHg atmospheric pressure and a fractional O_2 concentration (FiO_2) of about 0.21 (i.e., 21%). The latter represents the composition of the earth's atmosphere. For dry air, the normoxic partial pressure for O_2 (PO_2) is about 160 mmHg, or 760 mmHg × 0.21.

Hypobaric hypoxia refers to a lower than normoxic PO_2 due to lower ambient pressure as one rises through the air column with increasing altitude (Figure 11.1). Note that FiO_2 is constant with altitude. For example, the ambient PO_2 at 3000 m where Pb is ~525 mmHg is ~70% of the sea-level value, or 525 mmHg × 0.21 = 110 mmHg. This definition of hypoxia is not particularly helpful from a functional standpoint, and some distinction must be made between hypoxia per se and the level of hypoxia (or the ambient PO_2) where measurable effects on human physiology occur. The latter depends on the physiological system, the outcome measure, and the individual. For example, at the modest altitude of 1500 m, some very good athletes begin to experience decrements in aerobic capacity, and some visitors may note a transient hyperventilation on first exposure (Maresh et al., 1983). In this chapter, most of the populations under consideration live above 2500 m where there are clear effects on cellular metabolic processes underlying diverse human activities and human health. Some of these effects are briefly considered here to emphasize that hypobaric hypoxia is a formidable environmental stressor. This is a precondition for research approaches that focus on adaptive response.

Linear postnatal growth and development is nutrient and O_2 flow dependent (note: prenatal growth is discussed more fully in Section V below). Beginning with the studies of Paul Baker and his students in Nunoa, Peru (e.g., Frisancho, 1969), it has been suggested that hypoxia per se slows overall linear growth and delays sexual maturation. In the Andes, where the majority of growth studies have been conducted, much of the delay is probably established at birth as a result of intrauterine uterine growth retardation, with little catch-up growth thereafter (Greksa, 2006). At least

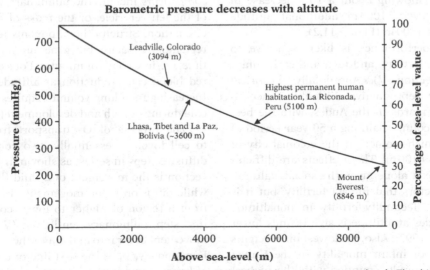

Barometric pressure decreases with altitude

11.1. Barometric pressure decreases with increasing altitude starting at sea-level. The highest permanent human habitation is likely the small town of La Riconada, in Peru (West, 2002).

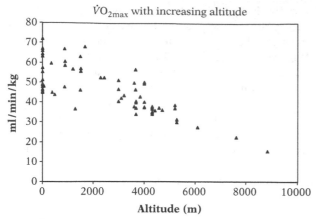

11.2. $\dot{V}O_{2max}$ decreases with altitude, approximately 10% per 1000 meters after 1500 meters. Data points are updated from Buskirk et al. (1967) and represent mean $\dot{V}O_{2max}$ values from studies of lowland native males who were exposed to hypobaric hypoxia. The length of exposure varies between studies (from acute exposure to several weeks of exposure), but $\dot{V}O_{2max}$ does not change much with ventilatory acclimatization to altitude (Bender et al., 1989). Some of the studies represented, particularly those at extreme altitude, were carried out in a hypobaric chamber.

some of the growth delay is attributable to poor health and nutrition in the research populations, but hypoxia likely has an independent effect evidenced by the delayed growth of European populations at altitude who generally live under relatively good socioeconomic conditions (Greksa et al., 1988).

Aerobic exercise is also an O_2 dependent process. One measure of performance is the maximal oxygen consumption ($\dot{V}O_{2max}$) reflecting the integrated functioning of pulmonary, cardiovascular, and muscle metabolic systems to obtain, deliver, and consume O_2 during maximal exertion. Buskirk et al. (1967) was the first to quantify the decrement in $\dot{V}O_{2max}$ ($\Delta\dot{V}O_{2max}$) with increasing altitude in lowland men of mostly European origins, showing about a 10% decrease in performance for every 1000 m additional altitude beginning at about 1500 m (Figure 11.2).

Reproductive performance is likely sensitive to hypoxia, including fertility, and fetal and early infant mortality (see Section V). Anecdotally, historical accounts describe reproductive difficulties encountered by early Spaniards in the Andes, with the best known of these accounts claiming a 50-year period of infertility for Spanish women in the colonial city of Potosí, Bolivia, at 4100 m! These effects are difficult to ascribe to hypoxia alone given the social, cultural, and nutritional factors that impact fertility, but it is worth noting that fecundity/fertility in nonaltitude native rodent species at altitude is significantly lower (Martin and Costa, 1992). Also, the lowest birthweights and highest rates of infant mortality in the United States occur in Colorado counties with the highest mean altitudes (Moore, 2003). A contributing factor

may be pre-eclampsia during pregnancy, a condition that increases mortality risk of both mother and fetus. Pre-eclampsia is ~3 times more common above 3000 m (Palmer et al., 1999; Keyes et al., 2003).

Various diseases are also associated with hypoxia, or exacerbated by hypoxic exposure. For example, initial exposure to altitude can lead to acute mountain sickness (AMS) in some individuals, with prevalence and severity depending on altitude and rate of ascent. Acute mountain sickness is usually transient, but often precedes or is associated with more serious and potentially fatal complications like pulmonary or cerebral edema. Lifelong or chronic hypoxic exposure is associated with a disease known as chronic mountain sickness (CMS), characterized by excessive polycythemia (high red blood cell level), severe hypoxemia (low blood O_2 levels), and in some cases moderate or severe pulmonary hypertension which may evolve to *cor pulmonale*, leading to congestive heart failure (Leon-Velarde et al., 2005).

SECTION II: THE PHYSIOLOGY OF O_2 TRANSPORT FROM ENVIRONMENT TO CELL

Given the O_2 poor environment of high altitude, the O_2 transport system is logically an area of major focus. This physiological system is frequently conceptualized as a linear sequence of functional and structural steps that brings oxygen into the body for delivery to cellular mitochondria (Figure 11.3). At the mitochondrial level, O_2 fulfills its role as the final electron acceptor in the process of oxidative phosphorylation to produce adenosine triphosphate (ATP). Examples of *functional components* of the O_2 transport system include the ventilation rate or the cardiac frequency (heart rate), both of which can be regulated physiologically to increase or decrease O_2 flux. Examples of *structural components* include the pulmonary volume, the mass of the left ventricle, or the mass of red blood cells in circulation. Structural components may possess some regulatory capacity, but generally not in physiological time, i.e., in seconds or minutes. For example, increasing red blood cell production at altitude requires weeks. Increasing total lung volume requires exposure to hypoxia during growth and development (see Section VI).

The process of O_2 transport from environment to cell involves, essentially, two convective and two diffusive steps in series, as shown in Figure 11.3. Convection is the movement of a fluid (like air or blood), while diffusion is the movement of a molecule or ion from a region of higher to lower concentration. The first step, pulmonary ventilation (V_E, L/min), or the movement of air into and out of the lung, is convective. Functionally, V_E is the first line of defense in the face of hypoxia and it has the purpose of maintaining constant values of the alveolar oxygen and carbon dioxide

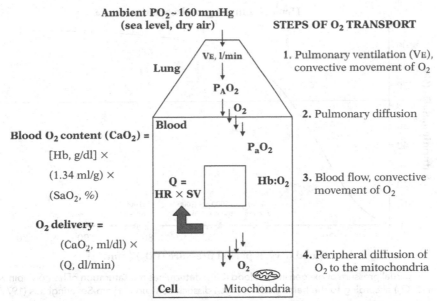

Ambient PO$_2$ ~ 160 mmHg
(sea level, dry air)

STEPS OF O$_2$ TRANSPORT

Lung   V$_E$, l/min

P$_A$O$_2$

O$_2$

Blood

P$_a$O$_2$

Q =
HR × SV   Hb:O$_2$

O$_2$

Cell   Mitochondria

Blood O$_2$ content (CaO$_2$) =

[Hb, g/dl] ×

(1.34 ml/g) ×

(SaO$_2$, %)

O$_2$ delivery =

(CaO$_2$, ml/dl) ×

(Q, dl/min)

1. Pulmonary ventilation (V$_E$),
   convective movement of O$_2$

2. Pulmonary diffusion

3. Blood flow, convective
   movement of O$_2$

4. Peripheral diffusion of
   O$_2$ to the mitochondria

11.3. The oxygen transport system has essentially four steps, two convective (pulmonary ventilation and blood flow), and two diffusive (pulmonary and peripheral, i.e., muscle diffusion). The structural/functional parameters that determine O$_2$ delivery to the cell are indicated including; VE, pulmonary ventilation; [Hb], hemoglobin concentration; SaO$_2$, arterial oxygen saturation; CaO$_2$, arterial oxygen content; SV, stroke volume; HR, heart rate; and Q, cardiac output. See text for further details.

partial pressures (P$_A$O$_2$ and P$_A$CO$_2$, respectively). The alveolar ventilation (V$_A$) is that proportion of V$_E$ that participates in gas exchange, taking into account the fact that the conducting airways of the lung are an "anatomic dead-space" where gas exchange cannot occur. The V$_A$ is not usually measured directly, but this chapter will make reference to the concept of an "effective V$_A$." Effective V$_A$ is assessed by consideration of the functional consequence of breathing which is to maintain/change gas partial pressures in the lung and thus circulation.

Note that P$_A$O$_2$ at sea-level is approximately 100 mmHg, or about 60 mmHg lower than the ambient PO$_2$, and that PO$_2$s fall at every subsequent step of the O$_2$ transport chain from environment to cell. For this reason, physiologists often refer to an "O$_2$ cascade" driven by the partial pressure gradients at every step. Importantly, P$_A$O$_2$ depends on ambient PO$_2$, and so the overall PO$_2$ gradient driving the diffusion of O$_2$ from lung to cell is reduced at altitude. V$_E$ itself is controlled in a complex manner by both central (brain) and peripheral chemoreceptors. The latter are known as the carotid and aortic bodies, and these sense changes in arterial blood gases (P$_a$O$_2$ and P$_a$CO$_2$) brought about through low ambient PO$_2$ or increased metabolic demand for O$_2$.

The first diffusive step (pulmonary diffusion) is from alveolar gas to blood across the pulmonary-capillary membrane. At rest and at sea-level, alveolar and capillary blood partial pressures equilibrate rapidly so that blood leaving the lung via the pulmonary

vein has a PO$_2$ similar to that of alveolar gas PO$_2$, i.e., ~100 mmHg. But, this is not always the case. At altitude, or sometimes during extremely strenuous exercise, full equilibration between alveolar gas and arterial blood PO$_2$ may not take place, a condition known as *diffusion limitation*. Diffusion limitation causes a widening of the alveolar-arterial partial pressure difference (A-a)DO$_2$ as P$_a$O$_2$ becomes less than P$_A$O$_2$. A widening of the (A-a)DO$_2$ may also occur from a condition known as ventilation-perfusion mismatch, also common at altitude, but without belaboring the details, it is important simply to realize that there are limits to the pulmonary gas exchange process. If one sees a small (A-a)DO$_2$ at altitude, this is a good indication of an efficient gas exchange process, and from an evolutionary or developmental perspective, one might hypothesize adaptations in either functional or structural components of the pulmonary system to account for it.

The ambient PO$_2$ and the process of V$_E$ together determine P$_a$O$_2$. In turn, P$_a$O$_2$ determines the saturation of hemoglobin with O$_2$ (SaO$_2$) according to the Hb-O$_2$ dissociation curve (Figure 11.4). The nonlinearity of the Hb-O$_2$ curve has several important physiological implications. Above a P$_a$O$_2$ of about 60, where the Hb-O$_2$ curve is relatively flat, large changes in P$_a$O$_2$ have little effect on SaO$_2$. However, below ~60 mmHg, or on the "steep" part of the Hb-O$_2$ curve, small changes in P$_a$O$_2$ produce large changes in SaO$_2$. In practical terms, this means that SaO$_2$ is fairly resistant to altitudes up to about 2500 m, but will rapidly

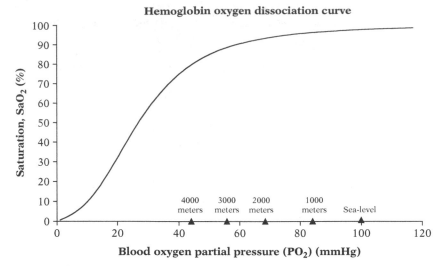

11.4. The partial pressure of oxygen in the blood (PO_2) determines the saturation of hemoglobin with oxygen (SaO_2) according to the hemoglobin-oxygen dissociation curve. From Severinghaus (1979).

decrease thereafter. SaO_2, hemoglobin concentration [Hb], and the O_2-binding capacity of hemoglobin (a constant at $1.34\,ml\,O_2/g$ hemoglobin) together determine the O_2 content of blood per unit volume, or CaO_2-ml O_2 per dl blood. CaO_2 multiplied by blood flow determines O_2 delivery. Blood flow, of course, is the second convective step of the O_2 transport process, and is determined by cardiac output (Q) and regulation of regional blood flow by vasodilation/vasoconstriction of arterioles to capillary beds. In turn, Q is determined by heart rate (HR) and stroke volume (SV). The final step of O_2 delivery is peripheral diffusion of O_2 into the cell down a diffusion gradient from capillary to mitochondria, where mitochondrial PO_2 may be just a few mmHG.

Given the importance of reproductive performance at altitude, it is important to also consider the physiology of O_2 delivery to the fetus. The fetus receives nutrients from uterine arteries that derive from the maternal arterial circulation. The uterine arteries branch out into a dense capillary network known as the placenta, which serves as a gas exchange organ between independent maternal and fetal circulations. Thus, O_2 delivery to the fetus involves an additional diffusive step across the placenta. Like O_2 delivery to skeletal muscle, O_2 delivery to the fetus is a function of uterine artery blood flow, [Hb], and SaO_2. The student of human evolutionary physiology should again consider the various structural or functional parameters that could be regulated (physiologically, developmentally, or across evolutionary time) to optimize O_2 delivery to the fetus under conditions of hypobaric hypoxia. For example, O_2 delivery could be upregulated by increasing uterine artery blood flow (a functional parameter), perhaps via

enlargement of the uterine artery (a structural parameter). Similarly, increased placental size (structure), analogous to increased pulmonary volume, would increase surface area for O_2 diffusion and thus increase O_2 delivery to the fetus. The physiology of O_2 delivery to the fetus in highland populations has been researched extensively by Moore and colleagues and is discussed in more detail in Section VI below.

SECTION III: ACCLIMATIZATION TO HYPOBARIC HYPOXIA

In general, an understanding of how acclimatization to high altitude affects a specific trait or feature is a prerequisite before analysis of how population trait distributions are conditioned by developmental exposure to hypoxia and/or by population genetic background. Some traits are very sensitive to acclimatization state, particularly ventilatory traits, and their analysis from a developmental and/or evolutionary perspective is therefore difficult. In other specific cases, populations may not show the expected acclimatization response. For example, Tibetan populations are characterized by [Hb] not different from sea-level normative values, counter to the expectation of hematological acclimatization (see below and Section VI for further discussion).

As described, the term acclimatization refers to a time dependent, short-term, and reversible physiological response to an environmental stress or stressors. To illustrate the concept, two text-book examples are considered here: (1) ventilatory acclimatization; and (2) hematological acclimatization. Ventilatory acclimatization involves the regulation of breathing by the nervous system including respiratory centers in the brain and peripheral chemoreceptors that are sensitive to changes in blood PCO_2, pH, and PO_2. On initial exposure to hypoxia, the peripheral chemoreceptors (primarily the carotid bodies) sense changes in blood gas partial pressures and signal an increase in V_E that manifests within seconds to minutes (hyperventilation). This hyperventilation is progressive and reaches a plateau after 5–6 days (Huang et al., 1984; Smith et al., 2001). Hyperventilation causes transient decreases in $PaCO_2$ and increases in blood pH to produce a respiratory alkalosis. Alkalosis may persist even after acclimatization, but over days there is some compensation to normalize blood chemistry, in part because of the action of the kidney which increases bicarbonate ion excretion. In humans, the end of ventilatory acclimatization is marked by a stability in V_E, $PaCO_2$, and pH. The functional result is that individuals maintain a higher V_A to partially offset the decreases in SaO_2 and CaO_2. It should be emphasized that this process does not return the body to a sea-level physiological state. Rather, SaO_2 levels remain low depending on the specific altitude. Hematological acclimatization involves the hormone erythropoietin which stimulates an increase in the production of red blood cells from precursor cells in the bone marrow. This process takes place over weeks with a resultant increase in the oxygen carrying capacity of blood.

SECTION IV: DO HIGH-ALTITUDE NATIVES HAVE ENHANCED WORK CAPACITY AT ALTITUDE?

Physical work is an important human activity, i.e., it is an adaptive domain that is dependent on O_2 availability. The question addressed here is whether high-altitude natives have higher upper limits of work capacity in hypoxia compared to their counterparts from sea-level? Work capacity is a general term, but specific aspects of work performance may be measured including the maximal oxygen consumption ($\dot{V}O_{2max}$, or aerobic capacity), the endurance capacity, and the work efficiency/work economy.

Aerobic capacity ($\dot{V}O_{2max}$)

$\dot{V}O_{2max}$ is usually measured on a treadmill or cycle ergometer by progressively increasing work-load over a 10–15-minute period in order to obtain a 1-minute average of maximal O_2 consumption. It is not surprising that $\dot{V}O_{2max}$ decreases with increasing altitude (Figure 11.2). But, there is substantial variation in the magnitude of the decrease between individuals

depending on age, gender, body-weight, body-composition, fitness state, and acclimatization state. Thus, partitioning the effects of genes, development, and/or environment on $\dot{V}O_{2max}$ is difficult. Compared to sea-level residents in hypoxia, many studies have documented a relatively high $\dot{V}O_{2max}$ (ml/min/kg) in Andean (Elsner et al., 1964; Kollias et al., 1968; Mazess, 1969a, 1969b; Frisancho et al., 1973; Baker, 1976) and Himalayan natives (Sun et al., 1990a; Ge et al., 1994b, 1995; Zhuang et al., 1996). Indigenous altitude natives also appear to experience smaller decreases in $\dot{V}O_{2max}$ ($\Delta\dot{V}O_{2max}$) when exposed to increasing hypoxia (Elsner et al., 1964; Velasquez, 1966; Baker, 1969; Frisancho et al., 1973; Vogel et al., 1974; Way, 1976; Hochachka et al., 1991; Brutsaert et al., 2003; Marconi et al., 2004). For example, $\Delta\dot{V}O_{2max}$ values in indigenous groups have been reported as between ~30% and 80% of the decrement in lowland comparison groups.

It is a difficult problem to explain the higher mean $\dot{V}O_{2max}$ values of indigenous high-altitude natives. Is the higher $\dot{V}O_{2max}$ due to population genetics, developmental exposure to hypoxia, or is it due to differences in lifestyle, particularly physical activity patterns? Consider the analyses shown in Figures 11.5. Figure 11.5 plots mean $\dot{V}O_{2max}$ values (ml/min/kg) by altitude for indigenous altitude-native males grouped by region. These mean values were taken from the literature based on studies conducted over the past ~50 years and they are superimposed on the mean values for acclimatized lowland-native males (sea-level reference) that were shown previously in Figure 11.2. Andean values across the range of altitude from 3000 to 5500 m are clearly and significantly above the sea-level reference line ($P = 0.002$). Unfortunately, statistical power is insufficient to test this hypothesis independently with natives from the Himalayas. In contrast, in a similar analysis, Figure 11.5 reveals no difference in mean $\dot{V}O_{2max}$ values for populations of lowland ancestry who were born and raised at altitude. The latter include populations of European ancestry living in Colorado and South America, and also populations of Han Chinese ancestry living on the Tibetan plateau. Therefore, a cautious interpretation of these results (given the small sample of studies) is that genes play a more important role than developmental adaptation in determining the high $\dot{V}O_{2max}$ of Andeans and Himalayans. Unfortunately, confounding factors cannot be excluded, particularly the high physical activity levels that are common among highland native groups (Kashiwazaki et al., 1995; Brutsaert, 1997).

Confounding is a difficult issue to resolve and limits the inference of genetic adaptation based on group differences using the comparative approach. This may perhaps explain the conflicting conclusions reached by different investigators and different studies.

(a)

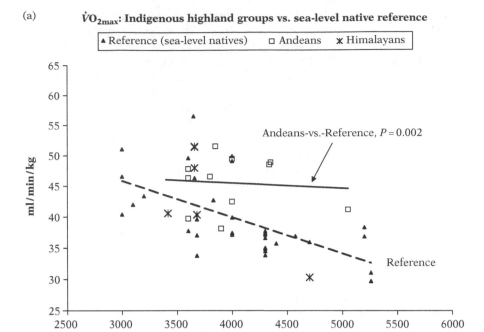

(b)

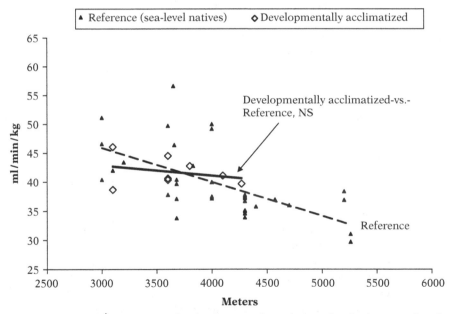

11.5. Published mean $\dot{V}O_{2max}$ values for developmentally acclimatized males living at altitude i.e., long-term European residents of Colorado and the Andes, and Han Chinese migrants to the Tibetan plateau. Mean values are superimposed on the sea-level native reference data (from Figure 11.2) across the altitude range from 3000–5500 meters. NS = nonsignificant for a test of the hypothesis that developmentally acclimatized males have higher altitude specific $\dot{V}O_{2max}$ compared to lowland native males (sea-level reference) using regression analysis.

One previous study in the Andes argued for a genetic basis to explain the high $\dot{V}O_{2max}$ of Aymara after controlling for physical activity level (Frisancho et al., 1995), while at least three others studies emphasized developmental and/or covariate effects (Greksa and

Haas, 1982; Greksa et al., 1985; Brutsaert et al., 1999b). A recent study by Brutsaert and colleagues (2003) argued for a genetic basis based on a negative correlation between $\Delta\dot{V}O_{2max}$ and the proportion of Native American ancestry in a sample of 32 lowland males of mixed Quechua-European ancestry who were transiently exposed to hypoxia. While admixture-based studies allow stronger inference regarding the action of genes, the problem of confounding remains. Thus, resolution of this issue will require a more direct interrogation of the genetic basis of human performance at altitude (see Sections VII and VIII) and/or a better understanding of how developmental exposure impacts oxygen transport capacity.

Work efficiency and the economy of locomotion

Work efficiency (WE) is defined as the ratio of external work (output) to metabolic cost (input), with external work typically measured on a treadmill or cycle ergometer and metabolic cost typically measured as O_2 cost, i.e., O_2 consumption (VO_2). Economy is a related but different construct, defined as the O_2 cost for a specific activity, like load carrying or running. In both cases, the evolutionary advantage of using O_2 efficiently to do work in an O_2 poor environment is self-evident, particularly considering the agricultural and pastoral labor demands that characterize life in the Andes (Kashiwazaki et al., 1995) and also on the Tibetan plateau. Alberto Hurtado, in his pioneering studies in Peru was the first to suggest that altitude natives have a higher metabolic WE (Hurtado, 1932, 1964). The subsequent literature in support of this hypothesis is conflicted. Several studies have reported higher WEs in Andeans versus lowland controls (Reynafarje and Velasquez, 1966; Haas et al., 1983; Hochachka et al., 1991), while other studies have shown no differences (Mazess, 1969a, 1969b; Brutsaert et al., 2004), or indeed lower WE in Andeans (Kollias et al., 1968). Two studies have reported significantly higher WE in Tibetans versus lowland controls at altitude (Ge et al., 1994b; Niu et al., 1995), while three other studies have reported no WE differences in Sherpas or Tibetans (Lahiri and Milledge, 1966; Lahiri et al., 1967; Kayser et al., 1994). One additional study reported higher WE in Tibetans resident at 4440 m versus Tibetans resident at 3658 m (Curran et al., 1998). Studies from Colorado show no differences in WE related to acclimatization state or growth and development at altitude (Dill et al., 1931; Balke, 1964; Grover et al., 1967). For economy, two recent studies have reported a lower O_2 cost for treadmill running (Bastien et al., 2005; Marconi et al., 2005) or load carrying (Bastien et al., 2005) in Sherpa. Both studies are difficult to interpret with respect to metabolic efficiency given differences in body size between study groups, and in the case of the Bastien et al. (2005) study, differences in study location. Thus, on balance, there is little compelling evidence to support the hypothesis that altitude natives use O_2 more efficiently in the performance of physical work.

Endurance performance

Hurtado's aforementioned classic study is the only report in the literature where a true endurance outcome was measured, i.e., time to exhaustion during a sustained bout of submaximal work. Hurtado reported a higher average tolerance time for treadmill running in 10 altitude natives of Morococha, Peru (4540 m) versus 10 sea-level residents of Lima, Peru (34.2 versus 59.4 minutes). The difference was impressive as each group was tested in their native environment! However, without information on subject fitness status and on the performance of both groups in the same environment, the study is also difficult to interpret.

Summary of work capacity studies

Numerous studies of indigenous altitude natives dating back nearly half a century reveal higher than expected values of $\dot{V}O_{2max}$ at a given altitude, but the genetic versus environmental origins of this difference remain obscure. Regarding energetic advantage during work or exercise, there is widespread conflict in the literature and no firm conclusions are possible. Regarding endurance capacity, this aspect of work performance has yet to be thoroughly evaluated.

SECTION V: DO HIGH-ALTITUDE NATIVES HAVE ENHANCED REPRODUCTIVE PERFORMANCE AT HIGH ALTITUDE?

Reproductive performance is central for any species. In this section, measures of reproductive success are considered, including fertility, fecundity, prenatal and postnatal mortality, birthweight, and various measures of maternal O_2 transport to the fetus.

Fertility and fecundity

As cited by Carlos Monge (1948), sixteenth-century Spanish historians were the first to note fertility/ fecundity impairments in Spanish women who settled in the Andean highlands. Monge proposed that hypobaric hypoxia lowers fertility, either by increasing fetal wastage or pregnancy loss (i.e., prenatal mortality), and/or by decreasing the ability of a woman to conceive (fecundity). These hypotheses are difficult to test in humans given myriad cultural and behavioral influences on fertility, and given the difficulty of directly assessing prenatal mortality or fecundity. However, the animal literature does give some evidence of reduced fertility/fecundity attributable to hypoxia when lowland native rats are raised in hypoxic conditions (Martin and Costa, 1992). In this context, what emerges as noteworthy is the apparently "normal" fertility and fecundity of indigenous highland populations in both the Andes and the Himalayas.

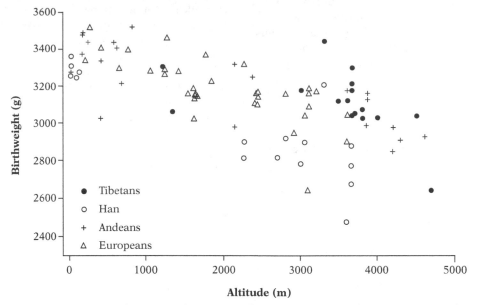

11.6. Mean birthweights by altitude from previously published studies of altitude residents from North and South America and the Tibetan plateau. Figure is from Moore et al. (2001). On average, birthweight (BW) decreases with altitude, approximately 100 grams per 1000 meters. See text for a discussion of group differences.

Although there is substantial interpopulation variability in fertility within altitude regions, the ranges observed are similar to those for natural fertility sea-level populations. Considering fecundity alone, only one study has intensively investigated this issue in an indigenous altitude group, with no evidence that the probability of conception was reduced in Andean Aymara women (Vitzthum and Wiley, 2003).

Prenatal and postnatal mortality

Prenatal, neonatal (birth to 28 days), and infant (birth to 1 year) mortality rates are generally high in highland populations, but this is not unexpected given the significant levels of poverty in most mountainous areas of the world. However, hypoxia may also have a direct effect. Consider that in Colorado, until the 1980s, high-altitude regions had higher neonatal and infant mortalities rates compared to lowland regions (Moore, 2003). Lowland migrants to altitude may have a higher mortality risk than indigenous altitude groups. Moore (2003) reported three-fold higher prenatal mortality and higher neonatal and infant mortality rates in Chinese migrants to the Tibetan plateau compared to Tibetan natives. Quantitative genetic studies by Beall et al. (2004) in Tibetans suggested that there may be genetic factors at work to explain these mortality-risk differences. In an original series of studies, these investigators measured resting SaO2 within families. A significant proportion of the age- and sex-adjusted

variance (from 21% to 39%) was attributed to additive genetic factors with the overall pattern of variance best explained by a major gene conferring a 5–6% point increase in resting SaO2 (Beall et al., 1997a, 1997b). In a follow-up study, Tibetan women with a high likelihood of possessing one to two of the putative alleles for the high SaO2 phenotype had more surviving children (Beall et al., 2004). This study provides evidence that hypoxia is acting as selective agent on the locus for SaO2 by the mechanism of higher infant survival of Tibetan women with high SaO2 genotypes. These results are also consistent with the physiological studies described below that suggest a pathway linking maternal O2 delivery, fetal growth, birthweight, and the probability of infant survival.

Birthweight

Birthweight has long been an important outcome to assess reproductive performance at altitude. Figure 11.6 shows the birthweight decrement with hypoxia (Δbirthweight), estimated to be about 100 g per 1000 m. Long-term resident populations appear to be buffered from the normal Δbirthweight, reminiscent of the pattern observed for $\Delta \dot{V}O_{2max}$. Specifically, on the basis of worldwide birthweight data, Moore and colleagues have argued that Δbirthweight as a consequence of hypoxia varies according to the duration of population exposure to hypoxia (Niermeyer et al., 2001; Moore, 2003; Julian et al., 2007). Populations with the shortest history at altitude, such as North Americans in Colorado or Han Chinese migrants to the Tibetan plateau, experience the greatest Δbirthweight. Populations with the longest exposure (i.e., ancestral exposure), such as Tibetans and Andeans, experience more modest Δbirthweight. Notably, altitude-specific birthweight is higher in indigenous populations compared to lowland controls despite socioeconomic differences that might otherwise predict lower birthweight. For example, Haas et al. (1980) found higher birthweight among Aymara women compared to European controls in Bolivia, despite the fact that the European women had better access to health care and better nutrition. Recently, a study by Bennett et al. (2008) provided evidence that genetic factors are involved to explain the higher birthweight of Andeans. Using an admixture approach based on surname analysis, this study shows a direct association of indigenous high-altitude ancestry with protection against hypoxia-associated fetal growth reduction in a cohort of 1343 singleton births in La Paz, Bolivia.

Whether there are differences in altitude-specific mean birthweights between Andeans and Tibetans is unclear. Moore (2000) has argued that Tibetans have higher mean birthweight compared to Andeans, but Beall (2000) has argued against this population difference and suggest that women from both populations are equally effective at supporting fetal development as measured by birthweight.

The placenta

How do indigenous altitude native women produce larger babies? One structural parameter to consider is the placenta which undergoes significant growth and remodeling during pregnancy to optimize gas and nutrient exchange between the maternal and fetal circulations. At altitude, it has consistently been observed that placentas are more vascularized, i.e., a higher density of blood vessels, perhaps to compensate for lower uteroplacental blood flow at altitude. There are other structural changes in the high-altitude placenta, but without going into detail, these generally operate to increase the diffusion capacity of this tissue. These changes are evident to some degree in both recent migrant- and native-altitude populations, and so it is not clear whether changes in placental architecture per se can account for the larger birthweight babies of Tibetans and Andeans. Indeed, only a few studies have directly compared placental morphology between highland native and lowland groups, and these have offered conflicting results (Zamudio, 2003).

Maternal O$_2$ transport

Another possibility to increase birthweight is enhanced maternal O$_2$ transport to the uteroplacental circulation.

Under normal conditions, maternal V_E increases during pregnancy as mediated by several reproductive hormones, but principally progesterone. Increased V_E serves to increase SaO$_2$, and this process could be especially important at altitude considering that CaO$_2$ may actually fall due to an expansion of blood volume during pregnancy without a concomitant expansion of the red blood cell mass, i.e., a kind of "anemia of pregnancy." Interestingly, McAuliffe et al. (2001) have shown that CaO$_2$ is stable during pregnancy in residents of Cerro de Pasco, Peru, with three or more generations of altitude ancestry, but falls during pregnancy in residents with less than three-generations exposure. This suggests that developmental effects accumulate across generations (i.e., maternal effects) affecting O$_2$ delivery to the fetus.

To emphasize the importance of CaO$_2$ on birthweight, consider the studies of Moore and colleagues conducted in Colorado, Peru, and the Himalayas (reviewed in Moore, 2003). These researchers have shown that larger birthweight babies are born to mothers who show higher V_E, greater increases in hypoxic ventilatory sensitivity during pregnancy, and higher CaO$_2$ during pregnancy. However, these are within group effects only, and CaO$_2$ differences per se probably do not explain the birthweight differences between groups. For example, in one study, Tibetan women had lower pregnancy [Hb] than Han Chinese, and thus lower pregnancy CaO$_2$. Despite lower CaO$_2$, Tibetan women produced babies nearly 0.5 kg heavier than Chinese residents at ~3600 m. This paradox led Moore and colleagues to consider the role of uteroplacental blood flow in determining birthweight, recalling that O$_2$ delivery (not CaO$_2$) is the important functional parameter depending on *both* CaO$_2$ and blood flow (Moore et al., 2001). Uterine artery blood flow velocity was higher in Tibetans, presumably increasing O$_2$ delivery to the uteroplacental circulation. The latter may be the simple result of structural adaptation, i.e., an enlargement in the diameter of the uterine artery permitting higher blood flow. Additional data consistent with this hypothesis are now emerging from the Andes. Compared to women of European ancestry resident at 3600 m, Andean women have greater uterine artery enlargement during pregnancy, increased uterine artery blood flow at week 36 of pregnancy, and thus a 1.6-fold greater uteroplacental O$_2$ delivery near term (Wilson et al., 2007). Unfortunately, comparisons have not yet been made with women of either Han Chinese or European ancestry who were born and raised at altitude. Thus, at present, little is known regarding the developmental and/or evolutionary origins of this structural change. Further, there are other poorly understood differences between highland populations in the physiological response to pregnancy. For example, Andean women have relatively high [Hb] and CaO$_2$ during pregnancy compared to Tibetan women. Andean women also have a different pattern of breathing during pregnancy to increase V_E compared to European women (Vargas et al., 2007).

Summary of reproductive performance studies

Most of the evidence in support of a fertility and/or mortality advantage among indigenous altitude groups is anecdotal. However, indigenous highland women clearly give birth to larger babies at altitude compared to lowland women. Generally, this is related to one or more mechanisms that operate to increase nutrient and oxygen flow across the placenta. Like the high exercise capacity of the indigenous altitude native, the developmental versus genetic origins of this difference are obscure. However, recent studies do provide some evidence that genetic factors are at work affecting various aspects of reproductive performance, both in Andean and Tibetan women (Beall et al., 2004; Bennett et al., 2008).

SECTION VI: SPECIFIC STRUCTURAL AND FUNCTIONAL TRAITS OF THE O_2 TRANSPORT SYSTEM

The focus of this section is on specific traits related to the O_2 transport system, rather than on broader adaptive domains like exercise or reproductive capacity. Following the organization of the O_2 transport system itself, the section begins at the lung with a consideration of V_E and then moves down the O_2 transport chain terminating at the muscle-metabolic level. A priori, two points are worth considering. Firstly, not all of the specific trait differences that are discussed below have obvious adaptive benefits. For example, high average [Hb] characterizes many populations at high altitude. While increased [Hb] certainly increases CaO_2, elevated red blood cell levels also increases blood viscosity which increases the work (afterload) on the heart. Thus, some investigators have argued that increases in [Hb] at altitude constitute a maladaptive response (Winslow et al., 1985, 1989). Secondly, some traits differ in interesting ways between highland native groups in the Andes, the Himalayas, and other regions. This means that there may be different patterns of adaptation in each region, offering different solutions to the same environmental problem of hypobaric hypoxia (Moore et al., 1992; Beall, 2000).

Resting ventilation

A study by Chiodi et al. (1957) in the Andes was the first to measure the relative resting V_A in a high-altitude native group. Compared to acclimatized lowland controls, lifelong residents of the Andean altiplano at 3990–4515 m showed lower "effective V_A" at rest. A full description of the concept of effective V_A is beyond the scope of this chapter, but functionally, low effective V_A implies lower ventilation (hypoventilation) to maintain a given value of the PaO_2. The basic finding of a relative hypoventilation in Andeans has been repeatedly confirmed (Hurtado, 1964; Severinghaus et al., 1966; Lahiri, 1968; Cudkowicz et al., 1972; Beall et al., 1997a; Moore, 2000), but has not generally been replicated in Tibetans or described in populations with lifelong developmental exposure to hypoxia (Weil et al., 1971; Moore, 2000). Thus, Andeans may be unique in showing a relative hypoventilation at rest in hypoxia, while Tibetans and developmentally exposed populations may have a "normal" resting V_A not different from the V_A of lowlanders after ventilatory acclimatization to hypoxia (Zhuang et al., 1993). The functional significance of differences in alveolar ventilation between populations is not known. One possibility for the Andean–Tibetan difference is that it relates to or explains the higher prevalence of chronic mountain sickness in the Andes, as suggested by Moore et al. (1998).

Ventilatory control and chemosensitivity

The low V_E and V_A in Andeans may have something to do with the ventilatory control system, which (as described previously) is both centrally and peripherally mediated by the brain respiratory center and carotid/aortic chemoreceptors, respectively. The earliest studies of ventilatory control were conducted at about the same time in the Andes, Colorado, and the Himalayas beginning in the late 1960s. In the Andes, a number of studies showed lower sensitivity of chemoreceptors (i.e., lower chemosensitivity) to hypoxia or a lower ventilatory response to hypoxia, with the latter termed the HVR, or the hypoxic ventilatory response (Severinghaus et al., 1966; Sorensen and Severinghaus, 1968b; Cudkowicz et al., 1972). In the Himalayas, early studies also suggested a lower HVR in Sherpas (Lahiri et al., 1967; Lahiri, 1968). Meanwhile, in Colorado, studies of Leadville residents also showed "blunted" chemosensitivity, with the implication being that blunting was acquired from lifelong exposure to hypobaric hypoxia (Forster et al., 1971; Weil et al., 1971; Byrne-Quinn et al., 1972). Thus, for a time there was apparent consensus in the literature that long-term hypoxic exposure resulted in desensitization of the ventilatory control system, and also that this was a universal human phenomenon that could explain the blunted chemosensitivity of disparate high-altitude groups worldwide. However, since the 1970s, additional studies have complicated this view somewhat.

While Andean studies are nearly uniform in showing a blunted chemosensitivity among Aymara/Quechua (Chiodi, 1957; Severinghaus et al., 1966; Sorensen and Severinghaus, 1968a; Lahiri et al., 1969; Cudkowicz et al., 1972; Leon-Velarde et al., 1996; Beall et al., 1997a; Moore, 2000; Gamboa et al., 2003; Brutsaert et al., 2005), in the Himalayas a different pattern seems evident. Most (Hackett et al., 1980; Huang et al., 1984; Zhuang et al., 1993; Ge et al., 1994a; Beall et al., 1997a) but not all (Santolaya et al., 1989) studies of Tibetans and Sherpa since the late 1960s have shown a normal HVR and high or normal resting V_E despite lifelong exposure to hypoxia. There is indirect evidence to support the idea that these traits are genetically determined in both Andean and Himalayan populations. For example, a study by Curran et al. (1997) showed lower HVR in admixed Chinese-Tibetans (Chinese fathers and Tibetan mothers) compared to nonadmixed Tibetans despite similar resting V_E between groups. In the Chinese admixed group only, HVR decreased with duration of altitude residence, suggesting that full Tibetan ancestry protected against hypoxic desensitization. In the Andes, Brutsaert et al. (2005) showed a strong negative association of HVR with the proportion of Native American ancestry, the latter assessed using a panel of 81 ancestry informative molecular markers. Finally, a study by Beall et al. (1997a) conducted at the same time in both the Andes and the Himalayas, could find no evidence of acquired blunting in either indigenous population, i.e., HVR did not decrease with age over time, at least from adolescence onward. However, in this study Andean HVR was clearly lower than Tibetan HVR.

V_E during exercise

At the onset of exercise, V_E increases commensurate to meet gas exchange requirements and metabolic demand. At altitude compared to sea-level, V_E (L/min) is higher for a given level of fixed work, depending on the specific altitude, and this is true for nonnatives and natives alike (Brutsaert et al., 2003; Marconi et al., 2004). Further, in lowland natives, depending on the severity of exercise and altitude, there is clear evidence of gas-exchange limitation, including diffusion limitation (Dempsey et al., 1995). Against this background, it is noteworthy that many exercise studies conducted at altitude report lower absolute V_E (L/min) and/or lower V_E relative to metabolic oxygen consumption (V_E/VO_2) in highland natives compared to lowland controls. These include one study of exercise in a Colorado group born and raised at altitude (Dempsey et al., 1971), nearly all studies in the Andes (Kollias et al., 1968; Schoene et al., 1990; Brutsaert et al., 2000; Wagner et al., 2002), and most (Lahiri and Milledge, 1966; Lahiri et al., 1967; Dua and Sen Gupta,

1980; Ge et al., 1994b; Zhuang et al., 1996) but not all (Sun et al., 1990a) studies in the Himalayas. It is not clear whether these are developmental or genetic effects. However, in support of the genetic hypothesis, one study by Brutsaert et al. (2005) shows a strong negative association of Quechua ancestry proportion with V_E and V_E/VO_2 in lowland-born subjects tested at 4338 m. This finding is consistent with better gas exchange in Quechua. For example, higher diffusion capacity or gas exchange efficiency could in theory allow more O_2 to enter the blood for the same or lower level of V_E. Indirectly, measures of increased pulmonary volume and/or increased diffusion capacity in highland natives (see below) also supports the idea that lower exercise V_E is made possible by better gas exchange. However, lower exercise V_E could also reflect a difference in ventilatory control that is independent of gas exchange.

Pulmonary volumes

Nearly all highland populations studied thus far have larger mean pulmonary volumes compared to sea-level controls, including total lung volume, vital capacity, and the residual volume. This includes both developmentally exposed populations and indigenous groups (Sun et al., 1990a; Frisancho et al., 1997; Brutsaert et al., 1999a). From studies of developmentally exposed populations, as well as from numerous animal studies (Johnson et al., 1985), it is clear that much of this effect is explained by lung growth during infant/child development in response to lifelong hypoxia. For example, Figure 11.7 is a comparison of lung volumes between two groups of Peruvian women who were matched for ancestry (i.e., genetics) but who differed by where they were born and raised (Lima, at sea-level, versus Cerro de Pasco at 4338 m). Cerro de Pasco women had ~15% larger total lung volume compared to Lima women. Whether "bigger is better" remains controversial, but at least one study in the Himalayas reported a positive correlation between forced vital capacity and $\dot{V}O_{2max}$ (Sun et al., 1990a). A similar study in the Andes failed to find this association within study groups (Brutsaert et al., 1999b, 2000).

Arterial O_2 saturation (SaO_2)

Numerous comparative studies show higher SaO_2s at rest and during submaximal and/or maximal exercise in indigenous high-altitude native populations (Sun et al., 1990a; Ge et al., 1994b, 1995; Favier et al., 1995; Zhuang et al., 1996; Chen et al., 1997; Brutsaert et al., 2000). It is important to note that these studies measured SaO_2 via pulse oximetry. Pulse oximetry is a non-invasive technique that correlates well with direct measures on whole blood, but there may be problems

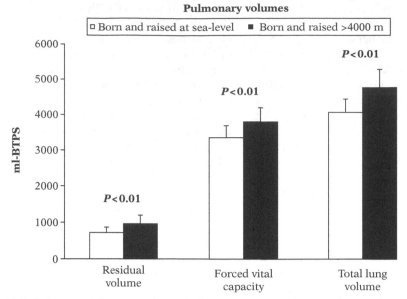

11.7. Pulmonary volumes are larger in Peruvian women who were born and raised above 4000 meters, compared to women born and raised at sea-level. Note: women in each group were matched on genetic background using a panel of ancestry informative molecular markers (see Brutsaert et al., 2003). BTPS refers to body temperature, pressure, and saturation.

with bias particularly during intense exercise (Yamaya et al., 2002). In the Andes, there is some conflict in the literature. Compared to fully acclimatized lowlanders, two studies show no difference in Aymara submaximal and maximal exercise SaO_2, one study shows comparable SaO_2s despite lower V_E in Aymara (Schoene et al., 1990), and three studies show higher SaO_2s during submaximal (Favier et al., 1995; Brutsaert et al., 2000) and/or maximal exercise (Frisancho et al., 1995). Two of these studies showed higher exercise SaO_2s in Aymara even when compared to Europeans who had been born and raised at 3600 m, suggesting a genetic effect (Favier et al., 1995; Brutsaert et al., 2000). In the Himalayas, nearly all studies show higher exercise SaO_2s at altitude in Tibetans or Sherpas compared to acclimatized lowland controls (Sun et al., 1990a; Ge et al., 1994b, 1995; Zhuang et al., 1996; Chen et al., 1997). Studies of resting SaO_2 in Tibetans indirectly support a genetic basis for the high exercise SaO_2s. Firstly, Tibetan-native neonates born at altitude have higher resting SaO_2s compared to neonates born to acclimatized lowland mothers (Niermeyer et al., 1995). Secondly, the quantitative genetic studies already described in detail (see Section V), suggest a major gene with a substantial phenotypic effect on resting SaO_2 (Beall et al., 1997a, 1997b). Finally, there is little evidence that SaO_2 is higher with developmental adaptation to altitude. In the few studies of developmentally exposed groups, SaO_2s were similar between acclimatized lowlanders and Europeans born and raised at altitude during submaximal exercise (Dempsey et al., 1971; Brutsaert et al., 2000) and at

$\dot{V}O_{2max}$ (Frisancho et al., 1995). A recent study from Qinghai China reports no differences in resting SaO_2 between large cohorts of Tibetans and Han Chinese who were both born and raised at altitude (Weitz and Garruto, 2007), although whether this is also the case for exercise SaO_2 cannot be determined from the resting data alone.

Blood gases and direct measures of pulmonary gas exchange

A handful of studies have measured arterial gas partial pressures at rest or during exercise in highland natives, and these have been very informative. In particular, evaluation of blood gases during exercise is useful given the additional demands placed on the pulmonary gas exchange system compared to rest. An early classic study by Dempsey et al. (1971) compared sojourners with residents of Leadville, Colorado at 3094 m. Colorado natives had smaller (A-a)DO_2 especially as exercise intensity increased. Recall that smaller (A-a)DO_2 can mean a better efficiency of gas exchange. Thus, from the Colorado work, it may be inferred that differences in (A-a)DO_2 between altitude natives and lowland controls are due at least in part to developmental adaptation to high altitude. However, this does not preclude the possibility of genetic effects. Zhuang et al. (1996) showed that Tibetans had lower V_E and about half the (A-a)DO_2 compared with acclimatized Han Chinese at 3658 m, again with the (A-a)DO_2 increasing between groups as VO_2 or power output increased. A recent study by Lundby et al. (2004) at 4100 m showed remarkably low (A-a)DO$_2$ (1–2 mmHg) at rest and during exercise to maximum in Aymara compared to Europeans with 8 weeks of acclimatization. A study by Wagner et al. (2002), at the relatively extreme altitude of 5260 m, also showed that Aymara natives have markedly lower V$_E$ and lower (A-a)DO$_2$, especially as exercise intensity increased. Improved gas exchange efficiency may have a simple structural basis in larger pulmonary volumes. Wagner et al. (2002) calculated that the O$_2$ diffusing capacity during maximal exercise was 40% higher in Aymara compared to acclimatized Europeans, and many other studies document higher diffusion capacities of highland natives at rest (Remmers and Mithoefer, 1969; Vincent et al., 1978).

The cardiovascular system

At the heart structural level, a small number of studies suggest changes in cardiovascular growth patterns at altitude, but there is no convincing evidence of meaningful population differences (Penaloza et al., 1963; Hulme et al., 2003). In persons born and raised at altitude, there tends to be a relative enlargement of the right ventricle, i.e., predominance. This is expected as hypoxia provokes a vasoconstriction of the pulmonary vasculature with a concomitant increase in pulmonary artery pressure (PAP). Presumably increased PAP is an adaptive response, perhaps serving to deliver a more uniform blood flow to the upper lung. On the other hand, persistent pulmonary hypertension is also associated with some of the acute and chronic problems of hypoxic exposure, including pulmonary edema and chronic mountain sickness. In this context, what is most interesting is the absence of pulmonary hypertension in many altitude-adapted species like the llama (Heath et al., 1974), and the very low prevalence of pulmonary hypertension in Tibetans (Groves et al., 1993). This is in some contrast with Andean groups and Han Chinese migrants to the Tibetan plateau who have higher prevalence of pulmonary hypertension and chronic mountain sickness (Sun et al., 1990b; Niermeyer et al., 2001).

At the heart functional level, cardiac output (Q) increases to match O$_2$ delivery to metabolic demand. Q is regulated by increases in the heart rate or stroke volume, and is also affected by the peripheral resistance of local capillary networks and the venous return of blood to the heart. Notwithstanding the significant hemodynamic changes that occur with altitude exposure and acclimatization (not discussed here), there is little to suggest that altitude natives differ with respect to cardiac output, although data are limited. One report suggests the possibility of higher stroke volume and lower peripheral resistance in Tibetans (Groves et al., 1993), but most reports from the Andes indicate normal Q at rest and during exercise (Banchero and Cruz, 1970; McKenzie et al., 1991), or indeed lower Q in residents of Cerro de Pasco, Peru (Vogel et al., 1974). In the terminal step of O$_2$ transport, oxygen must diffuse into working muscle, and a greater O$_2$ extraction at this level could obviate any need to increase Q. From limited data, Niermeyer and colleagues (2001) reported no differences in O$_2$ extraction between Andeans, Tibetans, and Colorado altitude natives, but at least one recent report showed paradoxically lower O$_2$ extraction at $\dot{V}O_{2max}$ in Andeans compared to acclimatized lowland controls (Lundby et al., 2006). The few studies in this area are potentially confounded by subject fitness status, which greatly affects stroke volume and other aspects of the hemodynamic response to exercise. Thus, more work is necessary before firm conclusions may be drawn.

One final trait considered in this section is exhaled pulmonary nitric oxide (NO), which may have some relevance to blood flow, both in the lung and at the systemic level. Beall and colleagues reported high values of exhaled NO in both Andean and Tibetan populations compared to sea-level standards (Beall et al., 2001; Hoit et al., 2005). Nitric oxide is a power vasodilator that regulates blood vessel diameter and local blood flow. Interestingly, Tibetans had nearly twice the mean exhaled NO of Andeans. Also, in Tibetans higher exhaled NO was associated with higher pulmonary blood flow. These authors suggest a beneficial role of NO in Tibetans allowing for higher pulmonary blood flow and O$_2$ delivery without the consequences of higher PAP.

Hemoglobin-O$_2$ affinity

Certain altitude-adapted species have high Hb-O$_2$ affinity, attributable to amino acid sequence variation in the hemoglobin molecule itself (Black and Tenney, 1980). However, this does not appear to be the case for human groups adapted to altitude. For both Andeans and Tibetans, Hb-O$_2$ affinity is similar to that of lowland groups as assessed on whole blood by the position of the Hb-O$_2$ dissociation curve (Samaja et al., 1979; Winslow et al., 1981). Further, at this time, there are no reports of hemoglobin genetic variants unique to altitude native populations. Similarly, for myoglobin, the muscle analog of hemoglobin, analysis of one exon in Tibetans does not show any novel polymorphism or selection for specific myoglobin alleles (Moore et al., 2002).

Hemoglobin concentration [Hb]

A number of large-scale surveys now make it clear that Tibetan populations have lower [Hb] compared to Andean, European, or Han Chinese populations resident at altitude (Beall et al., 1998; Moore et al., 2002;

Garruto et al., 2003; Wu et al., 2005). Indeed, Tibetan values at moderate altitude are not largely different from sea-level values, a paradoxical finding given the expected [Hb] increase with acclimatization. Hemoglobin production is regulated by the hormone erythropoietin, which is upregulated by hypoxemia. The low [Hb] in Tibetans suggests the absence of a hypoxic stimulus to increase erythropoietin, but how exactly this comes about is unknown. Also interesting is the emergent evidence from Ethiopia. A recent study of 236 Ethiopian native altitude residents by Beall et al. (2002) shows low [Hb], also within the ranges of sea-level populations. For myoglobin, one study by Gelfi et al. (2004) shows an upregulation of the myoglobin protein in Tibetans compared to lowland Nepali control subjects. However, the genetic, developmental, and/or environmental basis of this trait difference is unknown.

Muscle structure and metabolism

An early study in the Andes reported increased muscle myoglobin and oxidative enzyme concentration in Quechua compared to lowland controls (Reynafarje, 1962). However, this study has been criticized on the basis of training differences between the two comparison groups (Saltin et al., 1980), and subsequent studies have not replicated the findings. Indeed, prolonged exposure to hypoxia in lowlanders tends to decrease muscle oxidative capacity in both relative and absolute terms, i.e., decreased mitochondrial volume density and muscle mass. Hypoxia also decreases the activity of several key oxidative enzymes (Green et al., 1989; Hoppeler et al., 1990; Howald et al., 1990). Altitude natives from both the Andes and the Himalayas appear similar in this regard showing lower mitochondrial volume densities and/or oxidative enzyme activities (Kayser et al., 1991, 1996; Desplanches et al., 1996; Hoppeler et al., 2003). Further, in Andeans the muscle-training response is similar to that seen in lowlanders, including increases in capillary-to-fiber ratio, capillary density, the volume density of total mitochondria, and the activity of citrate synthase (Desplanches et al., 1996). Interestingly, Kayser et al. (1996) report lower mitochondrial volume density even in Tibetan migrants born at moderate altitude (1300 m), suggesting that this may be a fixed genetic trait. However, more work is needed in this area to replicate previous findings and special attention should be given to matching comparison groups on physical activity levels. Finally, there is no evidence that muscle fiber-type distributions are different in altitude natives, although again only two studies have addressed this question, one each in the Andes and in the Himalayas (Desplanches et al., 1996; Kayser et al., 1996).

At the muscle metabolic level, Hochachka et al. (1991) reported a persistent "lactate paradox" in Andeans transported and tested at sea-level. The lactate paradox refers to the observation that arterial lactate levels at a given level of work tend to be higher during exercise on acute exposure to hypoxia, but then return to near sea-level values after acclimatization time, despite continued hypoxia. Hochachka et al. (1991) reported persistently low lactate levels in Andeans even after six weeks at sea-level, and suggested this was part of a fundamental metabolic reorganization (i.e., adaptation) on the part of the altitude native subjects. According to Hochachka and colleagues, Andeans favor carbohydrate oxidation because glucose (glycogen) metabolism uses O_2 efficiently. The low lactate levels may be a reflection of a tight coupling between carbohydrate-based ATP synthesis and efficient pathways for ATP utilization. This hypothesis has yet to be confirmed and is at some variance with the recent study of Wagner et al. (2002) who showed similar lactate levels in Andeans and lowland controls at altitude. However, these authors did report an increased lactate acid buffering capacity in Andeans compared to acclimatized lowlanders on the basis of measured bicarbonate levels during exercise. In the Himalayas only a few studies have measured lactate levels. During exercise at 4700 m, Ge et al. (1994b) showed lower lactate levels, before and at the end of exercise, in Tibetans-versus-acclimatized Han Chinese. Two studies conducted in Kathmandu, Nepal (1300 m), show similar (Kayser et al., 1994) or lower (Marconi et al., 2005) lactate levels in Tibetans-versus-Nepali control populations. Unfortunately, the lactate response is highly dependent on subject fitness status and acclimatization state, and so studies conducted thus far are difficult to interpret regarding adaptive metabolic differences in lactate production/elimination in high-altitude natives.

SECTION VII: GENES AND ALTITUDE ADAPTATION

At the beginning of this chapter it was stated that no direct (genetic) evidence exists to support the hypothesis of natural selection in response to hypobaric hypoxia in a human population. What support exists for this hypothesis is by inference from trait differences between populations. Even the most directly comparative studies fall short of providing specific information on a genetic system that may have been modified by natural selection in an altitude native group. However, there is a growing library of candidate genes that are associated with the altitude response, and these may have relevance to the larger question of human adaptation.

Rupert and Koehle (2006) recently reviewed the literature on genetic associations with altitude disease, and much of it was centered on just a few candidate biochemical systems including polymorphisms in the pathway synthesizing nitric oxide, polymorphisms in the renin-angiotensin system that regulates cardiovascular homeostasis, and polymorphisms in the hypoxia inducible factor-1 (HIF-1) and erythropoiesis pathways. In a literature that currently numbers less than 20 independent studies, about half of the candidate genes tested against various altitude pathologies were statistically significant, and some of these are the focus of on-going current research.

One such genetic system, the insertion/deletion polymorphism of the angiotensin-converting enzyme (ACE), is considered here in some detail because it may prove to be paradigmatic of how gene-association studies are incorporated into our understanding of human adaptation to high altitude. The insertion (I) allele of the ACE gene is associated with lower tissue ACE activity, whereas the deletion (D) allele is associated with elevated serum ACE activity. In studies of altitude performance, the major focus has been on the possible benefit of the I allele as lower circulating ACE may attenuate the hypoxic pulmonary vasconstrictor response, attenuate pulmonary hypertension, and thus protect against AMS and high altitude pulmonary edema. There is some evidence in support of this hypothesis. In case control studies, the I allele was over-represented in a cohort of elite British climbers, and it has been associated with success in reaching the summit of Mt. Blanc (4807 m) (Woods and Montgomery, 2001; Tsianos et al., 2005). The I allele has also been associated with higher SaO_2 in relatively rapid but not slower ascents to 5000 m, and with a greater ventilatory response to exercise in hypoxia (Woods et al., 2002; Patel et al., 2003). In contrast, at least one study suggests an I-allele disadvantage at altitude. A study of Kyrgyz highlanders revealed a three-fold higher frequency of the I/I genotype in subjects with high altitude pulmonary hypertension (Aldashev et al., 2002). In addition, the highland Kyrgyz had lower I-allele frequency (0.56, $n = 87$) compared to a Bishkek lowland control group where the I-allele frequency was 0.65 ($n = 276$).

In any case, at the population level Andean Quechua have relatively high I-allele frequency (~0.72), with Tibetans showing slightly lower frequency (0.51–0.64) (Rupert et al., 1999; Gesang et al., 2002; Bigham et al., 2008). In most European populations, I-allele frequency is decidedly lower, ranging from ~0.15–0.55. Does this mean that the ACE I allele is an "altitude gene" that was driven to relatively high frequency by natural selection? The problem with this conclusion, as Rupert et al. (1999) first noted for Quechua, is that many other populations worldwide show comparable or higher I-allele frequency without a history of altitude exposure. For example, I-allele frequency is greater than 0.80 in a number of Native

American and Asian groups. Also, at a minimum, the evolutionary inference would require some demonstration of I-allele benefit on fertility/mortality in the population under consideration. A *phenotypic effect*, per se, is not always sufficient to make a compelling case for *phenotypic benefit* on population demography. Indeed, Bigham et al. (2008) have shown a strong I-allele effect determining higher resting and exercise SaO_2 in Peruvian Quechua ($P = 0.008$). However, it is unclear whether this is a common (within group) phenotypic effect of the ACE I allele, or whether the ACE gene has significance between groups as a locus of past natural selection.

SECTION VIII: FUTURE RESEARCH

There are certainly compelling physiological differences between highland and lowland populations. But, despite these differences, the hypothesis of natural selection cannot be adequately tested for a given trait until the genetic architecture of that trait is understood. Fortunately, the genomic information revolution has made it possible to interrogate the genetic basis of complex traits in new and powerful ways. Several approaches are currently being applied, including molecular studies of gene expression and genomic approaches that seek to identify the association of specific genes or genomic regions with traits of interest.

Genome-wide association (GWA) strategies have emerged as perhaps the most powerful and efficient means to dissect the genetic basis of complex traits (Risch and Merikangas, 1996; McCarthy et al., 2008), and the advent of high-density genotyping arrays has allowed a shift away from candidate gene studies. Using GWA, there have been many recent successes in the elucidation of genes involved in disease processes such as type II diabetes (Saxena et al., 2007; Scott et al., 2007; Sladek et al., 2007; Unoki et al., 2008; Yasuda et al., 2008), breast cancer (Easton et al., 2007; Hunter et al., 2007; Stacey et al., 2007, 2008; Gold et al., 2008), and prostate cancer (Yeager et al., 2007; Gudmundsson et al., 2007, 2008; Eeles et al., 2008; Thomas et al., 2008). Two key elements of these successes have been the collection of large sample sizes (i.e., thousands of individuals) with the consequent increase in study power to detect loci of modest effect, and the exponential advances in genotyping technologies that have dramatically improved genome coverage. Genome-wide association has not yet been applied to the study of the physiology of a highland native group, or to investigate any of the pathologies of high altitude, but several recent papers describe the potential utility of whole-genome approaches in this regard (Moore et al., 2004; Shriver et al., 2006).

At the molecular level, there has been intensive focus on the aforementioned HIF system. When intracellular O_2 levels fall, the HIF system is activated and HIF-1α works as a transcription factor to regulate cellular oxygen homeostasis via down-stream effects on numerous target genes, e.g., the erythropoietin gene which stimulates the production of red blood cells (Wenger and Gassmann, 1997). To date, two studies have examined sequence variation in the *HIF-1α* gene in Andeans (Hochachka and Rupert, 2003) and Sherpas (Suzuki et al., 2003). The latter study showed significant allele frequency differences in the Sherpa compared to lowland Japanese. Two other studies have investigated levels of gene expression of *HIF-1α* mRNA (i.e., the gene product) in leukocytes of Andeans (Appenzeller, 1998) and muscle cells of Sherpas (Gelfi et al., 2004). In both studies, compared to lowland controls, differences in gene expression or protein levels were detected. However, these studies are difficult to interpret in an evolutionary sense because no causal link has been established between genetic variation in the *HIF-1* gene, the levels of *HIF-1α* mRNA expression, and the ultimate effects on the phenotype. Nevertheless, gene-expression studies have already proven useful to accelerate gene discovery for complex traits that are related to chronic diseases in human and animal models, especially when integrated with genomic approaches (Farber and Lusis, 2008). Thus, the same approaches, used in the future to understand the genetic basis of highland native trait physiology, could greatly advance our understanding of the human evolutionary response to high altitude.

DISCUSSION POINTS

1. Distinguish between O_2 content, O_2 transport, O_2 delivery, O_2 extraction, and O_2 utilization.
2. Identify and distinguish between *structural* and *functional* components of the O_2 transport system. Speculate how these variables might best be regulated to increase O_2 delivery via acclimatization, or over developmental and evolutionary time frames.
3. Identify trait differences between Andean and Tibetan altitude native populations. Do these trait differences support the idea that Tibetans are "better adapted to altitude," or do these trait differences support the idea of "different, but equally effective patterns of adaptation to altitude?"
4. Why has birthweight been used so extensively to gauge population adaptation to hypoxia? Is birthweight a better outcome variable in this regard than $\dot{V}O_{2max}$?
5. Of all the complex traits discussed in this chapter, which, in your opinion, provides the best evidence of genetic adaptation to high altitude in a native group?
6. How important is developmental adaptation?

REFERENCES

Aldashev A. A., Sarybaev, A. S., Sydykov, A. S., et al. (2002). Characterization of high-altitude pulmonary hypertension in the Kyrgyz: association with angiotensin-converting enzyme genotype. *American Journal of Respiratory and Critical Care Medicine*, **166**, 1396–1402.

Appenzeller, O. (1998). Altitude and the nervous system. *Archive of Neurology*, **55**, 1007–1009.

Baker, P. T. (1969). Human adaptation to high altitude. *Science*, **163**, 1149–1156.

Baker, P. T. (1976). Work performance of highland natives. In *Man in the Andes: a Multidisciplinary Study of High-Altitude Quechua Natives*, P. T. Baker and M. A. Little (eds). Stroudsburg, PA: Wowden, Hutchinson, and Ross.

Balke, B. (1964). Work capacity and its limiting factors at high altitude. In *The Physiological Effects of Altitude*, W. H. Weihe (ed.). New York: Macmillan, pp. 233–247.

Banchero, N. and Cruz, J. C. (1970). Hemodynamic changes in the Andean native after two years at sea level. *Aerospace Medicine*, **41**, 849–853.

Bastien, G. J., Schepens, B., Willems, P. A., et al. (2005). Energetics of load carrying in Nepalese porters. *Science*, **308**, 1755.

Beall, C. M. (2000). Tibetan and Andean contrasts in adaptation to high-altitude hypoxia. In *Oxygen Sensing: Molecule to Man*, S. Lahiri (ed.). Springer, the Netherlands: Kluwer Academic/Plenum Publishers, pp. 63–74.

Beall, C. M., Strohl, K. P., Blangero, J., et al. (1997a). Ventilation and hypoxic ventilatory response of Tibetan and Aymara high altitude natives. *American Journal of Physical Anthropology*, **104**, 427–447.

Beall, C. M., Strohl, K. P., Blangero, J., et al. (1997b). Quantitative genetic analysis of arterial oxygen saturation in Tibetan highlanders. *Human Biology, an International Record of Research*, **69**, 597–604.

Beall, C. M., Brittenham, G. M., Strohl, K. P., et al. (1998). Hemoglobin concentration of high-altitude Tibetans and Bolivian Aymara. *American Journal of Physical Anthropology*, **106**, 385–400, erratum **107**, 421.

Beall, C. M., Laskowski, D., Strohl, K. P., et al. (2001). Pulmonary nitric oxide in mountain dwellers. *Nature*, **414**, 411–412.

Beall, C. M., Decker, M. J., Brittenham, G. M., et al. (2002). An Ethiopian pattern of human adaptation to high-altitude hypoxia. *Proceedings of the National Academy of Sciences of the United States of America*, **99**, 17215–17218.

Beall, C. M., Song, K., Elston, R. C., et al. (2004). Higher offspring survival among Tibetan women with high oxygen saturation genotypes residing at 4000 m. *Proceedings of the National Academy of Sciences of the United States of America*, **101**, 14300–14304.

Bender, P. R., McCullough, R. E., McCullough, R. G., et al. (1989). Increased exercise SaO_2 independent of ventilatory acclimatization at 4300 m. *Journal of Applied Physiology*, **66**, 2733–2738.

Bennett, A., Sain, S. R., Vargas, E., et al. (2008). Evidence that parent-of-origin affects birth-weight reductions at high altitude. *American Journal of Human Biology*, **20**, 592–597.

Bigham, A.W., Kiyamu, M., Leon-Velarde, F, et al. (2008). Angiotensin-converting enzyme genotype and arterial oxygen saturation at high altitude in Peruvian Quechua. *High Altitude Medicine and Biology*, **9**, 167–178.

Black, C.P. and Tenney, S.M. (1980). Oxygen transport during progressive hypoxia in high-altitude and sea-level waterfowl. *Respiration Physiology*, **39**, 217–239.

Brutsaert, T.D. (1997). Environmental, developmental, and ancestral (genetic) components of the exercise response of Bolivian high altitude natives. PhD thesis, Cornell University, Ithaca, NY.

Brutsaert, T.D., Soria, R., Caceres, E., et al. (1999a). Effect of developmental and ancestral high altitude exposure on chest morphology and pulmonary function in Andean and European/North American natives. *American Journal of Human Biology*, **11**, 383–395.

Brutsaert, T.D., Spielvogel, H., Soria, R., et al. (1999b). Effect of developmental and ancestral high-altitude exposure on VO_{2peak} of Andean and European/North American natives. *American Journal of Physical Anthropology*, **110**, 435–455.

Brutsaert, T.D., Araoz, M., Soria, R., et al. (2000). Higher arterial oxygen saturation during submaximal exercise in Bolivian Aymara compared to European sojourners and Europeans born and raised at high altitude. *American Journal of Physical Anthropology*, **113**, 169–181.

Brutsaert, T., Parra, E., Shriver, M., et al. (2003). Spanish genetic admixture is associated with larger $\dot{V}O_{2max}$ decrement from sea level to 4338 m in Peruvan Quechua. *Journal of Applied Physiology*, **95**, 519–528.

Brutsaert, T.D., Haas, J.D. and Spielvogel, H. (2004). Absence of work efficiency differences during cycle ergometry exercise in Bolivian Aymara. *High Altitude Medicine and Biology*, **5**, 41–59.

Brutsaert, T.D., Parra, E.J., Shriver, M.D., et al. (2005). Ancestry explains the blunted ventilatory response to sustained hypoxia and lower exercise ventilation of Quechua altitude natives. *American Journal of Physiology. Regulatory, Integrative and Comparative Physiology*, **289**(1), R225–R234.

Buskirk, E.R., Kollias, J., Akers, R.F., et al. (1967). Maximal performance at altitude and on return from altitude in conditioned runners. *Journal of Applied Physiology*, **23**, 259–266.

Byrne-Quinn, E., Sodal, I.E. and Weil, J.V. (1972). Hypoxic and hypercapnic ventilatory drives in children native to high altitude. *Journal of Applied Physiology*, **32**, 44–46.

Chen, Q.H., Ge, R.L., Wang, X.Z., et al. (1997). Exercise performance of Tibetan and Han adolescents at altitudes of 3,417 and 4,300 m. *Journal of Applied Physiology*, **83**, 661–667.

Chiodi, H. (1957). Respiratory adaptations to chronic high altitude hypoxia. *Journal of Applied Physiology*, **10**, 81–87.

Cudkowicz, L., Spielvogel, H. and Zubieta, G. (1972). Respiratory studies in women at high altitude (3600 m or 12200 ft and 5200 m or 17200 ft). *Respiration*, **29**, 393–426.

Curran, L.S., Zhuang, J., Sun, S.F., et al. (1997). Ventilation and hypoxic ventilatory responsiveness in Chinese-Tibetan residents at 3658 m. *Journal of Applied Physiology*, **83**, 2098–2104.

Curran, L.S., Zhuang, J., Droma, T., et al. (1998). Superior exercise performance in lifelong Tibetan residents of 4400 m compared with Tibetan residents of 3658 m. *American Journal of Physical Anthropology*, **105**, 21–31.

Dempsey, J.A., Reddan, W.G., Birnbaum, M.L., et al. (1971). Effects of acute through life-long hypoxic exposure on exercise pulmonary gas exchange. *Respiration Physiology*, **13**, 62–89.

Dempsey, J.A., Forster, H.V. and Ainsworth, D.M. (eds) (1995). *Regulation of Hyperpnea, Hyperventilation, and Respiratory Muscle Recruitment during Exercise*. New York: Marcel Dekker.

Desplanches, D., Hoppeler, H., Tuscher, L., et al. (1996). Muscle tissue adaptations of high-altitude natives to training in chronic hypoxia or acute normoxia. *Journal of Applied Physiology*, **81**, 1946–1951.

Dill, D.B., Edwards, H.T., Oberg, S.A., et al. (1931). Adaptations to the organism to changes in oxygen pressure. *Journal de physiologie*, **71**, 47–63.

Dua, G.L. and Sen Gupta, J. (1980). A study of physical work capacity of sea level residents on prolonged stay at high altitude and comparison with high altitude native residents. *Indian Journal of Physiology and Pharmacology*, **24**, 15–24.

Easton, D.F., Pooley, K.A., Dunning, A.M., et al. (2007). Genome-wide association study identifies novel breast cancer susceptibility loci. *Nature*, **447**, 1087–1093.

Eeles, R.A., Kote-Jarai, Z., Giles, G.G., et al. (2008). Multiple newly identified loci associated with prostate cancer susceptibility. *Nature Genetics*, **40**, 316–321.

Elsner, R.W., Blostad, A. and Forno, C. (1964). Maximum oxygen consumption of Peruvian Indians native to high altitude. In *The Physiological Effects of High Altitude*, W.H. Weihe (ed.). New York: Pergamon Press, pp. 217–223.

Farber, C.R. and Lusis, A.J. (2008). Integrating global gene expression analysis and genetics. *Advances in Genetics*, **60**, 571–601.

Favier, R., Spielvogel, H., Desplanches, D., et al. (1995). Maximal exercise performance in chronic hypoxia and acute normoxia in high-altitude natives. *Journal of Applied Physiology*, **78**, 1868–1874.

Forster, H.V., Dempsey, J.A., Birnbaum, M.L., et al. (1971). Effect of chronic exposure to hypoxia on ventilatory response to CO_2 and hypoxia. *Journal of Applied Physiology*, **31**, 586–592.

Frisancho, A.R. (1969). Human growth and pulmonary function of a high altitude Peruvian Quechua population. *Human Biology*, **41**, 365–379.

Frisancho, A.R., Martinez, C., Velasquez, T., et al. (1973). Influence of developmental adaptation on aerobic capacity at high altitude. *Journal of Applied Physiology*, **34**, 176–180.

Frisancho, A.R., Frisancho, H.G., Milotich, M., et al. (1995). Developmental, genetic, and environmental components of aerobic capacity at high altitude. *American Journal of Physical Anthropology*, **96**, 431–442.

Frisancho, A.R., Frisancho, H.G., Albalak, R., et al. (1997). Developmental, genetic and environmental components of lung volumes at high altitude. *American Journal of Human Biology*, **9**, 191–203.

Gamboa, A., Leon-Velarde, F., Rivera-Ch, M., et al. (2003). Selected contribution: acute and sustained ventilatory responses to hypoxia in high-altitude natives living at sea level. *Journal of Applied Physiology*, **94**, 1255–1262, discussion 1253–1254.

Garruto, R. M., Chin, C. T., Weitz, C. A., et al. (2003). Hematological differences during growth among Tibetans and Han Chinese born and raised at high altitude in Qinghai, China. *American Journal of Physical Anthropology*, **122**, 171–183.

Ge, R. L., Chen, Q. H. and He, L. G. (1994a). [Characteristics of hypoxic ventilatory response in Tibetan living at moderate and high altitudes.] *Chung Hua Chieh Ho Ho Hu Hsi Tsa Chih*, **17**, 364–366, 384.

Ge, R. L., Chen, Q. H., Wang, L. H., et al. (1994b). Higher exercise performance and lower $\dot{V}O_{2max}$ in Tibetan than Han residents at 4700 m altitude. *Journal of Applied Physiology*, **77**, 684–691.

Ge, R. L., He Lun, G. W., Chen, Q. H., et al. (1995). Comparisons of oxygen transport between Tibetan and Han residents at moderate altitude. *Wilderness and Environmental Medicine*, **6**, 391–400.

Gelfi, C., De Palma, S., Ripamonti, M., et al. (2004). New aspects of altitude adaptation in Tibetans: a proteomic approach. *FASEB Journal*, **18**, 612–614.

Gesang, L., Liu, G., Cen, W., et al. (2002). Angiotensin-converting enzyme gene polymorphism and its association with essential hypertension in a Tibetan population. *Hypertension Research*, **25**, 481–485.

Gold, B., Kirchhoff, T., Stefanov, S., et al. (2008). Genome-wide association study provides evidence for a breast cancer risk locus at 6q22.33. *Proceedings of the National Academy of Sciences of the United States of America*, **105**, 4340–4345.

Green, H. J., Sutton, J. R., Cymerman, A., et al. (1989). Operation Everest II: adaptations in human skeletal muscle. *Journal of Applied Physiology*, **66**, 2454–2461.

Greksa, L. P. (2006). Growth and development of Andean high altitude residents. *High Altitude Medicine and Biology*, **7**, 116–124.

Greksa, L. P. and Haas, J. D. (1982). Physical growth and maximal work capacity in preadolescent boys at high-altitude. *Human Biology*, **54**, 677–695.

Greksa, L. P., Spielvogel, H. and Paredes-Fernandez, L. (1985). Maximal exercise capacity in adolescent European and Amerindian high-altitude natives. *American Journal of Physical Anthropology*, **67**, 209–216.

Greksa, L. P., Spielvogel, H., Paz-Zamora, M., et al. (1988). Effect of altitude on the lung function of high altitude residents of European ancestry. *American Journal of Physical Anthropology*, **75**, 77–85.

Grover, R. F., Reeves, J. T., Grover, E. B., et al. (1967). Muscular exercise in young men native to 3100 m altitude. *Journal of Applied Physiology*, **22**, 555–564.

Groves, B. M., Droma, T., Sutton, J. R., et al. (1993). Minimal hypoxic pulmonary hypertension in normal Tibetans at 3658 m. *Journal of Applied Physiology*, **74**, 312–318.

Gudmundsson, J., Sulem, P., Manolescu, A., et al. (2007). Genome-wide association study identifies a second prostate cancer susceptibility variant at 8q24. *Nature Genetics*, **39**, 631–637.

Gudmundsson, J., Sulem, P., Rafnar, T., et al. (2008). Common sequence variants on 2p15 and Xp11.22 confer susceptibility to prostate cancer. *Nature Genetics*, **40**, 281–283.

Haas, J. D., Frongillo, E. J., Stepcik, C., et al. (1980). Altitude, ethnic, and sex differences in birth weight and length in Bolivia. *Human Biology*, **52**, 459–477.

Haas, J. D., Greksa, L. P., Leatherman, T. L., et al. (1983). Submaximal work performance of native and migrant preadolescent boys at high altitude. *Human Biology*, **55**, 517–527.

Hackett, P. H., Reeves, J. T., Reeves, C. D., et al. (1980). Control of breathing in Sherpas at low and high altitude. *Journal of Applied Physiology*, **49**, 374–379.

Heath, D., Smith, P., Williams, D., et al. (1974). The heart and pulmonary vasculature of the llama (*Lama glama*). *Thorax*, **29**, 463–471.

Hochachka, P. W. and Rupert, J. L. (2003). Fine tuning the HIF-1 "global" O_2 sensor for hypobaric hypoxia in Andean high-altitude natives. *Bioessays*, **25**, 515–519.

Hochachka, P. W., Stanley, C., Matheson, G. O., et al. (1991). Metabolic and work efficiencies during exercise in Andean natives. *Journal of Applied Physiology*, **70**, 1720–1730.

Hoit, B. D., Dalton, N. D., Erzurum, S. C., et al. (2005). Nitric oxide and cardio-pulmonary hemodynamics in Tibetan highlanders. *Journal of Applied Physiology*, **99**, 1796–1801.

Hoppeler, H., Kleinert, E., Schlegel, C., et al. (1990). Morphological adaptations of human skeletal muscle to chronic hypoxia. *International Journal of Sports Medicine*, **11**(suppl. 1), S3–S9.

Hoppeler, H., Vogt, M., Weibel, E. R., et al. (2003). Response of skeletal muscle mitochondria to hypoxia. *Experimental Physiology*, **88**, 109–119.

Howald, H., Pette, D., Simoneau, J. A., et al. (1990). Effect of chronic hypoxia on muscle enzyme activities. *International Journal of Sports Medicine*, **11**(suppl. 1), S10–S14.

Huang, S. Y., Alexander, J. K., Grover, R. F., et al. (1984). Hypocapnia and sustained hypoxia blunt ventilation on arrival at high altitude. *Journal of Applied Physiology*, **56**, 602–606.

Hulme, C. W., Ingram, T. E. and Lonsdale-Eccles, D. A. (2003). Electrocardiographic evidence for right heart strain in asymptomatic children living in Tibet – a comparative study between Han Chinese and ethnic Tibetans. *Wilderness and Environmental Medicine*, **14**, 222–225.

Hunter, D. J., Kraft, P., Jacobs, K. B., et al. (2007). A genome-wide association study identifies alleles in *FGFR2* associated with risk of sporadic postmenopausal breast cancer. *Nature Genetics*, **39**, 870–874.

Hurtado, A. (1932). Respiratory adaptation in the Indian natives of the Peruvian Andes. Studies at high altitude. *American Journal of Physical Anthropology*, **17**, 137–165.

Hurtado, A. (1964). Animals at high altitudes: resident man. In *Handbook of Physiology, Section 4, Adaptation and Environment*, D. B. Dill, E. F. Adolph and C. G. Wiber (eds). Washington, DC: American Physiological Society, pp. 843–860.

Johnson, R. L. Jr, Cassidy, S. S., Grover, R. F., et al. (1985). Functional capacities of lungs and thorax in beagles after prolonged residence at 3100 m. *Journal of Applied Physiology*, **59**, 1773–1782.

Julian, C. G., Vargas, E., Armaza, J. F., et al. (2007). High-altitude ancestry protects against hypoxia-associated reductions in fetal growth. *Archives of Disease in Childhood. Fetal and Neonatal Edition*, **92**, F372–F377.

Kashiwazaki, H., Dejima, Y., Orias-Rivera, J., et al. (1995). Energy expenditure determined by the doubly labeled water method in Bolivian Aymara living in a high altitude agropastoral community. *The American Journal of Clinical Nutrition*, **62**, 901–910.

Kayser, B., Hoppeler, H., Claassen, H., et al. (1991). Muscle structure and performance capacity of Himalayan Sherpas. *Journal of Applied Physiology*, **70**, 1938–1942.

Kayser, B., Marconi, C., Amatya, T., et al. (1994). The metabolic and ventilatory response to exercise in Tibetans born at low altitude. *Respiratory Physiology*, **98**, 15–26.

Kayser, B., Hoppeler, H., Desplanches, D., et al. (1996). Muscle ultrastructure and biochemistry of lowland Tibetans. *Journal of Applied Physiology*, **81**, 419–425.

Keyes, L. E., Armaza, J. F., Niermeyer, S., et al. (2003). Intrauterine growth restriction, preeclampsia, and intrauterine mortality at high altitude in Bolivia. *Pediatric Research*, **54**, 20–25.

Kollias, J., Buskirk, E. R., Akers, R. F., et al. (1968). Work capacity of long-time residents and newcomers to altitude. *Journal of Applied Physiology*, **24**, 792–799.

Lahiri, S. (1968). Alveolar gas pressures in man with lifetime hypoxia. *Respiratory Physiology*, **4**, 373–386.

Lahiri, S. and Milledge, J. S. (1966). Muscular exercise in the Himalayan high-altitude residents. *Federation Proceedings*, **25**, 1392–1396.

Lahiri, S., Milledge, J. S., Chattopadhyay, H. P., et al. (1967). Respiration and heart rate of Sherpa highlanders during exercise. *Journal of Applied Physiology*, **23**, 545–554.

Lahiri, S., Kao, F. F., Velasquez, T., et al. (1969). Irreversible blunted respiratory sensitivity to hypoxia in high altitude natives. *Respiratory Physiology*, **6**, 360–374.

Leon-Velarde, F., Vargas, M., Monge, C. C., et al. (1996). Alveolar P_{CO_2} and P_{O_2} of high-altitude natives living at sea level. *Journal of Applied Physiology*, **81**, 1605–1609.

Leon-Velarde, F., Maggiorini, M., Reeves, J. T., et al. (2005). Consensus statement on chronic and subacute high altitude diseases. *High Altitude Medicine and Biology*, **6**, 147–157.

Lundby, C., Calbet, J. A., van Hall, G., et al. (2004). Pulmonary gas exchange at maximal exercise in Danish lowlanders during 8 weeks of acclimatization to 4100 m and in high-altitude Aymara natives. *American Journal of Physiology. Regulatory, Integrative and Comparative Physiology*, **287**, R1202–R1208.

Lundby, C., Sander, M., van Hall, G., et al. (2006). Maximal exercise and muscle oxygen extraction in acclimatizing lowlanders and high altitude natives. *Journal of Physiology*, **573**, 535–547.

Lynch, T. F. (1990). Glacial-age man in South America? A critical review. *American Antiquity*, **55**, 12–36.

MacNeish, R. S. and Berger, R. P. (1970). Megafauna and man from Ayacucho, highland Peru. *Science*, **166**, 8975–8977.

Marconi, C., Marzorati, M., Grassi, B., et al. (2004). Second generation Tibetan lowlanders acclimatize to high altitude more quickly than Caucasians. *Journal of Physiology*, **556**, 661–671.

Marconi, C., Marzorati, M., Sciuto, D., et al. (2005). Economy of locomotion in high-altitude Tibetan migrants exposed to normoxia. *Journal of Physiology*, **569**, 667–675.

Maresh, C. M., Noble, B. J., Robertson, K. L., et al. (1983). Maximal exercise during hypobaric hypoxia (447 Torr) in moderate-altitude natives. *Medicine and Science in Sports and Exercise*, **15**, 360–365.

Martin, I. H. and Costa, L. E. (1992). Reproductive function in female rats submitted to chronic hypobaric hypoxia. *Archives internationales de physiologie, de biochimie et de biophysique*, **100**, 327–330.

Mazess, R. B. (1969a). Exercise performance at high altitude in Peru. *Federation Proceedings*, **28**, 1301–1306.

Mazess, R. B. (1969b). Exercise performance of Indian and white high altitude residents. *Human Biology*, **41**, 494–518.

Mazess, R. B. (ed.) (1978). *Adaptation: a Conceptual Framework*. The Hague, the Netherlands: Mouton.

McAuliffe, F., Kametas, N., Krampl, E., et al. (2001). Blood gases in pregnancy at sea level and at high altitude. *BJOG: An International Journal of Obstetrics and Gynaecology*, **108**, 980–985.

McCarthy, M. I., Abecasis, G. R., Cardon, L. R., et al. (2008). Genome-wide association studies for complex traits: consensus, uncertainty and challenges. *Nature Reviews. Genetics*, **9**, 356–369.

McKenzie, D. C., Goodman, L. S., Nath, C., et al. (1991). Cardiovascular adaptations in Andean natives after 6 week of exposure to sea level. *Journal of Applied Physiology*, **70**, 2650–2655.

Monge, C. (1948). *Acclimatization in the Andes*. Baltimore: The Johns Hopkins University Press.

Moore, L. G. (1990). Maternal O_2 transport and fetal growth in Colorado, Peru, and Tibet high-altitude residents. *American Journal of Human Biology*, **2**, 627–637.

Moore, L. G. (2000). Comparative human ventilatory adaptation to high altitude. *Respiratory Physiology*, **121**, 257–276.

Moore, L. G. (2001). Human genetic adaptation to high altitude. *High Altitude Medicine and Biology*, **2**, 257–279.

Moore, L. G. (2003). Fetal growth restriction and maternal oxygen transport during high altitude pregnancy. *High Altitude Medicine and Biology*, **4**, 141–156.

Moore, L. G., Curran-Everett, L., Droma, T. S., et al. (1992). Are Tibetans better adapted? *International Journal of Sports Medicine*, **13** (suppl. 1), S86–S88.

Moore, L. G., Asmus, L. and Curran, L. (1998). *Chronic Mountain Sickness: Gender and Geographic Variation: Progress in Mountain Medicine and High Altitude Physiology*. Japan: Matsumoto, pp. 114–119.

Moore, L. G., Zamudio, S., Zhuang, J., et al. (2001). Oxygen transport in Tibetan women during pregnancy at 3658 m. *American Journal of Physical Anthropology*, **114**, 42–53.

Moore, L. G., Zamudio, S., Zhuang, J., et al. (2002). Analysis of the myoglobin gene in Tibetans living at high altitude. *High Altitude Medicine and Biology*, **3**, 39–47.

Moore, L. G., Shriver, M., Bemis, L., et al. (2004). Maternal adaptation to high-altitude pregnancy: an experiment of nature – a review. *Placenta*, **25**, S60–S71.

Niermeyer, S., Yang, P., Shanmina, et al. (1995). Arterial oxygen saturation in Tibetan and Han infants born in Lhasa, Tibet. *New England Journal of Medicine*, **333**, 1248–1252.

Niermeyer, S., Zamudio, S. and Moore, L. G. (eds) (2001). *The People*. New York: Marcel Dekker.

Niu, W., Wu, Y., Li, B., et al. (1995). Effects of long-term acclimatization in lowlanders migrating to high altitude: comparison with high altitude residents. *European Journal of Applied Physiology*, **71**, 543–548.

Palmer, S. K., Moore, L. G., Young, D., et al. (1999). Altered blood pressure course during normal pregnancy and increased preeclampsia at high altitude (3100 m) in Colorado. *American Journal of Obstetrics and Gynecology*, **180**, 1161–1168.

Patel, S., Woods, D. R., Macleod, N. J., et al. (2003). Angiotensin-converting enzyme genotype and the ventilatory response to exertional hypoxia. *The European Respiratory Journal*, **22**, 755–760.

Penaloza, D., Banchero, N., Sime, F., et al. (1963). The heart in chronic hypoxia. *Biochemical Clinics*, **2**, 283–298.

Remmers, J. E. and Mithoefer, J. C. (1969). The carbon monoxide diffusing capacity in permanent residents at high altitude. *Respiratory Physiology*, **6**, 233–244.

Reynafarje, B. (1962). Myoglobin content and enzymatic activity of muscle and altitude adaptation. *Journal of Applied Physiology*, **17**, 301–305.

Reynafarje, B. and Velasquez, T. (1966). Metabolic and physiological aspects of exercise at high altitude. I. Kinetics of blood lactate, oxygen consumption and oxygen debt during exercise and recovery breathing air. *Federation Proceedings*, **25**, 1397–1399.

Risch, N. and Merikangas, K. (1996). The future of genetic studies of complex human diseases [see comments]. *Science*, **273**, 1516–1517.

Rupert, J. L. and Koehle, M. S. (2006). Evidence for a genetic basis for altitude-related illness. *High Altitude Medicine and Biology*, **7**, 150–167.

Rupert, J. L., Devine, D. V., Monsalve, M. V., et al. (1999). Angiotensin-converting enzyme (ACE) alleles in the Quechua, a high altitude South American native population. *Annals of Human Biology*, **26**, 375–380.

Saltin, B., Nygaard, E. and Rasmussen, B. (1980). Skeletal muscle adaptations in man following prolonged exposure to high altitude. *Acta Physiologica Scandinavica*, **109**, 31A.

Samaja, M., Veicsteinas, A. and Cerretelli, P. (1979). Oxygen affinity of blood in altitude Sherpas. *Journal of Applied Physiology*, **47**, 337–341.

Santolaya, R. B., Lahiri, S., Alfaro, R. T., et al. (1989). Respiratory adaptation in the highest inhabitants and highest Sherpa mountaineers. *Respiratory Physiology*, **77**, 253–262.

Saxena, R., Voight, B. F., Lyssenko, V., et al. (2007). Genome-wide association analysis identifies loci for type 2 diabetes and triglyceride levels. *Science*, **316**, 1331–1336.

Schoene, R. B., Roach, R. C., Lahiri, S., et al. (1990). Increased diffusion capacity maintains arterial saturation during exercise in the Quechua Indians of the Chilean Altiplano. *American Journal of Human Biology*, **2**, 663–668.

Scott, L. J., Mohlke, K. L., Bonnycastle, L. L., et al. (2007). A genome-wide association study of type 2 diabetes in Finns detects multiple susceptibility variants. *Science*, **316**, 1341–1345.

Severinghaus, J. W. (1979). Simple, accurate equations for human blood O_2 dissociation computations. *Journal of Applied Physiology*, **46**, 599–602.

Severinghaus, J. W., Bainton, C. R. and Carcelen, A. (1966). Respiratory insensitivity to hypoxia in chronically hypoxic man. *Respiratory Physiology*, **1**, 308–334.

Shriver, M. D., Mei, R., Bigham, A., et al. (2006). Finding the genes underlying adaptation to hypoxia using genomic scans for genetic adaptation and admixture mapping. *Advances in Experimental Medicine and Biology*, **588**, 89–100.

Sladek, R., Rocheleau, G., Rung, J., et al. (2007). A genome-wide association study identifies novel risk loci for type 2 diabetes. *Nature*, **445**, 881–885.

Smith, C., Dempsey, J. and Hornbein, T. (2001). Control of breathing at high altitude. In *High Altitude: an Exploration of Human Adaptation*, T. Hornbein and R. Schoene (eds). New York: Marcel Dekker, pp. 139–173.

Sorensen, S. C. and Severinghaus, J. W. (1968a). Irreversible respiratory insensitivity to acute hypoxia in man born at high altitude. *Journal of Applied Physiology*, **25**, 217–220.

Sorensen, S. C. and Severinghaus, J. W. (1968b). Respiratory sensitivity to acute hypoxia in man born at sea level living at high altitude. *Journal of Applied Physiology*, **25**, 211–216.

Stacey, S. N., Manolescu, A., Sulem, P., et al. (2007). Common variants on chromosomes 2q35 and 16q12 confer susceptibility to estrogen receptor-positive breast cancer. *Nature Genetics*, **39**, 865–869.

Stacey, S. N., Manolescu, A., Sulem, P., et al. (2008). Common variants on chromosome 5p12 confer susceptibility to estrogen receptor-positive breast cancer. *Nature Genetics*, **40**, 703–706.

Sun, S. F., Droma, T. S., Zhang, J. G., et al. (1990a). Greater maximal O_2 uptakes and vital capacities in Tibetan than Han residents of Lhasa. *Respiratory Physiology*, **79**, 151–161.

Sun, S. F., Huang, S. Y., Zhang, J. G., et al. (1990b). Decreased ventilation and hypoxic ventilatory responsiveness are not reversed by naloxone in Lhasa residents with chronic mountain sickness. *American Review of Respiratory Diseases*, **142**, 1294–1300.

Suzuki, K., Kizaki, T., Hitomi, Y., et al. (2003). Genetic variation in hypoxia-inducible factor 1alpha and its possible association with high altitude adaptation in Sherpas. *Medical Hypotheses*, **61**, 385–389.

Thomas, G., Jacobs, K. B., Yeager, M., et al. (2008). Multiple loci identified in a genome-wide association study of prostate cancer. *Nature Genetics*, **40**, 310–315.

Torroni, A., Miller, J. A., Moore, L. G., et al. (1994). Mitochondrial DNA analysis in Tibet: implications for the origin of the Tibetan population and its adaptation to high altitude. *American Journal of Physical Anthropology*, **93**, 189–199.

Tsianos, G., Eleftheriou, K. I., Hawe, E., et al. (2005). Performance at altitude and angiotensin I-converting enzyme genotype. *European Journal of Applied Physiology*, **93**, 630–633.

Unoki, H., Takahashi, A., Kawaguchi, T., et al. (2008). SNPs in KCNQ1 are associated with susceptibility to type 2 diabetes in East Asian and European populations. *Nature Genetics*, **40**, 1098–1102.

Vargas, M., Vargas, E., Julian, C. G., et al. (2007). Determinants of blood oxygenation during pregnancy in Andean and European residents of high altitude. *American Journal of Physiology*, **293**, R1303–R1312.

Velasquez, T. (1966). *Acquired Acclimatization to Sea-Level: Life at High Altitudes*. Washington, DC: Pan American Health Organization Scientific Publications, pp. 58–63.

Vincent, J., Hellot, M. F., Vargas, E., et al. (1978). Pulmonary gas exchange, diffusing capacity in natives and newcomers at high altitude. *Respiratory Physiology*, **34**, 219–231.

Vitzthum, V. J. and Wiley, A. S. (2003). The proximate determinants of fertility in populations exposed to chronic hypoxia. *High Altitude Medicine and Biology*, **4**, 125–139.

Vogel, J. A., Hartley, L. H. and Cruz, J. C. (1974). Cardiac output during exercise in altitude natives at sea level and high altitude. *Journal of Applied Physiology*, **36**, 173–176.

Wagner, P. D., Araoz, M., Boushel, R., et al. (2002). Pulmonary gas exchange and acid-base state at 5260 m in high-altitude Bolivians and acclimatized lowlanders. *Journal of Applied Physiology*, **92**, 1393–1400.

Way, A. B. (1976). Exercise capacity of high altitude Peruvian Quechua Indians migrant to low altitude. *Human Biology*, **48**, 175–191.

Weil, J. V., Byrne-Quinn, E., Sodal, I. E., et al. (1971). Acquired attenuation of chemoreceptor function in chronically hypoxic man at high altitude. *Journal of Clinical Investigation*, **50**, 186–195.

Weitz, C. A. and Garruto, R. M. (2007). A comparative analysis of arterial oxygen saturation among Tibetans and Han born and raised at high altitude. *High Altitude Medicine and Biology*, **8**, 13–26.

Wenger, R. H. and Gassmann, M. (1997). Oxygen(es) and the hypoxia-inducible factor-1. *Biological Chemistry*, **378**, 609–616.

Wilson, M. J., Lopez, M., Vargas, M., et al. (2007). Greater uterine artery blood flow during pregnancy in multi-generational (Andean) than shorter-term (European) high-altitude residents. *American Journal of Physiology*, **293**, R1313–R1324.

Winslow, R. M., Monge, C. C., Statham, N. J., et al. (1981). Variability of oxygen affinity of blood: human subjects native to high altitude. *Journal of Applied Physiology*, **51**, 1411–1416.

Winslow, R. M., Monge, C. C., Brown, E. G., et al. (1985). Effects of hemodilution on O_2 transport in high-altitude polycythemia. *Journal of Applied Physiology*, **59**, 1495–1502.

Winslow, R. M., Chapman, K. W., Gibson, C. C., et al. (1989). Different hematologic responses to hypoxia in Sherpas and Quechua Indians. *Journal of Applied Physiology*, **66**, 1561–1569.

Woods, D. R. and Montgomery, H. E. (2001). Angiotensin-converting enzyme and genetics at high altitude. *High Altitude Medicine and Biology*, **2**, 201–210.

Woods, D. R., Pollard, A. J., Collier, D. J., et al. (2002). Insertion/deletion polymorphism of the angiotensin I-converting enzyme gene and arterial oxygen saturation at high altitude. *American Journal of Respiratory and Critical Care Medicine*, **166**, 362–366.

Wu, T., Wang, X., Wei, C., et al. (2005). Hemoglobin levels in Qinghai-Tibet: different effects of gender for Tibetans vs. Han. *Journal of Applied Physiology*, **98**, 598–604.

Yamaya, Y., Bogaard, H. J., Wagner, P. D., et al. (2002). Validity of pulse oximetry during maximal exercise in normoxia, hypoxia, and hyperoxia. *Journal of Applied Physiology*, **92**, 162–168.

Yasuda, K., Miyake, K., Horikawa, Y., et al. (2008). Variants in *KCNQ1* are associated with susceptibility to type 2 diabetes mellitus. *Nature Genetics*, **40**, 1092–1097.

Yeager, M., Orr, N., Hayes, R. B., et al. (2007). Genome-wide association study of prostate cancer identifies a second risk locus at 8q24. *Nature Genetics*, **39**, 645–649.

Zamudio, S. (2003). The placenta at high altitude. *High Altitude Medicine and Biology*, **4**, 171–191.

Zhimin, A. (1982). Paleoliths and microliths from Shenja and Shuanghu, Northern Tibet. *Current Anthropology*, **23**, 493–499.

Zhuang, J., Droma, T., Sun, S., et al. (1993). Hypoxic ventilatory responsiveness in Tibetan compared with Han residents of 3658 m. *Journal of Applied Physiology*, **74**, 303–311.

Zhuang, J., Droma, T., Sutton, J. R., et al. (1996). Smaller alveolar-arterial O_2 gradients in Tibetan than Han residents of Lhasa (3658 m). *Respiratory Physiology*, **103**, 75–82.

12 Skin Coloration

Nina G. Jablonski

Human skin is functionally hairless and exhibits a wide range of natural colors from the most deeply saturated dark brown to pinkish off-white. Differences between people in skin color are readily perceived and have been used as the basis for classifying people into groups referred to as races or race-color identities (Harris et al., 1993). The array of colors observed in the skin of modern humans is greater than that of any other single mammalian species, and is the product of natural selection (Jablonski and Chaplin, 2000), despite some arguments to the contrary (Blum, 1961; Frost, 1988; Robins, 1991; Aoki, 2002). Skin pigmentation in humans evolved primarily to regulate the amount of ultraviolet radiation (UVR) penetrating the skin and, thus, modify its bioactive effects.

Color is imparted to skin by a variety of different substances, which are visible to varying degrees in different people. The most important of these substances is the pigment, melanin, which is produced in specialized cells called melanocytes within the skin. In people with very pale skin, the skin gets most of its color from the bluish-white connective tissue of the dermis and from oxyhemoglobin and deoxyhemoglobin associated with red blood cells circulating in the capillaries of the dermis. The red color produced by circulating hemoglobin becomes more obvious, especially on the face, when the arterioles dilate and become engorged with blood as a result of prolonged exercise or sympathetic nervous stimulation caused by embarrassment or anger (Jablonski, 2006). Variation in human skin color is due primarily to its melanin content, and so this chapter deals exclusively with the properties and evolutionary significance of melanin.

MELANIN AND MELANIN PRODUCTION

The remarkable range of brown hues seen in human skin is produced by melanin. "Melanin" is the collective term for a large family of molecules that are found in many types of organisms, including fungi, invertebrates, and vertebrates. Melanins are classified as pigments because they have a great capacity to absorb visible light. Much of their evolutionary value derives from their abilities to absorb more high-energy forms of electromagnetic radiation including UVR and ionizing radiation, and to neutralize the chemical by-products created when cells interact with these agents.

Eumelanin is the dominant form of melanin found in human skin. In its concentrated form, eumelanin is intensely dark because it absorbs broadly in the spectrum of visible light, but its protective effects on the body are due to its abilities to absorb more energetic and potentially damaging UVR. Eumelanin is a highly heterogeneous polymer consisting of 5,6-dihydroxyindole (DHI) and 5,6-dihydroxyindole-2-carboxylic acid (DHICA)-derived units bound to proteins (Prota, 1992b; Ito, 2003). Eumelanin polymers take on different physical conformations and are intractably stable, even when bombarded by high-energy radiation or reactive oxygen species (free radicals) (Fox and Vevers, 1960; Chedekel, 1995; Kollias, 1995a; Pathak, 1995; Sarna and Swartz, 1998; Meredith and Sarna, 2006). Ultraviolet radiation can break the chemical bonds that maintain the integrity of important molecules such as DNA and the constituents of cell membranes, causing a toxic cascade of events that produces reactive oxygen species (free radicals), which disrupt normal chemical reactions in cells (Caldwell et al., 1998; Hitchcock, 2001; Cleaver and Crowley, 2002). Eumelanin absorbs and scatters ultraviolet and visible light, and works chemically to prevent free radical formation and neutralize free radicals if they are formed (Mason et al., 1960; Meredith and Sarna, 2006). Eumelanin's superior antioxidant properties work to great advantage both in the skin and in the retina of the eye, where they prevent and quench free radical damage caused by incoming radiation (Zareba et al., 2006). Many forms of eumelanin with slightly different structures and colors exist, and the sheer heterogeneity of natural eumelanin contributes to its wide array of physical and chemical properties.

Eumelanin owes many of its physiological properties to the way in which it is packaged in cells.

Human Evolutionary Biology, ed. Michael P. Muehlenbein. Published by Cambridge University Press. © Cambridge University Press 2010.

Melanins are produced in specialized cells called melanocytes, which are found at the lowest level of the epidermis at the junction with the dermis. Melanocytes are considered one of the types of "immigrant cells" in the skin because their precursor cells migrate into the skin from the neural crest during early embryonic development. These cells establish connections with neighboring keratinocytes in a carefully orchestrated process of signaling and adhesion (Haass et al., 2005). Melanocytes produce melanin in the skin while the developing embryo is still very young, but the process ramps up slowly. Babies are born pale and their melanocytes do not produce melanin at full capacity until puberty (Robins, 1991).

Melanin is produced within melanocytes in small membrane-bound packages units called melanosomes. As melanosomes mature and become full of melanin they move into the dendrites of the melanocyte, and are from there transferred into adjacent keratinocytes. This process is precisely controlled by genetic and hormonal factors, but it can be accelerated in most people by exposure to UVR (Fitzpatrick et al., 1961; Prota, 1992a, 1992b; Aroca et al., 1993; Kollias, 1995b; Nordlund, 1995; Kollias et al., 1996; Li et al., 2001; Thong et al., 2003; Brenner and Hearing, 2008). One melanocyte supplies melanin to about 36 keratinocytes in a precisely and intricately choreographed process that is controlled by keratinocytes (Jimbow et al., 1976; Schallreuter, 2007). Loss of contact between melanocytes and keratinocytes allows the former to break free of their positions in the epidermis, start dividing rapidly, and leave the normal confines of the skin. This marks the beginning of the most serious type of skin cancer, melanoma (Haass and Herlyn, 2005; Haass et al., 2005). Within keratinocytes, melanin tends to be distributed in supranuclear caps that protect the cell nuclei from incoming UVR photons (Kobayashi et al., 1998; Gibbs et al., 2000). Cells with supranuclear melanin caps contained significantly less DNA photoproducts (cyclobutane pyrimidine dimers and 6–4 photoproducts) than those without.

People of different skin colors differ in the amounts and types of melanin they produce, and in the ways in which the melanins are packaged and distributed in the skin. The mechanisms controlling melanin production are genetically determined and involve the regulation of a series of chemical pathways in which enzyme tyrosinase plays a major part (Sturm et al., 2001; Alaluf et al., 2002; Sulaimon and Kitchell, 2003; Brenner and Hearing, 2008). The color and physical properties of skin are also caused by differences in the size and distribution of melanosomes in the skin, and in the types of melanin they contain. People with naturally darkly pigmented skin have melanosomes that are large, clumped, and filled with eumelanin (Szabo et al., 1969; Alaluf et al., 2002; Thong et al., 2003;

Tadokoro et al., 2005). Those with naturally light skin have smaller and more sparsely distributed melanosomes, which contain varying amounts and kinds of eumelanin and smaller amounts of its lighter-colored cousin, pheomelanin (Ancans et al., 2001; Thong et al., 2003; Lamason et al., 2005). Like eumelanin, pheomelanin is variable in its structure and its different forms vary in color from yellow to red (Ito, 2003). In humans, pheomelanin is most obvious in the hair of people from northernmost Europe (including the British Isles), but is present in small quantities in the skin of most people (Thody et al., 1991; Alaluf et al., 2002; Ito, 2003; Meredith and Sarna, 2006). Subtle variations in skin color between people that we easily detect with the eye – the reddish, yellowish, bluish, and other hues that are often remarked on in artistic descriptions of human skin – are due to different proportions of the different forms of eumelanin and pheomelanin in the skin (Thody et al., 1991; Alaluf et al., 2002; Hennessy et al., 2005; Kongshoj et al., 2006).

The complexity of the pigment production system is such that loss of pigmentation can occur in many different ways. Different types of albinism in humans are due to genetic mutations affecting different parts of the pigment production pathway or melanocytes in different parts of the body (Robins, 1991; Sulaimon and Kitchell, 2003; Hornyak, 2006; Sturm, 2006; Duffy et al., 2007). Melanocytes can be entirely absent or fail to produce melanin, or melanosomes can fail to mature and be transferred into keratinocytes (Bahadoran et al., 2003). The affliction vitiligo occurs when the ability to produce pigment in the skin is lost only in patches on the body. The absence of normal skin color on part or all of the body can have serious consequences for health and self-image, and so considerable research has been devoted to these afflictions. Individuals with albinism and vitiligo are at greater risk for skin cancer because of the absence of protective melanin in some or all of the skin (Johnson, 1998). Levels of serum folate and vitamin B_{12} are lower in vitiligo patients (Montes et al., 1992; Kim et al., 1999).

The distribution of pigment over the surface of the body is not always even, as in the case of freckles (ephelides or solar lentigines), which are small, flat spots of melanin that occur mostly in people with very lightly pigmented skin. They vary in color from yellow (predominantly pheomelanin-containing) to very dark brown (predominantly eumelanin-containing) and develop in a random pattern on the skin in response to repeated sun exposure. Ephelides are freckles which most commonly appear on the faces of children after sun exposure (Rhodes et al., 1991). The other type of freckle, solar lentigines ("liver spots"), populates the hands and faces of older people. These freckles tend to be darker than childhood freckles, and occur in people with a wide range of skin colors as the result of sun damage.

A person's genetically determined or baseline pigmentation is referred to as their constitutive pigmentation, in contrast to their facultative pigmentation, which is developed as the result of exposure to UVR. Over the last two centuries, the description of skin colors has developed from verbal portrayals of skin colors ("white," "yellow," "black," "brown," and "red") to color-matching methods such the von Luschan scale (von Luschan, 1897; Olivier, 1960) to reflectance spectrophotometry (Lasker, 1954a; Wassermann, 1974) as reviewed elsewhere (Jablonski, 2004; Parra, 2007). Reflectance spectrophotometry is the preferred method for the objective study of skin pigmentation because the incident light used and the distance between the light source and the subject are invariable (Wassermann, 1974). Constitutive skin pigmentation is measured on parts of the body not routinely exposed to sun, with the inner (medial) surface of the upper arm being preferred. Comparability of measurements made by color-matching techniques and reflectance spectrometry, and between skin reflectance measurements made by different types of reflectance spectrophotometers has been a problem in anthropology (Jablonski, 2004; Parra, 2007) that has still not been completely solved. Many human populations that were studied using color-matching methods or older types of reflectance spectrophotometers cannot be restudied because they are extinct or have become thoroughly admixed with neighboring populations.

Melanin is produced in the skin of most people following exposure to UVR through the tanning response. Nearly everyone can develop a "tan" when exposed to strong, UVR-containing sunlight, but some people can produce more melanin than others (Lasker, 1954b; Lee and Lasker, 1959; Agar and Young, 2005; Tadokoro et al., 2005). Dermatologists classify human skin by "phototype" or "sun reactive type" from Phototype (or Type) I (always burns, does not tan) to Type VI (never burns, always tans darkly) (Fitzpatrick et al., 1961; Fitzpatrick, 1988; Rubegni et al., 1999; Fitzpatrick and Ortonne, 2003). All people have similar numbers of melanocytes, but people with Types V or VI have about four times as much melanin in their skin as those with Type I skin, and will produce much more melanin and so develop deeper facultative pigmentation when exposed to UVR (Tadokoro et al., 2005). Higher melanin content affords more protection against damage to DNA and other biologically important molecules (Ortonne, 2002; Meredith and Sarna, 2006). The four-fold difference in melanin content in naturally dark people translates into a seven- to eight-fold difference in protection against damage to DNA (Tadokoro et al., 2005), but even the darkest skin does not protect completely against damage to DNA. Very low UVR exposures cause measurable damage to DNA in all people regardless of color, so there is no such thing as completely UVR-proof skin (Brenner and Hearing, 2008).

People with light constitutive pigmentation (Types I and II) make very little melanin in their melanocytes, and have no or negligible ability to produce melanin when exposed to UVR (Fitzpatrick and Ortonne, 2003). When exposed to UVB (290–320 nm), in particular, people with Types I or II skin react by mounting a strong inflammatory reaction resulting in sunburn. Erythema and pain are obvious and uncomfortable symptoms of sunburn, but the hidden damage done to the connective tissues and DNA of the skin is more serious because of the connection with premature aging and skin cancer, including melanoma (Cleaver and Crowley, 2002; Matsumura and Ananthawamy, 2004).

People with moderately or darkly pigmented skin (Types III–VI) produce melanin in their skin when they are exposed to UVR. The tanning reaction is complex and fully develops over the course of days and weeks, if exposure to the sun persists. In many people, the initial response of the skin to strong sunlight is the rapid development of blotchy grey-brown looking skin known as immediate pigment darkening (IPD). This mechanism appears to involve the redistribution and photo-oxidation of existing melanin in keratinocytes of the epidermis (Ortonne, 1990; Young, 2006). Normal tanning is also known as delayed tanning to distinguish it from IPD. Delayed tanning involves the stimulation of melanocytes into a program of long-term activity. The visible darkening that occurs within a week of sun exposure is as a result of upward movement of melanin in the epidermis, not because of the making of new pigment (Tadokoro et al., 2005; Nielsen et al., 2006b). Increases in melanin production occur later, and involve slower responses of melanocytes to modifications in the regulation of tyrosinase activity (Ortonne, 1990; Alaluf et al., 2002). Chronic exposure to UVR can result in a near doubling of melanin content in the skin relative to baseline amounts. The degree to which tans are protective against the harmful effects of UVR has been the subject of considerable debate. Heavy tans afford little protection against UVR-induced damage to DNA relative to the amount provided by naturally dark skin (Tadokoro et al., 2005; Nielsen et al., 2006a). For moderately pigmented skin, however, tanning affords some protection against seasonally varying intensities of UVR because melanin production increases slowly in relation to gradually rising UVR levels in the spring and so prevents bad sunburns from being experienced during the height of summer levels of UVB. This almost certainly accounts for the evolution of tanning abilities. Naturally dark skin affords great protection against UVR because of its higher eumelanin content, the superior UVR-absorbing abilities of large clumped melanosomes, and because the eumelanin can be mobilized faster from deep in the epidermis and brought to a position closer to the surface of the skin more quickly (Nielsen et al., 2006a).

Constitutive skin pigmentation gradually fades after early adulthood as part of the process called chronological aging. In people older than around 30 years, the number of active melanin-producing cells decreases on average by about 10–20% per decade (Quevedo, 1969; Ortonne, 1990; Fisher et al., 2002) in a pattern that is strongly correlated with the human reproductive career.

THE EVOLUTION OF HUMAN SKIN PIGMENTATION

The evolution of skin pigmentation in humans is inextricably connected to the evolution of hairlessness and enhanced sweating abilities (Jablonski, 2006). The probable ancestral condition of skin in the human lineage was pale skin covered by dark hair (Jablonski and Chaplin, 2000), and loss of functional body hair was associated with the evolution of an efficient whole-body cooling system based on sweating (Wheeler, 1985; Amaral, 1996; Jablonski and Chaplin, 2000; Jablonski, 2006). Modern humans have eccrine sweat glands distributed all over their bodies, and those on the forehead, back, and chest are especially quick to respond to heat and exertion (Sato and Dobson, 1970; Cotter et al., 1995; Shibasaki et al., 2006). Evaporation of sweat slightly reduces the temperature of the surface of the body, thereby cooling the blood flowing in the capillaries of the skin, including the scalp (Adams et al., 1975, 1992; Cabanac and Brinnel, 1985). The slightly cooled venous blood returning to the heart is oxygenated and then recirculated to the periphery, including the temperature-sensitive brain (Cabanac and Massonnet, 1977; Brinnel et al., 1987; Falk, 1990; Jablonski, 1993). Sweating is most effective in cooling the body when there is less hair on the surface to slow evaporation, hence the connection between an increased number of eccrine sweat glands and functionally naked skin. Naked skin is more vulnerable to environmental influences, and the naked skin of humans differs from that of close but hairier primate relatives in its greater water resistance and resistance to abrasion (Montagna, 1981). The genetic basis of these differences is just beginning to be understood (Chimpanzee Sequencing and Analysis Consortium, 2005), but genes related to the epidermal proteins that contribute to the barrier functions of the skin, the integrity of sweat glands, and the delicate nature of our body hair (Langbein et al., 2005) appear to be of particular importance. Lacking protective fur or hair also renders naked body skin much more vulnerable to damage from solar radiation, including UVR (Walsberg, 1988). Compensation for the loss of this protection came from evolution of increased thickness of the epidermis, especially of the most superficial layer, the *stratum corneum* (Montagna, 1971; Madison, 2003) and from evolution of permanent protective pigmentation in the skin to prevent the most energetic and damaging wavelengths of radiation from penetrating into the body.

Among modern humans, skin pigmentation as measured by reflectance spectrometry is highly correlated with latitude (Roberts and Kahlon, 1976; Tasa et al., 1985), but is even more highly correlated ($r^2 = -0.93$) with the annual average erythemal dose of UVR (the UVMED, or ultraviolet minimal erythemal dose, is the minimum dose of UVR, mostly UVB, necessary to produce a noticeable reddening of lightly pigmented skin) (Roberts, 1977; Jablonski and Chaplin, 2000; Chaplin, 2001, 2004). The strength of the correlation between UVMED and skin reflectance at all wavelengths is greater than with any other single environmental factor (Chaplin, 2001, 2004). The strength of this correlation strongly suggests the action of natural selection acting northward and southward to produce two reciprocal clines of skin pigmentation (Relethford, 1997).

Significant insight into the evolution of human skin pigmentation has come from studies of the *MC1R* (melanocortin 1 receptor) locus, one of several genes that contributes to skin, hair, and eye pigmentation in humans. In modern Africans, this gene exhibits no variation, but outside of Africa it is highly variable. The absence of variation in African forms of the gene provides evidence of strong positive selection or selective sweep occurring around 1.2 million years ago (Rogers et al., 2004) and the maintenance of a functional constraint on variation (purifying selection) in Africa thereafter (Rana et al., 1999; Rees, 2000; Makova and Norton, 2005). The ancestral form of *MC1R*, along with probable contributions from other pigmentation genes (Shriver et al., 2003; Norton et al., 2007), makes possible the production of large amounts of eumelanin in the melanocytes of the skin and appears to have been so effective in improving health and reproductive success that people carrying it quickly outnumbered and replaced those who did not. This evidence indicates that permanent, heavily melanized skin evolved *pari passu* with functionally hairless skin richly endowed with eccrine sweat glands, and was the ancestral condition for the genus *Homo* (Jablonski and Chaplin, 2000; Jablonski, 2006).

Natural selection and the evolution of dark skin pigmentation

What selective factors led to dark pigmentation being established by a selective sweep and being maintained by purifying selection in regions of high UVR? The main hypotheses that have been advanced to account for this have been lowered mortality due to skin cancer, enhanced fitness because of protection against the harmful effects of sunburn, the benefits of dark pigmentation with respect to predation avoidance or while hunting in poorly lit forested environments, enhanced fitness because of the antimicrobial properties of eumelanin, and enhanced fitness due to the protection of folate against its breakdown by UVR. These hypotheses will be discussed in turn.

Darkly pigmented, eumelanin-rich skin protects against considerable damage to DNA caused by UVR (Miyamura et al., 2007), and is associated with much lower rates of skin cancer than lightly pigmented skin (Barker et al., 1995; Armstrong and Kricker, 2001; Diepgen and Mahler, 2002; Soininen et al., 2002; Tadokoro et al., 2003; Saraiya et al., 2004; Agar and Young, 2005; Pfeifer et al., 2005; Rouzaud et al., 2005; Brenner and Hearing, 2008). The protective effects of eumelanin on DNA structure were established by an experimental study showing that heavily pigmented melanocytes resumed proliferation faster after UVB irradiation than can lightly pigmented ones, and that DNA from lightly pigmented melanocytes contained significantly higher numbers of cyclobutane pyrimidine dimers than did DNA from heavily pigmented melanocytes after irradiation with increasing doses of UVB (Barker et al., 1995; Cleaver, 2000; Cleaver and Crowley, 2002). In contrast, the production and presence of pheomelanin in lightly pigmented skin appears to increase the risk of oxidation stress in melanocytes. This, combined with the limited ability of pheomelanin to absorb UVR, may lead to an elevated skin cancer risk among light-skinned individuals (van Nieuwpoort et al., 2004). The damaging effects of UVR on DNA structure, especially those relating to the generation of carcinogenic cyclobutane pyrimidine dimers, are now widely recognized (Barker et al., 1995; Pflaum et al., 1998; Epel et al., 1999; Kielbassa and Epe, 2000; Cleaver and Crowley, 2002; Sinha and Hader, 2002; Pfeifer et al., 2005; Schreier et al., 2007). These are, however, mostly associated with the initiation of squamous and basal cell carcinomas (Dwyer et al., 2002; Christenson et al., 2005), which are skin cancers that mostly affect people toward the end or after their reproductive careers (Blum, 1961; Jablonski and Chaplin, 2000; Rigel, 2008). Of all the major types of skin cancer, cutaneous malignant melanoma is the only type with a high incidence rate among people of reproductive age, but overall incidence and mortality rates for melanoma prior to the mid-twentieth century were very low (<5 per 100 000) (Diepgen and Mahler, 2002). The low rates of mortality due to melanoma prior to 1970 (Jemal et al., 2001) argue that it was unlikely that melanoma was a significant driver of selection for darkly pigmented skin. Increases in the incidence of melanoma in the last 50 years are the result of lightly pigmented people being exposed to more intense or longer periods of sunlight and UVR (Jemal et al., 2001) and experiencing more painful sunburns (Kennedy et al., 2003; Veierod et al., 2003) because of migration to sunny places or involvement in recreational sun-tanning (Leiter and Garbe, 2008), and cannot be considered typical of our species prior to the twentieth century.

The protection conferred by eumelanin-rich skin against sunburn is raised (but rarely elaborated in writing) as a possible factor responsible for the selection of dark pigmentation in regions of high UVR. The prevalence and effects of serious sunburns have been studied mostly in relation to skin cancer, and strong links between repeated painful sunburns before the age of 20 years and cutaneous melanoma have been established (Kennedy et al., 2003). Serious sunburns alone are rarely linked to harmful immediate side-effects, however, despite the pain and discomfort they cause. There are no data to support the claim that sunburns cause damage sufficient to affect reproductive success. Only one study was found in the literature that examined the incidence of serious sunburns. This was a 1-year prospective study from an Irish hospital in which it was reported 4.7% of all burns (16 cases) treated in the hospital were caused by serious sunburn (Cronin et al., 1996). Two cases only required inpatient intravenous fluid replacement, and no deaths were reported. The absence of other reports in the medical literature on serious immediate consequences of sunburn suggests that the sunburn is not in itself a serious cause of morbidity and mortality and would have had only minor evolutionary potency. The malign effects of serious sunburns on the activity of eccrine sweat glands and thermoregulation have also been mooted in connection with the evolution of dark-skin pigmentation. To date, only one study has carefully evaluated the effect of sunburn on sweat rates and thermoregulation in humans (Pandolf et al., 1992). In this study, lightly pigmented subjects receiving artificially induced sunburns were able to maintain thermal homeostasis during vigorous exercise despite reduction of sweat rates 24 hours after UVB exposure (Pandolf et al., 1992). These results suggest that sunburn-induced damage to sweat glands does not adversely affect thermoregulation to the extent that was envisioned by some, and that damage to sweat glands was not a major selective force in the evolution of dark pigmentation in regions of high UVR.

Theories brought forth to account for the evolution of dark pigmentation based on the benefits of eumelanin other than those related to protection against solar radiation have not gained empirical support. The proposal that dark skin evolved because it provided superior concealment against visual detection in dark forest environments (Cowles, 1959) was not unreasonable when it was thought that much of human evolution took place in forests and not more open, well-lit environments. But this proposition is now untenable in light of the more than 40 years of paleoanthropological research demonstrating that most of the evolution of the human lineage took place in well- or brightly lit woodland or woodland-grassland environments. Another group of hypotheses have suggested that dark pigmentation evolved primarily because eumelanin-rich melanosomes and melanocytes confer strength to immune systems that are challenged by tropical infectious diseases and parasites (Wassermann, 1965, 1974; Mackintosh, 2001). Although eumelanin-rich melanosomes exhibit antimicrobial properties and may bolster the innate immune system, these benefits are of secondary importance compared to those they confer in direct connection with UVR protection. The ubiquity of eumelanin in nature appears to be due primarily to its role as a physical absorber of UVR and chemical neutralizer of the noxious by-products produced by UVR bombardment or, in other words, as a built-in sunscreen (Epel et al., 1999) not as an antimicrobial agent. The antimicrobial hypothesis also does not explain the evolution of deep-tanning abilities in human populations living remote from the tropics and tropical diseases (e.g., Inuit and native Tibetans), but exposed to high levels of environmental UVR.

If reduced fitness or mortality due to UVR-induced skin cancers, sunburns, a weakened immune system, or an absence of appropriate camouflage were not the main selective pressures driving the evolution of dark skin pigmentation in high UVR environments, then other agents capable of exerting these effects must be identified.

The effects of UVR on biological systems are wide-ranging, multifarious, and mostly destructive (Caldwell et al., 1998; Madronich et al., 1998; Rothschild, 1999). The deleterious effects of UVR on DNA have been emphasized because of the direct relationship between mutations of DNA in the skin and skin cancer. Ultraviolet radiation also breaks down other molecules of great biological importance, including the B vitamin, folate. The possibility that UVR might be implicated in the breakdown of folate in human blood was first mooted 30 years ago when it was recognized that folate levels in a small number of human patients were lowered significantly 1 hour after subjects were exposed to simulated strong sunlight (Branda and Eaton, 1978). The authors of the paper proposed that the light-induced breakdown of folate might be related to the evolution of skin color, but did not pursue research along this avenue probably because the importance of folate in normal development and cell proliferation was not then fully appreciated.

Folate is a water-soluble B vitamin that occurs naturally in food. Folate deficiency was long ago recognized as the primary cause of megaloblastic anemia but, by the late 1980s, folate's importance was enhanced by discovery of its role in nucleic acid synthesis (Green and Miller, 1999). Since then, studies of the interdiction of cell proliferation as a consequence of folate deficiency have had wide ramifications for understanding of normal development and birth defects, and normal cell division and neoplasia. Folate also participates in the formation of myelin and is important in the production of many neurotransmitters including serotonin (Djukic, 2007). Because the compound cannot be made by the body, humans get folate only from food or from supplements of folic acid, the synthetic form of folate. The best sources of natural food folates are green leafy vegetables (the word folate comes from the Latin "folium" for leaf), fruits, and dried beans and peas. Healthy levels of folate are difficult to maintain in the body because natural food folates are unstable, suffer from low bioavailability, and tend to break down when foods are boiled or stored (Gregory, 1995; McNulty and Scott, 2008). Natural folates are converted into various forms that are used immediately or stored in the liver. Folate deficiencies can be caused by insufficient intake of folate, improper absorption of the vitamin from the gut, or when serum folate is broken down by alcohol or UVR (Anonymous, 1983; Tamura and Halsted, 1983; Mastropaolo and Wilson, 1993; Komaromy-Hiller et al., 1997; Suh et al., 2001; Off et al., 2005; Steindal et al., 2006; Der-Petrossian et al., 2007). They are also influenced by genetic factors, notably by variations in the methylenetetrahydrofolate reductase locus (MTHFR) that affect DNA methylation and synthesis and homocysteine metabolism (Blom et al., 2006).

The essential connections between folate metabolism and the evolution of skin pigmentation are, firstly, the relationship between UVR exposure and folate breakdown and, secondly, the relationship between UVR-induced folate deficiency and reduced fitness due to failures of normal embryogenesis and spermatogenesis (Jablonski, 1992; Jablonski and Chaplin, 2000). Numerous epidemiological studies and meta-studies from the late 1980s onward have indicted folate deficiencies during pregnancy in the etiology of neural tube defects (NTDs) (Bower and Stanley, 1989, 1992; Minns, 1996; Copp et al., 1998; Fleming and Copp, 1998; Molloy et al., 1999; Lucock, 2000; Williams et al., 2005). These and other studies demonstrating the many important roles of folate were influential in the introduction in 1998 in the fortification with folic acid of the enriched flour used to make most breads and breakfast cereals in the United States and Canada. In the last 20 years, the relationship between folate deficiencies and NTDs has been thoroughly documented, and the importance of folate status is stressed for all women of reproductive age (Bailey, 1995; de Bree et al., 1997; Neuhouser et al., 1998; Caudill et al., 2001; Scott, 2007). Plentiful supplies of folate are essential to maintain the high rates of cell proliferation in the embryo, particularly in the developing central nervous system where high levels of folate-carrier protein are expressed (Djukic, 2007). Neural tube defects occur when the normal processes of cell division in the early nervous system are disrupted. In the fourth week of human intrauterine development, the neural tube closes like a two-ended zipper from the middle simultaneously toward the head and tail ends. At this time the embryo and its nervous system are particularly sensitive to low folate levels because rates of cell proliferation are high. Failure of the two edges of the tube to fuse securely can cause holes at the head end – leading to the fatal defect of anencephaly – or more distally, leading to the various forms and attendant disabilities of spina bifida. Folate is also important for normal sperm production, and folate status is increasingly being investigated as a reason for male infertility (Mathur et al., 1977; Ebisch et al., 2006). The emerging view is that folate status is crucial to reproductive success primarily because of its dual importance in embryonic differentiation and sperm production, but that it is unstable and subject to many environmental and genetic factors.

Evidence-based in vivo experiments for a causal relationship between UVR exposure and folate degradation in the human body has been slow to accumulate because the problem does not lend itself easily to experimental testing on human subjects. As a result, much of the research has involved exposure of human blood plasma outside of the body to UVR and the use of model systems in which folic acid was subjected to different wavelengths of UVR (Off et al., 2005; Nielsen et al., 2006b; Vorobey et al., 2006; Der-Petrossian et al., 2007). This research has shown that folate breaks down in the presence of UVR, and that the longer, more deeply penetrating UVA rays are particularly damaging. Ultraviolet A causes folate to degrade in three phases into a series of chemical intermediates, and some of these in turn act to sensitize folate to further degradation and also damage DNA. Photodegradation of the main form of folate in human plasma, 5-methyltetrahydrofolate (5-MTHF), involve reactive oxygen species generated by UVA and blue visible light and accelerated in the presence of riboflavin (Steindal et al., 2006). High concentrations of melanin significantly reduced folate destruction in vitro through absorption and scattering of UVA (Nielsen et al., 2006b). These studies support the theory that the major factor contributing to the evolution of dark skin pigmentation was breakdown of folate caused by UVR. This theory is also supported by the results of

epidemiological studies, in which the prevalence of NTDs relative to skin color has been examined (Buccimazza et al., 1994; Williams et al., 2005; Besser et al., 2007). Although these studies did not establish a definitive cause-and-effect relationship between skin color and NTDs, they indicated trends warranting further investigation. Darkly pigmented skin appears to contribute to the maintenance of healthy folate status by actively protecting circulating folate from UVR photolysis, resulting in fewer NTDs being observed in more darkly pigmented groups. Many factors probably account for the fact that the most darkly pigmented women suffer the lowest rates of NTDs, but the data suggest that high melanin concentrations in the skin have a protective effect and this warrants more explicit epidemiological investigation. The existence of a causal relationship between photodegradation of folate and increased incidence of NTDs in humans is also strengthened by the observation of a conception peak of May–June among NTD births in the Northern Hemisphere (Marzullo and Fraser, 2005).

The evolution of light skin pigmentation

The evidence that permanent dark skin pigmentation evolved as protection against the deleterious effects of UVR is overwhelming, and research is mounting that eumelanin confers photoprotection against both folate photolysis and indirect and direct damage to DNA. This accounts for the concentration of darkly pigmented indigenous peoples in areas of high UVR, mostly within the tropics (Jablonski and Chaplin, 2000; Jablonski, 2004), but it does not explain the clinal distribution of increasingly lightly pigmented skin outside of the tropics. As was the case with dark skin pigmentation, numerous hypotheses have been put forward to account for the evolution of light skin pigmentation. The three major hypotheses that have been put forward and that are discussed below in turn are: resistance against cold injury; loss of pigmentation through the probable mutation effect; and enhanced potential for production of vitamin D in the skin under conditions of reduced sunlight intensity.

According to the cold-injury hypothesis, darkly pigmented skin was actively selected against in colder and generally higher latitude environments because it was more susceptible to frostbite (Post et al., 1975). People afflicted with frostbite would be less able-bodied, less able to forage and hunt successfully, and more susceptible to possibly fatal secondary infections such as gangrene (Post et al., 1975). This hypothesis was based on observations of a slightly higher incidence of frostbite in twentieth-century US combat troops of African descent than those of European descent. The reasons for the slight difference in frostbite incidence revealed by the study are still not well understood, and there is good reason to suspect that variables relating to the equipment issued to soldiers were not adequately taken into account (Kittles, 1995). The difference in response to extreme cold probably has less to do with pigmentation than with other aspects of skin structure such as the distribution of fat and connective tissue (Steegmann, 1967; Kittles, 1995), or differences in the temperature responsiveness of peripheral capillaries.

The theory that lightly pigmented human skin evolved in the absence of selective pressure was first advanced within the framework of the probable mutation effect (Brace, 1963). The resulting "structural reduction" was seen as the main factor initiating the evolution of lightly pigmented skin outside of the tropical regions under the highest selective pressure for dark pigmentation (Brace, 1963). The subsequent spread of light pigmentation was then said to be promoted by assortative mating (Kittles, 1995), with sexual selection leading to even lighter pigmentation in females (Frost, 1988; Aoki, 2002). Doubt has been cast on the structural reduction hypothesis mainly because relaxation of selection on dark pigmentation would be expected to produce a more random pattern of skin pigmentation outside of high UVR regions, rather than the structured pattern characteristic of the action of purifying selection that is observed (Norton et al., 2007). The clinal distribution of skin pigmentation that is seen in the Eastern Hemisphere and, with lesser intensity, in the Western Hemisphere is one of the most significant characteristics of human skin and strongly suggests the operation of natural selection. A large proportion of global landmass is concentrated in regions that receive low UVR on an annual basis, and the distribution of skin pigmentation in modern humans is arranged such that increasingly lighter-skinned populations are distributed in areas of incrementally lower UVR (Relethford, 1997; Chaplin and Jablonski, 1998).

Mechanisms to account for the clinal distribution of light skin pigmentation in regions of increasingly low UVR must explain how this distribution was achieved during the process of hominid dispersal and maintained in the face of migration (gene flow) (Barton, 1999). The stable clinal arrangement of human skin pigmentation among indigenous peoples indicates the action of stabilizing selection over a spatially varying optimum condition. If dark pigmentation was the original condition for the genus *Homo* and has been maintained as an adaptation to high levels of UVR, then what must be elucidated is the selective pressure responsible for establishing and maintaining light pigmentation in regions of low UVR. Put another way, the question to be addressed surrounds the selective advantage of a continuously varying cline of UVR-attenuating pigment in the skin.

The strength of UVR, and of UVB in particular, declines greatly north of the Tropic of Cancer and south of the Tropic of Capricorn (Johnson et al., 1976, Relethford, 1997; Chaplin, 2001, 2004; Hitchcock, 2001; World Health Organization, 2002; Lucas et al., 2006). Humans living outside of the tropics face high UVR levels in the summer months, but extremely low levels, especially of UVB, in the fall and winter. In light of the many deleterious effects of UVR on biological systems, low levels of UVR might be regarded as universally beneficial, but they are not. The single overwhelmingly positive action of UVR is photosynthesis of vitamin D_3 (cholecalciferol) in the skin of land-living vertebrates (Coburn et al., 1974; Henry and Norman, 1984; Webb and Holick, 1988; Holick, 1997, 2003). Without the biologically active form of vitamin D, normal life and reproduction are not possible. The global disease burden linked to the vitamin D deficiencies caused by low UVR exposure now exceeds that connected with high UVR exposure (Lucas et al., 2008a).

Vitamin D_3 is made in the skin when UVR penetrates the skin and is absorbed by 7-dehydrocholesterol (7-DHC) in the epidermis and dermis to form previtamin D_3. This reaction only occurs at the Earth's surface in the presence of wavelengths of 290–315 nm in the UVB range, with peak conversion occurring at 295–297 nm. Photosynthesis of vitamin D_3 in the skin depends upon season and latitude, time of day, and on the amount of pigment and thickness of the skin (Mawer and Davies, 2001; Lips, 2006). This reaction also becomes less efficient with advancing adult age because of an age-dependent decrease in 7-DHC in the skin (Maclaughlin and Holick, 1985; Cerimele et al., 1990; Holick, 1995). Inherent limits on circulating levels of previtamin D_3 exist because continued sunlight exposure causes the photoisomerization of previtamin D_3 to lumisterol and tachysterol (Holick et al., 1981) regardless of skin color. In order to become biologically active, vitamin D_3 must undergo two successive hydroxylation steps, firstly in the liver to $25(OH)D_3$ (calcidiol) (also known as 25-hydroxyvitamin D (25(OH)D)) and then in the kidney under the influence of parathyroid hormone (PTH) into the active metabolite, $1,25(OH)_2D_3$ (calcitriol) (also known as 1,25-dihydroxyvitamin D_3 (1,25(OH)2D)). Extrarenal production of calcitriol occurs in several other tissues in humans, including breast, placenta, colon, and prostate, where it is used locally and does not enter the systemic circulation (Norman, 2008). (The naming conventions for vitamin D used here are those of the Joint Commission of Biochemical Nomenclature of the International Union of Pure and Applied Chemistry and International Union of Biochemistry [IUPAC-IUB Joint Commission on Biochemical Nomenclature, 1982].) It is calcitriol that acts as a steroid hormone through binding to its specific intranuclear receptor, the vitamin D-receptor (VDR), and subsequently modulates the transcription of responsive genes such as that of calcium binding protein, which regulates mineral ion homeostasis (Lips, 2006; Norman, 2008; St-Arnaud, 2008). Vitamin D is also available in low quantities in some foods, specifically vitamin D_3 in oily fish and liver, and vitamin D_2 in some plants including many fungi (Bjorn and Wang, 2000; Chen et al., 2007; van der Meer et al., 2008). Egg yolks contain vitamin D_3 only if the egg-producing chickens have been given vitamin-D-rich feed, and cow's milk contains it only if it has been specifically fortified at the time of processing (Lamberg-Allardt, 2006). Dietary sources of vitamin D must be converted into the biologically active form via the same hydroxylation steps undergone by cutaneously produced vitamin D_3. The most common clinical assays used to assess vitamin D status measure levels of calcidiol or 25(OH)D in the serum, hence the commonly used shorthand "serum 25(OH)D." Exposure to UVB does not automatically result in elevation of serum 25(OH)D or of serum 1,25(OH)2D levels because the hydroxylation steps occurring in the liver and kidney are under multiple hormonal influences.

The most obvious function of vitamin D in humans is in the building and maintenance of the bony skeleton. The essential connection between vitamin D status and bone health was established because serious vitamin D deficiency was linked to the highly visible and disfiguring bone disease, nutritional rickets. The classical view of vitamin D action has been that it exerts its effects on bone only indirectly, as a hormone, through regulation of absorption of calcium and phosphorus from the gut. This view is now being supplemented by the recognition that vitamin D directly modifies the activity of osteoblasts and chondrocytes, and many other nonclassical target tissues (Henry and Norman, 1984; Norman, 2008; St-Arnaud, 2008; Wolff et al., 2008). The discovery of VDRs in tissues of the brain, heart, stomach, pancreas, skin, gonads, in the activated T and B lymphocytes of the immune system, and in 28 other tissues has led to a growing appreciation of the varied and important roles vitamin D plays in the body (Henry and Norman, 1984; Norman, 2008). The active form, $1,25(OH)_2D_3$ (calcitriol), influences cell biology relevant to cancer through VDR-mediated gene transcription and inhibition of abnormal cell division (hyperproliferation) in several organs. For this reason, chronic deficiencies in vitamin D may be associated with breast, prostate, colon, ovarian, and possibly other cancers (Garland et al., 2006; Fleet, 2008; Grant, 2008). Strong correlations between hypovitaminosis D and a range of cardiovascular diseases also suggest a causal relationship between vitamin D status and the health of cardiac and smooth muscle (Chen et al., 2008; Kim et al., 2008). Low vitamin D status is also linked to impaired immune system activity, specifically Th1-mediated (T-helper-cell-type-1-mediated) autoimmunity and infectious immunity (Cantorna and Mahon, 2005; Cantorna et al., 2008). Hypovitaminosis D is correlated with a weakened immune response to influenza virus, and appears to predispose children to respiratory infections in general (Cannell et al., 2008). Developmental vitamin D deficiency in animal models has been linked to impairment of numerous activities of the brain (McGrath et al., 2004; Harms et al., 2008). Thus, the connection between adequate vitamin D status and fitness is not limited to development and maintenance of the strength of the musculoskeletal system, but to normal development and functioning of the immune system and brain.

The fact that UVA cannot initiate vitamin D photosynthesis is significant for understanding the evolution of human skin pigmentation. Most of the Earth is bathed in UVA and visible light for most of the year because their longer wavelengths can pass easily through the atmosphere. Shorter UVB wavelengths are more easily destroyed by atmospheric ozone or reflected by other molecules and dust in their path. The Earth's surface receives relatively little UVB (and none of the even more energetic and uniformly harmful UVC) because oxygen and ozone in the atmosphere are excellent filters of these kinds or radiation (Caldwell et al., 1998; Hitchcock, 2001; World Health Organization, 2002; Kimlin, 2004). The amount of UVB reaching the Earth's surface depends on the solar zenith angle. Ultraviolet B falls on the equator and within the tropics throughout the year because its path from the Sun through the atmosphere is short. Outside of the tropics the angular path taken by the sun's rays require that UVR pass through a thicker layer of atmosphere to reach the Earth's surface, resulting in the destruction or reflection of most UVB en route. Locations farther away from the equator receive less UVB on an annual basis and demonstrate less potential for cutaneous vitamin D biosynthesis (Jablonski and Chaplin, 2000; Chaplin, 2004; Chen et al., 2007). The major exception to this is the Tibetan Plateau, which receives higher UVB than other locations at its latitude (~34°N) because of the thinness of the atmosphere at its high altitude (average elevation 4500 m).

Once produced in the skin, vitamin D_3 can be broken down by UVA wavelengths (315–335 nm), even after exposures as short as 10 minutes in nontropical sunshine (Webb and Holick, 1988; Webb et al., 1989). The vitamin D_3 content of samples of lightly pigmented skin declined by 80% when exposed to 3 hours of sunlight in June in Boston (42.2°N) (Webb et al., 1989), and photolysis of vitamin D_3 was observed at lower rates during the nonsummer months. This mechanism is significant for two reasons. Firstly, it works – along with previtamin D_3 photoisomerization – to prevent vitamin D toxicity, which was considered in the past a causal explanation for the evolution of dark skin pigmentation (Loomis, 1967). Secondly, it means that UVA breaks down vitamin D_3 even during parts of the year when no UVB is present in the sunlight to catalyze previtamin D_3 photosynthesis (Webb and Holick, 1988).

The hypothesis that light skin pigmentation evolved outside of the tropics as an adaptation to lower levels of sunlight and diminished ability to produce vitamin D in the skin has been put forward with successive refinements for many years (Murray, 1934; Loomis, 1967; Jablonski and Chaplin, 2000). Today, it is the hypothesis that is supported by the largest and most convincing body of experimental and observational evidence. The many functions described above for vitamin D denote that hypovitaminosis D – comprising vitamin D deficiency and the less clearly defined vitamin D insufficiency – is a serious public health problem (Vieth, 1999, 2003; Hathcock et al., 2007; Kimball et al., 2008). Nutritional rickets in infants and children is the most obvious manifestation of the condition because the attendant problems of bone calcification in the developing skeleton lead to visible skeletal deformities such as bowing of the weight-bearing long bones. Deformities of the female pelvis associated with severe nutritional rickets impair and sometimes preclude normal childbirth, leading to mortality of the infant, the mother, or both (Vieth, 2003). In children and juveniles, the disease can also involve the less visible symptoms of dental agenesis and muscle weakness (sarcopenia or myopathy). New evidence suggests that depressed bone mineral accrual due to rickets in fetal development may affect prepubertal bone mass accumulation (Kimball et al., 2008). In adults, hypovitaminosis D is sinister because it usually betrays few or no obvious physical manifestations or acute symptoms. In the musculoskeletal system, the osteomalacia and sarcopenia associated with hypovitaminosis D are silent problems that can lead to increased morbidity from accidental falls (Wolff et al., 2008). In other systems, vitamin D deficiency and insufficiency – especially beginning in infancy or childhood – are associated with increased, but still not rigorously quantified, risk of certain cancers and autoimmune and infectious diseases, as described above. Prospective long-term studies involving the tracking of vitamin D status and the incidence of a wide range of infectious and chronic diseases across a wide range of human populations are needed.

Until the last decade, arguments for the evolution of light skin pigmentation outside of high UVR regions rested primarily on comparative physiological and epidemiological data from modern human populations that showed a convincing cause-and-effect relationship between hypovitaminosis D and reduced fitness caused primarily by the skeletal deformities of nutritional rickets, including those affecting the pelvic shape and childbirth in females (Neer, 1975; Holick et al., 1981; Clemens et al., 1982; Clements et al., 1987; Webb and Holick, 1988; Littleton, 1991; Matsuoka et al., 1991; Fogelman et al., 1995; Goor and Rubinstein, 1995; Brunvand et al., 1996; Mitra and Bell, 1997; Jablonski and Chaplin, 2000; Kreiter et al., 2000; Malvy et al., 2000; Holick, 2003; Shaw, 2003; Vieth, 2003; Calvo et al., 2005; Bouillon et al., 2006; Lips, 2006; Chen et al., 2007; Parra, 2007; Tran et al., 2008). The hypothesis that the light skin pigmentation is due to positive selection for depigmentation, that is, pigmentation lighter than the ancestral highly pigmented condition for the genus *Homo*, has been strengthened in the last decade by two further lines of evidence. The first connects hypovitaminosis vitamin D to the possible increased prevalence of certain cancers and compromised immune status, as described above. The evidence for these phenomena is strong, as judged by prodigious daily increases in the number of published studies available through major internet search engines. The second line of evidence comes from molecular genetic studies and points to lightly pigmented phenotypes in humans having been selected for multiple times (Lamason et al., 2005; Lalueza-Fox et al., 2007; Norton et al., 2007) and having been maintained by purifying selection (Makova and Norton, 2005; Norton and Hammer, 2007; Norton et al., 2007). In order to understand why positive selection for depigmentation occurred, it is useful to look in greater detail at the properties of eumelanin in human skin.

Eumelanin is such a good natural sunscreen that competes with 7-DHC for UVB photons in darkly pigmented skin (Types V and VI) to greatly slow production of previtamin D_3 (Clemens et al., 1982; Holick, 1987; Matsuoka et al., 1991; Goor and Rubinstein, 1995; Mitra and Bell, 1997; Jablonski and Chaplin, 2000; Kreiter et al., 2000; Malvy et al., 2000; Skull et al., 2003; Vieth, 2003; Webb, 2006; Chen et al., 2007; Cosman et al., 2007; Hathcock et al., 2007; Tran et al., 2008). When eumelanin is distributed throughout the entire thickness of the epidermis, within the melanosomes of keratinocytes and as "melanin dust" (disintegrated melanosomes), it absorbs most of the incident UVB. Penetration of UVR is related to the amount and the distribution of eumelanin, with large, more superficial, eumelanin-filled melanosomes being more effective in reducing previtamin D_3 production (Nielsen et al., 2006b). People with lightly pigmented (Type II) skin can produce previtamin D_3 in their skin at a rate 5–10 times faster than those with darkly pigmented (Type V) skin (Clemens et al., 1982; Matsuoka et al., 1991; Jablonski and Chaplin, 2000; Webb, 2006; Armas et al., 2007; Chen et al., 2007). This poses strict geographic limits on the distribution of darkly

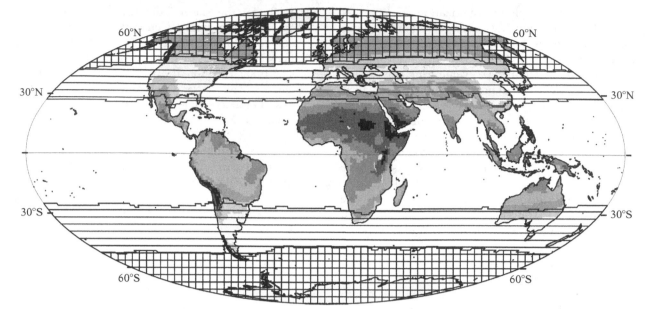

12.1. The geographic distribution of the potential for cutaneous vitamin D production, modified from a previous publication (Jablonski and Chaplin, 2000). For lightly pigmented skin, previtamin D_3 can be produced within the tropics during most of the year. Outside of the tropics, the absence of UVB in sunshine in some (or most) months prevents the vitamin D photosynthesis. Lightly pigmented people living within the zones denoted by horizontal lines experience at least one month during the year when they cannot produce vitamin D in their skin due to shortage of UVB. Lightly pigmented people living in the cross-hatched zones experience short bouts of UVB at the height of the summer only and cannot produce enough vitamin D from solar sources to satisfy their annual physiological requirements, and thus must supplement their diet with vitamin-D-rich or vitamin-D-fortified foods. For darkly pigmented people, the potential for cutaneous vitamin D production is considerably less under all UVR regimes due to the natural sun-screening effect of eumelanin. See text and a previous publication (Jablonski and Chaplin, 2000) for further discussion.

pigmented people outside of high UVB areas unless vitamin-D-rich foods or vitamin D supplements are consumed (Figure 12.1) (Jablonski and Chaplin, 2000; Jablonski, 2004; Chen et al., 2007).

The potential for cutaneous photosynthesis of previtamin D_3 is minimal in the winter at latitudes greater than 37°, and on an annual basis people living north of 50° cannot produce enough previtamin D_3 to satisfy their physiological needs (Jablonski and Chaplin, 2000; Chen et al., 2007; Holick et al., 2007). The problem is compounded by the facts that the main bioactive form of vitamin D_3, 1,25(OH)2D (calcitriol), has a half-life of approximately 15 hours in the circulation and can itself be broken down by UVA penetrating the skin, as described above (Holick et al., 1981; Webb and Holick, 1988; Webb et al., 1989; Jones, 2008). The half-life of serum 25(OH)D (calcidiol) is about two weeks. Storage of vitamin D as 25(OH)D (calcidiol) in human fat and skeletal muscle is possible (Mawer et al., 1971, 1972) and, in the absence of UVB or dietary vitamin D, the stores have a half-life of about two months (Jones, 2008). These stores are insufficient to provide adequate supplies of the vitamin in the absence of cutaneous production or dietary sources of the vitamin, especially in lean people. The persistent claim

that there was no selective pressure for loss of skin pigmentation in nontropical hominids because of the efficiency of cutaneous vitamin D production even in darkly pigmented skin and because of the body's capacity for long-term vitamin D storage (Robins, 1991) is not supported by any evidence. Depigmented skin was necessary for continuous human habitation at high latitudes. At the highest latitudes, rich dietary sources of vitamin D such as fatty fish, marine mammals, reindeer or caribou offal, and vitamin D-containing lichens were also essential for maintenance of health in the near-absence of UVB. These resources were and are heavily exploited by the indigenous inhabitants of the Arctic such as the Inuit and Saami.

Copious genetic information relevant to the question of the evolution of skin depigmentation in human evolution has been discovered in the last decade and has been summarized authoritatively elsewhere (Rees, 2003; Sturm, 2006; Lao et al., 2007). It is now recognized that as many as eight different genes contribute to variation in human skin pigmentation, including the *MC1R* locus (Sturm, 2006; Lao et al., 2007; Myles et al., 2007). Statistically significant correlations between the frequencies of four of these loci and specific skin pigmentation phenotypes have been recognized (Lao et al., 2007).

These studies are significant because they make possible the testing of hypotheses about the role of selection in determining variation in human pigmentation. As more genetic information on skin pigmentation is uncovered, it is salutary to remember that the pigmentation phenotypes, not genes, were the objects of natural selection. In the evolution of skin pigmentation, as in the evolution of any complex trait, different combinations of genes affecting different parts of pigment production pathways worked at particular times under particular environmental circumstances to produce reproductively successful phenotypes.

Variation in the *MC1R* locus of modern African peoples is minor and consists mostly of synonymous substitutions (John et al., 2003); the functional significance of the few nonsynonymous substitutions found has not been investigated. The apparent absence of functionally significant variation in this locus in modern native African peoples demonstrates the action of stabilizing or purifying selection that has worked to mostly eliminate variations in pigmentation that would not be able to survive and reproduce under high UVR regimes. In contrast, considerable variation at the *MC1R* locus is observed outside Africa, especially in northern Europe (Rana et al., 1999; Rees, 2003). High levels of polymorphism at the *MC1R* locus, and specific sets of polymorphisms at that locus, are associated with red hair and lightly pigmented skin, which has limited ability to produce melanin and which is highly susceptible to skin cancers (Rees, 2000, 2003, 2004, 2008). The contributions of other loci including *SLC25A5* (Lamason et al., 2005; Norton et al., 2007) and *SLC45A2* (MATP) (Graf et al., 2007) to skin pigmentation are now actively being investigated. Humans living at high latitudes in Paleolithic and Neolithic times with depigmented skin were probably at no significantly higher risk of developing lethal skin cancer because they were relatively short-lived and did not have the ability to migrate long distances into significantly higher UVR regions where they would be more prone to DNA-damaging sun exposure.

Variant forms of the *MC1R* gene associated with lightly pigmented skin also appear to have evolved independently in the Neanderthal lineage (Lalueza-Fox et al., 2007). The protein-coding sequence for the *MC1R* allele retrieved from the bone of a Neanderthal from Germany was a loss-of-function variant comparable in effect but different in sequence from any of those found in modern humans (Lalueza-Fox et al., 2007). This finding suggests that Neanderthals evolved a functional variant of the *MC1R* gene independently from modern humans as they dispersed into northerly latitudes, and thus supports the inference for the convergent evolution of depigmented skin in the Neanderthal lineage published prior to the elucidation of the gene sequence (Jablonski and Chaplin, 2000).

Variant forms of the *MC1R* gene are associated with red hair and pale skin in northern Europeans, but different variants of *MC1R* do not themselves cause skin color differences within this group (Sturm et al., 2001, 2003). Rather, these differences appear to be due to action and interaction of other loci including *SLC24A5*, *SLC45A2* (MATP), and *TYR* (Sturm et al., 2001; Lamason et al., 2005; Sturm, 2006; Norton et al., 2007). The particular importance of the *SLC24A5* locus in determining skin pigmentation in northern Europeans was demonstrated in an elegant series of experiments involving golden and wild-type zebrafish (Lamason et al., 2005). Golden zebrafish possess the *slc24a5* or *golden* gene, and exhibit fainter stripes and smaller and less dense pigment granules than those of the wild-type zebrafish. These differences parallel those that distinguish the melanosomes of lightly and darkly pigmented human skin. Using ingenious methodology and elegant experimental design, the research team established that the human ortholog of the variant *SLC24A5* gene that caused the golden zebrafish phenotype was probably responsible for the melanosomal structure and the light, pheomelanin-dominated pigmentation of northern European people (Lamason et al., 2005). This study showed that the European variant of *SLC24A5* was an important contributor to variability in human skin color, and that a single DNA base change (a single nucleotide polymorphism or SNP) that had undergone a selective sweep affected ancient European human populations (Norton and Hammer, 2007). The evolution and spread of the *SLC24A5* variant accounts for 25–38% of the difference in skin color between modern populations of Europeans and Africans (Lamason et al., 2005). The absence of the European variant of *SLC24A5* in Africans and in lightly pigmented East Asians implies that independent selection for depigmentation occurred in the populations leading to modern East Asians (Lamason et al., 2005; Myles et al., 2007; Norton and Hammer, 2007; Norton et al., 2007). When combined with the evidence of the Neanderthal *MC1R* polymorphism, this finding indicates that depigmented skin evolved independently three times in evolution through the action of positive selection on hominid populations living in areas receiving low UVB. The discovery of an early *Homo* fossil from Turkey with a pathological lesion probably caused by tuberculosis provides further evidence that maintenance of vitamin D status sufficient to protect against chronic infectious diseases has been a challenge since the first African dispersal event (Kappelman et al., 2008).

Skin pigmentation among the indigenous peoples of the New World follows a similar clinal pattern to that seen in the Old World, but is less pronounced (Jablonski and Chaplin, 2000). This may be due to the shorter length of time of habitation (15 000 years or less), the enhanced ability of the dispersing populations to buffer themselves culturally from the exigencies of new environments using clothing and shelter, or a combination of both factors. Most indigenous New World peoples have excellent tanning abilities (Lasker, 1954b; Lee and Lasker, 1959).

Sexual dimorphism in skin pigmentation

Adult human females are consistently lighter in pigmentation than males from the same population (Frost, 1988, 2007; Jablonski and Chaplin, 2000; Madrigal and Kelly, 2007). This fact has invited considerable speculation as to why this discrepancy evolved. Explanations based on the central role of sexual selection have dominated the literature and popular press and a reported global preference for lighter-than-average skin color in sexual partners offered as proof of male preference for lighter female mates (Frost, 1988, 2005, 2007; Aoki, 2002). The evolution of lighter pigmentation in females of lightly pigmented populations in low UVR regions, in particular, has been seen as evidence of the strength of sexual selection acting outside of the constraints of melanization maintained by purifying selection within the high-UVR tropics (Aoki, 2002). A recent study in which the degree of sexual dimorphism in skin pigmentation was examined relative to latitude revealed no evidence supporting this claim (Madrigal and Kelly, 2007).

A different perspective on sexual dimorphism in skin pigmentation comes from the recognition that human females require significantly higher amounts of calcium during pregnancy and lactation and, thus, must have lighter skin than males in the same environment in order to maximize their cutaneous vitamin D_3 production (Jablonski and Chaplin, 2000). The extra calcium needed for fetal and neonatal skeletal growth is insured by increased maternal calcium absorption, which is in turn facilitated by higher circulating levels of vitamin D_3 (Lucas et al., 2008b). The literature on the vitamin D status of pregnant and lactating women and their neonates has grown dramatically in recent years (Kreiter et al., 2000; Hollis and Wagner, 2004a, 2004b; Hollis, 2005; Kovacs, 2008; Lucas et al., 2008b). Controversies exist over the relationship between vitamin D status and calcium metabolism in adult women during pregnancy, lactation, and the months immediately following the cessation of lactation, and over the relationship between maternal and fetal/neonatal vitamin D status and calcium metabolism. Dark pigmentation and prolonged breast-feeding without supplementation are clear risk factors for vitamin D deficiencies in women and their neonates (Kreiter et al., 2000; Nozza and Rodda, 2001). To some extent, the healthy calcium status of neonates is insured against low maternal vitamin D status because calcium can be liberated from maternal bony stores through the action of parathyroid hormone-related protein (Kovacs, 2008). In the face of moderate maternal vitamin D deficiency, PTH concentrations rise in an apparent reflection of the need to maintain adequate plasma calcium concentrations through PTH-induced osteolysis (Okonofua et al., 1987). In cases of severe pre-existing maternal vitamin D deficiency, however, pregnancy precipitates osteomalacia in women and nutritional rickets in neonates (Kreiter et al., 2000; Nozza and Rodda, 2001; Kovacs, 2008). Few prospective studies have examined the vitamin D status and the course of calcium of women or neonates over the course of pregnancy and lactation (Hollis and Wagner, 2004a, 2004b; Kimball et al., 2008) and none have examined these parameters over the course of a relatively long lactation period such as that humans experienced in prehistory. We do not know the length of lactation in Paleolithic humans, but it probably was at least two years, based on averages for modern gathering and hunting people (Lee, 1980; Lunn, 1994). Despite the controversies and limitations of currently available data, the evidence now available suggests that chronically depressed vitamin D levels in actively reproducing women would compromise the female skeleton over successive pregnancies and lengthy periods of lactation, and would be associated with progressive hyperparathyroidism. It would also, in the most serious cases, compromise the integrity of the neonatal skeleton, and the future calcium status of children and juveniles so compromised as infants. Hypovitaminosis D in pregnant and lactating women and their neonates would also lead to depressed functioning of the immune system, with potentially deleterious or lethal consequences. Thus, strong clinical evidence continues to support the hypothesis that lighter skin pigmentation in females evolved primarily as a means to enhance the potential for cutaneous vitamin D production and maintain healthy long-term calcium status and skeletal health. Culturally based sexual selection probably acted to increase levels of sexual dimorphism in skin pigmentation in many populations (Jablonski and Chaplin, 2000; Jablonski, 2004, 2006). This effect has continued and may be increasing, as assortative mating is increasingly influenced by the widespread propagation of images of lightly pigmented females (Jones, 2000; Hill, 2002; Jablonski, 2006; Rondilla and Spickard, 2007).

SKIN PIGMENTATION AND HEALTH IN MODERN HUMAN POPULATIONS

One of the most biologically significant cultural differences between prehistoric and current populations of *Homo sapiens* is the potential for long-distance, high-speed migration. The ability of humans to migrate long distances through the use of domesticated animals, wheeled vehicles, watercraft, motorized transport, and air transportation has made it possible for people to move between regions with markedly different UVR regimes within single human lifetimes. This has resulted in many people living under UVR conditions very different in kind and intensity from those in which their ancestors evolved. The concomitant disease burden has been unexpected and high, and is visited both upon lightly pigmented people living under high UVR regimes and darkly pigmented people living under low UVR conditions. These problems are further exacerbated by cultural practices, from sun-bathing among the light-skinned to veil-wearing among the dark-skinned. Modern humans are excellent at making cultural adjustments to changed physical circumstances, but often the adjustments are insufficient to compensate for what has been lost and most are made in the absence of evolutionary knowledge. For example, the cultural encouragement of sun avoidance and sun protection as a means to reduce the incidence of skin cancer risk has been successful, but was launched in the absence of adequate cultural mechanisms for insuring adequate vitamin D levels in the absence of UVR exposure. Modern humans ignore their evolution history at their peril.

CONCLUSIONS

Our understanding of the evolution of human skin pigmentation has benefited greatly from new kinds of data and analytical methodologies being brought to bear on one of the oldest problems of anthropology and human biology. It is now possible for data on skin reflectances in modern indigenous human populations to be examined in conjunction with genetic and genomic studies of the nature and interaction of pigmentation genes. Experimental simulations of the reactivity of real or simulated human skin to the components of solar radiation are common. Direct measures of UVR and other physical characteristics of the environment are enabling powerful, geographically explicit explorations of the relationships between characteristics of the physical environment and the human organism. As a result of these new data and new kinds of analyses, we are gaining a much clearer understanding of the factors that have influenced the evolution of skin pigmentation during the history of the human lineage.

The human pigmentation phenotype has been determined by natural selection to maintain an optimum balance between photoprotection and photosynthesis over spatially varying conditions of ultraviolet irradiation. In regions of high UVR, including the regions of tropical Africa where the genus *Homo* and modern *Homo sapiens* emerged, this was achieved through the evolution of dark constitutive pigmentation maintained by purifying selection, and concomitant excellent tanning abilities. In populations dispersing to regions of lower UVR, incrementally lighter pigmentation evolved initially by positive selection for depigmentation and has been maintained by stabilizing selection. One of the great remaining challenges in the study of human skin color evolution is estimation of the intensity of selection pressure experienced by humans as they dispersed into different UVR regimes, and estimation of the time course for the evolutionary changes in pigmentation. This is an active area of research in which the author is now involved.

Our understanding of human skin color genetics is still incomplete, but comparative genomics is now providing evidence that skin color is a polygenic trait controlled by several genes that interact in complex ways. Skin, hair, and eye color are all affected by multiple genes, and their pleiotropic interactions. In some populations some variant forms of the genes account for more of the variation in skin color rather than in hair color and vice versa. Combinations of different forms of the genes have brought about the complex and continuous variation in skin coloration that we see in modern humans. With respect to evolutionary biology, what is important is that the human pigmentation phenotype has evolved to maintain an optimum balance of penetration of UVR over a spatially varying landscape of solar radiation. The genetic evidence demonstrating that light skin pigmentation was selected for three times independently in hominids in response to the selective pressure of low UVR regimes highlights the importance of skin pigmentation in maintaining physiological homeostasis and healthy reproductive status. It also denotes the lability of skin pigmentation in evolution and the unsuitability of skin reflectance as a character in cladistic analyses or in phenotypic sorting of human populations into color-based groups that are assumed to have genetic similarity or propinquity.

DISCUSSION POINTS

1. What functions does eumelanin play in human skin?
2. What hypotheses have been put forward to account for the evolution of darkly pigmented skin in humans inhabiting equatorial latitudes? What criteria need to be taken into account when assessing the validity of the statement, "skin pigmentation is adaptive"?
3. How is vitamin D made in the skin? What factors affect the body's ability to produce vitamin D?
4. What evidence supports the hypothesis that the evolution of lightly pigmented skin was promoted by positive selection?

5. What are some of the main health effects experienced by humans when they live under solar regimes different from those of their ancestors?
6. How does culture mitigate or intensify the effects of skin pigmentation on human physiology?

ACKNOWLEDGMENTS

I thank Michael Muehlenbein for inviting me to contribute to this volume, and for his patience. I am grateful to George Chaplin for numerous detailed discussions about the evolution of human skin color, for his constructive review of the first draft of this manuscript, and for preparing Figure 12.1. I also thank Tess Wilson for assisting in the gathering of reference material and for maintaining my bibliographic database. The constructive comments of two reviewers greatly improved the final draft of the manuscript. The financial support provided by an Alphonse Fletcher Sr. Fellowship is also gratefully acknowledged.

REFERENCES

Adams, W. C., Fox, R. H., Fry, A. J., et al. (1975). Thermoregulation during marathon running in cool, moderate, and hot environments. *Journal of Applied Physiology*, **38**(6), 1030–1037.

Adams, W. C., Mack, G. W., Langhans, G. W., et al. (1992). Effects of varied air velocity on sweating and evaporative rates during exercise. *Journal of Applied Physiology*, **73**, 2668–2674.

Agar, N. and Young, A. R. (2005). Melanogenesis: a photoprotective response to DNA damage? *Mutation Research*, **571**, 121–132.

Alaluf, S., Atkins, D., Barrett, K., et al. (2002). Ethnic variation in melanin content and composition in photoexposed and photoprotected human skin. *Pigment Cell Research*, **15**, 112–118.

Ancans, J., Tobin, D. J., Hoogduijn, M. J., et al. (2001). Melanosomal pH controls rate of melanogenesis, eumelanin/phaeomelanin ratio and melanosome maturation in melanocytes and melanoma cells. *Experimental Cell Research*, **268**, 26–35.

Amaral, L. Q. D. (1996). Loss of body hair, bipedality and thermoregulation. Comments on recent papers in the *Journal of Human Evolution*. *Journal of Human Evolution*, **30**(4), 357–366.

Anonymous (1983). Long-term effects of ethanol consumption on folate status of monkeys. *Nutrition Reviews*, **41**(7), 226–228.

Aoki, K. (2002). Sexual selection as a cause of human skin colour variation: Darwin's hypothesis revisited. *Annals of Human Biology*, **29**, 589–608.

Armas, L. A. G., Dowell, S., Akhter, M., et al. (2007). Ultraviolet-B radiation increases serum 25-hydroxyvitamin D levels: The effect of UVB dose and skin color. *Journal of the American Academy of Dermatology*, **57**, 588–593.

Armstrong, B. K. and Kricker, A. (2001). The epidemiology of solar radiation and skin cancer. In *Sun Protection in Man*, P. U. Giacomoni (ed.). Amsterdam: Elsevier Science, pp. 131–153.

Aroca, P., Urabe, K., Kobayashi, T., et al. (1993). Melanin biosynthesis patterns following hormonal stimulation. *Journal of Biological Chemistry*, **268**, 25650–25655.

Bahadoran, P., Ortonne, J.-P., King, R. A., et al. (2003). Albinism. In *Fitzpatrick's Dermatology in General Medicine*, I. M. Freedberg, A. Z. Eisen, K. Wolff, et al. (eds), 6th edn. New York: McGraw-Hill, pp. 826–835.

Bailey, L. B. (1995). Folate requirements and dietary recommendations. In *Folate in Health and Disease*, L. B. Bailey (ed.). New York: Marcel Dekker, Inc., pp. 123–151.

Barker, D., Dixon, K., Medrano, E. E., et al. (1995). Comparison of the responses of human melanocytes with different melanin contents to ultraviolet B irradiation. *Cancer Research*, **55**, 4041–4046.

Barton, N. H. (1999). Clines in polygenic traits. *Genetics Research*, **74**, 223–236.

Besser, L. M., Williams, L. J. and Cragan, J. D. (2007). Interpreting changes in the epidemiology of anencephaly and spina bifida following folic acid fortification of the US grain supply in the setting of long-term trends, Atlanta, Georgia, 1968–2003. *Birth Defects Research Part A: Clinical and Molecular Teratology*, **79**(11), 730–736.

Bjorn, L. O. and Wang, T. (2000). Vitamin D in an ecological context. *International Journal of Circumpolar Health*, **59**, 26–32.

Blom, H. J., Shaw, G. M., Den Heijer, M., et al. (2006). Neural tube defects and folate: case far from closed. *Nature Reviews Neuroscience*, **7**, 724–731.

Blum, H. F. (1961). Does the melanin pigment of human skin have adaptive value? *Quarterly Review of Biology*, **36**, 50–63.

Bouillon, R., Eelen, G., Verlinden, L., et al. (2006). Vitamin D and cancer. *Journal of Steroid Biochemistry and Molecular Biology*, **102**, 156–162.

Bower, C. and Stanley, F. J. (1989). Dietary folate as a risk factor for neural-tube defects: evidence from a case-control study in Western Australia. *Medical Journal of Australia*, **150**, 613–619.

Bower, C. and Stanley, F. J. (1992). The role of nutritional factors in the aetiology of neural tube defects. *Journal of Paediatrics and Child Health*, **28**, 12–16.

Brace, C. L. (1963). Structural reduction in evolution. *American Naturalist*, **97**, 39–49.

Branda, R. F. and Eaton, J. W. (1978). Skin color and nutrient photolysis: an evolutionary hypothesis. *Science*, **201**, 625–626.

Brenner, M. and Hearing, V. J. (2008). The protective role of melanin against UV damage in human skin. *Photochemistry and Photobiology*, **84**, 539–549.

Brinnel, H., Cabanac, M. and Hales, J. R. S. (1987). Critical upper levels of body temperature, tissue thermosensitivity, and selective brain cooling in hyperthermia. In *Heat Stress: Physical Exertion and Environment*, J. R. S. Hales and D. A. B. Richards (eds). Amsterdam: Excerpta Medica, pp. 209–240.

Brunvand, L., Quigstad, E., Urdal, P., et al. (1996). Vitamin D deficiency and fetal growth. *Early Human Development* **45**, 27–33.

Buccimazza, S. S., Molteno, C. D. and Viljoen, D. L. (1994). Prevalence of neural tube defects in Cape Town, South Africa. *Teratology*, **50**(3), 194–199.

Cabanac, M. and Brinnel, H. (1985). Blood flow in the emissary veins of the human head during hyperthermia. *European Journal of Applied Physiology and Occupational Physiology*, **54**(2), 172–176.

Cabanac, M. and Massonnet, B. (1977). Thermoregulatory responses as a function of core temperature in humans. *Journal of Physiology*, **265**, 587–596.

Caldwell, M. M., Bjorn, L. O., Bornman, J. F., et al. (1998). Effects of increased solar ultraviolet radiation on terrestrial ecosystems. *Journal of Photochemistry and Photobiology. B, Biology*, **46**, 40–52.

Calvo, M. S., Whiting, S. J. and Barton, C. N. (2005). Vitamin D intake: a global perspective of current status. *Journal of Nutrition*, **135**, 310.

Cannell, J. J., Zasloff, M., Garland, C. F., et al. (2008). On the epidemiology of influenza. *Virology Journal*, **5**, 29.

Cantorna, M. T. and Mahon, B. D. (2005). D-hormone and the immune system. *Journal of Rheumatology*, **76**, 11–20.

Cantorna, M. T., Yu, S. and Bruce, D. (2008). The paradoxical effects of vitamin D on type 1 mediated immunity. *Molecular Aspects of Medicine*, **29**, 369–375.

Caudill, M. A., Le, T., Moonie, S. A., et al. (2001). Folate status in women of childbearing age residing in southern California after folic acid fortification. *Journal of the American College of Nutrition*, **20**(2), 129–134.

Cerimele, D., Celleno, L. and Serri, F. (1990). Physiological changes in ageing skin. *British Journal of Dermatology*, **122**, 13–20.

Chaplin, G. (2001). The geographic distribution of environmental factors influencing human skin colouration. Thesis, Manchester Metropolitan University, Manchester, UK.

Chaplin, G. (2004). Geographic distribution of environmental factors influencing human skin coloration. *American Journal of Physical Anthropology*, **125**, 292–302.

Chaplin, G. and Jablonski, N. G. (1998). Hemispheric difference in human skin color. *American Journal of Physical Anthropology*, **107**(2), 221–224.

Chedekel, M. R. (1995). Photophysics and photochemistry of melanin. In *Melanin: its Role in Human Photoprotection*, L. Zeise, M. R. Chedekel and T. B. Fitzpatrick (eds). Overland Park, KS: Valdenmar Publishing Co., pp. 11–22.

Chen, S., Glenn, D. J., Ni, W., et al. (2008). Expression of the vitamin D receptor is increased in the hypertrophic heart. *Hypertension*, **52**, 1106–1112.

Chen, T. C., Chimeh, F., Lu, Z., et al. (2007). Factors that influence the cutaneous synthesis and dietary sources of vitamin D. *Archives of Biochemistry and Biophysics*, **460**, 213–217.

Chimpanzee Sequencing and Analysis Consortium (The) (2005). Initial sequence of the chimpanzee genome and comparison with the human genome. *Nature*, **437**, 69–87.

Christenson, L. J., Borrowman, T. A., Vachon, C. M., et al. (2005). Incidence of basal cell and squamous cell carcinomas in a population younger than 40 years. *Journal of the American Medical Association*, **294**, 681–690.

Cleaver, J. E. (2000). Common pathways for ultraviolet skin carcinogenesis in the repair and replication of defective groups of xeroderma pigmentosum. *Journal of Dermatological Science*, **23**(1), 1–11.

Cleaver, J. E. and Crowley, E. (2002). UV damage, DNA repair and skin carcinogenesis. *Frontiers in Bioscience*, **7**, 1024–1043.

Clemens, T. L., Henderson, S. L., Adams, J. S., et al. (1982). Increased skin pigment reduces the capacity of skin to synthesise vitamin D_3. *Lancet*, **1**, 74–76.

Clements, M. R., Johnson, L. and Fraser, D. R. (1987). A new mechanism for induced vitamin D deficiency in calcium deprivation. *Nature*, **325**, 62–65.

Coburn, J. W., Hartenbower, D. L. and Norman, A. W. (1974). Metabolism and action of the hormone vitamin D. Its relation to diseases of calcium homeostasis. *Western Journal of Medicine*, **121**(1), 22–44.

Copp, A. J., Fleming, A. and Greene, N. D. E. (1998). Embryonic mechanisms underlying the prevention of neural tube defects by vitamins. *Mental Retardation and Developmental Disabilities Research Reviews*, **4**, 264–268.

Cosman, F., Nieves, J., Dempster, D., et al. (2007). Vitamin D economy in blacks. *Journal of Bone and Mineral Research*, **22**, V34–V38.

Cotter, J. D., Patterson, M. J. and Taylor, N. A. S. (1995). The topography of eccrine sweating in humans during exercise. *European Journal of Applied Physiology*, **71**(6), 549–554.

Cowles, R. B. (1959). Some ecological factors bearing on the origin and evolution of pigment in the human skin. *American Naturalist*, **93**, 283–293.

Cronin, K. J., Butler, P. E. M., Mchugh, M., et al. (1996). A one-year prospective study of burns in an Irish paediatric burns unit. *Burns*, **22**, 221–224.

de Bree, A., van Dusseldorp, M., Brouwer, I. A., et al. (1997). Folate intake in Europe: recommended, actual and desired intake. *European Journal of Clinical Nutrition*, **51**(10), 643–660.

Der-Petrossian, M., Födinger, M., Knobler, R., et al. (2007). Photodegradation of folic acid during extracorporeal photopheresis. *British Journal of Dermatology*, **156**(1), 117–121.

Diepgen, T. L. and Mahler, V. (2002). The epidemiology of skin cancer. *British Journal of Dermatology*, **146**, 1–6.

Djukic, A. (2007). Folate-responsive neurologic diseases. *Pediatric Neurology*, **37**, 387–397.

Duffy, D. L., Montgomery, G. W., Chen, W., et al. (2007). A three-single-nucleotide polymorphism haplotype in intron 1 of OCA2 explains most human eye-color variation. *American Journal of Human Genetics*, **80**(2), 241–252.

Dwyer, T., Blizzard, L., Ashbolt, R., et al. (2002). Cutaneous melanin density of caucasians measured by spectrophotometry and risk of malignant melanoma, basal cell carcinoma, and squamous cell carcinoma of the skin. *American Journal of Epidemiology*, **155**, 614–621.

Ebisch, I. M. W., Pierik, F. H., De Jong, F. H., et al. (2006). Does folic acid and zinc sulphate intervention affect endocrine parameters and sperm characteristics in men? *International Journal of Andrology*, **29**(2), 339–345.

Epel, D., Hemela, K., Shick, M., et al. (1999). Development in the floating world: defenses of eggs and embryos against damage from UV radiation. *American Zoologist*, **39**, 271–278.

Eurocat Working Group. (1991). Prevalence of neural tube defects in 20 regions of Europe and the impact of prenatal diagnosis, 1980–1986. *Journal of Epidemiology and Community Health*, **45**(1), 52–58.

Falk, D. (1990). Brain evolution in *Homo*: the "radiator" theory. *Behavioral and Brain Sciences*, **13**, 333–381.

Fisher, G. J., Kang, S., Varani, J., et al. (2002). Mechanisms of photoaging and chronological skin aging. *Archives of Dermatological Research*, **138**(11), 1462–1470.

Fitzpatrick, T. B. (1988). The validity and practicality of sun reactive skin type I through VI. *Archives of Dermatology*, **124**, 869–871.

Fitzpatrick, T. B. and Ortonne, J.-P. (2003). Normal skin color and general considerations of pigmentary disorders. In *Fitzpatrick's Dermatology in General Medicine*, I. M. Freedberg, A. Z. Eisen, K. Wolff, et al. (eds), 6th edn. New York: McGraw-Hill, pp. 819–825.

Fitzpatrick, T. B., Seiji, M. and Mcgugan, A. D. (1961). Melanin pigmentation. *New England Journal of Medicine*, **265**, 328–332.

Fleet, J. C. (2008). Molecular actions of vitamin D contributing to cancer prevention. *Molecular Aspects of Medicine*, **29**, 388–396.

Fleming, A. and Copp, A. J. (1998). Embryonic folate metabolism and mouse neural tube defects. *Science*, **280**, 2107–2109.

Fogelman, Y., Rakover, Y. and Luboshitsky, R. (1995). High prevalence of vitamin D deficiency among Ethiopian women immigrants to Israel: exacerbation during pregnancy and lactation. *Israel Journal of Medical Sciences*, **31**, 221–224.

Fox, H. M. and Vevers, G. (1960). Melanin. In *The Nature of Animal Colours*. London: Sidgwick and Jackson Ltd, pp. 22–47.

Frost, P. (1988). Human skin color: a possible relationship between its sexual dimorphism and its social perception. *Perspectives in Biology and Medicine*, **32**, 38–59.

Frost, P. (2005). *Fair Women, Dark Men: the Forgotten Roots of Color Prejudice*. Christchurch, New Zealand: Cybereditions.

Frost, P. (2007). Human skin-color sexual dimorphism: a test of the sexual selection hypothesis. *American Journal of Physical Anthropology*, **133**, 779–780.

Garland, C. F., Garland, F. C., Gorham, E. D., et al. (2006). The role of vitamin D in cancer prevention. *American Journal of Public Health*, **96**, 252–261.

Gibbs, S., Murli, S., De Boer, G., et al. (2000). Melanosome capping of keratinocytes in pigmented reconstructed epidermis – effect of ultraviolet radiation and 3-isobutyl-1-methyl-xanthine on melanogenesis. *Pigment Cell Research*, **13**, 458–466.

Goor, Y. and Rubinstein, A. (1995). Vitamin D levels in dark-skinned people. *Israel Journal of Medical Sciences*, **31**, 237–238.

Graf, J., Voisey, J., Hughes, I., et al. (2007). Promoter polymorphisms in the *MATP* (*SLC45A2*) gene are associated with normal human skin color variation. *Human Mutation*, **28**, 710–717.

Grant, W. B. (2008). Solar ultraviolet irradiance and cancer incidence and mortality. *Advances in Experimental Medicine and Biology*, **624**, 16–30.

Green, R. and Miller, J. W. (1999). Folate deficiency beyond megaloblastic anemia: Hyperhomocysteinemia and other malfunctions of dysfunctional folate status. *Seminars in Hematology*, **36**, 47–64.

Gregory, J. F. III (1995). The bioavailability of folate. In *Folate in Health and Disease*, L. B. Bailey (ed.). New York: Marcel Dekker, Inc., pp. 195–235.

Haass, N. K. and Herlyn, M. (2005). Normal human melanocyte homeostasis as a paradigm for understanding melanoma. *Journal of Investigative Dermatology Symposium Proceedings*, **10**, 153–163.

Haass, N. K., Smalley, K. S. M., Li, L., et al. (2005). Adhesion, migration and communication in melanocytes and melanoma. *Pigment Cell Research*, **18**, 150–159.

Harms, L. R., Eyles, D. W., McGrath, J. J., et al. (2008). Developmental vitamin D deficiency alters adult behaviour in 129/SvJ and C57BL/6J mice. *Behavioural Brain Research*, **187**(2), 343–350.

Harris, M., Consorte, J. G., Lang, J., et al. (1993). Who are the whites? Imposed census categories and the racial demography of Brazil. *Social Forces*, **72**, 451–462.

Hathcock, J. N., Shao, A., Vieth, R., et al. (2007). Risk assessment for vitamin D. *American Journal of Clinical Nutrition*, **85**, 6–18.

Hennessy, A., Oh, C., Diffey, B., et al. (2005). Eumelanin and pheomelanin concentrations in human epidermis before and after UVB irradiation. *Pigment Cell Research*, **18**, 220–223.

Henry, H. L. and Norman, A. W. (1984). Vitamin D: metabolism and biological actions. *Annual Review of Nutrition*, **4**, 493–520.

Hill, M. E. (2002). Skin color and the perception of attractiveness among African Americans: does gender make a difference? *Social Psychology Quarterly*, **65**, 77–91.

Hitchcock, R. T. (2001). *Ultraviolet Radiation*, 2nd edn. Fairfax, VA: American Industrial Hygiene Association.

Holick, M. F. (1987). Photosynthesis of vitamin D in the skin: effect of environmental and life-style variables. *Federal Proceedings*, **46**, 1876–1882.

Holick, M. F. (1995). Environmental factors that influence the cutaneous production of vitamin D. *American Journal of Clinical Nutrition*, **61**, 638S–645S.

Holick, M. F. (1997). Photobiology of vitamin D. In *Vitamin D*, D. Feldman, F. H. Glorieux and J. W. Pike (eds). San Diego, CA: Academic Press, pp. 33–39.

Holick, M. F. (2003). Evolution and function of vitamin D. *Recent Results in Cancer Research*, **164**, 3–28.

Holick, M. F., Maclaughlin, J. A. and Doppelt, S. H. (1981). Regulation of cutaneous previtamin D_3 photosynthesis in man: skin pigment is not an essential regulator. *Science*, **211**, 590–593.

Holick, M. F., Chen, T. C., Lu, Z., et al. (2007). Vitamin D and skin physiology: a D-lightful story. *Journal of Bone and Mineral Research*, **22**, V28–V33.

Hollis, B. W. (2005). Circulating 25-hydroxyvitamin D levels indicative of vitamin D sufficiency: implications for establishing a new effective dietary intake recommendation for vitamin D. *Journal of Nutrition*, **135**, 317–322.

Hollis, B. W. and Wagner, C. L. (2004a). Assessment of dietary vitamin D requirements during pregnancy and lactation. *American Journal of Clinical Nutrition*, **79**, 717–726.

Hollis, B. W. and Wagner, C. L. (2004b). Vitamin D requirements during lactation: high-dose maternal supplementation as therapy to prevent hypovitaminosis D for both the mother and the nursing infant. *American Journal of Clinical Nutrition*, **80**, 1752S–1758S.

Hornyak, T. J. (2006). The developmental biology of melanocytes and its application to understanding human congenital disorders of pigmentation. *Advances in Dermatology*, **22**, 201–218.

Ito, S. (2003). A chemist's view of melanogenesis. *Pigment Cell Research*, **16**, 230–236.

IUPAC-IUB Joint Commission on Biochemical Nomenclature (1982). Nomenclature of vitamin D. *Molecular and Cellular Biochemistry*, **49**, 177–181.

Jablonski, N. G. (1992). Sun, skin and spina bifida: an exploration of the relationship between solar ultraviolet radiation, skin colour and neural tube defects. In *Proceedings of the Fifth Annual Conference of the Australasian Society for Human Biology*, N. W. Bruce (ed.). Perth: Centre for Human Biology, pp. 455–462.

Jablonski, N. G. (1993). Quaternary environments and the evolution of primates in East Asia, with notes on two new specimens of fossil cercopithecidae from China. *Folia Primatologica*, **60**, 118–132.

Jablonski, N. G. (2004). The evolution of human skin and skin color. *Annual Review of Anthropology*, **33**, 585–623.

Jablonski, N. G. (2006). *Skin: a Natural History*. Berkeley, CA: University of California Press.

Jablonski, N. G. and Chaplin, G. (2000). The evolution of human skin coloration. *Journal of Human Evolution*, **39**, 57–106.

Jemal, A., Devesa, S. S., Hartge, P., et al. (2001). Recent trends in cutaneous melanoma incidence among whites in the United States. *Journal of the National Cancer Institute*, **93**, 678–683.

Jimbow, K., Quevedo, W. C. Jr, Fitzpatrick, T. B., et al. (1976). Some aspects of melanin biology: 1950–1975. *Journal of Investigative Dermatology*, **67**, 72–89.

John, P. R., Makova, K., Li, W. H., et al. (2003). DNA polymorphism and selection at the melanocortin-1 receptor gene in normally pigmented southern African individuals. *Annals of the New York Academy of Sciences*, **994**, 299–306.

Johnson, B. L. Jr (1998). Vitiligo. In *Ethnic Skin: Medical and Surgical*, B. L. Johnson Jr, R. L. Moy and G. M. White (eds). St. Louis, MO: Mosby, pp. 187–194.

Johnson, F. S., Mo, T. and Green, A. E. S. (1976). Average latitudinal variation in ultraviolet radiation at the Earth's surface. *Photochemistry and Photobiology*, **23**, 179–188.

Jones, G. (2008). Pharmacokinetics of vitamin D toxicity. *American Journal of Clinical Nutrition*, **88**, 582S–586S.

Jones, T. (2000). Shades of brown: the law of skin color. *Duke Law Journal*, **49**, 1487–1557.

Kappelman, J., Alçiçek, M. C., Kazanci, N., et al. (2008). First *Homo erectus* from Turkey and implications for migrations into temperate Eurasia. *American Journal of Physical Anthropology*, **135**, 110–116.

Kennedy, C., Bajdik, C. D., Willemze, R., et al. (2003). The influence of painful sunburns and lifetime sun exposure on the risk of actinic keratoses, seborrheic warts, melanocytic nevi, atypical nevi, and skin cancer. *Journal of Investigative Dermatology*, **120**, 1087–1093.

Kielbassa, C. and Epe, B. (2000). DNA damage induced by ultraviolet and visible light and its wavelength dependence. *Methods in Enzymology*, **319**, 436–445.

Kim, D. H., Sabour, S., Sagar, U. N., et al. (2008). Prevalence of hypovitaminosis D in cardiovascular diseases (from the National Health and Nutrition Examination Survey 2001 to 2004). *American Journal of Cardiology*, **102**, 1540–1544.

Kim, S. M., Kim, Y. K. and Hann, S.-K. (1999). Serum levels of folic acid and vitamin B_{12} in Korean patients with vitiligo. *Yonsei Medical Journal*, **40**, 195–198.

Kimball, S., Fuleihan, G. H. and Vieth, R. (2008). Vitamin D: a growing perspective. *Critical Reviews in Clinical Laboratory Sciences*, **45**, 339–414.

Kimlin, M. G. (2004). The climatology of vitamin D producing ultraviolet radiation over the United States. *Journal of Steroid Biochemistry and Molecular Biology*, **89–90**(1–5), 479–483.

Kittles, R. (1995). Nature, origin, and variation of human pigmentation. *Journal of Black Studies*, **26**, 36–61.

Kobayashi, N., Nakagawa, A., Muramatsu, T., et al. (1998). Supranuclear melanin caps reduce ultraviolet induced DNA photoproducts in human epidermis. *Journal of Investigative Dermatology*, **110**, 806–810.

Kollias, N. (1995a). Melanin and non-melanin protection. In *Melanin: its Role in Human Photoprotection*, L. Zeise, M. R. Chedekel and T. B. Fitzpatrick (eds). Overland Park, KS: Valdenmar Publishing Co., pp. 233–237.

Kollias, N. (1995b). The physical basis of skin color and its evaluation. *Clinics in Dermatology*, **13**, 361–367.

Kollias, N., Malallah, Y. H., Al-Ajmi, H., et al. (1996). Erythema and melanogenesis action spectra in heavily pigmented individuals as compared to fair-skinned Caucasians. *Photodermatology, Photoimmunology and Photomedicine*, **12**, 183–188.

Komaromy-Hiller, G., Nuttall, K. L. and Ashwood, E. R. (1997). Effect of storage on serum vitamin B_{12} and folate stability. *Annals of Clinical and Laboratory Science*, **27**(4), 249–253.

Kongshoj, B., Thorleifsson, A. and Wulf, H. C. (2006). Pheomelanin and eumelanin in human skin determined by high-performance liquid chromatography and its relation to in vivo reflectance measurements. *Photodermatology, Photoimmunology and Photomedicine*, **22**(3), 141–147.

Kovacs, C. S. (2008). Vitamin D in pregnancy and lactation: maternal, fetal, and neonatal outcomes from human and animal studies. *American Journal of Clinical Nutrition*, **88**, 520S–528S.

Kreiter, S. R., Schwartz, R. P., Kirkman, H. N., et al. (2000). Nutritional rickets in African American breast-fed infants. *Journal of Pediatrics*, **137**, 153–157.

Lalueza-Fox, C., Rompler, H., Caramelli, D., et al. (2007). A melanocortin-1 receptor allele suggests varying pigmentation among Neanderthals. *Science*, **318**, 1453–1455.

Lamason, R. L., Mohideen, M.-A. P. K., Mest, J. R., et al. (2005). SLC24A5, a putative cation exchanger, affects pigmentation in zebrafish and humans. *Science*, **310**, 1782–1786.

Lamberg-Allardt, C. (2006). Vitamin D in foods and as supplements. *Progress in Biophysics and Molecular Biology*, **92**, 33–38.

Langbein, L., Rogers, M.A., Praetzel, S., et al. (2005). Characterization of a novel human type II epithelial keratin K1b, specifically expressed in eccrine sweat glands. *Journal of Investigative Dermatology*, **125**, 428–444.

Lao, O., De Gruijter, J.M., Van Duijn, K., et al. (2007). Signatures of positive selection in genes associated with human skin pigmentation as revealed from analyses of single nucleotide polymorphisms. *Annals of Human Genetics*, **71**(3), 354–369.

Lasker, G.W. (1954a). Photoelectric measurement of skin color in a Mexican Mestizo population. *American Journal of Physical Anthropology*, **12**, 115–122.

Lasker, G.W. (1954b). Seasonal changes in skin color. *American Journal of Physical Anthropology*, **12**, 553–558.

Lee, M.M.C. and Lasker, G.W. (1959). The sun-tanning potential of human skin. *Human Biology*, **31**, 252–260.

Lee, R.B. (1980). Lactation, ovulation, infanticide, and women's work: a study of hunter-gatherer population regulation. In *Biosocial Mechanisms of Population Regulation*, M.N. Cohen, R.S. Malpass and H.G. Klein (eds). New Haven, CT: Yale University Press, pp. 321–348.

Leiter, U. and Garbe, C. (2008). Epidemiology of melanoma and nonmelanoma skin cancer: the role of sunlight. *Advances in Experimental Medicine and Biology*, **624**, 89–103.

Li, D., Turi, T.G., Schuck, A., et al. (2001). Rays and arrays: the transcriptional program in the response of human epidermal keratinocytes to UVB illumination. *Molecular Biology and Evolution*, **15**, 2533–2535.

Lips, P. (2006). Vitamin D physiology. *Progress in Biophysics and Molecular Biology*, **92**, 4–8.

Littleton, J. (1991). Vitamin D deficiency rickets: prevalence in early societies. In *Living With Civilisation: Proceedings of the Australasian Society for Human Biology*, N.W. Bruce (ed.). Canberra, Australia: Centre for Human Biology, Australian National University, pp. 15–21.

Loomis, W.F. (1967). Skin-pigment regulation of vitamin-D biosynthesis in man. *Science*, **157**, 501–506.

Lucas, R.M., McMichael, A.J., Smith, W., et al. (2006). *Solar Ultraviolet Radiation: Global Burden of Disease from Solar Ultraviolet Radiation*. Geneva: World Health Organization.

Lucas, R.M., McMichael, A.J., Armstrong, B.K., et al. (2008a). Estimating the global disease burden due to ultraviolet radiation exposure. *International Journal of Epidemiology*, **37**, 654–667.

Lucas, R.M., Ponsonby, A.L., Pasco, J.A., et al. (2008b). Future health implications of prenatal and early-life vitamin D status. *Nutrition Reviews*, **66**, 710–720.

Lucock, M. (2000). Folic acid: nutritional biochemistry, molecular biology, and role in disease processes. *Molecular Genetics and Metabolism*, **71**(1–2), 121–138.

Lunn, P.G. (1994). Lactation and other metabolic loads affecting human reproduction. *Annals of the New York Academy of Sciences*, **709**, 77–85.

Mackintosh, J.A. (2001). The antimicrobial properties of melanocytes, melanosomes and melanin and the evolution of black skin. *Journal of Theoretical Biology*, **211**(2), 101–113.

Maclaughlin, J. and Holick, M.F. (1985). Aging decreases the capacity of human skin to produce vitamin D_3. *Journal of Clinical Investigation*, **76**, 1536–1538.

Madison, K.C. (2003). Barrier function of the skin: "*La raison d'être*" of the epidermis. *Journal of Investigative Dermatology*, **121**, 231–241.

Madrigal, L. and Kelly, W. (2007). Human skin-color sexual dimorphism: a test of the sexual selection hypothesis. *American Journal of Physical Anthropology*, **132**, 470–482.

Madronich, S., Mckenzie, R.L., Bjorn, L.O., et al. (1998). Changes in biologically active ultraviolet radiation reaching the Earth's surface. *Journal of Photochemistry and Photobiology B: Biology*, **46**, 5–19.

Makova, K. and Norton, H.L. (2005). Worldwide polymorphism at the *MC1R* locus and normal pigmentation variation in humans. *Peptides*, **26**, 1901–1908.

Malvy, D.J.-M., Guinot, C., Preziosi, P., et al. (2000). Relationship between vitamin D status and skin phototype in general adult population. *Photochemistry and Photobiology*, **71**, 466–469.

Marzullo, G. and Fraser, F.C. (2005). Similar rhythms of seasonal conceptions in neural tube defects and schizophrenia: a hypothesis of oxidant stress and the photoperiod. *Birth Defects Research Part A: Clinical and Molecular Teratology*, **73**, 1–5.

Mason, H.S., Ingram, D.J.E. and Allen, B. (1960). The free radical property of melanins. *Archives of Biochemistry and Biophysics*, **86**, 225–230.

Mastropaolo, W. and Wilson, M.A. (1993). Effect of light on serum B_{12} and folate stability. *Clinical Chemistry*, **39**(5), 913.

Mathur, U., Datta, S.L. and Mathur, B.B. (1977). The effect of aminopterin-induced folic acid deficiency on spermatogenesis. *Fertility and Sterility*, **28**(12), 1356–1360.

Matsumura, Y. and Ananthawamy, H.N. (2004). Toxic effects of ultraviolet radiation on the skin. *Toxicology and Applied Pharmacology*, **195**, 298–308.

Matsuoka, L.Y., Wortsman, J., Haddad, J.G., et al. (1991). Racial pigmentation and the cutaneous synthesis of vitamin D. *Archives of Dermatology*, **127**, 536–538.

Mawer, E.B. and Davies, M. (2001). Vitamin D nutrition and bone disease in adults. *Reviews in Endocrine and Metabolic Disorders*, **2**, 153.

Mawer, E.B., Schaefer, K., Lumb, G.A., et al. (1971). The metabolism of isotopically labelled vitamin D_3 in man: the influence of the state of vitamin D nutrition. *Clinical Science*, **40**, 39–53.

Mawer, E.B., Backhouse, J., Holman, C.A., et al. (1972). The distribution and storage of vitamin D and its metabolites in human tissues. *Clinical Science*, **43**, 413–431.

McGrath, J.J., Féron, F.P., Burne, T.H.J., et al. (2004). Vitamin D_3 – implications for brain development. *Journal of Steroid Biochemistry and Molecular Biology*, **89–90**, 557–560.

McNulty, H. and Scott, J.M. (2008). Intake and status of folate and related B-vitamins: considerations and challenges in achieving optimal status. *British Journal of Nutrition*, **99**(S3), S48–S54.

Meredith, P. and Sarna, T. (2006). The physical and chemical properties of eumelanin. *Pigment Cell Research*, **19**(6), 572–594.

Minns, R.A. (1996). Folic acid and neural tube defects. *Spinal Cord*, **34**, 460–465.

Mitra, D. and Bell, N.H. (1997). Racial, geographic, genetic, and body habitus effects on vitamin D metabolism. In *Vitamin D*, D. Feldman, F.H. Glorieux and J.W. Pike (eds). San Diego, CA: Academic Press, pp. 521–532.

Miyamura, Y., Coelho, S.G., Wolber, R., et al. (2007). Regulation of human skin pigmentation and responses to ultraviolet radiation. *Pigment Cell Research*, **20**, 2–13.

Molloy, A.M., Mills, J.L., Kirke, P.N., et al. (1999). Folate status and neural tube defects. *BioFactors*, **10**, 291–294.

Montagna, W. (1971). Cutaneous comparative biology. *Archives of Dermatology*, **104**, 577–591.

Montagna, W. (1981). The consequences of having a naked skin. *Birth Defects: Original Article Series*, **17**, 1–7.

Montes, L.F., Diaz, M.L., Lajous, J., et al. (1992). Folic acid and vitamin B_{12} in vitiligo: a nutritional approach. *Cutis*, **50**, 39–42.

Murray, F.G. (1934). Pigmentation, sunlight, and nutritional disease. *American Anthropologist*, **36**, 438–445.

Myles, S., Somel, M., Tang, K., et al. (2007). Identifying genes underlying skin pigmentation differences among human populations. *Human Genetics*, **120**, 613–621.

Neer, R.M. (1975). The evolutionary significance of vitamin D, skin pigment, and ultraviolet light. *American Journal of Physical Anthropology*, **43**, 409–416.

Neuhouser, M.L., Beresford, S.A.A., Hickok, D.E., et al. (1998). Absorption of dietary and supplemental folate in women with prior pregnancies with neural tube defects and controls. *Journal of the American College of Nutrition*, **17**(6), 625–630.

Nielsen, K.P., Zhao, L., Stamnes, J.J., et al. (2006a). The importance of the depth distribution of melanin in skin for DNA protection and other photobiological processes. *Journal of Photochemistry and Photobiology B: Biology*, **82**, 194–198.

Nielsen, K.P., Zhao, L., Stamnes, J.J., et al. (2006b). The importance of the depth distribution of melanin in skin for DNA protection and other photobiological processes. *Journal of Photochemistry and Photobiology B: Biology*, **82**, 194–198.

Nordlund, J.J. (1995). Melanin and melanocytes: their function and significance from a clinician's perspective. In *Melanin: its Role in Human Photoprotection*, L. Zeise, M.R. Chedekel and T.B. Fitzpatrick (eds). Overland Park, KS: Valdenmar Publishing, pp. 183–193.

Norman, A.W. (2008). From vitamin D to hormone D: fundamentals of the vitamin D endocrine system essential for good health. *American Journal of Clinical Nutrition*, **88**, 491S–499S.

Norton, H.L. and Hammer, M.F. (2007). Sequence variation in the pigmentation candidate gene *SLC24A5* and evidence for independent evolution of light skin in European and East Asian populations. *American Journal of Physical Anthropology*, **132**(S44), 179.

Norton, H.L., Kittles, R.A., Parra, E., et al. (2007). Genetic evidence for the convergent evolution of light skin in Europeans and East Asians. *Molecular Biology and Evolution*, **24**, 710–722.

Nozza, J.M. and Rodda, C.P. (2001). Vitamin D deficiency in mothers of infants with rickets. *Medical Journal of Australia*, **175**, 253–255.

Off, M.K., Steindal, A.E., Porojnicu, A.C., J, et al. (2005). Ultraviolet photodegradation of folic acid. *Journal of Photochemistry and Photobiology B: Biology*, **80**(1), 47–55.

Okonofua, F., Menon, R.K., Houlder, S., et al. (1987). Calcium, vitamin D and parathyroid hormone relationships in pregnant Caucasian and Asian women and their neonates. *Annals of Clinical Biochemistry*, **24**, 22–28.

Olivier, G. (1960). *Pratique Anthropologique*. Paris: Vigot Freres.

Ortonne, J.P. (1990). The effects of ultraviolet exposure on skin melanin pigmentation. *Journal of International Medical Research*, **18**, 8C–17C.

Ortonne, J.P. (2002). Photoprotective properties of skin melanin. *British Journal of Dermatology*, **146**, 7–10.

Pandolf, K.B., Gange, R.W., Latzka, W.A., et al. (1992). Human thermoregulatory responses during heat exposure after artificially induced sunburn. *American Journal of Physiology*, **262**, R610–R616.

Parra, E.J. (2007). Human pigmentation variation: evolution, genetic basis, and implications for public health. *American Journal of Physical Anthropology*, **134**, 85–105.

Pathak, M.A. (1995). Functions of melanin and protection by melanin. In *Melanin: its Role in Human Photoprotection*, L. Zeise, M.R. Chedekel and T.B. Fitzpatrick (eds). Overland Park, KS: Valdenmar Publishing Co., pp. 125–134.

Pfeifer, G.P., You, Y.H. and Besaratinia, A. (2005). Mutations induced by ultraviolet light. *Mutation Research*, **571**, 19–31.

Pflaum, M., Kielbassa, C., Garmyn, M., et al. (1998). Oxidative DNA damage induced by visible light in mammalian cells: extent, inhibition by antioxidants and genotoxic effects. *Mutation Research*, **408**, 137–146.

Post, P.W., Daniels, F. Jr and Binford, R.T. Jr (1975). Cold injury and the evolution of "white" skin. *Human Biology*, **47**, 65–80.

Prota, G. (1992a). Melanin-producing cells. In *Melanins and Melanogenesis*, G. Prota (ed.). San Diego, CA: Academic Press, pp. 14–33.

Prota, G. (1992b). Photobiology and photochemistry of melanogenesis. In *Melanins and Melanogenesis*, G. Prota (ed.). San Diego, CA: Academic Press, pp. 208–224.

Quevedo, W.C. (1969). Influence of age and UV on the populations of DOPA-positive melanocytes in human skin. *Journal of Investigative Dermatology*, **52**(3), 287–290.

Rana, B.K., Hewett-Emmett, D., Jin, L., et al. (1999). High polymorphism at the human melanocortin 1 receptor locus. *Genetics*, **151**, 1547–1557.

Rees, J.L. (2000). The melanocortin 1 receptor (*MC1R*): more than just red hair. *Pigment Cell Research*, **13**, 135–140.

Rees, J.L. (2003). Genetics of hair and skin color. *Annual Review of Genetics*, **37**, 67–90.

Rees, J.L. (2004). The genetics of sun sensitivity in humans. *American Journal of Human Genetics*, **75**, 739–751.

Rees, J.L. (2008). Melanoma: what are the gaps in our knowledge. *PLoS Medicine*, **5**, e122.

Relethford, J.H. (1997). Hemispheric difference in human skin color. *American Journal of Physical Anthropology*, **104**, 449–457.

Rhodes, A.R., Albert, L.S., Barnhill, R.L., et al. (1991). Sun-induced freckles in children and young adults: a correlation of clinical and histopathologic features. *Cancer*, **67**, 1990–2001.

Rigel, D. S. (2008). Cutaneous ultraviolet exposure and its relationship to the development of skin cancer. *Journal of the American Academy of Dermatology*, **58**, S129–S132.

Roberts, D. F. (1977). Human pigmentation: its geographical and racial distribution and biological significance. *Journal of the Society of Cosmetic Chemists*, **28**, 329–342.

Roberts, D. F. and Kahlon, D. P. S. (1976). Environmental correlations of skin colour. *Annals of Human Biology* **3** (1), 11–22.

Robins, A. H. (1991). *Biological Perspectives on Human Pigmentation*. Cambridge, UK: Cambridge University Press.

Rogers, A. R., Iltis, D. and Wooding, S. (2004). Genetic variation at the *MC1R* locus and the time since loss of human body hair. *Current Anthropology*, **45**, 105–124.

Rondilla, J. L. and Spickard, P. (2007). *Is Lighter Better? Skin-Tone Discrimination among Asian Americans*. Lanham, MD: Rowman and Littlefield.

Rothschild, L. J. (1999). The influence of UV radiation on protistan evolution. *Journal of Eukaryotic Microbiology*, **46**, 548–555.

Rouzaud, F., Kadekaro, A. L., Abdel-Malek, Z. A., et al. (2005). *MC1R* and the response of melanocytes to ultraviolet radiation. *Mutation Research/Fundamental and Molecular Mechanisms of Mutagenesis*, **571**, 133–152.

Rubegni, P., Cevenini, G., Barbini, P., et al. (1999). Quantitative characterization and study of the relationship between constitutive-facultative skin color and phototype in Caucasians. *Photochemistry and Photobiology*, **70**(3), 303–307.

Saraiya, M., Glanz, K., Briss, P. A., et al. (2004). Interventions to prevent skin cancer by reducing exposure to ultraviolet radiation: a systematic review. *American Journal of Preventive Medicine*, **27**, 422–466.

Sarna, T. and Swartz, H., M. (1998). The physical properties of melanins. In *The Pigmentary System, Physiology and Pathophysiology*, J. Nordlund, J. R. E. Boissy, V. J. Hearing, et al. (eds). New York: Oxford University Press, pp. 333–357.

Sato, K. and Dobson, R. L. (1970). Regional and individual variations in the function of the human eccrine sweat gland. *Journal of Investigative Dermatology* **54**(6), 443–449.

Schallreuter, K. U. (2007). Advances in melanocyte basic science research. *Dermatologic Clinics*, **25**, 283–291.

Schreier, W. J., Schrader, T. E., Koller, F. O., et al. (2007). Thymine dimerization in DNA is an ultrafast photoreaction. *Science*, **315**(5812), 625–629.

Scott, J. M. (2007). Reduced folate status is common and increases disease risk. It can be corrected by daily ingestion of supplements or fortification. *Novartis Foundation Symposium*, **282**, 105–117.

Shaw, N. J. (2003). Vitamin D deficiency rickets. In *Vitamin D and Rickets*, Z. E. Hochberg (ed.). Basel, Switzerland: Karger, pp. 93–104.

Shibasaki, M., Wilson, T. E. and Crandall, C. G. (2006). Neural control and mechanisms of eccrine sweating during heat stress and exercise. *Journal of Applied Physiology*, **100**(5), 1692–1701.

Shriver, M. D., Parra, E. J., Dios, S., et al. (2003). Skin pigmentation, biogeographical ancestry and admixture mapping. *Human Genetics*, 1–14.

Sinha, R. P. and Hader, D.-P. (2002). UV-induced DNA damage and repair: a review. *Photochemical and Photobiological Science*, **1**(4), 225–236.

Skull, S. A., Ngeow, J. Y. Y., Biggs, B. A., et al. (2003). Vitamin D deficiency is common and unrecognized among recently arrived adult immigrants from the horn of Africa. *International Medical Journal*, **33**, 47–51.

Soininen, L., Järvinen, S. and Pukkala, E. (2002). Cancer incidence among Sami in northern Finland, 1979–1998. *International Journal of Cancer*, **100**, 342–346.

St-Arnaud, R. (2008). The direct role of vitamin D on bone homeostasis. *Archives of Biochemistry and Biophysics*, **473**, 225–230.

Steegmann, A. T. Jr (1967). Frostbite of the human face as a selective force. *Human Biology*, **39**, 131–144.

Steindal, A. H., Juzeniene, A., Johnsson, A., et al. (2006). Photodegradation of 5-methyltetrahydrofolate: biophysical aspects. *Photochemistry and Photobiology*, **82**, 1651–1655.

Sturm, R. A. (2006). A golden age of human pigmentation genetics. *Trends in Genetics*, **22**, 464–468.

Sturm, R. A., Teasdale, R. D. and Fox, N. F. (2001). Human pigmentation genes: identification, structure and consequences of polymorphic variation. *Gene*, **277**, 49–62.

Sturm, R. A., Duffy, D. L., Box, N. F., et al. (2003). Genetic association and cellular function of *MC1R* variant alleles in human pigmentation. *Annals of the New York Academy of Sciences*, **994**, 348–358.

Suh, J. R., Herbig, A. K. and Stover, P. J. (2001). New perspectives on folate catabolism. *Annual Review of Nutrition*, **21**, 255–282.

Sulaimon, S. S. and Kitchell, B. E. (2003). The biology of melanocytes. *Veterinary Dermatology*, **14**, 57–65.

Szabo, G., Gerald, A. B., Pathak, M. A., et al. (1969). Racial differences in the fate of melanosomes in human epidermis. *Nature*, **222**, 1081–1082.

Tadokoro, T., Kobayashi, N., Zmudzka, B. Z., et al. (2003). UV-induced DNA damage and melanin content in human skin differing in racial/ethnic origin. *FASEB Journal*, **17**, 1177–1179.

Tadokoro, T., Yamaguchi, Y., Batzer, J., et al. (2005). Mechanisms of skin tanning in different racial/ethnic groups in response to ultraviolet radiation. *Journal of Investigative Dermatology*, **124**, 1326–1332.

Tamura, T. and Halsted, C. H. (1983). Folate turnover in chronically alcoholic monkeys. *Journal of Laboratory and Clinical Medicine*, **101**(4), 623–628.

Tasa, G. L., Murray, C. J. and Boughton, J. M. (1985). Reflectometer reports on human pigmentation. *Current Anthropology*, **26**(4), 511–512.

Thody, A. J., Higgins, E. M., Wakamatsu, K., et al. (1991). Pheomelanin as well as eumelanin is present in human epidermis. *Journal of Investigative Dermatology*, **97**, 340–344.

Thong, H. Y., Jee, S. H., Sun, C. C., et al. (2003). The patterns of melanosome distribution in keratinocytes of human skin as one determining factor of skin colour. *British Journal of Dermatology*, **149**, 498–505.

Tran, T. N. T., Schulman, J. and Fisher, D. E. (2008). UV and pigmentation: molecular mechanisms and social controversies. *Pigment Cell and Melanoma Research*, **21**, 509–516.

Van der Meer, I. M., Boeke, A. J. P., Lips, P., et al. (2008). Fatty fish and supplements are the greatest modifiable contributors to the serum 25-hydroxyvitamin D concentration in a multiethnic population. *Clinical Endocrinology*, **68**, 466–472.

Van Nieuwpoort, F., Smit, N. P. M., Kolb, R., et al. (2004). Tyrosine-induced melanogenesis shows differences in morphologic and melanogenic preferences of melanosomes from light and dark skin types. *Journal of Investigative Dermatology*, **122**, 1251–1255.

Veierod, M. B., Weiderpass, E., Thorn, M., et al. (2003). A prospective study of pigmentation, sun exposure, and risk of cutaneous malignant melanoma in women. *Journal of the National Cancer Institute*, **95**(20), 1530–1538.

Vieth, R. (1999). Vitamin D supplementation, 25-hydroxyvitamin D concentrations, and safety. *American Journal of Clinical Nutrition*, **69**, 842–856.

Vieth, R. (2003). Effects of vitamin D on bone and natural selection of skin color: how much vitamin D nutrition are we talking about? In *Bone Loss and Osteoporosis: an Anthropological Perspective*, S. C. Agarwal and S. D. Stout (eds). New York: Kluwer Academic/Plenum Press, pp. 135–150.

Von Luschan, F. (1897). *Beitrage zur Volkerkunde der Deutschen Schutzgebiete*. Berlin: D. Reimer.

Vorobey, P., Steindal, A. E., Off, M. K., et al. (2006). Influence of human serum albumin on photodegradation of folic acid in solution. *Photochemistry and Photobiology*, **82**(3), 817–822.

Walsberg, G. E. (1988). Consequences of skin color and fur properties for solar heat gain and ultraviolet irradiance in two mammals. *Journal of Comparative Physiology B: Biochemical, Systemic, and Environmental Physiology*, **158**, 213–221.

Wassermann, H. P. (1965). Human pigmentation and environmental adaptation. *Archives of Environmental Health*, **11**, 691–694.

Wassermann, H. P. (1974). *Ethnic Pigmentation*. New York: Elsevier Publishing.

Webb, A. R. (2006). Who, what, where and when: influences on cutaneous vitamin D synthesis. *Progress in Biophysics and Molecular Biology*, **92**, 17–25.

Webb, A. R. and Holick, M. F. (1988). The role of sunlight in the cutaneous production of vitamin D_3. *Annual Review of Nutrition*, **8**, 375–399.

Webb, A. R., Decosta, B. R. and Holick, M. F. (1989). Sunlight regulates the cutaneous production of vitamin D_3 by causing its photodegradation. *Journal of Clinical Endocrinology and Metabolism*, **68**, 882–887.

Wheeler, P. E. (1985). The loss of functional body hair in man: the influence of thermal environment, body form and bipedality. *Journal of Human Evolution*, **14**(1), 23–28.

Williams, L. J., Rasmussen, S. A., Flores, A., et al. (2005). Decline in the prevalence of spina bifida and anencephaly by race/ethnicity: 1995–2002. *Pediatrics*, **116**(3), 580–586.

Wolff, A. E., Jones, A. N. and Hansen, K. E. (2008). Vitamin D and musculoskeletal health. *Nature Clinical Practice Rheumatology*, **4**, 580–588.

World Health Organization (2002). Ultraviolet radiation: global solar UV index. http://www.who.int/mediacentre/factsheets/fs271/en/ (accessed 2008).

Young, A. R. (2006). Acute effects of UVR on human eyes and skin. *Progress in Biophysics and Molecular Biology*, **92**, 80–85.

Zareba, M., Szewczyk, G., Sarna, T., et al. (2006). Effects of photodegradation on the physical and antioxidant properties of melanosomes isolated from retinal pigment epithelium. *Photochemistry and Photobiology*, **82**, 1024–1029.

13 Classic Markers of Human Variation

Robert J. Meier

INTRODUCTION

The door to the study of genetically based variation in humans cracked open at the beginning of the twentieth century with the discovery by Landsteiner (1901) of ABO blood group substances whose Mendelian mode of inheritance was later established by Bernstein (1924). Needless to say, an early trickling of discoveries has led to a flood over the past century of new and important revelations concerning the nature and significance of human genetic variation. This chapter will cover approximately three-fourths of that history as it unfolded via the discovery and elucidation of a host of markers. A helpful review of genetic markers known as of the early 1970s and their role in the study of human evolution can be found in Crawford (1973). Another recent review of classic markers and their contribution toward understanding North American Native genetic variation appeared in O'Rourke (2006). Two current textbooks that provide substantial coverage of traditional markers, along with DNA markers and other topics relevant to human biological variation, are Mielke et al. (2006) and Molnar (2006).

Over the years markers have come to mean fairly consistently defined hereditary units. An early reference to the use of the term is found in Race and Sanger (1962) who discuss markers, in their case human blood groups, as characters that help to locate genes on chromosomes. This was, of course, early in any attempts to construct physical gene maps. Sometime later, genetic markers were associated with specific polymorphic loci that defined particular segments of chromosomes (Cavalli-Sforza et al., 1994). In essence, genetic markers today signify genes or other identifiable segments of DNA whose inheritance can be consistently documented and mapped.

For purposes of this chapter, classical markers will refer to those historically researched phenotypically observable variants that for the most part preceded DNA methodologies that presently deal with nucleotide sequences in various representations, such as mitochondrial DNA (mtDNA), variable number tandem repeats (VNTRs), single nucleotide polymorphisms (SNPs), and microsatellite polymorphisms. In short, these are DNA markers. Hence, the time frame for reviewing markers will encompass anthropological field and laboratory studies carried out beginning at the turn of the twentieth century roughly through the 1980s. This end point is just prior to the widespread use of polymerase chain reaction (PCR) and the developing application of DNA markers to population studies. As a timetable guide, Roychoudhury and Nei (1988) were referred to for selecting markers from those that had been identified by the date of that publication. The major categories of human markers will include: blood group polymorphisms, serum protein and enzyme variants, and an open-ended set of markers that are of microevolutionary and anthropological interest. Polymorphism is used here to signify two or more alleles at a given locus each exceeding 1% frequency.

Tables follow that list these traditional markers. It should be noted that marker identification pertains to an earlier time period and may not conform to presently known markers and their allele or haplotype (a combination of alleles at multiple linked loci) labels. However, chromosome locations for the markers are current and were extracted from the National Center for Biotechnology Information (NCBI) website.

Some sampling of the total number of classical markers was done so as to not overwhelm the chapter. The selection process was based on how much information was available for the marker, how widely it was studied across human populations, and how informative the marker was in bringing out important points pertaining to human variation. Several classic works are available for anyone interested in a more complete and in depth compilation of markers. These include; Race and Sanger (1962), Buettner-Janusch (1966), Giblett (1969), Mourant et al. (1976), Harris (1980), Mourant (1983). Tills et al. (1983), Livingstone (1985), Roychoudhury and Nei (1988), and Cavalli-Sforza et al. (1994).

Human Evolutionary Biology, ed. Michael P. Muehlenbein. Published by Cambridge University Press. © Cambridge University Press 2010.

The organization for describing classical markers centers on five main topics: basics of marker identification and their expressed phenotypic physiological function, historical application of markers in classifying human populations and races, application of markers to population studies and microevolutionary processes, markers and their relationships with diseases, and contemporary use and future prospects for classic markers.

IDENTIFYING RED BLOOD CELL MARKERS

An early method for detecting variants of hereditary expression, beyond that of parental selection and crossing as devised by Mendel, were serological reactions. Discovery of blood group markers arose through unintended consequences of transfusions, involving donor–recipient mismatches that led to adverse clinical outcomes for patients. Working off these unfortunate medical mishaps, Landsteiner's discovery in 1900 of the ABO system was based on laboratory tests of a person's blood cells against serum from a different person. Particular combinations of cells and sera produced visible agglutination reactions between red blood cell surface antigens and corresponding bivalent antibodies found in the serum. Following extensive cross matches of this sort, Landsteiner was able to identify persons who had either A or B antigens, both A and B, or neither. The success of his work partly depended on the fact that the ABO system has naturally occurring antibodies that form shortly after birth. This then, was the launch into a succession of subsequent discoveries of both improved medical procedures requiring blood transfusion, and in identifying additional blood group marker systems.

Table 13.1 provides a listing of antigens, their chromosome location and their marker notations, and year of discovery. Some of these markers, for example in the Rh system, were originally observed in a manner similar to that noted above, that is, due to adverse transfusion reactions. Others were found through deliberate laboratory procedures of injecting human blood cells into animals (often rabbits) and extracting any produced antiserum or agglutinin, which then could be tested against humans for positive or negative reactions. An example of this is anti-N in the MN system. Interestingly, plant extracts (lectins) also were found to differentially react with receptor sites of human blood cell antigens. Also in the MN system, anti-N reagent was extracted from *Vicea graminea*, a legume. In the case of the ABO system, anti-A_1 was made from *Dolichos biflorus*, and while *Ulex europaeus* (also a legume in the gorse bush family) differentiated A_2 from A_1 and also was used to establish that persons who lacked both A and B antigens, that

TABLE 13.1. Blood cell and secretion markers.

Locus	Chromosome	Markers	Discovered
ABO	9q34.1–q34.2	A_1,A_2,B,O	1900/30
Secretor	19p13.1–p13.11	Se,se	1932
Lewis	19p13.3	Le^a,Le^b,le	1946/54
Rh	1p36.11	*D,C,c,E,e	1940
MN/Ss	4q28–q31	MS,Ms,NS, Ns	1927/47
P	22q11.2–q13.2	P,p	1927
Luth./Auber.	19q13.2	Lu^a,Lu^b	1945/61
Kell/Sutter	7q33	K,k	1946/58
Duffy	1q21–q22	Fy^a,Fy^b,Fy^0	1950
Kidd	18q11–q12	Jk^a,Jk^b	1951
Diego	17q21–q22	Di^a,Di^b	1955
Xg	Xp22.23	Xg^a,Xg	1962
Hemoglobin	11p15.5 (β)	A,S,C,E	1947/49
HLA	6p21.3	*A,B,C,DQ, DR	1962/64/67

Note: *These closely linked marker loci have multiple alleles and haplotypes.

is the O phenotype, did in fact possess an antigen called H. This finding provided an alternative name for the system, ABH. The H antigen is in fact present in virtually all ABO phenotypes, in decreasing amount from O type to AB. One exception to this is the Bombay phenotype O that lacks red blood cell H antigen, and carries a corresponding anti-H, as well as anti-A and anti-B in the serum.

SOLUBLE ANTIGENS

In addition to blood cell surfaces, antigenic markers also can appear in water soluble form throughout body fluids, with particular reference here to saliva. Presence of salivary antigens in the ABH system were detected with the inhibition test, where known antisera were mixed with the tested saliva, and then checked against known red cell antigens to see which antisera had already reacted with or had been inhibited, and thus revealed the identity of the antigen.

ABH secretion (now designated as *FUT1*) of soluble antigens turned out to be just the first system that was later found to be among a linkage group on chromosome 19 that also included the Secretor locus (or *FUT2*) itself along with Lewis (or *FUT3*), Lutheran, and Auberger antigenic markers. Of historical significance, autosomal linkage of Secretor and Lutheran was the first of its kind to be shown (Mohr, 1951). Another linkage to note here is that between Kell and Sutter blood groups on chromosome 7. There has also been evidence that Kell is linked with the PTC trait, a marker to be discussed later. Linkage detail on Kell/Sutter and the secretor loci noted above was incomplete or not known at the time of major marker compilations, such as Roychoudhury and Nei (1988). Linkage between MN and Ss was known, and allele combinations of these two systems, which undergo little recombination, should probably be treated as haplotypes. At the molecular level of the red blood cell membrane, it is now established that the M and N antigens are bound to glycophorin A (GPA) while the S and s antigens are carried by glycophorin B (GPB). The final red blood antigen system to mention here is Xg, obviously so-named because it is located on the X chromosome. It was discovered through conventional serological methods in 1962. That year marks the end of the initial period of discovery of red blood cell polymorphisms, at least those that figured most prominently in anthropological field studies.

Modes of inheritance for red blood cell and secretor groups can be codominant (as in the MN group), dominant-recessive (as in the Rhesus group for the D antigen), a combination (as in the ABO group), or sex-linked (as in the Xg group). For some groups, detection of heterozygote phenotypes depends upon the specificity of the serological reagents. Molecular methods now make many of the earlier dominant-recessive designations obsolete or incomplete.

For a reasonably up-to-date compilation of red blood cell markers, dealing with those covered here and many more as well, the reader is referred to *The Blood Group Antigen FactsBook* (Reid and Lomas-Francis, 1997), wherein you will find descriptions and displays of the molecular basis of the markers along with additional categories of information befitting a complete reference source.

HEMOGLOBINS

Mode of inheritance: Two autosomal loci, segregating multiple codominant alleles, but depending upon which of the pleiotropic phenotypic expressions are considered, there can also be dominant and recessive conditions.

Table 13.1 also lists hemoglobins, which make up about 85% of the protein structure of red blood cells. Considering their early and continuing significance in microevolutionary studies, they could command a separate table. Electrophoresis was used in identifying hemoglobin variants. A primary function of hemoglobin is to bind oxygen molecules while blood has infused the lungs and transport this oxygen throughout the circulatory system where it is then released during metabolic activity. A very large number of hemoglobin variants have been found (Livingstone, 1985) but this review will focus on major variants found at the

β-chain locus on chromosome 11. This locus is of particular interest here due to its maintaining elevated marker frequencies in human groups that are at increased risk for contracting endemic malaria, a topic that will be covered later on.

HLA SYSTEMS

Mode of inheritance: Multiple autosomal loci and multiple codominant linkage groups or haplotypes.

Finally, Table 13.1 contains the HLA systems. Human leukocyte antigen (HLA) haplotypes are found on white blood cells and expressed at several closely linked loci on chromosome 6. In a broader context, HLA pertains to the major histocompatibility complex (MHC) as found throughout vertebrates. The MHC is of fundamental importance in defining an individual's immunological identity and consequently establishing a defense system against potential pathogens. The five HLA loci as listed in Table 13.1 contained markers for tracing human population relationships and for investigating associations with diseases. Three of these (HLA-A, -B, and -C) are tested through serological reactions, while the remaining two (HLA-DQ and -DR) are investigated through cytotoxic methods. Specific HLA haplotypes will be discussed later in the context of disease associations.

SERUM PROTEINS

The plasma or fluid portion of blood contains a large number of kinds of proteins, most of which were found to be polymorphic as well as variable among different human populations. Table 13.2 lists serum protein markers that will be reviewed here. The workhorse method for separating and identifying serum proteins was electrophoresis that utilized a variety of preparations, buffers, and media.

ALBUMINS

Mode of inheritance: Autosomal codominants with Al^A allele controlling the common albumin, and several variants, such as Al Naskapi, found in varying frequency in different populations.

Albumins make up about one-half of all serum proteins. Their genetic control is found on chromosome 4. One of their main functions is to bind and carry other serum constituents, such as fatty acids and steroids, and they also control fluid volume outside the cell. Albumin studies were regularly carried out by field researchers around the world and

TABLE 13.2. Serum protein, enzyme, and other markers.

Protein	Chromosome	Marker
Albumin	4q11–q13	Al^A,Al Naskapi and others
Gc	4q12–q13	Gc^1,Gc^2
Gm (IgG)	14.q32.33	Haplotypes
Inv (IgK)	2p12	Inv^1,Inv^2
Haptoglobin	16q22.1	Hp^1,Hp^2
Transferrin	3q22.1	Tf^C,Tf^B,Tf^D
Enzyme	**Chromosome**	**Marker**
Carbonic anhydrase	8q13–q22.1	CA II (B),CA II (C)
G6PD	Xq28	G6PD deficient
HEXA (Tay-Sachs)	15q23–q24	HEXA deficient
Lactase	2q21	Lactase persistent, deficient
PAH (PKU)	12q24.1	PAH deficient
Red cell acid phosphatase	2p25	P^a,P^b,P^c
Trait/condition	**Chromosome**	**Marker label**
Cerumen (ear wax)	16q12.1	Wet type, dry type
Cystic fibrosis (CF)	7q31.2	Affected
PTC tasting	7q34	Taster, nontaster

several variants were discovered. One of the first of these was Al Naskapi that was found in an Indian group in Quebec (Melartin and Blumberg, 1966). Al Naskapi was subsequently observed in other Canadian and US North American Indian samples, for example among the Dogrib Indians (Szathmáry et al., 1983).

GROUP SPECIFIC COMPONENT (GC)

Mode of inheritance: Two common autosomal codominant alleles.

Group specific component (Gc) is also found on chromosome 4 in close linkage with albumin. Its discovery was made by Hirschfeld (1959). Gc is well-understood to be a vitamin-D binding protein, with two common alleles, Gc^1 and Gc^2. From this function it might be expected there could be some interplay between Gc variants and the role of vitamin-D in blood cell formation, particularly in people subject to becoming anemic and in areas of reduced sunlight where human groups are at higher risk of rickets. With respect to the latter prediction, Gc^2 was thought to be more efficient in transporting vitamin D, and did show a higher frequency in some northern populations, but

with major exceptions, notably the Saami of Norway and Sweden (Roychoudhury and Nei, 1988).

IMMUNOGLOBULINS (GM AND INV)

Mode of inheritance: Multiple autosomal dominant/recessive and codominant alleles and linkage groups.

Immunoglobulins serve as the body's defense system by forming antibodies against foreign intruders such as bacteria and viruses. By the 1960s two types of globulins were identified, namely, Gm (IgG of the heavy chain of the antibody molecule) and Inv (IgK/Km of the light chain). Gm is located on chromosome 14, while Inv is mapped to chromosome 2. Marker variants and haplotypes segregating at these two loci were observed to differ by human population and region, probably as the immune responses were adaptively tailored to specific pathogenic threats. So while there are coding genes underlying Gm and Inv, their expression is mediated by environmental circumstances. Schanfield (1980) conducted a study of the anthropological usefulness of genetic markers in differentiating regional and continental populations and concluded that Gm haplotypes, along with HLA haplotypes and the Duffy blood group, were the leaders in carrying out this task when compared against a bank of red blood cell, serum protein and enzyme markers. Two essential components of usefulness were defined in terms of uniqueness of the marker and degree of polymorphism, and on both measures, Gm scored highly.

HAPTOGLOBINS

Mode of inheritance: Autosomal codominant alleles, Hp^1, Hp^2.

Haptoglobins (Hp) bind free hemoglobin (Hb) that is released from destroyed red blood cells. The Hp-Hb complex both prevents loss of hemoglobin from the body through excretion, and also apparently plays a role in reducing the risk of bacterial growth by hemoglobin (Eaton et al., 1982). Smithies (1955) was the first to demonstrate polymorphic variation in haptoglobins by using starch gel electrophoresis. Haptoglobin has since been mapped to the short arm of chromosome 16. As with other serum proteins, haptoglobin variants could be under selective forces that maintain polymorphic frequencies depending on environmental stressors. For example, Hp^1, which has a higher hemoglobin-binding capacity than Hp^2, generally reaches its highest frequency in tropically located African and Amazonian populations who face a high parasitic load and corresponding increased risk for anemia.

TRANSFERRIN

Mode of inheritance: Three autosomal codominant variants, Tf^C is common, and TF^B and Tf^D are rare.

Transferrin, as its name implies, is iron-binding protein that carries iron from the intestine and elsewhere, and delivers it to active tissues and dividing cells. As was the case for haptoglobins, Smithies (1958) discovered the polymorphic status of the transferrin locus, now mapped to chromosome 3. Could selection be maintaining the polymorphism? Transferrin variants might be implicated in persons or groups chronically stressed by iron-deficiency anemia or who are at high risk for red blood cell destruction. Also, transferrin may be involved with removing harmful allergens present in serum. While these are bases from which selection could operate, there was no clear evidence that this has been the case.

ENZYMES

Human variation in enzymes formed a vital area of research for anthropology/human biology, leading to field studies among non-Western populations in the 1960s. The most obvious interest, and of most clinical significance, were enzyme deficiencies commonly known as inborn errors of metabolism. Several traditional biochemical markers were identified, Table 13.2 lists a selected sample. A brief introduction to these is provided here, that will be followed later by a description of how these markers varied among different populations, and some discussion of possible bases for the variation. As in the case of serum proteins, electrophoresis was the then appropriate method of investigating enzyme variants in the 1960s. When first established as hereditary markers, enzymes were promoted as prime examples of the "one gene-one protein" notion that had to be modified after subsequent discoveries, as with the G6PD locus that has numerous variants all due to mutations of one structural gene.

CARBONIC ANHYDRASE

Mode of inheritance: Two linked loci on chromosome 8, CA I and CA II, each segregating dominant alleles, along with multiple recessive variants. Current status is that there now are at least 12 carbonic anhydrase loci, some linked and others on several different chromosomes.

Second only to hemoglobin, carbonic anhydrase forms a large portion of red blood cell protein. Its major function is to release carbon dioxide in the lungs in conjunction with the respiratory cycle. Carbonic anhydrases also play a role in bone resorption and calcification, and in maintaining an acid-base balance. Early population studies did not reveal very much variation except in Australia, with regard to CA I, and Africa, in terms of CA II, which showed polymorphisms (Roychoudhury and Nei, 1988). Given the many more recent discoveries of loci controlling the carbonic anhydrases, there is the potential of finding additionally interesting population variants.

GLUCOSE-6-PHOSPHATE DEHYDROGENASE (G6PD) DEFICIENCY

Mode of inheritance: Multiple codominant X-linked alleles.

Glucose-6-phosphate dehydrogenase (G6PD) is perhaps the most recognizable enzyme in anthropological study when it appears in one of several variants that result in reduced enzyme production or a deficiency. Its deficiency has received a high level of attention due to its interaction with malarial sensitivity and resistance, and hence, demonstrated increased frequencies in groups residing in endemic malarial regions. This topic will be discussed later. The normal functioning G6PD enzyme plays an important catalytic role in maintaining red blood cell membrane integrity. The enzyme is found throughout most of the body including skin and saliva. Its genetic control is located on the long arm of the X-chromosome.

HEXOSAMINIDASE A (HEXA) DEFICIENCY

Mode of inheritance: Multiple codominant autosomal alleles.

Hexosaminidase A (HEXA) is an example where an enzyme deficiency can have profound effects. The mutated *HEXA* gene causes a lethal condition known as Tay–Sachs disease. Persons having the classical form of Tay–Sachs disease experience developmental retardation and neurological degeneration in early infancy and in most cases die before reaching their third birthday. A normal functioning *HEXA* gene, located on chromosome 15, produces an enzyme that catalyzes the degradation of excess ganglioside (a constituent of cell membranes), whereas the mutated variant allows for the build up of ganglioside in neurons that causes the neurodegenerative disorder. Given the dire outcome for children with Tay–Sachs, it was surprising that the condition showed such a high frequency in Ashkenazi Jews of Eastern Europe. Homozygote recessives did not survive childhood so the variant marker would be expected to exist at a very low frequency. Initial thinking proposed that random drift had by chance elevated the mutant HEXA enzyme in the comparatively small and separated Jewish communities (Fraikor, 1977). Later, heterozygote advantage was invoked as a possible contributing explanation (Chakravarti and Chakraborty, 1978; Marks, 1995). It was argued that overcrowded urban ghettos posed severe risks for infectious diseases, for example, tuberculosis, but heterozygotes were somehow protected. A similar argument will be noted later with respect to cystic fibrosis. In a more recent report, the pendulum has swung back to explaining elevated *HEXA* gene frequencies as due to drift in the form of founder effect within a population experiencing rapid census growth (Frisch et al., 2004).

LACTASE DEFICIENCY

Mode of inheritance: Autosomal alleles with lactase persistence dominant over lactase deficiency.

Yet another example of an enzyme deficiency, but with comparatively low adverse consequences, involves lactase, a digestive enzyme of the milk sugar lactose. The lactase locus has been mapped to chromosome 2. Nearly all human babies produce sufficient amounts of this enzyme throughout their growing years, and then undergo a decline of enzyme output into maturity. Milk, and unfermented derived milk products, causes these adults to experience unpleasant digestive symptoms, including bloating and diarrhea. Yet adults in some parts of the world continue to produce higher amounts of lactase, and hence have none of the aforementioned symptoms. Population studies showed a strong association between cultures that had a long tradition of dairy farming and a persistence of lactase into adulthood. A genetic analysis based on family studies demonstrated that lactase deficiency was inherited as an autosomal recessive, meaning that heterozygotes and homozygotes possessing the dominant marker were lactose tolerant (Sahi, 1974). This was a classic example of a biocultural interaction. It also illustrated how environmentally dependent gene expression was, or that genetic predisposition required suitable conditions to become of significance to the organism.

PHENYLALANINE HYDROXLASE (PAH) DEFICIENCY

Mode of inheritance: Autosomal alleles, with PKU recessive.

Phenylalanine hydroxylase (PAH) is a catalytic enzyme that participates in the conversion of phenylalanine to tyrosine. A deficiency of PAH is an inborn error of metabolism that can lead to varying degrees of impaired mental functioning, and other pleiotropic effects, known as phenylketonuria (PKU). The mutated allele is located on chromosome 12. Phenylketonuria is

readily diagnosable and is routinely tested for as part of newborn screening, and is preventable through careful and consistent dietary management following a phenylalanine-free regimen at least through childhood. Like lactase deficiency noted above, PKU is another case of environmentally dependent or culturally mediated expression.

OTHER MARKERS

This section concludes with a description of variable human conditions or traits that were included occasionally in population studies (see Table 13.2). Of interest, investigations of phenylthiocarbamide (PTC) polymorphism were even extended to nonhuman primates. Methods of study were quite different ranging from visual and tactile examination for cerumen (ear wax) types, initially a host of clinical diagnosis and laboratory tests for cystic fibrosis that now include genetic analysis, and serial dilution or simple test paper strips for the PTC-tasting trait. They also show a range of consequences for the individual from being rather benign for cerumen types, to profoundly affecting the well-being of cystic fibrosis patients.

CERUMEN (EAR WAX) TYPES

Mode of inheritance: Allele for wet, sticky ear wax is autosomal dominant; dry ear wax is recessive.

Cerumen markers are expressed as wet (sticky, brown) and dry types (flakey, light colored) that are controlled by a locus on chromosome 16. There is human population distribution variability in these types along with implications of selection acting on ear wax type relative to climatic variables. The dry type is most often found in northern Asian populations, while the wet type is found in tropically located Asians, as well as in Africans and Europeans. Whatever adaptive significance there is for ear wax type polymorphism is yet to be determined.

CYSTIC FIBROSIS

Mode of inheritance: Multiple autosomal alleles, with CF recessive.

Cystic fibrosis (CF) is a debilitating condition that disrupts normal pancreatic, intestinal and respiratory functioning. After some intensive genetic research, the CF gene was mapped to chromosome 7 in 1985. Since affected individuals prior to more recent therapies were at high mortality risk as children and had reduced fertility as adults, it was puzzling why the condition had reached a high frequency in some European populations. A possible answer may be found in an association between the CF locus and risk for tuberculosis, paralleling heterozygote advantage explanations given for Tay–Sachs disease, and also for sickle cell anemia which will be discussed more fully later in the context of balancing selection and diseases.

PTC TASTING

Mode of inheritance: Taster allele is autosomal dominant; nontaster is recessive; variable expressivity in phenotypes.

The ability to taste the compound phenylthiocarbamide (PTC) is controlled by a major gene mapped to chromosome 7, with another locus likely involved as well. For an up-to-date confirmation of PTC chromosome mapping see Drayna et al. (2003) and for a complete historical review of this trait see Wooding (2006). Although a rigid bimodal distribution of tasters and nontasters is not observed, especially with applying the serial dilution procedure, there is a certain ease in collecting results, apparently so readily accessible that chimpanzees and rhesus monkeys became suitable subjects (Eaton and Gavan, 1965). Roychoudhury and Nei (1988) list nearly 80 human studies that had carried out PTC testing that virtually covered the world. Gene frequencies were highly variable both within and between continental samples, with no discernible patterns. There has been a suggestion of an interaction between dietary practices and thyroid function (Molnar, 2006). Phenylthiocarbamide, as a synthetic compound, serves as a proxy for a carbon-nitrogen-sulfur radical found in certain plant foods, particularly those of the cabbage family, that tasters perceive as bitter, and hence, to be avoided. This could be a protective behavior in that cabbage and its relatives may block the uptake of iodine, thereby reducing thyroid function, and resulting in depressed metabolism that in turn affected childhood growth and adult fertility. Conversely, nontasters have been shown to be more susceptible to developing nodular goiters, presumably due to a reaction of the thyroid gland to depressed amounts of iodine in the diet. Additional testing of the role selection and adaptation play in maintaining the PTC polymorphism seems warranted.

In concluding this section on basic marker identification, it should be pointed out that not all population studies utilized all of the markers described, or for that matter, had necessarily restricted their research to those that appear above. On the first point, research projects added markers as they were discovered and found to be anthropologically useful. As noted earlier, usefulness of markers was well investigated in

Schanfield and Fudenberg (1978) and Schanfield (1980), that dealt with the Gm and HLA systems and accompanying extensive tables of marker frequencies for world populations. Schanfield (1980) also notes a general problem in that certain markers could not be studied routinely because their reagents were not readily available, with particular reference to the HLA system.

Now that a set of classic markers has been introduced, the next section will offer a discussion of how these markers were applied in various contexts, the first being that of describing human biological diversity, including its most contentious application, that of classifying human races.

BLOOD GROUP MARKERS FOR CLASSIFYING HUMAN POPULATIONS

There is a long and tortuous history surrounding unsuccessful attempts to sort human populations into stable, mutually exclusive categories called races. Rather than extensively review that history here, the reader is referred to these works for that information (Montagu, 1964; Marks, 1995; Brace, 2005; Molnar, 2006). It is important, however, to trace the use of blood group markers as these became available to those choosing to carry out race classification. That story, as already mentioned, began with the discovery by Landsteiner in 1900 of the ABO blood group. A little more than a decade later, the ABO group was being studied by Ludwik Hirschfeld by conducting serological tests on thousands of persons, soldiers and civilians, from throughout Europe and even some from China, Japan, and Africa (Mourant, 1983). His results laid the groundwork for all subsequent studies showing serological distinctions across human populations, that is, the establishment of racial divisions. The premise applied was quite straightforward. First, accepting that the ABO blood group markers were inherited (which Hirschfeld helped to show), then frequencies of ABO blood group types (and later calculated gene frequencies) would indicate the degree of relationship between populations, the more similar they were the more closely they were related to each other, and vice versa. From there it was a matter of drawing lines between blocks of populations, a step that undoubtedly was greatly aided by geography and continental boundaries, and taken by Hirschfeld and his wife (Hirschfeld and Hirschfeld, 1919) in their defining of three ABO racial types, European, Asio-African and Intermediate (Marks, 1995). This was followed by other attempts at serological race classification (Ottenberg, 1925; Snyder, 1930; Wiener, 1948), but the effort that might have had a high potential for impacting anthropological thinking on races was that of Boyd. In his book, *Genetics and the Races of Man: an Introduction to Modern Physical Anthropology* (1950), Boyd set forth in highly explicit terms why he considered blood group markers to be more scientifically sound for racial classification than that any of the heretofore used methods utilizing morphological characters, including anthropometry.

When Boyd's work was published, blood group frequencies were available in large samples for the ABO, Rh, and MN systems. In addition, Boyd added PTC tasting and secretor status to his set of markers. His genetically defined races largely matched earlier classifications, particularly that of Wiener (1948). Not surprising then, Boyd's genetic races conformed closely with geography, a point that he seems to regard as confirmation of what he expected to find regarding human population descent histories and their patterns of separation and migration. His claims for the advantages of the genetic method over earlier classifications are that it is more simply done, completely objective, and that gene frequencies do not have the genetic uncertainty that is hidden in phenotypic traits, and gene frequencies provide quantitative rather than qualitative measures of population differences along with an assessment of admixture (Boyd, 1950). It should be noted that the erroneous claim of selective neutrality for blood group genes initially was accepted by Boyd (1950), except for maternal–fetal incompatibility in the Rh system, who then later abandoned it (Boyd, 1963a).

In this same year, Boyd (1963b) touted what he judged to be major accomplishments of the genetic method. He concluded that genetic methods had contributed to physical anthropology by: (a) confirming an Indian origin of Gypsies; (b) providing a quantitative assessment of white admixture in American Blacks; (c) establishing that Lapps were a distinctive European race; and (d) showing that Papuans of the New Guinea region were native to the South Pacific and had not migrated from Africa. With regard to one of these presumed feats there is recent caution expressed against the use of markers, sometimes single alleles, for calculating degree of admixture (O'Rourke, 2000).

By the time of Boyd's 1950 classification of serological races, it had already been reported (Boas, 1912; Shapiro and Hulse, 1940; Lasker, 1946) that head and body measurements were subject to modification in children of migrants who accommodated to new environmental conditions. Hence, this important finding would severely question the presumed stability of those variables, such as the cephalic index, that had been so heavily relied upon by race classifiers. However, by the end of the 1960s, race classification itself was on the wane, and genetic markers were not able to sustain efforts that sought to arbitrarily apportion human variation into discrete categories.

What helped to replace racial classification were attempts to discern the nature of human population relationships in terms of cultural historical and microevolutionary processes. An even more basic task was to be able to accurately analyze whatever biological differences existed between groups without any need to classify them. A study from Boyd's time period that illustrates this kind of endeavor was done by Sanghvi (1953). He included five endogamous Indian castes in an analysis of anthropometric versus genetic markers to discern their relationships. His list of markers, certainly short by subsequent standards, only consisted of ABO, MN, and Rh blood group phenotypes, taste reactions to PTC, and red-color-blindness. He concluded that either the genetic or morphological method could be more useful in reflecting biological relationships in certain cases, but more likely they will complement each other, and hence, both should be applied using many more measurements and markers than he did. We will see in the next section that this recommendation is indeed heeded within a decade with the launching of a number of major research projects.

Physical anthropology apparently was not so convinced of Boyd's approach not because it applied genetic markers, but because they were used to classify races. Two principal textbooks of roughly that time period perhaps best reflect the state of affairs. Montagu (1960) and Buettner-Janusch (1966) both are replete in their coverage of genetic markers, complete with tables of gene frequencies and allele distribution maps for the world. Beyond that they provided clear background information on the modes of inheritance and methods for identifying blood groups and serum proteins, and most importantly, what was then known about the selective basis of certain systems, such as the association of blood groups and diseases and the anthropological significance of hemoglobin variants at the sickle cell locus. Race classification utilizing genetic markers was seen as relatively unimportant and unproductive, in comparison with the study of selection and other microevolutionary processes that occurred within local populations. On a larger scale, research interest shifted to investigating how and when gene pools across and between continents came to differ from one another, again through microevolutionary processes. This state of affairs undoubtedly reflected the paradigm change that Washburn (1951) had proposed a decade or so earlier that the "new Physical Anthropology" should emphasize an understanding of function and process as opposed to an earlier focus on technique and description as a direct goal.

In opposition to race classification, a mid-twentieth century alternative was to view patterns of genetic variation expressed in terms of clinal distributions. Gene frequency clines joined the already recognized

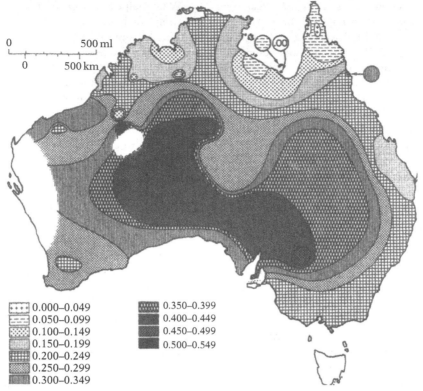

13.1. Frequency of the A_1 allele showing a clinal distribution in Australian groups. From Birdsell (1993). © 1993 Oxford University Press, reprinted with permission.

Legend:
- 0.000–0.049
- 0.050–0.099
- 0.100–0.149
- 0.150–0.199
- 0.200–0.249
- 0.250–0.299
- 0.300–0.349
- 0.350–0.399
- 0.400–0.449
- 0.450–0.499
- 0.500–0.549

gradients in human morphological variation with respect to body size and shape (Allen's and Bergmann's rules), and skin pigmentation (Golger's Rule). These so-called "ecogeographic" rules generally explained clinal variation in morphology as due to adaptive responses of populations residing in gradients of temperature, solar radiation, and other environmental conditions that occurred in latitudinal changes. It was reasoned that gene frequency clines, or genoclines, might also be the result of natural selection gradients, such as levels of disease stress, but could as well be explained by actions of gene flow, migration, and human mobility and settlement patterns. For here, a classic example of a genocline will be presented followed by a more recent application of genocline based on Australian data that had been collected decades earlier (Birdsell, 1993).

A textbook example of a genocline is the distribution of the *B* allele of the ABO system as its frequency was mapped from eastern Asia to the British Isles. Although it was not known why, the *B* allele had its maximum frequency in Asia at around 25% but then declined to less than 5% in much of Western Europe (Mourant, 1954). A likely explanation for the *B*-allele cline rested in historical migrations and invasions of peoples from Asia westward over the past couple of millennia. To be sure, the *B*-allele cline was not exactly

a smooth and steady transition across Eurasia. There were gaps in the big picture, and very likely if *B*-allele frequencies were filled in, a more detailed map composed of many local-level populations would show some breaks or even reversals of the general geographic trend. It is to be expected that there were historical episodes involving small founder groups that became isolated over sufficient amounts of time for genetic differentiation to have occurred. Lastly, gene flow and human population movements were not exclusively in a westerly direction.

For a more recently plotted example of a cline also from the ABO system, Figure 13.1 shows A_1-allele frequencies as isogenes (comparable to isotherm contour lines connecting points of equal temperature) for the Aboriginal Australian population (Birdsell, 1993). The overall range in frequency for A_1 is from a high of 0.53 near the center of the continent to a minimum of 0.03 at the coastal and northern island areas. As would be expected there are some steep declines along with more gradual gradients in the topographic display. Birdsell pointed out a generally recognized premise that single genes, such as the A_1 allele, more rapidly respond to microevolutionary processes than polygenic traits. Accordingly, in reviewing the history of human occupation of Australia, he attributed the gene frequency distribution shown in Figure 13.1 to be the product of settlement of small founding groups (that is, founder effect), successive major migration waves, and importantly, a population structure of local bands within a larger tribal population.

Clinal distributions of classic markers, such as those for the B and A_1 alleles just described, clearly demonstrated that race classification could not begin to capture the complexities and details of human population relationships and historical connections. In pursuit of that goal, the discussion now turns to population studies that set out to reconstruct history through an understanding of microevolution.

APPLICATION OF MARKERS TO POPULATION STUDIES AND MICROEVOLUTIONARY PROCESSES

Dynamic population study took precedence over static race labeling with the launching of a number of important human biology field research projects. One such effort was the Harvard Solomon Islands Project that was conceived by Albert Damon in the early 1960s (Friedlaender, 1987). This project served as a model of design for many more similar projects that were undertaken in roughly this time period. The Solomon Islands Project applied a multidisciplinary approach in which all four subfields of anthropology were represented, along with specialties from the biomedical sciences. It was reasoned that if population processes were to be adequately understood, it would be necessary to examine essential aspects of human behavior and decision-making. This meant that culture interacted with human biology, and in recognition of this connection a biocultural or biobehavioral approach was established. A clear illustration is to be found in population genetic measures of migration or gene flow, and even in selection and random drift. The strength of these processes very much depended upon human behavior and decision-making, such as cultural expressions in settlement patterns, mate choice and marriage customs, and culturally derived medical systems for diagnosis and treatment.

A major task within the sphere of population genetics in the Solomon Islands Project was to map biological variation among several groups on different islands with an aim to portray relationships of these groups in terms of microevolutionary processes, especially those pertaining to selection, random drift, and migration (Rhoads and Friedlaender, 1987). Among the markers included in that study were numerous blood polymorphisms, namely; eight red blood cell antigen systems, haptoglobins, transferrins, and Gm and Inv systems. Calculated allele frequencies from these markers were used in a distance analysis and other multivariate procedures that rendered

comparative findings for the Solomon Islands as well as with additional samples that had been obtained earlier from Bougainville (Friedlaender, 1975), and also samples from a broader Pacific Island context. The upshot of this aspect of the study that involved markers was a complex and not easily discernible pattern of genetic variation, but it seemed to indicate at each level from local groups to that of Pacific region and even beyond, that biological heterogeneity and variability extended deep into history and could not be explained simply by random drift of small, isolated groups.

Another aspect of the Solomon Islands Project, which also was carried out in many other research efforts in that time period including the earlier Bougainville study, was to incorporate genetic markers with multiple measures of distance as a test of correspondence between these measures for potentially realizing the same or similar outcomes of population relationships. In this regard, the Solomon Islands Project combined the distance measures of geography, language, anthropometry, odontometrics, dermatoglyphics, as well as the set of genetic markers noted earlier. One of the more enlightening results showed that genetic markers, along with anthropometrics and odontometrics, less closely corresponded with language and geography than did dermatoglyphic variation (Friedlaender, 1987).

Other studies have yielded varying results in these distance correspondence analyses. A brief review of this matter can be found in Meier (1980), who noted that incongruence between distance measures could be due to such factors as sample size and composition, number and kinds of markers used, and level at which the analysis is done, from local villages to large regions. For this discussion of genetic markers, it is perhaps best summarized with the appreciation that Mendelian traits could well be subjected to short-term and relatively rapid change in frequency via random drift and founder effect (particularly in small, semi-isolated groupings), but also undergo successive generational change due to selection processes. And hence, there is a great need to understand the nature and makeup of the sample upon which the marker frequencies are based, and to fully characterize samples even though most often there was little way to control sample makeup while conducting field studies. In the end, there remains considerable theoretical uncertainty whether the degree and rate of change in frequencies of markers are expected to correspond well with the other distance measures, such as anthropometric or language change. On this matter, Lewontin remarks in his Foreword to Friedlaender (1975) that linguistic distance at that time was too simply measured. However, Lewontin praised Friedlaender's work for its strong emphasis upon the historical perspective, that is, in reconstructing the action of evolution over time.

Several population studies in the late 1960s and into the 1970s paralleled portions of the Bougainville and Solomon Islands Project design, particularly for their application of the multidisciplinary, biocultural, and historical approaches. One set of such studies can be grouped under the International Biological Programme (IBP) Human Adaptability Projects. For a brief background, the IBP was composed of seven sections that directed a global effort toward measuring and understanding ecological productivity and its interaction with human welfare. One of these sections was that of Human Adaptability (HA) which got underway in the mid 1960s. Relevant to this discussion, methods for collecting specimens, such as blood from which markers could be determined, were presented in the IBP HA Handbook that first appeared in 1965 (Weiner and Lourie, 1969). This guide did not specify which markers were to be studied but rather set forth specifics of proven field methods for securing, storing, and transporting specimens so that they could be comparably analyzed, very often in a distantly located laboratory. A common problem was hemolysis during extended periods of travel, rupturing the red blood cell membrane and spilling out constituents that would have been used for serological testing. The IBP Handbook also detailed procedures for carrying out field testing of some markers, for example screening methods for G6PD and determining PTC taster status. A major concern that needed to be addressed was that of reliability of the serological results even when the specimens reached their destinations presumably intact. This matter had received some attention at the time.

Osborne (1958) had reported some major discrepancies for blood group testing when done at three well-established laboratories. Handling problems may have been an issue in another study documenting testing discrepancies (Livingstone et al., 1960). The least stable systems involved subtyping of A in the ABO system, and in the Duffy and P markers. Thus, it was imperative that blood specimens at the very least be handled with the utmost care to avoid degradation problems. One study that did a careful analysis of such problems was Neel et al. (1964) in which they had carried out field testing on blood specimens collected from the Xavante of Brazil, and then later retested them in their laboratory in Ann Arbor, Michigan. According to their full disclosure, discrepancies seemed to relate to different testing and laboratory conditions, and it was these problem areas that the IBP Handbook had hoped to rectify.

Under US IBP/HA auspices, multidisciplinary field studies that included a survey of genetic markers were carried out among human groups residing in Alaska (Inupiat Eskimos), Peru (Quechua), and Brazil (Yanomama and Makiritare), with the last cited having

TABLE 13.3. Selected population studies employing classical markers.

Study area/population	Year begun	Reference
Wales	Post-WWII	Harper and Sunderland (1986)
Australia	1952	Birdsell (1993)
Canada/Blackfeet Indians	1952	Chown and Lewis (1953)
Brazil/Xavante	1962	Neel et al. (1964)
South Africa/San, Herero, and others	1963	Jenkins et al. (1978)
Easter Island*	1964	Etcheverry (1967); Meier (1969)
Peruvian Andes/ Quechua	1965	Baker and Little (1976)
India/Gavdas	1966	Malhotra (1978)
Japan/Ainu	1966	Omoto (1978)
Bougainville	1966	Friedlaender (1975)
Solomon Islands	1966	Friedlaender (1987)
Alaska/Eskimos (Inupiat)*	1967	Jamison et al. (1978)
Southwestern United States/Papago	1967	Niswander et al. (1970)
Saharan Africa/Ideles	1968	Lefevre-Witier and Verges (1978)
Mexico/Tlaxcaltecan	1969	Crawford et al. (1974)
Central America/Black Caribs (Garifuna)	1975	Crawford (1984)
Canada/Dogrib Indians	1979	Szathmáry (1983)

Note: *Due to problems, complete serological testing could not be done on the Easter Island and Inupiat blood specimens. WWII, World War II.

the greatest emphasis on applying classic markers to population genetics questions (Neel and Ward, 1972). These along with a selection of additional field studies appear in Table 13.3.

Particular mention should be made here of a four-volume series published under the topic of *Anthropological Genetics* (Crawford and Workman, 1973; Mielke and Crawford, 1980; Crawford and Mielke, 1982; Crawford, 1984). (A fifth volume in this series – Crawford, 2007, presented an updating of the earlier volumes by focusing upon molecular genetics.) These works in general illustrated how useful classical markers were in population study, for example of the Black Caribs of Central America (Crawford, 1984). This volume contained several differently authored chapters devoted to marker description and frequency distributions and then went on to explore critical topics that employed these data in such matters as admixture estimates, fertility differentials (in the case of the sickle cell locus), and population structure.

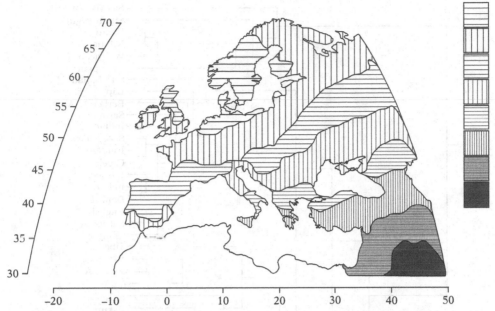

13.2. Synthetic map of Europe and western Asia based on first principal component (PC). The range between the maximum and minimum values of the PC has been divided into eight equal classes. From Cavalli-Sforza et al. (1994). © 1994 Princeton University Press, reprinted with permission.

It was these kinds of studies done on regionally demarked human groups for which genetic, biological, and cultural information could be combined that offer sharp insight to microevolutionary processes and population dynamics.

However, there was also the big picture to deal with, that is, the relationships of neighboring populations as well as those that were distant in both geography and in their historical connections. The work that epitomizes this effort was that of Cavalli-Sforza et al. (1994). If this tome can be described briefly, it is best depicted as a worldwide geography of human genes. As customary for physical geography, there are numerous maps that depict levels of gene frequencies for the major continents and Oceania. These are referred to as synthetic maps for their handling of an array of genetic markers by a multivariate procedure, namely, principal components (PC) analysis. An example of a synthetic map is found in Figure 13.2.

These maps then are interpreted in light of historical and microevolutionary processes whereby similarities and differences in PC values (seen as peaks and valleys on the maps) can represent migrational or selection patterns, sometimes according to gradients or clines, but possibly on a more local level show sharp breaks due to population isolation and random drift. Synthetic maps of this sort also were constructed from classic markers about a decade earlier for North American Native populations (Suarez et al., 1985) that assisted in sorting out population relationships and migration patterns.

A commonly applied procedure for depicting population relationships that was based on classic markers, and continues to be used with molecular data, was that of dendrograms or phylogenetic trees. An example of a dendrogram is shown in Figure 13.3. Various statistical methods were employed to generate graphically clear representations of genetic similarities or the opposite, genetic distance among populations. There generally was no unique solution in reconstructing trees; hence, multiple trees could lead to alternative interpretations. However, dendrograms, and also synthetic maps, could be viewed essentially as methods for reducing large data sets into manageable entities that might in fact partially answer questions concerning population affinities or perhaps even more importantly, point future research toward productive, new directions.

The final work to cover in this section on population study is a review that addressed the thorny question of peopling of the New World through a congruence of variables approach (Greenberg et al., 1986). They included linguistic, dental, and genetic lines of evidence in an attempt to reconstruct the timing and number of migrations. This work is cited because it stands at the transition between the use of classical markers and the then newly developing DNA technology, at that time devoted primarily to restriction fragment length polymorphisms (RFLPs). An extensive list of references can be found in the article. These are mostly dated from the late 1970s to the mid 1980s that include original study results for Native New World populations with respect to blood

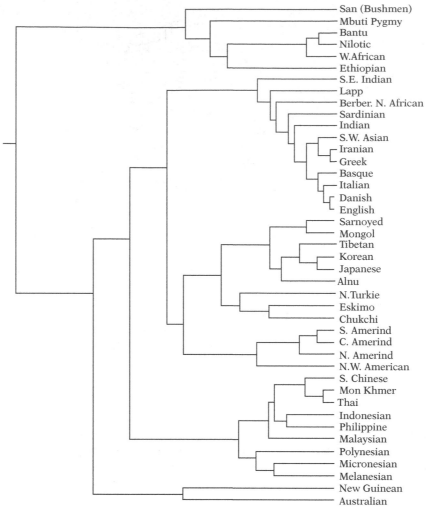

13.3. A dendrogram or phylogenetic tree based on 42 populations. From Cavalli-Sforza et al. (1994). © 1994 Princeton University Press, reprinted with permission.

group, serum protein, and enzyme polymorphisms. The review of these markers concluded that genetics could be complementary to the other two lines of evidence but could not stand alone in supporting a tripartite migration history of New World settlement. This work also provides a critical sense of how researchers viewed the claims of Greenberg et al. (1986) through multiple authored comments that directly followed the article.

GENETIC MARKERS AND SELECTION

This section will focus on the role that classical markers played in assessing natural selection as a major process for understanding genetic variation within and among human populations. To some degree this development was imbedded in the history of population genetics. During the 1930s Mendelian genetics interfaced with microevolutionary theory to form population genetics that at that time had a strong mathematical and theoretical emphasis. Later, by the 1940s, the modern synthesis of evolution was established with population genetics at its foundation.

While population genetics theory certainly received application to human groups over the next several decades as noted above in population studies, it achieved a substantial boost with respect to the academic realm in 1971 with the appearance of the initial edition of *The Genetics of Human Populations* (Cavalli-Sforza and Bodmer, 1999). Three chapters in this later edition are most relevant to note here for their in-depth treatment of classical markers with regard to maintaining polymorphisms (Chapter 4), possible adaptive relationships of blood group antigens and serum proteins to selective agents, particularly disease and incompatibility (Chapter 5), and marker frequency distributions across groups (parts of Chapter 11).

In this last cited chapter the authors continued to espouse a race concept, definitely not in the previous manifestations of racial typology or strict classification, but rather as an acknowledgment of biological differences and similarities they assumed to be based on race. It appears that their use of race is simply a matter of convenience for defining population units of study, which is to this day a vexing problem for anyone carrying out population studies. Who belongs within the sample and how will comparable study groups be defined? Island populations and semi-isolated villages offer fairly clearly demarked boundaries, and this may in fact underlie part of the attraction for the geographic selection in the field studies noted above.

In the manner in which Cavalli-Sforza and Bodmer (1999) had used race, it might have been more appropriate to have used a concept proposed by Montagu (1964), the "genogroup," which essentially defined population differences on the basis of gene frequencies. Then again, Montagu seemed to favor "ethnic group" as a substitute for race, which has gained fairly wide application. However, "ancestry" appears to be replacing all of these terms at the present time in the context of DNA markers. Aside from constructing tree diagrams to show partial conformity with prior race classifications, Cavalli-Sforza and Bodmer (1999) more importantly demonstrated how microevolutionary processes drove the course of population change. Their text, then, established a formal educational treatment from which to pursue interests in human population genetics. Needless to say, the genes upon which most of the principles rested were classical markers. And it is a consideration of these markers with respect to selection that is covered next.

ABO MARKERS AND DISEASE ASSOCIATIONS

As broader world surveys of the ABO blood-group distributions became available, it also became apparent that gene frequencies varied both within highly polymorphic loci as well as considerably across populations to a degree that could not be attributed to newly arising mutation nor to random drift or to gene flow, at least above local population levels. This, of course, pointed to selection as the active force in both changing and in maintaining gene frequency levels. There was ongoing debate as to whether ABO polymorphisms were in a stable state or transient and hence subject to eventual loss of alleles. The arguments for stability rested on the presence of the ABO blood group system in nonhuman primates – antigens analogous to A and B are present in Rhesus macaques (Duggleby, 1978), and the relatively narrow range of ABO gene frequencies within certain human groups. These lines of evidence indicated that ABO polymorphisms had a long history by virtue of their existence in nonhuman primates, and then varying

historical and environmental conditions for selection to operate on led to variation across different human groups. This section will cover the use of markers in trying to understand the nature of ABO polymorphisms with regard to selection.

The earliest work to implicate selection in the ABO distribution was faulty for its lack of statistical rigor. It simply amounted to collecting data showing associations between particular ABO blood types and a set of diseases, often using hospital patient records as the sampling source. However, flaws in the early research were in not having adequate control groups or unbiased samples that would allow any deviation from normal expectations to be properly ascertained. Once these study defects were corrected, many significant associations remained with respect to noninfectious diseases, but not nearly as clear as the results for infectious diseases, and these will be discussed first.

Armed with these findings, it was important to now offer possible selective mechanisms that would account for the blood group and disease associations, and that then would define directions for further investigation. Four categories of infectious disease had been identified as likely candidates for changing gene frequencies through natural selection (Vogel and Motulsky, 1997). These were: (1) acute infections, such as small pox, plague and cholera, that periodically spread as epidemics over large areas; (2) chronic infections that were highly contagious, such as tuberculosis and syphilis; (3) intestinal infections that afflicted all age groups but were likely fatal to infants and younger children; and (4) malarial infection.

These agents of selection would operate through differential mortality and fertility, although for some of them the former would be more likely in that many people would be stricken with the disease when they were past their prime reproductive years. In that situation, selection would not be acting directly on the reproductive success of the individual, but could have an impact on broader measures of fitness, e.g., inclusive fitness, due to the loss of support and resources provided by postreproductive members of the society.

Differential survival in connection with the ABO blood groups was presumed to be an increased susceptibility or resistance by virtue of certain markers a person possessed. Accordingly, it was found that individuals carrying the A antigen were more likely to contract smallpox (Vogel and Chakravarartti, 1966), while persons with the O blood type were at higher risk for cholera (Glass et al., 1985). It was presumed that these blood types were in fact more susceptible because of pathogen similarity to their own genetic makeup. Hence, there was a failure of their immune system to recognize the foreign invader, likely a bacterium or virus, and consequently did not initiate an appropriate defense response that then resulted in an

TABLE 13.4. Significant associations between ABO markers and noninfectious diseases.

Diagnosis	No. of series	No. of patients	No. of controls	Markers compared	Mean relative incidence*
Cancer, stomach	101	55 434	1 852 288	A:O	1.22
Cancer, colon	17	7 435	183 286	A:O	1.11
Cancer, pancreas	13	817	108 408	A:O	1.24
Duodenal ulcer	44	26 039	407 518	O:A	1.35
Gastric ulcer	41	22 052	448 354	O:A	1.17

Source: Data taken from Vogel and Motulsky (1997, p. 221).
*Relative incidence is determined by the ratio of, for example, A:O in patients compared with A:O in controls.

increased mortality risk. Highly relevant supporting evidence for this position came out of India for its comparatively high frequency of the *B* allele that could be explained by that country's long history of smallpox and cholera epidemics, that would render both the A and O markers at a selective disadvantage (Buchi, 1968). Conversely, over many generations persons carrying the B antigen had a proportionately higher survival rate that boosted the frequency of this marker.

Population and biomedical studies of these, and other blood group and infectious disease associations, have yielded varying and inconsistent results upholding the immunological hypothesis to explain ABO distributions. A troubling matter is that different disease associations have been found for the same locus, which raises a question of what the statistical associations actually demonstrate (Weiss, 1993). It does appear that selection acting in this manner through infectious disease probably does explain some of the worldwide marker frequency distributions. It has been reported that testing of the immunological hypothesis was discontinued sometime after the late 1970s (Vogel and Motulsky, 1997). Yet, there is at least one area of research along this line that remains very active. During the 1950s it was shown that persons with blood type O were more susceptible to having stomach ulcers (Table 13.4). It was subsequently discovered that nonsecretors of ABO substances were particularly vulnerable. This was followed by the highly significant finding that *Helicobacter pylori* was closely associated with stomach ulcers, and that the attachment of this bacillus (bacterial infection) to the gastric epithelium was mediated by blood group antigens. In particular, it was persons who carried the Lewis Le[b] antigen that appeared to be most likely infected and thereby experienced an ulceration process (Boren et al. 1993). However, a later study done in an outpatient clinic was not able to confirm this result (de Mattos et al., 2002), but it did convincingly match earlier findings of *H. pylori* infection predominantly in patients with O phenotypes. The pendulum continues to swing as a very recent report that studied monkeys concluded

that the ABO blood group and secretor status of individuals may in fact be part of an evolutionarily derived innate immunity against infectious diseases (Lindén et al., 2008).

When considering noninfectious disease associations with ABO, the list is much longer and statistically stronger but an understanding of the mechanism responsible was, and still is, virtually unknown. Table 13.4 provides a selected sampling of significant associations. These represent some of the largest samples of patients and controls from worldwide series, and undoubtedly established the statistical reality of ABO markers and disease associations. The selected set focuses on the digestive system, whereas a more thorough listing would also include malignant and nonmalignant conditions of the reproductive and vascular systems. For the gastrointestinal tract, initial speculation was that there were differential immune responses by persons with different ABO antigens, especially of the soluble form. This thinking would parallel the proposal noted above for ABO associations and infectious diseases. Testing of this hypothesis, however, is complicated due to the interplay in the gut between intestinal flora, highly variable dietary practices around the world, and ABO antigen specificities. It has been concluded that the contribution of ABO polymorphisms to the etiology of digestive ailments is quite small (Vogel and Motulsky, 1997), and further that, even though these markers cannot be considered neutral, the low level and uncertain direction of selection will not explain the maintenance of variation observed in the ABO distribution (Cavalli-Sforza and Bodmer, 1999).

INCOMPATIBILITY SELECTION AND BLOOD GROUP MARKERS

Another area of research interest that addressed the question of the persistence of blood group polymorphisms is that of incompatibility selection. Incompatibility refers to maternal/fetal situations where the mother would carry in her system antibodies against red blood cell antigens found in her developing fetus. For example in the ABO system, an O mother would be incompatible with a fetus having an AO or BO genotype. It is predicted under population genetics theory that ABO heterozygotes would be selected against, ultimately leading to a loss of genetic variability, unless there are counter selective forces that favor heterozygote survival. What these forces are for the ABO system is not yet known but probably do implicate the mother's immune system.

In the case of the Rh system, where incompatibility is prominent in some populations, a somewhat clearer picture has emerged. This discussion will be centered on the Dd locus, since it was this marker that directly caused transfusion problems that then led to the discovery of the system, and later was shown to be a cause of death of newborns in the situation of an Rh-negative mother (dd) whose infant was Rh-positive (Dd). The child of this incompatible Rh combination was at risk for having *erythroblastosis fetalis* or hemolytic disease of the newborn (HDN) due to Rh positive antibodies (the small 7S gamma G type) crossing the placenta and into the fetus.

Not all Rh-positive children born to Rh-negative mothers suffered from HDN. The explanation involves several mitigating circumstances. First of all, Rh antibodies are not naturally occurring, an Rh-negative mother has to be exposed to Rh antigens in order to stimulate her system to produce antibodies. This could have come through an improper blood transfusion, or much more likely, through carrying an incompatible fetus. Some of the fetus's red blood cells can enter the mother's system at the time of delivery, and thereby be present to stimulate her to later form antibodies. Obviously, an initial fetus is not affected because the mother has not yet been "immunized" or sensitized prior to the baby's birth. Subsequent incompatible conceptions increase the risk for HDN as the mother's antibody titer is raised. There appears to be some variation in women as to how many pregnancies are required to build antibody strength to the point of causing HDN. For some time now medical intervention has protected mothers from having HDN babies through the administration of a prepared antiserum that destroys any fetal blood cells that might have entered the mother's circulation during an incompatible pregnancy with the result that she is not sensitized. On rarer occasions, treatment has been in the form of exchange transfusions done either on the fetus or newborn. Prior to medical advancement severe HDN was often fatal, and obviously a source of much personal grief.

In the scientific realm a different kind of concern was brought out by the observation that the frequency of Rh-negative allele *d* was substantially higher than expected since all HDN children who were selected against were heterozygote *Dd*. This should mean that by losing *D* and *d* alleles equally, over time the rarer of the two eventually would be lost, and it was clear from many surveys the *d* allele was consistently at a lower frequency. On the other hand, surveys also showed that not only was the *d* allele not close to being selected out or anywhere near to a recurrent mutation rate, it seemed to be maintained at a much higher frequency than a simple selection against the heterozygote model would predict. There had to be circumstances that were countering selection.

Two of these circumstances were noted above in that the first-born children of Rh incompatibility are not affected and that later developing fetuses may not necessarily have HDN depending upon the state of sensitivity or antibody titer of the mother. Added to these is the segregation outcome of fathers who are either *DD* or *Dd*, and of course the latter have a 50% probability of contributing a *d* allele and thus a compatible Rh-negative child. There is also some evidence that parents who lost a child to HDN tended to overcompensate their loss by producing more homozygote *dd* children, thus increasing the *d* allele frequency. There is a final explanation for why the *d* allele is elevated beyond model prediction that interestingly involved mothers who had a double incompatibility with their fetus for both Rh and ABO systems.

Maternal/fetal incompatibility in the ABO system does result in HDN, but this occurs in a small percentage of overall potential circumstances. In another portion of the ABO incompatible cases, say with an O mother and an A fetus, as the baby's red blood cells enter her system they are quickly destroyed by her normally present anti-A antibody. If the blood cells also happen to carry the Rh-positive antigen and hence are Rh incompatible, then the cells are destroyed before there is time for the mother to build Rh antibody. In effect, double incompatibility serves to protect subsequent fetuses from HDN, and, of course, it would help to maintain a higher *d* allele frequency.

HLA AND DISEASE ASSOCIATIONS

From the early 1960s onward there has been increasing anthropological interest in the application of the HLA polymorphisms to investigating population relationships and to the study of microevolutionary processes, particularly selection. The HLA system has turned out to be highly useful in both regards. Certainly it is the most polymorphic system, and discovery of new alleles is still occurring. In Roychoudhury and Nei (1988) population data were compiled for 5 HLA loci and a total of 89 alleles. Updating of HLA polymorphisms can be found at an online website – *IMGT/HLA Database* – that currently reports nearly 2300 alleles for what is now known as Class I HLA alleles. Even going back to the late 1980s, allelic variation was high both within and between sampled regions. For example, the *B7* allele ranged from 0.066 to 0.144 in several European countries, 0.031 to 0.060 in Asian populations, and an African sample yielded a frequency of 0.115 (Roychoudhury and Nei, 1988). Much of the variation in HLA polymorphisms can be ascribed to populations that had undergone random drift and to population movements and resultant gene flow. With regard to selection, an early and striking association was found between *B27* allele and ankylosing spondylitis (AS) in a British study. It is now known that a vast majority of persons who have AS (which causes inflammation of the spine and other arthritic symptoms) possess the *B27* marker. However, the marker frequency varies across ethnic groups, and indeed AS can also occur even when the *B27* marker is absent. Ankylosing spondylitis is part of a group of autoimmune diseases, where the body's immune system fails and then makes antibodies against itself.

The *B27* marker, along with numerous other HLA alleles, probably predispose their hosts to autoimmune reactions as well as to infectious diseases at times as a single allele but often in combination with other alleles in the form of haplotypes. HLA haplotypes provide a clear example of linkage disequilibrium due to selection. In one study, haplotype A1 B8 DR$_3$ was present in nearly all hemophilic patients who showed a rapid course of developing AIDS after they were inadvertently treated with contaminated blood (Steel et al., 1988). Later research with HLA polymorphisms continued to find important marker associations with both susceptibility to and protection against disease, as will be referenced in the final section of the chapter.

SICKLE CELL LOCUS AND MALARIAL RESISTANCE

The marker that probably garnered the greatest amount of attention from anthropologists in the 1960s was a hemoglobin variant, Hb^S, the sickle cell allele. Hb^S is a point mutation that results from a single nucleotide substitution on the β-chain from the normal adult hemoglobin structure coded by the Hb^A allele. This mutation leads to one amino acid change in the hemoglobin molecule. There are a number of such variants classed as hemoglobinopathies, and two of these, Hb^C and Hb^E, will be discussed later.

Persons who are homozygote Hb^S/Hb^S episodically manifest sicklemia or sickle cell anemia due to multiple cascading effects of crescent-shaped red blood cells that are prone to hemolysis and also prevent blood from freely flowing through capillary beds. In the absence of medical attention, severe attacks are generally fatal.

Given this unfortunate outcome, it would be predicted that the Hb^S marker should decline in frequency ultimately to the level of a recurrent mutation rate at the locus. However, in endemic malarial regions of Africa it was proposed that there were counter-selection forces that were helping to maintain the Hb^S allele frequency in the population when it was combined with the normal Hb^A allele. Balancing selection operated against both classes of homozygotes, the Hb^S/Hb^S (from sickle cell anemia) and Hb^A/Hb^A (from malaria), and there was selection for the heterozygote Hb^A/Hb^S (protection from malarial morbidity).

The malarial parasite *Plasmodium falciparum*, carried by *Anopheles* mosquitoes, does not find a hospitable cellular environment in the heterozygote to complete its normal life cycle. This heterozygote advantage (also called overdominance) meant that both Hb^A and Hb^S alleles were being maintained at the same time that they were being selected out, a process that could lead to an equilibrium state, or balanced polymorphism. The alleles would be equal to each other in frequency only if selection against each class of homozygotes was at the same level, which it is not. Selection is much more severe against Hb^S/Hb^S than it is against Hb^A/Hb^A, so an equilibrium gene frequency would occur with the Hb^A allele proportionately much higher. This can be shown through the aid of some basic population genetics using the concept of relative fitness.

Relative reproductive fitness (w) is calculated with respect to specified genotypes Hb^A/Hb^A (AA), Hb^A/Hb^S (AS), and Hb^S/Hb^S (SS) and their expected frequencies, if there was Hardy–Weinberg Equilibrium at the hemoglobin locus. For a quick review, Hardy–Weinberg Equilibrium of genotypic frequencies depends on there being random mating at the locus in question and there being no evolution occurring at this locus. A further theoretical condition is that the population being studied be infinitely large. In spite of this last never-to-be-realized requirement, the Hardy–Weinberg Equilibrium was regularly found at a number of blood group loci, but not for the sickle cell locus due mainly to the action of selection.

Relative fitness for sickle cell-locus genotypes was calculated by Allison (1956) who derived his data from a number of African populations. He set the heterozygote fitness at $w_{AS} = 1.00$ (which would make it relatively the most fit), and found that fitness for one homozygote was approximately $w_{AA} = 0.80$ and for the other it was $w_{SS} = 0.20$. These values would translate into a 20% reduction in fertility for Hb^A/Hb^A parents and an 80% drop in fertility for Hb^S/Hb^S parents. Again through an application of population genetics calculations, over time it would be expected that the frequency of the Hb^S allele would decrease until it reached an equilibrium of about 0.20. Since this frequency was found in some African populations, it provided evidence for a balanced polymorphism.

However, some human population geneticists continued to question whether the sickle cell locus was in a balanced or transient state using the available evidence on differential fertility and mortality, along with computer simulation models to help predict the timing of equilibrium gene frequencies. In some African populations, Hb^S frequency appeared to be stable, as noted above, but in other regions it was subject to change (Livingstone, 1989). Complicating the picture was an interaction between Hb^S and other hemoglobin variants, especially Hb^C in West Africa, where one or the other marker was possibly being replaced. Added to this, the severity of malaria disease changed by region and altitude, and in fact very much depended upon the kind and intensity of certain horticultural practices.

Livingstone (1958) and other researchers proposed that as subsistence patterns of some African cultures shifted toward the clearing of once forested land for farming activities several thousand years ago, this established a more suitable environment for mosquito populations to flourish. As a consequence, the presence of the malarial parasite was promoted and malaria became a major disease stressor. This would set up the next step in the malaria hypothesis that involved the sickle cell locus, where probably the Hb^S allele was already present at a low frequency, and the malarial environment benefited the relatively better adapted heterozygote due to its resistance to the disease. It would be expected that this process was not strictly linear in the sense that selection for the heterozygote led to a lasting balanced polymorphism, but rather that as local populations adapted and increased in number, they also expanded in area by clearing more land and opening up broader opportunity for malarial disease, and another cycle of selection for the heterozygote. The dynamics between genetics and behavior in the sickle cell case can be considered a hallmark of biocultural interaction that is so important in understanding human population history.

The geographic distribution of the Hb^S marker, beyond Africa and into India and South Asia, has raised questions concerning the number of origins for the mutation. While Livingstone (1989) had argued for a single origin, others have proposed multiple mutations (Labie et al., 1986). Marks (1995) summarizes evidence showing that the sickle cell allele is found in five haplotypes corresponding to four African areas and one Indian source, that could be interpreted as independent origins or possibly also a single mutation that underwent successive later mutational and crossing-over events. A more complete resolution of this matter may help to answer the question as to whether or not the Hb^S marker had a single African origin and subsequently spread by gene flow beyond the continent

and into Asia and Europe. On a related research front, Volkman et al. (2001) have proposed that *P. falciparum* has a recent origin from a single common ancestor.

MALARIAL RESISTANCE FROM OTHER MARKERS

By the late 1980s, Hb^S was the hemoglobin marker for which most population data were available. Limited results for Hb^C and Hb^E were beginning to show a geographic distribution coinciding with malaria that suggested that these too were adaptive in heterozygotes in providing resistance to this disease. As noted earlier, Hb^C may have been replacing Hb^S in West Africa, while Hb^E showed a high frequency in southeastern Asia, where a hilly, forested habitat fostered a different mosquito vector of malaria. Southeast Asia presented its own set of intriguing research questions with respect to genetic protection from malaria. In this region, another kind of faulty hemoglobin, referred to as thalassemias, appeared to be interacting with Hb^E.

For some background, thalassemias are the result of point mutations, usually nucleotide substitutions or small deletions in regulatory genes that interfere with the normal synthesis of hemoglobin. They can occur on both the α- and β-globin chains. Clinical significance relates to the degree of hemoglobin reduction, and ranges from mildly affected to a fatal condition called *hydrops fetalis*. As in the case of hemoglobinopathies discussed above, heterozygotes for thalassemia offer protection against malaria, and once again due to balancing selection, thalassemia markers are maintained, but not necessarily balanced, polymorphisms.

Elevated frequencies for these markers were initially found in Mediterranean populations, particularly in Greece, Italy, and nearby islands where malaria is endemic. Later, polymorphic frequencies were recorded in New Guinea and Africa, as well as southeastern Asia, where the co-occurrence with Hb^E was noted above. A population genetics survey of this last region showed that after comparing gene frequencies for two populations, one had adapted via the thalassemia marker while the other gained malarial resistance from the Hb^E variant. Hence, population history would account for the different adaptive route (Vogel and Motulsky, 1997, p. 533). Additionally, the analysis went on to reveal that since the homozygote Hb^E/Hb^E was less deleterious than the homozygote thalassemia genotype, thalassemia was being replaced by Hb^E as the more effective and less costly balancing selection process.

Thalassemia was also found to co-occur geographically with an enzyme deficiency of G6PD, and it was presumed this was due to their both offering genetic resistance to malaria. The biochemical explanation for why G6PD deficiency was protective was proved by

Friedman and Trager (1981) who showed that the malarial falciparum parasite did not survive in G6PD-deficient cells due to a lack of potassium. The island of Sardinia provided a test of the malarial selection hypothesis for both thalassemia and G6PD deficiency. Frequencies for both markers closely correlated with altitude which largely determined the malarial parasitic load. Correspondingly, the lower the altitude the greater the selection pressure at the G6PD locus.

The remaining example pertaining to the malarial selection hypothesis involves the Duffy blood group system. A laboratory was the venue for testing the susceptibility of red blood cells of the different Duffy phenotypes to infection by the malaria parasite *Plasmodium vivax*, a benign form and probable predecessor to the malignant derivative *P. falciparum* (Mourant, 1983). It was found that phenotypes carrying at least one copy of Fy^a or Fy^b were readily infected whereas the Duffy negative Fy^0Fy^0 (the null Fy^0 allele also has been designated as Fy', Fy, and Fy^4) was highly resistant to infection. Based on these findings it was theorized that the Fy^0 mutation arose from either of the common alleles, and by chance increased in frequency enough to form a few homozygotes that were then at a selective advantage when *P. vivax* malaria became endemic. Livingstone (1984) countered this portion of the theory by concluding that the Duffy null allele had reached a sufficiently high enough frequency to prevent vivax malaria from becoming endemic in West Africa.

The theory continues that over time the Fy^0 marker progressively replaced Fy^a and Fy^b, and even reached fixation in some African populations, as observed today. The next phase was to theorize that tropical Africa was free of malaria until *P. falciparum* appeared and set off a new direction of selection operating on hemoglobin variants, including the sickle cell allele.

It is unlikely that historical details of this theory can ever be subjected to direct testing. However, considering the several examples cited above that overwhelmingly confirm the malarial selection hypothesis, it can be expected that portions of it will hold up to further scrutiny. Indeed, the selection hypothesis was favored more recently by Hamblin and Di Rienzo (2000) in their investigation of DNA sequence variation. In general, the early application of markers to a testing of crucial population genetic questions, especially those surrounding an understanding of how and why certain polymorphisms are maintained, must be viewed as highly successful.

ADDITIONAL AND PRACTICAL APPLICATIONS OF CLASSICAL MARKERS

Brief mention will be made here with respect to additional applications of markers. These are:

parentage questions and paternity exclusion, zygosity determination in twins, and bone and mummified tissue typing.

Race and Sanger (1962) devoted chapters to "Blood Groups" and "Problems of Parentage and Identity" and to "The Blood Groups of Twins." All of these applications relied on serological testing for red blood cell antigens from most of the common blood groups, and for identity determinations rare antigens or unusual combinations of common antigens were thought to be the most useful in assigning individuals to a race. On this last point, the examples that were provided indicated a low level of reliability for race identification. Race and Sanger (1962) did not attempt to cover forensic evidence in criminal cases, except to mention that at that time only the ABO system was applicable through specialized techniques for examining human blood and other fluid stains. Of course, present-day forensic science has an extensive array of DNA-based methods for analyzing evidence derived from criminal activities.

Little coverage of bone and tissue typing appeared in Race and Sanger (1962), which they saw as fraught with technical difficulties, some of which were inherent to the inhibition method that was being used. These problems were well documented in a study by Thieme and Otten (1957), who demonstrated that bacterial contamination could lead to false inhibition results, usually due to the transformation of A and B antigens into O. Their conclusion was that bone ABO typing was highly unreliable under certain conditions, especially if the bones had been recovered from damp soils. Later researchers made improvements in the preparation and testing procedures, and claimed to have achieved reliability from typing skeletal materials from an archaeological site in Israel (Micle et al., 1977), and mummified tissue from pre-Columbian sites in Peru (Allison, et al., 1978) and Chile (Llop and Rothhammer, 1988). Beyond the limited application and success of bone/tissue typing, the fledgling field of "paleoserology" was replaced by ancient DNA extraction and sequencing.

TAKING STOCK AND LOOKING AHEAD

Half a century after Boyd's bold claims for accomplishments of genetic analysis it probably is pretentious to make any similar such statements, but certainly there were significant contributions between 1950 when classic markers prevailed and prior to the dominance of molecular methods that is seen today. This section will offer an overview of the various applications of traditional markers as covered earlier but now discussed in terms of how they might continue to contribute to our understanding of human genetic variation.

POPULATION RELATIONSHIPS AND HISTORY

The first point to make under this topic is that the concept of biological race possessed little practical scientific reality or application among anthropologists/ human biologists investigating population affinities. To be sure, there is a research need to define units of study, but this requirement could not be met through race classification. In its place, population and variants such as deme or ancestral group or nonbiological linguistic and ethnic designations, have been used. A broad-based effort that employed language as its unit of study identifier was the Human Genome Diversity Project (HGDP). Growing out of the Human Genome Project in 1991, the HGDP had as its basic aim to map human DNA sequence diversity in order to deduce genetic history of our species. The HGDP became highly controversial due to ethical issues surrounding personal, civil, and legal rights of indigenous peoples from whom DNA would be obtained. Since its data set consists of DNA markers it is beyond the purview of this discussion, and the reader is directed to M'charek (2005) for a comprehensive review of the science behind the HGDP. Another global effort to study human genetic diversity through DNA markers is the Genographic Project. Under the direction of researchers from the National Geographic Society and IBM, the Genographic Project seeks to trace the deep migrational history of our species. Up-to-date accounts of the Project can be found at the National Geographic and IBM websites.

A recent study (Relethford, 2004) does exemplify a successful application of classical markers, along with microsatellite DNA markers and craniometrics, in a worldwide analysis of genetic variation in human populations. Relethford obtained frequencies of blood cell polymorphisms (blood groups, serum proteins and enzymes) from Roychoudhury and Nei (1988), much as he had done in a previous study (Relethford and Harpending, 1995). A major point to bring out here is that earlier databases of classical markers obviously can continue to serve contemporary research purposes. For the most part Birdsell (1993), in his study of microevolution on Australia, also followed this path in utilizing marker frequencies that he had obtained from his fieldwork of nearly a half-century earlier. These examples of data mining show that while DNA markers certainly rule the day, there still can be a place for already available classical marker frequencies to be used in investigating human population variation and genetic history. At an even more basic level than marker frequency

data sets, there are likely to be many instances of stored aliquots of human sera, saliva specimens, and hair samples that were collected during much earlier fieldwork projects that potentially could be subjected to laboratory analysis, but only after all ethical matters and human subjects concerns are satisfactorily resolved.

In sum, classical markers served their purpose well in describing human variation and in proposing plausible affinities between populations until such matters came under the finer-grained scrutiny of DNA markers. It seems fair to say that although they are no longer at the forefront of population study, classical markers will continue to contribute to these endeavors as evidenced in the examples cited above.

MARKERS AND DISEASE ASSOCIATIONS

It also seems appropriate to conclude that certain classical markers have been increasing over the years in their contribution to the study of selection and disease associations. For instance, the HLA system, with its numerous haplotype combinations, is highly noteworthy in this regard. Jackson's (2000) informative discussion is replete with relatively recent examples of interaction between HLA haplotypes and several infectious diseases, notably HIV/AIDS. Along this HIV research front, a very recent study has reported a connection between HIV susceptibility and Duffy antigen status. The presence of DARC (duffy antigen receptor for chemokines) appears to increase the susceptibility to infection by HIV. But following infection, the DARC-negative phenotype leads to a slowing down of the progression of the disease (He et al., 2008).

For another marker, a major breakthrough was reported by Allen et al. (1997) in that not only did thalassemia provide malaria resistance in Papua New Guinea children it also protected against other infectious diseases, and importantly this finding may apply to other malarial resistant genes, such as Hb^S, as well. Finally, two very recent studies can be cited to show how active this field remains. One found that a haptoglobin phenotype was at much higher risk for cardiovascular disease in individuals with diabetes mellitus (Levy et al., 2002), while the other demonstrated a complex interaction between one of the transferrin markers and an allele at another iron metabolism locus as these posed a prominent risk factor for developing Alzheimer's disease (Robson et al., 2004).

It probably is not surprising that these ongoing examples of classical marker associations with various diseases are still being discovered, and of course, just as in the earliest of such disease associations discovered decades ago, they are subject to further confirmation. There is every reason to predict that this area of research will continue to be a fruitful endeavor to pursue.

DISCUSSION POINTS

1. Which classical markers continue to be most useful in studying human variation and/or disease associations?
2. Discuss the pros and cons of Boyd's classification of human races based on blood groups.
3. What are the ways in which variation can be maintained at a polymorphic locus?
4. On what theoretical grounds can different measures of population distance (including genetic, anthropometric, odontometric, dermatoglyphic, linguistic, geographic) be expected to correspond to or, conversely, to not agree with one another?
5. Discuss the evidence for and against the malarial hypothesis regarding the frequency distributions of hemoglobin variants, transferrins, G6PD, and Duffy markers. Include in this discussion the role of cultural activities related to subsistence patterns.
6. What are the underlying reasons that classical markers can be expected to be associated with diseases?

REFERENCES

Allen, S. J., O'Donnell, A., Alexander, N. D. E., et al. (1997). α^{+}-Thalassemia protects children against disease caused by other infections as well as malaria. *Proceedings of the National Academy of Sciences of the United States of America*, **94**, 14736–14741.

Allison, A. C. (1956). The sickle and hemoglobin C-genes in some African populations. *Annals of Human Genetics*, **21**, 67–89.

Allison, M. J. J., Hossaini, A. A., Munizaga, J., et al. (1978). ABO blood groups in Chilean and Peruvian Mummies. *American Journal of Physical Anthropology*, **49**, 139–142.

Baker, P. T. and Little, M. A. (1976). *Man in the Andes: a Multidisciplinary Study of High-Altitude Quechua*. Stroudsburg, PA: Dowden, Hutchinson, and Ross.

Bernstein, F. (1924). Ergebnisse einer biostatischen zusammenfassenden betrachtung uber die erblichen blutstruckeren des Menschen. *Klinische Wochenschrift*, **3**, 1495–1497.

Birdsell, J. B. (1993). *Microevolutionary Patterns in Aboriginal Australia*. New York: Oxford University Press.

Boas, F. (1912). Changes in bodily form of descendents of immigrants. *American Anthropologist*, **14**, 530–562.

Boren, T., Falk, P., Roth, K. A., et al. (1993). Attachment of *Helicobactor pylori* to human gastric epithelium mediated by blood group antigens. *Science*, **262**, 1892–1895.

Boyd, W. C. (1950). *Genetics and the Races of Man: an Introduction to Modern Physical Anthropology*. Boston: Little, Brown, and Company.

Boyd, W. C. (1963a). Genetics and the human race. *Science*, **140**, 1057–1064.

Boyd, W. C. (1963b). Four achievements of the genetical method in physical anthropology. *American Anthropologist*, **65**, 243–252.

Brace, C. L. (2005). *"Race" Is A Four-Letter Word*. Oxford: Oxford University Press.

Buchi, E. C. (1968). Somatic groups composing the modern populations of India. In *Proceedings of Eighth International Congress of Anthropological and Ethnological Sciences*. Tokyo: Science Council of Japan, pp. 154–162.

Buettner-Janusch, J. (1966). *Origins of Man. Physical Anthropology*. New York: John Wiley and Sons.

Cavalli-Sforza, L. L. and Bodmer, W. F. (1999). *The Genetics of Human Populations*. Mineola, NY: Dover Publications.

Cavalli-Sforza, L. L., Menozzi, P. and Piazza, A. (1994). *The History and Geography of Human Genes*. Princeton: Princeton University Press.

Chakravarti, A. and Chakraborty, R. (1978). Elevated frequency of Tay–Sachs disease among Ashkenazic Jews unlikely by genetic drift alone. *American Journal of Human Genetics*, **30**, 256–261.

Chown, B. and Lewis, M. (1953). The ABO, MNSs, P, Rh, Lutheran, Kell, Lewis, Duffy and Kidd blood groups and the secretor status of the Blackfoot Indians of Alberta, Canada. *American Journal of Physical Anthropology*, **11**, 369–383.

Crawford, M. H. (1973). The use of genetic markers of the blood in the study of the evolution of human populations. In *Methods and Theories of Anthropological Genetics*, M. H. Crawford and P. L. Workman (eds). Albuquerque, NM: University of New Mexico Press, pp. 19–38.

Crawford, M. H. (ed.) (1984). *Current Developments in Anthropological Genetics*, vol. 3. *Black Caribs. A Case Study in Biocultural Adaptation*. New York: Plenum Press.

Crawford, M. H. (ed.) (2007). *Anthropological Genetics: Theory, Methods and Applications*. New York: Cambridge University Press.

Crawford, M. H. and Mielke, J. H. (ed.) (1982). *Current Developments in Anthropological Genetics*, vol. 2. New York: Plenum Press.

Crawford, M. H. and Workman, P. L. (ed.) (1973). *Methods and Theories of Anthropological Genetics*. Albuquerque, NM: University of New Mexico Press.

Crawford, M. H., Leyshon, W. C., Brown, K., et al. (1974). Human Biology in Mexico, part 2: a comparison of blood group, serum, and red cell enzyme frequencies, and genetic distances of the Indian populations of Mexico. *American Journal of Physical Anthropology*, **41**, 251–268.

De Mattos, L. C., Cintra, J. R., Sanches, F. E., et al. (2002). ABO, Lewis, secretor and non-secretor phenotypes in patients infected or uninfected by the *Helicobacter pylori* bacillus. *São Paulo Medical Journal*, **120**(2), doi: 10.1590/S1516–31802002000200006.

Drayna, D., Coon, H., Kim, U. K., et al. (2003). Genetic analysis of a complex trait in the Utah Genetic Reference Project: a major locus for PTC taste ability on chromosome 7q and a secondary locus on chromosome 16p. *Human Genetics*, **112**, 567–572.

Duggleby, C. R. (1978). Blood group antigens and the population genetics of *Macaca mulatta* on Cayo Santiago. *American Journal of Physical Anthropology*, **48**, 35–40.

Eaton, J. W. and Gavan, J. A. (1965). Sensitivity to P-T-C among primates. *American Journal of Physical Anthropology*, **23**, 381–388.

Eaton, J. W., Brandt, P., Mahoney, J. R., et al. (1982). Haptoglobin: a natural bacteriostat. *Science*, **215**, 691–693.

Etcheverry, R. (1967). Blood groups in natives of Easter Island. *Nature*, **216**, 690–691.

Fraikor, A. L. (1977). Tay–Sachs disease: genetic drift among the Ashkenazi Jews. *Social Biology*, **24**(2), 117–134.

Friedlaender, J. S. (1975). *Patterns of Human Variation: the Demography, Genetics, and Phenetics of Bougainville Islanders*. Cambridge, MA: Harvard University Press.

Friedlaender, J. S. (ed.) (1987). *The Solomon Islands Project*. Oxford: Clarendon Press.

Friedman, M. J. and Trager, W. (1981). The biochemistry of resistance to malaria. *Scientific American*, **244**(3), 154–164.

Frisch, A., Colombo, R., Michaelovsky, E., et al. (2004). Origin and spread of the 1278insTATC mutation causing Tay–Sachs disease in Ashkenazi Jews: genetic drift as a robust and parsimonious hypothesis. *Human Genetics*, **114**, 366–376.

Giblett, E. R. (1969). *Genetic Markers in Human Blood*. Oxford: Blackwell Scientific Publications.

Glass, R. I., Holmgren, J., Haley, C. E., et al. (1985). Predisposition for cholera of individuals with O blood group: possible evolutionary significance. *American Journal of Epidemiology*, **121**, 791–796.

Greenberg, J. H., Turner, C. G. and Zegura, S. L. (1986). The settlement of the Americas: a comparison of the linguistic, dental and genetic evidence. *Current Anthropology*, **27**, 477–497.

Hamblin, M. T. and Di Rienzo, A. (2000). Detection of the signature of natural selection in humans: evidence from the Duffy blood group locus. *American Journal of Human Genetics*, **66**, 1669–1679.

Harper, P. S. and Sunderland, E. (1986). *Genetic and Population Studies in Wales*. Cardiff: University of Wales Press.

Harris, H. (1980). *The Principles of Human Biochemical Genetics*, 3rd edn. Amsterdam: Elsevier/North-Holland Biomedical Press.

He, W., Neil, S., Kukarni, H., et al. (2008). Duffy antigen receptor for chemokines mediates trans-infection of HIV-1 from red blood cells to target cells and affects HIV-AIDS susceptibility. *Cell Host and Microbe*, **4**, 52–62.

Hirschfeld, J. (1959). Immuno-electrophoretic demonstration of qualitative differences in human sera and their relation to the haptoglobins. *Acta Pathologica et Microbiologica Scandinavica*, **47**, 160–168.

Hirschfeld, L. and Hirschfeld, H. (1919). Serological differences between the blood of different races. *Lancet*, **18**, 675–679.

Jackson, F. L. C. (2000). Human adaptations to infectious disease. In *Human Biology. An Evolutionary and Biocultural Perspective*, S. Stinson, B. Bogin, R. Huss-Ashmore, et al. (eds). New York: Wiley-Liss, pp. 273–293.

Jamison, P. L., Zegura, S. L. and Milan, F. A. (1978). *Eskimos of Northwestern Alaska: a Biological Perspective*. Stroudsburg, PA: Dowden, Hutchinson, and Ross.

Jenkins, T., Harpending, H. and Nurse, G. T. (1978). Genetic distances among certain Southern African populations. In *Evolutionary Models and Studies in Human Diversity*, R. J. Meier, C. M. Otten and F. Abdel-Hameed (eds). The Hague: Mouton Publishers, pp. 227–243.

Labie, D., Pagnier, J., Wajcman, H., et al. (1986). The genetic origin of the variability of the phenotypic expression of the Hb S gene. In *Genetic Variation and its Maintenance*, D. F. Roberts and G. F. DeStefano (eds). Cambridge: Cambridge University Press, pp. 149–156.

Landsteiner, K. (1901). Uber agglutinationserscheinungen normalen menschlichen blutes. *Wiener Klinische Wochenschrift*, **14**, 1132–1134.

Lasker, G. W. (1946). Migration and physical differentiation. *American Journal of Physical Anthropology*, **4**, 273–300.

Lefevre-Witier, P. and Verges, H. (1978). Genetic structure of Ideles. In *Evolutionary Models and Studies in Human Diversity*, R. J. Meier, C. M. Otten and F. Abdel-Hameed (eds). The Hague: Mouton Publishers, pp. 255–277.

Levy, A. P., Hochberg, I., Jablonski, K., et al. (2002). Haptoglobin phenotype is an independent risk factor for cardiovascular disease in individuals with diabetes: the strong heart study. *Journal of the American College of Cardiology*, **40**, 1984–1990.

Lindén, S., Mahdavi, J., Semino-Mora, C., et al. (2008). Role of ABO secretor status in mucosal innate immunity and *H. pylori* infection. *PLoS Pathogens* **4**(1), e2.

Livingstone, F. B. (1958). Anthropological implications of sickle cell gene distribution in West Africa. *American Anthropologist*, **60**, 533–562.

Livingstone, F. B. (1984). The Duffy blood groups, vivax malaria, and malaria selection in human populations. *Human Biology*, **56**, 413–425.

Livingstone, F. B. (1985). *Frequencies of Hemoglobin Variants*. New York: Oxford University Press.

Livingstone, F. B. (1989). Who gave whom hemoglobin S: the use of restriction site haplotype variation for the interpretation of the evolution of the β^s-globin gene. *American Journal of Human Biology*, **1**, 289–302.

Livingstone, F. B., Gershowitz, H., Neel, J. V., et al. (1960). The distribution of several blood group genes in Liberia, the Ivory Coast, and Upper Volta. *American Journal of Physical Anthropology*, **18**, 161–178.

Llop, E. and Rothhammer, F. (1988). A note on the presence of blood groups A and B in pre-Columbian South America. *American Journal of Physical Anthropology*, **75**, 107–111.

Malhotra, K. C. (1978). Microevolutionary dynamics among Gavdas of Goa. In *Evolutionary Models and Studies in Human Diversity*, R. J. Meier, C. M. Otten and F. Abdel-Hameed (eds). The Hague: Mouton Publishers, pp. 279–314.

Marks, J. (1995). *Human Biodiversity. Genes, Race, and History*. New York: Aldine de Gruyter.

M'charek, A. (2005). *The Human Genome Diversity Project. An Ethnography of Scientific Practice*. Cambridge: Cambridge University Press.

Meier, R. J. (1969). The Easter Islander: a study in human biology. PhD thesis, University of Wisconsin, Madison, WI, USA.

Meier, R. J. (1980). Anthropological dermatoglyphics: a review. *Yearbook of Physical Anthropology*, **23**, 147–178.

Melartin, L. and Blumberg, B. S. (1966). Albumin Naskapi: a new variant of serum albumin. *Science*, **153**, 1664–1666.

Micle, E., Kobilyansky, M., Nathan, M., et al. (1977). ABO-typing of ancient skeletons from Israel. *American Journal of Physical Anthropology*, **47**, 89–92.

Mielke, J. H. and Crawford, M. H. (eds) (1980). *Current Developments in Anthropological Genetics*, vol. **1**. New York: Plenum Press.

Mielke, J. H., Koningsberg, L. W. and Relethford, J. H. (2006). *Human Biological Variation*. New York: Oxford University Press.

Mohr, J. (1951). Search for linkage between Lutheran blood group and other hereditary characters. *Acta Pathologica et Microbiologica Scandinavica*, **28**, 207–210.

Molnar, S. (2006). *Human Variation: Races, Types and Ethnic Groups*, 6th edn. Upper Saddle River, NJ: Pearson/Prentice Hall.

Montagu, M. F. A. (1960). *An Introduction to Physical Anthropology*, 3rd edn. Springfield, IL: Charles C. Thomas.

Montagu, M. F. A. (1964). The concept of race. In *The Concept of Race*, M. F. A. Montagu (ed.). London: Collier-Macmillan, pp. 12–28.

Mourant, A. E. (1954). *Distribution of Human Blood Groups*. Oxford: Blackwell Scientific Publications.

Mourant, A. E. (1983). *Blood Groups and Anthropology*. London: Oxford University Press.

Mourant, A. E., Kopec, A. C. and Domaniewska-Sobczak, K. (1976). *The Distribution of the Human Blood Groups and Other Polymorphisms*. Oxford: Oxford University Press.

Neel, J. V. and Ward, R. H. (1972). The genetic structure of a tribal population, the Yanomama Indians, part six: analysis by *F*-statistics (including a comparison with Makiritare and Xavante). *Genetics*, **72**, 639–666.

Neel, J. V., Salzano, F. M., Junqueira, P. C., et al. (1964). Studies on the Xavante Indians of the Brazilian Mato Grosso. *American Journal of Human Genetics*, **16**, 52–140.

Niswander, J. D., Brown, K. S., Iba, B. Y., et al. (1970). Population studies on southwestern Indian tribes. I. History, culture, and genetics of the Papago. *American Journal of Human Genetics*, **22**, 7–23.

Omoto, K. (1978). Blood protein polymorphisms and the problem of genetic affinities of the Ainu. In *Evolutionary Models and Studies in Human Diversity*, R. J. Meier, C. M. Otten and F. Abdel-Hameed (eds). The Hague: Mouton Publishers, pp. 333–341.

O'Rourke, D. (2000). Genetics, geography, and human variation. In *Human Biology. An Evolutionary and Biocultural Perspective*, S. Stinson B. Bogin, R. Huss-Ashmore, et al. (eds). New York: Wiley-Liss, pp. 87–133.

O'Rourke, D. (2006). Blood groups, immunoglobulins, and genetic variation. In *Environment, Origins, and Population*, vol. **3**. D. H. Ubelaker (ed.). Washington, DC: Smithsonian Institution, pp. 762–776.

Osborne, R. H. (1958). Serology in physical anthropology. *American Journal of Physical Anthropology*, **16**, 187–195.

Ottenberg, R. (1925). A classification of human races based on geographic distribution of blood groups. *Journal of the American Medical Association*, **84**, 1393–1395.

Race, R. R. and Sanger, R. (1962). *Blood Groups in Man*, 4th edn. Philadelphia: F. A. Davis.

Reid, M. E. and Lomas-Francis, C. (1997). *The Blood Group Antigen FactsBook*. San Diego, CA: Harcourt Brace and Company.

Relethford, J. H. (2004). Global patterns of isolation by distance based on genetic and morphological data. *Human Biology*, **76**, 499–513.

Relethford, J. H. and Harpending, H. C. (1995). Ancient differences in population size can mimic a recent African origin of modern humans. *Current Anthropology*, **36**, 667–674.

Rhoads, J. G. and Friedlaender, J. S. (1987). Blood polymorphism variation in the Solomon Islands. In *The Solomon Islands Project*, J. S. Friedlaender (ed.). Oxford: Clarendon Press, pp. 125–154.

Robson, K. J. H., Lehmann, D. J., Wimhurst, V. L. C., et al. (2004). Synergy between the C2 allele of transferrin and the C282Y allele of the haemochromatosis gene (HFE) as risk factors for developing Alzheimer's disease. *Journal of Medical Genetics*, **41**, 261–265.

Roychoudhury, A. K. and Nei, M. (1988). *Human Polymorphic Genes. World Distribution*. New York: Oxford University Press.

Sahi, T. (1974). The inheritance of selective adult-type lactose malabsorption. *Scandinavian Journal of Gastroenterology*, **9**(suppl. **30**), 1–73.

Sanghvi, L. D. (1953). Comparison of genetical and morphological methods for a study of biological differences. *American Journal of Physical Anthropology*, **11**, 385–404.

Schanfield, M. S. (1980). The anthropological usefulness of polymorphic systems: HLA and immunological allotypes. In *Current Developments in Anthropological Genetics*, vol. **1**, J. H. Mielke and M. H. Crawford (eds). New York: Plenum Press, pp. 65–85.

Schanfield, M. S. and Fudenberg, H. H. (1978). The anthropological usefulness of IgA allotypic markers. In *Evolutionary Models and Studies in Human Diversity*, R. J. Meier, C. M. Otten and F. Abdel-Hameed (eds), The Hague: Mouton Publishers, pp. 343–351.

Shapiro, H. L. and Hulse, F. (1940). *Migration and Environment*. New York: Oxford University Press.

Smithies, O. (1955). Zone electrophoresis in starch gels: group variations in the serum proteins of normal human adults. *Biochemical Journal*, **61**, 629–641.

Smithies, O. (1958). Third allele at the serum β-globulin locus in humans. *Nature*, **181**, 1203–1204.

Snyder, L. H. (1930). The "laws" of serologic race-classification studies in human inheritance IV. *Human Biology*, **2**, 128–133.

Steel, C. M., Beatson, D., Cuthbert, R. J. G., et al. (1988). HLA haplotype A1 B8 DR$_3$ as a risk factor for HIV related disease. *Lancet*, **331**, 1185–1188.

Suarez, B. K., Crouse, J. D. and O'Rourke, D. H. (1985). Genetic variation in North American populations: the geography of gene frequencies. *American Journal of Physical Anthropology*, **67**, 217–232.

Szathmáry, E. J. E. (1983). Dogrib Indians of the NWT, Canada: genetic diversity and genetic relationship among subarctic Indians. *Annals of Human Biology*, **10**, 147–162.

Szathmáry, E. J. E., Ferrell, R. E. and Gershowitz, H. (1983). Genetic differentiation in Dogrib Indians: serum protein and erythrocyte enzyme variation. *American Journal of Physical Anthropology*, **62**, 249–254.

Thieme, F. P. and Otten, C. M. (1957). The unreliability of blood typing aged bone. *American Journal of Physical Anthropology*, **15**, 387–397.

Tills, D., Kopec, A. C. and Tills, R. E. (1983). *The Distribution of the Human Blood Groups and other Polymorphisms*. Oxford: Oxford University Press.

Vogel, F. and Chakravararti, M. R. (1966). ABO blood groups and smallpox in a rural population of Bengal and Bihar (India). *Human Genetics*, **3**, 166–180.

Vogel, F. and Motulsky, A. G. (1997). *Human Genetics: Problems and Approaches*, 3rd edn. Berlin, Springer.

Volkman, S. K., Barry, A. E., Lyons, E. J., et al. (2001). Recent origin of *Plasmodium falciparum* from a single progenitor. *Science*, **293**, 482–484.

Washburn, S. L. (1951). The new physical anthropology. *Transactions of the New York Academy of Sciences* Series II, **13**, 298–304.

Weiner, J. S. and Lourie, J. A. (1969). *Human Biology. A Guide to Field Methods. IBP Handbook No. 9*. Philadephia, PA: F. A. Davis Company.

Weiss, K. M. (1993). *Genetic Variation and Human Disease*. Cambridge: Cambridge University Press.

Wiener, A. S. (1948). Blood grouping tests in anthropology. *American Journal of Physical Anthropology*, **6**, 236–237.

Wooding, S. (2006). Phenylthiocarbamide: a 75-year adventure in genetics and natural selection. *Genetics*, **172**, 2015–2023.

14 DNA Markers of Human Variation

Michael E. Steiper

INTRODUCTION

Historically, questions relating to human genetics and variation have been addressed by the study of "classical" genetic markers (see Chapter 13 of this volume). Classical genetic markers are polymorphic proteins, which run the gamut from blood group antigens such as the ABO system to enzymes including G6PD. Each of these classical marker loci has characteristics that are useful for addressing questions about human genetic variation. These characteristics usually include an appreciable level of polymorphism (variation) and a methodological ability to consistently detect that polymorphism using techniques such as gel electrophoresis. Often, these variations make clear links between particular alleles and genetic diseases (e.g., the Hb^S allele of the β-globin gene with sickle cell anemia [Pauling et al., 1949], while at other times, these relationships are statistical (e.g., particular *ABO* blood type alleles and susceptibilities to diseases [see Chapter 13 of this volume])). The use of classical markers for understanding human genetic variation is reviewed in Chapter 13 of this volume.

In the current chapter, I review the application of more modern "DNA markers" to studies of human variation. A host of methodological advances coalesced in the 1980s that enabled scientists to investigate human variation directly at the level of the genetic material, hence the name "DNA markers." Because there are several advantages to assaying human genetic diversity directly at the DNA level, a transition to use of these markers followed. Importantly, throughout the transition from classical to DNA markers, many of the key evolutionary questions of human genetics and variation have remained constant. These studies address the processes that have shaped human genetic diversity; the origins, maintenance, and levels of human genetic diversity; the relationship between genetic diversity, diseases, and disease resistance; the effect of natural selection on the human genome; the role of genetics in human biocultural evolution; and the tracing of human migrations around the globe.

Although the questions addressed by different marker types are similar, it is critical to understand the relationship between the classical and DNA markers. Perhaps the most important feature of classical markers is that their allelic variation is due to amino acid level differences. Therefore, in classical marker studies, the genetic variation ascertained is based solely on DNA regions involved in transcription and/or translation. Genetically, these differences between classical marker alleles are caused either by variation in DNA that encodes that gene's amino acids or variation in its sequence that affects it expression. This has important consequences for understanding human variation through their lens because the number of DNA sites that contribute to variation at this level represents only a small portion of the genome (<5%) (International Human Genome Sequencing Consortium, 2001, 2004). Also, because they encode polypeptides, classical loci are likely to be under greater surveillance by natural selection than portions of the genome that do not encode for or regulate the expression of amino acids.

DNA markers, on the other hand, directly address the physical basis of heredity. As such, DNA markers can derive from anywhere within the entire nuclear genome or the mitochondrial DNA, whether the region is coding, regulatory, or noncoding. This is one of the key differences between classical markers and DNA markers. One of the greatest strengths of the direct study of DNA variation is that most of the genome is noncoding, and thus presumably not under natural selection. Consequently, DNA markers occurring in these regions are considered to be "neutral" in their effects. Neutral DNA markers are extremely useful for assessing the demographic history of humans, since these regions are expected to reflect mainly population level effects, such as drift, expansions, admixture and migration.

The ABO blood group system (reviewed in Chapter 13 of this volume) provides a useful example of the relationships between classical and DNA markers (Figure 14.1). The three main alleles of this system

Human Evolutionary Biology, ed. Michael P. Muehlenbein. Published by Cambridge University Press. © Cambridge University Press 2010.

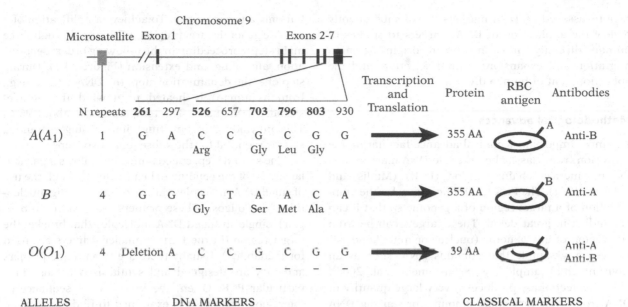

14.1. Schematic diagram of the relationship between DNA markers and classical markers in the ABO blood group system. At left are the three main alleles: $A(A_1)$, B, and $O(O_1)$ (there are many other subtypes of the alleles). The DNA markers associated with the different alleles follow to the right (adapted from Hosoi, 2008). There is a minisatellite DNA marker in the upstream region of the *ABO* gene (Irshaid et al., 1999). There are alleles with one repeat unit and others with four repeat units. There are a number of single nucleotide polymorphism (SNP) markers in the sixth and seventh exon of the different alleles. The numbers indicate the base pair location within the coding region, the base pairs are below the numbers, as well as the amino acids for nonsynonymous changes. The SNPs that change the amino acid (nonsynonymous) are shown in bold (e.g., 703). A single base pair deletion is present at position 261 on the O allele, which results in a frameshift that eventually leads to a premature stop codon. There are also synonymous SNPs in the different alleles (e.g., 930). When these alleles are transcribed and translated from DNA to amino acids, they form part of the basis for the ABO erythrocyte antigens (at right of large arrows). Note that individuals who have the genotype *AA* and *AO* are both phenotypically A with respect to their classical markers. The use of DNA markers, on the other hand, could type the base pair at position 261 or the number of minisatellite repeats and distinguish between these two genotypes.

are *A*, *B*, and *O*. The gene underlying this system, the *ABO* gene, is 18 kilobases long, comprising 7 exons, and located on human chromosome 9 (q34) (Cook et al., 1978; Yamamoto et al., 1995). The *A* and *B* alleles are codominant and *O* is recessive. The A and B antigens differ at four amino acids caused by nonsynonymous differences in the DNA sequence between the *A* and *B* alleles (Yamamoto et al., 1990). The *O* allele, which does not produce a functional antigen (more precisely, the H antigen is produced), contains a single base deletion. This deletion, a nonsense mutation that results in a frameshift (Yamamoto et al., 1990), causes a premature stop codon and a truncated amino acid chain.

The classical methodological approaches are able to distinguish these alleles based on different amino acid chains in the case of *A* and *B*, or their nonproduction, in the case of *O*. However, studies of the DNA sequence encoding the *A*, *B*, and *O* alleles have directly examined the nucleotides responsible for the amino acid differences. Perhaps more important than the DNA changes that result in amino acid chain

differences are the "silent" (or synonymous) DNA sequence differences between the *A*, *B*, and *O* alleles. These synonymous DNA differences, as well as changes within the introns, and the upstream and downstream regions, are not usually reflected in the amino acid chain and, therefore, are "invisible" to classical approaches. They are also likely to be invisible to natural selection.

However, the ascertainment of the complete set of differences in the DNA sequence is instrumental for obtaining a more precise and accurate picture of the population genetics of the ABO system. This information indicates the origin of the alleles, their evolutionary history, and their maintenance by drift and/or natural selection. For example, DNA marker analysis of the *ABO* gene suggests that the *O* allele has arisen three independent times in human evolution, suggesting it is a selectively advantageous allele (Calafell et al., 2008).

In this chapter, I review the methodological advances that have permitted the use of DNA markers in studies of human diversity, including the properties of different marker types and the different genomic regions assayed by DNA markers. I will subsequently review the application of DNA markers to studies of human diversity, modern human origins, human migration and expansion, natural selection, and the apportionment of human diversity.

Methodological advances

The most important technical advance facilitating the transition from classical markers to DNA markers was the polymerase chain reaction (PCR) (Mullis and Faloona, 1987). The PCR technique allows for the amplification of a target region of a genome so that it can be studied in great detail. These targets can be from any region of any genome (nuclear or mitochondrial) or even any sample that contains DNA, such as an environmental sample (e.g., Sebastianelli et al., 2008). The PCR technique produces a very large quantity of DNA from even very small amounts of starting DNA from the target DNA region. This PCR-amplified DNA can be examined by a host of other methods, which indicate its sequence or length. The PCR technique is critical to most subsequent applications in the ascertainment of population genetic variation. Indeed, the use of the PCR to examine genetic polymorphisms from a wide sample of humans is the essence of modern human population genetics with DNA markers. Importantly, PCR is relatively inexpensive, rapid, and standardized.

Obtaining DNA for use in PCR requires access to any human cells that carry genomic DNA. This is very different from classical markers, which are proteins, and can only be purified from samples in which the proteins are produced. For example, to study the classical protein marker locus albumin, the plasma portion of blood must be used to purify the albumin proteins. However, the DNA that underlies the production of the albumin proteins is located on human chromosome 4 and present in all nucleated human cells. Therefore, the direct study of an albumin gene can be conducted from DNA purified from any human cell sample: blood, cheek swabs, placenta, hair bulbs, or other tissues. Combined with very minimal requirements for the amount of genomic DNA to be used with the PCR, genetic analysis of DNA markers is simpler than analysis of classical markers. Obtaining tissue or DNA samples from a sufficient sample of humans for population genetics study can be difficult. Global sampling has been greatly facilitated by the collection and availability of cell lines from diverse humans (Cann et al., 2002).

Polymerase chain reaction is a technique that is an in vitro appropriation of an organism's mechanisms for replicating their own genomes using DNA polymerase enzymes. However, as noted above, during most PCRs, it is not an entire genome that is being replicated, but only a target region. To achieve amplification of a specific genomic target region, PCR goes through three main steps, proceeding in the following order: denaturation, annealing, and extension (Figure 14.2). During step one, the denaturation step, the DNA sample (e.g., from a human) is heated to unwind the double-stranded DNA helix. The temperature is about 95°C. This unwinding (or melting) into a single-stranded state is required for the subsequent two steps.

The second step, called annealing, allows a particular region of the genome to be targeted through the use of small DNA molecules called "primers," "oligonucleotides," or "oligos." These primers are small (~20 base pair), single-stranded DNA molecules that bracket the target region (in the figure, smaller primers are used for readability). Usually there are two unique primers and they are designed and synthesized for use in a particular PCR. Often, the known DNA sequence of the region of interest is examined to find appropriate locations for the primers. This has been greatly facilitated by the many DNA sequences publicly available (see GenBank at http://www.ncbi.nlm.nih.gov [Benson et al., 2004]). One primer has the complementary sequence to a region of the genome located at the most upstream (or 5′) margin of the target region, and the other primer has the complementary sequence of the most downstream (or 3′) margin of the target region.

The step is called annealing because the reaction is cooled to a temperature (~55°C) that allows these two primers to anneal or bind to the denatured DNA. These primers are able to bind to specific regions of the genome in the target sequence because they are generated from a known DNA sequence (and thus can be designed to match the sequence of the target region exactly, ensuring that they will anneal to the target). Also, their length (approximately 20 base pairs) usually precludes spurious matches elsewhere in the genome. Software programs for primer design are available (e.g., Untergasser et al., 2007).

The third step of PCR is called extension or elongation. In this step, the DNA polymerase enzyme elongates DNA molecules from the 3′-most-base of both primers using the genome in the sample as a template. The reaction is heated to 72°C which causes the polymerase to elongate a new DNA molecule along the template strand. This is done in the presence of additional free DNA bases (deoxy A, C, T, and Gs) that are incorporated by the polymerase during the elongation. DNA polymerase used in PCRs comes from a heat-adapted bacterium called *Thermus aquaticus* (or *Taq*) that lives in hot springs. This enzyme is used because it can withstand heating to denaturation temperatures.

In these initial steps, when the sample is the main template DNA, the elongated copies extend from one primer along the DNA molecule beyond the second primer. These three steps comprise a single PCR cycle.

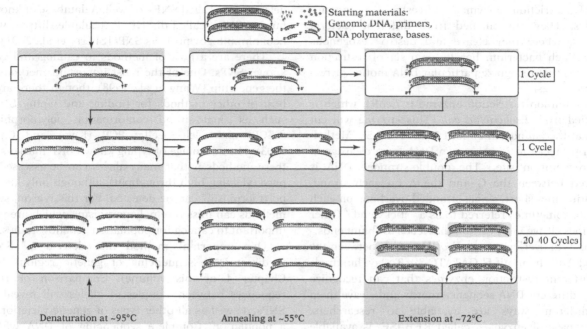

Starting materials:
Genomic DNA, primers,
DNA polymerase, bases.

1 Cycle

1 Cycle

20–40 Cycles

Denaturation at ~95°C   Annealing at ~55°C   Extension at ~72°C

14.2. A schematic of the polymerase chain reaction (PCR), which is described in the text. Polymerases are gray dots. Fine arrows at the end of DNA strands denote that the strand continues in this direction. The 5′ and 3′ denote the directionality of the different DNA strands; the format of the first cycle is followed throughout. The DNA strands given in black are present in the initial stage of the reaction and are the genomic DNA and primers. The DNA strands given in medium gray are those that have a primer anchoring one side of the strand, but extend beyond the target area. The DNA strands given in the lightest gray are only the target region, between the two primer annealing locations. It is these light gray DNA copies that are created in large amounts during PCR. These are highlighted by faint gray in the bottom right box.

At the end of this PCR cycle, there are now additional copies of the target region in addition to the genomic template. This cycle is then repeated, as it is a "chain reaction." In the second cycle, the newly polymerized DNA molecules are also denatured. Similarly, during annealing, the primers can bond to either the original template or the newly generated templates. Extension will also occur at the genomic DNA and the newly generated templates. During these subsequent cycles, when a primer anneals to a newly generated template, molecules will be created that only correspond to the target region between the primers. With each cycle, additional copies of the target region are made and these copies themselves can serve as templates in subsequent cycles. This leads to an exponential increase in the number of target DNA molecules. Most PCR protocols utilize between 20 and 45 cycles to generate large amounts of the target region. Indeed, PCR can be initiated from very low numbers of genomic DNA copies, even single genomes, making it useful for applications such as forensics and ancient DNA where samples may contain only small amounts of template DNA.

At the end of a PCR, a large amount of a target region of the genome is generated. Often, a sample of the target DNA generated in a PCR is then visualized with gel electrophoresis. Electrophoresis uses an electric charge to separate DNA by length across a matrix, such as agarose gel. It is at this stage that the level of genetic variation can be assessed by the examination of many different properties of the amplified DNA target. The two main properties of these amplified target regions are their size (or length) and their DNA sequence.

Types of DNA markers

Three of the most important types of DNA markers for studies of human variation will be addressed at length: microsatellites, single nucleotide polymorphisms, and restriction fragment length polymorphisms. Less frequently used DNA markers, such as *Alu* and LINE elements (Xing et al., 2007) and copy number variants (Jakobsson et al., 2008), are important for understanding human diversity, but will not be directly addressed here.

Though they are not used extensively as markers at present, restriction fragment length polymorphisms (RFLPs) (Botstein et al., 1980) are useful to understand as a methodology for both historical and practical purposes. This type of marker exposes DNA to enzymes that cut at short and specific DNA motifs of approximately 4–6 base pairs in length. These enzymes are called restriction enzymes or restriction endonucleases. They are purified from different bacteria species, where they cleave and destroy exogenous DNA. Each bacterium has its own set of restriction enzymes that recognize particular DNA motifs, or recognition sites.

A common restriction enzyme is *Eco*RI, which is purified from *Escherichia coli*. This enzyme will cut DNA at the palindromic motif 5′-GAATTC-3′. (Methylation protects an *E. coli*'s own DNA from cleavage at the recognition site.) The double-stranded DNA is cleaved between the G and the A on each strand, resulting in a short, single-stranded overhang on each strand, sometimes referred to as a "sticky end." Other enzymes cleave in a position that leads to "blunt ends," without an overhang (e.g., *Eco*RV cutting between the T and A in the motif 5′-GATATC-3′). Today, hundreds of different restriction enzymes that can recognize many different DNA sequence motifs and cleave them in different ways are available to researchers. A database of enzymes, called REBASE, is available (see http://rebase.neb.com/rebase/rebase.html; Roberts et al., 2005).

As a simple RFLP example, one can use PCR to amplify a target region 600 base pairs in length. At the 200th base pair of this region, the sequence motif 5′-GATATC-3′ is present. Nowhere else in this 600 base-pair region is this motif present. When this amplified DNA sample is exposed to *Eco*RV, it will be digested or restricted into two pieces, one that is approximately 200 base pairs and another that is 400 base pairs. When separated by size using gel electrophoresis, two DNA fragments will be present, one 200 and the other 400 base pairs in length. When this sequence motif is not present within the amplified fragment, only one DNA size will be present, 600 base pairs, even if it is exposed to the *Eco*RV restriction enzyme. If this motif is variable within a population (e.g., 5′-GATATC-3′ in one allele and 5′-GATACC-3′ in another allele), it can be assessed as a polymorphic genetic marker. These markers are used by assessing the variation in the fragment lengths generated by different restriction enzymes, hence, the term RFLPs. Because a relatively large amount of DNA is required to visualize RFLPs, the DNA examined is often PCR amplified DNA or purified mitochondrial DNA.

Single nucleotide polymorphisms (SNPs) are single base-pair differences between two alleles of a particular DNA region. For example, two alleles might differ at the 112th base pair in a gene sequence, with one having a "G" base and the other having an "A" base. Single nucleotide polymorphisms are found throughout the nuclear and mitochondrial genomes, and act as codominant markers. While the level of polymorphism at these positions varies, they are usually less variable than satellite markers. Sometimes small insertions and

deletions are called SNPs, as well. A database of known human SNPs, called dbSNP, is available (http://www.ncbi.nlm.nih.gov/projects/SNP/ [Sherry et al., 2001]).

There are a host of methods to catalog and type known SNPs. One of the most important methods is the gene chip (Wang et al., 1998), though there are a host of other methods for finding and typing SNPs, such as single-strand conformation polymorphism (SSCP) and allele-specific PCR (Kwok and Chen, 2003). The advantage of typing known SNP markers by these methods is that many alleles can be assessed at once (yielding high throughput), although only known polymorphisms can be detected. For this reason, such methods can miss rare polymorphisms, which are very important to catalog for population genetics studies.

One important method that detects all SNPs is direct DNA sequencing of genetic regions (see Chapter 4 of this volume). Comparison of DNA sequences between samples of alleles will reveal all SNPs, as well as all other types of genetic variation in a population. Complete sequencing of DNA within population samples is sometimes referred to as "resequencing" (e.g., Voight et al., 2006; Kelley et al., 2009). Resequencing has the advantage of finding novel polymorphisms, but it has the disadvantage of covering regions of DNA that are identical, which can be costly. Advances in automated DNA sequencing methods will enhance the ability to easily sequence stretches of DNA for population studies (Bentley, 2006; Stratton, 2008).

Microsatellite DNA regions are a class of repetitive DNA elements composed of tandemly repeated motifs of up to six base pairs in length. These regions are also known as short tandem repeat loci (STRs) or simple sequence repeats (SSRs). The STRs have motifs such as "GA" or "TAC" repeated a number of times, usually up to a few hundred base pairs in length. The alleles are designated by their repeat length, e.g., $(GA)_n$.

Microsatellites have a number of properties that make them very useful genetic markers for evolutionary studies (Schlötterer, 2000). Firstly, these regions are interspersed throughout the human nuclear genome. Because evolution can be a stochastic (i.e., random and nondeterministic) process, this allows markers to derive from many different independently evolving genomic regions. Also, markers can be chosen from particular regions of the genome (e.g., neutral noncoding regions, low recombination regions, particular chromosomes, or regions near a particular gene). Secondly, they are codominant and relatively simple to type, i.e., to ascertain the different alleles present at a specific locus. In brief, PCR primers are designed for regions flanking the STR region, and the region is amplified. Subsequent to PCR, a sensitive electrophoresis is used to decipher size differences of the alleles, distinguishing between alleles that even differ by a single repeat, e.g., $(CA)_{22}$ and $(CA)_{23}$ alleles.

This resolution is achieved through the use of fluorescently tagged primers and capillary electrophoresis-based machines. Thirdly, STRs are often highly variable due to their having a high mutation rate, about 10^{-6} to 10^{-2} per generation, due to slippage during DNA replication (Schlötterer, 2000). This means there are usually many alleles at a particular microsatellite locus, allowing for discrimination between individuals at very close levels. Fourthly, multiple microsatellite loci can often be amplified within the sample PCR, allowing these loci to be typed at the same time (referred to as multiplexing).

Overall, STRs are useful DNA markers for population genetics studies, relatedness studies (e.g., paternity or sibship analyses), molecular ecology applications, forensics, and genetic mapping. Some microsatellite markers are named for genes that they are associated with (e.g., FGA is a marker in the *fibrinogen α-chain* gene), while others are named for their location in the genome (e.g., D3S1358 is located on human chromosome 3). A database of known human microsatellites, called STRbase, is also available for researchers (http://www.cstl.nist.gov/biotech/strbase/ [Ruitberg et al., 2001]). A related kind of DNA marker is the minisatellite, which involves repeat motifs longer than microsatellites, up to about 100 base pairs.

In a sense, RFLP loci are similar to SNPs because they usually reflect single base-pair changes, but they are assessed in a manner similar to STRs, based on their length. The RFLP method is useful because it is relatively rapid, inexpensive and simple, and the markers behave in a codominant manner. A shortcoming is that only very few base-pair differences (SNPs) are assessed in a sequence by restriction enzymes, although use of multiple enzymes can result in a large proportion of total nucleotides being assayed. Also, absence of digestion can be caused by multiple different mutations, meaning that more alleles may be present than can be reliably assessed.

Regions of the human genome

For the purposes of assaying human genetic diversity using DNA markers, it is imperative to understand the properties of the different genomic regions that comprise "the human genome." Most important is the distinction between the mitochondrial genome and the nuclear genome. Both genomes are made from DNA and reside within most cells. Within each nucleated cell, there is one copy of the nuclear genome in the cell nucleus. However, within each cell's cytoplasm there are from a few to many thousands of mitochondria, which produce energy in the form of adenosine triphosphate (ATP) for the cell. Each mitochondrion possesses many copies of the mitochondrial genome

(mtDNA). Therefore, in most tissues, there are many more mitochondrial genomes present than nuclear genomes.

The mtDNA genome has a number of unique properties because of its evolutionary history (reviewed by Pakendorf and Stoneking, 2005). To begin with, it is a plasmid, or a circular molecule, of relatively small size, only 16 569 base pairs in length. The majority of its DNA consists of coding sequences or genes, although it uses a slightly different genetic code than nuclear DNA. In addition, it does not recombine, is haploid and maternally inherited, and mutates more rapidly than nuclear DNA. The first mitochondrial genome sequence was published in 1981 (Anderson et al., 1981) and has been further refined (Andrews et al., 1999).

Relative to the mtDNA genome, the nuclear genome is very different. A human nuclear genome draft sequence was first published in 2001 (International Human Genome Sequencing Consortium, 2001; Venter et al., 2001) and then finished several years later (International Human Genome Sequencing Consortium, 2004). Based on this work, we now know that a complete haploid human DNA sequence is about 3.08 billion base pairs long. Somewhat surprisingly, researchers discovered that less than 5% of the genome encodes genes and that there are between 20 000 and 40 000 protein-coding genes per genome (International Human Genome Sequencing Consortium, 2001, 2004). Many important features of the genome, such as the number of coding loci, the GC-content, and recombination rates, show considerable heterogeneity. Interestingly, about 50% of our nuclear genome comprised transposable elements (International Human Genome Sequencing Consortium, 2001). More information about the Human Genome Project, including access to "Landmark Papers," is found at the following website: http://www.ornl.gov/sci/techresources/ Human_Genome/home.shtml.

Within the nuclear genome, there are 22 autosomal chromosomes, each with biparental inheritance, and one set of sex chromosomes (X and Y). Males have an X and Y chromosome; females have two X chromosomes. The Y chromosome is haploid, relatively small with few genes, and is paternally inherited (i.e., passed from father to son). It also does not recombine for most of its length (except where required in male meiosis). The nonrecombining part of the Y chromosome (sometimes abbreviated as NRY for nonrecombining Y chromosome region) is often targeted for evolutionary studies. The X chromosome, on the other hand, is relatively large, has many more genes than the Y chromosome, and recombines in females (who have two X chromosomes), but not in males. The autosomes, X chromosome, Y chromosome, and mtDNA are sometimes referred to as different genetic compartments (e.g., Garrigan and Hammer, 2006).

Choosing DNA markers for the study of human variation

The three types of DNA markers discussed above, microsatellites, SNPs, and RFLPs, have different distributions in the genome. Loci that contain small single base-pair polymorphisms, SNPs and RFLPs, are distributed throughout the entire human genome. Due to the compact and mainly functional nature of the mtDNA genome, microsatellites are not found there. These repeat loci are instead scattered throughout the nuclear genome, usually outside of coding regions.

Markers have other properties that make them more or less useful for examining different population genetics questions. For fine-scale processes that have a recent time frame, such as recent migrations, recent expansions, and relationships between individuals, markers that harbor a great deal of diversity are used. These markers have higher mutation rates. For examining broader processes including those operating further in the past, more slowly evolving markers are preferable, since the action of subsequent mutations can sometimes overwrite these genetic signals. In general, microsatellite loci mutate more rapidly than the base substitutions that lead to SNPs and RFLPs (Schlötterer, 2000). Therefore, in general terms microsatellites are useful for assessing diversity within the nuclear genome and can trace fairly recent and fine-scale population genetics processes including relatedness, demography, and migration. Both SNPs and RFLPs are useful in this regard, but their slower mutation rate may not provide enough variation to resolve these questions. Also, in general, the SNP and RFLP loci are more frequently present within or near coding regions and, therefore, can help to answer direct questions about natural selection's role in shaping diversity at loci. DNA markers from the mtDNA, the Y, and the X chromosomes can also be used to trace sex-specific processes, such as migration of females, described further below.

HUMAN GENETIC DIVERSITY

Human genetic diversity can be assessed using these different markers. For example, an analysis of 1.42 million nuclear SNP markers in 24 diverse humans estimated nucleotide diversity (π) to be 0.075% (International SNP Map Working Group, 2001). Nucleotide diversity is the average number of pairwise differences in a sample of alleles (Nei, 1987). Studies that have examined nuclear DNA sequence data from larger samples of humans from across the globe usually find somewhat higher levels of polymorphism, despite examining fewer overall SNPs

(e.g., chr. 22 $\pi = 0.088\%$ [Zhao et al., 2000]; chr. 1 $\pi = 0.057\%$ [Yu et al., 2001]; 50 autosomal loci $\pi = 0.088\%$ [Yu et al., 2002a]; 20 autosomal loci $\pi = 0.116\%$ [Wall et al., 2008]; review of earlier studies $\pi = 0.116$ for autosomal loci [Przeworski et al., 2000]). These latter estimates were generated from "resequencing" studies that examined DNA sequences completely, not through the screening of samples for known SNP markers. However, use of SNP markers can introduce an ascertainment bias because the populations used to find variable SNPs are likely to show higher variation than SNPs recovered from a resequencing strategy (Wall et al., 2008). Although there is variation across the genome, a "rule of thumb" is that two random human alleles have about one difference per 1000 base pairs of autosomal DNA sequence (Tishkoff and Verrelli, 2003).

There are additional patterns in the genome. In the X chromosome, nucleotide diversity is somewhat lower than the autosomes (e.g., $\pi = 0.047\%$ [International SNP Map Working Group, 2001]; $\pi = 0.1\%$ [Wall et al., 2008]; $\pi = 0.065\%$ [Przeworski et al., 2000]) while Y chromosome nucleotide diversity is lowest of all (e.g., $\pi = 0.0135\%$ [Hammer et al., 2003]; $\pi = 0.0151\%$ [International SNP Map Working Group, 2001]). Mitochondrial DNA, on the other hand, has much higher polymorphism (based on 53 whole mtDNA genomes [not including D-Loop] $\pi = 0.28\%$ [Ingman et al., 2000]; based on 277 whole mtDNA genomes [protein-coding genes only] $\pi = 0.306\%$ [Kivisild et al., 2006]) because of its high mutation rate.

These levels of diversity can be closely examined in a number of ways to help understand human variation. Because DNA markers often reside in regions of the genome that are not subject to selection, they can be very informative about human demography and population history. Indeed, the level of neutral genetic diversity in a population or species is a reflection of demographic parameters. A simple formula, called, the population mutation parameter, or population mutation rate, expresses this simply as $4N_e\mu$. This formula means that the overall level of genetic diversity at a locus (θ) is equal to the effective size of a population (N_e) multiplied by the locus's neutral mutation rate per generation (μ) times 4 (4 for nuclear genes; 2 for mtDNA and Y chromosomes; 3 for X chromosomes; this reflects how many effective copies are present in individuals). The formula holds for a population that is at "equilibrium," when the role of mutation (μ) in adding new alleles to a gene pool has stabilized with effective population size (N_e), which determines the loss of alleles due to genetic drift. For example, a study of the same locus in two populations at equilibrium with equal mutation rates (μ), the population with more diversity (a larger θ) would be inferred to have a larger effective population size (N_e) than the population with the smaller θ (all other things being equal).

One useful estimator of θ is nucleotide diversity (π), which was summarized above for autosomes, sex chromosomes, and mtDNA. Using the equilibrium formula and estimators for the mutation rate, levels of human nucleotide diversity are consistent with a human effective population size (N_e) on the order of 10 000 individuals across loci (Garrigan and Hammer, 2006). Clearly, this is far fewer than our current census size (N) of 6 billion individuals. Although the effective population size is usually less than the census size (Frankham, 1995) due to factors such as age stratification in the census population, an effective population-size estimate of 10 000 is substantially different from our census size. This strongly suggests that humans are currently not in an equilibrium state and perhaps have an evolutionary history reflective of population fluctuations and sizes that were different than at present.

In addition to nucleotide diversity, DNA markers preserve a great deal of additional information such as the pattern and depth of branching relationship between alleles, patterns of linkage disequilibrium (nonrandom association of alleles), and recombination. More careful scrutiny of particular sets of these markers can therefore be informative about the past processes that have resulted in our current levels, patterns, and distribution of genetic diversity. Studies of these markers in humans have led to considerable insight in modern human origins, patterns of human migration, instances of natural selection, and other topics, which will now be reviewed.

DNA markers and the origin of modern humans

Our current patterns, levels, and distribution of genetic diversity are the product of evolutionary processes, including the ongoing effects of mutation, drift, gene flow, and selection. Because of this, anthropologists have examined genetic diversity to offer insights into the evolutionary history of our species. One critical issue is modern human origins, which specifically concerns questions about when, where, and how our species originated. Because much of our genome is noncoding, it is expected to primarily reflect our demographic history, instead of the effects of natural selection. DNA markers from neutral parts of the genome can therefore be especially useful to understanding human origins.

One of the pioneering studies that used DNA markers to address human evolution was that of Cann et al. (1987), which examined RFLPs from the purified mitochondria from 147 humans. This is the classic "Mitochondrial Eve" paper. In this study, Cann et al. (1987) generated a tree based on the mtDNA RFLPs from their sample of humans. An example of this type of tree, a coalescent tree, is shown in Figure 14.3 and is discussed further below. This human mtDNA tree had

a number of noteworthy features. Firstly, the tree was structured such that the deepest branch led to one set of individuals of African ancestry and a second set of individuals that were from Africa and elsewhere in the world. Parsimony suggested that the root of this tree, and thus the origin of modern humans, would have to have been in Africa. Secondly, Cann et al. (1987) used a mutation rate scaled to time to estimate a date for the deepest node in the study (referred to as the time to the most recent common ancestor, or TMRCA). This rate yielded a TMRCA estimate of 140–290 kya (thousands of years ago). These two aspects of the tree were interpreted to be consistent with a "Recent African Origin" model for anatomically modern *Homo sapiens* (AMHS) (this is also known as the "Rapid Replacement," "Mitochondrial Eve," or "Out of Africa" model).

There are a number of models for modern human origins. All models attempt to explain how AMHS became distributed throughout the globe within the last 100 kya. Prior to AMHS, there were other hominins present in the Old World. These included *H. erectus*, *H. neanderthalensis* (Neanderthals), and *H. heidelbergensis*. The earliest of these hominins (*H. erectus* and related species) are found throughout the Old World (Africa, Eurasia, and Southeast Asia) and date to about 1.8 million years ago (Feibel et al., 1989; Swisher et al., 1994; Gabunia et al., 2000).

Models for modern human origins attempt to explain how all of these species are related to one another in time and space. In general, the Recent African Origin model holds that AMHS originated approximately 100–200 kya within Africa (Stringer and Andrews, 1988). These AMHS individuals spread from Africa and rapidly replaced all other archaic populations of hominins without interbreeding. The main competing model is the "Multiregional Continuity" model (Wolpoff et al., 1984). This model generally argues for a degree of genetic continuity and ancestor-descendant relationships between the archaic and modern forms of humans. The continuity was maintained in some manner via gene flow, perhaps since the earliest members of *Homo*, such as *H. erectus* left Africa ~1.8 million years ago, up to our present gene pool.

Importantly, these two models make predictions for human genetic variation that can be tested with data from DNA markers. The Recent African Origin model predicts that the TMRCA for most loci should be distributed recently (Takahata, 1993; Ruvolo, 1996; Garrigan and Hammer, 2006), with the TMRCA being ~200 kya for mtDNA and Y chromosomes, and ~600 kya for X-linked DNA and nuclear DNA. Differences in age predictions among markers are due to the different effective population sizes of these different genomic regions based on their mode of inheritance. On the

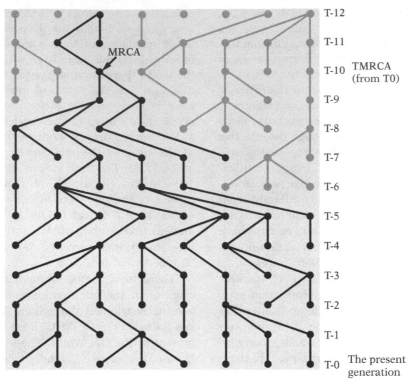

14.3. A simplified coalescent tree of alleles at a haploid locus in a "Wright–Fisher" population, which has a number of simplifying assumptions, e.g., nonoverlapping generations. Redrawn from a simulation at the website www.coalescent.dk. The present simulation is only one random genealogy with 8 individuals for 12 generations. Additional simulations would result in different genealogies. Additional simulations under different parameters can be conducted at the website listed above. T-0 denotes the present, and T-N corresponds to past generations. In general, when the only factors affecting mutation and the number of alleles that are transmitted are chance, this is the neutral model. In this case, the amount of genetic diversity is a result of the population size (drift) and the mutation rate. In this example, sampling all alleles at T-0 (black lines) and estimating a coalescent tree would recover an MRCA at the time T-10 (this is the most recent common ancestor [TMRCA]). Gray lines indicate individuals that do not have descendents that reach the present generation. A larger population size would likely mean that the TMRCA would be older. This can be investigated further at the simulation website. Note that at T-10, the TMRCA exists in a single individual, but that individual is part of a population of 8. This is analogous to the "Mitochondrial Eve": a single allele in a population of alleles. Also, note that in past generations, the TMRCA would be different. For example, if all alleles were sampled at a different time in the past, the TMRCA would be older. The same process works going forward, for example, one of the alleles at T-0 may be the TMRCA of alleles at a generation in the future. Also, genetic compartments with different effective population sizes, such as mitochondrial DNA (mtDNA) and autosomes, are predicted to have different TMRCAs.

other hand, the Multiregional Continuity model predicts that each gene's TMRCA could be substantially older, perhaps over 1.8 million years old.

The Recent African Origin and Multiregional Continuity models also make different predictions for the human effective population size, which can be estimated from genetic data using $\theta = 4N_e\mu$. The Recent African Origin model predicts a small effective population size due to AMHS having recently evolved from a restricted distribution. The Multiregional Continuity model predicts a larger effective population size, due to extensive migration across most of the globe among a large, fairly cohesive gene pool, which is likely to require many humans (Rogers and Harpending,

1992). The Recent African Origin model also predicts that most genes would be most diverse within African populations than elsewhere, and that trees of alleles will have their deepest node inferred to be placed in Africa (Takahata et al., 2001). The Multiregional Continuity model predicts a lack of patterning in these features.

It is also important to note that these predictions are based on very "strong" and simple versions of each model, i.e., these two models do not incorporate much complexity. While these "strong" models are useful historically and pedagogically, there are more sophisticated models of human evolution which contain elements of both (Relethford, 2001; Stringer, 2002).

Returning to Cann et al. (1987), based on the predictions of each model, this study is widely held to support the Recent African Origin model. Furthermore, subsequent studies of human mtDNA have confirmed this interpretation with a TMRCA of roughly 200 kya and the root node of the tree parsimoniously placed in Africa (e.g., TMRCA = 166–249 kya, 135 *control region* seqs [Vigilant et al., 1991]; TMRCA = 222 kya, 5 human *COII* seqs [Ruvolo et al., 1993]; TMRCA = 171.5 kya, 53 whole mtDNA seqs [Ingman et al., 2000]; TMRCA = 160 kya, 277 mtDNA genome seqs [Kivisild et al., 2006]; TMRCA = 194.3 kya, 320 mtDNA genome seqs [Gonder et al., 2007]; TMRCA = 203 kya, 624 mtDNA genome seqs [Behar et al., 2008], see original studies for confidence intervals). The levels of nucleotide diversity can also be compared to assess their fit to the different models. A recent, comprehensive study of mtDNA genomes found much higher diversity estimates for samples from Africa ($\pi = 0.392$) versus outside of Africa ($\pi = 0.181$) (Gonder et al., 2007), which is consistent with the Recent African Origin model.

Other interesting inferences can be drawn from mtDNA data sets. For example, the difference in estimates of our effective population size (10 000) and census size suggests that human population size has changed dramatically over our history, as a consequence of expanding or contracting through bottlenecks (i.e., dramatic population size reductions), during our 200 000-year history. Human mtDNA markers have other features that suggest population expansion, including an excess of rare sequence variants and a "star-shaped" phylogeny of alleles (Di Rienzo and Wilson, 1991; Merriwether et al., 1991).

A useful summary of these features is the distribution of differences when alleles are compared pairwise (the pairwise mismatch distribution) (Di Rienzo and Wilson, 1991; Rogers and Harpending, 1992; Harpending et al., 1993, 1998; Sherry et al., 1994). Population expansions result in a unimodel or "bell-shaped" mismatch distribution, while a multimodal (or "ragged") distribution is consistent with constant population size. Furthermore, expansion times and populations sizes can be estimated for unimodal distributions.

Human mtDNA marker data sets generally show evidence for population expansions in different human groups beginning about 100 kya (Rogers and Harpending, 1992; Harpending et al., 1993, 1998; Excoffier and Schneider, 1999; Ingman et al., 2000). These analyses also suggest that some populations did not experience expansions, while others experienced bottlenecks, and the timing of expansion differed in different parts of the world (Di Rienzo and Wilson, 1991; Rogers and Harpending, 1992; Harpending et al., 1993, 1998; Excoffier and Schneider, 1999; Ingman et al., 2000). For example, Excoffier and Schneider (1999) recovered evidence from mtDNA hypervariable sequences for the oldest expansion in East Africa (105 kya) and the youngest in Europe (42 kya). Ingman et al. (2000) examined 53 nearly whole genome mtDNA sequences from 14 different linguistic groups and detected evidence for an expansion outside of Africa dating to 38.5 kya.

These studies provided two interesting insights about modern human origins. Firstly, the estimated population size before the expansion was estimated to be small (<2000 individuals), which is difficult to reconcile with a Multiregional Continuity model that requires a contiguous population distributed throughout the Old World (Rogers and Harpending, 1992). Secondly, the timing of the expansion of human populations substantially postdates the TMRCA for the mtDNA trees suggesting that the "strong" Recent African Origin model may not strictly hold (Harpending et al., 1993). Human dispersals, differentiation, and isolation may have occurred between the mitochondrial TMRCA but before population expansion, leading to the proposal of the "Weak Garden of Eden" model (Harpending et al., 1993).

While these studies all support the Recent African Origin model, there is evidence that natural selection may have played a role in shaping human mtDNA diversity (e.g., Ruiz-Pesini et al., 2004; Kivisild et al., 2006; Balloux et al., 2009). Although the patterns cited above are not in dispute, the cause of them may be natural selection rather than demographic processes. Because mtDNA does not recombine, the action of selection will affect the entire genome due to genetic "hitchhiking." This makes the action of positive selection strong in its potential effects on reducing diversity. Additional work demonstrating the different relative roles that selection and demography have played on human mtDNA diversity would be welcome.

The Y chromosome has also been the target of a number of studies that address modern human origins. It has some features that are analogous to those of the mtDNA, such as its uniparental inheritance and its lack of recombination over a large part of its length. In other ways, it is quite different, in particular, because of its paternal inheritance, larger gene-poor sequence, and slower mutation rate.

In one of the first studies of the Y chromosome, Hammer (1995) examined DNA sequences from a 2.6 kilobase-pair region of the Y chromosome from 16 humans. The TMRCA of these alleles was estimated to be 188 kya and the effective population size was ~10 000 individuals. These aspects of the analysis were inconsistent with the Multiregional Continuity model, and the Y chromosome and mtDNA were shown to be in agreement with the Recent African Origin model.

Subsequent studies of Y chromosome diversity have included the analysis of microsatellite markers

(Hammer et al., 1998; Pritchard et al., 1999). Overall, this work has confirmed the consistency of the Y chromosome data with the Recent African Origin model, including a recent TMRCA, higher diversity within African populations compared to those from other parts of the world, and the node of the human Y chromosome tree being parsimoniously placed in Africa (e.g., Hammer et al., 1998, 2003; Pritchard et al., 1999; Thomson et al., 2000; Underhill et al., 2001). Microsatellite markers (Pritchard et al., 1999) and SNP markers (Thomson et al., 2000) on the Y chromosome also show evidence for a population expansion.

Because it is nonrecombining, the Y chromosome may be subject to natural selection in the same way that the mtDNA genome is. Some analyses support a role for selection in shaping Y chromosome diversity in humans (Jaruzelska et al., 1999). On the other hand, the TMRCA of the Y chromosome alleles has been shown to be consistently about half as old as for mtDNA, which may reflect demographic differences between males and females (e.g., differences in mating) (Wilder et al., 2004b). However, overall, with regard to the origin of modern humans, mtDNA and Y chromosomes are thought to yield generally similar scenarios (Underhill and Kivisild, 2007).

Often, because of its matrilineal nonrecombining inheritance, the most recent common ancestor (MRCA) of the human mtDNA alleles has often been referred to as "Mitochondrial Eve," as if this were a particular person. Although it is true that modern mtDNA alleles coalesced in an allele that existed in an individual at the base of the tree, this individual was not the original individual that marked the beginning of our species, nor was this individual alone during her lifetime (Ayala, 1995). Every region of the genome has its own history, including its own MRCA. Indeed, there is a male counterpart to the mtDNA MRCA – sometimes named "Y chromosome Adam." While regions that undergo recombination often have a more complicated tree structure of alleles, the most essential and critical idea is that every locus has its own history, including its own tree structure and its own TMRCA.

In neutral loci these features of a tree are a result of the forces of mutation, population size, recombination, linkage disequilibrium, and gene flow. Because these features are stochastic, even under the same demographic and evolutionary history different loci will have different TMRCAs and tree shapes. A simplified example of a coalescent tree is shown in Figure 14.3. Secondly, at the time of any gene's TMRCA, that particular allele exists within a population of alleles, i.e. group of humans. Given these caveats, and the fact that mtDNA and Y-chromosome studies each reflect the evolution of a single locus, it is essential to analyze tree shapes, levels of diversity, and TMRCAs from a number

of unlinked loci from across the genome in order to make more precise and accurate interpretations for the origins of modern humans, including differentiation between models including the Recent African Origin model, the Multiregional Continuity model, and others (Ruvolo, 1996; Garrigan and Hammer, 2006). This has been a major impetus for examining numerous DNA markers from the recombining nuclear genome for addressing modern human origins.

Early studies of DNA markers within the recombining nuclear genome (the 22 autosomes and the X chromosome) provided many additional windows onto human evolution. Goldstein et al. (1995) estimated a TMRCA of 156 kya and a tree rooted in Africa for 30 microsatellite markers. Tishkoff et al. (1996) examined two linked DNA markers within the *CD4* gene in 1600 humans and found a recent African origin for all non-African populations (~100 kya). Harding et al. (1997) examined 349 DNA sequences from the β-*globin* locus, and estimated the TMRCA of its alleles to be 800 kya and most parsimoniously placed in Africa, although there was a large amount of ancient diversity in Asia. By contrast, Harris and Hey (1999b) examined DNA sequences from the X-linked *PDHA1* gene, and estimated a TMRCA of 1860 kya, as well as extremely different patterns of variation between Africans (high diversity) and non-Africans (little diversity), and large sequence differences between these two sets of alleles (though this pattern did not hold with additional sampling (Yu and Li, 2000)).

In this sample of early studies, the different estimates for the TMRCA and other patterns in these nuclear data sets were clearly heterogeneous, a pattern that was noted at the time (Hey, 1997; Harris and Hey, 1999a). Another conundrum was that the recombining genomic regions did not all show the consistent evidence for the population expansions found in mtDNA and the Y chromosome (Harpending and Rogers, 2000; Wall and Przeworski, 2000). At this time, only nuclear microsatellite markers revealed evidence for such recent expansions (Reich and Goldstein, 1998; Kimmel et al., 1998; Zhivotovsky et al., 2000).

The heterogeneity in these studies was likely due to any of a number of reasons. Firstly, the effective population sizes of autosomes is four times higher than that of the mtDNA and the Y-chromosome, and this predicts a four-fold longer expected time to a MRCA (three-fold for X-linked genes due to the hemizygous state for males). The older TMRCA for these data sets may be in line with the expectation from the Recent African Origin model (Ruvolo, 1996), although the TMRCA for the *PDHA1* alleles was exceptionally old. Secondly, it may be the case that one set of the loci studied could be under some form of selection while others are reflecting primarily the demography of human groups (Harris and Hey, 1999a; Harpending and Rogers, 2000; Przeworski et al., 2000; Wall and Przeworski, 2000). Selection is not only a concern for mtDNA and Y-chromosome genes, but also for the other nuclear loci because they were largely located within or near coding regions. Because of the potential action of selection, it is difficult to resolve whether it is the nuclear genes or the mtDNA and Y chromosome genes that reflect demography at these loci. Thirdly, the cause of the discrepancy may be related to the mutation rate of the loci examined. Specifically, it could be the case that the rapid mutation rate of the mtDNA and the microsatellites provided the resolution required to track recent demographic events, while the slower rate of mutations at SNP markers allowed only the tracking of older events. Despite the heterogeneity in TMRCA and expansion estimates, the pattern of gene trees shapes being rooted mainly within Africa is consistent with an African origin for diversity and not a Multiregional Continuity model (Takahata et al., 2001).

Given these issues, subsequent DNA-sequencing projects targeted regions of the recombining nuclear genome that were very likely to be neutral, and therefore reflect demographic processes. In this light, Kaessmann et al. (1999) conducted a study of a 10 kilobase-pair intergenic, presumably neutral, X-linked region (Xq13.3) in 69 humans. They recovered a TMRCA at 535 kya with the highest levels of diversity being seen in African populations (Kaessmann et al., 1999). This region also showed evidence for a population expansion (Kaessmann et al., 2001).

There have been a number of subsequent DNA sequence studies of long (~10 kilobase-pair) mainly noncoding regions in humans. These include analyses of chromosome 22 (Zhao et al., 2000), chromosome 1 (Yu et al., 2001), the X chromosome (Yu et al., 2002b), and chromosome 6 (Zhao et al., 2006). These five data sets were summarized by Zhao et al. (2006) as tending to support an "Out-of-Africa" model for human origins rather than the Multiregional Continuity model. This included greater African diversity, African alleles having a closer proximity to the root, an autosomal TMRCA of 860 kya, and a low autosomal effective population size (Zhao et al., 2006). Importantly, particular aspects of the data sets suggested that these two models were both too simple to account for all aspects of observed data, such as the old TMRCA estimates for non-African alleles and intermediate frequency mutations outside of Africa (Zhao et al., 2006).

Recent analyses of a handful of data sets are especially suggestive that simple models like Recent African Origin and Multiregional Continuity hypothesis may not be sufficient to account for all aspects of the existing genetic data. These loci include the X-linked *RRM2P4* region (Garrigan et al., 2005b; Cox et al., 2008), the *Xp11.22* region (Shimada et al., 2007),

and the *microcephalin* locus (Evans et al., 2006). In these loci, a highly divergent allele is present and occurs most frequently outside of Africa, suggesting a non-African root node. In other cases a very divergent allele is present within Africa (Garrigan et al., 2005a). These data sets have a host of possible explanations. The *microcephalin* case may reflect the introgression of an "archaic" allele in the AMHS gene pool followed by locus-specific selective processes (Evans et al., 2006). One hypothesis is that the "*D*" allele, which may be from an archaic human population, confers a selective advantage related to brain size causing it to spread in the modern gene pool after introgression (Evans et al., 2006). However, while this locus does play a role in microcephaly, the precise phenotypic changes or selective advantage of extant human genetic variants is unknown (Mekel-Bobrov et al., 2007; Timpson et al., 2007). The other data sets have been explained by demographic models that include admixture between AMHS and archaic humans (Garrigan et al., 2005b; Garrigan and Hammer, 2006; Cox et al., 2008), and/or significant population structure (meaning that the population was not one large panmictic population) in the early AMHS gene pool (Shimada et al., 2007). Others question whether polymorphic chromosomal inversions and selection may be creating a false signal of admixture (Reed and Tishkoff, 2006). Studies on the patterns of linkage disequilibrium in modern human populations also suggest some admixture between humans and archaic populations (~5%) (Plagnol and Wall, 2006). More complex models that include multiple dispersals to and from Africa over the Pleistocene have also been forwarded to account for the variance in human population genetics data sets (Templeton, 2007).

Direct genetic evidence from other hominins would be very helpful for distinguishing between the different models of human evolution. Indeed, since the initial publication of a Neanderthal mtDNA sequence (Krings et al., 1997), sequences from these archaic humans have been a critical part of the modern human origins debate. The collection of these sequences (and others) is a testament to the power of PCR and other molecular genetics techniques to amplify and sequence DNA from ancient specimens (Pääbo et al., 2004). A Recent African Origin model suggests that Neanderthal mtDNA sequences would be significantly different from all AMHS sequences, while the Multiregional Continuity model could potentially show Neanderthal sequences to be nested within AMHS diversity.

Over the last 10 years, the data obtained from ancient Neanderthal mtDNA are relatively unequivocal about structure of genetic relationships between Neanderthals and AMHS. The hypervariable mtDNA sequence from the Neanderthal sample was phylogenetically situated outside of an AMHS group with a TMRCA about three times as deep as the deepest AMHS node (Krings et al., 1999). Furthermore, the Neanderthal sequence was about equally different from all AMHS, showing no specific affinities with modern European sequences (Krings et al., 1997, 1999). These sequences were augmented recently to six complete Neanderthal mtDNA genomes (Green et al., 2008). This data set shows that humans and Neanderthals form two distinct monophyletic clades separated from each other for at least 439 kya (Briggs et al., 2009).

Although this is interpreted as suggesting that Neanderthals did not contribute to the AMHS gene pool, it has alternatively been suggested that the two clade patterns at mtDNA is statistically compatible with a scenario where AMHS and Neanderthals formed a merged population in the past, but the Neanderthal mtDNA alleles were subsequently lost due to drift (Nordborg, 1998). Simulations suggest that direct investigations of large numbers of ancient sequences from both AMHS and Neanderthals would be required to refute a Neanderthal contribution to the modern mtDNA gene pool (Nordborg, 1998; Serre et al., 2004). While levels of admixture over 25% can be excluded, estimates below this number cannot despite the lack of direct evidence for admixture (Serre et al., 2004). These admixture estimates are very conservative, and more complex scenarios that model population interactions between humans and Neanderthals suggest that there was nearly no admixture (Currat and Excoffier, 2004).

As learned during the human origin debate, mtDNA represents a single locus and data from the nuclear genome is necessary to address questions more meaningfully. To this end, two large nuclear DNA sequence data sets have been published from Neanderthal material (~1 million base pairs [Green et al., 2006]; 65 250 base pairs [Noonan et al., 2006]). These studies came to vastly different conclusions on the relationship of Neanderthals to AMHS. Specifically, Noonan et al. (2006) supported significant divergence between humans and Neanderthals while Green et al. (2006) supported a model where Neanderthals were within the range of diversity found within modern humans. In a reanalysis of the data, Wall and Kim (2007) showed that the sequences of Green et al. (2006) reflected either the inclusion of contaminant DNA sequences or sequencing errors in the analysis. Considering only the Noonan et al. (2006) data, the TMRCA between humans and Neanderthals was 706 kya, a population split occurred at ~350 kya, and the most likely amount of admixture between humans and Neanderthals was estimated to be 0%, although the confidence interval spanned 0–20% (Noonan et al., 2006) or 0–39% (Wall and Kim,

2007). These findings are consistent with the mtDNA analyses.

A recent analysis of unlinked nuclear loci has recently been shown to strongly support a Recent African Origin model without admixture (Fagundes et al., 2007). This analysis examined sequences from 50 unlinked ~500 base-pair regions from three populations (African, Asian, and Native American) (Yu et al., 2002a; Fagundes et al., 2007). Using these data, a number of models were assessed including African replacement models, assimilation/admixture models, and Multiregional Continuity models. An African replacement model was strongly favored, with a "speciation time" of 141 kya for AMHS and a migration from Africa at 51 kya. Fagundes et al. (2007) also showed that under some versions of an African replacement model for human origins, some old TMRCAs are expected, including genes that have their root outside of Africa. These findings may indicate that admixture is not necessary to explain data sets like *RRM2P4* (Garrigan et al., 2005b) or *Xp21.1* (Garrigan et al., 2005a), whose patterns could be due to chance. Clearly, Fagundes et al. (2007) support a Recent African Origin model, but with additional model complexity in the form of large African population sizes and dramatic size changes outside of Africa.

In summary, the current wealth of data appears to be most consistent with some form of a Recent African Origin model for AMHS rather than a Multiregional Continuity model (Tishkoff and Verrelli, 2003). However, the totality of these studies suggests that a simple "single ancestry" Recent African Origin model cannot explain the variance in the loci sampled (Garrigan and Hammer, 2006). Instead of a simple Recent African Origin model, those that incorporate elements such as ancestral population structure in Africa, low levels of migration, and/or population expansions seem to fit the data best (Garrigan and Hammer, 2006; Fagundes et al., 2007; Campbell and Tishkoff, 2008). Models that incorporate some level of admixture between archaics and AMHS may also prove useful (Garrigan and Hammer, 2006; Wall and Hammer, 2006), although more data are required to assess this effect more fully (Wall, 2000). In the future, additional genetic data sets and analytical techniques, in conjunction with new material and interpretations from the paleontological record, will be required for further refining and advancing our understanding of modern human origins.

Global patterns of human migration and gene flow

Many aspects of human migration are implicit in the study of modern human origins, such as the movement of humans out of Africa and across the world. Studies of

DNA markers have been utilized extensively to directly examine the relationship between genetic diversity and geography. DNA markers have been used to explain the general, large-scale patterns of human migrations, such as the out of Africa migration event that is part of the Recent African Origin model. These markers are also used to explain more specific instances of migrations, such as the peopling of the Pacific or Iceland. These questions most specifically relate to the relationship between genetic diversity and geography.

However, at the lower levels "migration" is also synonymous with gene flow or admixture, the movement of alleles between populations. This has also been implicitly discussed, for example in the potential cases of admixture between archaics and AMHS, which constitutes a degree of gene flow between these populations. Studies of DNA markers have been utilized extensively to examine gene flow in the form of differences in patterns of gene flow between males and females and patterns of gene flow between human populations.

The relationship of genetic diversity and geography is critical for understanding fine-scale population genetic questions. At classical marker loci, it has long been noted that human polymorphisms generally behave in a pattern where genetic distance and geographic distance are positively correlated and that gene frequencies change clinally (Cavalli-Sforza et al., 1994). This pattern is generally referred to as "isolation by distance" which suggests that human migration patterns are related to genetic distance, and that matings predominantly occur over short distances. Interestingly, variation in cranial morphological data is in agreement with classical markers and microsatellites, all being consistent with isolation by distance models at global scales (Relethford, 2004).

A number of studies have recently re-examined the general patterns of human geography and genetic diversity using large sets of neutral DNA markers. Ramachandran et al. (2005) examined a global microsatellite marker data set from 1027 diverse humans (Cann et al., 2002; Rosenberg et al., 2002) and compared pairwise genetic distances between populations (using F_{ST}, an index of genetic differentiation between groups) using "great circle" distances between populations and "waypoint" distances, i.e., distances that used more realistic migration paths between points (e.g., requiring Egypt to connect points between Africa and Asia). These models produce high correlation coefficient (R^2) values of 0.59 and 0.78, respectively, showing that these genetic and geographic distances are closely related.

Given the nonequilibrium nature of the human populations, this relationship was interpreted as a "Serial Founder Effect," a model where humans originated from a single source, with successive migrations sampling from its most recent source populations.

Because it provides the most explanatory power in generating correlations, Africa was indicated as the likely source population. This Serial Founder Effect model explains up to 78% of the variation between genetic differences between populations (this value comes from the R^2 estimated for the model), suggesting that the remaining variation can be explained by population-specific factors, such as drift and selection (Ramachandran et al., 2005).

Models that estimate time in the context of this data set show a population-size expansion beginning at about 56 kya from a very small founding population (~1000 individuals) in East Africa (Liu et al., 2006). This date is generally consistent with the TMRCA estimates for the non-African portions of allele trees and the evidence from mismatch distributions summarized above. Similar studies based on different markers, estimates of genetic differentiation, and/or geographic modeling provide qualitatively similar results (Manica et al., 2005; Prugnolle et al., 2005; Ray et al., 2005; Jakobsson et al., 2008; Li et al., 2008; Tishkoff et al., 2009).

Interestingly, some analyses of large-scale microsatellite data suggest that humans form discrete clusters that corresponded to geography (Rosenberg et al., 2002). The cause of this pattern was quickly debated, perhaps due to the possible implication that this pattern was an affirmation of human "races" (e.g., Serre and Pääbo, 2004). One suggestion was that the discontinuous sampling regime of human groups in this study led spuriously to this result (Serre and Pääbo, 2004). Alternative explanations suggest that the relationship between genetic and geographic distance is largely clinal, yet has patterning that can be attributed to minor but real differences in geography (Rosenberg et al., 2005).

Hunley et al. (2009) recently crystallized these studies into four different models for the relationship between genetic and geographic distance in humans. These models include independent regions (i.e., each geographic region evolves independently), isolation by distance (described above), serial founders (described above), and nested regions. In a nested region model, non-African diversity is represented by an increasingly reduced set of alleles with increasing distance from Africa (Tishkoff et al., 1996) based on a mix of bottlenecks and expansions. Comparing simulations under each model with the pattern observed from a 783 microsatellite marker data set including 1032 humans (Cann et al., 2002; Rosenberg et al., 2005) showed that none of these models explained the totality of the data. Their preferred model includes a combination of serial founding, bottlenecks, migrations, and gene flow (Hunley et al., 2009). In other words, a migration scenario that is considerably more complex than any of the ones previously suggested. However, the overall pattern from these studies remains consistent with a Recent African Origin model for human origins and dispersal.

Finer-scale examination of DNA markers can assist in making further refinements of the migration of humans throughout the globe. Targets of study have included regional migration events, such as the original migration out of Africa. For example, similar to studies of human origins, the ages of particular nodes have been used to date the out-of-Africa migration, using the TMRCA of the non-African alleles within the trees. Based on the TMRCA of non-African mtDNA genomes, two estimates for an out of Africa migration are 94 kya (Gonder et al., 2007) and 52 kya (Ingman et al., 2000). The Y chromosome yields younger estimates for non-Africans, e.g., 40 kya, (Thomson et al., 2000). Autosomal studies generally agree with these estimates (Fagundes et al., 2007). Other studies have directly addressed the geography of the migration routes out of Africa (reviewed in Reed and Tishkoff, 2006).

Many other migration events also have been investigated. Two events that have received a great deal of study are the peopling of the Americas (reviewed in Dillehay, 2009) and the peopling of the Pacific (Friedlaender et al., 2008; Kayser et al., 2008a, 2008b). Other studies have focused on understanding the relationships of population movements within continents, e.g., Africa (Reed and Tishkoff, 2006; Tishkoff et al., 2009). Even more fine-scale studies have focused on the peopling of particular islands, such as Iceland (Helgason et al., 2009), and the origins of particular human populations, such as the Lemba, a South African group with ancestral ties to the Near East (Thomas et al., 2000). Lastly, genetic research using molecular markers has allowed for greater understanding of historical movements of people and admixture between human groups (e.g., Alves-Silva et al., 2000; Quintana-Murci et al., 2004; Tishkoff et al., 2009).

Sex-specific gene flow within and between populations plays an important role in shaping local patterns of human genetic diversity. In an analysis comparing the diversity between nuclear, mitochondrial, and Y-chromosome-linked DNA markers in humans, Seielstad et al. (1998) demonstrated a pattern in which Y-chromosome markers showed markedly reduced diversity within populations compared to autosomal markers and mtDNA (35.5% for Y chromosomes vs. ~81–86% for autosomes and mtDNA). Furthermore, there were more genetic differences between populations at Y-chromosomes DNA markers than mtDNA or autosomal markers (Seielstad et al., 1998). This means that there are greater differences at the Y chromosome between populations, i.e., the Y-linked diversity is more geographically restricted than autosomes and mtDNA markers.

Because the Y chromosome reflects a paternal inheritance pattern, this finding suggested differences in aspects of demography between males and females. Seielstad et al. (1998) showed that sex differences in migration rate best explained the data and reflect an overall patrilocal pattern for humans, i.e., that women moved further than men to mate and reproduce. This patrilocal tendency is supported by the ethnographic record (Murdock, 1967) and is consistent with other global studies of human Y-chromosome and mtDNA markers (Wilder et al., 2004b). However, a resequencing study of 389 humans from 10 populations showed no differences in within group variation at Y chromosomes versus mtDNA, calling this result into question (Wilder et al., 2004a). Markers from mtDNA and Y chromosome may not exclusively trace demography because of the potential confounding action of selection.

Examination of neutral markers on autosomes versus the X chromosome can potentially circumvent these problems (Ramachandran et al., 2004; Wilkins and Marlowe, 2006). This is because autosomes are found equally in males and females, while X chromosomes spend more time residing in females, on average. Also, both genome compartments recombine, mitigating the effect of selection on linked sites. In X chromosome versus autosome comparisons, examination of microsatellite markers demonstrated no differences between these compartments (Ramachandran et al., 2004) while SNP markers suggested a model in which genetic drift has actually been stronger on the X chromosome perhaps due to selection or high long-range male migration after leaving Africa (Keinan et al., 2009). In contrast, Hammer et al. (2008) used a resequencing strategy and recovered higher levels of X-chromosome diversity. This was interpreted by the authors to be caused by polygyny's effects on male reproductive variance, which can act to lower males' effective population size relative to that of females. However, this result is also consistent with higher rates of female migration (Hammer et al., 2008).

One of the main problems with ascertaining the effect of residence patterns in humans may be our flexible social structure. As Wilkins and Marlowe (2006) pointed out, while most current human groups are patrilocal, this may reflect recent changes. Over the course of human history our residence patterns may have followed that of current forager populations, who may be best represented as bilocal (Marlowe, 2004). This historical difference may help explain the differences in the global studies cited above (Wilkins and Marlowe, 2006), and further suggest that recent mating and residence patterns have a stronger effect on diversity.

Indeed, at the local level, the relationship between sex-specific migration/residence patterns and genetic diversity appears to have a stronger relationship to cultural practices. For example, Oota et al. (2001) examined mtDNA and Y-linked markers within three patrilocal and three matrilocal populations in Thailand. These populations fit the predictions of the sex-biased migration model, with the patrilocal populations having low diversity on Y markers and high diversity at mtDNA markers. This finding is supported by data from other populations (e.g., Chaix et al., 2004; Destro-Bisol et al., 2004; Bolnick et al., 2006).

Interestingly, further studies have shown differences between exogenous populations, in which marriage partners can be drawn from outside the population, and strictly endogenous populations, in which marriage partners are drawn only from within the population (Kumar et al., 2006). In endogenous marriages, differences between mtDNA and Y-chromosome diversity are not expected to accrue, resulting in no differences between patrilocal and matrilocal populations (Kumar et al., 2006). Analysis of endogenous groups from India reflect this expectation (Kumar et al., 2006). Local-scale studies have also used X chromosome versus autosome comparisons to demonstrate that patrilineal herders have more female migration and higher female effective population size than bilineal groups (Ségurel et al., 2008).

In summary, migration, gene flow, and geography all have had an effect on human genetic diversity. DNA markers show that genetic and geographic distances are strongly correlated globally (Ramachandran et al., 2005), although the exact model underlying this correlation is complex (Hunley et al., 2009). At the global scale, human residence patterns may play a role in patterning human diversity (Seielstad et al., 1998), although the effect is debated (Wilder et al., 2004a). However, at the local level human mating patterns appear to strongly influence genetic diversity (e.g., Oota et al., 2001).

DNA markers and natural selection

Natural selection is the evolutionary force responsible for generating adaptations. Natural selection works via differential reproductive success that covaries with genetic variation. Humans have many unique traits that are thought to be adaptations, such as language, encephalized brains, bipedal locomotion, and opposable thumbs with precision grip. Adaptive traits such as these are often held up to be what has "made us human" (Varki and Altheide, 2005) and, as such, have been the focus of considerable investigation. Humans also differ from our living primate relatives in other ways, for example, in disease susceptibility (Varki and Altheide, 2005). These adaptive traits were fixed by natural selection in the human lineage. Therefore, they have limited variance among humans and differentiate

us from our closest living relatives, the apes, and our fossil ancestors. The study of DNA markers has been used to investigate the molecular causes of these phenotypic differences between humans and nonhuman primates. Phylogenetic comparisons between human genes and their orthologs (genes that are identical by descent) in chimpanzees and other mammals have been widely employed in the search for the bases of human uniqueness. These interspecific comparisons search for two main classes of evolutionary changes: those in protein-coding regions (i.e., genes) and those in noncoding regions.

Studies on protein-coding genes have focused on uncovering instances of molecular adaptation along the human lineage. Positively selected genes are those that show evidence of more amino acid change than expected, deciphered from comparisons of the rates of change in nonsynonymous codon sites versus synonymous sites (commonly referred to as the K_a/K_s ratio, the d_N/d_S ratio, or ω) (Kimura, 1977; Miyata et al., 1980). These comparisons have been done in "candidate genes," i.e., those with a priori evidence for a relationship to the human phenotype. Candidate gene studies for human molecular adaptation have included genes relating to brain size (e.g., *ASPM* [Zhang, 2003; Evans et al., 2004; Kouprina et al., 2004]) and language (e.g., *FOXP2* [Enard et al., 2002]). In both of these genes positive selection was demonstrated along the human lineage. Scans for selection have also been done at the genomic level, comparing thousands of coding regions among mammalian species (Clark et al., 2003; Chimpanzee Sequencing and Analysis Consortium, 2005; Arbiza et al., 2006; Bakewell et al., 2007; Rhesus Macaque Genome Sequencing and Analysis Consortium, 2007; Kosiol et al., 2008). (Access to genome data is available at the UCSC Genome Browser http://genome.ucsc.edu [Kent et al., 2002].)

Some genomic analyses show that genes that are positively selected along the human lineage (called positively selected genes or PSGs) are enriched for biological mechanisms related to immune defense, sensory perception, reproduction, and cancer processes (apoptosis, cell cycle control, and tumor suppression) (Clark et al., 2003; Nielsen et al., 2005; Bakewell et al., 2007; Kosiol et al., 2008). However, there are a few interesting findings from these comparative genomic studies. Firstly, thus far, the processes that have been identified in these genomic scans do not seem to correspond to the phenotypic features classically considered to be the hallmarks of our species. Secondly, one study reported few PSGs exclusive to the human lineage (7 out of 17 489 orthologs [Kosiol et al., 2008]). While this may be related to the low power of the statistical tests used, it nevertheless is surprising that so few genes have undergone selection exclusively in humans. Thirdly, there are more PSGs found on the chimpanzee lineage than the human lineage (Bakewell et al., 2007). These results are surprising because the hominin lineage is a great deal more derived than the chimpanzee lineage. As of yet, there is no dramatic genomic signature in protein-coding genes documenting human uniqueness.

Other studies have comparatively examined noncoding regions of the genome for evidence of natural selection. The main rationale for these studies is the hypothesis that the majority of human phenotypic evolution is caused by regulatory change, not by protein change (King and Wilson, 1975). To this end, studies have looked for rapidly evolving noncoding regions on the human lineage (Prabhakar et al., 2006). An excess of human-specific accelerated noncoding regions have been found near genes involved in neuronal function, although the chimpanzee lineage has a similar pattern (Prabhakar et al., 2006). Indeed, human accelerated regions may have a function in gene expression (Bird et al., 2007). One particular region stands out as particularly interesting. This is human accelerated region 1 (HAR1), which is a 118 base-pair region that appears to be undergoing exceptionally accelerated change in the human lineage (Pollard et al., 2006). HAR1 is part of an RNA gene expressed in the developing human brain suggesting a role in human brain size (Pollard et al., 2006). Additional work on other human accelerated noncoding regions is likely to provide considerable insight into the evolution of the human phenotype.

Numerous studies have also looked for selection in the form of highly conserved noncoding regions, under the assumption that conserved regulatory elements are likely to be functionally important (Dermitzakis et al., 2003; Bejerano et al., 2004). Studies of these noncoding conserved regions demonstrate that they are indeed functional (Woolfe et al., 2005) and these regions are constrained within humans (Drake et al., 2006).

While these PSGs, human accelerated regions, and conserved noncoding regions are common to nearly all humans, other features that are hypothesized to be under selection are variable within our species. These include traits that vary in humans throughout the globe, such as body size and proportions (Katzmarzyk and Leonard, 1998), skin color (Jablonski and Chaplin, 2000), lactase persistence (Swallow, 2003), susceptibility to infectious disease (Hill, 2006), and others covered in this volume. In general, many of these are thought to be adaptive responses to different climatic, dietary, and infectious environments that occur throughout the globe. Furthermore, selection has also acted differently on humans over time, both before and after the shift to agriculture (Livingstone, 1958; Armelagos and Harper, 2005).

Understanding the role of selection at genes within a species is a population genetic question that is intimately tied to understanding the population genetics of neutral DNA markers, described above. Essentially, understanding the role of selection at particular genes often relies on different methodological comparisons to the diversity present in neutral genes, where diversity is due to demographic processes (Kreitman, 2000; Bamshad and Wooding, 2003; Harris and Meyer, 2006; Sabeti et al., 2006). For example, a locus that has two alleles under balancing selection is likely to be more diverse and have an older TMRCA than a neutral locus, because selection preserves these alleles in the gene pool, allowing them to build more diversity through mutations and a longer genealogical history than expected (human example: *CD209L* [Barreiro et al., 2005]). In contrast, an allele at a locus that is being swept rapidly through a population due to selection will have lower diversity, high linkage disequilibrium, and a shallower TMRCA than at neutral loci (human example: the *lactase* gene (*LCT*) [Bersaglieri et al., 2004; Tishkoff et al., 2007]). In genes that are undergoing local adaptation, selected loci will show more genetic differentiation between populations than will neutral loci (human example: pigmentation genes [Norton et al., 2007]). In each case, neutral loci set the expectations for diversity, which is why they are critical to understanding whether genes are under natural selection.

Similar to the genomic scans for interspecific cases of natural selection, there have been many studies mining genomic-scale human population genetics data sets for evidence of selection (a recent review examined 21 population genomics scans in humans [Akey, 2009]). Furthermore, there is currently a wealth of different methods that ascertain deviations in genomic data from the neutral expectation in different ways (Sabeti et al., 2006; Akey, 2009). For example, some scans detect selection by finding unusually long haplotypes (Sabeti et al., 2002b, 2006; Voight et al., 2006). When a positively selected allele spreads rapidly to fixation in a population, many linked sites extending in both directions from the selected site also spread in the population as an extended haplotype. Eventually recombination "breaks up" a long haplotype, but for a period of time it remains in the population as a signal of positive selection.

Sabeti et al. (2007) examined over three million SNP markers and found over 300 of these "candidate" regions. Further refinements revealed genes under selection were related to lactase persistence, skin color, and viral diseases (Sabeti et al., 2007). Akey (2009) recently collated results from 9 human population genomic scans for selected regions and found 5110 total regions under selection. Surprisingly, only 722 (14.1%) of these putatively selected regions were discovered in more than 1 of the 9 studies. While these 722 regions contained genes that are known to be under selection (e.g., *LCT*), there were also a considerable number of novel loci such as genes involved in cochlear function (Akey, 2009).

Before the advent of genomics, the ascertainment of natural selection relied on the "candidate gene" approach, which examined the population genetics of loci suspected to be under selection a priori. These hypotheses can be derived from genes suspected to be under selection based on classical markers, studies of disease distributions, and/or from suspected selective agents. One of the most well-known selective agents suspected to be acting on classical loci, DNA markers, and disease polymorphisms is malaria (Allison, 1954, 1961; Kwiatkowski, 2005). The disease has a tropical and subtropical distribution that currently includes 109 countries in Africa, the Middle East, Asia, and the New World (World Health Organization, 2008). Annually, about 247 million humans contract malaria and it causes 1 million deaths, mainly in children under 5 years of age (World Health Organization, 2008). The major parasite responsible for mortality is *Plasmodium falciparum*, though the *P. vivax* parasite is also responsible for a large number of cases of illness, especially outside of Africa (Mendis et al., 2001). (See Chapter 27 of this volume for a more complete discussion of human malaria.)

Livingstone (1958) suggested that malaria was a recent selective agent connected to agriculture, which caused changes in land use, increasing both mosquito habitat and population density. These features allowed malaria to become a selective agent, which subsequently gave rise to an increase in sickle cell anemia (Livingstone, 1958). For these reasons the population genetics of human malaria resistance has been extensively studied at candidate loci in many populations of humans around the globe. These studies provide substantial evidence for the action of natural selection on modern humans.

Based on a review of early classical marker and epidemiological studies, Allison (1961) provided evidence for a relationship between malaria and multiple β-*globin* gene alleles (*Hb^S*, *Hb^C*, and *Hb^E*), thalassemias (deficiencies in the production of the α- and β-*globin* genes), and G6PD enzyme deficiencies. Similar relationships were hypothesized between malaria and Southeast Asian ovalocytosis (the *SLC4A1* gene) (Serjeantson et al., 1977) and Duffy blood group negativity (the *FY* gene) (Miller et al., 1976; Livingstone, 1984). These studies, and others, provided a host of "candidate genes" for examination by DNA markers. It is worth noting that many of these genes have extreme medical importance, which has led to direct clinical investigations studying their effects (Kwiatkowski, 2005). Here, I will review the inferential evidence for natural selection acting on two erythrocyte related genes β-*globin* and α-*globin*, and a novel candidate locus from the immune system (*CD40L*).

The hemoglobin protein is a tetramer comprising two α- and two β-chains that carries oxygen and carbon dioxide to and from tissues. This tetramer fills red blood cells. Carriers for the *Hb^S* allele are semiprotected against death from malarial infection (Allison, 1954), although individuals homozygous for *Hb^S* can have sickle-shaped red blood cells. The *Hb^S* allele is common throughout Africa, southern Europe, and South Asia.

DNA marker studies of the β-*globin* gene have produced a number of interesting findings from a number of different parts of the world. Although all *Hb^S* alleles have the same mutation, which affects the sixth amino acid, this same change has occurred on a number of different haplotypes. Using RFLP markers, Pagnier et al. (1984) showed that the *Hb^S* mutation occurred on at least three different haplotypes in Africa, and Kulozik et al. (1986) found an additional origination in Asia. This shows that the mutation has occurred multiple times, and despite its apparent selective cost, each time increased in frequency due to its protective effects against malaria. Furthermore, at least one of the *Hb^S* mutations is apparently recent, having occurred less than 2100 years ago (Currat et al., 2002). Single nucleotide polymorphism studies of *Hb^S* alleles show that linkage disequilibrium extends for over 400 kilobases across a recombination hotspot (Hanchard et al., 2007).

Other hemoglobin alleles have more recent origins. The *Hb^C* allele is common in West Africa and is semiprotective against death from malarial infection (Modiano et al., 2001). This allele has an origin within 5000 years as well as extended linkage disequilibrium consistent with natural selection (Wood et al., 2005). The *Hb^E* allele, common in Southeast Asia, is protective against malaria (Chotivanich et al., 2002). The *Hb^E* allele is also inferred to have arisen more than once in Southeast Asia and Europe (Kazazian et al., 1984). A microsatellite analysis of a single *Hb^E* type showed extended linkage disequilibrium and an origin for the allele within the last 4400 years (Ohashi et al., 2004). The recent origin of all of these alleles is strongly consistent with Livingstone's (1958) "agriculture hypothesis" for malaria.

Mutations in the α-*globin* genes also show evidence for malarial selection. Humans have two identical α-*globin* genes arranged in tandem. The major mutations in α-*globin* are whole gene deletions, not amino acid replacements. These deletions cause a reduction in α-chain production, resulting in a disease called α-thalassemia. Flint et al. (1986) investigated the distribution of α-*globin* gene deletions and malaria in Melanesia using RFLPs. They found that the frequency of α-*globin* deletion alleles was higher in coastal Papua New Guinea (which has higher rates of malaria) than highland Papua New Guinea (which has low rates of malaria). Within Melanesia, a clinal latitudinal gradient was found both for malarial endemicity and the frequency of α-*globin* deletion alleles. These relationships did not hold for other markers. These findings strongly supported a link between malaria and α-thalassemia in Melanesia. The exact mechanism for protection, however, is not yet fully understood (Kwiatkowski, 2005).

CD40L is an X-linked gene that encodes the CD40 ligand protein and plays a crucial role in the immune system, facilitating immunoglobulin class switching. Mutations in this gene can lead to X-linked hypergammaglobulinemia (XHIM) (Allen et al., 1993) leaving individuals immune deficient and susceptible to recurrent infections from pathogens such as *Pneumocystis carinii* and *Cryptosporidium* (Levy et al., 1997). A SNP marker in the *CD40L* promoter region has been shown to ameliorate malaria in Gambian males (Sabeti et al., 2002a). Subsequent analysis of this SNP marker allele showed that the allele was present at high frequency in Africa and at much stronger linkage disequilibrium than other SNP markers (Sabeti et al., 2002b). This allele is inferred to be about 6500 years old (Sabeti et al., 2002b). Interestingly, the molecular mechanism of this protective allele is not currently known, and it remains to be shown what particular genes in this large haplotype have a direct role in malarial adaptation (Sabeti et al., 2002a).

Studies of DNA markers in human populations have also inferred selection acting on other candidate genes involved in malarial resistance. These include *SLC4A1* (the basis for the Diego blood group) (Wilder et al., 2009), *GYPA* (the basis for the MN blood group) (Baum et al., 2002), and *G6PD* (a red blood cell enzyme) (Tishkoff et al., 2001).

Natural selection is the force of evolution that creates adaptations. Analyses of humans and other primates have demonstrated the action of selection on candidate loci for uniquely human traits (e.g., Zhang, 2003; Evans et al., 2004; Kouprina et al., 2004). Comparative genomic scans, on the other hand, have found few genes under natural selection at the protein level in humans overall (Kosiol et al., 2008), and when compared to chimpanzees (Bakewell et al., 2007). Comparisons of noncoding regions are suggestive that some unique human traits may be affected by differences in gene regulation (Pollard et al., 2006), consistent with earlier hypotheses (King and Wilson, 1975).

Within the human species, genomic analyses have provided evidence that a number of regions are under natural selection (Akey, 2009). Some of the genes within these regions were previously known to be under selection, while others are novel. One of the strongest, most well-documented selective forces acting on humans is malaria. In response, natural selection has shaped the evolution of a number of genes in populations across the globe. Future research that links comparative genomics, population genetics, and functional studies are likely to provide considerable insight into the action of natural selection on many human traits, those that vary both interspecifically and intraspecifically.

THE APPORTIONMENT OF HUMAN VARIATION

One of the most interesting aspects of human population genetics relates to the apportionment of human diversity. These studies address what amount of the total variation present in the human species is apportioned at lower levels. In human studies, these levels have most often related to that of "race" and "population." Occasionally, "geographical region" (e.g., continent) has been used as a proxy for levels of diversity above population.

The most important analysis of the apportionment of human diversity remains Lewontin's (1972) seminal analysis of classical marker polymorphisms. His study compiled polymorphism data from 17 classical loci from a number of human studies to estimate the degree of variation within and between human populations, and races within the species as a whole. In this study, "population" was a group of individuals with a shared language or cultural identity, e.g., Navajo from North America, Luo from Africa, and Basque from Europe. The particular "race" classification chosen by Lewontin had seven categories. The results across loci showed that 85.4% of total human diversity was due to individual variation within a population. The differences between populations – but within a "race" category – accounted for an additional 8.3% of the variation. Together, these number account for 93.7% of total human diversity within race categories, apportioning only 6.3% of the variation to differences between race categories (see Chapter 15 of this volume for further discussion).

This finding was replicated in other classical marker studies (e.g., Ryman et al., 1983). This apportionment of human diversity strongly argued against typological models for human racial classification. Because classical marker loci can potentially be influenced by natural selection, subsequent DNA marker studies were critical for providing an additional window onto the apportionment of genetic diversity.

One of the first studies to employ DNA markers to assess the apportionment of human diversity examined mtDNA RFLPs (Excoffier et al., 1992). This study compiled data from 34 RFLP sites from 2 populations from 5 regions (akin to race categories) of the globe (10 total populations). Their hierarchical analysis found that between 75% and 81% of the diversity was found within the population variance component and between 16% and 22% was apportioned between the 5 regions. Subsequent analysis of DNA sequences from the mtDNA "hypervariable" regions recovered similar findings (Jorde et al., 2000). While widely used as a molecular marker, mtDNA has some properties that make it unique, as discussed above.

Study of the nuclear genome can examine larger numbers of independent loci, thus providing a more comprehensive view of the apportionment of human diversity. Barbujani et al. (1997) studied a total of 109 nuclear autosomal RFLP and microsatellite loci, reflecting a four-fold increase over the number of loci examined in classical marker studies. The analysis of these DNA marker loci partitioned the molecular variance into the hierarchical levels of "sample" (akin to population), within samples of the same continent (akin to race), and within all of the samples (akin to the total diversity). The results of this study were extremely similar to those using classical markers: the within sample variance component was 84.4% and the between continent variance component was 10.8% of the total genetic variance. This finding continues to be replicated in ever-larger studies of nuclear data from a range of marker types (e.g., Jorde et al., 2000; Rosenberg et al., 2002; Li et al., 2008; Long et al., 2009). It is exceptionally interesting to note the remarkable concordance of results based on classical markers and DNA markers, which range from examinations of 17 classical marker loci (Lewontin, 1972) to 650 000 SNP DNA markers (Li et al., 2008).

The apportionment of human diversity has been compared to that seen in our closest relatives, the hominoids. Ruvolo (1997) showed that the apportionment of genetic variation in humans was most similar to other ape subspecies (e.g., *Pan troglodytes schweinfurthii*), suggesting that our genus has a different genetic structure than other ape genera. This pattern underscores the recency of our evolutionary history and the importance of effects such as migration on our diversity. Chapter 15 of this volume places human diversity studies into their essential scientific, cultural, and historical context, including the relationship to and implications of these studies for race and racism.

DISCUSSION POINTS

1. What are the genes and DNA polymorphisms underlying the Rhesus and Duffy classical markers and their different alleles?

2. How would low levels of admixture between archaics and anatomically modern humans change our understanding of human origins? Would there be implications for human taxonomy? What about no admixture? Or high rates of admixture?

3. What are the expectations for comparative levels of mtDNA, X-linked, and Y-linked genetic diversity in groups for the different residence patterns and social structures commonly found in humans and other primates (i.e., patrilocal/virilocal, matrilocal/uxorilocal, bilocal, male philopatry, female philopatry, monogamy, polygyny, polyandry)?

4. How does the study of natural selection on simple genetic traits, e.g., β-*globin*, relate to the role of natural selection in shaping complex polygenetic traits, e.g., body size?

REFERENCES

Akey, J. M. (2009). Constructing genomic maps of positive selection in humans: where do we go from here? *Genome Research*, **19**, 711–722.

Allen, R. C., Armitage, R. J., Conley, M. E., et al. (1993). CD40 ligand gene defects responsible for X-linked hyper-IgM syndrome. *Science*, **259**, 990–993.

Allison, A. C. (1954). Protection afforded by sickle-cell trait against subtertian malarial infection. *British Medical Journal*, **1**, 290–294.

Allison, A. C. (1961). Genetic factors in resistance to malaria. *Annals of the New York Academy of Sciences*, **91**, 710–729.

Alves-Silva, J., Da Silva Santos, M., Guimaraes, P. E., et al. (2000). The ancestry of Brazilian mtDNA lineages. *American Journal of Human Genetics*, **67**, 444–461.

Anderson, S., Bankier, A. T., Barrell, B. G., et al. (1981). Sequence and organization of the human mitochondrial genome. *Nature*, **290**, 457–465.

Andrews, R. M., Kubacka, I., Chinnery, P. F., et al. (1999). Reanalysis and revision of the Cambridge reference sequence for human mitochondrial DNA. *Nature Genetics*, **23**, 147.

Arbiza, L., Dopazo, J. and Dopazo, H. (2006). Positive selection, relaxation, and acceleration in the evolution of the human and chimp genome. *PLoS Computational Biology*, **2**, e38.

Armelagos, G. J. and Harper, K. N. (2005). Genomics at the origins of agriculture, part two. *Evolutionary Anthropology*, **14**, 109–121.

Ayala, F. J. (1995). The myth of Eve: molecular biology and human origins. *Science*, **270**, 1930–1936.

Bakewell, M. A., Shi, P. and Zhang, J. (2007). More genes underwent positive selection in chimpanzee evolution than in human evolution. *Proceedings of the National Academy of Sciences of the United States of America*, **104**, 7489–7494.

Balloux, F., Handley, L. J., Jombart, T., et al. (2009). Climate shaped the worldwide distribution of human mitochondrial DNA sequence variation. *Proceedings of the Royal Society Biological Sciences*, **276**, 3447–3455.

Bamshad, M. and Wooding, S. P. (2003). Signatures of natural selection in the human genome. *Nature Reviews Genetics*, **4**, 99–111.

Barbujani, G., Magagni, A., Minch, E., et al. (1997). An apportionment of human DNA diversity. *Proceedings of the National Academy of Sciences of the United States of America*, **94**, 4516–4519.

Barreiro, L. B., Patin, E., Neyrolles, O., et al. (2005). The heritage of pathogen pressures and ancient demography in the human innate-immunity *CD209/CD209L* region. *American Journal of Human Genetics*, **77**, 869–886.

Baum, J., Ward, R. H. and Conway, D. J. (2002). Natural selection on the erythrocyte surface. *Molecular Biology and Evolution*, **19**, 223–229.

Behar, D. M., Villems, R., Soodyall, H., et al. (2008). The dawn of human matrilineal diversity. *American Journal of Human Genetics*, **82**, 1130–1140.

Bejerano, G., Pheasant, M., Makunin, I., et al. (2004). Ultraconserved elements in the human genome. *Science*, **304**, 1321–1325.

Benson, D. A., Karsch-Mizrachi, I., Lipman, D. J., et al. (2004). GenBank: update. *Nucleic Acids Research*, **32**, D23–D26.

Bentley, D. R. (2006). Whole-genome re-sequencing. *Current Opinion in Genetics and Development*, **16**, 545–552.

Bersaglieri, T., Sabeti, P. C., Patterson, N., et al. (2004). Genetic signatures of strong recent positive selection at the lactase gene. *American Journal of Human Genetics*, **74**, 1111–1120.

Bird, C. P., Stranger, B. E., Liu, M., et al. (2007). Fast-evolving noncoding sequences in the human genome. *Genome Biology*, **8**, R118.

Bolnick, D. A., Bolnick, D. I. and Smith, D. G. (2006). Asymmetric male and female genetic histories among Native Americans from eastern North America. *Molecular Biology and Evolution*, **23**, 2161–2174.

Botstein, D., White, R. L., Skolnick, M., et al. (1980). Construction of a genetic linkage map in man using restriction fragment length polymorphisms. *American Journal of Human Genetics*, **32**, 314–331.

Briggs, A. W., Good, J. M., Green, R. E., et al. (2009). Targeted retrieval and analysis of five Neanderthal mtDNA genomes. *Science*, **325**, 318–321.

Calafell, F., Roubinet, F., Ramirez-Soriano, A., et al. (2008). Evolutionary dynamics of the human *ABO* gene. *Human Genetics*, **124**, 123–135.

Campbell, M. C. and Tishkoff, S. A. (2008). African genetic diversity: implications for human demographic history, modern human origins, and complex disease mapping. *Annual Review of Genomics and Human Genetics*, **9**, 403–433.

Cann, R. L., Stoneking, M. and Wilson, A. C. (1987). Mitochondrial DNA and human evolution. *Nature*, **325**, 31–36.

Cann, H. M., de Toma, C., Cazes, L. et al. (2002). A human genome diversity cell line panel. *Science*, **296**, 261–262.

Cavalli-Sforza, L. L., Menozzi, P. and Piazza, A. (1994). *The History and Geography of Human Genes*. Princeton: Princeton University Press.

Chaix, R., Austerlitz, F., Khegay, T., et al. (2004). The genetic or mythical ancestry of descent groups: lessons from the Y chromosome. *American Journal of Human Genetics*, **75**, 1113–1116.

Chimpanzee Sequencing and Analysis Consortium (2005). Initial sequence of the chimpanzee genome and comparison with the human genome. *Nature*, **437**, 69–87.

Chotivanich, K., Udomsangpetch, R., Pattanapanyasat, K., et al. (2002). Hemoglobin E: a balanced polymorphism protective against high parasitemias and thus severe *P. falciparum* malaria. *Blood*, **100**, 1172–1176.

Clark, A. G., Glanowski, S., Nielsen, R., et al. (2003). Inferring nonneutral evolution from human-chimp-mouse orthologous gene trios. *Science*, **302**, 1960–1963.

Cook, P. J., Robson, E. B., Buckton, K. E., et al. (1978). Segregation of ABO, AK1 and ACONs in families with abnormalities of chromosome 9. *Annals of Human Genetics*, **41**, 365–377.

Cox, M. P., Mendez, F. L., Karafet, T. M., et al. (2008). Testing for archaic hominin admixture on the X chromosome: model likelihoods for the modern human *RRM2P4* region from summaries of genealogical topology under the structured coalescent. *Genetics*, **178**, 427–437.

Currat, M. and Excoffier, L. (2004). Modern humans did not admix with Neanderthals during their range expansion into Europe. *PLoS Biology*, **2**, e421.

Currat, M., Trabuchet, G., Rees, D., et al. (2002). Molecular analysis of the *β-globin* gene cluster in the Niokholo Mandenka population reveals a recent origin of the β^s Senegal mutation. *American Journal of Human Genetics*, **70**, 207–223.

Dermitzakis, E. T., Reymond, A., Scamuffa, N., et al. (2003). Evolutionary discrimination of mammalian conserved non-genic sequences (CNGs). *Science*, **302**, 1033–1035.

Destro-Bisol, G., Donati, F., Coia, V., et al. (2004). Variation of female and male lineages in sub-Saharan populations: the importance of sociocultural factors. *Molecular Biology and Evolution*, **21**, 1673–1682.

Di Rienzo, A. and Wilson, A. C. (1991). Branching pattern in the evolutionary tree for human mitochondrial DNA. *Proceedings of the National Academy of Sciences of the United States of America*, **88**, 1597–1601.

Dillehay, T. D. (2009). Probing deeper into first American studies. *Proceedings of the National Academy of Sciences of the United States of America*, **106**, 971–978.

Drake, J. A., Bird, C., Nemesh, J., et al. (2006). Conserved noncoding sequences are selectively constrained and not mutation cold spots. *Nature Genetics*, **38**, 223–227.

Enard, W., Przeworski, M., Fisher, S. E., et al. (2002). Molecular evolution of *FOXP2*, a gene involved in speech and language. *Nature*, **418**, 869–872.

Evans, P. D., Anderson, J. R., Vallender, E. J., et al. (2004). Adaptive evolution of *ASPM*, a major determinant of cerebral cortical size in humans. *Human Molecular Genetics*, **13**, 489–494.

Evans, P. D., Mekel-Bobrov, N., Vallender, E. J., et al. (2006). Evidence that the adaptive allele of the brain size gene *microcephalin* introgressed into *Homo sapiens* from an archaic *Homo* lineage. *Proceedings of the National Academy of Sciences of the United States of America*, **103**, 18178–18183.

Excoffier, L. and Schneider, S. (1999). Why hunter-gatherer populations do not show signs of pleistocene demographic expansions. *Proceedings of the National Academy of Sciences of the United States of America*, **96**, 10597–10602.

Excoffier, L., Smouse, P. E. and Quattro, J. M. (1992). Analysis of molecular variance inferred from metric distances among DNA haplotypes: application to human mitochondrial DNA restriction data. *Genetics*, **131**, 479–491.

Fagundes, N. J., Ray, N., Beaumont, M., et al. (2007). Statistical evaluation of alternative models of human evolution. *Proceedings of the National Academy of Sciences of the United States of America*, **104**, 17614–17619.

Feibel, C. S., Brown, F. H. and Mcdougall, I. (1989). Stratigraphic context of fossil hominids from the Omo group deposits: northern Turkana Basin, Kenya and Ethiopia. *American Journal of Physical Anthropology*, **78**, 595–622.

Flint, J., Hill, A. V. S., Bowden, D. K., et al. (1986). High frequencies of α-thalassaemia are the result of natural selection by malaria. *Nature*, **321**, 744–750.

Frankham, R. (1995). Effective population size/adult population size ratios in wildlife: a review. *Genetical Research*, **66**, 95–107.

Friedlaender, J. S., Friedlaender, F. R., Reed, F. A., et al. (2008). The genetic structure of Pacific Islanders. *PLoS Genetics*, **4**, e19.

Gabunia, L., Vekua, A., Lordkipanidze, D., et al. (2000). Earliest Pleistocene hominid cranial remains from Dmanisi, Republic of Georgia: taxonomy, geological setting, and age. *Science*, **288**, 1019–1025.

Garrigan, D. and Hammer, M. F. (2006). Reconstructing human origins in the genomic era. *Nature Reviews Genetics*, **7**, 669–680.

Garrigan, D., Mobasher, Z., Kingan, S. B., et al. (2005a). Deep haplotype divergence and long-range linkage disequilibrium at Xp21.1 provide evidence that humans descend from a structured ancestral population. *Genetics*, **170**, 1849–1856.

Garrigan, D., Mobasher, Z., Severson, T., et al. (2005b). Evidence for archaic Asian ancestry on the human X chromosome. *Molecular Biology and Evolution*, **22**, 189–192.

Goldstein, D. B., Ruiz Linares, A., Cavalli-Sforza, L. L., et al. (1995). Genetic absolute dating based on microsatellites and the origin of modern humans. *Proceedings of the National Academy of Sciences of the United States of America*, **92**, 6723–6727.

Gonder, M. K., Mortensen, H. M., Reed, F. A., et al. (2007). Whole-mtDNA genome sequence analysis of ancient African lineages. *Molecular Biology and Evolution*, **24**, 757–768.

Green, R. E., Krause, J., Ptak, S. E., et al. (2006). Analysis of one million base pairs of Neanderthal DNA. *Nature*, **444**, 330–336.

Green, R. E., Malaspinas, A. S., Krause, J., et al. (2008). A complete Neanderthal mitochondrial genome sequence determined by high-throughput sequencing. *Cell*, **134**, 416–426.

Hammer, M. F. (1995). A recent common ancestry for human Y chromosomes. *Nature*, **378**, 376–378.

Hammer, M. F., Karafet, T., Rasanayagam, A., et al. (1998). Out of Africa and back again: nested cladistic analysis of human Y chromosome variation. *Molecular Biology and Evolution*, **15**, 427–441.

Hammer, M. F., Blackmer, F., Garrigan, D., et al. (2003). Human population structure and its effects on sampling Y chromosome sequence variation. *Genetics*, **164**, 1495–1509.

Hammer, M. F., Mendez, F. L., Cox, M. P., et al. (2008). Sex-biased evolutionary forces shape genomic patterns of human diversity. *PLoS Genetics*, **4**, e1000202.

Hanchard, N., Elzein, A., Trafford, C., et al. (2007). Classical sickle β-globin haplotypes exhibit a high degree of long-range haplotype similarity in African and Afro-Caribbean populations. *BMC Genetics*, **8**, 52.

Harding, R. M., Fullerton, S. M., Griffiths, R. C., et al. (1997). Archaic African and Asian lineages in the genetic ancestry of modern humans. *American Journal of Human Genetics*, **60**, 772–789.

Harpending, H. and Rogers, A. (2000). Genetic perspectives on human origins and differentiation. *Annual Review of Genomics and Human Genetics*, **1**, 361–385.

Harpending, H. C., Sherry, S. T., Rogers, A. R., et al. (1993). Structure of ancient human populations. *Current Anthropology*, **34**, 483–496.

Harpending, H. C., Batzer, M. A., Gurven, M., et al. (1998). Genetic traces of ancient demography. *Proceedings of the National Academy of Sciences of the United States of America*, **95**, 1961–1967.

Harris, E. E. and Hey, J. (1999a). Human demography in the Pleistocene: do mitochondrial and nuclear genes tell the same story? *Evolutionary Anthropology*, **8**, 81–86.

Harris, E. E. and Hey, J. (1999b). X chromosome evidence for ancient human histories. *Proceedings of the National Academy of Sciences of the United States of America*, **96**, 3320–3324.

Harris, E. E. and Meyer, D. (2006). The molecular signature of selection underlying human adaptations. *American Journal of Physical Anthropology*, **43**(Suppl.), 89–130.

Helgason, A., Lalueza-Fox, C., Ghosh, S., et al. (2009). Sequences from first settlers reveal rapid evolution in Icelandic mtDNA pool. *PLoS Genetics*, **5**, e1000343.

Hey, J. (1997). Mitochondrial and nuclear genes present conflicting portraits of human origins. *Molecular Biology and Evolution*, **14**, 166–172.

Hill, A. V. S. (2006). Aspects of genetic susceptibility to human infectious diseases. *Annual Review of Genetics*, **40**, 469–486.

Hosoi, E. (2008). Biological and clinical aspects of ABO blood group system. *Journal of Medical Investigation*, **55**, 174–182.

Hunley, K. L., Healy, M. E. and Long, J. C. (2009). The global pattern of gene identity variation reveals a history of long-range migrations, bottlenecks, and local mate exchange: implications for biological race. *American Journal of Physical Anthropology*, **139**, 35–46.

Ingman, M., Kaessmann, H., Pääbo, S., et al. (2000). Mitochondrial genome variation and the origin of modern humans. *Nature*, **408**, 708–713.

International Human Genome Sequencing Consortium (2001). Initial sequencing and analysis of the human genome. *Nature*, **409**, 860–921.

International Human Genome Sequencing Consortium (2004). Finishing the euchromatic sequence of the human genome. *Nature*, **431**, 931–945.

International SNP Map Working Group (2001). A map of human genome sequence variation containing 1.42 million single nucleotide polymorphisms. *Nature*, **409**, 928–933.

Irshaid, N. M., Chester, M. A. and Olsson, M. L. (1999). Allele-related variation in minisatellite repeats involved in the transcription of the blood group *ABO* gene. *Transfusion Medicine (Oxford, England)*, **9**, 219–226.

Jablonski, N. G. and Chaplin, G. (2000). The evolution of human skin coloration. *Journal of Human Evolution*, **39**, 57–106.

Jakobsson, M., Scholz, S. W., Scheet, P., et al. (2008). Genotype, haplotype and copy-number variation in worldwide human populations. *Nature*, **451**, 998–1003.

Jaruzelska, J., Zietkiewicz, E. and Labuda, D. (1999). Is selection responsible for the low level of variation in the last intron of the ZFY locus? *Molecular Biology and Evolution*, **16**, 1633–1640.

Jorde, L. B., Watkins, W. S., Bamshad, M. J., et al. (2000). The distribution of human genetic diversity: a comparison of mitochondrial, autosomal, and Y-chromosome data. *American Journal of Human Genetics*, **66**, 979–988.

Kaessmann, H., Heißig, F., Von Haeseler, A., et al. (1999). DNA sequence variation in a non-coding region of low recombination on the human X chromosome. *Nature Genetics*, **22**, 78–81.

Kaessmann, H., Wiebe, V., Weiss, G., et al. (2001). Great ape DNA sequences reveal a reduced diversity and an expansion in humans. *Nature Genetics*, **27**, 155–156.

Katzmarzyk, P. T. and Leonard, W. R. (1998). Climatic influences on human body size and proportions: ecological adaptations and secular trends. *American Journal of Physical Anthropology*, **106**, 483–503.

Kayser, M., Choi, Y., Van Oven, M., et al. (2008a). The impact of the Austronesian expansion: evidence from mtDNA and Y chromosome diversity in the Admiralty Islands of Melanesia. *Molecular Biology and Evolution*, **25**, 1362–1374.

Kayser, M., Lao, O., Saar, K., et al. (2008b). Genome-wide analysis indicates more Asian than Melanesian ancestry of Polynesians. *American Journal of Human Genetics*, **82**, 194–198.

Kazazian, H. H. Jr, Waber, P. G., Boehm, C. D., et al. (1984). Hemoglobin E in Europeans: further evidence for multiple origins of the β_1-globin gene. *American Journal of Human Genetics*, **36**, 212–217.

Keinan, A., Mullikin, J. C., Patterson, N., et al. (2009). Accelerated genetic drift on chromosome X during the human dispersal out of Africa. *Nature Genetics*, **41**, 66–70.

Kelley, J. L., Turkheimer, K., Haney, M., et al. (2009). Targeted resequencing of two genes, *RAGE* and *POLL*, confirms findings from a genome-wide scan for adaptive evolution and provides evidence for positive selection in additional populations. *Human Molecular Genetics*, **18**, 779–784.

Kent, W. J., Sugnet, C. W., Furey, T. S., et al. (2002). The human genome browser at UCSC. *Genome Research*, **12**, 996–1006.

Kimmel, M., Chakraborty, R., King, J. P., et al. (1998). Signatures of population expansion in microsatellite repeat data. *Genetics*, **148**, 1921–1930.

Kimura, M. (1977). Preponderance of synonymous changes as evidence for the neutral theory of molecular evolution. *Nature*, **267**, 275–276.

King, M. C. and Wilson, A. C. (1975). Evolution at two levels in humans and chimpanzees. *Science*, **188**, 107–116.

Kivisild, T., Shen, P., Wall, D. P., et al. (2006). The role of selection in the evolution of human mitochondrial genomes. *Genetics*, **172**, 373–387.

Kosiol, C., Vinař, T., Da Fonseca, R. R., et al. (2008). Patterns of positive selection in six mammalian genomes. *PLoS Genetics*, **4**, e1000144.

Kouprina, N., Pavlicek, A., Mochida, G. H., et al. (2004). Accelerated evolution of the *ASPM* gene controlling brain size begins prior to human brain expansion. *PLoS Biology*, **2**, e126.

Kreitman, M. (2000). Methods to detect selection in populations with applications to the human. *Annual Review of Genomics and Human Genetics*, **1**, 539–559.

Krings, M., Stone, A., Schmitz, R. W., et al. (1997). Neandertal DNA sequences and the origin of modern humans. *Cell*, **90**, 19–30.

Krings, M., Geisert, H., Schmitz, R. W., et al. (1999). DNA sequence of the mitochondrial hypervariable region II from the Neanderthal type specimen. *Proceedings of the National Academy of Sciences of the United States of America*, **96**, 5581–5585.

Kulozik, A. E., Wainscoat, J. S., Serjeant, G. R., et al. (1986). Geographical survey of β^s-globin gene haplotypes: evidence for an independent Asian origin of the sickle-cell mutation. *American Journal of Human Genetics*, **39**, 239–244.

Kumar, V., Langstieh, B. T., Madhavi, K. V., et al. (2006). Global patterns in human mitochondrial DNA and Y-chromosome variation caused by spatial instability of the local cultural processes. *PLoS Genetics*, **2**, e53.

Kwiatkowski, D. P. (2005). How malaria has affected the human genome and what human genetics can teach us about malaria. *American Journal of Human Genetics*, **77**, 171–192.

Kwok, P. Y. and Chen, X. (2003). Detection of single nucleotide polymorphisms. *Current Issues in Molecular Biology*, **5**, 43–60.

Levy, J., Espanol-Boren, T., Thomas, C., et al. (1997). Clinical spectrum of X-linked hyper-IgM syndrome. *Journal of Pediatrics*, **131**, 47–54.

Lewontin, R. C. (1972). The apportionment of human diversity. *Evolutionary Biology*, **6**, 381–398.

Li, J. Z., Absher, D. M., Tang, H., et al. (2008). Worldwide human relationships inferred from genome-wide patterns of variation. *Science*, **319**, 1100–1104.

Liu, H., Prugnolle, F., Manica, A., et al. (2006). A geographically explicit genetic model of worldwide human-settlement history. *American Journal of Human Genetics*, **79**, 230–237.

Livingstone, F. B. (1958). Anthropological implications of sickle cell gene distributions in West Africa. *American Anthropologist*, **60**, 533–562.

Livingstone, F. B. (1984). The Duffy blood groups, vivax malaria, and malaria selection in human populations: a review. *Human Biology*, **56**, 413–425.

Long, J. C., Li, J. and Healy, M. E. (2009). Human DNA sequences: more variation and less race. *American Journal of Physical Anthropology*, **139**, 23–34.

Manica, A., Prugnolle, F. and Balloux, F. (2005). Geography is a better determinant of human genetic differentiation than ethnicity. *Human Genetics*, **118**, 366–371.

Marlowe, F. W. (2004). Marital residence among foragers. *Current Anthropology*, **45**, 277–284.

Mekel-Bobrov, N., Posthuma, D., Gilbert, S. L., et al. (2007). The ongoing adaptive evolution of *ASPM* and *Microcephalin* is not explained by increased intelligence. *Human Molecular Genetics*, **16**, 600–608.

Mendis, K., Sina, B. J., Marchesini, P., et al. (2001). The neglected burden of *Plasmodium vivax* malaria. *American Journal of Tropical Medicine and Hygiene*, **64**, 97–106.

Merriwether, D. A., Clark, A. G., Ballinger, S. W., et al. (1991). The structure of human mitochondrial DNA variation. *Journal of Molecular Evolution*, **33**, 543–555.

Miller, L. H., Mason, S. J., Clyde, D. F., et al. (1976). The resistance factor to *Plasmodium vivax* in blacks. The Duffy-blood-group genotype, *FyFy*. *New England Journal of Medicine*, **295**, 302–304.

Miyata, T., Yasunaga, T. and Nishida, T. (1980). Nucleotide sequence divergence and functional constraint in mRNA evolution. *Proceedings of the National Academy of Sciences of the United States of America*, **77**, 7328–7332.

Modiano, D., Luoni, G., Sirima, B. S., et al. (2001). Haemoglobin C protects against clinical *Plasmodium falciparum* malaria. *Nature*, **414**, 305–308.

Mullis, K. B. and Faloona, F. A. (1987). Specific synthesis of DNA in vitro via a polymerase-catalyzed chain reaction. *Methods in Enzymology*, **155**, 335–350.

Murdock, G. P. (1967). *Ethnographic Atlas*. Pittsburgh, PA: University of Pittsburgh.

Nei, M. (1987). *Molecular Evolutionary Genetics*. New York: Columbia University Press.

Nielsen, R., Bustamante, C., Clark, A. G. et al. (2005). A scan for positively selected genes in the genomes of humans and chimpanzees. *PLoS Biology* **3**, e170.

Noonan, J. P., Coop, G., Kudaravalli, S. et al. (2006). Sequencing and analysis of Neanderthal genomic DNA. *Science*, **314**, 1113–1118.

Nordborg, M. (1998). On the probability of Neanderthal ancestry. *American Journal of Human Genetics*, **63**, 1237–1240.

Norton, H. L., Kittles, R. A., Parra, E., et al. (2007). Genetic evidence for the convergent evolution of light skin in Europeans and East Asians. *Molecular Biology and Evolution*, **24**, 710–722.

Ohashi, J., Naka, I., Patarapotikul, J., et al. (2004). Extended linkage disequilibrium surrounding the hemoglobin E variant due to malarial selection. *American Journal of Human Genetics*, **74**, 1198–1208.

Oota, H., Settheetham-Ishida, W., Tiwawech, D., et al. (2001). Human mtDNA and Y-chromosome variation is correlated with matrilocal versus patrilocal residence. *Nature Genetics*, **29**, 20–21.

Pääbo, S., Poinar, H., Serre, D., et al. (2004). Genetic analyses from ancient DNA. *Annual Review of Genetics*, **38**, 645–679.

Pagnier, J., Mears, J. G., Dunda-Belkhodja, O., et al. (1984). Evidence for the multicentric origin of the sickle cell hemoglobin gene in Africa. *Proceedings of the National Academy of Sciences of the United States of America*, **81**, 1771–1773.

Pakendorf, B. and Stoneking, M. (2005). Mitochondrial DNA and human evolution. *Annual Review of Genomics and Human Genetics*, **6**, 165–183.

Pauling, L., Itano, H. A., Singer, S. J., et al. (1949). Sickle cell anemia: a molecular disease. *Science*, **110**, 543–548.

Plagnol, V. and Wall, J. D. (2006). Possible ancestral structure in human populations. *PLoS Genetics*, **2**, e105.

Pollard, K. S., Salama, S. R., Lambert, N., et al. (2006). An RNA gene expressed during cortical development evolved rapidly in humans. *Nature*, **443**, 167–172.

Prabhakar, S., Noonan, J. P., Pääbo, S, et al. (2006). Accelerated evolution of conserved noncoding sequences in humans. *Science*, **314**, 786.

Pritchard, J. K., Seielstad, M. T., Perez-Lezaun, A., et al. (1999). Population growth of human Y chromosomes: a study of Y chromosome microsatellites. *Molecular Biology and Evolution*, **16**, 1791–1798.

Prugnolle, F., Manica, A. and Balloux, F. (2005). Geography predicts neutral genetic diversity of human populations. *Current Biology*, **15**, R159–R160.

Przeworski, M., Hudson, R. R. and Di Rienzo, A. (2000). Adjusting the focus on human variation. *Trends in Genetics*, **16**, 296–302.

Quintana-Murci, L., Chaix, R., Wells, R. S., et al. (2004). Where West meets East: the complex mtDNA landscape of the Southwest and Central Asian corridor. *American Journal of Human Genetics*, **74**, 827–845.

Ramachandran, S., Rosenberg, N. A., Zhivotovsky, L. A., et al. (2004). Robustness of the inference of human population structure: a comparison of X-chromosomal and autosomal microsatellites. *Human Genomics*, **1**, 87–97.

Ramachandran, S., Deshpande, O., Roseman, C. C., et al. (2005). Support from the relationship of genetic and geographic distance in human populations for a serial founder effect originating in Africa. *Proceedings of the National Academy of Sciences of the United States of America*, **102**, 15942–15947.

Ray, N., Currat, M., Berthier, P., et al. (2005). Recovering the geographic origin of early modern humans by realistic and spatially explicit simulations. *Genome Research*, **15**, 1161–1167.

Reed, F. A. and Tishkoff, S. A. (2006). African human diversity, origins and migrations. *Current Opinion in Genetics and Development*, **16**, 597–605.

Reich, D. E. and Goldstein, D. B. (1998). Genetic evidence for a Paleolithic human population expansion in Africa. *Proceedings of the National Academy of Sciences of the United States of America*, **95**, 8119–8123.

Relethford, J. H. (2001). *Genetics and the Search for Modern Human Origins*. New York: Wiley-Liss.

Relethford, J. H. (2004). Global patterns of isolation by distance based on genetic and morphological data. *Human Biology*, **76**, 499–513.

Rhesus Macaque Genome Sequencing and Analysis Consortium (2007). Evolutionary and biomedical insights from the rhesus macaque genome. *Science*, **316**, 222–234.

Roberts, R. J., Vincze, T., Posfai, J., et al. (2005). REBASE – restriction enzymes and DNA methyltransferases. *Nucleic Acids Research*, **33**, D230–D232.

Rogers, A. R. and Harpending, H. (1992). Population growth makes waves in the distribution of pairwise genetic differences. *Molecular Biology and Evolution*, **9**, 552–569.

Rosenberg, N. A., Pritchard, J. K., Weber, J. L., et al. (2002). Genetic structure of human populations. *Science*, **298**, 2381–2385.

Rosenberg, N. A., Mahajan, S., Ramachandran, S., et al. (2005). Clines, clusters, and the effect of study design on the inference of human population structure. *PLoS Genetics*, **1**, e70.

Ruitberg, C. M., Reeder, D. J. and Butler, J. M. (2001). STRBase: a short tandem repeat DNA database for the human identity testing community. *Nucleic Acids Research*, **29**, 320–322.

Ruiz-Pesini, E., Mishmar, D., Brandon, M., et al. (2004). Effects of purifying and adaptive selection on regional variation in human mtDNA. *Science*, **303**, 223–226.

Ruvolo, M. (1996). A new approach to studying modern human origins: hypothesis testing with coalescence time distributions. *Molecular Phylogenetics and Evolution*, **5**, 202–219.

Ruvolo, M. (1997). Genetic diversity in hominoid primates. *Annual Review of Anthropology*, **26**, 515–540.

Ruvolo, M., Zehr, S., Von Dornum, M., et al. (1993). Mitochondrial COII sequences and modern human origins. *Molecular Biology and Evolution*, **10**, 1115–1135.

Ryman, N., Chakraborty, R. and Nei, M. (1983). Differences in the relative distribution of human gene diversity between electrophoretic and red and white cell antigen loci. *Human Heredity*, **33**, 93–102.

Sabeti, P., Usen, S., Farhadian, S., et al. (2002a). CD40L association with protection from severe malaria. *Genes and Immunity*, **3**, 286–291.

Sabeti, P. C., Reich, D. E., Higgins, J. M., et al. (2002b). Detecting recent positive selection in the human genome from haplotype structure. *Nature*, **419**, 832–837.

Sabeti, P. C., Schaffner, S. F., Fry, B., et al. (2006). Positive natural selection in the human lineage. *Science*, **312**, 1614–1620.

Sabeti, P. C., Varilly, P., Fry, B., et al. (2007). Genome-wide detection and characterization of positive selection in human populations. *Nature*, **449**, 913–918.

Schlötterer, C. (2000). Evolutionary dynamics of microsatellite DNA. *Chromosoma*, **109**, 365–371.

Sebastianelli, A., Sen, T. and Bruce, I. J. (2008). Extraction of DNA from soil using nanoparticles by magnetic bioseparation. *Letters in Applied Microbiology*, **46**, 488–491.

Ségurel, L., Martinez-Cruz, B., Quintana-Murci, L. et al. (2008). Sex-specific genetic structure and social organization in Central Asia: insights from a multi-locus study. *PLoS Genetics*, **4**, e1000200.

Seielstad, M. T., Minch, E. and Cavalli-Sforza, L. L. (1998). Genetic evidence for a higher female migration rate in humans. *Nature Genetics*, **20**, 278–280.

Serjeantson, S., Bryson, K., Amato, D., et al. (1977). Malaria and hereditary ovalocytosis. *Human Genetics*, **37**, 161–167.

Serre, D. and Pääbo, S. (2004). Evidence for gradients of human genetic diversity within and among continents. *Genome Research*, **14**, 1679–1685.

Serre, D., Langaney, A., Chech, M., et al. (2004). No evidence of Neanderthal mtDNA contribution to early modern humans. *PLoS Biology*, **2**, e57.

Sherry, S. T., Rogers, A. R., Harpending, H., et al. (1994). Mismatch distributions of mtDNA reveal recent human population expansions. *Human Biology*, **66**, 761–775.

Sherry, S. T., Ward, M. H., Kholodov, M., et al. (2001). dbSNP: the NCBI database of genetic variation. *Nucleic Acids Research*, **29**, 308–311.

Shimada, M. K., Panchapakesan, K., Tishkoff, S. A., et al. (2007). Divergent haplotypes and human history as revealed in a worldwide survey of X-linked DNA sequence variation. *Molecular Biology and Evolution*, **24**, 687–698.

Stratton, M. (2008). Genome resequencing and genetic variation. *Nature Biotechnology*, **26**, 65–66.

Stringer, C. (2002). Modern human origins: progress and prospects. *Philosophical Transactions of the Royal Society of London. Series B*, **357**, 563–579.

Stringer, C. B. and Andrews, P. (1988). Genetic and fossil evidence for the origin of modern humans. *Science*, **239**, 1263–1268.

Swallow, D. M. (2003). Genetics of lactase persistence and lactose intolerance. *Annual Review of Genetics*, **37**, 197–219.

Swisher, C. C. 3rd, Curtis, G. H., Jacob, T., et al. (1994). Age of the earliest known hominids in Java, Indonesia. *Science*, **263**, 1118–1121.

Takahata, N. (1993). Allelic genealogy and human evolution. *Molecular Biology and Evolution*, **10**, 2–22.

Takahata, N., Lee, S. H. and Satta, Y. (2001). Testing multiregionality of modern human origins. *Molecular Biology and Evolution*, **18**, 172–183.

Templeton, A. R. (2007). Genetics and recent human evolution. *Evolution*, **61**, 1507–1519.

Thomas, M. G., Parfitt, T., Weiss, D. A., et al. (2000). Y chromosomes traveling south: the cohen modal haplotype and the origins of the Lemba – the "Black Jews of

Southern Africa." *American Journal of Human Genetics*, **66**, 674–686.

Thomson, R., Pritchard, J. K., Shen, P., et al. (2000). Recent common ancestry of human Y chromosomes: evidence from DNA sequence data. *Proceedings of the National Academy of Sciences of the United States of America*, **97**, 7360–7365.

Timpson, N., Heron, J., Smith, G. D., et al. (2007). Comment on papers by Evans et al. and Mekel-Bobrov et al. on evidence for positive selection of *MCPH1* and *ASPM*. *Science*, **317**, 1036, author reply 1036.

Tishkoff, S. A. and Verrelli, B. C. (2003). Patterns of human genetic diversity: implications for human evolutionary history and disease. *Annual Review of Genomics and Human Genetics*, **4**, 293–340.

Tishkoff, S. A., Dietzsch, E., Speed, W., et al. (1996). Global patterns of linkage disequilibrium at the *CD4* locus and modern human origins. *Science*, **271**, 1380–1387.

Tishkoff, S. A., Varkonyi, R., Cahinhinan, N., et al. (2001). Haplotype diversity and linkage disequilibrium at human *G6PD*: recent origin of alleles that confer malarial resistance. *Science*, **293**, 455–462.

Tishkoff, S. A., Reed, F. A., Ranciaro, A., et al. (2007). Convergent adaptation of human lactase persistence in Africa and Europe. *Nature Genetics*, **39**, 31–40.

Tishkoff, S. A., Reed, F. A., Friedlaender, F. R., et al. (2009). The genetic structure and history of Africans and African Americans. *Science*, **324**, 1035–1044.

Underhill, P. A. and Kivisild, T. (2007). Use of Y chromosome and mitochondrial DNA population structure in tracing human migrations. *Annual Review of Genetics*, **41**, 539–564.

Underhill, P. A., Passarino, G., Lin, A. A., et al. (2001). The phylogeography of Y chromosome binary haplotypes and the origins of modern human populations. *Annals of Human Genetics*, **65**, 43–62.

Untergasser, A., Nijveen, H., Rao, X., et al. (2007). Primer3-Plus, an enhanced web interface to Primer3. *Nucleic Acids Research*, **35**, W71–W74.

Varki, A. and Altheide, T. K. (2005). Comparing the human and chimpanzee genomes: searching for needles in a haystack. *Genome Research*, **15**, 1746–1758.

Venter, J. C., Adams, M. D., Myers, E. W., et al. (2001). The sequence of the human genome. *Science*, **291**, 1304–1351.

Vigilant, L., Stoneking, M., Harpending, H., et al. (1991). African populations and the evolution of human mitochondrial DNA. *Science*, **253**, 1503–1507.

Voight, B. F., Kudaravalli, S., Wen, X., et al. (2006). A map of recent positive selection in the human genome. *PLoS Biology*, **4**, e72.

Wall, J. D. (2000). Detecting ancient admixture in humans using sequence polymorphism data. *Genetics*, **154**, 1271–1279.

Wall, J. D. and Hammer, M. F. (2006). Archaic admixture in the human genome. *Current Opinion in Genetics and Development*, **16**, 606–610.

Wall, J. D. and Kim, S. K. (2007). Inconsistencies in Neanderthal genomic DNA sequences. *PLoS Genetics*, **3**, 1862–1866.

Wall, J. D. and Przeworski, M. (2000). When did the human population size start increasing? *Genetics*, **155**, 1865–1874.

Wall, J. D., Cox, M. P., Mendez, F. L., et al. (2008). A novel DNA sequence database for analyzing human demographic history. *Genome Research*, **18**, 1354–1361.

Wang, D. G., Fan, J. B., Siao, C. J., et al. (1998). Large-scale identification, mapping, and genotyping of single-nucleotide polymorphisms in the human genome. *Science*, **280**, 1077–1082.

Wilder, J. A., Kingan, S. B., Mobasher, Z., et al. (2004a). Global patterns of human mitochondrial DNA and Y-chromosome structure are not influenced by higher migration rates of females versus males. *Nature Genetics*, **36**, 1122–1125.

Wilder, J. A., Mobasher, Z. and Hammer, M. F. (2004b). Genetic evidence for unequal effective population sizes of human females and males. *Molecular Biology and Evolution*, **21**, 2047–2057.

Wilder, J. A., Stone, J. A., Preston, E. G., et al. (2009). Molecular population genetics of *SLC4A1* and Southeast Asian ovalocytosis. *Journal of Human Genetics*, **54**, 182–187.

Wilkins, J. F. and Marlowe, F. W. (2006). Sex-biased migration in humans: what should we expect from genetic data? *Bioessays*, **28**, 290–300.

Wolpoff, M. H., Wu, X. and Thorne, A. G. (1984). Modern *Homo sapiens* origins: a general theory of hominid evolution involving the fossil evidence from East Asia. In *The Origins of Modern Humans: a World Survey of the Fossil Evidence*, F. H. Smith and F. Spencer (eds). New York: Liss, pp. 411–483.

Wood, E. T., Stover, D. A., Slatkin, M., et al. (2005). The β-globin recombinational hotspot reduces the effects of strong selection around HbC, a recently arisen mutation providing resistance to malaria. *American Journal of Human Genetics*, **77**, 637–642.

Woolfe, A., Goodson, M., Goode, D. K., et al. (2005). Highly conserved non-coding sequences are associated with vertebrate development. *PLoS Biology*, **3**, e7.

World Health Organization (2008). *World Malaria Report*. Geneva: World Health Organization.

Xing, J., Witherspoon, D. J., Ray, D. A., et al. (2007). Mobile DNA elements in primate and human evolution. *American Journal of Physical Anthropology*, **45**(Suppl.), 2–19.

Yamamoto, F., Clausen, H., White, T., et al. (1990). Molecular genetic basis of the histo-blood group ABO system. *Nature*, **345**, 229–233.

Yamamoto, F., Mcneill, P. D. and Hakomori, S. (1995). Genomic organization of human histo-blood group ABO genes. *Glycobiology*, **5**, 51–58.

Yu, N. and Li, W. (2000). No fixed nucleotide difference between Africans and non-Africans at the pyruvate dehydrogenase E1 α-subunit locus. *Genetics*, **155**, 1481–1483.

Yu, N., Zhao, Z., Fu, Y. X., et al. (2001). Global patterns of human DNA sequence variation in a 10-kb region on chromosome 1. *Molecular Biology and Evolution*, **18**, 214–222.

Yu, N., Chen, F. C., Ota, S., et al. (2002a). Larger genetic differences within Africans than between Africans and Eurasians. *Genetics*, **161**, 269–274.

Yu, N., Fu, Y. X. and Li, W. H. (2002b). DNA polymorphism in a worldwide sample of human X chromosomes. *Molecular Biology and Evolution*, **19**, 2131–2141.

Zhang, J. (2003). Evolution of the human *ASPM* gene, a major determinant of brain size. *Genetics*, **165**, 2063–2070.

Zhao, Z., Jin, L., Fu, Y. X., et al. (2000). Worldwide DNA sequence variation in a 10-kilobase noncoding region on human chromosome 22. *Proceedings of the National Academy of Sciences of the United States of America*, **97**, 11354–11358.

Zhao, Z., Yu, N., Fu, Y. X., et al. (2006). Nucleotide variation and haplotype diversity in a 10-kb noncoding region in three continental human populations. *Genetics*, **174**, 399–409.

Zhivotovsky, L. A., Bennett, L., Bowcock, A. M., et al. (2000). Human population expansion and microsatellite variation. *Molecular Biology and Evolution*, **17**, 757–767.

15 Ten Facts about Human Variation

Jonathan Marks

INTRODUCTION

The idea of race, so intrinsic a part of American social life, is a surprisingly ephemeral one. The ancient world conceptualized human diversity in purely local terms, and the idea that the human species could be naturally partitioned into a reasonably small number of reasonably discrete kinds of people does not seem to have been seriously entertained until the late seventeenth century (Hannaford, 1996; Hudson, 1996; Jahoda, 1999; Stuurman, 2000). The term "race" was introduced into biological discourse by Buffon in the eighteenth century, but he used the term in an entirely colloquial, not taxonomic, way. In this sense the term meant the equivalent of a "strain" or "variety" – a group of organisms linked by the possession of familial features. Buffon's rival Linnaeus, the founder of modern taxonomy, divided humans into four geographical subspecies – although he did not call them "races." The succeeding generation fused Buffon's word with Linnaeus's concept, and thus created the scientific term "race," used well into the twentieth century.

The Linnaean concept of race, however, was a Platonic or essentialist idea – describing not a reality (how organisms are), but a hyper-reality (the imaginary form they represent). Thus, Linnaeus (1758, p. 21) defined "*Homo sapiens Europaeus albus*" – that is to say, white European *Homo sapiens* – as having "long flowing blond hair" and "blue eyes" (*Pilis flavescentibus prolixis. Oculis caeruleis*). Even taking account of the fact that Linnaeus did not travel much outside of his native Sweden, it is difficult to imagine him being that naïve. Linnaeus clearly was describing an ideal type, a metaphysical form, not the actual indigenous inhabitants of Europe.

The essentialized race was not necessarily limited to the continents. Since it was not an empirical concept to begin with, it could be easily extended to any group of people with a distinct identity. Thus, one could just as readily talk about the "Aryan race," the "French race," or the "Jewish race," even though the terms technically applied to a linguistic group, national group, and religious group, respectively. Race was taken to inhere in an individual, as a group quality producing a specific identifiable form and expression in different people. Even so recent a scholar as the Harvard physical anthropologist Earnest Hooton (1926) could think of race as something to be diagnosed, on a medical analogy.

By the early 1930s, partly in response to the rise of racist ideologies in Europe, the concept of race underwent a revision. It became a group of people, a population, rather than an inner quality or spirit. This reversed the locus of race; instead of a race residing within a person, a person would now be a part of a race. Further, the laws of genetics did not seem to permit anything to be transmitted as race was thought to be. What would be "passing" to the previous generation – pretending to be a race you really were not – would be merely the facts of complex ancestry, or euphemistically, "gene flow," under the new concept. Finally, an appreciation for the significance of cultural distinctions in maintaining boundaries between human groups made it necessary to distinguish between ostensibly biological units of the human species, and culturally constituted group differences, and to juxtapose the latter category of human diversity against the study of race; it would come to be known as "ethnicity."

Physical anthropology, like human genetics, had to reinvent itself after World War II, since it was in fact not terribly easy to distinguish the good American science of race from its evil Nazi counterpart. Hooton had tried to do so as early as 1936, publishing an indignant review article in *Science* as "a physical anthropologist, who ... desires emphatically to dissociate the finding of his science from the acts of human injustice which masquerade as 'racial measures' or 'racial movements'" (Hooton, 1936, p. 512).

Hooton, however, was unsuccessful. A "new" physical anthropology (Washburn, 1951) would study human adaptation and microevolution, which was local, not continental. The human species would now be seen "as constituting a widespread network of

Human Evolutionary Biology, ed. Michael P. Muehlenbein. Published by Cambridge University Press. © Cambridge University Press 2010.

more-or-less interrelated, ecologically adapted and functional entities" (Weiner, 1957, p. 80).

The Civil Rights movement precipitated a second revision of the ontology of race for physical anthropology. If the units of the human species were indeed local populations, then higher-order clusters of populations could now be recognized as arbitrary and ephemeral (Thieme, 1952; Hulse, 1962; Johnston, 1966). Thus Frank Livingstone could epigrammatically declare, "There are no races, there are only clines" (1962, p. 279).

Of course, there was the embarrassment of having the President of the American Association of Physical Anthropologists (Carleton Coon of the University of Pennsylvania, Earnest Hooton's second doctoral student at Harvard) colluding with the segregationists in 1962, but Coon stood alone in defending the segregationist literature from censure by the American Association of Physical Anthropologists (Coon, 1981; Lasker, 1999). The "new physical anthropology" gave practitioners leeway to abandon race to the cultural anthropologists and sociologists on one side (as ethnicity), and to the population geneticists (as science) on the other. Thus, widely used biological anthropology texts, such as Frank Johnston's (1973) *Microevolution of Human Populations* and Jane Underwood's (1979) *Human Variation and Human Microevolution* could get by without even mentioning race in the index.

Population geneticists, however, were actually multivocal about race. On the one hand, Lewontin's (1972) famous "apportionment of human diversity" was able to quantify what fieldworkers had long known: there are all kinds of people, everywhere. Lewontin's discovery that there is approximately six times more within-group variation than between-group variation detectable in the human species seemed to put the lie to any possibility that the human species could be naturally divided into a small number of relatively discrete gene pools. On the other hand, other population geneticists would use races as natural categories quite unproblematically and unreflectively (Cavalli-Sforza, 1974; Nei and Roychoudhury, 1974).

By the 1990s, race was undergoing yet another transformation at the hands of population geneticists, from a geographically localized gene pool or population to the small amount of difference detectable among the most geographically separated peoples, after overlooking the major patterns of human variation – the cultural, polymorphic, clinal, and local. This is a new concept of race as a genetic residual, a successor to the race as population and the race as essence, and it is the idea of race employed by most contemporary defenders of race in physical anthropology and population genetics. Nevertheless, it would be largely unintelligible to scholars of earlier generations, who might otherwise be inclined to agree with the proposition that race is "real."

To understand race properly, however, we must appreciate that it is a biocultural category, the result of a negotiation between patterns of difference and perceptions of otherness. Old categories of identity are obliterated, or are relegated to "ethnicities" rather than to "races" (Ignatiev, 1996; Brodkin, 1999) and newer, more politically salient categories become racialized. Notably, the 2000 US Census separated the question of "Spanish/Hispanic/Latino" from that of "race" on the quite sensible grounds that "Spanish/Hispanic/Latino" designates a linguistic category, and thus cross-cuts race. One could, after all, reasonably fall within that category with mostly Native American Ancestry, mostly Filipino ancestry, mostly southern European ancestry, mostly Afro-Caribbean ancestry, and most especially, a mixture of several of those. Then the Census provided the familiar choices in the "race" question (White, Black, Native American, Asian, and Pacific Islander), but also included the option "Some other race." That choice, "Some other race," was checked by 42% of self-identified Hispanics, but by only a negligible amount of non-Hispanics. It seems that the Census Bureau over thought the matter: Hispanic has effectively become "some other race" – the cultural basis of its demarcation notwithstanding (Mays et al., 2003).

Population geneticists have not been able to resolve race because it is not a genetic category (Graves, 2004). Race is a human group which, like all human groups down to "family," is a coproduction of historical/cultural processes and of microevolutionary biological processes. It is not a question of whether humans differ, but of how they do so, and of how we concurrently make sense of it. And any scientific sense we make of human variation must ultimately by consistent with 10 empirical generalizations produced by anthropology and genetics over the last century and a half.

1. HUMAN GROUPS DISTINGUISH THEMSELVES PRINCIPALLY CULTURALLY

This is the singular discovery of anthropology. When E. B. Tylor (1871) separated biology or race from "culture," he described it as "that complex whole which includes knowledge, belief, art, morals, law, custom, and any other capabilities and habits acquired by man as a member of society" (Tylor, 1871, p. 1) – in other words, as the myriad things that we key on to differentiate "us" from our neighbors, "them." Today we would certainly expand the list to include the things that give us the earliest and most basic signals of who we are and who we're not: language, mode of dress or personal grooming, food preferences, body movement. This seems to be what humans evolved doing, and may well precede the emergence of our species itself.

In distinguishing our group from others, in these socially transmitted, historically constructed, and symbolically powerful ways, we structure most of our daily lives. What makes us group members also renders all of our sensory input and experience meaningful. We think and communicate using the metaphors and symbols of our group. We groom and dress ourselves according to the conventions of our group; indeed, the decisions we actually make during the course of our lives are rigidly constrained by the relatively meager options culturally available. In other words, the vast bulk of human behavioral and mental diversity is culturally constituted.[1]

It is of some significance that the strongest cultural distinctions are maintained between neighboring groups, who are nevertheless very closely related genetically to one another. This would constitute a paradox if there were a close and deterministic relationship between genetics and human behavior. Rather, however, if the bulk of human behavioral and cognitive diversity is of the sort that differentiates one group from another (culture), and this variation is social and historical in origin, then genetic variation can be invoked to explain at best a tiny part of human difference in thought and deed – presumably some of the differences identifiable among members of the same group.

Considered another way, an imaginary neuropeptide whose variant allele made someone a bit more aggressive, say, might be found both in a wealthy Parisian and in a poor Sri Lankan (and in many others). The variant allele might make its possessor slightly more aggressive, but a Sri Lankan sharing the allele with a Parisian would hardly have their lives thereby rendered significantly more similar. Their different lives would be shaped by their different cultural traditions and practices. Even the (culturally mediated) responses to their aggressive behavior would cause their personal experiences and perceptions to diverge. If the question, then, is to understand the major features of human behavioral diversity, a focus on behavioral genetics is manifestly a case of the tail wagging the dog.

---

[1] Biologized theories of human history have been put forward periodically from Arthur de Gobineau's (1853–1855) *The Inequality of Human Races* through C. D. Darlington's (1969) *The Evolution of Man and Society*. It is in this narrow historical sense, where (as Émile Durkheim famously noted) social facts are only explicable by prior social facts, that the analytic separation between biological (microevolutionary) and cultural phenomena has been most useful. In a broader sense, the interaction between the "natural" and the "cultural" is more complex and problematic. At very least, historical events often have biological consequences, which in turn engender different responses – as evidenced in the well-known relationships among agriculture, malaria, the human gene pool, and modern medicine.

2. HUMAN BIOLOGICAL VARIATION IS CONTINUOUS, NOT DISCRETE

In his 1749 discussion of human variation from *Natural History, General and Particular*, Buffon wrote, "On close examination of the peoples who compose each of these black races, we will find as many varieties as in the white races, and we will find all the shades from brown to black, as we have found in the white races all the shades from brown to white" (Buffon, 1749, p. 454).

That would seem, on the face of it, to preclude the possibility of taxonomically dividing people neatly into black and white; or even into black, brown, and white. Buffon's empiricism, unfortunately, had already lost the day to Linnaean idealism in the area of human taxonomy. Linnaeus's rigorous hierarchical approach to biological systematics was so obviously right in permitting us to understand the relationships among species, that it stood to reason that Linnaeus was correct in applying his ideas below the level of the human species as well. This created a paradox in the writings of Johann Friedrich Blumenbach a generation later.

Blumenbach was, like Buffon, an empiricist in matters of human variation; but he was also, like Linnaeus, a taxonomist. Thus, he famously wrote in 1775, "One variety of mankind does so sensibly pass into the other, that you cannot mark out the limits between them" (Bendyshe, 1865, pp. 98–99), and yet nevertheless proceeded to do just that. The same paradox inheres in the work of population geneticists over two centuries later (e.g., Cavalli-Sforza et al., 1994).

An alternative to the taxonomic approach in microevolutionary studies was suggested by Julian Huxley in 1938. Since a large component of the variation that exists within a species is structured as geographical gradients, he suggested, why not simply describe them that way, rather than trying to shoehorn the populations into taxonomic categories? In fact, Huxley did not mention humans among his examples; nor did he reject the establishment of subspecific taxa. In the 1950s other zoologists began to suggest rejecting the subspecies altogether (Wilson and Brown, 1953), and Livingstone (1962) was extending the argument to humans when he denied the very biological existence of human races.

Trying to explain clinal variation in human physical form from northern to southern Europe in taxonomic terms is what compelled William Z. Ripley (1899) to introduce a subdivision of "the races of Europe" into Teutonic (Nordic), Alpine, and Mediterranean. (Today, even the simple use of a plural in his title seems foreign to us.) Carleton Coon's (1939) revision of *The Races of Europe* identified over a dozen of them. Where no criteria exist other than "difference," certainly a broad cline of physical form could be subdivided in a pseudo-taxonomic fashion effectively without limit. It is simply a classic square-peg/round-hole problem.

This clinal pattern is evident for most human traits, extending from lactose persistence through to skin color. The reason for this pattern is two-fold: (1) natural selection, with environmental conditions varying gradually over space; and (2) gene flow, culturally mediated in humans. There are very few systems that do not show much in the way of geographical gradients. Yet even the genetic markers that permit full differentiation of disparate groups (almost all one allele in West Africans and almost all another allele in East Asia at the Duffy blood group locus on chromosome 2 exhibit clines of differing intensities in different regions (East Africa, West and South Asia).

It seems, then, that a division of the world into human races – reasonably discrete from one another and relatively few in number – was an aberration, derived from a peculiar view of human variation adopted by scientists from the seventeenth to the twentieth century. Scholars have differentiated the peoples they encounter according to diverse criteria, but human variation in nearly all times and places has been perceived on a local, not a continental/global, scale. This is because fundamental patterns of human difference are principally gradational, not discrete.

3. CLUSTERING POPULATIONS IS ARBITRARY

Human identities are culturally produced, and can assume a wide range of forms. Those that are principally geographic can be extensively subdivided; one can be Caucasian, Nordic, Slavic, Baltic, and Latvian simultaneously. All have been racialized by someone or another.

Approaching the issue from the bottom, so to speak, where the most basic human populations are local, how do they fit together into more inclusive entities?

We could try to cluster them genealogically, but as Frederick Hulse (1962) pointed out, there is no reason to think human populations are actually genealogically structured entities, and every reason to think they are not. Gene flow (both small-scale and long-term, and large-scale and short-term) is a pervasive feature of human history, and the horizontal modes of genetic transmission it produces are complementary to the vertical modes of genetic transmission depicted in genealogical trees (Fix, 2005). Consequently, the more accurate mode of representation of human populations is not as a tree, but as a trellis, capillary system, or rhizome (Moore, 1994; Pálsson, 2007; Arnold, 2009).

Further, the relationship between processes of human demographic history, and the products they have yielded at different times, is often far from clear. Patterns of relative genetic distinctiveness might be expected from several different demographic processes. Consequently, different clustering analyses applied to human populations by different researchers have often yielded different results. Clusters of populations may be produced as well simply by sampling discontinuously (Serre and Pääbo, 2004).

The idea that human populations fall naturally into genealogical clusters is itself the result of a gloss on the Biblical theory of human biogeography. Genesis 10 tells us that Noah's three sons (Ham, Shem, and Japheth) went out and populated the world after surviving the Deluge. Ham has sons named Cush, Mizraim, Phut, and Canaan – and is the ancestor of both the Babylonians (Babel) and the Egyptians (Mizraim). Shem has sons named Elam, Asshur, Arphaxad, Lud, and Aram – and is an ancestor of other local city-states. And Japheth sires Gomer, Magog, Madai, Javan, Tubal, Meshech, and Tiras – and once again, is an ancestor of a group of city-states. "These are the families of the sons of Noah," the Bible tells us, "after their generations, in their nations: and by these were the nations divided in the earth after the flood."

By the first century, the Jews understood this to explain the peopling of the three known continents. According to *The Antiquities of the Jews* by Flavius Josephus (Book I, Chapter 6), Ham heads south to beget the Egyptians, Ethiopians, and other Africans; Shem begets the Asians as far east as India (including the Hebrews themselves, through his son Heber); and Japheth is the ancestor of the European peoples, as far west as Spain.

In the nineteenth century, this story was embellished even further, as Noah curses his grandson Canaan for an ambiguous sexual deed perpetrated by his father Ham. Josephus had interpreted the curse in the context of Jewish origins, and the political/religious/military transformation of "Canaan" into "Judea." But to American physical anthropologists in the era of slavery, that curse became the Biblical justification for the modern enslavement of Africans.

Nevertheless, there was very little change in the biohistorical model explaining the human race. The three sons of Noah emigrate to the corners of the earth and populate it, becoming the pure progenitors of the people living there; and where their remote descendants encounter one another, impure races are found. The power of this model is such that it even underlies some genetic studies of the modern era. Thus, prominent population geneticists can casually write, as recently as 1993:

[H]uman populations can be subdivided into five major groups: (A) negroid (Africans), (B) Caucasoid (Europeans and their related populations), (C) mongoloid (East Asians and Pacific Islanders), (D) Amerindian (including Eskimos), and (E) australoid (Australians and Papuans). (There are intermediate populations, which are apparently products of gene admixture of these major groups, but they are ignored here.) (Nei and Roychoudhury, 1993, pp. 936–937)

Of course, there was never a time when people lived only in Lagos, Oslo, and Seoul; indeed, the most ancient representatives of *Homo sapiens sapiens* are right there in the middle. That raises a crucial question about the statistical clustering of populations: What do the clusters actually represent? What is their connection to human history? While most population geneticists readily acknowledge that the clusters are statistical reifications (Templeton, 1998), it is not too difficult to find them naïvely interpreted as clado-genetic events, with that occasional rare admixture. And indeed, philosopher Robin Andreasen (2000, 2004) misunderstands the evolutionary meanings of those trees in precisely that fashion, as a series of literal, historical bifurcations that produced – you guessed it – races.

4. POPULATIONS ARE BIOLOGICALLY REAL, NOT RACES

Gilmour and Gregor (1939) coined the word "deme" to refer to the local population that exists as an ecological and social unit in nature. The focus on the population genetics of human demes is what permitted biological anthropologists of the 1970s to avoid "race" altogether.

The application of this concept to human diversity revolutionized the study of physical anthropology in the years following World War II. The genetical processes described in the evolutionary synthesis were measurable and meaningful at the local level; Sewall Wright's work showed that local populations were effectively the units of general microevolution. That is consequently where the study of human population genetics would have to focus.

Larger units than the deme lack cohesion or time depth. Their evolutionary meaning is consequently not obvious. To adopt a unit of analysis of human biology larger than that of the local population or deme, then, is what requires some justification today. Perhaps the most interesting question in this vein is that of representation: Can local populations "stand for" anything other than themselves? In one famous study, geneticists used 94 African pygmies, 64 "Chinese … living in the San Francisco Bay Area," 110 samples from "individuals of European origin from ongoing studies in our laboratories or reported in the literature," and concluded sweepingly that "ancestral Europeans are

estimated to be an admixture of 65% ancestral Chinese and 35% ancestral Africans" (Bowcock et al., 1991, p. 839). That is, the samples were intended to represent larger categories assumed to be natural and separate.

5. POPULATIONS ALSO HAVE A CONSTRUCTED COMPONENT

"Population" is a term that is notoriously difficult to define rigorously. The usage above is intended to juxtapose the "local" against the "global" – or ontologically real "demes" against reified human mega-populations. And yet, local human populations, as previously noted, tend to distinguish themselves by features such as language, dress, religion, and dietary prohibitions or preferences. These are not biological attributes, but they help circumscribe an entity that is to some extent biological, namely the local human population or deme.

The boundaries being nonbiological, they are consequently porous to biological input, in the form of gene flow (e.g., Hunley and Long, 2005). This can take place through social practices, such as exogamy and adoption; economic practices, such as trade and subsistence; and political practices, such as warfare, slave raids, and forced migrations.

Unfortunately, a large class of population genetics models have tended to work best for populations in isolation from one another, which in turn necessitates a high degree of "purity" for the populations under study. This assumption was raised during the public discussion over the Human Genome Diversity Project in the 1990s, as the Project itself continually talked of "isolated" populations. But this had in fact been highlighted as a problem half-a-century earlier, as Boston University's anthropological geneticist William C. Boyd had proclaimed the purity of the Navajo group he was studying. But cultural anthropologist Clyde Kluckhohn knew the specific community and its ethnohistory, and knew of its extensive interbreeding, with Walapai, Apache, Laguna, and Anglo/Spanish contributions. "In spite of all this, [they] conclude from their blood group data that the Ramah Navaho represent an 'unusually pure' Indian group" (Kluckhohn and Griffith, 1950, p. 401). The implication was clear that the population in question would actually have their complex history erased by the geneticists, and would be falsely simplified and reified into one in which they were more-or-less "pure."

The myth that non-European peoples are "pure" and "unmixed," and have more or less always been where (and as) we find them today, was comprehensively refuted by Eric Wolf (1982) in *Europe and the People without History*. That it complicates some population genetic analyses is unfortunate (Moore, 1994;

Templeton, 1998), but human populations are biocultural units, connected economically, socially, and genetically; and with complex histories intertwined with those of their neighbors (Lasker and Crews, 1996).

6. THERE IS MUCH MORE VARIATION WITHIN GROUPS (POLYMORPHISM) THAN BETWEEN GROUPS (POLYTYPY)

Lewontin's (1972) calculation that there is six times more within-group variation than between-group variation in the gene pool of *Homo sapiens* has been the subject of periodic criticism, but the results have proved remarkably robust to the kinds of genetic data analyzed. Barbujani et al. (1997) found a similar result for nuclear DNA, as did Rosenberg et al. (2002).

Indeed, the recognition that variation within human groups vastly exceeds that between human groups was noted explicitly in the second (1951) UNESCO statement on race. Now, however, with genetic data, the observation could be quantified. The most obvious conclusion is that the human species does not come naturally partitioned into reasonably discrete gene pools, which had been the predominant theory of race for most of the twentieth century.

A. W. F. Edwards (2003) has recently criticized the invocation of these numbers against the race concept as "Lewontin's fallacy," on the grounds that a proportion of the diversity detectable in the human gene pool is indeed correlated with geography, and thus can be used to sort people into large groups, if one focuses upon it closely enough. The argument here is not with the data, but with the meaning of the data and its relation to human races. Geographical correlations are far weaker hypotheses than genetically discrete races, and they obviously exist in the human species (whether studied somatically or genetically). What is unclear is what this has to do with "race" as that term has been used through much of the twentieth century – the mere fact that we can find groups to be different and can reliably allot people to them is trivial. Again, the point of the theory of race was to discover large clusters of people that are principally homogeneous within, and heterogeneous between, contrasting groups. Lewontin's analysis shows that such groups do not exist in the human species, and Edwards's critique does not contradict that interpretation.

Moreover, the Lewontin numbers show that patterns of human genetic diversity simply do not map well onto the patterns of human behavioral or cognitive diversity. The latter kinds of differences tend to be localized at the borders of human groups, as noted above, and are of the sort we call cultural (Peregrine et al., 2003; Bell et al., 2009). To the extent that genetic diversity is structured quite differently (mostly polymorphism and clines), it seems unlikely that genetic differences could play a significant role in understanding the major patterns of human behavior, unless variation in the hypothetical genes involved were structured quite differently from the rest of the known human gene pool.

7. PEOPLE ARE SIMILAR TO THOSE NEARBY AND DIFFERENT FROM THOSE FAR AWAY

The primary factor governing between-group variation in our species is geography, a fact known even to the ancients. This allows us grossly to predict patterns of relatedness: a Dane will tend be more similar to an Italian than to a Hopi. This, however, only allows us to classify the Dane and the Italian *in relation to the Hopi*; it does not tell us whether Danes and Italians themselves belong to the same group or to different ones. There are indeed geographical patterns in the human gene pool, and they can indeed be used to allot people into groups (Witherspoon et al., 2007); the groups simply do not correspond to "races," in any previously or generally understood sense of that term. The ability to discriminate Swedes from Nigerians genetically does not tell you what to do with Moroccans. The existence of genetic variation over space is thus disconnected from race as theory of human groups and their classification – a point sufficiently important, yet subtle, as to be lost on some geneticists! In fact, one needs neither statistics nor genetics to tell an Inca from a Dinka.

In general, the most geographically proximate peoples are the most genetically similar. In rare cases, a (permeable) barrier of language, politics, or ethnicity might serve to reinforce a genetic distinction between one people and their neighbors (Hulse, 1957); these differences are nevertheless often genetically subtle, arbitrary, and discordant. If the Ainu of Hokkaido are more hirsute than other Japanese, can one be a glabrous Ainu? Likewise, can one be an Rh$^+$ Basque, or a tall pygmy?

The answer is presumably "yes" to all of those, although perhaps with varying degrees of aspersion cast upon one's ancestry, in proportion to the degree of purity ascribed to the group itself. Once again, however, this is hardly meaningful in the context of races; but rather, only in the context of local populations.

Perhaps the most celebrated confusion of geographic difference for race followed the publication of Genetic Structure of Human Populations (Rosenberg et al., 2002). The authors studied genetic variation in 1052 people from 52 populations and then asked a computer program called Structure to group the samples. When they asked it to produce two groups, Structure gave them EurAfrica and East Asia–Oceania–America. When asked for three groups, Structure gave Europe, Africa, and East Asia–Oceania–America. When asked for four, it gave Europe, Africa,

East Asia–Oceania, and America. When asked for five, it gave roughly the continents. And when asked for six, it gave the continents and the Kalash people of Pakistan. When asked for more (up to twenty groups), it gave more (Bolnick, 2008).

This was more or less what population geneticists had been doing with the human gene pool since the pioneering work of Cavalli-Sforza and Edwards (1965). On the face of it, once again, this would seem to have little relevance for race. The user specifies the number of groups, and geographic proximity is the strongest predictor of similarity, so asking the computer to break the human species into five groups might reasonably be expected to yield groups roughly corresponding to the continents. And the Kalash people of Pakistan certainly do not have green skin and square heads; nor do they constitute a "natural" contrast against Europeans or Africans.

Nevertheless, a headline in the New York Times announced, "Gene study identifies five main human populations, linking them to geography" and quoted Marcus Feldman[2], the principal author of the study, to the effect that "the finding essentially confirmed the popular conception of race" (Wade, 2002).

Of course the popular conception of race as a classification system applies not just to the more-or-less indigenous peoples surveyed by the geneticists, but as well to the entire admixed urban populations of the modern world, especially the Americas. This raises an important criticism of genetic "racial" studies: their focus on a mythological past rather than on a real present (Cartmill, 1998). What biological relevance does an exercise like this have, after all, for the peoples of New York, Chicago, Los Angeles, Mexico City, Rio de Janeiro, or Johannesburg? It is indeed an odd and perverse approach to history, geography, and genetics that would cast a blind eye to the centuries of colonial contact and demographic reconfiguration that have constructed the human gene pool.

In modern American populations, it is certainly reasonable to expect people who look "black" to tend to cluster genetically with Africans when examined with carefully selected genetic markers (Bamshad et al., 2003), but the vagaries of Mendelian genetics and the complexities of human history will combine to place an increasing amount of weight on the phrase "tend to." Further, given nontrivial amounts of polymorphism and admixture, there is always a nontrivial possibility that a particular person may have the "wrong" racial marker at a specific locus. That is ultimately why a racialized pharmacopoeia is a very poor and risky substitute for an individualized one, which will have to be predicated on the direct assessment of individual genotypes.

[2] Feldman (personal communication) said it was a misquotation.

8. RACIAL CLASSIFICATION IS HISTORICAL AND POLITICAL, AND DOES NOT REFLECT NATURAL BIOLOGICAL PATTERNS

The contemporary racialization of Hispanics in the United States (see above) is certainly prima facia evidence for the political embeddedness of racial classifications. In classic anthropological fashion, the cultural aspects of race are revealed most clearly when we contrast the classifications and their uses from place to place and time to time. Thus, while "Black" in the United States has effectively meant "possessing any recent African ancestry," that category in the United Kingdom traditionally referred to South Asian ancestry (meaningful in the context of the colonial relationships between Britain and India), and only recently has the category "Afro-Caribbean" emerged there to designate what Americans mean by "Black." People of South Asian ancestry in the United Kingdom are now commonly regarded as "Asian" in the United Kingdom, but in the United States the term instead tends to connote people of East Asian ancestry.

Central and South American classifications have tended to incorporate more categories, based on actual variation in skin shade, in contrast to the "one drop of blood" rule prevalent in the United States. While there is commonly status differential associated with skin color, it is nevertheless quite different from the binary racial system of the United States.

The point is that biological or genetic difference can be studied and quantified, but it is not race. Race is a sense-making system imposed upon the facts of difference. Races are not merely human divisions, they are politically salient human divisions. All classifications exist to serve a purpose; the purpose of a racial classification is to naturalize human differences – that is, to establish important categories and make their distinctions appear to be rooted in nature, rather than in history or politics.

The pervasive tendency for racial classifications to see sub-Saharan Africans as a single group, for example, has far more to do with the politics and history of slavery than with the gene pool of Africans. After all, fieldworkers like Seligman (1930) and Hiernaux (1975) consistently emphasized the physical diversity of Africans. Julian Huxley could write, "It is a commonplace of anthropology that many single territories of tropical Africa, such as Nigeria or Kenya, contain a much greater diversity of racial type than all Europe" (Huxley, 1931, p. 379). Today, their genetic diversity is generally considered to harbor the ancestral gene pool of the rest of the world. Sub-Saharan Africans thus encompass more genetic diversity than other "races," and more significantly, constitute a paraphyletic category, and are thus not even taxonomically comparable to other "races" (Marks, 1995). So if the empirical data have long been known to contradict it, how then do we account for the presentation of sub-Saharan Africans as consistently monolithic in racial classifications as late as those of Campbell (1962) and Boyd (1963)? [3]

9. HUMANS HAVE LITTLE GENETIC VARIATION

Ferris et al. (1981) found a much greater degree of heterogeneity in the mitochondrial DNA of chimpanzees and gorillas than in humans. This finding was soon extended to nuclear DNA by Deinard (1997) and Kaessmann et al. (2001). Stone et al. (2002) found very different patterns of diversity in chimpanzees and humans as well, chimpanzees having deeper coalescences, and more between-group variation (which is especially striking, given their considerably more restricted range), than humans. At some loci where humans are variable, apes turn out to be less variable, but this is the result of a statistical bias – if we try to identify variation in apes where humans are already known to vary, then we miss the many loci at which apes vary but humans do not.

Although it must be noted that there is a conservation-driven push towards "taxonomic inflation" in the apes, the levels and degrees of genetic differentiation in our closest relatives seem to be considerably different from our own. Ape subspecies appear to cluster strongly together with mitochondrial DNA, for example, while human races do not. To the extent that they have traditionally been divided into subspecies, then, these great ape taxa represent very different entities than human races.

One interesting consequence of finding such high levels of genetic diversity in the apes is the difficulty it imposes upon phylogenetic reconstruction (Ruano et al., 1992; O'hUigin et al., 2002). Very high levels of homoplasy and ancestral polymorphism undermine the assumption of parsimony in molecular phylogenetics (Marks, 1994; Satta et al., 2000; Chen and Li, 2001), and contribute to the relatively large statistical errors associated with the calculation of divergence times of human and ape species (Stauffer et al., 2001; Glazko and Nei, 2003; Kumar et al., 2005). This in turn suggests the need for models of ape-human ancestry more complex than just a sequence of simple bifurcations (Chaline et al., 1991; Deinard, 1997; Barbulescu et al., 2001; Marks, 2002; Patterson et al., 2006; Arnold, 2009).

Thus, the temptation to represent evolutionary history as a series of cladogenetic events seems to be nearly as problematic just above the human species as just below it. Clearly, the demographic histories of these populations made the patterns of genetic difference we see today more difficult to interpret than earlier generations of scholars appreciated.

10. RACIAL ISSUES ARE SOCIAL–POLITICAL–ECONOMIC, NOT BIOLOGICAL

The most important aspect of the study of race is its connection to racism, a political ideology in which humans are ranked according to group membership. It has occasionally been argued that the absence of taxa equivalent to zoological subspecies in humans invalidates racism, as if all we had do to disband the Ku Klux Klan would be to teach them some population genetics.

This view, however, misrepresents the basis of racism, for it takes racism to be predicated on science. In fact racism is independent of science, and is simply one of many anti-democratic political discourses that function to rationalize social inequalities. Sexism, anti-Semitism, and homophobia are quite real, in spite of the fact that the groups constituted by women, Jews, and homosexuals possess varying degrees of "naturalness." In other words, it is the social ranking and prejudice, not the biology, which comprise the salient features of racism.

Race is thus paradoxically of minor relevance to racism. The "race" in "racism" is the first – the essentialist – version of race, in which any group can possess its own innate qualities, and individual people can be relied upon to embody those qualities. The categories are still real and experienced, however, despite how little they may correspond to biology (Smedley and Smedley, 2005).

If "white" and "black" denote intractably large, highly heterogeneous, extensively overlapping populations, then, as Lewontin (1972) recognized, there can be little justification for ascribing great biological meaning to the perceived discontinuities between them. On the other hand, if: (1) considerable social inequality is mapped onto the categories; and (2) phenotypes are coconstructions of genotypes and the cultural conditions under which the genotypes are expressed, then it follows that; (3) significant perceived differences between the two groups, particularly etiologically complex ones like odor (Classen, 1995), body form (Bogin, 1988), or intelligence (Lewontin et al., 1984), are simply more likely to be attributable to their different social statuses (especially class and ethnicity) than to their gene pools.

[3] Coon's (1962) *The Origin of Races* followed Gates' (1948) *Human Ancestry* in splitting the Khoisan peoples of southern Africa off from other Africans, thus doubling the number of African races.

This conclusion, obviously, is not value-neutral. The ascription of inequality to biological causes is a political position that minimizes the role of political–economic factors in producing and maintaining that social inequality. The implication is that biological causes require biological remedies, or at least, not remedies involving significant expenditures on social programs. Obviously there is considerable harmony between this ostensibly scientific conclusion and a political agenda of social conservatism, often explicitly so. Indeed, this is what links the reasoning of the social Darwinists, eugenicists, and segregationists of earlier eras with works like *The Bell Curve* (Herrnstein and Murray, 1994) in the modern era. Consequently they necessitate a higher degree of scrutiny than the ordinary run of scientific work, and generally, they do not stand up well to it (Boas, 1911; Hogben, 1931; Merton and Ashley-Montagu, 1940; Dobzhansky, 1962, 1963; Gould, 1981; Lieberman, 2001; Marks, 2005).

In the case of health care, for example, it is quite uncontroversial that identifiers such as ancestry, age, and occupation carry different statistical health risks and that knowledge of them can aid in producing a proper diagnosis. Being born white carries a risk of 1 in 2500 of having cystic fibrosis; being born black carries a risk of 1 in 15 000. Nevertheless, one needs to guard carefully against misdiagnosing the presentation of symptoms in a black child, say, on the grounds that cystic fibrosis is a white child's disease, since that act puts lives directly at risk (Garcia, 2003). Further, race itself is a red herring here: being Ashkenazi Jewish, Pennsylvania Amish, "not northern European," a football player, a primary school teacher, or a computer hacker puts one at higher risk for familial dysautonomia, Ellis–van Creveld syndrome, lactose intolerance, knee problems, mild viral infections, and carpal tunnel syndrome, respectively, but those labels do not designate groups we would identify as races. And more importantly, since the social inequality associated with race is a significant variable affecting many aspects of life and health care (Sankar et al., 2004), it should not be surprising the some of the most well-known racialized medical issues – low birthweight and hypertension – also do not stand up well under scrutiny as innate differences (David and Collins, 1997; Kaufman and Hall, 2003).

Most significantly, the modern context of racial science involves another player, in addition to science and politics – the economics of health care, in which "racial pharmacogenomics" is being positioned as a source of new markets for the pharmaceutical industry (Duster, 2005; Bibbins-Domingo and Fernandez, 2007). With such conflicting interests, it becomes harder than ever to evaluate the merits of scientific research on the genetics of race. A broad perspective on what we already know about science and human difference is consequently often quite valuable. After all, the newest work is hardly carried out in an intellectual, historical, or cultural vacuum.

Earnest Hooton almost understood this, trying to differentiate his own ostensibly benign physical anthropology from that of the Nazis, while nevertheless remaining a eugenicist long after it fell out of fashion in American academia. He warned, somewhat poignantly,

There is a rapidly growing aspect of physical anthropology which is nothing less than a malignancy. Unless it is excised, it will destroy the science. I refer to the perversion of racial studies and of the investigation of human heredity to political uses and to class advantage ... [T]he output of physical anthropology may become so suspect that it is impossible to accept the results of research without looking behind them for a political motive (Hooton, 1937, pp. 217–218).

CONCLUSIONS

Both human beings, and the scientific study of human beings, are coproductions of nature and culture. Human biologists are very familiar with the manifold processes by which "culture" is inscribed upon the human organism, and is ultimately not separable from the biology, or the human phenotype – "nature." It has proven more difficult to accept the idea that science itself – despite being a human activity, taking place in a cultural context, and being subject to conflicting interests of various kinds – produces conclusions about nature that are ultimately also not separable from culture. The idea that you can separate the natural from the cultural with a high degree of confidence, however, is an Aristotelian survival (Goodman et al., 2003).

The most significant aspect of the study of human diversity is that it consists of natural–cultural facts. These facts emanate from the kinds of questions framed, the manner in which categories are envisioned and established, the applications that assign people to the categories, the meanings attributed to group membership, and of course, the program of the investigator. Certainly there is a base of data that can inform us about the patterns of diversity that exist in our species, both somatic and genetic. The problem lies in the presumptions: (1) that the biological data on human variation are fundamentally separable from their cultural context and values, and from the interests of the scientists producing them; and (2) that the data themselves are meaningful independently of a stream of Euro-American ideas about difference, heredity, and hierarchy. That is why the problem of race has never been resolved by genetics; its domain is anthropological, rather than biological.

DISCUSSION POINTS

1. What are the incompatibilities among the three concepts of race discussed in this essay?
2. Why can't we separate facts of nature from culture?
3. Are Hispanics a race?
4. Old anthropology books used to show maps of the races of the world, with, for example, no presence of Europeans, Asians, or Africans in America. What are the merits of, and problems with, that?
5. What are the major patterns of human genetic variation and the major patterns of human cognitive variation, and how do they relate to one another? What implications can be drawn from that?

REFERENCES

Andreasen, R. O. (2000). Race: biological reality or social construct? *Philosophy of Science*, **76**, 653–666.

Andreasen, R. O. (2004). The cladistic race concept: a defense. *Biology and Philosophy*, **19**, 425–442.

Arnold, M. (2009). *Reticulate Evolution and Humans: Origins and Ecology*. New York: Oxford University Press.

Bamshad, M. J., Wooding, S., Watkins, W. S., et al. (2003). Human population genetic structure and inference of group membership. *American Journal of Human Genetics*, **72**, 578–589.

Barbujani, G., Magagni, A., Minch, E., et al. (1997). An apportionment of human DNA diversity. *Proceedings of the National Academy of Sciences of the United States of America*, **94**, 4516–4519.

Barbulescu, M., Turner, G., Su, M., et al. (2001). A HERV-K provirus in chimpanzees, bonobos, and gorillas, but not humans. *Current Biology*, **11**, 779–783.

Bell, A., Richerson, P. and McElreath, R. (2009). Culture rather than genes provides greater scope for the evolution of large-scale human prosociality. *Proceedings of the National Academy of Sciences of the United States of America*, **106**, 17671–17674.

Bendyshe, T. (1865). *The Anthropological Treatises of Johann Friedrich Blumenbach*. London: Longman, Green, Longman, Roberts and Green.

Bibbins-Domingo, K. and Fernandez, A. (2007). Focus on the right health fight. *San Francisco Chronicle*, March 5, D9.

Boas, F. (1911). *The Mind of Primitive Man*. New York: Macmillan.

Bogin, B. (1988). *Patterns of Human Growth*. New York: Cambridge University Press.

Bolnick, D. A. (2008). Individual ancestry inference and the reification of race as a biological phenomenon. In *Revisiting Race in a Genomic Age*, B. A. Koenig, S. S. J. Lee and S. Richardson (eds). Piscataway, NJ: Rutgers University Press, pp. 70–85.

Bowcock, A. M., Kidd, J. R., Mountain, J. L., et al. (1991). Drift, admixture, and selection in human evolution: a study with DNA polymorphisms. *Proceedings of the National Academy of Sciences of the United States of America*, **88**, 839–843.

Boyd, W. C. (1963). Genetics and the human race. *Science*, **140**, 1057–1065.

Brodkin, K. (1999). *How Jews Became White Folks and What That Says About Race in America*. Piscataway, NJ: Rutgers University Press.

Buffon, Comte de (1749) Variétés dans l'espèce humaine. In *Histoire Naturelle, Générale et Particuliére*, vol. **3**. Paris: L'Imprimerie Royale, pp. 371–530.

Campbell, B. (1962). The systematics of man. *Nature*, **194**, 225–232.

Cartmill, M. (1998). The status of the race concept in physical anthropology. *American Anthropologist*, **100**, 651–660.

Cavalli-Sforza, L. L. (1974). The genetics of human populations. *Scientific American*, **231**, 81–89.

Cavalli-Sforza, L. L. and Edwards, A. W. F. (1965). Analysis of human evolution. In *Genetics Today: Proceedings of the XI International Congress of Genetics*, S. J. Geerts (eds.). Oxford: Pergamon, pp. 923–933.

Cavalli-Sforza, L. L., Menozzi, P. and Piazza, A. (1994). *The History and Geography of Human Genes*. Princeton: Princeton University Press.

Chaline, J., Dutrillaux, B., Couturier, J., et al. (1991). Un modèle chromosomique et paléobiogéographique d'évolution des primates supérieurs. *Geobios*, **24**, 105–110.

Chen, F.-C. and Li, W.-H. (2001). Genomic divergences between humans and other hominoids and the effective population size of the common ancestor of humans and chimpanzees. *American Journal of Human Genetics*, **68**, 444–456.

Classen, C. (1995). *Worlds of Sense: Exploring the Senses in History and Across Cultures*. New York: Routledge.

Coon, C. S. (1939). *The Races of Europe*. New York: Macmillan.

Coon, C. S. (1962). *The Origin of Races*. New York: Knopf.

Coon, C. S. (1981). *Adventures and Discoveries*. Englewood Cliffs, NJ: Prentice-Hall.

Darlington, C. D. (1969). *The Evolution of Man and Society*. London: Allen and Unwin.

David, R. J. and Collins, J. W. Jr (1997). Differing birth weights among infants of US-born blacks, African-born blacks, and US-born whites. *New England Journal of Medicine*, **337**, 1209–1214.

Deinard, A. (1997). The evolutionary genetics of the chimpanzees. PhD dissertation, Department of Anthropology, Yale University, New Haven, CT.

Dobzhansky, T. (1962). Genetics and equality. *Science*, **137**, 112–115.

Dobzhansky, T. (1963). Probability that *Homo sapiens* evolved independently five times is vanishingly small. *Current Anthropology*, **4**, 360, 364–366.

Duster, T. (2005). Race and reification in science. *Science*, **307**, 1050–1051.

Edwards, A. W. F. (2003). Human genetic diversity: Lewontin's fallacy. *Bioessays*, **25**, 798–801.

Ferris, S. D., Brown, W. M., Davidson, W. S., et al. (1981). Extensive polymorphism in the mitochondrial DNA of apes. *Proceedings of the National Academy of Sciences of the United States of America*, **78**, 6319–6323.

Fix, A. G. (2005). *Migration and Colonization in Human Microevolution*. New York: Cambridge University Press.

Garcia, R. S. (2003). The misuse of race in medical diagnosis. *Chronicle of Higher Education*, **49**, B15.

Gates, R. R. (1948). *Human Ancestry*. Cambridge, MA: Harvard University Press.

Gilmour, J. S. L. and Gregor, J. W. (1939). Demes: a suggested new terminology. *Nature*, **144**, 333–334.

Glazko, G. V. and Nei, M. (2003). Estimation of divergence times for major lineages of primate species. *Molecular Biology and Evolution*, **20**, 424–434.

Gobineau, J. A. (1853–1855). *Essai sur l'Inégalité des Races Humaines*. Paris: Firmin-Didot.

Goodman, A. H., Heath, D. and Lindee, M. S. (2003). *Genetic Nature/Culture: Anthropology and Science beyond the Two-Culture Divide*. Berkeley, CA: University of California Press.

Gould, S. J. (1981). *The Mismeasure of Man*. New York: W. W. Norton.

Graves, J. (2004). *The Race Myth: Why We Pretend Race Exists in America*. New York: Dutton.

Hannaford, I. (1996). *Race: the History of an Idea in the West*. Baltimore: Johns Hopkins University Press.

Herrnstein, R. and Murray, C. (1994). *The Bell Curve*. New York: Free Press.

Hiernaux, J. (1975). *The People of Africa*. New York: Scribner.

Hogben, L. (1931). *Genetic Principles in Medicine and Social Science*. London: Williams and Norgate.

Hooton, E. A. (1926). Methods of racial analysis. *Science*, **63**, 75–81.

Hooton, E. A. (1936). Plain statements about race. *Science*, **83**, 511–513.

Hooton, E. A. (1937). *Apes, Men, and Morons*. New York: G. P. Putnam's Sons.

Hudson, N. (1996). From "nation" to "race": the origin of racial classification in eighteenth-century thought. *Eighteenth-century Studies*, **29**, 247–264.

Hulse, F. S. (1957). Linguistic barriers to gene-flow; the blood-groups of the Yakima, Okanagon and Swinomish Indians. *American Journal of Physical Anthropology*, **15**, 235–246.

Hulse, F. S. (1962). Race as an evolutionary episode. *American Anthropologist*, **64**, 929–945.

Hunley, K. and Long, J. (2005). Gene flow across linguistic boundaries in native North American populations. *Proceedings of the National Academy of Sciences of the United States of America*, **102**, 1312–1317.

Huxley, J. S. (1931). *Africa View*. London: Chatto and Windus.

Huxley, J. S. (1938). Clines: an auxiliary taxonomic principle. *Nature*, **142**, 219–220.

Ignatiev, N. (1996). *How the Irish Became White*. New York: Routledge.

Jahoda, G. (1999). *Images of Savages: Ancient Roots of Modern Prejudice in Western Culture*. New York: Routledge.

Jennings, H. S. (1930). *The Biological Basis of Human Nature*. New York: W. W. Norton.

Johnston, F. E. (1966). The population approach to human variation. *Annals of the New York Academy of Sciences*, **134**, 507–515.

Johnston, F. E. (1973). *Microevolution of Human Populations*. Englewood Cliffs, NJ: Prentice-Hall.

Kaessmann, H., Wiebe, V. and Paabo, S. (2001). Great ape DNA sequences reveal a reduced diversity and an expansion in humans. *Nature Genetics*, **27**, 155–156.

Kaufman, J. and Hall, S. (2003). The slavery hypertension hypothesis: dissemination and appeal of a modern race theory. *Epidemiology and Society*, **14**, 111–126.

Kluckhohn, C. and Griffith, C. (1950). Population genetics and social anthropology. *Cold Spring Harbor Symposium on Quantitative Biology*, **15**, 401–408.

Kumar, S., Filipski, A., Swarna, V., et al. (2005). Placing confidence limits on the molecular age of the human-chimpanzee divergence. *Proceedings of the National Academy of Sciences of the United States of America*, **102**, 18842–18847.

Lasker, G. W. (1999). *Happenings and Hearsay: Reflections of a Biological Anthropologist*. Detroit, MI: Savoyard Books.

Lasker, G. and Crews, D. (1996). Behavioral influences on human genetic diversity. *Molecular Phylogenetics and Evolution*, **5**, 232–240.

Lewontin, R. C. (1972). The apportionment of human diversity. *Evolutionary Biology*, **6**, 381–398.

Lewontin, R. C., Rose, S. and Kamin, L. J. (1984). *Not in Our Genes*. New York: Pantheon.

Lieberman, L. (2001). How Caucasoids got such big crania and why they shrank: from Morton to Rushton. *Current Anthropology*, **42**, 63–85.

Linnaeus, C. (1758). *Systema Naturae*, 10th edn. Stockholm: Laurentius Salvius.

Livingstone, F. B. (1962). On the non-existence of human races. *Current Anthropology*, **3**, 279–281.

Marks, J. (1994). Blood will tell (won't it?): a century of molecular discourse in anthropological systematics. *American Journal of Physical Anthropology*, **94**, 59–80.

Marks, J. (1995). *Human Biodiversity: Genes, Race, and History*. New York: Aldine de Gruyter.

Marks, J. (2002). Genes, bodies, and species. In *Physical Anthropology: Original Readings in Method and Practice*, P. N. Peregrine, C. R. Ember and M. Ember (eds). Englewood Cliffs, NJ: Prentice-Hall, pp. 14–28.

Marks, J. (2005). Anthropology and the Bell Curve. In *Why America's Top Pundits are Wrong: Anthropologists Talk Back*, C. Besteman and H. Gusterson (eds). Berkeley, CA: University of California Press, pp. 206–227.

Mays, V. M., Ponce, N. A., Washington, D. L., et al. (2003). Classification of race and ethnicity: implications for public health. *Annual Review of Public Health*, **24**, 83–110.

Merton, R. K. and Ashley-Montagu, M. F. (1940). Crime and the anthropologist. *American Anthropologist*, **42**, 384–408.

Moore, J. H. (1994). Putting anthropology back together again: the ethnogenetic critique of Cladistic theory. *American Anthropologist*, **96**, 925–948.

Nei, M. and Roychoudhury, A. (1974). Genetic variation within and between the three major races of man, Caucasoids, Negroids, and Mongoloids. *American Journal of Human Genetics*, **26**, 421–443.

Nei, M. and Roychoudhury, A. (1993). Evolutionary relationships of human populations on a global scale. *Molecular Biology and Evolution*, **10**, 927–943.

O'hUigin, C., Satta, Y., Takahata, N., et al. (2002). Contribution of homoplasy and of ancestral polymorphism to the evolution of genes in anthropoid primates. *Molecular Biology and Evolution*, **19**, 1501–1513.

Pálsson, G. (2007). *Anthropology and the New Genetics*. New York: Cambridge University Press.

Patterson, N., Richter, D. J., Gnerre, S., et al. (2006). Genetic evidence for complex speciation of humans and chimpanzees. *Nature*, **441**, 1103–1108.

Peregrine, P., Ember, C. and Ember, M. (2003). Cross-cultural evaluation of predicted associations between race and behavior. *Evolution and Human Behavior*, **24**, 357–364.

Ripley, W. Z. (1899). *The Races of Europe*. New York: D. Appleton.

Rosenberg, N. A., Pritchard, J. K., Weber, J. L., et al. (2002). Genetic structure of human populations. *Science*, **298**, 2181–2185.

Ruano, G., Rogers, Jeffrey A., et al. (1992). DNA sequence polymorphism within hominoid species exceeds the number of phylogenetically informative characters for a HOX2 locus. *Molecular Biology and Evolution*, **9**, 575–586.

Sankar, P., Cho, M., Condit, C. M., et al. (2004). Genetic research and health disparities. *JAMA: Journal of the American Medical Association*, **291**, 2985–2989.

Satta, Y., Klein, J. and Takahata, N. (2000). DNA archives and our nearest relative: the trichotomy problem revisited. *Molecular Phylogenetics and Evolution*, **14**, 259–275.

Seligman, C. (1930). *Races of Africa*. Oxford: Oxford University Press.

Serre, D. and Pääbo, S. (2004). Evidence for gradients of human genetic diversity within and among continents. *Genome Research*, **14**, 1679–1685.

Smedley, A. and Smedley, B. D. (2005). Race as biology is fiction, racism as a social problem is real. *American Psychologist*, **60**, 16–26.

Stauffer, R. L., Walker, A., Ryder, O. A., et al. (2001). Human and ape molecular clocks and constraints on paleontological hypotheses. *Journal of Heredity*, **92**, 469–474.

Stone, A. C., Griffiths, R. C., Zegura, S. L., et al. (2002). High levels of Y-chromosome nucleotide diversity in the genus *Pan*. *Proceedings of the National Academy of Sciences of the United States of America*, **99**, 43–48.

Stuurman, S. (2000). François Bernier and the invention of racial classification. *History Workshop Journal*, **50**, 1–21.

Templeton, A. R. (1998). Human races: a genetic and evolutionary perspective. *American Anthropologist*, **100**, 632–650.

Thieme, F. P. (1952). The population as a unit of study. *American Anthropologist*, **54**, 504–509.

Tishkoff, S., Reed, F., Friedlaender, F., et al. (2009). The genetic structure and history of Africans and African Americans. *Science*, **324**, 1035–1044.

Tylor, E. B. (1871). *Primitive Culture: Researches into the Development of Mythology, Philosophy, Religion, Art, and Custom*. London: John Murray.

Underwood, J. H. (1979). *Human Variation and Human Microevolution*. Englewood Cliffs, NJ: Prentice-Hall.

Wade, N. (2002). Gene study identifies five main human populations. *New York Times*, December 20.

Washburn S. L. (1951). The new physical anthropology. *Transactions of the New York Academy of Sciences, Series II*, **13**, 298–304.

Weiner, J. S. (1957). Physical anthropology: an appraisal. *American Scientist*, **45**, 79–87.

Wilson, E. O. and Brown, W. L. Jr (1953). The sub-species concept and its taxonomic application. *Systematic Zoology*, **2**, 97–111.

Witherspoon, D., Wooding, S., Rogers, A., et al. (2007). Genetic similarities within and between human populations. *Genetics*, **176**, 351–359.

Wolf, E. R. (1982). *Europe and the People without History*. Berkeley, CA: University of California Press.

16 The Evolution and Endocrinology of Human Behavior: a Focus on Sex Differences and Reproduction

Peter B. Gray

INTRODUCTION

The aim of this chapter is to highlight some of the core concepts and empirical findings concerning the evolution and endocrinology of human behavior. To do this, we first review some basic principles in the evolution and endocrinology of behavior. These principles have been derived from studies with various taxa rather than humans alone. Indeed, nonhuman theoretical and empirical findings have inspired some of the human research, and the research on humans should be viewed in comparative contexts. Next, we investigate a series of examples illustrating empirical research on the evolution and endocrinology of human behavior with a focus on sex differences and reproductive behavior. These examples have been chosen because they illustrate well the relationships between human hormones and behavior and they offer enough data to enable drawing some conclusions. For some of these examples, cross-cultural data are also available, helping show the ways endocrine mechanisms underlie human behavior across variable sociocultural environments.

A comprehensive literature review of human findings, much less nonhuman behavioral endocrinology, would involve a very long book rather than a book chapter. Yet in the course of reviewing some well-documented examples, readers may gain an appreciation for the excitement of this research niche as well as the types of research questions remaining to be addressed. For the unavoidably hooked student or researcher, please see Ellison and Gray (2009), Carter et al. (2005), Adkins-Regan (2005), Nelson (2005), Sapolsky (2004), Becker et al. (2002), and Pfaff et al. (2002) for excellent overviews of hormones and behavior. In these latter volumes, readers can find full treatments of topics such as the relationships between hormones and diet, sleep, exercise, aggression, and stress that are largely outside the scope of the present chapter.

BACKGROUND

Overarching theoretical frameworks

There is more than one way to answer a "why" question in biology (such as why do sex differences exist). In fact, we can distinguish four complementary approaches to answering any biological question: (1) proximate (mechanism); (2) adaptation (function); (3) phylogenetic (evolutionary history); and (4) development (ontogeny) (Tinbergen, 1963; Bolhuis and Giraldeau, 2005). A focus on the evolution and endocrinology of human behavior most clearly integrates two of these approaches – endocrine mechanisms of behavior within a functional perspective – but may also consider the ontogeny of behavioral endocrine mechanisms and phylogenetic context.

Adoption of a functional, or adaptive, perspective has several advantages. For one, it stimulates the development of testable hypotheses. For example, suppose one recognizes that human males face a life history allocation problem of optimizing investment in mating (male–male competition and mate seeking) and parenting effort. One then wonders about the neuroendocrine mechanisms that modulate this allocation challenge. Another advantage is that an adaptive perspective helps organize what might otherwise be a pile of facts about endocrinology and behavior; it suggests some underlying organizational principles for why mechanisms work the way they do (e.g., why male–male competition and mate seeking are both commonly linked to elevations in male testosterone – both behaviors facilitate reproductive access to mates).

Adoption of a focus on the proximate, or mechanistic, understanding of behavior may have several different advantages (Panksepp et al., 2002; Piersma and Drent, 2003). For one, an understanding of mechanism helps constrain adaptationist thinking. Neuroendocrine mechanisms underlying behavior tend to evolve through "tinkering" of existing mechanisms (e.g., alterations in neurotransmitter or hormone

Human Evolutionary Biology, ed. Michael P. Muehlenbein. Published by Cambridge University Press. © Cambridge University Press 2010.

receptor distribution; changes in neural connections), casting doubt on novel, modular mechanisms of human behavior arising de novo during recent human evolution (Quartz and Sejnowski, 2002). Another advantage of focusing on mechanisms is that they suggest, based on principles of homology, mechanisms of behavior that may operate in humans (like in other species). In turn, this can yield predictions about the neuroendocrine bases of human behavior based on findings from nonhuman animals. For example, if research on rodents implicates the hormone oxytocin in pair bonding, then this knowledge may stimulate research in humans investigating the role of this same hormone in human pair bonding. A further benefit of studying mechanisms in a comparative approach is that the types of invasive, experimental designs conducted in, say, laboratory rodents are often not ethically or logistically feasible in humans, meaning that we are forced, to some degree, to rely on evidence from other species for making causal inferences on hormone-behavior relationships. The nonhuman work may converge with clinical findings from atypical human development (e.g., behavioral effects of a mutated hormone receptor or enzyme such as 5-α reductase). Though largely outside the scope of this chapter, another important aspect of mechanisms is that an understanding of them may have implications for the development of clinical treatments for behavioral problems (e.g., use of hormone replacement therapy for treating depression or sexual problems).

In this chapter, I highlight the interplay between functional and mechanistic approaches to human behavior. In sections below, brief summaries of the relevant concepts are provided. It also bears emphasis that any behavior has a developmental course to it (e.g., the onset of sex-differentiated behavior), even if less emphasis is given to development in this chapter. Moreover, phylogeny may have important implications, especially when considering whether mechanisms of human behavior may operate similarly or differently compared with other species. The most straightforward approach is to assume that mechanisms of human behavior may be similar to those of other taxa, especially the more closely related to us (e.g., rhesus monkeys more relevant than rats, in turn more relevant than fruit flies or nematodes). However, interesting exceptions to homologous mechanisms exist and should be appreciated; for example the masculinizing effects of hormones in rodent brains appear, unlike in primates, to entail the conversion of testosterone to estradiol (which does the masculinizing!) (Nelson, 2005). Also showing the importance of phylogeny, effects of hormonal changes across the menstrual cycle are much more pronounced in rodents and New World monkeys than in Old World monkeys and apes (Dixson, 1998).

Evolution of human behavior

Evolutionary theory recognizes that selection can act on multiple levels (gene, individual, group) but that selection on individuals tends to be the most powerful. Consequently, adaptive accounts of human behavior typically consider the fitness (relative reproductive success) costs and benefits to individual behavior. The overarching question from this adaptationist approach is: how do individuals behave in ways maximizing reproductive success?

In maximizing reproductive success, behaviors can arise that benefit others like kin at cost to oneself. Kin selection theory shows how heritable tendencies to benefit relatives can spread. Co-operative behavior among unrelated individuals can also evolve in various ways, including via processes of reciprocal altruism (I scratch your back now and you scratch my back later) or indirect reciprocity (I behave in ways leading to you wanting to co-operate with me). Fitness costs to behavior are also contingent on a variety of variables: reproductive value (likelihood of future reproductive output), social status, and physical condition of both actors and recipients of behavior impact whether selection will favor certain behaviors.

How individuals maximize reproductive success will depend on the social context. Hence, historical and cross-cultural variation in human sociocultural environments is important (Low, 2000; Richerson and Boyd, 2005). This sociocultural variation presents different opportunities and constraints to individuals: with kin to assist with allocare, a woman may have more children; in small-scale bellicose societies in which male–male bonds enable defense against neighbors, weaker husband–wife and father–child relationships may emerge; when formal education systems arise, children may reduce their care of younger siblings to advance their own education.

Several additional complications remain. A school of thought, championed by evolutionary psychologists, questions whether humans would be expected to behave adaptively in contemporary environments (Barkow et al., 1992; Irons, 1998). The psychological mechanisms underlying human behavior evolved in different environments from those in which we typically find ourselves (like a large city populated by strangers). Thus, we may behave maladaptively in our novel worlds, yet understandably according to scenarios of human evolution. Other reasons for humans behaving in nonoptimal ways include pleiotropic effects, trade-offs, and constraints. As an illustration, testosterone has multiple effects (pleiotropy), some of which may work against one another (e.g., a trade-off between maintaining elevated testosterone levels in response to social challenges that may compromise immune function). For review of these general principles concerning the evolution of human behavior, a number of clear accounts exist (Betzig, 1997; Pinker, 1997; Hrdy, 1999; Cartwright, 2000; Barrett et al., 2002; Konner, 2002; Buss, 2003, 2005; Gaulin and McBurney, 2004).

Some implications from evolutionary principles of human behavior can be drawn with respect to the underlying endocrine mechanisms. One implication is that males and females have faced different selective environments over evolutionary history, and thus we might expect to find endocrine mechanisms playing an important role in the development of sex differences. Human reproduction typically occurs within contexts of long-term pair bonds; consequently, we should ask about mechanisms underlying pair bonds and parental care in nonhuman animals to gain potential insights into these human behaviors. Kinship has occupied a central role in our ancestral social organization, even if it sometimes seems less important in large cities replete with strangers and friends rather than extended family members. We might expect that degree of relatedness between individuals (or proxies thereof, such as proximity during early childhood development) will have predictable impacts on endocrine-mediated behavior. Finally, we must locate the development and expression of behavior, and underlying mechanisms, in particular sociocultural contexts. The specific environment in which behavior develops and is expressed will shape how mechanisms work (see Chapter 24 of this volume).

ENDOCRINOLOGY OF HUMAN BEHAVIOR

As a complementary approach to functional questions concerning human behavior, a focus on endocrine mechanisms highlights some of the proximate ways in which human behavior occurs. Endocrine mechanisms play important roles in the expression of human behavior. The endocrine system helps co-ordinate behavior by serving as a key system of communication within the body (see Chapter 8 of this volume). The endocrine system incorporates information from both within and outside the body to help generate appropriate responses. Hormonal systems can guide an individual toward behaviors appropriate for its condition (e.g., age, nutritional status, disease status) and social context (e.g., availability of mates, infant caretaking).

The endocrine system consists of ductless glands that release hormones into the body to effect responses at specific tissues elsewhere in the body (Reed Larsen et al., 2003; Nelson, 2005). While this classic view of hormone action has many exceptions, this straightforward view captures key points of it. In the course of hormone release, complex feedback loops emerge. For example, the hypothalamus releases gonadotropin-releasing hormone (GnRH), in turn facilitating the pituitary to release gonadotropins that in turn lead to steroid release by the gonads. These feedback loops are usually negative, but occasionally positive. Moreover, they incorporate information from other structures and systems too; intricate co-ordination with the nervous system, immune system, and other endocrine axes allows behavioral responses appropriate for an individual's physical condition and social context.

The causal arrow between hormones and behavior points in both directions. Elevations or decreases in hormones may alter behavior; you may have noticed this if you have a neutered or spayed pet since these operations have altered sex steroid secretions. Behavior can also affect hormones. If you have felt an acute rise in arousal when giving a presentation to a large audience, it is quite possible that this behavior increased your cortisol levels. The bidirectional causal effects of hormones and behavior reflect the complex feedback loops through which these systems operate.

In this chapter, I will focus on a limited set of hormones. The hypothalamus-pituitary-gonadal axis alluded to above will feature in discussions of human sex differences and sex-typed behavior. Prolactin is a peptide hormone released, largely under negative control by dopamine, from the anterior pituitary gland. Oxytocin and vasopressin are small peptide hormones released from the posterior pituitary. Each of these hormones passes through the circulatory system until reaching target tissues where they are bound by the appropriate hormone receptor (e.g., the androgen receptor binds testosterone). Once bound to the appropriate receptor, physiological changes can be induced, including both rapid (e.g., progesterone altering the inhibitory neurotransmitter γ-aminobutyric [GABA] activity) and longer-acting (e.g., estradiol-facilitating gene expression) effects.

HUMAN HORMONES AND BEHAVIOR EXAMPLES

With this general background on the evolution and endocrinology of behavior in mind, let us now turn to a series of human behavioral endocrinology examples. These are examples focused on sex differences and reproductive behavior. This focus thus emphasizes examples closely tied to survival and reproductive success, the ultimate currency in a Darwinian world. For each of these examples, I begin with a question. In these studies, I present primarily human data. My phylogenetic bias may be excusable here because this volume focuses on *human* evolutionary biology. Moreover, the literature on hormones and behavior consists of elegant field and experimental research on species like voles and rhesus monkeys that researchers commonly extrapolate to humans. Here, we skip the extrapolation and proceed directly to humans.

WHAT IS THE HORMONAL BASIS OF SEX DIFFERENCES IN HUMAN BEHAVIOR?

Sex differences exist because they have been favored by natural selection, including sexual selection. Over evolutionary time, females and males have faced different selective pressures to maximize survival and reproductive success, leading to the kinds of sex differences in behavior described below. A complementary question to the functional origins of sex differences in behavior is to ask about the proximate mechanisms underpinning them. The fundamental mechanisms underlying human sex differences have been elegantly shown (Migeon and Wisniewski, 1998; Vilain and McCabe, 1998; Nelson, 2005). Centuries of experience castrating male domesticated animals and twentieth-century endocrinology experiments (e.g., castration and hormone replacement studies on laboratory rodents) laid the nonhuman groundwork for understanding the role of hormones in sex differences. The story of human hormonal bases to sex differences is yet another remarkable story of biological conservation: the mechanisms differentiating human males and females look much like those of a typical mammal.

The process of differentiating males from females begins with the sex chromosomes. Males possess an X and Y chromosome and females two X chromosomes. A gene on the Y chromosome – the *SRY* gene (for sex determining region of the Y chromosome) – encodes a protein that causes the initially undifferentiated fetal gonads to become testes. As the testes develop, they begin secreting testosterone and Müllerian inhibiting hormone (MIH) at around week 12 of gestation. Testosterone "tells" other undifferentiated tissues to develop male phenotypes (e.g., prostate gland, penis). In many cases, testosterone is first converted to a more potent androgen, dihydrotestosterone (DHT), which binds to the androgen receptor in appropriate tissues and thereby leads to the development of male phenotypes. The MIH secreted by the testes induces the Müllerian duct system to regress.

So in the absence of testicular hormones promoting male phenotypes, undifferentiated structures will develop as female phenotypes. Such observations have led to the concept of females as the default sex (an undifferentiated individual will become female in the absence of substances promoting maleness). This characterization largely holds for the role of hormones early in life. However, at puberty, hormones play central roles in promoting female and male phenotypes respectively (Ellison, 2001). At puberty, estrogens in women promote secondary sexual characteristics like sex-specific regional fat deposition but also neural mechanisms underlying behavior. Androgens in males also propel male secondary sexual characteristics, including those involved in behavior. In both sexes,

androgens promote development of secondary sexual characteristics such as underarm hair and sebaceous glands (which can manifest as acne).

Against this backdrop of human sex differentiation, roles for other genes, both on sex and autosomal chromosomes, have been identified (Arnold, 2002). The involvement of some of these other genes suggests slight modifications to the picture above about sex differentiation. Moreover, species differences among mammals exist (e.g., masculinization of the mouse brain, unlike that of an Old World monkey or ape brain, involves estradiol bound to estrogen receptors). Nonetheless, we do quite well employing the standard scenario above to understand the fundamental ways human sexes differ.

One other central concept to the role of hormones in human behavioral sex differences should be highlighted: the distinction between organizational and activational effects (Nelson, 2005). Organizational effects typically refer to permanent effects of hormones occurring early in life, whereas activational effects refer to more transient behavioral effects of changes in hormone levels. To illustrate the distinction, if we were to inject testosterone into a mouse fetus during a sensitive period early in life, we might "organize" its brain as masculine; injecting testosterone into that same male's adult brain might demonstrate short-lived "activational" effects too on male behavior.

This background on human sex differentiation, documenting the crucial roles played by sex steroids, underlies many of the sex differences in human behavior that have been observed. Consider, for a moment, some of the best-documented sex differences in behavior. Taken from several exhaustive reviews of human sex differences (Maccoby, 1998; Archer and Lloyd, 2002; Lippa, 2002; Hines, 2004; Cohen-Bendahan et al., 2005; Nelson, 2005), these are displayed in Table 16.1. The directionality and effect sizes given in Table 16.1 are drawn from these reviews, with effect sizes in some cases estimated by Cohen's d statistic (number of standard deviations differentiating males and females) and in other cases more subjective means. These patterns represent distributions (e.g., a given female may not differ from a given male) that can also be situated in a broader social context (e.g., the magnitude of the sex difference in openness to sexual behavior varies according to demographic and other factors [Schmitt, 2005]).

Several types of evidence convincingly demonstrate that the human behavioral sex differences shown in Table 16.1 can largely be traced to hormonal effects. There are five key types of evidence. The first is that experimental research on closely related organisms like mice, rats, and rhesus macaques commonly shows that comparable sex differences can be traced to organizational and activational effects of hormones

TABLE 16.1. Human sex differences in behavior.

Behavior	Male/female patterning	Effect size
Physical aggression	M > F	Large
Direct care of children	F > M	Large
Verbal fluency	F > M	Large
Toy preferences (e.g., males and balls; females and dolls)	–	Large
Rough-and-tumble play	M > F	Very large
Preference for boys as playmates	M > F	Very large
Preference for girls as playmates	F > M	Very large
Risk-taking behavior	M > F	Small
Socioemotional behavior in groups	F > M	Medium
Decoding nonverbal behavior	F > M	Medium
Task-oriented behavior in groups	M > F	Medium
Active nonverbal behavior/ body movement	M > F	Large
Openness to sexual behavior	M > F	Large
Orientation toward same-sex dominance hierarchy	M > F	Large
Sexual coercion (e.g., rape)	M > F	Large
Sexual attraction to males	F > M	Very large
Sexual attraction to females	M > F	Very large

(Nelson, 2005; Wallen, 2005). For example, rhesus female play patterns can be masculinized with androgen administration early in life. The second line of evidence is correlational human research. Here, as an example, weak, yet positive correlations between adult male testosterone levels and physical aggression have been observed (Book and Quinsey, 2005), suggesting activational differences in testosterone play some role in the sex difference in physical aggression. The third line of evidence involves clinical cases of atypical human development (Cohen-Bendahan et al., 2005). For example, XY individuals with a nonfunctional androgen receptor appear and behave largely like females, implicating androgens in phenotypic masculinization. A fourth line of evidence refers to human hormonal interventions leading to unintended consequences. As an example, women taking progestins in the 1960s and 1970s to improve pregnancy symptoms unintentionally exposed their fetuses to androgenic effects of these; the exposed offspring reported a higher likelihood of using physical aggression than controls, indicative of organizing influences of androgens on physical aggression. A fifth line of evidence refers to human hormonal interventions, usually undertaken for clinical, cosmetic, or quality-of-life reasons.

Providing vivid testimony of such effects, changes in cognition, mood, and behavior experienced by transsexuals while taking hormones to change their sex have been described (Roughgarden, 2004).

To illustrate these various lines of evidence implicating hormones in human behavioral sex differences, take rough-and-tumble play as an example. Experiments with rhesus monkeys support the masculinizing, and organizational, effects of prenatal androgens. So do human cases of atypical development. Some girls presenting with congenital adrenal hyperplasia, characterized by steroid enzyme abnormalities, have unusually high androgen levels; some of these same girls display more masculine play patterns as children. Hence, several lines of evidence like these highlight the roles of hormones in accounting for sex differences in human behavior (Hines, 2004; Nelson, 2005).

WHAT ARE THE HORMONAL CORRELATES OF HUMAN PAIR BONDING AND PARENTAL CARE?

Long-term bonds between adult mates (pair bonds) appear to be a derived feature of human behavior arising, perhaps in the genus *Homo*, within the past 1.8 mya (Gray et al., 2004b). Fossil evidence suggests a reduction in body size dimorphism consistent with a tendency toward reduced male–male contest competition and rise in pair bonding (Flinn et al., 2005), though new fossil evidence questions whether these trends began around 1.8 mya or more recently with archaic *Homo sapiens* around 0.5 mya (Lordkipanidze et al., 2007). Regardless of the exact timing of the origin of human pair bonds, however, these are nonetheless noteworthy because they occur in only about 5% of mammalian species, and are absent in our closest living relatives, common chimpanzees and bonobos (Reichard and Boesch, 2003). Changes in male parental investment (e.g., provisioning and/or direct care of children) may have arisen at this same juncture, or perhaps more recently in our lineage.

If asking about the hormonal bases of human pair bonding and parental care, we can turn to research on sheep, rats and voles – where most of the relevant and experimental research has been conducted. This body of research has focused on roles of the peptide hormones oxytocin and vasopressin (Carter, 1998; Young and Wang, 2004). In nonhuman animals, oxytocin and prolactin facilitates maternal care and, less consistently, adult pair bonds; effects of oxytocin and prolactin are more clearly recognized in females, but also appear sometimes in males. In nonhuman animals, vasopressin is involved in courtship, male–male competition, and mate and offspring guarding, with its effects more pronounced among males (Storm and

Tecott, 2005). Here, we draw inspiration from nonhuman animal functional and proximate considerations to briefly review the hormonal correlates of human pair bonding and parental care. We concentrate on whether oxytocin and vasopressin play roles in human pair bonding and parental care, but also take brief tours to investigate the roles of cortisol and prolactin.

Oxytocin has classically recognized roles in smooth muscle contraction, including that involved in uterine contractions and milk release (Sanchez et al., 2009). Orgasm also stimulates oxytocin release (Kruger et al., 2003), as do touch and nipple stimulation (Uvnas-Moberg, 1998). Psychological effects of oxytocin tend to be anxiolytic. Overall, effects of oxytocin have been characterized as those of a "relaxing, feel-good" hormone.

These psychological effects have important functional effects, particularly with respect to promoting affiliative social behavior. Increased oxytocin can condition rewarding association between sex, touch, and other interactions with a partner. Neural pathways activating dopamine facilitate these rewarding effects in social memory circuits (Young and Wang, 2004). Moreover, oxytocin tends to downregulate stress responsiveness (Uvnas-Moberg, 1998). This may be advantageous by immunizing an individual against stressors in favor of focused, calm attention on a social partner.

Oxytocin may play a role in human maternal care (Uvnas-Moberg, 1998). Rushes of oxytocin occur during uterine contractions, both facilitating birth and setting the stage for rewarding interactions between a mother and her newborn. Placing a newborn on a mother's chest increases the mother's oxytocin levels (Matthiesen et al., 2001). In addition to facilitating maternal–offspring bonding, these maternal contexts of oxytocin release appear to dampen a mother's stress responsiveness.

Mothers appear to have downregulated hypothalamus-pituitary-adrenal (HPA) activity compared with nonmothers, at least for the first days or weeks postpartum (Uvnas-Moberg, 1998). In fact, elevated cortisol levels among mothers of newborns appear positively associated with maternal interest and care of offspring (Fleming et al., 1997; Fleming, 2005). Aside from this initial window, maternal interactions with an infant, including breast-feeding and touch, increase oxytocin which, in turn, may help reduce stress reactivity. Breast-feeding mothers showed less HPA activity than control women on an exercise stressor test (Altemus et al., 1995) as well as a psychosocial stressor (Heinrichs et al., 2001), suggesting downregulatory effects of maternal care (and oxytocin) on stress responsiveness. The reduced stress responsiveness of mothers may facilitate more relaxing, positive interactions with her offspring, while blunting her responsiveness to stressors that might otherwise interfere with maternal care.

While oxytocin plays a role in human maternal care, it also seems to be involved in human pair bonding. In fact, some of the neuroendocrine mechanisms involved in parental care appear homologous to those involved in pair bonding (Bartels and Zeki, 2004). Several studies have examined oxytocin levels of men and women involved in relationships like marriage to ask whether individual differences in relationship quality are associated with baseline differences in oxytocin levels. Women reporting more frequent hugs in their marital relationships showed higher oxytocin levels and lower cardiovascular activity (Light et al., 2005).

Other research methods entailing brief interactions between male and female partners, measuring oxytocin levels before and after these interactions, have shown that changes in oxytocin levels appear positively related to pair bonding outcomes. For example, Grewen et al. (2005) found that both men and women reporting higher partner support had higher oxytocin levels before and after a brief behavioral interaction with their partner. Cardiovascular effects were more pronounced in women than men in this study, however. Furthermore, Chen and colleagues observed positive correlations between increases in female oxytocin levels and researcher ratings of partner support during brief interactions among couples (Sanchez et al., 2009).

Given the expectation that oxytocin will have stronger relationships with female behavior, it remains interesting that links between oxytocin and male behavior have also been shown. Several experimental interventions relying on intranasal oxytocin have observed increased trust (Kosfeld et al., 2005) and effects on social memory (Kirsch et al., 2005) in men, further showing that links between oxytocin and behavior occur in males too. Even if oxytocin may have a longer-standing role in promoting maternal behavior (e.g., in mammals), then its effects can carry over not only to females pair bonded to a mate, but also pair bonded males. This is particularly intriguing because estrogen is known to have facilitatory effects on oxytocin, another reason to expect relationships between oxytocin and behavior to be more prominent in women (Carter, 2007).

Although closely related to oxytocin, vasopressin has quite different physiological and behavioral effects. Its classical roles in vasoconstriction and water retention have long been recognized (and given rise to its alter-ego name of antidiuretic hormone). Psychological effects have also been recognized for decades, including increased anxiety (Landgraf and Holsboer, 2005). Research on nonhuman animals suggests vasopressin is more likely to be associated with male than female human behavior since its effects are potentiated by testosterone (Delville et al., 1996) and the vasopressin system is sexually dimorphic. Generally, we should expect vasopressin to facilitate adaptive anxiety and rapid behavioral responsiveness within contexts of male–male competition, courtship, and mate- and offspring defense.

There are remarkably few data on vasopressin and human male social behavior to address whether behavioral correlates of vasopressin are consistent with expectations. Only three studies in humans have examined links between vasopressin and male aggression. Positive associations between cerebrospinal fluid (CSF) vasopressin levels and aggression have been observed (Coccaro et al., 1998). Male subjects randomized to receive an intranasal spray of vasopressin displayed increased interest in angry faces compared to males given a placebo spray (Thompson et al., 2004). A follow-up study by Thompson et al. (2006) suggested sexually dimorphic effects of vasopressin administration. Both women and men reporting increased anxiety associated with vasopressin administration, but women (unlike men) channeled that anxiety into more socially affiliative cognition. That is, women appeared drawn toward positive interactions with others when given vasopressin, while males were more primed toward agonistic responses.

If vasopressin plays a role in male–male aggression, we might also expect from nonhuman data elevated vasopressin during courtship and in contexts of mate- and offspring-guarding. However, no data, to my knowledge, address these expectations in humans.

Roles for at least two other hormones involved in paternal care have been raised: prolactin and testosterone. Prolactin stimulates milk production, and is responsive to nipple stimulation, orgasm, and psychological stressors, among other stimuli. In a variety of taxa, elevated prolactin is associated with paternal care (Ziegler, 2000; Nelson, 2005). In humans, three studies have considered prolactin and paternal care. Two studies of highly invested Canadian fathers revealed increases in prolactin levels associated with responses to infant cries or interactions with a child (Berg and Wynne-Edwards, 2001; Fleming et al., 2002), although men's prolactin responses differed according to experience and recency of infant interactions (Delahunty et al., 2007). In a sample of 43 Jamaican men, fathers maintained relatively flat prolactin profiles across a 20 minute session interacting with an adult female partner and youngest child, whereas prolactin levels of single men sitting alone during this time declined (Gray et al., 2007a). In the next section, we consider the evidence that lower testosterone levels may be associated with paternal care.

To summarize the hormonal correlates of human pair bonding and parental care, oxytocin appears to facilitate human maternal care as well as both male and female involvement in adult pair bonds. Vasopressin appears linked to male–male competition, but it is premature to draw conclusions regarding a role for vasopressin in human pair bonding and parental care given the paucity of data. In mothers, cortisol is positively associated with initial engagement in intensive offspring care, then cortisol responsiveness appears downregulated with maternal care and effects of oxytocin. Prolactin levels appear positively related with human paternal care.

Do male testosterone levels differ according to relationship status?

Testosterone may be adaptively linked with variation in male reproductive effort by promoting male–male competition in reproductive contexts, but being lower in contexts of affiliative pair bonding and paternal care (Wingfield et al., 1990; Ketterson and Nolan, 1999). Such theoretical research has considerable empirical support in many, but certainly not all, vertebrates. This body of research provides one rationale for thinking human male testosterone levels too might differ according to differential reproductive effort – and we can use pair bonding status and paternal care as indicators of variation in reproductive effort.

The first human study addressing this question was published by Booth and Dabbs (1993) based on a sample of several thousand US Army veterans. They found that married men had significantly lower testosterone levels than their unmarried counterparts. A rapidly growing human literature on the topic of human male pair bonding, parenting, and testosterone has extended these initial findings to consider pair bonding status (not just marital status), paternal care, and societies outside the United States (reviewed in van Anders and Watson, 2006a; Gray and Campbell, 2009).

Studies in North America have consistently found that men involved in long-term committed relationships such as marriage have lower testosterone levels than unpaired men. Initially, the focus had been on marital relationships, with married US men found to have lower testosterone levels than single men (Booth and Dabbs, 1993; Mazur and Michalek, 1998; Gray et al., 2004a). Three other studies of business school students (Burnham et al., 2003) and undergraduate students (Gray et al., 2004b; McIntyre et al., 2006) found that similar differences held for men involved in relationships outside of marriage; that is, involvement in a committed, romantic relationship, whether married or not, was associated with lower testosterone levels. Moreover, lower testosterone levels of partnered men were observed among men regardless of whether they lived in the same city or were engaged in long-distance relationships with their partners (van Anders and Watson, 2007).

Extending this body of research to consider paternal care, two studies observed lower, though nonsignificant, testosterone levels among paired fathers compared with married men without children (Gray et al., 2002; Burnham et al., 2003). Three other studies of Canadian fathers showed a role for testosterone in fatherhood. Two of these found lower testosterone levels among fathers compared with control nonfathers (Berg and Wynne-Edwards, 2001; Fleming et al., 2002). Storey et al. (2000) found that fathers had lower testosterone levels shortly after birth compared to men whose babies were about to be born. Moreover, Fleming et al. (2002) found that paternal responsiveness to infant stimuli such as infant cries was inversely related to men's testosterone levels.

Several other North American studies have observed finer points in the links between testosterone, pair bonding and paternal care. Among both Harvard and University of New Mexico undergraduate students, those pair bonded men less interested in seeking extrapair mates tended to have lower testosterone levels than paired men more interested in extrapair mating (McIntyre et al., 2006). In examining links between testosterone and marital interactions among a recently wed Pennsylvanian sample, husbands and wives with similarly low testosterone levels were found to exhibit more supportive interactions when observed (Cohan et al., 2003). A Canadian study extended these findings by considering sexual orientation and both men and women: men paired with women and women paired with women had lower testosterone levels than their unpaired counterparts, but men paired with men and women paired with men did not show differences in testosterone levels according to partner status (Van Anders and Watson, 2006b).

Of course, what happens in North America may not be representative of what happens elsewhere in the world. Human pair bonding and paternal behavior vary cross-culturally. In some societies, husbands and wives may be highly supportive of each other; in others, as measured by indices such as cosleeping, coeating, spending leisure time together, and providing emotional support, husbands and wives may be quite aloof (Whiting and Whiting, 1975). Human paternal care also exhibits considerable variation both in terms of direct care (e.g., holding and carrying a child) and indirect care (e.g., defending and provisioning) (Marlowe, 2000).

Perhaps not surprisingly, the cross-cultural research on this topic is less consistent than that from North America (Gray and Campbell, 2009). In some societies, monogamously married men have lower testosterone levels than unmarried men (Ariaal of northern Kenya: Gray et al., 2007b), but in others they do not

(Kenyan Swahili: Gray, 2003; Beijing, China: Gray et al., 2006; Bangladesh: Magid et al., 2006; Japan: Sakaguchi et al., 2006). In the two societies in which male testosterone levels have been considered with respect to polygyny, in one society men with two wives had higher testosterone levels than other men in the sample (Kenyan Swahili: Gray, 2003), whereas in the other study men with multiple wives had lower testosterone levels (Ariaal of Kenya: Gray et al., 2007b). In Beijing, China (Gray et al., 2006), a rural village in Dominica (Gangestad et al., 2005a) and in urban Jamaica (Gray et al., 2007a), fathers had lower testosterone levels than nonfathers, although this was not true among Bangladeshi men (Magid et al., 2006).

Several factors may help account for the less consistent links between testosterone and relationship status outside of North America. If men's behavior changes little with marriage, and men are uninvolved in direct paternal care, then we would be less likely to find lower testosterone levels associated with pair bonding and fatherhood. If acquisition of an additional wife in a polygynous society is better predicted by male age, social status, or wealth, then links between polygynous marriage and elevated testosterone may be dependent on such considerations (and may help account for the discrepant Kenyan results). Even the context of extrapair mating may be differentially linked to finding elevated testosterone (Mazur and Michalek, 1998) or not (Gray et al., 2006); depending on variation in how socially sanctioned the pursuit of extrapair mates is, including the availability of access to mistresses or prostitutes, then perhaps elevated testosterone associated with affairs will only arise where males must engage in overt male–male competition and courtship. These considerations suggest that the use of relationship status as a proxy for human male mating and parenting effort has less validity in some sociocultural contexts than others, and may in turn help account for the less consistent findings cross-culturally.

While summarizing the human male testosterone, pair bonding and paternal care literature, it is important to bear in mind several caveats. Almost all of these studies have relied on cross-sectional research designs in which the causal relationships between hormones and behavior remains undefined. In one of the rare longitudinal designs, men's testosterone levels increased around the time of divorce (Mazur and Michalek, 1998), and we saw above that men's baseline testosterone predicted paternal responsiveness in a Canadian sample; these studies suggest, consistent with general hormone-behavior principles, that testosterone and men's social relationships are likely reciprocally related to each other. The sample sizes of most of these studies are modest, at least compared with epidemiological studies. The effect sizes (as measured by percentage differences in testosterone between groups of men, correlation coefficients, and *p* values) are also modest. Many things happen in different male relationships, and variation in testosterone levels commonly appears to be one of these things – especially in North America. All this said, male testosterone levels do appear to commonly differ according to relationship status – as one might have anticipated.

Does female behavior change across the menstrual cycle?

A tremendous amount of research has recently addressed potentially adaptive changes in human female mating psychology, including sexual behavior, across the menstrual cycle (reviewed in Gangestad et al., 2005b). As one illustration of this tantalizing literature, females prefer the scent of more symmetrical males around the time of ovulation (Gangestad and Thornhill, 1998). Here, we begin with functional logic predicting changes in several facets of female behavior including, but not limited to, sexual behavior across the menstrual cycle. The logic predicting changes in female behavior includes behavioral shifts viewed as adaptive as well as behavioral shifts viewed as by-products of other traits. Relevant empirical evidence is briefly considered below to determine whether a priori expectations are supported or not given the available data. Last, a case is made for more research directed at changes in female behavior studied across the so-called "reproductive cycle" (pregnancy, postpartum amenorrhea [suppression of ovulation after birth], cycling).

An adaptive perspective suggests human female behavior should change across the menstrual cycle. We might anticipate shifts in key behavioral domains such as sexual behavior, relationship dynamics, dietary intake, display activities, and sleep. Because the fertile window is short (1 day), and the life span of sperm in the reproductive tract also short (about 3–5 days), we would expect selection to have favored female behaviors resulting in successful mating behavior around the time of ovulation (Pillsworth et al., 2004). The most straightforward psychological and behavioral evidence of successful mating behavior would be increases in female libido, to motivate behavior, and sexual behavior (i.e., vaginal intercourse) itself around the time of ovulation. As other correlates of successful mating behavior around the time of ovulation, we might anticipate changes in relationship dynamics among pair bonded women. Women attached to their partners may be more likely to spend time around their mates; their mates may also be more likely to mate guard them. If women are unpaired or seeking additional mating opportunities, they may engage in more public display (e.g., more likely to flirt or spend time in places where they could meet mates) around the time of ovulation. Finally, the behavioral shift to mating may come at a sacrifice to other activities like eating and sleeping. We would expect such behavioral shifts around the time of ovulation to be due to effects of increased estrogen and/or increased androgen levels.

Since, after ovulation, physiological mechanisms initially act as if fertilization has occurred, we should expect the latter half of the menstrual cycle, or luteal phase, to support behaviors appropriate to pregnancy like increased eating, sleeping, and lower energy expenditure (fewer public displays). Moreover, sexual behavior may decrease relative to an expected periovulatory peak. Effects during the luteal phase would presumably be due largely to rising concentrations of progesterone.

If fertilization has not occurred, a consequence will be short-circuiting the endocrine mechanisms that had been acting as if the female were pregnant; behavioral effects at this time should be viewed as by-products rather than functionally designed. During the late luteal phase, a transient increase in sexual behavior could be expected from the rapidly diminishing physiological support for pregnancy; decreases in caloric intake might also arise. At the time, both estrogen and progesterone levels plummet, suggesting that they would be implicated in behavioral shifts associated with menstruation. Because of the potentiating effects of estrogen on oxytocin, premenstrual declines in oxytocin may also have behavioral effects indirectly through decreased oxytocin (Brizendine, 2006).

So do the expected changes in female behavior occur across the menstrual cycle? With respect to sexual behavior, drawing primarily on North American and Western European samples, results have been mixed (reviewed in Brewis and Meyer, 2005). In some studies, fluctuations in sexual activity have been observed, but not consistently. However, arguably two of the best designed studies involving North American pair bonded (but not maternal) females, including urine collection for hormone assay and cycle phase assignment, did find increases in intercourse frequency in the days prior to and around ovulation (Bullivant et al., 2004; Wilcox et al., 2004). In expanding the cross-cultural scope of this research, however, we often encounter more evolutionary-relevant contexts in which to address this question – contexts in which females already have children. In a sample of pair bonded, reproductive-age women from 13 countries (e.g., Guinea, Haiti, Peru) relying on questionnaire responses to identify cycle phase and sexual activity, no increases in sexual behavior appeared around the time of ovulation (Brewis and Meyer, 2005). The only patterned sexual behavior across the cycle was a reduction during menses, a pattern also reported in a different cross-cultural survey (Ford and

Beach, 1951). Interestingly, several negative predictors of sexual behavior observed in Brewis and Meyer's study were being older, breast-feeding, having more children, and being involved in a polygynous (rather than monogamous) union.

Several studies have investigated whether relationship dynamics (e.g., time spent with a partner, felt commitment to a partner) change across the menstrual cycle. In a diary study of 38 US women (in which women reported self and partner's behavior daily), women reported more mate guarding by their partner around the time of ovulation (Haselton and Gangestad, 2006). Similar patterns emerged from behavioral observations in a small, Caribbean village (Gangestad et al., 2005a). In a study of UK women, subjects reported higher commitment to their relationship on cycle days associated with high progesterone levels (Jones et al., 2005).

Fessler (2003) has recently reviewed both nonhuman and human data assessing potential changes in dietary intake and female activity patterns across the menstrual cycle. In his review, he observed that dietary intake decreased in 10 of 16 studies around the time of ovulation – evidence, he noted, that could be viewed as suggesting adaptive changes in the salience of female motivation at that time. There was mixed evidence whether female activity patterns shifted across the cycle. For example, one study observed an increase in volunteerism around the time of ovulation. In a recent exhaustive review of sleep research, no clear, consistent patterns were observed in sleep patterns across the menstrual cycle (Moline et al., 2003).

What should we conclude about variation in female behavior across the menstrual cycle? Behavioral effects tend to be modest, if they can be measured at all. This observation is consistent with the view that human behavior tends to be quite flexible – that conserved endocrine mechanisms operate but with greater cortical control over our behavior than, say, laboratory rodents (Panksepp et al., 2002; Keverne, 2005). Another conclusion is that patterns in female behavior across the menstrual cycle may be contingent upon the contexts in which they are measured. In samples of young, nulliparous women, we may be more apt to observe periovulatory peaks in sexual behavior than in samples of older women with multiple children present in the household. As Fessler (2003) reminds us, studies of dietary intake and activity are lacking in societies where food availability is constrained and where women walk as part of subsistence activities – perhaps we are underestimating these latter effects in places where these are more meaningful determinants of female reality and reproductive success.

From an evolutionary perspective, a focus on female behavioral changes across the menstrual cycle overlooks the more dramatic life history shifts, orchestrated by hormones, that occur across a "reproductive cycle" (menstrual cycle, pregnancy, postpartum amenorrhea). Women residing in human hunter-gatherer and other natural fertility populations spend the bulk of their reproductive years pregnant or lactating – not experiencing decades of continuous menstrual cycles (Ellison, 2001; Eaton et al., 1994). Consequently, in addition to examining behavioral shifts across the menstrual cycle, we should be devoting greater attention to the shifts in female behavior across the reproductive cycle. Toward this end, I briefly consider the female behavioral changes we might expect to occur with pregnancy and postpartum amenorrhea.

While pregnant, we might expect decreased sexual behavior (already pregnant), increased food consumption, and decreased activity levels, especially near the time of birth. Hormonal changes associated with pregnancy, including steady rises in estrogens, androgens, progesterone, dehydroepiandrosterone sulfate (DHEAS), and cortisol, may facilitate some of these expectations but also yield by-product behavioral effects (Buckwalter et al., 2001; Greenspan and Gardner, 2001). For example, the rising estrogen levels (especially estriol) and androgen levels during pregnancy may maintain libido and in turn promote sexual behavior more than might otherwise be expected during a time when a female cannot conceive. After birth, rapid decreases in pregnancy hormones may play a role in postpartum depression (Bloch et al., 2003; Nelson, 2005). Such changes should be viewed as by-products of pregnancy rather than adaptive. The suppressed gonadal steroid levels during postpartum amenorrhea prevent a new pregnancy during a time when a mother is subject to the metabolic challenges of breast-feeding, with maternal condition and energy balance playing additional roles in the resumption of menstrual cycles (Ellison, 2001). The dramatic changes in hormone levels may interact with perceived social support to help account for variation in mood and depression postpartum. Reduced sexual activity postpartum thus may be expected, but dietary intakes increased.

Focusing on sexual behavior, we observe both some consistent patterns as well as cross-cultural variation in female behavior across the reproductive cycle. Both in the United States (Laumann et al., 1994; Rathus et al., 2005) and cross-culturally (Ford and Beach, 1951), pregnant women tend to continue engaging in sexual activity. However, the frequency of sexual activity tends to diminish the longer the pregnancy. Intercourse is commonly avoided in the initial weeks postpartum, with resumption of sexual activity quite variable cross-culturally; in some societies, it is avoided for months to years, but this tends to be among societies practicing polygyny, meaning that a husband has alternative sexual outlets (Ford and Beach, 1951). Several reviews primarily of North American studies show that the frequency of sexual behavior tends to be less common postpartum than before pregnancy (von Sydow, 1999; De Judicibus and McCabe, 2002).

Breast-feeding tends to be negatively associated with postpartum sexual behavior, perhaps due to direct hormone effects (e.g., elevated prolactin helping suppress sex steroid levels) or other reasons (e.g., a mother feeling less desire for physical contact from her adult partner) (De Judicibus and McCabe, 2002). Fatigue, physical health concerns (e.g., after episiotomy), and changes in body image also tend to play common roles predicting postpartum sexual behavior. Generally, then, it does appear that patterning of sexual activity is consistent with expectations of highest frequencies among cycling women, but diminished frequencies with pregnancy (especially advanced pregnancy) and postpartum. Of course, these are quite general conclusions, calling for more behavioral data on women across this important transition in life.

CONCLUSIONS

This chapter has provided a taste of the evolution and endocrinology of human behavior. Beginning with some general principles concerning the evolution of behavior and endocrinology, we turned to a set of four examples to flesh out some of the best empirical evidence documenting the links between human hormones and behavior. The work on hormones and human sexual differentiation demonstrates elegant, well-documented roles for the role of hormones in generating sex differences in human behavior. A nascent literature on the hormonal correlates of human pair bonding and parenting has, in several studies, implicated the peptide oxytocin in maternal care as well as both male and female pair bonding. A review of the human male testosterone, pair bonding, and parenting literature revealed that pair bonded and paternal men commonly exhibit lower testosterone levels than their single counterparts, although these effects were more consistent in North American samples. Some evidence suggests changes in female behavior across the menstrual cycle, although that evidence is not especially strong and overlooks the more dramatic, and evolutionary relevant, changes in female behavior occurring across the reproductive cycle.

There are clinical and epidemiological implications of existing human research on the evolution and endocrinology of behavior. When parents face the decision of what gender to raise a child of ambiguous sex, knowledge of the biological basis of sex differences may inform their decision. Oxytocin may facilitate some of the health benefits of social relationships, in turn raising the possibility of oxytocin-based

interventions for treating social deficits. If testosterone interferes with social relationships, might those individuals taking high doses of exogenous androgens be compromising their intimate social relationships? If fluctuations in female sex steroids across the reproductive cycle account for variation in sexual behavior, clinicians might advise patients accordingly.

There are many exciting paths that research on the evolution and endocrinology of human behavior might take. The further links to functional imaging (e.g., Bartels and Zeki [2004] on the neural correlates of maternal and romantic love, which showed that neural structures rich in oxytocin and vasopressin receptors were activated by subjects viewing photographs of a romantic partner or one's child) may provide direct insight into neural actions of hormones (otherwise rarely accessible in humans). The availability of minimally invasive hormone measures (in saliva, urine, and finger-prick blood spots) opens the door to cross-cultural and epidemiological research that would otherwise be prohibitive in clinical settings (e.g., measuring hormonal correlates of allocare or aggression in naturalistic settings). Greater attention to the developmental course of neuroendocrine mechanisms will illuminate the ways early social experiences "get under the skin"; these may include effects of social deprivation (Fries et al., 2005), but also more typical early social experiences in various cultural settings (see Chapter 27 of this volume).

Indeed, the variable cross-cultural expression of human behavior effectively provides a remarkable set of "natural social experiments" unprecedented for any other vertebrate. International human migrations and travel increasingly bring this variation close to home (e.g., a doctor treating a diverse patient population; students, business people, and politicians interacting with people of diverse backgrounds). Many research questions could be addressed based on this variation: What are the hormonal correlates of sibling or grandparental allocare? What are the hormonal correlates of human to nonhuman animal interactions (see Odendaal and Meintjes, 2003)? Do adoptive parents show similar hormonal bases of allocare compared with biological parents? In the world of the evolution and endocrinology of human behavior, we have learned much recently, but fascinating questions remain.

DISCUSSION POINTS

1. Why do boys engage in more rough-and-tumble play than girls? Provide answers to this question employing each of the four perspectives highlighted in Tinbergen's framework.
2. Among hunter-gatherer societies, women spend most of their reproductive years pregnant or

lactating. Infants are typically nursed for two to four years. Contrast that reproductive context with one in which women commonly gives birth by cesarean section, often aided by administration of synthetic oxytocin (e.g., pitocin), to one or two children over the course of their lives, and which they nurse for a few months postpartum. What are some potential behavioral impacts of this latter reproductive scenario, both on the mother and off-spring, compared with that among our ancestors and more recently studied hunter-gatherers?

3. Suppose you were designing a clinical trial to determine whether testosterone supplementation affected family relationships in older men. Why might you expect (or not) that increased testosterone (say from the "low normal" to "high normal" range) would affect family relationships? How would you measure these behavioral effects?

4. Researchers have pointed out that species differences in pair bonding and paternal care of closely related species of voles (small rodents) appear to be underpinned by the oxytocin and vasopressin systems. What evidence is there that human female involvement in long-term pair bonds and maternal care is facilitated by oxytocin? Similarly, what evidence is there that human male involvement in long-term pair bonds and paternal care is facilitated by vasopressin?

5. Imagine you were designing a two-year field research study to investigate potential shifts in female behavior across the menstrual cycle and reproductive cycle in a small-scale, rural, population of horticulturalists (growing crops such as manioc and plantains) living in the Amazon Basin. Which behaviors would you seek to examine? How would you propose measuring them?

6. The causal arrow between hormones and behavior points in both directions. Suggest several potential behavioral effects of the "endocrine intervention" undertaken by women on the "The Pill" (e.g., hormonal contraception consisting of synthetic progesterone and estrogen that yields relatively constant, rather than the normal fluctuations, progesterone and estrogen levels across the menstrual cycle, apart from menses). Conversely, how might participating in competitive team sports alter hormone levels?

REFERENCES

Adkins-Regan, E. (2005). *Hormones and Animal Social Behavior*. Princeton: Princeton University Press.

Altemus, M., Deuster, P. A., Galliven, E., et al. (1995). Suppression of hypothalamic-pituitary-adrenal axis responses to stress in lactating women. *Journal of Clinical Endocrinology and Metabolism*, **80**, 2954–2959.

Archer, J. and Lloyd, B. (2002). *Sex and Gender*, 2nd edn. Cambridge: Cambridge University Press.

Arnold, A. P. (2002). Concepts of genetic and hormonal induction of vertebrate sexual differentiation in the twentieth century, with special reference to the brain. In *Hormones, Brain and Behavior*, vol. 4, D. W. Pfaff, A. P. Arnold, et al. (eds). San Diego, CA: Academic Press, pp. 105–135.

Barkow, J. H., Cosmides, L. and Tooby, J. (eds) (1992). *The Adapted Mind: Evolutionary Psychology and the Generation of Culture*. Oxford: Oxford University Press.

Barrett, L., Dunbar, R. and Lycett, J. (2002). *Human Evolutionary Psychology*. Princeton: Princeton University Press.

Bartels, A. and Zeki, S. (2004). The neural correlates of maternal and romantic love. *Neuroimage*, **21**, 1155–1166.

Becker, J. B., Breedlove, S. M., Crews, D., et al. (eds) (2002). *Behavioral Endocrinology*, 2nd edn. Cambridge, MA: MIT Press.

Berg, S. J. and Wynne-Edwards, K. E. (2001). Changes in testosterone, cortisol, and estradiol levels in men becoming fathers. *Mayo Clinic Proceedings*, **76**, 582–592.

Betzig, L. (ed.) (1997). *Human Nature: a Critical Reader*. Oxford: Oxford University Press.

Bloch, M., Daly, R. C. and Rubinow, D. R. (2003). Endocrine factors in the etiology of postpartum depression. *Comparative Psychiatry*, **44**, 234–246.

Bolhuis, J. J. and Giraldeau, L.-A. (2005). *The Behavior of Animals: Mechanisms, Function, and Evolution*. Malden, MA: Blackwell.

Book, A. S. and Quinsey, V. L. (2005). Re-examining the issues: a response to Archer et al. *Aggressive and Violent Behavior*, **10**, 637–646.

Booth, A. and Dabbs, J. M. (1993). Testosterone and men's marriages. *Social Forces*, **72**, 463–477.

Brewis, A. and Meyer, M. (2005). Demographic evidence that human ovulation is undetectable (at least in pair bonds). *Current Anthropology*, **46**, 465–471.

Brizendine, L. (2006). *The Female Brain*. New York: Morgan Road Books.

Buckwalter, J. G., Buckwalter, D. K., Bluestein, B. W., et al. (2001). Pregnancy and post partum: changes in cognition and mood. *Progress in Brain Research*, **133**, 303–319.

Bullivant, S. B., Sellergren, S. A., Stern, K., et al. (2004). Women's sexual experience during the menstrual cycle: identification of the sexual phase by noninvasive measurement of luteinizing hormone. *Journal of Sex Research*, **41**, 82–93.

Burnham, T. C., Chapman, J. F., Gray, P. B., et al. (2003). Men in committed, romantic relationships have lower testosterone. *Hormones and Behavior*, **44**, 119–122.

Buss, D. M. (2003). *Evolutionary Psychology: the New Science of the Mind*, 2nd edn. New Needham, MA: Allyn and Bacon.

Buss, D. M. (ed.) (2005). *Handbook of Evolutionary Psychology*. Hoboken, NJ: Wiley.

Carter, C. (1998). Neuroendocrine perspectives on social attachment and love. *Psychoneuroendocrinology*, **23**, 779–818.

Carter, C. S. (2007). Sex differences in oxytocin and vasopressin: implications for autism spectrum disorders? *Behavioural Brain Research*, **176**, 170–186.

Carter, C. S., Ahnert, L., Grossman, K. E., et al. (eds) (2005). *Attachment and Bonding: a New Synthesis*. Cambridge, MA: MIT Press.

Cartwright, J. (2000). *Evolution and Human Behavior*. Cambridge, MA: MIT Press.

Coccaro, E., Kavoussi, R., Hauger, R., et al. (1998). Cerebrospinal fluid vasopressin levels: correlates with aggression and serotonin function in personality-disordered subjects. *Archives of General Psychiatry*, **55**, 708–714.

Cohan, C. L., Booth, A. and Granger, D. A. (2003). Gender moderates the relationship between testosterone and marital interaction. *Journal of Family Psychology*, **17**, 29–40.

Cohen-Bendahan, C. C. C., van de Beek, C. and Berenbaum, S. A. (2005). Prenatal sex hormone effects on child and adult sex-typed behavior: methods and findings. *Neuroscience and Biobehavioral Reviews*, **29**, 353–384.

De Judicibus, M. A. and McCabe, M. (2002). Psychological factors and the sexuality of pregnant and postpartum women. *Journal of Sex Research*, **39**, 94–103.

Delahunty, K. M., McKay, D. W., Noseworthy, D. E., et al. (2007). Prolactin responses to infant cues in men and women: effects of parental experience and recent infant contact. *Hormones and Behavior*, **51**, 213–220.

Delville, Y., Mansour, K. and Ferris, C. (1996). Testosterone facilitates aggression by modulating vasopressin receptors in the hypothalamus. *Physiology and Behavior*, **60**, 25–29.

Dixson, A. F. (1998). *Primate Sexual Behavior*. Oxford: Oxford University Press.

Eaton, S. B., Pike, M. C., Short, R. V., et al. (1994). Women's reproductive cancers in evolutionary context. *Quarterly Review of Biology*, **69**, 353–367.

Ellison, P. T. (2001). *On Fertile Ground*. Cambridge, MA: Harvard University Press.

Ellison, P. T. and Gray, P. B. (eds) (2009). *Endocrinology of Social Relationships*. Cambridge, MA: Harvard University Press.

Fessler, D. M. T. (2003). No time to eat: an adaptationist account of periovulatory behavioral changes. *Quarterly Review of Biology*, **78**, 3–21.

Ford, C. S. and Beach, F. A. (1951). *Patterns of Sexual Behavior*. New York: Ace.

Fleming, A. S. (2005). Plasticity of innate behavior: experiences throughout life affect maternal behavior and its neurobiology. In *Attachment and Bonding: a New Synthesis*, C. S. Carter, L. Ahnert, K. E. Grossman, et al. (eds). Cambridge, MA: MIT Press, pp. 137–168.

Fleming, A. S., Ruble, D., Krieger, H., et al. (1997). Hormonal and experiential correlates of maternal responsiveness during pregnancy and the peuerperium in human mothers. *Hormones and Behavior*, **31**, 145–158.

Fleming, A., Corter, C., Stallings, J., et al. (2002). Testosterone and prolactin are associated with emotional responses to infant cries in new fathers. *Hormones and Behavior*, **42**, 399–413.

Flinn, M. V., Ward, C. V. and Noone, R. V. (2005). Hormones and the human family. In *Handbook of Evolutionary Psychology*, D. Buss (ed.). Hoboken, NJ: Wiley, pp. 552–583.

Fries, A. B. W., Ziegler, T. E., Kurian, J. R., et al. (2005). Early experience in humans is associated with changes in neuropeptides critical for regulating social behavior. *Proceedings of the National Academy of Sciences of the United States of America*, **102**, 17237–17240.

Gangestad, S. W. and Thornhill, R. (1998). Menstrual cycle variation in women's preference for the scent of symmetrical men. *Proceedings of the Royal Society of London. Series B*, **262**, 727–733.

Gangestad, S. W., Thornhill, R., Flinn, M. V., et al. (2005a). Men's testosterone and life history in a Caribbean rural village. Paper presented at the Human Behavior and Evolution Society meeting, June 2005, Austin, TX.

Gangestad, S. W., Thornhill, R. and Garver-Apgar, C. E. (2005b). Adaptations to ovulation. In *Handbook of Evolutionary Psychology*, D. Buss (ed.). New York: Wiley, pp. 344–371.

Gaulin, S. J. C. and McBurney, D. H. (2004). *Evolutionary Psychology*, 2nd edn. Upper Saddle River, NJ: Prentice Hall.

Gray, P. B. (2003). Marriage, parenting and testosterone variation among Kenyan Swahili men. *American Journal of Physical Anthropology*, **122**, 279–286.

Gray, P. B. and Campbell, B. C. (2009). Human male testosterone, pair bonding and fatherhood. In *Endocrinology of Social Relationships*, P. T. Ellison and P. B. Gray (eds). Cambridge, MA: Harvard University Press, pp. 270–292.

Gray, P. B., Kahlenberg, S. M., Barrett, E. S., et al. (2002). Marriage and fatherhood are associated with lower testosterone in males. *Evolution and Human Behavior*, **23**, 193–201.

Gray, P. B., Campbell, B. C., Marlowe, F. W., et al. (2004a). Social variable predict between-subject but not day-to-day variation in the testosterone of US men. *Psychoneuroendocrinology*, **29**, 1153–1162.

Gray, P. B., Chapman, J. F., Burnham, T. C., et al. (2004b). Human male pair bonding and testosterone. *Human Nature*, **15**, 119–131.

Gray, P. B., Yang, C. J. and Pope, H. G. Jr (2006). Fathers have lower salivary testosterone levels than unmarried men and married non-fathers in Beijing, China. *Proceedings of the Royal Society of London. Series B*, **273**, 333–339.

Gray, P. B., Parkin, J. C. and Samms-Vaughan, M. E. (2007a). Hormonal correlates of human paternal interactions: a hospital-based investigation in urban Jamaica. *Hormones and Behavior*, **52**, 499–507.

Gray, P. B., Ellison, P. T. and Campbell, B. C. (2007b). Testosterone and marriage among Ariaal men of northern Kenya. *Current Anthropology*, **48**, 750–755.

Greenspan, F. S. and Gardner, D. G. (eds) (2001). *Basic and Clinical Endocrinology*. New York: McGraw Hill.

Grewen, K., Girdler, S., Amico, J., et al. (2005). Effects of partner support on resting oxytocin, cortisol, norepinephrine, and blood pressure before and after warm partner contact. *Psychosomatic Medicine*, **67**, 531–538.

Heinrichs, M., Meinlschmidt, G., Neumann, I., et al. (2001). Effects of suckling on hypothalamic-pituitary-adrenal axis responses to psychosocial stress in postpartum lactating women. *Journal of Clinical Endocrinology and Metabolism*, **86**, 4798–4804.

Hines, M. (2004). *Brain Gender*. New York: Oxford University Press.

Hrdy, S. (1999). *Mother Nature*. New York: Pantheon.

Irons, W. (1998). Adaptively relevant environments versus the environment of evolutionary adaptedness. *Evolutionary Anthropology*, **6**, 194–204.

Jones, B. C., Little, A. C., Boothroyd, L., et al. (2005). Commitment to relationships and preferences for femininity and apparent health in faces are strongest on days of the menstrual cycle when progesterone level is high. *Hormones and Behavior*, **48**, 283–290.

Ketterson, E. D. and Nolan, V. Jr (1999). Adaptation, exaptation, and constraint: a hormonal perspective. *American Naturalist*, **154S**, S4–S25.

Keverne, E. B. (2005). Neurobiological and molecular approaches to attachment and bonding. In *Attachment and Bonding: A New Synthesis*, C. S. Carter, L. Ahnert, K. E. Grossman, et al. (eds). Cambridge, MA: MIT Press, pp. 101–117.

Kirsch, P., Esslinger, C., Chen, Q., et al. (2005). Oxytocin modulates neural circuitry for social cognition and fear in humans. *Journal of Neuroscience*, **25**, 11489–11493.

Konner, M. (2002). *The Tangled Wing: Biological Constraints on the Human Spirit*, 2nd edn. New York: Times Books.

Kosfeld, M., Heinrichs, M., Zak, P. J., et al. (2005). Oxytocin increases trust in humans. *Nature*, **435**, 673–676.

Kruger, T., Haake, P., Chereath, D., et al. (2003). Specificity of the neuroendocrine response to orgasm during sexual arousal in men. *Journal of Endocrinology*, **177**, 57–64.

Landgraf, R. and Holsboer, F. (2005). The involvement of neuropeptides in evolution, signaling, behavioral regulation and psychopathology: focus on vasopressin. *Drug Development Research*, **65**, 185–190.

Laumann, E. O., Gagnon, J. H., Michael, R. T., et al. (1994). *Social Organization of Sexuality: Sexual Practices in the United States*. Chicago: University of Chicago Press.

Light, K. C., Grewen, K. M. and Amico, J. A. (2005). More frequent partner hugs and higher oxytocin levels are linked to lower blood pressure and heart rate in premenopausal women. *Biological Psychology*, **69**, 5–21.

Lippa, R. A. (2002). *Gender, Nature and Nurture*. Mahwah, NJ: Erlbaum.

Lordkipanidze, D., Lordkipanidze, D., Vekua, A., et al. (2007). Postcranial evidence from early *Homo* from Dmanisi, Georgia. *Nature*, **449**, 305–310.

Low, B. (2000). *Why Sex Matters*. Princeton: Princeton University Press.

Maccoby, E. E. (1998). *The Two Sexes: Growing up Apart, Coming Together*. Cambridge: Harvard University Press.

Magid, K. W., Chatterton, R. T., Uddin Ahamed, F., et al. (2006). No effect of marriage or fatherhood on salivary testosterone levels in Bangladeshi men. Poster presented at the Human Behavior and Evolution Society meeting, June 2006, Philadelphia, PA.

Marlowe, F. (2000). Paternal investment and the human mating system. *Behavioural Processes*, **51**, 45–61.

Matthiesen, A. S., Ransjo-Arvidson, A. B. and Nissen, E. (2001). Postpartum maternal oxytocin release by newborns: effects of infant hand massage and sucking. *Birth*, **28**, 13–19.

Mazur, A., and Michalek, J. (1998). Marriage, divorce, and male testosterone. *Social Forces*, **77**, 315–330.

McIntyre, M. H, Gangestad, S. W., Gray, P. B., et al. (2006). Romantic involvement often reduces men's testosterone levels – but not always: the moderating roles of extrapair sexual interest. *Journal of Personality and Social Psychology*, **91**, 642–651.

Migeon, C. J. and Wisniewski, A. B. (1998). Sexual differentiation: from genes to gender. *Hormone Research*, **50**, 245–251.

Moline, M. L., Broch, L., Zak, R., et al. (2003). Sleep in women across the life cycle from adulthood through menopause. *Sleep Medicine Review*, **7**, 155–177.

Nelson, R. J. (2005). *An Introduction to Behavioral Endocrinology*, 3rd edn. Sunderland, MA: Sinauer.

Odendaal, J. and Meintjes, R. (2003). Neurophysiological correlates of affiliative behaviour between humans and dogs. *Veterinary Journal*, **165**, 296–301.

Panksepp, J., Moskal, J. R., Panksepp, J. B., et al. (2002). Comparative approaches in evolutionary psychology: molecular neuroscience meets the mind. *Neuroendocrinology Letters*, **23**, 105–115.

Pfaff, D. W., Arnold, A. P., Etgen, A. M., et al. (eds) (2002). *Hormones, Brain and Behavior*. San Diego, CA: Academic Press.

Piersma, T. and Drent, J. (2003). Phenotypic flexibility and the evolution of organismal design. *Trends in Evolution and Ecology*, **18**, 228–233.

Pillsworth, E. G., Haselton, M. G. and Buss, D. M. (2004). Ovulatory shifts in female sexual desire. *Journal of Sex Research*, **41**, 55–65.

Pinker, S. (1997). *How the Mind Works*. New York: Norton.

Quartz, S. R. and Sejnowski, T. J. (2002). *Liars, Lovers, and Heroes: What the New Brain Science Reveals about How We Become Who We Are*. New York: William Morrow.

Rathus, S. A., Nevid, J. S. and Fichner-Rathus, L. (2005). *Human Sexuality in a World of Diversity*, 6th edn. Boston, MA: Allyn and Bacon.

Reed Larsen, P., Kronenberg, H. M., Melmed, S., et al. (eds) (2003). *Williams Textbook of Endocrinology*, 10th edn. New York: Elsevier.

Reichard, U. H. and Boesch, C. (eds) (2003). *Monogamy: Mating Strategies and Partnerships in Birds, Humans and Other Mammals*. Cambridge: Cambridge University Press.

Richerson, P. J. and Boyd, R. (2005). *Not by Genes Alone: How Culture Transformed Human Evolution*. Chicago: University of Chicago Press.

Roughgarden, J. (2004). *Evolution's Rainbow: Diversity, Gender, and Sexuality in Nature and People*. Berkeley, CA: University of California Press.

Sakaguchi, K., Oki, M., Hasegawa, T., et al. (2006). Influence of relationship status and personality traits on salivary testosterone among Japanese men. *Personality and Individual Differences*, **41**, 1077–1087.

Sanchez, R., Parkin, J. C., Chen, J., et al. (2009). Oxytocin, vasopressin and human social behavior. In *Endocrinology of Social Relationships*, P. T. Ellison and P. G. Gray (eds). Cambridge, MA: Harvard University Press, pp. 319–338.

Sapolsky, R. (2004). *Why Zebras Don't Get Ulcers*, 3rd edn. New York: Holt.

Schmitt, D. P. (2005). Sociosexuality from Argentina to Zimbabwe: a 48-nation study of sex, culture, and strategies of human mating. *Behavioral and Brain Sciences*, **28**, 247–275.

Storm, E. E. and Tecott, L. H. (2005). Social circuits: peptidergic regulation of mammalian social behavior. *Neuron*, **47**, 483–486.

Storey, A. E., Walsh, C. J., Quinton, R. L., et al. (2000). Hormonal correlates of paternal responsiveness in new and expectant fathers. *Evolution and Human Behavior*, **21**, 79–95.

Thompson, R., Gupta, S., Miller, K., et al. (2004). The effects of vasopressin on human facial responses related to social communication. *Psychoneuroendocrinology*, **29**, 35–48.

Thompson, R. R., George, K., Walton, J. C., et al. (2006). Sex-specific influences of vasopressin on human social cognition. *Proceedings of the National Academy of Sciences of the United States of America*, **103**, 7889–7994.

Tinbergen, N. (1963). On aims and methods of ethology. *Zeitschrift fur Tierpsychologie*, **20**, 410–433.

Uvnas-Moberg, K. (1998). Oxytocin may mediate the benefits of positive social interaction and emotions. *Psychoneuroendocrinology*, **23**, 819–835.

Van Anders, S. and Watson, N. (2006a). Social neuroendocrinology: effects of states and behaviours on human neuroendocrine function. *Human Nature*, **17**, 212–237.

Van Anders, S. and Watson, N. (2006b). Relationship status and testosterone in North American heterosexual and non-heterosexual men and women: cross-sectional and longitudinal data. *Psychoneuroendocrinology*, **31**, 715–723.

Van Anders, S. and Watson, N. V. (2007). Testosterone levels in women and men who are single, in long-distance relationships, or same-city relationships. *Hormones and Behavior*, **51**, 286–291.

Vilain, E. and McCabe, R. B. (1998). Minireview: mammalian sex determination: from gonads to brain. *Molecular Genetics and Metabolism*, **65**, 74–84.

Von Sydow, K. (1999). Sexuality during pregnancy and after childbirth: a metacontent analysis of 59 studies. *Journal of Psychosomatic Research*, **47**, 27–49.

Wallen, K. (2005). Hormonal influences on sexually differentiated behavior in nonhuman primates. *Frontiers in Neuroendocrinology*, **26**, 7–26.

Whiting, J. and Whiting, B. (1975). Aloofness and intimacy of husbands and wives: a cross-cultural study. *Ethos*, **3**, 183–207.

Wilcox, A. J., Baird, D. D., Dunson, D. B., et al. (2004). On the frequency of intercourse around ovulation: evidence for biological influences. *Human Reproduction*, **19**, 1539–1543.

Wingfield, J. C., Hegner, R. E., Dufty, A. M., et al. (1990). The "challenge hypothesis": theoretical implications for patterns of testosterone secretion, mating systems, breeding strategies. *American Naturalist*, **136**, 829–846.

Young, L. J. and Wang, Z. (2004). The neurobiology of pair bonding. *Nature Neuroscience*, **7**, 1048–1054.

Ziegler, T. E. (2000). Hormones associated with non-maternal infant care: a review of mammalian and avian studies. *Folia Primatologia*, **71**, 6–21.

Part III

Reproduction

"The reproduction of mankind is a great marvel and mystery. Had God consulted me in the matter, I should have advised him to continue the generation of the species by fashioning them out of clay."

Martin Luther (1483–1546)

17 Human Mate Choice

David P. Schmitt

From an evolutionary perspective, animal mate choice depends in large part on the natural mating system of a species. The natural mating system of humans, however, seems at first glance to contain internal contradictions. On the one hand, humans show several signs of having a monogamous mating system. For example, humans are highly altricial – we have prolonged childhoods and rely heavily on extended families throughout our life spans (Alexander and Noonan, 1979). We also appear designed to form romantic pair bonds, having a dedicated neurochemistry of attachment associated with monogamy when comparisons are made across mammalian species (Fisher, 1998; Young, 2003).

On the other hand, humans seem to possess evolved design features associated with multimale/multifemale or "promiscuous" mating systems. For example, humans possess psychological and physiological adaptations to sperm competition (Baker and Bellis, 1995; Shackelford and LeBlanc, 2001), such as women's adaptive timing of extrapair copulations (Gangestad and Thornhill, 1998; Haselton and Miller, 2006), men's specialized expressions of sexual jealousy (Buss, 2000; Schützwohl, 2006), and the physical structure of the human penis serving as a semen displacement device (Gallup et al., 2003). Among men, causal sex with multiple partners is often viewed as desirable (Symons and Ellis, 1989; Oliver and Hyde, 1993), with most men agreeing to have sex with complete strangers when asked in field experiments (Clark and Hatfield, 1989). Adaptive patterns of premarital sex, extramarital sex, and mate poaching by both men and women have been documented across cultures (Broude and Greene, 1976; Schmitt et al., 2004a).

Moreover, there is further evidence that humans are designed, at least in part, for polygynous mating. For example, men and women have sexually differentiated life history traits such as men's tendencies to be more physically aggressive, to die much earlier, and to physically mature much later than women across all known cultures (Archer and Lloyd, 2002; Kaplan and Gangestad, 2005). Such sex differences are usually not seen among truly monogamous species, especially primates (Alexander et al., 1979). Moreover, across foraging cultures – the predominantly polygynous cultures in which humans have spent most of our evolutionary history (Frayser, 1985; Brown, 1991; Pasternak et al., 1997) – there are ethnographically pervasive links among men's status, polygynous marriage, and reproductive success (Turke and Betzig, 1985; Low, 2000). In contrast, very few cultures (<1%) have polyandrous marriage systems (Broude and Greene, 1976).

EVOLUTIONARY THEORIES OF MATE CHOICE

Evolutionary psychologists tend to reconcile these seemingly contradictory findings by acknowledging that humans, like several other species, are probably designed and adapted for more than one mating strategy (Mealey, 2000; Barash and Lipton, 2001). Specifically, most evolutionary psychologists view humans as coming equipped with specialized mate choice adaptations for both *long-term mating* (i.e., marriage and extended pair bonding) and *short-term mating* (i.e., promiscuity and infidelity; see Buss and Schmitt, 1993). Not all people pursue both mating strategies at all times. Instead, humans possess adaptive desires, preferences, and behavioral tactics that are differentially activated depending on whether a long-term or short-term mating strategy is actively being pursued at the time (Schmitt et al., 2003; Schmitt, 2005a).

Most evolutionary theories of human mating argue that such a flexible mating design – comprising both long-term monogamous adaptations and short-term promiscuous adaptations – would have provided important reproductive benefits to humans in our ancestral past, allowing individuals to functionally respond to a wide range of familial, cultural, and ecological contexts (Pedersen, 1991; Buss and Schmitt, 1993; Lancaster, 1994; Belsky, 1999; Gangestad and Simpson, 2000). Evolutionary theories further acknowledge that humans can benefit from shifting between long-term and short-term mating strategies during

Human Evolutionary Biology, ed. Michael P. Muehlenbein. Published by Cambridge University Press. © Cambridge University Press 2010.

their life span, when in different stages of romantic relationships, and across the ovulatory cycle (Gangestad, 2001; Klusmann, 2002; Schmitt et al., 2002). Thus, humans have evolved the capacity to follow a mix of different strategies – both long-term and short-term – depending on fitness-related circumstances.

Most evolutionary psychology approaches further postulate that men and women possess design features that cause sex differences in mate choice within long-term and short-term mating contexts. For example, when men seek short-term mates they appear motivated by adaptive desires for sexual variety – desires that lead them to functionally pursue numerous mating partners and to consent to sex relatively quickly compared to women (Clark and Hatfield, 1989; Symons and Ellis, 1989; Okami and Shackelford, 2001; Schmitt et al., 2003). Women's short-term mating motivations appear not to be rooted in the desire for numerous sexual partners and seem focused, instead, on other factors such as obtaining select men who display dominance, intelligence, or creativity (i.e., show high genetic quality; see Gangestad and Thornhill, 1997; Regan et al., 2000; Penton-Voak et al., 2003). As a consequence, evolutionary approaches predict men's and women's mate choices and attraction tactics will differ in important ways, especially within the context of short-term mating. Most evolutionary theories of human mate choice are based on the assumption that the sexes will differ in some ways, an assumption that can be traced to the logic of Parental Investment Theory (Trivers, 1972).

Parental Investment Theory

According to Parental Investment Theory (Trivers, 1972), the relative proportion of parental investment – the time and energy devoted to the care of individual offspring (at the expense of other offspring) – varies across the males and females of different species. In some species, males provide more parental investment than females. In other species, females possess the heavy-investing burdens (e.g., most mammals; Clutton-Brock, 1991). Sex differences in parental investment burdens are systematically linked to processes of sexual selection (Darwin, 1871). The sex that invests less in offspring is *intrasexually* more competitive, especially over gaining reproductive access to members of the opposite sex. That is, the lesser-investing sex is reliably more aggressive with their own sex, tends to die earlier, tends to mature later, and generally competes for mates with more vigor than the heavier-investing sex (Alcock, 2001). Furthermore, the lesser-investing sex of a species is *intersexually* less discriminating in mate choice than the heavier-investing sex. The lesser-investing sex is willing to mate more quickly, at lower cost, and will initiate relationships with more partners than the heavier-investing parent (Bateson, 1983).

Much of the evidence in favor of Parental Investment Theory (Trivers, 1972) has come from species where females happen to be the heavy-investing sex (see Clutton-Brock, 1991). In such species, Parental Investment Theory leads to the prediction that sexual selection has been more potent among males. Upon empirical examination, males of these species tend to display more competitiveness with each other over sexual access to heavier-investing females, and to exhibit more intrasexual competition through greater aggressiveness, riskier life history strategies, and earlier death than females (Trivers, 1985; Archer and Lloyd, 2002). Lesser-investing males also exhibit relatively indiscriminate intersexual mate choice, often seeking multiple partners and requiring less time before initiating sex than females do (Geary, 1998).

Perhaps the most compelling support for Parental Investment Theory (Trivers, 1972), however, has come from "sex-role reversed" species. In species where males are the heavy-investing parent, the processes of sexual selection are thought to have been more potent among females. Females of these species vie more ferociously for sexual access to heavy-investing males and require little from males before consenting to sex. Evidence of this form of sexual differentiation has been documented among such "sex-role reversed" species as the red-necked phalarope, the Mormon cricket, katydids, dance flies, water bugs, seahorses, and a variety of fish species (Alcock, 2001). Parental Investment Theory, therefore, is not a theory about males always having more interest in indiscriminate sex than females. Instead, it is a theory about differences in parental investment *obligations* systematically relating to sex differences in mate choice and relationship initiation.

Among humans, many men invest heavily in their children by teaching social skills, emotionally nurturing children, and investing in their offspring both resources and prestige. Nevertheless, men incur much lower levels of obligatory or "minimum" parental investment in offspring than women do (Symons, 1979). Women are obligated, for example, to incur the costs of internal fertilization, placentation, and gestation in order to reproduce. The minimum physiological obligations of men are considerably less – requiring only the contribution of sperm. Furthermore, all female mammals, including ancestral women, carried the obligatory investments associated with lactation. Lactation can last several years in human foraging environments (Kelly, 1995), years during which it is harder for women than men to reproduce and invest in additional offspring (Blurton Jones, 1986). Finally, across all known cultures human males typically invest less in active parenting effort than females (Low, 1989; Munroe and Munroe, 1997).

This asymmetry in parental investment should affect human mate choice, with the lesser-investing sex (i.e., men) displaying greater intrasexual competitiveness and lower intersexual "choosiness" in mate preferences. Numerous studies have shown that men exhibit greater physical size and competitive aggression (Archer and Lloyd, 2002), riskier life history strategies (Daly and Wilson, 1988), relatively delayed maturation (Geary, 1998), and earlier death than women do across cultures (Alexander and Noonan, 1979). In addition, men's mate preferences are, as predicted, almost always less "choosy" or discriminating than women's, especially in the context of short-term mating (Kenrick et al., 1990; Regan et al., 2000).

Because men are the lesser-investing sex of our species, they also should be more inclined toward initiating low-cost, short-term mating than women. Human sex differences in the desire for short-term sex have been observed in studies of sociosexuality (Simpson and Gangestad, 1991; Jones, 1998; Schmitt, 2005a), motivations for and prevalence of extramarital mating (Seal et al., 1994; Wiederman, 1997), quality and quantity of sexual fantasies (Ellis and Symons, 1990), quality and quantity of pornography consumption (Malamuth, 1996), motivations for and use of prostitution (McGuire and Gruter, 2003), willingness to have sex without commitment (Townsend, 1995), willingness to have sex with strangers (Clark and Hatfield, 1989; Clark, 1990), and in the fundamental differences between the short-term mating psychology of gay males and lesbians (Bailey et al., 1994). Clearly, sex differences in parental investment obligations have an influence on men's and women's fundamental mate choices and basic reproductive strategies.

Sexual Strategies Theory

Buss and Schmitt (1993) expanded on Parental Investment Theory (Trivers, 1972) by proposing Sexual Strategies Theory (SST). According to SST, men and women have evolved a pluralistic repertoire of mating strategies. One strategy within this repertoire is "long-term" mating. Long-term mating is usually marked – for both men and women – by extended courtship, heavy investment, pair bonding, the emotion of love, and the dedication of resources over a long temporal span to the mating relationship and any offspring that ensue. Another strategy within the human mating repertoire is "short-term" mating, defined as a relatively fleeting sexual encounter such as a brief affair, a hook-up, or one-night stand. Which sexual strategy or mix of strategies an individual pursues is predicted to be contingent on factors such as opportunity, personal mate value, sex ratio in the relevant mating pool, parental

influences, regnant cultural norms, and other features of social and personal context (see also Gangestad and Simpson, 2000; Schmitt, 2005a, 2005b).

EVOLUTION OF SEX DIFFERENCES IN MATE CHOICE

Sex differences in long-term mating

Although SST views both sexes as having long-term and short-term mating strategies within their repertoire, men and women are predicted to psychologically differ in what they desire (i.e., mate choice) and in how they tactically pursue (i.e., initiate) romantic relationships. In long-term mate choice, the sexes are predicted to differ in several respects. Men are hypothesized to possess long-term mate choice adaptations that place a greater premium on signals of fertility and reproductive value, such as a woman's youth and physical appearance (Symons, 1979; Buss, 1989; Kenrick and Keefe, 1992; Singh, 1993; Jones, 1995). Men also prefer long-term mates who are sexually faithful and are capable of good parenting (Buss and Schmitt, 1993). Women, in contrast, are hypothesized to place a greater premium when long-term mating on a man's status, resources, ambition, and maturity – cues relevant to his *ability* for long-term provisioning – as well as his kindness, generosity, emotional openness – cues to his *willingness* to provide for women and their children (Ellis, 1992; Feingold, 1992; Cashdan, 1993; Townsend and Wasserman, 1998; Buunk et al., 2001; Brase, 2006).

Conversely, men who display cues to long-term provisioning, and women who display youthfulness, tend to be the ones that are most effective at initiating, enhancing, and preserving monogamous mating relationships (Buss, 1988; Tooke and Camire, 1991; Walters and Crawford, 1994; Landolt et al., 1995; Hirsch and Paul, 1996; Schmitt, 2002). From an evolutionary perspective, the differing qualities that men and women preferentially respond to are thought to help solve the adaptive problems that men and women had to overcome throughout human evolutionary history (Schmitt and Buss, 1996). Of course, in our ancestral past men and women also faced similar problems of mate choice, leading to little or no sex differences in some domains (Buss and Schmitt, 1993).

Numerous survey and meta-analytic studies have confirmed many of the major tenets of SST, including the fact that men and women seeking long-term mates desire different attributes in potential partners (e.g., Cunningham et al., 1995; Jensen-Campbell et al., 1995; Graziano et al., 1997; Regan, 1998a, 1998b; Li et al., 2002; Kruger et al., 2003; Urbaniak and Kilmann, 2003). Several investigators have replicated or confirmed SST-related findings using nationally representative, cross-cultural, or multicultural samples (Feingold, 1992; Knodel et al., 1997; Sprecher et al., 1994; Walter, 1997; Schmitt et al., 2003). Other investigators have validated key SST hypotheses concerning sex differences in long-term mate choice using nonsurvey techniques such as studying actual mate attraction, online dating services, speed dating contexts, personal advertisements, actual marital choices, spousal conflicts, and which traits lead to divorce (Betzig, 1989; Wiederman, 1993; Kenrick et al., 1994; Townsend and Wasserman, 1998; Salmon and Symons, 2001; Schmitt et al., 2001; Dawson and McIntosh, 2006; Hitsch et al., 2006). These experimental, behavioral, and naturalistic methodologies suggest evolutionary-supportive findings are not merely stereotype artifacts or social desirability biases limited to self-reported mate choice.

Hitsch et al. (2006) recently examined 22 000 members of an online dating service and tested for sex differences in the attributes of members who were sought out and contacted by other members as prospective partners versus those who were not. For female members, physical attractiveness (as judged from a photograph) accounted for 30% of the variance in whether men contacted them. For men, physical attractiveness accounted for 18% of the variance. In contrast, women's income accounted for 4% of the variance in whether men contacted them. For men, income accounted for 7% of the variance. When combining physical attractiveness and income as predictors, researchers found for women that having high income could not make up for having relatively low or even average levels of physical attractiveness. However, a man of average attractiveness was considered just as desirable as the most attractive men if he made at least $205 500 per year (Hitsch et al., 2006).

Kenrick et al. (1994) demonstrated a double dissociation of "contrast effects" in that experimental exposure to physically attractive women tended to lessen a man's commitment to his current relationship partner. However, exposure to physically attractive men had no effect on women's commitment to their current partners. Conversely, when women were exposed to targets who had high status and resource-related attributes, this lessened women's (but not men's) commitment to their current romantic partners. Kenrick et al. (1994) argued that this indirect research method not only confirms self-reported mate preference findings, but further shows that men's and women's evolved mate preferences unconsciously influence men's and women's satisfaction and commitment over the long-term course of relationships (see also Buss and Shackelford, 1997; Little and Mannion, 2006).

Another indirect effect of sex-differentiated mating desires can be found in the context of relationship initiation and romantic attraction. According to Sexual Selection Theory (Darwin, 1871), the evolved mate preferences of one sex should have a substantive impact on the effectiveness of attraction tactics used by the opposite sex. If men possess an evolved preference for physical attractiveness, the argument goes, women should be more engaged than men at using mate initiation and attraction tactics that manipulate physical attractiveness. Conversely, if women prefer resources more than men do, men should be more engaged than women at using resource-related tactics of initiation and attraction. Empirical evaluations have supported this aspect of sexual selection in humans. For example, Buss (1988), Tooke and Camire (1991), and Walters and Crawford (1994) all demonstrated that women are judged more effective than men when using appearance-related tactics of initiation and attraction, whereas men are judged more effective than women when using resource-related tactics of romantic initiation and attraction (for a meta-analysis of mate attraction results, see Schmitt, 2002).

Perceived sex differences in physical appearance and resource-related tactic effectiveness have also been documented within more specialized rating contexts of romantic attraction. Buss (1988) found sex differences in effectiveness ratings of appearance and resource-related tactics when used by men and women to both attract and *retain* a long-term marital partner (see also Flinn, 1985; Bleske-Rechek and Buss, 2001). Schmitt and Buss (1996) documented sex differences in perceived tactic effectiveness across both self-promotion and *competitor derogation* forms of mate attraction (see also Greer and Buss, 1994; Walters and Crawford, 1994). Schmitt and Buss (2001) found sex differences in perceived appearance and resource-related mate attraction within the specialized context of obtaining a long-term mating partner who is already in a relationship, what they called the context of *mate poaching* (see also Bleske and Shackelford, 2001; Schmitt and Shackelford, 2003). Whether researchers ask people directly, observe their real-life behavior, or subtly look for indirect effects, the pervasive range of sex differences in long-term mating psychology supports the evolutionary perspective on mate choice.

Sex differences in short-term mating

According to SST, both sexes are hypothesized to pursue short-term mateships in certain contexts, but for different reproductive reasons that reflect sex-specific adaptive problems (Buss and Schmitt, 1993). For women, the asymmetry in obligatory parental investment (Trivers, 1972; Symons, 1979) leaves them little to gain in reproductive output by engaging in indiscriminate, short-term sex with high numbers of partners. Women can reap evolutionary benefits from short-term mating (Hrdy, 1981; Greiling and Buss, 2000); however, women's psychology of short-term mate choice appears to center on obtaining men of high genetic quality rather than numerous men in high-volume quantity (Smith, 1984; Gangestad and Thornhill, 1998; Banfield and McCabe, 2001; Li and Kenrick, 2006).

For men, the potential reproductive benefits from short-term mating with numerous partners can be profound. A man can produce as many as 100 offspring by mating with 100 women over the course of a year, whereas a man who is monogamous will tend to have only one child with his partner during that same time period. In evolutionary currencies, this represents a strong selective pressure – and a potent adaptive problem – for men's short-term mating strategy to center on obtaining large numbers of partners (Schmitt et al., 2003). Obviously, 100 instances of only one-time mating would rarely produce precisely 100 offspring; however, a man mating with 100 women over the course of a year – particularly repeated matings when the women are nearing ovulation and are especially interested in short-term mating (Gangestad, 2001) – would likely have significantly more offspring than a woman mating repeatedly with 100 interested men over the course of a year.

According to SST, three of the specific design features of men's short-term mating psychology are: (1) men possess a greater desire than women do for a variety of sexual partners; (2) men require less time to elapse than women do before consenting to sexual intercourse; and (3) men tend to more actively seek short-term mateships than women do (Buss and Schmitt, 1993). This suite of hypothesized sex differences has been well supported empirically. For example, Schmitt et al. (2003) documented these fundamental sex differences across 10 major regions of the world. When people from North America were asked "Ideally, how many different sexual partners would you like to have in the next month?" over 23% of men, but only 3% of women, indicated that they would like *more than one* sexual partner in the next month. This finding confirmed that many men, and few women, desire sexual variety in the form of multiple sexual partners over short time intervals. Similar degrees of sexual differentiation were found in South America (35.0% vs. 6.1%), Western Europe (22.6% vs. 5.5%), Eastern Europe (31.7% vs. 7.1%), Southern Europe (31.0% vs. 6.0%), the Middle East (33.1% vs. 5.9%), Africa (18.2% vs. 4.2%), Oceania (25.3% vs. 5.8%), South/Southeast Asia (32.4% vs. 6.4%), and East Asia (17.9% vs. 2.6%). These sex differences also persisted across a variety of demographic statuses, including age, socioeconomic status, and sexual orientation. Moreover, when men and women who reported actively pursuing a short-term mating strategy were asked whether they wanted more than one partner in the next month, over 50% of men,

but less than 20% of women, expressed desires for multiple sexual partners (Schmitt et al., 2003). This finding supports the key SST hypothesis that men's short-term mating strategy is very different from women's and is based, in part, on obtaining large numbers of sexual partners.

Other findings from the cross-cultural study by Schmitt et al. (2003) documented that men universally agree to have sex after less time has elapsed than women do, and that men from all world regions expend more effort on seeking brief sexual relationships than women do. For example, across all cultures nearly 25% of married men, but only 10% of married women, reported that they are actively seeking short-term, extramarital relationships (see also Wiederman, 1997). These culturally universal findings support the view that men evolved to seek large numbers of sex partners when they pursue a short-term mating strategy. Some women also pursue short-term mates. However, when women seek short-term mates they are more selective and tend to seek out men who are physically attractive, intelligent, and otherwise possess high-quality genes (Buss and Schmitt, 1993; Gangestad and Thornhill, 1997, 2003).

EVOLUTION OF INDIVIDUAL DIFFERENCES IN MATE CHOICE

The previous section addressed the evolutionary psychology of *how* men and women choose and initiate short-term and long-term mating relationships. Another important question is *when* and *why* an individual man or woman would choose to pursue a long-term mateship versus a short-term mateship. Several theories have suggested that personal circumstances – including stage of life, phenotypic quality, local ecology – play an adaptive role in shaping or evoking people's strategic mating choices (Buss and Schmitt, 1993; Gangestad and Simpson, 2000). Among the more important sex-specific personal characteristics that affect mating strategies are men's overall mate value and women's ovulatory status.

Mating differences within men

According to SST (Buss and Schmitt, 1993), men possess a menu of alternative mating strategies that they can follow. Whether a man chooses to pursue a short-term or long-term mating strategy (or both) may depend, in part, on his status and prestige. In foraging cultures, men with higher status and prestige tend to possess multiple wives (Betzig, 1986; Borgerhoff Mulder, 1987, 1990; Cronk, 1991; Heath and Hadley, 1998), and in so doing polygynous men are able to satisfy aspects of both their long-term pair bonding desires and short-term "numerous partner" desires. In most modern cultures, men with high status are unable to legally marry more than one woman. However, they are more likely to have extramarital affairs and to practice de facto or "effective" polygyny in the form of serial divorce and remarriage (Brown and Hotra, 1988; Fisher, 1992; Buss, 2000). Given an equal sex ratio of men and women in a given culture, this results in other men – namely those with low status and prestige – being limited to monogamy in the form of one wife. Some low-status men are left with no wives at all and may be forced to resort to coercive, promiscuous-mating strategies (Thornhill and Palmer, 2000). Consequently, an important source of individual variation in men's mate choice and relationship initiation tactics is status and prestige.

Whether a man follows a more short-term or long-term oriented mating strategy depends on other factors as well, many of which relate to the man's overall value in the mating marketplace (Gangestad and Simpson, 2000). A man's "mate value" is determined, in part, by his status and prestige. It is also affected by his current resource holdings, long-term ambition, intelligence, interpersonal dominance, social popularity, sense of humor, reputation for kindness, maturity, height, strength, and athleticism (Chagnon, 1988; Ellis, 1992; Pierce, 1996; Miller, 2000; Nettle, 2002).

Most studies of men in modern cultures find that, when they are able to do so as a result of high mate value, men choose to engage in multiple mating relationships. For example, Lalumiere et al. (1995) designed a scale to measure overall mating opportunities. The scale, similar to overall mate value, included items such as "relative to my peer group, I can get dates with ease." They found that men with higher mate value tended to have sex at an earlier age, to have a larger number of sexual partners, and to follow a more promiscuous mating strategy overall (see also Landolt et al., 1995; James, 2003).

Another potential indicator of mate value is the social barometer of self-esteem (Kirkpatrick et al., 2002). Similar to the results with mating opportunities, men who score higher on self-esteem scales tend to choose and to successfully engage in more short-term mating relationships (Walsh, 1991; Baumeister and Tice, 2001). Indeed, in a recent cross-cultural study by Schmitt (2005b), this revealing trend was evident across several world regions. The same relationship was usually not evident, and was often reversed, among women in modern nations (see also Mikach and Bailey, 1999). That is, women with high self-esteem were more likely to pursue monogamous, long-term mating strategies. These findings would seem to support Parental Investment Theory (Trivers, 1972), in that when mate value is high and people are given a choice, men prefer short-term mating

(sometimes in addition to long-term mating) whereas women strategically prefer a single monogamous mateship. An important determinant of individual mate choice, therefore, is overall mate value in the mating marketplace, with men of high mate value and women of low mate value more likely to pursue short-term mating strategies.

According to Strategic Pluralism Theory (Gangestad and Simpson, 2000), men should also be more likely to engage in short-term mating when they exhibit the physical characteristics most preferred by women who desire a short-term mate, especially those traits indicative of high genetic quality. Higher facial symmetry, for example, is indicative of low genetic mutation load in men, and women adaptively prefer facial symmetry when pursing short-term mates (Gangestad and Thornhill, 1997). This is because one of the key benefits women can reap from short-term mating is to gain access to high-quality genes that they might not be able to secure from a long-term partner (Gangestad, 2001).

Evidence that physically attractive men adaptively respond to women's desires and become more promiscuous comes from other sources, as well. For example, men who possess broad and muscular shoulders, a physical attribute preferred by short-term oriented women (Frederick et al., 2003), tend toward short-term mating as reflected in an earlier age of first intercourse, more sexual partners, and more extrapair copulations (Hughes and Gallup, 2003). In numerous studies, Gangestad and his colleagues have shown that women who seek short-term mates place special importance on the physical attractiveness of their partners, and that physically attractive men are more likely to pursue short-term mating strategies (Thornhill and Gangestad, 1994, 1999; Gangestad and Thornhill, 1997; Gangestad and Cousins, 2001).

Some research suggests that genetic and hormonal predispositions may affect men's mate choice and relationship initiation strategies (Bailey et al., 2000). Much of this research focuses on the moderating effects of testosterone (Dabbs and Dabbs, 2000). For example, married men, compared to their same-age single peers, tend to have lower levels of testosterone (Burnham et al., 2003), though this is not true among married men who are also interested in concurrent extrapair copulations or short-term mateships (McIntyre et al., 2006). Men who are expectant fathers and hope to have children only with their current partner have relatively low testosterone (Gray et al., 2002), whereas men possessing high testosterone tend to have more sexual partners, to start having sex earlier, to have higher sperm counts, to be more interested in sex, to divorce more frequently, and are more likely to have affairs (Alexander and Sherwin, 1991; Udry and Campbell, 1994; Mazur and Booth, 1998; Manning,

2002). The root cause of this mate choice variability may lie in early testosterone exposure and its effects on the activation of men's short-term mating psychology. Exposure to high testosterone levels in utero causes increased masculinization of the human brain and increased testosterone in adulthood (Manning, 2002; Ridley, 2003). If men's brains are programmed for greater short-term mating in general (Trivers, 1972; Symons, 1979), this would lead to the hypothesis that those who are exposed to higher testosterone in utero would be more likely to develop short-term mating strategies in adulthood. In women, though, other factors appear to adaptively influence mating strategy choice.

Mating differences within women

Women's desires for engaging in sexual intercourse tend to vary across their ovulatory cycles. On average, women's desires for sex peak during the late follicular phase, just before ovulation when the odds of becoming pregnant would be maximized (Regan, 1996). It was once thought that this shift in sexual desire evolved because it increased the probability of having conceptive intercourse in our monogamous female ancestors. However, several studies have now documented that women's short-term desires for men with high-quality genes actually peak in the highly fertile days just before ovulation (Gangestad and Thornhill, 1997; Gangestad, 2001; Haselton and Miller, 2006).

For example, women who are interested in short-term mating tend to prefer men who are high in dominance and masculinity (Buss and Schmitt, 1993), as indicated by testosterone-related attributes such has prominent brows, large chins, and other features of facial masculinity (Mueller and Mazur, 1998; Perrett et al., 1998; Penton-Voak and Chen, 2004). Short-term oriented women may prefer these attributes because facial markers of testosterone are honest indicators of immunocompetence quality in men (Gangestad and Thornhill, 2003). During the late follicular phase, women's preferences for men with masculine faces conspicuously increase (Johnston et al., 2001; Penton-Voak et al., 2003), as do their preferences for masculine voices (Puts, 2006), precisely as though women are shifting their mating psychology to follow a more short-term oriented strategy around ovulation.

A similar ovulatory shift can be seen in women's preference for symmetrical faces. Women who generally pursue a short-term mating strategy express strong preferences for male faces that are symmetrical, perhaps because facial symmetry is indicative of low mutation load (Gangestad and Thornhill, 1997). During the late follicular phase, women's preference for symmetrical faces increases even further (Gangestad and Cousins, 2001), again as though they have shifted

their psychology to that of a short-term mating strategist. It has also been shown that women who are nearing ovulation find the pheromonal smell of symmetrical men more appealing than when women are less fertile (Gangestad and Thornhill, 1998; Rikowski and Grammer, 1999), that women who mate with more symmetrical men have more frequent and intense orgasms (Thornhill et al., 1995), and that men with attractive faces have qualitatively better health (Shackelford and Larsen, 1999) and semen characteristics (Soler et al., 2003). Finally, women appear to dress more provocatively when nearing ovulation (Grammer et al., 2004), though women near ovulation also reduce risky behaviors associated with being raped, especially if they are not taking contraception (Bröder and Hohmann, 2003).

Overall, there is compelling evidence that women's mating strategies shift, from a long-term mating psychology to a more short-term oriented mating psychology, precisely when they are the most fertile. It is possible that these shifts reflect women seeking high-quality genes from extrapair copulations while maintaining a long-term relationship with a heavily investing partner (Gangestad, 2001; Haselton and Miller, 2006).

Additional individual differences and personal situations may be linked to adaptive variability in women's mate choices and relationship initiation strategies. For example, short-term mating strategies are more likely to occur when in adolescence, when one's partner is of low mate value, when one desires to get rid of a mate, and when one has recently divorced – all situations where short-term mating may serve specific adaptive functions (Cashdan, 1996; Greiling and Buss, 2000). In some cases, short-term mating seems to emerge as an adaptive reaction to early developmental experiences within the family (Michalski and Shackelford, 2002). For example, short-term mating strategies are more likely occur among women growing up in father-absent homes (Moffit et al., 1992; Quinlan, 2003) especially in homes where a stepfather is present (Ellis and Garber, 2000). In these cases, the absence of a father, and presence of a stepfather, may indicate to young women that mating-age men are unreliable. In such environments, short-term mating may serve as the more viable mating strategy choice once in adulthood (see also Belsky, 1999).

Finally, some have argued that frequency-dependent or other forms of selection have resulted in different heritable tendencies toward long-term versus short-term mating (Gangestad and Simpson, 1990). There is behavioral genetic evidence that age at first intercourse, lifetime number of sex partners, and sociosexuality – a general trait that varies from restricted long-term mating to unrestricted short-term mating – are somewhat heritable (Bailey et al., 2000;

Rowe, 2002). However, most findings suggest heritability in mate choice and mating strategy is stronger in men than in women (Dunne et al., 1997).

EVOLUTION OF CULTURAL DIFFERENCES IN MATE CHOICE

Sex ratios and human mating

In addition to sex and individual differences in mating strategies, mate choices and attraction behaviors appear to vary in evolutionary-relevant ways across cultures (Frayser, 1985; Kelly, 1995; Pasternak et al., 1997). Pedersen (1991) has speculated that the relative number of men versus women in a given culture should influence mating behavior. Operational sex ratio can be defined as the relative balance of marriage-age men versus marriage-age women in the local mating pool (Secord, 1983). Sex ratios are considered "high" when the number of men significantly outsizes the number of women in a local culture. Sex ratios are considered "low" when there are relatively more women than men in the mating market. In most cultures women tend to slightly outnumber men, largely because of men's polygyny-related tendency to have a higher mortality rate (Daly and Wilson, 1988). Nevertheless, significant variation often exists in sex ratios across cultures, and within cultures when viewed over historical time (Guttentag and Secord, 1983; Grant, 1998).

Pedersen (1991) argued that a combination of Sexual Selection Theory (Darwin, 1871) and Parental Investment Theory (Trivers, 1972) leads to a series of predictions concerning the effects of sex ratios on human mating strategies. According to sexual selection theory, when males desire a particular attribute in potential mating partners, females of that species tend to respond by competing in the expression and provision of that desired attribute. Among humans, when sex ratios are especially low and there are many more women than men, men should become an especially scarce resource that women compete for with even more intensity than normal (see also Guttentag and Secord, 1983).

When combined with the parental investment notion described earlier in which men tend to desire short-term mating (Trivers, 1972), this leads to the hypothesis that humans in cultures with lower sex ratios (i.e., more women than men) should possess more short-term oriented mating strategies. Conversely, when sex ratios are high and men greatly outnumber women, men must enter into more intense competition for the limited number of potential female partners. Women's preferences for long-term monogamous relationships become the key desires that must be responded to if men are to remain competitive in the courtship marketplace.

Using data from sex ratio fluctuations over time within the United States, Pedersen (1991) marshaled a compelling case for a causal link between sex ratios and human mating strategies (see also Guttentag and Secord, 1983). For example, high sex ratio fluctuations have been historically associated with increases in monogamy, as evidenced by lower divorce rates and men's greater willingness to invest in their children. Low sex ratios have been historically associated with indexes of short-term mating, such as an increase in divorce rates and a reduction in what he termed female "sexual coyness." In a recent cross-cultural study (Schmitt, 2005a), national sex ratios were correlated with direct measures of basic human mating strategies across 48 nations in an attempt to test Pedersen's (1991) theory. As expected, cultures with more men than women tended toward long-term mating, whereas cultures with more women than men tended toward short-term mating (see also Barber, 2000).

Attachment and human mating

Several combinations of life history theory (Low, 1998) and attachment theory (Bowlby, 1969) have suggested that certain critical experiences during childhood play a role in the development of human mating strategies (Belsky, 1999). Perhaps the most prominent of these theories is a life span model developed by Belsky et al. (1991). According to this model, early social experiences adaptively channel children down one of two reproductive pathways. Children who are socially exposed to high levels of stress – especially insensitive/inconsistent parenting, harsh physical environments, and economic hardship – tend to develop insecure attachment styles. These children also tend to physically mature earlier than those children who are exposed to less stress. According to Belsky and his colleagues, attachment insecurity and early physical maturity subsequently lead to the evolutionary-adaptive development of what is called an "opportunistic" reproductive strategy in adulthood (i.e., short-term mating). In cultures with unpredictable social environments, it is therefore argued, children adaptively respond to stressful cues by developing the more viable strategy of short-term mating.

Conversely, those children exposed to lower levels of stress and less environmental hardship tend to be more emotionally secure and to physically mature later. These children are thought to develop a more "investing" reproductive strategy in adulthood (i.e., long-term mating) that pays evolutionary dividends in low-stress environments. Although the causal mechanisms that influence strategic mating are most prominently located within the family, this model also suggests that certain aspects of culture may be related to mating strategy variation (see also Belsky, 1999).

A closely related theory has been proposed by Chisholm (1996). Chisholm argues that local mortality rates – presumably related to high stress and inadequate resources – act as cues that facultatively shift human mating strategies in evolutionary-adaptive ways. In cultures with high mortality rates and unpredictable resources, the optimal mating strategy is to reproduce early and often, a strategy related to insecure attachment, short-term temporal orientations, and promiscuous mating strategies. In cultures that are physically safe and have abundant resources, mortality rates are lower and the optimal strategy is to invest heavily in fewer numbers of offspring. In safer environments, therefore, one should pursue a long-term strategy associated with more monogamous mating. Collectively, the Belsky et al. (1991) and Chisholm (1996) theories can be referred to as a "developmental-attachment theory" of human mating strategies.

Numerous studies have provided support for developmental-attachment theory (Moffit et al., 1992; Belsky, 1999; Ellis and Garber, 2000; Barber, 2003; Quinlan, 2003). In a recent attempt to test developmental-attachment theory, Schmitt et al. (2004b) measured the romantic attachment styles of over 17 000 people from 56 nations. They related insecure attachment styles to various indexes of familial stress, economic resources, mortality, and fertility. They found overwhelming support for developmental-attachment theory. For example, nations with higher fertility rates, higher mortality rates, higher levels of stress (e.g., poor health and education), and lower levels of resources tended to have higher levels of insecure romantic attachment. Schmitt (2005a) also found that short-term mating was related to insecure attachment across cultures. As expected, the dismissing form of insecure attachment was linked to short-term mating in men and fearful/preoccupied forms of insecure attachment were linked to short-term mating in women. These findings support the view that stressful environments cause increases in insecure romantic attachment, increases presumably linked to short-term mating strategies (see also Kirkpatrick, 1998).

CONCLUSIONS

From an evolutionary perspective, humans seem to possess psychological adaptations related to monogamy, polygyny, and promiscuity. It appears that humans evolved a pluralistic mating repertoire, organized in terms of basic long-term and short-term mating psychologies. The activation and pursuit of these mating psychologies – including concomitant patterns of mate choice and relationship initiation – differ in adaptive ways across sex, individual circumstance, and cultural context.

The sexes differ significantly in their adaptations for short-term mate choice. Men's short-term mating strategy is based primarily on obtaining large numbers of partners, being quick to consent to sex, and more actively seeking brief sexual encounters. Women's short-term strategy is more heavily rooted in obtaining partners of high genetic quality, including men who possess masculine and symmetrical features. Both sexes desire long-term monogamous partners who are kind and understanding, but men place more emphasis on youth, and women on social status, when considering a long-term mate.

Individual differences in mate choice within each sex appear to emerge as adaptive responses to key personal circumstances. Men high in social status and mate value, for example, tend to pursue more short-term oriented mating strategies than other men, and highly valued men strive for polygynous marriages where possible (or serial marriages where polygyny is illegal). Women nearing ovulation tend to manifest desires indicative of their short-term mating psychology, expressing more potent mate choice for masculine and dominant men and being more sensitive to the pheromones of symmetrical men.

Features of culture and local ecology influence the differential pursuit of long-term versus short-term mating strategies. In cultures with high stress levels and high fertility rates, insecure attachment and resulting short-term mating psychologies in men and women may be more common. As a result, in these cultures evolutionary psychologists expect men to emphasize obtaining large numbers of partners and women to emphasize physical features associated with masculinity and symmetry in potential mates (see Schmitt, 2005a). Finally, the relative sex ratio of men versus women in the local mating pool may play a causal role in generating differences in mate choice behavior both over historical time and across the many diverse forms of human culture.

DISCUSSION POINTS

1. What features of mate choice and romantic attraction in humans show the most sexual differentiation, and what features show a high degree of sexual similarity? What adaptive problems faced by humans over the ancestral past explain sexual differentiations in mate choice?
2. How would a shift toward higher levels of short-term mating influence a culture? For example, if women were to engage in increasingly more short-term mating over time, how would women's

evolved short-term mate preferences come to more greatly influence a culture? How would men's evolved short-term mate preferences influence a culture experiencing a shift toward a short-term mating system?

3. Can short-term mating that results from insecure parent–child attachment serve adaptive functions? Describe those functions.

4. How would a bias in sex ratios due to warfare or high rates of male imprisonment (i.e., more women than men) come to influence a culture's mating behavior? Discuss an example from recent history.

REFERENCES

Alcock, J. (2001). *Animal Behavior*, 7th edn. Sunderland, MA: Sinauer.

Alexander, G. M. and Sherwin, B. B. (1991). The association between testosterone, sexual arousal, and selective attention for erotic stimuli in men. *Hormones and Behavior*, **25**, 367–381.

Alexander, R. D. and Noonan, K. M. (1979). Concealment of ovulation, parental care, and human social interaction. In *Evolutionary Biology and Human Social Behavior: an Anthropological Perspective*, N. A. Chagnon and W. Irons (eds). North Scituate, MA: Duxbury, pp. 436–453.

Alexander, R. D., Hoogland, J. L., Howard, R. D., et al. (1979). Sexual dimorphism and breeding systems in pinnipeds, ungulates, primates, and humans. In *Evolutionary Biology and Human Social Behavior: an Anthropological Perspective*, N. A. Chagnon and W. Irons (eds). North Scituate, MA: Duxbury, pp. 402–435.

Archer, J. and Lloyd, B. B. (2002). *Sex and Gender*, 2nd edn. New York: Cambridge University Press.

Bailey, J. M., Gaulin, S., Agyei, Y., et al. (1994). Effects of gender and sexual orientation on evolutionary relevant aspects of human mating psychology. *Journal of Personality and Social Psychology*, **66**, 1081–1093.

Bailey, J. M., Kirk, K. M., Zhu, G., et al. (2000). Do individual differences in sociosexuality represent genetic or environmentally contingent strategies? Evidence from the Australian twin registry. *Journal of Personality and Social Psychology*, **78**, 537–545.

Baker, R. R. and Bellis, M. A. (1995). *Human Sperm Competition*. London: Chapman and Hall.

Banfield, S. and McCabe, M. P. (2001). Extra relationship involvement among women: are they different from men? *Archives of Sexual Behavior*, **30**, 119–142.

Barash, D. P. and Lipton, J. E. (2001). *The Myth of Monogamy*. New York: W. H. Freeman.

Barber, N. (2000). On the relationship between country sex ratios and teen pregnancy rates: a replication. *Cross-Cultural Research*, **34**, 26–37.

Barber, N. (2003). Paternal investment prospects and cross-national differences in single parenthood. *Cross-Cultural Research*, **37**, 163–177.

Baumeister, R. F. and Tice, D. M. (2001). *The Social Dimension of Sex*. Needham Heights, MA: Allyn and Bacon.

Bateson, P. (ed.) (1983). *Mate Choice*. Cambridge: Cambridge University Press.

Belsky, J. (1999). Modern evolutionary theory and patterns of attachment. In *Handbook of Attachment*, J. Cassidy and P. R. Shaver (eds). New York: Guilford, pp. 141–161.

Belsky, J., Steinberg, L. and Draper, P. (1991). Childhood experience, interpersonal development, and reproductive strategy: an evolutionary theory of socialization. *Child Development*, **62**, 647–670.

Betzig, L. (1986). *Despotism and Differential Reproduction: a Darwinian View of History*. New York: Aldine de Gruyter.

Betzig, L. (1989). Causes of conjugal dissolution: a cross-cultural study. *Current Anthropology*, **30**, 654–676.

Bleske, A. L. and Shackelford, T. K. (2001). Poaching, promiscuity, and deceit: combating mating rivalry in same-sex friendships. *Personal Relationships*, **8**, 407–424.

Bleske-Rechek, A. L. and Buss, D. M. (2001). Opposite-sex friendship: sex differences and similarities in initiation, selection, and dissolution. *Personality and Social Psychology Bulletin*, **27**, 1310–1323.

Blurton Jones, N. (1986). Bushman birth spacing: a test for optimal interbirth intervals. *Ethology and Sociobiology*, **7**, 91–105.

Borgerhoff Mulder, M. (1987). Cultural and reproductive success: Kipsigis evidence. *American Anthropologist*, **89**, 617–634.

Borgerhoff Mulder, M. (1990). Kipsigis women's preferences for wealthy men: evidence for female choice in mammals. *Behavioral Ecology and Sociobiology*, **27**, 255–264.

Bowlby, J. (1969). *Attachment and Loss*, vol. 1. New York: Basic Books.

Brase, G. L. (2006). Cues of parental investment as a factor in attractiveness. *Evolution and Human Behavior*, **27**, 145–157.

Bröder, A. and Hohmann, N. (2003). Variations in risk taking over the menstrual cycle: an improved replication. *Evolution and Human Behavior*, **24**, 391–398.

Broude, G. J. and Greene, S. J. (1976). Cross-cultural codes on 20 sexual attitudes and practices. *Ethnology*, **15**, 403–409.

Brown, D. E. (1991). *Human Universals*. New York: McGraw-Hill.

Brown, D. E. and Hotra, D. (1988). Are prescriptively monogamous societies effectively monogamous? In *Human Reproductive Behavior: a Darwinian Perceptive*, L. Betzig, M. Borgerhoff Mulder and P. Turke (eds). Cambridge: Cambridge University Press, pp. 153–159.

Burnham, T. C., Chapman, J. F., Gray, P. B., et al. (2003). Men in committed, romantic relationships have lower testosterone. *Hormones and Behavior*, **44**, 119–122.

Buss, D. M. (1988). From vigilance to violence: tactics of mate retention in American undergraduates. *Ethology and Sociobiology*, **9**, 291–317.

Buss, D. M. (1989). Sex differences in human mate preferences: evolutionary hypotheses tested in 37 cultures. *Behavioral and Brain Sciences*, **12**, 1–49.

Buss, D. M. (2000). *The Dangerous Passion*. New York: The Free Press.

Buss, D. M. and Schmitt, D. P. (1993). Sexual strategies theory: an evolutionary perspective on human mating. *Psychological Review*, **100**, 204–232.

Buss, D. M. and Shackelford, T. K. (1997). From vigilance to violence: mate retention tactics in married couples. *Journal of Personality and Social Psychology*, **72**, 346–361.

Buunk, A. P., Dijkstra, P., Kenrick, D. T., et al. (2001). Age preferences for mates as related to gender, own age, and involvement level. *Evolution and Human Behavior*, **22**, 241–250.

Cashdan, E. (1993). Attracting mates: effects of paternal investment on mate attraction strategies. *Ethology and Sociobiology*, **14**, 1–24.

Cashdan, E. (1996). Women's mating strategies. *Evolutionary Anthropology*, **5**, 134–143.

Chagnon, N. A. (1988). Life histories, blood revenge, and warfare in a tribal population. *Science*, **239**, 985–992.

Chisholm, J. S. (1996). The evolutionary ecology of attachment organization. *Human Nature*, **7**, 1–38.

Clark, R. D. (1990). The impact of AIDS on gender differences in willingness to engage in casual sex. *Journal of Applied Social Psychology*, **20**, 771–782.

Clark, R. D. and Hatfield, E. (1989). Gender differences in receptivity to sexual offers. *Journal of Psychology and Human Sexuality*, **2**, 39–55.

Clutton-Brock, T. H. (1991). *The Evolution of Parental Care*. Princeton, NJ: Princeton University Press.

Cronk, L. (1991). Wealth, status, and reproductive success among the Mukogodo of Kenya. *American Anthropologist*, **93**, 345–360.

Cunningham, M. R., Roberts, R., Barbee, A. P., et al. (1995). Their ideas of attractiveness are, on the whole, the same as ours: consistency and variability in the cross-cultural perception of female attractiveness. *Journal of Personality and Social Psychology*, **68**, 261–279.

Dabbs, J. M. and Dabbs, M. G. (2000). *Heroes, Rogues, and Lovers: Testosterone and Behavior*. New York: McGraw-Hill.

Daly, M. and Wilson, M. (1988). *Homicide*. New York: Aldine de Gruyter.

Darwin, C. R. (1871). *The Descent of Man and Selection in Relation to Sex*. London: Murray.

Dawson, B. L. and McIntosh, W. D. (2006). Sexual strategies theory and internet personal advertisements. *Cyber Psychology and Behavior*, **9**, 614–617.

Dunne, M. P., Martin, N. G., Statham, D. J., et al. (1997). Genetic and environmental contributions to variance in age at first intercourse. *Psychological Science*, **8**, 211–216.

Ellis, B. J. (1992). The evolution of sexual attraction: evaluative mechanisms in women. In *The Adapted Mind*, J. H. Barkow, L. Cosmides and J. Tooby (eds). New York: Oxford University Press, pp. 267–288.

Ellis, B. J. and Garber, J. (2000). Psychosocial antecedents of variation in girl's pubertal timing: maternal depression, stepfather presence, and marital and family stress. *Child Development*, **71**, 485–501.

Ellis, B. J. and Symons, D. (1990). Sex differences in sexual fantasy: an evolutionary psychological approach. *Journal of Sex Research*, **27**, 527–556.

Feingold, A. (1992). Gender differences in mate selection preferences: a test of the parental investment model. *Psychological Bulletin*, **112**, 125–139.

Fisher, H. E. (1992). *Anatomy of Love: the Natural History of Monogamy, Adultery, and Divorce*. New York: W. W. Norton.

Fisher, H. E. (1998). Lust, attraction, and attachment in mammalian reproduction. *Human Nature*, **9**, 23–52.

Flinn, M. V. (1985). Mate guarding in a Caribbean village. *Ethology and Sociobiology*, **9**, 1–28.

Frayser, S. (1985). *Varieties of Sexual Experience: an Anthropological Perspective*. New Haven, CT: HRAF Press.

Frederick, D., Haselton, M. G., Buchanan, G. M., et al. (2003). Male muscularity as a good-genes indicator: Evidence from women's preferences for short-term and long-term mates. Paper presentation to the 15th Annual Meeting of Human Behavior and Evolution Society, Lincoln, NE.

Gallup, G. G., Burch, R. L., Zappieri, M. L., et al. (2003). The human penis as a semen displacement device. *Evolution and Human Behavior*, **24**, 277–289.

Gangestad, S. W. (2001). Adaptive design, selective history, and women's sexual motivations. In *Evolutionary Psychology and Motivation*, J. A. French, A. C. Kamil and D. W. Leger (eds). Lincoln, NE: University of Nebraska Press, pp. 37–74.

Gangestad, S. W. and Cousins, A. J. (2001). Adaptive design, female mate preferences, and shifts across the menstrual cycle. *Annual Review of Sex Research*, **12**, 145–185.

Gangestad, S. W. and Simpson, J. A. (1990). Toward an evolutionary history of female sociosexual variation. Special Issue: biological foundations of personality: evolution, behavioral genetics, and psychophysiology. *Journal of Personality*, **58**, 69–96.

Gangestad, S. W. and Simpson, J. A. (2000). The evolution of human mating: trade-offs and strategic pluralism. *Behavioral and Brain Sciences*, **23**, 573–644.

Gangestad, S. W. and Thornhill, R. (1997). The evolutionary psychology of extrapair sex: the role of fluctuating asymmetry. *Evolution and Human Behavior*, **18**, 69–88.

Gangestad, S. W. and Thornhill, R. (1998). Menstrual cycle variation in women's preferences for the scent of symmetrical men. *Proceedings of the Royal Society of London. Series B*, **265**, 927–933.

Gangestad, S. W. and Thornhill, R. (2003). Facial masculinity and fluctuating asymmetry. *Evolution and Human Behavior*, **24**, 231–241.

Geary, D. C. (1998). *Male, Female: the Evolution of Human Sex Differences*. Washington, DC: American Psychological Association.

Grammer, K., Renninger, L. and Fischer, B. (2004). Disco clothing, female sexual motivation, and relationship status: is she dressed to impress? *Journal of Sex Research*, **41**, 66–74.

Grant, V. J. (1998). *Maternal Personality, Evolution and the Sex Ratio*. London: Routledge.

Gray, P. B., Kahlenberg, S. M., Barrett, E. S., et al. (2002). Marriage and fatherhood are associated with lower testosterone in males. *Evolution and Human Behavior*, **23**, 193–201.

Graziano, W. G., Jensen-Campbell, L. A., Todd, M., et al. (1997). Interpersonal attraction from an evolutionary perspective: women's reactions to dominant and prosocial men. In *Evolutionary Social Psychology*, J. A. Simpson and D. T. Kenrick (eds). Mahwah, NJ: Lawrence Erlbaum, pp. 141–167.

Greer, A. E. and Buss, D. M. (1994). Tactics for promoting sexual encounters. *Journal of Sex Research*, **31**, 185–201.

Greiling, H. and Buss, D. M. (2000). Women's sexual strategies: the hidden dimension of short-term mating. *Personality and Individual Differences*, **28**, 929–963.

Guttentag, M. and Secord, P. F. (1983). *Too Many Women? The Sex Ratio Question*. Beverly Hills, CA: Sage.

Haselton, M. G. and Miller, G. F. (2006). Women's fertility across the cycle increases the short-term attractiveness of creative intelligence. *Human Nature*, **17**, 50–73.

Heath, K. M. and Hadley, C. (1998). Dichotomous male reproductive strategies in a polygynous human society: mating versus parental effort. *Current Anthropology*, **39**, 369–374.

Hirsch, L. R. and Paul, L. (1996). Human male mating strategies, I. Courtship tactics of the "quality" and "quantity" alternatives. *Ethology and Sociobiology*, **17**, 55–70.

Hitsch, G. J., Hortacsu, A. and Ariely, D. (2006). What makes you click? Mate preferences and matching outcomes in online dating. *MIT Sloan Research Paper*, No. 4603–4606.

Hrdy, S. B. (1981). *The Woman That Never Evolved*. Cambridge, MA: Harvard University Press.

Hughes, S. M. and Gallup, G. G. (2003). Sex differences in morphological predictors of sexual behavior: shoulder to hip and waist to hip ratios. *Evolution and Human Behavior*, **24**, 173–178.

James, J. (2003). Sociosexuality and self-perceived mate value: A multidimensional approach. Poster presentation to the 15th Annual Meeting of Human Behavior and Evolution Society, Lincoln, NE.

Jensen-Campbell, L. A., Graziano, W. G. and West, S. G. (1995). Dominance, prosocial orientation, and female preferences: do nice guys really finish last? *Journal of Personality and Social Psychology*, **68**, 427–440.

Johnston, V., Hagel, R., Franklin, M., et al. (2001). Male facial attractiveness: evidence for hormone-mediated adaptive design. *Evolution and Human Behavior*, **22**, 251–267.

Jones, D. (1995). Sexual selection, physical attractiveness, and facial neoteny: cross-cultural evidence and implications. *Current Anthropology*, **36**, 723–748.

Jones, M. (1998). Sociosexuality and motivations for romantic involvement. *Journal of Research in Personality*, **32**, 173–182.

Kaplan, H. S. and Gangestad, S. W. (2005). Life history theory and evolutionary psychology. In *The Handbook of Evolutionary Psychology*, D. M. Buss (ed.). Hoboken, NJ: Wiley, pp. 68–95.

Kelly, R. L. (1995). *The Foraging Spectrum: Diversity in Hunter-Gatherer Lifeways*. Washington, DC: Smithsonian Institution Press.

Kenrick, D. T. and Keefe, R. C. (1992). Age preferences in mates reflect sex differences in human reproductive strategies. *Behavioral and Brain Sciences*, **15**, 75–133.

Kenrick, D. T., Sadalla, E. K., Groth, G., et al. (1990). Evolution, traits, and the stages of human courtship: qualifying the parental investment model. Special issue: biological foundations of personality: evolution, behavioral genetics, and psychophysiology. *Journal of Personality*, **58**, 97–116.

Kenrick, D. T., Neuberg, S. L., Zierk, K. L., et al. (1994). Evolution and social cognition: contrast effects as a function of sex, dominance, and physical attractiveness. *Personality and Social Psychology Bulletin*, **20**, 210–217.

Kirkpatrick, L. A. (1998). Evolution, pair-bonding, and reproductive strategies: a reconceptualization of adult attachment. In *Attachment Theory and Close Relationships*, J. A. Simpson and W. S. Rholes (eds). New York: Guilford, pp. 353–393.

Kirkpatrick, L. A., Waugh, C. E., Valencia, A., et al. (2002). The functional domain specificity of self-esteem and the differential prediction of aggression. *Journal of Personality and Social Psychology*, **82**, 756–767.

Klusmann, D. (2002). Sexual motivation and the duration of partnership. *Archives of Sexual Behavior*, **31**, 275–287.

Knodel, J., Low, B., Saengtienchai, C., et al. (1997). An evolutionary perspective on Thai sexual attitudes and behavior. *Journal of Sex Research*, **34**, 292–303.

Kruger, D. J., Fisher, M. and Jobling, I. (2003). Proper and dark heroes as dads and cads: alternative mating strategies in British romantic literature. *Human Nature*, **14**, 305–317.

Lalumiere, M. L., Seto, M. C. and Quinsey, V. L. (1995). Self-perceived mating success and the mating choices of males and females. Unpublished manuscript. Queen's University at Kingston, Ontario, Canada.

Lancaster, J. B. (1994). Human sexuality, life histories, and evolutionary ecology. In *Sexuality across the Life Course*, A. S. Rossi (ed.). Chicago: University of Chicago Press, pp. 39–62.

Landolt, M. A., Lalumiere, M. L. and Quinsey, V. L. (1995). Sex differences in intra-sex variations in human mating tactics: an evolutionary approach. *Ethology and Sociobiology*, **16**, 3–23.

Li, N. P. and Kenrick, D. T. (2006). Sex similarities and differences in preferences for short-term mates: what, whether, and why. *Journal of Personality and Social Psychology*, **90**, 468–489.

Li, N. P., Bailey, M. J., Kenrick, D. T., et al. (2002). The necessities and luxuries of mate preferences: testing the tradeoffs. *Journal of Personality and Social Psychology*, **82**, 947–955.

Little, A. C. and Mannion, H. (2006). Viewing attractive or unattractive same-sex individuals changes self-rated attractiveness and face preferences in women. *Animal Behavior*, **72**, 981–987.

Low, B. S. (1989). Cross-cultural patterns in the training of children: an evolutionary perspective. *Journal of Comparative Psychology*, **103**, 313–319.

Low, B. S. (1998). The evolution of human life histories. In *Handbook of Evolutionary Psychology*, C. Crawford and D. L. Krebs (eds). Mahwah, NJ: Erlbaum, pp. 131–161.

Low, B. S. (2000). *Why Sex Matters*. Princeton, NJ: Princeton University Press.

Malamuth, N. M. (1996). Sexually explicit media, gender differences, and evolutionary theory. *Journal of Communication*, **46**, 8–31.

Manning, J.T. (2002). *Digit Ratio: a Pointer to Fertility, Behavior, and Health*. New Brunswick, NJ: Rutgers University Press.

Mazur, A. and Booth, A. (1998). Testosterone and dominance in men. *Behavioral and Brain Sciences*, **21**, 353–397.

McGuire, M. and Gruter, M. (2003). Prostitution: an evolutionary perspective. In *Human Nature and Public Policy: an Evolutionary Approach*, A. Somit and S. Peterson (eds). New York: Palgrave Macmillan, pp. 29–40.

McIntyre, M., Gangestad, S.W., Gray, P.B., et al. (2006). Romantic involvement often reduces men's testosterone levels – but not always: the moderating role of extrapair sexual interest. *Journal of Personality and Social Psychology*, **91**, 642–651.

Mealey, L. (2000). *Sex Differences: Developmental and Evolutionary Strategies*. San Diego, CA: Academic Press.

Michalski, R.L. and Shackelford, T.K. (2002). Birth order and sexual strategy. *Personality and Individual Differences*, **33**, 661–667.

Mikach, S.M. and Bailey, J.M. (1999). What distinguishes women with unusually high numbers of sex partners? *Evolution and Human Behavior*, **20**, 141–150.

Miller, G.F. (2000). *The Mating Mind*. New York: Doubleday.

Moffit, T.E., Caspi, A., Belsky, J., et al. (1992). Childhood experience and the onset of menarche: a test of a sociobiological model. *Child Development*, **63**, 47–58.

Mueller, U. and Mazur, A. (1998). Facial dominance in *Homo sapiens* as honest signaling of male quality. *Behavioral Ecology*, **8**, 569–579.

Munroe, R.L. and Munroe, R.H. (1997). A comparative anthropological perspective. In *Handbook of Cross-Cultural Psychology*, vol. 1, J.W. Berry, Y.H. Poortinga and J. Pandey (eds), 2nd edn. Boston: Allyn and Bacon, pp. 171–213.

Nettle, D. (2002). Height and reproductive success in a cohort of British men. *Human Nature*, **13**, 473–491.

Okami, P. and Shackelford, T.K. (2001). Human sex differences in sexual psychology and behavior. *Annual Review of Sex Research*, **12**, 186–241.

Oliver, M.B. and Hyde, J.S. (1993). Gender differences in sexuality: a meta-analysis. *Psychological Bulletin*, **114**, 29–51.

Pasternak, B., Ember, C. and Ember, M. (1997). *Sex, Gender, and Kinship: a Cross-Cultural Perspective*. Upper Saddle River, NJ: Prentice Hall.

Pedersen, F.A. (1991). Secular trends in human sex ratios: their influence on individual and family behavior. *Human Nature*, **2**, 271–291.

Penton-Voak, I.S. and Chen, J.Y. (2004). High salivary testosterone is linked to masculine male facial appearance in humans. *Evolution and Human Behavior*, **25**, 229–241.

Penton-Voak, I.S., Little, A.C., Jones, B.C., et al. (2003). Female condition influences preferences for sexual dimorphism in faces of male humans (*Homo sapiens*). *Journal of Comparative Psychology*, **117**, 264–271.

Perrett, D.I., Lee, K.J., Penton-Voak, I.S., et al. (1998). Effects of sexual dimorphism on facial attractiveness. *Nature*, **394**, 884–887.

Pierce, C.A. (1996). Body height and romantic attraction: a meta-analytic test of the male-taller norm. *Social Behavior and Personality*, **24**, 143–149.

Puts, D.A. (2006). Cyclic variation in women's preferences for masculine traits: potential hormonal causes. *Human Nature*, **17**, 114–127.

Quinlan, R.J. (2003). Father absence, parental care, and female reproductive development. *Evolution and Human Behavior*, **24**, 376–390.

Regan, P.C. (1996). Rhythms of desire: the association between menstrual cycle phases and female sexual desire. *Canadian Journal of Human Sexuality*, **5**, 145–156.

Regan, P.C. (1998a). Minimum mate selection standards as a function of perceived mate value, relationship context, and gender. *Journal of Psychology and Human Sexuality*, **10**, 53–73.

Regan, P.C. (1998b). What if you can't get what you want? Willingness to compromise ideal mate selection standards as a function of sex, mate value, and relationship context. *Personality and Social Psychology Bulletin*, **24**, 1294–1303.

Regan, P.C., Levin, L., Sprecher, S., et al. (2000). Partner preferences: what characteristics do men and women desire in their short-term and long-term romantic partners? *Journal of Psychology and Human Sexuality*, **12**, 1–21.

Ridley, M. (2003). *Nature via Nurture*. New York: Harper Collins.

Rikowski, A. and Grammer, K. (1999). Human body odor, symmetry, and attractiveness. *Proceedings of the Royal Academy of London. Series B*, **266**, 869–874.

Rowe, D.C. (2002). On genetic variation in menarche and age at first intercourse: a critique of the Belsky–Draper hypothesis. *Evolution and Human Behavior*, **23**, 365–372.

Salmon, C. and Symons, D. (2001). *Warrior Lovers: Erotic Fiction, Evolution, and Female Sexuality*. London: Weidenfeld and Nicolson.

Schmitt, D.P. (2002). A meta-analysis of sex differences in romantic attraction: do rating contexts affect tactic effectiveness judgments? *British Journal of Social Psychology*, **41**, 387–402.

Schmitt, D.P. (2005a). Sociosexuality from Argentina to Zimbabwe: a 48-nation study of sex, culture, and strategies of human mating. *Behavioral and Brain Sciences*, **28**, 247–275.

Schmitt, D.P. (2005b). Is short-term mating the maladaptive result of insecure attachment? A test of competing evolutionary perspectives. *Personality and Social Psychology Bulletin*, **31**, 747–768.

Schmitt, D.P. and Buss, D.M. (1996). Strategic self-enhancement and competitor derogation: sex and context effects on the perceived effectiveness of mate attraction tactics. *Journal of Personality and Social Psychology*, **70**, 1185–1204.

Schmitt, D.P. and Buss, D.M. (2001). Human mate poaching: tactics and temptations for infiltrating existing mateships. *Journal of Personality and Social Psychology*, **80**, 894–917.

Schmitt, D.P. and Shackelford, T.K. (2003). Nifty ways to leave your lover: the tactics people use to entice and disguise the process of human mate poaching. *Personality and Social Psychology Bulletin*, **29**, 1018–1035.

Schmitt, D. P., Couden, A. and Baker, M. (2001). Sex, temporal context, and romantic desire: an experimental evaluation of Sexual Strategies Theory. *Personality and Social Psychology Bulletin*, **27**, 833–847.

Schmitt, D. P., Shackelford, T. K., Duntely, J., et al. (2002). Is there an early-30s peak in female sexual desire? Cross-sectional evidence from the United States and Canada. *Canadian Journal of Human Sexuality*, **11**, 1–18.

Schmitt, D. P., Alcalay, L., Allik, J., et al. (2003). Universal sex differences in the desire for sexual variety: tests from 52 nations, 6 continents, and 13 islands. *Journal of Personality and Social Psychology*, **85**, 85–104.

Schmitt, D. P., Alcalay, L., Allik, J., et al. (2004a). Patterns and universals of mate poaching across 53 nations: the effects of sex, culture, and personality on romantically attracting another person's partner. *Journal of Personality and Social Psychology*, **86**, 560–584.

Schmitt, D. P., Alcalay, L., Allensworth, M., et al. (2004b). Patterns and universals of adult romantic attachment across 62 cultural regions: are models of self and other pancultural constructs? *Journal of Cross-Cultural Psychology*, **35**, 367–402.

Schützwohl, A. (2006). Sex difference in jealousy: information search and cognitive preoccupation. *Personality and Individual Differences*, **40**, 285–292.

Seal, D. W., Agostinelli, G. and Hannett, C. A. (1994). Extra-dyadic romantic involvement: moderating effects of sociosexuality and gender. *Sex Roles*, **31**, 1–22.

Secord, P. F. (1983). Imbalanced sex ratios: the social consequences. *Personality and Social Psychology Bulletin*, **9**, 525–543.

Shackelford, T. K. and Larsen, R. J. (1999). Facial attractiveness and physical health. *Evolution and Human Behavior*, **20**, 71–76.

Shackelford, T. K. and LeBlanc, G. J. (2001). Sperm competition in insects, birds, and humans: insights from a comparative evolutionary perspective. *Evolution and Cognition*, **7**, 194–202.

Simpson, J. A. and Gangestad, S. W. (1991). Individual differences in sociosexuality: evidence for convergent and discriminant validity. *Journal of Personality and Social Psychology*, **60**, 870–883.

Singh, D. (1993). Adaptive significance of female physical attractiveness: role of waist-to-hip ratio. *Journal of Personality and Social Psychology*, **65**, 293–307.

Smith, R. L. (1984). Human sperm competition. In *Sperm Competition and the Evolution of Animal Mating Systems*, R. L. Smith (ed.). New York: Academic Press, pp. 601–659.

Soler, C., Nunez, M., Gutierrez, R., et al. (2003). Facial attractiveness in men provides clues to semen quality. *Evolution and Human Behavior*, **24**, 199–207.

Sprecher, S., Sullivan, Q. and Hatfield, E. (1994). Mate selection preferences: gender differences examined in a national sample. *Journal of Personality and Social Psychology*, **66**, 1074–1080.

Symons, D. (1979). *The Evolution of Human Sexuality*. New York: Oxford University Press.

Symons, D. and Ellis, B. J. (1989). Human male–female differences in sexual desire. In *Sociobiology of Sexual and Reproductive Strategies*, A. Rasa, C. Vogel and E. Voland (eds). London: Chapman and Hall, pp. 131–146.

Thornhill, R. and Gangestad, S. W. (1994). Human fluctuating asymmetry and sexual behavior. *Psychological Science*, **5**, 297–302.

Thornhill, R. and Gangestad, S. W. (1999). The scent of symmetry: a human sex pheromone that signals fitness? *Evolution and Human Behavior*, **20**, 175–201.

Thornhill, R. and Palmer, C. T. (2000). *A Natural History of Rape*. Cambridge, MA: MIT Press.

Thornhill, R., Gangestad, S. W. and Comer, R. (1995). Human female orgasm and mate fluctuating asymmetry. *Animal Behaviour*, **50**, 1601–1615.

Tooke, W. and Camire, L. (1991). Patterns of deception in intersexual and intrasexual mating strategies. *Ethology and Sociobiology*, **12**, 345–364.

Townsend, J. M. (1995). Sex without emotional involvement: an evolutionary interpretation of sex differences. *Archives of Sexual Behavior*, **24**, 173–205.

Townsend, J. M. and Wasserman, T. (1998). Sexual attractiveness: sex differences in assessment and criteria. *Evolution and Human Behavior*, **19**, 171–191.

Trivers, R. (1972). Parental investment and sexual selection. In *Sexual Selection and the Descent of Man: 1871–1971*, B. Campbell (ed.). Chicago: Aldine de Gruyter, pp. 136–179.

Trivers, R. (1985). *Social Evolution*. Menlo Park, CA: Benjamin/Cummings.

Turke, P. and Betzig, L. (1985). Those who can do: wealth, status, and reproductive success on Ifaluk. *Ethology and Sociobiology*, **6**, 79–87.

Udry, J. R. and Campbell, B. C. (1994). Getting started on sexual behavior. In *Sexuality over the Life Course*, A. S. Rossi (ed.). Chicago: University of Chicago Press, pp. 187–207.

Urbaniak, G. C. and Kilmann, P. R. (2003). Physical attractiveness and the "nice guy paradox": do nice guys really finish last? *Sex Roles*, **49**, 413–426.

Walsh, A. (1991). Self-esteem and sexual behavior: exploring gender differences. *Sex Roles*, **25**, 441–450.

Walter, A. (1997). The evolutionary psychology of mate selection in Morocco: a multivariate analysis. *Human Nature*, **8**, 113–137.

Walters, S. and Crawford, C. B. (1994). The importance of mate attraction for intrasexual competition in men and women. *Ethology and Sociobiology*, **15**, 5–30.

Wiederman, M. W. (1993). Evolved gender differences in mate preferences: evidence from personal advertisements. *Ethological and Sociobiology*, **14**, 331–352.

Wiederman, M. W. (1997). Extramarital sex: prevalence and correlates in a national survey. *Journal of Sex Research*, **34**, 167–174.

Young, L. J. (2003). The neural basis of pair bonding in a monogamous species: a model for understanding the biological basis of human behavior. In *Offspring: Human Fertility Behavior in Biodemographic Perspective*, K. W. Wachter and R. A. Bulatao (eds). Washington, DC: The National Academies Press, pp. 91–103.

18 Mate Choice, the Major Histocompatibility Complex, and Offspring Viability

Claus Wedekind and Guillaume Evanno

INTRODUCTION

Human offspring become independent relatively late in life and fitness in our species is therefore critically linked to the amount and quality of the parental care that children receive. This explains, in evolutionary terms, why humans generally have a resource-based mating system where both sexes invest substantially into offspring (Trivers, 1972). Women are expected to prefer men who are likely to provide much parental care in the form of active parenting and/or in the amount of resources they can provide. These men are, in turn, expected to be choosy about the women they select as mates (Trivers, 1972; Johnstone et al., 1996). The present chapter is not about this rather obvious link between mutual mate choice and offspring survival. Instead, we focus on probably less obvious aspects of mate selection that are likely to influence offspring viability. These include various genetic aspects that may play a role, e.g., inbreeding avoidance or preference for characteristics that are linked to heritable viability or to genes that may be complementary to the chooser's own genes. Genes of the major histocompatibility complex (MHC) have been intensively studied in this context. The MHC genes play a central role in controlling immunological self- and nonself recognition (Apanius et al., 1997). The MHC is also one of the most polymorphic regions of the genome, and it has been found to influence sexual selection. We will outline the various concepts that have been proposed about possible genetic aspects of sexual selection, and we will summarize a number of experimental studies on humans and other organisms that provide support for one or the other model. We will also briefly discuss the possibility of differential parental investment, i.e., of life history strategies that take the perceived quality of a present partner and the perceived future reproductive success into account. Differential parental investment has been intensely studied in various animal models. If such life history strategies

exist in humans, they are likely to also affect offspring viability.

Let us start with a more general question, namely, why is there sex at all? Or: why do we need males in our species? Asexual reproduction is common in many plants and animals. Thinking harder about the evolutionary costs and benefits of sex may help us to better understand why mate choice may indeed be strongly linked to genetic aspects, and why mate choice for "high genetic quality" partners may in turn help maintaining sex as an, in evolutionary terms, costly way of reproduction.

SEX, SEXUAL SELECTION, AND INFECTIOUS DISEASES

For evolutionary biology, explaining the success of sexual reproduction has been a challenge because sex involves a number of enormous disadvantages compared to asexual reproduction. The major disadvantage is called "the cost of meiosis" (Williams, 1975; Maynard Smith, 1978): a female that reproduces sexually is only 50% related to her offspring while an asexual female transmits 100% of her genes to each of her daughters. Hence, gene transmission is about twice as efficient in asexuals as it is in sexuals. The other evolutionary disadvantages of sex are, for example, cellular-mechanical costs, damages through recombination, exposure to risks, mate choice, mate competition, etc. (Andersson, 1994). Therefore, if asexuals would have a similar survival probability as sexuals, a mutation causing a female to produce only asexual daughters would, when introduced into a sexually reproducing population, rapidly increase in frequency and outcompete sexuals in numbers within few generations (Williams, 1975; Maynard Smith, 1978). Obviously, this does not seem to happen very often, so sex must come with large evolutionary benefits. What might

Human Evolutionary Biology, ed. Michael P. Muehlenbein. Published by Cambridge University Press. © Cambridge University Press 2010.

these benefits be? And what are the disadvantages of asexual reproduction?

One serious disadvantage of asexual clones is that they are likely to die out after some hundred or thousand generations because of a fatal mechanism called "Muller's ratchet" (Muller, 1932). Muller's ratchet predicts that slightly deleterious mutations are accumulated in asexuals from generation to generation until the genome does not code for a viable organism any longer and the population goes extinct (Andersson, 1994). On a first glance it may therefore seem that sex must be so successful because recombination followed by selection can result in the efficient removal of damaged genes from generation to generation. However, such a benefit of sex is likely to have a significant effect only when it is too late, i.e., after an asexual population has outcompeted all its sexual conspecifics. Without any further benefits of sex, asexuals are expected to outcompete sexuals within only few generations because of the "cost of meiosis" and the other costs mentioned above, i.e., sexuals would die out before Muller's ratchet drove the asexual population to extinction (Michod and Levin, 1987; Kondrashov, 1993; Howard and Lively, 1994). Muller's ratchet itself, therefore, cannot explain the evolutionary success of sex.

Another set of hypotheses suggests that sex enables the spread or even the creation of advantageous traits. This second category of hypotheses on the evolutionary significance of sex requires that the direction of selection is continuously changing, i.e., the main sources of mortality are short-term environmental changes. The condition is especially fulfilled in coevolving systems such as parasite–host communities. The idea is that host resistance genes that are advantageous today might become disadvantageous in the near future because parasites evolve to overcome them. Therefore, hosts must continuously change gene combinations, and sex is an efficient means to do so. This idea is called the "Red Queen hypothesis" (Jaenike, 1978; Hamilton, 1980). Of course, the argument can be turned around to predict that any form of sex and recombination in parasite populations is maintained as a diversity-generating mechanism to counteract the rapidly changing selection imposed by the host's immune system (Gemmill et al., 1997; Wedekind and Rüetschi, 2000). Hence, the evolutionary conflict between pathogens and hosts may select for diversity-generating mechanisms on both sides and in turn prevents both parties from dying out as a direct or indirect consequence of Muller's ratchet.

Sexual reproduction is a very powerful diversity-generating mechanism. But there is more to sex than just creating diversity, even in evolutionary terms. As soon as we have two different mating types (males and females), different reproductive strategies can evolve. This usually starts with anisogamy, i.e., an unequal

investment into gametes as the result of disruptive selection: smaller gametes can be produced in greater number, and larger gametes provide the later zygote with more resources to survive better (Parker et al., 1972). Such anisogamy automatically leads to a race among males to ensure that their sperm find and fertilize eggs. This is when sexual selection comes into play. It is expected already in the most primitive sexual organisms, and although the difference in egg and sperm size can be trivial in more complex organisms with lactation or intense sexual advertisment, sexuals will probably never get rid of sexual selection (Kokko et al., 2006). In order to propagate their genes, individuals must not only survive natural selection, i.e., all the lethal threats imposed by harsh climates, predators, pathogens, and competitors, but they must also be able to find a mate and withstand competition from rivals of the same sex for access to potential mates. They have to be successful in sexual selection. Sometimes, females have to compete for access to males and are chosen by males. However, sexual selection is usually stronger on the male sex, because males usually have a much higher potential rate of reproduction than females (Cluttonbrock and Vincent, 1991; Cluttonbrock and Parker, 1992). Regardless of the direction, sexual selection shapes the evolution of a species, and it might even help to maintain sexual reproduction itself (Agrawal, 2001; Siller, 2001).

Mate choice (intersexual selection) and competition for mates (intrasexual selection) are two large categories of sexual selection. They both cause a number of immediately negative aspects, including the waste of resources for being attractive (the peacock's tail as a classical example in animals, various forms of status-signaling in humans) or the risk of injury or distraction from predator surveillance during intrasexual struggles or courtship behavior. Furthermore, transmission, maintenance, and growth of many pathogens are increased during the courtship phase of their host, either directly by sexual behavior itself (transmittance of ectoparasites or sexually transmitted microparasites) or indirectly by a reduction of immunocompetence of the host (Grossman, 1985; Folstad and Karter, 1992), which may be a consequence of an adaptive resource reallocation during the courtship phase, mating, and reproduction (Wedekind and Folstad, 1994; Gustafson et al., 1995). These disadvantages that are associated to sexual selection add to the evolutionary costs of sex mentioned above.

Sexual selection does not need to be over when mating has occurred. There are various postcopulatory selection levels possible, especially in mammals with their internal fertilization and their intense mother–embryo interaction (see points 2–8 in Figure 18.1). Postcopulatory inter-sexual selection (also called

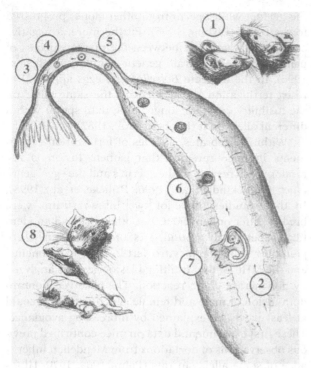

18.1. The levels at which females or female reproductive tissue could select for or invest differentially into different offspring genotypes: (1) male choice; (2) selection on sperm with the female reproductive tract; (3) egg choice for sperm; (4) the second meiotic division in the egg (that only starts some time after the genetic material of the sperm has entered the egg); (5) selection on the early embryo by the oviduct; (6) implantation; (7) maternal supply to the developing embryo and spontaneous or induced abortion; (8) maternal care after birth. Figure reproduced from Wedekind (1994).

"cryptic female choice") may include selection on sperm within the female reproductive tract, nonrandom gamete fusion, a nonrandom second meiotic division in the egg, and/or selective support of the embryo or the offspring. Preferences for certain kind of sperm have been observed in vertebrates and invertebrates (reviewed in Birkhead, 1996, 2000). In many plants, growth of the pollen tube is often affected by the stigma and depends on the combination of male and female alleles on the self-incompatibility locus (Jordan et al., 2000). In mice (*Mus musculus*), gamete fusion and the second meiotic division of the egg is not fully random with respect to genetics (Pomiankowski and Hurst, 1993; Wedekind et al., 1996; Rülicke et al., 1998). Analogous results have been found in other species (Eberhard, 1996; Zeh and Zeh, 1997), but Stockley (1997) could not find it in common shrews (*Sorex araneus*). Last but not least, in many mammals the outcome of pregnancy may often depend on the dam's perception of the sire's attractiveness or other aspects that may be fitness relevant (Bruce, 1954; Rülicke et al., 2006). Such differential maternal investment has been observed in some mammals and

is hence a possibility that is discussed for humans too (see below).

INBREEDING AND INBREEDING DEPRESSION

If sex evolved as a diversity-generating mechanism and sexual selection as a consequence of conflicts between and within the sexes, it may not be surprising that mate choice often takes at least the degree of kinship between two individuals into account. Mating between kin not only reduces the efficiency of sex as a diversity-generating mechanism but also gives rise to inbreeding depression. Inbreeding depression is the immediately reduced fitness of inbred offspring that results from the manifestation of deleterious traits. The more closely the breeding pair is related, the more homozygous the offspring will be. Therefore, recessive deleterious mutations will be expressed. The significance of inbreeding depression has been studied in many systems. Pedigree and/or molecular data from natural populations that allow for identifying kinship and quantifying inbreeding have revealed that inbreeding depression is usually substantial enough to affect individual performance, but it varies across taxa, populations, and environments. In bird and mammal populations, inbreeding depression usually affects birthweight, survival, reproduction, tolerance or resistance to diseases, predators, and other stress factors (Keller and Waller, 2002). Analogously, in plants it affects seed set, germination, survival, and tolerance or resistance to stress (Keller and Waller, 2002). Quantitative estimates of the effects of inbreeding depression are, however, often difficult to obtain, because there are a number of potentially confounding factors that need to be taken into account.

In the case of humans, matings between first-degree cousins are frequent in various cultures, and various quantitative estimates of inbreeding depression exist. The estimates are quite consistent: Descendants of such pair bondings have on average a 4–5% increased probability of not reaching their 10th year of life (Bittles and Neel, 1994; Modell and Darr, 2002). One could carefully project this estimate to other kinds of kin matings, for example brother–sister matings. The average degree of relatedness between first-degree cousins is 12.5%, for half-sibs it is 25%, and for full sibs it is 50%. Inbreeding depression might then rise up to a roughly 20% increased probability of inbred children not reaching their 10th year of life. Other authors estimate this increased mortality to 25% (Aoki, 2005). This is obviously a significant reduction in offspring viability and fitness that may, however, still be widespread in many human cultures because of the significant social benefits that consanguineous marriages often provide (Modell and Darr, 2002). In captive animals (e.g., in zoos), full-sib matings lead on average to about a 33% increased mortality in the offspring (Ralls et al., 1988).

Inbreeding depression is a genetic effect, but it is not directly heritable. If an inbred individual mates with an unrelated one, the resulting offspring is not likely to suffer from a reduced viability as compared to all other offspring of outbreeding pairs. However, the immediate viability and fitness reduction that is linked to inbreeding depression is expected to select for mechanisms that lead to kin recognition and inbreeding avoidance. Indeed, there are many examples of inbreeding avoidance in animals and humans (Pusey and Wolf, 1996).

Incestuous matings are in many human societies a cultural taboo, but these taboos may often be politically motivated (e.g., to avoid too much accumulation of power), and there is much variation among the various human societies with regard to this taboo (Thornhill, 1991; Scheidel, 1996). Beside such social factors that influence the rate of inbreeding, there also exists a number of more biological factors. Avoidance of kin as mates often relies on recognition of familiarity (Clarke and Faulkes, 1999; Penn and Potts, 1999). In some species, odors that reveal information about highly polymorphic loci seem to play a crucial role in this recognition of familiarity and hence inbreeding avoidance (Brown and Eklund, 1994; Penn and Potts, 1999). In humans, the so-called "Westermarck effect" (Westermarck, 1891) may also play an important role in avoiding inbreeding. Westermarck postulated that men and women who were raised together in some sort of family-like structure are usually not sexually attracted to each other, regardless of whether they are genetically related or not. Shepher (1983), for example, found that marriages among communally reared children in an Israelian kibbutz were virtually absent. Analogous observations in other cultures followed (Wolf, 1995), and more recent studies (Lieberman et al., 2003; Fessler and Navarrete, 2004) found that a history of cosocialization with an opposite-sex individual is linked with a decreased tolerance of other's incestuous behaviors. Such effects are stronger in females than in males (Lieberman et al., 2003; Fessler and Navarrete, 2004), and the effect seems to have, among others, an olfactory basis (Schneider and Hendrix, 2000; Weisfeld et al., 2003).

Inbreeding avoidance does not need to be confined to mating decisions. Nonrandom gamete fusion with respect to kinship has been observed in various taxa. The ascidian *Diplosoma listeranum*, for example, is a colonial, sessile, marine filter-feeder that disperses sperm into surrounding water. Sperm are taken up and pass via the oviduct to reach oocytes within the ovary. Autoradiography of labeled sperm revealed that sperm from the same clone were normally stopped in the oviduct while sperm from other clones progressed to the ovary (Bishop, 1996). Furthermore, a negative correlation was found between the mating success of pairs and their overall genetic similarity (Bishop, 1996). In the tunicate *Botryllus* sp., eggs appeared to resist fertilization by sperm with the same allele on the fusibility locus for longer time than sperm with a different allele on it (Scofield et al., 1982).

Within vertebrates, a series of fertilization experiments in mice revealed that gamete fusion is not random with respect to the sperm's and the eggs' genotypes (Wedekind et al., 1996; Rülicke et al., 1998). In these studies, mice of two inbred strains were bred to differ only in MHC but otherwise had an identical genetic background. F1s of both strains were paired or used for in vitro fertilization experiments, and the MHC of the resulting blastocysts was analyzed by polymerase chain reaction. The observed nonrandom fusion of male and female genetic material could at least partly be explained by inbreeding avoidance. These first experimental data on mice confirmed previous observations of deviations from Mendelian inheritance of some alleles in rats (Palm, 1969, 1970; Hings and Billingham, 1981, 1983, 1985), mice (Potts et al., 1991), and lizards (Olsson et al., 1996). However, the data could also be explained in the light of hypotheses that compete to the inbreeding avoidance hypothesis. We will therefore come back to these examples later.

"GOOD-GENES" AND "COMPATIBLE-GENES" SEXUAL SELECTION

Apart from inbreeding avoidance, there are a number of more sophisticated forms of sexual selection, based on general phenotypic appearance or on sexual signals such as odors, certain forms of behavior, fluctuating asymmetry, or secondary sexual characteristics. The literature usually groups criteria for mate choice into three classes (Andersson, 1994): (1) criteria that reveal direct benefits, such as parental care or nuptial gifts; (2) "Fisherian traits", i.e., traits that are attractive to members of the other sex but do not reveal anything else (Fisher, 1930); and (3) indicators of "good genetic quality" (Zahavi, 1975; Hamilton and Zuk, 1982; Grafen, 1990). In the following we concentrate on this last class of criteria, because a preference for "good genetic quality" is expected to increase offspring survival while a preference for Fisherian traits is not.

"Genetic quality" is, in the context of sexual selection, an umbrella term. It includes "good genes" and "compatible genes" (Neff and Pitcher, 2005). "Good genes" are alleles that increase fitness directly and, in principle, independently of the rest of the genome. If variation in such loci exists within a population, i.e., when there is variation in fitness that can be linked to such loci, population studies will show additive genetic variance for viability, and populations are expected to evolve in response to directed selection. "Compatible genes" are defined as alleles that increase fitness only when in combination with other alleles, i.e., only when in a specific genotype. Such fitness effects can be due to epistasis, i.e., due to interactions with alleles on other loci, or due to any sort of heterozygote advantage (e.g., overdominance) when the allele is paired with a specific homolog. In population studies, "compatible-gene" effects will be revealed in significant nonadditive genetic variance for viability: some male–female combinations will then have a higher fitness than other combinations, regardless of possible "good-genes" effects. In principle, alleles that contribute to any kind of "good genetic quality" can code for all sort of traits. However, there are good reasons to assume that most of these alleles play a role in the coevolution between hosts and their pathogens, i.e., that sexual selection for "good genetic quality" is expected to be pathogen driven.

The most important aspect of the interaction between a pathogen and its host is the virulence an infection is associated with, here defined as the loss of fitness the pathogen causes to the host. Although this definition sounds as if virulence were a specific trait of a pathogen, it is not. Rather, it is the result of the interaction between this pathogen and its host (Bull, 1994; Read, 1994; Ebert and Hamilton, 1996). Sexual reproduction and the different forms of pathogen-driven sexual selection have the potential to increase or decrease virulence. Firstly, as indicated above, the ability of the host to reproduce sexually plays a significant if not the most important role here (Ebert and Hamilton, 1996). Sexual reproduction (especially in the case of outcrossing) results in a rearrangement of host genes. Pathogen populations that have adapted to one host line have to readapt to the next one and so on. Hence, host reproduction by sex leads to the existing pathogen population being suboptimally adapted to their current host population. This could typically mean that the pathogens are less virulent than would be expected if they were optimally adapted to their host (Ebert, 1994). This benefit of sexual reproduction could be amplified by certain forms of mate choice.

Hamilton and Zuk (1982) were the first who suggested that individuals in good health and vigor are preferred in mate choice because they are the ones that are likely to possess heritable resistance to the predominant pathogens. By preferring healthy individuals one may thereby produce resistant progeny. This could result in subsequent generations of hosts that are better adapted to the local pathogens, i.e., that are less susceptible (Grahn et al., 1998). In a meta-analysis on the available empirical evidence, Møller and Alatalo (1999) concluded that sexual selection for "good genes" is widespread across vertebrate taxa, but its effect on offspring survival varies. Their meta-analysis revealed that male sexual characters chosen by females on average accounted for only 1.5% of the variance in offspring viability, but the authors stressed that many of the studies included in their analysis may only partly estimate the full fitness consequences of mate choice for offspring survival. The effects were generally stronger for studies where the target of selection had been identified than those with an unknown target of selection. Indeed, more recent experimental studies demonstrated that "good genes effect" of mate choice can be very strong: optimal mate choice in a whitefish (*Coregonus* sp.), for example, would reduce pathogen-correlated embryo mortality by 67% as compared to random mating (Wedekind et al., 2001). Hence, the potential effects of pathogen-driven sexual selection on offspring viability can be, at least in some vertebrate taxa, very significant.

A potential indicator of genetic quality that has been intensely studied is fluctuating asymmetry (FA): the random deviations from perfect bilateral symmetry sometimes observed on various body parts. Developmental instability reveals the tendency of an organism to be influenced by various stress factors (environmental, e.g. pathogens, or genetic, e.g. mutations) during development (Thornhill and Gangestad, 2006). Fluctuating asymmetry can then be considered as a measurement of developmental instability. Many studies provide evidence for a link between FA and mate attractiveness (reviewed in Møller and Thornhill, 1998; more recent studies include Scheib et al., 1999; Brown et al., 2005; Little and Jones, 2006). An up-to-date meta-analysis of studies on humans reported a statistically significant global effect size of 0.30 (Pearson's r) between facial symmetry and mate attractiveness (Rhodes and Simmons, 2007). In these studies, attractiveness was either measured as self-reported reproductive success or by rating. The authors also analyzed the attractiveness of body symmetry separately from face symmetry and found a lower but significant effect size of 0.14 ($P < 0.001$). It is, however, less clear whether there exists a relationship between FA and mate quality. Rhodes and Simmons (2007) also made a meta-analysis of studies looking at symmetry and health measured as self-reported health status. They found low and not significant effect sizes both for facial and body symmetry. The link between parental FA and offspring survival has been rarely investigated (Waynforth, 1998) but one study revealed a higher FA in mothers of preterm infants (Livshits et al., 1998). Overall, while the relationship between FA and mate attractiveness is well established in humans, further investigations are needed to demonstrate a potential link between FA and mate quality.

A preference for indicators of health and vigor will lead to populations where all individuals of one sex have the same mate preference, i.e., the members of the opposite sex can be ranked in a universally valid order of attractiveness, and less attractive individuals would only be taken as mates if the more attractive ones are not available for some reasons. However, in some species this prediction does not appear to be fulfilled (Wedekind, 1994, 2002a; Neff and Pitcher, 2005). This may be, at least partly, due to the fact that an offspring's level of resistance depends on both its mother's and father's genetic contribution. At loci that are important for the parasite–host interaction (e.g., immunogenes), certain combinations of alleles may be more beneficial than others. If individuals choose their mates in order to produce such beneficial allele combinations, their preferences would have to depend on their own genotypes as well as their partners' ones (Wedekind, 1994). As a consequence, individuals with different alleles would show different preferences, and there would be no universally valid order of sexual attractiveness with respect to signals that reveal heritable disease resistance. Sexual selection for such "complementary genes" is therefore expected to lead to very different dynamics as compared to sexual selection for "good genes."

The "compatible-genes" sexual selection hypothesis usually predicts that dissimilar genotypes are more attractive to each other, e.g., because mating with dissimilar genotypes leads to heterozygous offspring that may benefit from some sort of heterozygote advantage, as has often been proposed in the case of the MHC (Thursz et al., 1997; Carrington et al., 1999; McClelland et al., 2003). It has to be stressed, however, that MHC heterozygotes are sometimes less resistant to infections than expected from the average response of homozygotes (McClelland et al., 2003; Wedekind et al., 2005).

Preferences for mates or for sperm of genetically dissimilar types have been observed in different vertebrate taxa. Olsson et al. (1996) found in a population of sand lizards (*Lacerta agilis*), where most females mate with more than one male, that the male's genetic similarity to the female correlates with the proportion of her offspring that he sires: more dissimilar males sire more offspring, both in the field and in the laboratory. They concluded that the female reproductive tissue actively selects from genetically dissimilar sperm. So far, the best example of mate preferences that depend on the chooser's own genotype is the mouse and the effects of genes on the MHC. Mice base their mate choice to a large extent on odors. These odors reveal some of the allelic specificity of the MHC (Yamazaki et al., 1979, 1983a, 1983b, 1994), and this information is often used in mate choice by males and females (Yamazaki et al., 1976, 1988; Egid and Brown, 1989;

18.2. Participant of the study by Wedekind et al. (1995) where the odors of six T-shirts worn by men were evaluated for pleasantness, sexiness, and intensity. One T-shirt was unused. Three of the T-shirts had been worn by men who shared no or very few major histocompatibility complex (MHC) antigens with the woman who judged the odors, while three other T-shirts had been worn by men who shared as many MHC antigens with the woman as possible in the study sample. It turned out that women who were not using the contraceptive pill at the time of the experiments preferred the odors of MHC-dissimilar men (as predicted from mouse studies [Penn and Potts, 1999]), while women on the pill preferred the odors of MHC-similar men. This preference change could be due to hormones that simulate aspects of pregnancy. Pregnancy is known to change MHC-linked odor preferences in mice (Manning et al., 1995).

Potts et al., 1991; Roberts and Gosling, 2003). They chose according to their own MHC-types, apparently to reach certain allele combinations or to avoid certain allele combinations in the progeny.

Humans can distinguish the odors of two congenic inbred mouse strains that differ only in their MHC (Gilbert et al., 1986), and rodents seem to be able to recognize human MHC-types (Ferstl et al., 1992). Wedekind et al. (1995) found that women's preferences for male odors correlated with the degree of similarity of their own and the men's MHC type. T-shirt odors were judged as more pleasant when they were worn by men whose MHC genotype was different from that of the judging woman (Figure 18.2). The difference in odor assessment was reversed when the women were taking oral contraceptives (studies by Thorne et al., 2002 and Roberts et al. 2008, revealed further effects of oral contraceptives on odor perception). Furthermore, the odors of MHC-dissimilar men more frequently reminded the women of their own present or former partners than did the odors of MHC-similar men. These memory associations suggested that the MHC or other linked genes influence human mate choice. Later experiments (Wedekind and Füri, 1997; Thornhill et al., 2003; Santos et al., 2005; Roberts et al., 2008) with new combinations of T-shirt wearers and smellers provided further support of a link between MHC and body odors (but see Jacob et al. 2002, and

Wedekind's 2002b, comments on this). When men and women sniffed at male and female odors, there was no significant effect of gender in the correlation between pleasantness and MHC similarity (Wedekind and Füri, 1997). Carol Ober and her colleagues then found in a large study on American Hutterites that married couples were less likely to share MHC loci than expected by chance, even after incest taboos were statistically controlled for (Ober et al., 1997), but Hedrick and Black (1997) could not confirm the effect in South Amerindians. See Roberts and Little (2008) for a recent review on the subject.

It is still not yet clear whether MHC-correlated mate preferences in mice or in humans optimize the offspring's immunogenetics or whether the MHC merely serves as a marker of kinship to avoid inbreeding (Potts and Wakeland, 1993; Apanius et al., 1997). If the first variant holds, this would further improve host defense against pathogen populations, i.e., it would further reduce the observable level of virulence. However, even if the first variant holds, it remains unclear what kind of MHC-complementarity is preferred, i.e., whether individuals simply prefer other types to ensure a higher proportion of MHC-heterozygous offspring (Brown, 1997) because heterozygosity at the MHC appears to be beneficial on average (Hedrick and Thomson, 1983; Thursz et al., 1997; Carrington et al., 1999; but see Lipsitch et al., 2003; Wedekind et al., 2005), or whether mate choice aims to reach specific allele combinations that are more beneficial under given environmental conditions than others (Wedekind and Füri, 1997). At the moment, there are only few indications that such beneficial allele combinations exist. The strong linkage disequilibrium observed between some alleles of the MHC could be explained by long-term epistatic fitness effects (Klein, 1986; Maynard Smith, 1989), but there is still much need for research on the beneficial or deleterious aspects of various homo- or heterozygous combinations of MHC-haplotypes under given environmental conditions, i.e., under given pathogen pressures.

Choice for complementary alleles would be most efficient, i.e., it would result in the highest fitness return, if individuals were able to choose their mate conditionally (Wedekind, 1999). A well-tuned condition-dependent mate selection could have evolved in some species because of a nontrivial fitness advantage. Conditional choice would take the present pathogen pressure into account and would promote allele combinations in the host that ensure the optimal defense against these pathogens. However, this requires physiological achievements that have not been demonstrated so far.

Two studies on mice suggest that sexual selection takes not only the male and female MHC-genotypes into account, but is also conditional since it responds to at least one external factor that can vary over time. In an in vitro experiment with two congenic mouse strains, Wedekind et al. (1996) tested whether: (1) eggs select for sperm according to their MHC; and (2) whether the second meiotic division of the egg is influenced by the MHC type of the fertilizing sperm (Figure 18.1). They found that neither egg–sperm fusion nor the second meiotic division is random, but that both processes depend on the MHC of both the egg and the sperm. However, these selection levels did not simply select for heterozygous MHC-combinations. Sometimes the eggs appeared to prefer homozygous combinations, and sometimes they appeared to prefer heterozygous combinations. This effect of time was statistically significant. It seemed that the external factor that influenced sexual selection was an uncontrolled epidemic by mouse hepatitis virus (MHV). The presence of MHV appeared to stimulate a preference for heterozygous combinations, while when absent, the mice seemed to prefer homozygous variants. To test this hypothesis, Rülicke et al. (1998) repeated the experiment in vivo with two groups of mice: some females were experimentally exposed to MHV while the others were sham exposed. When the authors typed the blastocysts of these mice for the MHC, they found again that infected mice had more MHC-heterozygous embryos than noninfected mice. This time, the finding was the outcome of an experiment designed to test an a priori hypothesis.

Several authors have previously searched for deviations from the expected ratios of MHC-heterozygosity in the progeny of controlled matings (Palm, 1969, 1970; Hings and Billingham, 1981, 1983, 1985; Potts et al., 1991). They reported a significant variability of MHC-heterozygote frequencies which, however, remained poorly understood because there appeared to be a general inability to replicate previous findings (see discussion in Hings and Billingham, 1985). Rülicke et al.'s (1998) results could lead to an explanation for the apparently controversial findings published before, since they could perform the experiments under defined hygienic conditions with a selective and monitored viral infection, i.e., they could control for the factor infection that was not fully controlled for in the previous studies and that could have influenced their outcome. However, it is still not clear whether MHC-heterozygous offspring of the mice strains used in the above experiments have a greater resistance to MHV-infection than homozygous variants, and it is still not known whether homozygous offspring have higher survival in the absence of MHV. If so, nonrandom fusion of egg and sperm with regard to their respective MHC and to the presence or absence of MHV could improve the health of the progeny, i.e., it would decrease the observed level of virulence in a locally adapted host–pathogen system.

DIFFERENTIAL PARENTAL INVESTMENT

Evolutionary theory predicts that parents invest in each offspring according to the potential fitness return of that offspring (Fisher, 1930). If, for example, the relative reproductive value of sons and daughters differs for different females or different males, sex allocation theory predicts that females adjust the sex ratio of their offspring according to their own condition or according to their mate's attractiveness (Trivers and Willard, 1973). Life history theory also predicts that parental investment depends on the perceived mate quality (Williams, 1966). Such conditional parental investment was first demonstrated in experiments and field studies that showed that females increase their investment in the current brood when mated with a preferred male (Burley, 1982; Delope and Møller, 1993; Petrie and Williams, 1993). Increased parental effort may lower one's own survival and future reproductive potential (Saino et al., 1999).

Recent studies on birds have identified the mechanisms of these life history decisions. Some female birds lay more eggs (Petrie and Williams, 1993) or larger eggs (Cunningham and Russell, 2000) after copulating with preferred males. In the latter case the females produced offspring of better body condition when paired with preferred males. Gil et al. (1999) found that females deposit higher amounts of testosterone and 5α-dihydrotestosterone in their eggs when mated to attractive males. In kestrels, maternal hormones influence offspring survival (Sockman and Schwabl, 1999), and in canaries, chick social rank is positively correlated with the concentration of yolk testosterone in the eggs from which they hatched, suggesting that the development of aggressive behavior of offspring may be modified by maternal testosterone (Schwabl, 1993; Schwabl et al., 1997). There is evidence that the effect exists also in somewhat more primitive taxa. The tapeworm *Schistocephalus solidus*, for example, produced large eggs if given the opportunity to outbreed, but relatively small ones if forced to reproduce by selfing (Wedekind et al., 1998).

Some data suggest that there is differential parental investment in humans, too. However, in contrast to many of the studies on animals, most evidence that suggests human differential investment is correlational, i.e., causes and effects are not clear. As mentioned above, the degree of MHC-similarity could be shown to influence mate preferences or aspects of mate attractiveness in some human populations. Life history theory may therefore predict that differential maternal investment is linked to the degree of sharing of MHC antigens between a couple. Indeed, this degree of sharing is linked to the rate of successful pregnancies after in vitro fertilization or tubal embryo transfer (Weckstein et al., 1991; Ho et al., 1994), to the interbirth interval (Ober et al., 1988), to the placenta and baby birthweight (Reznikoff Etievant et al., 1991), and to the likelihood of otherwise unexplained spontaneous abortions in the first three month of pregnancy, as found in the United States (Beer et al., 1985), Germany (Karl et al., 1989), Japan (Koyama et al., 1991), Finland (Laitinen, 1993), or a Chinese population in Taiwan (Ho et al., 1990).

POSSIBLE GENETIC CONSEQUENCES OF FREE MATE CHOICE

There are many examples of human interference with animal and plant reproduction. Free mate choice is usually circumvented in many farm animals and plants, and it is often rather restricted in zoo animals. Even in our own species there are cases in which free mate choice is at least partly prevented, e.g., for some cultural reasons. From a genetical point of view, this can also be the case as a result of infertility treatment with some forms of assisted reproductive technology, especially so in donor insemination and in egg or embryo donation (donors are usually anonymous), but possibly also in intracytoplasmic sperm injection where potential egg choice is not allowed for.

It may be too early to speculate about the evolutionary consequences of such interference in our own species. The implication of sexual selection on parasite–host coevolution is not well understood in natural systems, and it is even less understood in a culturally shaped species like our own one. Moreover, while the evidence for cryptic female choice (Figure 18.1) is increasingly convincing in some plants and animals, the evidence for it in humans is only correlational, i.e., cause and effects are unclear. The existing data can therefore be interpreted in various ways (Hedrick, 1988; Verrell and McCabe, 1990; Wedekind, 1994).

Assisted reproductive technology is now responsible for tens of thousands of new births annually. Gosgen et al. (1999) discussed the possibility of genetic costs of assisted reproductive technology in humans. The authors found that the incidence of birth defects in these children is not higher than in those conceived naturally, and that any other possible negative consequence of circumventing free mate choice is at least not obvious in the context discussed here. However, the success of assisted reproductive technology may depend on the respective MHC types of the genetic parents of an embryo (Weckstein et al., 1991; Ho et al., 1994). Gosgen et al. (1999) called for more research on the impact of new reproductive technologies on individuals and the population, and whether or not donor insemination programs should reflect female choice.

Many studies on sexual selection have suggested that genetically dissimilar mates are sexually preferred (see examples above), probably because high genetic diversity in the offspring is beneficial. It should, however, be noted that a preference for genetic dissimilarity could not always be observed (e.g., Yamazaki et al., 1976; Wedekind et al., 1996; Hedrick and Black, 1997; Rülicke et al., 1998), and it is not clear whether these exceptions would have led to higher viability in the offspring under given environmental conditions.

CONCLUSION

Sexual mixing of genes has two main evolutionary advantages, namely that recombination followed by selection results in the efficient removal of deleterious mutations, and that it creates genetic diversity which is important in evolutionary arms races, especially in host–pathogen coevolution. It may therefore not be surprising that mating in nature is often not random with respect to genetics, and that it may often be linked to host–pathogen coevolution. Different kinds of sexual selection could have different potential consequences on offspring viability. A preference for unrelated individuals (i.e., inbreeding avoidance) is already a very simple form of sexual selection. Several more sophisticated kinds of sexual selection that can be relevant for parasite–host coevolution have been proposed, some of them even after mate choice has occurred, i.e., before, during, and after gamete fusion. These possibilities have been investigated in some model species, especially with respect to the MHC, i.e., to a set of genes that are crucial in the parasite–host interaction. Moreover, sexual selection is often connected to life history decisions about the use of resources, as for example in peahens that lay more eggs if mated to a more attractive peacock. In humans, mate attractiveness has been found to be linked to the degree of sharing of MHC alleles, and this degree of sharing is linked to the rate of successful pregnancies after in vitro fertilization or tubal embryo transfer, the interbirth interval, placenta and baby birthweight, and the likelihood of otherwise unexplained spontaneous abortions in the first three months of pregnancy. These correlations to a genetic aspect of mate attractiveness may indicate two points, namely that: (1) some sort of "compatible-genes" sexual selection may still influence mate attractiveness and hence mate choice in our species; and (2) that genetic aspects of mate attractiveness may sometimes lead to differential maternal investment. Such differential maternal investment obviously influences zygote and embryo survival. It remains unclear, however, whether MHC-linked sexual selection itself affects offspring viability in modern human societies, at least as long as it is not used as a mechanism for inbreeding avoidance.

DISCUSSION POINTS

1. What are the genetic and social costs and benefits of various degrees of inbreeding in humans?
2. The idea of "good-genes" sexual selection assumes that potential mates differ in their heritable viability. What maintains variation in heritable viability over time in our species?
3. The idea of "compatible-genes" sexual selection predicts that some combinations of alleles lead to fitter offspring. If choice for "compatible-genes" leads to offspring that are heterozygous on some loci, does this mean that heterozygote individuals are generally more attractive in mate choice?
4. If differential maternal investment really exists in humans, what does it say about the fitness consequences of mate choice based on attractiveness traits?
5. How important are odors in mate choice today?
6. Is it possible that some forms of pregnancy problems in humans may be induced by body odors? If so, what would that mean for therapy?

ACKNOWLEDGMENTS

We thank Michael Muehlenbein and the two anonymous reviewers for suggestions and comments on the manuscript, and the Swiss National Science Foundation for financial support.

REFERENCES

Agrawal, A. F. (2001). Sexual selection and the maintenance of sexual reproduction. *Nature*, **411**, 692–695.

Andersson, M. (1994). *Sexual Selection*. Princeton: Princeton University Press.

Aoki, K. (2005). Avoidance and prohibition of brother–sister sex in humans. *Population Ecology*, **47**, 13–19.

Apanius, V., Penn, D., Slev, P. R., et al. (1997). The nature of selection on the major histocompatibility complex. *Critical Reviews in Immunology*, **17**, 179–224.

Beer, A. E., Semprini, A. E., Zhu, X. Y., et al. (1985). Pregnancy outcome in human couples with recurrent spontaneous abortions: HLA antigen profiles, HLA antigen sharing, female serum MLR blocking factors, and parental leukocyte immunization. *Experimental and Clinical Immunogenetics*, **2**, 137–153.

Birkhead, T. R. (1996). In it for the eggs: a review of "Female Control: Sexual Selection by Cryptic Female Choice," by W. G. Eberhard. *Nature*, **382**, 772.

Birkhead, T. (2000). *Promiscuity*. London: Faber and Faber.

Bishop, J. D. D. (1996). Female control of paternity in the internally fertilizing compound ascidian *Diplosoma listerianum*. 1. Autoradiographic investigation of sperm movements in the female reproductive tract. *Proceedings of the Royal Society of London. Series B*, **263**, 369–376.

Bittles, A. H. and Neel, J. V. (1994). The costs of human inbreeding and their implications for variations at the DNA level. *Nature Genetics*, **8**, 117–121.

Brown, J. L. (1997). A theory of mate choice based on heterozygosity. *Behavioral Ecology*, **8**, 60–65.

Brown, J. L. and Eklund, A. (1994). Kin recognition and the major histocompatibility complex – an integrative review. *American Naturalist*, **143**, 435–461.

Brown, W. M., Cronk, L., Grochow, K., et al. (2005). Dance reveals symmetry especially in young men. *Nature*, **438**, 1148–1150.

Bruce, H. M. (1954). An exertoceptive block to pregnancy in the mouse. *Nature*, **184**, 105.

Bull, J. J. (1994). *Virulence. Evolution*, **48**, 1423–1437.

Burley, N. (1982). Reputed band attractiveness and sex manipulation in zebra finches. *Science*, **215**, 423–424.

Carrington, M., Nelson, G. W., Martin, M. P., et al. (1999). HLA and HIV-1: heterozygote advantage and B*35-Cw*04 disadvantage. *Science*, **283**, 1748–1752.

Clarke, F. M. and Faulkes, C. G. (1999). Kin discrimination and female mate choice in the naked mole-rat *Heterocephalus glaber*. *Proceedings of the Royal Society of London. Series B*, **266**, 1995–2002.

Cluttonbrock, T. H. and Parker, G. A. (1992). Potential reproductive rates and the operation of sexual selection. *Quarterly Review of Biology*, **67**, 437–456.

Cluttonbrock, T. H. and Vincent, A. C. J. (1991). Sexual selection and the potential reproductive rates of males and females. *Nature*, **351**, 58–60.

Cunningham, E. J. A. and Russell, A. F. (2000). Egg investment is influenced by male attractiveness in the mallard. *Nature*, **404**, 74–77.

Delope, F. and Møller, A. P. (1993). Female reproductive effort depends on the degree of ornamentation of their mates. *Evolution*, **47**, 1152–1160.

Eberhard, W. G. (1996). *Female Control: Sexual Selection by Cryptic Female Choice*. Princeton: Princeton University Press.

Ebert, D. (1994). Virulence and local adaptation of a horizontally transmitted parasite. *Science*, **265**, 1084–1086.

Ebert, D. and Hamilton, W. D. (1996). Sex against virulence: the coevolution of parasitic diseases. *Trends in Ecology and Evolution*, **11**, 79–82.

Egid, K. and Brown, J. L. (1989). The major histocompatibility complex and female mating preferences in mice. *Animal Behaviour*, **38**, 548–550.

Ferstl, R., Eggert, F., Westphal, E., et al. (1992). MHC-related odors in human. In *Chemical Signals in Vertebrates VI*, R. L. Doty (ed.). New York: Plenum, pp. 205–211.

Fessler, D. M. T. and Navarrete, C. D. (2004). Third-party attitudes toward sibling incest – evidence for Westermarck's hypotheses. *Evolution and Human Behavior*, **25**, 277–294.

Fisher, R. A. (1930). *The Genetical Theory of Natural Selection*. Oxford: Clarendon Press.

Folstad, I. and Karter, A. J. (1992). Parasites, bright males, and the immunocompetence handicap. *American Naturalist*, **139**, 603–622.

Gemmill, A. W., Viney, M. E. and Read, A. F. (1997). Host immune status determines sexuality in a parasitic nematode. *Evolution*, **51**, 393–401.

Gil, D., Graves, J., Hazon, N., et al. (1999). Male attractiveness and differential testosterone investment in zebra finch eggs. *Science*, **286**, 126–128.

Gilbert, A. N., Yamazaki, K., Beauchamp, G. K., et al. (1986). Olfactory discrimination of mouse strains (*Mus musculus*) and major histocompatibility types by humans (*Homo sapiens*). *Journal of Comparative Psychology*, **100**, 262–265.

Gosgen, R. G., Dunbar, R. I. M., Haig, D., et al. (1999). Evolutionary interpretation of the diversity of reproductive health and disease. In *Evolution in Health and Disease*, S. C. Stearns (ed.). Oxford: Oxford University Press, pp. 108–121.

Grafen, A. (1990). Biological signals as handicaps. *Journal of Theoretical Biology*, **144**, 517–546.

Grahn, M., Langesfors, A. and Von Schantz, T. (1998). The importance of mate choice in improving viability in captive populations. In *Behavioral Ecology and Conservation Biology*, T. M. Caro (ed.). Oxford: Oxford University Press, pp. 341–363.

Grossman, C. J. (1985). Interactions between the gonadal steroids and the immune system. *Science*, **227**, 257–261.

Gustafson, L., Qarnström, A. and Sheldon, B. (1995). Trade-offs between life-history traits and a secondary sexual character in male collared flycatcher. *Nature*, **375**, 311–313.

Hamilton, W. D. (1980). Sex versus non-sex versus parasite. *Oikos*, **35**, 282–290.

Hamilton, W. D. and Zuk, M. (1982). Heritable true fitness and bright birds – a role for parasites. *Science*, **218**, 384–387.

Hedrick, P. W. (1988). HLA-sharing, recurrent spontaneous-abortion, and the genetic hypothesis. *Genetics*, **119**, 199–204.

Hedrick, P. W. and Black, F. L. (1997). HLA and mate selection: no evidence in South Amerindians. *American Journal of Human Genetics*, **61**, 505–511.

Hedrick, P. W. and Thomson, G. (1983). Evidence for balancing selection at HLA. *Genetics*, **104**, 449–456.

Hings, I. M. and Billingham, R. E. (1981). Splenectomy and sensitization of Fischer female rats favors histoincompatibility of R2 backcross progeny. *Transplantation Proceedings*, **13**, 1253–1255.

Hings, I. M. and Billingham, R. E. (1983). Parity-induced changes in the frequency of Rt1 heterozygotes in an R2-backcross. *Transplantation Proceedings*, **15**, 900–902.

Hings, I. and Billingham, R. E. (1985). Maternal fetal immune interactions and the maintenance of major histocompatibility complex polymorphism in the rat. *Journal of Reproductive Immunology*, **7**, 337–350.

Ho, H. N., Gill, T. J., Nsieh, R. P., et al. (1990). Sharing of human leukocyte antigens in primary and secondary recurrent spontaneous abortions. *American Journal of Obstetrics and Gynecology*, **163**, 178–188.

Ho, H. N., Yang, Y. S., Hsieh, R. P., et al. (1994). Sharing of human-leukocyte antigens in couples with unexplained infertility affects the success of in vitro fertilization and tubal embryo transfer. *American Journal of Obstetrics and Gynecology*, **170**, 63–71.

Howard, R. S. and Lively, C. M. (1994). Parasitism, mutation accumulation and the maintenance of sex. *Nature*, **367**, 554–557.

Jacob, S., Mcclintock, M. K., Zelano, B., et al. (2002). Paternally inherited HLA alleles are associated with women's choice of male odor. *Nature Genetics*, **30**, 175–179.

Jaenike, J. (1978). An hypothesis to account for the maintenance of sex within populations. *Evolutionary Theory*, **3**, 191–194.

Johnstone, R. A., Reynolds, J. D. and Deutsch, J. C. (1996). Mutual mate choice and sex differences in choosiness. *Evolution*, **50**, 1382–1392.

Jordan, N. D., Ride, J. P., Rudd, J. J., et al. (2000). Inhibition of self-incompatible pollen in *Papaver rhoeas* involves a complex series of cellular events. *Annals of Botany*, **85**, 197–202.

Karl, A., Metzner, G., Seewald, H. J., et al. (1989). HLA compatibility and susceptibility to habitual abortion. Results of histocompatibility testing of couples with frequent miscarriages. *Allergie und Immunologie*, **35**, 133–140.

Keller, L. F. and Waller, D. M. (2002). Inbreeding effects in wild populations. *Trends in Ecology and Evolution*, **17**, 230–241.

Klein, J. (1986). *Natural History of the Major Histocompatibility Complex*. New York: Wiley and Sons.

Kokko, H., Jennions, M. D. and Brooks, R. (2006). Unifying and testing models of sexual selection. *Annual Review of Ecology and Systematics*, **37**, 43–66.

Kondrashov, A. S. (1993). Classification of hypotheses on the advantage of amphimixis. *Journal of Heredity*, **84**, 372–387.

Koyama, M., Saji, F., Takahashi, S., et al. (1991). Probabilistic assessment of the HLA sharing of recurrent spontaneous-abortion couples in the Japanese population. *Tissue Antigens*, **37**, 211–217.

Laitinen, T. (1993). A set of MHC haplotypes found among Finnish couples suffering from recurrent spontaneous-abortions. *American Journal of Reproductive Immunology*, **29**, 148–154.

Lieberman, D., Tooby, J. and Cosmides, L. (2003). Does morality have a biological basis? An empirical test of the factors governing moral sentiments relating to incest. *Proceedings of the Royal Society of London. Series B*, **270**, 819–826.

Lipsitch, M., Bergstrom, C. T. and Antia, R. (2003). Effect of human leukocyte antigen heterozygosity on infectious disease outcome: the need for allele-specifc measures. *BMC Medical Genetics*, **4**, 2.

Little, A. C. and Jones, B. C. (2006). Attraction independent of detection suggests special mechanisms for symmetry preferences in human face perception. *Proceedings of the Royal Society of London. Series B*, **273**, 3093–3099.

Livshits, G., Davidi, L., Kobyliansky, E., et al. (1998). Decreased developmental stability as assessed by fluctuating asymmetry of morphometric traits in preterm infants. *American Journal of Medical Genetics*, **29**, 793–805.

Manning, C. J., Dewsbury, D. A., Wakeland, E. K., et al. (1995). Communal nesting and communal nursing in house mice, *Mus musculus domesticus*. *Animal Behaviour*, **50**, 741–751.

Maynard Smith, J. (1978). *The Evolution of Sex*. Cambridge: Cambridge University Press.

Maynard Smith, J. (1989). *Evolutionary Genetics*. Oxford: Oxford University Press.

McClelland, E. E., Penn, D. and Potts, W. K. (2003). Major histocompatibility complex heterozygote superiority during coinfection. *Infection and Immunity*, **71**, 2079–2086.

Michod, R. E. and Levin, B. R. (eds) (1987). *The Evolution of Sex. An Examination of Current Ideas*. Sunderland, MA: Sinauer Associates.

Modell, B. and Darr, A. (2002). Genetic counselling and customary consanguineous marriage. *Nature Reviews Genetics*, **3**, 225–229.

Møller, A. P. and Alatalo, R. (1999). Good-genes effects in sexual selection. *Proceedings of the Royal Society of London. Series B*, **266**, 85–91.

Møller, A. P. and Thornhill, R. (1998). Bilateral symmetry and sexual selection: a meta-analysis. *American Naturalist*, **151**, 174–192.

Muller, H. J. (1932). Some genetic aspects of sex. *American Naturalist*, **66**, 118–138.

Neff, B. D. and Pitcher, T. E. (2005). Genetic quality and sexual selection: an integrated framework for good genes and compatible genes. *Molecular Ecology*, **14**, 19–38.

Ober, C., Elias, S., Obrien, E., et al. (1988). HLA sharing and fertility in Hutterite couples – evidence for prenatal selection against compatible fetuses. *American Journal of Reproductive Immunology and Microbiology*, **18**, 111–115.

Ober, C., Weitkamp, L. R., Cox, N., et al. (1997). HLA and mate choice in humans. *American Journal of Human Genetics*, **61**, 497–504.

Olsson, M., Shine, R., Madsen, T., et al. (1996). Sperm selection by females. *Nature*, **383**, 585.

Palm, J. (1969). Association of maternal genotype and excess heterozygosity for Ag-B histocompatibility antigens among male rats. *Transplantation Proceedings*, **1**, 82–84.

Palm, J. (1970). Maternal–fetal interaction and histocompatibility antigens polymorphisms. *Transplantation Proceedings*, **2**, 162–173.

Parker, G. A., Smith, V. G. F. and Baker, R. R. (1972). The origin and evolution of gamete dimorphism and the male–female phenomenon. *Journal of Theoretical Biology*, **36**, 181–198.

Penn, D. J. and Potts, W. K. (1999). The evolution of mating preferences and major histocompatibility complex genes. *American Naturalist*, **153**, 145–164.

Petrie, M. and Williams, A. (1993). Peahens lay more eggs for peacocks with larger trains. *Proceedings of the Royal Society of London. Series B*, **251**, 127–131.

Pomiankowski, A. and Hurst, L. D. (1993). Evolutionary genetics – Siberian mice upset Mendel. *Nature*, **363**, 396–397.

Potts, W. K. and Wakeland, E. K. (1993). Evolution of MHC genetic diversity – a tale of incest, pestilence and sexual preference. *Trends in Genetics*, **9**, 408–412.

Potts, W. K., Manning, C. J. and Wakeland, E. K. (1991). Mating patterns in seminatural populations of mice influenced by MHC genotype. *Nature*, **352**, 619–621.

Pusey, A. and Wolf, M. (1996). Inbreeding avoidance in animals. *Trends in Ecology and Evolution*, **11**, 201–206.

Ralls, K., Ballou, J. D. and Templeton, A. (1988). Estimates of lethal equivalents and the cost of inbreeding in mammals. *Conservation Biology*, **2**, 185–193.

Read, A. F. (1994). The evolution of virulence. *Trends in Microbiology*, **73**, 73–76.

Reznikoff Etievant, M. F., Bonneau, J. C., Alcalay, D., et al. (1991). HLA antigen-sharing in couples with repeated spontaneous abortions and the birthweight of babies in successful pregnancies. *American Journal of Reproduction and Immunology*, **25**, 25–27.

Rhodes, G. and Simmons, L. W. (2007). Symmetry, attractiveness, and sexual selection. In *The Oxford Handbook of Evolutionary Psychology*, R. I. M. Dunbar and L. Barrett (eds). Oxford: Oxford University Press, pp. 333–364.

Roberts, S. C. and Gosling, L. M. (2003). Genetic similarity and quality interact in mate choice decisions by female mice. *Nature Genetics*, **35**, 103–106.

Roberts, S. C. and Little, A. C. (2008). Good genes, complementary genes and human mate preferences. *Genetics*, **132**, 309–321.

Roberts, S. C., Gosling, L. M., Carter, V., et al. (2008). MHC-correlated odour preferences in humans and the use of oral contraceptives. *Proceedings of the Royal Society of London. Series B*, **275**, 2715–2722.

Rülicke, T., Chapuisat, M., Homberger, F. R., et al. (1998). MHC-genotype of progeny influenced by parental infection. *Proceedings of the Royal Society of London. Series B*, **265**, 711–716.

Rülicke, T., Guncz, N. and Wedekind, C. (2006). Early maternal investment in mice: no evidence for compatible-genes sexual selection despite hybrid vigor. *Journal of Evolutionary Biology*, **19**, 922–928.

Saino, N., Calza, S., Ninni, P., et al. (1999). Barn swallows trade survival against offspring condition and immunocompetence. *Journal of Animal Ecology*, **68**, 999–1009.

Santos, P. S. C., Schinemann, J. A., Gabardo, J., et al. (2005). New evidence that the MHC influences odor perception in humans: a study with 58 Southern Brazilian students. *Hormones and Behavior*, **47**, 384–388.

Scheib, J. E., Gangestad, S. W. and Thornhill, R. (1999). Facial attractiveness, symmetry and cues of good genes. *Proceedings of the Royal Society of London. Series B*, **266**, 1913–1917.

Scheidel, W. (1996). Brother–sister and parent–child marriage outside royal families in ancient Egypt and Iran: a challenge to the sociobiological view of incest avoidance? *Ethology and Sociobiology*, **17**, 319–340.

Schneider, M. A. and Hendrix, L. (2000). Olfactory sexual inhibition and the Westermarck effect. *Human Nature – An Interdisciplinary Biosocial Perspective*, **11**, 65–91.

Schwabl, H. (1993). Yolk is a source of maternal testosterone for developing birds. *Proceedings of the National Academy of Sciences of the United States of America*, **90**, 11446–11450.

Schwabl, H., Mock, D. W. and Gieg, J. A. (1997). A hormonal mechanism for parental favouritism. *Nature*, **386**, 231.

Scofield, V. L., Schlumpberger, J. M., West, L. A., et al. (1982). Protochordate allorecognition is controlled by a MHC-like gene system. *Nature*, **295**, 499–502.

Shepher, J. (1983). *Incest, a Biosocial View*. New York: Academic Press.

Siller, S. (2001). Sexual selection and the maintenance of sex. *Nature*, **411**, 689–692.

Sockman, K. W. and Schwabl, H. (1999). Female kestrels hormonally regulate the survival of their offspring. *American Zoologist*, **39**, 369.

Stockley, P. (1997). No evidence of sperm selection by female common shrews. *Proceedings of the Royal Society of London. Series B*, **264**, 1497–1500.

Thorne, F., Neave, N., Scholey, A., et al. (2002). Effects of putative male pheromones on female ratings of male attractiveness: influence of oral contraceptives and the menstrual cycle. *Neuroendocrinology Letters*, **23**, 291–297.

Thornhill, N. W. (1991). An evolutionary analysis of rules regulating human inbreeding and marriage. *Behavioral and Brain Sciences*, **14**, 247–260.

Thornhill, R. and Gangestad, S. W. (2006). Facial sexual dimorphism, developmental stability, and susceptibility to disease in men and women. *Evolution and Human Behavior*, **27**, 131–144.

Thornhill, R., Gangestad, S. W., Miller, R., et al. (2003). Major histocompatibility complex genes, symmetry, and body scent attractiveness in men and women. *Behavioral Ecology*, **14**, 668–678.

Thursz, M. R., Thomas, H. C., Greenwood, B. M., et al. (1997). Heterozygote advantage for HLA class-II type in hepatitis B virus infection. *Nature Genetics*, **17**, 11–12.

Trivers, R. L. (1972). Parental investment and sexual selection. In *Sexual Selection and the Descent of Man 1871–1971*, B. Campbell (ed.). Chicago: Aldine de Gruyter, pp. 136–179.

Trivers, R. L. and Willard, D. E. (1973). Natural selection of parental ability to vary the sex ratio of offspring. *Science*, **179**, 90–91.

Verrell, P. A. and McCabe, N. R. (1990). Major histocompatibility antigens and spontaneous abortion – an evolutionary perspective. *Medical Hypotheses*, **32**, 235–238.

Waynforth, D. (1998). Fluctuating asymmetry and human male life-history traits in rural Belize. *Proceedings of the Royal Society of London. Series B*, **265**, 1497–1501.

Weckstein, L. N., Patrizio, P., Balmaceda, J. P., et al. (1991). Human leukocyte antigen compatibility and failure to achieve a viable pregnancy with assisted reproductive technology. *Acta Europaea Fertilitatis*, **22**, 103–107.

Wedekind, C. (1994). Mate choice and maternal selection for specific parasite resistances before, during and after fertilization. *Philosophical Transactions of the Royal Society of London. Series B*, **346**, 303–311.

Wedekind, C. (1999). Pathogen-driven sexual selection and the evolution of health. In *Evolution in Health and Disease*, S. C. Stearns (ed.). Oxford: Oxford University Press.

Wedekind, C. (2002a). Sexual selection and life-history decisions: implications for supportive breeding and the management of captive populations. *Conservation Biology*, **16**, 1204–1211.

Wedekind, C. (2002b). The MHC and body odors: arbitrary effects caused by shifts of mean pleasantness. *Nature Genetics*, **31**, 237.

Wedekind, C. and Folstad, I. (1994). Adaptive or nonadaptive immunosuppression by sex-hormones. *American Naturalist*, **143**, 936–938.

Wedekind, C. and Füri, S. (1997). Body odour preferences in men and women: do they aim for specific MHC combinations or simply heterozygosity? *Proceedings of the Royal Society of London. Series B*, **264**, 1471–1479.

Wedekind, C. and Rüetschi, A. (2000). Parasite heterogeneity affects infection success and the occurrence of within-host competition: an experimental study with a cestode. *Evolutionary Ecology Research*, **2**, 1031–1043.

Wedekind, C., Seebeck, T., Bettens, F., et al. (1995). MHC-dependent mate preferences in humans. *Proceedings of the Royal Society of London. Series B*, **260**, 245–249.

Wedekind, C., Chapuisat, M., Macas, E., et al. (1996). Nonrandom fertilization in mice correlates with the MHC and something else. *Heredity*, **77**, 400–409.

Wedekind, C., Strahm, D. and Schärer, L. (1998). Evidence for strategic egg production in a hermaphroditic cestode. *Parasitology*, **117**, 373–382.

Wedekind, C., Müller, R. and Spicher, H. (2001). Potential genetic benefits of mate selection in whitefish. *Journal of Evolutionary Biology*, **14**, 980–986.

Wedekind, C., Walker, M. and Little, T. J. (2005). The course of malaria in mice: major histocompatibility complex (MHC) effects, but no general MHC heterozygote advantage in single-strain infections. *Genetics*, **170**, 1427–1430.

Weisfeld, G. E., Czilli, T., Phillips, K. A., et al. (2003). Possible olfaction-based mechanisms in human kin recognition and inbreeding avoidance. *Journal of Experimental Child Psychology*, **85**, 279–295.

Westermarck, E. (1891). *The History of Human Marriage*. London: Macmillan.

Williams, G. C. (1966). Natural selection, the cost of reproduction, and a refinement of Lack's principle. *American Naturalist*, **100**, 687–690.

Williams, G. C. (1975). *Sex and Evolution*. Princeton: Princeton University Press.

Wolf, A. P. (1995). *Sexual Attraction and Childhood Association: a Chinese Brief for Edward Westermarck*. Stanford, CA: Stanford University Press.

Yamazaki, K., Boyse, E. A., Mike, V., et al. (1976). Control of mating preference in mice by genes in the major histocompatibility complex. *Journal of Experimental Medicine*, **144**, 1324–1335.

Yamazaki, K., Yamaguchi, M., Baranoski, L., et al. (1979). Recognition among mice. Evidence from the use of a Y-maze differentially scented by congenic mice of different major histocompatibility types. *Journal of Experimental Medicine*, **150**, 755–760.

Yamazaki, K., Beauchamp, G. K., Egorov, I. K., et al. (1983a). Sensory distinction between H-2b and H-2bm1 mutant mice. *Proceedings of the National Academy of Sciences of the United States of America*, **80**, 5685–5688.

Yamazaki, K., Beauchamp, G. K., Wysocki, C. J., et al. (1983b). Recognition of H-2 types in relation to the blocking of pregnancy in mice. *Science*, **221**, 186–188.

Yamazaki, K., Beauchamp, G. K., Kupniewski, D., et al. (1988). Familial imprinting determines H-2 selective mating preferences. *Science*, **240**, 1331–1332.

Yamazaki, K., Beauchamp, G. K., Shen, F. W., et al. (1994). Discrimination of odor types determined by the major histocompatibility complex among outbred mice. *Proceedings of the National Academy of Sciences of the United States of America*, **91**, 3735–3738.

Zahavi, A. (1975). Mate selection – a selection for a handicap. *Journal of Theoretical Biology*, **53**, 205–214.

Zeh, J. A. and Zeh, D. W. (1997). The evolution of polyandry. 2. Post-copulatory defences against genetic incompatibility. *Proceedings of the Royal Society of London. Series B*, **264**, 69–75.

19 Why Women Differ in Ovarian Function: Genetic Polymorphism, Developmental Conditions, and Adult Lifestyle

Grazyna Jasienska

VARIATION IN LEVELS OF REPRODUCTIVE HORMONES

High levels of ovarian steroid hormones in menstrual cycles are crucial for successful pregnancy (Lipson and Ellison, 1996; Venners et al., 2006) and, as such, are important determinants of female reproductive success and evolutionary fitness. However, there are substantial differences in mean levels of estradiol and progesterone between populations, among women within a single population, and among menstrual cycles of a single woman (Figure 19.1) (Ellison et al., 1993; Jasienska and Jasienski, 2008). For example, urban women in the United States have progesterone levels that are on average 65% higher than those of women from the Democratic Republic of Congo (Ellison et al., 1993). In a rural population from Poland, as much as 46% of the among-cycle variation in salivary progesterone is due to differences among individual women, while the remaining 54% of variation is due to differences among cycles of individual women (Jasienska and Jasienski, 2008). Such high intercycle variation is probably caused by a seasonality of agricultural workload and is much higher than in nonseasonal, industrial populations. However, even in urban women from the United States and the United Kingdom, where lifestyle is less influenced by seasons, progesterone levels vary from cycle to cycle (Lenton et al., 1983; Sukalich et al., 1994; Gann et al., 2001).

The present chapter reviews recent findings about variation in human female ovarian function, and more specifically, the levels of two primary female reproductive hormones: 17-β estradiol and progesterone. These steroids are involved in processes leading to ovulation, fertilization, and implantation of the fertilized egg. Their levels are important for successful completion of these processes and as such are directly responsible for the establishment of pregnancy. Estradiol levels positively correlate with follicle size and egg quality, and are related to morphology and thickness of

the endometrium (Eissa et al., 1986; Dickey et al., 1993). The endometrium is the lining of the uterus in which the fertilized egg implants and begins its further development. Progesterone is essential for maturation of the endometrium and a dose–response relationship has been described between progesterone levels and transformation of endometrium during the second half of the menstrual cycle (Santoro et al., 2000). Cycles that vary in levels of these hormones show corresponding variation in the chance of pregnancy. In a group of Caucasian women from the United States who were attempting pregnancy, women were more likely to achieve pregnancy during menstrual cycles with higher levels of estradiol in the follicular phase (Lipson and Ellison, 1996). In this group of women, average estradiol levels were associated with a 12% probability of conception, while a 37% rise in estradiol levels was associated with a 35% probability of conception. In healthy Chinese women who were trying to conceive, cycles characterized by higher levels of estradiol resulted in higher rates of conception (Venners et al., 2006). Progesterone levels, especially during the mid-luteal phase, are also positively correlated with the probability of successful conception (Lu et al., 1999).

While it is well established that, among women from the same population, lower levels of ovarian hormones lead to a lower probability of conception, it is less clear if differences in average levels of these hormones described for different population can be interpreted the same way. That is, between populations, do lower average levels of estradiol and progesterone lead to lower probability of conception in one population versus another? It has been suggested that ovarian function may have different set-points, depending on lifestyle conditions (Vitzthum, 2001). Women with chronically poor energetic conditions would have lower levels of ovarian hormones, but that these levels would be sufficient for conception to occur. For example, rural Bolivian women conceived during cycles with mean progesterone levels approximately

Human Evolutionary Biology, ed. Michael P. Muehlenbein. Published by Cambridge University Press. © Cambridge University Press 2010.

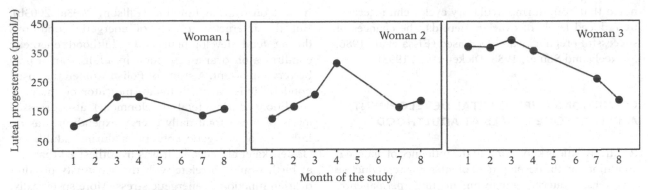

19.1. Inter-individual and intra-individual differences in mean luteal progesterone levels among healthy Polish rural women with regular menstrual cycles. Progesterone was measured for each woman in daily collected saliva samples during six menstrual cycles.

40% lower than the levels measured during conception cycles of urban US women (Vitzthum et al., 2004). It is argued that women living in chronically poor energetic conditions are not expected to have levels of hormones equally high as women having good energetic conditions (Ellison, 1990; Lipson, 2001). Lower levels of hormones during the cycle observed in women from chronically poor environments lead to a longer waiting time to conception, but conception is still possible.

It is important to note that comparisons of mean levels of ovarian hormones among populations or among individual women are difficult. One needs to make sure that women were in a similar age range, were selected for participation in a study using similar criteria, and that differences in laboratory procedures do not contribute to variation in hormone levels. In addition, hormonal values from blood or serum are not directly comparable with values obtained from saliva or urine (see Chapter 8 of this volume). In consequence, reliable comparisons of hormone levels can be made, at present, only for very few populations (Ellison et al., 1993; Ellison, 1994; Jasienska and Thune, 2001a).

WHY DO WOMEN AND POPULATIONS DIFFER IN LEVELS OF HORMONES?

Levels of hormones are influenced by many factors: genes, developmental conditions during fetal and childhood growth, and adult lifestyle (Figure 19.2). It is also well established that levels of reproductive hormones change with age. The lowest levels are observed in the years past menarche and before the menopause, and the highest levels for women between 25 and 35 years of age. For a review of age-related variation in ovarian function see Ellison (1994).

This chapter does not review variation in hormonal levels due to various diseases or anatomical/physiological pathologies, but rather focuses on variation in ovarian hormone levels in healthy women. It is

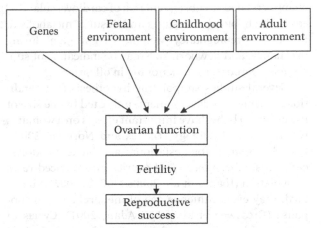

19.2. Ovarian function is influenced by genes, developmental conditions during fetal and childhood growth, and adult lifestyle. Ovarian function, and more specifically, levels of estradiol and progesterone, are important determinants of female fecundity, and as such directly related to her reproductive success.

important to emphasize that considerable variation in the levels of estradiol and progesterone exists among cycles in such women, and that substantial variation is still present even among cycles that are regular and ovulatory.

Ovarian function is sensitive to changes in lifestyle and can respond with reproductive suppression, especially to changes in energy availability. Reproductive suppression is understood here as any change in reproductive function which lowers the probability of pregnancy. When reproductive suppression results from the presence of environmental stressors, it usually occurs in gradual fashion (Prior, 1985; Ellison, 1990): a small reduction in energy intake may cause low progesterone production in the luteal phase of the menstrual cycle. A more serious energy limitation may result in the absence of ovulation or even total suppression of cycling (amenorrhea). The probability of pregnancy is, of course, reduced to zero when menstrual cycles are anovulatory or absent. However, it must be noted that even during ovulatory cycles characterized by reduced levels of ovarian steroids, the chances of successful pregnancy are decreased (Eissa et al., 1986; McNeely and Soules, 1988; Dickey et al., 1993).

CONDITIONS DURING FETAL DEVELOPMENT, AND HORMONE LEVELS AT ADULTHOOD

Relatively little is known about the influence of maternal conditions on the development of offspring reproductive functions. Maternal environment may permanently influence numerous aspects of fetal physiology (see Chapters 2 and 30 of this volume), but most research projects have been concerned with physiological changes related to subsequent risk of cardiovascular diseases, high blood pressure, and insulin metabolism. However, prepregnancy weight, weight gain during pregnancy, and newborn size may be indicators of subsequent reproductive functioning in offspring.

Several studies support the hypothesis that conditions during fetal development, indicated by the size of the newborn baby, have important effects on a woman's reproductive physiology (Davies and Norman, 2002). Girls born small for gestational age have, as adolescents, a smaller uterus and ovaries and reduced rates of ovulation (Ibanez et al., 2000a, 2000b, 2002). Size at birth may even influence age of menarche and menopause (Cresswell et al., 1997; Adair, 2001). Cycles in women with early menarche may also be characterized by higher levels of ovarian steroids (Apter, 1996).

While the studies listed above focused on small size for gestational age (usually below 2500 g), variation in "normal" size at birth may also be related to ovarian function. In healthy Polish women with regular menstrual cycles, variation in levels of estradiol was related to their size at birth (Jasienska et al., 2006e). The ponderal index (calculated as body weight/body length^3) describes fatness at birth and, therefore, can be used as an indicator of a newborn's nutritional status. This measure significantly predicted levels of estradiol in menstrual cycles: mean estradiol was 16.4 pmol/L in the group of women with low ponderal index, 17.3 pmol/L in the group with moderate ponderal index, and 19.6 pmol/L in the group with high ponderal index. These results suggest that conditions during fetal life influence physiological mechanisms responsible for production of reproductive hormones in adulthood, and may be one of the central causes of interindividual variation in reproductive function observed in adult women.

Of course, the levels of ovarian hormones in adult women result not only from impact of the fetal environment, but also from the influences of the adult environment. As discussed later in this chapter, metabolic conditions during adult life have a well established

impact on ovarian function (Ellison, 2003a, 2003b). But the interactive effects of energetic conditions during fetal development and adulthood energetic conditions on ovarian function in adult women may be very important. A study of Polish women that used ponderal index at birth (as an indicator of energetic conditions during fetal development) also assessed levels of mean total daily energy expenditure (as an indicator of energetic conditions during adulthood) (Jasienska et al., 2006d). It was hypothesized that size at birth would correlate with the sensitivity of adult ovarian function to energetic stress. More specifically, women born with poorer nutritional status should have higher sensitivity to energetic stress and lower response thresholds in adulthood than women born in better nutritional status. Higher sensitivity may be indicated by a reduction in the levels of estradiol in menstrual cycles at a lower threshold of energetic stress.

Results of this study indicated that nutritional status at birth is a predictor of the sensitivity of adult ovarian function to energetic stress. Physical activity may suppress ovarian function. However, women who had a high ponderal index at birth did not exhibit ovarian suppression in response to moderate levels of physical activity at adulthood. They responded with suppression only to higher levels of physical activity, while, in contrast, women who had a lower ponderal index at birth showed ovarian suppression even under mild energetically stressful conditions (Figure 19.3).

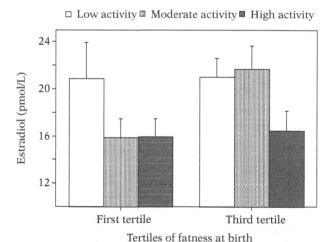

□ Low activity ▨ Moderate activity ■ High activity

19.3. Mean (with 95% confidence intervals) levels of estradiol in women from the first and the third tertiles of fatness at birth (as quantified by the ponderal index) at three different levels of mean daily physical activity. At low physical activity, both tertiles of fatness at birth had similar levels of estradiol. These tertiles of fatness also did not differ at high levels of physical activity. Such activity was associated with reduced estradiol in both groups. Moderate levels of physical activity, however, were sufficient to suppress estradiol levels in women who were relatively skinny at birth (the first tertile), but not in women relatively fat at birth (the third tertile).

The predictive adaptive response hypothesis was developed by Gluckman and Hanson (2005) in order to explain how physiological adjustments made by a fetus may result from developmental constraints or, alternatively, are adaptive responses to future environment. It was suggested that the developing fetus receives information about past environmental conditions that serves as cues for making predictions about future environmental conditions (Ellison, 2005; Gluckman et al., 2005b; Kuzawa, 2005; Jasienska, 2009). Therefore, women who develop under poor in utero conditions may receive a signal predicting poor environmental conditions in adulthood. In these women, ovarian sensitivity to energetic stress may be more acute, so that relatively mild energetic stressors in adulthood (a possible signal of deteriorating environmental conditions) would result in reproductive suppression. Such ovarian suppression, associated with a reduced probability of conception (Lipson and Ellison, 1996), may represent a developmentally plastic, predictive adaptive response, lowering the probability of conception when prospects for successful gestation are poor (Ellison, 2003a; Jasienska, 2003).

Studies of body asymmetry, finger length ratios (2D:4D), and ovarian hormones also reveal an influence of the uterine environment on adult ovarian function. Fluctuating asymmetry (FA), or random deviation from perfect symmetry expected in bilateral structures of bilaterally symmetric organisms, may result from the influences of environmental stressors operating during fetal development (Moller and Swaddle, 1997; Jones et al., 2001). Results pointing to a relationship between FA and reproductive physiology in women are preliminary, but such a relationship was indicated between breast symmetry and fertility-related traits, since women with higher breast symmetry had a higher number of offspring (Moller et al., 1995). Another study confirmed this finding and, in addition, documented an earlier age at first birth for more symmetric women (Manning et al., 1997). A relationship between FA and ovarian function was also documented in Polish women: mean mid-cycle estradiol levels in more symmetric urban women were almost 30% higher than estradiol levels in asymmetric women from the same population (Jasienska et al., 2006b).

Digit length ratio (index to ring fingers; 2D:4D) may also be another marker of uterine conditions, particularly the influence of androgens on prenatal sex differentiation (Manning et al., 2002). In women of reproductive age, estradiol measured in a single serum sample was positively correlated with 2D:4D (Manning et al., 1998). McIntyre et al. (2007) also identified a positive correlation between right hand 2D:4D and estradiol, but not progesterone, levels. While the latter study was advantageous by measuring hormones in daily collected saliva samples, neither study controlled for the effects of potentially confounding factors, like sports ability (Paul et al., 2006). Since energy expenditure resulting from sports participation is one of the most important factors influencing levels of ovarian hormones, such a variable should be controlled for (e.g., as a covariate) when investigating a relationship between 2D:4D and hormone levels.

CONDITIONS DURING CHILDHOOD AND OVARIAN HORMONES

Nutritional conditions during childhood can influence the age at menarche: girls with poor nutritional status mature later than those with good nutritional status (Ellison, 1982, 1990; Vermeulen, 1993). In those that mature early, ovarian hormone levels are higher, in comparison with those that mature late, for at least several years past menarche (Vihko and Apter, 1984). It is unclear if early menarche somehow changes the future trajectory of reproductive physiology or if both age at menarche and hormone levels are independently influenced by nutritional conditions.

Recently, more direct evidence pointed to the impact of childhood conditions on adult ovarian function. Nunez-De La Mora et al. (2007) analyzed levels of ovarian hormones in Bangladeshi women of reproductive age living in Bangladesh with Bangladeshi women who migrated to the United Kingdom at different points of their lives, and also with Caucasian women living in the United Kingdom. Bangladeshi women who spent their childhood in Bangladesh had lower levels of progesterone than Bangladeshi women who migrated to the United Kingdom as children, or those who were born in the United Kingdom. Those women who migrated as young children (between zero and eight years of age) had significantly higher levels of progesterone than those who migrated at later ages. They also exhibited earlier age at menarche and taller adult size. For women who migrated as children (prior to menarche), age at migration predicted age at menarche and average levels of luteal progesterone: early migration was associated with earlier maturation and higher hormone levels at adulthood. Energetic stressors resulting from maintaining immune responses to high infectious disease burden in Bangladesh was suggested to be responsible for lower levels of ovarian function in women who remained in Bangladesh and those who emigrated at later ages.

ENERGY METABOLISM DURING ADULT LIFE: ENERGY INTAKE AND PHYSICAL ACTIVITY

The influence of adult lifestyle on reproductive function has been studied intensely (Ellison, 2003b). Levels of ovarian hormones are often reduced due to poor nutritional condition. Negative energy balance, resulting from either high levels of energy expenditure or from low levels of energy intake, correlates with reduced levels of ovarian hormones.

Intense exercise and very negative energy balance are associated with increased incidence of menstrual disturbances and even total amenorrhea (absence of cycles) (Prior et al., 1982; Prior, 1985; Broocks et al., 1990; Rosetta, 1993, 2002; Rosetta et al., 1998). High incidence of menstrual disturbances in female endurance athletes has been well documented (Feicht et al., 1978; Prior et al., 1992; Rosetta, 1993). Such disruptions can be induced in previously untrained women who are subjected to demanding regimes of aerobic exercise (Bullen et al., 1985).

More moderate variation in energy expenditure (e.g., women engaged in various forms of recreational exercise) is associated with less dramatic changes in ovarian function, including lower levels of ovarian hormone, but without the disruption of menstrual patterns, and often even without any change in menstrual cycle regularity (Pirke et al., 1989; Broocks et al., 1990; Ellison, 1990; De Souza et al., 1998; Jasienska and Ellison, 1998, 2004; Jasienska et al., 2006f). For example, weekly running, even at very low distances, causes changes in the length of the luteal phase of the menstrual cycle (Shangold et al., 1979). Recreational joggers who run on average less than 24 km per week have suppressed levels of salivary progesterone (Ellison and Lager, 1985, 1986; Bledsoe et al., 1990). The incidence of anovulatory cycles in adolescent girls also shows a dose–response relationship to weekly energy expenditure (Bernstein et al., 1987).

Negative energy balance resulting from low energy intake also influences ovarian function. During the Dutch famine in the winter of 1944–1945, severe caloric restrictions resulted in reduced ability to conceive, as indicated by birth records (Vigersky et al., 1977; Painter et al., 2005). Women who lost 15% of initial weight exhibited amenorrhea and disturbances in the release of pituitary gonadotropins (Vigersky et al., 1977). In young women, caloric restriction (i.e., dieting) is associated with increased incidence of menstrual disturbances and suppressed ovarian hormone profiles (Pirke et al., 1985). Eating disorders, such as anorexia nervosa and bulimia nervosa are often associated with serious menstrual and hormonal disorders (Becker et al., 1999). Women do not need to be thin, however, to develop ovarian suppression in response to weight loss. Lower salivary progesterone profiles have been observed in women who lose only moderate amounts of mass through caloric restriction (Lager and Ellison, 1990). In these women, suppression of ovarian steroid levels was even more pronounced in the cycle following that in which the weight loss occurred. Suppression of ovarian function

is stronger when weight loss occurs in combination with exercise (Bullen et al., 1985), or when it occurs in young women (Schweiger et al., 1989).

Variation in ovarian function caused by energetic factors is not restricted to cases of voluntary exercise or dieting in urban women. It has also been described as a result of workload or seasonal food shortages in women with more traditional lifestyles. Farm women in rural Poland have profiles of salivary progesterone that vary with the intensity and duration of their workloads (Jasienska, 1996; Jasienska and Ellison, 1998, 2004). Women in Zaire and Nepal show seasonally suppressed levels of ovarian steroids associated with changes in workload and energy balance (Ellison et al., 1986; Panter-Brick et al., 1993). In Zaire this seasonal variation in ovarian function has been implicated as the main cause of observed seasonality in conceptions (Bailey et al., 1992).

Energy expenditure due to sport participation or occupational work may influence ovarian function independently of negative energy balance. Bullen et al. (1985) investigated the effects of intense physical exercise on ovarian function in previously untrained women. While women in one group were losing weight during eight weeks of intense training regime, a second group of women were receiving controlled diets with enough additional calories to maintain prestudy body weight. Even though the suppression of ovarian function was more pronounced in the weight-loss group, women who did not lose weight when training also showed evidence of moderate ovarian suppression. Ovarian suppression has also been documented in women who were maintaining stable body weight while running on average 20 km per week (Ellison and Lager, 1986). Even though the runners' menstrual cycles were of similar length as those in the control group, they were characterized by lower levels (and shorter profiles) of luteal progesterone. In Polish women, variation in the amount of habitual physical activity (as measured by average total daily energy expenditure) corresponds to variation in estradiol levels (Jasienska et al., 2006f) (Figure 19.4). Women from the low-activity group had 30% higher estradiol concentration than women from the high-activity group.

Occupational work that does not necessarily cause negative energy balance may still result in ovarian suppression (Jasienska, 1996; Jasienska and Ellison, 1998, 2004; Jasienska et al., 2006f). Polish women working on their own farms had salivary progesterone levels reduced by almost 25% during the months of intense harvest-related activities in comparison to months with less demanding work. These women were in good nutritional status (mean body mass index [BMI] of 24.4 kg/m², mean body fat percentage of 27.5) and did not exhibit negative energy balance as a result of

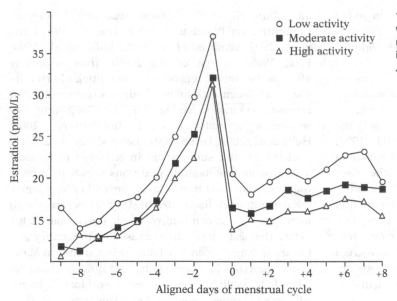

19.4. Profiles of mean estradiol for groups of women with low, moderate, and high levels of mean daily habitual physical activity. Confidence intervals were omitted for clarity. Reprinted from Jasienska et al. (2006f).

intense work. Results of these studies clearly indicate that negative energy balance is not a necessary condition for the occurrence of physical activity-induced reproductive suppression in women.

IS BODY FAT RELATED TO FECUNDITY?

Women with very low and very high body fat levels often experience menstrual and hormone level irregularities, and thus are less likely to conceive. Several studies have described a positive or an inverse U-shape relationship between BMI or body fat percentage and estradiol levels in women (Bruning et al., 1992; Barnett et al., 2002; Furberg et al., 2005; Ziomkiewicz, 2006). None of these studies, however, controlled for the effects of energy balance. It is possible that variation in BMI or body fat is correlated with variation in energy balance. Many women are able to achieve low BMI or low fat percentages due to conscious control of body size through low energy diet or exercise. These women are more likely, on average, to be in a state of negative energy balance or have high levels of energy expenditure. As discussed earlier, both of these factors have well established suppressive effects on ovarian function. In further analyses, Ziomkiewicz et al. (2008) suggested a positive relationship between body fat percentage and estradiol levels, but only in women with positive energy balance, while in women with negative energy balance, body fat did not correlate with ovarian function.

An excess of body fat seems to be detrimental to fecundity. Women who are obese or overweight have a lower likelihood of conceiving and suffer more complications during pregnancy than women of healthy body weight (Baird et al., 2006; Homan et al., 2007). For example, in a study which examined women who were patients of fertility clinics from the United States and Canada, obese women (BMI ≥ 27) were three times less likely to conceive than women with a BMI between 20 and 25 (Grodstein et al., 1994). Similarly, obese British women reported more menstrual problems and were less likely to conceive than women with lower body weight (Lake et al., 1997). These obese women often experienced pregnancy complications, including hypertension. In the United States, the risk of infertility increases from 1.1 among women with a BMI of 22.0–23.9 to 2.7 in women with BMI ≥ 32 (Rich-Edwards et al., 1994). Obese women who are able to lose a modest amount of weight (even less than 10%) often observe improved menstrual regularity, higher rate of ovulation, higher ovarian hormone levels, and, consequently, a higher chance of conception (Falsetti et al., 1992; Clark et al., 1995; Galletly et al., 1996; Norman and Clark, 1998). Therefore, weight loss is recommended for obese women who are unable to become pregnant. Some authors even suggest that weight loss programs should be offered to women before any other treatments for infertility (Norman et al., 2004).

OVARIAN SUPPRESSION AS AN EVOLUTIONARY ADAPTIVE PHENOMENON

Ellison (1990, 2003a) proposed that physiological responses of the reproductive system of contemporary women to energetic factors are not pathological, but rather reflect important features of human biology during the evolution of modern humans, particularly throughout the late Paleolithic era. Human physiology has remained basically unchanged since this time (Eaton et al., 1988, 2002), and energetic stressors have been, and still are, salient features of life in many traditional populations (Roberts et al., 1982; Lawrence and Whitehead, 1988; Panter-Brick, 1993; Adams, 1995; Benefice et al., 1996; Sellen, 2000).

Human female reproduction is energetically expensive. Relatively little energy is required to maintain ovarian and uterine functions. However, resting metabolic rate rises by 6–12% for several days during the luteal phase (Bisdee et al., 1989; Meijer et al., 1992; Howe et al., 1993; Curtis et al., 1996), and women tend to increase their caloric intake during the days following ovulation, perhaps compensating for additional energetic expenses of hormone production (Johnson et al., 1994). Such evidence points to the fact that some additional energy beyond regular maintenance needs is required to support regular menstrual function (Strassmann, 1996), although these costs are negligible in comparison with the energetic demands associated with pregnancy and lactation.

Frisch (1984) proposed that, due to the high energetic costs of pregnancy and lactation, ovarian function should respond in an on/off manner to the body fat stores. A woman with body fat insufficient to support the energetic costs of pregnancy and lactation should be unable to conceive. However, while fat stores may indeed be necessary to support the energetic costs of milk production (McNamara, 1995), such stores are insufficient to support the energetic costs rendered by both pregnancy and lactation.

Data demonstrating that women in developing countries often have very low fat reserves (Lawrence et al., 1987; Little et al., 1992; Panter-Brick, 1996) led to the hypothesis that, in the developing world, the function of fat stores is to serve as an emergency resource. Maternal fat stores may become useful when environmental conditions decline, but they cannot be used to steadily support milk production (Prentice and Prentice, 1990; Lunn, 1994). Even in well-nourished Western women, body fat stores used to support just half of lactation costs would last not longer than 11 months. In poorly nourished Gambian women, fat reserves used for this function would last for only 4 months (Prentice and Prentice, 1990; Lunn, 1994).

It should be stressed that women are capable of supporting a reproductive episode on a limited energy supply, but this does entail substantial long-term costs. Reproduction in women in poor energetic conditions is often associated with diminished reproductive outcomes, reflected by poor newborn health conditions (Lechtig et al., 1975; Roberts et al., 1982; Kusin et al., 1992; Siniarska, 1992; Pike, 2000). These children are, as adults, at increased risk of several metabolic diseases (Barker, 1994; Gluckman and Hanson, 2004; Gluckman et al., 2005a). Reproduction during poor energetic conditions also worsens maternal nutritional status ("the maternal depletion syndrome") (Merchant

and Martorell, 1988; National Academy of Sciences Committee on Population, 1989; Tracer, 1991; Little et al., 1992; Winkvist et al., 1992; Miller et al., 1994; Pike, 1999; George et al., 2000), thus negatively affecting her future reproductive potential. Shorter life spans in women with many children also suggests that reproduction may indeed have negative long-term consequences for maternal health (Doblhammer, 2000; Helle et al., 2002; Dribe, 2004; Jasienska et al., 2006c).

Reproductive suppression in response to temporary, poor environmental conditions serves to protect the maternal condition and optimize lifetime reproductive output (Ellison, 2003b). With longer interbirth intervals, women can improve their nutritional status before the next energy drain caused by pregnancy and lactation. Negative energy balance is a transient state, often occurring just seasonally, and quickly changing when food availability improves or workload lightens. Ellison's hypothesis elegantly explains why reproductive suppression is adaptive in women with a negative energy balance. However, suppression occurring in response to high levels of physical activity in women with a neutral or positive energy balance (i.e., not losing or gaining body weight) requires additional explanation.

The "constrained downregulation" hypothesis proposed that intense workload compromises a woman's ability to allocate sufficient energy to reproduce (Jasienska, 2001, 2003). Women who, as a result of increased physical activity, remain in a state of high-energy flux (high-energy expenditure compensated by high-energy intake), may have an impaired ability to downregulate their own metabolism when faced with increasing energetic needs of pregnancy and lactation. Lowered basal metabolism serves as one mechanism allowing women from traditional subsistence populations to allocate more energy to reproduction (Poppitt et al., 1993, 1994; Sjodin et al., 1996). On the other hand, increased basal metabolism is frequently observed in individuals who exhibit increased physical activity (Sjodin et al., 1996). It is possible that these hard-working women with elevated basal metabolism have constrained ability to redirect energy for reproductive processes. In this case, temporary suppression of ovarian function may be adaptive even in individuals who are still sustaining positive energy balance.

THE HOURGLASS FIGURE: RELATIONSHIPS BETWEEN BODY SHAPE AND HORMONE LEVELS

According to human evolutionary psychology, physical characteristics like breast size and waist-to-hip ratio (WHR) are used by human males to assess female attractiveness (Singh, 1993; Tovee et al., 1999). Males may pay attention to these features because they may serve as cues to fecundity and health. However, only a few studies have reported significant relationships between a low WHR (narrow waist, wider hips) or large breast size and increased fecundity. In addition, in most such studies, women with high WHR were also obese (Evans et al., 1983; Moran et al., 1999), or were patients of fertility clinics (Zaadstra et al., 1993). Low fecundity in such women may be the direct effect of obesity and not a high WHR.

One study investigated the relationship between body shape and fecundity in nonobese, fecund women (Jasienska et al., 2004). Women with a higher breast-to-underbreast ratio (larger breasts) and women with a relatively low waist-to-hip ratio (narrower waists) had significantly higher fecundity as assessed by estradiol and progesterone levels. Even more interesting was the finding that women who were characterized by both a narrow waist and large breasts had 26% higher mean estradiol and 37% higher mean mid-cycle estradiol levels than women with other combinations of body shape variables (i.e., low WHR with small breasts and high WHR with either large or small breasts). In this study, breast sizes and WHR were large or small, in a relative sense; population mean values were used as criteria for categorizing women into groups of large or small sizes. Therefore, a relationship between body shape and reproductive potential can be detected even within a group of otherwise "average" women.

PSYCHOLOGICAL STRESS AND FECUNDITY

We know that infertility often causes psychological stress, but there is limited evidence that psychological stress causes infertility (Campagne, 2006). Many women with fertility problems report high levels of psychological stress, and these results are often interpreted as evidence that psychological stress influences fecundity. Despite great interest in the potential impact of psychological stress on the ability to conceive, or on the risk of early pregnancy loss (Domar et al., 2000; Smeenk et al., 2001; Boivin, 2003; Cwikel et al., 2004; Anderheim et al., 2005; Nepomnaschy et al., 2006), it is not clear if moderate psychological stress may indeed have such detrimental effects.

Studies investigating relationships between psychological stress and fertility in women are plagued by several methodological problems (Ellison et al., 2007). Frequently, stress levels of women at fertility clinics are compared to stress levels in women who do not exhibit fertility problems (Harrison et al., 1986; Domar et al., 1990). It is not surprising if the former group experiences more severe psychological stress. Cognitive-behavioral therapy may improve some aspects of fertility in otherwise infertile couples

(Tuschen-Caffier et al., 1999), but these results cannot be used to suggest that psychological stress was the primary factor causing infertility.

"Psychological stress" is a broad concept, and studies use various ways to qualify and quantify psychological stress, focusing on "stressors," "perceived stress," or physiological "stress responses." Cortisol levels are often used as markers of psychological stress, and while they indeed become elevated in response to stressful stimuli, cortisol is also elevated in response to energetic stress. Mobilization of energy due to, for example, physical activity, infection, or low ambient temperature is associated with high levels of cortisol (Hackney and Viru, 1999). Therefore, a relationship between cortisol levels and ovarian hormones cannot be used as the evidence that psychological stress suppresses ovarian function. To complicate things further, psychological stress and energetic stress often coincide. For example, in traditional agricultural populations, famine may lead to both weight loss and high anxiety levels.

A potential impact of moderate, acute psychological stress on levels of ovarian steroid hormones was addressed in a study of United States college women taking the Medical College Admission Test (MCAT) (Ellison et al., 2007). Women who took the MCAT had significantly higher scores of anxiety during the months preceding the MCAT exams than several months after the exams. During the exam period, their anxiety was also higher than in women from the control group. Despite the differences in levels of self-reported anxiety reported by women, no statistically significant differences were observed in the levels of cortisol, estradiol, or progesterone. In addition, the relationship between chronic anxiety levels and ovarian function was examined in women 27–41 years of age. There was no statistically significant relationship between chronic anxiety and levels of either cortisol or ovarian hormones. In total, the study did not identify any suppressive effects of moderate anxiety, either acute or chronic, on ovarian function.

In general, it is too early to conclude if psychological stress has a suppressive effect on ovarian function. Carefully controlled studies on the effects of energetic stress on ovarian function in women from traditional, non-Western populations may help resolve this problem.

IS VARIATION IN HORMONE LEVELS EXPLAINED BY GENETIC VARIATION?

Levels of ovarian hormones within and between populations may be influenced by genetic polymorphisms, and there are several candidate genes. For example, some studies suggest that polymorphisms of the estrogen receptor genes are related to variation in levels of androgens in premenopausal women (Westberg et al., 2001) and variation in levels of estradiol in postmenopausal women (Peter et al., 2008). However, a large multiethnic study has found only weak (explaining < 4% of variation) and inconsistent associations between receptor genetic polymorphism and estradiol levels (Sowers et al., 2006).

Genetic variants of *CYP17*, *CYP19*, *CYPA1A*, and *CYPB1B* are involved in the steroid metabolic pathway and code for enzymes involved in steroid production and metabolism. These genetic variants can therefore influence levels of circulating steroid hormones. The most intensely studied of all genes involved in steroid metabolism are *CYP19* and *CYP17*.

CYP19 encodes aromatase, a key enzyme in the synthesis of estrogens. Several studies have documented a relationship between polymorphism of this gene and estrogen levels in postmenopausal women (Tworoger et al., 2004; Haiman et al., 2007; Peter et al., 2008), but not in the menstrual cycles of younger women (Garcia-Closas et al., 2002). *CYP17* encodes cytochrome P450c17α, which mediates activities of the enzymes 17α-hydroxylase and 17,20-lyase, both involved in the biosynthesis of estrogen (Small et al., 2005). A single nucleotide polymorphism in the 5′-untranslated region of *CYP17* is relatively common, and presence of the *A2* allele is thought to increase transcription rates. In other words, having a simple mutation in *CYP17* (e.g., having the *A2* allele as part of the *CYP17* genotype) increases enzyme production rates needed for the synthesis of estrogens.

The relationship between *CYP17* polymorphism and estradiol levels in premenopausal women has been the focus of several studies (Feigelson et al., 1998; Garcia-Closas et al., 2002; Travis et al., 2004; Hong et al., 2004; Lurie et al., 2005; Small et al., 2005), although results are inconsistent. Estradiol levels measured around day 11 of the menstrual cycle have been reported to be 11% and 57% higher among women with genotypes *A1/A2* and *A2/A2*, respectively, compared to *A1/A1* women (Feigelson et al., 1998). Estradiol levels during the luteal phase, around day 22 of the cycle, were reported to be 7% and 28% higher for women with *A1/A2* and *A2/A2*, respectively. Women with the *A2/A2* genotype had 42% (and heterozygotes *A1/A2* 19%) higher estradiol than the *A1/A1* homozygotes, but this was true only for women with BMI values not greater than 25 kg/m^2 (Small et al., 2005). It is important to note that both of these studies were based on only one or two estradiol values measured per woman.

Another study of premenopausal women did not identify any differences in estradiol levels among *CYP17* genotypes, but rather documented significant differences in dehydroepiandrosterone levels (Hong et al., 2004). This study, in addition to two others

(Garcia-Closas et al., 2002; Travis et al., 2004) which also did not identify significant differences in estradiol levels among *CYP17* genotypes, used a single blood sample per woman for hormone measurements.

Only two studies have controlled for within-cycle variability in estradiol levels by measuring hormone levels via multiple samples. Over a 2-year period, Lurie et al. (2005) sampled women, on average, at 4.4 time points and did not identify any significant relationships between estradiol levels and *CYP17* genotype. In contrast, Jasienska et al. (2006a) measured ovarian hormone levels in saliva samples collected daily across an entire menstrual cycle. In this case, variation in estradiol levels was partially explained by polymorphism at the *CYP17* locus: women with *A2/A2* genotypes had 54% higher mean estradiol levels than women with *A1/A1* genotypes, and 37% higher mean estradiol levels than women who had only one *A2* allele. Heterozygous *A1/A2* women had 13% higher estradiol levels than homozygous *A1/A1* women. Levels of estradiol during the preovulatory day were 72% higher in *A2/A2* compared to *A1/A1*, and 52% higher compared to *A1/A2*.

While variation in levels of hormones in response to energetic factors, such as energy balance or energy expenditure, can be explained as adaptive, the existence of genetic variation at the loci responsible for hormone production is not immediately clear. Since high levels of hormones are important determinants of female fecundity, one would expect a strong and consistent selective pressure promoting the alleles encoding high levels of reproductive hormones. These alleles should be most prevalent in all contemporary populations. Instead, in all studied populations there is a considerable polymorphism in genes involved in steroid production and metabolism. Why are the low-steroid alleles present in modern populations at all?

Females with alleles that code for high production of reproductive hormones could have a clear selective advantage over females with allelic variants coding for lower steroid levels. However, those with high-level alleles could potentially exhibit lower lifetime fitness compared to other females. While estrogens are in general beneficial for fecundity and health in women, high estrogen levels are also responsible for increased risk of hormone-dependent cancers, including breast cancer (Pike et al., 1983, 1993; Bernstein and Ross, 1993). Therefore, women with high levels of estradiol would have the advantage of having higher potential fertility, but at the risk of dying from reproductive cancers. Reproductive cancers are, however, most prevalent in postmenopausal women.

An alternative explanation is that low-level alleles are simply sufficient for normal physiological functions, but that lifestyle changes associated with the introduction of agriculture led to the consumption of larger quantities of food which contain high concentrations of phytoestrogens. These chemicals bind to estrogen receptors and influence activity of enzymes involved in the steroid metabolism (Kuiper et al., 1998; Xu et al., 1998). When consumed in large quantities, these chemicals may reduce levels of endogenous estrogens (Xu et al., 1998; Kapiszewska et al., 2006). Phytoestrogens exhibit much more potent estrogen receptor binding abilities when diet is rich in carbohydrates, a common feature in agricultural populations (Setchell and Cassidy, 1999).

In ancestral populations that lacked high consumption rates of phytoestrogens, the levels of hormones coded by low-level alleles could have been optimal for ovarian function, ensuring sufficiently high probability of conception. But because increased phytoestrogen consumption may cause lower fecundity in women, consuming large quantities of phytoestrogens after the spread of agriculture may have placed a selective advantage on high-level genotypes.

HORMONE LEVELS AND GENE–ENVIRONMENT INTERACTIONS

In healthy women of comparable ages, variation in hormone levels can be explained, to a large extent, by variation in energetic factors, which in turn are related to environmental conditions. Energy availability to a developing fetus, to a growing girl, and to an adult woman influences the levels of hormones produced during the menstrual cycles. But what about gene–environment interactions? Do women with genetically influenced low levels of hormones have comparable responses to environmental energetic factors as do women with genetically influenced high levels? Would physical exercise of exactly the same duration and intensity have the same effect on women of different genotypes? No studies have so far addressed these questions. If the suppression of reproductive function is an adaptive response to low availability of metabolic energy, should similar responses be observed in all women, regardless of whether they are genetically low or high hormone "producers?"

Furthermore, as suggested before, genetic variation in hormone levels may be a relatively new evolutionary phenomenon, appearing after the origin of agriculture. If so, it may be that an inadequate amount of time has elapsed for different genotype-specific responses to environmental challenges to have evolved.

DETERMINANTS OR MERELY CORRELATES OF HORMONE LEVELS?

While some of the factors described here clearly influence levels of hormones, others just show correlations with hormonal levels. Results of cross-sectional, observational studies allow us to only conclude that two variables are correlated. For example, causality cannot be determined between the correlation of WHR ratio and estrogen levels. Estrogen influences the pattern of fat distribution so that women with higher estrogen levels may have higher fat deposition in the hip region, or lower fat deposition in the waist region, or both. At the same time, body fat influences estrogen levels, and women with high abdominal fat may have reduced levels of estradiol. Causality in this example is likely complex and bidirectional. The relationship between physical activity, ovarian function, and estrogen levels seems to be much more straightforward.

ARE HIGH LEVELS OF HORMONES IN WESTERN WOMEN A PHYSIOLOGICAL NORM?

Medical sciences have assumed that urban women from industrialized countries have physiology operating at optimal levels. Therefore, high levels of ovarian hormones in menstrual cycles are considered a physiological norm. Lower hormone levels in women from non-Western populations are not discussed much in medical literature. Many physicians have and still do consider low ovarian hormonal levels in these women as pathological. However, as pointed out by Ellison (2003b), hormone levels in Western women actually appear to be abnormally high. Abundant availability of energy during fetal and childhood development and during adult life contributes to the presence of high levels of ovarian hormones. Such energetic conditions were unlikely features throughout the majority of human evolution.

In Western populations, women also have high numbers of cycles during their lives (Strassmann, 1997; Eaton and Eaton III, 1999). Early age at menarche and late age at menopause expand the range of years during which cycles occur. With fewer pregnancies, a woman spends more time cycling. With fewer pregnancies, a woman also spends less time breast-feeding. Even while breast-feeding, a woman may resume her cycles early, because of infrequent nursing episodes and good maternal nutritional status (Valeggia and Ellison, 2004). Combined, this may expose her to higher lifetime levels of estrogens.

In women, many aspects of health and disease are estrogen-dependent, including cardiovascular function, bone density, psychological well-being, and reproductive cancers (Key and Pike, 1988; Barrett-Conor and Bush, 1991; Pike et al., 1993; Nguyen et al., 1995; Jasienska et al., 2000; Jasienska and Thune, 2001a, 2001b; Jasienska, 2002). In premenopausal women of reproductive age, those with higher estrogen levels may also be in better health, since estrogen at normal physiological levels can be immunostimulatory (Kovacs et al., 2002; Jacobson and Ansari, 2004). However, women who have high estrogen levels during their reproductive years may suffer detrimental health effects in their postmenopausal years. High lifetime levels of reproductive hormones are related to increased risk of hormone-dependent cancers (Key and Pike, 1988; Pike et al., 1993; Jasienska et al., 2000; Jasienska and Thune, 2001a, 2001b; Yue et al., 2003). The knowledge of factors capable of influencing the levels of hormones is crucial for understanding determinants of health and disease, and for designing effective programs of disease prevention for women.

DISCUSSION POINTS

1. What factors influence levels of ovarian steroid hormones in adult women?
2. What is the relationship between conditions experienced during fetal development and ovarian function during adulthood?
3. Reproduction in women is energetically costly. Would you expect a linear, positive relationship between energy stored as body fat and levels of ovarian hormones in women? Why, or why not?
4. Reduced levels of ovarian steroid hormones lead to reduced ability of conception, yet a hypothesis suggests that "ovarian suppression" in response to physical activity or weight loss can be adaptive (i.e., is evolutionarily advantageous). Can you explain this apparent paradox?
5. Are high levels of estrogen beneficial for women?

REFERENCES

Adair, L. S. (2001). Size at birth predicts age at menarche. *Pediatrics*, **107**, e59.

Adams, A. M. (1995). Seasonal variations in energy balance among agriculturalists in central Mali: compromise or adaptation? *European Journal of Clinical Nutrition*, **49**, 809–823.

Anderheim, L., Holter, H., Bergh, C., et al. (2005). Does psychological stress affect the outcome of in vitro fertilization? *Human Reproduction*, **20**, 2969–2975.

Apter, D. (1996). Hormonal events during female puberty in relation to breast cancer risk. *European Journal of Cancer Prevention*, **5**, 476–482.

Bailey, R. C., Jenike, M. R., Ellison, P. T., et al. (1992). The ecology of birth seasonality among agriculturalist in central Africa. *Journal of Biosocial Science*, **24**, 393–412.

Baird, D. T., Cnattingius, S., Collins, J., et al. (2006). Nutrition and reproduction in women. *Human Reproduction Update*, **12**, 193–207.

Barker, D. J. P. (1994). *Mothers, Babies, and Disease in Later Life*. London: BMJ Publishing.

Barnett, J. B., Woods, M. N., Rosner, B., et al. (2002). Waist-to-hip ratio, body mass index and sex hormone levels associated with breast cancer risk in premenopausal Caucasian women. *Journal of Medical Sciences*, **2**, 170–176.

Barrett-Conor, E. and Bush, T. L. (1991). Estrogen and coronary heart disease in women. *Journal of the American Medical Association*, **265**, 1861–1867.

Becker, A. E., Grinspoon, S. K., Klibanski, A., et al. (1999). Current concepts – eating disorders. *New England Journal of Medicine*, **340**, 1092–1098.

Benefice, E., Simondon, K. and Malina, R. M. (1996). Physical activity patterns and anthropometric changes in Senegalese women observed over a complete seasonal cycle. *American Journal of Human Biology*, **8**, 251–261.

Bernstein, L. and Ross, R. K. (1993). Endogenous hormones and breast cancer risk. *Epidemiological Reviews*, **15**, 48–65.

Bernstein, L., Ross, R. K., Lobo, R. A., et al. (1987). The effects of moderate physical activity on menstrual cycle patterns in adolescence: implications for breast cancer prevention. *British Journal of Cancer*, **55**, 681–685.

Bisdee, J., James, W. and Shaw, M. (1989). Changes in energy expenditure during the menstrual cycle. *British Journal of Nutrition*, **61**, 187–199.

Bledsoe, R. E., O'Rourke, M. T. and Ellison, P. T. (1990). Characterization of progesterone profiles of recreational runners. *American Journal of Physical Anthropology*, **81** (abstract), 195–196.

Boivin, J. (2003). A review of psychosocial interventions in infertility. *Social Science and Medicine*, **57**, 2325–2341.

Broocks, A., Pirke, K. M., Schweiger, U., et al. (1990). Cyclic ovarian function in recreational athletes. *Journal of Applied Physiology*, **68**, 2083–2086.

Bruning, P. F., Bonfrer, J. M. G., Hart, A. A. M., et al. (1992). Body measurements, estrogen availability and the risk of human breast cancer: a case-control study. *International Journal of Cancer*, **51**, 14–19.

Bullen, B. A., Skrinar, G. S., Beitins, I. Z., et al. (1985). Induction of menstrual disorders by strenuous exercise in untrained women. *New England Journal of Medicine*, **312**, 1349–1353.

Campagne, D. M. (2006). Should fertilization treatment start with reducing stress? *Human Reproduction*, **21**, 1651–1658.

Clark, A. M., Ledger, W., Galletly, C., et al. (1995). Weight-loss results in significant improvement in pregnancy and ovulation rates in anovulatory obese women. *Human Reproduction*, **10**, 2705–2712.

Cresswell, J. L., Egger, P., Fall, C. H. D., et al. (1997). Is the age of menopause determined in-utero? *Early Human Development*, **49**, 143–148.

Curtis, V., Henry, C. J. K., Birch, E., et al. (1996). Intraindividual variation in the basal metabolic rate of women: effect of the menstrual cycle. *American Journal of Human Biology*, **8**, 631–639.

Cwikel, J., Gidron, Y. and Sheiner, E. (2004). Psychological interactions with infertility among women. *European Journal of Obstetrics, Gynecology and Reproductive Biology*, **117**, 126–131.

Davies, M. J. and Norman, R. J. (2002). Programming and reproductive functioning. *Trends in Endocrinology and Metabolism*, **13**, 386–392.

De Souza, M. J., Miller, B. E., Loucks, A. B., et al. (1998). High frequency of luteal phase deficiency and anovulation in recreational women runners: blunted elevation in follicle-stimulating hormone observed during luteal-follicular transition. *Journal of Clinical Endocrinology and Metabolism*, **83**, 4220–4232.

Dickey, R. P., Olar, T. T., Taylor, S. N., et al. (1993). Relationship of endometrial thickness and pattern to fecundity in ovulation induction cycles: effect of clomiphene citrate alone and with human menopausal gonadotropin. *Fertility and Sterility*, **59**, 756–760.

Doblhammer, G. (2000). Reproductive history and mortality later in life: a comparative study of England and Wales and Austria. *Population Studies: a Journal of Demography*, **54**, 169–176.

Domar, A. D., Seibel, M. M. and Benson, H. (1990). The Mind Body Program for Infertility – a new behavioral treatment approach for women with infertility. *Fertility and Sterility*, **53**, 246–249.

Domar, A. D., Clapp, D., Slawsby, E. A., et al. (2000). Impact of group psychological interventions on pregnancy rates in infertile women. *Fertility and Sterility*, **73**, 805–811.

Dribe, M. (2004). Long-term effects of childbearing on mortality: evidence from pre-industrial Sweden. *Population Studies: a Journal of Demography*, **58**, 297–310.

Eaton, S. B., Konner, M. and Shostak, M. (1988). Stone agers in the fast lane: chronic degenerative diseases in evolutionary perspective. *American Journal of Medicine*, **84**, 739–749.

Eaton, S. B. and Eaton S. B. III, (1999). Breast cancer in evolutionary context. In *Evolutionary Medicine*, W. R. Trevathan, E. O. Smith and J. J. Mckenna (eds). New York: Oxford University Press, pp. 429–442.

Eaton, S. B., Strassman, B. I., Nesse, R. M., et al. (2002). Evolutionary health promotion. *Preventive Medicine*, **34**, 109–118.

Eissa, M. K., Obhrai, M. S., Docker, M. F., et al. (1986). Follicular growth and endocrine profiles in spontaneous and induced conception cycles. *Fertility and Sterility*, **45**, 191–195.

Ellison, P. T. (1982). Skeletal growth, fatness and menarcheal age: a comparison of two hypotheses. *Human Biology*, **54**, 269–281.

Ellison, P. T. (1990). Human ovarian function and reproductive ecology: new hypotheses. *American Anthropologist*, **92**, 933–952.

Ellison, P. T. (1994). Salivary steroids and natural variation in human ovarian function. *Annals of the New York Academy of Sciences*, **709**, 287–298.

Ellison, P. T. (2003a). Energetics and reproductive effort. *American Journal of Human Biology*, **15**, 342–351.

Ellison, P. T. (2003b). *On Fertile Ground*. Cambridge, MA: Harvard University Press.

Ellison, P. T. (2005). Evolutionary perspectives on the fetal origins hypothesis. *American Journal of Human Biology*, **17**, 113–118.

Ellison, P. T. and Lager, C. (1985). Exercise-induced menstrual disorders. *New England Journal of Medicine*, **313**, 825–826.

Ellison, P. T. and Lager, C. (1986). Moderate recreational running is associated with lowered salivary progesterone profiles in women. *American Journal of Obstetrics and Gynecology*, **154**, 1000–1003.

Ellison, P. T., Peacock, N. R. and Lager, C. (1986). Salivary progesterone and luteal function in two low-fertility populations of northeast Zaire. *Human Biology*, **58**, 473–483.

Ellison, P. T., Lipson, S. F., O'Rourke, M. T., et al. (1993). Population variation in ovarian function. *Lancet*, **342**, 433–434.

Ellison, P. T., Lipson, S. F., Jasienska, G., et al. (2007). Moderate anxiety, whether acute or chronic, is not associated with ovarian suppression in healthy, well-nourished, Western women. *American Journal of Physical Anthropology*, **134**, 513–519.

Evans, D. J., Hoffmann, R. G., Kalkhoff, R. K., et al. (1983). Relationship of androgenic activity to body-fat topography, fat-cell morphology, and metabolic aberrations in premenopausal women. *Journal of Clinical Endocrinology and Metabolism*, **57**, 304–310.

Falsetti, L., Pasinetti, E., Mazzani, M. D., et al. (1992). Weight-loss and menstrual-cycle – clinical and endocrinologic evaluation. *Gynecological Endocrinology*, **6**, 49–56.

Feicht, C. B., Johnson, T. S., Martin, B. J., et al. (1978). Secondary amenorrhoea in athletes. *Lancet*, **26**, 1145–1146.

Feigelson, H. S., Shames, L. S., Pike, M. C., et al. (1998). Cytochrome p450c17α gene (*CYP17*) polymorphism is associated with serum estrogen and progesterone concentrations. *Cancer Research*, **58**, 585–587.

Frisch, R. E. (1984). Body-fat, puberty and fertility. *Biological Reviews of the Cambridge Philosophical Society*, **59**, 161–188.

Furberg, A. S., Jasienska, G., Bjurstam, N., et al. (2005). Metabolic and hormonal profiles: HDL cholesterol as a plausible biomarker of breast cancer risk. The Norwegian EBBA study. *Cancer Epidemiology, Biomarkers and Prevention*, **14**, 33–40.

Galletly, C., Clark, A., Tomlinson, L., et al. (1996). Improved pregnancy rates for obese, infertile women following a group treatment program – an open pilot study. *General Hospital Psychiatry*, **18**, 192–195.

Gann, P. H., Giovanazzi, S., Van Horn, L., et al. (2001). Saliva as a medium for investigating intra- and interindividual differences in sex hormone levels in premenopausal women. *Cancer Epidemiology, Biomarkers and Prevention*, **10**, 59–64.

Garcia-Closas, M., Herbstman, J., Schiffman, M., et al. (2002). Relationship between serum hormone concentrations, reproductive history, alcohol consumption and genetic polymorphisms in pre-menopausal women. *International Journal of Cancer*, **102**, 172–178.

George, D., Everson, P., Stevenson, J., et al. (2000). Birth intervals and early childhood mortality in a migrating Mennonite community. *American Journal of Human Biology*, **12**, 50–63.

Gluckman, P. D. and Hanson, M. A. (2004). Living with the past: evolution, development, and patterns of disease. *Science*, **305**, 1733–1736.

Gluckman, P. D. and Hanson, M. A. (2005). *The Fetal Matrix: Evolution, Development and Disease.* Cambridge: Cambridge University Press.

Gluckman, P. D., Cutfield, W., Hofman, P., et al. (2005a). The fetal, neonatal, and infant environments – the long-term consequences for disease risk. *Early Human Development*, **81**, 51–59.

Gluckman, P. D., Hanson, M. A. and Spencer, H. G. (2005b). Predictive adaptive responses and human evolution. *Trends in Ecology and Evolution*, **20**, 527–533.

Grodstein, F., Goldman, M. B. and Cramer, D. W. (1994). Body-mass index and ovulatory infertility. *Epidemiology*, **5**, 247–250.

Hackney, A. C. and Viru, A. (1999). Twenty-four-hour cortisol response to multiple daily exercise sessions of moderate and high intensity. *Clinical Physiology*, **19**, 178–182.

Haiman, C. A., Dossus, L., Setiawan, V. W., et al. (2007). Genetic variation at the CYP19A1 locus predicts circulating estrogen levels but not breast cancer risk in postmenopausal women. *Cancer Research*, **67**, 1893–1897.

Harrison, R. F., Omoore, R. R. and Omoore, A. M. (1986). Stress and fertility – some modalities of investigation and treatment in couples with unexplained infertility in Dublin. *International Journal of Fertility*, **31**, 153–159.

Helle, S., Lummaa, V. and Jokela, J. (2002). Sons reduced maternal longevity in preindustrial humans. *Science*, **296**, 1085.

Homan, G. F., Davies, M. and Norman, R. (2007). The impact of lifestyle factors on reproductive performance in the general population and those undergoing infertility treatment: a review. *Human Reproduction Update*, **13**, 209–223.

Hong, C. C., Thompson, H. J., Jiang, C., et al. (2004). Association between the T27C polymorphism in the cytochrome P450c17α (*CYP17*) gene and risk factors for breast cancer. *Breast Cancer Research and Treatment*, **88**, 217–230.

Howe, J. C., Rumpler, W. V. and Seale, J. L. (1993). Energy expenditure by indirect calorimetry in premenopausal women: Variation within one menstrual cycle. *Journal of Nutritional Biochemistry*, **4**, 268–273.

Ibanez, L., Potau, N. and De Zegher, F. (2000a). Ovarian hyporesponsiveness to follicle stimulating hormone in adolescent girls born small for gestational age. *Journal of Clinical Endocrinology and Metabolism*, **85**, 2624–2626.

Ibanez, L., Potau, N., Enriquez, G., et al. (2000b). Reduced uterine and ovarian size in adolescent girls born small for gestational age. *Pediatric Research*, **47**, 575–577.

Ibanez, L., Potau, N., Ferrer, A., et al. (2002). Reduced ovulation rate in adolescent girls born small for gestational age. *Journal of Clinical Endocrinology and Metabolism*, **87**, 3391–3393.

Jacobson, J. D. and Ansari, M. A. (2004). Immunomodulatory actions of gonadal steroids may be mediated by gonadotropin-releasing hormone. *Endocrinology*, **145**, 330–336.

Jasienska, G. (1996). Energy expenditure and ovarian function in Polish rural women. PhD thesis, Harvard University, Cambridge, MA.

Jasienska, G. (2001). Why energy expenditure causes reproductive suppression in women. An evolutionary and bioenergetic perspective. In *Reproductive Ecology and Human Evolution*, P. T. Ellison (ed.). New York: Aldine de Gruyter, pp. 59–85.

Jasienska, G. (2002). Are sex steroids suitable as biomarkers of breast cancer risk? Variation and repeatability of estimates. *International Journal of Cancer*, **13**(suppl.), 129.

Jasienska, G. (2003). Energy metabolism and the evolution of reproductive suppression in the human female. *Acta Biotheoretica*, **51**, 1–18.

Jasienska, G. (2009). Low birth weight of contemporary African Americans: intergenerational effect of slavery? *American Journal of Human Biology*, **21**, 16–24.

Jasienska, G. and Ellison, P. T. (1998). Physical work causes suppression of ovarian function in women. *Proceedings of the Royal Society of London. Series B*, **265**, 1847–1851.

Jasienska, G. and Ellison, P. T. (2004). Energetic factors and seasonal changes in ovarian function in women from rural Poland. *American Journal of Human Biology*, **16**, 563–580.

Jasienska, G. and Jasienski, M. (2008). Interpopulation, interindividual, intercycle, and intracycle natural variation in progesterone levels: a quantitative assessment and implications for population studies. *American Journal of Human Biology*, **20**, 35–42.

Jasienska, G. and Thune, I. (2001a). Lifestyle, hormones, and risk of breast cancer. *British Medical Journal*, **322**, 586–587.

Jasienska, G. and Thune, I. (2001b). Lifestyle, progesterone and breast cancer. *British Medical Journal*, **323**, 1002.

Jasienska, G., Thune, I. and Ellison, P. T. (2000). Energetic factors, ovarian steroids and the risk of breast cancer. *European Journal of Cancer Prevention*, **9**, 231–239.

Jasienska, G., Ziomkiewicz, A., Ellison, P. T., et al. (2004). Large breasts and narrow waists indicate high reproductive potential in women. *Proceedings of the Royal Society of London. Series B*, **271**, 1213–1217.

Jasienska, G., Kapiszewska, M., Ellison, P. T., et al. (2006a). CYP17 genotypes differ in salivary 17-β estradiol levels: a study based on hormonal profiles from entire menstrual cycles. *Cancer Epidemiology Biomarkers and Prevention*, **15**, 2131–2135.

Jasienska, G., Lipson, S. F., Ellison, P. T., et al. (2006b). Symmetrical women have higher potential fertility. *Evolution and Human Behavior*, **27**, 390–400.

Jasienska, G., Nenko, I. and Jasienski, M. (2006c). Daughters increase longevity of fathers, but daughters and sons equally reduce longevity of mothers. *American Journal of Human Biology*, **18**, 422–425.

Jasienska, G., Thune, I. and Ellison, P. T. (2006d). Fatness at birth predicts adult susceptibility to ovarian suppression: an empirical test of the Predictive Adaptive Response hypothesis. *Proceedings of the National Academy of Sciences of the United States of America*, **103**, 12759–12762.

Jasienska, G., Ziomkiewicz, A., Lipson, S. F., et al. (2006e). High ponderal index at birth predicts high estradiol levels in adult women. *American Journal of Human Biology*, **18**, 133–140.

Jasienska, G., Ziomkiewicz, A., Thune, I., et al. (2006f). Habitual physical activity and estradiol levels in women of reproductive age. *European Journal of Cancer Prevention*, **15**, 439–445.

Johnson, W. G., Corrigan, S. A., Lemmon, C. R., et al. (1994). Energy regulation over the menstrual cycle. *Physiology and Behavior*, **56**, 523–527.

Jones, B. C., Little, A. C., Penton-Voak, I. S., et al. (2001). Facial symmetry and judgements of apparent health – support for a "good genes" explanation of the attractiveness-symmetry relationship. *Evolution and Human Behavior*, **22**, 417–429.

Kapiszewska, M., Miskiewicz, M., Ellison, P. T., et al. (2006). High tea consumption diminishes salivary 17 β-estradiol concentration in Polish women. *British Journal of Nutrition*, **95**, 989–995.

Key, T. J. A. and Pike, M. C. (1988). The role of oestrogens and progestagens in the epidemiology and prevention of breast cancer. *European Journal of Clinical Oncology*, **24**, 29–43.

Kovacs, E. J., Messingham, K. A. N. and Gregory, M. S. (2002). Estrogen regulation of immune responses after injury. *Molecular and Cellular Endocrinology*, **193**, 129–135.

Kuiper, G., Lemmen, J. G., Carlsson, B., et al. (1998). Interaction of estrogenic chemicals and phytoestrogens with estrogen receptor β. *Endocrinology*, **139**, 4252–4263.

Kusin, J. A., Kardjati, S., Renqvist, U., et al. (1992). Reproduction and maternal nutrition in Madura, Indonesia. *Tropical and Geographical Medicine*, **44**, 248–255.

Kuzawa, C. W. (2005). Fetal origins of developmental plasticity: are fetal cues reliable predictors of future nutritional environments? *American Journal of Human Biology*, **17**, 5–21.

Lager, C. and Ellison, P. T. (1990). Effect of moderate weight loss on ovarian function assessed by salivary progesterone measurements. *American Journal of Human Biology*, **2**, 303–312.

Lake, J. K., Power, C. and Cole, T. J. (1997). Women's reproductive health: the role of body mass index in early and adult life. *International Journal of Obesity*, **21**, 432–438.

Lawrence, M. and Whitehead, R. G. (1988). Physical activity and total energy expenditure in child-bearing Gambian women. *European Journal of Clinical Nutrition*, **42**, 145–160.

Lawrence, M., Coward, W. A., Lawrence, F., et al. (1987). Fat gain during pregnancy in rural African women: the effect of season and dietary status. *American Journal of Clinical Nutrition*, **45**, 1442–1450.

Lechtig, A., Yarbrough, C., Delgado, H., et al. (1975). Influence of maternal nutrition on birth weight. *American Journal of Clinical Nutrition*, **28**, 1223–1233.

Lenton, E. A., Lawrence, G. F., Coleman, R. A., et al. (1983). Individual variation in gonadotrophin and steroid concentration and in lengths of the follicular and luteal phases in women with regular menstrual cycles. *Clinical Reproduction and Fertility*, **2**, 143–150.

Lipson, S. F. (2001). Metabolism, maturation, and ovarian function. In *Reproductive Ecology and Human Evolution*, P. T. Ellison (ed.). New York: Aldine de Gruyter, pp. 235–248.

Lipson, S. F. and Ellison, P. T. (1996). Comparison of salivary steroid profiles in naturally occurring conception and non-conception cycles. *Human Reproduction*, **11**, 2090–2096.

Little, M. A., Leslie, P. W. and Campbell, K. L. (1992). Energy reserves and parity of nomadic and settled Turkana women. *American Journal of Human Biology*, **4**, 729–738.

Lu, Y. C., Bentley, G. R., Gann, P. H., et al. (1999). Salivary estradiol and progesterone levels in conception and non-conception cycles in women: evaluation of a new assay for salivary estradiol. *Fertility and Sterility*, **71**, 863–868.

Lunn, P. G. (1994). Lactation and other metabolic loads affecting human reproduction. *Annals of the New York Academy of Sciences*, **709**, 77–85.

Lurie, G., Maskarinec, G., Kaaks, R., et al. (2005). Association of genetic polymorphisms with serum estrogens measured multiple times during a two-year period in premenopausal women. *Cancer Epidemiology Biomarkers and Prevention*, **14**, 1521–1527.

Manning, J. T., Scutt, D., Whitehouse, G. H., et al. (1997). Breast asymmetry and phenotypic quality in women. *Evolution and Human Behavior*, **18**, 223–236.

Manning, J. T., Scutt, D., Wilson, J., et al. (1998). The ratio of second to fourth digit length: a predictor of sperm numbers and concentrations of testosterone, luteinizing hormone and oestrogen. *Human Reproduction*, **13**, 3000–3004.

Manning, J. T., Bundred, P. E. and Flanagan, B. F. (2002). The ratio of second to fourth digit length: a proxy for transactivation activity of the androgen receptor gene? *Medical Hypotheses*, **59**, 334–336.

McIntyre, M. H., Chapman, J. F., Lipson, S. F., et al. (2007). Index-to-ring finger length ratio (2D:4D) predicts levels of salivary estradiol, but not progesterone, over the menstrual cycle. *American Journal of Human Biology*, **19**, 434–436.

McNamara, J. P. (1995). Role and regulation of metabolism in adipose tissue during lactation. *Journal of Nutritional Biochemistry*, **6**, 120–129.

McNeely, M. J. and Soules, M. R. (1988). The diagnosis of luteal phase deficiency: a critical review. *Fertility and Sterility*, **50**, 1–9.

Meijer, G. A. L., Westerterp, K. R., Saris, W. H. M., et al. (1992). Sleeping metabolic rate in relation to body composition and the menstrual cycle. *American Journal of Clinical Nutrition*, **55**, 637–640.

Merchant, K. S. and Martorell, R. (1988). Frequent reproductive cycling: does it lead to nutritional depletion of mothers? *Progress in Food and Nutrition Science*, **12**, 339–369.

Miller, J. E., Rodriguez, G. and Pebley, A. R. (1994). Lactation, seasonality, and mother's postpartum weight change in Bangladesh: an analysis of maternal depletion. *American Journal of Human Biology*, **6**, 511–524.

Moller, A. P. and Swaddle, J. P. (1997). *Asymmetry, Developmental Stability, and Evolution*. Oxford: Oxford University Press.

Moller, A. P., Soler, M. and Thornhill, R. (1995). Breast asymmetry, sexual selection, and human reproductive success. *Ethology and Sociobiology*, **16**, 207–219.

Moran, C., Hernandez, E., Ruiz, J. E., et al. (1999). Upper body obesity and hyperinsulinemia are associated with anovulation. *Gynecologic and Obstetric Investigation*, **47**, 1–5.

National Academy of Sciences Committee on Population (1989). *Contraception and Reproduction: Health*

Consequences for Women and Children in the Developing World. Washington, DC: National Academy Press.

Nepomnaschy, P. A., Welch, K. B., Mcconnell, D. S., et al. (2006). Cortisol levels and very early pregnancy loss in humans. *Proceedings of the National Academy of Sciences of the United States of America*, **103**, 3938–3942.

Nguyen, T. V., Jones, G., Sambrook, P. N., et al. (1995). Effects of estrogen exposure and reproductive factors on bone mineral density and osteoporotic fractures. *Journal of Clinical Endocrinology and Metabolism*, **80**, 2709–2714.

Norman, R. J. and Clark, A. M. (1998). Obesity and reproductive disorders: a review. *Reproduction Fertility and Development*, **10**, 55–63.

Norman, R. J., Noakes, M., Wu, R. J., et al. (2004). Improving reproductive performance in overweight/obese women with effective weight management. *Human Reproduction Update*, **10**, 267–280.

Nunez-De La Mora, A., Chatterton, R. T., Choudhury, O. A., et al. (2007). Childhood conditions influence adult progesterone levels. *PLoS Medicine*, **4**, e167.

Painter, R. C., Roseboom, T. J. and Bleker, O. P. (2005). Prenatal exposure to the Dutch famine and disease in later life: an overview. *Reproductive Toxicology*, **20**, 345–352.

Panter-Brick, C. (1993). Seasonality and levels of energy expenditure during pregnancy and lactation for rural Nepali women. *American Journal of Clinical Nutrition*, **57**, 620–628.

Panter-Brick, C. (1996). Physical activity, energy stores, and seasonal energy balance among men and women in Nepali households. *American Journal of Human Biology*, **8**, 263–274.

Panter-Brick, C., Lotstein, D. S. and Ellison, P. T. (1993). Seasonality of reproductive function and weight loss in rural Nepali women. *Human Reproduction*, **8**, 684–690.

Paul, S. N., Kato, B. S., Hunkin, J. L., et al. (2006). The big finger: the second to fourth digit ratio is a predictor of sporting ability in women. *British Journal of Sports Medicine*, **40**, 981–983.

Peter, I., Kelley-Hedgepeth, A., Fox, C. S., et al. (2008). Variation in estrogen-related genes associated with cardiovascular phenotypes and circulating estradiol, testosterone, and dehydroepiandrosterone sulfate levels. *Journal of Clinical Endocrinology and Metabolism*, **93**, 2779–2785.

Pike, I. L. (1999). Age, reproductive history, seasonality, and maternal body composition during pregnancy for nomadic Turkana of Kenya. *American Journal of Human Biology*, **11**, 658–672.

Pike, I. L. (2000). Pregnancy outcome for nomadic Turkana pastoralists of Kenya. *American Journal of Physical Anthropology*, **113**, 31–45.

Pike, M. C., Krailo, M. D., Henderson, B. E., et al. (1983). "Hormonal" risk factors, "breast tissue age" and the age-incidence of breast cancer. *Nature*, **303**, 767–770.

Pike, M. C., Spicer, D. V., Dahmoush, L., et al. (1993). Estrogens, progestogens, normal breast cell proliferation, and breast cancer risk. *Epidemiological Reviews*, **15**, 17–35.

Pirke, K. M., Schweiger, V., Lemmel, W., et al. (1985). The influence of dieting on the menstrual cycle of healthy young women. *Journal of Clinical Endocrinology and Metabolism*, **60**, 1174–1179.

Pirke, K. M., Wuake, W. and Schweiger, U. (eds) (1989). *The Menstrual Cycle and its Disorders*. Berlin: Springer-Verlag.

Poppitt, S. D., Prentice, A. M., Jequier, E., et al. (1993). Evidence of energy sparing in Gambian women during pregnancy: a longitudinal study using whole-body calorimetry. *American Journal of Clinical Nutrition*, **57**, 353–364.

Poppitt, S. D., Prentice, A. M., Goldberg, G. R., et al. (1994). Energy-sparing strategies to protect human fetal growth. *American Journal of Obstetrics and Gynecology*, **171**, 118–125.

Prentice, A. M. and Prentice, A. (1990). Maternal energy requirements to support lactation. In *Breastfeeding, Nutrition, Infection and Infant Growth in Developed and Emerging Countries*, S. A. Atkinson, L. A. Hanson and R. K. Chandra (eds). St. John's, Newfoundland, Canada: ARTS Biomedical Publishers and Distributors, pp. 69–86.

Prior, J. C. (1985). Luteal phase defects and anovulation: adaptive alterations occurring with conditioning exercise. *Seminars in Reproductive Endocrinology*, **3**, 27–33.

Prior, J. C., Cameron, K., Yuen, B. H., et al. (1982). Menstrual cycle changes with marathon training: anovulation and short luteal phase. *Canadian Journal of Applied Sport Sciences*, **7**, 173–177.

Prior, J. C., Vigna, Y. M. and Mckay, D. W. (1992). Reproduction for the athletic woman: new understandings of physiology and management. *Sports Medicine*, **14**, 190–199.

Rich-Edwards, J. W., Goldman, M. B., Willett, W. C., et al. (1994). Adolescent body-mass index and infertility caused by ovulatory disorder. *American Journal of Obstetrics and Gynecology*, **171**, 171–177.

Roberts, S. B., Paul, A. A., Cole, T. J., et al. (1982). Seasonal changes in activity, birth weight and lactational performance in rural Gambian women. *Transactions of the Royal Society for Tropical Medicine and Hygiene*, **76**, 668–678.

Rosetta, L. (1993). Female reproductive dysfunction and intense physical training. *Oxford Reviews in Reproductive Biology*, **15**, 113–141.

Rosetta, L. (2002). Female fertility and intensive physical activity. *Science and Sports*, **17**, 269–277.

Rosetta, L., Harrison, G. A. and Read, G. F. (1998). Ovarian impairments of female recreational distance runners during a season of training. *Annals of Human Biology*, **25**, 345–357.

Santoro, N., Goldsmith, L. T., Heller, D., et al. (2000). Luteal progesterone relates to histological endometrial maturation in fertile women. *Journal of Clinical Endocrinology and Metabolism*, **85**, 4207–4211.

Schweiger, U., Tuschl, R. J., Laessle, R. G., et al. (1989). Consequences of dieting and exercise on menstrual function in normal weight women. In *The Menstrual Cycle and its Disorders*, K. M. Pirke, W. Wuttke and U. Schweiger (eds). Berlin: Springer-Verlag, pp. 142–149.

Sellen, D. W. (2000). Seasonal ecology and nutritional status of women and children in a Tanzanian pastoral community. *American Journal of Human Biology*, **12**, 758–781.

Setchell, K. D. R. and Cassidy, A. (1999). Dietary isoflavones: biological effects and relevance to human health. *Journal of Nutrition*, **129**, 758S–767S.

Shangold, M., Freeman, R., Thysen, B., et al. (1979). The relationship between long-distance running, plasma progesterone, and luteal phase length. *Fertility and Sterility*, **31**, 130–133.

Singh, D. (1993). Body shape and women's attractiveness – the critical role of waist-to-hip ratio. *Human Nature – an Interdisciplinary Biosocial Perspective*, **4**, 297–321.

Siniarska, A. (1992). Socio-economic conditions of the family and somatic and physiological properties of parents and offspring. *Studies in Human Ecology*, **10**, 139–154.

Sjodin, A. M., Forslund, A. H., Westerterp, K. R., et al. (1996). The influence of physical activity on BMR. *Medicine and Science in Sports and Exercise*, **28**, 85–91.

Small, C. M., Marcus, M., Sherman, S. L., et al. (2005). *CYP17* genotype predicts serum hormone levels among pre-menopausal women. *Human Reproduction*, **20**, 2162–2167.

Smeenk, J. M. J., Verhaak, C. M., Eugster, A., et al. (2001). The effect of anxiety and depression on the outcome of in-vitro fertilization. *Human Reproduction*, **16**, 1420–1423.

Sowers, M. R., Jannausch, M. L., Mcconnell, D. S., et al. (2006). Endogenous estradiol and its association with estrogen receptor gene polymorphisms. *American Journal of Medicine*, **119**, 16–22.

Strassmann, B. I. (1996). Energy economy in the evolution of menstruation. *Evolutionary Anthropology*, **5**, 157–164.

Strassmann, B. I. (1997). The biology of menstruation in *Homo sapiens*: total lifetime menses, fecundity, and nonsynchrony in a natural-fertility population. *Current Anthropology*, **38**, 123–129.

Sukalich, S., Lipson, S. F. and Ellison, P. T. (1994). Intra and interwomen variation in progesterone profiles. *American Journal of Physical Anthropology*, **18**(suppl.), 191.

Tovee, M. J., Maisey, D. S., Emery, J. L., et al. (1999). Visual cues to female physical attractiveness. *Proceedings of the Royal Society of London. Series B*, **266**, 211–218.

Tracer, D. P. (1991). Fertility-related changes in maternal body composition among the Au of Papua New Guinea. *American Journal of Physical Anthropology*, **85**, 393–406.

Travis, R. C., Churchman, M., Edwards, S. A., et al. (2004). No association of polymorphisms in *CYP17*, *CYP19*, and *HSD17-B1* with plasma estradiol concentrations in 1090 British women. *Cancer Epidemiology Biomarkers and Prevention*, **13**, 2282–2284.

Tuschen-Caffier, B., Florin, I., Krause, W., et al. (1999). Cognitive-behavioral therapy for idiopathic infertile couples. *Psychotherapy and Psychosomatics*, **68**, 15–21.

Tworoger, S. S., Chubak, J., Aiello, E. J., et al. (2004). Association of *CYP17*, *CYP19*, *CYP1B1*, and *COMT* polymorphisms with serum and urinary sex hormone concentrations in postmenopausal women. *Cancer Epidemiology Biomarkers and Prevention*, **13**, 94–101.

Valeggia, C. and Ellison, P. T. (2004). Lactational amenorrhoea in well-nourished Toba women of Formosa, Argentina. *Journal of Biosocial Science*, **36**, 573–595.

Venners, S. A., Liu, X., Perry, M. J., et al. (2006). Urinary estrogen and progesterone metabolite concentrations in menstrual cycles of fertile women with non-conception, early pregnancy loss or clinical pregnancy. *Human Reproduction*, **21**, 2272–2280.

Vermeulen, A. (1993). Environment, human reproduction, menopause, and andropause. *Environmental Health Perspectives*, **2**, 91–100.

Vigersky, R. A., Anderson, A. E., Thompson, R. H., et al. (1977). Hypothalamic dysfunction in secondary amenorrhea associated with simple weight loss. *New England Journal of Medicine*, **297**, 1141–1145.

Vihko, R. and Apter, D. (1984). Endocrine characteristics of adolescent menstrual cycles – impact of early menarche. *Journal of Steroid Biochemistry and Molecular Biology*, **20**, 231–236.

Vitzthum, V. J. (2001). Why not so great is still good enough. Flexible responsiveness in human reproductive functioning. In *Reproductive Ecology and Human Evolution*, P. T. Ellison (ed.). New York: Aldine de Gruyter, pp. 179–202.

Vitzthum, V. J., Spielvogel, H. and Thornburg, J. (2004). Interpopulational differences in progesterone levels during conception and implantation in humans. *Proceedings of the National Academy of Sciences of the United States of America*, **101**, 1443–1448.

Westberg, L., Baghaei, F., Rosmond, R., et al. (2001). Polymorphisms of the androgen receptor gene and the estrogen receptor β gene are associated with androgen levels in women. *Journal of Clinical Endocrinology and Metabolism*, **86**, 2562–2568.

Winkvist, A., Rasmussen, K. M. and Habicht, J. P. (1992). A new definition of maternal depletion syndrome. *American Journal of Public Health*, **82**, 691–694.

Xu, X., Duncan, A. M., Merz, B. E., et al. (1998). Effects of soy isoflavones on estrogen and phytoestrogen metabolism in premenopausal women. *Cancer Epidemiology Biomarkers and Prevention*, **7**, 1101–1108.

Yue, W., Santen, R. J., Wang, J. P., et al. (2003). Genotoxic metabolites of estradiol in breast: potential mechanism of estradiol induced carcinogenesis. *Journal of Steroid Biochemistry and Molecular Biology*, **86**, 477–486.

Zaadstra, B. M., Seidell, J. C., Vannoord, P. A. H., et al. (1993). Fat and female fecundity – prospective study of effect of body fat distribution on conception rates. *British Medical Journal*, **306**, 484–487.

Ziomkiewicz, A. (2006) [Anthropometric correlates of the concentration of progesterone and estradiol in menstrual cycles of women age 24–37 living in rural and urban area of Poland]. PhD thesis, Jagiellonian University, Krakow, Poland.

Ziomkiewicz, A., Ellison, P. T., Lipson, S. F., et al. (2008). Body fat, energy balance and estradiol levels: a study based on hormonal profiles from complete menstrual cycles. *Human Reproduction*, **23**, 2555–2563.

20 Pregnancy and Lactation

Ivy L. Pike and Lauren A. Milligan

In this chapter, we draw on life history theory to create a comparative framework for exploring pregnancy and lactation as coevolved elements of the primate reproductive strategy, broadly defined. We draw on the physiological evidence to create a picture of how pregnancy and lactation represent a highly integrative reproductive strategy. Life history theory offers several advantages for understanding this integrative primate reproductive strategy. Firstly, by comparing life history characteristics across nonhuman and human primates we can examine benefits and costs to this energetically expensive strategy and how natural selection may have tailored aspects of the larger strategy toward more species specific ends. Secondly, life history theory emphasizes trade-offs between survival and current reproduction, current and future reproduction, and number, size, and sex of offspring among many others (Stearns, 1992). We take the position that these trade-offs must be examined from the dual vantage point of the mother and the fetus/infant since the two will not always have completely overlapping interests.

The chapter unfolds across four sections. In the first section, we explore the primate reproductive strategy by comparing life history parameters across broad phylogenetic groupings, making note of similarities and differences. The next section examines gestation from the perspective of maternal constraint, maternal investment in reproduction, and physiological mechanisms that might shift the cost of any given pregnancy. Mechanisms that might enhance fetal survival and offer developmental cues are also explored. In the lactation section, lactation is viewed as a highly conserved part of the reproductive strategy but with considerable biological and behavioral flexibility. We conclude by arguing that an evolutionary approach that draws on both biocultural and life span perspectives to examine pregnancy and lactation has important policy implications.

THE PRIMATE REPRODUCTIVE STRATEGY: SITUATING HUMANS IN A BROADER CONTEXT

Life history theory asserts that life history traits have evolved as a suite, and the target of selection is fitness over an individual's lifetime rather than fitness at one point in time or the maximization of one particular life history trait (Alberts and Altmann, 2003). Using this theoretical framework, a species' reproductive strategy is predicted to maximize the mother's fitness, or reproductive success, over her lifetime. However, pregnancy and lactation also affect the fitness of the offspring. The offspring is predicted to maximize its own lifetime fitness, which may or may not be at odds with that of the mother (Trivers, 1974). Given the importance of maternal transfer of nutrients during gestation and that milk is the only source of nutrition for neonates and developing infants, selection is expected to favor a strategy that balances the interests of maternal *and* offspring survival. The pattern and rate of growth and development of the offspring, established in utero, requires the production of sufficient milk quality and quantity. In turn, maternal energy balance and body size determine the mother's ability to provide the necessary resources for growth and development during lactation (Lee, 1999).

Life history traits are products of natural selection and therefore a species' life history strategy is intimately tied to that species' evolutionary history. Natural selection elaborates, rather than innovates, leading to similarities in life history traits among closely related species. Across the mammalian class considerable physiological similarities exist with the important exception of reproduction (Smith, 2007). For example, variation on how placental mammals proceed to parturition varies widely (Smith, 2007). Only anthropoid primates produce corticotrophin-releasing hormone (CRH), a neuropeptide found in amphibians and mammals (Denver, 1997), in the placenta (Smith et al., 1999; Power and Schulkin, 2006). Also, only gorillas, chimps, and humans experience increasing levels of CRH across pregnancy, apparently playing a role in timing parturition (Smith et al., 1999; Bowman et al., 2001; Power et al., 2006). The wide range in age at weaning among primates (e.g., Ross, 2003) indicates that the duration of lactation has been modified by selection over the course of primate evolution. Such data suggest different components of pregnancy and

Human Evolutionary Biology, ed. Michael P. Muehlenbein. Published by Cambridge University Press. © Cambridge University Press 2010.

lactation also would be likely targets of selection over the course of primate evolution. However, similarity of life history traits within genera, families, and even superfamilies indicate a phylogenetic ordering to this variation. A species' reproductive strategy therefore may carry phylogenetic baggage, both in regard to the genetic makeup of inherited trait and the range of variation (phenotypic plasticity) in this expression.

Like other life history traits, the primate reproductive strategy also may be an adaptive response to ecological variation (Morbeck, 1997a, 1997b; Hill and Kaplan, 1999; Kaplan et al., 2000). Morbeck (1997b) argues that external factors such as climate, the quality and quantity of available food, disease ecology, and group structure can affect the timing, duration, and energetic effort of an individual's life history. As an example, mechanisms that modify fecundity and conception have been documented and suggest the ability to delay conception when conditions are poor (Wood, 1994; Ellison, 2001, 2005). Primate mothers may choose to lengthen lactation if ecological conditions are poor, or shorten lactation when conditions are good (Lee, 1996). The extent to which they are able to do this is part of their evolutionary history, however, and the degree of plasticity is itself a product of natural selection (Morbeck, 1997a).

Gestation represents a considerable energetic investment but the daily costs of fetal growth are lower than might be expected among primates when examined across mammals (Dufour and Sauther, 2002; Ulijaszek, 2002) and this may be achieved in part by having longer gestational durations (Martin, 1996). When compared to great apes, humans invest more heavily in fetal tissue, with fetal body fat comprising approximately 61% of the total cost of pregnancy (Dufour and Sauther, 2002; Ulijaszek, 2002). There is considerable flexibility in meeting the energetic demands of gestation (Dufour and Sauther, 2002), with examples of lower basal metabolic costs for females who experience marginally adequate nutritional levels (Durnin, 1987; Prentice and Goldberg, 2000). Ultimately, fetal growth and body composition reflect a compromise between maternal and fetal interests (Wells, 2003).

Lactation is the most energetically expensive component of a mammalian female's reproductive strategy (Pond, 1984, 1997; Gittleman and Thompson, 1988), with energetic costs and micronutrient requirements shifting to reflect infant growth and development trajectories (Picciano, 2003). Like gestation, the cost of lactation is argued to be related to maternal body size and to the evolutionary history of the species (Martin and MacLarnon, 1985). Lee (1996) argues that the energy cost of lactation is inescapable; mothers must convert maternal nutrients or body reserves to milk. Although the conversion of maternal stores or dietary intake to lactation is relatively efficient, most mammalian mothers still need to increase their non-reproductive energy intake above normal levels to sustain infant growth (Lee, 1996). For example, baboon mothers are estimated to require 1.5 times and human females 1.3 times their normal energy intake (non-pregnant and nonlactating) (Lee, 1996, 1999).

The lactation strategy can be divided into four, interrelated traits: the composition of the milk produced, the volume of milk produced, the frequency at which the mother feeds the infant throughout the day, and the total duration of lactation per reproductive cycle. All these elements must work together to deliver the necessary nutrients and tactile stimulation to the infant for its growth and development without irreversibly compromising maternal health. The duration of lactation is considered a life history trait of the species; a trait that exhibits considerable flexibility in timing within and across species. Milk composition, volume of milk produced, and nursing frequency can also be modeled as life history traits. As such, each of the four components of the lactation strategy is subject to natural selection, as each interacts with the environment and other life history traits to maximize the fitness of the individual.

FAT STORAGE, GESTATION, AND LACTATION – IS IT ALL ABOUT THE BRAIN?

The human brain is approximately one-third lipid, all of which must be supplied to fetuses and young infants by the mother in utero and in milk, respectively. A relatively larger brain with a longer period of postnatal growth may have been linked to the selection of an increased ability in human females to store energy and later mobilize this stored energy for lactation. The human brain has "obligatory and inflexible requirements" (Kuzawa, 1998, citing Armstrong, 1983) across a range of nutrients, with some less "inflexible" than others. Development of significant fat stores may be a physiologically linked adaptation to changes in rate and pattern of brain growth (Ulijaszek, 2002). Although large fat stores are found in many female mammals, including nonhuman primates, human female fat stores are argued to be relatively larger than (Dufour and Sauther, 2002) and different from (Pond, 1997) the female primate pattern. Further, the degree of difference in fat mass between males and females is greater in humans than nonhuman primates. While human males are heavier than females, females are fatter, with 34% more fat mass on average than males (Wells, 2006). Wells (2006) argues that the greater fat mass of human females relative to human males and to nonhuman primate females indicates the importance of reproductive energetics in human evolution. Specifically, he proposes that the evolution of fatness in human females is at least partly the result of encephalization.

Dufour and Sauther (2002) report on four non-Western human populations, in which females have greater than 20% body fat. There is little comparative data on nonhuman apes, but data on captive lowland gorillas (Zihlman and McFarland, 2000) show that one captive female gorilla had the potential to store similar amounts of body fat. It is unknown if wild living apes are capable of maintaining the percent body fat of human females. Such data would be invaluable to determine if adult female fat storage is indeed a unique-derived human adaptation to support brain growth.

Another possibility is that the cost of growing a larger brain was not met through alterations to the reproductive strategy at all. Modern human infants have a precocious condition of adipose development at birth (Kuzawa, 1998; Ulijaszek, 2002) and are the fattest species on record at birth (Kuzawa, 1998). While variation in human fatness at birth exists across a range of pregnancy-related circumstances, including maternal weight gain and fat stores during pregnancy, fatness is argued to be an adaptation to the higher lipid needs of newborns for rapid brain growth during the first year of life (Kuzawa, 1998; Ulijaszek, 2002). Like maternal depot fat stores, infant fat stores (deposited primarily during the third trimester) would act as a buffer against disruptions in energy transfer during lactation (Kuzawa, 1998). As reviewed by Kuzawa (1998), this hypothesis is indirectly supported by data. Without comparative data from nonhuman primates, it is not possible to say whether this is a unique human adaptation or a nonadaptive consequence of other ontogenetic changes (such as increased fat storage by the mother).

PREGNANCY

Gestation long has been considered a critical selective point in reproductive success. In addition to low fecundability (Wood, 1994), fetal loss during the first 4 weeks of gestation are estimated at 40–50% (Wood, 1994), with fetal deaths across pregnancy representing a considerable modifier of birth intervals and thus total reproductive potential (Wood, 1994; Holman and Wood, 2001). Methodological constraints focused attention on pregnancy outcomes, such as gestational duration and weight and length at birth, as proxies for the adequacy of the fetal experience. Yet more recent advances in physiology, genetics, endocrinology, and gestational age assessments, have provided opportunities to refine our understanding of the selective pressures and evolutionary relationships between the maternal–fetal unit.

From the maternal perspective, reproduction represents a balance of shorter- and longer-term trade-offs (Ellison, 2001, 2005). Given the energetic cost of pregnancy, maternal investment in any given pregnancy only makes sense if the demands do not threaten survival. If maternal condition, defined as the adequacy of her growth, health, and nutritional experience across the life span, is marginal then investing in an individual pregnancy is weighed against future reproductive opportunity. If opportunity is low, for example due to older maternal age, then investing in gestation now may be worth the risk and may benefit overall reproductive success. If future reproductive opportunity is high, delaying conception until maternal condition or environmental circumstances improve offers a reasonable trade-off. Strategies look different if maternal condition is poor as a result of an acute insult versus chronic energetic deficiency across the life span. In the presence of chronic stress, waiting for maternal condition to improve may be a higher risk strategy. Instead, relying on mechanisms that might limit the cost of individual pregnancies may be a more viable strategy (Prentice and Goldberg, 2000; Pike, 2005).

Several different mechanisms have been proposed for lowering the energetic cost of pregnancy. One well-documented physiological mechanism is to lower basal metabolic rate during gestation, particularly via reduction in diet-induced thermogenesis (Durnin, 1987; Dufour et al.,1999). Another widely examined mechanism is to reduce energy expenditure, particularly during the third trimester when the energy demands of even basic movements are higher (Lawrence et al., 1985; Lawrence and Whitehead, 1988; Panter-Brick, 1989, 1992; Dufour et al., 1999). Although humans exhibit slower fetal growth across longer gestations when compared to other nonhuman primates (Dufour and Sauther, 2002; Ulijaszek, 2002), several lines of research suggest limiting fetal growth may incur energetic savings. Drawing on studies of genomic imprinting and Trivers's (1974) ideas regarding parent–offspring conflict, Haig (1993) suggests there may be maternally derived fetal genes that limit fetal growth, with predictions that paternally derived genes might encourage fetal growth. Recent evidence suggests that maternal neurohormonal cues that signal a poor or insufficient intrauterine environment can accelerate fetal developmental trajectories (McClean et al., 1995; Pike, 2005), creating a "thriftier phenotype" (Hales and Barker, 2001). A shortened gestational duration once the fetus is viable may offer important energetic savings for the mother by limiting the higher costs of fetal fat deposition and the active transfer of maternal nutrients (Peacock, 1991). Such strategies may limit the cost of an individual pregnancy but come with important risks for perinatal and longer-term survival for the infant.

The developmental experiences of the mother, strongly patterned by her environmental context, also set important constraints on fetal growth and development. If a mother experienced intrauterine growth restriction as a fetus, she is more likely to mature at an earlier age (Adair, 2001), is smaller overall in size at maturity (Simondon et al., 1998), and may have a smaller pelvic size, and smaller uterus and ovaries (Ibanez et al., 2000, 2006). Such responses to a poor intrauterine nutritional environment may place limits on adequate fetal growth. There is additional evidence that chronic psychosocial stress across a mother's life span alters her developmental trajectory. In contexts of high psychosocial stress an earlier age at menarche has been documented (Ellis and Garber, 2000; Coall and Chisholm, 2003). These responses may limit the overall cost of a pregnancy and thus offer increased opportunity for reproduction but come with intergenerational consequences that have important long-term public health implications.

From the fetal perspective, one body of literature, in particular, has generated considerable interest in possible mechanisms that allow fetal adaptations to a poor or insufficient maternal environment. This research suggests fetal growth restriction, combined with the modifying effects of early childhood growth and development, may result in permanent alterations in organ and metabolic function that place individuals at greater risk for adult-onset diseases (Barker et al., 1993) (see Chapter 30 of this volume for more details). One closely related body of literature suggests maternal stress, as measured by elevated glucocorticoids and CRH, relays information of a poor or insufficient fetal environment across the placenta to the fetus, initiating a hormonal cascade that accelerates fetal maturation and results in preterm delivery (defined as <37 weeks from last menstrual period). Such research suggests developmental trade-offs that respond to signals of a poor or insufficient environment, resulting in enhanced short-term survival with long-term consequences and risks. For example, in addition to the many consequences of preterm delivery such as increased risk for mortality and increased vulnerability to respiratory diseases, fetal exposure to elevated maternal stress hormones appears to increase reactivity to stress in early childhood and even into adulthood (Brouwers et al., 2001; Weinstock, 2001).

The influence of environment on development has been appreciated since the earliest studies of growth (Tanner, 1998). While research on pregnancy outcomes highlights the importance of the maternal environment on fetal growth, more recent research suggests the embryo/fetus monitors information about the adequacy of the intrauterine environment (Wadhwa et al., 1993, 2001; McLean et al., 1995). Such information may serve as necessary cues that guide environmentally specific fetal developmental trajectories (Wadhwa et al., 2001). Signals of an insufficient environment, in turn, trigger a cascade of facultative responses in the fetus. These responses include an acceleration of fetal organ maturation (Dodic et al., 1999; McGrath and Smith, 2001), maximization of blood flow to the fetus (McLean et al., 1995; Smith et al., 2002), asymmetrical growth resulting from preferential allocation of resources to important growth centers (Fowden, 2001), and when extreme threats are present then a shift in the endocrine cascade leading to parturition may occur allowing early expulsion from the stressful environment (McLean et al., 1995). The full range of potential responses and some of the precise mechanisms remain vague but some intriguing pathways have been illuminated.

Additional evidence for the importance of environmental cues for fetal development comes from links between placental CRH and the role it plays in fetal neurodevelopment (Wadhwa et al., 2001). The central nervous system develops over 11–12 years but the fetal period remains critical to normal development (Rodier et al., 1994). The sensitivity of neurotransmitters in the central nervous system are set during critical periods of development and affect the organisms response to all subsequent experiences (Rodier et al., 1994). Environmental perturbations in utero and the subsequent response in the central nervous system are poorly understood (Wadhwa et al., 2002). However, experimental studies using a series of vibroacoustic stimuli over the fetal head suggest the fetus can detect and habituate to external stimuli by the third trimester (Sandman et al., 1997). From the same study, mother–fetus pairs with elevated CRH levels were tested using the same stimuli and measures of responsiveness (fetal heart rate). The researchers found a blunted response to novel external stimuli in fetuses with high CRH levels (Sandman et al., 1997). If subtly stressful stimuli impact fetal responsiveness there is considerable evidence that maternal anxiety and experiences of traumatic events also have lasting effects on the fetus and infant (Brouwers et al., 2001; Tagle et al., 2007). Thus, the prenatal environment appears to play an important role in appraisal, responsiveness, and habituation in postnatal life (Wadhwa et al., 2002).

The maternal–fetal unit simultaneously experiences competing and overlapping interests. Wells (2003) suggests fetal size and body composition at birth reflect a compromise between maternal and fetal strategies. Constrained fetal growth and altered metabolism may represent a series of facultative responses manipulated by the mother in response to the quality of the environment (Adair, 2001; Wells, 2003). Lampl (2005), however, reviews the evidence for maternal versus fetal adaptations to an insufficient environment and concludes the opposite, that the fetus controls developmental responses to an insufficient environment.

In either case, selective pressure to meet competing evolutionary demands between mother and fetus seems likely and highlights the importance of examining reproduction from a maternal/offspring perspective.

As further evidence of the physiological interdependency between mother and offspring, parturition reflects a transition from active transport of nutrients across the placenta to more passive transport via breast milk. The process of lactogenesis begins in mid-pregnancy as mammary glands differentiate and prepare for secretion (Neville, 2001; Neville et al., 2001). The successful production of milk, defined as lactogenesis stage II, occurs at approximately day four postpartum and is mediated by the neurophysiological experience of labor and delivery (Chen et al., 1999; Neville and Morton, 2001). For example, stressful births that result in cesarian section deliveries are commonly associated with delayed expression of colostrum (Dewey, 2001; Lau, 2001). Early initiation of breastfeeding, defined as suckling immediately following delivery, is aided by maternal breast odor (Porter and Winberg, 1999; Winberg, 2005) and offers considerable benefit to the newborn including aiding the postnatal gastrointestinal-tract transition. The skin-to-skin contact associated with suckling appears to help regulate temperature and blood glucose levels (Winberg, 2005 citing Christenssen et al., 1992). In turn, the close contact and suckling creates physiological responses in the mother. Suckling appears to improve the efficiency of maternal energy conversion through improved use of ingested calories as a result of increases in gastrointestinal-tract hormone release (Winberg, 2005). Suckling also releases higher levels of oxytocin, a hormone thought to have an impact on the maternal brain, especially spatial learning and memory (Kinsley et al., 1999; Monks et al., 2003). These very early interactions between newborn and mother help regulate a more mutual physiological interaction but also appear to play a critical role in the process of mother/infant attachment (Porter and Winberg, 1999; Insel and Young, 2001) and thus reproductive success.

LACTATION

Lactation represents an energetically expensive phase of reproduction for all mammalian mothers (Gittleman and Thompson, 1988; Oftedal and Iverson, 1995). It requires mothers to mobilize and transfer large quantities of nutrients in milk, placing nutrient demands on the mother (Oftedal, 2000). Mammalian mothers meet these demands by increasing the energy in their diet (Altmann, 1983; Forsum et al., 1992; Sauther, 1994; Butte et al., 1999; Oftedal, 2000), reducing energy output (Roberts et al., 1985; Panter-Brick, 1993; Piperata and Dufour, 2007), or by storing nutrients during periods

of abundance (Oftedal, 2000; Pond, 1984). The demands of lactation are met in a variety of ways depending on factors such as the condition of the environment, the species' ontogenetic priorities, and genetically programmed physiological traits, such as fat storage and metabolism (Oftedal, 1984, 2000).

Parent–offspring conflict theory (Trivers, 1974) predicts that mothers and infants should have *behavioral* conflicts over the allocation of parental investment, such as length of lactation. Much empirical work has been done on the issue of weaning conflict, that is, the conflict between mother and infant on the scheduling of when the infant must become nutritionally independent of the mother (see Maestripieri, 2002, for review of literature). Because milk composition represents a large investment by the mother, an extension of Trivers's theory would predict that milk composition would reflect the *physiological* conflict, or compromise, between what the infant wants and what the mother is able to give. Infants may grow faster if mothers produce milk with higher fat concentrations, but mothers may be limited by their physiological energy stores. The amount of fat in the milk of extant mammals, including humans, is thus a compromise between infant energy needs for growth and development and maternal abilities to access fat in the diet, or store fat on her body and mobilize those fat stores during lactation (Oftedal, 2000).

Among primates, the duration of lactation is relative to body size and among many species, including apes, exceeds gestation in energy requirements and length (Harvey et al., 1987; Ross, 1998, 2003). Indeed, the life history of primates is considered one of the "slowest" among mammals (Harvey et al., 1987; Charnov and Berrigan, 1993). Primate milk composition is argued to reflect this derived life history pattern (Oftedal, 1984; Oftedal and Iverson, 1995; Sellen, 2007). Relative to other mammalian orders, primates produce milks that are low in fat, protein, and dry matter, and high in lactose. Within the primate order, however, significant interspecific (Oftedal, 1984; Oftedal and Iverson, 1995; Milligan, 2007; Milligan et al., 2008a; Milligan and Bazinet, 2008) and intraspecific (Lönnerdal et al., 1984; Tilden and Oftedal, 1997; Power et al., 2002, 2008; Hinde, 2007a, 2007b; Milligan et al., 2008b) variation has been identified. Primates diverge from other mammalian orders in the lack of correlation between body size and energy in milk (Power et al., 2002). Variation may instead relate to the rate of somatic growth, with faster-growing primate species producing milks with more energy from protein (Oftedal, 1984; Power et al., 2002; Milligan, 2007). Energy from fat also is highly variable both within and among nonhuman primate species (Milligan, 2007). One explanation for this variability may be infant sex. Primiparous rhesus macaque mothers produce significantly higher fat milks, and thus higher energy milks, for sons compared to daughters (Hinde, 2007b). Variation may also relate to maternal energy balance. Old and New World monkey mothers in positive energy balance may be able to convert excess energy intake into milk energy, producing milks with significantly more energy than predicted for a primate (Milligan, 2007). As is true in humans, milk fat is the most variable component of nonhuman primate milk composition and the most malleable with respect to ecological variation (Oftedal, 1984; Oftedal and Iverson, 1995; Power et al., 2002, 2008; Milligan et al., 2008a). These results suggest significant flexibility among nonhuman primate mothers in milk composition. Intraspecific variation in age at weaning (Harvey et al., 1987; Ross, 1998, 2003) further supports the picture of a primate lactation strategy that is extremely flexible and responsive to current ecological conditions and future reproductive events.

Like nonhuman primate milks, human milk is dilute, low in fat and protein, and high in lactose (Jenness, 1979; Lönnerdal and Atkinson, 1995; Oftedal and Iverson, 1995; Prentice, 1996). Humans also exhibit variability in milk composition (Prentice, 1995), flexibility in the duration of lactation (Dettwyler, 2004; Kennedy, 2005) and flexibility in meeting the costs of lactation (Forsum et al., 1992; Panter-Brick, 1993; Butte et al., 1999; Piperata and Dufour, 2007). As argued by Dufour and Sauther (2002), the human lactation strategy most likely differs from nonhuman primates in degree, rather than kind.

Several aspects of human milk composition and the larger human lactation strategy have been argued to be derived traits in humans. These include the concentration and/or composition of long-chain polyunsaturated fatty acids (LCPUFA) (Martin, 1983; Vasey and Walker, 2001), the use of complementary foods during weaning (Sellen, 2007), and immunological components (Slade and Schwartz, 1987; Goldman, 1993; Goldman et al., 1998).

In a test of Martin's hypothesis for unique LCPUFA composition of human milk, Milligan et al. (2008a) and Milligan and Bazinet (2008) found that the range of LCPUFA composition of anthropoid milks, including the highly encephalized *Cebus* monkey and several species of hominoid, was identical to cross-cultural findings on human milk LCPUFA. Like human milk, nonhuman primate milks are highly variable in docosahexaenoic acid (DHA) composition with respect to dietary intake of DHA. Higher levels of milk DHA and other LCPUFA are predicted in any primate population that consumes foods with preformed sources of DHA (Milligan and Bazinet, 2008).

Sellen (2007) argues that a flexible complementary feeding strategy is a derived feature of human lactation. Complementary feeding describes the transition from exclusive breast-feeding to the inclusion of other food items. This adaptation is tied to timing and flexibility of weaning. Taken together, these adaptations allow human mothers to resolve trade-offs between the high energetic cost of lactation and infant morbidity and mortality. While complementary feeding may be a critical aspect of the human lactation strategy, it may not be a unique attribute. Among nonhuman primates, the transition from milk to nutritional independence is a process rather than a particular point in time, with infants nursing long after the inclusion of other food items into the diet (Lee et al., 1991; Lee, 1996, 1999).

The high concentration of immune factors in human milk is well documented but little comparative work has been done to determine if this is species specific, or a primitive aspect of primate lactation (but see Lönnerdal et al., 1984). Research on rhesus macaque milk (Milligan, 2005) suggests that human milk may be unique among nonhuman primates in its high concentration of secretory immunoglobulin A (sIgA). The cultural changes associated with agriculture, including increased population density and a more sedentary lifestyle, promoted an increase in infectious diseases, thereby creating a novel ecological setting for human populations (Cohen, 1989; Barrett et al., 1998). Increases in number of pathogens and pathogen virulence would have placed strong selective pressure on the human immune system, particularly the immune system of infants and children and may have selected for increased immune factors in milk to increase neonatal and infant survival (Milligan, 2005).

Milk is generally buffered from maternal condition, with some exceptions

Supplementation experiments are used to determine the effect of diet on milk composition (see Prentice, 1995, for review of literature). Providing malnourished women (determined by a body mass index [BMI] < 18.5) in The Gambia with biscuits containing protein, calcium, and/or carbohydrate resulted in little to no change in milk composition or milk volume (Prentice, 1995). Nutritionally compromised women show similar concentrations of milk protein to well-nourished women (Prentice, 1995; Prentice and Prentice, 1995), suggesting that even low levels of dietary protein intake allow for production of milk with the necessary concentration of protein for optimal human infant (somatic) growth. Indeed, Prentice and Prentice (1995) found the macronutrient content of human breast milk (protein, fat, lactose) to be surprisingly insensitive to maternal dietary differences.

Changes were observed when supplementation involved fatty acids and vitamin B_{12} (Prentice and Prentice, 1995); supplementation of the maternal diet with these factors increases their composition in milk.

Milk fatty acid profiles are most affected by maternal diet, as dietary fat is one source for milk fat. The other two sources of milk fat are de novo synthesis within the mammary gland and depot fat transferred through the maternal bloodstream (Stini et al., 1980; Prentice, 1996). Human milk fatty acid profiles vary within and between populations because they are so intimately tied to dietary habits and maternal condition (Koletzko et al., 1992, 2001; Sanders, 1999). This is especially true for LCPUFA, such as DHA, which are supplied directly from the maternal plasma (from the diet or depot fat stores) or as metabolites of precursor fatty acids in the maternal diet (e.g., α-linolenic acid, ALA). Variation in LCPUFA concentration is attributed directly to dietary differences in foods containing these fatty acids. For example, in a study of 9 human populations, Yuhas et al. (2006) found the percent composition of milk DHA to range from 0.17% to 0.99% of total fatty acids. Those populations with the highest DHA values had higher fish consumption, the primary dietary source for DHA. Vegetarian mothers produced milk that contained little to no fatty acids derived from animal fat and more fatty acids derived from dietary vegetable fat (Finley and Lönnerdal 1985; Dettwyler and Fishman, 1992). Agostoni (2005) offers an interesting perspective on LCPUFA fatty acids in milk. He proposes that fetuses may become accustomed to the supply of fatty acids from their mother during intrauterine life. Infants are "imprinted" with a specific fatty acid pattern during gestation, and their fatty acid metabolism may be "programmed" to the maternal environment and the infant's genetic background.

Alternative sources of fat in milk production allows for contingencies or compensatory actions when environmental conditions do not permit adequate dietary intake of fat. Most important for humans may be the use of depot fat stores during lactation. Human females gain weight during pregnancy by way of fat deposition (McFarland, 1997). Fat that is stored in the hip and the thigh is more metabolically active during lactation than fat in other depots in the body and is also highly resistant to weight loss (McFarland, 1997). When dietary fat is low, lactating women are able to use fat from their depot fat stores as a source for milk fat. Indeed, lactating women on low fat diets produced more milk fat from subcutaneous depot fat rather than from dietary fat (Emmett and Rogers, 1997) or as a result of decreased physical activity (Schutz et al., 1980). Fat content in milk is of particular importance because the human brain is approximately one-third lipid, all of which must be supplied to fetuses and young infants by the mother in utero and in milk, respectively. The importance of maternal transfer of stored fat to the infant during lactation therefore is an extension of the pattern observed during gestation. Placental transfer of fat includes the mobilization of up

to 3–5% of the maternal brain volume during the last trimester (Holdcroft et al., 1997), probably because of the high concentration of essential fatty acids, particularly DHA, in this tissue (Vasey and Walker, 2001). After birth, fat stored in the mother's hips and thighs can be converted into milk; fuel for the metabolic requirements of the growing infant brain.

Investigations on the relationship between maternal body composition and milk composition indicate that there is not a strong relationship between energy balance and milk quality. Although human populations vary with respect to maternal energy balance, human milk fat (and energy) is relatively conserved and appears to be as variable within females (and populations) as between females (and populations) (Jensen et al., 1995; Prentice, 1995). Women who have large stores of depot fat (BMI > 26) were found to produce milk with higher fat concentration than women with low BMIs (<18). However, the higher BMI group produced less milk than the lower BMI group (Barbosa et al., 1997). As a result of this inverse relationship between milk fat concentration and milk volume, the total amount of fat secreted into milk had no significant association with maternal body fat (Butte et al., 1984; Barbosa et al., 1997). Even when milk volume is ignored, the difference in milk fat production between these two groups was small (2.73 ± 0.37% vs. 2.89 ± 0.35%) and well within the range of variation reported for populations of well-nourished women (Jensen et al., 1995). Probably as a consequence, these authors (Barbosa et al., 1997) reported that infant growth velocities were not significantly different between the low and high BMI groups.

Because milk fat is the main energy source for human infants, supplying more than 50% of the calories (Jenness, 1979; Jensen et al., 1995), consistent production of essential milk fat from diverse sources may have been critical to fitness of the infant and the mother. The reliance of reproductive females on stored fat also is established. If dietary intake of fat during lactation is low, human females compensate by utilizing fat from their depot stores (Emmett and Rogers, 1997). Indeed, it is believed that women in undernourished populations are able to continue to lactate through reliance on fat reserves (Jelliffe and Jelliffe, 1978).

It has been suggested that despite evidence that diet and nutritional status have only a small effect on milk composition and yield, there may be a threshold below which the quality and quantity of milk may be compromised (Lönnerdal, 1986; Emmett and Rogers, 1997). Lönnerdal (1986) argues that there must be a lower limit of dietary nutrient intake below which it would be inadequate for milk production. However, of the human populations investigated thus far (Prentice, 1995), breast milk composition and volume seem to be very "well-buffered" (Jensen et al., 1995) against ecological variability (Prentice et al., 1994). For example, Butte et al. (1984) found no relationship between milk quantity or milk quality and maternal anthropometric indices such as maternal weight, height, and body fat. The lack of cross-cultural variation in human milk composition and the relatively small role played by the environment in altering this composition seem to support the view that milk composition in humans is highly conserved.

MILK AS CUES TO DISEASE ECOLOGY

Milk provides more than nutrition to the developing infant. Nonnutritive immune factors that are passed from mother to infant (passive immunity) are necessary both for immediate immunocompetence of the infant, and for enhancement and development of the infant's immune system (Goldman et al., 1982, 1998; Van de Perre, 2003). Additionally, breast milk contains immune factors that stimulate the development of the neonate's intestine and play a critical role in the neonate's development of intestinal host immune defenses (Bines and Walker, 1991; Cruz et al., 1991). The immune factors present in milk are well adapted to the environment in which they are needed: SIgA, lactoferrin, and lysozyme are resistant to digestive enzymes and can persist in the gut without degradation (Cruz et al., 1991; Goldman and Goldblum, 1995; Lönnerdal, 1996). Most micro-organisms enter the neonate through the mucosal tissues, and antibodies present in milk are thus able to react with and provide immunity against pathogens of mucosal surfaces, especially enteric bacteria (Hoshower, 1992; Goldman et al., 1998; Goldman, 2001; Van de Perre, 2003). Indeed, the majority of sIgA antibodies are directed against enteric pathogens (Cruz et al., 1991). What is especially important about these transformed maternal antibodies is that they are specific to antigens that would be recognized by antibodies in the gastrointestinal tract of the mother (Goldman, 2001). In this respect, sIgA antibodies ingested and utilized by the infant through the breast milk will be directed against pathogens encountered by the mother (they are her memory B cells) and therefore, pathogens the infant is likely to encounter in the environment. Passive immunity conferred through breast milk is an example of "inheritance" of an acquired characteristic, maintained by a genetic component subject to natural selection; the mother passes her antigen experience to her offspring via milk.

CONCLUSIONS AND POLICY IMPLICATIONS

As our understanding of the importance of environments for gene expression expands, a life span perspective becomes a necessary starting point for asking questions about fitness and reproductive success. Pregnancy offers an essential window of opportunity to examine how a woman's previous biological experiences impact pregnancy outcomes and simultaneously allow us to examine the earliest of environments and longer-term impacts of variable intrauterine experiences. Lactation, argued here as a more fluid continuation of the same reproductive strategy, also demands a life span perspective. The evidence for metabolic programming in utero and the links to the early postnatal environment also argue for a broader more inclusive framework. For example, early postnatal growth trajectories are strongly patterned by gestational duration and intrauterine growth and lactation, via suckling bouts and the composition of milk, helps facilitate this growth (Jochum et al., 2005; Kovacs et al., 2005; De Blasio et al., 2007). Agostoni's (2005) work on programmed fatty acid metabolism offers similar support. Such data suggest we should follow previous pleas (e.g., Dufour and Sauther, 2002; Ulijaszek, 2002; Sellen, 2007) and encourage future studies that link nonhuman primate and human physiological flexibility and reproductive strategies for a better understanding of how reproduction has been conserved and modified across evolutionary time.

Behavioral flexibility is a critical and potentially integral part of the human life history strategy. Given the importance of behavioral flexibility in meeting the demands of reproduction across human and nonhuman primate species and our long-standing awareness of the importance of contexts/environments shaping biology, a biosocial perspective offers another useful framework. The example of how fetal stress reactivity develops in response to maternal stress and anxiety highlights the importance of social contexts. Examining how social structures encourage or discourage mother/infant attachment and broader social support has long-term implications for infant/child development and overall health. Flexibility in patterns of supplemental feeding, social support during reproductive events, and balancing maternal work demands with infant feeding all require a nuanced understanding of how social contexts shape biological variation.

Human evolutionary biology and social epidemiology are increasingly compatible and offer new opportunities to examine how environmental contexts, including social environments, shape local biologies (Lock and Kaufert, 2001; Pike, 2005; Worthman and Khort, 2005). Trends suggest a stronger presence of evolutionary interpretations for biomedical research (e.g., Wells, 2003; Bateson et al., 2004; Gluckman et al., 2007; Gluckman and Hanson, 2007; Smith, 2007, among many others). Pregnancy and lactation offer important opportunities to simultaneously examine mechanisms that enhance reproductive success and fitness and offer insights on pressing public health concerns. A host of examples can be cited including preterm delivery (Pike, 2005; Gluckman and Hanson, 2007), sudden infant death syndrome (McKenna et al., 1997; Mosko et al., 1997), complications associated with birth (Rosenberg and Trevathan, 2002), complementary feeding strategies and infant survival (Sellen, 2007), and the implications of the evolution of fatty acid content of milk and infant formula (Uauy et al., 1996; Carlson, 1999, 2001; Gibson and Makrides, 1999; Agostoni et al., 2001) among so many others. We hope this chapter, and the volume more generally, further encourages the exploration of evolutionary underpinnings of pressing public health concerns.

DISCUSSION POINTS

1. How does life history theory enhance our ability to understand the evolution of primate reproduction?
2. The authors propose the idea that the primate reproductive strategy is best examined as an integrative strategy that crosses pregnancy, birth, and lactation. What are the strengths and weaknesses of such an integrative approach?
3. How does our understanding of pregnancy and lactation benefit from a comparative perspective across primate groups? Provide examples of both conserved traits across primate groups and a trait specific to human reproduction.
4. Provide evidence for the importance of reproductive energetics for primate females.
5. How might a better understanding of milk composition inform health policy?

REFERENCES

Adair, L. S. (2001). Size at birth predicts age at menarche. *Pediatrics*, **107**, 59–66.

Agostoni, C. (2005). LC-PUFA content in human milk: is it always optimal? *Acta Paediatrica*, **94**, 1532–1537.

Agostoni, C., Marangoni, F., Giovannini, M., et al. (2001). Prolonged breast-feeding (six months or more) and milk fat content at six months are associated with higher developmental scores at one year of age within a breast-fed population. In *Bioactive Components of Human Milk*, D. S. Newburg (ed.). New York: Kluwer Academic/Plenum Publishers, pp. 137–141.

Alberts, S. C. and Altmann, J. (2003). Matrix models for primate life history analysis. In *Primate Life Histories and Socioecology*, P. M. Kappeler and M. E. Pereira (eds). Chicago: University of Chicago Press, pp. 66–102.

Altmann, J. (1983). Costs of reproduction in baboons (*Papio cynocephalus*). In *Behavioral Energetics: the Cost of Survival in Vertebrates*, W. Aspe and S. Lustick (eds). Columbus, OH: Ohio State University Press, pp. 67–88.

Armstrong, E. (1983). Relative brain size and metabolism in mammals. *Science*, **220**, 1302–1304.

Barbosa, L., Butte, N. F., Villalpando, S., et al. (1997). Maternal energy balance and lactation performance of Meso-amerindians as a function of body mass index. *American Journal of Clinical Nutrition*, **66**, 575–583.

Barker, D., Gluckman, P., Godfrey, K., et al. (1993). Fetal nutrition and cardiovascular disease in adult life. *Lancet*, **341**, 938–1001.

Barrett, R., Kuzawa, C. W., McDade, T., et al. (1998). Emerging and re-emerging infectious diseases: the third epidemiological transition. *Annual Review of Anthropology*, **27**, 247–271.

Bateson, P., Barker, D., Clutton-Brock, T., et al. (2004). Developmental plasticity and human health. *Nature*, **430**, 419–421.

Bines, J. E. and Walker, W. A. (1991). Growth factors and the development of the neonatal host defense. *Advances in Experimental Medicine and Biology*, **310**, 31–39.

Bowman, M. E., Lopata, A., Jaffe, R. B., et al. (2001). Corticotropin-releasing hormone-binding protein in primates. *American Journal of Primatology*, **53**, 123–130.

Brouwers, E. P. M., van Baar, A. L. and Pop, V. J. M. (2001). Maternal anxiety during pregnancy and subsequent infant development. *Infant Behavior and Development*, **24**, 95–106.

Butte, N. F., Garza, C., Stuff, J. E., et al. (1984). Effect of maternal diet and body composition on lactational performance. *American Journal of Clinical Nutrition*, **39**, 296–306.

Butte, N. F., Hopkinson, J. M., Mehta, N., et al. (1999). Adjustments in energy expenditure and substrate utilization during late pregnancy and lactation. *American Journal of Clinical Nutrition*, **69**, 299–307.

Carlson, S. E. (1999). Long-chain polyunsaturated fatty acids and development of human infants. *Acta Paediatrica*, **88**, 72–77.

Carlson, S. E. (2001). Docosahexaenoic acid and arachidonic acid in infant development. *Seminars in Neonatology*, **6**, 437–449.

Chang, C. L. and Hsua, S. Y. T. (2004). Ancient evolution of stress-regulating peptides in vertebrates. *Peptides*, **25**, 1681–1688.

Charnov, E. L. and Berrigan, D. (1993). Why do female primates have such long lifespans and so few babies? "or" Life in the slow lane. *Evolutionary Anthropology*, **1**, 191–194.

Chen, D. C., Nommsen-Rivers, L., Dewey, K. G., et al. (1999). Stress during labor and delivery and early lactation performance. *Obstetrical and Gynecological Survey*, **54**, 81–82.

Christensson, K., Siles, C., Moreno, L., et al. (1992). Temperature, metabolic adaptation, and crying in healthy, full-term newborns cared for skin-to-skin or in a cot. *Acta Paediatrica*, **81**, 488–493.

Coall, D. A. and Chisholm, J. S. (2003). Evolutionary perspectives on pregnancy: maternal age at menarche and infant birth weight. *Social Science and Medicine*, **57**, 1771–1781.

Cohen, M. N. (1989). *Health and Rise of Civilization*. New Haven: Yale University Press.

Cruz, J., Cano, F. and Caceres, P. (1991). Association of human milk sIgA antibodies with maternal intestinal exposure to microbial antigens. In *Immunology of Milk and the Neonate*, J. Mestecky, C. Blair and P. L. Ogra (eds). New York: Plenum Press, pp. 193–200.

De Blasio, M. J., Gatford, K. L., Robinson, J. S., et al. (2007). Placental restriction of fetal growth reduces size at birth and alters postnatal growth, feeding activity, and adiposity in the young lamb. *American Journal of Physiology – Regulatory Integrative and Comparative Physiology*, **292**, R875–R886.

Denver, R. (1997). Environmental stress as a developmental cue: corticotropin-releasing hormone is a proximate mediator of adaptive phenotypic plasticity in amphibian metamorphosis. *Hormones and Behavior*, **31**, 169–179.

Dettwyler, K. A. (2004). When to wean: biological versus cultural perspectives. *Clinical Obstetrics and Gynecology*, **47**, 712–723.

Dettwyler, K. A. and Fishman, C. (1992). Infant feeding practices and growth. *Annual Review of Anthropology*, **21**, 171–204.

Dewey, K. G. (2001). Maternal and fetal stress are associated with impaired lactogenesis in humans. *Journal of Nutrition*, **131**, 3012S–3015S.

Dodic, M., May, C. N., Wintour, E. M., et al. (1999). An early prenatal exposure to excess glucocorticoid leads to hypertensive offspring in sheep. *Clinical Science*, **94**, 149–155.

Dufour, D. L. and Sauther, M. L. (2002). Comparative and evolutionary dimensions of the energetics of human pregnancy and lactation. *American Journal of Human Biology*, **14**, 584–602.

Dufour, D., Reina, J. and Spurr, G. (1999). Energy intake and expenditure of free-living pregnant Columbian women in an urban setting. *American Journal of Clinical Nutrition*, **70**, 269–276.

Durnin, J. V. G. A. (1987). Energy-requirements of pregnancy – an integration of the longitudinal data from the five-country study. *Lancet*, **2**, 1131–1133.

Ellis, B. J. and Garber, J. (2000). Psychosocial antecedents of variation in girls' pubertal timing: maternal depression, stepfather presence, and marital and family stress. *Child Development*, **71**, 485–501.

Ellison, P. T. (2001). *On Fertile Ground: a Natural History of Human Reproduction*. Cambridge, MA: Harvard University Press.

Ellison, P. T. (2005). Evolutionary perspectives on the fetal origins hypothesis. *American Journal of Human Biology*, **17**, 113–118.

Emmett, P. M. and Rogers, I. S. (1997). Properties of human milk and their relationship with maternal nutrition. *Early Human Development*, **49**, S7–S28.

Finley, D.-A. C. and Lönnerdal, B. (1985). Fattty acid composition of breast milk from vegetarian and non-vegetarian lactating women. In *Composition and Physiological Properties of Human Milk*, J. Schaub (ed.). Amsterdam: Elsevier Science Publishers, pp. 203–211.

Forsum, E., Kabir, N., Sadurskis, A., et al. (1992). Total energy expenditure of healthy Swedish women during pregnancy and lactation. *American Journal of Clinical Nutrition*, **56**, 334–342.

Fowden, A. (2001). Growth and metabolism. In *Fetal Growth and Development*, R. Harding and A. D. Bocking (eds). Cambridge: Cambridge University Press, pp. 44–69.

Gibson, R. A. and Makrides, M. (1999). Long-chain polyunsaturated fatty acids in breast milk: are they essential? In *Bioactive Components of Human Milk*, D. Newburg (ed.). New York: Plenum Publishers, pp. 375–383.

Gittleman, J. L. and Thompson, S. D. (1988). Energy allocation in mammalian reproduction. *American Zoologist*, **28**, 863–875.

Gluckman, P. D. and Hanson, M. A. (2007). Developmental plasticity and human disease: research directions. *Journal of Internal Medicine*, **261**, 461–471.

Gluckman, P. D., Hanson, M. A. and Beedle, A. S. (2007). Early life events and their consequences for later disease: a life history and evolutionary perspective. *American Journal of Human Biology*, **19**, 1–19.

Goldman, A. S. (1993). The immune system of human milk: antimicrobial, antiinflammatory and immunomodulating properties. *Pediatric Infectious Disease Journal*, **12**, 664–671.

Goldman, A. S. (2001). The immunological system in human milk: the past: a pathway to the future. *Advances in Nutritional Research*, **10**, 15–31.

Goldman, A. S. and Goldblum, R. M. (1995). Defense agents in milk. A. Defense agents in human milk. In *Handbook of Milk Composition*, R. G. Jensen (ed.). San Diego, CA: Academic Press, pp. 727–745.

Goldman, A. S., Garza, C., Nichols, B. L., et al. (1982). Immunological factors in human-milk during the first year of lactation. *Journal of Pediatrics*, **100**, 563–567.

Goldman, A. S., Chheda, S. and Garofalo, R. (1998). Evolution of immunologic functions of the mammary gland and the postnatal development of immunity. *Pediatric Research*, **43**, 155–162.

Haig, D. (1993). Genetic conflicts of pregnancy. *Quarterly Review of Biology*, **68**, 495–532.

Hales, C. N. and Barker, D. J. P. (2001). The thrifty phenotype hypothesis. *British Medical Bulletin*, **60**, 5–20.

Harvey, P. H., Martin, R. D. and Clutton-Brock, T. H. (1987). Life histories in comparative perspective. In *Primate Societies*, B. B. Smuts, D. L. Cheney, R. M. Seyfarth, et al. (eds). Chicago: University of Chicago Press, pp. 181–196.

Hill, K. and Kaplan, H. (1999). Life history traits in humans: theory and empirical studies. *Annual Review of Anthropology*, **28**, 397–430.

Hinde, K. J. (2007a). Milk composition varies in relation to the presence and abundance of *Balantidium coli* in the mother in captive rhesus macaques (*Macaca mulatta*). *American Journal of Primatology*, **69**, 625–634.

Hinde, K. J. (2007b). First-time macaque mothers bias milk composition in favor of sons. *Current Biology*, **17**, R958–R959.

Holdcroft, A., Oatridge, A., Hajnal, J. V., et al. (1997). Changes in brain size in normal human pregnancy. *Journal of Physiology*, **498P**, 80P–81P.

Holman, D. and Wood, J. (2001). Pregnancy loss and fecundability in women. In *Reproductive Ecology and Human Evolution*, P. T. Ellison (ed.). New York: Aldine de Gruyter, pp. 15–38.

Hoshower, L. M. (1992). Brief communication: immunologic aspects of human colostrum and milk – a misinterpretation. *American Journal of Physical Anthropology*, **94**, 421–425.

Ibanez, L., Potau, N., Enriquez, G., et al. (2000). Reduced uterine and ovarian size in adolescent girls born small for gestational age. *Pediatric Research*, **47**, 575–577.

Ibanez, L., Jimenez, R. and de Zegher, F. (2006). Early puberty-menarche after precocious pubarche: relation to prenatal growth. *Pediatrics*, **117**, 117–121.

Insel, T. R. and Young, L. J. (2001). The neurobiology of attachment. *Nature Reviews Neuroscience*, **2**, 129–136.

Jelliffe, D. B. and Jelliffe, E. F. P. (1978). The volume and composition of human milk in poorly nourished communities. A review. *American Journal of Clinical Nutrition*, **31**, 492–515.

Jenness, R. (1979). The composition of human milk. *Seminars in Perinatology*, **3**, 225–239.

Jensen, R. G., Bitman, J., Carlson, S. E., et al. (1995). Human milk lipids. In *Handbook of Milk Composition*, R. G. Jensen (ed.). San Diego, CA: Academic Press, pp. 495–542.

Jochum, F., Loui, A., Weber, A., et al. (2005). Low soluble Fas (sFas) and sFas ligand (sFasL) content in breast milk after preterm as opposed to term delivery. *Acta Paediatrica*, **94**, 143–146.

Kaplan, H., Hill, K., Lancaster, J., et al. (2000). A theory of human life history evolution: diet, intelligence, and longevity. *Evolutionary Anthropology*, **9**, 156–185.

Kennedy, G. E. (2005). From the ape's dilemma to the weanling's dilemma: early weaning and its evolutionary context. *Journal of Human Evolution*, **48**, 123–145.

Kinsley, C. H., Madonia, L., Gifford, G. W., et al. (1999). Motherhood improves learning and memory. *Nature*, **402**, 137–138.

Koletzko, B., Thiel, I. and Abiodum, P. O. (1992). The fatty acid composition of human milk in Europe and Africa. *Journal of Pediatrics*, **120**, S62–S70.

Koletzko, B., Agostoni, C., Carlson, S. E., et al. (2001). Long chain polyunsaturated fatty acids (LC-PUFA) and perinatal development. *Acta Paediatrica*, **90**, 460–464.

Kovacs, A., Funke, S., Marosvolgyi, T., et al. (2005). Fatty acids in early human milk after preterm, and full-term delivery. *Journal of Pediatric Gastroenterology and Nutrition*, **41**, 454–459.

Kuzawa, C. W. (1998). Adipose tissue in human infancy and childhood: an evolutionary perspective. *Yearbook of Physical Anthropology*, **43**, 177–209.

Lampl, M. (2005). Cellular life histories and bow tie biology. *American Journal of Human Biology*, **17**, 66–80.

Lau, C. (2001). Effects of stress on lactation. *Pediatric Clinics of North America*, **48**, 221–234.

Lawrence, M. and Whitehead, R. (1988). Physical activity and total energy expenditure of child-bearing Gambian village women. *European Journal of Clinical Nutrition*, **42**, 1442–1450.

Lawrence, M., Singh, J., Lawrence, F., et al. (1985). The energy cost of common daily activities in African women: increased expenditure in pregnancy? *American Journal of Clinical Nutrition*, **42**, 753–763.

Lee, P. C. (1996). The meanings of weaning: growth, lactation and life history. *Evolutionary Anthropology*, **5**, 87–98.

Lee, P. C. (1999). Comparative ecology of postnatal growth and weaning among haplorhine primates. In *Comparative Primate Socioecology*, P. C. Lee (ed.). Cambridge: Cambridge University Press, pp. 111–136.

Lee, P. C., Majluf, P. and Gordon, I. J. (1991). Growth, weaning and maternal investment from a comparative perspective. *Journal of Zoology*, **225**, 99–114.

Lock, M. and Kaufert, P. (2001). Menopause, local biologies, and cultures of aging. *American Journal of Human Biology*, **13**, 494–504.

Lönnerdal, B. (1986). Effects of maternal nutrition on human lactation. In *Human Lactation 2: Maternal and Environmental Factors*, M. Hamosh and A. S. Goldman (eds). New York: Plenum Press, pp. 301–324.

Lönnerdal, B. (1996). Recombinant human milk proteins – an opportunity and a challenge. *American Journal of Clinical Nutrition*, **63**, 622S–626S.

Lönnerdal, B. and Atkinson, S. (1995). Nonprotein nitrogen factors in human milk. In *Handbook of Milk Composition*, R. G. Jensen (ed.). San Diego, CA: Academic Press. pp. 351–387.

Lönnerdal, B., Keen, C. L., Glazier, C. E., et al. (1984). A longitudinal study of rhesus monkey (*Macaca mulatta*) milk composition: trace elements, minerals, protein, carbohydrate, and fat. *Pediatric Research*, **18**, 911–914.

Maestripieri, D. (2002). Parent–offspring conflict in primates. *International Journal of Primatology*, **23**, 923–951.

Martin, R. D. (1983). *Human Brain Evolution in an Ecological Context*. New York: American Museum of Natural History.

Martin, R. D. (1996). Scaling of the mammalian brain: the maternal energy hypothesis. *News in Physiological Sciences*, **11**, 149–156.

Martin, R. D. and MacLarnon, A. (1985). Gestation period, neonatal size and maternal investment in placental mammals. *Nature*, **313**, 220–223.

McClean, M., Bisits, A., Davies, J., et al. (1995). A placental clock controlling the length of human pregnancy. *Nature Medicine*, **1**, 460–463.

McFarland, R. (1997). Female primates: fat or fit? In *The Evolving Female*, M. E. Morbeck, A. Galloway and A. L. Zilhman (eds). Princeton: Princeton University Press, pp. 163–175.

McGrath, S. and Smith, R. (2001). Corticotrophin-releasing hormone and parturition. *Clinical Endocrinology*, **55**, 593–595.

McKenna, J. J., Mosko, S. S. and Richard, C. A. (1997). Bed-sharing promotes breastfeeding. *Pediatrics*, **100**, 214–219.

Milligan, L. A. (2005). Concentration of sIgA in the milk of *Macaca mulatta* [abstract]. *American Journal of Physical Anthropology*, **128**, 153.

Milligan, L. A. (2007). Nonhuman primate milk composition: relationship to phylogeny, ontogeny, and ecology. PhD thesis, University of Arizona, Tucson, AZ.

Milligan, L. A. and Bazinet, R. P. (2008). Evolutionary modifications on human milk composition – evidence from long-chain polyunsaturated fatty acid composition of anthropoid milks. *Journal of Human Evolution*, **55**, 1086–1095.

Milligan, L. A., Rapoport, S. I., Cranfield, M. R., et al. (2008a). Fatty acid composition of wild anthropoid milks. *Comparative Biochemistry and Physiology, Part B.*, **149**, 74–82.

Milligan, L. A., Gibson, S. V., Williams, L. E., et al. (2008b). The composition of milk from Bolivian squirrel monkeys

(*Saimiri bolivienses bolivienses*). *American Journal of Primatology*, **70**, 35–43.

Monks, D. A., Lonstin, J. S. and Breedlove, M. (2003). Got milk? Oxytocin triggers hippocampal plasticity. *Nature Neuroscience*, **6**, 327–328.

Morbeck, M. E. (1997a). Life history, the individual, and evolution. In *The Evolving Female*, M. E. Morbeck, A. Galloway and A. L. Zihlman (eds). Princeton: Princeton University Press, pp. 4–14.

Morbeck, M. E. (1997b). Reading life history in teeth, bones, and fossils. In *The Evolving Female*, M. E. Morbeck, A. Galloway and A. L. Zihlman (eds). Princeton: Princeton University Press, pp. 117–131.

Mosko, S., Richard, C. and McKenna, J. (1997). Infant arousals during mother–infant bed sharing: implications for infant sleep and sudden infant death syndrome research. *Pediatrics*, **100**, 841–849.

Neville, M. C. (2001). Anatomy and physiology of lactation. *Pediatric Clinics of North America*, **48**, 13–34.

Neville, M. C. and Morton, J. (2001). Physiology and endocrine changes underlying human lactogenesis II. *Journal of Nutrition*, **131**, 3005S–3008S.

Neville, M. C., Morton, J. and Umemura, S. (2001). Lactogenesis: the transition from pregnancy to lactation. *Pediatric Clinics of North America*, **48**, 35–52.

Oftedal, O. T. (1984). Milk composition, milk yield and energy output at peak lactation: a comparative review. *Symposia of the Zoological Society of London*, **51**, 33–85.

Oftedal, O. T. (2000). Use of maternal reserves as a lactation strategy in large mammals. *Proceedings of the Nutrition Society*, **59**, 99–106.

Oftedal, O. T. and Iverson, S. J. (1995). Comparative analysis of nonhuman milks: phylogenetic variation in the gross composition of milks. In *Handbook of Milk Composition*, R. G. Jensen (ed.). San Diego, CA: Academic Press, pp. 749–788.

Panter-Brick, C. (1989). Motherhood and subsistence work: the Tamang of rural Nepal. *Human Ecology*, **17**, 205–228.

Panter-Brick, C. (1992). The energy cost of common tasks in rural Nepal: levels of energy expenditure compatible with sustained physical activity. *European Journal of Applied Physiology and Occupational Physiology*, **64**, 477–484.

Panter-Brick, C. (1993). Seasonality of energy expenditure during pregnancy and lactation for rural Nepali women. *American Journal of Clinical Nutrition*, **57**, 620–628.

Peacock, N. (1991). An evolutionary perspective on the patterning of maternal investment in pregnancy. *Human Nature*, **2**, 351–385.

Picciano, M. F. (2003). Pregnancy and lactation: physiological adjustments, nutritional requirements and the role of dietary supplements. *Journal of Nutrition*, **133**, 1997S–2002S.

Pike, I. L. (2005). Maternal stress and fetal responses: evolutionary perspectives on preterm delivery. *American Journal of Human Biology*, **17**, 55–65.

Piperata, B. A. and Dufour, D. L. (2007). Diet, energy expenditure, and body composition among lactating Ribeirinha women in the Brazilian Amazon. *American Journal of Human Biology*, **19**, 722–734.

Pond, C. M. (1984). Physiological and ecological importance of energy storage in the evolution of lactation: evidence for a common pattern of anatomical organization of adipose tissue in mammals. *Symposia of the Zoological Society of London*, **51**, 32.

Pond, C. M. (1997). The biological origins of adipose tissue in humans. In *The Evolving Female*, M. E. Morbeck, A. Galloway and A. L. Zihlman (eds). Princeton: Princeton University Press, pp. 147–162.

Porter, R. H. and Winberg, J. (1999). Unique salience of maternal breast odors for newborn infants. *Neuroscience and Biobehavioral Reviews*, **23**, 439–449.

Power, M. L. and Schulkin, J. (2006). Functions of corticotropin-releasing hormone in anthropoid primates: from brain to placenta. *American Journal of Human Biology*, **18**, 431–447.

Power, M. L., Oftedal, O. T. and Tardif, S. D. (2002). Does the milk of callitrichid monkeys differ from that of larger anthropoids? *American Journal of Primatology*, **56**, 117–127.

Power, M. L., Bowman, M. E., Smith, R., et al. (2006). Pattern of maternal serum corticotropin-releasing hormone concentration during pregnancy in the common marmoset (*Callithrix jacchus*). *American Journal of Primatology*, **68**, 181–188.

Power, M. L., Verona, C., Ruiz-Miranda, C., et al. (2008). The composition of milk from free-living common marmosets (*Callithrix jacchus*) in Brazil. *American Journal of Primatology*, **70**, 78–83.

Prentice, A. (1995). Regional variations in the composition of human milk. In *Handbook of Milk Composition*, R. G. Jensen (ed.). San Diego, CA: Academic Press, pp. 155–221.

Prentice, A. (1996). Constituents of human milk. *Food and Nutrition Bulletin*, **17**, 305–312.

Prentice, A. M. and Goldberg, G. R. (2000). Energy adaptations in human pregnancy: limits and long-term consequences. *American Journal of Clinical Nutrition*, **71**, 1226S–1232S.

Prentice, A. M. and Prentice, A. (1995). Evolutionary and environmental influences on human lactation. *Proceedings of the National Academy of Sciences of the United States of America*, **54**, 391–400.

Prentice, A. M., Goldberg, G. R. and Prentice, A. (1994). Body mass index and lactation performance. *European Journal of Clinical Nutrition*, **48**, S86–S89.

Rodier, P., Cohen, I. and Buelke-Sam, J. (1994). Developmental nerotoxicology: neuroendocrine manifestations of CNS insult. In *Developmental Toxicology*, C. A. Kimmel and J. Buelke-Sam (eds). New York: Raven Press, pp. 65–92.

Roberts, S. B., Cole, T. J. and Coward, W. A. (1985). Lactational performance in relation to energy-intake in the baboon. *American Journal of Clinical Nutrition*, **41**, 1270–1276.

Rosenberg, K. and Trevathan, W. (2002). Birth, obstetrics and human evolution. *BJOG – An International Journal of Obstetrics and Gynaecology*, **109**, 1199–1206.

Ross, C. (1998). Primate life histories. *Evolutionary Anthropology*, **6**, 54–63.

Ross, C. (2003). Life history, infant care strategies, and brain size in primates. In *Primate Life Histories and Socioecology*,

P. M. Kappeler and M. E. Pereira (eds). Chicago: University of Chicago Press, pp. 266–284.

Sanders, T. A. B. (1999). Essential fatty acid requirements of vegetarians in pregnancy, lactation, and infancy. *American Journal of Clinical Nutrition*, 70, 555S–559S.

Sandman, C. A., Wadhwa, P. D., Chicz-DeMet, A., et al. (1997). Maternal stress, HPA activity, and fetal/infant outcome. *Annals of the New York Academy of Sciences*, 814, 266–275.

Sauther, M. L. (1994). Wild plant use by pregnant and lactating lemurs, with implications for early hominid foraging. In *Eating on the Wild Side*, N. L. Etkin (ed.). Tucson: University of Arizona Press, pp. 240–256.

Schutz, Y., Lechtig, A. and Bradfield, R. B. (1980). Energy expenditures and food intakes of lactating women in Guatemala. *American Journal of Clinical Nutrition*, 33, 892–902.

Sellen, D. W. (2007). Evolution of infant and young child feeding: implications for contemporary public health. *Annual Review of Nutrition*, 27, 123–148.

Simondon, K. B., Simondon, F., Simon, I., et al. (1998). Preschool stunting, age at menarche and adolescent height: a longitudinal study in rural Senegal. *European Journal of Clinical Nutrition*, 52, 412–418.

Slade, H. B. and Schwartz, S. A. (1987). Mucosal immunity: the immunology of breast milk. *Journal of Allergy and Clinical Immunology*, 80, 348–358.

Smith, R. (2007). *Parturition. New England Journal of Medicine*, 356, 271–283.

Smith, R., Wickings, E., Bowman, M., et al. (1999). Corticotropin-releasing hormone in chimpanzee and gorilla pregnancies. *Journal of Clinical Endocrinology and Metabolism*, 84, 2820–2825.

Smith, R., Mesiano, S., and McGrath, S. (2002). Hormone trajectories leading to human birth. *Regulatory Peptides*, 108, 159–164.

Stearns, S. C. (1992). *The Evolution of Life Histories*. Oxford: Oxford University Press.

Stini, W. A., Weber, C. W., Kemberling, S. R., et al. (1980). Bioavailability of nutrients in human breast milk as compared to formula. *Studies in Physical Anthropology*, 6, 32–35.

Tagle, N. M., Neal, C., Glover, V., et al. (2007). Antenatal stress and long-term effects on child neurodevelopment: how and why? *Journal of Child Psychology and Psychiatry*, 48, 245–261.

Tanner, J. (1998). A brief history of the study of human growth. In *The Encyclopedia of Human Growth*, S. J. Ulijaszek, F. E. Johnston and M. Preece (eds). Cambridge: Cambridge University Press, pp. 2–7.

Tilden, C. D. and Oftedal, O. T. (1997). Milk composition reflects pattern of maternal care in prosimian primates. *American Journal of Primatology*, 41, 195–211.

Trivers, R. (1974). Parent–offspring conflict. *American Zoologist*, 14, 249–264.

Uauy, R., Peirano, P., Hoffman, D., et al. (1996). Role of essential fatty acids in the function of the developing nervous system. *Lipids*, 31, 167S–176S.

Ulijaszek, S. J. (2002). Comparative energetics of primate fetal growth. *American Journal of Human Biology*, 14, 603–608.

Van de Perre, P. (2003). Transfer of antibody via mother's milk. *Vaccine*, 21, 3374–3376.

Vasey, N. and Walker, A. (2001). Neonate body size and hominid carnivory. In *Meat-eating and Human Evolution*, C. B. Stanford and H. T. Bunn (eds). Oxford: Oxford University Press, pp. 332–349.

Wadhwa, P., Sandman, C., Porto, M., et al. (1993). The association between prenatal stress and infant birth-weight and gestational age at birth – a prospective investigation. *American Journal of Obstetrics and Gynecology*, 169, 858–865.

Wadhwa, P. D., Culhane, J. F., Rauh, V., et al. (2001). Stress, infection and preterm birth: a biobehavioural perspective. *Paediatric and Perinatal Epidemiology*, 15, 17–29.

Wadhwa, P. D., Glynn, L., Hobel, C. J., et al. (2002). Behavioral perinatology: biobehavioral processes in human fetal development. *Regulatory Peptides*, 108, 149–157.

Weinstock, M. (2001). Alterations induced by gestational stress in brain morphology and behaviour of the offspring. *Progress in Neurobiology*, 65, 427–451.

Wells, J. C. K. (2003). The thrifty phenotype hypothesis: thrifty offspring or thrifty mother? *Journal of Theoretical Biology*, 221, 143–161.

Wells, J. C. K. (2006). The evolution of human fatness and susceptibility to obesity: an ethological approach. *Biological Reviews*, 81, 183–205.

Winberg, J. (2005). Mother and newborn baby: mutual regulation of physiology and behavior – a selective review. *Developmental Psychobiology*, 47, 217–229.

Wood, J. (1994). *Dynamics of Human Reproduction: Biology, Biometry, Demography*. Hawthorne, NY: Aldine de Gruyter.

Worthman, C. M. and Khort, B. (2005). Receding horizons of health: biocultural approaches to public health paradoxes. *Social Science and Medicine*, 61, 861–878.

Yuhas, R., Pramuk, K. and Lien, E. L. (2006). Milk fatty acid composition from nine countries varies most in DHA. *Lipids*, 41, 851–858.

Zilhman, A. L. and McFarland, R. K. (2000). Body mass in lowland gorillas: a quantitative analysis. *American Journal of Physical Anthropology*, 113, 61–78.

21 Male Reproduction: Physiology, Behavior, and Ecology

Michael P. Muehlenbein and Richard G. Bribiescas

INTRODUCTION

Women and men exhibit vast differences in reproductive physiologies, behaviors, and ecologies. The present chapter aims to illustrate these fundamental differences by utilizing recent theoretical and empirical developments in addition to clinical and anthropological data that help clarify the evolutionary bases for human male reproductive functions. We begin with a discussion of the various aspects of male reproductive effort, including the physiology and behaviors associated with seeking, attracting, and choosing mates, competing for and controlling mates, paternal behaviors, and the proximate determinants of spermatogenesis, libido, and erection. For the specific purpose of better understanding male reproductive ecology, we next provide readers with a description of male reproductive physiology, including major aspects of development, endocrinology, and senescence. Testosterone may be synonymous with maleness; however, it may impose a number of costs, including negative energy balance, immunosuppression, and prostate cancer.

A discussion on male reproductive ecology ends our chapter because it draws from our discussions on reproductive effort and physiology. Readers are encouraged to utilize the information presented here to further discussion and research on human male reproduction, both the proximate mechanisms involved in addition to the evolutionary explanations for the ways we function and behave.

MALE REPRODUCTIVE EFFORT

Reproductive effort refers to investments of both time and energy into reproduction. Optimization of reproductive effort is of central importance, especially for capital breeding, iteroparous organisms that must budget time and energy over a number of reproductive events within a lifetime. Males and females differ greatly in the amount of time and energy they invest in reproduction. For mammals, this is the direct result of contrasting offspring investment requirements that stem from internal gestation. Mammalian females produce relatively small quantities of large, energy-rich gametes, and must energetically commit to gestation and lactation in addition to other postnatal activities that increase the likelihood of offspring survival. Lifetime reproductive success for females is therefore largely limited by access to resources (Trivers, 1972). Females set the pace of childbearing, and population fertility is determined by the length of interval between successive gestations. This interbirth interval is determined largely through robusticity of ovarian function, length of lactational amenorrhea, likelihood of fetal loss, and coital frequency (Wood, 1994b). The male contribution to this is relatively low.

Sexual reproduction and the exchange of genes followed by mutation, recombination, and independent assortment function largely to create variation in a population (Fisher, 1932), which, among other things, provides some advantage against pathogens also evolving in the environment (Van Valen, 1973). Humans are not parthenogenic, and males may make some contributions to population fertility and the interbirth interval, albeit much less so than that of females.

Unlike women, men are not constrained by the direct metabolic costs of gestation, lactation, childbirth, or menstruation. Men, like other male mammals, are characterized by the continuous production of spermatozoa with the potential for inseminating many females. Greater mate access can therefore increase offspring production. Some males will be more successful than others in the reproductive race, resulting in high levels of reproductive variance. For example, some men may not have any children whereas the Sultan of Morocco likely produced more than 800 (Low, 2000). For all men, enhancement of evolutionary fitness is possible through increased coital opportunities. Beyond sexual performance, male reproductive effort can include seeking, choosing, and attracting mates, competing for and controlling mates, as well as protection and provisioning of offspring and mates. Below we discuss each of these concepts in detail.

Human Evolutionary Biology, ed. Michael P. Muehlenbein. Published by Cambridge University Press. © Cambridge University Press 2010.

Given the scope of this text, homosexual preference will not be considered here, but evolutionary and developmental causes of homosexuality have been considered elsewhere (Roughgarden, 2009).

Seeking and attracting mates

Males of some species will put so much effort into seeking a mate that they forgo eating during the breeding season, which can have serious effects on survivorship (McMillin et al., 1980; Bobek et al., 1990). Mate-seeking behaviors can compromise survivorship by increasing day-range lengths and energy expenditure as well as exposure to predators. Sexual promiscuity also increases the risk of sexually transmitted infections. Despite this, a significant number of men surveyed are willing to consent to sexual intercourse with a potential partner they have only recently met (Buss and Schmidt, 1993).

Animal courtship displays can be quite complex, with ornate and brightly colored or decorated males vying for the attention of onlookers. The same can be said for humans, in which men may attempt to advertise their resource potential by displaying resources such as expensive cars and jewelry. Men, particularly adolescent males, may participate in dangerous sporting behaviors to gain attraction. This does not imply that such behaviors are consciously performed for the purpose of mate attraction. Rather, psychological mechanisms have certainly been naturally selected to maximize lifetime reproductive success. These may encompass dangerous behaviors that attract attention and demonstrate survivability.

A man may also use his physique and/or intellect (i.e., education level) to attract a mate. The former is augmented through the anabolic actions of testosterone and other androgens, and thus elevated testosterone levels and muscle anabolism may be considered direct investments in male reproductive effort (see below). At the same time, high doses of testosterone can compromise survivorship (also see below). Despite this, rates of anabolic-androgenic steroid and other ergogenic drug use for both athletic enhancement and improvement of appearance are increasing in the United States. Adolescents are exposing themselves to these substances at an earlier age, with an estimated 10% of high school male athletes taking illegal steroids (Calfee and Fadale, 2006). Anabolic-androgenic steroid use can cause physical and psychological damages, including liver failure, depression, psychosis, and rage (Hall and Hall, 2005). Future mortality data from men taking androgen supplements will likely provide much-needed data on this interesting experimental endocrine intervention.

Female reproductive value peaks during the 20s regardless of population and declines gradually with age (Hill and Hurtado, 1996; Kuhnert and Nieschlag, 2004). Since there is no evidence of an analogous decline in male fertility, it may be that males simply do not have access to fertile females due to female choice. Compromised female partner fertility (menopause) is also a factor in declining male reproductive value. Declination in attractiveness in addition to decreased ability to compete with conspecifics and acquire resources may reflect a type of "social andropause" in elderly men (Bribiescas, 2006). Such shifts in reproductive possibilities perhaps encourage men to dedicate more time and energy into offspring and mate investment.

Choosing a mate

Because of the high costs of reproduction, females should be under selection to choose carefully when and with whom they mate (Williams, 1966; Trivers, 1972). Female mate choice may be expressed to maximize resources, protection, and paternal care in addition to beneficial genetic variation that could be passed to offspring. For example, females may favor mates that display honest indications of disease resistance (Hamilton and Zuk, 1982) or other attributes of survivability (Zahavi, 1975). Human women of many sociocultural groups consistently exhibit preference for older, intelligent, emotionally sensitive men who are successful at their careers and are willing to invest financial and other resources, including parental care, in a long-term relationship (Buss and Schmidt, 1993). In fact, income and education (i.e., resources in general) are the best predictors of lifetime reproductive success in men (Mace, 1996; Waynforth, 1998). While women tend to prefer fewer short-term mates to long-term ones, they can also gain financial and "genetic" resources by employing a short-term reproductive strategy (see Chapter 17 of this volume for a complete discussion about human mating strategies).

If afforded access to fecund females, male animals may actually employ some mate choice. This is in contrast to the long-standing "Bateman's principle" in which a male, because of the minimal costs of sperm production, should be physiologically and behaviorally "ready and willing" to mate at all times and be indiscriminate about his choice of mates. We now recognize male mate choice as a real phenomenon in various species. For example, male chimpanzees prefer to consort and mate with older, parous females with proven maternal skills (Muller et al., 2006). Estrus females of various species are also more attractive to males (Beach, 1976).

Mammalian males may exhibit mate choice because the number of quality ejaculates is limited (Dewsbury, 1982). That is, the delivery system (ejaculation) can outpace production (spermatogenesis), and it may therefore pay for a male to choose who receives the ejaculate. Sperm counts do decline in some animals following repeated ejaculations, which is especially the case in insects (Odendaal et al., 1989). Furthermore, the refractory period between consecutive ejaculations increases in a number of species, including primates (Small, 1988). The refractory period between consecutive ejaculations can range from minutes to hours depending on physical condition (Aversa et al., 2000). An exception of this is known as the "Coolidge effect" in which the refractory periods between ejaculations diminish with the presentation of novel receptive females (Beamer et al., 1969). In this regard, males can partition ejaculates between females. Such phenomena have not been definitively demonstrated in humans. Human male sperm counts do decline after each consecutive ejaculation, but not to a point significantly low enough to be considered infertile (Nnatu et al., 1991; Cooper et al., 1993).

As with females, human males exhibit preference for both long-term and short-term mating partners. In men, these choices are based largely on fecundity, the likelihood of fidelity, and maternal ability (Buss and Schmidt, 1993; Gangestad et al., 2005; Sugiyama, 2005). On average, men consistently exhibit preference for younger mates and more of them, reflecting possible selection for promiscuity towards highly fecund women (Buss and Schmidt, 1993). Visual signs of fecundity in women include a high breast-to-underbreast ratio (large breasts) and a low waist-to-hip ratio (narrow waist) (Reynolds, 1991; Tovee et al., 1999). Women with these characteristics tend to have higher reproductive potential as measured through estradiol and progesterone levels (Jasienska et al., 2004).

Unlike females of some other species (Domb and Pagel, 2001), women do not exhibit sexual swellings that advertise fecundity or reveal the time of ovulation to males. However, scent may play some role in human mate choice. Pheromones are chemical signals released from apocrine glands of the skin to induce various physiological and behavioral responses (Albone and Shirley, 1984). For example, in rodents the presence of an adult male's odor can cause females to enter into estrus (the "Whitten effect"; Whitten, 1956), miscarriages in pregnant females (the "Bruce effect"; Bruce, 1959), and cause females to enter puberty earlier (the "Vandenbergh effect"; Vandenbergh, 1967, 1973). The importance of these chemicals in facilitating human reproduction is debated, but they are likely to be more than just vestigial (Rodriguez et al., 2000; Kohl et al., 2001). In women, female preference for male scents changes across the ovulatory cycle (Wedekind et al., 1995; Gangestad and Thornhill, 1998). Additionally, the pheromone androstenol may enhance attractiveness of males to females (Filsinger et al., 1985). The role of pheromones in male mate choice and detection of fecundity has not yet been extensively studied.

Competing for mates

Males may rely on various physical attributes in order to increase reproductive fitness, including body size, weaponry (e.g., canines), and sperm volume. The existence of sperm competition implies that by increasing the volume of spermatozoa and ejaculate, males can outcompete each other for successful fertilizations (Harcourt et al., 1981). Therefore, species with multi-male groups have larger testes relative to body size than those with single-male polygynous mating systems (Harvey and Harcourt, 1984).

The existence of sperm competition in humans has been debated, and it appears that sperm volumes in coital ejaculates can vary as a function of duration of separation from their committed partners (Baker and Bellis, 1989). That said, the role of sperm competition in our own species is reduced compared to chimpanzees; short-term bonding and intensive mating promiscuity have selected for greater investment in sperm quantity and accessory gland development (i.e., seminal vesicles) in chimpanzees (Dixson, 1999). Humans exhibit only mild sexual dimorphism (Smith and Leigh, 1998), but this may still be indicative of the large role polygyny has played in our evolutionary past. In fact, approximately 80% of all human populations may currently practice polygyny (Buss and Schmidt, 1993).

Intrasexual competition is an important mechanism of sexual selection (Trivers, 1972), and male–male competition over access to fecund females is a common phenomenon, even in humans. One way to compete for access to mates is to dominate or enhance one's status over conspecifics, and testosterone is related to dominance-seeking behaviors in males in interesting ways. For example, testosterone increases in preparation for competitions (Booth et al., 1989; Mazur, 1992). This may function to augment muscle tissue (especially upper body strength), which itself may have been selected to facilitate the production and use of weapons to settle these male–male conflicts (Bercovitch, 2001). Elevated testosterone levels prior to competition may also improve co-ordination and cognitive performance, increase self-confidence and motivation, as well as possibly interfere with self-control. Results of competition also alter hormone levels: testosterone levels remain high in winners and decline in losers, possibly to facilitate further successful competition in winners, and temper provocative actions in losers (Booth et al., 1989). Testosterone levels in men respond to, and are determined by, dominance activity and status change (Mazur and Booth, 1998).

Aggressive behavior with the intention of inflicting physical or mental damage is one means of attaining and maintaining dominance status. Although testosterone levels have been reported to correlate with aggressive behaviors in many species (Dabbs and Dabbs, 2000), associations between testosterone and aggression in humans have proved relatively inconsistent. Testosterone levels have been significantly correlated with self-reports of diverse antisocial behaviors, including marital disruption and violence, in military personnel (Booth and Dabbs, 1993). Testosterone levels have also correlated with violence of crimes and peer ratings of toughness in prison inmates (Dabbs et al., 1987, 1995). However, short-term administration of supraphysiological doses of testosterone in otherwise healthy men (eugonadal, noncriminal) is unrelated to various measures of anger and aggressive behavior (Tricker et al., 1996; Pope et al., 2000). Therefore, changing levels of testosterone may potentiate pre-existing patterns of aggression, but fluctuations in testosterone levels in normal individuals do not predict changes in aggressive behavior. Violent aggression and antisocial behaviors are more consistently associated with serotonergic dysfunction, such as defect in the neurotransmitter-metabolizing enzyme monoamine oxidase A gene, rather than high testosterone levels (Morell, 1993; Krakowski, 2003).

Fluctuations in testosterone levels are consistently related to the frequency and intensity of reproductive aggression in most species examined to date. That is, testosterone is most consistently associated with aggressive behaviors during challenges with conspecifics, mate guarding, the formation of dominance relationships, or establishment of territorial boundaries (the "challenge hypothesis"; Wingfield et al., 1990). Testosterone variation may therefore involve behavioral and physiological manipulation in accordance with the needs to compete with conspecifics for reproductive opportunities.

Testosterone can reduce the refractory period between action potentials running down the stria terminalis between the hypothalamus and amygdala, potentiating an aggressive response following the correct stimuli (Kendrick and Drewett, 1979). Testosterone also reduces the amount of hypothalamic stimulation needed to generate an aggressive response (Bermond et al., 1982). Testosterone increases glucose uptake and metabolic rate of muscle cells (Bhasin et al., 1996; Tsai and Sapolsky, 1996), and stimulates fat catabolism and muscle anabolism (Welle et al., 1992). In total, testosterone can prepare the body, physically and mentally, for competition. The challenge hypothesis and its context of reproductive aggression may be less applicable to humans because acute variation in androgens does not necessarily trigger behavioral changes. However, changes in testosterone level before and during competition could still function to prime a man's mind and body.

Controlling mates

Another general strategy commonly employed by males of a number of species to gain access to females is sexual coercion. One frequently noted variation of this strategy is infanticide, the killing of an infant sired by another male in order to gain reproductive access to the mother (Hausfater and Hrdy, 1984; Smuts and Smuts, 1993). The existence of such adaptive behaviors in humans is debatable. Certainly the likelihood of being fatally abused is much higher for stepchildren compared to children living with both biological parents (Daly and Wilson, 1988).

Various aspects of modern humans demonstrate relative degrees of male control/coercion over female reproduction. A number of modern human cultures attempt to exert male-biased control over female reproduction. Religious and other institutional, judicial, and social systems have attempted to control human sexual morality and behaviors, particularly that of women (Kinsey et al., 1948, 1953). Arranged marriages, strictures on divorce, anti-abortion campaigns, regulation or illegalization of prostitution, physical sequestration, female genital mutilation, and dress codes (i.e., veils, chuddars, chastity belts, foot binding, etc.) are all modern examples of institutionalized control of female sexual expression (Potts and Short, 1999; Gruenbaum, 2001; Hendrix, 2004). Human male control over female reproduction has also materialized through sexual jealousy and spousal abuse, sexual harassment, forced copulation, male adultery, asymmetry in parental investment, and emphasis on female virginity (Potts and Short, 1999; Thornhill and Palmer, 2000).

Interestingly, synchronized estrous in some groups of female primates may have been selected to prevent males from monopolizing receptive and ovulatory moments (Zinner et al., 1994), increase paternal investment (Alexander and Noonan, 1979), and reduce the likelihood of infanticide by confusing paternity (Hrdy, 1979; Altmann, 1990). The phenomenon of synchronized ovulation in group-living females may or may not exist in humans (McClintock, 1971; Strassmann, 1999). Women, like females of many different species, also exhibit concealed ovulation with no monthly sexual swellings.

Sperm

Although healthy human males are capable of continuous production of gametes throughout a lifetime, there are notable variations in sperm quality and quantity between individuals. Healthy sperm counts can vary between 1 and 120 million per milliliter of semen, with clinically defined minimal limits between 10 and 60 million per milliliter (Glass, 1991; World Health Organization, 1999; Guzick et al., 2001). There may even be a significant amount of variation in sperm counts between populations, with non-Western populations manifesting lower sperm counts compared to American males (Fisch et al., 1996). The highest reported sperm counts are among French men (102.9 million/ml) while the Thai seem to be at the low end of the spectrum (52.9 million/ml) (Fisch et al., 1996). However these numbers should be viewed with caution due to interlaboratory and counting method variability (Matson, 1995). Large-scale longitudinal sperm and semen assessments from other countries indicate some variation but none to suggest differences in fertility (Tortolero et al., 1999; Seo et al., 2000). In general, there is little evidence that any significant differences in sperm quantity between individuals or populations would correlate with differences in fecundity (Polansky and Lamb, 1988; Fisch et al., 1996). Sperm morphology may be a better predictor of male fertility (Guzick et al., 2001), although ultimately the contribution of sperm quality and quantity to couple fertility is uncertain at this time.

In some species, particularly arthropods, spermatogenesis is energetically expensive (van Voorhies, 1992; Gems and Riddle, 1996). The act of mating alone can compromise survivorship in male fruit flies (Partridge and Farquhar, 1981). In contrast, energetic investment in mammalian spermatogenesis is negligible, accounting for less than 1% of basal metabolic rate in human males (Elia, 1992). It is therefore not surprising that spermatogenesis can withstand extremely taxing energetic circumstances. Exercise may affect sperm quality and quantity only under endurance-training circumstances (Arce et al., 1993), but still not to a point where compromised fecundity becomes evident (Roberts et al., 1993). Other factors like diet composition (Wong et al., 2000, 2003), photoperiodicity (Levine et al., 1992), and body temperature (Mieusset and Bujan, 1995) may be influential. Furthermore, variation in testosterone levels does not correlate well with variation in spermatogenesis (Weinbauer and Nieschlag, 1990). High follicle-stimulating hormone and inhibin B levels exhibit modest links with lower sperm counts (Meeker et al., 2007). Genetic variation in luteinizing hormone (LH) and follicle-stimulating hormone (FSH) production is evident in some individuals, sometimes leading to reproductive disorders and compromised fertility (Berger et al., 2005; Lofrano-Porto et al., 2007). Variation in FSH levels within the common range of variation is not associated with differences in spermatogenesis. Indeed, spermatogenesis is tolerant of a broad range of FSH exposure (Kumar et al., 1997; Tapanainen et al., 1997). However, severe FSH deficiencies can result in oligospermia (low sperm count) or azoospermia (total lack of sperm).

Interestingly, some data suggest a significant decline in sperm counts, independent of age, in North American and European populations over the past century (Becker and Berhane, 1997; Swan et al., 2000) in addition to increased rates of testicular cancer and genital abnormalities (Giwercman et al., 1993). Other data suggest that testosterone levels have decreased over the last two decades and that these changes are independent of age, diagnosed illnesses, smoking, body fat percentage, or other lifestyle factors (Travison et al., 2007). However, the results of such studies have been contested based on sampling biases (Saidi et al., 1999), changes in clinical norms (Bromwich et al., 1994), and inappropriate statistical methods (Olsen et al., 1995; Emanuel et al., 1998). We may also now be more likely to identify, record, and treat these conditions than we have been in the past.

One popular hypothesis is that environmental pollutants with endocrine activity ("endocrine disruptors") could interfere with normal endocrine function, resulting in developmental abnormalities, infertility, and pathology (Sharpe and Skakkebaek, 1993). These chemicals include phytoestrogens (natural endocrine disruptors with estrogenic activity), xenoestrogens (synthetic endocrine disruptors), and industrial contaminants such as dioxins, biphenyls, heavy metals, pesticides, and others. It is possible that exposure to these chemicals could cause adverse developmental and reproductive events, especially following high levels of occupational exposure. Furthermore, high levels of exposure could cause germline mutations that result in transgenerational effects, such as decreased sperm count in subsequent male generations (Anway et al., 2005). However, there is still little direct evidence that normal exposure to these chemicals cause male reproductive malfunction in the general population (Safe, 2000; Joffe, 2003; Pflieger-Bruss et al., 2004).

Libido

A fear of diminished libido is likely to be one of the primary reasons why a hormonal male contraceptive may never achieve widespread adoption (Martin et al., 2000). But the relationship between hormones and sexual motivation is equivocal anyway. Male reproductive endocrinology is sensitive to perceived coital opportunities. For example, male rhesus macaques demonstrate increased testosterone levels in both the breeding and nonbreeding seasons when provided access to females (Bernstein et al., 1977; Glick, 1984). Increased androgen levels would be beneficial to support increased frequency of sexual activity (Rose et al., 1972) and actively establish and defend mating partners (Bernstein et al., 1977; Glick, 1984).

Testosterone levels in men also increase with anticipation of a sexual encounter. For example, a classic study describes how a lone male conducting research on a remote island weighed his daily beard clippings as a gross androgen bioassay and found that his beard growth rate increased immediately prior to leaving the island for female companionship (Anonymous, 1970). More recently, a sample of heterosexual men exhibited increased testosterone levels during short conversations with women (Roney et al., 2003). There was also a positive relationship between change in testosterone and males' ratings of female romantic potential.

Testosterone levels also change in response to copulation, with significantly increased androgen levels following the coital event (Dabbs and Mohammed, 1992). Likewise, exposure to erotic sensory stimuli is correlated with increases in testosterone levels in heterosexual men (Carani et al., 1990). It is difficult, however, to identify any consistent relationships between variation in a man's testosterone levels and actual sexual motivation. In hypogonadal men, testosterone injections at high doses are related to increased frequencies of erotic thoughts, erections, and sexual activity (Davidson et al., 1978). In contrast, testosterone supplementation in healthy, eugonadal men is unrelated to intensity of sexual feelings or activity (Buena et al., 1993). Therefore, testosterone levels appear to be associated with libido only in men with low testosterone to begin with, such as during clinical deficiencies or prolonged abstinence.

Variation in testosterone levels does not correlate well with sexual motivation in healthy men. Rather, sexual motivation depends largely on previous sexual experiences. Testosterone and other hormones may increase the likelihood that a sexual stimulus will elicit sexual behaviors, but only in the context of appropriate stimuli. Similarly, androgens can reduce the amount of stimuli required for an erection. They are, however, not required to produce or maintain an erection.

Erection

When afforded the capability of generating and maintaining an erection followed by successful ejaculation, an elderly man can fertilize an ovum. Pablo Picasso became a father at age 68, as did Charlie Chaplin at age 73. The major rate-limiting steps at this age are attracting a fecund mate and generating/maintaining an erection. Causes of erectile dysfunction (ED) can be psychological, neuroendocrinological, and pharmacological in origin. Degeneration of the vascular system due to stress, disease, or lifestyle (diet and smoking) can interfere with the normal process of erection (Caretta et al., 2006). The prevalence of ED ranges from 10–22% among various countries (Rosen et al., 2004), and the risk and degree of ED increases with age, around 8% per year in men aged 45–60 (Kratzik et al., 2005). At age 70, more than 50% of men will exhibit some form (minimal to severe) of ED (Kratzik

et al., 2005). Even in remote populations, ED is common among older men (Gray and Campbell, 2005). Furthermore, the incidence of ED appears to be increasing worldwide, with more than 300 million adult men expected to be afflicted with it in the year 2025 (McKinlay, 2000). The number of cases reported may be increasing due to the recent acceptance of discussing and diagnosing such ailments in the elderly.

A considerable amount of research efforts have now gone into the treatment of ED, from penile implants and prostheses to pharmacotherapy (Fabbri et al., 1997). One of the most significant advances has been the development of sildenafil citrate (Viagra®, Pfizer Inc.), an oral phosphodiesterase inhibitor that inhibits the breakdown of nitric oxide-induced cyclic guanosine monophosphate (cGMP), mimicking the effects of nitric oxide by stimulating the relaxation of muscles around the corpora cavernosa (Bivalacqua et al., 2000; Argiolas and Melis, 2003).

Paternal behaviors

A direct consequence of internal fertilization is that men cannot be absolutely certain about their paternity. The rate of human extrapair copulations may be as high as 30% (Baker and Bellis, 1995), and the worldwide misappropriated paternity rate is around 2%, although it can be as high as 30% in cases of low paternity confidence (Anderson, 2006). It is therefore not surprising that natural selection has favored the production of behavioral mechanisms that predispose men (in general) to invest more heavily in mating effort than parental effort. However, humans are among the 5% of mammalian species that do exhibit some type of paternal care (Clutton-Brock, 1991). Basic paternal care is characterized not only by defense, but also by holding, carrying, provisioning, and social interaction.

The level of paternal investment is directly correlated with both paternity confidence (Anderson, 2006) and assessment of a wife's fidelity (Apicella and Marlowe, 2004). Physical resemblance between father and child can increase paternal care and resource investment (Christenfeld and Hill, 1995). Phenotype matching, or the recognition of common phenotypic traits such as appearance or odor, certainly plays some role in kin recognition in primates and other animals (Alberts, 1999; Parr and de Waal, 1999). For example, adult male baboons can differentiate between their own offspring and unrelated juveniles and support their own offspring more frequently during agonistic disputes (Buchan et al., 2003). Physical resemblance between a human father and child is actually not more common than that between mother and child, and misidentification of fathers of young children from photographs is quite random, averaging around 50% (Brédart and French, 1999).

Unlike women, men are not required by their biology to provide any care for offspring beyond the contribution of sperm. Human male biology does however respond in interesting manners to both pair bonding and fatherhood. Such behaviors are characterized by elevated prolactin and suppressed testosterone levels in men (Storey et al., 2000; Gray et al., 2004). Combined, these may function to decrease interest and effort in acquiring new mates as well as facilitate interest in paternal behaviors (Schoech et al., 1998) (see Chapter 16 of this volume for a more complete discussion).

MALE REPRODUCTIVE PHYSIOLOGY

Taking a step back for a moment, we wish to provide readers with some basic background on male reproductive physiology, in part because this may clarify some of the above-mentioned discussion, as well as prepare the reader for our discussion of male reproductive ecology to follow. From a life history perspective, male reproductive physiology has evolved to optimize energetic allocation decisions and to maximize lifetime reproductive success. Investment in reproductive function and somatic tissue reflective of reproductive effort are primary targets of hormonal action and therefore subject to significant selection. That is, males with reproductive systems that were most efficient at allocating energy likely had higher lifetime reproductive success. Such efficiency can emerge from variation in hormone levels, receptor number, receptor sensitivity, and perhaps genetic differences in the coding and transcription of receptor and hormone proteins. It is important to remember that the evolution of male reproductive physiology was shaped by paternal uncertainty and sex differences in offspring investment resulting from internal gestation.

Endocrinology and development

Male endocrine function is based on the hypothalamic-pituitary-testicular (HPT) system. The preoptic area of the hypothalamus, a small collection of neurons at the base of the brain, secretes gonadotropin-releasing hormone (GnRH) into a conduit called the hypophyseal portal leading to the pituitary gland. Gonadotropin-releasing hormone stimulates the production of two gonadotropins, luteinizing hormone (LH) and follicle-stimulating hormone (FSH). Both of these gonadotropins are transcribed from specific genes and are composed of a common alpha dimer and a specific beta dimer (Kronenberg et al., 2007). LH and FSH are therefore similar in molecular structure but impose unique physiological functions. LH is primarily responsible for stimulating the production of testosterone by Leydig cells in the testes and FSH aids in the maturation and production of sperm by Sertoli cells. In general terms,

the HPT axis acts as a negative feedback loop. As sex hormones such as testosterone rise, it increases the likelihood of decreasing the production of GnRH and LH. Estradiol also appears to have a significant effect on the negative feedback process on the hypothalamus (Hayes et al., 2000; Rochira et al., 2006). As testosterone levels drop, the production of GnRH and LH increases. Gonadotropin-releasing hormone and all downstream hormones are produced in a pulsatile fashion due to the initial neuronal impulses that trigger and maintain GnRH secretion. Similar negative feedback loops govern spermatogenesis although in a more attenuated fashion. As FSH rises, production of inhibin A and B rise, which also have negative feedback effects on GnRH. Activin, another peptide, promotes FSH production (Kronenberg et al., 2007).

The effects of androgens on male somatic, reproductive, and behavioral development begin in utero (Knickmeyer and Baron-Cohen, 2006). Müllerian-inhibiting factor, testosterone, and dihydrotestosterone promote defeminization and masculinization of the genitalia (Grumbach and Conte, 1998). Testosterone levels rise during mid-gestation and then fall prior to birth (Siiteri and Wilson, 1974). There is a second rise in testosterone level that falls again prior to the first year of age (Forest et al., 1974). The functions of these surges in androgen levels are not completely understood, although they may play important roles in sexual differentiation of the central nervous system as well as priming of androgen target tissues (De Moor et al., 1973; Waldhauser et al., 1981; Wilson, 1982; Corbier et al., 1992; Davies and Norman, 2002). Several factors, including genetic polymorphisms of pathway synthesis and clearance enzymes, nutritional status, immunological stress, and social stress during development may play important roles in "programming" baseline testosterone secretion for later adulthood (Allen et al., 2001; Bribiescas, 2001; Zitzmann and Nieschlag, 2001; Muehlenbein and Bribiescas, 2005; Jakobsson et al., 2006; Muehlenbein, 2008). For example, populations experiencing chronic energetic stress have lower insulin levels that may inhibit gonadotropin secretion, adolescent hormone priming, and subsequently lower adult testosterone levels (Bruning et al., 2000).

Differential testosterone levels between individuals and populations (discussed in detail below) may be caused by these early priming effects on receptor number and sensitivity as well as HPT function in general. It is clear that priming effects of exogenous testosterone in boys with delayed puberty enhances the effects of growth hormone (Muller et al., 2004), and that individuals with idiopathic hypogonadotropic hypogonadism, a condition characterized by a congenital deficiency of GnRH, can respond to exogenous GnRH if they have prior exposure to testosterone, suggesting a priming effect (Spratt and Crowley, 1988). Among chronically undernourished Indian men, administration of human chorionic gonadotropin (hCG, a potent stimulator of testosterone production) resulted in a substantially muted testosterone increase compared to well-fed controls (Smith et al., 1975). Testosterone variation between adult individuals and populations may be the result of similar effects prior to reproductive maturation.

Puberty is a pivotal point in the life history of a male, marked by a shift in reproductive and somatic investment from survivorship to reproductive effort. Androgens (particularly the reduction of testosterone to dihydrotestosterone) control hair, vocal cord and genitalia development, fat catabolism, and skeletal muscle anabolism in boys. Increases in adrenal androgens anticipate the pubertal rise in sex steroids. The process, known as adrenarche, promotes the desensitization of the hypothalamus to androgen levels and contributes to the onset of puberty. Growth of the adrenal gland increases levels of adrenal androgens that in turn desensitize the hypothalamic negative feedback response, resulting in greater tolerance for higher levels of androgens such as testosterone (Conley et al., 2004; Campbell, 2006). See Chapter 8 in this volume for a longer discussion of adrenarche.

Timing of the onset of pubertal maturation may be affected by environmental factors such as activity, lifestyle, and nutrition. Kulin et al. (1984) reported lower levels of urinary LH in malnourished Kenyan boys. Later puberty in boys in association with chronic energetic stress is also evident among the Turkana of Kenya and the Tonga of Zambia (Campbell et al., 2004, 2005a). The effects on fertility are unknown although well-nourished boys tend to be larger, have greater body mass, and higher testosterone levels as adults. There is such an association between larger body size and higher fertility in Aché men and women of Paraguay (Hill and Hurtado, 1996).

Social factors such as group composition may also influence variation in developmental timing. For example, in rodents, exposure of female pups to adult males is associated with accelerated maturation whereas exposure to adult females can cause delayed sexual maturation (Vandenbergh, 1967, 1973). In response to social cues, males from a variety of species can exhibit alternative adult phenotypes with different reproductive strategies. Such alternative male phenotypes/strategies have been demonstrated in lizards, reef fish, and primates. An excellent example includes orangutans for which there are two morphotypes of adult males. "Flanged" males are those with fully developed secondary sexual characteristics (e.g., large cheek pouches) that usually maintain a territorial range and mate with resident females (Mitani, 1985). The "unflanged" males are much smaller, lacking fully

developed secondary sexual characteristics. They do not normally maintain a territorial range, are usually tolerated by flanged males in the area, and have been observed force copulating with females (Mitani, 1985). Adults of each phenotype may be of reproductive age but exhibit different hormone profiles. The unflanged males appear to be developmentally stunted with lower androgen, gonadotropin, growth hormone, and thyroid-stimulating hormone levels compared to flanged males, yet both are reproductively capable (Maggioncalda et al., 1999, 2000). By maintaining a smaller body size, unflanged males are tolerated by flanged males and do not pay the costs of increased energetic expenditure and immunosuppression caused by higher androgen levels, although they still gain access to some reproductive opportunities. The development of this alternative phenotype in other males may be triggered by harassment from the fully flanged male, the release of pheromones from the flanged male, or even via their vocalizations. The hypothalamus does possess numerous auditory connections, and thus an auditory signal could theoretically affect GnRH secretion from the hypothalamus, altering the developmental process of the animal (Ronnekliev and Resko, 1990). Either way, these alternative male phenotypes are viewed as adaptations to the social environment.

Social factors may also influence variation in human developmental timing. For example, Ellis and Garber (2000) have reported that girls who are raised in families characterized by relatively high levels of stress (e.g., history of maternal mood disorders, stressful interpersonal relationships, lack of resources, biological father absence, stepfather presence, etc.) may mature more quickly, although the relevant effect size appears to be very small. Presence of an alternative father figure may trigger earlier development in peripubertal women in order to facilitate mating between the female and unrelated male. Alternatively, the presence of a stepfather may contribute to family discord, resulting in psychopathology, and earlier maturation in females may facilitate dispersal from the unstable/dangerous environment. Future studies should be designed to investigate similar effects in males. It is likely that the energetic determinants of menarche and pubarche far outweigh any social determinants.

Given the amount of phenotypic variation in humans, it is not possible to categorize men into alternative reproductive phenotypes. Different men certainly practice different reproductive strategies. Men also consistently develop later than women, and it has been suggested that earlier maturation in females may allow women to acquire important parenting skills while still being subfecund whereas delayed maturation in males may postpone direct competition with other males for access to mates and allow for attainment of larger body size (Bogin, 1999). In this manner, human pubertal timing for both males and females may be an adaptation to maximize lifetime reproductive success in response to environmental cues.

Spermatogenesis and erectile function

Spermatogenesis is the process of the production of male gametes. It is a 64 day cycle in which haploid germ cells (spermatogonia) eventually develop into mature sperm. Spermatogenesis takes place within the seminiferous tubules in the testes. Separating the seminiferous tubules is interstitial tissue which contains Leydig cells, the primary source of testosterone. The stages of sperm development are quite distinct. Germ cells, known as spermatogonia, initially divide mitotically to form spermatocytes. Spermatocytes slowly migrate through interstitial space between Sertoli cells that aid in their maturation and development. Leydig cells also provide endocrine support for spermatocyte maturation. An initial process of meiosis results in haploid secondary spermatocytes. During a second meiotic division, spermatids are formed. During the final stages of maturation, the acrosome and tail develop before they pass into the lumen of the seminiferous tubule. In total, about 300–600 sperm per gram of testicular tissue are produced (Nussey and Whitehead, 2001). Unlike oogenesis, some daughter spermatogonia do not complete the maturation cycle and revert to a primordial stage, therefore allowing a virtually inexhaustible number of sperm to be produced. However, most daughter cells undergo several divisions and ultimately produce spermatids and complete cell differentiation.

Sperm consist of three basic components, the head, tail, and nucleus. The head consists of the nucleus as well as the acrosome, which contains enzymes that allow penetration of the ova and deposition of its genetic material. The tail contains numerous mitochondria and flagellates vigorously to propel the sperm towards the ova. Some sperm are however defective and do not swim correctly or have multiple heads or tails (Guzick et al., 2001). Because so many sperm are produced, gamete quality control is less rigorous than in oogenesis. Various hormones such as FSH, inhibin, and testosterone are all involved in the maturation process. Immature sperm are passed from the rete testes, to the vasa efferentia, and ultimately stored in the epididymis where they are mixed with seminal fluid (semen) produced by the seminal vesicle and prostate gland. At this stage, the sperm become fully motile. During ejaculation sperm and semen are ejected through the vas deferens and out through the urethra.

In order to complete coital intromission, the penis must increase its rigidity through the production and maintenance of an erection. In a flacid state, blood circulation through the penis is unrestricted. Controlled by the parasympathetic aspect of the autonomic nervous system, blood enters the base of the penis, circulates down the shaft, to the head (glans penis) and back out to the body. In a healthy man, sexual stimuli induce changes in neurotransmitter levels, specifically elevations in nitric oxide levels in nerve terminals of the penis which then activates formation of the enzyme guanylate cyclase. This enzyme causes increased formation of cGMP with consequent decreases in intracellular calcium levels in corpora cavernosa (erectile body) muscle cells. Relaxation of this tissue allows for increased blood flow and subsequent engorgement (Andersson and Wagner, 1995; Wagner and Mulhall, 2001).

Senescence

Human senescence can be broken down into two basic components, somatic and reproductive. In life history theory, senescence is unique from aging in that senescence is the manner, timing, and rate at which physiological aspects lose functional capacity and efficiency, ultimately leading to death (see Chapter 31 of this volume), whereas aging can be viewed as the passage of time. This is a key conceptual difference in life history theory since patterns of senescence reveal much about the energetic and time constraints that molded the evolution of a species. Endothermy, ectothermy, body size and metabolic rate, and extrinsic mortality (i.e., predation) all contribute to the evolution of senescence. For example, a tortoise and a mouse may be of equal age, say three years old, but exhibit very different patterns of senescence. Mice, who are endothermic and small, have high metabolic rates and spend a relatively great amount of energy on early sexual maturation and reproduction due to high extrinsic mortality, hence their short life spans. Interestingly, rates of senescence differ significantly between humans and chimpanzees living in the wild and in captivity (Hill et al., 2001), suggesting that decreases in extrinsic mortality may have been an important aspect of human evolution (Kaplan et al., 2000).

Reproductive senescence in human males is distinct from menopause in women. Unlike menopause, there is no abrupt cessation of reproductive function and onset of permanent infertility. In men, declines in reproductive function are more subtle and exhibit a wide range of population variation. Male reproductive senescence may involve declines in fertility although further evidence is needed (de La Rochebrochard et al., 2006). Various studies have reported deterioration of sperm morphology and motility with age in otherwise healthy men (Schwartz et al., 1983; Kidd et al., 2001), whereas other studies have not (Andolz et al., 1999;

Tortolero et al., 1999). Genetic abnormalities are more common in sperm of older men (Plas et al., 2000), and older men are at increased risk of producing children with autism (Reichenberg et al., 2006) and schizophrenia (Malaspina et al., 2001).

Sex hormone levels also tend to change with age, although such changes are not ubiquitous across populations. In a meta-analysis of several populations in which all laboratory analyses were conducted using the same assay protocol, Ellison et al. (2002) showed that while Boston men exhibited the standard decline in testosterone, other populations showed a more modest decline and others none at all. More refined differences among Japanese men included a testosterone decrease from the second decade of life until their 40s. However, testosterone then remained unchanged well into their 70s (Uchida et al., 2006). Increases in sex hormone binding globulin (SHBG) contribute to declines in free testosterone (Gray et al., 1991; Harman et al., 2001). Changes in estradiol are sometimes also evident with age, although this may be the result of increased adiposity and aromatization (Vermeulen et al., 2002), and decreases have also been reported (Orwoll et al., 2006).

In most well-nourished populations testosterone exhibits a modest steady decline after the age of 40 while LH and FSH exhibit a steady rise, most likely due to decreased testicular receptor sensitivity. There is little or no variation between populations in gonadotropin changes with age. Among Aché men of Paraguay, testosterone levels were significantly lower compared to US men and exhibited no decline with age. However, FSH and LH were positively associated with age, as is common in other populations suggesting that this aspect of male reproductive senescence is not subject to environmental variability (Bribiescas, 2005). Of further interest is the decline in hypothalamic sensitivity to energetic status in older men. Fasted older men fail to exhibit any improvement in LH pulse number or amplitude in response to GnRH administration compared to younger men (Bergendahl et al., 1998). Perhaps a previously unrecognized aspect of male reproductive senescence is the compromised ability to adjust reproductive function in response to energetic availability (Bribiescas, 2006).

Somatic changes with age are also common. Muscle mass declines and adiposity increases (Guo et al., 1999). Free testosterone is associated with muscle mass and negatively associated with fat mass in elderly men in Western countries (van den Beld et al., 2000). Basal metabolic rate decreases with age in tandem with lower testosterone (Fukagawa et al., 1990). Strength also declines with age, a characteristic of male aging that is shared across populations (Walker and Hill, 2003). Interestingly, peak hunting efficiency occurs decades after the height of physical strength and performance in Aché men, suggesting a decoupling between foraging efficiency and strength as well as an increased reliance on skill, planning, and experience (Walker et al., 2002).

Costs of testosterone

Although most readers might assume that testosterone and other androgens are synonymous with maleness in most cultures, the fact is that male physiology imposes costs to survivorship beginning at a young age. Male fetuses and newborns exhibit higher mortality than females (Kellokumpu-Lehtinen and Pelliniemi, 1984; Byrne and Warburton, 1987; Ingemarsson, 2003). The average life span of a woman in almost all populations is significantly longer than a man's (Buettner, 1995). In order to maintain some equilibrium between the sexes in a population, the sex-ratio at birth needs to be slightly male-biased, which is typically the case (Parazzini et al., 1998). In young adulthood, the chances of a male being murdered rises dramatically compared to a female, and men account for over 85% of violent crime committed in the United States (Daly and Wilson, 1983). In fact, it is safe to say that high mortality (relative to females) is a male characteristic.

Energetic costs

As discussed above, the energetic burden of spermatogenesis varies by species. For some, like *Drosophila* and *Caenorhabditis*, spermatogenesis can be expensive, whereas for mammals it is not (Elia, 1992; Gems and Riddle, 1996). In men, sperm quality or quantity are relatively unaffected by even long-term, high energetic output (Bagatell and Bremner, 1990). However, attaining and maintaining adequate musculoskeletal function (e.g., skeletal muscle mass, red blood cells, cortical bone density, etc.) are energetically taxing. This would reflect an investment in male reproductive effort by augmenting inter- and intrasexual competition (i.e., male–male conflict and female sexual coercion), mate attraction, and protection of mates and offspring. Given the energetic demands of musculoskeletal function, such an investment could compromise survivorship under conditions of resource restriction (Bribiescas, 2001).

Muscle anabolism is enabled through the actions of testosterone and other androgens (Bhasin et al., 1996; Tsai and Sapolsky, 1996). In addition to anabolic effects, testosterone stimulates adipose tissue redistribution (Marin et al., 1992a, 1992b; Welle et al., 1992). Altered somatic composition and increased energetic costs due to elevated androgen levels can result in a negative energy balance that can compromise survivability. This has been demonstrated in various species of birds and reptiles (Marler and Moore, 1988; Ketterson et al., 1992; Marler et al., 1995). In men, skeletal muscle tissue accounts for approximately

20% of basal metabolic rate (Elia, 1992), and this measure surely increases during periods of high activity. Elevated testosterone levels and metabolic rates could not only contribute to elevated production of free oxygen radicals (Zirkin and Chen, 2000) but also reduced tissue and organ maintenance in humans (Bribiescas, 2001; Muehlenbein and Bribiescas, 2005).

Immunosuppression

Elevated testosterone levels could compromise survivorship by causing suppression of immune functions. These immunoregulatory actions of testosterone are based on: (1) comparing male and female differences in immunocompetence; (2) examining associations between circulating endogenous testosterone levels and measurements of immune function, such as size of immune organs or lymphocyte counts in healthy or parasitized animals; (3) experimentally manipulating testosterone levels through castration or supplementation and observing subsequent effects on immunocompetence; and (4) performing in vitro analyses of immune-endocrine interactions (for a complete review, see Muehlenbein and Bribiescas, 2005).

In brief, males tend to be more susceptible to a variety of diseases, and both prevalence and intensity of infection is often higher in males than in females (Poulin, 1996; Zuk and McKean, 1996; Fedigan and Zohar, 1997; Moore and Wilson, 2002). For example, the rate of primary and secondary syphilis is greater in American men than women, and the incidence of syphilis in men continues to increase (CDC, 2009). However, male biases in disease may be caused by a number of factors besides dimorphism in endocrine function. Several factors such as exposure rates, social and sexual behaviors, habitat, diet, and hormone levels may account for some of these differences (Klein, 2000).

Only a few studies have investigated immune-endocrine correlates in human males. In a large sample of healthy military men, Granger et al. (2000) found no association between testosterone and T or B lymphocytes. In contrast, testosterone levels were positively associated with parasitemia of *Plasmodium vivax* malaria in Honduran men (Muehlenbein et al., 2005).

Testosterone replacement therapy is currently being evaluated as a means to improve physical function as well as general health perceptions in a number of cohorts, including elderly, obese, and HIV-infected men (Marin et al., 1992a, 1992b; Wilson et al., 2000). However, the effects of testosterone supplementation on immune parameters in these individuals are not well investigated. It appears that varying doses of testosterone do not significantly alter C-reactive protein levels in healthy eugonadal males (Singh et al., 2002). Similarly, testosterone supplementation does not cause changes in absolute and percentage $CD4^+$ and $CD8^+$ cell counts and plasma HIV RNA copy number in HIV-infected men (Bhasin et al., 2000). Given the popularity of testosterone supplementation, both legal and illegal, research investigations should be specifically designed to more adequately determine any negative effects on immune functions. In vitro analyses clearly indicate that testosterone can inhibit lymphocyte proliferation and activity, and antibody and cytokine production (Weinstein and Bercovich, 1981; Daynes and Araneo, 1991; Grossman et al., 1991; Grossman, 1995; Olsen and Kovacs, 1996; Giltay et al., 2000; Straub and Cutolo, 2001; Burger and Dayer, 2002; Wunderlich et al., 2002).

In addition to causing immunosuppression, elevated testosterone levels could also compromise survivorship by decreasing the amount of energy and nutrients available for somatic repair and the maintenance and activation of immune responses (Wedekind and Folstad, 1994; Sheldon and Verhulst, 1996; Muehlenbein and Bribiescas, 2005). Investment in male reproductive effort generates a significant energetic demand that will theoretically trade-off with the competing energetic demands of immunocompetence, which would increase susceptibility to disease or otherwise negatively affect convalescence (Sheldon and Verhulst, 1996; Raberg et al., 1998, 2002; Verhulst et al., 1999; Owens, 2002; Schmid-Hempel, 2003; Muehlenbein and Bribiescas, 2005). Such effects may be more exaggerated in energetically limited populations, and more muted in well-nourished populations.

Immunocompetence is energetically expensive. Prolonged energy restriction can lead to immune suppression, increasing the risk of infection from opportunistic pathogens (Kramer et al., 1997; Klasing, 1998; Lin et al., 1998; Shephard et al., 1998; Ing et al., 2000; Koski and Scott, 2001). Likewise, strenuous exercise or participation in energetically demanding tasks, such as reproduction, can also compromise immune functions (Norris et al., 1994; Richner et al., 1995; Deerenberg et al., 1997; Nordling et al., 1998; Nelson et al., 2002; Bonneaud et al., 2003). Changes in metabolism following infection as well as in vivo assessments of energy consumption by the immune system also suggest that immunocompetence is energetically expensive (Newsholme and Newsholme, 1989; Demas et al., 1997, 2003; Lochmiller and Deerenberg, 2000). Resting metabolic rates increase in infants during upper respiratory tract infections (Fleming et al., 1994), as well as in East African children during acute episodes of the measles (Duggan et al., 1986). Compared to other species, the energetic costs of mounting an immune response in adult humans have not been well described (but see Muehlenbein et al., in press), and the clinical medicine and evolutionary biology communities would benefit from further clarification of these costs.

Prostate cancer

The prostate gland surrounds the urethra just below the bladder. Its primary function is to produce, store, and secrete a fluid that protects sperm when ejaculated into the vaginal tract. Prostate cancer is now the most common malignancy of men in the United States: approximately one in six men will develop malignant prostate cancer in their lifetime (Crawford, 2003). The incidence of prostate cancer has varied over the past century, with a significant increase following the widespread use of digital rectal examination combined with assays for prostate-specific antigen in the 1980s and early 1990s (Crawford, 2003). Incidence also varies widely between populations with rates much higher in industrialized regions than in less developed areas (Quinn and Babb, 2002), although rates in Asia are very low compared to the rest of the world (Kurihara et al., 1989). Within the United States, African-American men have an incidence 60% greater than that of Caucasians (Crawford, 2003; Reddy et al., 2003).

There are likely multiple causative factors for prostate cancer. One risk factor is a high fat diet (Whittemore et al., 1995), as evidenced by increased risk in Asian populations following immigration into the United States and a switch from traditional diets to one with greater fat and meat intake (Cook et al., 1999). There is also some genetic predisposition to prostate cancer (Shibata and Whittemore, 1997; Amundadottir et al., 2006), with at least seven inherited susceptibility genes (Simard et al., 2002) and a number of other somatic gene mutations (Nelson et al., 2003).

Although the relationship between lifetime exposure to androgens and risk of prostate cancer remains unclear (Eaton et al., 1999; Hsing, 2001), treatment of prostate enlargement or cancer often involves the use of androgen suppressors (Anderson, 2003). Testosterone and dihydrotestosterone bind to prostate cells to induce proliferation and thus may contribute to prostate carcinoma (Carter et al., 1995). Maintaining high androgen levels would function to increase male reproductive effort, which would just outweigh the costs of increased risk of prostate cancer later in life. These physiological trade-offs are central to a modern understanding of male reproductive ecology.

MALE REPRODUCTIVE ECOLOGY

Life history theory and reproductive ecology

An organism's interconnected adaptations for development, survival and reproduction combine to form its life history strategy, and its life history traits describe age- and size-specific schedules of development, mortality, and fertility (Stearns, 1992). Life history theory is that part of evolutionary ecology which attempts to explain phenotypic evolution, and the elements of demography, trade-offs, bet-hedging, lineage-specific effects, quantitative genetics, and reaction norms help us to explain variation in life history traits and strategies between organisms. This suite of methods can be used for studying processes common to all species, including humans.

Reproduction is obviously central to the process of evolution. Reproductive physiologies and behaviors are evolved response systems, shaped by natural selection to adapt individuals to changing environments. This allows for a variable response in which a genotype can produce a range of phenotypes depending on environmental conditions (a "reaction norm"). However, this phenotypic plasticity is limited through lineage-specific effects (canalization of certain traits) as well as trade-offs (Stearns, 1989). Trade-offs are, in fact, central to life history theory, which predicts that selection will act on physiological and behavioral mechanisms that efficiently regulate the allocation of energy and time between general competing functions, specifically reproduction, maintenance (i.e., survival), and growth (Stearns, 1992; Hill, 1993; Kaplan et al., 2000). Because time and energy used for one purpose cannot be used for another, organisms often face trade-offs, particularly under conditions of resource restriction.

Reproductive physiologies and behaviors therefore represent compromised adaptations that have been designed by natural and sexual selection to maximize lifetime reproductive success. Suppression of current reproduction in order to increase the likelihood of successful future reproduction should function to maximize lifetime reproductive success, particularly in unpredictable and stochastic natural and social environments (Wasser and Barash, 1983). The trade-off between current and future reproduction is a common one in iteroparous organisms, as is the trade-off between reproduction and survivorship. Because of this, reproductive physiologies and behaviors are flexible and responsive to environmental cues, such as diet, stress, disease, and the availability of mates.

The field of reproductive ecology examines the effects of ecological factors on reproductive effort within the context of evolutionary and life history theories (Ellison, 2001). Recent detailed studies of human female reproductive ecology have clarified how female fecundity represents an adaptive reaction norm that can respond to changes in energy flux, disease, and psychological stress (Ellison, 1990, 1994, 2003; Ellison et al., 1993; Jasienska and Ellison, 1998, 2004). No longer viewed by most researchers as pathological, decreased ovarian function in response to ecological stressors likely represents an adaptive mechanism to lower the likelihood of a reproductive outcome if environmental circumstances are not optimal, allowing the female to wait and invest in a future reproductive event with a better likelihood of success.

Energetic investment in female mammalian reproduction is obviously expensive and very different from that of males. Until recently, the study of human male physiology has typically been conducted by clinicians and other medical investigators, not biological anthropologists or evolutionary biologists. This has changed in recent years with increased clarification of the adaptive functions of male reproductive variation (Campbell and Leslie, 1995; Bribiescas, 2001; Muehlenbein and Bribiescas, 2005).

Causes of variation in male testosterone levels

There are enormous amounts of variation in hormone levels within and between men. Moment-to-moment variation in testosterone level is caused by the pulsatile release of GnRH from the hypothalamus (Spratt and Crowley, 1988). Testosterone production can be influenced by many factors, including genetics (Jameson, 1996; Beranova et al., 2001; Ring et al., 2005) and diet. For example, high alcohol consumption (Muller et al., 2003), low zinc intake (Abbasi et al., 1980), and a low carbohydrate diet (Anderson et al., 1987) may all cause decreased testosterone synthesis, although Key et al. (1990) and Deslypere and Vermeulen (1984) have reported no difference in testosterone levels in omnivorous versus vegan/vegetarian (low fat) diets. Others have suggested low testosterone levels are associated with a low protein diet (Christiansen, 1991b). Obese individuals also typically have lower testosterone levels, due in part to aromatization of testosterone into estrogens in adipose tissue (Kley et al., 1981). However, as discussed earlier, relative changes in energy balance have little effect on male reproductive function.

Testosterone levels may vary according to energy balance, but only under the most taxing circumstances. For example, complete fasting decreases testosterone and hypothalamic-pituitary functioning, although these effects appear to be quickly reversible (Klibanski et al., 1981; Röjdmark, 1987; Bergendahl et al., 1991; Opstad, 1992; Veldhuis et al., 1993). Acute changes in food availability also do not cause significant changes in testosterone levels in human males (Bentley et al., 1993; Ellison and Panter-Brick, 1996). Similar results have been identified in nonhuman primates: testosterone levels do not significantly decrease in wild male orangutans or chimpanzees during periods of low food availability (Knott, 1999; Muller and Wrangham, 2005).

Variation in energy expenditure also appears to have little impact on male reproductive function.

Recall that spermatogenesis is relatively insensitive to variation in energy balance. Variation in normal or seasonal workload and exercise also appears to have little effect on testosterone levels (Ellison and Panter-Brick, 1996; Bribiescas, 2001). In fact, energy expenditure appears to negatively affect testosterone levels only after long-term high output (Bagatell and Bremner, 1990; Roberts et al., 1993; Gomez-Merino et al., 2003, 2005). Acute, anabolic exercise can actually cause increases in testosterone levels (Cumming et al., 1986).

Depressed testosterone levels can result from acute or chronic exposure to psychological stressors, including skydiving (Chatterton et al., 1997) and military training (Gomez-Merino et al., 2003, 2005). These effects are likely caused by the actions of glucocorticoids and endogenous opioids on the hypothalamic-pituitary-gonadal system. For example, glucocorticoids like cortisol can directly suppress Leydig cell function (Gao et al., 2002; Hardy et al., 2005) and downregulate testicular LH receptors (Aakvaag et al., 1978; Bambino and Hsueh, 1981). Glucocorticoids can also suppress the production and secretion of gonadotropins from the hypothalamus and pituitary (Doerr and Pirke, 1976; Attardi et al., 1997; Kalantaridou et al., 2004; Mitchell et al., 2005; Breen and Karsch, 2006). Endogenous opioids and cytokines can similarly affect the hypothalamus (Gilbeau and Smith, 1985; Sapolsky and Krey, 1988; Isseroff et al., 1989; Bonavera et al., 1993; Oktenli et al., 2004).

These proximate mechanisms, in addition to others, are also responsible for inhibited reproductive processes, including hypogonadism and hypogonadotropism, following injury, infection and immune activation. The degree of response is often associated with the degree of disrupted somatic injury (Spratt et al., 1993; Cernak et al., 1997). Some infections can cause direct testicular pathology in humans (Nelson, 1958; Iturregui-Pagán et al., 1976) in addition to diminished libido (Yirmiya et al., 1995). Men infected with HIV frequently exhibit low testosterone levels, testicular pathology, and azoospermia, possibly through direct infection of the testes and/or hypothalamus-pituitary (Poretsky et al., 1995; Lo and Schambelan, 2001; Dobs and Brown, 2002). Honduran men infected with *Plasmodium vivax* exhibit significantly lower testosterone levels during infection compared to postrecovery samples as well as age-matched healthy controls (Muehlenbein et al., 2005). Similar results have been identified with African sleeping sickness (*Trypanosoma brucei* ssp.) (Ikede et al., 1988; Hublart et al., 1990; Soudan et al., 1992; Reincke et al., 1998), toxoplasmosis (*Toxoplasma gondii*) (Stahl et al., 1995; Antonios et al., 2000; Oktenli et al., 2004), filaria (*Loa loa* and *Mansonella perstans*) (Landsoud-Soukate et al., 1989), and schistosomiasis (Saad et al., 1999).

Summary of male reproductive ecology

It is clear that acute immunological and psychological stress in addition to chronic negative energy balance are all potential factors that can affect male testosterone levels. Based on evidence presented throughout this chapter, human male reproductive neuroendocrine function therefore represents a reaction norm in which a number of factors interact to produce a phenotypically plastic response to environmental stimuli. The ability to alter testosterone levels in response to stimuli, such as availability of resources and/or injury and illness, likely represents an adaptive mechanism to augment either male reproductive effort or survivorship. As described above, augmenting mammalian male reproductive effort is largely accomplished through elevated testosterone levels and increased musculoskeletal performance (Bribiescas, 2001), which aids in work capacity, intersexual competition, intrasexual coercion, and mate choice. Spermatogenesis is energetically inexpensive and relatively insensitive to variation in testosterone levels, whereas musculoskeletal function is energetically expensive and is under tight physiological control by androgens. Elevated testosterone levels may enhance male reproductive effort, but at significant metabolic and immunological costs (Bribiescas, 2001; Muehlenbein and Bribiescas, 2005). These costs likely account for the functional significance of high variability in testosterone levels within and between individuals.

In environments characterized by high extrinsic mortality, unpredictable resources and/or high disease risk (i.e., periods of negative energy balance and immune activation), depressed testosterone levels and muscle atrophy would function to avoid some of the metabolic and immunological costs caused by otherwise higher testosterone levels. In this case, energy would be allocated to processes that maximize survivorship, such as adipocyte deposition or immunocompetence. In environments characterized by lower extrinsic mortality, high access to resources and low disease burden, testosterone levels could be elevated. These hypotheses may explain why, compared to more industrialized populations in the United States and Europe, testosterone levels are typically lower in forager, horticultural, and pastoral populations, including the Aché of Paraguay (Bribiescas, 1996), Ariaal of Kenya (Campbell et al., 2003), Aymara of Bolivia (Beall et al., 1992), Efe and Lese of the Democratic Republic of Congo (Ellison et al., 1989; Bentley et al., 1993), !Kung San of Namibia (Worthman and Konner, 1987; Christiansen, 1991a), Tamang and Kami of Nepal (Ellison and Panter-Brick, 1996), Gainj of New Guinea (Campbell, 1994), Turkana of Kenya (Campbell et al., 2006), men from Harare, Zimbabwe (Lukas et al., 2004), and the Okavango Delta of Namibia (Christiansen, 1991b). In fact, testosterone levels are about twice as high in North American men compared to other populations around the world, and their range of variation is about 10-fold greater. We see a similar pattern in chimpanzees: captive animals with high food abundance throughout their entire lives are characterized by relatively higher testosterone levels compared to natural populations with lower food abundance (Muller and Wrangham, 2005).

The relative contributions of food availability and disease risk in accounting for population variation in human testosterone levels are not yet tested. However, it is likely the case that ecological differences exert some developmental effects. Nutritional and immunological differences during development may play important roles in "programming" baseline testosterone synthesis and secretion as well as variability in later adulthood. Given its interconnected regulatory relationship with the reproductive, metabolic, and immune systems, testosterone levels in adulthood could then function as an important information transducer, regulating the differential investment in competing trade-offs according to the availability of energy (Bribiescas, 2001) and disease risk in the environment (Muehlenbein, 2008). Like females, human males maintain some ability to alter the expression of reproductive hormones, and therefore manage the life history trade-off of reproduction versus survival, in response to environmental stimuli.

DISCUSSION POINTS

1. Describe in detail the various aspects of male reproductive effort.
2. What are the costs and benefits of testosterone for males?
3. How does hormone variation affect sperm production?
4. How is male reproductive ecology different from that of females?
5. Discuss all of the major sources of variation in testosterone levels within and between men.
6. How would you test the hypothesis that differences in energy availability and disease/injury risk account for differences in baseline hormone levels and patterns of variation between people and populations?
7. What are some of the primary characteristics of male reproductive senescence?
8. From the standpoint of sexual selection, why would some males choose to "enhance" themselves through use of anabolic-androgenic steroid use? Why is this a poor decision?
9. Search the World Health Organization website for country-specific data on sex differences in mortality rates. In what countries, if any, do men outlive women on average? Why, according to this chapter, would these men tend to outlive women?

REFERENCES

Aakvaag, A., Sand, T., Opstad, P. K., et al. (1978). Hormonal changes in serum in young men during prolonged physical strain. *European Journal of Applied Physiology and Occupational Physiology*, **39**, 283–291.

Abbasi, A. A., Prasad, A. S., Rabbani, P., et al. (1980). Experimental zinc deficiency in man. Effect on testicular function. *Journal of Laboratory and Clinical Medicine*, **96**(3), 544–550.

Alberts, S. C. (1999). Paternal kin discrimination in wild baboons. *Proceedings of the Royal Society of London. Series B*, **266**, 1501–1506.

Albone, E. S. and Shirley, S. G. (1984). *Mammalian Semiochemistry: the Investigation of Chemical Signals between Mammals*. Chichester: John Wiley and Sons.

Alexander, R. D. and Noonan, K. M. (1979). Concealment of ovulation, parental care, and human social evolution. In *Evolutionary Biology and Human Social Behavior*, N. A. Chagnon and W. Irons (eds). North Scituate, MA: Duxbury Press, pp. 402–435.

Allen, N. E., Forrest, M. S. and Key, T. J. (2001). The association between polymorphisms in the *CYP17* and 5α-reductase (*SRD5A2*) genes and serum androgen concentrations in men. *Cancer Epidemiology, Biomarkers and Prevention*, **10**, 185–189.

Altmann, J. (1990). Primate males go where the females are. *Animal Behavior*, **39**, 193–195.

Amundadottir, L. T., Sulem, P., Gudmundsson, J., et al. (2006). A common variant associated with prostate cancer in European and African populations. *Nature Genetics*, **38**, 652–658.

Anderson, J. (2003) The role of antiandrogen monotherapy in the treatment of prostate cancer. *British Journal of Urology International*, **91**(5), 455–461.

Anderson, K. E., Rosner, W., Khan, M. S., et al. (1987). Diethormone interactions: protein/carbohydrate ratio alters reciprocally the plasma levels of testosterone and cortisol and their respective binding globulins in man. *Life Science*, **40**, 1761–1768.

Anderson, K. G. (2006). How well does paternity confidence match actual paternity? *Current Anthropology*, **47**, 513–520.

Andersson, K. E. and Wagner, G. (1995). Physiology of penile erection. *Physiological Review*, **75**, 191–236.

Andolz, P., Bielsa, M. A. and Vila, J. (1999). Evolution of semen quality in north-eastern Spain: a study in 22 759 infertile men over a 36 year period. *Human Reproduction*, **14**, 731–735.

Anonymous (1970). Effects of sexual activity on beard growth in man. *Nature*, **226**, 869–870.

Antonios, S. N., Ismail, H. I. and Essa, T. (2000). Hypothalamic origin of reproductive failure in chronic experimental toxoplasmosis. *Journal of Egyptian Society of Parasitology*, **30**(2), 593–599.

Anway, M. D., Cupp, A. S., Uzumcu, M., et al. (2005). Epigenetic transgenerational actions of endocrine disruptors and male fertility. *Science*, **3**(308), 1466–1469.

Apicella, C. and Marlowe, F. (2004). Perceived mate fidelity and paternal resemblance predict men's investment in children. *Evolution and Human Behavior*, **25**, 371–378.

Arce, J. C., De Souza, M. J., Pescatello, L. S., et al. (1993). Subclinical alterations in hormone and semen profile in athletes. *Fertility and Sterility*, **59**, 398–404.

Ardawi, M. S. M. and Newsholme, E. A. (1985). Metabolism in lymphocytes and its importance in the immune response. *Essays in Biochemistry*, **21**, 1–44.

Argiolas, A. and Melis, M. R. (2003). The neurophysiology of the sexual cycle. *Journal of Endocrinological Investment*, **26**(3), 20–22.

Attardi, B., Klatt, B., Hoffman, G. E., et al. (1997). Facilitation or inhibition of the estradiol-induced gonadotropin surge in the immature rat by progesterone: regulation of GnRH and LH messenger RNAs and activation of GnRH neurons. *Journal of Neuroendocrinology*, **9**, 589–599.

Aversa, A., Mazzilli, F., Rossi, T., et al. (2000). Effects of sildenafil (Viagra™) administration on seminal parameters and post-ejaculatory refractory time in normal males. *Human Reproduction*, **15**, 131–134.

Bagatell, C. J. and Bremner, W. J. (1990). Sperm counts and reproductive hormones in male marathoners and lean controls. *Fertility and Sterility*, **53**, 688–692.

Baker, R. R. and Bellis, M. A. (1989). Number of sperm in human ejaculates varies in accordance with sperm competition theory. *Animal Behaviour*, **37**, 867–869.

Baker, R. R. and Bellis, M. A. (1995). *Human Sperm Competition*. London: Chapman and Hall.

Bambino, T. H. and Hsueh, A. J. (1981). Direct inhibitory effect of glucocorticoids upon testicular luteinizing hormone receptor and steroidogenesis in vivo and in vitro. *Endocrinology*, **108**, 2142–2148.

Beach, F. A. (1976). Sexual attractivity, proceptivity, and receptivity in female mammals. *Hormones and Behavior*, **7**, 105–138.

Beall, C. M., Worthman, C. M., Stallings, J., et al. (1992). Salivary testosterone concentration of Aymara men native to 3600 m. *Annals of Human Biology*, **19**, 67–78.

Beamer, W., Bermant, G. and Clegg, M. (1969). Copulatory behavior in the ram, *Ovis aries*, II: factors affecting copulatory satiety. *Animal Behavior*, **17**, 706–711.

Becker, S. and Berhane, K. (1997). A meta-analysis of 61 sperm count studies revisited. *Fertility and Sterility*, **67**, 1103–1108.

Bentley, G. R., Harrigan, A. M., Campbell, B., et al. (1993). Seasonal effects on salivary testosterone levels among Lese males of the Ituri Forest, Zaire. *American Journal of Human Biology*, **5**, 711–717.

Beranova, M., Oliveira, L. M. B., Bédécarrats, G. Y., et al. (2001). Prevalence, phenotypic spectrum, and modes of inheritance of gonadotropin-releasing hormone receptor mutations in idiopathic hypogonadotropic hyprogonadism. *Journal of Clinical Endocrinology and Metabolism*, **86**, 1580–1588.

Bercovitch, F. B. (2001). Reproductive ecology of Old World monkeys. In *Reproductive Ecology and Human Evolution*, P. T. Ellison (ed.). New York: Aldine de Gruyter, pp. 369–396.

Bergendahl, M., Perheentupa, A. and Huhtaniemi, I. (1991). Starvation-induced suppression of pituitary-testicular function in rats is reversed by pulsatile gonadotropin-releasing hormone substitution. *Biology of Reproduction*, **44**, 413–419.

Bergendahl, M., Aloi, J. A., Iranmanesh, A., et al. (1998). Fasting suppresses pulsatile luteinizing hormone (LH) secretion and enhances orderliness of LH release in young but not older men. *Journal of Clinical Endocrinology and Metabolism*, **83**, 1967–1975.

Berger, K., Souza, H., Brito, V. N., et al. (2005). Clinical and hormonal features of selective follicle-stimulating hormone (FSH) deficiency due to FSH beta-subunit gene mutations in both sexes. *Fertility and Sterility*, **83**, 466–470.

Bermond, B., Mos, J., Meelis, W., et al. (1982). Aggression induced by stimulation of the hypothalamus: effects of androgens. *Pharmacology Biochemistry and Behavior*, **16**, 41–45.

Bernstein, I. S., Rose, R. M. and Gordon, T. P. (1977). Behavioural and hormonal responses of male rhesus monkeys introduced to females in the breeding and non-breeding seasons. *Animal Behavior*, **25**, 609–614.

Bhasin, S., Storer, T. W., Berman, N., et al. (1996). The effects of supraphysiologic doses of testosterone on muscle size and strength in normal men. *New England Journal of Medicine*, **335**, 1–7.

Bhasin, S., Storer, T. W., Javanbakht, M., et al. (2000). Testosterone replacement and resistance exercise in HIV-infected men with weight loss and low testosterone levels. *Journal of the American Medical Association*, **283**, 763–770.

Bivalacqua, T. J., Champion, H. C., Mehta, Y. S., et al. (2000). Adenoviral gene transfer of endothelial nitric oxide synthase (eNOS) to the penis improves age-related erectile dysfunction in the rat. *International Journal of Impotence Research*, **12**, S8–S17.

Bobek, B., Perzanowski, K. and Weiner, J. (1990). Energy expenditure for reproduction in male red deer. *Journal of Mammalogy*, **71**, 230–232.

Bogin, B. (1999). Evolutionary perspective on human growth. *Annual Review of Anthropology*, **28**, 109–153.

Bonavera, J. J., Kalra, S. P. and Kalra, P. S. (1993). Mode of action of interleukin-1 in suppression of pituitary LH release in castrated male rats. *Brain Research*, **612**, 1–8.

Bonneaud, C., Mazuc, J., Gonzalez, G., et al. (2003). Assessing the cost of mounting an immune response. *American Naturalist*, **161**(3), 367–379.

Booth, A., and Dabbs, J. M. (1993). Testosterone and men's marriages. *Social Forces*, **72**, 463–477.

Booth, A. G., Shelley, A., Mazur, G., et al. (1989). Testosterone and winning and losing in human competition. *Hormones and Behavior*, **23**, 556–571.

Brédart, S. and French, R. M. (1999). Do babies resemble their fathers more than their mothers? A failure to replicate Christenfeld and Hill (1995). *Evolution and Human Behavior*, **20**, 129–135.

Breen, K. M. and Karsch, F. J. (2006). Does season alter responsiveness of the reproductive neuroendocrine axis to the suppressive actions of cortisol in ovariectomized ewes? *Biology of Reproduction*, **74**, 41–45.

Bermond, B., Mos, J., Meelis, W., et al. (1982). Aggression induced by stimulation of the hypothalamus: effects of androgens. *Pharmacology, Biochemistry and Behavior*, **16**, 41–45.

Bribiescas, R. G. (1996). Testosterone levels among Aché hunter/gatherer men: a functional interpretation of population variation among adult males. *Human Nature*, **7**, 163–188.

Bribiescas, R. G. (2001). Reproductive ecology and life history of the human male. *Yearbook of Physical Anthropology*, **44**, 148–176.

Bribiescas, R. G. (2005). Age-related differences in serum gonadotropin (FSH and LH), salivary testosterone, and 17-β estradiol levels among Aché Amerindian males of Paraguay. *American Journal of Physical Anthropology*, **127**, 114–121.

Bribiescas, R. G. (2006). On the evolution, life history, and proximate mechanisms of human male reproductive senescence. *Evolutionary Anthropology*, **15**, 132–141.

Bromwich, P., Cohen, J., Stewart, I., et al. (1994). Decline in sperm counts: an artifact of changed reference range of "normal"? *British Medical Journal, Clinical Research*, **309**, 19–20.

Bruce, H. M. (1959). An exteroceptive block to pregnancy in the mouse. *Nature*, **184**, 105.

Bruning, J. C., Gautam, D., Burks, D. J., et al. (2000). Role of brain insulin receptor in control of body weight and reproduction. *Science*, **289**, 2122–2125.

Buchan, J. C., Alberts, S. C., Silk, J. B., et al. (2003). True paternal care in a multi-male primate society. *Nature*, **425**, 179–181.

Buena, F., Swerdloff, R. S., Steiner, B. S., et al. (1993). Sexual function does not change when serum testosterone levels are pharmacologically varied within the normal male range. *Fertility and Sterility*, **59**, 1118–1123.

Buettner, T. (1995). Sex differentials in old-age mortality. *Population Bulletin of the United Nations*, **39**, 18–44.

Burger, D. and Dayer, J. M. (2002). Cytokines, acute-phase proteins, and hormones: IL-1 and TNF-α production in contact-mediated activation of monocytes by T lymphocytes. *Annals of the New York Academy of Sciences*, **966**, 464–473.

Buss, D. M. and Schmidt, D. P. (1993). Sexual strategies theory: an evolutionary perspective on human mating. *Psychological Review*, **100**, 204–232.

Byrne, J. and Warburton, D. (1987). Male excess among anatomically normal fetuses in spontaneous abortions. *American Journal of Medical Genetics*, **26**, 605–611.

Calfee, R. and Fadale, P. (2006). Popular ergogenic drugs and supplements in young athletes. *Pediatrics*, **117**, e577–e589.

Campbell, B. C. (2006). Adrenarche and the evolution of human life history. *American Journal of Human Biology*, **18**, 569–589.

Campbell, B. C. and Leslie, P. W. (1995). Reproductive ecology of human males. *Yearbook of Physical Anthropology*, **38**, 1–26.

Campbell, B. C., O'Rourke, M. and Lipson, S. F. (2003). Salivary testosterone and body composition among Ariaal males. *American Journal of Human Biology*, **15**, 697–708.

Campbell, B. C., Gillett-Netting, R. and Meloy, M. (2004). Timing of reproductive maturation in rural versus urban Tonga boys, Zambia. *Annals of Human Biology*, **31**, 213–227.

Campbell, B. C., Leslie, P. W., Little, M. A., et al. (2005a). Pubertal timing, hormones, and body composition among adolescent Turkana males. *American Journal of Physical Anthropology*, **128**, 896–905.

Campbell, B. C., Prossinger, H. and Mbzivo, M. (2005b). Timing of pubertal maturation and the onset of sexual behavior among Zimbabwe school boys. *Archives of Sexual Behavior*, **34**, 505–516.

Campbell, B. C., Gray, P. B. and Ellison, P. T. (2006). Age-related patterns of body composition and salivary testosterone among Ariaal men of Northern Kenya. *Aging Clinical and Experimental Research*, **18**, 470–476.

Campbell, K. L. (1994). Blood, urine, saliva, and dip-sticks: experiences in Africa, New Guinea, and Boston. *Annals of the New York Academy of Sciences*, **709**, 312–330.

Carani, C., Bancroft, J., Del Rio, G., et al. (1990). The endocrine effects of visual erotic stimuli in normal men. *Psychoneuroendocrinology*, **15**, 207–216.

Carani, C., Scuteri, A., Marrama, P., et al. (1990). The effects of testosterone administration and visual erotic stimuli on nocturnal penile tumescence in normal men. *Hormones and Behavior*, **24**, 435–441.

Caretta, N., Palego, P., Roverato, A., et al. (2006). Age-matched cavernous peak systolic velocity: a highly sensitive parameter in the diagnosis of arteriogenic erectile dysfunction. *International Journal of Impotence Research*, **18**, 306–310.

Carter, H. B., Pearson, J. D., Metter, E. J., et al. (1995). Longitudinal evaluation of serum androgen levels in men with and without prostate cancer. *Prostate*, **27**, 25–31.

Centers for Disease Control and Prevention (CDC) (2009). *Sexually Transmitted Disease Surveillance 2007 Supplement, Syphilis Surveillance Report*. Atlanta, GA: US Department of Health and Human Services, Centers for Disease Control and Prevention.

Cernak, I., Savic, J. and Lazarov, A. (1997). Relations among plasma prolactin, testosterone, and injury severity in war casualties. *World Journal of Surgery*, **21**, 240–245.

Chatterton, R. T. Jr, Vogelsong, K. M., Lu, Y., et al. (1997). Hormonal responses to psychological stress in men preparing for skydiving. *Journal of Clinical Endocrinology and Metabolism*, **82**, 2503–2509.

Christenfeld, N. J. S. and Hill, E. A. (1995). Whose baby are you? Scientific correspondence. *Nature*, **378**, 669.

Christiansen, K. (1991a). Sex hormone levels, diet and alcohol consumption in Namibian Kavango men. *Homo*, **42**, 43–62.

Christiansen, K. H. (1991b). Serum and saliva sex hormone levels in !Kung San men. *American Journal of Physical Anthropology*, **86**, 37–44.

Clutton-Brock, T. H. (1991). *The Evolution of Parental Care*. Princeton, NJ: Princeton University Press.

Conley, A. J., Pattison, J. C. and Bird, I. M. (2004). Variations in adrenal androgen production among (nonhuman) primates. *Seminars in Reproductive Medicine*, **22**, 311–326.

Cook, L. S., Goldoft, M., Schwartz, S. M., et al. (1999). Incidence of adenocarcinoma of the prostate in Asian immigrants to the United States and their descendants. *Journal of Urology*, **161**, 152–155.

Cooper, T. G., Keck, C., Oberdieck, U., et al. (1993). Effects of multiple ejaculations after extended periods of sexual abstinence on total, motile and normal sperm numbers, as well as accessory gland secretions, from healthy normal and oligozoospermic men. *Human Reproduction*, **8**, 1251–1258.

Corbier, P., Edwards, D. A. and Roffi, J. (1992). The neonatal testosterone surge: a comparative study. *Archives Internationales de Physiologie*, **100**, 127–131.

Crawford, E. D. (2003). Epidemiology of prostate cancer. *Urology*, **62**, 3–12.

Cumming, D. C., Brunsting, L. A., Strich, G., et al. (1986). Reproductive hormone increases in response to acute exercise in men. *Medicine and Science in Sports and Exercise*, **18**, 369–373.

Dabbs, J. M. Jr and Dabbs, M. G. (2000). *Heroes, Rogues, and Lovers: Testosterone and Behavior*. New York: McGraw-Hill.

Dabbs, J. M. Jr and Mohammed, S. (1992). Male and female salivary testosterone concentrations before and after sexual activity. *Physiological Behavior*, **52**, 195–197.

Dabbs, J. M. Jr, Frady, R. L., Carr, T. S., et al. (1987). Saliva testosterone and criminal violence in young adult prison inmates. *Psychosomatic Medicine*, **49**, 174–182.

Dabbs, J. M., Carr, T. S., Frady, R. L., et al. (1995). Testosterone, crime, and misbehavior among 692 male prison inmates. *Personality and Individual Differences*, **18**(5), 627–633.

Daly, M. and Wilson, M. (1983). *Sex, Evolution, and Behavior*. Belmont, CA: Wadsworth.

Daly, M. and Wilson, M. (1988). *Homicide*. New York: Aldine de Gruyter.

Davidson, J. M., Camargo, C. A. and Smith, E. R. (1978). Effects of androgen on sexual behavior in hypogonadal men. *Journal of Clinical Endocrinology and Metabolism*, **48**, 955–958.

Davies, M. J. and Norman, R. J. (2002). Programming and reproductive functioning. *Trends in Endocrinology and Metabolism*, **13**, 386–392.

Daynes, R. A. and Araneo, B. A. (1991). Regulation of T-cell function by steroid hormones. In *Cellular and Cytokine Networks in Tissue Immunity*, M. A. Meltzer and A. Mantovani (eds). New York: Wiley-Liss, pp. 77–82.

Deerenberg, C., Arpanius, V., Daan, S., et al. (1997). Reproductive effort decreases antibody responsiveness. *Proceedings of the Royal Society of London. Series B*, **264**, 1021–1029.

de La Rochebrochard, E., de Mouzon, J., Thepot, F., et al. (2006). Fathers over 40 and increased failure to conceive: the lessons of in vitro fertilization in France. *Fertility and Sterility*, **85**, 1420–1424.

De Moor, P., Verhoeven, G. and Heyns, W. (1973). Permanent effects of fetal and neonatal testosterone secretion on steroid metabolism and binding. *Differentiation*, **1**, 241–253.

Demas, G. E., Chefer, V., Talan, M. I., et al. (1997). Metabolic costs of mounting an antigen-stimulated immune response in adult and aged C57BL/6J mice. *American Journal of Physiology: Regulatory, Integrative and Comparative Physiology*, **42**, R1331–R1367.

Demas, G. E., Drazen, D. L. and Nelson, R. J. (2003). Reductions in total body fat decrease humoral immunity. *Proceedings of the Royal Society of London. Series B*, **270**, 905–911.

Deslypere, J. P. and Vermeulen, A. (1984). Leydig cell function in normal men: effect of age, life style, residence, diet and activity. *Journal of Clinical Endocrinology and Metabolism*, **59**, 955–962.

Dewsbury, D. A. (1982). Ejaculate cost and male choice. *American Naturalist*, **1**, 601–610.

Dixson, A. F. (1999). *Primate Sexuality: Comparative Studies of the Prosimians, Monkeys, Apes, and Human Beings*. New York: Oxford University Press.

Dobs, A. and Brown, T. (2002). Metabolic abnormalities in HIV disease and injection drug use. *Journal of Acquired Immune Deficiency Syndromes*, **31**, S70–S77.

Doerr, P. and Pirke, K. M. (1976). Cortisol-induced suppression of plasma testosterone in normal adult males. *Journal of Clinical Endocrinology and Metabolism*, **43**, 622–629.

Domb, L. G. and Pagel, M. (2001). Sexual swellings advertise female quality in wild baboons. *Nature*, **410**, 204–206.

Duggan, M. B., Alwar, J. and Milner, R. D. (1986). The nutritional cost of measles in Africa. *Archives of Disease in Childhood*, **61**, 61–66.

Eaton, N. E., Reeves, G. K., Appleby, P. N., et al. (1999). Endogenous sex hormones and prostate cancer: a quantitative review of prospective studies. *British Journal of Cancer*, **80**, 930–934.

Elia, M. (1992). Organ and tissue contribution to metabolic rate. In *Energy Metabolism: Tissue Determinants and Cellular Corollaries*, J. Kinney and H. N. Tucker (eds). New York: Raven Press, pp. 61–79.

Ellis, B. J. and Garber, J. (2000). Psychosocial antecedents of variation in girls' pubertal timing: maternal depression, stepfather presence, and marital and family stress. *Child Development*, **71**, 485–501.

Ellison, P. T. (1990). Human ovarian function and reproductive ecology: new hypotheses. *American Anthropology*, **92**, 933–952.

Ellison, P. T. (1994). Advances in human reproductive ecology. *Annual Review of Anthropology*, **23**, 255–275.

Ellison, P. T. (2001). *On Fertile Ground: A Natural History of Human Reproduction*. Cambridge, MA: Harvard University Press.

Ellison, P. T. (2003). Energetics and reproductive effort. *American Journal of Human Biology*, **15**, 342–351.

Ellison, P. T. and Panter-Brick, C. (1996). Salivary testosterone levels among Tamang and Kami males of central Nepal. *Human Biology*, **68**, 955–965.

Ellison, P. T., Lipson, S. F. and Meredith, M. D. (1989). Salivary testosterone levels in males from the Ituri forest of Zaire. *American Journal of Human Biology*, **1**, 21–24.

Ellison, P. T., Panter-Brick, C., Lipson, S. F., et al. (1993). The ecological context of human ovarian function. *Human Reproduction*, **8**, 2248–2258.

Ellison, P. T., Bribiescas, R. G., Bentley, G. R., et al. (2002). Population variation in age-related decline in male salivary testosterone. *Human Reproduction*, **17**, 3251–3253.

Emanuel, E. R., Goluboff, E. T. and Fisch, H. (1998). MacLeod revisited: sperm count distributions in 374 fertile men from 1971 to 1994. *Urology*, **51**, 86–88.

Fabbri, A., Aversa, A. and Isidori, A. (1997). Erectile dysfunction: an overview. *Human Reproduction Update*, **3**, 455–466.

Fedigan, L. M. and Zohar, S. (1997). Sex differences in mortality of Japanese macaques. *American Journal of Physical Anthropology*, **102**, 161–165.

Filsinger, E. E, Braun, J. J. and Monte, W. C. (1985). An examination of the effects of putative pheromones on human judgments. *Ethology and Sociobiology*, **6**, 227–236.

Fisch, H., Ikeguchi, E. F. and Goluboff, E. T. (1996). Worldwide variations in sperm counts. *Urology*, **48**, 909–911.

Fisher, R. A. (1932). The genetical theory of natural selection. *Nature*, **66**, 118–138.

Fleming, P. J., Howell, T., Clements, M., et al. (1994). Thermal balance and metabolic rate during upper respiratory tract infection in infants. *Archives of Disease in Childhood*, **70**, 187–191.

Forest, M. G., Sizonenko, P. C., Cathiard, A. M., et al. (1974). Hypophyso-gonadal function in humans during the first year of life. I: Evidence for testicular activity in early infancy. *Journal of Clinical Investigation*, **53**, 819–828.

Fukagawa, N. K., Bandini, L. G. and Young, J. B. (1990). Effect of age on body composition and resting metabolic rate. *American Journal of Physiology*, **259**, E233–E238.

Gangestad, S. W. and Thornhill, R. (1998). Menstrual cycle variation in women's preference for the scent of symmetrical men. *Proceedings of the Royal Society of London. Series B*, **265**, 927–933.

Gangestad, S. W., Thornhill, R. and Garver-Apgar, C. E. (2005). Women's sexual interests across the ovulatory cycle depend on primary partner developmental instability. *Proceedings of the Royal Society of London. Series B*, **272**, 2023–2027.

Gao, H. B., Tong, M. H., Hu, Y. Q., et al. (2002). Glucocorticoid induces apoptosis in rat Leydig cells. *Endocrinology*, **143**, 130–138.

Gems, D. and Riddle, D. L. (1996). Longevity in *Caenorhabditis elegans* reduced by mating but not gamete production. *Nature*, **379**, 723–725.

Gilbeau, P. M. and Smith, C. G. (1985). Naloxone reversal of stress-induced reproductive effects in the male rhesus monkey. *Neuropeptides*, **5**, 335–338.

Giltay, E. J., Fonk, J. C., von Blomberg, B. M., et al. (2000). In vivo effects of sex steroids on lymphocyte responsiveness and immunoglobulin levels in humans. *Journal of Clinical Endocrinology and Metabolism*, **85**, 1648–1657.

Giwercman, A., Carlsen, E., Keiding, N., et al. (1993). Evidence for increasing incidence of abnormalities of the human testis: a review. *Environmental Health Perspectives*, **101**, 65–71.

Glass, R. H. (1991). Infertility. In *Reproductive Endocrinology: Physiology, Pathophysiology, and Clinical Management*, S. Samuel, C. Yen and R. B. Jaffe (eds). Philadelphia: W.B. Saunders, pp. 689–709.

Glick, B. B. (1984). Male endocrine responses to females: effects of social cues in cynomolgus macaques. *American Journal of Primatology*, **6**, 229–239.

Gomez-Merino, D., Chennaouia, M, Burnat, P., et al. (2003). Immune and hormonal changes following intense military training. *Military Medicine*, **168**, 1034–1038.

Gomez-Merino, D., Drogoua, C., Chennaouia, M., et al. (2005). Effects of combined stress during intense training on cellular immunity, hormones and respiratory infections. *Neuroimmunomodulation*, **12**, 164–172.

Granger, D. A., Booth, A. and Johnson, D. R. (2000). Human aggression and enumerative measures of immunity. *Psychosomatic Medicine*, **62**, 583–590.

Gray, A., Feldman, H. A., McKinlay, J. B., et al. (1991). Age, disease, and changing sex hormone levels in middle-aged men: results of the Massachusetts Male Aging Study. *Journal of Clinical Endocrinology and Metabolism*, **73**, 1016–1025.

Gray, P. B. and Campbell, B. C. (2005). Erectile dysfunction and its correlates among the Ariaal of northern Kenya. *Internal Journal of Impotence Research*, **17**, 445–449.

Gray, P. B., Chapman, J. F., Burnham, T. C., et al. (2004). Human male pair bonding and testosterone. *Human Nature*, **15**, 119–131.

Griffin, J. E. (2004). Male reproductive function. In *Textbook of Endocrine Physiology*, J. E. Griffin and S. R. Ojeda (eds), 5th edn. New York: Oxford University Press, pp. 243–264.

Grossman, C. J. (1995). The role of sex steroids in immune system regulation. In *Bilateral Communication between the Endocrine and Immune Systems*, C. J. Grossman (ed.). New York: Springer-Verlag, pp. 1–11.

Grossman, C. J., Roselle, G. A. and Mendenhall, C. L. (1991). Sex steroid regulation of autoimmunity. *Journal of Steroid Biochemistry and Molecular Biology*, **40**, 649–659.

Gruenbaum, E. (2001). *The Female Circumcision Controversy: an Anthropological Perspective*. Philadelphia, PA: University of Pennsylvania Press.

Grumbach, M. M. and Conte, F. A. (1998). Disorders of sex differentiation. In *Williams Textbook of Endocrinology*, P. R. Larsen, H. M. Kronenberg, S. Melmed et al. (eds), 9th edn. Philadelphia: Saunders, pp. 1303–1425.

Guo, S. S., Zeller, C., Chumlea, W. C., et al. (1999). Aging, body composition, and lifestyle: the Fels Longitudinal Study. *American Journal of Clinical Nutrition*, **70**, 405–411.

Guzick, D. S., Overstreet, J. W., Factor-Litvak, P., et al. (2001). Sperm morphology, motility, and concentration in fertile and infertile men. *New England Journal of Medicine*, **345**, 1388–1393.

Hall, R. C. and Hall, R. C. (2005). Abuse of supraphysiologic doses of anabolic steroids. *Southern Medical Journal*, **98**(5), 550–555.

Hamilton, W. D. and Zuk, M. (1982). Heritable true fitness and bright birds: a role for parasites? *Science*, **218**, 384–387.

Harcourt, A. H., Harvey, P. H., Larson, S. G., et al. (1981). Testis weight, body weight and breeding system in primates. *Nature*, **293**, 55–57.

Hardy, M. P., Gao, H. B., Dong, Q., et al. (2005). Stress hormone and male reproductive function. *Cell and Tissue Research*, **322**, 147–153.

Harman, S. M., Metter, E. J., Tobin, J. D., et al. (2001). Longitudinal effects of aging on serum total and free testosterone levels in healthy men. Baltimore Longitudinal Study of Aging. *Journal of Clinical Endocrinology and Metabolism*, **86**, 724–731.

Harvey, P. H. and Harcourt, A. H. (1984). Sperm competition, testes size, and breeding system in primates. In *Sperm Competition and the Evolution of Animal Mating Systems*, R. L. Smith (ed.). New York: Academic Press, pp. 573–587.

Hausfater, G. and Hrdy, S. B. (1984). *Infanticide in Comparative and Evolutionary Perspectives*. New York: Aldine de Gruyter.

Hayes, F. J., Seminara, S. B., Decruz, S., et al. (2000). Aromatase inhibition in the human male reveals a hypothalamic site of estrogen feedback. *Journal of Clinical Endocrinology and Metabolism*, **85**, 3027–3035.

Hendrix, C. T. (2004). The history of violence against women in three acts. *Violence Against Women*, **10**, 1383–1394.

Hill, K. (1993). Life history theory and evolutionary anthropology. *Evolutionary Anthropology*, **2**, 78–88.

Hill, K. and Hurtado, A. M. (1996). *Ache Life History: the Ecology and Demography of a Foraging People*. New York: Aldine de Gruyter.

Hill, K., Boesch, C., Goodall, J., et al. (2001). Mortality rates among wild chimpanzees. *Journal of Human Evolution*, **40**, 437–450.

Hrdy, S. B. (1979). Infanticide among animals: a review, classification, and examination of the implications for the reproductive strategies of females. *Ethology and Sociobiology*, **1**, 13–40.

Hsing, A. W. (2001). Hormones and prostate cancer: what's next? *Epidemiological Review*, **23**, 42–58.

Hublart, M., Tetaert, D., Croix, D., et al. (1990). Gonadotropic dysfunction produced by *Trypanosoma brucei brucei* in the rat. *Acta Tropica*, **47**, 177–184.

Ikede, B. O., Elhassan, E. and Akpavie, S. O. (1988). Reproductive disorders in African trypanosomiasis: a review. *Acta Tropica*, **45**, 5–10.

Ing, R., Su, Z., Scott, M. E., et al. (2000). Suppressed T helper 2 immunity and prolonged survival of a nematode parasite in protein-malnourished mice. *Proceedings of the National Academy of Sciences of the United States of America*, **97**, 7078–7083.

Ingemarsson, I. (2003). Gender aspects of preterm birth. *BJOG: an International Journal of Obstetrics and Gynecology*, **110**, 34–38.

Isseroff, H., Sylvester, P. W., Bessette, C. L., et al. (1989). Schistosomiasis: role of endogenous opioids in suppression of gonadal steroid secretion. *Comparative Biochemistry and Physiology*, **94**, 41–45.

Iturregui-Pagán, J. R., Fortuno, R. F. and Noy, M. A. (1976). Genital manifestation of filariasis. *Urology*, **8**, 207–209.

Jakobsson, J., Ekstrom, L., Inotsume, N., et al. (2006). Large differences in testosterone excretion in Korean and Swedish men are strongly associated with a UDP-glucuronosyl transferase 2B17 polymorphism. *Journal of Clinical Endocrinology and Metabolism*, **91**, 687–693.

Jameson, J. L. (1996). Inherited disorders of the gonadotropin hormones. *Molecular and Cellular Endocrinology*, **125**, 143–149.

Jasienska, G. and Ellison, P. T. (1998). Physical work causes suppression of ovarian function in women. *Proceedings of the Royal Society of London. Series B*, **265**, 1847–1851.

Jasienska, G. and Ellison, P. T. (2004). Energetic factors and seasonal changes in ovarian function in women from rural Poland. *American Journal of Human Biology*, **16**, 563–580.

Jasienska, G., Ziomkiewicz, A., Ellison, P. T., et al. (2004). Large breast and narrow waist indicate high reproductive potential in women. *Proceedings of the Royal Society of London. Series B*, **271**, 1213–1217.

Joffe, M. (2003). Invited commentary: the potential for monitoring of fecundity and the remaining challenges. *American Journal of Epidemiology*, **157**, 89–93.

Kalantaridou, S. N., Makrigiannakis, A., Zoumakis, E., et al. (2004). Reproductive functions of corticotropin-releasing hormone. Research and potential clinical utility of antalarmins (CRH receptor type 1 antagonists). *American Journal of Reproductive Immunology*, **51**, 269–274.

Kaplan, H., Hill, K., Lancaster, J., et al. (2000). A theory of human life history evolution: diet, intelligence, and longevity. *Evolutionary Anthropology*, **9**, 156–185.

Kellokumpu-Lehtinen, P. and Pelliniemi, L. J. (1984). Sex ratio of human conceptuses. *Obstetrics and Gynecology*, **64**, 220–222.

Kendrick, K. M. and Drewett, R. F. (1979). Testosterone reduces refractory period of stria terminalis neurons in the rat brain. *Science*, **2004**, 877–879.

Ketterson, E. D. and Nolan, V. Jr (1992). Hormones and life histories: an integrative approach. *American Naturalist*, **140**, 33–62.

Ketterson, E. D., Nolan, V., Wolf, L., et al. (1992). Testosterone and avian life histories: effects of experimentally elevated testosterone on behavior and correlates of fitness in the dark eyed junco (*Junco hyemalis*). *American Naturalist*, **140**, 980–999.

Key, T. J., Roe, L., Thorogood, M., et al. (1990). Testosterone, sex hormone-binding globulin, calculated free testosterone, and oestradiol in male vegans and omnivores. *British Journal of Nutrition*, **64**, 111–119.

Kidd, S. A., Eskenazi, B. and Wyrobek, A. J. (2001). Effects of male age on semen quality and fertility: a review of the literature. *Fertility and Sterility*, **75**, 237–248.

Kinsey, A. C., Pomeroy, W. R. and Martin, C. E. (1948). *Sexual Behavior in the Human Male*. Bloomington, IN: Indiana University Press.

Kinsey, A. C., Pomeroy, W. R., Martin, C. E., et al. (1953). *Sexual Behavior in the Human Female*. Bloomington, IN: Indiana University Press.

Klasing, K. C. (1998). Nutritional modulation of resistance to infectious diseases. *Poultry Science*, **77**, 1119–1125.

Klein, S. L. (2000). The effects of hormones on sex differences in infection: from genes to behavior. *Neuroscience and Biobehavioral Reviews*, **24**, 627–638.

Kley, H. K., Deselaers, T. and Peerenboom, H. (1981). Evidence for hypogonadism in massively obese males due to decreased free testosterone. *Hormone and Metabolic Research*, **11**, 639–641.

Klibanski, A., Beitins, I. Z., Badger, T., et al. (1981). Reproductive function during fasting in men. *Journal of Clinical Endocrinology and Metabolism*, **53**, 258–263.

Knickmeyer, R. C. and Baron-Cohen, S. (2006). Fetal testosterone and sex differences. *Early Human Development*, **82**, 755–760.

Knott, C. D. (1999). *Reproductive, Physiological and Behavioral Responses of Orangutans in Borneo to Fluctuations in Food Availability (Dissertation)*. Cambridge, MA: Harvard University Press.

Kohl, J. V., Atzmueller, M., Fink, B., et al. (2001). Human pheromones: integrating neuroendocrinology and ethology. *Neuroendocrinology Letters*, **22**, 309–321.

Koski, K. G. and Scott, M. E. (2001). Gastrointestinal nematodes, nutrition and immunity: breaking the negative spiral. *Annual Review of Nutrition*, **21**, 297–321.

Krakowski, M. (2003). Violence and serotonin: influence of impulse control, affect regulation, and social functioning.

Journal of Neuropsychiatry and Clinical Neuroscience, **15**, 294–305.

Kramer, T. R., Moore, R. J., Shippee, R. L., et al. (1997). Effects of food restriction in military training on T-lymphocyte responses. *International Journal of Sports Medicine*, **18**, S84–S90.

Kratzik, C., Schatzl, G., Lunglmayr, G., et al. (2005). The impact of age, body mass index and testosterone on erectile dysfunction. *Journal of Urology*, **174**, 240–243.

Kronenberg, H. M., Melmed, S., Polonsky, K. S., et al. (eds) (2007). *Williams Textbook of Endocrinology*, 9th edn. Philadelphia: Saunders.

Kuhnert, B. and Nieschlag, E. (2004). Reproductive functions of the ageing male. *Human Reproduction Update*, **10**, 327–339.

Kulin, H. E., Bwibo, N., Mutie, D., et al. (1984). Gonadotropin excretion during puberty in malnourished children. *Journal of Pediatrics*, **105**, 325–328.

Kumar, T. R., Wang, Y., Lu, N., et al. (1997). Follicle stimulating hormone is required for ovarian follicle maturation but not male fertility. *Nature Genetics*, **15**, 201–204.

Kurihara, M., Aoki, K., and Hisamichi, S. (eds) (1989). *Cancer Mortality Statistics in the World 1950–1985*. Nagoya, Japan: University of Nagoya Press.

Levine, R. J., Brown, M. H., Bell, M., et al. (1992). Air-conditioned environments do not prevent deterioration of human semen quality during the summer. *Fertility and Sterility*, **57**, 1075–1083.

Lin, E., Kotani, J. G. and Lowry, S. F. (1998). Nutritional modulation of immunity and the inflammatory response. *Nutrition*, **14**, 545–550.

Lo, J. C. and Schambelan, M. (2001). Reproductive function in human immunodeficiency virus infection. *Journal of Clinical Endocrinology and Metabolism*, **86**, 2338–2343.

Lochmiller, R. L. and Deerenberg, C. (2000). Trade-offs in evolutionary immunology: just what is the cost of immunity? *Oikos*, **88**, 87–98.

Lofrano-Porto, A., Barra, G. B., Giacomini, L. A., et al. (2007). Luteinizing hormone beta mutation and hypogonadism in men and women. *New England Journal of Medicine*, **357**, 897–904.

Low, B. S. (2000). *Why Sex Matters: a Darwinian Look at Human Behavior*. Princeton, NJ: Princeton University Press.

Lukas, W. D., Campbell, B. C. and Ellison, P. T. (2004). Testosterone, aging, and body composition in men from Harare, Zimbabwe. *American Journal of Human Biology*, **16**, 704–712.

Mace, R. (1996). Biased parental investment and reproductive success in Gabbra pastoralists. *Behavioral Ecology and Sociobiology*, **38**, 75–81.

Maggioncalda, A. N., Sapolsky, R. M. and Czekala, N. M. (1999). Reproductive hormone profiles in captive male orangutans: implications for understanding developmental arrest. *American Journal of Physical Anthropology*, **109**, 19–32.

Maggioncalda, A. N., Czekala, N. M. and Sapolsky, R. M. (2000). Growth hormone and thyroid stimulating hormone concentrations in captive male orangutans: implications for understanding developmental arrest. *American Journal of Primatology*, **50**, 67–76.

Malaspina, D., Harlap, S., Fennig, S., et al. (2001). Advancing paternal age and risk of schizophrenia. *Archives of General Psychiatry*, **58**, 361–367.

Marin, P., Holmang, S., Jonsson, L., et al. (1992a). The effects of testosterone treatment on body composition and metabolism in middle-aged obese men. *International Journal of Obesity Related Metabolic Disorders*, **16**, 991–997.

Marin, P., Krotkiewski, M. and Bjorntorp, P. (1992b). Androgen treatment of middle-aged, obese men: effects on metabolism, muscle and adipose tissues. *European Journal of Medicine*, **1**, 329–336.

Marler, C. A. and Moore, M. C. (1988). Evolutionary costs of aggression revealed by testosterone manipulations in free-living lizards. *Behavior Ecology and Sociobiology*, **23**, 21–26.

Marler, C. A., Walsberg, G., White, M. L., et al. (1995). Increased energy expenditure due to the increased territorial defense in male lizards after phenotypic manipulation. *Behavioral Ecology and Sociobiology*, **37**, 225–231.

Martin, C. W., Riley, S. C., Everington, D., et al. (2000). Dose-finding study of oral desogestrel with testosterone pellets for suppression of the pituitary–testicular axis in normal men. *Human Reproduction*, **15**, 1515–1524.

Matson, P. L. (1995). External quality assessment for semen analysis and sperm antibody detection: results of a pilot scheme. *Human Reproduction*, **10**, 620–625.

Mazur, A. (1992). Testosterone and chess competition. *Social Psychology Quarterly*, **55**, 70–77.

Mazur, A. and Booth, A. (1998). Testosterone and dominance in men. *Behavior and Brain Science*, **21**, 353–364.

McClintock, M. K. (1971). Menstrual synchrony and suppression. *Nature*, **291**, 244–245.

McKinlay, J. B. (2000). The worldwide prevalence and epidemiology of erectile dysfunction. *International Journal of Impotence Research*, **12**, S6–S11.

McMillin, J. M., Seal, U. S. and Karns, P. D. (1980). Hormonal correlates of hypophagia in white-tailed deer. *Federation Proceedings*, **39**, 2964–2968.

Meeker, J. D., Godfrey-Bailey. L. and Hauser, R. (2007). Relationships between serum hormone levels and semen quality among men from an infertility clinic. *Journal of Andrology*, **28**, 397–406.

Mieusset, R. and Bujan, L. (1995). Testicular heating and its possible contributions to male infertility: a review. *International Journal of Andrology*, **18**, 169–184.

Mitani, J. (1985). Mating behavior of male orangutans in Kutai Game Reserve, Indonesia. *Animal Behavior*, **33**, 392–402.

Mitchell, J. C., Li, X. F., Breen, L., et al. (2005). The role of the locus coeruleus in corticotrophin-releasing hormone and stress-induced suppression of pulsatile luteinizing hormone secretion in the female rat. *Endocrinology*, **146**, 323–331.

Moore, S. L. and Wilson, K. (2002). Parasites as a viability cost of sexual selection in natural populations of mammals. *Science*, **297**, 2015–2018.

Morell, V. (1993). Evidence found for a possible "aggression gene." *Science*, **262**, 321.

Muehlenbein, M. P., Jordan, J. L., Bonner, J. Z., et al. (in press). Towards quantifying the usage costs of human immunity: altered metabolic rates and hormone levels during acute immune activation in men. *American Journal of Human Biology*.

Muehlenbein, M. P. (2008). Adaptive variation in testosterone levels in response to immune activation: empirical and theoretical perspectives. *Social Biology*, **53**, 13–23.

Muehlenbein, M. P. and Bribiescas, R. G. (2005). Testosterone-mediated immune functions and male life histories. *American Journal of Human Biology*, **17**, 527–558.

Muehlenbein, M. P., Algier, J., Cogswell, F., et al. (2005). The reproductive endocrine response to *Plasmodium vivax* infection in Hondurans. *American Journal of Tropical Medicine and Hygiene*, **73**, 178–187.

Muller, G., Keller, A., Reich, A., et al. (2004). Priming with testosterone enhances stimulated growth hormone secretion in boys with delayed puberty. *Journal of Pediatric Endocrinology and Metabolism*, **17**, 77–83.

Muller, M. N. and Wrangham, R. W. (2005). Testosterone and energetics in wild chimpanzees (*Pan troglodytes schweinfurthii*). *American Journal of Primatology*, **66**, 119–130.

Muller, M., den Tonkelaar, I., Thijssen, J. H., et al. (2003). Endogenous sex hormones in men aged 40–80 years. *European Journal of Endocrinology*, **149**, 583–589.

Muller, M., Thompson, M. and Wrangham, R. (2006). Male chimpanzees prefer mating with old females. *Current Biology*, **16**, 2234–2238.

Nelson, G. S. (1958). *Schistosoma mansoni* infection in the West Nile District of Uganda. I. The incidence of *S. mansoni* infection. *East African Medical Journal*, **35**, 311–319.

Nelson, R. J., Demas, G. E., Klein, S. L., et al. (2002). *Seasonal Patterns of Stress, Immune Function, and Disease.* New York: Cambridge University Press.

Nelson, W. G., De Marzo, A. M. and Isaacs, W. B. (2003). Prostate cancer. *New England Journal of Medicine*, **349**, 366–381.

Newsholme, P., and Newsholme, E. A. (1989). Rates of utilization of glucose, glutamine and oleate and formation of end-products by mouse peritoneal macrophages in culture. *Biochemistry Journal*, **261**, 211–218.

Nnatu, S. N., Giwa-Osagie, O. F. and Essien, E. E. (1991). Effect of repeated semen ejaculation on sperm quality. *Clinical and Experimental Obstetrics*, **18**, 39–42.

Nordling, D., Andersson, M., Zohari, S., et al. (1998). Reproductive effort reduces specific immune response and parasite resistance. *Proceedings of the Royal Society of London. Series B*, **265**, 1291–1298.

Norris, K., Anwar, M. and Read, A. F. (1994). Reproductive effort influences the prevalence of haematozoan parasites in great tits. *Journal of Animal Ecology*, **63**, 601–610.

Nussey, S. and Whitehead, S. A. (2001). *Endocrinology: an Integrated Approach.* Oxford: Taylor and Francis.

Odendaal, F. J., Jones, K. N. and Stermitz, F. R. (1989). Mating behavior and male investment in euphydryas-anicia lepidoptera nymphalidae. *Journal of Research on the Lepidoptera*, **28**, 1–13.

Oktenli, C., Doganci, L., Ozgurtas, T., et al. (2004). Transient hypogonadotrophic hypogonadism in males with acute toxoplasmosis: suppressive effect of interleukin-1 on the secretion of GnRH. *Human Reproduction*, **19**, 859–866.

Olsen, G. W., Bodner, K. M., Ramlow, J. M., et al. (1995). Have sperm counts been reduced 50 percent in 50 years? A statistical model revisited. *Fertility and Sterility*, **63**, 887–893.

Olsen, N. J. and Kovacs, W. J. (1996). Gonadal steroids and immunity. *Endocrinological Review*, **17**, 369–384.

Opstad, P. K. (1992). Androgenic hormones during prolonged physical stress, sleep, and energy deficiency. *Journal of Clinical Endocrinology and Metabolism*, **74**, 1176–1183.

Orwoll, E., Lambert, L. C., Marshall, L. M., et al. (2006). Testosterone and estradiol among older men. *Journal of Clinical Endocrinology and Metabolism*, **91**, 1336–1344.

Owens, I. P. (2002). Sex differences in mortality rate. *Science*, **297**, 2008–2009.

Parazzini, F., La Vecchia, C., Levi, F., et al. (1998). Trends in male:female ratio among newborn infants in 29 countries from five continents. *Human Reproduction*, **13**, 1394–1396.

Parr, L. A. and de Waal, F. B. M. (1999). Visual kin recognition in chimpanzees. *Nature*, **399**, 647–648.

Partridge, L. and Farquhar, M. (1981). Sexual activity reduces lifespan of male fruit flies. *Nature*, **294**, 580–582.

Pflieger-Bruss, S., Schuppe, H. C. and Schill, W. B. (2004). The male reproductive system and its susceptibility to endocrine disrupting chemicals. *Andrologia*, **36**, 337–345.

Plas, E., Berger, P., Hermann, M, et al. (2000). Effects of aging on male fertility? *Experimental Gerontology*, **35**, 543–551.

Polansky, F. F. and Lamb, E. J. (1988). Do the results of semen analysis predict future fertility? A survival analysis study. *Fertility and Sterility*, **49**, 1059–1065.

Pope, H. G., Kouri, E. M. and Hudson, J. I. (2000). Effects of supraphysiologic doses of testosterone on mood and aggression in normal men: a randomized controlled trial. *Archives of General Psychiatry*, **57**, 133–140.

Poretsky, L., Can, S. and Zumoff, B. (1995). Testicular dysfunction in human immunodeficiency virus-infected homosexual men. *Metabolism*, **44**, 946–953.

Potts, M. and Short, R. (1999). *Ever Since Adam and Eve*. Cambridge, MA: Cambridge University Press.

Poulin, R. (1996). Sexual inequalities in helminth infections: a cost of being a male? *American Naturalist*, **147**, 287–295.

Quinn, M. and Babb, P. (2002). Patterns and trends in prostate cancer incidence, survival, prevalence and mortality. Part I: international comparisons. *British Journal of Urology International*, **90**, 162–173.

Raberg, L., Grahn, M., Hasselquist, D., et al. (1998). On the adaptive significance of stress-induced immunosuppression. *Proceedings of the Royal Society of London. Series B*, **265**, 1637–1641.

Raberg, L., Vestberg, M., Hasselquist, D., et al. (2002). Basal metabolic rate and the evolution of the adaptive immune system. *Proceedings of the Royal Society of London. Series B*, **269**, 817–821.

Raven, P. H. and Johnson, G. B. (1988). *Understanding Biology*. St. Louis: Times Mirror/Mosby College Publishing.

Reddy, S., Shapiro, M., Morton, R. Jr, et al. (2003). Prostate cancer in black and white Americans. *Cancer and Metastasis Reviews*, **22**, 83–86.

Reichenberg, A., Gross, R., Weiser, M., et al. (2006). Advancing paternal age and autism. *Archives of General Psychiatry*, **63**, 1026–1032.

Reincke, M., Arlt, W., Heppner, C., et al. (1998). Neuroendocrine dysfunction in African Trypanosomiasis: the role of cytokinesa. *Annals of the New York Academy of Sciences*, **840**, 809–821.

Reynolds, V. (1991). The biological basis of human patterns of mating and marriage. In *Marriage and Mating*, V. Reynolds and J. Kellett (eds). Oxford: Oxford University Press, pp. 46–90.

Richner, H., Christe, P. and Oppliger, A. (1995). Paternal investment affects prevalence of malaria. *Proceedings of the Royal Society of London. Series B*, **92**, 1192–1194.

Ring, H. Z., Lessov, C. N., Reed, T., et al. (2005). Heritability of plasma sex hormones and hormone binding globulin in adult male twins. *Journal of Clinical Endocrinology and Metabolism*, **90**, 3653–3658.

Roberts, A. C., McClure, R. D., Weiner, R. I., et al. (1993). Overtraining affects male reproductive status. *Fertility and Sterility*, **60**, 686–692.

Rochira, V., Zirilli, L., Genazzani, A. D., et al. (2006). Hypothalamic-pituitary-gonadal axis in two men with aromatase deficiency: evidence that circulating estrogens are required at the hypothalamic level for the integrity of gonadotropin negative feedback. *European Journal of Endocrinology*, **155**, 513–522.

Rodriguez, I., Greer, C. A., Mok, M. Y., et al. (2000). Putative pheromone receptor gene expressed in human olfactory mucosa. *Nature Genetics*, **26**, 18–19.

Röjdmark, S. (1987). Influence of short-term fasting on the pituitary-testicular axis in normal men. *Hormone Research*, **25**, 140–146.

Roney, J. R., Mahler, S. V. and Maestripieri, D. (2003). Behavioral and hormonal responses of men to brief interactions with women. *Evolution and Human Behavior*, **24**, 365–375.

Ronnekliev, O. and Resko, J. (1990). Ontogeny of gonadotropin-releasing hormone-containing neurons in early fetal development of rhesus macaques. *Endocrinology*, **126**, 498–511.

Rose, R. M., Kreuz, L. E., Holaday, J. W., et al. (1972). Diurnal variation of plasma testosterone and cortisol. *Journal of Endocrinology*, **54**, 177–178.

Rosen, R. C., Fisher, W. A., Eardley, I., et al. (2004). The multinational Men's Attitudes to Life Events and Sexuality (MALES) study: I. Prevalence of erectile dysfunction and related health concerns in the general population. *Current Medical Research and Opinion*, **20**, 607–617.

Roughgarden, J. (2009). *Evolution's Rainbow: Diversity, Gender, and Sexuality in Nature and People*, 2nd edn. Los Angeles: University of California Press.

Saad, S. H., Abdelbaky, A., Osman, A. M., et al. (1999). Possible role of *Schistosoma mansoni* infection in male hypogonadism. *Journal of the Egyptian Society of Parasitology*, **29**, 307–323.

Safe, S. H. (2000). Endocrine disruptors and human health: is there a problem? An update. *Environmental Health Perspectives*, **108**(6), 487–493.

Saidi, J. A., Chang, D. T., Goluboff, E. T., et al. (1999). Declining sperm counts in the United States? A critical review. *Journal of Urology*, **161**, 460–462.

Sapolsky, R. M. and Krey, L. C. (1988). Stress-induced suppression of luteinizing hormone concentrations in wild baboons: role of opiates. *Journal of Clinical Endocrinology and Metabolism*, **66**, 722–726.

Schmid-Hempel, P. (2003). Variation in immune defence as a question of evolutionary ecology. *Hormone Research*, **270**, 357–366.

Schoech, S. J., Ketterson, E. D., Nolan, V. Jr, et al. (1998). The effect of exogenous testosterone on parental behavior, plasma prolactin and prolactin binding sites in dark-eyed juncos. *Hormones and Behavior*, **34**, 1–10.

Schwartz, D., Mayaux, M. J., Spira, M. L., et al. (1983). Semen characteristics as a function of age in 833 fertile men. *Fertility and Sterility*, **39**, 530–535.

Seo, J. T., Rha, K. H., Park, Y. S., et al. (2000). Semen quality over a 10-year period in 22 249 men in Korea. *International Journal of Andrology*, **23**, 194–198.

Sharpe, R. M. and Skakkebaek, N. E. (1993). Are oestrogens involved in falling sperm counts and disorders of the male reproductive tract? *Lancet*, **341**, 1392–1395.

Sheldon, B. C. and Verhulst, S. (1996). Ecological immunology: costly parasite defenses and trade-offs in evolutionary ecology. *Trends in Ecological Evolution*, **11**, 317–321.

Shephard, R. J., Castellani, J. W. and Shek, P. N. (1998). Immune deficits induced by strenuous exertion under adverse environmental conditions: manifestations and countermeasures. *Critical Reviews in Immunology*, **18**, 545–568.

Shibata, A. and Whittemore, A. S. (1997). Genetic predisposition to prostate cancer: possible explanations for ethnic differences in risk. *The Prostate*, **32**, 65–72.

Siiteri, P. K. and Wilson, J. D. (1974). Testosterone formation and metabolism during male sexual differentiation in the human embryo. *Journal of Clinical Endocrinology and Metabolism*, **38**, 113–125.

Simard, J., Dumont, M., Soucy, P., et al. (2002). Perspective: prostate cancer susceptibility genes. *Endocrinology*, **143** (6), 2029–2040.

Singh, A. B., Hsia, S., Alaupovic, P., et al. (2002). The effects of varying doses of testosterone on insulin sensitivity, plasma lipids, apolipoproteins, and C-reactive protein in healthy young men. *Journal of Clinical Endocrinology and Metabolism*, **87**, 136–143.

Small, M. F. (1988). Female primate sexual behavior and conception: are there really sperm to spare? *Current Anthropology*, **29**, 81–100.

Smith, R. J. and Leigh, S. R. (1998). Sexual dimorphism in primate neonatal body mass. *Journal of Human Evolution*, **34**, 173–201.

Smith, S. R., Chhetri, M. K., Johanson, J., et al. (1975). The pituitary-gonadal axis in men with protein-calorie malnutrition. *Journal of Clinical Endocrinology and Metabolism*, **41**, 60–69.

Smuts, B. B. and Smuts, R. W. (1993). Male aggression and sexual coercion of females in nonhuman primates and other mammals: evidence and theoretical implications. *Advances in the Study of Behavior*, **22**, 1–63.

Soudan, B., Tetaert, D., Racadot, A., et al. (1992). Decrease of testosterone level during an experimental African trypanosomiasis: involvement of a testicular LH receptor desensitization. *Acta Endocrinology (Copenhagen)*, **127**, 86–92.

Spratt, D. I. and Crowley, W. Jr (1988). Pituitary and gonadal responsiveness is enhanced during GnRH-induced puberty. *American Journal of Physiology*, **254**, E652–E657.

Spratt, D. I., Cox, P., Orav, J., et al. (1993). Reproductive axis suppression in acute illness is related to disease severity. *Journal of Clinical Endocrinology and Metabolism*, **76**, 1548–1554.

Stahl, W., Kaneda, Y., Tanabe, M., et al. (1995). Uterine atrophy in chronic murine toxoplasmosis due to ovarian dysfunction. *Journal of Parasitology Research*, **81**, 109–113.

Stearns, S. C. (1989). Trade-offs in life-history evolution. *Functional Ecology*, **3**, 259–268.

Stearns, S. C. (1992). *The Evolution of Life Histories*. Oxford: Oxford University Press.

Storey, A. E., Walsh, C. J., Quinton, R. L., et al. (2000). Hormonal correlates of paternal responsiveness in new and expectant fathers. *Evolution and Human Behavior*, **21**, 79–95.

Strassmann, B. I. (1999). Menstrual synchrony pheromones: cause for doubt. *Human Reproduction*, **14**, 579–580.

Straub, R. H. and Cutolo, M. (2001). Involvement of the hypothalamic–pituitary–adrenal/gonadal axis and the peripheral nervous system in rheumatoid arthritis: viewpoint based on a systemic pathogenetic role. *Arthritis and Rheumatism*, **44**, 493–507.

Sugiyama, L. S. (2005). Physical attractiveness in adaptationist perspective. In *The Handbook of Evolutionary Psychology*, D. M. Buss (ed.). New York: Wiley.

Swan, S. H., Elkin, E. P. and Fenster, L. (2000). The question of declining sperm density revisited: an analysis of 101 studies published 1934–1996. *Environmental Health Perspectives*, **108**, 961–966.

Tapanainen, J. S., Aittomaki, K., Min, J., et al. (1997). Men homozygous for an inactivating mutation of the follicle-stimulating hormone (FSH) receptor gene present variable suppression of spermatogenesis and fertility. *Nature Genetics*, **15**, 205–266.

Thornhill, R. and Palmer, C. (2000). *A Natural History of Rape: Biological Basis of Sexual Coercion*. Cambridge, MA: MIT Press.

Tortolero, I., Arata, G., Lozano, R., et al. (1999). Semen analysis in men from Merida, Venezuela, over a 15-year period. *Archives of Andrology*, **42**, 29–34.

Tovee, M. J., Maisey, D. S., Emery, J. L., et al. (1999). Visual cues to female physical attractiveness. *Proceedings of the Royal Society of London. Series B*, **266**, 211–218.

Travison, T. G., Araujo, A. B., O'Donnell, A. B., et al. (2007). A population-level decline in serum testosterone levels in American men. *Journal of Clinical Endocrinology and Metabolism*, **92**, 196–202.

Tricker, R., Casaburi, R., Storer, T. W., et al. (1996). The effects of supraphysiological doses of testosterone on angry behavior in healthy eugonadal men – a clinical research center study. *Journal of Clinical Endocrinology and Metabolism*, **81**, 3754–3758.

Trivers, R. L. (1972). Parental investment and sexual selection. In *Sexual Selection and the Descent of Man*, B. Campbell (ed.). Chicago: Aldine de Gruyter, pp. 136–179.

Tsai, L. W. and Sapolsky, R. M. (1996). Rapid stimulatory effects of testosterone upon myotubule metabolism and sugar transport, as assessed by silicon microphysiometry. *Aggressive Behavior*, **22**, 357–364.

Uchida, A., Bribiescas, R. G., Ellison, P. T., et al. (2006). Age related variation of salivary testosterone values in healthy Japanese males. *Aging Male*, **9**, 207–213.

van den Beld, A. W., de Jong, F. H., Grobbee, D. E., et al. (2000). Measures of bioavailable serum testosterone and estradiol and their relationships with muscle strength, bone density, and body composition in elderly men. *Journal of Clinical Endocrinology and Metabolism*, **85**, 3276–3282.

Van Valen, L. (1973). A new evolutionary law. *Evolutionary Theory*, **1**, 1–30.

Van Voorhies, W. A. (1992). Production of sperm reduces nematode lifespan. *Nature*, **360**, 456–458.

Vandenbergh, J. C. (1967). Effect of the presence of a male on the sexual maturation of female mice. *Endocrinology*, **81**, 345–349.

Vandenbergh, J. G. (1973). Acceleration and inhibition of puberty in female mice by pheromones. *Journal of Reproduction and Fertility*, **19**, 411–419.

Veldhuis, J. D., Wilkowski, M. J., Zwart, A. D., et al. (1993). Evidence for attenuation of hypothalamic gonadotropin-releasing hormone (GnRH) impulse strength with preservation of GnRH pulse frequency in men with chronic renal failure. *Journal of Clinical Endocrinology and Metabolism*, **76**, 648–654.

Verhulst, S., Dieleman, S. J. and Parmentier, H. K. (1999). A tradeoff between immunocompetence and sexual ornamentation in domestic fowl. *Proceedings of the National Academy of Sciences of the United States of America*, **96**, 4478–4481.

Vermeulen, A., Kaufman, J. M., Goemaere, S., et al. (2002). Estradiol in elderly men. *Aging Male*, **5**, 98–102.

Wagner, G. and Mulhall, J. (2001). Pathophysiology and diagnosis of male erectile dysfunction. *British Journal of Urology International*, **88**, 3–10.

Waldhauser, F., Weissenbacher, G., Frisch, H., et al. (1981). Pulsatile secretion of gonadotropins in early infancy. *European Journal of Pediatrics*, **137**, 71–74.

Walker, R. and Hill, K. (2003). Modeling growth and senescence in physical performance among the Aché of eastern Paraguay. *American Journal of Human Biology*, **15**, 196–208.

Walker, R., Hill, K., Kaplan, H., et al. (2002). Age-dependency in hunting ability among the Aché of eastern Paraguay. *Journal of Human Evolution*, **42**, 639–657.

Wasser, S. K. and Barash, D. P. (1983). Reproductive suppression among female mammals: implications for biomedicine and sexual selection theory. *Quarterly Review of Biology*, **58**, 513–538.

Waynforth, D. (1998). Fluctuating asymmetry and human male life history traits in rural Belize. *Proceedings of the Royal Society of London. Series B*, **265**, 1497–1501.

Wedekind, C. and Folstad, I. (1994). Adaptive or nonadaptive immunosuppression by sex-hormones. *American Naturalist*, **143**, 936–938.

Wedekind, C., Seebeck, T., Bettens, F., et al. (1995). MHC-dependent mate preferences in humans. *Proceedings of the Royal Society of London. Series B*, **260**, 245–249.

Weinbauer, G. F. and Nieschlag, E. (1990). The role of testosterone in spermatogenesis. In *Testosterone: Action, Deficiency, Substitution*, E. Nieschlag and H. M Behre (eds). Berlin: Springer-Verlag, pp. 23–50.

Weinstein, Y. and Bercovich, Z. (1981). Testosterone effects on bone marrow, thymus and suppressor T cells in the (NZB × NZW) F1 mice: its relevance to autoimmunity. *Journal of Immunology*, **126**, 998–1002.

Welle, S., Jozefowicz, R., Forbes, G., et al. (1992). Effect of testosterone on metabolic rate and body composition in normal men and men with muscular dystrophy. *Journal of Clinical Endocrinology and Metabolism*, **74**, 332–335.

Whittemore, A. S., Wu, A. H., Kolonel, L. N., et al. (1995). Family history and prostate cancer risk in black, white, and Asian men in the United States and Canada. *American Journal of Epidemiology*, **141**, 732–740.

Whitten, W. K. (1956). The effect of removal of the olfactory bulbs on the gonads of mice. *Journal of Endocrinology*, **14**, 160–163.

Williams, G. C. (1966). *Adaptation and Natural Selection*. Princeton, NJ: Princeton University Press.

Wilson, I. B., Roubenoff, R., Knox, T. A., et al. (2000). Relation of lean body mass to health-related quality of life in persons with HIV. *Journal of Acquired Immune Deficiency Syndrome*, **24**, 137–146.

Wilson, J. D. (1982). Gonadal hormones and sexual behavior. In *Clinical Neuroendocrinology*, G. M. Besser and L. Martini (eds). New York: Academic Press, pp. 1–29.

Wingfield, J. C., Hegner, R. E., Dufty, A. M. Jr, et al. (1990). The "challenge hypothesis": theoretical implications for patterns of testosterone secretion, mating systems, and breeding strategies. *American Naturalist*, **136**, 829–846.

Wong, W. Y., Thomas, C. M. G., Merkus, H. M. W. M., et al. (2000). Cigarette smoking and the risk of male factor subfertility: minor association between cotinine in seminal plasma and semen morphology. *Fertility and Sterility*, **74**, 930–935.

Wong, W. Y., Zielhuis, G. A., Thomas, C. M. G., et al. (2003). New evidence of the influence of exogenous and endogenous factors on sperm count in man. *European Journal of Obstetrics and Gynecology and Reproductive Biology*, **110**, 49–54.

Wood, J. W. (1994b). *Dynamics of Human Reproduction: Biology, Biometry, Demography*. New York: Aldine de Gruyter.

World Health Organization (1999). *Laboratory Manual for Examination of Human Semen*. Cambridge, MA: Cambridge University Press.

Worthman, C. M. and Konner, M. J. (1987). Testosterone levels change with subsistence hunting effort in !Kung San men. *Psychoneuroendocrinology*, **12**, 449–458.

Wunderlich, F., Benten, W. P., Lieberherr, M., et al. (2002). Testosterone signaling in T cells and macrophages. *Steroids*, **67**, 535–538.

Yirmiya, R., Avitsur, R., Donchin, O., et al. (1995). Interleukin-1 inhibits sexual behavior in female but not in male rats. *Brain, Behavior and Immunity*, **9**, 220–233.

Zahavi, A. (1975). Mate selection – a selection for a handicap. *Journal of Theoretical Biology*, **53**, 205–214.

Zinner, D. P., Schwibbe, M. H. and Kaumanns, W. (1994). Cycle synchrony and probability of conception in female hamadryas baboons, *Papio hamadryas*. *Behavioral Ecology and Sociobiology*, **35**, 175–184.

Zirkin, B. R. and Chen, H. (2000). Regulation of Leydig cell steroidogenic function during aging. *Biology of Reproduction*, **63**, 977–981.

Zitzmann, M. and Nieschlag, E. (2001). Testosterone levels in healthy men and the relation to behavioural and physical characteristics: facts and constructs. *European Journal of Endocrinology*, **144**, 183–197.

Zuk, M. and McKean, K. A. (1996). Sex differences in parasite infections: patterns and processes. *International Journal of Parasitology*, **26**, 1009–1023.

Part IV

Growth and Development

"Man is an intelligence, not served by, but in servitude to his organs."

Aldous Huxley (1894–1963)

22 Evolution of Human Growth

Barry Bogin

INTRODUCTION

Why should a human evolutionary biologist study growth of the human body? The reason is because for multicellular organisms, most major evolutionary change proceeds by alterations in the pattern of growth, development, and maturation. The human species is no exception. In this chapter we review both classic and recent research on the evolution of the human pattern of growth. The major points of this review are:

1. Humans have four stages of growth and development between birth and adulthood. These are infancy, childhood, juvenile, and adolescent.
2. The infancy and juvenile stages are shared with most nonhuman primates, social carnivores, elephants, and many cetaceans. The childhood and adolescence stages are human species-specific features.
3. Human childhood and adolescence evolved because they confer reproductive advantages, increasing the fertility of the parents and reducing the mortality of their offspring. This is classic natural selection.
4. Adolescence may have evolved by both natural selection and sexual selection. Adolescents may contribute significant amounts of food and labor to their families and this enhances reproduction by the parents and survival of their offspring (natural selection). The sex-specific features of adolescent girls and boys enhances opportunities for an apprenticeship-type of learning and practice of the wide variety of economic, social, political, and sexual skills needed for their own adulthood and successful reproduction (sexual selection).

GROWTH AND EVOLUTION

Growth may be defined as a quantitative increase in size or mass. Measurements of height in centimeters or weight in kilograms indicate how much growth has taken place in a child. Development is defined as a progression of changes, either quantitative or qualitative, that lead from an undifferentiated or immature state to a highly organized, specialized, and mature state. Physical maturation is measured by functional capacity, for example, the maturation of bipedal walking results from changes with age in skeletal, muscular, and motor skills of the infant and child. Human growth, development, and maturation have evolved, sometimes as discrete processes, but more often as an integrated series of biological events.

Dobzhansky (1973) said that, "nothing in biology makes sense except in the light of evolution." Human growth, development, and maturation, which follow a unique pattern among the mammals and, even, the primates, are no exceptions to Dobzhansky's admonition. A consideration of the chimpanzee and the human, two closely related (genetically) extant primates, shows the value of taking an evolutionary perspective on growth. Huxley (1863) demonstrated many anatomical similarities between chimpanzees and humans. By the mid 1970s, King and Wilson (1975) established that these anatomical similarities are due to a near identity of the structural DNA of the two species. The recent specifications of both the human and chimpanzee genome show a greater than 99% identity in the coding DNA (Shankar et al., 2007). Nevertheless, there are fundamental differences between the chimpanzee and human species in anatomy, including cranial shape, brain size, and relative length of arms versus legs.

Variation in the expression of common genetic sequences by regulatory DNA (Davidson, 2001; Mattick, 2004), acting in concert with neuroendocrine products (Finch and Rose, 1995; Cobourne and Sharpe, 2003; Shibaguchi et al., 2003), is likely to alter growth rates of shared human–chimpanzee anatomical structures to produce structural and functional differences. D'Arcy Wentworth Thompson showed in 1917 (Thompson, 1917, 1942) that the differences in form between the adults of various species may be accounted for by differences in growth rates from an initially identical – one might better say "similar" – form. Thompson's transformational grid system (Figure 22.1) for the growth of the chimpanzee and human skull from birth

Human Evolutionary Biology, ed. Michael P. Muehlenbein. Published by Cambridge University Press. © Cambridge University Press 2010.

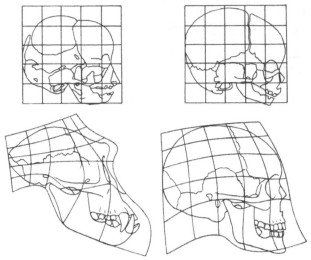

22.1. Transformation grids for the chimpanzee (left) and human (right) skull during growth. Fetal skull proportions are shown above for each species. The relative amount of distortion of the grid lines overlying the adult skull proportions indicate the amount of growth of different parts of the skull (inspired by the transformational grid method of D'Arcy Thompson, 1942, and redrawn from Lewin, 1993).

to maturity, show how both may be derived from a common neonatal form. Different patterns of growth of the cranial bones, maxilla, and mandible are all that are required to produce the adult differences in skull shape. Of course, the differences in skull growth are related to size and shape of the brain, and size of the dentition (both species have the same number and classes of teeth). In a similar manner, the differences in the postcranial anatomy between chimpanzee and human being result from unequal rates of growth for common skeletal and muscular elements.

The evolution of the human pattern of growth may be understood by the study of growth and development in fossil species and by comparison of growth patterns in living species, especially the primates. It is now clear that no living species of nonhuman primate has all of the characteristics of human growth, in particular the human childhood and adolescent stages of life (these are defined and discussed below). This strongly suggests that the human pattern could only have evolved within the taxonomic group of the hominins, a group that includes the human species and extinct forms that were obligated to bipedal locomotion (members of the Genus *Homo*) or facultatively capable of bipedality (e.g., genera such as *Australopithecus, Pranthropus, Sahelanthropus, Orronin*, etc). Hominin primates appear by six to seven million years ago in Africa. The locomotor, behavioral, ecological, and reproductive isolation of Homininae from the Paninae (chimpanzees, gorillas, bonobos, and their extinct ancestors) over the past few million years provides the time depth for the evolutionary differences between

species. In this chapter we explore the selective pressures that appear to have caused ape–human differences relating to physical growth and development and reproductive success.

EVOLUTION OF HUMAN GROWTH AND REPRODUCTION

Why the juxtaposition of physical growth with reproduction in the heading for this section? As Charles Darwin so astutely observed, evolution by natural selection occurs due to differential reproductive success, that is, differences in fertility and mortality. So, evolution is dependent on the process of reproduction. Consider the following. There are about 6.5 billion people alive today (http://www.census.gov). Our closest evolutionary relative, the chimpanzee is an endangered species. Indeed the total population of great apes – orangutans, gorillas, bonobos, and chimpanzees – can be seated in any three modern football stadiums. By this or any other measure, human reproductive success stands in sharp contrast to that of almost all other species of nondomesticated mammals, even nonhuman species of primates.

Comparison of human women (*Homo sapiens*) and wild-living chimpanzee (*Pan troglodytes*) females illustrates the difference in fertility. Women delay the onset of reproduction (first birth) to a relatively late age, about 19 years on average for a worldwide sample of human societies (Bogin, 2001). Chimpanzees in the wild have an average age at first birth of 13.1 years of age (Littleton, 2005, based on research at the Gombe, Mahale, Tai, and Bossou research sites in Africa). Chimpanzees continue to reproduce until death, which usually occurs by 35 years of age (Kaplan et al., 2000). Women end reproduction in their 40s, but live for one or more decades after menopause (cessation of menstrual cycles subsequent to the loss of ovulation). Some female chimpanzees in the wild live into their 40s, making their total reproductive life span six to eight years longer than women (Gage, 1998). Even so, people can produce as many or more newborns than chimpanzees in the shorter time. The total fertility rate (TFR) for wild chimpanzees averages 5.9 offspring and the TFR human women in rural Costa Rica averages 6.4 (Gage, 1998). Average TFR for human women varies by society, but the rural Costa Rica example is fairly typical of human agricultural-based societies.

The documented maximum offspring production in wild chimpanzees is for a high-ranking female named Fifi from the Gombe site. In the year 2003 Fifi, at 44 years of age, became a mother for the 9th time (Goodall, 2003). In 2006 both Fifi and her two-year-old daughter Furaha were reported missing and presumed dead (http://www.janegoodall.org/news). One chimpanzee researcher describes Fifi's fertility as "exceptional"

(Dr. Anne Pusey, personal communication). The known maximum TFR for human societies is for the Hutterites (a religious isolate in Canada) at 9.83 during the time period 1946–1965 (Nonaka et al., 1994; Figure 22.2). In contrast with the exceptional case of Fifi, the Hutterite average is for 1682 Hutterite women of the Dariusleut population of Canada, living in 125 colonies with a total population of 10 000 individuals (Sato et al., 1994).

The TFR is based on another fertility statistic, the age-specific fertility rate (ASFR). Figure 22.2 illustrates ASFR for captive chimpanzees (Littleton, 2005), for the Hutterites, and several other human societies. The numbers given on the y-axis indicate the number of births per female of a given age, per year, per 1000 females in the population. The captive chimpanzees reside at the Taronga Zoo in Australia where they have been allowed natural breeding (no contraception) for the past 50 years. Excellent records by zoo staff ensure the quality of the fertility data. Controlled diets and health care allow for optimal fertility of the females. We may assume that the ASFR for these captive chimpanzees represent the near maximum values for the species. Estimated ASFR for wild-living chimpanzees average about 25–30% lower than rates for the captive chimpanzees until age 30 years. After age 30 years the ASFRs are similar.

The human groups illustrated in Figure 22.2 are so-called "natural fertility" populations. These are defined as societies without conscious family size limitations due to contraception or induced abortion. Only married women are represented in this figure, because in natural fertility populations it is more likely that married women will reproduce than unmarried women. The Hutterites are a religious order who prize high fertility of women. They practiced agriculture and were generally well-nourished and healthy. Adherents marry after age 15 years and, traditionally, did not practice any form of birth control. The curve labeled "Henry's 13" represents 13 natural fertility small societies studied by the French historical demographer Louis Henry (1961), who first noted the relatively constant shape of the age specific fertility rate in different populations. The "Chinese farmers" represent a group of rural villagers studied in the 1930s and the "!Kung" are a society of hunters and gathers living in the Kalahari desert of Botswana in southern Africa studied in the 1960s and 1970s. While the rural Chinese farmers and the !Kung are considered to be "natural fertility," note that their levels of fertility are lower at all ages than the other two groups. In fact, !Kung and Chinese women are known to use herbal medicines and other means to prevent or terminate pregnancies. Lower fertility among the rural Chinese and the !Kung may also be due to the stress of inadequate nutrition (Howell, 1979).

One reason that the Hutterites and "Henry's 13" have greater ASFRs and greater TFR than chimpanzees is due to relatively short birth intervals. For Hutterite women the interval between live-births was 22.1 months at age 30 and 31.2 months at age 40. These intervals included 6.5 months of postpartum amenorrhea (Larsen et al., 2003). Women living in other agricultural societies may reduce the birth interval to less than 2.0 years, but if

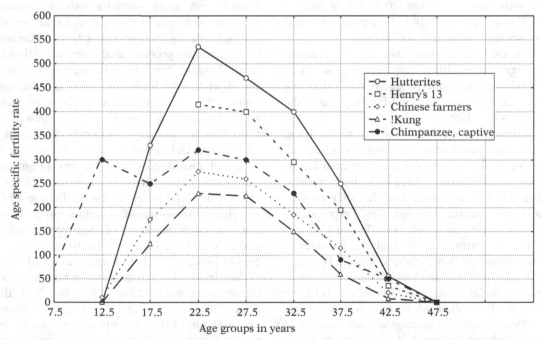

22.2. Age-specific fertility rates (per 1000 females in the population) for captive chimpanzees and for married women in natural fertility populations. The captive chimpanzee data are from Littleton (2005). The human data are redrawn from Ellison and O'Rourke (2000).

the mother is malnourished and/or ill this often has negative health consequences to the infant and the mother (Bogin, 2001). In several human forager societies, including the Aché, Hadza, Hiwi, and !Kung, the interval between successful births averages 3.4 years (40.8 months) for women (Kaplan et al., 2000). The longer birth interval for these forager women may be due to marginal nutrition, heavy physical labor, and an extended period of lactation, all of which drain energy resources away from ovulation (Ellison and O'Rourke, 2000).

In contrast to these human examples, the interval between successful births for wild-living chimpanzee females ranges from 5.2 years to 5.6 years at the Gombe and Mahale research sites in Tanzania (Teleki et al., 1976; Goodall, 1983; Nishida et al., 1990). At one of the Kibale Forest research sites in Uganda, chimpanzees average 7.0 years between successful births (Pusey, 2001). High-ranking female chimpanzees can reproduce successfully every 4.0 years and have high offspring survival (Pusey et al., 1997), but they still lag behind human women in the production of new births.

The human advantage in reduced birth intervals is compounded by greater survival of the infant to adulthood, when the offspring begin their own reproduction. Typically, only two offspring that a female chimpanzee produces live to adulthood (Pusey, 2001). This means that only about 32% of live-born chimpanzee infants survive to maturity (age 15 years). Even in captivity the mortality rates for infant and juvenile chimpanzees is high relative to human beings (Littleton, 2005). In contrast, more than 50% of live-born human infants survive to age 20 years in forager societies (Kaplan et al., 2000). In agricultural and industrial societies the rate of human survival is greater, reaching more than 90% in the wealthy nations today (Gage, 1998; http://www.census.gov). Human survival to reproductive age is the best of any animal species, and chimpanzee infant survival is the second best (Bogin, 1999).

The net reproductive result of these contrasts in age at reproduction, reproductive life span, birth interval, and infant survival to adulthood is that human beings out reproduce chimpanzees. Studies of birth, deaths, and migrations of chimpanzees find that their populations are, at best, stable. Goodall (1983) reports that between the years 1965 and 1980, 51 births and 49 deaths occurred in one community of wild chimpanzees at the Gombe Stream National Park, Tanzania. During a 10-year period, Nishida et al. (1990) observed 74 births, 74 deaths, 14 immigrations, and 13 emigrations in one community at the Mahale Mountains National Park, Tanzania. In contrast, the average annual increase of the human population is about 1.2%. By some estimates, the world population will reach nine billion people by the year 2050 (Bogin, 2001).

EARLY WEANING AND CHILDHOOD

The relative rapidity and absolutely greater fitness (survival to adulthood) of human reproduction depends on many factors, and two of the most important are the early weaning of infants, relative to other species, and the co-operative social care of children. Human infants are fed by lactation, and this is true for all mammalian infants. Mammalian mothers must nurse their infants until the young are capable of independent feeding. This usually coincides with the eruption of the first permanent molar (M1), which is needed so that the infant can eat an adult-type diet (Smith, 1991). The infant must also be able to forage for itself, which usually coincides with M1 eruption in the majority of mammalian species. The chimpanzee infant's M1 erupts at a mean age of 3.1 years (Smith et al., 1994; Anemone et al., 1996), but the mother continues to nurse for about another 2 years as the infant learns how to acquire and process foods (Pusey, 1983). Because of the chimpanzee infant's dependency on the mother, the average period between successful births is delayed relative to M1 eruption.

Human infants are weaned early relative to M1 eruption, at a median age of 30–36 months in traditional human societies (Dettwyler, 1995) and as early as 6.5 months in the Hutterites. The human M1 erupts at about six years of age. Even at age six years, human offspring have much to learn before they can survive on their own. The relatively early weaning of human infants is therefore quite unexpected when compared with other primates and mammals. However, the short birth interval gives women a distinct advantage over the apes: women can produce and rear two offspring through infancy in the time it takes chimpanzees or orangutans to produce and rear one offspring. The weaned infant, though, must survive to adulthood if the short human birth spacing is to result in a true reproductive advantage. How is it possible that human beings trade-off early weaning for increased reproductive frequency and still ensure offspring survival?

Short birth intervals entail a compromise between maternal investments in a current infant and in a future infant. A mother who stops nursing her current infant leaves the infant in the predicament of how to eat. Human three-year-olds cannot move on to feeding independence; they cannot forage for themselves. Even if they could get hold of food from others, children cannot process the diet of juveniles or adults because of their fragile deciduous dentition and the small size of the digestive tract (Behar, 1977; Smith et al., 1994). Sellen (2006) explains that, "... digestion of some foods and absorption of nutrients is constrained by small stomach size and short intestinal length throughout childhood (Hamosh, 1995)" – meaning until seven years of age.

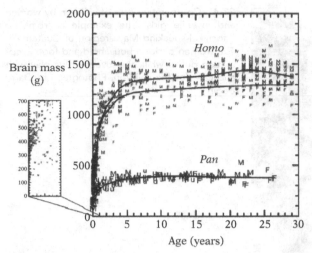

22.3. Brain-mass growth data for humans (*Homo sapiens*) and chimpanzees (*Pan troglodytes*). Brain mass increases during the postnatal period in both species. Lines represent best-fit Lowess regressions through the data points. "M," males; "F," females; "U," sex unidentified (Vrba, 1998). The human regressions separate into male (upper) and female (lower) curves. The inset shows brain-mass growth for each species during the first postnatal year. Reproduced from Leigh (2004) with kind permission of the author.

Another biological constraint on children is that they have relatively large, fast-growing brains that are energy demanding. Infancy and childhood are the times of the most rapid, postnatal brain growth in human beings. The high rate of brain growth is energetically expensive. The human newborn uses 87% of its resting metabolic rate (RMR) for brain growth and function. By the age of 5 years, the percent of RMR usage is still high, at 44%, whereas in the adult human, the figure is between 20% and 25% of RMR. At birth, chimpanzees have smaller brains than humans, and the difference in size between the species increases rapidly (Leigh, 2004; Figure 22.3). Consequently, the RMR values for the chimpanzee are only about 45% at birth, 20% at age 5 years, and 9% at adulthood (Leonard and Robertson, 1994).

Several studies of human forager societies show that children cannot produce enough food to meet their energy and nutrient needs. Indeed, food production remains below food requirements until age 15–20 years in these societies (Kaplan et al., 2000; Kramer, 2002; Gurven and Walker, 2006; Robinson et al., 2008).

CO-OPERATIVE CHILD CARE IN HUMAN SOCIETIES

The child, then, needs to be supplied with foods. These foods need to be specially chosen and prepared so that they are easy to chew and swallow because of the small and fragile deciduous teeth of the child. The foods also must be nutrient dense due to the small digestive system of the child. Essentially the diet for a weaned three-year-old must consist of what we call today "baby foods." Human mothers, however, do not have to provide 100% of nutrition and care directly to their children. Because they are weaned, dependent children may be fed and protected by any older individuals in the social group. Indeed, traditional societies deal with the problem of childcare by spreading the responsibility among many individuals, including older juveniles, adolescents, grandmothers, and other adults.

Co-operative childcare seems to be a human universal (Hrdy, 1999). For example, in Hadza society (African hunters and gatherers), grandmothers and great aunts supply a significant amount of food and care to children (Hawkes et al., 1997; Blurton Jones, 2006). In Agta society (Philippine hunter-gatherers), women hunt large game animals but still retain primary responsibility for childcare (Estioko-Griffin, 1986). They accomplish this dual task by living in extended family groups – two or three brothers and sisters, their spouses, children, and parents – and sharing the childcare. Among the Maya of Guatemala (horticulturists and agriculturists), many people live together in extended family compounds. Women of all ages work together in food preparation, clothing manufacture, and childcare (Bogin field notes, 1988–1993; Figure 22.4). In some societies including the Agta and the Aka pygmies, hunter-gatherers of central Africa (Hewlett, 1991), fathers provide significant childcare.

Death of the chimpanzee mother almost always results in death of her infant. This is because chimpanzee females usually provide 100% of infant care to their offspring (Tutin, 1994). Moreover, female chimpanzees are generally in competition with each other for access to critical resources and antagonistic toward each other and each others offspring. Adoptions of orphaned infants do occur occasionally in chimpanzee social groups, but only infants that are older than four years and able to forage for themselves survive more than a few weeks (Goodall, 1983; Tutin, 1994). Goodall noted deterioration in the health and behavior of infant chimpanzees whose mothers had died. The behavioral changes include symptoms of clinical depression, such as listlessness, whimpering, refusing to eat or interact with others, and less play. Health changes, such as loss of weight, were observed. Goodall reported that even those older infants who survived the death of their mothers were affected by delays in physical growth and maturation.

Because of human co-operative care and kinship organization, family members may adopt orphaned infants and children. It is well known that human infants and children also show physical and behavioral pathology after the death of one or both parents (Bowlby, 1969). Humans, however, usually overcome this and both survive and thrive under the care of adoptive

22.4. Co-operative care of children by women and juvenile girls. The example is from the Kaqchikel-speaking Maya region of Guatemala. The women perform household and food preparation duties while the juvenile girls play with and care for the children. Photograph by Barry Bogin.

parents. It seems that the human infant and child can more easily make new attachments to other caretakers than can the chimpanzee infant (Chisholm, 1999). The ability of a variety of human caretakers to attach to one or several human infants may also be an important factor. The psychological and social roots of this difference between human and nonhuman species in attachment behavior are not well understood. The flexibility in attachment behavior evolved by hominin ancestors may have contributed, in part, to the evolution of childhood and the reproductive efficiency of the human species.

Summarizing the data from many human societies, Lancaster and Lancaster (1983) call this kind of co-operative childcare and feeding "the hominid adaptation" because no other primate or mammal does all of this. The evolutionary reward is that by reducing the length of the infancy stage of life (i.e., shortening the period of lactation) and by developing the special features of the human childhood stage (i.e., flexibility in attachment), humans have the potential for greater lifetime fertility than any ape.

LIFE HISTORY THEORY

How did human beings come to have this unusual fertility and co-operative childcare? Life history theory is the study of the evolution and function of life stages and behaviors related to these stages (Stearns, 1992). The life history of a species may be defined as the evolutionary adaptations used to allocate limited resources and energy toward growth, maintenance, reproduction, raising offspring to independence, and avoiding death. Life history patterns of species are often a series of trade-offs between growth versus reproduction, quantity versus quality of offspring, and possibilities given limited time and resources. Figure 22.5 illustrates the amount of growth, or distance, and rate of growth, or velocity, of healthy human beings from birth until adulthood. The velocity changes in growth correspond with stages of human life history.

The infancy stage begins at birth and lasts until about 3.0 years of age. Postnatal infant growth is rapid, as is its rate of deceleration. Human childhood encompasses the ages of about 3.0–7.0 years. Body growth during childhood proceeds at a steady rate of 5–6 cm per year. Brain growth is rapid and the difference in growth rates is an example of a life history trade-off given limited energy. Many children experience a transient and small "spurt" in growth rate as they transition into the juvenile period. Juvenile mammals are sexually immature, but physically and mentally capable of providing for much of their own care. In many human societies, juveniles perform important work including food production and the care of children (i.e., "babysitting"). Juvenile growth rate declines until puberty, representing another trade-off between current growth versus building a higher quality body and behavioral repertoire over a longer period of time. Puberty is a short-term event of the central nervous system, which initiates sexual maturation and the adolescent life stage. In humans, the hormones responsible for sexual maturation also cause the adolescent growth spurt in stature and other skeletal dimensions. The growth spurt, which is a notable feature of the adolescent growth stage, but not the only defining characteristic, begins at about 10.0 years for girls and 12.0 for boys. The adolescent spurt and growth of the skeleton ends at about 18–19 years for girls and 20–22 years for boys, and with this the adulthood, or reproductive stage of life history, begins.

There are also significant changes in dentition, motor control, muscular development, cognitive function, and emotions associated with human infancy, childhood, juvenile, and adolescent development. The integration

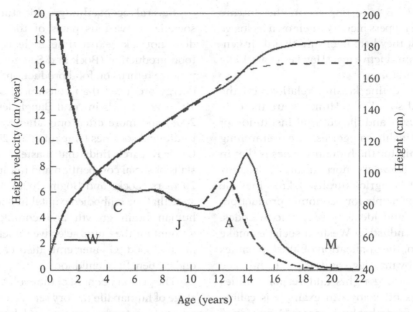

22.5. Average velocity and distance curves of growth in height for healthy girls (dashed lines) and boys (solid lines), showing the postnatal stages of human growth. Values for the distance curves are on the left y-axis; values for the velocity curve are on the right y-axis. In the velocity curves, note the spurts in growth rate at mid-childhood and adolescence for both girls and boys. The postnatal stages: I, infancy; C, childhood; J, juvenile; A, adolescence; M, mature adult. In traditional human societies, weaning (W) of infants from any breast-feeding occurs at an average age of 30 months, with a range of 6–60 or more months (Dettwyler, 1995; Sellen, 2006). Figure based on Bogin (1999).

of these separate domains of function also takes place with age and maturation. These have been much studied by developmental biologists, psychologists and others, but only a few key features of human childhood and adolescence are discussed below (Bogin, 1999 provides further discussion and references).

THE EVOLUTION OF HUMAN CHILDHOOD

The origins of human childhood have fascinated scholars from many disciplines. Some researchers argue that childhood is an invention of recent human societies. Philippe Ariès (1965) proposed that the concept of childhood came into existence in sixteenth-century Europe. Prior to that time, youngsters were considered to be "miniature adults," meaning that once weaned, and less dependent on their mother, children entered the adult world of work. Children were thought to have adult-like understanding of the physical and social world, an idea that was overturned with the developmental psychology studies of Piaget (1954).

A biological hypothesis for human childhood is that the human patterns of growth and development evolved by heterochrony, an evolutionary process that alters the timing of growth stages from ancestors to their descendants. An early twentieth-century version of heterochrony claimed that human childhood evolved by neoteny, meaning that human beings become sexually mature

while still in the juvenile stage of our ancestors. Neoteny implies that human beings are, in a sense, "permanent children." Another version of heterochrony suggests the opposite, which is that human beings, and childhood, evolved by extending the developmental stages of our ancestors. Childhood, in this view, is just a prolongation of infancy (Bogin, 1997 critiques these notions).

The evolutionary evidence, however, favors the hypothesis that childhood evolved as a new stage in hominin life history (hominins or hominids are living humans and our bipedal ancestors), first appearing about two million years ago, during the time of *Homo habilis* (Thompson et al., 2003). Darwinian evolution proceeds by adaptations that increase the fertility of adults and decrease the mortality of offspring prior to their own reproductive age. As explained above, the evolutionary value of childhood is that it allows a woman to give birth to new offspring while allowing her existing dependent offspring to receive care and feeding from close kin and other members of the social group. With this assistance, human women may reproduce every three years without sacrificing the health or life of their previous offspring.

This type of co-operative breeding is found in some species of birds and other mammals (e.g., wolves and hyenas) and it works to increase net reproductive output (Bergmuller et al., 2007). In those species, and in many but not all human groups, the co-operative breeders are close genetic relatives of the mother

(Clutton-Brock, 2002). By assisting the mother to care for her offspring, the helpers increase their own inclusive fitness, meaning that they help to ensure that their genetic kin survive to reproductive age (Hawkes et al., 1998; Hawkes and Paine, 2006).

Human societies define kinship relations on the basis of genetic and social ties. Humans are the only species to use language and the cultural institution of marriage to define kinship categories. The overarching importance of kinship for the human species is that in traditional societies (foragers, horticulturalist, pastoralists, and preindustrial agriculturalists), kinship is the central organizing principle for economic production, social organization, and ideology (e.g., moral codes, religious behavior). Industrial Western societies make use of fictive kinship, the application of kinship names to people unrelated by marriage or descent, to enhance social relations, including rights and responsibilities towards each others offspring. An example is calling the close friend of one's mother by the name "Aunt Maria" instead of Mrs. Smith. "Aunt Maria" may provide food, supervision, protection, gifts, and other types of parental investment to her "niece" and the "niece" is expected to behave in accordance with the rules of interaction between family members. Human co-operative breeding, therefore, is biocultural in nature – explained by both genetic and fictive kinship. Human biocultural co-operative breeding enhances the social, economic, political, religious, and ideological "fitness" of the group as much or more than it contributes to genetic fitness.

CHILDHOOD AND LEARNING

Another explanation for childhood is called the "embodied capital hypothesis" (Kaplan et al., 2000; see Chapter 26 of this volume). By "capital" these authors mean both the quality of the human physical body in terms of strength, skill, immune function, and co-ordination as well as the size and quality of the brain in terms of knowledge, social networks, strategies for resource acquisition, mating competition, parenting style, and social dominance.

Observations of nonhuman primates, elephants, social carnivores, and other mammals show that all of the feeding, social, and reproductive behavior they need to learn and practice can be accomplished during their infant and juvenile stages of life (Pereira and Fairbanks, 1993). Human beings have more to learn, such as symbolic language, kinship systems, and the use of complex technology. The embodied capital hypothesis suggests that it may require 20 years or more to develop the brain and body needed to acquire it all. Human childhood may have coevolved with these behavioral complexes to provide the extra time needed.

Several recent ethnographic studies with traditional societies in various parts of the world show that it does not take extra time to learn the intricacies of food production (Bock and Sellen, 2002). However, net positive returns on food production, meaning more food energy produced than consumed, occur only after age 10 years for girls in rural Bangladesh (Robinson et al., 2008) and more often only after age 15 years in other traditional societies (Kaplan et al., 2000; Kramer, 2002). Other research finds that mastery of language and the skills of social competition and mating do require nearly 20 years (Locke and Bogin, 2006). Perhaps it is best to state that the embodied capital hypothesis for prolonged human brain growth and complex learning cannot account for the initial selective impetus for the evolution of childhood. Greater embodied capital may be a secondary benefit of childhood.

The primary benefit of the evolution of the childhood stage of human life history seems to be the reproductive advantages to the mother and her close genetic and social kin. The result of the human type of reproduction, with childhood and biocultural co-operative breeding, means that human woman can successfully produce two infants in the time it takes a chimpanzee to produce one. The investments in childcare by family members translate into higher survival for human children than for the, generally, unassisted chimpanzee juvenile. In modern humans, the childhood stage, along with much embodied capital, results in the potential for more rapid reproduction, the greatest survival to adulthood, and greater longevity than any other primate species.

JUVENILE TO ADOLESCENT

Following the childhood life history stage the individual enters the juvenile stage, which may be defined as the time from about 7–10 years old for girls and 7–12 years old for boys. It is a stage of slowing growth rate prior to the onset of sexual maturation (Figure 22.5). Juveniles have sufficient maturity of many body systems, such as the dental, locomotor, and cognitive systems, to allow for self-feeding capability. Much important learning of economic and social skills takes place during the juvenile stage. Juvenility ends with puberty, which is an event of short duration (days or a few weeks) that takes place in the brain, within or near the hypothalamic-pituitary axis, and results in a reactivation within the central nervous system of sexual development. This includes a dramatic increase in secretion of sex hormones, which is one marker of the onset of adolescence.

The adolescent life history stage of growth, development, and maturation lasts 5–10 years after the onset of puberty: Other notable features of adolescence include a growth spurt in height and weight, the completion of permanent tooth eruption, development of secondary sexual characteristics (fat and muscle typical of each sex), and the intensification of interest in and practice of adult social, economic, and sexual activities leading to sociosexual maturation. Adolescence ends with the cessation of skeletal growth in length (the closing of the epiphyses of the long bones), the completion of dental development (eruption of the third molar, if it is present), and sexual maturation (measured for women as the age at first reproduction). On a worldwide basis, including living and historical societies, the age of onset of adulthood averages 19 years for women and 21–25 years for men (Bogin, 2001).

ADOLESCENCE AS APPRENTICESHIP

Some life history theorists hypothesize that an adolescent stage of human growth evolved to provide the time to learn and practice complex economic, social, and sexual skills required for effective food production and reproduction and parenting (reviewed in Bogin, 1993, 1999; also see Chapter 26 of this volume). In this perspective, adolescence is a time for an apprenticeship, working and learning alongside older and more experienced members of the social group. The benefits of the skills acquired during adolescence are lower mortality of both first-time mothers and their offspring. This places the "apprenticeship hypothesis" for the learning and practice value of adolescence firmly within Darwinian natural selection theory. There is much human ethnographic and demographic evidence to support the apprenticeship hypothesis and it is likely that the learning and practice of adult skills play an important role in human growth, development, and maturation.

However, apprenticeship cannot be the primary cause for the evolution of adolescence. Learning for childcare is an example. The ethnographic literature documents that in human societies juvenile girls often are expected to provide significant amounts of childcare for their younger siblings (Weisner, 1987, 1996). Human girls enter adolescence with considerable knowledge of the needs of young children. Learning about childcare, then, is not the reason why human girls experience adolescence.

Just as childhood evolved so that the mother could resume reproduction more quickly, adolescence is likely to have evolved as a reproductive adaptation for older individuals. The reason for this is that natural selection works on differential fertility and differential mortality between individuals. An additional 5–10 years of infertility associated with adolescence could not evolve for all humans, since those individuals who "cheated" by terminating growth at an earlier age would begin reproducing sooner and would be at a reproductive advantage. All other primates do, in fact, begin reproducing at earlier ages than humans, and none of the nonhuman primates has a human-like adolescent growth spurt, nor many of the other biological and behavioral features of the human adolescent stage of life history (Hamada and Udono, 2002; Bogin, 2006; but see Leigh, 2001 for evidence of growth spurts in body weight). Clearly, a juvenile primate does not need to pass through a lengthy period of adolescence, with apprenticeship type learning, just to be reproductively successful.

What factors, then, could give rise to adolescence and further delays in reproduction. The answer, it seems, lies in a combination of natural selection for the type of biocultural co-operative breeding that characterizes human beings and sexual selection. Darwin proposed sexual selection as an independent and complementary process to natural selection. Sexual selection was defined by Darwin as an evolutionary process in animals that, "… depends on the advantage which certain individuals have over other individuals of the same sex and species, in exclusive relation to reproduction" (Darwin, 1871, vol. 1, p. 276). Today we would replace the word *reproduction* with *mating*, as not all mating opportunities are intended to result in a pregnancy. Darwin also wrote that:

There are many other structures and instincts which must have been developed through sexual selection – such as the weapons of offence and the means of defence possessed by the males for fighting with and driving away their rivals – their courage and pugnacity – their ornaments of many kinds – their organs for producing vocal or instrumental music – and their glands for emitting odours; most of these latter structures serving only to allure or excite the female (Darwin 1871, vol. 1, pp. 257–258).

It is known today that sexual selection also works for females, meaning that female specific physical and behavioral traits may evolve via competition between the females for mating opportunities with males.

NATURAL SELECTION FOR ADOLESCENCE

The natural selection case for adolescence is derived from the fact that the evolution of childhood afforded hominid females the opportunity to give birth at shorter intervals. Producing offspring, however, is only a small part of reproductive fitness. Rearing the young to their own reproductive maturity is a surer indicator of success. Studies of yellow baboons (Altmann, 1980), toque macaques (Dittus, 1977), and chimpanzees (Teleki et al., 1976) show that between 50% and 60% of first-born offspring die in infancy. By contrast, in hunter-gatherer human societies, such as the Hadza of eastern Africa, 39% of offspring die in infancy (Blurton Jones et al., 1992). For the !Kung of southern Africa the figure is 44% (Howell, 1979). Just for comparison, it may be noted that in the United States, in the year 1960, about 2.5% of all live, first-born children died before the age of 1 year (Vavra and Querec, 1973).

The infants of nonhuman primates are fed and cared for by either the mother or, in a few species, the father and other close kin. As was mentioned earlier, infant and childcare in human societies is different because many members of the social group provide care and protection. The contribution of food and other resources by many group members helps to insure infant survival.

Kaplan et al. (2000) have documented the number of calories of food produced by juveniles, adolescents, and adults in several human foraging societies (Aché, Hiwi, !Kung, and Hadza). Kramer (2002) performed a similar analysis for a village of traditional Maya horticulturalists in the Yucatan, Mexico, and Robinson et al. (2008) did the same for Bangladeshi children, juveniles, adolescents, and adults in a traditional farming village. These studies show that rates of food production increase most steeply after age 15 years, that is, in mid adolescence. While adolescent rates of energy production fall well below those of men and women over 20 years of age, they are still significant and by age 15 years average between 1500 and 2000 kcal per day. That is nearly enough energy to meet the needs of the adolescents, which means that these adolescents are no longer a drain on older individuals. Extra food foraged by older people can be shared with still dependent infants, children, and juveniles. This helps to ensure the survival of the dependent young. In reality, of course, the foods gathered by adolescents may be combined with those foraged and hunted by adults, so that all members of the group receive a balanced diet. The main point to stress, however, is that adolescents contribute to the reproductive success for their parents and other close kin. The net reproductive gain from this co-operative breeding may offset the reproductive loss to the individual adolescent who must delay her own first birth by several more years. However, this still does not rule out "cheating" by reproducing at age 13 years rather than at 19 years or later. As will be explained below, such "cheating" has harmful consequences for the young mother and her infant, and this maintains natural selection for the average age at first birth at 19 years.

SEXUAL SELECTION FOR ADOLESCENCE

There is still much the adolescent girl needs to acquire in order to negotiate adult life besides caring for infants. She must gain physical size, strength, and fat stores to support pregnancy and lactation. She must attain proficiency in economic productivity. Finally, she must learn and practice skills in social co-operation–competition and in sexual politics if she is to successfully contend with other women for desirable mating opportunities. The pattern of physical growth and development during adolescence allows girls to achieve all of these needs. The adolescent growth spurt serves as a signal of maturation. Early in the spurt, before peak height velocity is reached, girls develop pubic hair and fat deposits on breasts, buttocks, and thighs. They appear to be maturing sexually. About a year after peak height velocity, girls experience menarche, an unambiguous external signal of internal reproductive system development.

These changes give adolescent girls the physical characteristics they need to appear adult-like and enter the sociosexual world of adults. The adolescent gains skill in economic productivity and some proficiency in sexual politics because they look mature sexually, and are treated as such, several years before they actually become fertile. Most adolescent girls experience one to three years of anovulatory menstrual cycles after menarche. Nevertheless, the dramatic changes of adolescence stimulate both the girls and the adults around them to participate in adult social, sexual, and economic behavior. For the postmenarchial adolescent girl, this participation is "risk free" in terms of pregnancy for several years. Even after ovulation begins it takes about 5 years for the adolescent to achieve the adult frequency of monthly ovulations, which is defined as 65% of menstrual cycles (Worthman, 1993). This means that if menarche occurs between ages 12 and 13 years (as is the case for well-nourished populations) the typical healthy and sexually active adolescent has a reduced possibility of becoming pregnant until she is 17–18 years old. Indeed the worldwide median age at first birth is 19 years (Bogin, 2001).

Although adolescents younger than 17 years old can and do have babies, both the teenage mothers and the infants are at risk because of the reproductive immaturity of the mother. Risks include a low-birthweight infant, premature birth, high blood pressure in the mother, and death for the fetus, the mother, or both. Much of the reason for these risks is that adolescent girls are still growing and competing for resources against their own fetus. The likelihood of these risks declines and the chance of successful pregnancy and birth increases markedly after age 18. There is, therefore, considerable natural selection pressure working in concert with the sexual selection for the adolescent female phenotype of sexual maturity combined with infertility to delay pregnancy until the end of adolescence.

UNUSUAL GROWTH OF THE HUMAN PELVIS

A fundamental feature of human growth that delays female fertility until the late teenage years is the growth of the pelvis. Ellison (1982) and Worthman (1993) found that age at menarche is best predicted by bi-iliac width, the distance between the iliac crests of the pelvis. A median width of 24 cm is needed for menarche in

American girls living in Berkeley, California, Kikuyu girls of East Africa, and Bundi girls of highland New Guinea. The pelvic width constant occurs at different ages in these three cultures, about 13, 16, and 17 years old, respectively. The later ages for menarche are due to chronic malnutrition and disease in Kenya and Bundi.

Moerman (1982, now known as Marquisa LaVelle) also reported a special human relationship between growth in pelvic size and reproductive maturation. She found that the crucial variable for successful first birth is size of the pelvic inlet, the bony opening of the birth canal. Moerman measured pelvic X-rays from a longitudinal sample of healthy, well-nourished American girls who achieved menarche between 12 and 13 years. These girls did not attain adult pelvic inlet size until 17–18 years of age. Quite unexpectedly, the adolescent growth spurt, which occurs before menarche, does not influence the size of the pelvis in the same way as the rest of the skeleton. Rather, the female pelvis has its own slow pattern of growth, which continues for several years after adult stature is achieved.

Why the pelvis follows this unusual pattern of growth is not clearly understood. Perhaps bipedal walking, another special human attribute, is a factor. Bipedalism is known to have changed the shape of the human pelvis from the basic ape-like shape. Apes have a cylindrical-shaped pelvis, but humans have a bowl-shaped pelvis. The human shape is more efficient for bipedal locomotion but less efficient for reproduction because it restricts the size of the birth canal. Human women may need a longer period of pelvic growth to compensate for the restriction. Whatever the reason, cross-cultural studies of reproductive behavior shows that human societies acknowledge (consciously or not) this special pattern of pelvic growth. Marriage, a uniquely human cultural behavior, usually precedes first childbirth. Age at marriage clusters around 18 years for women from such diverse cultures as the Aché of Paraguay, the Hadza of Tanzania, the !Kung of Botswana, the Kikuyu of Kenya, Mayans of Guatemala, Copper Eskimos of Canada, and both the colonial and early twentieth-century United States (Bogin, 1999; Kaplan et al., 2000).

Extra widening of the pelvic inlet late in adolescence may assist successful birth for human women. But, even with this widening human births are difficult. In virtually all cultures older, experienced women assist the mother to deliver her infant (Trevathan, 1987, 1996). This assistance by both biological and social kinswomen is another example of biocultural co-operative breeding in the human species.

The special human pattern of pelvic growth helps explain the delay from menarche to full reproductive maturity. That time of waiting seems to be the result of both natural and sexual selection and provides the adolescent girls with many opportunities to be perceived as more physically mature than they in fact are. Grafted onto this biological delay in fertility is the embodied capital that comes from learning and practicing important adult sexual, social, economic, and political behaviors that lead to increase reproductive fitness in later life.

SEXUAL DEVELOPMENT IN HUMANS AND CHIMPANZEES

Some features of the sequence of adolescent events for girls described above are illustrated in Figure 22.6. The human female pattern of physical, behavioral, and sexual development after puberty is quite different from our closest cousin, the chimpanzee. A key difference is that chimpanzee females have an estrus cycle, not a menstrual cycle. Chimpanzee females advertise their fertility via anogenital swelling. In the wild, females have their first "small irregular swellings of the clitoris between eight and nine years" (Pusey, 1990, p. 208). The swellings enlarge with time and by about 11 years of age they have their first swelling that is clearly visible to all members of the social group (Wallis, 1997). At this point, adult males show sexual interest in the females when they are swollen (Pusey, 1990). Menarche takes place a few months after the first mature swelling cycle, and ovulation does not occur for an average of two years after menarche, although the range of variation is five months to three years (Pusey, 1990).

Once female chimpanzees achieve mature estrus cycles they swell for 13 of the 36 days of their cycle (Pusey, 2001). Ovulation occurs near the end of the swelling period. Male chimpanzees seem to calibrate their sexual interest in the females based on the phases of estrus, and the most dominant males concentrate their reproductive efforts toward the likely time of ovulation (Goodall, 1986).

Human adolescent girls and adult women have no anogenital swellings. While human breast development and menstrual bleeding may be signs of impending fertility, they are not tightly correlated with the onset of ovulation or the timing of ovulation during the menstrual cycle. Even in mature women, the exact timing of human ovulation is "hidden" from other members of the social group. Women are often not certain of their own ovulation (at least in the United States where books to help women time their ovulatory cycles are "best sellers"). As a probable consequence of this uncertainty, human males show sexual interest in women during most phases of the menstrual cycle. In some cultures, sexual relations are prohibited during and just after menstruation, but these are the least likely times for ovulation.

There are other important differences, and a few similarities, between people and chimpanzees in sociosexual development. Female chimpanzees remain in their natal

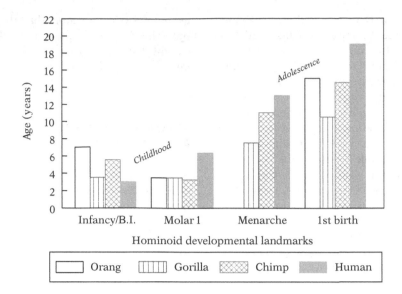

22.6. Hominoid female developmental landmarks. Data based on observations of wild-living individuals of the ape species *Pongo pygmaeus* (labeled "Orang"); *Gorilla gorilla* (labeled "Gorilla"); *Pan troglodytes* (labeled "Chimp"); and *Homo sapiens* (labeled "human"). For humans, the data comprise healthy individuals from various cultures. The label "Infancy/B.I." is the period of dependency on the mother for survival, usually coincident with the mean age at weaning and/or a new birth ("B.I." means "birth interval"); "Molar 1" is the mean age at eruption of the first permanent molar; "Menarche" is the mean age at first estrus or menstrual bleeding; "1st birth" is the mean age of females at first offspring delivery. The label "Childhood" denotes the period of time of human life history from about age 3.0 to 6.0 years between age at weaning and age of eruption of the first permanent molar. Childhood "fills the gap" between feeding by lactation and semi-independent feeding of the juvenile life history stage (maturation of dention, motor skills, and cognitive behaviors required for self-feeding). The label "Adolescence" denotes the period of time of human female life history from menarche to adulthood (first reproduction). In a strict biological sense, adolescence begins at puberty, which takes place two to three years before menarche. However, many human societies ascribe special biocultural meaning to menarche, including rites of passage from "girlhood" to "womanhood." The biocultural events surrounding menarche and the years following it assist girls to become women in the economic, social, and reproductive spheres of life. Similar rites of passage are reserved for boys and are tied to biocultural events of their growth, development, and maturation (e.g., the adolescent growth spurt, growth of the penis and pubic hair, development of greater muscle mass). Figure modified from Bogin (1999).

social group until they begin estrus cycles. As they cycle, the young females leave their natal groups to copulate with males from neighboring groups. These visits eventually lead to permanent emigration to a new group. Once pregnant the female chimpanzee continues to have one or two estrus cycles and many copulations (Pusey, 2001). Human women may or may not emigrate from their natal group. An analysis based on the "World Ethnographic Sample" database (Ember et al., 2002) shows that 15% of human societies are matrilocal (married couple resides with the wife's social group), 4% are avunculocal (living near or with the wife's mother's brother), 67% are patrilocal (couple resides with the husband's group), 7% are bilocal (living with either the bride's or groom's family) and 5% are neolocal (married couple establish an independent residence). In this respect and many others, human societies follow cultural rules of behavior that work with and against biology to produce a myriad of sexual and reproductive behaviors. Human cultural behavior is more complex and varied than the social behavior of chimpanzee communities. A final comparison is that human pregnancy usually terminates menstrual cycling; however, a single menstruation (as opposed to "spotting" and pathological bleeding) following fertilization can occur if the pregnancy takes place close to the expected start of the next period. As for chimpanzees, copulation may continue during pregnancy.

ADOLESCENCE IS ONLY FOR HUMANS

The sexual development of human adolescent girls often is associated with a series of sociocultural events that are not reported for any other species. Breast development and menarche may promote rites of passage, that is, initiation ceremonies or events that mark the transition between girlhood and womanhood. This transition often takes years to complete. Even when marriage takes place before or near the time of menarche it may take years until the bride makes a complete transference from her natal family to live with her husband. A detailed description of the variety and significance of human rites of passage is beyond the scope of this chapter. Interested readers may consult the extensive ethnographic literature (e.g., Schlegel and Barry, 1991; Schlegel, 1995).

The physiological, anatomical, growth, developmental, maturational, and behavioral differences between people and chimpanzees lead to the conclusion that chimpanzees do not have adolescence. Some time after the divergence from their common ancestor, chimpanzees and humans evolved different life history patterns for sexual development. Physiologically, chimpanzee females seem to have evolved estrus swelling as something new, because gibbons and orangutans do not have them (de Waal, 2001). Human females evolved permanent breasts, the adolescent growth spurt, and menarche followed by hidden ovulation as a unique set of traits.

WHY DO BOYS HAVE ADOLESCENCE?

Natural and sexual selection for adolescence applies to both girls and boys. The forces of natural selection are the same for both sexes but the particulars of sexual selection are different. The adolescent development of boys is quite distinct from that of girls. Boys become fertile well before they assume the size and the physical characteristics of men. Analysis of urine samples from boys 11–16 years old show that they begin producing sperm at a median age of 13.4 years (Muller et al., 1989). Yet cross-cultural evidence indicates that few boys successfully father children until they are into their third decade of life. In the United States, for example, only 3.09% of live-born infants in 1990 were fathered by men under 20 years of age. In Portugal, for years 1990, 1994, and 1999, the percentage of fathers under 20 years of age was always below 3% (Instituto Nacional de Estatística, 1999). In 2001, Portugal stopped presenting results concerning the percentage of fathers below 20 because there were too few of them (Instituto Nacional de Estatística, 2001). Among the traditional Kikuyu of East Africa, men do not marry and become fathers until about age 25 years, although they become sexually active after their circumcision rite at around age 18 (Worthman, 1993).

The explanation for the lag between sperm production and fatherhood is not likely to be a simple one of sperm performance, such as not having the endurance to swim to an egg cell in the woman's fallopian tubes. More likely is the fact that the average boy of 13.4 years is only beginning his adolescent growth spurt (Figure 22.5). Growth researchers have documented that in terms of physical appearance, physiological status, psychosocial development, and economic productivity, the 13-year-old boy is still more a juvenile than an adult. Anthropologists working in many diverse cultural settings report that few women (and more important from a cross-cultural perspective, few prospective in-laws) view the teenage boy as a biologically, economically, and socially viable husband and father.

The delay between sperm production and reproductive maturity is not wasted time in either a biological or social sense. The obvious and the subtle psychophysiological effects of testosterone and other androgen hormones that are released after gonadal maturation may "prime" boys to be receptive to their future roles as men. Alternatively, it is possible that physical changes provoked by the endocrines, such as deepening of the voice and appearance of pubic hair, provide a social stimulus toward adult behaviors. The following is an example of the interaction between biology and behavior. In 2001, a research team measured and interviewed Portuguese and Cape Verde boys, ages 10–15 years old, living near Lisbon, Portugal (Bogin and Varela Silva, 2003). We assessed pubic hair development and voice "breaking." The older, more mature boys told us that in school they "speak and act like men" because they have pubic hair, but at home they speak like boys to show respect for their parents and older siblings. Such are the social pressures of male adolescence!

Whether the influences are primarily physical or social, early in adolescence, sociosexual feelings including guilt, anxiety, pleasure, and pride intensify. At the same time, adolescent boys become more interested in adult activities, adjust their attitude to parental figures, and think and act more independently. In short, they begin to behave like men.

However – and this is where the survival advantage may lie – they still look like boys. One might say that a healthy, well-nourished 13.5-year-old human male, at a median height of 160 cm (62 inches) "pretends" to be more childlike than he really is. Because their adolescent growth spurt occurs late in sexual development, young males can practice behaving like adults before they are actually perceived as adults. The sociosexual antics of young adolescent boys are often considered to be more humorous than serious. Yet, they provide the experience to fine-tune their sexual and social roles before their lives or those of their offspring depend on them. For example, competition between men for women favors the older, more experienced man. Because such competition may be fatal, the childlike appearance of the immature but hormonally and socially primed adolescent male may be life-saving as well as educational.

Adolescent boys do not begin to look like men until their spurt in muscle development, which takes place at about age 17 years. Prior to this, adolescent boys are fertile but still look like juveniles. This is a type of reverse sexual selection when compared with girls.

Adolescent girls learn and practice adult behaviors when they are infertile but look like women.

Language development in adolescent boys is another influence on social and sexual success. Indeed, vocal and verbal performance is an essential aspect of sexual/ mating behavior in human beings. Adolescent speech becomes more complex in vocabulary, including slang, more rapid in speaking rate, and assumes more rhythmic fluency (Locke and Bogin, 2006). Boys and men in many societies engage in vocal and verbal duels, use of riddles, and other complex patterns of language. These duels develop sequentially during adolescence. Those who can handle this complex vocal and verbal material are considered intelligent by members of the social group. Anthropologists find that most oral societies, and many literate societies, promote verbal skill for attention, power, prestige, and success (Locke and Bogin, 2006). Vocal and verbal dueling is almost always performed in front of an audience and is often used to attract mating opportunities.

Girls and women also engage in vocal and verbal contests, but less often in highly public displays. Girls and younger women focus relatively more on social talking, including gossip, deceiving, mollifying, negotiating, and persuading (Locke and Bogin, 2006). As for boys, this use of language for girls does not reach adult levels of complexity and effectiveness until the later teenage years. Human language, in this sense, conforms closely to Darwin's examples of sexual selection for "... organs for producing vocal or instrumental music ..." which influence opportunities for mating.

WHEN DID ADOLESCENCE EVOLVE?

In contrast to the relatively ancient evolution of hominin childhood, an adolescent life stage may be relatively recent. Based on skeletal and dental development there is little or no evidence of human adolescence in *Homo erectus* or any earlier hominin (Dean et al., 2001; Antón, 2003). Moreover, the fossil evidence indicates that the adolescent growth stage, and the adolescent growth spurt, evolved only in the lineage leading directly to modern *Homo sapiens*.

There is a possibility that adolescence may first appear in *Homo antecessor*, a hominin from Spain that is about 800 000 years old (Bermudez de Castro et al., 1999). The evidence for this is based on patterns of tooth formation, which while not directly linked to the presence of an adolescent growth spurt is suggestive. Possible descendants of *H. antecessor* are the Neanderthals. There is one fossil of a Neanderthal youth in which the associated dental and skeletal remains needed to assess adolescent growth are preserved. It is called Le Moustier 1, most likely the remains of a male, and it was found in 1908 in Western France. The specimen is

dated at between 42 000 and 37 000 years BP. Using modern human development reference data, Thompson and Nelson (1997) estimate that Le Moustier 1 has a dental age of 15.5 years and a stature age of 11–12 years based on the length of his femur. The dental age and the stature age are in poor agreement, and indicate that Le Moustier 1 may not have followed a human pattern of adolescent growth (Bogin, 1999). Alternatively, Nelson and Thompson (2002) suggest that Neanderthals, both the young and adults, may have reduced limb growth due to cold adaptation (Allen's rule). There is also the suggestion that Neanderthals of Western Europe suffered iodine deficiency (Dobson, 1998; Bogin and Rios, 2003). Cold adaptation and/or nutrient deficiency may obscure evidence for an adolescent spurt in limb-length growth.

Given the uncertainties of the evidence derived from *H. antecessor* and the Neanderthals, it is quite likely adolescence is no older than the appearance of archaic *H. sapiens* in Africa at about 125 000 years ago – possibly even more recently at about 60 000 years ago. With the evolution of adolescence the modern pattern of human life history was established. With the addition of adolescence to human life history sociocultural behaviors and forms, such as marriage and the family, could come into being. When all of these came to exist is not known. From the ancient roots for the evolution of childhood to the more recent emergence of adolescence our hominin ancestors were transformed. Only with a life history including both childhood and adolescence could the human species, in most essential biocultural aspects, have evolved.

DISCUSSION POINTS

1. How does the pattern of human brain growth during infancy and childhood help us to understand the biological and behavioral development of human beings?

2. In what ways does human parental investment in offspring differ from that of other primates, especially the great apes? What value does human co-operative care of infants and children have for human reproduction and infant-child survival?

3. Boys and girls proceed through puberty and adolescence in different ways regarding the development of sexual characters and adult behaviors. What may be the biological and social value of the two distinct paths of development?

4. Bogin proposes that human childhood is fundamentally a feeding and reproductive adaptation to assist the mother's reproductive success. Find other explanations for the value of human childhood and discuss the evidence in favor and against Bogin's hypothesis.

REFERENCES

Altmann, J. (1980). *Baboon Mothers and Infants*. Cambridge, MA: Harvard University Press.

Anemone, R. L., Mooney, M. P. and Siegel, M. I. (1996). Longitudinal study of dental development in chimpanzees of known chronological age: implications for understanding the age at death of Plio-Pleistocene hominids. *American Journal of Physical Anthropology*, **99**, 119–133.

Antón, S. C. (2003). Natural history of *Homo erectus*. *American Journal of Physical Anthropology*, **37**(suppl.), 126–170.

Ariès, P. (1965). *Centuries of Childhood: a Social History of Family Life*, transl. R. Baldich. New York: Vintage Books.

Behar, M. (1977). Protein-calorie deficits in developing countries. *Annals of the New York Academy of Sciences*, **300**, 176–187.

Bergmuller, R., Johnstone, R. A., Russell, A. F. and Bshary, R. (2007). Integrating cooperative breeding into theoretical concepts of cooperation. *Behavioural Processes*, **76**, 61–72.

Bermudez de Castro, J. M., Rosas, A., Carbonell, E., et al. (1999). A modern human pattern of dental development in Lower Pleistocene hominids from Atapuerca-TD6 (Spain). *Proceedings of the National Academy of Sciences of the United States of America*, **96**, 4210–4213.

Blurton Jones, N. G. (2006). Contemporary hunter-gatherers and human life history evolution. In *The Evolution of Human Life History*, K. Hawkes and R. L. Paine (eds). Santa Fe, NM: School of American Research Press, pp. 231–266.

Blurton Jones, N. G., Smith, L. C., O'Connel, J. F., et al. (1992). Demography of the Hadza, an increasing and high density population of savanna foragers. *American Journal of Physical Anthropology*, **89**, 159–181.

Bock, J. and Sellen, D. W. (2002). Childhood and the evolution of the human life course. *Human Nature*, **B13**, 153–159.

Bogin, B. (1993). Why must I be a teenager at all? *New Scientist*, **137**, 34–38.

Bogin, B. (1997). Evolutionary hypotheses for human childhood. *Yearbook of Physical Anthropology*, **40**, 63–89.

Bogin, B. (1999). *Patterns of Human Growth*, 2nd edn. Cambridge: Cambridge University Press.

Bogin, B. (2001). *The Growth of Humanity*. New York: Wiley-Liss.

Bogin, B. (2006). Modern human life history: the evolution of human childhood and adult fertility. In *The Evolution of Human Life History*, K. Hawkes and R. R. Paine (eds). Santa Fe, NM: School of American Research Press, pp. 197–230.

Bogin, B. and Rios, L. (2003). Rapid morphological change in living humans: implications for modern human origins. *Comparative Biochemistry and Physiology, Part A*, **136**, 71–84.

Bogin, B. and Varela Silva, M. I. (2003). Anthropometric variation and health: a biocultural model of human growth. *Journal of Children's Health*, **1**, 149–172.

Bowlby, R. (1969). *Attachment and Loss*. New York: Basic Books.

Chisholm, J. S. (1999). *Sex, Hope, and Death*. Cambridge: Cambridge University Press.

Clutton-Brock, T. (2002). Breeding together: kin selection and mutualism in cooperative vertebrates. *Science*, **296**, 69–72.

Cobourne, M. T. and Sharpe, P. T. (2003). Tooth and jaw: molecular mechanisms of patterning in the first branchial arch. *Archives of Oral Biology*, **48**, 1–14.

Darwin, C. (1871). *The Descent of Man and Selection in Relation to Sex*. London: John Murray.

Davidson, E. (2001). *Genomic Regulatory Systems: Development and Evolution*. San Diego, CA: Academic Press.

de Waal, F. B. M. (2001). Introduction. In *Tree of Origin*, F. B. M. de Waal (ed.). Cambridge: Harvard University Press, pp. 1–9.

Dean, C., Leakey, M. G., Reid, D., et al. (2001). Growth processes in teeth distinguish modern humans from *Homo erectus* and earlier hominins. *Nature*, **414**, 628–631.

Dettwyler, K. A. (1995). A time to wean: the hominid blueprint for the natural age of weaning in modern human populations. In *Breastfeeding: Biocultural Perspectives*, P. Stuart-Macadam and K. A. Dettwyler (eds). New York: Aldine de Gruyter, pp. 39–74.

Dittus, W. P. J. (1977). The social regulation of population density and age-sex distribution in the Toque Monkey. *Behaviour*, **63**, 281–322.

Dobson, J. E. (1998). The iodine factor in health and evolution. *Geographical Review*, **88**, 1–28.

Dobzhansky, T. (1973). Nothing in biology makes sense except in the light of evolution. *American Biology Teacher*, **35**, 125–129.

Ellison, P. T. (1982). Skeletal growth, fatness, and menarcheal age: a comparison of two hypotheses. *Human Biology*, **54**, 269–281.

Ellison, P. T. and O'Rourke, M. T. (2000). Population growth and fertility regulation. In *Human Biology: an Evolutionary and Biocultural Perspective*, S. Stinson, B. Bogin, R. Huss-Ashmore, et al. (eds). New York: Wiley, pp. 553–586.

Ember, C. R., Ember, M. and Peregrine, P. N. (2002). *Anthropology*, 10th edn. Upper Saddle River, NJ: Prentice Hall.

Estioko-Griffin, A. (1986). Daughters of the forest. *Natural History*, **95**, 36–43.

Finch, C. E. and Rose, M. R. (1995). Hormones and the physiological architecture of life history evolution. *Quarterly Review of Biology*, **70**, 1–52.

Gage, T. B. (1998). The comparative demography of primates: with some comments on the evolution of life histories. *Annual Review of Anthropology*, **27**, 197–221.

Goodall, J. (1983). Population dynamics during a 15-year period in one community of free-living chimpanzees in the Gombe National Park, Tanzania. *Zietschrift fur Tierpsychologie*, **61**, 1–60.

Goodall, J. (1986). *The Chimpanzees of Gombe: Patterns of Behavior*. Cambridge, MA: Harvard University Press.

Goodall, J. (2003). Fifi fights back. *National Geographic Magazine*, **203**, 76–89.

Gurven, M. and Walker, R. (2006). Energetic demand of multiple dependents and the evolution of slow human growth. *Proceedings of the Royal Society of London. Series B*, **273**, 835–841.

Hamada, Y. and Udono, T. (2002). Longitudinal analysis of length growth in the chimpanzee (*Pan troglodytes*). *American Journal of Physical Anthropology*, **118**, 268–284.

Hamosh, M. (1995). Lipid metabolism in pediatric nutrition. *Pediatric Clinics of North America*, **42**, 839–859.

Hawkes, K. and Paine, R. R. (eds) (2006). *The Evolution of Human Life History*. Santa Fe, NM: School of American Research Press.

Hawkes, K., O'Connell, J. F. and Blurton Jones, N. G. (1997). Hadza women's time allocation, offspring provisioning, and the evolution of post-menopausal lifespans. *Current Anthropology*, **38**, 551–578.

Hawkes, K., O'Connell, J. F., Blurton Jones, N. G., et al. (1998). Grandmothering, menopause, and the evolution of human life histories. *Proceedings of the National Academy of Sciences of the United States of America*, **95**, 1336–1339.

Henry, L. (1961). Some data on natural fertility. *Eugenics Quarterly*, **8**, 81–91.

Hewlett, B. (1991). *Intimate Fathers: the Nature and Context of Aka Pygmy Paternal Infant Care*. Ann Arbor, MI: University of Michigan Press.

Howell, N. (1979). *Demography of the Dobe !Kung*. New York: Academic Press.

Hrdy, S. B. (1999). *Mother Nature: a History of Mothers, Infants, and Natural Selection*. New York: Pantheon.

Huxley, T. H. (1863). *Evidence as to Man's Place in Nature*. London: Williams and Norwood.

Instituto Nacional de Estatística (1999). *Resultados definitivos: a natalidade em Portugal, 1998 – Informação à Comunicação Social – Destaque de 16 de Setembro de 1999*. Lisboa: Instituto Nacional de Estatística.

Instituto Nacional de Estatística (2001). *Resultados definitivos: a natalidade em Portugal, 2001 – Informação à Comunicação Social – Destaque de 4 de Julho de 2001*. Lisboa: Instituto Nacional de Estatística.

Kaplan, H., Hill, K., Lancaster, J., et al. (2000). A theory of human life history evolution: diet, intelligence, and longevity. *Evolutionary Anthropology*, **9**, 156–185.

King, M. C. and Wilson, A. C. (1975). Evolution at two levels: molecular similarities and differences between humans and chimpanzees. *Science*, **188**, 107–116.

Kramer, K. L. (2002). Variation in juvenile dependence: helping behavior among Maya children. *Human Nature*, **13**, 299–325.

Lancaster, J. B. and Lancaster, C. S. (1983). Parental investment: the hominid adaptation. In *How Humans Adapt*, D. J. Ortner (ed.), Washington, DC: Smithsonian Institution Press, pp. 33–65.

Larsen, U., Yan, S. and Yashin, A. (2003). Controlling for postpartum amenorrhea and heterogeneity in the analysis of fecundability using birth interval data: a simulation study with application to Hutterite reproductive histories. *Population Review*, **42**, http://muse.jhu.edu/demo/population_review/v042/42.1larsen.pdf.

Leigh, S. R. (2001). The evolution of human growth. *Evolutionary Anthropology*, **10**, 223–236.

Leigh, S. R. (2004). Brain growth, life history, and cognition in primate and human evolution. *American Journal of Primatology*, **62**, 139–164.

Leonard, W. R. and Robertson, M. L. (1994). Evolutionary perspectives on human nutrition: the influence of brain and body size on diet and metabolism. *American Journal of Human Biology*, **6**, 77–88.

Lewin, R. (1993). *Human Evolution: an Illustrated Introduction*. Oxford: Blackwell.

Littleton, J. (2005). Fifty years of chimpanzee demography at Taronga Park Zoo. *American Journal of Primatology*, **67**, 281–298.

Locke, J. L. and Bogin, B. (2006). Language and life history: a new perspective on the development and evolution of human language. *Behavioral and Brain Sciences*, **29**, 259–325.

Mattick, J. S. (2004). The hidden genetic program of complex organisms. *Scientific American*, **291**, 60–67.

Moerman, M. L. (1982). Growth of the birth canal in adolescent girls. *American Journal of Obstetrics and Gynecology*, **143**, 528–532.

Muller, J., Nielsen, C. T. and Skakkebaek, N. E. (1989). Testicular maturation and pubertal growth and development in normal boys. In *The Physiology of Human Growth*, J. M. Tanner and M. A. Preece (eds). Cambridge: Cambridge University Press, pp. 201–207.

Nelson, A. J. and Thompson, J. L. (2002). Adolescent postcranial growth in *Homo neanderthalensis*. In *Human Evolution through Developmental Change*, N. Minugh-Purvis and K. J. McNamara (eds). Baltimore, MD: Johns Hopkins University Press, pp. 442–463.

Nishida, T., Takasaki, H. and Takahata, Y. (1990). Demography and reproductive profiles. In *The Chimpanzees of the Mahale Mountains: Sexual and Life History Strategies*, T. Nishida (ed.). Tokyo: University of Tokyo Press, pp. 63–97.

Nonaka, K., Miura, T. and Peter, K. (1994). Recent fertility decline in Dariusleut Hutterites: an extension of Eaton and Meyer's Hutterite fertility study. *Human Biology*, **66**, 411–420.

Pereira, M. E. and Fairbanks, L. A. (1993). *Juvenile Primates: Life History, Development, and Behavior*. New York: Oxford University Press.

Piaget, J. (1954). *The Construction of Reality in the Child*. New York: Basic Books.

Pusey, A. (1983). Mother–offspring relationships in chimpanzees after weaning. *Animal Behavior*, **31**, 363–377.

Pusey, A. (1990). Behavioral changes at adolescence in chimpanzees. *Behaviour*, **115**, 203–246.

Pusey, A. (2001). Of genes and apes: chimpanzee social organization and reproduction. In *Tree of Origin*, F. B. M. de Waal (ed.). Cambridge, MA: Harvard University Press, pp. 10–37.

Pusey, A., Williams, J. and Goodall, J. (1997). The influence of dominance rank on the reproductive success of female chimpanzees. *Science*, **227**, 828–831.

Robinson, R. S., Lee, R. D. and Kramer, K. L. (2008). Counting women's labour: a reanalysis of children's net production using Cain's data from a Bangladeshi village. *Population Studies (Cambridge)*, **62**, 25–38.

Sato, T., Nonaka, K., Miura, T., et al. (1994). Trends in cohort fertility of the Dariusleut Hutterite population. *Human Biology*, **66**, 421–432.

Schlegel, A. (ed.) (1995). Special issue on adolescence. *Ethos*, **23**, 3–103.

Schlegel, A. and Barry, H. (1991). *Adolescence: an Anthropological Inquiry*. New York: Free Press.

Sellen, D. W. (2006). Lactation, complementary feeding and human life history. In *The Evolution of Human Life History*, K. Hawkes and R. R. Paine (eds). Santa Fe, NM: School of American Research Press, pp. 155–196.

Shankar, R., Chaurasia, A., Ghosh, B., et al. (2007). Non-random genomic divergence in repetitive sequences of human and chimpanzee in genes of different functional categories. *Molecular Genetics and Genomics*, **277**, 441–455.

Shibaguchi, T., Kato, J., Abe, M., et al. (2003). Expression and role of Lhx8 in murine tooth development. *Archives of Histology and Cytology*, **66**, 95–108.

Smith, B. H. (1991). Age at weaning approximates age of emergence of the first permanent molar in non-human primates [abstract]. *American Journal of Physical Anthropology*, **12**(suppl.), 163–164.

Smith, B. H., Crummett, T. L. and Brandt, K. L. (1994). Ages of eruption of primate teeth: a compendium for aging individuals and comparing life histories. *Yearbook of Physical Anthropology*, **37**, 177–231.

Stearns, S. C. (1992). *The Evolution of Life Histories*. Oxford: Oxford University Press.

Teleki, G. E., Hunt, E. and Pfifferling, J. H. (1976). Demographic observations (1963–1973) on the chimpanzees of the Gombe National Park, Tanzania. *Journal of Human Evolution*, **5**, 559–598.

Thompson, D. W. (1917). *On Growth and Form*. Cambridge: Cambridge University Press.

Thompson, D. W. (1942). *On Growth and Form*, revised edition. Cambridge: Cambridge University Press.

Thompson, J. L. and Nelson, A. J. (1997). Relative postcranial development of Neandertals. *Journal of Human Evolution*, **32**, A23–A24.

Thompson, J. L., Krovitz, G. E. and Nelson, A. J. (2003). *Patterns of Growth and Development in the Genus Homo*. Cambridge: Cambridge University Press.

Trevathan, W. R. (1987). *Human Birth: an Evolutionary Perspective*. New York: Aldine de Gruyter.

Trevathan, W. R. (1996). The evolution of bipedalism and assisted birth. *Medical Anthropology Quarterly*, **10**, 287–298.

Tutin, C. E. G. (1994). Reproductive success story: variability among chimpanzees and comparisons with gorillas. In *Chimpanzee Cultures*, R. W. Wrangham, W. C. McGrew, F. B. M. de Waal, et al. (eds). Cambridge, MA: Harvard University Press, pp. 181–193.

Vavra, H. M. and Querec, L. J. (1973). *A Study of Infant Mortality from Linked Records by Age of Mother, Total-birth Order, and other Variables*. DHEW Publication No. (HRA). 74–1851. Washington, DC: US Government Printing Office.

Wallis, A. (1997). A survey of reproductive parameters in free-ranging chimpanzees of Gombe National Park. *Journal of Reproduction*, **109**, 297–307.

Weisner, T. S. (1987). Socialization for parenthood in sibling caretaking societies. In *Parenting Across the Life Span: Biosocial Dimensions*, J. B. Lancaster, J. Altmann, A. S. Rossi, et al. (eds). New York: Aldine de Gruyter, pp. 237–270.

Weisner, T. S. (1996). The 5–7 transition as an ecocultural project. In *Reason and Responsibility: the Passage through Childhood*, A. Samaroff and M. Haith (eds.). Chicago: University of Chicago Press, pp. 295–326.

Worthman, C. M. (1993). Biocultural interactions in human development. In *Juvenile Primates: Life History, Development, and Behavior*, M. E. Perieira and L. A. Fairbanks (eds). New York: Oxford University Press, pp. 339–357.

23 Variation in Human Growth Patterns due to Environmental Factors

Stanley J. Ulijaszek

INTRODUCTION

The human growth pattern is characterized by rapid growth in infancy, followed by an extensive period of childhood, and a relatively intense adolescent spurt (see Chapter 22 of this volume). The extended period of biological immaturity relative to other mammalian species is associated with high environmental sensitivity and growth plasticity (Johnston, 1998), illustrated by the processes of stunting and wasting in response to poor nutrition and infection (Waterlow, 1988) and of catch-up growth during environmental improvements following episodes of environmental stress (Prader et al., 1963).

Known environmental factors that influence growth, body size, and body composition of children postnatally include nutrition (Barclay and Weaver, 2006), infection (Bhan et al., 2001), interactions between the two (Ruel, 2001), psychosocial stress (Powell et al., 1967), food contaminants (Gong et al., 2004), pollution (Schell, 1991a), and hypoxia (Frisancho, 1977). Most of these factors are conditioned by poverty and socioeconomic status (Martorell et al., 1988). They are also conditioned historically, culturally, and politically (Ulijaszek, 2006), interacting with each other, but also with individual genotypes in the production of growth, body size, and body composition.

Diet, nutrition, disease, hypoxia, pollution, contamination, behavioral toxicants, deprivation, and psychosocial stress can be clustered as proximate environmental agents that can influence growth (Figure 23.1). They vary in importance according to circumstance and the age and stage in infancy, childhood, and adolescence. Culture, society, behavior, socioeconomic status, poverty, and political economy can also be clustered as structurally powerful but distal agents in the production of growth and body size outcomes, at all ages and stages of childhood and adolescence. This latter cluster is conditioned historically. In the developing world, the risks associated with poverty include low income, poor food security, inadequate health infrastructure, and environmental hazards. In the industrialized world, the risks for impaired child growth that are associated with

low socioeconomic status include single parenthood, overcrowding, low disposable income, paternal ill health, dependence on social welfare, and parental abuse of alcohol and drugs (Schell, 1991b).

In both developed and developing worlds there are similar associations between stature and socioeconomic status, height correlating positively with wealth (Bogin, 1999). Weight, however, does not always relate positively with wealth, overweight and obesity being the provenance of low socioeconomic status in most industrialized nations (Sobal, 1991), and becoming increasingly so among emerging nations undergoing the health transition (Xu et al., 2005). Growth stunting in association with overweight was first identified in Peruvian children (Trowbridge et al., 1987) and more recently in children aged between three and nine years in four nations viewed to be undergoing nutrition transition: Russia, Brazil, Republic of South Africa, and China (Popkin et al., 1996).

Early life experiences involving environmental stress, intrauterine growth retardation, poor growth in early childhood, and subsequent catch-up growth can impact on growth, body composition, and health outcomes later in life (Henry and Ulijaszek, 1996). Catch-up growth is an acceleration of child growth-rate following either medical or environmental intervention or environmental improvement, such that body size approaches or reaches normality, as defined by appropriate growth references (Prader et al., 1963). It can take place at all stages of child growth, including adolescence (Golden, 1994). However, when the factors responsible for growth faltering or failure are ubiquitous, this process is constrained and individuals fail to reach their potential for maximal growth and optimal body size (Martorell et al., 1988).

Human growth and body size responds with sensitivity to environmental quality. The term "secular trend" is used to describe marked changes in growth and development of successive generations of human populations living in the same territories. Positive secular trends in increased stature and weight, and earlier timing of the adolescent growth spurt, have been documented among European, European-origin,

Human Evolutionary Biology, ed. Michael P. Muehlenbein. Published by Cambridge University Press. © Cambridge University Press 2010.

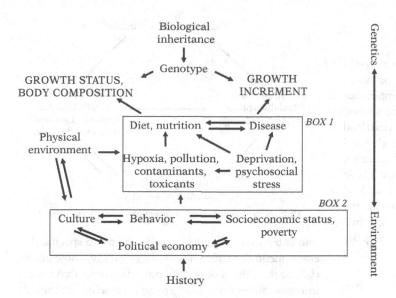

23.1. Proximate and distal agents influencing child growth. Adapted from Ulijaszek (2006).

and Asian populations (Ulijaszek, 2001). Negative secular trends have been identified among populations in Africa (Henneberg and van den Berg, 1990), Papua New Guinea (Ulijaszek, 1993), and Central and Latin America (Bogin, 1999). Positive secular trends have largely been attributed to improved social, political, nutritional, and health conditions, while negative secular trends are often seen as outcomes of environmental, social, or political deterioration (Bogin, 1999). The best example of a positive secular trend is that of the Netherlands (Van Wieringen, 1986), where mean stature has increased from 165cm in 1860 to 181cm in 1990, and 184cm in 1997 (Cole, 2000). In largest part, this has been attributed to improved nutrition (in terms of both quantity and quality) that came with economic improvements across the twentieth century, as well as the control of, and decline in, infectious disease morbidity (Van Wieringen, 1986). However, although most populations in Europe have experienced a positive secular trend across the twentieth century, there is evidence that some experienced a negative secular trend in the late eighteenth century, due largely to poor harvests, high grain prices, and the poor infant and child nutrition that followed (Komlos, 1985; Floud et al., 1990).

The vast majority of research on environmental influences on human growth has focused on birthweight (Wharton, 1989), infancy and infant feeding (Frongillo, 2001), and early childhood, up to the age of five years (Waterlow, 1988). Later childhood is taken to be from five years of age until onset of puberty. In contrast, the environmental influences on preadolescent growth have attracted relatively little attention (Stoltzfus, 2001). Growth and body composition in adolescence has been researched to a greater extent than in preadolescence, but to a much smaller extent than among children of preschool age (Stoltzfus, 2001).

While adolescent growth may be under stronger genetic control than growth in childhood (Hauspie and Susanne, 1998), environment can influence both of these measures of adolescent growth and maturation, but to a lesser extent than genetics (Bogin, 1999). This chapter describes important environmental influences on the growth of children from birth to adulthood, focusing primarily on measures of height and weight, the primary anthropometric measures used in child health surveillance, screening, and monitoring.

INFANCY AND EARLY CHILDHOOD

Human infancy is the period when the mother provides all or some nourishment to her offspring by way of lactation (Bogin, 1998). Human infancy ends when the child is weaned from the breast, which in preindustrialized societies usually continues to beyond two years of age (Sellen, 2001). Exclusive breast-feeding usually provides adequate nutrition to support good child growth until 6 months of age (Butte et al., 2002), and dietary supplementation of the infant often begins around or before that time (Sellen, 2001). Of the various environmental factors influencing growth of children in developing countries, diet, nutrition, and infection are particularly powerful in infancy and childhood. While infants are breast-fed, they are usually protected from the broad disease environment nutritionally, immunologically (Ulijaszek, 1990), and behaviorally. Where on-demand breast-feeding is usual, infants are often kept close to their mother and are buffered from contact with objects or foods contaminated with pathogens. In developing countries, interactions between undernutrition and infection usually lead to growth faltering from about six months of age (Waterlow, 1988; Lunn, 2000).

Interactions between undernutrition and infection

Undernutrition–infection interactions can be initiated in two ways (Figure 23.2). The first involves poor nutritional status leading to impaired immunocompetence and reduced resistance to infection, while the second involves an exposure to infectious disease which can lead to appetite loss and anorexia, malabsorption, elevated basal metabolic rate, as well as gut mucosal damage, and protein catabolism in order to fuel acute phase protein production. Delayed supplementation may lead to growth faltering and undernutrition, leaving the infant more suspectible to infectious diseases, while earlier dietary supplementation may provide adequate nutrient intake, but concomitantly introduce the child to diarrheal agents.

Once started, the interactions between these two major environmental stressors become increasingly complex, with the nature of the disease ecology influencing the duration and severity of infection, and adaptive immunity, its effect on subsequent disease experience, and the extent, if any, of anorexia, fever, and malabsorption during infectious episodes, which impact on nutritional status. Specific nutritional deficiencies can subsequently influence immune status and responsiveness, as well as adaptive immunity. In addition, cultural factors and poverty can influence patterns of disease management and sickness behavior, which can in turn affect the incidence, severity, and duration of infection, and their effects on nutritional status.

Infections that influence nutritional status and linear growth are either acute and invasive, provoking a systemic response (such as dysentery and pneumonia), or chronic, affecting the host over a sustained period (including gut helminth infections). Infections can diminish linear growth by affecting nutritional status, by way of decreased food intake, impaired nutrient absorption, direct nutrient losses, increased metabolic requirements, catabolic losses of nutrients, and/or impaired transport of nutrients to target tissues. In addition, induction of the acute phase response and host elaboration of pro-inflammatory cytokines (Friedman et al., 2003) may contribute to growth faltering because they directly inhibit the process of bone remodeling that is needed for long-bone growth (Stephensen, 1999).

In Table 23.1, the nutrition – infection processes associated with growth faltering of children are reduced to four types of effect. These are: (1) the diseases and disease categories that are known to affect nutritional status; (2) the diseases and disease categories known to be influenced by nutritional status; (3) the nutritional deficiencies that inhibit immune system function; and (4) the diseases known to inhibit immune system function. The general disease categories of diarrhea, intestinal parasitic infestation, and upper respiratory tract infections fall

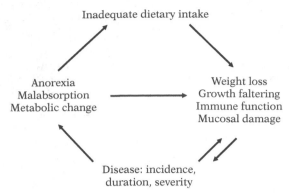

23.2. Nutrition–infection interactions in early childhood.

into categories (1) and (2), as do the more specific diseases pneumonia, measles, malaria, and typanosomiasis (Ulijaszek, 2006). The nutritional deficiencies that inhibit immune system include energy, protein, vitamin A, pyridoxine, iron, and zinc (Tomkins, 2002). However, in the absence of overt infection, deficiencies of zinc and iron have only a small effect on linear growth, while vitamin A is unlikely to have any important effect (Bhandari et al., 2001). Diseases known to inhibit immune system function include AIDS, measles, leprosy, and malaria. The ways in which growth faltering is associated with the interaction between undernutrition and infection are thus manifold, and varies with specific local disease and nutritional ecologies. Gut damage due to infection also varies across ecologies (Lunn, 2000). While the growth outcomes of undernutrition–infection interactions may look similar across the developing world, they are in fact different in specific causation.

Growth faltering due to undernutrition and infection may continue for months or years, depending on the severity of the disease environment, and the abundance and quality of the nutritional environment. In most populations, the process of growth faltering is complete by the age of two years, after which the shorter, stunted child may follow a parallel trajectory to the Western growth references (Eveleth and Tanner, 1990). This period of departure from the growth references derived from measures of Western populations can be regarded in one sense as an accommodation to the disease and nutritional environment. However, this accommodation is usually associated with high rates of mortality and morbidity, reduced energy for play, and compromised intellectual development, and cannot therefore be regarded as desirable.

Poverty

Childhood undernutrition and infection are associated with poverty. Farmer (2004) has described the extensive ways in which poverty and social inequality are embodied as differential risks for infection with HIV and tuberculosis in developing countries, while Walls

TABLE 23.1. Nutrition–infection processes associated with growth faltering of children.

Diseases and disease categories known to affect nutritional status:

Diarrhea; upper respiratory tract infections; pneumonia; measles; malaria; intestinal parasites; AIDS

Diseases and disease categories known to be influenced by nutritional status:

Diarrhea; cholera; leprosy; pertussis; upper respiratory tract infections; pneumonia; measles; malaria; intestinal parasites; trypanosomiasis

Nutritional deficiencies that inhibit immune system function:

Energy; protein; vitamin A; pyridoxine; iron; zinc

Diseases known to inhibit immune system function:

AIDS; measles; leprosy; malaria

and Shingadia (2004) identified overcrowding, poverty, and the HIV epidemic as contributing to the resurgence of tuberculosis globally. Bates et al. (2004) identified poverty as a key factor operating at individual, household, and community levels in increasing vulnerability to malaria, tuberculosis, and HIV infection. Poverty also underpins the nutrition and infection interactions that impact on child growth. Tuberculosis, associated with household crowding and poverty, may not be directly associated with growth faltering, but it is associated with nutritional status. Vitamin A can cause growth delay when combined with infection. Children born to HIV-infected women who are vitamin A deficient during pregnancy are more likely to experience growth failure (Semba et al., 1997). Furthermore, HIV infection, malaria, and diarrheal disease adversely affect growth of preschool-age children, and are associated with increased prevalence of vitamin A deficiency (Villamor et al., 2002).

Environmental contamination

Exposure to aflatoxin contamination, particularly at the time of weaning, has been shown to inhibit early childhood growth in West Africa (Gong et al., 2003). Aflatoxins are mold metabolites produced by toxigenic strains of *Aspergillus* species, a number of which are hepatotoxic and immunotoxic. Primary commodities susceptible to aflatoxin contamination include corn, peanuts, cottonseed, and animal-derived foods such as milk when the animal is fed aflatoxin-contaminated feed. Although excessive aflatoxin contamination is not global, significant dietary contamination has been demonstrated in many parts of West Africa, Asia, and South America (Ulijaszek, 2006). Risks associated with aflatoxin-contaminated foods can be reduced through the use of multiple processing and decontamination procedures, including physical cleaning and separation procedures (Park, 2002), but not with simple cooking procedures available to poor mothers. Although aflatoxins are partially destroyed during nixtamalization, the alkaline cooking procedure employed to prepare tortillas in

Mexico (Mendez-Albores et al., 2004), high levels of aflatoxin often remain in the food (Plasencia, 2004).

Levels of pollutants that human populations are exposed to vary markedly, depending partly on the degree of, and proximity to, industrialization (Schell, 1991b). Generalized air pollution has been identified as an environmental risk factor for poor growth of children in communities in Silesia and Belgium (Schell, 1991b), pollution from hazardous waste sites having similar effects (Paigen et al., 1987). Exposure to polychlorinated biphenyls (PCB) at very high doses can affect child growth, while exposure to lead can affect growth at moderate to low levels. The risk of environmental pollutant exposure to child growth is increasing as developing countries industrialize. China has seen pronounced increases in anthropogenic lead level during the past two decades (Huh and Chen, 1999), while in Taiwan, high serum lead levels are associated with lead exposure from drinking water sources and residential proximity to factories, as well as occupational lead exposure (Chu et al., 1998).

Psychological stress

The idea that psychological stress causes growth faltering in some children was first put forward by Widdowson (1951), who published evidence that the presence of a sadistic schoolteacher caused child growth in an orphanage to falter, despite a concurrent increase in the amount of food eaten. Furthermore, family conflict has been shown to be associated with short stature in childhood as well as short adult height in the British 1958 cohort study (Montgomery et al., 1997). The dominant mechanism by which short stature emerges is nutritional, by way of appetite loss and anorexia (Skuse, 1998).

Altitude

Relative to lowland populations, humans living at high altitude are more likely to be born at low birthweight and undergo postnatal growth which is slow and prolonged (Frisancho, 1993). The poor growth in infancy and early childhood among some high altitude populations is associated with poverty, poor food security and exposure to infectious disease, in addition to hypoxia. Ethiopians living at high altitude have better nutrition, growth, and socioeconomic conditions than low altitude populations, and this is reflected in their better growth rates (Clegg et al., 1972), illustrating the importance of non-hypoxic factors influencing the growth of high altitude populations.

GROWTH IN THE LATER CHILDHOOD, JUVENILE AND ADOLESCENCE STAGES

Most of the environmental factors associated with growth in adolescence are common to growth in pre-adolescence, the most important of which are nutrition, infection, and the interactions between the two. Additionally, birthweight, catch-up growth, breast-feeding, and early adiposity rebound have impacts on growth and/or body composition into puberty. Growth in later childhood and adolescence continues to show great biological plasticity. Across populations, Stoltzfus (2001) characterized patterns of growth that deviate greatly from normative patterns, as represented by growth references. These include populations that display: (1) prepubertal catch-up growth; (2) prepubertal stunting combined with catch-up growth in puberty; and (3) prepubertal stunting with no catch-up growth in puberty. She concludes that between-population variation in growth among school-age children and adolescents is as great as among children in early childhood. Pattern three is the most common across the developing world (Waterlow, 1988), and is usually associated with poverty and low socioeconomic status (Martorell et al., 1988). It is also associated with infant failure to thrive in industrialized nations, where growth-retarded children in infancy have been shown to remain height and weight deficit at the age of six years (Tomkins, 1994). Patterns one and two can occur as a result of: (1) different types of infant feeding and weaning behavior (Frongillo, 2001); (2) varying illness management practices (Tomkins, 1986); (3) dietary manipulation (Steckel, 1987); and (4) changing environments during childhood (Golden, 1994). The latter pattern has been observed to take place across secular trends (Proos, 1993), and during nutrition transition (Sawaya et al., 2003).

The first two of Stoltzfus's (2001) growth patterns, representing prepubertal catch-up growth, catch-up growth in puberty, and prepubertal stunting combined with catch-up growth in puberty, respectively, are associated with critical periods of development which can have long-term implications for health, and body composition in later childhood (Barker et al., 2001). Body fatness reaches a postinfancy low level typically between the ages of five and seven years, followed by increased body fatness, a phenomenon termed adiposity rebound by Rolland-Cachera et al. (1984). Early adiposity rebound has been associated with earlier age at menarche (Barker et al., 2001) and increased relative weight and obesity later in life, including during adolescence (Cameron and Demerath, 2002). Early growth restriction followed by catch-up growth is also associated with the development of abdominal obesity (Dulloo, 2006), while higher growth velocity in early childhood, prior to adiposity rebound, has been shown to be associated with greater fatness and obesity in subsequent years (Monteiro et al., 2003). The combination of small size at birth and during infancy, followed by accelerated weight gain from age 3 to 11 years, predicts large differences in the cumulative incidence of coronary heart disease, non-insulin dependent diabetes, and hypertension in later life (Barker et al., 2001). In the United States, low levels of vigorous physical activity and high levels of television viewing have been associated with fatness in children during the adiposity rebound period (Janz et al., 2002), which expose American children to an increased risk of obesity and chronic disease in adult life.

In addition to undernutrition and infection, child neglect and abuse, exposure to industrial pollutants, food contaminants, behavioral toxicants, single parenthood, overcrowding, and parental ill health are often important contributors to growth outcomes, according to circumstance (Schell, 1991a). Seasonality of growth is found among populations of both developed and developing countries. In the former, effects are quite subtle climatic ones, while in the latter, they are largely due to seasonal variation in food availability and infectious disease exposure (Cole, 1993). If infants are breast-fed, they are usually shielded from most of the stresses associated with seasonality of infection and nutrition. However, such seasonal factors can become significant after weaning, and persist across childhood and adolescence.

The relationships between proximate and distal influences on child growth (Figure 23.1) have been examined at macro-level. Blakely et al. (2005) identified strong relationships between poverty and childhood malnutrition, access to unsafe water and sanitation, and exposure to indoor air pollution, while Frongillo et al. (1997) found the most important determinants of stunting in children below the age of five years to be dietary energy availability, female literacy, and gross national product. In the developed world, relationships between poverty and child growth persist but perhaps to a lesser extent than prior to positive secular trends that took place from the late nineteenth and across the twentieth century (Ulijaszek, 2001).

Diet

Both dietary quantity and quality can influence growth and body composition in later childhood, as can infant feeding patterns. While there appears to be no difference in weight and height between breast-fed and formula-fed children by the time they reach school-age, breast-fed infants are less likely to become obese as adults (Frongillo, 2001). Deficiencies of energy, protein, and zinc have been implicated in growth faltering, while diets high in fat have been associated with obesity (Ulijaszek, 2006). Vegetarian and vegan diets in both developed and developing countries have been shown to be deficient in micronutrients. In a study from the Netherlands, children aged 0–10 years consuming macrobiotic diets with adequate protein and energy intakes and protected from bacterial contamination were shown to have growth patterns similar to those of poor children in developing countries (Dagnelie et al., 1994).

While broad descriptive studies of adolescent growth are plentiful, there are few that have critically evaluated the relative importance of specific environmental factors in this process. This is because of the great variation both within- and between-populations in the maturation rate and the timing and magnitude of peak weight and height velocities. In general, adolescent growth is sensitive to nutritional deficit and surfeit (Eveleth and Tanner, 1990). For many industrialized nations, there have been secular trends in the timing and size of the pubertal growth spurt which have been taken as evidence for nutritional improvement (Eveleth and Tanner, 1990). Furthermore, many of the socioeconomic differences between groups in growth in adolescence have been attributed to nutritional differences (Eveleth and Tanner, 1990). Despite this, there appears to be only one longitudinal study that demonstrates direct nutritional effects on growth in adolescence (Berkey et al., 2000), among girls in Boston, in the United States. Those who consumed more dietary energy and animal protein than average two years before peak growth were shown to experience both earlier age at peak growth velocity, and higher peak height velocity than average. The latter is likely to be conditional upon the former.

Infection

The impacts of infection on adolescent growth are of much lesser importance than nutrition. This is because the immune system has matured and adaptive immunity is largely in place by adolescence (Ulijaszek, 1998), and as a consequence of this, contraction of most common infections of early childhood is much reduced. However, perinatal HIV-1 infection has been shown to interfere with sexual maturation in children surviving this infection into adolescence (Buchacz et al., 2003). Other infections that persist in their effects on physical growth and development into adolescence are tuberculosis and *Helicobacter pylori*. Furthermore, a number of chronic diseases can delay onset of puberty and reduce the size of the pubertal growth spurt (Simon, 2002).

Pollution

Exposure of adolescents to environmental pollutants is different from earlier life, as the likelihood of occupational exposure increases, at least in the developing world. Perinatal exposure to polybrominated biphenyl (PBB) has been shown to be associated with earlier age at menarche (Blanck et al., 2000), while early exposure to PCB is associated with delayed sexual maturation in both males and females (den Hond et al., 2002). In the one study in which the concurrent effects of most common pollutants to which children might be exposed was examined, Denham et al. (2005) found attainment of menarche to be sensitive to relatively low levels of lead and certain PCB congeners. Dichlorodiphenyltrichloroethan (DDT) is a chemical once used widely in agriculture but which is now limited largely to public health use, especially in malarial vector control programs in nations where equally effective and affordable alternatives are not locally available. The one study in the United States which rigorously examined the possible effects of prenatal DDT exposure on pubertal growth and development, found no effect (Gladen et al., 2004).

Altitude

Adolescent growth and development among populations living at high altitude is often characterized by slow growth and delayed puberty, resulting in smaller adult body size (Weitz and Garruto, 2004). Slower growth at high altitude is a consequence of hypoxia (Frisancho, 1993), poor economic conditions, and nutritional inadequacy (Weitz and Garruto, 2004). A study of European children of higher socioeconomic status migrating to high altitude in the Andes has shown them to be slightly shorter and lighter than their peers of same socioeconomic status living at sea-level, indicating that high altitude hypoxia has a small but independent impact on growth (Stinson, 1982). There is delayed sexual maturation among many, but not all, high latitude populations relative to lower altitude populations, as well as a late and poorly defined adolescent growth spurt, at least among Andean populations (Frisancho and Baker, 1970).

CONCLUSIONS

This chapter reviews the range of environmental agents known to influence growth in weight and stature across infancy, childhood, and adolescence. Diet, nutrition, disease, hypoxia, pollution, contamination, behavioral toxicants, deprivation, and psychosocial stress are clustered as proximate environmental agents that can influence growth, which vary in importance according to circumstance and the age and stage in infancy, childhood, and adolescence. Culture, society, behavior, socioeconomic status, social status, poverty, and political economy are clustered as structurally powerful but distal agents in the production of growth and body size outcomes, at all ages and stages of childhood and adolescence. Early life experiences involving environmental stress, intrauterine growth retardation, poor growth in early childhood, and subsequent catch-up growth can impact on growth, body composition, and health outcomes later in life.

The extent to which human growth and body size responds with sensitivity to environmental quality is demonstrated in the secular trends in growth and body size that have taken place across successive generations of human populations living in the same territories. These have been mostly positive, resulting from environmental improvements, but negative trends have also been demonstrated. Developmental plasticity is an adaptive property of humans that allows them to survive and thrive in very varied environmental circumstances.

DISCUSSION POINTS

1. In what ways does plasticity in human development manifest itself, and what are its longer-term consequences?
2. What is important about the secular trend in growth and body size?
3. How does growth and development in early childhood differ from that in later childhood?
4. How do environmental factors that influence child growth cluster together?
5. In what ways do nutrition and infection influence patterns of growth and development from birth and into adolescence?

REFERENCES

Barclay, A. and Weaver, L. (2006). Feeding the normal infant, child and adolescent. *Medicine*, **34**, 551–556.

Barker, D. J. P., Forsen, T., Uutela, A., et al. (2001). Size at birth and resilience to effects of poor living conditions in adult life: longitudinal study. *British Medical Journal*, **323**, 1273–1276.

Bates, I., Fenton, C., Gruber, J., et al. (2004). Vulnerability to malaria, tuberculosis, and HIV/AIDS infection and disease. Part I: Determinants operating at individual and household level. *Lancet Infectious Disease*, **4**, 267–277.

Berkey, C. S., Gardner, J. D., Frazier, A. L., et al. (2000). Relation of childhood diet and body size to menarche and adolescent growth in girls. *American Journal of Epidemiology*, **152**, 446–452.

Bhan, M. K., Bahl, R. and Bhandari, N. (2001). Infection: how important are its effects on child nutrition? In *Nutrition and Growth*, R. Martorell and F. Haschke (eds). Philadelphia: Lippincott Williams and Wilkins, pp. 197–217.

Bhandari, N., Bahl, R. and Taneja, S. (2001). Effect of micronutrient supplementation on linear growth of children. *British Journal of Nutrition*, **85**(suppl. 2), S131–S137.

Blakely, T., Hales, S., Kieft, C., et al. (2005). The global distribution of risk factors by poverty level. *Bulletin of the World Health Organization*, **83**, 118–126.

Blanck, H. M., Marcus, M., Tolbert, P. E., et al. (2000). Age at menarche and Tanner stage in girls exposed in utero and postnatally to polybrominated biphenyl. *Epidemiology*, **11**, 641–647.

Bogin, B. (1998). Patterns of human growth. In *Encyclopedia of Human Growth and Development*, S.J. Ulijaszek, F.E. Johnston and M.A. Preece (eds). Cambridge: Cambridge University Press, pp. 91–95.

Bogin, B. (1999). *Patterns of Human Growth*, 2nd edn. Cambridge: Cambridge University Press.

Buchacz, K., Rogol, A. D., Lindsey, J. C., et al. (2003). Delayed onset of pubertal development in children and adolescents with perinatally acquired HIV infection. *Journal of Acquired Immune Deficiency Syndromes*, **33**, 56–65.

Butte, N. F., Lopez-Alarcon, M. G. and Garza, C. (2002). *Nutrient Adequacy of Exclusive Breastfeeding for the Term Infant during the First Six Months of Life*. Geneva: World Health Organization.

Cameron, N. and Demerath, E. W. (2002). Critical periods in human growth and their relationship to diseases of aging. *Yearbook of Physical Anthropology*, **45**, 159–184.

Chu, N. F., Liou, S. H., Wu, T. N., et al. (1998). Risk factors for high blood lead levels among the general population in Taiwan. *European Journal of Epidemiology*, **14**, 775–781.

Clegg, E. J., Pawson, I. G., Ashton, E. H., et al. (1972). The growth of children at different altitudes in Ethiopia. *Philosophical Transactions of the Royal Society of London. Series B*, **264**, 403–437.

Cole, T. J. (1993). Seasonal effects on physical growth and development. In *Seasonality and Human Ecology*, S. J. Ulijaszek and S. S. Strickland (eds). Cambridge: Cambridge University Press, pp. 89–106.

Cole, T. J. (2000). Secular trends in growth. *Proceedings of the Nutrition Society*, **59**, 317–324.

Dagnelie, P. C., van Dusseldorp, M., van Staveren, W. A., et al. (1994). Effects of macrobiotic diets on linear growth in infants and children until 10 years of age. *European Journal of Clinical Nutrition*, **48**(suppl. 1), S103–S111.

den Hond, E., Roels, H. A., Hoppenbrouwers, K., et al. (2002). Sexual maturation in relation to polychlorinated aromatic hydrocarbons: Sharpe and Skakkebaek's hypothesis revisited. *Environmental Health Perspectives*, **110**, 771–776.

Denham, M., Schell, L. M., Deane, G., et al. (2005). Relationship of lead, mercury, mirex, dichlorodiphenyldichloroethylene, hexachlorobenzene, and polychlorinated biphenyls to timing of menarche among Akwesasne Mohawk girls. *Pediatrics*, **115**, e127–e134.

Dulloo, A. G. (2006). Regulation of fat storage via suppressed thermogenesis: A thrifty phenotype that predisposes individuals with catch-up growth to insulin resistance and obesity. *Hormone Research*, **65**(suppl. 3), 90–97.

Eveleth, P. B. and Tanner, J. M. (1990). *Worldwide Variation in Human Growth*, 2nd edn. Cambridge: Cambridge University Press.

Farmer, P. (2004). Sidney W. Mintz lecture for 2001. An anthropology of structural violence. *Current Anthropology*, **45**, 305–325.

Floud, R., Wachter, K. and Gregory, A. (1990). *Height, Health and History*. Cambridge: Cambridge University Press.

Friedman, J. F., Kurtis, J. D., Mtalib, R., et al. (2003). Malaria is related to decreased nutritional status among male adolescents and adults in the setting of intense perennial transmission. *Journal of Infectious Disease*, **188**, 449–457.

Frisancho, A. R. (1977). Developmental adaptation to high altitude hypoxia. *International Journal of Biometeorology*, **21**, 135–146.

Frisancho, A. R. (1993). *Human Adaptation and Accommodation*. Ann Arbor, MI: University of Michigan Press.

Frisancho, A. R. and Baker, P. T. (1970). Altitude and growth: a study of the patterns of physical growth of a high altitude Peruvian Quechua population. *American Journal of Physical Anthropology*, **32**, 279–292.

Frongillo, E. A. (2001). Growth of the breast-fed child. In *Nutrition and Growth*, R. Martorell and F. Haschke (eds). Philadelphia: Lippincott Williams and Wilkins, pp. 37–49.

Frongillo, E. A., de Onis, M. and Hanson, K. M. P. (1997). Socioeconomic and demographic factors are associated with worldwide patterns of stunting and wasting of children. *Journal of Nutrition*, **127**, 2302–2309.

Gladen, B. C., Klebanoff, M. A., Hediger, M. L., et al. (2004). Prenatal DDT exposure in relation to anthropometric and pubertal measures in adolescent males. *Environmental Health Perspectives*, **112**, 1761–1767.

Golden, M. H. N. (1994). Is complete catch-up possible for stunted malnourished children? *European Journal of Clinical Nutrition*, **48**(suppl. 1), S58–S71.

Gong, Y. Y., Egal, S., Hounsa, A., et al. (2003). Determinants of aflatoxin exposure in young children from Benin and Togo, West Africa: the critical role of weaning. *International Journal of Epidemiology*, **32**, 556–562.

Gong, Y. Y., Hounsa, A., Egal, S., et al. (2004). Postweaning exposure to aflatoxin results in impaired child growth: a longitudinal study in Benin, West Africa. *Environmental Health Perspectives*, **112**, 1334–1338.

Hauspie, R. C. and Susanne, C. (1998). Genetics of child growth. In *Cambridge Encyclopedia of Human Growth and Development*, S. J. Ulijaszek, F. E. Johnston and M. A. Preece (eds). Cambridge: Cambridge University Press, pp. 124–128.

Henneberg, M. and van den Berg, E. R. (1990). Test of socio-economic causation of secular trend: stature changes among favored and oppressed South Africans are parallel. *American Journal of Physical Anthropology*, **83**, 459–465.

Henry, C. J. K. and Ulijaszek, S. J. (1996). *Long-term Consequences of Early Environment. Growth, Development and the Lifespan Developmental Perspective*. Cambridge: Cambridge University Press.

Huh, C. A. and Chen, H. Y. (1999). History of lead pollution recorded in East China Sea sediments. *Marine Pollution Bulletin*, **38**, 545–549.

Janz, K. F., Levy, S. M., Burns, T. L., et al. (2002). Fatness, physical activity, and television viewing in children during the adiposity rebound period: the Iowa bone development study. *Preventive Medicine*, **35**, 563–571.

Johnston, F. E. (1998). The ecology of post-natal growth. In *Cambridge Encyclopedia of Human Growth and Development*, S. J. Ulijaszek, F. E. Johnston and M. A. Preece (eds). Cambridge: Cambridge University Press, pp. 315–319.

Komlos, J. (1985). Stature and nutrition in the Habsburg monarchy: the standard of living and economic development. *American Historical Review*, **90**, 1149–1161.

Lunn, P. G. (2000). The impact of infection and nutrition on gut function and growth in childhood. *Proceedings of the Nutrition Society*, **59**, 147–154.

Martorell, R., Mendoza, F. and Castillo, R. (1988). Poverty and stature in children. In *Linear Growth Retardation in Less Developed Countries*, J. C. Waterlow (ed.). New York: Raven Press, pp. 57–70.

Mendez-Albores, J. A., Arambula-Villa, G., Loarca-Pina, M. G., et al. (2004). Aflatoxins' fate during the nixtamalization of contaminated maize by two tortilla-making processes. *Journal of Stored Products Research*, **40**, 87–94.

Monteiro, P. O. A., Victora, C. G., Barros, F. C., et al. (2003). Birth size, early childhood growth, and adolescent obesity in a Brazilian birth cohort. *International Journal of Obesity*, **27**, 1274–1282.

Montgomery, S. M., Bartely, M. J. and Wilkinson, R. G. (1997). Family conflict and slow growth. *Archives of Disease in Childhood*, **77**, 326–330.

Paigen, B., Goldman, L. R., Magnant, M. M., et al. (1987). Growth of children living near the hazardous waste site, Love Canal. *Human Biology*, **59**, 489–508.

Park, D. L. (2002). Effect of processing on aflatoxin. *Advances in Experimental Medicine and Biology*, **504**, 173–179.

Plasencia, J. (2004). Aflatoxins in maize: a Mexican perspective. *Journal of Toxicology – Toxin Reviews*, **23**, 155–177.

Popkin, B. M., Richards, M. K. and Monteiro, C. A. (1996). Stunting is associated with overweight in children of four nations that are undergoing the nutrition transition. *Journal of Nutrition*, **126**, 3009–3016.

Powell, G. F., Brasel, J. A., Raiti, S., et al. (1967). Emotional deprivation and growth retardation simulating idiopathic hypopituitarism: I. Clinical evaluation of the syndrome. *The New England Journal of Medicine*, **276**, 1271–1278.

Prader, A., Tanner, J. M. and von Harnack, G. A. (1963). Catch-up growth following illness or starvation. An example of developmental canalization in man. *Journal of Pediatrics*, **62**, 646–659.

Proos, L. A. (1993). Anthropometry in adolescence – secular trends, adoption, ethnic and environmental differences. *Hormone Research*, **39**(suppl. 3), 18–24.

Rolland-Cachera, M.-F., Deheeger, M., Bellisle, F., et al. (1984). Adiposity rebound in children: a simple indicator for predicting obesity. *American Journal of Clinical Nutrition*, **39**, 129–135.

Ruel, M. T. (2001). The natural history of growth failure: importance of intrauterine and postnatal periods. In *Nutrition and Growth*, R. Martorell and F. Haschke (eds). Philadelphia: Lippincott Williams and Wilkins, pp. 123–153.

Sawaya, A. L., Martins, P., Hoffman, D., et al. (2003). The link between childhood undernutrition and risk of chronic diseases in adulthood: a case study of Brazil. *Nutrition Reviews*, **61**, 168–175.

Schell, L. M. (1991a). Effects of pollutants on human prenatal and postnatal growth: noise, lead, polychlorobiphenyl compounds, and toxic wastes. *Yearbook of Physical Anthropology*, **34**, 157–188.

Schell, L. M. (1991b). Risk focusing: an example of biocultural interaction. In *Health and Lifestyle Change*, R. Huss-Ashmore, J. Schall and M. Hediger (eds), vol. **9**. Philadelphia, PA: MASCA Research Papers in Science and Archaeology, pp. 137–144.

Sellen, D. W. (2001). Comparison of infant feeding patterns reported for nonindustrial populations with current recommendations. *Journal of Nutrition*, **131**, 2707–2715.

Semba, R. D., Miotti, P., Chiphangwi, J. D., et al. (1997). Maternal vitamin A deficiency and child growth failure during human immunodeficiency virus infection. *Journal of Acquired Immune Deficiency Syndrome and Human Retrovirology*, **14**, 219–222.

Simon, D. (2002). Puberty in chronically diseased patients. *Hormone Research*, **57**, 53–56.

Skuse, D. H. (1998). Growth and psychosocial stress. In *Cambridge Encyclopedia of Human Growth and Development*, S. J. Ulijaszek, F. E. Johnston and M. A. Preece (eds). Cambridge: Cambridge University Press, pp. 341–342.

Sobal, J. (1991). Obesity and socio-economic status: a framework for examining relationships between physical and social variables. *Medical Anthropology*, **13**, 231–247.

Steckel, R. H. (1987). Growth depression and recovery: the remarkable case of American slaves. *Annals of Human Biology*, **14**, 111–132.

Stephensen, C. B. (1999). Burden of infection on growth failure. *Journal of Nutrition*, **374**, 534S–538S.

Stinson, S. (1982). The effect of high altitude on the growth of children of high socioeconomic status in Bolivia. *American Journal of Physical Anthropology*, **59**, 61–73.

Stoltzfus, R. J. (2001). Growth of school-age children. In *Nutrition and Growth*, R. Martorell and F. Haschke (eds). Philadelphia: Lippincott Williams and Wilkins, pp. 257–276.

Tomkins, A. (1996). Growth monitoring, screening and surveillance in developing countries. In *Anthropometry: the Individual and the Population*, S. J. Ulijaszek and C. G. N. Mascie-Taylor (eds). Cambridge: Cambridge University Press, pp. 108–116.

Tomkins, A. (2002). Nutrition, infection and immunity: public health implications. In *Nutrition and Immune Function*, P. C. Calder, C. J. Field and H. S. Gill (eds). Wallingford, Oxfordshire: CABI Publishing.

Trowbridge, F. L., Marks, J. S., Deromana, G. L., et al. (1987). Body composition of Peruvian children with short stature and high weight-for-height. 2. Implications for the interpretation of weight-for-height as an indicator of nutritional status. *American Journal of Clinical Nutrition*, **46**, 411–418.

Ulijaszek, S. J. (1990). Nutritional status and susceptibility to infectious disease. In *Diet and Disease*, G. A. Harrison and J. C. Waterlow (eds). Cambridge: Cambridge University Press, pp. 137–154.

Ulijaszek, S. J. (1993). Evidence for a secular trend in heights and weights of adults in Papua New Guinea. *Annals of Human Biology*, **20**, 349–355.

Ulijaszek, S. J. (1998). Immunocompetence. In *Cambridge Encyclopedia of Human Growth and Development*, S. J. Ulijaszek, F. E. Johnston and M. A. Preece (eds). Cambridge: Cambridge University Press, p. 340.

Ulijaszek, S. J. (2001). Ethnic differences in patterns of human growth in stature. In *Nutrition and Growth*, R. Martorell and F. Haschke (eds). Philadelphia: Lippincott Williams and Wilkins, pp. 1–15.

Ulijaszek, S. J. (2006). The International Growth Reference for Children and Adolescents Project: environmental influences on preadolescent and adolescent growth in weight and height. *Food and Nutrition Bulletin*, **27**(suppl.), S279–S294.

Van Wieringen, J. C. (1986). Secular growth changes. In *Human Growth: A Comprehensive Treatise*, vol. **3**, F. Falkner and J. M. Tanner (eds). New York: Plenum Press, pp. 307–331.

Villamor, E., Msamanga, G., Spiegelman, D., et al. (2002). Effect of multivitamin and vitamin A supplements on weight gain during pregnancy among HIV-1-infected women. *American Journal of Clinical Nutrition*, **76**, 1082–1090.

Walls, T. and Shingadia, D. (2004). Global epidemiology of paediatric tuberculosis. *Journal of Infection*, **48**, 13–22.

Waterlow, J. C. (1988). Observations on the natural history of stunting. In *Linear Growth Retardation in Less Developed Countries*, J. C. Waterlow (ed.). New York: Raven Press, pp. 1–12.

Weitz, C. A. and Garruto, R. M. (2004). Growth of Han migrants at high altitude in Central Asia. *American Journal of Human Biology*, **16**, 405–419.

Wharton, B. (1989). Causes of low birthweight in developing countries. In *Intrauterine Growth Retardation*, J. Senterre (ed.). New York: Raven Press, pp. 143–155.

Widdowson, E. M. (1951). Mental contentment and physical growth. *Lancet*, **1**, 1316–1318.

Xu, F., Yin, X. M., Zhang, M., et al. (2005). Family average income and body mass index above the healthy weight range among urban and rural residents in regional mainland China. *Public Health Nutrition*, **8**, 47–51.

24 Evolutionary Biology of Hormonal Responses to Social Challenges in the Human Child

Mark V. Flinn

Human children are remarkably tuned-in to their social environments. They are informational sponges, absorbing bits of knowledge from others at a phenomenal pace, equipped with life's most sophisticated and creative communication system (human language). This sensitivity to social interactions is interwoven with the ontogeny of flexible cognitive skills – including empathy, self awareness, social-scenario building, and theory of mind (ToM) – that are the foundation of human relationships. In this chapter I examine the neuroendocrine systems that facilitate the development of these distinctively human sociocognitive adaptations.

Neuroendocrine systems may be viewed as complex sets of mechanisms designed by natural selection to communicate information among cells and tissues. Steroid and peptide hormones, associated neurotransmitters, and other chemical messengers guide behaviors of mammals in many important ways (Ellison, 2001; Lee et al., 2009; Panksepp, 2009). Analysis of patterns of hormone levels in naturalistic contexts can provide important insights into the evolutionary functions of the neuroendocrine mechanisms that guide human behaviors. Here I focus on the apparent evolutionary paradox of neuroendocrine response to psychosocial stressors.

Acute and chronic stressful experiences are associated with a variety of negative health outcomes in humans, including susceptibility to upper respiratory infections (Cohen et al., 2003), anxiety and depression (Heim and Nemeroff, 2001), and coronary heart disease (McEwen, 1998). The effects of psychosocial stress can be substantial: in the rural community of Bwa Mawego, Dominica, where I have studied child health for the past 22 years, overall morbidity among children for the 3–6 days following an acute stress event is more than double the normal rate (Flinn and England, 2003). Studies of populations within the United States indicate that chronic stress is similarly associated with a long-term three-fold increase in adverse health conditions (Cohen et al., 1993, 2007). Exposure to stressful events early in development, moreover, appears to have lifelong effects (Heim et al., 2002; Fox et al., 2007; Kolassa and Elbert, 2007; Meaney et al., 2007; Champagne, 2008; Seckl, 2008). Stress endocrinology is suspected to have an important role in the links between social environment and health. Chronic release of stress hormones such as cortisol in response to psychosocial challenges is posited to have incidental deleterious effects on immune and metabolic regulatory functions (Ader et al., 2001; Sapolsky, 2005). Release of androgens such as testosterone, dehydroepiandrosterone (DHEA), and dehydroepiandrosterone sulfate (DHEAS) are also influenced by social conditions (see Chapter 16 of this volume), and can affect immunocompetence (Muehlenbein and Bribiescas, 2005; Muehlenbein, 2008).

This importance of the social environment for a child's physical and mental health presents an evolutionary puzzle. Why, given the apparent high cost to human health of psychosocial stress, would natural selection have favored links between the psychological mechanisms that assess social challenges, and the neuroendocrine mechanisms that regulate stress and reproductive physiology and downstream immune functions? I approach this question from the integrative evolutionary paradigm of Niko Tinbergen (1963), who emphasized the importance of linking proximate physiological explanations with ontogeny (development), phylogeny (ancestry), and adaptive function (natural selection). My basic argument is that hormonal stress response to psychosocial challenges facilitates the neural remodeling and potentiation that is necessary to adapt to the dynamic informational arms race of the human sociocultural environment.

Human Evolutionary Biology, ed. Michael P. Muehlenbein. Published by Cambridge University Press. © Cambridge University Press 2010.

WHY IS THE HUMAN CHILD SO SENSITIVE TO THE SOCIAL ENVIRONMENT?

The human child is a social creature, motivated by and highly sensitive to interpersonal relationships (Gopnik et al., 1999). The life history stage of human childhood enables the development of necessary social skills (Alexander, 1987; Joffe, 1997; Bogin, 1999; Geary and Bjorklund, 2000; Flinn, 2004), including emotional regulation. Learning, practice, and experience are imperative for social success. The information processing capacity used for human social interactions is considerable, and perhaps significantly greater than that involved with foraging skills (Roth and Dicke, 2005).

The child needs to master complex dynamic tasks such as learning the personalities, social biases, relationships, and so forth of peers and adults in the local community, and developing appropriate cognitive and emotional responses to these challenges (Bugental, 2000). The learning environments that facilitate and channel these astonishing aspects of human mental phenotypic plasticity appear to take on a special importance. Much of the data required for the social behavior necessary to be successful as a human cannot be "preprogrammed" into specific, detailed, fixed responses. Social cleverness in a fast-paced, cumulative cultural environment must contend with dynamic, constantly shifting strategies of friends and enemies, and hence needs information from experiential social learning (Flinn, 1997, 2006a). The links among psychosocial stimuli, emotions, and physiological stress response may guide both the acute and long-term neurological plasticity necessary for adapting to the dynamic aspects of human sociality.

HUMAN SOCIALITY: KEY EVOLUTIONARY PUZZLES

Humans are characterized by a distinctive set of traits, including: (1) large brains; (2) long periods of juvenile dependence; (3) extensive biparental care including large transfers of information; (4) multigenerational bilateral kin networks; (5) habitual bipedal locomotion; (6) use of the upper limbs for tool use including projectile weapons; (7) concealed or "cryptic" ovulation; (8) menopause; (9) culture including language; and (10) lethal competition among kin-based coalitions. A few other species exhibit several of these traits; only humans, however, are characterized by the entire combination (Alexander, 2005). This suite of traits presents several questions or puzzles that are key to understanding human evolutionary biology. Here I first briefly describe these puzzles, and suggest a common resolution based on the importance of social competition during human evolution. I then return to

the paradox of hormonal response to social challenge, hypothesizing that glucocorticoids and androgens help facilitate neural remodeling and long-term potentiation necessary for dynamic social cognition.

Paternal care in multimale groups

Mammals that live in groups with multiple males – such as chimpanzees (*Pan troglodytes*) – usually have little or no paternal care, because the nonexclusivity of mating relationships obscures paternity (Alexander, 1974; Clutton-Brock, 1991). Chimpanzee males appear to lack reliable cues for identifying their offspring. In contrast, it is common for human fathers to provide protection, information, food, and social status for their children. Paternal care in humans appears to be facilitated by relatively stable pair bonds, which not only involves co-operation between mates that often endures over the life span, but which requires an unusual type of co-operation among coresiding males – respect for each other's mating relationships.

The relatively exclusive mating relationships that are characteristic of humans generate natural factions within the group. Mating relationships also can create alliances in human groups, linking two families or clans together. By way of comparison, in chimpanzee communities it is difficult for even the most dominant male to monopolize an estrous female; most of the males in a community mate with most of the females (Goodall, 1986). Although dominant males sire a higher proportion, chimpanzee males in effect "share" a common interest in the community's females and their offspring. Human groups, in contrast, are composed of family units, each with distinct reproductive interests. Human males do not typically share mating access to all the group's females; consequently, there are usually reliable cues identifying which children are their genetic offspring, and which are those of other males (for variations see Flinn, 1981; Beckerman and Valentine, 2002). Because humans live in multimale groups, yet typically maintain fairly stable mating relationships, the potential for fission along family lines is high. Still, human groups overcome this inherent conflict between family units to form large, stable coalitions (Chapais, 2008).

This unusual tolerance among coresidential males and females stands in contrast to the norm of polygamous mate competition in nonhuman primates. Selection pressures favoring such tolerance are uncertain, but likely involve the importance of both male parental investment (Alexander, 1990b; Geary and Bjorklund, 2000) and male coalitions for intraspecific conflict (Alexander, 1989, 2006; Wrangham, 1999; Geary and Flinn, 2001; Bernhard et al., 2006). The hormonal mechanisms that enable these unusual aspects of human male relationships are uncertain.

Analysis of patterns of levels of candidate hormones – such as vasopressin, testosterone, DHEA/S and cortisol – in natural social conditions may provide useful clues as to the evolved functions of male coalitions and pair bonding among humans.

An extended period of juvenile dependence and child development

The human baby is unusually altricial (helpless). Infants must be carried, fed, and protected for a long period in comparison with other primates. Human childhood and adolescence are also lengthy (Smith, 1994; Bogin, 1999; Leigh, 2004). This extension of the juvenile period that delays reproduction for much longer than the other hominoids appears costly in evolutionary terms. Parental and other kin investment continues for an unusually long time, often well into adulthood and perhaps even after the death of the parents (Alexander, 1987; Coe, 2003; Hrdy, 2009).

The selective pressures responsible for this unique suite of life history characteristics appear central to understanding human evolution (Alexander, 1990a, 1990b; Kaplan et al., 2000; Bjorklund and Pellegrini, 2002; Rosenberg, 2004). The normal delay of reproduction until at least 15 years of age involves prolonged exposure to extrinsic causes of mortality and longer generation intervals. What advantages of an extended childhood could have outweighed the heavy costs of reduced fecundity and late reproduction (Williams, 1966; Stearns, 1990) for our hominin ancestors?

Intelligence, information, and social power

The human brain is an astonishing organ. Its cortex comprises about 30 billion neurons of 200 different types, each of which are interlinked by about a thousand synapses, resulting in a million billion connections working at rates of up to 10 billion interactions per second (Williams and Herrup, 1988; Koch, 1999; Edelman, 2006). Quantifying the transduction of these biophysical actions into specific cognitive activities – e.g., thoughts and emotions – is difficult, but it is likely that humans have more information processing capacity than any other species (Roth and Dicke, 2005).

The human brain evolved at a rapid pace: hominin cranial capacity tripled (from an average of about 450 to 1350 cc) in less than 2 million years (Lee and Wolpoff, 2003) – roughly 100 000 neurons and supportive cells per generation. Structural changes such as increased convolutions, thickly myelinated cortical neurons, lateral asymmetries, increased von Economo neurons, expansion of the neo-cortex,

and enhanced integration of the cerebellum also appear significant (Allman, 1999; Amodio and Frith, 2006). In comparison with most other parts of the human genome, selection on genes involved with brain development was especially intense (Gilbert et al., 2005).

The human brain has high metabolic costs: about 50% of an infant's, and 20% of an adult's, energetic resources are used to support this neurological activity (Aiello and Wheeler, 1995). Although the increase in energetic resources allocated to the brain was accompanied by a corresponding decrease in digestive tissue, this does not explain what the selective pressures were for enhanced information processing, or why the resources were not reallocated to direct reproductive function. The obstetric difficulties associated with birthing a large-headed infant generate additional problems (Rosenberg and Trevathan, 2002). The selective advantages of increased intelligence must have been high to overcome these costs.

The human brain, in short, is a big evolutionary puzzle. It is developmentally and metabolically expensive, evolved rapidly, enables uniquely human cognitive abilities such as language, empathy, foresight, consciousness, and ToM, and generates unusual levels of novelty. Advantages of a larger brain may include enhanced information processing capacities to contend with ecological pressures that involve sexually dimorphic activities such as hunting and complex foraging (Kaplan and Robson, 2002). There is little evidence, however, of sufficient domain-specific enlargement of those parts of the brain associated with selective pressures from the physical environment (Geary and Huffman, 2002; Adolphs, 2003). Indeed, human cognition has little to distinguish itself in the way of specialized ecological talents. A large brain may have been sexually selected because it was an attractive trait for mate choice (Miller, 2000; Gavrilets and Vose, 2006). However, there is little sexual dimorphism in encephalization quotient or intelligence psychometrics (Jensen, 1998), nor is there a clear reason why brains would have been a target for sexual selection driven by mate choice uniquely among hominins.

One area in which humans are truly extraordinary is sociality. Humans are able to mentally represent the feelings and thoughts of others. Humans have unusually well-developed mechanisms for ToM (Leslie et al., 2004; Amodio and Frith, 2006), and associated specific pathologies in this domain (Baron-Cohen, 1995; Gilbert, 2001). We have exceptional linguistic abilities for transferring information from one brain to another (Pinker, 1994), enabling complex social learning. Social and linguistic competencies are roughly equivalent in both males and females, although human mothers appear to have especially important roles in the development of their offspring's sociocognitive development (Simons et al., 2001; Deater-Deckard et al., 2004).

Human coalitionary dynamics appear to have become increasingly based on information and social skills. Intense intergroup competition created pressure for within-group social cohesion (Alexander, 1990a; Flinn et al., 2005a) that required not only fighting abilities, but complex social strategies.

Kin networks and multiple caretakers

All human societies recognize kinship as a key organizational principle (Brown, 1991). All languages have kinship terminologies and concomitant expectations of nepotism (Murdock, 1949; Fortes, 1969). Human kinship systems appear unique in the consistency of both bilateral (maternal and paternal) and multigenerational structure, with a general trend for coresidence of male kin. These aspects of human kinship link families into broader co-operative systems, and provide additional opportunities for alloparental care during the long social childhood. Human grandparents stand out as unusually important in this regard (Hrdy, 2005; Flinn and Leone, 2006, 2009).

Grandparents and grand-offspring share 25% of their genes identical by descent, a significant opportunity for kin selection. Few species, however, live in groups with multiple overlapping generations of kin. Fewer still have significant social relationships among individuals two or more generations apart. Humans appear rather exceptional in this regard. Grandparenting is cross-culturally ubiquitous and pervasive (Murdock, 1967; Sear et al., 2000). Our life histories allow for significant generational overlaps, including an apparent extended postreproductive stage facilitated by the unique human physiological adaptation of menopause (Alexander, 1974, 1987; Hawkes, 2003).

The significance of emotional bonding between grandparents and grandchildren is beyond doubt. The evolved functions are uncertain, but likely involve the exceptional importance of long-term extensive and intensive investment for the human child. The emotional and cognitive processes that guide grand-relationships must have evolved because they enhanced survival and eventual reproductive success of grandchildren. In addition to the physical basics of food, protection, and hygienic care, development of the human child is strongly influenced by the dynamics of the social environment (Konner, 1991; Hetherington, 2003a, 2003b; Dunn, 2004). Grandparents may have knowledge and experience that are important and useful for helping grandchildren and other relatives succeed in social competition (Coe, 2003). Humans are unusual in the role of kin in alloparental care and group coalitions (Hrdy, 2009).

THE SOCIAL ENVIRONMENT AS A KEY SELECTIVE PRESSURE

Information processing is a core human adaptation

Children are especially tuned to their social worlds and the information that it provides. The social world is a rich source of useful information for cognitive development. The human brain appears designed by natural selection to take advantage of this bonanza of data (Tooby and Cosmides, 1992; Bjorklund and Pellegrini, 2002; Belsky, 2005). "Culture" may be viewed as a highly dynamic information pool that coevolved with the extensive information processing abilities associated with our flexible communicative and sociocognitive competencies (Alexander, 1979). With the increasing importance and power of information in hominin social interaction, culture and tradition may have become an arena of social co-operation and competition (Coe, 2003; Flinn, 2004, 2006a; Baumeister, 2005).

The key issue is *novelty*. One of the most difficult challenges to understanding human cognitive evolution, and its handmaiden culture, is the unique informational arms race that underlies human behavior. The reaction norms posited by evolutionary psychology to guide evoked culture within specific domains may be necessary but insufficient (Chiappe and MacDonald, 2005). The mind does not appear limited to a predetermined Pleistocene set of options – such as choosing mate A if in environment X, but choose mate B if in environment Y – analogous to examples of simple phenotypic plasticity (MacDonald and Hershberger, 2005).

Keeping up in the hominin social chess game requires imitation. Getting ahead favors creativity to produce new solutions to beat the current winning strategies. Random changes, however, are risky and ineffective. Hence the importance of cognitive abilities to hone choices among imagined innovations in ever more complex social scenarios. The theater of the mind that allows humans to "understand other persons as intentional agents" (Tomasello, 1999, p. 526) provides the basis for the evaluation and refinement of creative solutions to the never-ending novelty of the social arms race. This process of filtering the riot of novel information generated by the creative mind favored the cognitive mechanisms for recursive pattern recognition in the "open" domains of both language (Pinker, 1994, 1997; Nowak et al., 2001) and social dynamics (Geary, 2005; Flinn, 1997, 2006a). Cultural "traditions" passed down through the generations also help constrain the creative mind (Coe, 2003; Flinn and Coe, 2007). The evolutionary basis for these psychological mechanisms underlying the importance of social learning and culture appears rooted in a process of "runaway social selection" (Alexander, 2005; Flinn and Alexander, 2007).

Runaway social selection

Darwin (1871) recognized that there could be important differences between: (1) selection occurring as a consequence of interaction with ecological factors such as predators, climate, and food; and (2) selection occurring as a consequence of interactions among conspecifics, i.e., members of the same species competing with each other over resources such as nest sites, food, and mates. The former is termed "natural selection" and the latter "social selection" of which sexual selection may be considered a special subtype (West-Eberhard, 1983). The pace and directions of evolutionary changes in behavior and morphology produced by these two types of selection – natural and social – can be significantly different (Fisher, 1930; West-Eberhard, 2003).

Selection that occurs as a consequence of interactions between species can be intense and unending – for example with parasite-host red queen evolution (Hamilton et al., 1990) and other biotic arms races. Intraspecific social competition may generate selective pressures that cause even more rapid and dramatic evolutionary changes. Relative to natural selection, social selection has the following characteristics (West-Eberhard, 1983): (1) The intensity of social selection (and consequent genetic changes) can be very high because competition among conspecifics can have especially strong effects on differential reproduction. (2) Because the salient selective pressures involve competition among members of the same species, the normal ecological constraints are often relaxed for social selection. Hence traits can evolve in seemingly extreme and bizarre directions before counter-balancing natural selection slows the process. (3) Because social competition involves *relative* superiority among conspecifics, the bar can be constantly raised in a consistent direction generation after generation in an unending arms race. (4) Because social competition can involve multiple iterations of linked strategy and counter-strategy among interacting individuals, the process of social selection can become autocatalytic, its pace and directions partly determined from within, generating what might be termed "secondary red queens." For example, reoccurrence of social competition over lifetimes and generations can favor flexible phenotypic responses such as social learning that enable constantly changing strategies. Phenotypic flexibility of learned behavior to contend with a dynamic target may benefit from enhanced information processing capacities, especially in regard to foresight and scenario-building.

Human evolution appears characterized by these circumstances generating a process of runaway social selection (Alexander, 2005; Flinn and Alexander, 2007). Humans, more so than any other species, appear to have become their own most potent selective pressure via social competition involving coalitions (Alexander, 1989; Geary and Flinn, 2002), and dominance of their ecologies involving niche construction (Laland et al., 2000). The primary functions of the most extraordinary and distinctive human mental abilities – language, imagination, self-awareness, ToM, foresight, and consciousness – involve the negotiation of social relationships (Siegal and Varley, 2002; Tulving, 2002; Flinn et al., 2005a). The multiple-party reciprocity and shifting nested subcoalitions characteristic of human sociality generate especially difficult information processing demands for these cognitive facilities that underlie social competency. Hominin social competition involved increasing amounts of novel information and creative strategies. Culture emerged as an intensive selective pressure on the evolving brain.

Evolution of the cultural brain

As noted above, the human brain is a big evolutionary paradox. It has high metabolic costs, takes a long time to develop, evolved rapidly, enables behavior to change quickly, has unique linguistic and social aptitudes, and generates unusual levels of informational novelty. Its primary functions include dealing with other human brains (Adolphs, 2003; Alexander, 2005; Amodio and Frith, 2006). The currency is not footspeed or antibody production, but the generation and processing of data in the social worlds of the human brain's own collective and historical information pools. Some of the standout features of the human brain that distinguish us from our primate relatives are asymmetrically localized in the prefrontal cortex, including especially the dorsolateral prefrontal cortex and frontal pole (Ghazanfar and Santos, 2004; for review see Geary, 2005). These areas appear to be involved with "social scenario building" or the ability to "see ourselves as others see us so that we may cause competitive others to see us as we wish them to" (Alexander, 1990b, p. 7), and are linked to specific social abilities such as understanding sarcasm (Shamay-Tsoory et al., 2005) and morality (Moll et al., 2005). An extended childhood seems to enable the development of these necessary social skills (Joffe, 1997).

Evolution of the human family as a nest for the child's social mind

To summarize, the human family is the nexus for the suite of extraordinary and unique human traits. Humans are the only species to live in large multimale groups with complex coalitions and extensive paternal and alloparental care, and the altricial infant is indicative of a protective environment provided by intense parenting and alloparental care in the context of kin groups (Chisholm, 1999). The human baby does not need to be physically precocial, instead the brain continues rapid growth, and the corresponding cognitive competencies largely direct attention toward the social environment. Plastic neural systems adapt to the nuances of the local community, such as its language (Alexander, 1990a; Geary and Bjorklund, 2000; Bjorklund and Pellegrini, 2002; Fisher, 2005). In contrast to the slow development of ecological skills of movement, fighting, and feeding, the human infant rapidly acquires skill with the complex communication system of human language (Pinker, 1994; Sakai, 2005). The extraordinary information-transfer abilities enabled by linguistic competency provide a conduit to the knowledge available in other human minds. This emergent capability for intensive and extensive communication potentiates the social dynamics characteristic of human groups (Deacon, 1997; Dunbar, 1998) and provides a new mechanism for social learning and culture.

An extended childhood appears useful for acquiring the knowledge and practice to hone social skills and to build coalitional relationships necessary for successful negotiation of the increasingly intense social competition of adolescence and adulthood. Ecologically related play and activities (e.g., exploration of the physical environment) are also important (e.g., Geary et al., 2003), but appear similar to that of other primates. The unusual scheduling of human reproductive maturity, including an "adrenarche" (patterned increases in adrenal activities preceding puberty) and a delay in direct mate competition among males appears to extend the period of practicing social roles and extends social ontogeny (Campbell, 2006; Del Giudice 2009; Flinn et al., 2009).

The advantages of intensive parenting, including paternal protection and other care, require a most unusual pattern of mating relationships: moderately exclusive pair bonding in multiple-male groups. No other primate (or mammal) that lives in large, co-operative multiple-reproductive-male groups has extensive male parental care, although some protection by males is evident in baboons (Buchan et al., 2003). Competition for females in multiple-male groups usually results in low confidence of paternity (e.g., chimpanzees). Males forming exclusive "pair bonds" in multiple-male groups would provide cues of nonpaternity to other males, and hence place their offspring in great danger of infanticide (Hrdy, 1999). Paternal care is most likely to be favored by natural selection in conditions where males can identify their offspring with sufficient probability to offset the costs of investment, although reciprocity with mates is also likely to be involved (Smuts and Smuts, 1993; Geary and Flinn, 2001; Chapais, 2008).

Humans exhibit a unique "nested family" social structure, involving complex reciprocity among males and females to restrict direct competition for mates among group members.

It is difficult to imagine how this system of pair bonds and male coalitions could be maintained in the absence of another unusual human trait: concealed or "cryptic" ovulation (Alexander and Noonan, 1979). Human groups tend to be male philopatric (males tending to remain in their natal groups), resulting in extensive male kin alliances, useful for competing against other groups of male kin (Wrangham and Peterson, 1996; LeBlanc, 2003). However, unlike chimpanzees, human groups and communities are often composed of several bilateral kin factions, interwoven by pair bond relationships among them. Human females also have complex alliances, but usually are not involved directly in the overt physical aggression characteristic of intergroup relations (Campbell, 2002; Geary and Flinn, 2002). Parents and other kin may be especially important for the child's mental development of social and cultural maps because they can be relied upon as landmarks who provide relatively honest information. From this perspective, the evolutionary significance of the human family in regard to child development is viewed more as a nest from which social skills may be acquired than just as an economic unit centered on the sexual division of labor (Flinn et al., 2005b).

To summarize my argument to this point, human childhood is viewed as a life history stage that appears necessary and useful for acquiring the information and practice to build and refine the mental algorithms critical for negotiating the social coalitions that are key to success in our species. Mastering the social environment presents special challenges for the human child. Social competence is difficult because the target is constantly changing and similarly equipped with ToM and other cognitive abilities. The family environment, including care from fathers and grandparents, is a primary source and mediator of the ontogeny of social competencies. Human biology has been profoundly affected by our evolutionary history as unusually social creatures, including, perhaps, a special reliance upon smart mothers, co-operative fathers, and helpful grandparents. Indeed, the mind of the human child may have design features that enable its development as a group project, guided by the multitudinous informational contributions of its ancestors and codescendants (Coe, 2003; Hrdy, 2009). Studies of the patterns of hormonal responses to these complex components of human sociality may provide important clues about the selective pressures that guided human evolution.

NEUROENDOCRINE RESPONSE TO THE SOCIAL ENVIRONMENT

The constellation of behaviors associated with the human family and the dynamics of social competition described in previous sections are enabled by complex regulatory systems. In this section, I first briefly review the potential mechanisms for human pair bonding, maternal and paternal attachment to offspring, kin attachment, and male coalitions. Much of the research on the basic mechanisms has been done with nonhuman models and is not easily applied directly to some aspects of human psychology. I then turn to a more detailed analysis of how the neuroendocrine stress response system functions to enable acquisition of social competencies during childhood in the context of the human family environment.

The chemical messenger systems that orchestrate the ontogeny and regulation of sexual differentiation, metabolism, neurogenesis, immune function, growth, and other complex somatic processes, tend to be evolutionarily conservative among primates and more generally among mammals. Hence rodent and nonhuman primate models provide important comparative information about the functions of specific human neuroendocrine systems, for which we often have little direct empirical research. It is the particular balance of human mechanisms and abilities that is unique and reflects the history of selection for complex social interactions that shaped the human lineage.

The chemistry of affection

Some of the most precious of all our human feelings are stimulated by close social relationships: a mother holding her newborn infant for the first time, brothers reunited after a long absence, or lovers entangled in each other's arms. Natural selection has designed our neurobiological mechanisms, in concert with our endocrine systems, to generate potent sensations in our interactions with these most evolutionarily significant individuals. We share with our primate relatives the same basic hormones and neurotransmitters that underlie these mental gifts. But our unique evolutionary history has modified us to respond to different circumstances and situations; we are rewarded and punished for somewhat different stimuli than our phylogenetic cousins. Chimpanzees and humans share the delight – the sensational reward – when biting into a ripe, juicy mango. But the endocrine, neurological, and associated emotional responses of a human father to the birth of his child (e.g., Storey et al., 2000) are likely to be quite different from those of a chimpanzee male. Happiness for a human (Buss, 2000) has many

unique designs, such as romantic love (Fisher et al., 2002), that involve shared endogenous messengers from our phylogenetic heritage.

Attachments or bonding are central in the lives of the social mammals. Basic to survival and reproduction, these interdependent relationships are the fabric of the social networks that permit individuals to maintain co-operative relationships over time. Although attachments can provide security and relief from stress, close relationships also exert pressures on individuals to which they continuously respond. It should not be surprising, therefore, that the neuroendocrine mechanisms underlying attachment and stress are intimately related to one another. And although at the present time a good deal more is known about the stress response systems than the affiliative systems, some of the pieces of the puzzle are beginning to fall into place (Panksepp, 2004).

The mother–offspring relationship is at the core of mammalian life, and it appears that some of the biochemistry at play in the regulation of this intimate bond was also selected to serve in primary mechanisms regulating bonds between mates, paternal care, the family group, and even larger social networks (Hrdy, 1999; Fisher et al., 2002). Although a number of hormones and neurotransmitters are involved in attachment and other components of relationships, the two peptide hormones, oxytocin (OT) and arginine-vasopressin (AVP), appear to be primary (Carter, 2002; Young and Insel, 2002; Curtis and Wang, 2003; Lim et al., 2004; Heinrichs and Domes, 2008; Lee et al., 2009), with dopamine, cortisol, and other hormones and neurotransmitters having mediating effects.

The hypothalamus is the major brain site where OT and AVP (closely related chains of nine amino acids) are produced. From there they are released into the central nervous system (CNS) as well as transported to the pituitary where they are stored until secreted into the bloodstream. Oxytocin and AVP act on a wide range of neurological systems, and their influence varies among mammalian species and stage of development. The neurological effects of OT and AVP appear to be key mechanisms (e.g., Bartels and Zeki, 2004) involved in the evolution of human family behaviors. The effects of OT and AVP in humans are likely to be especially context dependent, because of the variable and complex nature of family relationships.

Parental care

Along with OT and AVP, prolactin, estrogen, and progesterone are involved in parental care among mammals (Insel and Young, 2001). The roles of these hormones vary across species and between males and females. The effects of these hormones are influenced by experience and context. Among rats, for example, estrogen and progesterone appear to prime the brain during pregnancy for parental behavior. Estrogen has been found to activate the expression of genes that increase the receptor density for OT and prolactin, thus increasing their postnatal influence (Young and Insel, 2002).

Oxytocin is most well known for its role in regulating birth and lactation, but along with AVP, it has also been found to play a central role in maternal care and attachment (Fleming et al., 1999). Just prior to birth, an increase in OT occurs, which is seen as priming maternal care. An injection of OT to virgin rats has been found to induce maternal care, while an OT antagonist administered to pregnant rats interferes with the development of maternal care (Carter, 2002).

The new rat mother requires hormonal activation to initially stimulate maternal behavior. Once she has begun to care for her pups, however, hormones are not required for maternal behavior to continue. Olfactory and somatosensory stimulation from interactions between pups and mother are, however, required for the parental care to continue (Fleming et al., 1999). The stimulation from suckling raises OT levels in rodents and breast-feeding women, which then results in not only milk letdown but also a decrease in limbic hypothalamic-anterior pituitary-adrenal cortex system (HPA) activity and a shift in the autonomic nervous system (ANS) from a sympathetic tone to a parasympathetic tone. This results in a calmness seen as conducive to remaining in contact with the infant. It also results in a shift from external-directed energy toward the internal activity of nutrient storage and growth (Uvnas-Moberg, 1998).

Experience also affects the neuroendocrine systems involved in the expression of maternal care. The HPA system of offspring during development is influenced by variation in maternal care, which then influences their maternal behavior as adults. Such changes involve the production of, and receptor density for, stress hormones and OT (Fleming et al., 1999; Champagne and Meaney, 2001).

The HPA-modulated hormones and maternal behavior are related in humans during the postpartum period (Fleming et al., 1997). During this time, cortisol appears to have an arousal effect, focusing attention on infant bonding. Mothers with higher cortisol levels were found to be more affectionate, more attracted to their infant's odor, and better at recognizing their infant's cry during the postpartum period.

Functional magnetic resonance imaging (fMRI) studies of brain activity involved in maternal attachment in humans indicate that the activated regions are part of the reward system and contain a high density of receptors for OT and AVP (Bartels and Zeki, 2004;

Fisher et al., 2006). These studies also demonstrate that the neural regions involved in attachment activated in humans are similar to those activated in non-human animals. Among humans, however, neural regions associated with social judgment and assessment of the intentions and emotions of others exhibited some deactivation during attachment activities, suggesting possible links between psychological mechanisms for attachment and management of social relationships. Falling in love with a mate and affective bonds with offspring may involve temporary deactivation of psychological mechanisms for maintaining an individual's social "guard" in the complex reciprocity of human social networks. Dopamine levels are likely to be important for both types of relationship but may involve some distinct neural sites. It will be interesting to see what fMRI studies of attachment in human males indicate because that is where the most substantial differences from other mammals would be expected. Similarly, fMRI studies of attachment to mothers, fathers, and alloparental-care providers in human children may provide important insights into the other side of parent–offspring bonding.

Paternal care

Paternal care is not common among mammals. For evolutionary reasons noted earlier, it is found among some rodent and primate species, including humans. The extent and types of paternal care vary among species. The hormonal influence in parental care among males appears to differ somewhat from that found among females. Vasopressin (AVP) appears to function as the male addition to OT (Young and Insel, 2002). Along with prolactin and OT, AVP prepares the male to be receptive to and care for infants (Bales et al., 2004).

Paternal care is more common in monogamous than polygamous mammals and is often related to hormonal and behavioral stimuli from the female. In the monogamous California mouse, disruption of the pair bond does not affect maternal care but does diminish paternal care (Gubernick, 1996). In other species with biparental care, however, paternal care is not as dependent on the presence of the female (Young and Insel, 2002). Experience also plays a role in influencing hormonal activation and paternal behavior. Among tamarins, experienced fathers have higher levels of prolactin than first-time fathers (Ziegler and Snowdon, 1997).

Androgens including testosterone also appear to be involved in the regulation of paternal behavior. For example, human fathers tend to have lower testosterone levels when they are involved in childcare activities (Berg and Wynne-Edwards, 2002; Fleming et al., 2002;

Gray and Campbell, 2009; also see Chapter 16 of this volume), although the relation with the key paternal role of offspring protection is uncertain. Human males stand out as very different from our closest relatives the chimpanzees in the areas of paternal attachment and investment in offspring. Investigation of the neuroendocrine mechanisms that underpin male parental behavior may provide important insights into these critical evolutionary changes.

Pair bonding

Like male parental care, bonding between mates is also uncommon among mammals but has been selected for when it has reproductive advantages for both parents (Clutton-Brock, 1991; Carter, 2002; Young et al., 2002). Monogamy is found across many mammalian taxa, but most of the current knowledge related to the neuroendocrine basis of this phenomenon has been obtained from the comparative study of two closely related rodent species. The prairie vole (*Microtus ochrogaster*) mating pair nest together and provide prolonged biparental care, while their close relatives, the meadow vole (*Microtus pennsylvanicus*), do not exhibit these behaviors (Young et al., 2002). As with other social behaviors in rodents, OT and AVP have been found to be central in the differences these related species exhibit with respect to pair bonding.

Pair bonding occurs for the prairie vole following mating. Vagino-cervical stimulation results in a release of OT and the development of a partner preference for the female (Carter, 2002). For the male, it is an increase in AVP following mating and not just OT that results in partner preference. Exogenous OT injected in the female and exogenous AVP in the male prairie vole result in mate preference even without mating. This does not occur with meadow voles (Young et al., 2002).

The receptor density for OT and AVP in specific brain regions might provide the basis for mechanisms underlying other social behaviors. Other neurotransmitters, hormones, and social cues also are likely to be involved, but slight changes in gene expression for receptor density, such as those found between the meadow and prairie voles in the ventral palladium (located near the nucleus accumbens, an important component of the brain's reward system), might demonstrate how such mechanisms could be modified by selection (Lim et al., 2004). The dopamine D2 receptors in the nucleus accumbens appear to link the affiliative OT and AVP pair bonding mechanisms with positive rewarding mental states (Aragona et al., 2003; Curtis and Wang, 2003). The combination results in the powerful addiction that parents have for their offspring.

Given the adaptive value of extensive biparental care and prolonged attachment found in the mating pair and larger family network, it is not surprising that similar neurohormonal mechanisms active in the maternal–offspring bond would also be selected to underlie these other attachments. Though there is some variation among species and between males and females, the same general neurohormonal systems active in pair bonding in other species are found in the human (Wynne-Edwards, 2003; Panksepp, 2004; Lee et al., 2009). Androgen response to pair bonding appears complex (e.g. van der Meij et al., 2008), but similar to parent–offspring attachment in that pair bonded males tend to have lower testosterone levels in nonchallenging conditions (Alvergne et al., 2009; Gray and Campbell, 2009). Moreover, males actively involved in caretaking behavior appear to have temporarily diminished testosterone levels (Gray et al., 2007).

The challenge before human evolutionary biologists and psychologists is to understand how these general neuroendocrine systems have been modified and linked with other special human cognitive systems (e.g., Allman et al., 2001; Blakemore et al., 2004) to produce the unique suite of human family behaviors. Analysis of hormonal responses to social stimuli may provide important insights into the selective pressures that guided the evolution of these key aspects of the human mind.

The chemistry of stress, family, and the social mind

The evolutionary scenario proposed in previous sections posits that the family is of paramount importance in a child's world. Throughout human evolutionary history, parents and close relatives provided calories, protection, and information necessary for survival, growth, health, social success, and eventual reproduction. The human mind, therefore, is likely to have evolved special sensitivity to interactions with family care providers, particularly during infancy and early childhood (Bowlby, 1969; Baumeister and Leary, 1995; Daly and Wilson, 1995; Belsky, 1997, 1999; Geary and Flinn, 2001; Flinn et al., 2009).

The family and other kin provide important cognitive "landmarks" for the development of a child's understanding of the social environment. The reproductive interests of a child overlap with those of its parents more than with any other individuals. Information (including advice, training, and incidental observation) provided by parents is important for situating oneself in the social milieu and developing a mental model of its operations. A child's family environment may be an especially important source and mediator of stress, with consequent effects on health.

Psychosocial stressors are associated with increased risk of infectious disease (Cohen et al., 2003) and a variety of other illnesses (Ader et al., 2001). Physiological stress responses regulate the allocation of energetic and other somatic resources to different bodily functions via a complex assortment of neuroendocrine mechanisms. Changing, unpredictable environments require adjustment of priorities. Digestion, growth, immunity, and sex are irrelevant while being chased by a predator (Sapolsky, 1994). Stress hormones help shunt blood, glucose, and so on to tissues necessary for the task at hand. Chronic and traumatic stress can diminish health, evidently because resources are diverted away from important health functions. These costs can be referred to as "allostatic load" (McEwen, 1995). Such diversions of resources may have special significance during childhood because of the additional demands of physical and mental growth and development and possible long-term ontogenetic consequences.

Stress response mechanisms and theory

Physiological response to environmental stimuli perceived as stressful is modulated by the limbic system (amygdala and hippocampus) and basal ganglia. These components of the CNS interact with the sympathetic and parasympathetic nervous systems and two neuroendocrine axes, the sympathetic-adrenal medullary system (SAM) and the HPA. The SAM and HPA systems affect a wide range of physiological functions in concert with other neuroendocrine mechanisms and involve complex feedback regulation. The SAM system controls the catecholamines norepinephrine and epinephrine (adrenalin). The HPA system regulates glucocorticoids, primarily cortisol (for reviews, see Weiner, 1992; McEwen, 1995; Sapolsky et al., 2000).

Cortisol is a key hormone produced in response to physical and psychosocial stressors (Mason, 1968; Selye, 1976). It is produced and stored in the adrenal cortex. Release into the plasma is primarily under the control of pituitary adrenocorticotropic hormone (ACTH). The free or unbound portion of the circulating cortisol may pass through the cell membrane and bind to a specific cytosolic glucocorticoid receptor. This complex may induce genes coding for at least 26 different enzymes involved with carbohydrate, fat, and amino acid metabolism in brain, liver, muscle, and adipose tissue (Yuwiler, 1982).

Cortisol modulates a wide range of somatic functions, including: (1) energy release (e.g., stimulation of hepatic gluconeogenesis in concert with glucagon and inhibition of the effects of insulin); (2) immune activity (e.g., regulation of inflammatory response and the cytokine cascade); (3) mental activity (e.g., alertness, memory, and learning); (4) growth (e.g., inhibition of growth hormone and somatomedins); and (5) reproductive function (e.g., inhibition of gonadal steroids, including testosterone). These complex multiple effects of cortisol muddle understanding of its adaptive functions. The demands of energy regulation must orchestrate with those of immune function, attachment bonding, and so forth. Mechanisms for localized targeting (e.g., glucose uptake by active versus inactive muscle tissues and neuropeptide-directed immune response) provide fine-tuning of the preceding general physiological effects. Cortisol regulation allows the body to respond to changing environmental conditions by preparing for *specific* short-term demands (Mason, 1971; Munck et al., 1984; Weiner, 1992).

These temporary beneficial effects of glucocorticoid stress response, however, are not without costs. Persistent activation of the HPA system is associated with immune deficiency, cognitive impairment, inhibited growth, delayed sexual maturity, damage to the hippocampus, and psychological maladjustment (Glaser and Kiecolt-Glaser, 1994; Dunn, 1995; McEwen, 1995; Ader et al., 2001). Chronic stress may diminish metabolic energy (Ivanovici and Wiebe, 1981; Sapolsky, 1991) and produce complications from autoimmune protection (Munck and Guyre, 1991). Stressful life events – such as divorce, death of a family member, change of residence, or loss of a job – are associated with infectious disease and other health problems (Herbert and Cohen, 1993; Maier et al., 1994).

Current psychosocial stress research suggests that cortisol response is stimulated by uncertainty that is perceived as significant and for which behavioral responses will have unknown effects (Weiner, 1992; Kirschbaum and Hellhammer, 1994). That is, important events are going to happen; the child does not know how to react but is highly motivated to figure out what should be done. Cortisol release is associated with unpredictable, uncontrollable events that require full alert readiness and mental anticipation. In appropriate circumstances, temporary moderate increases in stress hormones (and associated neuropeptides) may enhance mental activity for short periods in localized areas, potentially improving cognitive processes for responding to social challenges (Beylin and Shors, 2003; Lupien, 2009). Other mental processes may be inhibited, perhaps to reduce external and internal "noise" (Servan-Schreiber et al., 1990; cf. Newcomer et al., 1994).

Relations between cortisol production and emotional distress, however, are difficult to assess because of temporal and interindividual variation in HPA response (Kagan, 1992; Nachmias et al., 1996). Habituation may occur to repeated events for which a child acquires an effective mental model. Attenuation and below-normal levels of cortisol may follow a day or more after emotionally charged events. Chronically stressed children may develop abnormal cortisol response, possibly via changes in binding globulin levels and/or reduced affinity or density of glucocorticoid or corticotrophin-releasing hormone (CRH)/vasopressin receptors in the brain (Fuchs and Flugge, 1995). Early experience – such as perinatal stimulation of rats (Meaney et al., 1991), prenatal stress of rhesus macaques (Schneider et al., 1992; Clarke, 1993), and sexual abuse among humans (de Bellis et al., 1994) – may permanently alter HPA response. Personality may also affect HPA response (and vice versa) because children with inhibited temperaments tend to have higher cortisol levels than extroverted children (Kagan et al., 1988; cf. Gunnar et al., 1995; Hertsgaard et al., 1995; Nachmias et al., 1996).

Further complications arise from interaction between the HPA stress response and a wide variety of other neuroendocrine activities, including modulation of catecholamines, melatonin, testosterone, serotonin, β-endorphins, cytokines, and enkephalins (de Kloet, 1991; Sapolsky, 1992; Saphier et al., 1994). Changes in cortisol for energy allocation and modulation of immune function may be confused with effects of psychosocial stress. As reviewed in the previous section, OT and vasopressin intracerebral binding sites are associated with familial attachment in mammals and may influence distress involving caretaker–child relationships. Other components of the HPA axis such as CRH and melanocyte-stimulating hormone have effects that are distinct from cortisol.

Stress response and family environment

Composition of the family or caretaking household may have important effects on child development (Kagan, 1984; Whiting and Edwards, 1988). For example, in Western cultures, children with divorced parents may experience more emotional tension or "stress" than children living in a stable two-parent family (Wallerstein, 1983; Pearlin and Turner, 1987; Gottman and Katz, 1989).

Investigation of physiological stress responses in the human family environment has been hampered by the lack of noninvasive techniques for measurement of stress hormones. Frequent collection of plasma samples to assess temporal changes in endocrine function is not feasible in nonclinical settings. The development of saliva immunoassay techniques, however, presents new opportunities for stress research. Saliva is relatively easy to collect and store, especially under adverse field conditions faced by anthropologists (Ellison, 1988). In this section I review results from a longitudinal, 20-year study of child stress and health in a rural community on the island of Dominica (for reviews see Flinn and England, 1995, 1997, 2003; Flinn, 1999, 2006b). The research design uses concomitant monitoring of a child's daily activities, stress hormones, and psychological conditions to investigate the

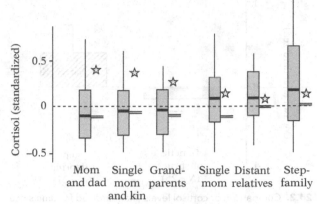

24.1. Cortisol levels and household composition of children living in Bwa Mawego, Dominica. Box and whisker plots are for children's mean values of cortisol standardized for time since awakening (for descriptions of methods see Flinn, 2006b). Double lines represent average cortisol levels when an absence of stressful events were observed or reported. Stars indicate average cortisol levels during holidays (day before Christmas and August holiday weekends). Data include 21 673 salivary cortisol samples from 268 children collected from 1989 to 2001.

effects of naturally occurring psychosocial events in the family environment.

Associations between average cortisol levels of children and household composition indicate that children living with nonrelatives, stepfathers, and half-siblings (stepfather has children by the stepchild's mother), or single parents without kin support had higher average levels of cortisol than children living with both parents, single mothers with kin support, or grandparents (Figure 24.1). Note, however, that these differences in cortisol levels are diminished when comparisons are made during nonstressed conditions (double line bars in Figure 24.1). Moreover, the pattern is reversed during the excitement of holidays and other apparently hedonic emotional circumstances (stars in Figure 24.1). Hence cortisol appears to be elevated during "positive" as well as "negative" social challenges.

A further test of the hypothesis that difficult family environments are stressful is provided by comparison of step- and genetic children residing in the same households. Stepchildren had higher average cortisol levels than their half-siblings residing in the same household who were genetic offspring of both parents (Figure 24.2).

Several caveats need emphasis. Firstly, not all children in difficult family environments have elevated cortisol levels. Secondly, household composition is not a uniform indicator of family environment. Some single-mother households, for example, appear more stable, affectionate, and supportive than some two-parent households. Thirdly, children appear

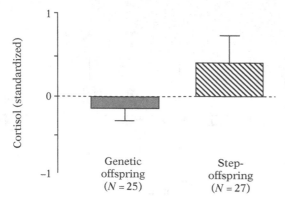

24.2. Comparison of cortisol levels (standardized for time since awakening) of children (maternal half-siblings) living in the same household that were either genetic offspring or step-offspring of the resident adult male.

differentially sensitive to different aspects of their caretaking environments, reflecting temperamental and other individual differences.

These caveats, however, do not invalidate the general association between household composition and childhood stress. There are several possible reasons underlying this result. Children in difficult caretaking environments may experience chronic stress resulting in moderate-high levels of cortisol (i.e., a child has cortisol levels that are above average day after day). They may experience more acute stressors that substantially raise cortisol for short periods of time. They may experience more frequent stressful events (e.g., parental chastisement or marital quarreling – see Wilson et al., 1980; Flinn, 1988; Finkelhor and Dzuiba-Leatherman, 1994) that temporarily raise cortisol. There may be a lack of reconciliation between parent and child. And they may have inadequate coping abilities, perhaps resulting from difficult experiences in early development.

The events in children's lives that are associated with elevated cortisol are not always traumatic or even "negative." Activities such as eating meals, hard physical work, routine competitive play (e.g., cricket, basketball, and "king of the mountain" on ocean rocks), return of a family member who was temporarily absent (e.g., father returning from a job in town for the weekend), and holiday excitement (stars in Figure 24.1) were associated with temporary moderate increases (from about 10% to 100%) in cortisol among healthy children. These moderate stressors usually had rapid attenuation (<1 hour) of cortisol levels (some stressors had characteristic temporal "signatures" of cortisol level and duration).

High-stress events (cortisol increases from 100% to 2000%), however, most commonly involved trauma from family conflict or change (Flinn et al., 1996; Flinn and England, 2003). Punishment, quarreling, and residence change substantially increased cortisol levels,

whereas calm, affectionate contact was associated with diminished (–10% to –50%) cortisol levels. Of all cortisol values that were more than two standard deviations (2 SD) above mean levels (i.e., indicative of substantial stress), 19.2% were temporally associated with traumatic family events (residence change of child or parent/caretaker, punishment, "shame," serious quarreling, and/or fighting) within a 24-hour period. In addition, 42.1% of traumatic family events were temporally associated with substantially elevated cortisol (i.e., at least one of the saliva samples collected within 24 hours was >2 SD above mean levels). Chronic elevations of cortisol levels sometimes occurred among children in difficult family environments, but this was difficult to assess quantitatively (Flinn, 2009).

There was considerable variability among children in cortisol response to family disturbances. Not all individuals had detectable changes in cortisol levels associated with family trauma. Some children had significantly elevated cortisol levels during some episodes of family trauma but not during others. Cortisol response is not a simple or uniform phenomenon. Numerous factors, including preceding events, habituation, specific individual histories, context, and temperament, might affect how children respond to particular situations.

Nonetheless, traumatic family events were associated with elevated cortisol levels for all ages of children more than any other factor that we examined. These results suggest that family interactions were a critical psychosocial stressor in most children's lives, although the sample collection during periods of intense family interaction (early morning and late afternoon) may have exaggerated this association.

Although elevated cortisol levels are associated with traumatic events such as family conflict, long-term stress may result in diminished cortisol response. In some cases, chronically stressed children had blunted response to physical activities that normally evoked cortisol elevation. Comparison of cortisol levels during "nonstressful" periods (no reported or observed crying, punishment, anxiety, residence change, family conflict, or health problem during the 24-hour period before saliva collection) indicates a striking reduction and, in some cases, reversal of the family environment–stress association (see double bars in Figure 24.1). Chronically stressed children sometimes had subnormal cortisol levels when they were not in stressful situations. For example, cortisol levels immediately after school (walking home from school) and during noncompetitive play were lower among some chronically stressed children (cf. Long et al., 1993). Some chronically stressed children appeared socially "tough" or withdrawn and exhibited little or no arousal to the novelty of the first few days of the saliva collection procedure.

Relations between family environment and cortisol stress response appear to result from a combination of factors including frequency of traumatic events, frequency of positive "affectionate" interactions, frequency of negative interactions such as irrational punishment, frequency of residence change, security of "attachment," development of coping abilities, and availability or intensity of caretaking attention. Probably the most important correlate of household composition that affects childhood stress is maternal care. Mothers in socially "secure" households (i.e., permanent amiable coresidence with mate and/or other kin) appeared more able and more motivated to provide physical, social, and psychological care for their children. Mothers without mate or kin support were likely to exert effort attracting potential mates and may have viewed dependent children as impediments to this. Hence coresidence of father may provide not only direct benefits from paternal care but also may affect maternal care (Lamb et al., 1987; Belsky et al., 1991; Flinn, 1992; Hurtado and Hill, 1992). Young mothers without mate support usually relied extensively on their parents or other kin for help with childcare.

Children born and raised in household environments in which mothers have little or no mate or kin support were at greatest risk for abnormal cortisol profiles and associated health problems. Because socioeconomic conditions influence family environment, they have consequences for child health that extend beyond direct material effects. Also, because health in turn may affect an individual's social and economic opportunities, a cycle of poor health and poverty may be perpetuated generation after generation.

CONCLUDING REMARKS

People in difficult or inequitable social environments tend to be less healthy in comparison with their more fortunate peers (e.g., Flinn, 1999; Hertzman, 1999; Dressler and Bindon, 2000; Wilkinson, 2001; Cohen et al., 2003). Social support can have reproductive consequences in group-living species (e.g., Silk et al., 2003; Cheney and Seyfarth, 2007). If the brain evolved as a social tool, then the expenditure of somatic resources (e.g., glucose) to resolve psychosocial problems makes sense. Relationships, especially family relationships, are of paramount importance. They are likely to have been a key factor affecting human reproductive success at least for over half a million years, and selection may have shaped our hormonal, neural, and psychological mechanisms to respond to this critical selective

pressure. In Bwa Mawego, and perhaps in most human societies, children elevate their stress hormone (cortisol) levels more frequently and extensively in response to psychosocial stimuli than to challenges associated with the physical environment. The adaptive effects of the major stress hormones (Huether, 1996, 1998; Koolhaas et al., 2006; Fox et al., 2007) and affiliative neurotransmitters on neural reorganization appear consistent with observations of sensitivity to the social world (Flinn, 2006b).

Social competence is extraordinarily difficult because the competition is constantly changing and similarly equipped with ToM and other cognitive abilities. The sensitivity of the stress-response and affiliative systems to the social environment may enable adaptive neural reorganization to this most salient and dynamic puzzle. Childhood appears necessary and useful for acquiring the information and practice to build and refine the mental algorithms critical for negotiating the social coalitions that are the key to success in our species. The human family provides critical support for the child to develop sociocognitive skills. Traumatic early environments may result in diminished abilities to acquire social competencies as a consequence of glucocorticoid hypersensitivity disrupting neurogenesis, particularly in the hippocampus and other components of the limbic system (Mirescu et al., 2004; Weaver et al., 2004). An improved understanding of the hormonal and neurological mechanisms that facilitate the intensive and extensive relationships involved with human families and broader kin coalitions, including comparisons between humans and our close primate relatives, may provide important insights into the selective pressures that shaped key features of human biology.

DISCUSSION POINTS

1. What happened to our hominin ancestors? Why are we the only species left? If humans suddenly went extinct, would another life form eventually evolve high intelligence? How?
2. Do you think chimpanzee fathers love their offspring? Why or why not? And chimpanzee grandparents?
3. Do school exams make you sick? Collect data from your classmates to test your hypotheses.
4. What events cause you to become "stressed?" Why do you think natural selection produced psychological mechanisms that result in this sensitivity?
5. How are the events that cause stress for you similar and/or different from the events that were stressful for your parents, your grandparents, and your distant hominin ancestors?

REFERENCES

Ader, R., Felten, D. L. and Cohen, N. (2001). *Psychoneuroimmunology*, 3rd edn. San Diego, CA: Academic Press.

Adolphs, R. (2003). Cognitive neuroscience of human social behavior. *Nature Reviews, Neuroscience*, 4(3), 165–178.

Aiello, L. C. and Wheeler, P. (1995). The expensive-tissue hypothesis: the brain and the digestive system in human and primate evolution. *Current Anthropology*, 36, 199–221.

Alexander, R. D. (1974). The evolution of social behavior. *Annual Review of Ecology and Systematics*, 5, 352–383.

Alexander, R. D. (1979). *Darwinism and Human Affairs*. Seattle: University of Washington Press.

Alexander, R. D. (1987). *The Biology of Moral Systems*. Hawthorne, NY: Aldine de Gruyter.

Alexander, R. D. (1989). Evolution of the human psyche. In *The Human Revolution*, P. Mellars and C. Stringer (eds). Chicago: University of Chicago Press, pp. 455–513.

Alexander, R. D. (1990a). Epigenetic rules and Darwinian algorithms: the adaptive study of learning and development. *Ethology and Sociobiology*, 11, 1–63.

Alexander, R. D. (1990b). *How Humans Evolved: Reflections on the Uniquely Unique Species. Museum of Zoology (Special Publication No. 1)*. Ann Arbor, MI: The University of Michigan.

Alexander, R. D. (2005). Evolutionary selection and the nature of humanity. In *Darwinism and Philosophy*, V. Hosle and C. Illies (eds). South Bend, IN: University of Notre Dame Press, pp. 301–348.

Alexander, R. D. (2006). The challenge of human social behavior. *Evolutionary Psychology*, 4, 1–32.

Alexander, R. D. and Noonan, K. M. (1979). Concealment of ovulation, parental care, and human social evolution. In *Evolutionary Biology and Human Social Behavior: an Anthropological Perspective*, N. A. Chagnon and W. Irons (eds). North Scituate, MA: Duxbury Press, pp. 436–453.

Allman, J. (1999). *Evolving Brains*. New York: Scientific American Library.

Allman, J., Hakeem, A., Erwin, J. M., et al. (2001). The anterior cingulate cortex: the evolution of an interface between emotion and cognition. *Annals of the New York Academy of Sciences*, 935, 107–117.

Alvergne, A., Faurie, C. and Raymond, M. (2009). Variation in testosterone levels and male reproductive effort: insight from a polygynous human population. *Hormones and Behavior*, 56(5), 491–497.

Amodio, D. M. and Frith, C. D. (2006). Meeting of minds: the medial frontal cortex and social cognition. *Nature Reviews Neuroscience*, 7(4), 268–277.

Aragona, B. J., Liu, Y., Curtis, J. T., et al. (2003). A critical role for nucleus accumbens dopamine in partner-preference formation in male prairie voles. *Journal of Neuroscience*, 23(8), 3483–3490.

Bales, K. L., Kim, A. J., Lewis-Reese, A. D., et al. (2004). Both oxytocin and vasopressin may influence alloparental behavior in male prairie voles. *Hormones and Behavior*. 45(5), 354–361.

Baron-Cohen, S. (1995). *Mindblindness: an Essay on Autism and Theory of Mind*. Boston, MA: MIT/Bradford.

Bartels, A. and Zeki, S. (2004). The neural correlates of maternal and romantic love. *NeuroImage*, 21, 1155–1166.

Baumeister, R. F. (2005). *The Cultural Animal: Human Nature, Meaning, and Social Life*. New York: Oxford University Press.

Baumeister, R. F. and Leary, M. R. (1995). The need to belong: desire for interpersonal attachment as a fundamental human motive. *Psychological Bulletin*, 117, 497–529.

Beckerman, S. and Valentine, P. (eds) (2002). *Cultures of Multiple Fathers: the Theory and Practice of Partible Paternity in South America*. Gainesville, FL: University of Florida Press.

Belsky, J. (1997). Attachment, mating, and parenting: an evolutionary interpretation. *Human Nature*, 8, 361–381.

Belsky, J. (1999). Modern evolutionary theory and patterns of attachment. In *Handbook of Attachment: Theory, Research, and Clinical Applications*. J. Cassidy and P. R. Shaver (eds). New York: Guilford Press, pp. 141–161.

Belsky, J. (2005). Differential susceptibility to rearing influence: an evolutionary hypothesis and some evidence. In *Origins of the Social Mind: Evolutionary Psychology and Child Development*, B. J. Ellis and D. F. Bjorklund (eds). New York: Guilford Press, pp. 139–163.

Belsky, J., Steinberg, L. and Draper, P. (1991). Childhood experience, interpersonal development, and reproductive strategy: an evolutionary theory of socialization. *Child Development*, 62, 647–670.

Berg, S. J. and Wynne-Edwards, K. E. (2002). Salivary hormone concentrations in mothers and fathers becoming parents are not correlated. *Hormones and Behavior*, 42, 424–436.

Bernhard, H., Fischbacher, U. and Fehr, E. (2006). Parochial altruism in humans. *Nature*, 442(7105), 912–915.

Beylin, A. V. and Shors, T. J. (2003). Glucocorticoids are necessary for enhancing the acquisition of associative memories after acute stressful experience. *Hormones and Behavior*, 43, 124–131.

Bjorklund, D. F. and Pellegrini, A. D. (2002). *The Origins of Human Nature: Evolutionary Developmental Psychology*. Washington, DC: APA Press.

Blakemore, S.-J., Winston, J. and Frith, U. (2004). Social cognitive neuroscience: where are we heading? *Trends in Cognitive Neurosciences*, 8(5), 216–222.

Bogin, B. (1999). *Patterns of Human Growth*, 2nd edn. Cambridge: Cambridge University Press.

Bowlby, J. (1969). *Attachment and Loss:* vol. 1. *Attachment*. London: Hogarth.

Brown, D. E. (1991). *Human Universals*. Philadelphia: Temple University Press.

Buchan, J. C., Alberts, S. C., Silk, J. B., et al. (2003). True paternal care in a multi-male primate society. *Nature*, 425(6954), 179–181.

Bugental, D. B. (2000). Acquisition of the algorithms of social life: a domain-based approach. *Psychological Bulletin*, 26, 187–209.

Buss, D. M. (2000). The evolution of happiness. *American Psychologist*, 55, 15–23.

Campbell, A. (2002). *A Mind of Her Own: the Evolutionary Psychology of Women*. Oxford: Oxford University Press.

Campbell, B. C. (2006). Adrenarche and the evolution of human life history. *American Journal of Human Biology* **18**, 569–589.

Carter, C. S. (2002). Neuroendocrine perspectives on social attachment and love. In *Foundations in Social Neuroscience*, J. T. Cacioppo, G. G. Berntson, R. Adolphs, et al. (eds). Cambridge, MA: MIT Press, pp. 853–890.

Champagne, F. A. (2008). Epigenetic mechanisms and the transgenerational effects of maternal care. *Frontiers in Neuroendocrinology*, **29**, 386–397.

Champagne, F. A. and Meaney, M. J. (2001). Like mother, like daughter: evidence for non-genomic transmission of parental behavior and stress responsivity. *Progress in Brain Research*, **133**, 287–302.

Chapais, B. (2008). *Primeval Kinship: How Pair-bonding Gave Birth to Human Society*. Cambridge, MA: Harvard University Press.

Cheney, D. L. and Seyfarth, R. M. (2007). *Baboon Metaphysics: the Evolution of a Social Mind*. Chicago: University of Chicago Press.

Chiappe, D. and MacDonald, K. (2005). The evolution of domain-general mechanisms in intelligence and learning. *Journal of General Psychology*, **132**(1), 5–40.

Chisholm, J. S. (1999). *Death, Hope, and Sex*. Cambridge: Cambridge University Press.

Clarke, A. S. (1993). Social rearing effects on HPA axis activity over early development and in response to stress in rhesus monkeys. *Developmental Psychobiology*, **26**(8), 433–446.

Clutton-Brock, T. H. (1991). *The Evolution of Parental Care*. Princeton, NJ: Princeton University Press.

Coe, K. (2003). *The Ancestress Hypothesis: Visual Art as Adaptation*. New Brunswick, NJ: Rutgers University Press.

Cohen, S., Doyle, W. J., Turner, R. B., et al. (2003). Emotional style and susceptibility to the common cold. *Psychosomatic Medicine*, **65**(4), 652–657.

Cohen, S., Kessler, R. C. and Underwood, G. L. (1993). Negative life events, perceived stress, negative effect, and susceptibility to the common cold. *Journal of Personality and Social Psychology*, **64**, 131–140.

Cohen, S., Janicki-Deverts, D., Miller, G. E. (2007). Psychological stress and disease. *Journal of the American Medical Association*, **298**(14), 1685–1687.

Curtis, T. J. and Wang, Z. (2003). The neurochemistry of pair bonding. *Current Directions in Psychological Science*, **12**(2), 49–53.

Daly, M. and Wilson, M. (1995). Discriminative parental solicitude and the relevance of evolutionary models to the analysis of motivational systems. In *The Cognitive Neurosciences*, M. S. Gazzaniga (ed.). Cambridge, MA: MIT Press, pp. 1269–1286.

Darwin, C. R. (1871). *The Descent of Man and Selection in Relation to Sex*. London: John Murray.

de Bellis, M., Chrousos, G. P., Dorn, L. D., et al. (1994). Hypothalamic-pituitary-adrenal axis dysregulation in sexually abused girls. *Journal of Clinical Endocrinology and Metabolism*, **78**, 249–255.

de Kloet, E. R. (1991). Brain corticosteroid receptor balance and homeostatic control. *Frontiers in Neuroendocrinology*, **12**(2), 95–164.

Deacon, T. W. (1997). *The Symbolic Species: the Co-Evolution of Language and the Brain*. New York: Norton.

Deater-Deckard, K., Atzaba-Poria, N. and Pike, A. (2004). Mother- and father–child mutuality in Anglo and Indian British families: a link with lower externalizing behaviors. *Journal of Abnormal Child Psychology*, **32**(6), 609–620.

Del Giudice, M. (2009). Sex, attachment, and the development of reproductive strategies. *Behavioral and Brain Sciences*, **32**, 1–21.

Dressler, W. and Bindon, J. R. (2000). The health consequences of cultural consonance: cultural dimensions of lifestyle, social support, and arterial blood pressure in an African American community. *American Anthropologist*, **102**(2), 244–260.

Dunbar, R. I. M. (1998). The social brain hypothesis. *Evolutionary Anthropology*, **6**, 178–190.

Dunn, A. J. (1995). Interactions between the nervous system and the immune system: implications for psychopharmacology. In *Psychopharmacology: the Fourth Generation of Progress*, F. R. Bloom and D. J. Kupfer (eds). New York: Raven Press.

Dunn, J. (2004). Understanding children's family worlds: family transitions and children's outcome. *Merrill-Palmer Quarterly*, **50**(3), 224–235.

Edelman, G. M. (2006). *Second Nature: Brain Science and Human Knowledge*. New Haven: Yale University Press.

Ellison, P. (1988). Human salivary steroids: methodological considerations and applications in physical anthropology. *Yearbook of Physical Anthropology*, **31**, 115–142.

Ellison, P. T. (2001). *On Fertile Ground, a Natural History of Human Reproduction*. Cambridge, MA: Harvard University Press.

Finkelhor, D. and Dzuiba-Leatherman, J. (1994). Victimization of children. *American Psychologist*, **49**(3), 173–183.

Fisher, H. E., Aron, A., Mashek, D., et al. (2002). Defining the brain systems of lust, romantic attraction and attachment. *Archives of Sexual Behavior*, **31**(5), 413–419.

Fisher, H. E., Aron, A. and Brown, L. L. (2006). Romantic love: a mammalian brain system for mate choice. *Philosophical Transactions of the Royal Society of London. Series B*, **361**, 2173–2186.

Fisher, R. A. (1930). *The Genetical Theory of Natural Selection*. Oxford: Clarendon Press.

Fisher, S. E. (2005). On genes, speech, and language. *New England Journal of Medicine*, **353**, 1655–1657.

Fleming, A. S., Steiner, M. and Corter, C. (1997). Cortisol, hedonics, and maternal responsiveness in human mothers. *Hormones and Behavior*, **32**(2), 85–98.

Fleming, A. S., O'Day, D. H. and Kraemer, G. W. (1999). Neurobiology of mother–infant interactions: experience and central nervous system plasticity across development and generations. *Neuroscience and Biobehavioral Reviews*, **23**, 673–685.

Fleming, A. S., Corter, C., Stallings, J., et al. (2002). Testosterone and prolactin are associated with emotional responses to infant cries in new fathers. *Hormones and Behavior*, **42**, 399–413.

Flinn, M. V. (1981). Uterine and agnatic kinship variability. In *Natural Selection and Social Behavior: Recent Research and New Theory*, R. D. Alexander and D. W. Tinkle (eds). New York: Blackwell Press, pp. 439–475.

Flinn, M. V. (1988). Step and genetic parent/offspring relationships in a Caribbean village. *Ethology and Sociobiology*, **9**(3), 1–34.

Flinn, M. V. (1992). Paternal care in a Caribbean village. In *Father–Child Relations: Cultural and Biosocial Contexts*, B. Hewlett (ed.). Hawthorne, NY: Aldine de Gruyter, pp. 57–84.

Flinn, M. V. (1997). Culture and the evolution of social learning. *Evolution and Human Behavior*, **18**(1), 23–67.

Flinn, M. V. (1999). Family environment, stress, and health during childhood. In *Hormones, Health, and Behavior*, C. Panter-Brick and C. Worthman (eds). Cambridge: Cambridge University Press, pp. 105–138.

Flinn, M. V. (2004). Culture and developmental plasticity: evolution of the social brain. In *Evolutionary Perspectives on Child Development*, K. MacDonald and R. L. Burgess (eds). Thousand Oaks, CA: Sage, pp. 73–98.

Flinn, M. V. (2006a). Cross-cultural universals and variations: the evolutionary paradox of informational novelty. *Psychological Inquiry*, **17**, 118–123.

Flinn, M. V. (2006b). Evolution and ontogeny of stress response to social challenge in the human child. *Developmental Review*, **26**, 138–174.

Flinn, M. V. (2009). Are cortisol profiles a stable trait during child development? *American Journal of Human Biology*, **21**(6), 769–771.

Flinn, M. V. and Alexander, R. D. (2007). Runaway social selection. In *The Evolution of Mind*, S. W. Gangestad and J. A. Simpson (eds). New York: Guilford Press, pp. 249–255.

Flinn, M. V. and Coe, K. C. (2007). The linked red queens of human cognition, coalitions, and culture. In *The Evolution of Mind*, S. W. Gangestad and J. A. Simpson (eds). New York: Guilford Press, pp. 339–347.

Flinn, M. V. and England, B. G. (1995). Family environment and childhood stress. *Current Anthropology*, **36**(5), 854–866.

Flinn, M. V. and England, B. G. (1997). Social economics of childhood glucocorticoid stress response and health. *American Journal of Physical Anthropology*, **102**, 33–53.

Flinn, M. V. and England, B. G. (2003). Childhood stress: endocrine and immune responses to psychosocial events. In *Social and Cultural Lives of Immune Systems*, J. M. Wilce (ed.). London: Routledge Press, pp. 107–147.

Flinn, M. V. and Leone, D. V. (2006). Early trauma and the ontogeny of glucocorticoid stress response: grandmother as a secure base. *Journal of Developmental Processes*, **1**(1), 31–68.

Flinn, M. V. and Leone, D. V. (2009). Alloparental care and the ontogeny of glucocorticoid stress response among stepchildren. In *Substitute Parents*, G. Bentley and R. Mace (eds). Oxford: Berghahn Books, pp. 266–286.

Flinn, M. V., Quinlan, R., Turner, M. T., et al. (1996). Male–female differences in effects of parental absence on glucocorticoid stress response. *Human Nature*, **7**(2), 125–162.

Flinn, M. V., Geary, D. C. and Ward, C. V. (2005a). Ecological dominance, social competition, and coalitionary arms races: why humans evolved extraordinary intelligence. *Evolution and Human Behavior*, **26**(1), 10–46.

Flinn, M. V., Ward, C. V. and Noone, R. (2005b). Hormones and the human family. In *Handbook of Evolutionary Psychology*, D. Buss (ed.). New York: Wiley, pp. 552–580.

Flinn, M. V., Muehlenbein, M. P. and Ponzi, D. (2009). Evolution of neuroendocrine mechanisms linking attachment and life history: the social neuroendocrinology of middle childhood. *Behavioral and Brain Sciences*, **32**(1), 27–28.

Fortes, M. (1969). *Kinship and the Social Order*. Chicago, IL: Aldine de Gruyter.

Fox, N. A., Hane, A. A. and Pine, D. S. (2007). Plasticity for affective neurocircuitry: how the environment affects gene expression. *Current Directions in Psychological Science*, **16**, 1–5.

Fuchs, E. and Flugge, G. (1995). Modulation of binding sites for corticotropin-releasing hormone by chronic psychosocial stress. *Psychoneuroendocrinology*, **30**(1), 33–51.

Gavrilets, S. and Vose, A. (2006). The dynamics of Machiavellian intelligence. *Proceedings of the National Academy of Sciences of the United States of America*, **103**(45), 16823–16828.

Geary, D. C. (2005). *The Origin of Mind: Evolution of Brain, Cognition, and General Intelligence*. Washington, DC: American Psychological Association.

Geary, D. C. and Bjorklund, D. F. (2000). Evolutionary developmental psychology. *Child Development*, **71**(1), 57–65.

Geary, D. C. and Flinn, M. V. (2001). Evolution of human parental behavior and the human family. *Parenting: Science and Practice*, **1**, 5–61.

Geary, D. C. and Flinn, M. V. (2002). Sex differences in behavioral and hormonal response to social threat. *Psychological Review*, **109**(4), 745–750.

Geary, D. C. and Huffman, K. J. (2002). Brain and cognitive evolution: forms of modularity and functions of mind. *Psychological Bulletin*, **128**(5), 667–698.

Geary, D. C., Byrd-Craven, J., Hoard, M. K., et al. (2003). Evolution and development of boys' social behavior. *Developmental Review*, **23**, 444–470.

Ghazanfar, A. A. and Santos, L. R. (2004). Primate brains in the wild: the sensory bases for social interactions. *Nature Reviews Neuroscience*, **5**(8), 603–616.

Gilbert, P. (2001). Evolutionary approaches to psychopathology: the role of natural defences. *Australian and New Zealand Journal of Psychiatry*, **35**(1), 17–27.

Gilbert, S. L., Dobyns, W. B. and Lahn, B. T. (2005). Genetic links between brain development and brain evolution. *Nature Reviews Genetics*, **6**(7), 581–590.

Glaser, R. and Kiecolt-Glaser, J. K. (eds) (1994). *Handbook of Human Stress and Immunity*. New York: Academic Press.

Goodall, J. (1986). *The Chimpanzees of Gombe: Patterns of Behavior*. Cambridge, MA: Belknap Press of Harvard University Press.

Gopnik, A., Meltzoff, A. N. and Kuhl, P. K. (1999). *The Scientist in the Crib: Minds, Brains, and How Children Learn.* New York: William Morrow and Co.

Gottman, J. M. and Katz, L. F. (1989). Effects of marital discord on young children's peer interaction and health. *Developmental Psychology*, **25**(3), 373–381.

Gray, P. B. and Campbell, B. C. (2009). Human male testosterone, pair bonding and fatherhood. In *Endocrinology of Social Relationships*, P. B. Gray and P. T. Ellison (eds). Cambridge: Harvard University Press.

Gray, P. B., Parkin, J. C. and Samms-Vaughan, M. E. (2007). Hormonal correlates of human paternal interactions: a hospital-based investigation in urban Jamaica. *Hormones and Behavior*, **52**, 499–507.

Gubernick, D. (1996). A natural family system. *Family Systems*, **3**, 109–124.

Gunnar, M., Porter, F. L., Wolf, C. M., et al. (1995). Neonatal stress reactivity: predictions to later emotional temperament. *Child Development*, **66**, 1–13.

Hamilton, W. D., Axelrod, R. and Tanese, R. (1990). Sexual reproduction as an adaptation to resist parasites (a review). *Proceedings of the National Academy of Sciences of the United States of America*, **87**, 3566–3573.

Hawkes, K. (2003). Grandmothers and the evolution of human longevity. *American Journal of Human Biology*, **15**, 380–400.

Heim, C., Newport, D., Wagner, D., et al. (2002). The role of early adverse experience and adulthood stress in the prediction of neuroendocrine stress reactivity in women: a multiple regression analysis. *Depression and Anxiety*, **15**, 117–125.

Heim, C. and Nemeroff, C. (2001). The role of childhood trauma in the neurobiology of mood and anxiety disorders: preclinical and clinical studies. *Society of Biological Psychiatry*, **49**, 1023–1039.

Heinrichs, M. and Domes, G. (2008). Neuropeptides and social behaviour: effects of oxytocin and vasopressin in humans. *Progress in Brain Research*, **170**, 337–350.

Henry, J. P. and Wang, S. (1998). Effect of early stress on adult affiliative behavior. *Psychoneuroendocrinology*, **23**(8), 863–875.

Herbert, T. B. and Cohen, S. (1993). Stress and immunity in humans: a meta-analytic review. *Psychosomatic Medicine*, **55**, 364–379.

Hertsgaard, L., Gunnar, M., Erickson, M. F., et al. (1995). Adrenocortical responses to the strange situation in infants with disorganized/disoriented attachment relationships. *Child Development*, **66**, 1100–1106.

Hertzman, C. (1999). The biological embedding of early experience and its effects on health in adulthood. *Annals of the New York Academy of Sciences*, **896**, 85–95.

Hetherington, E. M. (2003a). Intimate pathways: changing patterns in close personal relationships across time. *Family Relations: Interdisciplinary Journal of Applied Family Studies*, **52**(4), 318–331.

Hetherington, E. M. (2003b). Social support and the adjustment of children in divorced and remarried families. *Childhood: a Global Journal of Child Research*, **10**(2), 217–236.

Hrdy, S. B. (1999). *Mother Nature: a History of Mothers, Infants, and Natural Selection.* New York: Pantheon.

Hrdy, S. B. (2005). Evolutionary context of human development: the cooperative breeding model. In *Attachment and Bonding: a New Synthesis*, C. S. Carter and L. Ahnert (eds). Cambridge, MA: MIT Press, pp. 9–32.

Hrdy, S. B. (2009). *Mothers and Others: The Evolutionary Origins of Mutual Understanding.* Cambridge, MA: Harvard University Press.

Huether, G. (1996). The central adaptation syndrome: psychosocial stress as a trigger for adaptive modifications of brain structure and brain function. *Progress in Neurobiology*, **48**, 568–612.

Huether, G. (1998). Stress and the adaptive self organization of neuronal connectivity during early childhood. *International Journal of Developmental Neuroscience*, **16**(3/4), 297–306.

Hurtado, A. M. and Hill, K. R. (1992). Paternal effect on offspring survivorship among Ache and Hiwi hunter-gatherers: implications for modeling pair-bond stability. In *Father–Child Relations: Cultural and Biosocial Contexts*, B. Hewlett (ed.). Hawthorne, NY: Aldine de Gruyter, pp. 31–55.

Insel, T. R. and Young, L. R. (2001). The neurobiology of attachment. *Nature Reviews Neuroscience*, **2**, 129–136.

Ivanovici, A. M. and Wiebe, W. J. (1981). Towards a working "definition" of "stress": a review and critique. In *Stress Effects on Natural Ecosystems*, G. W. Barrett and R. Rosenberg (eds). New York: Wiley, pp. 13–17.

Jensen, A. R. (1998). *The G Factor: the Science of Mental Ability.* New York: Praeger.

Joffe, T. H. (1997). Social pressures have selected for an extended juvenile period in primates. *Journal of Human Evolution*, **32**, 593–605.

Kagan, J. (1984). *The Nature of the Child.* New York: Basic Books.

Kagan, J. (1992). Behavior, biology, and the meanings of temperamental constructs. *Pediatrics*, **90**, 510–513.

Kagan, J., Resnick, J. S. and Snidman, N. (1988). The biological basis of childhood shyness. *Science*, **240**, 167–171.

Kaplan, H. S. and Robson, A. J. (2002). The emergence of humans: the coevolution of intelligence and longevity with intergenerational transfers. *Proceedings of the National Academy of Sciences of the United States of America*, **99**(15), 10221–10226.

Kaplan, H., Hill, K., Lancaster, J., et al. (2000). A theory of human life history evolution: diet, intelligence and longevity. *Evolutionary Anthropology*, **9**, 156–183.

Kirschbaum, C. and Hellhammer, D. H. (1994). Salivary cortisol in psychneuroendocrine research: recent developments and applications. *Psychoneuroendocrinology*, **19**, 313–333.

Koch, C. (1999). *Biophysics of Computation. Information Processing in Single Neurons.* New York: Oxford University Press.

Kolassa, I.-T. and Elbert, T. (2007). Structural and functional neuroplasticity in relation to traumatic stress. *Current Directions in Psychological Science*, **16**, 321–325.

Konner, M. (1991). *Childhood.* Boston, MA: Little, Brown and Co.

Koolhaas, J. M., de Boer, S. F. and Buwalda, B. (2006). Stress and adaptation: toward ecologically relevant animal models. *Current Directions in Psychological Science*, **15**, 109–112.

Korte, S. M., Koolhaas, J. M., Wingfield, J. C., et al. (2005). The Darwinian concept of stress: benefits of allostasis and costs of allostatic load and the trade-offs in health and disease. *Neuroscience and Biobehavioral Reviews*, **29**(1), 3–38.

Laland, K. N., Odling-Smee, J. and Feldman, M. W. (2000). Niche construction, biological evolution, and cultural change. *Behavioral and Brain Sciences*, **23**, 131–175.

Lamb, M., Pleck, J., Charnov, E., et al. (1987). A biosocial perspective on paternal behavior and involvement. In *Parenting Across the Lifespan: Biosocial Dimensions*, J. B. Lancaster, J. Altmann, A. Rossi, et al. (eds). Hawthorne, NY: Aldine de Gruyter, pp. 111–142.

LeBlanc, S. A. (2003). *Constant Battles: the Myth of the Peaceful, Noble Savage*. New York: St. Martin's Press.

Lee, H.-J., Macbeth, A. H., Pagani, J. H., et al. (2009). Oxytocin: the great facilitator of life. *Progress in Neurobiology*, **88**(2), 127–151.

Lee, S. H. and Wolpoff, M. H. (2003). The pattern of evolution in Pleistocene human brain size. *Paleobiology*, **29**, 186–196.

Leigh, S. R. (2004). Brain growth, cognition, and life history in primate and human evolution. *American Journal of Primatology*, **62**, 139–164.

Leslie, A. M., Friedmann, O. and German, T. P. (2004). Core mechanisms in "theory of mind." *Trends in Cognitive Sciences*, **8**(12), 529–533.

Lim, M. M., Wang, Z., Olazabal, D. E., et al. (2004). Enhanced partner preference in a promiscuous species by manipulating the expression of a single gene. *Nature*, **429**, 754–757.

Long, B., Ungpakorn, G. and Harrison, G. A. (1993). Home-school differences in stress hormone levels in a group of Oxford primary school children. *Journal of Biosocial Sciences*, **25**, 73–78.

Lupien, S. J. (2009). Brains under stress. *Canadian Journal of Psychiatry – Revue Canadienne de Psychiatrie*, **54**(1), 4–5.

MacDonald, K. and Hershberger, S. L. (2005). Theoretical issues in the study of evolution and development. In *Evolutionary Perspectives on Human Development*, R. L. Burgess and K. MacDonald (eds). Thousand Oaks, CA: Sage, pp. 21–72.

Maier, S. F., Watkins, L. R. and Fleschner, M. (1994). Psychoneuroimmunology: the interface between behavior, brain, and immunity. *American Psychologist*, **49**, 1004–1007.

Mason, J. W. (1968). A review of psychoendocrine research on the pituitary-adrenal cortical system. *Psychosomatic Medicine*, **30**, 576–607.

Mason, J. W. (1971). A re-evaluation of the concept of "non-specificity" in stress theory. *Journal of Psychosomatic Research*, **8**, 323–334.

McEwen, B. S. (1995). Stressful experience, brain, and emotions: developmental, genetic, and hormonal influences. In *The Cognitive Neurosciences*, M. S. Gazzaniga (ed.). Cambridge, MA: MIT Press, pp. 1117–1135.

McEwen, B. S. (1998). Protective and damaging effects of stress mediators. *New England Journal of Medicine*, **338**, 171–179.

Meaney, M. J., Mitchell, J., Aitken D., et al. (1991). The effects of neonatal handling on the development of the adrenocortical response to stress: implications for neuropathology and cognitive deficits later in life. *Psychoneuroendocrinology*, **16**, 85–103.

Meaney, M. J., Szyf, M. and Seckl, J. R. (2007). Epigenetic mechanisms of perinatal programming of hypothalamic-pituitary-adrenal function and health. *Trends in Molecular Medicine*, **13**(7), 269–277.

Miller, G. E. (2000). *The Mating Mind: How Sexual Choice Shaped the Evolution of Human Nature*. New York: Doubleday.

Mirescu, C., Peters, J. D. and Gould, E. (2004). Early life experience alters response of adult neurogenesis to stress. *Nature Reviews: Neuroscience*, **7**(8), 841–846.

Moll, J., Zahn, R., de Oliveira-Souza, R., et al. (2005). The neural basis of human moral cognition. *Nature Reviews: Neuroscience*, **6**(10), 799–809.

Muehlenbein, M. P. (2008). Adaptive variation in testosterone levels in response to immune activation: empirical and theoretical perspectives. *Social Biology*, **53**, 13–23.

Muehlenbein, M. P. and Bribiescas, R. G. (2005). Testosterone-mediated immune functions and male life histories. *American Journal of Human Biology*, **17**, 527–558.

Munck, A. and Guyre, P. M. (1991). Glucocorticoids and their immune function. In *Psychoneuroimmunology*, R. Ader, D. L. Felten, and N. Cohen (eds). New York: Academic Press, pp. 447–474.

Munck, A., Guyre, P. M. and Holbrook, N. J. (1984). Physiological functions of glucocorticoids in stress and their relation to pharmacological actions. *Endocrine Reviews*, **5**, 25–44.

Murdock, G. P. (1949). *Social Structure*. New York: Macmillan.

Murdock, G. P. (1967). *Ethnographic Atlas*. Pittsburgh, PA: University of Pittsburgh Press.

Nachmias, M., Gunnar, M., Mangelsdorf, S., et al. (1996). Behavioral inhibition and stress reactivity: the moderating role of attachment security. *Child Development*, **67**, 508–522.

Newcomer, J. W., Craft, S., Hershey, T., et al. (1994). Glucocorticoid-induced impairment in declarative memory performance in adult humans. *Journal of Neuroscience*, **14**(4), 2047–2053.

Nowak, M. A., Komarova, N. L. and Niyogi, P. (2001). Evolution of universal grammar. *Science*, **291**, 114–118.

Pande, H., Unwin, C. and Haheim, L. L. (1997). Factors associated with the duration of breastfeeding. *Acta Paediatrica*, **86**, 173–177.

Panksepp, J. (2004). *Affective Neuroscience: the Foundations of Human and Animal Emotions*. New York: Oxford University Press.

Panksepp, J. (2009). Carving "natural" emotions: "kindly" from bottom-up but not top-down. *Journal of Theoretical and Philosophical Psychology*, **28**(2), 395–422.

Pearlin, L. I. and Turner, H. A. (1987). The family as a context of the stress process. In *Stress and Health: Issues in Research Methodology*, S. V. Kasl and C. L. Cooper (eds). New York: Wiley, pp. 143–165.

Pinker, S. (1994). *The Language Instinct*. New York: William Morrow.

Pinker, S. (1997). *How the Mind Works*. New York: Norton.

Rosenberg, K. (2004). Living longer: information revolution, population expansion, and modern human origins. *Proceedings of the National Academy of Sciences of the United States of America*, **101**(30), 10847–10848.

Rosenberg, K. and Trevathan, W. (2002). Birth, obstetrics and human evolution. *BJOG: An International Journal of Obstetrics and Gynecology*, **109**(11), 1199–1206.

Roth, G. and Dicke, U. (2005). Evolution of the brain and intelligence. *Trends in Cognitive Sciences*, **9**(5), 250–257.

Sakai, K. L. (2005). Language acquisition and brain development. *Science*, **310**, 815–819.

Saphier, D., Welch, J. E., Farrar, G. E., et al. (1994). Interactions between serotonin, thyrotropin-releasing hormone and substance P in the CNS regulation of adrenocortical secretion. *Psychoneuroendocrinology*, **19**, 779–797.

Sapolsky, R. M. (1991). Effects of stress and glucocorticoids on hippocampal neuronal survival. In *Stress: Neurobiology and Neuroendocrinology*, M. R. Brown, G. F. Koob and C. Rivier (eds). New York: Dekker, pp. 293–322.

Sapolsky, R. M. (1992). *Stress, the Aging Brain, and the Mechanisms of Neuron Death*. Cambridge, MA: MIT Press.

Sapolsky, R. M. (1994). *Why Zebras Don't Get Ulcers*. New York: W. H. Freeman and Co.

Sapolsky, R. M. (2005). The influence of social hierarchy on primate health. *Science*, **308**(5722), 648–652.

Sapolsky, R. M., Romero, L. M. and Munck, A. U. (2000). How do glucocorticoids influence stress responses? *Endocrine Reviews*, **21**(1), 55–89.

Schneider, M. L., Coe, C. L. and Lubach, G. R. (1992). Endocrine activation mimics the adverse effects of prenatal stress on the neuromotor development of the infant primate. *Developmental Psychobiology*, **25**, 427–439.

Sear, R., Mace, R. and McGregor, I. A. (2000). Maternal grandmothers improve the nutritional status and survival of children in rural Gambia. *Proceedings of the Royal Society of London. Series B*, **267**, 1641–1647.

Seckl, J. R. (2008). Glucocorticoids, developmental "programming" and the risk of affective dysfunction. *Progress in Brain Research*, **167**, 17–34.

Selye, H. (1976). *The Stress of Life*. New York: McGraw-Hill.

Servan-Schreiber, D., Printz, H. and Cohen, S. D. (1990). A network model of catecholamine effects: gain, signal-to-noise ratio, and behavior. *Science*, **249**, 892–895.

Shamay-Tsoory, S. G., Tomer, R. and Aharon-Peretz, J. (2005). The neuroanatomical basis of understanding sarcasm and its relationship to social cognition. *Neuropsychology*, **19**(3), 288–300.

Siegal, M. and Varley, R. (2002). Neural systems involved with "Theory of Mind." *Nature Reviews Neuroscience*, **3**, 463–471.

Silk, J. S., Alberts, S. C. and Altmann, J. (2003). Social bonds of female baboons enhance infant survival. *Science*, **302**, 1231–1234.

Simons, K., Paternite, C. E. and Shore, C. (2001). Quality of parent/adolescent attachment and aggression in young adolescents. *Journal of Early Adolescence*, **21**, 182–203.

Smith, B. H. (1994). Patterns of dental development in homo, Australopithecus, pan, and gorilla. *American Journal of Physical Anthropology*, **94**(3), 307–325.

Smuts, B. B. and Smuts, R. W. (1993). Male aggression and sexual coercion of females in nonhuman primates and other mammals: evidence and theoretical implications. *Advances in the Study of Behavior*, **22**, 1–63.

Stearns, S. C. (1992). *The Evolution of Life Histories*. Oxford: Oxford University Press.

Storey, A. E., Walsh, C. J., Quinton, R. L., et al. (2000). Hormonal correlates of paternal responsiveness in new and expectant fathers. *Evolution and Human Behavior*, **21**(2), 79–95.

Tinbergen, N. (1963). On the aims and methods of ethology. *Zeitschrift für Tierpsychologie*, **20**, 410–463.

Tomasello, M. (1999). *The Cultural Origins of Human Cognition*. Cambridge, MA: Harvard University Press.

Tooby, J. and Cosmides, L. (1992). The psychological foundations of culture. In *The Adapted Mind*, J. H. Barkow, L. Cosmides and J. Tooby (eds). Oxford: Oxford University Press, pp. 19–36.

Tulving, E. (2002). Episodic memory: from mind to brain. *Annual Review of Psychology*, **53**, 1–25.

Uvnas-Moberg, K. (1998). Oxytocin may mediate the benefits of positive social interaction and emotions. *Psychoneuroendocrinology*, **23**, 819–835.

van Anders, S. M. and Gray, P. B. (2007). Hormones and human partnering. *Annual Review of Sex Research*, **18**, 60–93.

van der Meij, L., Buunk, A. P., van de Sande, J. P., et al. (2008). The presence of a woman increases testosterone in aggressive dominant men. *Hormones and Behavior*, **54**, 640–644.

Wallerstein, J. S. (1983). Children of divorce: stress and developmental tasks. In *Stress, Coping, and Development in Children*, N. Garmezy and M. Rutter (eds). New York: McGraw-Hill, pp. 265–302.

Weaver, I. C. G., Cervoni, N., Champagne, F. S., et al. (2004). Epigenetic programming by maternal behavior. *Nature Reviews: Neuroscience*, **7**(8), 847–854.

Weiner, H. (1992). *Perturbing the Organism*. Chicago: University of Chicago Press.

West-Eberhard, M. J. (1983). Sexual selection, social competition, and speciation. *Quarterly Review of Biology*, **58**, 155–183.

West-Eberhard, M. J. (2003). *Developmental Plasticity and Evolution*. Oxford: Oxford University Press.

Whiting, B. B. and Edwards, C. (1988). *Children of Different Worlds*. Cambridge, MA: Harvard University Press.

Wilkinson, R. G. (2001). *Mind the Gap: Hierarchies, Health, and Human Evolution*. New Haven, CT: Yale University Press.

Williams, G. C. (1966). *Adaptation and Natural Selection*. Princeton: Princeton University Press.

Williams, R. W. and Herrup, K. (1988). The control of neuron number. *Annual Review of Neuroscience*, **11**, 423–453.

Wilson, M. I., Daly, M. and Weghorst, S. J. (1980). Household composition and the risk of child abuse and neglect. *Journal of Biosocial Sciences*, **12**, 333–340.

Wrangham, R. W. (1999). Evolution of coalitionary killing. *Yearbook of Physical Anthropology*, **42**, 1–30.

Wrangham, R. W. and Peterson, D. (1996). *Demonic Males*. New York: Houghton Mifflin Company.

Wynne-Edwards, K. E. (2003). From dwarf hamster to daddy: The intersection of ecology, evolution, and physiology that produces paternal behavior. In *Advances in the Study of Behavior*, P. J. B. Slater, J. S. Rosenblatt, C. T. Snowden et al. (eds). San Diego, CA: Academic Press, **32**, pp. 207–261.

Young, L. J. and Insel, T. R. (2002). Hormones and parental behavior. In *Behavioral endocrinology*, J. B. Becker, S. M. Breedlove, D. Crews, et al. (eds). Cambridge, MA: MIT Press, pp. 331–369.

Young, L., Wang, Z. and Insel, T. R. (2002). Neuroendocrine bases of monogamy. In *Foundations in Social Neuroscience*, J. T. Cacioppo, G. G. Berntson, R. Adolphs, et al. (eds). Cambridge, MA: MIT Press, pp. 809–816.

Yuwiler, A. (1982). Biobehavioral consequences of experimental early life stress: effects of neonatal hormones on monoaminergic systems. In *Critical Issues in Behavioral Medicine*, L. J. West and M. Stein (eds). Philadelphia: J. P. Lippincott, pp. 59–78.

Ziegler, T. E. and Snowdon, C. T. (1997). Role of prolactin in paternal care in a monogamous New World primate, *Saguinus oedipus*. The integrative neurobiology of affiliation. *Annals of the New York Academy of Sciences*, **807**, 599–601.

25 Human Biology, Energetics, and the Human Brain

Benjamin C. Campbell

INTRODUCTION

Human biology has made great progress in applying modern techniques to the understanding of human biological variation at the genetic, physiological, developmental, and phenotypic levels. For instance, rather than only asking about disease symptoms, human biologists measure immune makers which represent an underlying element of health (McDade et al., 2005; Muehlenbein et al., 2005; Snodgrass et al., 2007); rather than simply measuring body fat, they determine leptin levels as a signal of energy stores (Bribiescas, 2005). In addition to collecting self-reports of stress, they assay for salivary cortisol (Nepomnaschy et al., 2006; Pike and Williams, 2006). Along with collecting reproductive histories, they measure gonadal steroids including estrogen, progesterone (Lipson and Ellison, 1996), and testosterone (Campbell et al., 2006).

At the same time, human biologists have also become more evolutionarily sophisticated as they have adopted a life history perspective (Hill, 1993; Kuzawa, 2007). There is a growing understanding that much of the variation in the phases of the life cycle, starting in utero, and moving through childhood, adolescence, adulthood, and aging represents an inter-related response to energetic availability. The impact of nutrition during fetal development, in particular may have important implications for the rest of the life cycle (Jasienska et al., 2006; Kuzawa, 2007). Furthermore, the understanding that growth and development, immune function, and reproductive function are responsive to energetics means that human biologists can use energy as a currency through which the basic functions of growth, maintenance, and reproduction can be traded off (Hill, 1993).

Yet human biologists have tended to avoid studying the brain (see Leonard et al., 2003, 2007 for an exception). Such shyness surely reflects the Cartesian dichotomy, in which the mind and body are fundamentally different spheres and hence must be studied using different techniques. At the same time, it may also reflect the shear complexity of the brain. As the organ of behavior, the brain seems to have an almost infinitely flexible phenotypic expression. In contrast, the standard methods of human biology focus on the measurement of relatively simple phenotypes such as the size and shape of anatomical features and variation in their developmental timing. These can be easily quantified using methods suitable for the field and compared across populations.

However, in terms of the brain, simple and robust phenotypes are hard to come by. Measures of brain size and shape are not accessible in the samples of living human beings favored by human biologists, even if we knew how to interpret them. On the other hand, proxy measures such as skull size and shape are eminently measurable, but variation across human populations reflect climatic variation, not brain function (Beals et al., 1984).

Thus some other simple and robust metric for measuring brain function is necessary. An energetic approach has at least two advantages. Firstly, the brain is energetically expensive with 2% of body mass using 20% of energy among humans (Elia, 1992), increasing the probability of selection based on energetic constraints. These include trade-offs within the brain itself as well as somatic maintenance and growth as outlined by life history theory. Secondly, an energetic perspective is consistent with an emerging focus on energetics in human evolution more generally (Aiello and Wheeler, 1995; Leonard and Ulijaszek, 2002; Leonard et al., 2007).

In what follows I attempt to provide a sufficient outline of human brain metabolism, including energy utilization, storage, and substrate availability, and its association with brain development to be able to consider its implications for human evolution. In the first part of the paper, I set the stage with a brief review of earlier work on brain metabolism and evolution. I then review recent advances in assessing energy utilization in the brain which highlight the intrinsic and intimate interplay between energetic utilization and neuronal function. I argue that the elevated energetic costs of the human brain are balanced against the benefit

Human Evolutionary Biology, ed. Michael P. Muehlenbein. Published by Cambridge University Press. © Cambridge University Press 2010.

of increased neuroplasticity. I suggest that glucose utilization in the human brain is directly tied to synaptic plasticity through a process referred to as glutamate cycling, and that a high glutamate flux is important in human neuroplasticity.

In the second part of the paper I review the recent literature showing changes in the energetic utilization of the human brain during development. I argue that changes in energy utilization in the developing brain are closely tied to on-going processes of brain development. More specifically, I suggest that age-related changes in glucose utilization (Chugani, 1998) are consistent with development patterns of neuroplasticity, social context, and behavioral development in humans. I hope to show that rather than representing a release from energetic constraints, the development of the human brain becomes a focal point for the selective power of energetic constraints during evolution.

HISTORICAL BACKGROUND

Early interest in human brain metabolism and evolution grew out of attempts to understand the basis for the allometric relationship between brain and body across mammals (Martin, 1981; Armstrong, 1983, 1985). When the scaling coefficient of brain to body size was thought to be around 0.67, it was suggested to reflect the relationship between neural sensory input/output function and body surface area. However, Armstrong (1983) demonstrated that variation in brain size across mammals is related to the relative amount of energy utilized by the brain, i.e., its scales with the amount of energy reserves (Armstrong, 1985), clearly suggesting the important of metabolism in brain evolution. Interestingly Crile (1941) had anticipated the importance of metabolism in brain evolution and argued that the importance of basal metabolism to brain size must have coevolved with the thyroid function.

When other analyses indicated that the scaling coefficient was closer to 0.75, than 0.67, arguments about body surface area no longer made sense. Instead, Martin (1981) suggested that the underlying relationship might reflect constraints on maternal metabolic investment in fetal brain development. More specifically, in terms of human evolution, Martin (1989) has argued that australopithecines demonstrate increased brain to body size ratios relative to the great apes. He then suggests that a reduction in gut along with an increase in high quality food would have been necessary to allow for the increased energetic demands of the brain. It is this idea that was taken up by Aiello and Wheeler (1995) and dubbed the "expensive tissue hypothesis." In what follows I explore in more detail why the brain is expensive tissue.

OVERVIEW OF ENERGETICS AND THE BRAIN

It is well known to human biologists that the brain is an energetically expensive organ, consuming 20% of the body's energy, despite being only 2% of the body's weight (Elia, 1992). But this figure only superficially reflects how pervasive the effects of energetic factors are in brain function. The brain depends almost entirely on glucose as a metabolic fuel (Van Itallie and Nufert, 2003) and accounts for 50% of total body glucose utilization (Fehm et al., 2006). Furthermore, brain structures such as the hippocampus and amygdala known for their role in basic functions of memory and emotional regulation are also important in regulation of somatic energy use and food intake (Fehm et al., 2006). Thus, brain activity is only tied to energy consumption, but with regulating its own energy status (Peters et al., 2004).

In order to understand the role of metabolism in brain function, it is useful to first characterize the brain on a more global level. Neurobiologists list several qualities of the brain that may be important. These include: (1) the brain is very plastic, i.e., capable of changing; (2) it has little storage, i.e., there is little evidence for glycogen storage in neurons; (3) is substrate specific, i.e., it will only metabolize glucose; (4) is separated from the body by the blood–brain barrier reducing its exposure to the general circulation (Peters et al., 2004). To these I would add: (5) the brain is energetically demanding (Armstrong, 1985; Fehm et al., 2006), i.e., it is always active and can not survive long if its energy supply is interrupted.

Brain function is a metabolically dynamic process, with specific brain activity dependent on increased energy utilization (Shulman et al., 2004). Neural plasticity, the alteration of neural connections based on experience, enhances the ability of the brain to adapt to the environmental by directing that energy toward the most commonly used neuronal pathways and neural circuits. In contrast, the other four qualities listed above; energetically demanding, little storage, substrate specificity, and the blood–brain barrier, all act to limit the physiological range in which the brain can operate. Thus the trade-off between energy and brain function in humans can been seen as a basic function of energy utilization and the maintenance of neural plasticity.

The cellular basis of both energy utilization and neural plasticity has come into clearer focus in recent years. Though still controversial (Pellerin et al., 2007), recent findings have been taken to suggest a metabolic cycle between astrocytes and neurons in which glucose is used by astrocytes to convert glutamate to glutamine. In the process glucose is transformed into lactate which is consumed by neurons (Magistretti, 2006). This tight metabolic coupling of neurons and astrocytes, referred to as the astrocyte-lactate neuronal shuttle appears central to understanding brain plasticity. Glutamate is critical to neuronal plasticity and its association with glucose directly links energetic processes and neural plasticity throughout the life cycle (Magistretti, 2006). In fact, Ullian et al. (2004) have suggested that astrocytes should be understood as regulators of synaptic plasticity.

Furthermore, it is now clear on the basis of both animal models (Gruetter, 2003; Brown, 2004) and human studies (Oz et al., 2007) that the brain does in fact have important glycogen stores, albeit at much lower levels than found in other tissues such as muscle. Such stores appear to vary across regions of the brain and are most abundant in more metabolically active parts of the brain such as the hippocampus (Dalsgaard et al., 2007). Thus glycogen is more likely to represent a local energy buffer during normal use (Brown, 2004; Brown and Ransom, 2007) than storage against pathological conditions such as global ischemia, as previously assumed.

While the blood–brain barrier remains an important reality and glucose is the major energy source transported across the blood–brain barrier (Peters et al., 2004), other substrates such as ketones and lactate can cross the blood–brain barrier (Emery, 2005), and may serve physiologically important functions in addition to their role as energy substrates. For instance, ketones appear to be important during early development when they are an important substrate in the production of lipids by oligodendrocytes (Edmond, 1992; Nehlig, 2004), including lipids associated with myelinization, so important to early brain development. In another example, transport of lactate into the brain is particularly important during exercise (Dalsgaard, 2006), suggesting that the brain does not always have total priority for glucose utilization and physiological mechanisms other than glucose metabolism may play a role in adjusting the allocation of energy between the brain and body.

Finally, there is growing evidence that the brain has cells that sense glucose (Rao et al., 2006) and react accordingly, allocating energy to the brain depending on its needs. Peters et al. (2004) have argued that because the brain has priority of allocation and controls other parts of the body, it is in control of its own energetic requirements. From a life history perspective, such a "selfish brain" should include potential mechanisms for shifting allocation of energy to brain versus the body as the importance of the brain relative to reproduction shifts over the life course.

ENERGY CONSUMPTION AND BRAIN ACTIVITY

Before considering the implications of specific aspects of energy metabolism for brain function, it is important to outline the processes that contribute to the total energetic costs of the brain. The development of functional magnetic resonance imaging (fMRI) has focused attention on energy utilization during the activation of particular brain regions (Shulman et al., 2004). However, such functional activation appears to represent a remarkably small fraction of the total energy consumed by the adult human brain (Raichle and Mintun, 2006). Based on both glucose and blood utilization studies it is estimated that activation of neuronal processes accounts for some 1% of energy utilization in the brain (Raichle and Mintun, 2006), the other 99% of the brain energy consumption is related to baseline or "resting" activity levels. Only 15% of total brain energy consumption is thought to be related to the maintenance of resting action potentials and glial cell activity (Attwell and Laughlin, 2001), leaving approximately 80% of brain energy consumption devoted to something else (Raichle and Gusnard, 2002).

That something else has been referred to as intrinsic brain activity (Raichle and Mintun, 2006). Intrinsic brain activity is not well defined, but is thought to represent a default brain system (Raichle et al., 2001) that is attenuated but not deactivated during a conscious task. Furthermore, this brain system appears to show spontaneous cycles of energy usage suggesting it is dynamically active (Fox et al., 2005; Mantini et al., 2007). The function of such a brain system may reflect stimulus independent thought (Mason et al., 2007) and may involve somatic self-monitoring (Raichle et al., 2001) and the processing of episodic memory (Greicius et al., 2003; Greicius and Menon, 2004). Similar brain activity appears to occur during sleep, making intrinsic brain activity on-going and independent of wakefulness (Horovitz et al., 2007).

High levels of on-going brain activity suggest that the energetic demands of the brain are both constant and dynamic. I argue that such high energy utilization and glucose dependence is directly related to synaptic plasticity at the cellular level, through the process of glutamate cycling (Shulman et al., 2004; Magistretti, 2006). Glutamate cycling appears to run at a higher rate in the brain of humans and great apes relative to other primate relatives (Burki and Kaessmann, 2004), thus allowing for higher local energy utilization and potentially greater neural plasticity. I elaborate the physiological details of this argument below.

The astrocyte-neuron-lactate shuttle

Recent work has suggested that most neurons do not metabolize glucose directly, but obtain their energy through an interaction with surrounding glial cells, a process referred to as the astrocyte-neuron-lactate shuttle (Bittar et al., 1996; Pellerin et al., 1998). As part of this cycle, glutamate released into the synapse is taken up by a glutamate transporter (EAAT2) and brought into astrocytes surrounding the synapse. There glucose is used to convert glutamate to glutamine through the act of glutamine synthetase (Daikhan and Yudkoff, 2000). The glutamine produced in the astrocyte is then taken up by the neuron. In the process, lactate is created within the astrocyte which is later taken up by the neuron and used to reconvert the glutamine to glutamate for release in the synapse (Magistretti, 2006). In addition to glutamate, a small fraction of the glutamine is converted into gamma-butyric acid (GABA) the major inhibitory transmitter in the brain (Patel et al., 2005).

As mentioned previously, the function of glutamate cycling has not been fully established, though it is been suggested as the basis for neural plasticity (Magistretti, 2006). It is generally thought that a build up of glutamate from the synapse into the extracellular space decreases the signal-to-noise ratio in the synapse and at extreme levels can lead to axonal depolarization and the death of neurons. By removing glutamate from the synapse and metabolizing it within the astrocyte neurons are less susceptible to glutamate excitotoxicity, allowing for increased strengthening of synaptic connections on the basis of neuronal firing and experience.

Rocher et al. (2003) show that levels of synaptophysin, a marker of synaptic density, are related to regional glucose utilization in baboons, supporting the idea that glucose metabolism is associated with neural connectivity. Thus the high rate of glucose utilization of the human brain may translate directly into the benefits of synaptic plasticity.

At the cellular level, synaptic plasticity is known to be associated with the growth and development of dendrites and dendritic spines (Calabrese et al., 2006). A greater proliferation of dendrites allows more room for dendritic spines. Since many synapses form on dendrite spines, more dendrites spines allow for more synapses as well. However, it is only recently that the dynamic nature of dendritic spine development has become clear. Small filaments on the dendrites emerge and start to grow to larger filaments (Ziv and Smith, 1996). Some of these larger filaments appear to persist and take the form of dendritic spines (Goda and Davis, 2003). If stabilized these spines then become synapses (Calabrese et al., 2006). Small filaments that do not become large filaments presumably die off, forming an ongoing cycle of production, growth, and loss.

The basis for the development of new synapses is not fully understood. However, N-methyl-D-aspartate (NMDA) receptors appear to be involved. In the case of dopaminergic neurons, NMDA receptors are important in trapping D_1 dopamine receptors into synapses (Scott et al., 2002), thus creating a dopaminergic synapse. Since NMDA is a glutamate receptor, increased glutamate flux may act to promote the stabilization of dendritic dopaminergic synapses. In addition, NMDA receptors are thought to be particularly important in the growth of dendritic spines during long-term potentiation (LTP) (Park et al., 2006), a basic neural mechanism underlying learning.

Regardless of the exact cellular mechanisms involved in neural plasticity, it is clear that glutamate cycling is very important in terms of brain energetics. Ninety percent of synapses release glutamate (Abeles, 1991; Braitenberg and Schuz, 1998), a figure that reflects the corelease of neurotransmitters by individual neurons (Trudeau and Gutiérrez, 2007). Eighty percent of the utilization of glucose is linearly related to glutamate cycling (Sibson et al., 1998; Shen et al., 1999). In addition, it has been estimated that about 20% of the energy consumed by glutamate cycling is used in the production of GABA (Patel et al., 2005), giving this neurotransmitter a smaller, but still potentially important, role in the relationship between energy consumption and brain function.

The importance of glutamate in human brain metabolism is illustrated by the recent discovery of a hominoid specific version of the glutamate dehydrogenase gene (GLUD). GLUD2 is a variant of GLUD found only in humans and the great apes (Burki and Kaessman, 2004) resulting in an increased capacity to oxidize glutamate under low oxygen conditions (Plaitakis et al., 2003), indicative of increased glutamate turnover. Since neural plasticity is based on the maintenance of synaptic connections, increased glutamate cycling among both humans and the great apes may be associated with a greater capacity for neural plasticity than other primates.

Energetics in specific parts of the brain

Given that 90% of all neurons express glutamate, glutamate cycling can be expected to play a role in maintaining neural plasticity throughout the brain. However, given its association with glucose utilization, the impact of glutamate cycling may be particularly apparent in more metabolically active parts of the brain, such as the hippocampus.

The existence of high levels of glycogen stores (Pelligri et al., 1996) in the hippocampus is consistent with a high local energy flux. Furthermore, in humans, there is a substantial literature demonstrating that glucose ingestion improves performance on memory-dependent, but not other, cognitive tasks (Meikle et al., 2004; Riby et al., 2006), a finding that is thought to reflect the effects of glucose on hippocampal function.

In terms of neuroplasticity, Segovia et al. (2006) report that environmental enrichment has an impact on the hippocampus through elevated levels of glutamate. In a sample of rats, environmental enrichment promoted both neurogenesis and increased levels of glutamate and GABA in the CA3 area of the hippocampus. While this finding is intriguing more research is needed to confirm a role for glutamate cycling in hippocampal function.

In addition to the hippocampus, the anterior cingulate cortex may also be particularly metabolically active as part of its role in executive function (Posner and Rothbart, 1998), the default brain network (Margulies et al., 2007) and as a way station between emotion and cognition (Allman et al., 2001). Even when the brain is not engaged in a focused task, the anterior cingulate cortex may be active in maintaining generalized attention to both external and internal signals. Recent work in humans has linked glutamate levels in the anterior cingulate cortex and hippocampus to sensation-seeking (Gallinat et al., 2007), consistent with an important role for glutamate cycling in these two brain structures. Again more research is needed to determine if glutamate cycling is associated with higher glucose consumption in the anterior cingulate cortex.

ENERGY STORAGE

It was once thought that the brain had practically no energy storage leaving it critically and globally dependent on an immediate supply of blood glucose. However, recent research has emphasized the existence of glycogen stores throughout the brain (Gruetter, 2003; Brown, 2004). The primary storage of glycogen is in astrocytes (Phelps, 1972), though some may be stored in neurons (Brown, 2004). Glycogen stores in the brain had been previously overlooked largely because of their low levels. At 0.1% of total brain weight (Brown, 2004) they are much lower than those of other tissues, for instance 20 times lower than that found in muscle (Oz et al., 2007).

It is widely thought that glycogen stores are sufficient to support brain function for only a few minutes (Clarke and Sokoloff, 1999). More recent work has extended that estimate to 100 minutes (Gruetter, 2003). This is rather surprising given the sensitivity of the brain to hypoxia and suggests that ultimately the limiting factor in brain function is not energy but oxygen. Global brain oxygen depletion, however, is not a normal condition. On the other hand,

hypoglycemia within local areas of the brain is a potential outcome of reduced glucose availability. Given the high demand for glucose throughout the brain hypoglycemia could occur on a very short time scale. Thus the intensity of short-term activity of specific regions of the brain may be directly related to local glycogen availability.

In the only human study to date, Oz et al. (2007) report that total glycogen stores represent three to four times the energy equivalents from glucose in brain circulation, suggesting that in terms of normal function glycogen stores are in fact rather substantial, and may play a key role in brain metabolism. On the other hand, Oz et al. (2007) also report negligible consumption of glycogen from neurons in the visual cortex during a 20 minute visual task, suggesting that glycogen stores may not be important in the course of normal brain activation and/or function.

In contrast, animal models indicate that glycogen is depleted under hypoglycemic conditions (Gruetter, 2003; Brown, 2004). Furthermore, during brain activation oxygen consumption does not initially increase, which has been taken to indicate that other sources of energy, i.e., glycogen are being utilized rather than direct oxidation of glucose (Raichle and Mintun, 2006). It is estimated that during brain activation in the rat glycogen stores are depleted by about 15%, suggesting that glycogen stores within astrocytes are used to support the initial costs of brain activation (Schurr et al., 1999; Shulman et al., 2001).

Furthermore, the more metabolically active parts of the brain have higher levels of glycogen storage and deplete those stores faster than less metabolically active regions (Brown, 2004). For instance, mouse cerebral cortex has a higher glycogen content than do deeper layers (Folbergrová et al., 1970) and the dentate gyrus of the hippocampus, the only area with active neurogenesis, has twice the glycogen content of the rest of the hippocampus (Lipton, 1989). These findings clearly suggest that glycogen stores act as an active buffer against glucose utilization during regional brain metabolism.

Energy storage and sleep

Sleep, with its drastic behavioral inhibition, provides one way of considering the role of energy metabolism, including that of glycogen, in brain function (Brown, 2004). Among humans, overall brain energy consumption as measured by oxygen uptake has been estimated to decrease by 3–11% during light sleep, and by 25–44% during slow wave sleep, but very little during REM sleep (Madsen and Virstrup, 1991; Madsen et al., 1991). The lack of decline during REM sleep emphasizes that the brain is in fact metabolically active, rather than quiescent, during dreaming.

Detailed studies in rats suggest that glycogen synthesis is dramatically increased during slow wave sleep relative to waking (Karnovsky et al., 1983). Furthermore, sleep deprivation leads to a decrease in glycogen in the frontal cortex of rats (Djuricic et al., 1977; Kong et al., 2002; though see Gip et al., 2002). This has led to the suggestion that the function of sleep may be the replenishment of glycogen (Bennington and Heller, 1995).

More recent work demonstrating variation in glycogen content in response to sleep deprivation in different strains of mice has been taken to suggest that accumulation of glycogen is not the primary function of sleep (Franken et al., 2003). However, to the extent that glycogen represents a short-term buffer against increased energy demands of neuronal activity, interference with glycogen metabolism may be an important intermediate in explaining the impact of sleep disruption on brain function (McEwen, 2006), and ultimately the function of sleep.

SUBSTRATE SPECIFICITY

As noted earlier, the brain was long considered to be entirely dependent on glucose as a fuel. From an evolutionary perspective, such dependence suggests that glucose metabolism may have been an important limiting factor in the evolution of the large human brain. While glucose can be produced from protein in the liver by the process of gluconeogenesis, the process is rather slow and glucose is most easily derived from the consumption of carbohydrates. As has been argued for polyunsaturated fatty acids (PUFAs) (Broadhurst et al., 2002, though see Carlson and Kingston, 2007 for an opposing perspective) or protein (Kennedy, 2005) the availability of carbohydrates could have been an important constraint in the evolution of the human brain. Thus consumption of underground storage organs, with relatively high starch composition, might have played a role in the evolution of the human brain (Yeakel et al., 2007).

However, in addition to glucose, the brain can also utilize ketones for fuel, and ketones have long been considered a back-up supply in the case of starvation when glucose is in short supply (Emery, 2005). While peripheral fat produces fatty acids which can be utilized by muscle, ketones are produced from abdominal fat suggesting that the development of a substantial abdominal fat depot in humans may be an important buffer for the human brain (Peters et al., 2004). This may be particularly important in infants (Kuzawa, 1998; Cunnane and Crawford, 2003).

The discovery of the neuronal-lactate shuttle suggests the possibility of a more finely tuned facultative use of alternative substrates by the human brain. While the brain can not run on lactate by itself, the utilization of lactate by neurons suggests that lactate can be substituted for glucose to sustain neurons in the short term. Such substitution is potentially important since lactate is a by-product of other energy processes in the body.

In fact, during vigorous exercise global brain glucose uptake declines, while the uptake of lactate by the brain increases, in inverse proportion (Kemppainen et al., 2005). At the same time lactate levels in the brain do not increase, suggesting utilization of lactate in neurons (Dalsgaard et al., 2004). The brain appears to utilize lactate produced by somatic effort to directly fuel the metabolism of neurons, by-passing the conversion of glucose or glycogen within astrocytes. Furthermore, glycogen stores increase during recovery from exercise (Dalsgaard, 2006), leading to the conclusion that they are depleted with the onset of neural activity associated with exercise, but not afterwards. Thus it appears that during strenuous exercise, glucose may be preferentially allocated to somatic energy uses over those of the brain, which is forced to deplete glycogen stores and utilize somatic metabolic wastes for its energetic requirements.

Such reallocation of energy substrates may have important implications for human brain evolution. Based on anatomical features, humans have been argued to have adopted endurance running as a basic strategy for hunting (Bramble and Lieberman, 2005; Lieberman and Bramble, 2007) and hence energy acquisition. If so, running in the tropical heat would have placed additional energy demands on both the brain and the body. Increased use of lactate from muscle would decrease the need for the immediate metabolism of glucose, increasing the duration of exercise possible without endangering brain function, and linking the potential for increased brain size with the success of endurance running as part of a subsistence strategy.

BRAIN DEVELOPMENT

In the first part of this paper I argued that the energetics of the human brain could be seen as a trade-off between the constraints of energetic cost, demand, substrate specificity, and blood–brain barrier versus the benefits of neuroplasticity. Such trade-offs will be present at all times, but may be particularly evident during conditions of high energy demand such as high activity levels. Brain development is another, more extended, period of elevated brain energy utilization. Thus the trade-off between energy and neuroplasticity should be evident during the development of the human brain as well.

At the broadest level, the overall energetic cost of brain development may be subsidized by additional somatic energy stores or traded-off against somatic growth and development by temporally offsetting periods of elevated brain and somatic energy utilization. For instance, the high level of adiposity seen in human infants has been argued to represent energy stores put on in utero to support continued brain growth for the first year postnatally (Kuzawa, 1998; Cunnane and Crawford, 2003). Similarly, very slow growth during human childhood (Bogin, 1999) may be a reduction of somatic energy costs in favor of brain development and neuroplasticity during a period of high brain metabolism from 4 to 10 years of age (Chugani, 1998).

In addition to overall energy consumption, the specific substrates which the brain uses to meet its energy requirements change during development. For instance, despite the importance of glucose to brain function, it is well known that infants utilize ketones and lactate as fuel, thus potentially sparing glucose for other functions (Nehlig, 2004). However, the capacity of the brain to metabolize ketones declines as the child grows, leading to an increasing demand for glucose. In what follows we consider brain development and its implications for the allocation of energy to the brain and body starting with infancy.

Infant brain development

Brain metabolism is estimated to represent some 50–80% of the energy budget of the human infant (Holliday, 1986). This reflects the large size of the brain relative to the body, as well as rapid brain growth during the first two years of life (Leigh, 2004), a period of extensive synaptogenesis and high levels of synaptic density, as well as the myelinization of major nerve tracts (Carmody et al., 2004). In addition, the hippocampus and amgydala are developing during this period.

It is important to note that breast milk, the primary food in early infancy is rich in lipids, but a relatively poor source of glucose. However, lactate and ketone bodies, not glucose appear to be important energy substrates for the brain during this period (Edmond, 1992; Nehlig, 2004; Medina and Tabernero, 2005). In particular, ketones can be used by oligodendrocytes to make the lipids that are part of myelin, sparing glucose for other pathways (Nehlig, 2004). In addition, ketones appear to shunt glutamine toward the production of GABA rather than glutamate (Melø et al., 2006), thus potentially reducing the risk of glutamate excitotoxicity.

In fact, glucose may be in relatively short supply in the infant brain. Glucose utilization rates during infancy start relatively low, reaching adult levels by around the age of two years (Chugani, 1998). Furthermore, levels of EAAT2, the transporter that removes glutamate from the synapse to astrocytes are relatively low from one to two months to two years (Lauriat et al., 2007). Together reduced glucose availability and low levels of EAAT2 may leave the infant brain particularly susceptible to glutamate excitotoxicity, as suggested by the relatively high rate of seizures during this period (Lauriat et al., 2007).

During infancy it is thought that experience plays a key role in the development of fundamental emotional circuits in the brain. Alan Schore (1994, 1997, 2002) has suggested that negative affect associated with trauma and neglect during the first two years of life results in a hypermetabolic state of arousal, including elevated cortisol levels and the activation of the sympathetic nervous system. Schore suggests that if sufficiently prolonged, such arousal may lead to elevated levels of glutamate and resulting excitotoxicity, particularly in the orbitofrontal cortex, anterior cingulate cortex, and amygdala, which are developing during this period.

Glutamate exotoxic shaping of emotional circuits appears to reflect a close association of nutrition and emotional conditions during infancy. Schore's argument is focused on early childhood trauma and neglect and its effects on the development of psychopathology. But such an argument is applicable to infant brain development under conditions of nutritional as well as emotional stress. Given the potentially low levels of glucose availability during infancy, as well as the risk of infectious disease in traditional societies, many infants may exist in a state of metabolic brain arousal, leading to glutamate excitotoxicity during periods of acute undernutrition. In support of this, acute undernutrition during infancy has been associated with cerebral atrophy (Hazin et al., 2007) and disruption of dendrite development (Benitez-Bribiesca et al., 1999).

Thus it has been suggested that the rather remarkable degree of adiposity exhibited by human infants may represent a metabolic buffer for brain development (Kuzawa, 1998; Cunnane and Crawford, 2003). Adipose tissue itself can not be metabolized by the brain. However, the visceral component of adiposity can be metabolized to produce ketones for the brain, which are an important source of energy and mylenization for the infant brain as outlined above. On the other hand, subcutaneous fat stores may be used to support the energetic requirements of the immune system during infancy, which would free up glucose to be used by the brain (Kuzawa, 1998).

Childhood brain development

Important changes in human brain metabolism are evident around the age of four to eight years, when glucose utilization rates reach their peak, at about twice the level observed in adults (Chugani, 1998). The executive attention system, which integrates attention and executive function, is undergoing rapid development between three and seven years of age (Posner, 2005), including the development of the thalamus, the parietal lobe, the orbitofrontal cortex, and the anterior cingulate cortex. Thus the early increase in glucose utilization may be directly related to the growth and development to these parts of the brain.

Glucose utilization rates, however, remain elevated until around the age of 11 when they begin to decline to adult rates (Chugani, 1998). Brain growth is 95% complete by the age of seven (Caviness et al., 1996), suggesting that the high rate of glucose utilization after the age of seven is not a function of increasing brain size. Instead, the high glucose demand of the brain during this period is presumably related to enhanced synaptic plasticity (Chugani, 1998), which I have argued is maintained by the energetic costs of glutamate cycling.

In fact, the juvenile period is associated with synaptic pruning, i.e., the loss of established neuronal connections, a process known to continue in adolescence (Huttenlocker, 1984; Huttenlocher and Dabholkar, 1997). Such pruning is based on experience; neuronal connections that are used are strengthened and those that are not are lost. Recent studies indicating cortical maturation associated with declines in gray matter starting around the age of six (Gogtay et al., 2004) are consistent with the decreased number of synaptic connections during this period.

Thus, the age pattern of glucose utilization rates is suggestive of an important shift in brain function during childhood linking social context, nutrition, and behavior. The onset of peak levels at three to four years of age is roughly coincident with the natural end of lactation (Martin, 2007). Weaning represents a switch away from breast milk to other sources of food that are generally higher in carbohydrate content and may provide a greater source of glucose for the brain. In addition, weaning marks a change in the child's social environment as well, as children start to spend increasing time in play with sibling and other children, and less time in such close proximity with their mother. In other words, children between the ages of approximately 4–11 years of age soak up an increasing variety of environmental influences, all of which may promote the development of synaptic connections with relatively little additional reinforcement.

The period from 4–11 years of age, roughly the period between weaning and puberty, is also a period of very slow and decreasing somatic growth (Bogin, 1999). It has been argued that somatic growth is suppressed during this stage to allow for the increased energetic costs of growing a large brain. However, given that growth in brain volume is almost entirely completed by the age of seven, it seems more accurate to say that the slow rate of somatic growth during the juvenile period represents a priority in the use of glucose for brain *development* over somatic growth.

In fact, the energetic demands of brain development may be quite a bit higher than usually calculated, since they include the costs of physical activity critical to brain development as well as the metabolic costs of brain tissue. For instance, it has been calculated that 17% of the total energy budget for children six years of age is consumed by physical activity (Dufour, 1997). Much of the physical activity during childhood clearly serves multiple purposes and can not be attributed solely to a specific function. However, the costs of physical activity, which helps shape brain development and promotes learning through the release of brain-derived neurotrophic factor (BDNF) (Winter et al., 2007), must be taken into account in calculating the full energetic cost of brain development.

Adolescence

Compared to childhood, during which elevated glucose utilization rates suggest a dynamic process of synaptic formation and pruning, adolescence is thought to be associated with the pruning of synaptic connections (Huttenlocher and Dabholkar, 1997; Gogtay et al., 2004). If I am correct, such synaptic pruning is associated with reduction of glucose-fueled glutamate cycling and reflected in declining rates of cerebral glucose utilization throughout the brain. In fact, there does appear to be a loss of generalized neuroplasticity, as indicated by declining language acquisition ability, around the age of 12 (Sakai, 2005), roughly the same time that that glucose utilization rates start to decline. However, localized synaptic plasticity may be maintained, particularly in the prefrontal cortex, still maturing throughout adolescence (Sowell et al., 1999; Gogtay et al., 2004).

In addition to declining brain plasticity, the pubertal drop in glucose utilization by the brain appears to reflect an increased allocation of energy to somatic growth associated with the pubertal growth spurt. The onset of puberty has been linked to abdominal fat stores in both boys and girls (Vizmanos and Marti-Henneberg, 2000), the same fat stores that are thought to serve as a reserve source of ketones for the brain (Peters et al., 2004), potentially placing reproductive maturation and brain development in direct energetic competition.

Boys appear to utilize prepubertal adipose storage as the basis for increased size and muscularity in association with increased testosterone production (Campbell and Mbzivo, 2006). Behaviorally these changes are presumably associated with male–male competition for mates (Hilton et al., 2000). For girls, puberty is associated with increased estrogen production and the deposition of fat stores that play an important role in the modulation of reproductive function, including ovarian cycling, and potential pregnancy and lactation (Ellison, 2001).

The same gonadal steroids that promote secondary sexual characteristics and fuel somatic growth also act on the brain to change behavior during adolescence (Sisk and Zehr, 2005). The continued maturation of the prefrontal cortex involves the experience-dependent maturation of the cortico-thalamic-striatal circuit (Chambers et al., 2003; Crews et al., 2007). Testosterone may play a role in this process by modulating the dopaminergic reward system (Wood, 2004; Frye, 2007) while estrogen may have a role through its effects on the serotonergic system (Lasiuk and Hegadoren, 2007), thus linking reproductive hormones and the reinforcement of behavioral predispositions.

From an energetic point of view, the timing of hormonal changes and their behavioral consequences during adolescence is a function of energy availability. Greater energy availability promotes faster childhood growth with earlier pubertal onset, including reproductive hormones (Ellison, 2001). Reproductive hormones promote secondary characteristics which attract the attention of others (see Waylen and Wolke, 2004, for a recent review), while at the same time acting on the developing circuits in the brain to organize behavior, including libido (Sisk and Foster, 2004; Sisk and Zehr, 2005). Thus earlier maturers not only stand out physically from their peers, but their brains are being shaped by the experience of being more advanced sexually than their peers, which may reinforce sexual behavior during adulthood as well (Ostovich and Sabini, 2005).

Young adulthood

Even after puberty is complete, and the production of reproductive hormones has leveled off, the brain continues to develop finishing with the maturation of the prefrontal cortex some time in the early 20s (Gogtay et al., 2004). Glucose utilization rates in the anterior cingulate cortex, a central structure in monitoring emotional impulses from the amygdala (Pezawas et al., 2005; Etkin et al., 2006) have been shown to increase until the middle of the 20s (Van Bogaert et al., 1998). The anterior cingulate cortex appears particularly affected by the adrenal hormone dehydro-epiandrosterone (DHEA) (Alhaj et al., 2006), which continues to increase into the 20s as well (Orentreich et al., 1984; Sulcova et al., 1997). Together, these findings suggest a young-adult period characterized by continued brain maturation in the absence of further somatic and reproductive maturation.

However, such a characterization may be misleading given that favorable energetic conditions in industrialized populations has led to early reproductive maturation. In many societies with low food availability somatic growth continues into the early 20s

(see Campbell et al., 2005, for an example). Thus in the natural human life course, rather than being a distinct postpubertal period, young adulthood may represent a final stage of development in which reproductive, somatic, and brain maturation all converge to produce a fully functioning reproductive adult.

SUMMARY

Recent advances in neuroscience and brain imaging have greatly enlarged our understanding of human brain metabolism and made the brain much more amenable to an energetic and evolutionary analysis. In addition to the well-known overall energetic cost of the brain and its dependence on glucose as a substrate, more recent work has emphasized the fact that brain is energetically demanding and in a position to regulate its own energy supply. The human brain is also highly plastic, and can utilize more than one substrate, both during development and adulthood. It is this plasticity (in addition to the large size) that gives the human brain its amazing flexibility and may play a central role in the cognitive powers that we hold so crucial to the nature of our species.

I argue that much of the energetic cost of the human brain can be linked to the use of glucose in glutamate cycling and its role in maintaining synaptic plasticity. Glutamate cycling involves the neuron-lactate-astrocyte shuttle, an on-going process, which may underlie intrinsic brain activity, thus accounting for the energetically expensive and demanding properties of the human brain. Furthermore, though glucose remains the preferred fuel for brain activity, the capacity of neurons to metabolize lactate and ketones appears to provide additional mechanisms for trade-offs between brain and somatic energy utilization. Such mechanisms are important in physical exercise and may have provided a means to circumvent energetic constraints on physical activity associated with subsistence strategies underwriting the evolution of a large human brain.

Developmentally, changes in the energetic costs of the brain appear to map onto important neurological and behavioral stages of human development. The infant brain appears to be buffered from the effects of potential low energy availability during its rapid growth by a store of adipose tissue. On the other hand, elevated brain glucose utilization between the ages of 4 and 11 years appears to coincide with the onset of adrenarche, suggesting that the neuroprotective effects of dehydroepiandrosterone sulfate (DHEAS) may be important in maintaining synaptic plasticity, thus promoting learning and socialization in prepubescent children.

During adolescence the timing of puberty and the rise of reproductive hormones reflects energy availability. Thus those who physically mature early will not only show earlier brain maturation, but such maturation will tend to develop in social circumstances favoring sexual behavior and thus shape their brain to expect similar circumstances during adulthood. Though it appears that this integrative developmental process is centered on adolescence in our society, in subsistence societies where somatic growth is often slower it may also include young adulthood.

Finally, the understanding of the dynamic nature of brain energetics has fundamental philosophical implications for the brain as a self-regulating homeostatic system. While it has long been held that information is the key element of energy flow through the brain, the recent characterization of intrinsic brain activity suggests that such a statement is not a metaphor, but a literal truth. The consumption of glucose by the brain is directly related to the dynamic neural processes that allow for on-going processing in the brain, a large part of which appears to reflect the integration of the response to incoming environmental information and with the vast amount of information already stored in the brain.

DISCUSSION POINTS

1. Why is human brain energetics central to an understanding of human evolutionary physiology?
2. What two other substrates, besides glucose, can the human brain utilize? Under what conditions does is each of these substrates take on particular importance?
3. What role are astrocytes thought to play in the mobilization of glycogen stores for neuronal metabolism?
4. What is meant by the term "neuroplasticity?" Why should neuroplasticity be energetically costly?
5. What part of the human brain appears to be most sensitive to changes in glucose availability?
6. What is meant by "intrinsic" brain activity? What implications does it have for evolutionary constraints on the evolution of the large human brain?
7. At what point during the human life cycle does the rate of brain glucose consumption peak?
8. What are the potential benefits/costs of prolonged juvenile brain development in humans?
9. How is the timing of human brain development related to reproductive maturation? To somatic growth?

ACKNOWLEDGMENTS

I would like to thank colleagues with whom I have discussed elements of this topic over the past two years, including Peter Ellison, Dan Eisenberg, Peter Gray, and the Tea and Hormones group. I would also like to thank Robert Campbell for his continued encouragement to pursue this topic. All errors are my own.

REFERENCES

Abeles, M. (1991). *Corticonics: Neural Circuits of the Cerebral Cortex*. New York: Cambridge University Press.

Aiello, L. and Wheeler, P. (1995). The expensive-tissue hypothesis; the brain and digestive system in human and primate evolution. *Current Anthropology*, **36**, 199–221.

Alhaj, H. A., Massey, A. E. and McAllister-Williams, R. H. (2006). Effects of DHEA administration on episodic memory, cortisol and mood in healthy young men: a double-blind, placebo-controlled study. *Psychopharmacology (Berlin)*, **188**, 541–551.

Allman, J. M., Hakeem, A., Erwin, J. M., et al. (2001). The anterior cingulate cortex. The evolution of an interface between emotion and cognition. *Annals of the New York Academy of Science*, **935**, 107–117.

Armstrong, E. (1983). Relative brain size and metabolism in mammals. *Science*, **220**, 1302–1304.

Armstrong, E. (1985). Allometric considerations of the adult mammalian brain, with emphasis on primates. In *Size and Scaling in Primate Biology*, W. L. Jungers (ed.). New York: Plenum, pp. 115–146.

Attwell, D. and Laughlin, S. B. (2001). An energy budget for signaling in the grey matter of the brain. *Journal of Cerebral Blood Flow and Metabolism*, **21**, 1133–1145.

Beals, K. L., Smith, C. L. and Dodd, S. M. (1984). Brain size, cranial morphology, climate, and time machines. *Current Anthropology*, **25**, 301–328.

Benitez-Bribiesca, L., De la Rosa-Alvarez, I. and Mansilla-Olivares, A. (1999). Dendritic spine pathology in infants with severe protein-calorie malnutrition. *Pediatrics*, **104**, e21.

Bennington, J. H. and Heller, H. C. (1995). Restoration of brain energy metabolism as the function of sleep. *Progress in Neurobiology*, **45**, 347–360.

Bittar, P. G., Charnay, Y., Pellerin, L., et al. (1996). Selective distribution of lactate dehydrogenase isoenzymes in neurons and astrocytes of human brain. *Journal of Cerebral Blood Flow and Metabolism*, **16**, 1079–1089.

Bogin, B. (1999). Evolutionary perspective on human growth. *Annual Review of Anthropology*, **28**, 109–153.

Braitenberg, V. and Schuz, A. (1998). *Cortex: Statistics and Geometry of Neuronal Connectivity*. New York: Springer.

Bramble, D. M. and Lieberman, D. E. (2005). Endurance running and the evolution of *Homo*. *Nature*, **432**, 345–352.

Bribiescas, R. G. (2005). Serum leptin levels in Aché Amerindian females with normal adiposity are not significantly different from American anorexia nervosa patients. *American Journal of Human Biology*, **17**, 207–210.

Broadhurst, C. L., Wang, Y., Crawford, M. A., et al. (2002). Brain-specific lipids from marine, lacustrine, or terrestrial food resources: potential impact on early African *Homo sapiens. Comparative Biochemistry and Physiology B: Biochemistry and Molecular Biology*, **131**, 653–673.

Brown, A. M. (2004). Brain glycogen re-awakened. *Journal of Neurochemistry*, **89**, 537–552.

Brown, A. M. and Ransom, B. R. (2007). Astrocyte glycogen and brain energy metabolism. *Glia*, **55**, 1263–1271.

Burki, F. and Kaessmann, H. (2004). Birth and adaptive evolution of a hominoid gene that supports high neurotransmitter flux. *Nature Genetics*, **36**, 1061–1063.

Calabrese, B., Wilson, M. S. and Halpain, S. (2006). Development and regulation of dendritic spine synapses. *Physiology (Bethesda)*, **21**, 38–47.

Campbell, B. C. and Mbzivo, M. T. (2006). Testosterone, reproductive maturation and somatic growth among Zimbabwe boys. *Annals of Human Biology*, **33**, 17–25.

Campbell, B. C., Leslie, P. W., Little, M. A., et al. (2005). Pubertal timing, hormones and body composition among adolescent Turkana males. *American Journal of Physical Anthropology*, **128**, 896–905.

Campbell, B. C., Gray, P. B. and Ellison, P. T. (2006). Age-related changes in body composition and salivary testosterone among Ariaal males. *Aging: Clinical and Experimental Research*, **18**, 470–476.

Carlson, B. A. and Kingston, J. D. (2007). Docosahexaenoic acid, the aquatic diet, and hominin encephalization: difficulties in establishing evolutionary links. *American Journal of Human Biology*, **19**, 132–141.

Carmody, D. P., Dunn, S. M., Boddie-Willis, A. S., et al. (2004). A quantitative measure of myelination development in infants, using MR images. *Neuroradiology*, **46**, 781–786.

Caviness, V. S. Jr, Kennedy, D. N., Richelme, C., et al. (1996). The human brain age 7–11 years: a volumetric analysis based on magnetic resonance images. *Cerebral Cortex*, **6**, 726–736.

Chambers, R. A., Taylor, J. R. and Potenza, M. N. (2003). Developmental neurocircuitry of motivation in adolescence: a critical period of addiction vulnerability. *American Journal of Psychiatry*, **160**, 1041–1052.

Chugani, H. T. (1998). Critical period of brain development: studies of cerebral glucose utilization with PET. *Preventative Medicine*, **27**, 184–188.

Clarke, D. D. and Sokoloff, L. (1999). Circulation and energy metabolism of the brain. In *Basic Neurochemistry: Molecular, Cellular, and Medical Aspects*, G. J. Siegel, B. W. Agranoff, R. W. Albers, et al. (eds), 6th edn. Philadelphia: Lippincott-Raven.

Crews, F., He, J. and Hodge, C. (2007). Adolescent cortical development: a critical period of vulnerability for addiction. *Pharmacology, Biochemistry and Behavior*, **86**, 189–199.

Crile, G. W. (1941). *Intelligence, Power and Personality*. New York: Wittlesy.

Cunnane, S. C. and Crawford, M. A. (2003). Survival of the fattest: fat babies were the key to evolution of the large human brain. *Comparative Biochemistry and Physiology A: Molecular and Integrative Physiology*, **136**, 17–26.

Daikhan, Y. and Yudkoff, M. (2000). Compartmentation of brain glutamate metabolism in neurons and glia. *Journal of Nutrition*, **130**, 1026S–1031S.

Dalsgaard, M. K. (2006). Fueling cerebral activity in exercising man. *Journal of Cerebral Blood Flow and Metabolism*, **26**, 731–750.

Dalsgaard, M. K., Quistorff, B., Danielsen, E. R., et al. (2004). A reduced cerebral metabolic ratio in exercise reflects metabolism and not accumulation of lactate within the human brain. *Journal of Physiology*, **554**, 571–578.

Dalsgaard, M. K., Madsen, F. F., Secher, N. H., et al. (2007). High glycogen levels in the hippocampus of patients with epilepsy. *Journal of Cerebral Blood Flow and Metabolism*, **27**, 1137–1141.

Djuricic, B., Masirevic, G. and Susic, V. (1977). Paradoxical sleep deprivation: effects on brain energy metabolism. *Archives Internationale de Physiologie et de Biochemie*, **85**, 213–219.

Dufour, D. L. (1997). Nutrition, activity, and health in children. *Annual Review of Anthropology*, **26**, 541–565.

Edmond, J. (1992). Energy metabolism in developing brain cells. *Canadian Journal of Physiology and Pharmacology*, **70**, S118–S129.

Elia, M. (1992). Organ and tissue contribution to metabolic rate. In *Energy Metabolism: Tissue Determinants and Cellular Corollaries*, J. M. Kinner and H. N. Tucker (eds). New York: Raven Press, pp. 61–79.

Ellison, P. T. (2001). *On Fertile Ground: a Natural History of Reproduction*. Cambridge, MA: Harvard University Press.

Emery, P. W. (2005). Metabolic changes in malnutrition. *Eye*, **19**, 1029–1034.

Etkin, A., Egner, T., Peraza, D. M., et al. (2006). Resolving emotional conflict: a role for the rostral anterior cingulate cortex in modulating activity in the amygdala. *Neuron*, **51**, 871–882.

Fehm, H. L., Kern, W. and Peters, A. (2006). The selfish brain: competition for energy resources. *Progress in Brain Research*, **153**, 129–140.

Folbergrová, J., Lowry, O. H. and Passonneau, J. V. (1970). Changes in metabolites of the energy reserves in individual layers of mouse cerebral cortex and subjacent white matter during ischemia and anaesthesia. *Journal of Neurochemistry*, **17**, 1155–1162.

Fox, M. D., Snyder, A. Z., Vincent, J. L., et al. (2005). The human brain is intrinsically organized into dynamic, anticorrelated functional networks. *Proceedings of the National Academy of Sciences of the United States of America*, **102**, 9673–9678.

Franken, P., Gip, P., Hagiwara, G., et al. (2003). Changes in brain glycogen after sleep deprivation vary with genotype. *American Journal of Physiology. Regulatory, Integrative, and Comparative Physiology*, **285**, R413–R419.

Frye, C. A. (2007). Some rewarding effects of androgens may be mediated by actions of its 5α-reduced metabolite 3α-androstanediol. *Pharmacology, Biochemistry and Behavior*, **86**, 354–367.

Gallinat, J., Kunz, D., Lang, U. E., et al. (2007). Association between cerebral glutamate and human behaviour: the sensation seeking personality trait. *NeuroImage*, **34**, 671–678.

Gip, P., Hagiwara, G., Ruby, N. F., et al. (2002). Sleep deprivation decreases glycogen in the cerebellum but not in the cortex of young rats. *American Journal of Physiology. Regulatory, Integrative, and Comparative Physiology*, **283**, R54–R59.

Goda, Y. and Davis, G. W. (2003). Mechanisms of synapse assembly and disassembly. *Neuron*, **40**, 243–264.

Gogtay, N., Giedd, J. N., Lusk, L., et al. (2004). Dynamic mapping of human cortical development during childhood through early adulthood. *Proceedings of the National Academy of Sciences of the United States of America*, **101**, 817–819.

Greicius, M. D. and Menon, V. (2004). Default-mode activity during a passive sensory task: uncoupled from deactivation but impacting activation. *Journal of Cognitive Neuroscience*, **16**, 1484–1492.

Greicius, M. D., Krasnow, B., Reiss, A. L., et al. (2003). Functional connectivity in the resting brain: a network analysis of the default mode hypothesis. *Proceedings of the National Academy of Sciences of the United States of America*, **100**, 253–258.

Gruetter, R. (2003). Glycogen: the forgotten cerebral energy store. *Journal of Neuroscience Research*, **74**, 179–183.

Hazin, A. N., Alves, J. G. and Rodrigues Falbo, A. (2007). The myelination process in severely malnourished children: MRI findings. *International Journal of Neuroscience*, **117**, 1209–1214.

Hill, K. (1993). Life history theory and evolutionary anthropology. *Evolutionary Anthropology*, **2**, 78–88.

Hilton, N. Z., Harris G. T. and Rice, M. E. (2000). The functions of aggression by male teenagers. *Journal of Personality and Social Psychology*, **79**, 988–994.

Holliday, M. A. (1986). Body composition and energy needs during growth. In *Human Growth: a Comprehensive Treatise*, F. Falkner and J. M. Tanner (eds). New York: Plenum Press.

Horovitz, S. G., Fukunaga, M., de Zwart, J. A., et al. (2007). Low frequency BOLD fluctuations during resting wakefulness and light sleep: A simultaneous EEG-fMRI study. *Human Brain Mapping*, **29**, 671–682.

Huttenlocher, P. R. (1984). Synapse elimination and plasticity in developing human cerebral cortex. *American Journal of Mental Deficiency*, **88**, 488–496.

Huttenlocher, P. R. and Dabholkar, A. S. (1997). Regional differences in synaptogenesis in human cerebral cortex. *Journal of Comparative Neurology*, **387**, 167–178.

Jasienska, G., Thune, I. and Ellison, P. T. (2006). Fatness at birth predicts adult susceptibility to ovarian suppression: an empirical test of the Predictive Adaptive Response hypothesis. *Proceedings of the National Academy of Sciences of the United States of America*, **103**, 12759–12762.

Karnovsky, M. L., Reich, P., Anchors, J. M., et al. (1983). Changes in brain glycogen during slow-wave sleep in the rat. *Journal of Neurochemistry*, **41**, 1498–1501.

Kemppainen, J., Aalto, S., Fujimoto, T., et al. (2005). High intensity exercise decreases global brain glucose uptake in humans. *Journal of Physiology*, **568**, 323–332.

Kennedy, G. E. (2005). From the ape's dilemma to the weanling's dilemma: early weaning and its evolutionary context. *Journal of Human Evolution*, **48**, 123–145.

Kong, J., Shepel, P. N., Holden, C. P., et al. (2002). Brain glycogen decreases with increased periods of wakefulness:

implications for homeostatic drive to sleep. *Journal of Neuroscience*, **22**, 5581–5587.

Kuzawa, C. W. (1998). Adipose tissue in human infancy and childhood: an evolutionary perspective. *Yearbook of Physical Anthropology*, **41**, 177–206.

Kuzawa, C. W. (2007). Developmental origins of life history: growth, productivity, and reproduction. *American Journal of Human Biology*, **19**, 654–661.

Lasiuk, G. C. and Hegadoren, K. M. (2007). The effects of estradiol on central serotonergic systems and its relationship to mood in women. *Biological Research for Nursing*, **9**, 147–160.

Lauriat, T. L., Schmeidler, J. and McInnes, L. A. (2007). Early rapid rise in EAAT2 expression follows the period of maximal seizure susceptibility in human brain. *Neuroscience Letters*, **412**, 89–94.

Leigh, S. R. (2004). Brain growth, life history, and cognition in primate and human evolution. *American Journal of Primatology*, **62**, 139–164.

Leonard, W. R. and Ulijaszek, S. J. (2002). Energetics and evolution: an emerging research domain. *American Journal of Human Biology*, **14**, 547–550.

Leonard, W. R., Robertson, M. L., Snodgrass, J. J., et al. (2003). Metabolic correlates of hominid brain evolution. *Comparative Biochemistry and Physiology A: Molecular and Integrative Physiology*, **136**, 5–15.

Leonard, W. R., Snodgrass, J. J. and Robertson, M. L. (2007). Effects of brain evolution on human nutrition and metabolism. *Annual Review of Nutrition*, **27**, 311–327.

Lieberman, D. E. and Bramble, D. M. (2007). The evolution of marathon running capabilities in humans. *Sports Medicine*, **37**, 288–290.

Lipson, S. F. and Ellison, P. T. (1996). Comparison of salivary steroid profiles in naturally occurring conception and non-conception cycles. *Human Reproduction*, **11**, 2090–2096.

Lipton, P. (1989). Regulation of glycogen in the dentate gyrus of the in vitro guinea pig hippocampus; effect of combined deprivation of glucose and oxygen. *Journal of Neuroscience Methods*, **28**, 147–154.

Madsen, P. L. and Vorstrup, S. (1991). Cerebral blood flow and metabolism during sleep. *Cerebrovascular and Brain Metabolism Reviews*, **3**, 281–296.

Madsen, P. L., Schmidt, J. F., Wildschiodtz, G., et al. (1991). Cerebral O_2 metabolism and cerebral blood flow in humans during sleep and rapid-eye-movement sleep. *Journal of Applied Physiology*, **70**, 2597–2601.

Magistretti, P. J. (2006). Neuron-glia metabolic coupling and plasticity. *Journal of Experimental Biology*, **209**, 2304–2311.

Mantini, D., Perrucci, M. G., Del Gratta, C., et al. (2007). Electrophysiological signatures of resting state networks in the human brain. *Proceedings of the National Academy of Sciences of the United States of America*, **104**, 13170–13175.

Margulies, D. S., Kelly, A. M., Uddin, L. Q., et al. (2007). Mapping the functional connectivity of anterior cingulate cortex. *NeuroImage*, **37**, 579–588.

Martin, R. D. (1981). Relative brain size and basal metabolic rate in terrestrial vertebrates. *Nature*, **293**, 57–60.

Martin, R. D. (1989). Evolution of the brain in early hominids. *Ossa*, **4**, 49–62.

Martin, R. D. (2007). The evolution of human reproduction: a primatological perspective. *American Journal of Physical Anthropology*, **134**, 59–84.

Mason, M. F., Norton, M. I., Van Horn, J. D., et al. (2007). Wandering minds: the default network and stimulus-independent thought. *Science*, **315**, 393–395.

McDade, T. W., Leonard, W. R., Burhop, J., et al. (2005). Predictors of C-reactive protein in Tsimane' 2 to 15 year-olds in lowland Bolivia. *American Journal of Physical Anthropology*, **128**, 906–913.

McEwen, B. S. (2006). Sleep deprivation as a neurobiologic and physiologic stressor: allostasis and allostatic load. *Metabolism*, **55**, S20–S23.

Medina, J. M. and Tabernero, A. (2005). Lactate utilization by brain cells and its role in CNS development. *Journal of Neuroscience Research*, **79**, 2–10.

Meikle, A., Roby, L. M. and Stollery, B. (2004). The impact of glucose ingestion and gluco-regulatory control on cognitive performance: a comparison of young and middle aged adults. *Human Psychopharmacology*, **19**, 523–535.

Melø, T. M., Nehlig, A. and Sonnewald, U. (2006). Neuronal-glial interactions in rats fed a ketogenic diet. *Neurochemistry International*, **48**, 498–507.

Muehlenbein, M. P., Algier, J., Cogswell, F., et al. (2005). The reproductive endocrine response to *Plasmodium vivax* infection in Hondurans. *American Journal of Tropical Medicine and Hygiene*, **73**, 178–187.

Nehlig, A. (2004). Brain uptake and metabolism of ketone bodies in animal models. *Prostaglandins, Leukotrienes, and Essential Fatty Acids*, **70**, 265–275.

Nepomnaschy, P. A., Welch, K. B., McConnell, D. S., et al. (2006). Cortisol levels and very early pregnancy loss in humans. *Proceedings of the National Academy of Sciences of the United States of America*, **103**, 3938–3942.

Orentreich, N., Brind, J. L., Rizer, R. L., et al. (1984). Age changes and sex differences in serum dehydroepiandrosterone sulfate concentrations throughout adulthood. *Journal of Clinical Endocrinology and Metabolism*, **59**, 551–555.

Ostovich, J. M. and Sabini, J. (2005). Timing of puberty and sexuality in men and women. *Archives of Sexual Behavior*, **34**, 197–206.

Oz, G., Seaquist, E. R., Kumar, A., et al. (2007). Human brain glycogen content and metabolism: implications on its role in brain energy metabolism. *American Journal of Physiology. Endocrinology and Metabolism*, **292**, E946–E951.

Park, M., Salgado, J. M., Ostroff, L., et al. (2006). Plasticity-induced growth of dendritic spines by exocytic trafficking from recycling endosomes. *Neuron*, **52**, 817–830.

Patel, A. B., de Graaf, R. A., Mason, G. F., et al. (2005). The contribution of GABA to glutamate/glutamine cycling and energy metabolism in the rat cortex in vivo. *Proceedings of the National Academy of Sciences of the United States of America*, **102**, 5588–5593.

Pellerin, L., Pellegri, G., Bittar, P. G., et al. (1998). Evidence supporting the existence of an activity-dependent astrocyte-neuron lactate shuttle. *Developmental Neuroscience*, **20**, 291–299.

Pellerin, L., Bouzier-Sore, A. K., Aubert, A., et al. (2007). Activity-dependent regulation of energy metabolism by astrocytes: an update. *Glia*, **55**, 1251–1262.

Pelligri, G., Rossier, C., Magistretti, P. J., et al. (1996). Cloning, localization, and induction of mouse brain glycogen synthase. *Brain Research and Molecular Brain Research*, **38**, 191–199.

Peters, A., Schweiger, U., Pellerin, L., et al. (2004). The selfish brain: competition for energy resources. *Neuroscience and Biobehavioral Reviews*, **28**, 143–180.

Pezawas, L., Meyer-Lindenberg, A., Drabant, E. M., et al. (2005). 5-HTTLPR polymorphism impacts human cingulate-amygdala interactions: a genetic susceptibility mechanism for depression. *Nature Neuroscience*, **8**, 828–834.

Phelps, C. H. (1972). Barbiturate-induced glycogen accumulation in brain. An electron microscopic study. *Brain Research*, **39**, 225–234.

Pike, I. L. and Williams, S. R. (2006). Incorporating psychosocial health into biocultural models: preliminary findings from Turkana women of Kenya. *American Journal of Human Biology*, **18**, 729–740.

Plaitakis, A., Spanaki, C., Mastorodemos, V., et al. (2003). Study of structure-function relationships in human glutamate dehydrogenases reveals novel molecular mechanisms for the regulation of the nerve tissue-specific (GLUD2) isoenzyme. *Neurochemistry International*, **43**, 401–410.

Posner, M. I. (2005). Genes and experience shape brain networks of conscious control. *Progress in Brain Research*, **150**, 173–183.

Posner, M. I. and Rothbart, M. K. (1998). Attention, self-regulation and consciousness. *Proceedings of the Royal Society of London. Series B*, **353**, 1915–1927.

Raichle, M. E. and Gusnard, D. A. (2002). Appraising the brain's energy budget. *Proceedings of the National Academy of Sciences of the United States of America*, **99**, 10237–10239.

Raichle, M. E. and Mintun, M. A. (2006). Brain work and brain imaging. *Annual Review of Neuroscience*, **29**, 449–476.

Raichle, M. E., MacLeod, A. M., Snyder, A. Z., et al. (2001). A default mode of brain function. *Proceedings of the National Academy of Sciences of the United States of America*, **98**, 676–682.

Rao, J., Oz, G. and Sequist, E. R. (2006). Regulation of cerebral glucose metabolism. *Minerva Endocrinologica*, **31**, 149–158.

Riby, L. M., McMurtrie, H., Smallwood, J., et al. (2006). The facilitative effects of glucose ingestion on memory retrieval in younger and older adults: is task difficulty or task domain critical? *British Journal of Nutrition*, **95**, 414–420.

Rocher, A. B., Chapon, F., Blaizot, X., et al. (2003). Resting-state brain glucose utilization as measured by PET is directly related to regional synaptophysin levels: a study in baboons. *NeuroImage*, **20**, 1894–1898.

Sakai, K. L. (2005). Language acquisition and brain development. *Science*, **310**, 815–819.

Schore, A. (1994). *Affect Regulation and the Origin of the Self: the Neurobiology of Emotional Development*. Hillsdale, NJ: Erlbaum.

Schore, A. (1997). Early origins of the nonlinear right brain and the development of a predisposition to psychiatric disorders. *Development and Psychopathology*, **9**, 595–631.

Schore, A. (2002). Advances in neuropsychoanalysis, attachment theory, and trauma research: implications for self psychology. *Psychoanalytic Inquiry*, **22**, 433–484.

Schurr, A., Miller, J. J., Payne, R. S., et al. (1999). An increase in lactate output by brain tissue serves to meet the energy needs of glutamate-activated neurons. *Journal of Neuroscience*, **19**, 34–39.

Scott, L., Kruse, M. S., Forssberg, H., et al. (2002). Selective up-regulation of dopamine D_1 receptors in dendritic spines by NMDA receptor activation. *Proceedings of the National Academy of Sciences of the United States of America*, **99**, 1661–1664.

Segovia, G., Yague, A. G., Garcia-Verdugo, J. M., et al. (2006). Environmental enrichment promotes neurogenesis and changes in extracellular concentrations of glutamate and GABA in the hippocampus of aged rats. *Brain Research Bulletin*, **70**, 8–14.

Shen, N. R., Petersen, K., Behar, K. L., et al. (1999). Determination of the rate of the glutamate/glutamine cycle in the human brain by in vivo 13C NMR. *Proceedings of the National Academy of Sciences of the United States of America*, **96**, 8235–8240.

Shulman, R. G., Hyder, F. and Rothman, D. L. (2001). Cerebral energetics and the glycogen shunt: neurochemical basis of functional imaging. *Proceedings of the National Academy of Sciences of the United States of America*, **98**, 6417–6422.

Shulman, R. G., Rothman, D. L., Behar, K. L., et al. (2004). Energetic basis of brain activity: implications for neuroimaging. *Trends in Neuroscience*, **27**, 489–495.

Sibson, N. R., Dhankhar, A., Mason, G. F., et al. (1998). Stoichiometric coupling of brain glucose metabolism and glutamatergic neuronal activity. *Proceedings of the National Academy of Sciences of the United States of America*, **95**, 316–321.

Sisk, C. L. and Foster, D. L. (2004). The neural basis of puberty and adolescence. *Nature Neuroscience*, **7**, 1040–1047.

Sisk, C. L. and Zehr, J. L. (2005). Pubertal hormones organize the adolescent brain and behavior. *Frontiers of Neuroendocrinology*, **26**, 163–174.

Snodgrass, J. J., Leonard, W. R., Tarskaia, L. A., et al. (2007). Anthropometric correlates of C-reactive protein among indigenous Siberians. *Journal of Physiological Anthropology*, **26**, 241–246.

Sowell, E. R., Thompson, P. M., Holmes, C. J., et al. (1999). In vivo evidence for post-adolescent brain maturation in frontal and striatal regions. *Nature Neuroscience*, **2**, 859–861.

Sulcova, J., Hill, M., Hampl, R., et al. (1997). Age and sex related differences in serum levels of unconjugated dehydroepiandrosterone and its sulfate in normal subjects. *Journal of Endocrinology*, **154**, 57–62.

Trudeau, L. E. and Gutiérrez, R. (2007). On cotransmission and neurotransmitter phenotype plasticity. *Molecular Interventions*, **7**, 138–146.

Ullian, E. M., Christopherson, K. S. and Barres, B. A. (2004). Role for glia in synaptogenesis. *Glia*, **47**, 209–216.

Van Bogaert, P., Wikler, D., Damhaut, P., et al. (1998). Regional changes in glucose metabolism during brain development from the age of 6 years. *NeuroImage*, **8**, 62–68.

Van Itallie, T. B. and Nufert, T. H. (2003). Ketones: metabolism's ugly duckling. *Nutrition Review*, **61**, 327–341.

Vizmanos, B. C. and Marti-Henneberg, C. (2000). Puberty begins with a characteristic body fat mass in each sex. *European Journal of Clinical Nutrition*, **54**, 203–208.

Waylen, A. and Wolke, D. (2004). Sex 'n' drugs 'n' rock 'n' roll: the meaning and social consequences of pubertal timing. *European Journal of Endocrinology*, **151**, U151–U159.

Winter, B., Breitenstein, C., Mooren, F. C., et al. (2007). High impact running improves learning. *Neurobiology of Learning and Memory*, **87**, 597–609.

Wood, R. I. (2004). Reinforcing aspects of androgens. *Physiology and Behavior*, **83**, 279–289.

Yeakel, J. D., Bennett, N. C., Koch, P. L., et al. (2007). The isotopic ecology of African mole rats informs hypotheses on the evolution of human diet. *Proceedings of the Royal Society of London. Series B*, **274**, 1723–1730.

Ziv, N. E. and Smith, S. J. (1996). Evidence for a role of dendritic filopodia in synaptogenesis and spine formation. *Neuron*, **17**, 91–102.

26 Embodied Capital and Extra-somatic Wealth in Human Evolution and Human History

Jane B. Lancaster and Hillard S. Kaplan

INTRODUCTION

This chapter presents a theory of brain and life span evolution and applies it to both primates in general, and to the hominid line, in particular. To address the simultaneous effects of natural selection on the brain and on the life span, it extends the standard life history theory in biology which organizes research into the evolutionary forces shaping age-schedules of fertility and mortality. This extension, the embodied capital theory, integrates existing models with an economic analysis of capital investments and the value of life.

The chapter begins with a brief introduction to embodied capital theory, and then applies it to understanding major trends in primate evolution and the specific characteristics of humans. The evolution of brain size, intelligence, and life histories in the primate order are addressed first. The evolution of the human life course is then considered, with a specific focus on the relationship between cognitive development, economic productivity, and longevity. It will be argued that the evolution of the human brain entailed a series of coevolutionary responses in human development and aging.

The second section on embodied capital and extra-somatic wealth discusses humans in a comparative context, beginning with the hunting and gathering lifestyle because of its relevance to the vast majority of human evolutionary history. However, in the past 10 000 years human history traced a series of behavioral adaptations based on ecology and individual condition. The introduction of extra-somatic capital, first in the form of livestock and later in land and other types of wealth and power, radically changed the shape of human life history parameters and produced new patterns of fertility, parental investment, and reproductive regimes as access to extra-somatic capital became a focus of life history strategies.

Finally, modern skills-based, competitive labor markets, combined with reduced fertility during the nineteenth century, mark a returning focus on embodied capital in the form of skills, education, and training. In past civilizations, going back to Babylonia in the third millennia BC, literacy and numeracy were known but exceedingly rare skills. This pattern continued worldwide until 1800 in Western Europe, including England, where these skills went from rarity to the norm in under a century (Clark, 2007). Labor markets with a particular demand for embodied capital in their workers place new demands on human life history and reproductive strategies in terms of mate choice, fertility, investment in children, and the timing of reproduction in the life course. Once again, human life history radically changed in shape to a new emphasis on the acquisition of skills through training and education, postponement of reproduction to the late 20s, and radically reduced completed family size with the reproductive part of the life course compressed into less than a decade.

EMBODIED CAPITAL AND THE COEVOLUTION OF INTELLIGENCE, DIET, AND LONGEVITY

According to the theory of evolution by natural selection, organic evolution is the result of a process in which variant forms compete to harvest energy from the environment and convert that energy into replicates of those forms. Forms that can capture more energy and convert that energy more efficiently into replicates of themselves become more prevalent through time. This simple issue of harvesting energy and converting energy into offspring generates many complex problems that are time-dependent (Gadgil and Bossert, 1970).

Two fundamental trade-offs determine the action of natural selection on life history strategies. The first trade-off is between current and future reproduction. By growing, an organism can increase its energy capture capacities in the future and thus increase its future fertility. For this reason, organisms typically have a juvenile phase in which fertility is zero until

Human Evolutionary Biology, ed. Michael P. Muehlenbein. Published by Cambridge University Press. © Cambridge University Press 2010.

they reach a size at which some allocation to reproduction increases lifetime fitness more than does growth. Similarly, among organisms that engage in repeated bouts of reproduction (humans included), some energy during the reproductive phase is diverted away from reproduction and allocated to maintenance so that they can live to reproduce again. Natural selection is expected to optimize the allocation of energy to current reproduction and to future reproduction (via investments in growth and maintenance) at each point in the life course so that genetic descendents are maximized (Gadgil and Bossert, 1970). Variation across taxa and across conditions in optimal energy allocations is shaped by ecological factors, such as food supply, disease, access to mates, and predation rates.

A second fundamental life history trade-off is between offspring number (quantity) and offspring fitness (quality). This trade-off occurs because parents have limited resources to invest in offspring and each additional offspring produced necessarily reduces average investment per offspring. Most biological models operationalize this trade-off as number versus survival of offspring (Lack, 1954; Smith and Fretwell, 1974; Lloyd, 1987). However, parental investment may not only affect survival to adulthood, but also the adult productivity and fertility of offspring. This is especially true of humans. Thus, natural selection is expected to shape investment per offspring and offspring number so as to maximize offspring number times their average lifetime fitness.

The embodied capital theory generalizes existing life history theory by treating the processes of growth, development, and maintenance as investments in stocks of somatic, or embodied, capital. In a physical sense, embodied capital is organized somatic tissue – muscles, digestive organs, immune competence, brains, etc. In a functional sense, embodied capital includes strength, speed, immune function, skill, knowledge, and other qualities such as social networks and status. Since such stocks tend to depreciate with time, allocations to maintenance can also be seen as investments in embodied capital. Thus, the present–future reproductive trade-off can be understood in terms of optimal investments in own embodied capital versus reproduction, and the quantity–quality trade-off can be understood in terms of investments in the embodied capital of offspring versus their number.

The brain as embodied capital

The brain is a special form of embodied capital. Neural tissue is involved in monitoring the organism's internal and external environments and organizing physiological and behavioral adjustments to those stimuli (Jerison, 1976). Portions (particularly the cerebral cortex) are also involved in transforming past and present experience into future performance. Cortical expansion among higher primates, along with enhanced learning abilities, reflects increased investment in transforming present experience into future performance (Armstrong and Falk, 1982; Fleagle, 1999).

The action of natural selection on neural tissue involved in learning and memory should depend on costs and benefits realized over the organism's lifetime. Three kinds of costs are likely to be of particular importance. Firstly, there are the initial energetic costs of growing the brain. Among mammals, those costs are largely born by the mother during pregnancy and lactation. Secondly, there are the energetic costs of maintaining neural tissue. Among infant humans, about 65% of all resting energetic expenditure supports maintenance and growth of the brain (Holliday, 1978). Thirdly, certain brain abilities may actually decrease performance early in life. Specifically, the capacity to learn and increased behavioral flexibility may entail reductions in "preprogrammed" behavioral routines. The incompetence with which human infants and children perform many motor tasks is an example.

Some allocations to investments in brain tissue may provide immediate benefits (e.g., perceptual abilities, motor co-ordination). Other benefits of brain tissue are only realized as the organism ages. The acquisition of knowledge and skills has benefits that, at least in part, depend on their impact on future productivity. Consider two alternative cases, using as an example, the difficulty and learning-intensiveness of the organism's foraging niche. In the easy-feeding niche where there is little to learn and information to process, net productivity (excess energy above and beyond maintenance costs of brain and body) reaches its asymptote early in life. There is a relatively small impact of the brain on productivity late in life (because there has been little to learn), but there are higher costs of the brain early in life. Unless the life span is exceptionally long, natural selection will favor the smaller brain.

In the difficult-feeding niche, the large-brain creature is slightly worse off than the small-brain one early in life (because the brain is costly and learning is taking place), but much better off later in life. The effect of natural selection will depend upon the probabilities of reaching the older ages. If those probabilities are sufficiently low, the small brain will be favored, and if they are sufficiently high, the large brain will be favored. Thus, selection on learning-based neural capital depends not only on its immediate costs and benefits, but also upon mortality schedules which affect the expected gains in the future.

The human adaptive complex

The human adaptive complex is a coadapted complex of traits, including: (1) the life history of development, aging and longevity; (2) diet and dietary physiology;

(3) energetics of reproduction; (4) social relationships among men and women; (5) intergenerational resource transfers; and (6) co-operation among related and unrelated individuals (Kaplan, 1997; Kaplan et al., 2000, 2001, 2003, 2005, 2007; Kaplan and Robson, 2002; Robson and Kaplan, 2003; Gurven and Kaplan, 2006; Gurven and Walker, 2006). It describes a very specialized niche, characterized by: (1) the highest-quality, most nutrient-dense, largest package size, food resources from both plants and animals; (2) learning-intensive, sometimes technology-intensive, and often co-operative, food acquisition techniques; (3) a large brain to learn and store a great deal of context-dependent environmental information and to develop creative food acquisition techniques; (4) a long period of juvenile dependence to support brain development and learning; (5) low juvenile and even lower adult mortality rates, generating a long productive life span and a population age structure with a high ratio of adult producers to juvenile dependents; (6) a three-generational system of downward resource flows from grandparents to parents, to children; (7) biparental investment with men specializing in energetic support and women combining energetic support with direct care of children; (8) marriage and long-term reproductive unions; (9) co-operative arrangements among kin and unrelated individuals to reduce variance in food availability through sharing and to more effectively acquire resources in group pursuits.

The publications cited above show that the majority of the foods consumed by contemporary hunter-gatherers worldwide are calorically dense, hunted, and extracted (taken from an embedded or protected matrix – underground, in shells, etc.) resources, accounting for 60% and 35% of calories, respectively. Extractive foraging and hunting proficiency generally does not peak until the mid-30s, because they are learning – and technique – intensive. Hunting, in particular, demands great skill and knowledge that takes years to learn, with the amount of meat acquired per unit time more than doubling from age 20 to age 40, even though strength peaks in the early 20s. This learning-intensive foraging niche generates large calorie deficits until age 20, and great calorie surpluses later in life. This life history profile of hunter-gatherer productivity is only economically viable with a long expected adult life span.

LIFE HISTORIES OF WILD CHIMPANZEES AND HUMAN FORAGERS

To appreciate the implications of the human adaptive complex for the life histories of foragers, it is useful to compare humans with the chimpanzee, another large-bodied, long-lived mammal, and our closest-living

TABLE 26.1. Life history characteristics and diet of human foragers and chimpanzees (after Lancaster and Kaplan, 2008).

Life history characteristics	Human foragers		Chimpanzees
Maximum life span	~100		~60
Probability of survival to age 15	0.6		0.35
Expected age of death at 15 (years)	54.1		29.7
Mean age first reproduction (years)	19.7		14.3
Mean age last reproduction (years)	39		27.7**
Interbirth interval* (months)	41.3		66.7
Mean weight at age 5 (kg)	15.7		10
Mean weight at age 10 (kg)	24.9		22.5
Composition of diet (%)			
Collected	9		94
Extracted	31		4
Hunted	60		2
Contributions by sex (%)	**Men**	**Women**	
Adult calories	68	32	Sexes independent
Adult Protein	88	12	
Caloric support for offspring	97	3	
Protein support for offspring	100	0	

Notes: *Mean interbirth interval following a surviving infant.
**Age of last reproduction for chimpanzee females was estimated as two years prior to the mean adult life expectancy.

relative in phylogenetic terms. Table 26.1 presents major differences in five critical parameters of life history: (1) survivorship to age of first reproduction; (2) life expectancy at the beginning of the reproductive period; (3) absolute and relative length of the post-reproductive period; (4) spacing between births of surviving offspring; and (5) growth during the juvenile period (Kaplan et al., 2000; Lancaster et al., 2000). The data for these analyses are based on published data sets on the only four forager groups for which full demographic data are available as well as food consumption and production throughout the year for all age and sex categories (Aché, Hadza, Hiwi, and !Kung). The data on chimpanzees are based on studies at the African field sites of Bossou, Gombe, Kibale, Mahale, and Tai. The data and full citation list are presented in Kaplan et al. (2000, Table 1, p. 158).

Human and chimpanzee life history parameters based on data from these extant groups of hunter-gatherers and wild chimpanzees indicate that forager children experience higher survival to age 15 (60% vs. 35%) and higher growth rates during the first 5 years of life (2.6 kg/year vs. 1.6 kg/year). Chimpanzees, however, grow faster both in absolute and proportional weight gain between the ages of 5–10 years. The early high-weight gain in humans may be the result of the earlier weaning age (2.5 years vs. 5 years) followed by provisioning of highly processed and nutritious foods, foods that juvenile chimpanzees could never collect to any extent. Fast growth and weight gain during infancy and the early juvenile period may also represent an adaptation to support the energetic demands of brain growth development, since a significant portion of this weight gain is in the form of fat.

The chimpanzee juvenile period is shorter than that for humans with age at first birth by chimpanzee females about five years earlier than among forager women. This is followed by a dramatically shorter adult life span for chimpanzees. At age 15, chimpanzee life expectancy is an additional 15 years, whereas foragers can expect to live an additional 38 years having survived to age 15. Importantly, women spend more than a third of their adult lives in a postreproductive phase, whereas few chimpanzee females spend any time as postreproductives. The differences in overall survival probabilities and life span of the two species are striking: less than 10% of chimpanzees ever born survive to age 40 and virtually none survive past 50, whereas 45% of foragers reach 40 and more than 15% of foragers born survive to age 70!

Finally, despite the fact that human juvenile and adolescent periods take longer and that human infants are larger than chimpanzees at birth, forager women are characterized by higher fertility. The mean inter-birth interval between offspring, when the first survives to the birth of the second, is 1.6 times longer among wild chimpanzees than among modern forager populations.

To summarize, human foragers show a juvenile period 1.4 times longer and a mean adult life span 2.5 times longer than chimpanzees. They experience higher survival at all ages postweaning, but slower growth rates during mid childhood. Despite a long juvenile period, slower growth, an equal length reproductive period, and a long postreproductive life span, forager women achieve higher fertility than do chimpanzees.

CONSUMPTION AND PRODUCTIVITY THROUGH THE LIFE COURSE

Table 26.1 also demonstrates the overlap in component categories of the diets of foraging societies and chimpanzee communities as well as wide differences in relative proportions (Kaplan et al., 2000; Lancaster et al., 2000). For example, hunted meat makes up about 2% of chimpanzee but 60% of forager diets. Chimpanzees rely on collected foods for 94% of their nutrition, especially ripe fruits. Such foods are nutritious and are neither hard to harvest nor learning intensive, at least relative to human resource pursuits. Humans depend on extracted or hunted foods for 91% of their diet. The data suggest that humans specialize in rare but nutrient dense resources (meat, roots, nuts) whereas chimpanzees specialize in ripe fruit and fibrous plant parts. These fundamental differences in diet are reflected in gut morphology and food passage times in which chimpanzees experience rapid passage of bulky, fibrous meals processed in the large intestine whereas humans process nutritionally dense, lower volume meals amenable to slow digestion in the small intestine (Milton and Demment, 1988).

Table 26.1 also summarizes the relative contributions of both sexes to the nutritional support of group members through food sharing, one of the critical features of the human adaptive pattern. This table is based on contributions by sex in 10 modern forager societies (Onge, Anbarra, Arnhm, Aché, Nukak, Hiwi, !Kung (2), Gwi and Hadza) where daily adult caloric production of meat, roots, fruits, and other has been documented (Kaplan et al., 2000, Table 2, p. 162). Generally, women produce virtually no animal protein and the carbohydrate calories they produce help to support themselves and male hunters. As described in the next paragraph, calories and protein consumed by children mostly comes from the large surpluses supplied by adult males.

Figure 26.1 presents survivorship and net food production through the life course of humans and chimpanzees (Kaplan et al., 2000). Humans consume more than they produce for the first third of their life course. In contrast chimpanzees are self-supporting by the age of five. Thus, human juveniles, unlike chimpanzee juveniles, have an evolutionary history of dependency on adults to provide their daily energy needs. Even more striking is the steady increase in productivity over consumption among humans into their 30s and early 40s. Forager males begin to produce more than they consume in their late teens, but their peak productivity builds slowly from their early 20s until their mid-to-late 30s and then is sustained for 20 or more years at a level of approximately 6500 kcals per day. In contrast forager women vary greatly from group to group in energy production, depending upon the

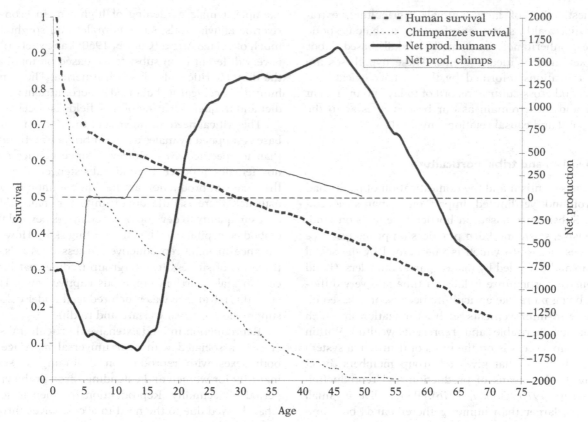

26.1. Survivorship and net food production through the life course of humans and chimpanzees. After Kaplan and Lancaster (2003).

demands of intensive childcare (Hurtado and Hill, 1990). In some groups, they consume more than they produce until sometime after menopause, when they are finally freed from childcare demands; whereas in others, such as the Aché, they remain nutritionally dependent on men throughout their lives. The provisioning of reproductive women and children has a powerful effect on the production of children by humans by reducing the energy cost and health risk of lactation to the mother and by lifting the burden of self-feeding from the juvenile, thus permitting a shortened interbirth interval without an increase in maternal or juvenile mortality (Hawkes et al., 1998).

The human adaptive complex is both broad and flexible, in one sense, and very narrow and specialized in another. It is broad in the sense that as foragers, humans have existed successfully in virtually all of the Earth's major habitats. It is narrow and specialized in that it is based on a diet composed of nutrient-dense, difficult-to-acquire foods and a life history with a long, slow development, a heavy commitment to learning and intelligence, and an age-profile of production shifted towards older ages. In order to achieve this diet, humans are very unproductive as children, have very costly brains, are extremely productive as adults, and engage in extensive food sharing both within and among age- and sex-classes.

EMBODIED CAPITAL AND EXTRA-SOMATIC WEALTH IN THE PAST 10 000 YEARS

For most of human history from perhaps 2 million years until 10 000 years ago, humans depended on investments embodied in their brains and bodies to survive and reproduce. They invested in themselves and their offspring through patterns of behavior that emphasized accessing high energy, hard to acquire foods that demanded skilled, learned performances, food sharing, the feeding of juveniles, and a complementary division of labor between men and women. However, about 10 000 years ago at the end of the last Ice Age, the distribution of resources that humans depended on and the means to access them began to change as a result of climate change and an increase in population density. At the beginning these changes had little effect except to promote population growth. Later their effects were so profound that patterns of marriage, investment in children, and social organization appeared to reinvent themselves.

In the following sections we will evaluate the impact that changes in subsistence base and social organization made on the division of labor, family formation strategies, fertility, and investment in children in response to sedentism, horticulture, the domestication of large animals, agriculture, extra-somatic wealth, social stratification, archaic despotic states, inheritance, and modern skills-based labor markets and political systems. These hypotheses are generalizations informed by the archaeological, historic, and cross-cultural record of today and the recent past and must remain as our best guesses as to the temporal and causal relations involved.

Sedentism and tribal horticulture

Village sedentism and the domestication of plants had a profound, yet limited, impact on human socioecology. Subsistence based on horticulture rests on land-extensive, slash-and-burn practices on prime resource patches, access to which is maintained by the social group and defended by males against outsiders. (Land intensive horticulture is later in time and very different, being more like agriculture because it is based on long-term improvements such as irrigation in which fields are heritable and represent wealth.) Within the group, access is on the basis of usufruct, a system of land tenure that gives all group members direct rights to the means of production and reproduction (Boserup, 1970; Goody, 1976). People live in small villages, larger than hunter-gatherer bands but similarly scaled in terms of face-to-face, kinship-laden interactions.

There is evidence that sedentism brought a reduction in child mortality compared to hunter-gatherers, as well as higher female fertility, although it is unclear whether the strongest effects are in reduced birth spacing or in higher rates of child survival. Bentley et al. (2001), in comparing the fertility changes associated with the prehistoric transition to agriculture report that, when comparing subsistence modes and fertility rates, forager, horticultural, and pastoral groups had similar fertility rates whereas increases were strongly associated with a higher dependence on agriculture. The potential of deaths from chronic intergroup warfare and raiding increased. Using the archaeological and historic record, Keeley (1996) found that for males the percentage mobilized in war often reached 35–40% and male deaths ranged from 10% to nearly 60%.

Among contemporary horticulturalists, complementarity in the male and female division of labor is complex because of its link to local ecology. Garden production by women using the digging stick and hoe provides the carbohydrate and caloric base of the diet and is easily combined with childcare (Boserup, 1970; Goody, 1976). Males contribute their labor in clearing fields, in animal protein through hunting and fishing, and in protection of the village resource base through defense. The relative contribution, type, and imperative of male help varies by ecological context. For example, female gardening of high-protein crops on riverine alluvial soils, such as millet and sorghum in much of village Africa (Colson, 1960; Lancaster, 1981), is very different from subsistence based on manioc in the thin, lateritic soils of South America. There, male hunting is critical to balanced macronutrients in the diet and frequent clearing of new fields is necessary.

The critical need for defense of the village resource base is supplied by males as an umbrella benefit rather than to specific wives. However, since neither males nor females produce beyond subsistence needs and the means of production are held in common through usufruct, there is little opportunity for major differences in quality to develop between males beyond their embodied capital (age, health, hunting skill). However, variance in male reproductive success does arise on the basis of success in intergroup raiding that brings certain male warriors numerous captive wives. However, this advantage was much reduced under colonial suppression of tribal warfare and raiding.

Reproduction in land-extensive, horticultural societies is associated with near universal marriage for both sexes with reproduction beginning at sexual maturity for women and extending through the entire period of fecundity. Reproduction for men is somewhat delayed due to the need to access wives through either bride service (local group) or bride capture (outside group); the first being a personal cost in labor contributed to the bride's family and the second a cost in risk. However, the possibility of polygyny extends the male reproductive period as new and younger wives can be added through time.

The high frequency of polygynous husbands associated with horticulture is likely because each wife is essentially able to support herself and her children through her own labor (Murdock, 1967; Lancaster and Kaplan, 1992). Males do not have to ponder whether they can afford additional wives and children, only how they get and keep them. As White and Burton (1988) found, polygyny is most associated with fraternal interest groups, warfare for the capture of women, absence of constraints on expansion into new lands and, especially for horticulturalists, environmental quality and homogeneity. The frequent practice of widow inheritance by husband's kin also increases the frequency of polygyny (Kirwin, 1979). Sororal polygyny (the marriage of sisters) is at its highest frequency among horticulturalists perhaps due to the ease of closely related women forming collaborative, horticultural work groups and childcare (White and Burton, 1988).

Parental investment in horticultural societies focuses on raising healthy children without concern for their marriage market endowments of extra-somatic wealth or inheritance of resources. Birth into a social group provides all the inheritance a child needs to access the means of production and reproduction. Such concepts as bastardy or disinheritance do not play a formal role in family dynamics. Child labor is valuable to families since horticulture provides a number of relatively low-skilled tasks that older children can perform (Bock, 2002b). In fact Kramer (2005b) demonstrated that among Maya horticulturalists older children contribute at the level of "helpers-at-the-nest," significantly increasing their parents' fertility and without whose help their parents could not add additional offspring to the family.

Variance in reproductive success is relatively low for women because marriage is universal, and female fertility and fecundity depend on their own health, productivity, and work effort (Prentice and Whitehead, 1987; Jasienska, 2000; Ellison, 2001). Greater variance among men is possible on the basis of raiding and bride capture but the social system itself is not stratified and individual men cannot amass or control access to resources relative to other men or pass them on to their sons.

Tribal pastoralism and extra-somatic wealth

For most of human history, humans depended on somatic wealth or embodied capital to fund growth and reproduction. However, the domestication of animals, particularly large herd animals such as cattle, camels, and horses, had a profound effect on human social and reproductive patterns. Large, domesticated livestock have intrinsic qualities that affected human social relationships, marriage patterns, and investment in children. For the first time in human history, men could control a form of extra-somatic wealth that could be held by individuals, thus increasing the variance in male quality based on the resources each can control.

Secondly, herds are the basis of a domestic economy through their products of meat, milk, and hides. There are advantages to dependence on such a resource supply: (1) improvements in diets rich in animal protein; (2) stability of diet since animals are stored hedges against fluctuation in annual or seasonal climatic effects; and (3) flexibility due to the divisibility of herds into smaller units that can be moved about the landscape on the basis of the richness and concentration of local resources (Barth, 1961). This improvement in diet may result in higher survivorship of women and children compared to foraging and horticulture, but also results in higher mortality for males due to endemic conflict.

Large-animal herding demands a high degree of complementarity between female processing and childcare and male risk-taking in herd management and defense. The products of herds require intensive processing of meat, milk, and hides, labor provided by women. In contrast, the very existence of extra-somatic wealth in large stock in such a readily divisible and moveable form (as opposed to agricultural land) puts a high premium on males as defenders and raiders. We find the warrior complex full-blown, with chronic internal warfare, blood feuds, social segregation of a male warrior age class, fraternal interest groups, a geographic flow of women from subordinate to dominant groups through bride capture, and expansionist, segmentary lineages based on the male line (Sahlins, 1961; DiVale and Harris, 1976; White and Burton, 1988; Low, 2000). Men with strong social alliances are more likely to find at least some of their wives from within their own social groups, whereas men from small or subordinate lineages are less likely to be offered brides and are willing to take more risks in lieu of performing bride service (Lancaster, 1981; Chagnon, 1988, 2000).

The original distinction made by Orians (1969) between resource defense polygyny and harem defense polygyny is relevant here. The chronic warfare of pastoralists (White and Burton, 1988; Manson and Wrangham, 1991; Keeley, 1996) can be understood as resource defense polygyny, as opposed to harem defense polygyny described earlier for horticultural societies. Both types of societies raid to capture women to form polygynous unions (harems), but pastoralists also raid to capture resources that can be used to acquire and maintain new wives and their children. In later socially stratified societies, successful male resource holders do not have to do bride service, pay bride wealth, or capture brides; brides will flock to them and their families will even pay for the opportunity for their daughters to marry such a quality male. In a study of 75 traditional societies, the principal cause of warfare was either to capture women (45% of cases) or steal material resources to use to obtain them (39% of the cases), particularly in pastoral societies where bride wealth must be paid (Manson and Wrangham, 1991).

Resource defense polygyny means that males will compete to control the resources that females must have for successful reproduction. A male's ability to successfully control more resources translates directly into more wives and children (Borgerhoff Mulder, 1985, 1988b, 1989). One extraordinary result of extra-somatic wealth, particularly readily partible wealth, is the institution of a new pawn on the marriage market table, bride wealth. Women and their families come to marriage negotiations with their traditional offers of embodied capital (youth, health, fecundity, and female labor). Men, however, now have to come up with a significant payment of extra-somatic resources in the form of bride wealth as a preferred substitute for bride service. Men who depend on bride service are limited in their polygyny because of the years of service each bride's family requires. Men who inherit resources can start their families early and marry often.

Bride wealth among pastoralists consists of horses, cattle, or camels with sheep or goats as supplements or lower-valued substitutes. Among African pastoralists the close male kin of the groom help him with his first bride-wealth payment, but the acquisition of subsequent wives is his own responsibility. Livestock used for bride wealth has interesting attributes: (1) it creates conflicts of interest between fathers and sons and among brothers for its use to obtain a bride (Borgerhoff Mulder, 1988a); (2) men from poor families will be more willing to take risks to obtain bride wealth or brides though capture (Dunbar, 1991); and (3) livestock can be inherited.

Investment in children takes a novel form under a pastoralist system. The payment of bride wealth improves health and survivorship among young girls because their marriages bring in resources that can be used by their fathers and brothers to acquire more wives (Borgerhoff Mulder, 1998). Sub-Saharan Africa is notable for the fact that in spite of the patrilineal bias in so many societies, neither a survival nor a nutritional advantage is found for boys over girls (Svedberg, 1990). Furthermore, among the Kipsigis, who are agro-pastoralists, early maturing (and presumably better fed and healthier) women have higher lifetime reproductive success than late-maturing women. As a result, they command higher bride wealth and hence constitute a higher return on parental investment for their upbringing (Borgerhoff Mulder, 1989). They also represent a better investment for a husband's bride-wealth payment because of a higher return in fertility.

Furthermore, children are able to provide child care of younger siblings as well as low-skilled labor in stock care and the processing of animal products, so they are able to substantially but not completely offset the costs of their rearing compared to foragers (Bock, 2002a, 2002b). Child labor plays an important role in the economies of both pastoral and agricultural societies because their contributions through simple tasks such as carrying water contribute to food production by freeing mothers to become more effective producers (Blurton Jones et al., 1994; Kramer, 2005a). However, this reduction in cost of rearing is countered by the fact that the parents of sons now have a new cost to meet; the balloon payment (bride wealth) needed to establish sons on the marriage market. The flow of stock through families who are both bride-wealth receivers and givers helps maintain the system, at the same time that it creates problems for families with unfavorable ratios of sons to daughters (Borgerhoff Mulder, 1998).

Finally and most significantly, there is suggestive evidence that for the first time humans begin to reproduce at levels that may not maximize the number of descendents in association with the appearance of extra-somatic wealth and its inheritance. Among modern East African pastoralists men appear to marry

fewer wives than they could afford in the interests of providing each child with a greater endowment. In other words, male pastoralists may pit quality against quantity of children to preserve a lineage status and resource base and rather than simply maximizing the immediate number of descendents (Luttbeg et al., 2000; Mace, 2000).

Social stratification, states, and despotism

The rise of civilizations, beginning about 6000 years ago in Mesopotamia and occurring at different times and places around the world (for example, Egypt in the Near East, the Aztec and Inca in the Americas, and India and China in Asia) marked a critical shift in how humans organized themselves in social systems and in relation to the environment (Goody, 1976; Betzig, 1993; Summers, 2005). These civilizations appear to have developed independently in response to local conditions without being the products of either conquest or diffusion. In spite of this historical independence, they evidence significant similarities: (1) the presence of large, stratified social groupings settled on particularly large and productive resource patches; and (2) the appearance of social despots, men who use coercive political power to defend their wealth and reproduction and practice warfare to acquire more resource patches and slaves (Betzig, 1986). These two major effects flow from the nature of the resource patches.

The patches upon which the first civilizations were settled had special qualities: (1) they were highly productive but set in environments where there was a rapid fall off to unproductive lands such as desert or forest; and (2) these productive patches could not be intensively utilized without complex political organization as in regional irrigation systems. Political control and organization rested on the power of men. Although female primates often form alliances with their female kin to protect and control access to the resources necessary for their reproduction (Isbell, 1991; Sterck et al., 1997), the reproductive benefits of extra-somatic resources are much greater for men than for women, because of their impact on polygyny. The end result of these environmental conditions associated with early social stratification was that men competed for control of the resources necessary for reproduction, formed despotic hierarchies involving social alliances and stratification, with low-status men "agreeing" to live under political despotism because they could not readily move to another resource base.

The increased reliability of food resources, the costs of warfare, and the concentrations of large populations into small and sometimes urban areas each had impacts on mortality and morbidity. A cross-cultural analysis of fertility and mode of subsistence found that, for a 10% increase in dependence on agriculture between two related cultures one of which moves towards agriculture, there is a fertility increase of approximately 0.2 live births per woman (Sellen and Mace, 1997). Bentley et al. (2001), in reviewing the cross-cultural and archaeological evidence, suggest a series of multidirectional effects: higher fertility due to more consistent food supply and earlier maturation; increased infectious diseases with regular visitations as well as endemic diseases (malaria and tuberculosis) due to long-distance trade and large urban populations; and a shift in peak mortality from infancy to middle childhood. Furthermore, warfare continues to reduce the numbers of young men in the mate pool.

With social stratification comes a complex division of labor with specialists in war, farming, crafts for the production of goods and services, and war captives and slaves for the hardest manual labor, as well as long-distance trade in luxury goods and slaves. The introduction of the plow in Eurasia, perhaps as early as the sixth century BC, and the need for food production beyond simple subsistence to service urban markets led to significant changes in the division of labor (Goody, 1976; Ember, 1983) and extremely high complementarity between male labor and resource acquisition and female labor and childcare. There is evidence of increased workloads for women in spite of the fact that men assume more responsibility for farm labor, because of increased demands for women to process grains or secondary animal products such as milk, hides, and wool (Bentley et al., 2001).

Variance in male fertility in these first civilizations in the Near East, Central and South America, and Asia was probably the greatest it has ever been before or after in human history (Betzig, 1986, 1992a, 1992b, 1993; Summers, 2005). The reason for this is that despotic males had enormous political and social control with the ability to eliminate rivals and their entire families through despotic edict, to wage war to increase personal and state resource bases, to acquire slaves and war captives for labor and reproduction, and to determine political succession for favored sons. This extreme variance in male resource holding inevitably produces social and political instability due to the creation of too many potential heirs (sons of many wives) and too many males (slaves) without access to the means of reproduction. The great wealth to be gained from domination also motivated expansion and intergroup conflict among would-be despots.

Despotic males are an extreme example of resource defense polygyny (Orians, 1969); that is, as individuals they control access to the resource base for reproduction that females require and, with few competitors, polygynous marriages to them become the only family formation strategy option for many women. The mating markets of despotic systems are characterized by historic extremes in male variance in resource holding and power. As Betzig (1993) notes, the extreme sizes of royal harems ranging from 4000–16 000 women are associated with smaller but still impressive numbers of wives and concubines for the royal relatives and supporters. In the case of the Inca the size of a man's harem was regulated by law and in direct relationship to his social/political rank (Betzig, 1993). Among the Inca there were nine levels of political rankings with polygyny ceilings for each except the topmost. These harems were exclusive holdings of large numbers of young, fecund women with their children and sexual access to them was restricted to their mate and regulated with some sophistication to optimize female fertility. Many of these wives and concubines were collected as tribute or war booty; but others, as principal wives, probably represented important political alliances with their male relatives.

Variance in male quality and the marriage market

There are two clear outcomes of such extreme variance in male quality. The first is that many men remain unmated or have only one wife, so that male celibacy or at least nonmarital sex is prominent. In the words of Dickemann (1981, p. 427), polygyny in the context of extreme social stratification is "characterized not only by arbitrary sexual rights of lords and rulers but by large numbers of masculine floaters and promiscuous semi-floaters, beggars, bandits, outlaws, kidnappers, militia, and resentful slaves and serfs." Nevertheless, these early despotic states lasted for thousands of years. A second outcome of variance in male resource holding and male mating success is that there tends to be universal marriage for women with only those most severely compromised by health or other personal qualities being unlikely to find a role as secondary wife or in a minor union. For access to the mating market men must bring extra-somatic wealth, power, and land in order to be favorably placed or else get wives as high-risk booty in state warfare (Low, 2000; Clarke and Low, 2001).

Women, too, bring their traditional embodied capital qualities of youth, health, and fecundity along with their labor for access to the marriage market. However, there was a historic shift in how women and their families approached marriage negotiations that has been richly described by Dickemann (1979a, 1979b, 1981) in a series of papers on hypergyny, dowry, female infanticide, and paternity confidence. The extreme variance in male quality created by despotism and harem polygyny forces the families of women to put down more and more value on the mate market table to access a desirable groom or to move a daughter up in the social hierarchy. These extra payments include actual wealth, in the form of dowry, and guarantees of paternity confidence (bridal virginity and wifely chastity). Guarantees of a daughter's virginity and chastity (a prerequisite for a bride destined to produce heirs to a male lineage holding a reproductive estate) are costly forms of embodied capital, involving female seclusion (special women's quarters, harem guards, chaperones), and female incapacitation (foot-binding and corseting) that bars their daughters from the outside world of productive labor.

Parental concern over the ability of their children to access reproductive estates transformed the nature of the marriage market. Parental investment in these systems varies in relation to the power and wealth of the male's family. As is to be expected, under such conditions where male access to and control of resources is the basis of social stratification, patrilineal descent and patrilocal residence are highly favored since males are the principal resource holders (Hartung, 1982). Resource-holding parents commit to a "balloon payment" in launching their children in marriage. This balloon payment takes the form of endowments and promised inheritance for sons and dowry as anticipatory inheritance for daughters (Goody, 1973, 1976; Dickemann, 1979a, 1979b). For resource-holding families then, the marriage market formed by stratified social systems proved costly in terms of parental investment and forced a focus on endowments for both sexes at the age of marriage. Poor parents, on the other hand, attempted to balance labor demands with fertility, since in agricultural systems children can be productive at low skill tasks or childcare and add to the family economy. Thus, they might try to regulate birth spacing to optimize the productivity of already born children before another mouth to feed is added to the family.

Finally, a notable characteristic of the premodern period in many parts of the world is evidence for a growing rural population resulting from higher fertility and an associated growing concern regarding saturation of the resource environment. This is often associated with urban growth, empire building, and expansionism, providing opportunities for migration by noninheriting or low status children to areas of both higher mortality and risk but also with the potential for the acquisition of land, or wealth and power. It also generated a new concern about keeping the family estate intact and about the management of inheritance.

Premodern states and narrowing the pool of inheritors

With population growth and increased saturation of arable lands, parents adopted patterns of restricted and differential inheritance in order to keep the family estate intact and maintain the concentration of wealth, or in the case of the poor, to balance food supply with family size. This trend, although it occurred in response to population pressure on existing resource bases all over the world at different time periods, is particularly well documented in premodern Europe. Human evolutionary ecologists in collaboration with historical demographers provide us with a unique record of the relationships between fertility, family formation strategies, and socioecological context during the premodern and early modern periods of European history (Voland, 2000). Their studies, based on heraldic or parish records of births, marriages, deaths, and inheritance of estates, can be used to directly link reproductive strategies with resource holdings. This time period witnessed developments that had began centuries earlier but occurred without the benefit of quantifiable documentation. Boone (1986a, 1986b), for example, traces the historic process of parental investment among Portuguese elites during the late medieval/early modern periods of the fifteenth and sixteenth centuries. Saturation and resource stress are evident with a progressive narrowing of the numbers of claimants to an inheritance, first through monogamy to create a single bloodline of inheritors and bastardy to disenfranchise offspring who are not the product of a legitimate union (Goody, 1976, 1983), followed by a preference for sons over daughters as inheritors, and finally by birth order effects with preference for primogeniture within each sex for access to resources and the creation of celibate children to live as priests, nuns, bachelors, and spinsters (Hrdy and Judge, 1993). For the first time in human history mating and reproduction is no longer a universal for women and siblings of the same sex are pitted against each other in competition for access to reproductive estates. With survival through child- and young adult-hood still quite problematic, ancillary practices develop in which both sons and daughters would be held in reserve in monasteries and nunneries for inheritance and reproduction should their older same-sex sibling die (Goody, 1976, 1983; Boone, 1986a, 1986b). Within the scope of these restrictions that limit half-sibling and sibling competition, parents with wealth raise as many children as they can but endow a select number at adulthood.

During most of this historic period there is a strong correlation between wealth, probability of marriage, younger age at marriage, and completed fertility (Voland, 2000). However, restricted inheritance decreases the reproductive benefits of polygyny. The desire to concentrate wealth also limits the reproductive success of noninheriting sons and daughters. This is a second striking example in which reproductive and parental investment behavior in response to extrasomatic wealth results in outcomes that did not maximize parental fitness. In fact, towards the end of the period, as life expectancy improved and economic structures became saturated, resource holding groups delayed marriage into the late 30s and early 40s for men and mid 20s for women (Szreter and Garrett, 2000; Voland, 2000).

The family reconstruction studies document very different reproductive strategies according to class[1]. Generally, wealth brings higher probability of marriage, at a younger age, to a younger spouse, and more children. However, as environments become more saturated, local resource competition among siblings differentially affected resource-holding families, as opposed to day laborers, and increased the likelihood of dispersal of later-born children (Clarke and Low, 1992; Voland and Dunbar, 1997; Towner, 1999, 2001). With saturation, the benefits to resource holders of having an above average number of children was offset by more and more intense sibling competition for access to inheritance (Voland, 2000). Parents without resources had no need to manipulate their offspring and were more likely to benefit from opportunistic strategies by their children (Voland and Dunbar, 1995).

Wet-nursing presents a fascinating example of how differentiation in parental investment strategies develops into extreme forms for both the highest and the lowest status groups of women. Throughout human history there has always existed a conflict between production (acquisition of food) and reproduction (lactation and childcare) for women, a conflict that in fact troubles female mammals in general. Human women are especially caught in this conflict because they have multiple, dependent young of differing ages and needs (Draper, 1992), which means that true respite never occurs until all children are reared. Cross-culturally women's work is organized by its compatibility with childcare (Brown, 1970); however, this compatibility is never complete – only more or less so (Hurtado and Hill, 1990; Lancaster, 1997; Lancaster et al., 2000). As a result of social stratification high status women are able to subvert the physiological capacity of lactation of other women to serve their own reproductive ends. Since intense breast-feeding lowers the likelihood of ovulation, a wet-nurse, even if paid, sacrifices her own fecundity to another (Hrdy, 1994, 1999). Typically high status women did not breast-feed their own children but used wet-nurses. This increased the fertility of high status women,

whose main function was to produce heirs, to a nearly annual birth rate (among the highest for any group of women in human history). In contrast, the birth spacing for wet-nurses was closer to four years (Hrdy, 1994). A second group of women also used wet-nurses, especially towards the end of this historic period. These were single women working in urban centers or the wives of poor tradesman who found themselves in positions of servitude or trade where the incompatibility between breast-feeding and work was complete. To the great detriment of their infants' survival, these women placed their children with commercial wet-nurses at baby farms (Hrdy, 1994). In these cases the demands of maternal work far outweighed the needs of infant growth, perhaps to improve the development of weaned, older children.

The past 10 000 years of human history brought many changes to what was originally the forager adaptive niche. As the last glaciers withdrew, humans began to intensify their extraction of resources from the environment by domesticating plants and animals. At first, land extensive horticulture combined with hunting did little to alter the human experience of small groups, face-to-face social networks, and subsistence economies. Family formation practices continued the relatively low rate of polygyny, nearly universal marriage, bride service and bride capture, and the production of children regulated only by the health and well-being of the mother and each child.

The first transformation in human experience followed from the appearance of extra-somatic wealth in the form of large domesticated animals and later land. Extra-somatic wealth has an intrinsic quality, it can be taken by force and stronger individuals and groups can amass or control access to it. This necessarily creates much wider variance in male quality than occurs in forager men dependent on embodied capital investments of health, vigor, and skill. Family formation strategies responded by turning the old rules upside down. Instead of men paying for access to wives with bride service, bride wealth, or the dangers of bride capture, payments in the marriage market reversed direction. As social groups became stratified and wealth differentials increased, women and their families began to bring more to the bargaining table by offering both dowry and paternity confidence in the form of virginity and chastity. During the final phase of the premodern period, societies became obsessed with the preservation of the family's reproductive estate by successively narrowing the possible number of inheritors. First the line of inheritance went only to the children of the principal wife with others labeled with bastardy, then daughters could only inherit via a dowry lower in value than a son's inheritance, next only the first or a selected son could be endowed with the family estate and the rest had to find other niches

---

[1] The behaviors of nobility are documented by Boone in Portugal (Boone, 1986a; Kramer, 1998) and Dickemann in Europe, the Middle East, China, and India (Dickemann, 1979b, 1981), and for gentry and land-holding peasants as well as day laborers by Voland and colleagues in Germany (Voland, 1990; Voland and Engel, 1990; Voland et al., 1991, 1997; Voland and Dunbar, 1995, 1997; Voland and Chasiotis, 1998; Voland, 2000), Low in Sweden (Low, 1990, 1991, 1994), Towner in the United States, and Hughes (1986) and Scott and Duncan (1999) in England.

in the society or migrate. Just before the onset of modernization, the world had become full of bachelor and spinster noninheriting children with no guaranteed access to the right or means of reproduction, qualities of life that were part and parcel of the original human adaptive niche.

The modern world and embodied capital

Given rural reproductive and survival rates, the restricted inheritance system discussed in the previous section produced excess adults without access to land and the means of production. Colonization through conquest was one response by males to this situation, especially later-born sons (Boone, 1986a, 1986b; Curtin, 1989). Another response by both men and women was to provide services for others, and migration to cities in search of employment. This supply of labor and of consumers helped fuel the growth of a skills-based, mercantile economy that was to gradually supplant the power- and land-based hierarchies of the premodern period based almost entirely on extra-somatic wealth.

Those conditions set the stage for dramatic changes in reproductive and parental investment strategies. In the early 1800s, changes in the relationship between humans and their economies began in a small part of the world, Western Europe, including England (Clark, 2007). This change has been labeled the "demographic transition." For nearly all of human history, available evidence suggests that human populations responded to greater resource availability with increased fertility, and reduced fertility when resources were scarce. Women's reproductive physiology appears well designed to make adaptive responses to increases and decreases in energy availability (Ellison, 2001, 2003).

However, after 1800 with the demographic transition, the size of human families began to shrink, first among the wealthier segments of society, even as standards of living and energy availability were increasing. Unlike before when individuals in good condition had more progeny than individuals in poor condition (just as is the case with other species), higher status humans began having fewer children than did the poor.

This trend only lasted for a time. Today there are remarkably few differences between classes or even ethnicities in completed family size. For example, in the United States today the average US woman produces 1.9 children, considerably below the replacement level of 2.2 children. When broken down by ethnicity, the numbers are 1.7 for Asian Americans, 1.8 for non-Hispanic Whites, 2.0 for Blacks and 2.3 for Hispanics (US Census Bureau Report, 2008). Although the range between the highest and the lowest is three-fifths of a child, the main message from this data is consensus: two children are enough.

Not all the world today has experienced the demographic transition, but completed family size of replacement level or less is typical of modern economies with skills-based labor markets as in Western Europe, North America, Japan, China, and parts of Latin America (Cuba, Chile, Costa Rica, Puerto Rico, and Trinidad and Tobago) (Population Reference Bureau, 2008). Furthermore, for the first time the world population is evenly divided between rural and urban areas, and by 2050 urban residents are likely to make up 70% of the world's population (Population Reference Bureau, 2008). This reversal in family reproductive strategies from having as many children as possible to only two is related to a strategic shift from quantity to quality, in which quality is most often expressed in education and training to be used for access to resources, not inheritance.

Modern skills-based labor markets and the expenditure of extra-somatic wealth to embody human capital

Changes in the nature of resource production and the economic forces that determine wages in labor markets appear to underlie these changes in reproductive and parental investment strategies, and explain their patterning over time and space. The directional change in the nature of labor markets towards greater wage premiums for skill- and education-based capital over the last two centuries is well documented (Newcomer, 1955; Burck, 1976; Herrnstein and Murray, 1994; Vinovskis, 1994; Clark, 2007). As the extent of the labor and consumer markets grew, along with advances in production technologies, there was a concomitant increase in both private and public investments in education. In a sense, the relationship between embodied capital and production in modern skills-based labor markets is more similar to the foraging life way than to its agricultural predecessor. Rather than generating wealth through control of land, people now invest in learning to increase productivity, and individuals are free to move through the environment in search of economic opportunities because they carry their embodied capital with them.

These increases in educational capital investment and the nature of labor markets were accompanied by improvements in the "technology" of disease prevention and treatment, and by increased public and private investments in health and mortality reduction. During the nineteenth century, there were large changes in the scientific understanding of disease (Preston and Haines, 1991). This led to a dramatic decline in infant, child, and adult mortality rates that continued for close to a century. As scientific advances enabled reductions in mortality rates, there was strong pressure to increase public investments in health and disease prevention from the protection of the water supply to the development of vaccines and public access to medical care. As a result, infant and child mortality rates reduced dramatically, greatly increasing the probability that investments in children will be realized in terms of productive adulthoods. The length of the productive adult life span, especially when time lost to morbidity is taken into account, also increased significantly. Together, the two shifts in production processes and mortality rates favor increased human embodied capital investment in a way that is reminiscent of the initial dietary shift leading to the hominid specialization discussed above (Kaplan et al., 2002).

This historical process also resulted in much greater labor force participation by women. During the initial demographic transition in the developed world, the breadwinner–homemaker family structure was dominant. With increased demand for labor that requires skill as opposed to strength and with growth in the service sector of the economy, wage-earning opportunities for women increased. At the same time the payoffs to "home" production decreased with labor-saving devices, such as washing machines and refrigerators, and smaller family size reduced the number of years spent caring for small children. Over time the value of male strength through labor and the time women spent caring for small children was reduced; thus leading to a trend from greater to lesser complementarity between men and women so that men and women are now closer to being interchangeable units in work effort.

Although the shift towards an education-based wage structure has been largely monotonic, those changes occurred at different times in the developed and developing worlds and the details of the supply and demand for labor of different levels of human capital have been both historically and regionally variable. Moreover, both within and among societies, there appears to be a great deal of variation in rates of return on investments in educational capital.

The production of human capital is also human-capital intensive (Becker and Barro, 1988) and associated with a reduction of the value of children's labor as their time is taken up with education and training. To see this, it is useful to think of an "education production function." In each year of a child's life the amount a child learns, and the changes in his or her knowledge, reading, writing, logic, and mathematical skills, will depend upon many different inputs, such as the child's time, prior abilities, parents' time, and teachers' time. The value of those inputs, in terms of the educational capital produced, depends on the quality of in those inputs. First, consider inputs of parents' time. There is significant evidence that the nature of parent–child interaction varies with the educational level of parents (Hart and Risley, 1995; Hoff-Ginsberg and Tardiff,

1995). For example, Hart and Risley report that, by the age of three, children have heard six million words if their mothers are professionals, three million words if their mothers are "working class," and only one million words if their mothers are on welfare. By the time children enter the public education system there are clear differences among them in school-related skills, and those differences are related to socioeconomic status.

Second, the rate at which a child learns may depend on the knowledge and skills she already possesses. Much of the education offered in schools is based upon the premise that knowledge is cumulative (Cromer, 1993). Basic skills are acquired first, and those skills are used as a foundation for the acquisition of the next set of skills. This implies that the impact of the child's time inputs would depend upon skills already in place. It also means that the net increase in embodied capital at each age is a function of both the quality of inputs, and the embodied capital acquired at younger ages.

Moreover, those qualities tend to be correlated across inputs. Children with more educated parents also attend better schools with better teachers and better fellow students. At the other extreme, children in developing nations often come from families in which neither parent has had formal schooling and attend schools with very large class sizes, almost no library resources, and teachers with only primary education themselves. Under those conditions, much less is learned per year spent in schooling. For example, in a study of a predominantly Black township school in Cape Town, South Africa, Anderson, Kaplan and Lam (unpublished manuscript) found that on average, it took children 15 years to complete 12 grades of schooling. By that age (20–21), only about 10% of students have passed the final matriculation exam and earned a high school diploma. The variance in those inputs leads to an increasing differentiation in educational capital with age.

This within-population heterogeneity in the costs of embodying capital in children means that the environment does not determine diminishing returns to parental investment as it would be in primary production economies, but will be frequency-dependent. Individuals with low levels of human capital are more likely to be unemployed as well as having a lower income, when employed. This is especially true in urban areas in the developing world. The massive rural to urban migration over the last four decades has resulted in very large populations of people with low levels of education competing for a limited number of low skill jobs in the economy. In many places, male unemployment can be as high as 70% or more. This variability in educational capital, along with its impacts on income variation both across individuals and within individuals over time, has profound effects on family formation and reproduction.

In fact, the link between education and income increased in intensity during the second half of the twentieth century. For example, real wages actually dropped from 1958 to 1990 among men without high school degrees in the United States. In 1958, men with graduate education earned about 2.3 times as much as men with elementary education; by 1990, they earned more than 3.5 times as much. Wage differentials among men with some college education, bachelor's degrees, and graduate degrees also increased substantially. For women, wage differentials among educational attainment levels increased substantially in the 1980s (Kaplan et al., 2002).

THE HUMAN ADAPTATION: SOMATIC AND EXTRA-SOMATIC INVESTMENTS

Human history is based on a remarkable coevolved pattern of investment in a large brain, slow growth, long life, and access to resources based on skills-based performances. This pattern, along with marriage, a complementary division of labor between the sexes, food-sharing, and the support of offspring well into adulthood, has allowed humans to people the world and control the top of the food chain wherever they go. Most of the human history of the past two million years depended on capital investments in mind and body; embodied capital in the form of skills, experience, immune function, and social networks created the opportunity for adaptations to highly variable socioecological environments. Then, only 10 000 years ago the domestication of animals and plants shifted the value of such investments to the point that social organization, family structure, marriage patterns, status of women, and investment in children all seem obsessed with access, control, and defense of extra-somatic wealth in the form of animals or productive land. It is only in the last 200 years that the pendulum, linked to the appearance of skills-based, competitive labor markets demanding a new form of embodied capital based on education and training for access to resources, has reversed its swing. Today humans are moving into a new, more flexible productive niche based on the investment of extra-somatic wealth to embody human capital in the form of education and training.

DISCUSSION POINTS

1. Why was the division of labor between males and females such a critical feature in human evolution? How does it differ from feeding and parental investment patterns typical of most mammals? How did it alter how humans access and distribute food?

2. What is the impact of the socioecological context on human marriage markets and family formation strategies in terms of the distribution of resources and the means to access them?

3. More and more modern societies are experiencing a reduction of completed family size to replacement level (2.2 offspring) or below. Why should this be so when these societies have the highest standard of living known in human history? How might such small family size impact parental investment and family formation patterns?

4. Why does variation in male quality impact the marriage market? What features in male quality have been important in different kinds of economies and social organization? Are these qualities inherent or acquired? How so?

5. What are the factors that have led to a division of labor in which female work is closely linked to compatibility with childcare? Are these factors as salient today as in the past?

6. Embodied capital has been critical to human affairs in both the simplest and most modernized societies. What are the similarities and differences in embodied capital in these two contexts?

REFERENCES

Armstrong, E. and Falk, D. (ed.) (1982). *Primate Brain Evolution: Methods and Concepts*. New York: Plenum Press.

Barth, F. (1961). *Nomads of South Persia*. Boston: Little, Brown and Co.

Becker, G. S. and Barro, R. J. (1988). A reformulation of the economic theory of fertility. *Quarterly Journal of Economics*, **103**, 1–25.

Bentley, G. R., Paine, R. R. and Boldsen, J. L. (2001). Fertility changes with the prehistoric transition to agriculture. In *Reproductive Ecology and Human Evolution*, P. T. Ellison (ed.). Hawthorne, NY: Aldine de Gruyter, pp. 203–232.

Betzig, L. (1986). *Despotism and Differential Reproduction: a Darwinian View of History*. Hawthorne, NY: Aldine de Gruyter.

Betzig, L. (1992a). Roman monogamy. *Ethology and Sociobiology*, **13**, 351–383.

Betzig, L. (1992b). Roman polygyny. *Ethology and Sociobiology*, **13**, 309–349.

Betzig, L. (1993). Sex, succession and stratification in the first six civilizations: how powerful men reproduced, passed power on to their sons, and used their power to defend their wealth, women and children. In *Social Stratification and Socioeconomic Inequality*, L. Ellis (ed.). New York: Praeger, pp. 37–74.

Blurton Jones, N., Hawkes, K. and Draper, P. (1994). Foraging returns of !Kung adults and children: why didn't !Kung children forage? *Journal of Anthropological Research*, **50**(4), 217–228.

Bock, J. (2002a). Evolutionary theory and the search for a unified theory of fertility. *American Journal of Human Biology*, **14**, 145–148.

Bock, J. (2002b). Learning, life history, and productivity: children's lives in the Okavango Delta, Botswana. *Human Nature*, **13**, 161–198.

Boone, J. (1986a). Parental investment and elite family structure in preindustrial states: a case study of late medieval–early modern Portuguese genealogies. *American Anthropologist*, **88**, 859–878.

Boone, J. L. (1986b). Parental investment, social subordination and population processes among the fifteenth and sixteenth century Portuguese nobility. In *Human Reproductive Behavior: a Darwinian Perspective*, L. Betzig, P. Turke and M. Borgerhoff Mulder (eds). Cambridge: Cambridge University Press, pp. 201–220.

Borgerhoff Mulder, M. (1985). The polygyny threshold: a Kipsigis case study. *National Geographic Research Reports*, **21**, 33–39.

Borgerhoff Mulder, M. (1988a). Kipsigis bridewealth payments. In *Human Reproductive Behavior*, L. Betzig, M. Borgerhoff Mulder and P. Turke (eds). Cambridge: Cambridge University Press, pp. 65–82.

Borgerhoff Mulder, M. (1988b). Reproductive success in three Kipsigi cohorts. In *Reproductive Decisions*, T. Clutton Brock (ed.). Cambridge: Cambridge University Press, pp. 419–435.

Borgerhoff Mulder, M. (1989). Early maturing Kipsigis women have higher reproductive success than late maturing women and cost more to marry. *Behavioral Ecology and Sociobiology*, **24**, 145–153.

Borgerhoff Mulder, M. (1998). Brothers and sisters: how sibling interactions affect optimal parental investment. *Human Nature*, **9**, 119–162.

Boserup, E. (1970). *Women's Role in Economic Development*. London: Allen and Unwin.

Brown, J. K. (1970). A note on the division of labor by sex. *American Anthropologist*, **72**, 1073–1078.

Burck, C. G. (1976). A group profile of the Fortune 500 child executives. *Fortune*, May, 173–177.

Chagnon, N. A. (1988). Life histories, blood revenge, and warfare in a tribal population. *Science*, **239**, 985–992.

Chagnon, N. A. (2000). Manipulating kinship rules: a form of male Yanomamo reproductive competition. In *Adaptation and Human Behavior: an Anthropological Perspective*, L. Cronk, N. Chagnon and W. Irons (eds). Hawthorne, NY: Aldine de Gruyter, pp. 115–132.

Clark, G. (2007). *A Farewell to Alms: a Brief Economic History of the World*. Princeton, NJ: Princeton University Press.

Clarke, A. L. and Low, B. S. (1992). Ecological correlates of human dispersal in nineteenth century Sweden. *Animal Behavior*, **44**, 677–693.

Clarke, A. L. and Low, B. S. (2001). Testing evolutionary hypotheses with demographic data. *Population and Development Review*, **27**, 633–660.

Colson, E. (1960). *Social Organization of the Gwembe Tonga*. Manchester, UK: Rhodes–Livingstone Institute.

Cromer, A. (1993). *Uncommon Sense: the Heretical Nature of Science*. New York: Oxford University Press.

Curtin, P. D. (1989). *Death by Migration: Europe's Encounter with the Tropical World in the Nineteenth Century*. Cambridge: Cambridge University Press.

Dickemann, M. (1979a). The ecology of mating systems in hypergynous dowry societies. *Social Science Information*, **18**, 163–195.

Dickemann, M. (1979b). Female infanticide, reproductive strategies, and social stratification: a preliminary model. In *Evolutionary Societies and Human Social Behavior*, N. A. Chagnon and W. Irons (eds). North Scituate, MA: Duxbury, pp. 321–368.

Dickemann, M. (1981). Paternal confidence and dowry competition: a biocultural analysis of purdah. In *Natural Selection and Social Behavior*, R. D. Alexander and D. W. Tinkle (eds). New York: Chiron Press, pp. 417–438.

DiVale, W. and Harris, M. (1976). Population, warfare, and the male supremacist complex. *American Anthropologist*, **80**, 21–41.

Draper, P. (1992). Room to manuever: !Kung women cope with men. In *Sanctions and Sanctuary: Cultural Perspectives on the Beating of Wives*, D. Counts, J. K. Brown and J. Campbell (eds). Boulder, CO: Westview Press, pp. 43–63.

Dunbar, R. I. M. (1991). Sociobiological theory and the Cheyenne case. *Current Anthropology*, **32**, 169–173.

Ellison, P. T. (ed.) (2001). *Reproductive Ecology and Human Evolution*. Hawthorne, NY: Aldine de Gruyter.

Ellison, P. T. (2003). Energetics and reproductive effort. *American Journal of Human Biology*, **52**, 342–351.

Ember, C. R. (1983). The relative decline in women's contributions to agriculture with intensification. *American Anthropologist*, **85**, 285–304.

Fleagle, J. G. (1999). *Primate Adaptation and Evolution*. New York: Academic Press.

Gadgil, M. and Bossert, W. H. (1970). Life historical consequences of natural selection. *American Naturalist*, **104**, 1–24.

Goody, J. (1973). Bride wealth and dowry in Africa and Eurasia. In *Bridewealth and Dowry*, J. R. Goody and S. J. Tambiah (eds). Cambridge: Cambridge University Press.

Goody, J. (1976). *Production and Reproduction: a Comparative Study of the Domestic Domain*. Cambridge: Cambridge University Press.

Goody, J. (1983). *The Development of the Family and Marriage in Europe*. Cambridge: Cambridge University Press.

Gurven, M. D. and Kaplan, H. S. (2006). Determinants of time allocation to production across the lifespan among the Machiguenga and Piro Indians of Peru. *Human Nature*, **17**, 1–49.

Gurven, M. and Walker, R. (2006). Energetic demand of multiple dependents and the evolution of slow human growth. *Proceedings of the Royal Society of London. Series B*, **273**, 835–841.

Hart, B. and Risley, T. (1995). *Meaningful Differences in the Everyday Experience of Young American Children*. Baltimore, MD: Brookes.

Hartung, J. (1982). Polygyny and the inheritance of wealth. *Current Anthropology*, **23**, 1–12.

Hawkes, K., O'Connell, J. F., Blurton Jones, N. G., et al. (1998). Grandmothering, menopause, and the evolution of human life histories. *Proceedings of the National Academy of Sciences of the United States of America*, **95**, 1226–1339.

Herrnstein, R. J. and Murray, C. (1994). *The Bell Curve: Intelligence and Class Structure in American Life*. New York: Free Press.

Hoff-Ginsberg, E. and Tardiff, T. (1995). Socioeconomic status and parenting. In *The Handbook of Parenting*, M. Bornstein (ed.). Hillsdate, NJ: Erlbaum.

Holliday, M. A. (1978). Body composition and energy needs during growth. In *Human Growth*, F. Falker and J. M. Tanner (eds). New York: Plenum Press, pp. 117–139.

Hrdy, S. B. (1994). Fitness tradeoffs in the history and evolution of delegated mothering with special reference to wet-nursing, abandonment, and infanticide. In *Infanticide and Parental Care*, S. Parmigiani and F. S. vom Saal (eds). Chur, Switzerland: Harwood Academic Publishers, pp. 3–42.

Hrdy, S. B. (1999). *Mother Nature: a History of Mothers, Infants, and Natural Selection*. New York: Pantheon.

Hrdy, S. B. and Judge, D. (1993). Darwin and the puzzle of primogeniture: an essay on biases in parental investment after death. *Human Nature*, **4**, 1–45.

Hughes, A. L. (1986). Reproductive success and occupational class in eighteenth-century Lancashire, England. *Social Biology*, **33**, 109–115.

Hurtado, A. M. and Hill, K. (1990). Seasonality in a foraging society: variation in diet, work effort, fertility, and the sexual division of labor among the Hiwi of Venezuela. *Journal of Anthropological Research*, **46**, 293–345.

Isbell, L. A. (1991). Contest and scramble competition: patterns of female aggression and ranging behavior among primates. *Behavioral Ecology*, **2**, 143–155.

Jasienska, G. (2000). Why energy expenditure causes reproductive suppression in women. In *Reproductive Ecology and Human Evolution*, P. Ellison (ed.). Hawthorne, NY: Aldine de Gruyter, pp. 59–84.

Jerison, H. J. (1976). Paleoneurology and the evolution of the mind. *Scientific American*, **234**, 90–101.

Kaplan, H. S. (1997). The evolution of the human life course. In *Between Zeus and Salmon: the Biodemography of Aging*, K. Wachter and C. Finch (eds). Washington, DC: National Academy of Sciences, pp. 175–211.

Kaplan, H. S. and Lancaster, J. B. (2003). An evolutionary and ecological analysis of human fertility, mating patterns, and parental investment. In *Offspring: Human Fertility Behavior in Biodemographic Perspective*, K. W. Wachter and R. A. Bulatao (eds). Washington, DC: National Academies Press, pp. 170–223.

Kaplan, H. S. and Robson, A. (2002). The emergence of humans: the coevolution of intelligence and longevity with intergenerational transfers. *Proceedings of the National Academy of Sciences of the United States of America*, **99**, 10221–10226.

Kaplan, H., Hill, K., Lancaster, J. B., et al. (2000). A theory of human life history evolution: diet, intelligence and longevity. *Evolutionary Anthropology*, **9**, 156–185.

Kaplan, H. S., Hill, K., Hurtado, A. M., et al. (2001). The embodied capital theory of human evolution. In *Reproductive Ecology and Human Evolution*, P. T. Ellison (ed.). Hawthorne, NY: Aldine de Gruyter, pp. 293–318.

Kaplan H. S., Lancaster, J. B., Tucker, W. T., et al. (2002). An evolutionary approach to below replacement fertility. *American Journal of Human Biology*, **14**, 1–24.

Kaplan, H., Mueller, T., Gangestad, S., et al. (2003). Neural capital and lifespan evolution among primates and humans. In *Brain and Longevity*, C. E. Finch, J. M. Robine and Y. Christen (eds). New York: Springer-Verlag, pp. 69–98.

Kaplan, H., Gurven, M., Hill, K., et al. (2005). The natural history of human food sharing and cooperation: a review and a new multi-individual approach to the negotiation of norms. In *Moral Sentiments and Material Interests: on the Foundations of Cooperation in Economic Life*, H. Gintis, S. Bowles, R. Boyd, et al. (eds). Cambridge, MA: MIT Press, pp. 75–113.

Kaplan, H. S., Gurven, M. and Lancaster, J. B. (2007). Brain evolution and the human adaptive complex: an ecological and social theory. In *The Evolution of Mind: Fundamental Questions and Controversies*, S. W. Gangestad and J. A. Simpson (eds). New York: Guilford Press, pp. 269–279.

Keeley, L. H. (1996). *War before Civilization: the Myth of the Peaceful Savage*. Oxford: Oxford University Press.

Kirwin, M. (1979). *African Widows*. Maryknoll, NY: Orbis.

Kramer, K. L. (1998). Variation in children's work among modern Maya subsistence agriculturalists. PhD thesis, University of New Mexico, Albuquerque, NM.

Kramer, K. L. (2005a). Children's help and the pace of reproduction: Cooperative breeding in humans. *Evolutionary Anthropology*, **14**, 224–237.

Kramer, K. L. (2005b). *Maya Children: Helpers at the Farm*. Cambridge, MA: Harvard University Press.

Lack, D. (1954). *The Natural Regulation of Animal Numbers*. Oxford: Oxford University Press.

Lancaster, C. S. (1981). *The Goba of the Zambezi: Sex Roles, Economics, and Change*. Norman, OK: University of Oklahoma Press.

Lancaster, J. B. (1997). The evolutionary history of human parental investment in relation to population growth and social stratification. In *Feminism and Evolutionary Biology*, P. A. Gowaty (ed.). New York: Chapman and Hall, pp. 466–489.

Lancaster, J. B. and Kaplan, H. (1992). Human mating and family formation strategies: the effects of variability among males in quality and the allocation of mating effort and parental investment. In *Topics in Primatology: Human Origins*, T. Nishida, W. C. McGrew, P. Marler, et al. (eds). Tokyo: University of Tokyo Press, pp. 21–33.

Lancaster, J. B. and Kaplan, H. S. (2008). The endocrinology of the human adaptive complex. *Endocrinology of Social Relationships*, P. T. Ellison and P. G. Gray (eds). Cambridge, MA: Harvard University Press, pp. 95–119.

Lancaster, J. B., Kaplan, H., Hill, K., et al. (2000). The evolution of life history, intelligence and diet among chimpanzees and human foragers. In *Perspectives in Ethology: Evolution, Culture and Behavior*, F. Tonneau and N. S. Thompson (eds). New York: Plenum, pp. 47–72.

Lloyd, D. G. (1987). Selection of offspring size at independence and other size-versus-number strategies. *American Naturalist*, **129**, 800–817.

Low, B. S. (1990). Occupational status, land ownership, and reproductive behavior in nineteenth century Sweden: Tuna Parrish. *American Anthropologist*, **92**, 457–468.

Low, B. S. (1991). Reproductive life in nineteenth century Sweden: an evolutionary perspective on demographic phenomena. *Ethology and Sociobiology*, **12**, 411–448.

Low, B. S. (1994). Men in the demographic transition. *Human Nature*, **5**, 223–253.

Low, B. S. (2000). *Why Sex Matters: a Darwinian Look at Human Behavior*. Princeton, NJ: Princeton University Press.

Luttbeg, B., Borgerhoff Mulder, M. and Mangel, M. S. (2000). To marry again or not: a dynamic model for the demographic transition. In *Adaptation and Human Behavior: an Anthropological Perspective*, L. Cronk, N. A. Chagnon and W. Irons (eds). Hawthorne, NY: Aldine de Gruyter, pp. 345–368.

Mace, R. (2000). An adaptive model of human reproductive rate where wealth is inherited: why people have small families. In *Adaptation and Human Behavior: an Anthropological Perspective*, L. Cronk, N. Chagnon and W. Irons (eds). Hawthorne, New York: Aldine de Gruyter, pp. 261–282.

Manson, J. and Wrangham, R. (1991). Intergroup aggression in chimpanzees and humans. *Current Anthropology*, **32**, 369–390.

Milton, K. and Demment, M. (1988). Digestive and passage kinetics of chimpanzees fed high and low fiber diets and comparison with human data. *Journal of Nutrition*, **118**, 107.

Murdock, G. P. (1967). Ethnographic atlas: a summary. *Ethnology*, **6**, 109–236.

Newcomer, M. (1955). *The Big Business Executive: the Factors That Made Him, 1890–1950*. New York: Columbia University Press.

Orians, G. H. (1969). On the evolution of mating systems in birds and mammals. *American Naturalist*, **103**, 589–603.

Population Reference Bureau (2008). World population highlights. *Population Bulletin*, **63**(3), 1–12.

Prentice, A. M. and Whitehead, R. G. (1987). The energetics of human reproduction. *Symposium of the Zoological Society of London*, **57**, 275–304.

Preston, S. and Haines, M. (1991). *Fatal Years: Child Mortality in the Late Nineteenth Century America*. Princeton, NJ: Princeton University Press.

Robson, A. and Kaplan, H. (2003). The co-evolution of longevity and intelligence in hunter-gatherer economies. *American Economic Review*, **93**, 150–169.

Sahlins, M. D. (1961). The segmentary lineage: an organization of predatory expansion. *American Anthropologist*, **63**(2), 322–345.

Scot, S. and Duncan, C. J. (1999). Reproductive strategies and sex-biased investment. *Human Nature*, **10**, 85–108.

Sellen, D. W. and Mace, R. (1997). Fertility and mode of subsistence: a phylogenetic analysis. *Current Anthropology*, **38**, 878–889.

Smith, C. C. and Fretwell, S. D. (1974). The optimal balance between size and number of offspring. *American Naturalist*, **108**, 499–506.

Sterck, E. H. M., Watts, D. P. and van Schaik, C. P. (1997). The evolution of female social relationships in nonhuman primates. *Behavioral Ecology and Sociobiology*, **41**, 291–309.

Summers, K. (2005). The evolutionary ecology of despotism. *Evolution and Human Behavior*, **26**, 106–135.

Svedberg, P. (1990). Undernutrition in sub-Saharan Africa: is there a gender bias? *Journal of Development Studies*, **26**, 469–486.

Szreter, S. and Garrett, E. (2000). Reproduction, compositional demography, and economic growth in early modern England. *Population and Development Review*, **26**, 45–80.

Towner, M. (1999). A dynamic model of human dispersal in a land-based economy. *Behavioral Ecology and Sociobiology*, **46**, 82–94.

Towner, M. (2001). Linking dispersal and resources in humans: life history data from Oakham, Massachusetts (1750–1850). *Human Nature*, **12**, 321–350.

Vinovskis, M. A. (1994). Education and the economic transformation of nineteenth century America. In *Age and Structural Lag: Society's Failure to Provide Meaningful Opportunities in Work, Family, and Leisure*, M. W. Riley, R. L. Kahn and A. Foner (eds). New York: Wiley and Sons, pp. 171–196.

Voland, E. (1990). Differential reproductive success within the Krummhorn population (Germany, eighteenth and nineteenth centuries). *Behavioral Ecology and Sociobiology*, **26**, 65–72.

Voland, E. (2000). Contributions of family reconstitution studies to evolutionary reproductive ecology. *Evolutionary Anthropology*, **9**, 134–146.

Voland, E. and Dunbar, R. I. M. (1995). Resource competition and reproduction: the relationship between economic and parental strategies in the Krummhorn population (1720–1874). *Human Nature*, **6**, 33–49.

Voland, E. and Dunbar, R. I. M. (1997). The impact of social status and migration on female age at marriage in an historical population in north-west Germany. *Journal of Biosocial Science*, **29**, 355–360.

Voland, E. and Engel, C. (1990). Female choice in humans: a conditional mate selection strategy of the Krummhorn women (Germany, 1720–1874). *Ethology*, **84**, 144–154.

Voland, E., Stegelkow, E. and Engel, C. (1991). Cost/benefit oriented parental investment by high status families: the Krummhorn case. *Ethology and Sociobiology*, **12**, 105–118.

Voland, E., Dunbar, R. I. M., Engel, C., et al. (1997). Population increase and sex-biased parental investment: evidence from eighteenth and nineteenth century Germany. *Current Anthropology*, **38**, 129–135.

White, D. R. and Burton, M. L. (1988). Causes of polygyny: ecology, economy, kinship, and warfare. *American Anthropologist*, **90**, 871–887.

Part V

Health and Disease

"The deviation of man from the state in which he was originally placed by nature seems to have proved to him a prolific source of diseases."

Edward Jenner (1749–1823), An Inquiry into the Causes and Effects of the Variolae Vaccine. (1798)

27 Evolutionary Medicine, Immunity, and Infectious Disease

Michael P. Muehlenbein

The purpose of this chapter is to provide readers with introductions to several topics central to a modern understanding of human evolutionary biology. Infectious pathogens have placed critical selective constraints on the evolution of our hominin ancestors, and our own species continues to coevolve with infectious organisms today. Our understanding of the processes that shaped this evolutionary struggle have changed, and now an adaptationist perspective offered by the discipline of evolutionary medicine helps to shed light on our vulnerabilities to infectious diseases and noninfectious degenerative diseases. It also aids in our understanding of the purpose and outcomes of our coevolutionary conflicts with the microscopic predators that parasitize us. So as to compete in these interactions, we have developed a marvelously complex immune system capable of dynamic, varied responses. Insight into these mechanisms provides fascinating examples of real-time Darwinian processes of survival and fitness maximization in the face of invading competitors within the human host. Interestingly the ontogeny and deployment of these responses are dependent on several factors, including genetic and ecological constraints.

The discussion offered below provides an introduction to evolutionary medicine with the specific purpose of better understanding human–pathogen coevolution and the development of human immune responses. A current, detailed description of human immunity is included, but this discussion is far from complete. Comparative aspects of evolutionary immunology are emphasized, as are the genetic and ecological sources of variation in these responses. Immune functions have played important roles in the evolution of organismal life histories and so our discussion includes how immune mechanisms could be selected for via natural and sexual selections, develop according to environmental exposure and genetic factors, and then be maintained through trade-offs and constraints. Basic aspects of epidemiology are also introduced but with particular emphasis on host–parasite coevolution and its consequences for the evolution of antibiotic resistance

and pathogen virulence. Finally, the evolutionary histories of several key pathogens are discussed, specifically to illustrate the human host adaptations, both biological and behavioral, to disease emergence and evolution. Review of such a diverse topic produces an admittedly large reading, but it is hoped that readers may use this as a source for further discussion and development.

EVOLUTIONARY MEDICINE

The central role of disease in human evolution was queried decades ago by John Burdon Sanderson Haldane (1892–1964). He hypothesized that much of the biochemical diversity found in serological studies of humans likely played important roles in disease resistance (Haldane, 1949). Today, the integration of medicine and evolutionary biology forms the basis of the discipline of evolutionary, or Darwinian, medicine. This field recognizes that medical research can benefit significantly from a priori understanding of adaptation by natural selection and its role in the causation of health-related outcomes. This might include understanding the various adaptations we use to combat pathogens as well as the adaptations that pathogens use to counter our own defense mechanisms. This also includes understanding the necessary costs imposed by our current adaptations against disease, and that there are mismatches between our current form, which evolved in the past, and the present environments in which we find ourselves (Williams and Nesse, 1991). Whereas current medial research and practices focus on describing how we become ill with the purpose of identifying cures and preventions, evolutionary medicine utilizes the adaptationist perspective in evolutionary biology to describe why some people get sick in different environments. Caution must be used in over-applying an adaptationist paradigm in evolutionary medicine (Marks, 2008), but we must still ask what aspects of our physiology, morphology, and behavior make us vulnerable to disease, why have these traits

Human Evolutionary Biology, ed. Michael P. Muehlenbein. Published by Cambridge University Press. © Cambridge University Press 2010.

TABLE 27.1. Some manifestations of disease benefit the host whereas others benefit the invading pathogens. Adapted from Nesse and Williams (1996).

Observations	Examples	Primary beneficiary
Hygienic measures	Killing mosquitoes, avoiding detritus	Host
Host defenses	Immune responses, sneezing*, vomiting*	Host
Repair of damage	Regeneration of tissue, inflammation*	Host
Compensation for damage	Chewing on opposite side to avoid tooth pain*	Host
Damage to host tissue	Liver damage from hepatitis*	Neither
Impairment of host	Decreased detoxification, lameness*	Neither
Evasion of host defenses	Molecular mimicry, change in antigens	Pathogen
Attack on host defenses	Destruction of host cells*	Pathogen
Uptake and use of nutrients by pathogen	Growth and proliferation of pathogens	Pathogen
Dispersal of pathogen	Transfer of blood parasite to new host by mosquito	Pathogen

Note: *Some manifestations could benefit both the host and pathogen.

evolved, and why do they persist today? Evolutionary medicine attempts to emphasize the ultimate or evolutionary explanations for medical phenomena rather than just the proximate causes of morbidity and mortality.

For a more complete review of evolutionary medicine, readers are encouraged to utilize the following texts: Nesse and Williams (1996), Ewald (1996), Trevathan et al. (2007), Elton and O'Higgins (2008), Stearns and Koella (2008). In brief, evolutionary explanations for disease can be organized into several categories, including defense mechanisms, design compromises, and conflict between organisms (Williams and Nesse, 1991; Nesse and Williams, 1996). Defenses are mechanisms that have evolved to prevent, limit, or eliminate disease, and we must recognize which are signs and symptoms of disease that materialize to benefit the invading pathogen, which are manifestations of disease that are actually beneficial host responses shaped by natural selection to defend against infection, which are side effects of infection/disease, and which mechanisms benefit both the host and pathogen (Table 27.1). Although they may be unpleasant, if they are actually beneficial host responses, then blocking these mechanisms may cause more harm than good. Fever, nausea, and vomiting during pregnancy, revulsion towards odors associated with bacterial decomposition, and mental conditions like fear of snakes and heights have all been proposed to be defensive mechanisms that evolved to increase host fitness. For example, fever may make it more difficult for certain pathogens to multiply within the host, and elevated temperatures may also optimize immune cell activity of the host (Kluger et al., 1998). Inhibiting or reducing fever during infection may not be very beneficial in certain circumstances (Doran et al., 1989; Graham et al., 1990; Kramer et al., 1991). Of course, high fevers can cause much more than just discomfort, including sterility and death.

Some traits may have evolved as design compromises. In this case, a trait may have been beneficial in an ancestral environment, but is no longer beneficial in the current environment, and disease can result from this mismatch (the "discordance hypothesis"). For example, a preference for fat, carbohydrates, and salt in the ancestral hominin diet was beneficial, but may now result in cardiovascular disease and metabolic disorders in modern, developed environments with less physical activity (Eaton and Eaton, 1999). Some traits may be evolutionary legacies that evolved in the past, but can predispose individuals to illness in the present. Immunoglobulin E (IgE), eosinophils, mast cells, and other proximate mediators of allergic responses likely evolved to protect us against helminth and other extracellular infections. Improvements in sanitation and the absence of these infections in most people in developed countries may produce an imbalanced Th-cell immune phenotype (see below), and our systems overreact to novel and largely unimportant allergens, like dust mites (Yazdanbakhsh et al., 2002). Thus our evolutionary legacy may be a hypersensitivity response that now predisposes us to atopy.

The propensity for drug abuse is also likely a design compromise and the result of a mismatch between our bodies and modern environments: humans are vulnerable to abuse of psychoactive drugs because these substances mimic the activation of neural mechanisms involved in regulating feelings of pleasure and other necessary emotions (Nesse and Berridge, 1997). We have been programmed by natural selection to pursue behaviors that produce such chemical incentives. Unfortunately, these substances are now available in forms, concentrations, and delivery systems that are all too potent and readily available. Evolutionary considerations suggest that "we cannot reasonably expect to win the war on drug abuse, but we can use our knowledge to develop sensible strategies for prevention, treatment, and public policy to manage a problem that is likely to persist because it is rooted in the fundamental design of the human nervous system" (Nesse and Berridge, 1997, p. 65). Furthermore, psychoactive substances that inhibit all negative emotions can also impose fitness costs. Nonpathological low mood and anxiety may be beneficial in some situations, forcing individuals to remove themselves from potentially harmful circumstances (Nesse, 2006).

Despite contributing to morbidity and mortality, some genetic and phenotypic traits can be maintained in a population because they are beneficial in some forms or in certain environments. Of course, some genes can be maintained in a population due to novel mutations, genetic drift and other random processes, or because they are very rare disorders. Many disease conditions are caused by organismal defects, and these may be transmitted to future generations and maintained in the population because they do not affect people until after reproduction (Haldane, 1941; Medawar, 1952). Nonetheless, there are trade-offs in which traits can exhibit antagonistic pleiotropic functions, including some which are beneficial, particularly in the genetic heterozygous state. For example, individuals that are heterozygous for alleles on chromosome 7 (e.g., *δF508*, *W1282X*, *1677delTA*) which code for cystic fibrosis may better resist disease from *Vibrio cholerae*, *Salmonella typhi*, *Escherichia coli*, and even *Mycobacterium tuberculosis* (Poolman and Galvani, 2007).

Organisms are shaped by evolution in ways that can make disease almost inevitable. Autoimmune diseases, such as type I diabetes, rheumatoid arthritis, and multiple sclerosis, can result from overactivation of immune responses which become targeted towards an individual's own antigens. Increased risk of leukemia may even be a price we pay for maintaining intricate and powerful immune responses that would help to control microbial infections (Greaves, 2006). Maintaining high testosterone in mammalian males can augment male reproductive effort by increasing musculoskeletal performance (which aids in work capacity, intersexual competition, intrasexual coercion, and mate choice) but can also compromise survivorship by increasing risk of prostate cancer, production of oxygen radicals, risk of injury due to hormonally augmented behaviors such as aggression, violence and risk taking, reduced tissue and organ maintenance, negative energy balance from adipose tissue catabolism, and suppression of immune functions (Muehlenbein and Bribiescas, 2005; Muehlenbein, 2008a). These are compromises that our genotypes and phenotypes have made through evolutionary processes, and understanding this can be insightful for medical research and practices. This is especially the case in considering infectious diseases, their evolution, and our immune responses to them.

HUMAN IMMUNITY

Immunity is obviously vital for defense against invading pathogens, cellular maintenance, and renewal and protection against cancer. No discussion of infectious diseases would be complete without an introduction to the immune system and its responses. For a comprehensive description of immunology and human immunity see Paul (2008), and for a brief review see Delves and Roitt (2000a, 2000b). Here I provide readers with an introductory review of the human immune system and its complex responses. Comparative aspects of evolutionary immunology are emphasized as are the genetic and ecological sources of variation in these responses.

Innate immunity

The mammalian immune system is usually organized into two primary components: innate (constitutive) and adaptive (acquired) immune responses (Figure 27.1). Innate immunity consists of primary nonspecific, generalized mechanisms that block or eliminate foreign particles from invasion of the host. Such defenses include anatomical barriers (mucus membranes, skin), resident flora (nonpathogenic bacteria), humoral factors (lysozyme, complement, and other acute phase proteins), and cells (phagocytic cells like neutrophils, monocytes, and macrophages; inflammatory mediators produced by basophils, mast cells, and eosinophils; and natural killer cells) (Delves and Roitt, 2000a). Note that white blood cells (leukocytes) include macrophages, dendritic cells, granulocytes (eosinophils, basophils, and neutrophils) and lymphocytes (T cells and B cells). All leukocytes originate from hematopoietic stem cells in bone marrow.

Macrophages are mononuclear phagocytes that perform many tasks, such as phagocytosis, cytokine secretion, chemotaxis, and antigen processing and presentation (Hume et al., 2002). Interdigitating dendritic cells function in antigen presentation (Medzhitov and Janeway, 1997). Toll-like receptors on macrophages and dendritic cells bind to foreign antigens and initiate the cascade of innate and adaptive immune effector mechanisms (Visintin et al., 2001). Eosinophils attack extracellular parasites by the release of various chemical mediators (Wardlaw et al., 1995). Basophils and mast cells facilitate atopic, inflammatory reactions (Abraham and Arock, 1998). Natural killer cells attack the membranes of infected or malignant target cells through a variety of processes, including antibody-dependent cellular cytotoxicity (Herberman et al.,

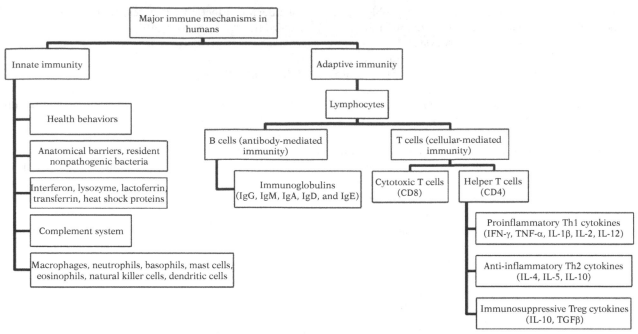

27.1. Many of the major immune mechanisms in humans.

1986). Interferon is produced by virally infected cells to signal apoptosis and prevent infection of adjacent cells (Samuel, 2001). Heat shock proteins are intracellular molecules that, among other functions, aid in antigen presentation and stimulation of proinflammatory responses (Robert et al., 2003). Lactoferrin, transferrin, and other proteins bind circulating iron, limiting availability during bacterial infections (Baker et al., 2002). Innate immunity also includes resident flora in the gut and other tissues that prevent pathogen colonization. Health behaviors that decrease the likelihood of illness, such as handwashing and avoidance of detritus, might also be included in the category of innate immunity, although most of these behaviors arguably have a learned component.

The complement system includes enzymes that function to eliminate micro-organisms by promoting inflammatory responses, such as changes in local vascular permeability and entry of immune cells into infection sites. Complement also functions in lysis of foreign cells through the formation of membrane attack complexes, and in mediation of phagocytosis through the coating (opsonization) of pathogens and infected cells (Carroll, 1998). Complement is also important for stimulating adaptive immune responses (Dempsey et al., 1996). Within mammals, the presence of foreign molecules activates one of three different complement pathways (innate = alternative and lectin; adaptive = classical). In most cases, the C3 component binds to foreign molecules and activates phagocytes and other complement components (approximately 30 different proteins) that produce the membrane

attack complex and a cytolytic response (Law and Reid, 1995).

Adaptive immunity

Because of the extreme diversity and short multiplication times of most pathogens, hosts are under intense selection pressure to produce variable defensive responses. This is accomplished through the high diversification and rapid responses of acquired, specific immunity. In this case, effector mechanisms allow fast, secondary responses during subsequent exposures. First, foreign antigens are recognized by immune cells in the tonsils, adenoids, and Peyer's patches after inhalation or ingestion. Foreign substances are transported in the lymph and trapped in the lymph nodes or transported in the blood and filtered by the spleen. In these locations, lymphocytes react with specific pathogen antigens and facilitate an adaptive immune response. Peripheral circulation of lymphocytes also allows for continual monitoring of infection and injury in strategic sites throughout the body.

Lymphocytes come in two main forms: B cells and T cells. All lymphocytes are produced in bone marrow and the fetal liver, but the maturation of T cells takes places in the thymus. During development, those T and B cells that autoreact to self-antigens (estimated at 98% of all T cells) are eliminated in the process of thymic education to minimize self-reactivity (Sprent, 1993). Positive selection for developing T cells involves downregulating signals that would otherwise induce apoptosis if autoreactivity occurs, resulting in tolerance of self-antigens. Negative selection involves induction of apoptosis in T and B cells that autoreact with self-molecules on antigen-presenting cells (macrophages and interdigitizing dendritic cells) in the thymus (Rathmell and Thompson, 1999; Sebzda et al., 1999). Without the ability to distinguish between self and nonself tissues, immunopathology due to autoimmune disease would rampantly consume the individual. The ontogeny of lymphocytes therefore invokes Darwin's principles of natural selection in evolutionary processes that occur within an organism.

The development of the immune system begins early in gestation (Remington and Klein, 1990). Acquired immunological characteristics (particularly immunoglobulins G and A) are transferred from mother to offspring via the placenta and breast milk (Keller, 1992; Goldman, 1993). As offspring encounter new antigens, as many as 10 million different lymphocytes react to these antigens to cause subsequent proliferation of particular lymphocytes lineages. The entire process of antigen recognition and lymphocyte activation and proliferation is known as "clonal selection." Clonal selection for lymphocytes is a process of somatic evolution in which antigen receptor diversity is maximized during a critical stage of development when offspring are continuously encountering new infections. The process of clonal selection produces a diverse pool of lymphocytes with millions of different antigen-binding possibilities, all from simple progenitor cells (Burnet, 1959). Throughout this process, thymus volume regresses, lymphocytes diversify, and immunoglobulin levels increase (Hannet et al., 1992; George and Ritter, 1996). Despite this so-called thymic involution, T cells still develop in this structure during adult life (Jamieson et al., 1999).

Antibody-mediated immunity

B cells represent antibody-mediated (humoral) immunity. Antigen recognition by lymphocytes promotes cellular proliferation and the differentiation of B cells into plasma cells, which secrete antibodies or "immunoglobulins," and memory cells, which function in immunosurveillance. B cells have the capacity to bind "native" or free antigen. T cells, on the other hand, recognize "processed" antigen. Receptors on antigen-presenting cells (including dendritic cells and B cells) bind to antigens which are internalized, degraded, and presented onto the cell surface via major histocompatibility complex (MHC; or human leukocyte antigen, HLA) class I (HLA-A, -B, and -C) and II (HLA-DP, DQ, and DR) molecules. The MHC class I molecules present their antigens to killer T cells (CD8), which initiate a cytotoxic response, whereas MHC class II molecules present their antigens to helper T cells (CD4), which produce cytokines that facilitate the clonal expansion of other T and B cells (Meyer and

Thomson, 2001). The HLA superlocus is located on chromosome 6 (Fischer and Mayr, 2001).

Antibodies are glycoproteins that neutralize pathogens and their products, block binding of parasites to host cells, induce complement activation, promote cellular migration to sites of infection, and enhance phagocytosis, among other actions. Antibodies in mammals are composed of two light chains (with single variable [V] and joining [J] elements) and two heavy chains (with single variable [V] and joining [J] elements, and multiple diversity [D] elements) bound together by disulfide bonds (Edelman, 1973). The variable end interacts with antigens whereas the constant region determines the class and subclass of antibody: IgG, IgM, IgA, IgD, and IgE. In general, IgG is the most predominant in circulation and some of its functions include opsonization of bacteria and infected cells, activation of the complement cascade and antibody-dependent cellular cytotoxicity. Immunoglobulin M activates the complement cascade and is the first antibody produced during a response. Immunoglobulin A protects mucosal surfaces from infection. Immunoglobulin E mediates allergic reactions (immediate-type hypersensitivity) and functions to clear extracellular helminth infections. The functions of the secreted form (nonmembrane bound) of IgD are not yet known (Wallace Taylor, 2002).

The variable region of the antibody, produced by four bounded polypeptide chains, confers high specificity for the target molecule, while the constant region of the antibody binds the appropriate immune effector cells for activation. Genes for antigen-binding regions on lymphocytes as well as antibodies themselves are randomly assorted from different gene segments or clusters (variable [V], diversity [D], joining [J] and constant [C]) on chromosomes 2, 14, and 22. This random process is also susceptible to splicing errors and additional nucleotide insertions, and B cells can undergo further receptor editing in their V gene in secondary lymphoid organs (Schatz et al., 1992; Radic and Zouali, 1996). The results are millions of different unique antibodies and antigen-binding sites on lymphocytes (Tonegawa, 1983). In fact, this rearrangement of mini-gene segments allows for much greater diversity in antigen binding (estimated 10^{15} variable regions on T and B cells) than there are genes in the human body that could otherwise produce each individual antibody.

Cellular-mediated immunity

T cells represent cellular immunity, and different subsets are identified by their surface markers that regulate cellular activation and adhesion (CD number refers to "cluster of differentiation"). Cytotoxic T cells (CD8) destroy infected host cells via perforin and lysis. Cytotoxic T cells and natural killer cells are particularly important for eliminating intracellular pathogens (Berke, 1997). Suppressor T cells downregulate T-cell responses after infection (see below).

Helper T cells (CD4) secrete cytokines and activate B cells to secrete antibodies. Cytokines are glycoproteins that perform a variety of functions such as regulation of cell growth and development (Snapper, 1996). Cytokines have several striking features; most importantly, they perform pleiotropic actions and interact in different complex ways with each another. Cytokines have pleiotropic, redundant, and epistatic (synergistic and antagonistic) actions. That is, single cytokines can have multiple functions, multiple cytokines can have similar functions, some cytokines work together to facilitate single functions, and some cytokines have opposite functions to one another. There may even be significant heritable variation in cytokine levels (Williams-Blangero et al., 2004; Curran et al., 2005).

CD4 helper T cells are generally differentiated into two major subsets depending on the type of cytokine produced: type 1 (Th1) and type 2 (Th2) phenotypes (Mosmann et al., 1986; Mosmann and Coffman, 1989a, 1989b; Mosmann, 1991a, 1991b; Coffman and Mosmann, 1991; O'Garra 1998; Reiner and Seder, 1999). Th1 cytokines include, among others, interferon-gamma (IFN-γ), tumor necrosis factor-alpha and -beta (TNF-α, -β), and various interleukins (IL-1β, IL-2, IL-3, IL-12, etc.). These cytokines activate macrophages, neutrophils and natural killer (NK) cells, mediate inflammatory responses and cellular immunity (T cells), promote cytotoxicity toward tumor cells, and enhance chemotaxis of leukocytes (Kobayashi et al., 1989; Gazzinelli et al., 1993; Dinarello, 2000; Burger and Dayer, 2002; Trinchieri, 2003).

The Th2, anti-inflammatory cytokines include many interleukins (IL-4, IL-5, IL-10, etc.) that induce humoral immunity and antibody production (B cells), eosinophil activation, mast cell degranulation, goblet cell hyperplasia, mucin production, and intestinal mastocytosis (resulting in histamine release). This cytokine phenotype is important for protection against intestinal infections as well as the facilitation of allergic reactions (Barrett et al., 1988; Rothwell, 1989; Cox and Liew, 1992; King and Nutman, 1992; Sher and Coffman, 1992; Urban et al., 1992; Allen and Maizels, 1996; Else and Finkelman, 1998; Dinarello, 2000; MacDonald et al., 2002). Despite the fact that Th1 and Th2 cytokines act antagonistically to one another, both are usually present within the host at any given time, although during infection one phenotype usually predominates.

Newborn humans tend to have a dominant Th2 phenotype, whereas the Th1 phenotype develops later with age (Jones et al., 2000). In the absence of a balanced helper T-cell phenotype, atopic disease can become more pronounced. Lack of early exposure to allergens and pathogens may predispose individuals to a dominant Th2 response with high IgE levels and subsequent allergies and asthma later in life (the "hygiene hypothesis") (Strachan, 1989; Cookson and Moffatt, 1997; Matricardi et al., 2000; Wills-Karp et al., 2001; Yazdanbakhsh et al., 2002). Lack of persistent exposure to allergens and pathogens may not allow for proper programming of an anti-inflammatory regulatory response, such as IL-10 production (Wills-Karp et al., 2001). In fact, nematode infection is associated with fewer allergies and less asthma in some populations (Wilson and Maizels, 2004). Improvements in sanitation may therefore explain the disproportionate increase in asthma morbidity and mortality over the past 25 years, despite improvements in medications (Von Mutius et al., 1992; Braun-Fahrlander et al., 1999; Yazdanbakhsh et al., 2002). Clearly, early life events produce various physiological effects in later adulthood (Barker et al., 2002; Barker, 2007). Childhood environments likely program the development of immune responses, and differences in local disease exposure can explain differences in immune development between populations (Mohammed et al., 1973; Lisse et al., 1997; McDade et al., 2004).

Other Th cell types include Th17 cells that produce IL-17, IL-6, and related cytokines, and Tregs, or induced regulatory T cells, that express Foxp3 (a forkhead winged-helix transcription factor). The absence of Tregs has been implicated in autoimmune diseases like inflammatory bowel disease (Shevach, 2008). The CD25[+]CD4[+] Tregs produce IL-10 and transforming growth factor beta (TGF-β) and are important for mucosal immunity (Kiyono et al., 2008). Th3 (which produces high levels of TGF-β), Tr1 and CD8[+]CD28[-] suppressor T cells are other subsets of suppressor T cells that may be different from Tregs (Kiyono et al., 2008).

Comparative aspects

Susceptibility to host infection depends on many factors, including whether or not the pathogen has been encountered before, inducing acquired immunity. Nutritional status also plays an important role since immune responses generate significant energetic burdens (see below). Genetic predisposition also accounts for differences in disease outcome in various species. There is an incredible amount of polymorphism in the genes that code for immune responses (Trowsdale and Parham, 2004). Several HLA alleles have been associated with resistance and susceptibility to a variety of diseases, including malaria, tuberculosis, and HIV (Hill, 1998). There is also population variation in the alleles that code for Toll-like receptors on the surfaces of leukocytes that bind to antigens and trigger important innate inflammatory responses (Lazarus et al., 2002).

The Sm1 locus on chromosome 5 has been associated with susceptibility to *Schistosoma mansoni* in Brazilian and Senegalese populations (Marquet et al., 1996), and Sm2 on chromosome 6 has been associated with a high risk of liver fibrosis caused by *S. mansoni* in a Sudanese population (Dessein et al., 1999). These alleles likely cause dysfunctions in cytokine and lymphocyte proliferation responses to infection (Rodrigues et al., 1999). Research also suggests significant heritable variation in hookworm infection intensity in a Zimbabwe population (Williams-Blangero et al., 1997) as well as heritable variation in *Ascaris lumbricoides* and *Trichuris trichiura* infection intensities in a Nepalese population (Williams-Blangero et al., 1999, 2002a). Several loci on chromosomes 1, 8, 9, 11, 12, 13, and 18 may alter cytokine and immunoglobulin responses to these infection, resulting in heritable variation in immune responses in these populations (Williams-Blangero et al., 2002b, 2008a, 2008b). While there are certainly important genetic components to variation in immune responses, even adaptive ones, to human pathogens, the utility of heritability estimates obtained from studies not using monozygotic twins is equivocal (Vitzthum, 2003). Several studies have reported no significant heritable variation in susceptibility to *Schistosoma haematobium* in a Kenyan population (King et al., 2004) or in infection intensity from *Strongyloides fuelleborni* in a population of Papua New Guinea (Smith et al., 1991).

Despite these disagreements, it is well accepted that all organisms (examined to date) have some immune responses, including phagocytic abilities and the ability to recognize self from nonself. Notwithstanding these similarities, there is extraordinary variation in the immune responses within the Animal Kingdom (Flajnik and du Pasquier, 2004). Invertebrate immunity closely resembles innate immunity in vertebrates (Hoffmann and Reichhart, 2002). Phagocytic cells circulate in the hemolymph and encapsulate pathogens. Lectins opsonize foreign antigens and interference RNA (RNAi) inhibits viral gene function. Proteins, such as fibrinogen-related proteins (FREPs) in the snail *Biomphalaria glabrata* and the products of Down's syndrome cell-adhesions molecule (*DSCAM*) genes in *Drosophila melanogaster*, bind invading molecules (Adema et al., 1997; Yu and Kanost, 2002; Meister, 2004; Loker et al., 2004).

Plants exhibit hypersensitivity responses that cause cellular apoptosis and exhibit characteristics of localized and systemic acquired resistance via antimicrobial peptides and pathogen-associated molecular patterns (PAMPs) (Dangl and Jones, 2001; Innes, 2004; Chisholm et al., 2006). Some aspects of the complement system (like the C3 component) are shared between vertebrates, invertebrate deuterostomes (e.g., sea urchins), and even some protostomes (e.g., nematodes)

(Nonaka and Yoshizaki, 2004). However, adaptive immunity with the ability to generate diverse antigen receptors on lymphocytes appears to have evolved only in vertebrates. The adaptive immune system in vertebrates may have evolved in response to selection pressures by diverse parasitic flatworms, viruses, and bacteria as well as longer life spans, higher metabolic rates, and bigger genomes in the vertebrate hosts (Rolff, 2007). A more complex immune response may also have been helpful in differentiating potential pathogens from symbiotic microflora (Pancer and Cooper, 2006).

Within vertebrates, the process of producing the lymphocyte receptor repertoire is different in the jawed gnathostomes compared to the jawless agnathans, with conventional rearrangeable immunoglobulin gene segments in the former and rearrangement of leucine-rich repeat-encoding modules in the latter (Flajnik and Kasahara, 2001; Pancer et al., 2004; Alder et al., 2005; Cooper and Alder, 2006; Pancer and Cooper, 2006). Jawed vertebrates adapting a predatory lifestyle may have needed more complex and robust immune responses, particularly in their gastrointestinal tracts (Matsunaga and Rhaman, 1998). In these jawed vertebrates, the adaptive immune responses likely evolved when a transposable element (mobile DNA) produced an immunoglobulin-like gene (*RAGs*, recombination activating genes). The transposon inserted itself into an Ig superfamily (IgSF) gene of the variable (V) type, leaving behind the machinery for the original receptor gene to cut and paste independent loci which allowed for somatic gene rearrangement for the various V(D)J mini-gene combinations (Agrawal et al., 1998; Fugmann et al., 2000). The results are the production of highly polymorphic combinations of α/β- and γ/δ-chains of T-cell antigen receptors and the heavy/light chains of immunoglobulin B-cell receptors. This mechanism is estimated to have evolved more than 500 million years ago (Flajnik, 2002).

Immune responses within all jawed vertebrates appear to be rather conserved, with significant homogeneity in the defense mechanisms of different species (Marchalonis and Schluter, 1994; Litman et al., 2005). In fact, there are very few biochemical and genetic differences in the immune systems of humans and common chimpanzees (Muchmore, 2001). This is in spite of the fact that chimpanzees are less susceptible to some infections, like immunodeficiency virus, and more susceptible to others, like pneumonia. The comparative study of primate immunology will prove quite insightful for human evolutionary biologists.

Evolutionary and ecological immunology

Research on immunological stress has traditionally focused primarily on understanding somatic maintenance, repair, and defense against pathogens. From an evolutionary perspective, immune functions are critical for maximizing survivorship, and an optimized immune system should always be highly selected for. Immune functions have therefore played important roles in the evolution of organismal life histories. Life history strategies are complex adaptations for survival and reproduction via the co-ordinated evolution of somatic and reproductive developmental processes (Stearns, 1992). There exists an enormous amount of variation in organismal life history strategies, and much of the physiological variation between and within organisms can be explained using several concepts of life history theory, most notably trade-offs and reaction norms (Stearns, 1992; Ricklefs and Wikelski, 2002).

Immunity as an evolved reaction norm

Both somatic and reproductive physiologies are evolved response systems, shaped by natural and sexual selections to adapt individuals to changing environments. This allows for a variable physiological response (a "reaction norm") in which a genotype can produce a range of phenotypes (through short-term changes, such as acclimatization to altitude, and long-term adaptations) depending on environmental conditions. However, this phenotypic plasticity is limited through lineage-specific effects (or the canalization of certain traits) as well as trade-offs. Assuming a limited supply of energy and time, organisms are required to allocate physiological resources between a number of competing functions, particularly reproduction, maintenance (i.e., survival) and growth (Stearns, 1989). Organisms will therefore be under selection to develop and maintain physiological systems that allow for the efficient regulation of resources between these functions. In a stochastic environment, those organisms that can most efficiently regulate the allocation of resources between competing traits will likely exhibit increased lifetime reproductive success.

An individual's immune system is an excellent example of a reaction norm that allows for short- and long-term phenotypic plasticity in response to environmental signals such as pathogens, allergens, and injury. Immunocompetence, or the ability to mount an effective immune response, is an integral component of organismal life histories precisely because: (1) it is crucial for maximizing evolutionary fitness; and (2) it is energetically expensive to produce, maintain, and activate. This important function can be characterized as an energetic burden subject to allocation mechanisms (Sheldon and Verhulst, 1996; Demas et al., 1997; Raberg et al., 1998, 2002; Verhulst et al., 1999; Owens, 2002; Schmid-Hempel, 2003; Muehlenbein and Bribiescas, 2005; Muehlenbein, 2008a). For example, prolonged energy restriction can lead to immune suppression in humans and nonhuman animals alike

(Rosenberg and Bowman, 1984; Hadju et al., 1995; Kramer et al., 1997; Klasing, 1998; Shephard et al., 1998; Lin et al., 1998; Koski et al., 1999; Ing et al., 2000). In turn, acute infection in adult humans can cause high amount of protein loss, greater than 1 g/kg of body weight per day (Scrimshaw, 1992). Strenuous exercise or participation in energetically demanding tasks, such as migration, breeding, or molting, can also compromise immune functions (Nelson et al., 2002). For example, increased brood size is associated with reduced antibody response against Newcastle disease virus and increased *Haemoproteus* infection intensity in collared flycatchers (*Ficedula albicollis*) (Nordling et al., 1998), reduced antibody response against sheep red blood cells in zebra finches (*Taeniopygia guttata*) (Deerenberg et al., 1997) and female tree swallows (*Tachycineta bicolor*) (Ardia et al., 2003), and increased prevalence of *Plasmodium* in male great tits (*Parus major*) (Richner et al., 1995).

Metabolic rate, oxygen consumption and thermogenic activity also frequently increase following immune stimulation (Newsholme and Newsholme, 1989; Spurlock, 1997; Lockmiller and Deerenberg, 2000). For example, house mice (*Mus musculus*) injected with the antigen keyhole limpet hemocyanin show a 20–30% increase in oxygen consumption (Demas et al., 1997). Likewise, blue tits (*Parus caeruleus*) immunized with diphtheria-tetanus vaccine exhibit an 8–13% increase in resting metabolic rate (Svensson et al., 1998), and great tits (*Parus major*) and collared doves (*Streptopelia decaocto*) injected with sheep red blood cells exhibit a 9% increase in resting metabolic rate (Ots et al., 2001; Eraud et al., 2005). In humans, fever typically results in a 7–15% increase in resting metabolic rate for every 1°C rise in body temperature (Barr et al., 1922; Roe and Kinney, 1965; Elia, 1992). In a population of adult male college students sampled during and after acute upper respiratory tract infection, resting metabolic rate was, on average, 8% higher during infection compared to samples taken after complete recovery (Muehlenbein, 2008b; Muehlenbein et al., in press). It is interesting that these metabolic values were elevated even in the absence of fever, likely reflecting increased energetic demands despite mild infection. Further research should investigate changes in metabolic rates of adult humans during illnesses of varying severities and in individuals with different states of energy balance.

Trade-offs with immunity

"Every trait must be analyzed in terms of the costs and benefits of the trade-offs in which it is involved" (Stearns et al., 2008, p. 11). As immune responses are energetically expensive, optimized immune functions should trade-off with other critical life history functions, like growth. Nutrient deficiencies can have significant, long-term negative effects on the human immune system (Lunn, 1991; Gershwin et al., 2000), and these effects may begin early in life. For example, infants in the Philippines born small-for-gestational age exhibit slower growth rates and produce less thymopoietin as adolescents (McDade et al., 2001b). These individuals are also less likely to produce antibodies in response to typhoid vaccination (McDade et al., 2001a). Elevated concentrations of α-1 antichymotrypsin (an acute phase protein produced by the liver during inflammation) are also associated with growth faltering (lower height-for-age) in Nepalese adolescents (Panter-Brick et al., 2000). Activation of proinflammatory immune responses (as in the case of inflammatory bowel disease) is associated with delayed puberty in even adequately nourished individuals (Ballinger et al., 2003), and elevated C-reactive protein levels are associated with reduced gains in height across three months in Tsimane children of Amazonian Bolivia (McDade et al., 2008). Continued work on comparative developmental immunology within and between populations will prove interesting. This will include continued efforts to qualify and quantify trade-offs between immunity and reproductive effort.

Optimization of reproductive effort is of central importance, especially for capital breeding, iteroparous organisms that must budget time and stored energy over a number of reproductive events within a lifetime. Under conditions of resource restriction, a trade-off between current and future reproduction is predicted: investments in current reproductive events may negatively affect future reproductive returns (the "cost of reproduction" argument). Investments in reproduction should also compromise survivorship through depressed immune functions. Conversely, investment in immune activation should compromise reproductive effort. In mosquitoes (*Anopheles gambiae*), injection with lipopolysaccharide is associated with reduced egg production (Ahmed and Hurd, 2006). Deerenberg et al. (1997) have shown that only 47% of breeding zebra finches (*Taeniopygia guttata*) produced antibodies in response to infection with sheep red blood cells whereas all nonbreeding birds produced antibodies.

In humans, hypogonadism (decreased levels of hormones from the testes or ovaries) and hypogonadotropism (decreased levels of gonadotropins from the hypothalamus and pituitary glands) are common physiological responses to somatic injury. For example, in men, serum testosterone decreases during sepsis, burns, myocardial infarction, and surgery (Spratt et al., 1993; Spratt, 2001). Honduran men infected with *Plasmodium vivax* exhibit significantly lower testosterone levels than age-matched healthy controls (Muehlenbein et al., 2005). Similarly, experimental Venezuelan Equine Encephalitis virus infection in captive male macaques (*Macaca fasicularis*) is associated with significant declines in serum testosterone levels (Muehlenbein et al., 2006). Hypogonadism has also been reported in association with African sleeping sickness (*Trypanosoma brucei*) (Reincke et al., 1998), toxoplasmosis (*Toxoplasma gondii*) (Oktenli et al., 2004), schistosomiasis (*Schistosoma mansoni*) (Saad et al., 1999), and filarial infection (*Loa loa* and *Mansonella perstans*) (Landsoud-Soukate et al., 1989). Changes in testosterone levels throughout the range of physiological variation may function as a basic aspect of male phenotypic plasticity and an adaptive response that facilitates the allocation of metabolic resources according to available energy and disease risk in a stochastic environment. Assuming testosterone's immunomodulatory actions are primarily suppressive (for review see Muehlenbein and Bribiescas, 2005), depressed testosterone levels during illness or injury could function to prevent immunosuppression by otherwise higher testosterone levels (Wedekind and Folstad, 1994). In addition, depressed testosterone levels could function to limit metabolic investment in energetically expensive anabolic functions (Muehlenbein, 2008a).

Testosterone increases energetic costs through direct actions on muscle tissue and metabolism (Welle et al., 1992; Bhasin et al., 1996), and this may decrease survivorship in resource-limited environments (Ketterson et al., 1992; Bribiescas, 2001). The problem would become exacerbated in pathogen-rich environments because of the immunosuppressive actions of testosterone and because investment in muscle anabolism generates a significant energetic demand that will theoretically trade-off with the competing energetic demands of immunocompetence. An evolutionary and ecological perspective on immunity would suggest that natural and sexual selections favor individuals that can best balance the trade-offs so as to maximize reproductive effort and survivorship (i.e., immunity) given different ecological conditions. In lower pathogen-risk environments (e.g., higher latitudes), less of a premium may be placed on immunity, and it may pay to select for less robust immune responses (Muehlenbein, 2008a).

Immunity and mate choice

It is also predicted that individuals should develop honest signals of survivorship (i.e., immunocompetence) in an effort to maximize mate choice (Zahavi, 1975; Hamilton and Zuk, 1982). Animals should be under selective pressure to evolve preferences for those mates that possess reliable indicators of pathogen resistance by scrutinizing characteristics that honestly reflect health or the ability to resist pathogens. A number of morphological and behavioral characteristics appear to be honest sexual signals of immunocompetence in avian and other species. For example, tail length was positively associated with cell-mediated immune function in male barn swallows (*Hirundo rustica*) (Saino et al., 2002). Male barn swallows with longer outermost tail feathers also exhibited stronger primary antibody responses following an immunization (Saino et al., 2003b), had higher testosterone levels (Saino and Moller, 1994), and were preferred by females, both as social mates and extrapair copulation partners (Saino et al., 1999). Fluctuating asymmetry of antlers in male reindeer (*Rangifer tarandus*) was associated with immune parameters during the rut, suggesting that low fluctuating asymmetry in sexually selected ornaments may also signal the ability to resist parasites (Lagesen and Folstad, 1998).

Some primates may exhibit signals that honestly indicate health and survivability, such as coloration in the facial, scrotal, and perianal regions. Examples may include sexual colorations in adult male vervet monkeys (*Cercopithecus aethiops sabaeus*) (Gerald, 2001) and mandrills (*Mandrillus sphinx*) (Setchell and Dixson, 2001). However, there have been no published studies to date that have investigated relationships between immunocompetence and degree of sexual coloration in primates. For humans, muscle mass may be an honest indicator of survivorship due to the significant costs of anabolic steroids, including increased energetic costs and the risk of negative energy balance, increased risk of prostate cancer, production of oxygen radicals, increased risk of injury due to hormonally augmented behaviors such as aggression, violence, and risk taking, reduced tissue (especially adipose) and organ maintenance, and suppression of immune functions (Muehlenbein and Bribiescas, 2005; Muehlenbein 2008a). It may also be the case that morphological symmetry is a potential indicator of immunological status in humans. Morphologically symmetric humans are frequently judged as more attractive and are preferred as potential mates (Grammer and Thornhill, 1994; Gangestad and Thornhill, 1998; Perrett et al., 1998). More research should clearly be conducted with humans in this area, particularly given the interesting sexual selection behaviors that humans use and the myriad of pathogens that can infect us.

INFECTIOUS DISEASE

The initial description that diseases have natural rather than supernatural causes is attributed to Hippocrates of Cos (460–370 BC). However, several others further developed the concept of "contagion," including Marcus Terentius Varro (116–27 BC), Abū Alī ibn Sīna (980–1037), Girolamo Fracastoro (1478–1553) Francesco Redi (1626–1697), Anton van Leeuwenhoek

(1632–1723), Ignaz Philipp Semmelweis (1818–1865), and John Snow (1813–1858) (see Kiple, 1993, 2003; Porter, 2001, 2006). The actual germ theory of disease is attributed to Agostino Bassi (1773–1856) and Louis Pasteur (1822–1895) (see Kiple, 1993, 2003; Porter, 2001, 2006). Pasteur demonstrated that micro-organisms do not arise spontaneously, but are the products of reproduction by existing micro-organisms. Around the same time, Heinrich Hermann Robert Koch (1843–1910), along with his mentor Friedrich Gustav Jakob Henle (1809–1885) and associate Friedrich August Johannes Loeffler (1852–1915), provided the scientific community with the "postulates" or experimental criteria to establish a causal relationship between these micro-organisms and disease (see Kiple, 1993, 2003; Porter, 2001, 2006). In brief, the agent or micro-organism must be found in all cases of the diseased, but not healthy, hosts. The organism must be isolated from diseased hosts and cultured. Inoculation of the culture into susceptible, healthy hosts must reproduce the disease, and the agent must be reisolated from the newly infected host. This form of deductive reasoning, although criticized by many, uses observable, empirical evidence to test hypotheses about the cause of infectious disease.

Infectious organisms include thousands of species of viruses (and bacteriophages), bacteria (including rickettsiae), parasitic protozoa and helminthes (nematodes, cestodes, and trematodes), and fungi. These parasitic organisms live all or part of their lives in or on a host from which biological necessities are derived. This state of metabolic dependence usually results in host energy loss, lowered survival, and reduced reproductive potential. Disease or illness is the impairment of host body function done by a pathogen.

There is fantastic variation in the transmission dynamics of infectious (communicable) organisms (Anderson and May, 1992; Combes, 2004; Poulin, 2006). The primary infection transmission routes include fecal-oral (ingestion of contaminated food, water, or other objects), respiratory, vector-borne (e.g., mosquitoes, ticks, flies, etc.), blood-borne, sexually transmitted, vertical transmission (congenital; mother to offspring), and nosocomial (hospital-acquired). Zoonotic infections are those acquired from nonhuman animals. A reservoir is the biotic or abiotic source where a pathogen normally lives and reproduces. Latency is the period of inactivity of the pathogen inside the host, or the period between initial host infection and the subsequent ability of the infected host to infect new hosts. The incubation period is the interval between initial infection and onset of clinical illness (with signs and symptoms). This period can vary depending on dose of exposure, host susceptibility, pathogenicity of the infectious agent, and other factors. A carrier is an infected host that does not present any signs or symptoms of infection, but is still capable of infecting other hosts.

Epidemiology is the study of the distribution (frequency and pattern) and determinants (etiology) of health-related events within populations. Endemic diseases are those that occur regularly at low to moderate frequency, epidemics are outbreaks that occur above endemic levels, and pandemics are epidemics that affect a large proportion of the world's population. Specifics of transmission dynamics, epidemiologic study designs, and how health-related conditions are specifically associated with exposures will not be discussed here (see Sattenspiel, 2000, for an introduction to epidemiologic study designs).

Coevolution

Infectious organisms offer a near-ubiquitous selective evolutionary force (Levin, 1996). Many of these pathogens have infected humans and our hominin ancestors for a large part of our evolutionary history (Hoberg et al., 2001; Goncalves et al., 2003; Van Blerkom, 2003). For example, hominins have certainly been infected with herpesviruses for millions of years (Sharp et al., 2008). *Helicobacter pylori* has likely infected humans for as much as 60 000 years (Linz et al., 2007), and *Salmonella typhi*, the causative agent of typhoid fever, has likely affected humans for 30 000 years (Roumagnac et al., 2006).

Conflict between organisms is ubiquitous, even between parents and offspring (Trivers, 1974; Haig, 1993). Just as predators and prey coevolve in an escalating cycle of complexity with predators' improved hunting techniques countered by prey's improved armor or defensive adaptations, hosts and parasites must evolve in order to maintain current levels of adaptation (Dawkins and Krebs, 1978). Hosts and pathogens coevolve together in a constant state of flux, with reciprocal modification of evolutionary strategies producing evolutionary change in host traits in response to evolutionary change in pathogen traits, and vice versa (Thompson, 1994). The result is oscillations in levels of host resistance and pathogen invasiveness, and a subsequent arms race of specializations over time.

The struggle for existence does not get easier, no matter how well a species may adapt to its current environment because competitors and enemies are also adjusting, causing significant change in the adaptive landscape (the "Red Queen hypothesis," Van Valen, 1973). Sexual reproduction may have evolved specifically as a mechanism to combat against high pathogen evolutionary rates and infectious disease potential (Howard and Lively, 1994). However, host recombination is often met with genetic recombination in pathogens (e.g., recombination and mutation, lateral gene combination via conjugation between bacteria or transduction from bacteriophages, sexual reproduction in malaria, etc.). Coevolution of organisms, and the conflict that often ensues, provides significant selection pressure for the evolution of traits, and understanding this can assist in the explanation of health-related traits. Understanding the conflicts between hosts and pathogens can aid in our understanding of why antibiotic resistance evolves and why some diseases are very deadly and others less so.

Virulence and antibiotic resistance

The traditional view of host–pathogen coevolution suggested that evolutionary processes should theoretically lead to reduced antagonism or symbiosis since this would be in the best interest for both the pathogen and host (Smith, 1934; Swellengrebel, 1940; Allee et al., 1949). Longer coexistence between hosts and pathogens would theoretically lead to attenuated infections (Burnet and White, 1972). This "group selectionist" reasoning implies some type of co-operation between hosts and parasites. However, a more modern view is that natural selection will favor the increase of fitness in both hosts and parasites, which may not lead to obligate evolution towards benign interactions.

Virulent pathogens are characterized by high levels of host exploitation, producing high morbidity and mortality. Virulence can evolve when the benefits of host exploitation are outweighed by the costs to the pathogen from host damage (Galvani, 2003). High virulence is more common when host immobility does not disrupt pathogen transfer. Examples include vector-borne, water-borne, and nosocomial infections in which intense host exploitation comes at little cost to the pathogen because it is not dependent on host mobility for transmission to new hosts (Ewald, 1996). Such is the case for malaria and cholera. Infectious organisms that can survive in the external environment for lengthy periods of time also often evolve high levels of virulence, as is the case of anthrax (Walther and Ewald, 2004). Vertical transmission (mother to offspring) is more often associated with less virulent infections because the well-being of both the host and parasite are linked (Bull et al., 1991; Messenger et al., 1999).

New, accidental infections in dead-end hosts may result in high virulence, as in the case of rabies (Ebert and Bull, 2008). Colonization of a new host population following successful host switching may also be associated with initial increases in virulence. Level of pathogen virulence may also be a function of host recovery rate, the geographic and temporal distribution of the host population, and even host age (Koella and Turner, 2008). Pathogen–pathogen competition over host resources also plays an important role (Nowak and May, 1994). Within the host, pathogens must evolve strategies to either coexist or outcompete

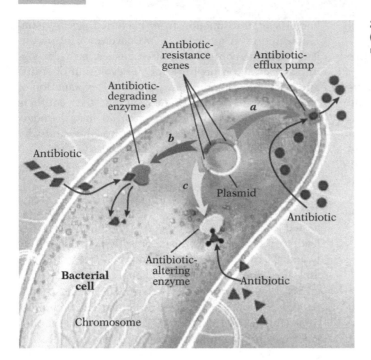

27.2. Modes of antibiotic resistance. Bacteria can "eject" **(a)**, degrade **(b)**, or even inactivate **(c)** antibiotics. Figure reprinted with permission from Laurie Grace.

other members of its own or separate species (Kim, 1985). If within-host genetic relatedness of such pathogens is great, the likelihood of evolution towards avirulence is increased as these pathogens should tend to "co-operate" with each other, analogous to "kin selection." Alternatively, high within-host genetic relatedness of pathogens may lead to increased virulence via kin selection if the benefits associated with increased virulence are shared across the pathogen population (Ewald and Cochran, 2004).

In contrast, high within-host genetic variability could favor increased competition with an escalation towards high virulence. A more exploitive variant of a pathogen may be more successful at reproduction, further increasing within-host pathogen genetic variability and favoring parasite–parasite competition (between different strains or species) and increased virulence. One example is the high virulent outcomes during coinfection with multiple malaria strains (Conway et al., 1991).

Imperfect vaccines designed to limit pathogen growth and toxicity could contribute to increased pathogen virulence (Gandon et al., 2001; Mackinnon et al., 2008). A host population imperfectly protected against morbidity and mortality could generate selection pressure for the evolution of more virulent organisms. Such strains could develop to increase competitive advantage over other strains during coinfection as well as increase the likelihood of successful transmission, assuming transmission is still possible from vaccinated hosts. This would place the unvaccinated host population at increased risk of death and disability from the new dangerous strains.

High pathogen variability within hosts can also contribute to antibiotic resistance. Antibiotics are chemotherapeutic substances that kill or inhibit bacterial growth by disrupting cell wall, nucleic acid, and protein synthesis, and altering metabolic pathways and cell membrane integrity (Levy, 1998). Broad-spectrum antibiotics, like tetracyclines, are useful in controlling several types of bacteria, like chlamydias, rickettsias, gram-positive and gram-negative bacteria. They are often used before the pathogens are identified or antibiotic susceptibility has been tested for. However, they are also more likely to negatively affect normal flora in the body.

Pathogens previously susceptible to antimicrobials are becoming increasingly resistant. The antibacterial effects of penicillin, a β-lactam derived from *Penicillium* mold, were described by John Tyndall (1820–1893), Ernest Duchesne (1874–1912), Clodomiro Picado Twight (1887–1944), and Alexander Fleming (1881–1955). Within a very short time of penicillin's mass production and clinical use, resistant strains of *Staphylococcus aureus* were identified (Spink and Ferris, 1945). Now over half of all such strains are resistant to penicillin and its derivatives (Chambers, 1997). Microbes are accomplishing resistance by blocking entry of antimicrobials into cells or removing (drug efflux), degrading or otherwise altering the antimicrobials (Figure 27.2). This does occur naturally, but humans have facilitated these processes by inappropriate use of antimicrobials through self-medication, lack of patient compliance (i.e., premature termination of treatments), demands on physicians to over-prescribe, and overuse in livestock (Austin et al., 1999). Millions of kilograms of antibiotics are produced per year, and the majority of them are used in livestock (Mellon et al., 2001).

Random mutations, recombination, reassortment, and lateral gene transfer (including transduction from bacteriophages, uptake of naked DNA and transposons, and conjugation or plasmid exchange between bacteria) can all produce resistant microbes (Levy, 1998; Lipsitch, 2001; Levy and Marshall, 2004). The selection pressures we impose on them provides ample impetus for rapid evolution and proliferation of resistant species. Fortunately not all pathogen populations develop resistance because resistant bacteria appear to compete poorly against sensitive bacteria in the absence of antibiotics, possibly because of energetic costs associated with carrying nonessential plasmids (Courvalin, 2008). Future chemotherapeutic agents will benefit from directing actions towards virulent pathogens, making them less competitive against benign counterparts. Simultaneous treatment with several different antibiotics also creates a more heterogeneous environment for bacterial populations to overcome (Bergstrom et al., 2004).

Case studies in human evolutionary biology

I conclude this chapter with detailed accounts from three of the most notable bacterial, viral, and protozoan pathogens in humans: tuberculosis, human immunodeficiency virus (HIV), and malaria. Each has its own unique evolutionary history and all offer insight into human–pathogen coevolution. Finally, emerging infectious diseases, their causes, and consequences are considered so as to illustrate the importance of identifying cultural as well as biological adaptations against human disease.

Tuberculosis

Tuberculosis is caused by one of several bacteria of the genus *Mycobacterium*. These gram-positive bacteria are aerobic, nonmotile, and posses a thick hydrophobic cell wall. The *Mycobacterium* "complex" includes *M. tuberculosis*, *M. africanum* and *M. canettii* of humans, *M. microti* of rodents, *M. caprae* of various mammals (including humans), *M. pinnipedii* of seals, and *M. bovis* of bovine and many other mammalian species. Many other atypical species are found in soil, water, and other animals (from fish to birds to monkeys), and several of these species can cause disease in humans, particularly in immunocompromised individuals. The causative agent of leprosy (*Mycobacterium leprae*) naturally infects humans, chimpanzees, armadillos, and several species of monkeys, but is only distantly genetically related to tuberculosis (Grosskinsky et al., 1989).

Mycobacterium tuberculosis is one of the most ubiquitous pathogens in humans, with an estimated worldwide prevalence (number of total cases) as great as two billion people, resulting in as many as three million human fatalities annually (Bloom and Murray, 1992). The highest incidence (number of new cases within a given time period) of tuberculosis are presently in India, China, Indonesia, Bangladesh, Pakistan, and several countries in Africa (Dye et al., 1999). Use of the Bacille Calmette-Guérin (BCG) vaccine is widespread, but of limited efficacy in adults (Ellner, 1997). Rifamycin-, isoniazid- and multidrug-resistant strains have evolved and spread rapidly (Drobniewski et al., 2002).

Tuberculosis infection is spread via the inhalation of infected respiratory secretions. Macrophages attempt to phagocytose, and then kill via reactive oxygen and nitrogen intermediates, the tubercle bacilli following inhalation. Those mycobacteria which escape this destruction will trigger a proinflammatory cellular immune response (effector T cells with subsequent production of proinflammatory cytokines, like TNF-α, IL-12, and IFN-γ), which are vital for controlling infection (Lenzini et al., 1977; Havlir et al., 1991; Ellner, 1997; van Crevel et al., 2002). The active form of vitamin D (1,25 dihydrocholecalciferol) impairs growth of *M. tuberculosis* inside activated macrophages (Rook, 1988). Vitamin D deficiency, even due to seasonal variation in food resources, may therefore increase the risk of disease from tuberculosis (Douglas et al., 1996; Wilkinson et al., 2000).

In most people, infection becomes stabilized as a solid fibrous granuloma inhibits further growth of the pathogen. However, diminished Th1 proinflammatory cytokine responses, caused in part by the overproduction of TGF-β (Toossi et al., 1995), can result in hematogenous dissemination and spread of infection throughout the lungs and other tissues (van Crevel et al., 2002). Slow progression of disease is an excellent mechanism by which the likelihood of transmission to susceptible hosts is increased. The mycobacteria can exist within the phagosomes of macrophages in human lungs for years.

People can exhibit a wide spectrum of immune responses to tuberculosis infection (Lenzini et al., 1977), and not surprisingly resistance and susceptibility to infection has been associated with some genetic polymorphisms, particularly in the human leukocyte antigens that present mycobacterial proteins to effector cells (Goldfeld et al., 1998). The *HLA-DQB1*0503* allele is significantly associated with susceptibility to clinical tuberculosis (Goldfeld et al., 1998) as are members of the HLA-DR2 serotype (Bothamley et al., 1989; Brahmajothi et al., 1991; Rani et al., 1993). Variants of the solute carrier family 11, member 1 gene (*NRAMP1*) have been associated with susceptibility to tuberculosis in a variety of human populations, including Japanese and West Africans (Bellamy et al., 1998; Gao et al., 2000).

Mutations in the Toll-interleukin-1 receptor domain and Toll-like receptor-2 genes are also associated with severe disease outcome (Hawn et al., 2006; Thuong et al., 2007). A single-point mutation in IFN-γ receptor 1 gene is also associated with progressive infection due to downregulated cell-mediated inflammatory responses (Newport et al., 1996). There may be other genetic variants that affect susceptibility and resistance to tuberculosis.

Mycobacteria have certainly been infecting human populations for a long time. The last common ancestor of all mycobacteria is estimated at 2.6–2.8 million years before present (Gutierrez et al., 2005). As there is relatively low variation in the housekeeping genes within *M. tuberculosis*, Sreevatsan et al. (1997) have concluded that the organism experienced an evolutionary bottleneck around 15 000–20 000 years before present. More sensitive analyses revealed genetic variation that allowed others to conclude that the *M. africanum*, *M. microti*, and *M. bovis* lineages likely diverged from the *M. tuberculosis* ancestor, not that *M. tuberculosis* evolved from *M. bovis* (Brosch et al., 2002). *Mycobacterium bovis* was long considered ancestral to *M. tuberculosis* because the former has a wide range of hosts, whereas the latter is human-specific (Cole et al., 1998). Genetic analyses now confirms the opposite, that *M. tuberculosis* is likely ancestral to the other members of the *Mycobacterium* complex. This complex was likely African in origin, and the youngest strains causing modern endemic human tuberculosis were introduced into the various geographic regions. *Mycobacterium bovis* is actually the youngest of the complex species (Smith et al., 2006).

Although many have argued against a bovine origin for *M. tuberculosis* (Mostowy et al., 2002), 20 000 years may not have been enough time for the diverse strains to develop in their respective hosts (Brosch et al., 2002). Others estimate the most recent last common ancestor of the complex at around 35 000 years before present (Hughes et al., 2002). *Mycobacterium canettii* may have even diverged before this bottleneck (Gutierrez et al., 2005). Also prior to the bottleneck of the complex members, *M. tuberculosis* likely acquired several virulence genes (e.g., Rv0986–8 that inhibit macrophage functions) via horizontal transfer of a plasmid from a gammaproteobacterium donor species, like *Agrobacterium* (Rosas-Magallanes et al., 2006). Several other virulence genes have now been identified in *M. tuberculosis* (Ernst et al., 2007).

Tuberculosis certainly infected humans in North America prior to European contact (Braun et al., 1998) (Figure 27.3). *Mycobacterium tuberculosis* was present in North American bovids, including bison, bighorn sheep, and musk ox, at least 17 000 years before present (Rothschild et al., 2001). Presently, the various strains of *M. tuberculosis* are rather

27.3. A Pre-Columbian human specimen infected with tuberculosis. The photograph shows a fused lumbar spine and sacrum from a 19–22-year-old human female recovered from the Middle Mississippian Schild Cemetery, Greene County, Illinois, excavated by Gregory Perino in 1962. The specimen dates to 1020 AD (±110 years) and is genetically diagnosed with *Mycobacterium tuberculosis* (Braun et al., 1998). Photograph courtesy of Della Cook.

geographically distinct. There is strong evidence for phylogeographic relationships between these different strains and their human hosts. For example, East Asian individuals seem to be more susceptible to East Asian strains of the mycobacteria, even if the infection is acquired outside of East Asia, suggesting some coevolution of the parasites with the hosts (Gagneux et al., 2006).

Human immunodeficiency virus

Human immunodeficiency virus (HIV) is a double-stranded RNA virus of the family Retroviridae, subfamily Lentivirus (see Hutchinson, 2001, and Rambaut et al., 2004, for review). Type 1 HIV (HIV-1) includes groups N and O, found primarily in Gabon and Cameroon, and the pandemic group M, for which there are 11 different subtypes or viral clusters (A–K). There are eight subtypes (A–H) of HIV-2 which are mostly found in West Africa and India. Infection by HIV-2 is characterized by much slower disease progression compared to HIV-1 (Marlink et al., 1994).

Transmission of HIV is through fluid exchange, primarily sexual contact, inoculation with blood or blood products, and perinatal transmission. Upon entry into the body, HIV's envelope protein (gp120) binds to CD4 and chemokine coreceptors (e.g., CCR5 and CXCR4) on the surface of helper T cells. The virus fuses to the host cell surface and inserts its viral core. The viral genome is reverse transcribed and integrated into the host genomic DNA. Viral RNA is then transcribed, viral proteins are translated, viruses are assembled and bud from the infected cell (Rambaut et al., 2004). Due to the imperfect reverse transcription process, many errors accumulate, resulting in a massive amount of HIV genetic diversity in a host at any given time (Rodrigo, 1999).

Initial infection is followed by high viral replication and mild, 'flu-like illness followed by an asymptomatic period of approximately 10 years. During this time, the virus typically exhibits low replication rates and is often sequestered out of peripheral circulation and into gut-associated lymphoid tissue (Chun et al., 2008). Acquired immune deficiency syndrome (AIDS) results when the density of CD4 T cells drops below 200 cells per microliter of blood or when one of several indicator conditions are present (e.g., Kaposi's sarcoma, Burkitt's lymphoma, infection with Toxoplasma, Pneumocystis, Cryptosporidium, Cytomegalovirus, etc.). Death results from opportunistic infection, cancers, wasting, and neurological complications.

The estimated global prevalence of HIV is approximately 33 million people with the highest prevalence rates found in sub-Saharan Africa (Joint United National Programme on HIV/AIDS, 2008). Approximately 2.7 million people are infected each year, with approximately 2 million deaths annually. Within the United States, the majority of those infected are young males (Hall et al., 2008). Although the majority of those infected in the United States are white non-Hispanics, death rates are highest in the non-Hispanic black populations. However, life expectancy of HIV-infected individuals within the United States has generally increased due to the use of highly active anti-retroviral therapy (HAART), which includes reverse transcriptase blockers like azidothymidine (AZT), integrase blockers, and protease inhibitors. Still the most effective means at prevention is to eliminate high-risk behaviors. Because of the long latency period of infection, a significant proportion of infected individuals are unaware of their infection. Approximately one-quarter of all those infected with HIV in the United States are unaware of their HIV infection (Hall et al., 2008).

The modern HIV epidemic was first recognized as a cluster of Pneumocystis carinii pneumonia cases in homosexual men in Los Angeles, California in 1981.

27.4. Chimpanzees are often used for bushmeat. Photograph courtesy of David Watts.

The source of HIV was not biological warfare or divine retribution against this community. Rather, HIV originated from nonhuman primate simian immunodeficiency viruses (SIVs). Based on overwhelming genetic similarities (genome structure and protein homology), HIV-1 groups M and N likely originated from chimpanzees (Pan troglodytes troglodytes) in Cameroon (Figure 27.4), and group O from Western lowland gorillas (Gorilla gorilla) in Cameroon. HIV-2 originated from sooty mangabeys (Cercocebus atys) in Côte d'Ivoire (Gao et al., 1999; Hahn et al., 2000; Peeters et al., 2002; Bailes et al., 2003; Apetrei et al., 2005; Santiago et al., 2005; Keele et al., 2006; Van Heuverswyn et al., 2006).

There are currently 18 described strains of SIV found in 38 species of nonhuman primates, including vervets, mangabeys, guenons, colobus, talapoins, mandrills, patas, baboons, chimpanzees, and gorillas (Gao et al., 1999; Hahn et al., 2000; Peeters et al., 2002; Bailes et al., 2003; Apetrei et al., 2005; Santiago et al., 2005; Keele et al., 2006; Van Heuverswyn et al., 2006). Simian immunodeficiency viruses likely entered into the human population in West Africa due to cutaneous or mucous membrane exposure to infected nonhuman primates. Direct exposure to nonhuman primate blood through hunting and butchering is common (Wolfe et al., 2004), and SIV has been identified in nonhuman primate bushmeat and pet animals in West Africa (Peeters et al., 2002; Apetrei et al., 2005). Simian immunodeficiency viruses have also been reported in bushmeat hunters in Cameroon (Kalish et al., 2005).

Through the hunting and butchering of nonhuman primates in West Africa, SIV likely entered into the human population many times and became established within the human population as HIV around 1900 in what is now the Democratic Republic of Congo (Worobey et al., 2008). Despite infecting humans for only just over 100 years, several allelic variants appear to provide protective responses against HIV infection. Historic selection pressure through high mortality rates may account for the presence of these restriction genes within human populations, but the influence of genetic drift and gene flow should not be discounted. Alterations in the morphology of the chemokine co-receptors that the virus uses to bind to and enter host cells can confer resistance against HIV infection. Several of these AIDS restriction genes that delay the onset of AIDS include *CCR2-64I*, *HLA-B*27* and *B*57* and others (Dean et al., 2002; Winkler et al., 2004). Individuals homozygous for *CCR5-δ32* can have complete resistance against infection (Zimmerman et al., 1997).

The *CCR5-δ32* allele is found at high frequencies in many European populations (e.g., 15% in Scandinavia, 5% in Italy) (Stephens et al., 1998). Galvani and Slatkin (2003) hypothesized that the huge amount of deaths due to plague (*Yersinia pestis*) and smallpox (*Variola major*) caused selection pressure to increase the frequency of the mutation. However, while there is evidence that *CCR5*-deficient macrophages have reduced uptake of *Y. pestis* in vitro, the mutation appears to have no protective effect in mice artificially infected with *Y. pestis* (Mecsas et al., 2004). Furthermore, recent data suggest that the *CCR5* mutation is approximately 5000 years old, which predates the outbreaks of plague and smallpox in Europe (Duncan et al., 2005; Hummel et al., 2005; Sabeti et al., 2005). With current selection pressures, this HIV-resistance genotype may increase to over 50% in South Africans within 100 years (Galvani and Slatkin, 2003). Such a change would not come without an evolutionary cost though: individuals homozygous for the *CCR5-δ32* allele are at increased risk of fatal outcome from West Nile virus due to altered leukocyte trafficking to the brain (Glass et al., 2006). As discussed earlier, health-related adaptations are constrained by trade-offs.

Malaria

Malaria is presently endemic in most tropical regions of the world, with approximately one billion people at risk of acquiring malaria (Guerra et al., 2008). The global incidence is estimated to be more than 300 million new clinical cases each year (Trigg and Kondrachine, 1998) with millions of deaths from malaria annually, a majority occurring in children (World Health Organization, 1999). Five to ten percent of children born in tropical Africa will likely die from malaria before the age of five (World Health Organization, 1999; Carter and Mendis, 2002). The economic and social impacts of this disease are enormous (Gallup and Sachs, 2001).

Malaria is a mosquito-borne disease caused by protozoa of the genus *Plasmodium* (phylum Apicomplexa, suborder Haemosporidiidea, family Plasmodiidae), with 172 named species that parasitize reptiles, birds, and mammals (Garnham, 1966; Coatney et al., 1971; Levine, 1988). Of these, only 4 usually infect humans (*P. falciparum*, *vivax*, *ovale*, and *malariae*), 19 infect nonhuman primates, and 19 infect various other mammals. The majority of *Plasmodium* species infect birds and reptiles. The parasites are transmitted by over 50 species of female *Anopheles* mosquitoes (Kiszewski et al., 2004). During the *Plasmodium* life cycle (Figure 27.5), sporozoites are injected from the mosquito's salivary glands into the vertebrate host. The sporozoites penetrate the parenchymal cells of the liver where they remain for a variable period of time and asexually reproduce ("exo-erythrocytic schizogony"). Schizonts develop, which rupture to release merozoites into the vertebrate blood stream. These merozoites invade red blood cells (erythrocytes), in which they develop into trophozoites, the form of the parasite that essentially feeds off the nutrient supply of the erythrocyte. The trophozoites undergo erythrocytic schizogony to produce either more merozoites (which reinfect surrounding red blood cells) or the sexual gametocytes (macro and micro). The gametocytes are ingested by the *Anopheles* mosquito, after which they escape the erythrocyte ("exflagellation"). Within the gut of the mosquito, the male and female gametes fuse and form a fertilized ookinete. The ookinete develops into an oocyst which is implanted into the stomach wall of the mosquito. There it undergoes sporogony. The oocyst eventually ruptures to release many sporozoites which migrate to the mosquito's salivary glands. The sporozoites are released into the vertebrate host during the next blood-meal (see http://www.dpd.cdc.gov/DPDx/HTML/malaria.htm).

In humans, malaria causes 'flu-like nausea with headache and muscle pain. The paroxysms of fever and shivering followed by sweating and fatigue correspond with the length of shizogony and the synchronous rupture of schizonts: either every 48 hours (tertian: *vivax*, *ovale*, *falciparum*) or 72 hours (quartan: *malariae*). Malaria can cause liver, spleen, and kidney failure, and cardiovascular and placental damage. *Plasmodium falciparum*-infected red blood cells are capable of adhering to one another, to noninfected cells, and to cerebral microvasculature, causing seizure and coma (Taylor et al., 2004). Immune responses include cell- and antibody-mediated ones, with primary activation of proinflammatory cytokines, including interleukin 12, TNF-α, and IFN-γ (Stevenson and Riley, 2004). Acquisition of effective immune responses to malaria requires repeated exposure and inoculation (Carter and Mendis, 2002).

The evolutionary history of human malaria is an interesting one. These parasites are likely descended from a coccidian ancestor that first parasitized the intestinal tract of either a reptile host (Garnham, 1966; Coatney et al., 1971) or aquatic invertebrate

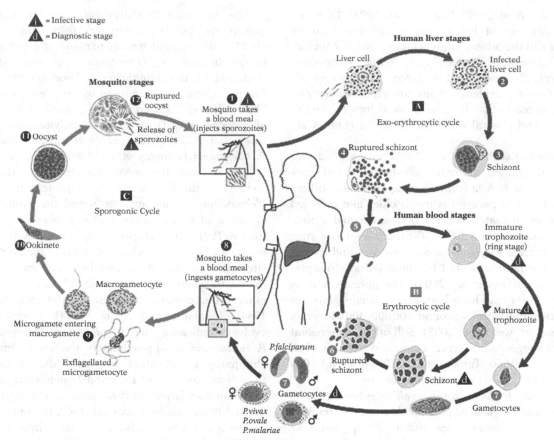

27.5. Life cycle of *Plasmodium*. The malaria parasite life cycle involves two hosts. During a blood meal, a malaria-infected female *Anopheles* mosquito inoculates sporozoites into the human host (1). Sporozoites infect liver cells (2) and mature into schizonts (3), which rupture and release merozoites (4). (Of note, in *P. vivax* and *P. ovale* a dormant stage [hypnozoites] can persist in the liver and cause relapses by invading the bloodstream weeks, or even years later.) After this initial replication in the liver (exo-erythrocytic schizogony [A]), the parasites undergo asexual multiplication in the erythrocytes (erythrocytic schizogony [B]). Merozoites infect red blood cells (5). The ring stage trophozoites mature into schizonts, which rupture releasing merozoites (6). Some parasites differentiate into sexual erythrocytic stages (gametocytes) (7). Blood stage parasites are responsible for the clinical manifestations of the disease. The gametocytes, male (microgametocytes) and female (macrogametocytes), are ingested by an *Anopheles* mosquito during a blood meal (8). The parasites' multiplication in the mosquito is known as the sporogonic cycle [C]. While in the mosquito's stomach, the microgametes penetrate the macrogametes generating zygotes (9). The zygotes in turn become motile and elongated (ookinetes) (10), which invade the midgut wall of the mosquito where they develop into oocysts (11). The oocysts grow, rupture, and release sporozoites (12), which make their way to the mosquito's salivary glands. Inoculation of the sporozoites (1) into a new human host perpetuates the malaria life cycle. Figure and caption reprinted with permission from the US Centers for Disease Control and Prevention: http://www.cdc.gov/malaria/biology/life_cycle.htm.

(Wilson and Williamson, 1997). Members of the genus *Plasmodium* likely diverged from the other Haemosporidiidea around 500 mya, perhaps around the time of the Cambrian explosion (Escalante and Ayala, 1994). Biting dipteran insects were later introduced into the parasite's life cycle, possibly as early as 200 million years ago (Carter and Mendis, 2002). The *Plasmodium* radiation is thought to have occurred around 150 million years ago, which parallels the diversification of their vector's lineages (Escalante and Ayala, 1995; Ayala et al., 1998).

Within the subgenus *P. (Plasmodium)*, three malarial parasites naturally infect humans: *vivax*, *ovale*, and *malariae*. *Plasmodium (Plasmodium) vivax* (Grassi and Feletti, 1890) is a malarial parasite with worldwide distribution (Mendis et al., 2001). It possesses a tertian periodicity in which the process of schizogony lasts approximately 48 hours. Because hypnozoites can remain dormant in the liver for years, relapse of illness can occur. A number of biomolecular investigations suggest that *P. (Plasmodium) vivax* is closely related to *P. (Plasmodium) cynomolgi* of macaques (McCutchan et al., 1984; White, 1993; Waters et al., 1993). There are also a number of life history similarities between *P. vivax* and the other tertian primate malarias, including *P. schwetzi*, *P. cynomolgi*, *P. youngi*, *P. hylobati*, *P. pitheci*, *P. eyelesi*, and *P. jefferyi* (Coatney et al., 1971). Genetic evidence from the circumsporozoite protein suggests that *P. vivax* evolved into *P. simium* in South and Central American monkeys (Lim et al., 2005).

Whereas Escalante et al. (2005) have argued that *P. vivax* originated approximately 46 000–82 000 years before present in Asia from a macaque malaria, others suggest that the parasite is much older and diverged from other primate malarias between 5 and 7 mya (Jongwutiwes et al., 2005). *Plasmodium vivax* may have infected early *Homo erectus* in Asia and spread to Africa sometime around 1 million years before present (Jongwutiwes et al., 2005). The parasite and its hominin hosts may have both experienced significant population bottlenecks around 200 000–300 000 years ago (Jongwutiwes et al., 2005). Still other more recent analyses suggest a most recent common ancestor of all *P. vivax* at 600 000 (Tanabe et al., 2007) or even 10 000 (Leclerc et al., 2004) years before present.

Plasmodium (Plasmodium) ovale (Stephens, 1922) is a human tertian malarial parasite that is most common in West Africa, New Guinea, and the Philippines. Like *Plasmodium vivax*, hypnozoites of *Plasmodium ovale* can remain dormant in the liver for years, relapse of illness can occur. This species is morphologically very similar to *P. vivax*, although *P. (Plasmodium) ovale* possesses very characteristic oocyst and exoerythrocytic morphology with very large nuclei (Garnham, 1966). *Plasmodium (Plasmodium) ovale* also bears some morphological resemblance to *P. simium* of howler monkeys (Garnham, 1966; Coatney et al., 1971). There appear to be no close extant genetic relatives of *P. ovale* (Ayala et al., 1998).

Plasmodium (Plasmodium) malariae (Grassi and Feletti, 1890) is the malarial parasite responsible for causing quartan fever (72 hour schizogony) in humans. This parasite can remain in peripheral circulation of human hosts for years. Some investigators have suggested that there is a phylogenetic connection between *P. malariae* of humans, *P. hylobati* of gibbons, and *P. brasilianum* of New World monkeys since all of the parasites exhibit quartan periodicity, coarse pigment and dense cytoplasm (White, 1993). Escalante et al. (1995) analyzed the conserved regions of the gene coding for the circumsporozoite protein (a surface protein expressed at the sporozoite stage) in various *Plasmodium* species and found that *P. malariae* was indistinguishable from *P. brasilianum*. The genetic similarity between the two species indicates that host switching or host sharing may have recently occurred.

The subgenus *P. (Laverania)* contains only two known species: *P. (Laverania) falciparum* (Welch, 1897) is the virulent tertian parasite which naturally infects humans; *P. (Laverania) reichenowi* (Sluiter et al., 1922) is the chimpanzee and gorilla counterpart. The two species are diagnosed by the presence of ring-like trophozoites and crescentric gametocytes. Analysis of mitochondrial DNA sequence polymorphism indicates that *P. falciparum* is most closely related to *P. reichenowi* (Conway et al., 2000). Escalante et al. (1995) analyzed the conserved regions of the gene coding for the circumsporozoite protein in various *Plasmodium* species and confirmed the evolutionary closeness of *P. falciparum* and *P. reichenowi*. In fact, the two *P. (Laverania)* species are more closely related to one another than either are to other malarial parasites (Qari et al., 1996), and both are more closely related to rodent and avian malarial parasites than to other primate malarial parasites (McCutchan et al., 1984). Escalante and Ayala (1994) present rRNA evidence indicating that *P. falciparum* diverged from *P. reichenowi* at approximately the same time that chimpanzees and other hominins did, 5–10 mya. In contrast, analysis of *Dhfr* and *Ts* genes suggest that all extant populations of *P. falciparum* may be recently derived from a single ancestral stock, the most recent common ancestor of which may have lived between 25 000 and 58 000 years ago (Rich et al., 1998). Extremely low nucleotide polymorphism is consistent with a recent origin for *P. falciparum* (Rich et al., 1998; Conway et al., 2000; Volkman et al., 2001; Carter and Mendis, 2002; Hartl, 2004). Worldwide climatic changes throughout the last glaciation and the advent of agriculture would have facilitated the spread of the *Anopheles gambiae* vector responsible for the radiation of *Plasmodium* (Coluzzi, 1999). During this agrarian revolution, higher density, sedentary human populations provided mosquitoes with necessary bloodmeals and potential mosquito breeding sites (Livingstone, 1958; Hartl, 2004).

Still other analyses using mitochondrial DNA suggest an origin of more than 100 000 years before present for *P. falciparum* (Hughes and Verra, 2001). Joy et al. (2003) have predicted, based on mitochondrial DNA evidence, that *P. falciparum* really began to spread within the human populations in Africa around 10 000 years ago. There is obviously great discrepancy in these dates, and it will be important to continue to elucidate the phylogenetic history of these parasites.

As malaria has been such a large evolutionary driving force for humans, we have developed several resistance genetic polymorphisms (for review, see Carter and Mendis, 2002). Some of the major hemoglobinopathies, enzymopathies, and erythrocyte variants common today include sickle cell disorder, the O blood group, hemoglobins C and E, various glycophorins and human leukocyte antigens, ovalocytosis, α- and β-thalassemias, glucose-6-phosphate-dehydrogenase (G6PD) deficiency, and Duffy antigen receptor negativity (Livingstone, 1971; Weatherall and Clegg, 2001; Carter and Mendis, 2002; Kwiatkowski, 2005).

Sickle cell disorder is an autosomal recessive condition caused by a single-point mutation from glutamate to valine in the sixth position of the β-chain of hemoglobin. The condition is found throughout Africa, India, the Middle East, and the Mediterranean (Allison, 1954; Livingstone, 1967). In the homozygous state, sickle cell anemia causes significant morbidity and mortality (Weatherall and Clegg, 2001). Although the exact protective mechanism against malaria is equivocal, in the heterozygous state this trait could confer resistance through increased clearance or reduced resetting of infected cells, reduced parasite growth in sickled cells, and enhanced immune responses against infected cells (Williams et al., 2005; Williams, 2006). The α- and β-thalassemias, caused by deletions or mutations on chromosomes 16 and 11 with resultant alterations of the α- and β-chains of hemoglobin, are also likely characterized by greater immune responses towards malaria-infected erythrocytes (Allen et al., 1997).

Hemoglobin C is found in West Africa and results from a single-point mutation from glutamate to lysine in the sixth position of the β-chain of hemoglobin. In the homozygous state, there is a significant protective affect against death from *P. falciparum* (Modiano et al., 2001). In contrast, the homozygous state of ovalocytosis appears to always cause prenatal mortality (Genton et al., 1995). Ovalocytosis is caused by a 27-base-pair deletion in the erythrocyte band 3 (*AE1*) gene (Allen et al., 1999). In the heterozygous state, erythrocytes become oval-shaped which may inhibit merozoite invasion as well as decrease cytoadherence of infected erythrocytes to cerebral microvessels. Individuals with blood group O also exhibit significantly less binding between infected and uninfected erythrocytes (Rowe et al., 2007). Blood group O may have evolved as a mutation of blood group A in response to selective pressure from *P. falciparum* just prior to the major migration of anatomically modern *Homo sapiens* out of African around 100000 years ago (Cserti and Dzik, 2007).

More recent protective genotypic polymorphisms include G6PD deficiency and the Duffy-negative allele. The former is caused by several hundred mutational variations of the X-linked *G6PD* gene (Ruwende et al., 1995). Again, the exact protective mechanism is unknown, but infected cells may be more susceptible to phagocytosis or hemolysis. The Duffy-negative allele results in a mutation at the FY locus of the Duffy antigen receptor for chemikines (DARC). This mutation eliminates a binding site on the surface of erythrocytes which *P. vivax* requires for entry into host cells

(Tournamille et al., 1995). Both G6PD deficiency and the Duffy-negative allele are thought to have evolved approximately 10000 years before present, corresponding with the spread of agriculture (Carter and Mendis, 2002; Tishkoff et al., 2001). Interestingly, a Duffy-negative genotype (−46C/C) is now associated with increased susceptibility to HIV-1 infection, but slower disease progression, in African-American individuals (He et al., 2008).

Emerging infectious diseases

"Emerging infectious diseases" are those that have recently increased in incidence, expanded in geographic range, moved into a new host population, or are caused by newly evolved pathogens (Morens et al., 2004; Weiss and McMichael, 2004). Today, examples include dengue hemorrhagic fever, West Nile virus, Nipah virus, and H5N1 avian influenza (Figure 27.6). However, just because new, potentially deadly diseases are emerging and evolving does not necessarily mean that the entire modern human species will be wiped out by some exotic airborne Ebola-like virus, as the cinema likes to portray. As illustrated above, humans have been very good at developing genotypic and phenotypic adaptations to combat infections. Interestingly, the primary causes of emerging infectious diseases in human populations have been through anthropogenic modification of the physical and social environments (Daily and Ehrlich, 1996; Patz et al., 2004). In the short-term, remediation of the effects of these pathogens will be primarily accomplished by human behavioral changes rather than genotypic adaptations.

Human and livestock populations continue to grow rapidly, increasing the number of hosts potentially susceptible to novel infections. Mass transportation of people, products, livestock, and vectors of disease bring each of these closer to one another, and more quickly at that (Kimball et al., 2005). Population movements due to war, social disruption, and rural-to-urban migration in addition to general urbanization increase the densities of nonimmune human hosts and pose significant sanitation problems. Natural disasters and bioterrorism may destroy public-health infrastructure and other resources (Watson et al., 2007). Sex tourism, intravenous drug abuse, the reuse of injectable medical equipment, ("iatrogenic") and improper disinfection or ineffective protective measures in hospitals ("nosocomial") all contribute to the rapid evolution of resistant and deadly pathogens.

Human encroachment into previously undisturbed areas increases remote area accessibility and introduces more vectors and reservoirs of infection to new hosts. Encroachment, extensification of agricultural land, urban sprawl, and habitat fragmentation all alter population densities and distributions of wildlife,

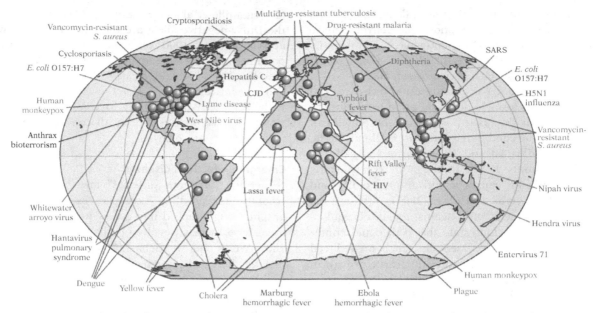

27.6. Examples of emerging infectious diseases. Figure reprinted with permission from Morens et al. (2004).

which changes disease dynamics (Patz et al., 2004). Deforestation results in loss of plant species and subsequent loss of undiscovered therapeutic drugs. Changes in water usage, such as during the construction of dams, culverts, and irrigation systems, can increase the potential breeding sites of vector species like mosquitoes and snails (Keiser et al., 2005; Steinmann et al., 2006). Biodiversity loss due to global climate change, deforestation, the spread of invasive species, overexploitation, and other causes reduces the buffer of hosts in an environment, increasing the likelihood of cross-species transmission (Maillard and Gonzalez, 2006).

As discussed previously, the use of bushmeat has been the source of HIV. In fact, the majority of all emerging pathogens in humans are zoonotic in origin (Jones et al., 2008), and population, ecological, and behavioral changes that increase contact with wildlife exacerbate emergence of these pathogens (Daszak et al., 2000). Another example is the severe acute respiratory syndrome caused by a coronavirus (SARS-CoV). It is transmitted through aerosolized particles, and symptoms include fever and pneumonia. Between November 2002 and April 2004, SARS resulted in 774 confirmed deaths from 8098 cases in 29 countries (Hughes, 2004; Morens et al., 2004). The outbreak originated in farmers and animal workers in the Foshan Municipality of the Guangdong Province of China (Yu et al., 2003). Later, the virus was identified in Himalayan palm civets (*Paguna larvata*), raccoon dogs (*Nyctereutes procuyoinboides*), and Chinese ferret badgers (*Melongdale moschata*) sold in wet markets for consumption. The actual reservoir of the virus is the horseshoe bat (genus *Rhinolophus*) (Li et al., 2005). These animals likely came into contact with civet farms, possibly through the feeding of bats to farmed civets. Unfortunately, it took close to a year to contain the outbreak, in no small part because of the lag of case reporting and under-reports of the actual casualties in China (Parry, 2003). Fear of decreased tourism, travel, and trade eventually cost the region billions of dollars and hundreds of lives. This is an excellent example of how emphasis on national sovereignty over global solidarity can produce significant negative health effects for the global population (Heymann, 2004, 2006).

Humans have also caused altered vector and reservoir distributions through the production of greenhouse gases and global climate change. The *Aedes aegypti* mosquito has expanded its geographic range in response to increased global temperatures, resulting in increased risk of Dengue virus in subtropical and temperate climates (Hales et al., 2002). Likewise, the risk of Lyme disease (*Borrelia burgdorferi*) has increased with elevated global temperatures (Brownstein et al., 2005). The outbreak of Hantavirus pulmonary syndrome in the southwestern United States in the early 1990s was caused by heavy summer rains associated with the El Niño Southern Oscillation effect and the subsequent proliferation of pine nuts and deer mice (*Peromyscus maniculatus*), the natural reservoir of the Hantaan virus (family Bunyaviridae) (Engelthaler et al., 1999).

Nipah virus is another excellent example of how human-induced ecological changes have altered the risk of emerging infectious diseases. Nipah is a single-stranded RNA virus of the family Paramyxoviridae that causes severe acute febrile encephalitis in humans. Its natural reservoir is the flying fox (genus *Pteropus*), and the virus has been identified in Malaysia, Singapore, and Bangladesh (Epstein et al., 2006). Between October 1997 and February 2000, Nipah virus resulted in 105 confirmed human fatalities in Malaysia. The cause of the outbreak is now attributed to a complex interaction of human-induced environmental changes. Specifically, fire-mediated deforestation for the expansion of oil palm plantations produced significant air pollution in the region. Combined with a drought produced by the El Niño Southern Oscillation, the availability of flowering and fruiting forest trees was reduced (Chua et al., 2002). Bats (*Pteropus vampyrus* and *Pteropus hypomelanus*) began feeding in human orchards that had been strategically planted next to pig farms so as to use pig waste as fertilizer for the orchards. Bat waste entered the pigsties and the virus amplified within the swine. Despite producing severe respiratory disease (but low mortality), the infected pigs were distributed throughout Malaysia. Economic losses were enormous, ultimately because we were less mindful about the impacts of environmental change on animal ecology and human health than we should have been.

Widespread, irreversible modification and overuse of our environments have been characteristics of most human populations over the past few thousand years, but particularly the past century. Such rapid modification of the physical and social environments has increased our reliance on cultural adaptations against disease over biological adaptations that take generations to proliferate. The current avian influenza outbreak illustrates the need for this realization. Influenza is a single-stranded RNA virus of the family Orthomyxoviridae, of three main types (A, B, and C), and classified into subgroups according to its glycosylated surface antigens: hemagglutinin (H) and neuraminidase (N). The current avian influenza epidemic that began in 2003 is caused by a type A, H5N1 virus. Influenza viruses are common in wild migratory waterfowl, particularly ducks, geese, swans, gulls, and terns (Chen et al., 2005; Olsen et al., 2006). Infection with the low pathogenic forms is typically asymptomatic in the birds. However, the high pathogen forms can cause significant mortality is some species, like bar-headed geese. Influenza experts are concerned that such a virus could cause a devastating pandemic. The 1918 Spanish influenza (H1N1) pandemic that killed as many as 50 million people was caused by an avian influenza virus that reassorted (i.e., antigenic shift) within swine (Taubenberger et al., 1997, 2005; Gamblin et al., 2004).

Despite the fact that the recent avian influenza outbreak has attracted a great deal of attention, avian influenza has likely affected humans (although to a much lesser degree) for hundreds of years. Due to the huge amount of commercial poultry produced today, the likelihood of viral entry into the poultry and human populations from wild birds has increased. Although human-to-human transmission of the virus has been suggested in several cases (Ungchusak et al., 2005), the avian influenza virus appears to bind to receptors deep in the human respiratory tract, which would limit direct human transmission and decrease the likelihood of a pandemic (Shinya et al., 2006). However, reliance on genetic adaptations against this epidemic is clearly unwise; human genotypic adaptations to such diseases will be delayed compared to rapid pathogen evolution. The mis-governance of epidemics with decentralized testing, misdiagnoses, and underreporting of cases has resulted in slow public health responses (Cyranoski, 2005; Normile, 2005). A future pandemic could produce considerable economic burden due to destruction of commercial poultry, high human health-care costs and loss of productivity (Meltzer et al., 1999). An understanding of human evolution against infectious disease is incomplete and inapplicable for today's epidemics without consideration of the social, political, and environmental causes of morbidity and mortality, and the behaviors we might employ to ameliorate some of these challenges.

DISCUSSION POINTS

1. What contributions can evolutionary biology make to medicine?
2. How can environmental conditions during child development affect immune functions?
3. How does lymphocyte development resemble Darwinian evolution by natural selection?
4. How do some infectious pathogens evolve high levels of virulence whereas others do not?
5. What genotypic and phenotypic adaptations have humans developed to combat malaria infection?
6. What are some of the social and environmental causes of emerging infectious diseases?

ACKNOWLEDGMENTS

Paul Ewald, Lisa Becker, Sean Prall, Laurah Jones, and two anonymous reviewers assisted with the production of this manuscript.

REFERENCES

Abraham, S. N. and Arock, M. (1998). Mast cells and basophils in innate immunity. *Seminars in Immunology*, **10**, 373–381.

Adema, C. M., Hertel, L. A., Miller, R. D., et al. (1997). A family of fibrinogen-related proteins that precipitates parasite-derived molecules is produced by an invertebrate after infection. *Proceedings of the National Academy of Sciences of the United States of America*, **94**, 8691–8696.

Agrawal, A., Eastman, Q. M. and Schatz, D. G. (1998). Transposition mediated by RAG1 and RAG2 and its implications for the evolution of the immune system. *Nature*, **394**, 744–751.

Ahmed, A. M. and Hurd, H. (2006). Immune stimulation and malaria infection impose reproductive costs in *Anopheles gambiae* via follicular apopstosis. *Microbes and Infection*, **8**, 308–315.

Alder, M. N., Rogozin, I. B., Lakshminarayan, M. I., et al. (2005). Diversity and function of adaptive immune receptors in a jawless vertebrate. *Science*, **310**, 1970–1973.

Allee, W. C., Park, O., Emerson, A. E., et al. (1949). *Principles of Animal Ecology*. Philadelphia: W.B. Saunders.

Allen, J. E. and Maizels, R. M. (1996). Immunology of human helminth infection. *International Archives of Allergy and Immunology*, **109**, 3–10.

Allen, S. J., O'Donnell, A., Alexander, N. D., et al. (1997). Alpha-thalassaemia protects children against disease caused by other infections as well as malaria. *Proceedings of the National Academy of Sciences of the United States of America*, **94**, 14736–14741.

Allen, S. J., O'Donnell, A., Alexander, N. D., et al. (1999). Prevention of cerebral malaria in children in Papua New Guinea by Southeast Asian ovalocytosis band 3. *American Journal of Tropical Medicine and Hygiene*, **60**, 1056–1060.

Allison, A. C. (1954). Protection afforded by sickle-cell trait against subtertian malarial infection. *British Medical Journal*, **1**, 290–294.

Anderson, R. M. and May, R. M. (1992). *Infectious Diseases of Humans: Dynamics and Control*. New York: Oxford University Press.

Apetrei, C., Metzger, M. J., Richardson, D., et al. (2005). Detection and partial characterization of simian immunodeficiency virus SIVsm strains from bush meat samples from rural Sierra Leone. *Journal of Virology*, **79**, 2631–2636.

Ardia, D. R., Schat, K. A. and Winkler, D. W. (2003). Reproductive effort reduces long-term immune function in breeding tree swallows (*Tachycineta bicolor*). *Proceedings of the Royal Society of London. Series B*, **270**, 1679–1683.

Austin, D. J., Kristinsson, K. G. and Anderson, R. M. (1999). The relationships between the volume of antimicrobial consumption in human communities and the frequency of resistance. *Proceedings of the National Academy of Sciences of the United States of America*, **96**, 1152–1156.

Ayala, F. J., Escalante, A. A., Lal, A. A., et al. (1998). Evolutionary relationships of human malaria parasites. In *Malaria: Parasite Biology, Pathogenesis and Protection*, I. W. Sherman (ed.). Washington, DC: ASM Press, pp. 285–300.

Bailes, E., Gao, F., Bibollet-Ruche, F., et al. (2003). Hybrid origin of SIV in chimpanzees. *Science*, **300**, 1713.

Baker, E. N., Baker, H. M. and Kidd, R. D. (2002). Lactoferrin and transferrin: functional variations on a common structural framework. *Biochemistry and Cell Biology*, **80**, 27–34.

Ballinger, A. B., Savage, M. O. and Sanderson, I. R. (2003). Delayed puberty associated with inflammatory bowel disease. *Pediatric Research*, **53**, 205–210.

Barker, D. J. (2007). The origins of the developmental origins theory. *Journal of Internal Medicine*, **261**, 412–417.

Barker, D. J., Eriksson, J. G., Forsen, T., et al. (2002). Fetal origins of adult disease: strength of effects and biological basis. *International Journal of Epidemiology*, **31**, 1235–1239.

Barr, D. P., Russell, M. D., Cecil, L., et al. (1922). Clinical calorimetry XXXII: temperature regulation after the intravenous injections of protease and typhoid vaccine. *Archives of Internal Medicine*, **29**, 608–634.

Barrett, K. E., Neva, F. A., Gam, A. A., et al. (1988). The immune response to nematode parasites: modulation of mast cell numbers and function during *Strongyloides stercoralis* infections in nonhuman primates. *American Journal of Tropical Medicine and Hygiene*, **38**, 574–581.

Bellamy, R., Ruwende, C., Corrah, T., et al. (1998). Variations in the *NRAMP1* gene and susceptibility to tuberculosis in West Africans. *New England Journal of Medicine*, **338**, 640–644.

Berke, G. (1997). Killing mechanisms of cytotoxic lymphocytes. *Current Opinion in Hematology*, **4**, 32–40.

Bergstrom, C. T., Lo, M. and Lipsitch, M. (2004). Ecological theory suggests that antimicrobial cycling will not reduce antimicrobial resistance in hospitals. *Proceedings of the National Academy of Sciences of the United States of America*, **101**, 13285–13290.

Bhasin, S., Storer, T. W., Berman, N., et al. (1996). The effects of supraphysiological doses of testosterone on muscle size and strength in normal men. *New England Journal of Medicine*, **335**, 1–7.

Bloom, B. R. and Murray, C. J. (1992). Tuberculosis: commentary on a reemergent killer. *Science*, **257**, 1055–1064.

Bothamley, G. H., Beck, J. S., Schreuder, G. M., et al. (1989). Association of tuberculosis and *M. tuberculosis*-specific antibody levels with HLA. *Journal of Infectious Diseases*, **159**, 549–555.

Brahmajothi, V., Pitchappan, R. M., Kakkanaiah, V. N., et al. (1991). Association of pulmonary tuberculosis and HLA in South India. *Tubercle*, **72**, 123–132.

Braun, M., Cook, D. C. and Pfeiffer, S. (1998). DNA from *Mycobacterium tuberculosis* complex identified in North American, pre-Columbian human skeletal remains. *Journal of Archaeological Science*, **25**, 271–277.

Braun-Fahrlander, C., Gassner, M., Grize, L., et al. (1999). Prevalence of hay fever and allergic sensitization in farmer's children and their peers living in the same rural community. *Clinical and Experimental Allergy*, **29**, 28–34.

Bribiescas, R. G. (2001). Reproductive ecology and life history of the human male. *Yearbook of Physical Anthropology*, **44**, 148–176.

Brosch, R., Gordon, S. V., Marmiesse, M., et al. (2002). A new evolutionary scenario for the *Mycobacterium tuberculosis* complex. *Proceedings of the National Academy of Sciences of the United States of America*, **99**, 3684–3689.

Brownstein, J. S., Holford, T. R. and Fish, D. (2005). Effect of climate change on Lyme disease risk in North America. *EcoHealth*, **2**, 38–46.

Bull, J. J., Molineux, I. J. and Rice, W. R. (1991). Selection of benevolence in a host-parasite system. *Evolution*, **45**, 875–882.

Burger, D. and Dayer, J. M. (2002). Cytokines, acute-phase proteins, and hormones: IL-1 and TNF-α production in contact-mediated activation of monocytes by T lymphocytes. *Annals of the New York Academy of Sciences*, **966**, 464–473.

Burnet, F. M. (1959). *The Clonal Selection Theory of Acquired Immunity*. Nashville: Vanderbilt University Press.

Burnet, F. M. and White, D. O. (1972). *Natural History of Infectious Disease*, 4th edn. Cambridge: Cambridge University Press.

Carroll, M. C. (1998). The role of complement and complement receptors in induction and regulation of immunity. *Annual Review of Immunology*, **16**, 545–568.

Carter, R. and Mendis, K. N. (2002). Evolutionary and historical aspects of the burden of malaria. *Clinical Microbiology Reviews*, **15**, 564–594.

Chambers, H. F. (1997). Methicillin resistance in Staphylococci: genetics and mechanisms of resistance. *Clinical Microbiology Reviews*, **10**, 781–791.

Chen, H., Smith, G. J. D., Zhang, S. Y., et al. (2005). H5N1 virus outbreak in migratory waterfowl. *Nature*, **436**, 191–192.

Chisholm, S. T., Coaker, G., Day, B., et al. (2006). Host-microbe interactions: shaping the evolution of the plant immune response. *Cell*, **124**, 803–814.

Chua, K. B., Chua, B. H. and Wang, C. W. (2002). Anthropogenic deforestation, El Niño and the emergence of Nipah virus in Malaysia. *Malaysian Journal of Pathology*, **24**, 15–21.

Chun, T. W., Nickle, D. C., Justement, J. S., et al. (2008). Persistence of HIV in gut-associated lymphoid tissue despite long-term antiretroviral therapy. *Journal of Infectious Diseases*, **197**, 714–720.

Coatney, R. G., Collins, W. E., Warren, M., et al. (1971). *The Primate Malarias*. Bethesda, MD: National Institutes of Health.

Coffman, R. L. and Mosmann, T. R. (1991). CD4$^+$ T-cell subsets: regulation of differentiation and function. *Research in Immunology*, **142**, 7–9.

Cole, S. T., Brosch, R., Parkhill, J., et al. (1998). Deciphering the biology of *Mycobacterium tuberculosis* from the complete genome sequence. *Nature*, **393**, 537–544.

Coluzzi, M. (1999). The clay feet of the malaria giant and its African roots: hypotheses and inferences about origin, spread and control of *Plasmodium falciparum*. *Parassitologia*, **41**, 277–283.

Combes, C. (2004). *Parasitism: the Ecology and Evolution of Intimate Interactions*. Chicago: University of Chicago Press.

Conway, D. J., Greenwood, B. M. and McBride, J. S. (1991). The epidemiology of multiple-clone *Plasmodium falciparum* infections in Gambian patients. *Parasitology*, **103**, 1–6.

Conway, D. J., Fanello, C., Lloyd, J. M., et al. (2000). Origin of *Plasmodium falciparum* malaria is traced by mitochondrial DNA. *Molecular and Biochemical Parasitology*, **111**, 163–171.

Cookson, W. O. C. M. and Moffatt, M. F. (1997). Asthma: an epidemic in the absence of infection. *Science*, **275**, 41–42.

Cooper, M. and Alder, M. (2006). The evolution of adaptive immune systems. *Cell*, **124**, 815–822.

Courvalin, P. (2008). Predictable and unpredictable evolution of antibiotic resistance. *Journal of Internal Medicine*, **264**, 4–16.

Cox, F. E. and Liew, F. Y. (1992). T-cell subsets and cytokines in parasitic infections. *Immunology Today*, **13**, 445–448.

Cserti, C. M. and Dzik, W. H. (2007). The ABO blood group system and *Plasmodium falciparum* malaria. *Blood*, **110**, 2250–2258.

Curran, J. E., Jowett, J. B. M., Elliott, K. S., et al. (2005). Genetic variation in selenoprotein S influences inflammatory response. *Nature Genetics*, **37**, 1234–1241.

Cyranoski, D. (2005). Tests in Tokyo reveal flaws in Vietnam's bird flu surveillance. *Nature*, **433**, 787.

Daily, G. C. and Ehrlich, P. R. (1996). Global change and human susceptibility to disease. *Annual Review of Energy and the Environment*, **21**, 125–144.

Dangl, J. L. and Jones, J. D. (2001). Plant pathogens and integrated defence responses to infection. *Nature*, **411**, 826–833.

Daszak, P., Cunningham, A. A. and Hyatt, A. D. (2000). Emerging infectious diseases of wildlife – threats to biodiversity and human health. *Science*, **287**, 443–449.

Dawkins, R. and Krebs, J. R. (1978). Arms races between and within species. *Proceedings of the Royal Society of London. Series B*, **205**, 489–511.

Dean, M., Carrington, M. and O'Brien, S. J. (2002). Balanced polymorphism selected by genetic versus infectious human disease. *Annual Review of Genomics and Human Genetics*, **3**, 263–292.

Deerenberg, C., Arpanius, V., Daan, S., et al. (1997). Reproductive effort decreases antibody responsiveness. *Proceedings of the Royal Society of London. Series B*, **264**, 1021–1029.

Delves, P. J. and Roitt, I. M. (2000a). Advances in immunology: the immune system I. *New England Journal of Medicine*, **343**, 37–49.

Delves, P. J. and Roitt, I. M. (2000b). Advances in immunology: the immune system II. *New England Journal of Medicine*, **343**, 108–117.

Demas, G. E., Chefer, V., Talan, M. I., et al. (1997). Metabolic costs of mounting an antigen-stimulated immune response in adult and aged C57BL/6J mice. *American Journal of Physiology – Regulatory, Integrative and Comparative Physiology*, **42**, R1331–R1367.

Dempsey P. W., Allison, M. E., Akkaraju, S., et al. (1996). C3d of complement as a molecular adjuvant: bridging innate and acquired immunity. *Science*, **271**, 348–350.

Dessein, A. J., Hillaire, D., Elwali, N. E., et al. (1999). Severe hepatic fibrosis in *Schistosoma mansoni* infection is controlled by a major locus that is closely linked to the interferon-gamma receptor gene. *American Journal of Human Genetics*, **65**, 709–721.

Dinarello, C. A. (2000). Proinflammatory cytokines. *Chest*, **118**, 503–508.

Doran, T. F., DeAngelis, C., Baumgardner, R. A., et al. (1989). Acetaminophen: more harm than good for chickenpox? *Journal of Pediatrics*, **114**, 1045–1048.

Douglas, A. S., Strachan, D. P. and Maxwell, J. D. (1996). Seasonality of tuberculosis: the reverse of other respiratory diseases in the UK. *Thorax*, **51**, 944–946.

Drobniewski, F., Balabanova, Y., Ruddy, M., et al. (2002). Rifampin- and multidrug-resistant tuberculosis in Russian civilians and prison inmates: dominance of the Beijing strain family. *Emerging Infectious Diseases*, **8**, 1320–1326.

Duncan, S. R., Scott, S. and Duncan, C. J. (2005). Reappraisal of the historical selective pressures for the CCR5-D32 mutation. *Journal of Medical Genetics*, **42**, 205–208.

Dye, C., Scheele, S., Dolin, P., et al. (1999). Global burden of tuberculosis: estimated incidence, prevalence, and mortality by country. *Journal of the American Medical Association*, **282**, 677–686.

Eaton, S. B. and Eaton, S. B. (1999). The evolutionary context of chronic degenerative diseases. In *Evolution in Health and Disease*, S. C. Stearns (ed.). New York: Oxford University Press, pp. 251–259.

Ebert, D. and Bull, J. J. (2008). The evolution and expression of virulence. In *Evolution in Health and Disease*, S. C. Stearns and J. C. Koella (eds), 2nd edn. New York: Oxford University Press, pp. 153–168.

Edelman, G. M. (1973). Antibody structure and molecular immunology. *Science*, **180**, 830–840.

Elia, M. (1992). Energy expenditure to metabolic rate. In *Energy Metabolism: Tissue Determinants and Cellular Corollaries*, J. M. McKinney and H. N. Tucker (eds). New York: Raven Press, pp. 19–49.

Ellner, J. J. (1997). Review: the immune response in human tuberculosis – implications for tuberculosis control. *Journal of Infectious Diseases*, **176**, 1351–1359.

Else, K. J. and Finkelman, F. D. (1998). Intestinal nematode parasites, cytokines and effector mechanisms. *International Journal for Parasitology*, **28**, 1145–1158.

Elton, S. and O'Higgins, P. (2008). *Medicine and Evolution: Current Applications, Future Prospects*. New York: CRC Press.

Engelthaler, D. M., Mosley, D. G., Cheek, J. E., et al. (1999). Climatic and environmental patterns associated with Hantavirus pulmonary syndrome, Four Corners region, United States. *Emerging Infectious Diseases*, **5**, 87–94.

Epstein, J., Field, H. E., Luby, S., et al. (2006). Nipah virus: impact, origins, and causes of emergence. *Current Infectious Disease Reports*, **8**, 59–65.

Eraud, C., Duriez, O., Chastel, O., et al. (2005). The energetic cost of humoral immunity in the collared dove, *Streptopelia decaocto*: is the magnitude sufficient to force energy-based trade-offs? *Functional Ecology*, **19**, 110–118.

Ernst, J. D., Trevejo-Nuñez, G. and Banaiee, N. (2007). Genomics and the evolution, pathogenesis, and diagnosis of tuberculosis. *Journal of Clinical Investigation*, **117**, 1738–1745.

Escalante, A. A. and Ayala, F. J. (1995). Phylogeny of the malarial genus *Plasmodium*, derived from rRNA gene sequences. *Proceedings of the National Academy of Sciences of the United States of America*, **91**, 11373–11377.

Escalante, A. A., Barrio, E. and Ayala, F. J. (1995). Evolutionary origins of human and primate malarias: evidence from the circumsporozoite protein gene. *Molecular Biology and Evolution*, **12**, 616–626.

Escalante, A. A., Cornejo, O. E., Freeland, D. E., et al. (2005). A monkey's tale: the origin of *Plasmodium vivax* as a human malaria parasite. *Proceedings of the National Academy of Sciences of the United States of America*, **102**, 1980–1985.

Ewald, P. W. (1996). *Evolution of Infectious Disease*. New York: Oxford University Press.

Ewald, P. W. and Cochran, G. M. (2004). Units of selection and the evolution of virulence. In *The Evolution of Population Biology*, R. Singh and M. Uyenoyama (eds). New York: Cambridge University Press, pp. 377–390.

Fischer, G. F. and Mayr, W. R. (2001). Molecular genetics of the HLA complex. *Wiener Klinische Wochenschrift*, **113**, 814–824.

Flajnik, M. F. (2002). Comparative analyses of immunoglobulin genes: surprises and portents. *Nature Reviews Immunology*, **2**, 688–698.

Flajnik, M. F. and du Pasquier, L. (2004). Evolution of innate and adaptive immunity: can we draw a line? *Trends in Immunology*, **25**, 640–644.

Flajnik, M. F. and Kasahara, M. (2001). Comparative genomics of the MHC: glimpses into the evolution of the adaptive immune system. *Immunity*, **15**, 351–362.

Fugmann, S. D., Lee, A. I., Shockett, P. E., et al. (2000). The RAG proteins and V(D)J recombination: complexes, ends, and transposition. *Annual Review of Immunology*, **18**, 495–527.

Gagneux, S., DeRiemer, K., Van, T., et al. (2006). Variable host–pathogen compatibility in *Mycobacterium tuberculosis*. *Proceedings of the National Academy of Sciences of the United States of America*, **103**, 2869–2873.

Gallup, J. L. and Sachs, J. D. (2001). The economic burden of malaria. *American Journal of Tropical Medicine and Hygiene*, **64**(suppl.), 85–96.

Galvani, A. P. (2003). Epidemiology meets evolutionary ecology. *Trends in Ecology and Evolution*, **18**, 132–139.

Galvani, A. P. and Slatkin, M. (2003). Evaluating plague and smallpox as historical selective pressures for the *CCR5-Δ32* HIV-resistance allele. *Proceedings of the National Academy of Sciences of the United States of America*, **100**, 15276–15279.

Gamblin, S. J., Haire, L. F., Russell, R. J., et al. (2004). The structure and receptor binding properties of the 1918 influenza hemagglutinin. *Science*, **303**, 1838–1842.

Gandon, S., Mackinnon, M. J., Nee, S., et al. (2001). Imperfect vaccines and the evolution of pathogen virulence. *Nature*, **414**, 751–756.

Gangestad, S. W. and Thornhill, R. (1998). Menstrual cycle variation in women's preferences for the scent of symmetrical men. *Proceedings of the Royal Society of London. Series B*, **265**, 927–933.

Gao, F., Bailes, E., Robertson, D. L., et al. (1999). Origin of HIV-1 in the chimpanzee *Pan troglodytes troglodytes*. *Nature*, **397**, 436–441.

Gao, P. S., Fujishima, S., Mao, X. Q., et al. (2000). Genetic variants of NRAMP1 and active tuberculosis in Japanese populations. International Tuberculosis Genetics Team. *Clinical Genetics*, **58**, 74–76.

Garnham, P. C. C. (1966). *Malaria Parasites and other Haemosporidia*. Oxford: Blackwell Scientific Publications.

Gazzinelli, R. T., Hieny, S., Wynn, T. A., et al. (1993). Interleukin 12 is required for the T-lymphocyte-independent induction of interferon gamma by an intracellular parasite and induces resistance in T-cell-deficient hosts. *Proceedings of the National Academy of Sciences of the United States of America*, **90**, 6115–6119.

Genton, B., Al-Yaman, F., Mgone, C. S., et al. (1995). Ovalocytosis and cerebral malaria. *Nature*, **378**, 564–565.

Gerald, M. S. (2001). Primate colour predicts social status and aggressive outcome. *Animal Behavior*, **61**, 559–566.

George, A. J. T. and Ritter, M. A. (1996). Thymic involution with ageing: obsolescence or good housekeeping? *Immunology Today*, **17**, 267–272.

Gershwin, M. E., German, J. B. and Keen, C. L. (2000). *Nutrition and Immunology*. Totowa: Humana Press.

Glass, W. G., McDermott, D. H., Lim, J. K., et al. (2006). CCR5 deficiency increases risk of symptomatic West Nile virus infection. *Journal of Experimental Medicine*, **203**, 35–40.

Goldfeld, A. E., Delgado, J. C., Thim, S., et al. (1998). Association of an HLA-DQ allele with clinical tuberculosis. *Journal of the American Medical Association*, **279**, 226–228.

Goldman, A. S. (1993). The immune system of human milk: antimicrobial, anti-inflammatory and immunomodulating properties. *Journal of Pediatric Infectious Diseases*, **12**, 664–671.

Goncalves, M. L., Araujo, A. and Ferreira, L. F. (2003). Human intestinal parasites in the past: new findings and a review. *Memorias do Instituto Oswaldo Cruz*, **98**, 103–118.

Graham, N. M., Burrell, C. J., Douglas, R. M., et al. (1990). Adverse effects of aspirin, acetaminophen, and ibuprofen on immune function, viral shedding, and clinical status in rhinovirus-infected volunteers. *Journal of Infectious Diseases*, **162**, 1277–1282.

Grammer, K. and Thornhill, R. (1994). Human (*Homo sapiens*) facial attractiveness and sexual selection: the role of symmetry and averageness. *Journal of Comparative Psychology*, **108**, 233–242.

Grassi, B. and Feletti, R. (1890). Parasites malariques chez les oiseaux. *Archives Italiennes de Biologie*, **13**, 297–300.

Greaves, M. (2006). Infection, immune responses and the aetiology of childhood leukemia. *Nature Reviews Cancer*, **6**, 193–203.

Grosskinsky, C. M., Jacobs, W. R., Clark-Curtiss, J. E., et al. (1989). Genetic relationships among *Mycobacterium leprae*, *Mycobacterium tuberculosis*, and candidate leprosy vaccine strains determined by DNA hybridization: identification of an *M. leprae*-specific repetitive sequence. *Infection and Immunity*, **57**, 1535–1541.

Gutierrez, M. C., Brisse, S., Brosch, R., et al. (2005). Ancient origin and gene mosaicism of the progenitor of *Mycobacterium tuberculosis*. *PLoS Pathogens*, **1**, 55–61.

Guerra, C. A., Giandi, P. W., Tatem, A. J., et al. (2008). The limits and intensity of *Plasmodium falciparum* transmission: implications for malaria control and elimination worldwide. *PLoS Medicine*, **5**, e38.

Hadju, V., Abadi, K., Stephenson, L. S., et al. (1995). Intestinal helminthiasis, nutritional status, and their relationship; a cross-sectional study in urban slum school children in Indonesia. *Southeast Asian Journal of Tropical Medicine and Public Health*, **26**, 719–729.

Hahn, B. H., Shaw, G. M., De Cock, K. M., et al. (2000). AIDS as a zoonosis: scientific and public health implications. *Science*, **287**, 607–614.

Haig, D. (1993). Genetic conflict in human pregnancy. *Quarterly Review of Biology*, **68**, 495–532.

Haldane, J. B. S. (1941). *New Paths in Genetics*. London: Allen and Unwin.

Haldane, J. B. S. (1949). Disease and evolution. *La Ricerca Scientifica Supplement A*, **19**, 68–76.

Hales, S., de Wet, N., Maindonald, J., et al. (2002). Potential effect of population and climate changes on global distribution of dengue fever: an empirical model. *Lancet*, **360**, 830–834.

Hall, H. I., Song, R., Rhodes, P., et al. (2008). Estimation of HIV incidence in the United States. *Journal of the American Medical Association*, **300**, 520–529.

Hamilton, W. D. and Zuk, M. (1982). Heritable true fitness and bright birds: a role for parasites? *Science*, **218**, 384–387.

Hannet, I., Erkeller-Yuksel, F., Lydyard, P., et al. (1992). Developmental and maturational changes in human blood lymphocyte subpopulations. *Immunology Today*, **13**, 215–218.

Hartl, D. L. (2004). The origin of malaria: mixed messages from genetic diversity. *Nature Reviews Microbiology*, **2**, 15–22.

Havlir, D. V., Wallis, R. S., Boom, W. H., et al. (1991). Human immune response to *Mycobacterium tuberculosis* antigens. *Infection and Immunity*, **59**, 665–670.

Hawn, T. R., Dunstan, S. J., Thwaites, G. E., et al. (2006). A polymorphism in Toll-interleukin 1 receptor domain containing adaptor protein is associated with susceptibility to meningeal tuberculosis. *Journal of Infectious Diseases*, **194**, 1127–1134.

He, W., Neil, S., Kulkarni, H., et al. (2008). Duffy antigen receptor for chemokines mediates trans-infection of HIV-1 from red blood cells to target cells and affects HIV-AIDS susceptibility. *Cell Host and Microbe*, **4**, 52–62.

Herberman, R. B., Reynolds, C. W. and Ortaldo, J. R. (1986). Mechanisms of cytotoxicity by natural killer (NK) cells. *Annual Review of Immunology*, **4**, 651–680.

Heymann, D. L. (2004). The international response to the outbreak of SARS in 2003. *Philosophical Transactions of the Royal Society of London. Series B*, **1447**, 1127–1129.

Heymann, D. L. (2006). SARS and emerging infectious diseases: a challenge to place global solidarity above national sovereignty. *Annals Academy of Medicine Singapore*, **35**, 350–353.

Hill, A. V. S. (1998). The immunogenetics of human infectious diseases. *Annual Review of Immunology*, **16**, 593–617.

Hoberg, E. P., Alkire, N. L., de Queiroz, A., et al. (2001). Out of Africa: origins of the *Taenia* tapeworm in humans. *Proceedings of the Royal Society of London. Series B*, **268**, 781–787.

Hoffmann, J. A. and Reichhart, J. M. (2002). *Drosophila* innate immunity: an evolutionary perspective. *Nature Immunology*, **3**, 121–126.

Howard, R. S. and Lively, C. M. (1994). Parasitism, mutation accumulation and the maintenance of sex. *Nature*, **367**, 554–557.

Hughes, A. L. and Verra, F. (2001). Very large long-term effective population size in the virulent human malaria parasite *Plasmodium falciparum*. *Proceedings of the Royal Society of London. Series B*, **268**, 1855–1860.

Hughes, A. L., Friedman, R. and Murray, M. (2002). Genome-wide pattern of synonymous nucleotide substitution in two complete genomes of *Mycobacterium tuberculosis*. *Emerging Infectious Disease*, **8**, 1342–1346.

Hughes, J. M. (2004). SARS: an emerging global microbial threat. *Transactions of the American Clinical and Climatological Association*, **115**, 361–374.

Hume, D. A., Ross, I. L., Himes, S. R., et al. (2002). The mononuclear phagocyte system revisited. *Journal of Leukocyte Biology*, **72**, 621–627.

Hummel, S., Schmidt, D., Kremeyer, B., et al. (2005). Detection of the *CCR5-Δ32* HIV resistance gene in Bronze-Age skeletons. *Genes and Immunity*, **6**, 371–374.

Hutchinson, J. F. (2001). The biology and evolution of HIV. *Annual Review of Anthropology*, **30**, 85–108.

Ing, R., Su, Z., Scott, M. E., et al. (2000). Suppressed T helper 2 immunity and prolonged survival of a nematode parasite in protein-malnourished mice. *Proceedings of the National Academy of Sciences of the United States of America*, **97**, 7078–7083.

Innes, R. W. (2004). Guarding the goods. New insights into the central alarm system of plants. *Plant Physiology*, **135**, 695–701.

Jamieson, B. D., Douek, D. C., Killian, S., et al. (1999). Generation of functional thymocytes in the human adult. *Immunity*, **10**, 569–575.

Joint United National Programme on HIV/AIDS (2008). *Report on the Global AIDS Epidemic*. Geneva: World Health Organization.

Jones, C. A., Holloway, J. A. and Warner, J. O. (2000). Does atopic disease start in foetal life? *Allergy*, **55**, 2–10.

Jones, K. E., Patel, N. G., Levy, M. A., et al. (2008). Global trends in emerging infectious diseases. *Nature*, **451**, 990–993.

Jongwutiwes, S., Putaporntip, C., Iwasaki, T., et al. (2005). Mitochondrial genome sequences support ancient population expansion in *Plasmodium vivax*. *Molecular and Biological Evolution*, **22**, 1733–1739.

Joy, D. A., Feng, X., Mu, J., et al. (2003). Early origin and recent expansion of *Plasmodium falciparum*. *Science*, **300**, 318–321.

Kalish, M. L., Wolfe, N. D., Ndongmo, C. B., et al. (2005). Central African hunters exposed to simian immunodeficiency virus. *Emerging Infectious Diseases*, **11**, 1928–1930.

Keele, B. F., Van Heuverswyn, F., Li, Y., et al. (2006). Chimpanzee reservoirs of pandemic and nonpandemic HIV-1. *Science*, **313**, 523–526.

Keiser, J., De Castro, M. C., Maltese, M. F., et al. (2005). Effect of irrigation and large dams on the burden of malaria on a global and regional scale. *American Journal of Tropical Medicine and Hygiene*, **72**, 392–406.

Keller, M. A. (1992). Immunology of lactation. In *Immunological Obstetrics*, C. B. Coulam, W. P. Faulk and J. A. McIntyre (eds). New York: W. W. Norton, pp. 315–330.

Ketterson, E. D., Nolan, V., Wolf, L., et al. (1992). Testosterone and avian life histories: effects of experimentally elevated testosterone on behavior and correlates of fitness in the darkeyed junco (*Junco hyemalis*). *American Naturalist*, **140**, 980–999.

Kim, K. C. (1985). Parasitism and coevolution: epilogue. In *Coevolution of Parasitic Arthropods and Mammals*, K. C. Kim (ed.). New York: John Wiley and Sons, pp. 661–682.

Kimball, A. M., Arima, Y. and Hodges, J. R. (2005). Trade related infections: farther, faster, quieter. *Globalization and Health*, **1**, 3.

King, C. H., Blanton, R. E., Muchiri, E. M., et al. (2004). Low heritable component of risk for infection intensity and infection-associated disease in urinary schistosomiasis among Wadigo village populations in Coast Province, Kenya. *American Journal of Tropical Medicine and Hygiene*, **70**, 57–62.

King, C. L. and Nutman, T. B. (1992). Biological role of helper T-cell subsets in helminth infections. *Chemical Immunology*, **54**, 136–165.

Kiple, K. F. (1993). *The Cambridge World History of Human Disease*. New York: Cambridge University Press.

Kiple, K. F. (2003). *The Cambridge Historical Dictionary of Disease*. New York: Cambridge University Press.

Kiszewski, A., Mellinger, A., Spielman, A., et al. (2004). A global index representing the stability of malaria transmission. *American Journal of Tropical Medicine and Hygiene*, **70**, 486–498.

Kiyono, H., Kunisawa, J., McGhee, J. R., et al. (2008). The mucosal immune system. In *Fundamental Immunology*, W. E. Paul (ed.), 6th edn. New York: Lippincott, Williams and Wilkins, pp. 983–1030.

Klasing, K. C. (1998). Nutritional modulation of resistance to infectious diseases. *Poultry Science*, **77**, 1119–1125.

Kluger, M. J., Kozak, W., Conn, C. A., et al. (1998). Role of fever in disease. *Annals of the New York Academy of Sciences*, **856**, 224–233.

Kobayashi, M., Fitz, L., Ryan, M., et al. (1989). Identification and purification of natural killer cell stimulatory factor (NKSF), a cytokine with multiple biologic effects on human lymphocytes. *Journal of Experimental Medicine*, **170**, 827–845.

Koella, J. C. and Turner, P. (2008). Evolution of parasites. In *Evolution in Health and Disease*, S. C. Stearns and J. C. Koella (eds), 2nd edn. New York: Oxford University Press, pp. 229–238.

Koski, K. G., Su, Z. and Scott, M. E. (1999). Energy deficits suppress both systemic and gut immunity during infection. *Biochemical and Biophysical Research Communications*, **264**, 796–801.

Kramer, M. S., Naimark, L. E., Roberts-Brauer, R., et al. (1991). Risks and benefits of paracetamol antipyresis in young children with fever of presumed viral origin. *Lancet*, **337**, 591–594.

Kramer, T. R., Moore, R. J., Shippee, R. L., et al. (1997). Effects of food restriction in military training on T-lymphocyte responses. *International Journal of Sports Medicine*, **18**, 84–90.

Kwiatkowski, D. P. (2005). How malaria has affected the human genome and what human genetics can teach us about malaria. *American Journal of Human Genetics*, **77**, 171–192.

Lagesen, K. and Folstad, I. (1998). Antler asymmetry and immunity in reindeer. *Behavioral Ecology and Sociobiology*, **44**, 135–142.

Landsoud-Soukate, J., Dupont, A., De Reggi, M. L., et al. (1989). Hypogonadism and ecdysteroid production in *Loa loa* and *Mansonella perstans* filariasis. *Acta Tropica*, **46**, 249–256.

Law, S. K. A. and Reid, K. B. M. (1995). *Complement*, 2nd edn. Oxford: IRL Press.

Lazarus, R., Vercelli, D., Palmer, L. J., et al. (2002). Single nucleotide polymorphisms in innate immunity genes: abundant variation and potential role in complex human disease. *Immunological Reviews*, **190**, 9–25.

Leclerc, M. C., Durand, P., Gauthier, C., et al. (2004). Meager genetic variability of the human malaria agent *Plasmodium vivax*. *Proceedings of the National Academy of Sciences of the United States of America*, **101**, 14455–14460.

Lenzini, L., Rottoli, P. and Rottoli, L. (1977). The spectrum of human tuberculosis. *Clinical and Experimental Immunology*, **27**, 230–237.

Levin, B. R. (1996). Evolution and maintenance of virulence in microparasites. *Emerging Infectious Diseases*, **2**, 93–102.

Levine, N. D. (1988). *The Protozoan Phylum Apicomplexa*, vols 1 and 2. Boca Raton: CRC.

Levy, S. B. (1998). The challenge of antibiotic resistance. *Scientific American*, **278**, 46–53.

Levy, S. B. and Marshall, B. (2004). Antibacterial resistance worldwide: causes, challenges and responses. *Nature Medicine*, **10**, S122–S129.

Li, W., Shi, Z., Yu, M., et al. (2005). Bats are natural reservoirs of SARS-like coronaviruses. *Science*, **310**, 676–679.

Lim, C. S., Tazi, L. and Ayala, F. J. (2005). *Plasmodium vivax*: recent world expansion and genetic identity to *Plasmodium simium*. *Proceedings of the National Academy of Sciences of the United States of America*, **102**, 15523–15528.

Lin, E., Kotani, J. G. and Lowry, S. F. (1998). Nutritional modulation of immunity and the inflammatory response. *Nutrition*, **14**, 545–550.

Linz, B., Balloux, F., Moodley, Y., et al. (2007). An African origin for the intimate association between humans and *Helicobacter pylori*. *Nature*, **445**, 915–918.

Lipsitch, M. (2001). The rise and fall of antimicrobial resistance. *Trends in Microbiology*, **9**, 438–444.

Lisse, I. M., Aaby, P., Whittle, H., et al. (1997). T-lymphocyte subsets in West African children: impact of age, sex, and season. *Journal of Pediatrics*, **130**, 77–85.

Litman, G. W., Cannon, J. P. and Dishaw, L. J. (2005). Reconstructing immune phylogeny: new perspectives. *Nature Reviews Immunology*, **5**, 866–879.

Livingstone, F. B. (1958). Anthropological implications of sickle cell gene distribution in West Africa. *American Anthropologist*, **60**, 533–562.

Livingstone, F. B. (1967). *Abnormal Hemoglobins in Human Populations*. New York: Aldine de Gruyter.

Livingstone, F. B. (1971). Malaria and human polymorphisms. *Annual Review of Genetics*, **5**, 33–64.

Lockmiller, R. L. and Deerenberg, C. (2000). Trade-offs in evolutionary immunology: just what is the cost of immunity? *Oikos*, **88**, 87–98.

Loker, E. S., Adema, C. M., Zhang, S. M., et al. (2004). Invertebrate immune systems – not homogeneous, not simple, not well understood. *Immunological Reviews*, **198**, 10–24.

Lunn, P. G. (1991). Nutrition, immunity and infection. In *The Decline of Mortality in Europe*, R. Schofield, D. S. Reher and A. Bideau (eds). New York: Oxford University Press, pp. 131–145.

MacDonald, A. S., Araujo, M. I. and Pearce, E. J. (2002). Immunology of parasitic helminth infections. *Infection and Immunity*, **70**, 427–433.

Mackinnon, M. J., Gandon, S. and Read, A. F. (2008). Virulence evolution in response to vaccination: the case of malaria. *Vaccine*, **26S**, C42–C52.

Maillard, J. C. and Gonzalez, J. P. (2006). Biodiversity and emerging diseases. *Annals of the New York Academy of Sciences*, **1081**, 1–16.

Marchalonis, J. J. and Schluter, S. F. (1994). Development of an immune system. *Annals of the New York Academy of Sciences*, **712**, 1–12.

Marks, J. (2008). Would Darwin recognize himself here? In *Medicine and Evolution: Current Applications, Future Prospects*, S. Elton and P. O'Higgins (eds). New York: CRC Press, pp. 273–288.

Marlink, R., Kanki, P., Thior, I., et al. (1994). Reduced rate of disease development after HIV-2 infection as compared to HIV-1. *Science*, **265**, 1587–1590.

Marquet, S., Abel, L., Hillaire, D., et al. (1996). Genetic localization of a locus controlling the intensity of infection by *Schistosoma mansoni* on chromosome 5q31–q33. *Nature Genetics*, **14**, 181–184.

Matricardi, P. M., Rosmini, F., Riondino, S., et al. (2000). Exposure to foodborne and orofecal microbes versus airborne viruses in relation to atopy and allergic asthma: an epidemiological study. *British Medical Journal*, **320**, 412–417.

Matsunaga, T. and Rhaman, A (1998). What brought the adaptive immune system to vertebrates? The jaw hypothesis and the seahorse. *Immunological Reviews*, **166**, 177–186.

McCutchan, T. F., Dame, J. B., Miller, L. H., et al. (1984). Evolutionary relatedness of *Plasmodium* species as determined by the structure of DNA. *Science*, **225**, 808–811.

McDade, T. W., Beck, M. A., Kuzawa, C. W., et al. (2001a). Prenatal undernutrition, postnatal environments, and antibody response to vaccination in adolescence. *American Journal of Clinical Nutrition*, **74**, 543–548.

McDade, T. W., Beck, M. A., Kuzawa, C. W., et al. (2001b). Prenatal undernutrition and postnatal growth are associated with adolescent thymic function. *Journal of Nutrition*, **131**, 1225–1235.

McDade, T. W., Kuzawa, C. W., Beck, M. A., et al. (2004). Prenatal and early postnatal environments are significant predictors of total IgE concentration in Filipino adolescents. *Clinical and Experimental Allergy*, **34**, 44–50.

McDade, T. W., Reyes-Garcia, V., Tanner, S., et al. (2008). Maintenance versus growth: investigating the costs of immune activation among children in lowland Bolivia. *American Journal of Physical Anthropology*, **136**, 478–484.

Mecsas, J., Franklin, G., Kuziel, W. A., et al. (2004). *CCR5* mutation and plague protection. *Nature*, **427**, 606.

Medawar, P. B. (1952). *An Unsolved Problem of Biology*. London: HK Lewis and Company.

Medzhitov, R. and Janeway, C. A. (1997). Innate immunity: impact on the adaptive immune response. *Current Opinion in Immunology*, **9**, 4–9.

Meister, M. (2004). Blood cells of Drosophila: cell lineages and role in host defense. *Current Opinion in Immunology*, **16**, 10–15.

Mellon, M., Benbrook, C. and Benbrook, K. L. (2001). *Hogging It: Estimates of Antimicrobial Abuse in Livestock*. Cambridge: UCS Publications.

Meltzer, M. I., Cox, N. J. and Fukuda, K. (1999). The economic impact of pandemic influenza in the United States: priorities for intervention. *Emerging Infectious Diseases*, **5**, 659–671.

Mendis, K. N., Sina, B. J., Marchesini, P., et al. (2001). The neglected burden of *Plasmodium vivax* malaria. *American Journal of Tropical Medicine and Hygiene*, **64** (suppl.), 97–106.

Messenger, S. L., Molineux, I. J. and Bull, J. J. (1999). Virulence evolution in a virus obeys a trade-off. *Proceedings of the Royal Society of London. Series B*, **266**, 397–404.

Meyer, D. and Thomson, G. (2001). How selection shapes variation of the human major histocompatibility complex: a review. *Annals of Human Genetics*, **65**, 1–26.

Modiano, D., Luoni, G., Sirima, B. S., et al. (2001). Haemoglobin C protects against clinical *Plasmodium falciparum* malaria. *Nature*, **414**, 305–308.

Mohammed, I., Tomkins, A. M. and Greenwood, B. M. (1973). Normal immunoglobulin's in the tropics. *Lancet*, **1**, 481.

Morens, D. M., Folkers, G. K. and Fauci, A. S. (2004). The challenge of emerging and re-emerging infectious diseases. *Nature*, **430**, 242–249.

Mosmann, T. R. (1991a). Cytokine secretion phenotypes of TH cells: how many subsets, how much regulation? *Research in Immunology*, **142**, 9–13.

Mosmann, T. R. (1991b). Cytokine secretion patterns and cross-regulation of T cell subsets. *Immunologic Research*, **10**, 183–188.

Mosmann, T. R. and Coffman, R. L. (1989a). Heterogeneity of cytokine secretion patterns and functions of helper T cells. *Advances in Immunology*, **46**, 111–147.

Mosmann, T. R. and Coffman, R. L. (1989b). TH1 and TH2 cells: different patterns of lymphokine secretion lead to different functional properties. *Annual Review of Immunology*, **7**, 145–173.

Mosmann, T. R., Cherwinski, H., Bond, M. W., et al. (1986). Two types of murine helper T cell clone. I. Definition according to profiles of lymphokine activities and secreted proteins. *Journal of Immunology*, **136**, 2348–2357.

Mostowy, S., Cousins, D., Brinkman, J., et al. (2002). Genomic deletions suggest a phylogeny for the *Mycobacterium tuberculosis* complex. *Journal of Infectious Diseases*, **186**, 74–80.

Muchmore, E. A. (2001). Chimpanzee models for human disease and immunobiology. *Immunological Reviews*, **183**, 86–93.

Muehlenbein, M. P. (2008a). Adaptive variation in testosterone levels in response to immune activation: empirical and theoretical perspectives. *Social Biology*, **53**, 13–23.

Muehlenbein, M. P. (2008b). Human immune functions are energetically costly. *American Journal of Physical Anthropology*, **135**, 158–159.

Muehlenbein, M. P. and Bribiescas, R. G. (2005). Testosterone-mediated immune functions and male life histories. *American Journal of Human Biology*, **17**, 527–558.

Muehlenbein, M. P., Algier, J., Cogswell, F., et al. (2005). The reproductive endocrine response to *Plasmodium vivax* infection in Hondurans. *American Journal of Tropical Medicine and Hygiene*, **73**, 178–187.

Muehlenbein, M. P., Cogswell, F., James, M., et al. (2006). Testosterone correlates with Venezuelan Equine Encephalitis virus infection in macaques. *Virology Journal*, **3**, 19–27.

Muehlenbein, M. P., Jordan, J. L., Bonner, J. Z. (in press). Towards quantifying the usage costs of human immunity: altered metabolic rates and hormone levels during acute immune activation in men. *American Journal of Human Biology*.

Nelson, R. J., Demas, G. E., Klein, S. L., et al. (2002). *Seasonal Patterns of Stress, Immune Function, and Disease*. New York: Cambridge University Press.

Nesse, R. M. (2006). Evolutionary explanations for mood and mood disorders. In *American Psychiatric Publishing Textbook of Mood Disorders*, D. J. Stein, D. J. Kupfur and A. F. Schatzberg (eds). Washington, DC: American Psychiatric Publishing, pp. 159–175.

Nesse, R. M. and Berridge, K. C. (1997). Psychoactive drug use in evolutionary perspective. *Science*, **278**, 63–66.

Nesse, R. M. and Williams, G. C. (1996). *Why We Get Sick: the New Science of Darwinian Medicine*. New York: Vintage Books.

Newport, M. J., Huxley, C. M., Huston, S., et al. (1996). A mutation in the interferon-γ-receptor gene and susceptibility to mycobacterial infection. *New England Journal of Medicine*, **335**, 1941–1949.

Newsholme, P. and Newsholme, E. A. (1989). Rates of utilization of glucose, glutamine and oleate and formation of end-products by mouse peritoneal macrophages in culture. *Biochemistry*, **261**, 211–218.

Nonaka, M. and Yoshizaki, F. (2004). Evolution of the complement system. *Molecular Immunology*, **40**, 897–902.

Nordling, D., Andersson, M., Zohari, S., et al. (1998). Reproductive effort reduces specific immune response and parasite resistance. *Proceedings of the Royal Society of London. Series B*, **265**, 1291–1298.

Normile, D. (2005). WHO faults China for lax outbreak response. *Science*, **309**, 684.

Nowak, M. A. and May, R. M. (1994). Superinfection and the evolution of parasite virulence. *Proceedings of the Royal Society of London. Series B*, **255**, 81–89.

O'Garra, A. (1998). Cytokines induce the development of functionally heterogeneous T helper cell subsets. *Immunity*, **8**, 275–283.

Oktenli, C., Doganci, L., Ozguratas, T., et al. (2004). Transient hypogonadotrophic hypogonadism in males with acute toxoplasmosis: suppressive effect of interleukin-1β on the secretion of GnRH. *Human Reproduction*, **19**, 859–866.

Olsen, B., Munster, V. J., Wallensten, A., et al. (2006). Global patterns of influenza A virus in wild birds. *Science*, **312**, 384–388.

Ots, I., Kerimov, A. B., Ivankina, E. V., et al. (2001). Immune challenge affects basal metabolic activity in wintering great tits. *Proceedings of the Royal Society of London. Series B*, **268**, 1175–1181.

Owens, I. P. (2002). Ecology and evolution. Sex differences in mortality rate. *Science*, **297**, 2008–2009.

Pancer, Z. and Cooper, M. D. (2006). The evolution of adaptive immunity. *Annual Review of Immunology*, **24**, 497–518.

Pancer, Z., Amrmiya, C. T., Ehrhardt, G. R. A., et al. (2004). Somatic diversification of variable lymphocyte receptors in the agnathan sea lamprey. *Nature*, **430**, 174–180.

Panter-Brick, C., Lunn, P. G., Baker, R., et al. (2000). Elevated acute-phase protein in stunted Nepali children reporting low morbidity: different rural and urban profiles. *British Journal of Nutrition*, **85**, 1–8.

Parry, J. (2003). WHO is worried that China is underreporting SARS. *British Medical Journal*, **326**, 1110.

Patz, J. A., Daszak, P., Tabor, G. M., et al. (2004). Unhealthy landscapes: policy recommendations on land use change and infectious disease emergence. *Environmental Health Perspectives*, **112**, 1092–1098.

Patz, J. A., Gibbs, H. K., Foley, J. A., et al. (2007). Climate change and global health: quantifying a growing ethical crisis. *EcoHealth*, **4**, 397–405.

Paul, W. E. (2008). *Fundamental Immunology*, 6th edn. New York: Lippincott Williams and Wilkins.

Peeters, M., Courgnaud, V., Abela, B., et al. (2002). Risk to human health from a plethora of simian immunodeficiency viruses in primate bushmeat. *Emerging Infectious Diseases*, **8**, 451–457.

Perrett, D. I., Lee, K. J., Penton-Voak, I. S., et al. (1998). Effects of sexual dimorphism on facial attractiveness. *Nature*, **394**, 884–887.

Poolman, E. M. and Galvani, A. P. (2007). Evaluating candidate agents of selective pressure for cystic fibrosis. *Journal of the Royal Society Interface*, **4**, 91–98.

Porter, R. (2001). *The Cambridge Illustrated History of Medicine*. New York: Cambridge University Press.

Porter, R. (2006). *The Cambridge History of Medicine*. New York: Cambridge University Press.

Poulin, R. (2006). *Evolutionary Ecology of Parasites*, 2nd edn. Princeton: Princeton University Press.

Qari, S. H., Shi, Y. P., Pieniazek, N. J., et al. (1996). Phylogenetic relationship among the malaria parasites based on small subunit rRNA gene sequences: monophyletic nature of the human malaria parasite, *Plasmodium falciparum*. *Molecular Phylogenetics and Evolution*, **6**, 157–165.

Raberg, L., Grahn, M., Hasselquist, D., et al. (1998). On the adaptive significance of stress-induced immunosuppression. *Proceedings of the Royal Society of London. Series B*, **265**, 1637–1641.

Raberg, L., Vestberg, M., Hasselquist, D., et al. (2002). Basal metabolic rate and the evolution of the adaptive immune system. *Proceedings of the Royal Society of London. Series B*, **269**, 817–821.

Radic, M. Z. and Zouali, M. (1996). Receptor editing, immune diversification, and self-tolerance. *Immunity*, **5**, 505–511.

Rambaut, A., Posada, D., Crandall, K. A., et al. (2004). The causes and consequences of HIV evolution. *Nature Reviews Genetics*, **5**, 52–61.

Rani, R., Fernandez-Vina, M. A., Zaheer, S. A., et al. (1993). Study of HLA class II alleles by PCR oligotyping in leprosy patients from north India. *Tissue Antigens*, **42**, 133–137.

Rathmell, J. C. and Thompson, C. B. (1999). The central effectors of cell death in the immune system. *Annual Review of Immunology*, **17**, 781–828.

Reincke, M., Arlt, W., Heppner, C., et al. (1998). Neuroendocrine dysfunction in African trypanosomiasis. The role of cytokines. *Annals of the New York Academy of Sciences*, **840**, 809–821.

Reiner, S. L. and Seder, R. A. (1999). Dealing from the evolutionary pawnshop: how lymphocytes make decisions. *Immunity*, **11**, 1–10.

Remington, J. S. and Klein, J. O. (1990). *Infectious Diseases of the Fetus and Newborn Infant*. Philadelphia: W. B. Saunders.

Rich, S. M., Licht, M. C., Hudson, R. R., et al. (1998). Malaria's eve: evidence of a recent population bottleneck throughout the world populations of *Plasmodium falciparum*. *Proceedings of the National Academy of Sciences of the United States of America*, **95**, 4425–4430.

Richner, H., Christie, P. and Oppliger, A. (1995). Paternal investment affects prevalence of malaria. *Proceedings of the National Academy of Sciences of the United States of America*, **92**, 1192–1194.

Ricklefs, R. E. and Wikelski, M. (2002). The physiology/life history nexus. *Trends in Ecology and Evolution*, **17**, 462–468.

Robert, J., Cohen, N., Maniero, G. D., et al. (2003). Evolution of the immunomodulatory role of the heat shock protein gp96. *Cellular and Molecular Biology*, **49**, 263–275.

Rodrigo, A. G. (1999). HIV evolutionary genetics. *Proceedings of the National Academy of Sciences of the United States of America*, **96**, 10559–10561.

Rodrigues, V. Jr, Piper, K., Couissinier-Paris, P., et al. (1999). Genetic control of schistosome infections by the SM1 locus of the 5q21–q33 region is linked to differentiation of type 2 helper T lymphocytes. *Infection and Immunity*, **67**, 4689–4692.

Roe, C. F. and Kinney, J. M. (1965). The caloric equivalent of fever: II. Influence of major trauma. *Annals of Surgery*, **161**, 140–147.

Rolff, J. (2007). Why did the acquired immune system of vertebrates evolve? *Developmental and Comparative Immunology*, **31**, 476–482.

Rook, W. G. A. (1988). Role of activated macrophages in the immunopathology of tuberculosis. *British Medical Bulletin*, **44**, 611–623.

Rosas-Magallanes, V., Deschavanne, P., Quintana-Murci, L., et al. (2006). Horizontal transfer of a virulence operon to the ancestor of *Mycobacterium tuberculosis*. *Molecular Biology and Evolution*, **23**, 1129–1135.

Rosenberg, I. H. and Bowman, B. B. (1984). Impact of intestinal parasites on digestive function in humans. *Federation Proceedings*, **43**, 246–250.

Rothschild, B. M., Martin, L. D., Lev, G., et al. (2001). *Myco-bacterium tuberculosis* complex DNA from an extinct bison dated 17 000 years before the present. *Clinical Infectious Diseases*, **33**, 305–311.

Rothwell, T. L. (1989). Immune expulsion of parasitic nematodes from the alimentary tract. *International Journal for Parasitology*, **19**, 139–168.

Roumagnac, P., Weill, F. X., Dolecek, C., et al. (2006). Evolutionary history of *Salmonella typhi*. *Science*, **314**, 1301–1304.

Rowe, J. A., Handel, I. G., Thera, M. A., et al. (2007). Blood group O protects against severe *Plasmodium falciparum* malaria through the mechanism of reduced resetting. *Proceedings of the National Academy of Sciences of the United States of America*, **104**, 17471–17476.

Ruwende, C., Khoo, S. C., Snow, R. W., et al. (1995). Natural selection of hemi- and heterozygotes for G6PD deficiency in Africa by resistance to severe malaria. *Nature*, **376**, 246–249.

Saad, A. H., Abdelbaky, A., Osman, A. M., et al. (1999) Possible role of *Schistosoma mansoni* infection in male hypogonadism. *Journal of the Egyptian Society of Parasitology*, **29**, 307–323.

Sabeti, P. C., Walsh, E., Schaffner, S. F., et al. (2005). The case for selection at *CCR5-Δ32*. *PLoS Biology*, **3**, e378.

Saino, N. and Moller, A. P. (1994). Secondary sexual characters, parasites and testosterone in the barn swallow, *Hirundo rustica*. *Animal Behaviour*, **48**, 1325–1333.

Saino, N., Stradi, R., Ninni, P., et al. (1999). Carotenoid plasma concentration, immune profile, and plumage ornamentation of male barn swallows (*Hirundo rustica*). *American Naturalist*, **154**, 441–448.

Saino, N., Incagli, M., Martinelli, R., et al. (2002). Immune response of male barn swallows in relation to parental effort, corticosterone plasma levels, and sexual ornamentation. *Behavioral Ecology*, **13**, 169–174.

Samuel, C. E. (2001). Antiviral actions of interferon's. *Clinical Microbiology Reviews*, **14**, 778–809.

Santiago, M. L., Range, F., Keele, B. F., et al. (2005). Simian immunodeficiency virus infection in free-ranging sooty mangabeys (*Cercocebus atye atys*) from the Tai Forest, Cote d'Ivoire: implications for the origin of epidemic human immunodeficiency virus type 2. *Journal of Virology*, **79**, 12515–12527.

Sattenspiel, L. (2000). The epidemiology of human disease. In *Human Biology: an Evolutionary and Biocultural Approach*, S. Stinson, B. Bogin, R. Huss-Ashmore, et al. (eds). New York: Wiley-Liss, pp. 225–271.

Schatz, D. G., Oettinger, M. A. and Schlissel, M. S. (1992). V(D)J recombination: molecular biology and regulation. *Annual Review of Immunology*, **10**, 359–383.

Schmid-Hempel, P. (2003). Variation in immune defence as a question of evolutionary ecology. *Proceedings of the Royal Society of London. Series B*, **270**, 357–366.

Scrimshaw, N. S. (1992). Effect of infection on nutritional status. *Proceedings of the National Science Council Republic of China. Part B Life Sciences*, **16**, 46–64.

Sebzda, E., Mariathasan, S., Ohteki, T., et al. (1999). Selection of the T cell repertoire. *Annual Review of Immunology*, **17**, 829–874.

Setchell, J. M. and Dixson, A. F. (2001). Changes in the secondary sexual adornments of male mandrills (*Mandrillus sphinx*) are associated with gain and loss of alpha status. *Hormones and Behavior*, **39**, 177–184.

Sharp, P. M., Bailes, E. and Wain, L. V. (2008). Evolutionary origins of diversity in human viruses. In *Evolution in Health and Disease*, S. C. Stearns and J. C. Koella (eds), 2nd edn. New York: Oxford University Press, pp. 169–184.

Sheldon, B. C. and Verhulst, S. (1996). Ecological immunology: costly parasite defenses and trade-offs in evolutionary ecology. *Trends in Ecology and Evolution*, **11**, 317–321.

Shephard, R. J., Castellani, J. W. and Shek, P. N. (1998). Immune deficits induced by strenuous exertion under adverse environmental conditions: manifestations and countermeasures. *Critical Reviews in Immunology*, **18**, 545–568.

Sher, A. and Coffman, R. L. (1992). Regulation of immunity to parasites by T cells and T cell-derived cytokines. *Annual Review of Immunology*, **10**, 385–409.

Shevach, E. M. (2008). Regulatory/suppressor T cells. In *Fundamental Immunology*, W. E. Paul (ed.), 6th edn. New York: Lippincott, Williams and Wilkins, pp. 943–982.

Shinya, K., Ebina, M., Yamada, S., et al. (2006). Avian flu: influenza virus receptors in the human airway. *Nature*, **440**, 435–436.

Sluiter, C. P., Swellengrebel, N. H. and Ihle, J. E. (1922). *De dierlijke parasieten van den mensch en van onze huisdieren*, 3rd edn. Amsterdam: Scheltema and Holkema.

Smith, N. H., Gordon, S. V., de la Rua-Domenech, R., et al. (2006). Bottlenecks and broomsticks: the molecular evolution of *Mycobacterium bovis*. *Nature Reviews Microbiology*, **4**, 670–681.

Smith, T. (1934). *Parasitism and Disease*. Princeton: Princeton University Press.

Smith, T., Bhatia, K., Branish, G., et al. (1991). Host genetic factors do not account for variation in parasite loads in *Strongyloides fuelleborni kellyi*. *Annals of Tropical Medicine and Parasitology*, **5**, 533–537.

Snapper, C. M. (1996). *Cytokine Regulation of Humoral Immunity: Basic and Clinical Aspects*. New York: Wiley.

Spink, W. W. and Ferris, V. (1945). Quantitative action of penicillin inhibitor from penicillin-resistant strain of staphylococcus. *Science*, **102**, 102–221.

Spratt, D. I. (2001). Altered gonadal steroidogenesis in critical illness: is treatment with anabolic steroids indicated? *Best Practice and Research Clinical Endocrinology and Metabolism*, **15**, 479–494.

Spratt, D. I., Cox, P., Orav, J., et al. (1993). Reproductive axis suppression in acute illness is related to disease severity. *Journal of Clinical Endocrinology and Metabolism*, **76**, 1548–1554.

Sprent, J. (1993). The thymus and T-cell tolerance. *Annals of the New York Academy of Sciences*, **681**, 5–15.

Spurlock, M. E. (1997). Regulation of metabolism and growth during growth challenge: an overview of cytokine function. *Journal of Animal Science*, **75**, 1773–1783.

Sreevatsan, S., Pan, X., Stockbauer, K. E., et al. (1997). Restricted structural gene polymorphism in the *Mycobacterium tuberculosis* complex indicates evolutionarily recent global dissemination. *Proceedings of the National*

Academy of Sciences of the United States of America, **94**, 9869–9874.

Stearns, S. (1989). Trade-offs in life-history evolution. *Functional Ecology*, **3**, 259–268.

Stearns, S. (1992). *The Evolution of Life Histories*. New York: Oxford University Press.

Stearns, S. C. and Koella, J. C. (2008). *Evolution in Health and Disease*, 2nd edn. New York: Oxford University Press.

Stearns, S. C., Nesse, R. M. and Haig, D. (2008). Introducing evolutionary thinking for medicine. In *Evolution in Health and Disease*, S. C. Stearns and J. C. Koella (eds), 2nd edn. New York: Oxford University Press, pp. 3–15.

Steinmann, P., Keiser, J., Bos, R., Tanner, et al. (2006). Schistosomiasis and water resources development: systematic review, meta-analysis, and estimates of people at risk. *Lancet Infectious Diseases*, **6**, 411–425.

Stephens, J. C., Reich, D. E., Goldstein, D. B., et al. (1998). Dating the origin of the *CCR5-Δ32* AIDS-resistance allele by the coalescence of haplotypes. *American Journal of Human Genetics*, **62**, 1507–1515.

Stephens, J. W. W. (1922). A new malaria parasite of man. *Annals of Tropical Medicine*, **16**, 383–388.

Stevenson, M. M. and Riley, E. M. (2004). Innate immunity to malaria. *Nature Reviews Immunology*, **4**, 169–180.

Strachan, D. P. (1989). Hay fever, hygiene, and household size. *British Medical Journal*, **299**, 1259–1260.

Svensson, E., Raberg, L., Koch, C., et al. (1998). Energetic stress, immunosuppression and the costs of an antibody response. *Functional Ecology*, **12**, 912–919.

Swellengrebel, N. H. (1940). The efficient parasite. In *Proceedings of the Third International Congress of Microbiology*. Baltimore: Waverly, pp. 119–127.

Tanabe, K., Escalante, A., Sakihama, N., et al. (2007). Recent independent evolution of msp1 polymorphism in *Plasmodium vivax* and related simian malaria parasites. *Molecular and Biochemical Parasitology*, **156**, 74–79.

Taubenberger, J. K., Reid, A. H., Krafft, A. E., et al. (1997). Initial genetic characterization of the 1918 "Spanish" influenza virus. *Science*, **275**, 1793–1796.

Taubenberger, J. K., Reid, A. H., Lourens, R. M., et al. (2005). Characterization of the 1918 influenza virus polymerase genes. *Nature*, **437**, 889–893.

Taylor, T. E., Fu, W. J., Carr, R. A., et al. (2004). Differentiating the pathologies of cerebral malaria by postmortem parasite counts. *Nature Medicine*, **10**, 143–145.

Thompson, J. N. (1994). *The Coevolutionary Process*. Chicago: Chicago University Press.

Thuong, N. T., Hawn, T. R., Thwaites, G. E., et al. (2007). A polymorphism in human TLR2 is associated with increased susceptibility to tuberculous meningitis. *Genes and Immunity*, **8**, 422–428.

Tishkoff, S. A., Varkonyi, R., Cahinhinan, N., et al. (2001). Haplotype diversity and linkage disequilibrium at human G6PD: recent origin of alleles that confer malarial resistance. *Science*, **293**, 455–462.

Tonegawa, S. (1983). Somatic generation of antibody diversity. *Nature*, **302**, 575–581.

Toossi, Z., Gogate, P., Shiratsuchi, H., et al. (1995). Enhanced production of TGF-β by blood monocytes from patients with active tuberculosis and presence of TGF-β in tuberculous granulomatous lung lesions. *Journal of Immunology*, **154**, 465–473.

Tournamille, C., Colin, Y., Cartron, J. P., et al. (1995). Disruption of a GATA motif in the Duffy gene promoter abolishes erythroid gene expression in Duffy-negative individuals. *Nature Genetics*, **10**, 224–228.

Trevathan, W. R., Smith, E. O. and McKenna, J. (2007). *Evolutionary Medicine and Health: New Perspectives*. New York: Oxford University Press.

Trigg, P. I. and Kondrachine, A. V. (1998). The current global malaria situation. In *Malaria: Parasite Biology, Pathogenesis and Protection*, I. W. Sherman (ed.). Washington, DC: ASM Press, pp. 11–22.

Trinchieri, G. (2003). Interleukin-12 and the regulation of innate resistance and adaptive immunity. *Nature Reviews Immunology*, **3**, 133–146.

Trivers, R. L. (1974). Parent–offspring conflict. *American Zoologist*, **14**, 249–264.

Trowsdale, J. and Parham, P. (2004). Defense strategies and immunity-related genes. *European Journal of Immunology*, **34**, 7–17.

Ungchusak, K., Auewarakul, P., Dowell, S. F., et al. (2005). Probably person-to-person transmission of avian influenza A (H5N1). *New England Journal of Medicine*, **352**, 333–340.

Urban, J. F. Jr, Madden, K. B., Svetic, A., et al. (1992). The importance of Th2 cytokines in protective immunity to nematodes. *Immunological Reviews*, **127**, 205–220.

Van Blerkom, L. M. (2003). Role of viruses in human evolution. *Yearbook of Physical Anthropology*, **46**, 14–46.

Van Crevel, R., Ottenhoff, T. H. M. and van der Meer, J. W. M. (2002). Innate immunity to *Mycobacterium tuberculosis*. *Clinical Microbiology Reviews*, **15**, 294–309.

Van Heuverswyn, F., Li, Y., Neel, C., et al. (2006). SIV infection in wild gorillas. *Nature*, **444**, 164.

Van Valen, L. (1973). A new evolutionary law. *Evolutionary Theory*, **1**, 1–30.

Verhulst, S., Dieleman, S. J. and Parmentier, H. K. (1999). A tradeoff between immunocompetence and sexual ornamentation in domestic fowl. *Proceedings of the National Academy of Sciences of the United States of America*, **96**, 4478–4481.

Visintin, A., Mazzoni, A., Spitzer, J. H., et al. (2001). Regulation of Toll-like receptors in human monocytes and dendritic cells. *Journal of Immunology*, **166**, 249–255.

Vitzthum, V. J. (2003). A number no greater than the sum of its parts: the use and abuse of heritability. *Human Biology*, **75**, 539–558.

Volkman, S. K., Barry, A. E., Lyons, E. J., et al. (2001). Recent origin of *Plasmodium falciparum* from a single progenitor. *Science*, **293**, 482–484.

Von Mutius, E., Fritzsch, C., Weiland, S. K., et al. (1992). Prevalence of asthma and allergic disorders among children in united Germany: a descriptive comparison. *British Medical Journal*, **305**, 1395–1399.

Wallace Taylor, D. (2002). The inducible defense system: antibody molecules and antigen-antibody reactions. In *Infection, Resistance, and Immunity*, J. P. Kreier (ed.). New York: Taylor and Francis, pp. 105–130.

Walther, B. A. and Ewald, P. W. (2004). Pathogen survival in the external environment and the evolution of virulence. *Biological Reviews*, **79**, 849–869.

Wardlaw, A. J., Moqbel, R. and Kay, A. B. (1995). Eosinophils: biology and role in disease. *Advances in Immunology*, **60**, 151–266.

Waters, A. P., Higgins, D. G. and McCutchan, T. F. (1993). Evolutionary relatedness of some primate models of *Plasmodium*. *Molecular Biology and Evolution*, **10**, 914–923.

Watson, J. T., Gayer, M. and Connolly, M. A. (2007). Epidemics after natural disasters. *Emerging Infectious Diseases*, **13**, 1–5.

Weatherall, D. J. and Clegg, J. B. (2001). Inherited haemoglobin disorders: an increasing global health problem. *Bulletin of the World Health Organization*, **79**, 704–712.

Wedekind, C. and Folstad, I. (1994). Adaptive or nonadaptive immunosuppression by sex-hormones. *American Naturalist*, **143**, 936–938.

Weiss, R. A. and McMichael, A. J. (2004). Social and environmental risk factors in the emergence of infectious diseases. *Nature Medicine*, **10**, S70–S76.

Welch, W. H. (1897). Malaria: definitions, synonyms, history and parasitology. In *Systemic Practical Medicine*, A. L. Loomis and W. G. Thompson (eds). Philadelphia, PA: Lea Brothers and Company, pp. 1–17.

Welle, S., Jozefowicz, R., Forbes, G., et al. (1992). Effect of testosterone on metabolic rate and body composition in normal men and men with muscular dystrophy. *Journal of Clinical Endocrinology and Metabolism*, **74**, 332–335.

White, N. J. (1993). Malaria parasites do ape. *Lancet*, **341**, 793.

Wilkinson, R. J., Llewelyn, M., Toossi, Z., et al. (2000). Influence of vitamin D deficiency and vitamin D receptor polymorphisms on tuberculosis among Gujarati Asians in west London: a case-control study. *Lancet*, **355**, 618–621.

Williams, G. C. and Nesse, R. M. (1991). The dawn of Darwinian medicine. *Quarterly Review of Biology*, **66**, 1–22.

Williams, T. N. (2006). Human red blood cell polymorphisms and malaria. *Current Opinions in Microbiology*, **9**, 388–394.

Williams, T. N., Mwangi, T. W., Roberts, D. J., et al. (2005). An immune basis for malaria protection by the sickle cell trait. *PLoS Medicine*, **2**, e128.

Williams-Blangero, S., Blangero, J. and Bradley, M. (1997). Quantitative genetic analysis of susceptibility to hookworm infection in a population from rural Zimbabwe. *Human Biology*, **69**, 201–208.

Williams-Blangero, S., Subedi, J., Upadhayay, R. P., et al. (1999). Genetic analysis of susceptibility to infection with *Ascaris lumbricoides*. *American Society of Tropical Medicine and Hygiene*, **60**, 921–926.

Williams-Blangero, S., McGarvey, S. T., Subedi, J., et al. (2002a). Genetic component to susceptibility to *Trichuris trichiura*: evidence from two Asian populations. *Genetic Epidemiology*, **22**, 254–264.

Williams-Blangero, S., VandeBerg, J. L., Subedi, J., et al. (2002b). Genes on chromosomes 1 and 13 have significant affects on *Ascaris* infection. *Proceedings of the National Academy of Sciences of the United States of America*, **99**, 5533–5538.

Williams-Blangero, S., Correa-Oliveira, R., Vandenberg, J. L., et al. (2004). Genetic influences on plasma cytokine variation in a parasitized population. *Human Biology*, **76**, 515–525.

Williams-Blangero, S., Vandeberg, J. L., Subedi, J., et al. (2008a). Localization of multiple quantitative trait loci influencing susceptibility to infection with *Ascaris lumbricoides*. *Journal of Infectious Diseases*, **197**, 66–71.

Williams-Blangero, S., Vandeberg, J. L., Subedi, J., et al. (2008b). Two quantitative trait loci influence whipworm (*Trichuris trichiura*) infection in a Nepalese population. *Journal of Infectious Diseases*, **197**, 1198–1203.

Wills-Karp, M., Santeliz, J. and Karp, C. L. (2001). The germless theory of allergic disease: revisiting the hygiene hypothesis. *Nature Reviews Immunology*, **1**, 69–75.

Wilson, M. S. and Maizels, R. M. (2004). Regulation of allergy and autoimmunity in helminth infection. *Clinical Reviews of Allergy and Immunology*, **26**, 35–50.

Wilson, R. J. M. and Williamson, D. H. (1997). Extrachromosomal DNA in the Apicomplexa. *Microbiology and Molecular Biology Reviews*, **61**, 1–16.

Winkler, C., An, P. and O'Brien, S. J. (2004). Patterns of ethnic diversity among the genes that influence AIDS. *Human Molecular Genetics*, **13**, R9–R19.

Wolfe, N. D., Prosser, A. T., Carr, J. K., et al. (2004). Exposure to nonhuman primates in rural Cameroon. *Emerging Infectious Diseases*, **10**, 2094–2099.

World Health Organization (1999). *World Health Report*. Geneva: World Health Organization.

Worobey, M., Gemmel, M., Teuwen, D. E., et al. (2008). Direct evidence of extensive diversity of HIV-1 in Kinshasa by 1960. *Nature*, **455**, 661–664.

Yazdanbakhsh, M., Kremsner, P. G. and van Ree, R. (2002). Allergy, parasites, and the hygiene hypothesis. *Science*, **296**, 490–494.

Yu, D., Li, H., Xu, R., et al. (2003). Prevalence of IgG antibody to SARS-associated coronavirus in animal traders – Guangdong Province, China, 2003. *Morbidity and Mortality Weekly Report*, **52**, 986–987.

Yu, X. Q. and Kanost, M. R. (2002). Binding of hemolin to bacterial lipopolysaccharide and lipoteichoic acid. An immunoglobulin superfamily member from insects as a pattern-recognition receptor. *European Journal of Biochemistry*, **269**, 1827–1834.

Zahavi, A. (1975). Mate selection – a selection for a handicap. *Journal of Theoretical Biology*, **53**, 205–214.

Zimmerman, P. A., Buckler-White, A., Alkhatib, G., et al. (1997). Inherited resistance to HIV-1 conferred by an inactivating mutation in CC chemokine receptor 5: studies in populations with contrasting clinical phenotypes, defined racial background, and quantified risk. *Molecular Medicine*, **3**, 23–36.

28 Complex Chronic Diseases in Evolutionary Perspective

S. Boyd Eaton

From an evolutionary point of view, most complex chronic diseases appear to be the result of imbalance, mismatch, between our genetic makeup and the conditions of life in Westernized twenty-first century nations. The basic contentions (Eaton and Eaton, 1999a; Eaton et al., 2002a) are that:

1. The contemporary human genome was selected over thousands of millennia during which our ancestral line existed as prehuman primates who became increasingly human-like until, during the period between 100 and 50 thousand years ago, they became behaviorly modern and lived as near equivalents of hunter-gatherers studied during the last century.
2. Our genetic makeup, especially that regarding our core metabolic and physiologic characteristics, has changed very little between the emergence of agriculture, roughly 10 000 years ago, and the present.
3. On the other hand, cultural change during these past 10 millennia has progressed at an ever-accelerating rate. The resulting dissonance between what amount to Stone Age genes and Space Age lives fosters development of multiple health disorders ranging from the potentially life threatening (cancer, pulmonary emphysema, heart attacks) to the more mundane, but still costly and uncomfortable conditions such as acne, high frequency hearing loss, myopia (near sightedness), and dental caries.

These contentions make up the "discordance hypothesis" and its corollary is that the "afflictions of affluence" might best be prevented by reincorporating the essentials of our ancestral living pattern into our contemporary lives – ideally blending the best from the past with best from the present.

OBJECTIONS TO THE EVOLUTIONARY APPROACH

On initially encountering these propositions, educated health-conscious individuals usually raise one or more of three potential reservations (Eaton et al., 2002b):

Firstly, what is the evidence that we are still genetic Stone Agers? Secondly, if our ancestral lifestyle was so much healthier, why do we now live longer? And thirdly, chronic illnesses ordinarily become apparent late in life. Isn't it just because we live longer that we develop these disorders?

Ten thousand years is plenty of time for at least some genetic evolution to occur and, indeed, scientists have identified a number of genetic changes that have emerged since agriculture initiated the watershed alteration in human life conditions which has characterized the last 10 millennia. The majority of these deal with defenses against infectious diseases, the nature of which has changed greatly since our ancestors lived as nomadic hunter-gatherers. Settled communities (with attendant sanitation problems), increased population density, and intimate association with domesticated animals promoted epidemic "crowd" diseases which had previously been of far lesser importance. Genetic evolutionary responses included the sickle cell condition and similar hemoglobinopathies (defenses against malaria), and also alterations in the genetics of our immune system. The latter explain why, after initial contact with Westerners, Native Americans, Hawaiians, and other such groups experienced disastrous death rates from "childhood" infectious illnesses which, in Old World cultures, had come to be infrequently fatal. Those populations who adopted dairying and milk consumption developed persistence of intestinal lactase into adulthood. Almost all mammals make the enzyme lactase during infancy because it allows digestion of lactose, the sugar in mothers' milk. After weaning, ancestral humans, like other mammals, no longer needed lactase because they did not consume dairy products. Animal husbandry made milk available to adults; hence lactase persistence became selectively beneficial.

As our ancestors migrated from Africa to other parts of the world, "races," as we know them today, differentiated. However, it is important to remember that the identifying features that distinguish current humans from varying geographical regions are largely superficial. To our eyes, Finns and Australian Aborigines look much more unlike than do mountain and lowland gorillas, but there is far greater genetic dissimilarity between the gorillas. It appears that cheetahs are the only mammalian species whose individual members are more genetically similar to each other than are humans.

A few other examples of "recent" genetic evolution have been proposed: a decreased ability to smell, changes in genes affecting the liver's capacity to detoxify natural poisons, and, most controversially, genes affecting brain size and/or function. Some of the latter are analogous to sickle cell condition: when a person has only one such gene, the situation may be beneficial, however, when two copies of the mutant gene are present, one from each parent, illness develops. An example is Tay–Sachs disease, which is caused by a single mutation and inherited in a fairly straightforward Mendelian pattern. Tay–Sachs and conditions like it are, apparently, unrelated to the common chronic degenerative diseases like diabetes and atherosclerosis, which are influenced by many relevant genes, and whose inheritance is much more complicated.

These genetic changes that have evolved during the past 10 000 years are exceptions which prove a general rule: they are few in number and have little effect on chronic illness susceptibility. Even though some modifications have occurred, biologists, evolutionary theorists, paleontologists, and human geneticists concur that, genetically, we remain almost identical to our late Paleolithic ancestors. If agriculture and industrialization had produced substantial alteration, there should be categorical, "taxonomic" genetic differences between peoples whose ancestors have been farmers for thousands of years (e.g. Near Easterners, Mayans, Chinese, etc.) and groups whose ancestors have been foragers until the past generation or two (e.g. !Kung San Bushmen, Paraguayan Aché, Australian Aborigines, etc.). Apart from genes affecting resistance to infectious disease, no such markers have been identified.

The second question, regarding life expectancy now versus the Stone Age, disturbs nearly everyone on initial consideration of the discordance hypothesis. The best available estimates are that until about 1800 average human life expectancy was relatively constant at 30 to 40 years for as far back as can be determined. The figure was likely less in urban centers until the past century: estimates for London in the 1660s range between 18 and 27 years! Infectious illnesses were the major killers, especially those affecting children. Individuals who reached age 40 are thought to have had an expectation of further life only moderately less than that of contemporary humans at the same age.

In affluent Westernized nations, average life expectancy has nearly doubled over the past two centuries, from around 40 in 1800 to almost 80 in 2000. This unprecedented increase results from societal affluence (better housing, more affordable food, cheaper clothing), which has increased host resistance to infectious agents, and from effective public health measures (immunizations, sanitary engineering, safer food) which decrease microbial transmission. Both factors work against contagious illnesses which were the major causes of human mortality until the last century. (In 1900, infectious diseases caused more deaths in the United States than did cancer and heart disease combined.) Increasingly sophisticated medical care has also played a role, although less than might be assumed.

On the other hand, personal lifestyle choices affecting health – those targeted by health promotion advocates – have actually had a negative impact. Compared with the early 1900s, per capita cigarette smoking has increased 10-fold; daily physical activity, previously over 1000 kcal/day, has decreased to 500 kcal/day. For adult men, average body mass indices have increased from 23 to 28, 25 being considered the upper limit of normal. The failure rate among candidates taking the Army's induction physical fitness test has become so high that the Pentagon has seriously considered relaxing standards so that more candidates can achieve eligibility. Intake of dietary fiber has decreased while intake of sugars and sweeteners has doubled. The interval between menarche and first full-term pregnancy has risen from 5 years to 15 and breast-feeding at 12 months (19.4%) is now one third what it was in 1912–1919 (58%). All these changes increase risk for one or more chronic illnesses.

The effect of increasing life expectancy on age-related chronic diseases is another potential stumbling block. A population with average life expectancy of 40 years will inevitably have much lower mortality from cancer, heart disease, diabetes, and stroke than will a population with an average life expectancy of 75. To this extent, comparisons between recently studied hunter-gatherers (imperfect, but the best available surrogates for prehistoric Stone Agers) and citizens of affluent Western nations are invalid. Of course diseases whose frequency increases with age will cause more deaths in a society with a greater proportion of older individuals.

This coin has another side, however. While chronic degenerative diseases generally produce mortality in later life, they begin much earlier, often in childhood. This allows comparison between age-matched younger members of industrial and technologically primitive societies. Biomarkers of developing abnormality such as obesity, rising blood pressure, nonobstructive coronary atherosclerosis, and insulin resistance are common among the former, but rare in the latter.

Measurements of muscular strength and aerobic power (endurance) reveal similar discrepancies, again favoring individuals whose lives more closely resemble the ancestral pattern. About 20% of hunter-gatherers reach age 60 or beyond, but even in this age-bracket, members of foraging and other technologically primitive cultures appear almost completely free from manifestations of most chronic degenerative diseases (osteoarthritis is an exception). Together, these observations strongly suggest that it is current Western lifestyle rather than age alone that promotes those "afflictions of affluence" the prevention of which is a major goal of contemporary health promotion efforts.

WHAT ARE COMPLEX CHRONIC DEGENERATIVE DISEASES?

Illnesses in this category account for the overwhelming majority of deaths and disability in Westernized nations. Such disorders are generally characterized by an uncertain etiology, multiple risk factors, a long latency period, a protracted course of illness, a non-contagious origin, functional impairment or disability, and, in many cases, incurability. Examples include asthma, arthritis, cancer, cardiovascular diseases (including stroke and heart attack), and diabetes. Of course, generalizations of this nature are subject to many exceptions and qualifications. Fortunately, many cancers are curable. There is increasing evidence that chronic, low-grade infections may actually be a prerequisite for several "noncontagious" chronic illnesses, including both malignancies and atherosclerosis. The relationship appears to be that infection is a necessary, but not sufficient, causal factor. Sudden death is the initial clinical manifestation for many individuals with coronary heart disease. The underlying arterial plaque formation may have taken decades to reach the critical level, but from the patient's subjective viewpoint, the disorder is hardly protracted and there has been no premonitory functional impairment.

One almost invariable attribute of chronic degenerative diseases is the necessary interaction of multiple causative influences – risk factors. Tobacco abuse, heavy alcohol ingestion, and inadequate dietary antioxidant intake are all implicated for esophageal cancer. These elements potentiate each other's effects in a multiplicative rather than a simple additive fashion. Similarly, abnormal body composition (hyperadiposity, sarcopenia), excessive intake of high glycemic-load foods, and physical inactivity combine to induce insulin resistance and, increasingly often, type II diabetes. That multiple risk factors affect chronic disease incidence implies the existence of numerous genes which are involved in the development of, or resistance

to, chronic diseases. This expectation has been well-substantiated: dozens to hundreds of genes have been found to play a role for most illnesses in this category: 250 and counting for obesity alone. There are exceptions such as hemoglobinopathies (e.g., sickle cell anemia), retinoblastoma, and phenylketonuria, but the rule holds for the most widespread, numerically important chronic conditions.

CATEGORIES OF DISCORDANCE

It is through their impact on risk factors that the cultural changes which have accumulated since agriculture's emergence influence chronic disease propagation. These cultural novelties fall into several categories, but in all cases the new conditions appear to adversely affect human health.

Nutrition

There is surprisingly little overlap between current foods and those of the Paleolithic (Cordain et al., 2002, 2005; Eaton, 2006). We get most of our calories from grains, domesticated livestock, dairy products, and refined sugars, but preagricultural humans ate naturally occurring plant foods and wild game. They used almost no cereal grains, no dairy foods, no separated oils, no commercial processing, and they had no sources of "empty calories." People in the Stone Age consumed about twice the animal protein current Westerners do. The proportion of total fat in Paleolithic diets was roughly equal to that for contemporary Americans – about 35%; however, intake of serum-cholesterol raising fat was nearly always far less than at present, and there were more dietary long-chain (C20 and above) polyunsaturated fatty acids. These are thought to be the most biologically significant of the fatty acids because they are necessary components of neurons and required for eicosanoid synthesis. The preagricultural essential fatty acid ratio (ω-6:ω-3) was perhaps 3:1; for average Americans it approximates 10:1 or more – an imbalance likely to promote atherogenesis. Dietary cholesterol content exceeded current US levels because both lean wild game and fatty supermarket steak contain cholesterol. Carbohydrate consumption was less than at present, but came almost exclusively from fruits and vegetables, not from cereals, refined sugars, and dairy products. The latter provide more readily absorbable "insulinogenic" glucose – thus increasing dietary glycemic load. That ancestral humans ate thrice the fruits and vegetables most affluent Westerners do should have enhanced their antioxidant capacity relative to ours. Compared with the typical American pattern, Paleolithic diets generally provided less sodium, but more potassium, fiber (soluble and insoluble), micronutrients, and, probably, phytochemicals. Before agriculture the net metabolic impact of human foods was slightly alkalotic, tending to raise bodily pH. Contemporary foods, especially dairy products and those made from grains, have the opposite – acidotic – effect and tend to slightly lower bodily pH (Sebastian, 2005). Stabilizing homeostatic biochemical mechanisms largely correct for this influence, but over decades, the result is excessive skeletal calcium loss and, consequently, osteoporosis.

These differences are pertinent to several areas of current nutrition-related research: e.g., ω-3 fatty acids and depression; ω-6:ω-3 ratios and coronary heart disease; fruits, vegetables, and phytochemicals as cancer preventive agents; optimal versus minimal requirements of vitamins and minerals; dietary sodium, hypertension, and overall mortality; and the appropriate contribution of fats to dietary energy.

Physical exertion

Through nearly all of human evolution, physical exertion and food procurement have been inextricably linked; in the past people had to work, physically, in order to eat. Hierarchical social stratification uncoupled this relationship for elites; industrialization and mechanization have completed the dissociation for practically everyone. Prior to the industrial era humans are estimated to have required a total of about 3000 kcal (12 MJ) daily; for current affluent populations comparable estimates are 2000 kcal (8 MJ) or less. This change has resulted from decreased energy expenditure through physical exertion: about 20 kcal/kg/day (84 kJ) for hunter-gatherers versus <5 kcal/kg/day (21 kJ) for sedentary Westerners – a four-fold differential (Eaton and Eaton, 2003).

Exercise has important effects on aerobic power, muscular strength, and skeletal robusticity, all of which were substantially greater for ancestral populations. Exercise likely affects the incidence of age-related fractures, some cancers, and atherosclerosis. Obligatory exertion promoted greater lean body mass while attenuating adipose tissue, thereby reducing type II diabetes risk for our ancestors.

Reproduction

Studies of women in foraging and other traditional settings suggest substantial differences between patterns of ancestral and modern reproduction (Eaton and Eaton, 1999b Ellison, 2001). For preindustrial women menarche was later (16 vs. 12.5 years) and first birth earlier (~19 years) so that the nubility (menarche to first birth) interval was only 3 years, versus about 12 years for average Americans and Europeans. Foragers who lived through their full reproductive span had high parity: typically 6 live births vs. 1.8 for Americans. Nursing was obligatory, intensive (on demand, not on schedule), and commonly lasted three years. Only about 50% of American babies are nursed at all and mean nursing duration is barely 3 months. Age at menopause is hard to ascertain for forager women, but menses apparently cease somewhat earlier than in affluent societies.

New reproductive patterns and the associated ovulatory differential (three times as many ovulations for Westerners not using oral contraceptives) are associated with increased risk for cancers of the breast, endometrium, and ovary. For example, immature breast lobules form at puberty; their rapidly dividing cells are relatively susceptible to natural mutation, genotoxic carcinogens, and clonal promotion. At first full-term pregnancy most lobules differentiate into mature forms whose cells divide more slowly and are hence more resistant. Prolonged nubility thus extends a period of high susceptibility to carcinogenesis.

Infection

Relationships between humans and microbes were altered by the rise of agriculture. Higher population density, frequent long-distance contacts, settled living, and interactions with domesticated animals vastly increased pathogen transmission. As a result, certain infections assumed greater importance, becoming selective forces that have subsequently affected the human genome (e.g., malaria, typhoid fever). More recently, improved sanitation has reduced transmission, a pivotal contribution to the past two centuries' increase in average life-expectancy. Discovery of antibiotics had dramatic impact, but intensive usage, including incorporation into animal feeds, has led to the emergence of resistant organisms. Consequently, "preventive" anti-infective chemotherapy must now aim at minimizing resistance as well as attaining clinical efficacy. To this end, mathematical models integrating classic pharmacological approaches with the principles of evolutionary biology may help optimize treatment protocols. Attempts to reduce pathogen virulence may also benefit from Darwinian considerations. For example, vaccines directed against virulence-enhancing microbial antigens might disproportionately affect dangerous strains and promote their displacement by milder variants (Eaton et al., 2002a).

While adequate food, public health measures, and medical interventions have lowered infectious disease mortality during the past century, the megapolitan crowding and unparalleled mobility in current affluent nations have probably increased transmission of certain organisms, especially those spread by sexual and respiratory contact. This phenomenon could affect chronic disease prevalence: there are well-established relationships between viral infections and certain cancers as well as intriguing hints of a causal link between microbes and atherosclerosis. Epidemiological correlation between infectious exposure rates and incidence of chronic "noninfectious" degenerative diseases might ultimately open new avenues for preventive intervention via evolution-based antibiotic prophylaxis and/or vaccine development. Immunization against human papillomavirus, which is strongly linked to development of cervical cancer, is an example of this approach.

The impact of public health measures and personal cleanliness may not have been altogether positive: the *hygiene hypothesis* posits that allergic conditions and the so-called autoimmune diseases are undesirable side effects (Yazdanbaksh et al., 2002). For our ancestors, genetically determined host defenses – white blood cells, antibodies, etc. (see Chapter 27 of this volume) – resisted the near-continuous onslaught of bacteria, parasites, helminths, and viruses. The human immune system has been designed through evolutionary selection for ongoing battle with our microscopic adversaries. But now the terms of engagement have changed so that, in a very real sense, our immune system has become under-employed. Previously prevailing circumstances (repetitive, unrelenting exposure to infectious agents) fostered an immunoregulatory atmosphere which maintained focus on deleterious micro-organisms while protecting against injury to our own tissues. In contrast, contemporary conditions apparently fail to activate appropriate immunoregulation so the immune system has become prone to attacking our lungs (asthma), paranasal sinuses (sinusitis, rhinitis), joints (various arthridides), bone marrow (childhood leukemias), intestines (Crohn's disease, ulcerative colitis), brain (multiple sclerosis), and pancreas (type I, "childhood" diabetes). These allergic or autoimmune disorders occur disproportionately among affluent, urban, and perhaps otherwise optimally hygienic populations for whom exposure to micro-organisms is (presumably) especially reduced as compared with prior human experience.

For immunologists, rheumatologists, and others concerned with diseases of this nature, the aim is to isolate those microbial constituents (antigens) that previously acted to induce appropriate immunoregulation and to create from meticulously selected components vaccines that will guard against development of allergic/autoimmune diseases while maintaining adequate host defense against infection.

Growth and development

In Western populations, less frequent and severe childhood infection, sharply reduced exercise requirements, unprecedented caloric availability, and epigenetic influences on maternal–fetal relationships (which are only now becoming understood) result in rapid bodily growth and early sexual maturation. Average adult height is asymptotically approaching a maximum while age at menarche has fallen to about 12.5 years, probably near the population's genetic limit. Most recent hunter-gatherers have been short-statured, reflecting the nutritional stress of foraging in marginal environments, but average height for Paleolithic humans (who lived in more fertile regions) appears to have roughly equaled that at present. Nevertheless, maturation may have been slower, as it is for athletic young women in Western nations. Traditional North African pastoralists – who have sufficient dietary protein, little access to empty calories, and high levels of physical exertion – may simulate the ancestral standard. They experience later puberty and slower growth in height than do Westerners, attaining full stature only in their early 20s; still, their average adult height equals that of Europeans.

Rapid growth is usually interpreted as a sign of societal health, but maximal is not necessarily optimal (Eaton et al., 2002a). The current experience of puberty three years earlier than the hunter-gatherer average may result in dissociation between psychological and sexual maturation, thus contributing to unwanted teenage pregnancies as well as suboptimal parenting. Both early menarche and youthful attainment of adult stature are associated with increased breast cancer risk. Rapid bodily growth may also affect blood pressure regulation if renal development is unable to keep pace allometrically, thus requiring compensatory blood pressure elevation to maintain homeostasis, and possibly establishing a pathophysiological trajectory towards subsequent hypertension. Also, in laboratory animals at least, slower growth during adolescence and early adulthood is associated with increased longevity – apparently independent of any effect on chronic disease susceptibility.

Psychosocial factors

Genes affecting human behavior are ancient and probably coevolved with our life history characteristics. For example, prolongation of childhood during hominid evolution may have facilitated learning and correlated with brain expansion occurring over the same period. But, like current sedentism and diet, the social circumstances of contemporary existence are novel. Many factors believed to exert important influence on psychological development and interpersonal relations are profoundly different from what they are thought to have been during our evolutionary past (Eaton et al., 2002a). Average birth spacing is now closer, while nursing and physical contact between infants and adults is much reduced. In most affluent societies, babies do not sleep with their mothers – a break from general primate experience dating back many million years. Ancestral childhood and adolescence were almost certainly characterized by multiage play groups, less restrictive supervision, and intense small group interpersonal dynamics quite different from the age-segregated, more structured routines of contemporary schools and little leagues. Based on what we know about hunter-gatherers, Paleolithic teenagers had relatively clear societal expectations, not the exciting-but-daunting array of life choices which confronts young people today. For adults, a global society has advantages, but it differs radically from the more human-scale experience of our ancestors who lived, found their roles, and developed self-esteem in bands of 15–50 people, most of whom were relatives. We have little concrete evidence, but it seems likely that these differences and others – frequent contact with strangers, conflicting social roles, wage labor, working in bureaucracies, reduced support from kin, socio-economic societal stratification, and education that questions social beliefs and ideologies – may contribute to syndromes such as attention deficit/hyperactivity, depression, anxiety disorders, and substance abuse.

REPRESENTATIVE ILLNESSES

Table 28.1 lists many chronic disorders, all of which can be considered discordance-related conditions. Discussing each of these in detail would greatly exceed the length allowed for this chapter; consequently, four disparate ailments, whose ties to discordance vary from the simple and straightforward to complicated and indirect, will be analyzed as illustrative examples.

Hypertension

For decades blood pressure (BP) readings of 120/80 mmHg were considered normal, but recent studies have shown that adverse health effects begin at pressures exceeding 115/75 mmHg. Accordingly, in 2003, a National Institutes of Health (NIH) select committee redefined "normal" as *under* 120/80 mmHg (Joint National Committee, 2003). Blood pressures between 120/80 and 139/89 mmHg are now considered prehypertensive, while BPs of 140/90 mmHg or above indicate that hypertension (HT), or high BP is present. Hypertension is rife in Westernized nations: 25% of North American adults and an even higher proportion

TABLE 28.1. A partial list of discordance-related diseases.

Acne
Age-related fractures
Alcoholism
Asthma
Cancer
"Childhood" infections
Cirrhosis
Coronary heart disease
Dental caries
Dental malocclusion (impacted teeth)
Depression
Diabetes
Emphysema
Epidemic infectious disease
Gingivitis (gum disease)
High frequency hearing loss
Hypertension
Lactose intolerance
Low back pain syndrome
Myopia (near sightedness)
Obesity
Osteoporosis
Peripheral vascular disease
Sickle cell anemia
Stroke

of Europeans have BPs above 140/90 mmHg or are on anti-hypertensive medication. Also HT is becoming increasingly common: in the United States, its prevalence rose 15% between 1990 and 2000.

Across the globe, many cultures studied during the last century were found to have no HT (Table 28.2). Their BPs averaged 110/70 mmHg – within the range newly designated as normal by the NIH. Members of normotensive cultures were not immune from HT because, when they adopted a Western lifestyle, either by migration or by acculturation within their homeland, their BP experience soon began to parallel that of Americans and Europeans. These normal BP cultures existed in varied climatic circumstances – in the arctic, the rainforest, the desert, and the savanna – but they shared a number of essential similarities, each of which reprised, or continued, the experience of our late Paleolithic ancestors and which were the reciprocal of postulated causal factors for HT.

High BP is a treacherous condition because, of itself, it produces no symptoms. It has become known as the "silent killer" because its complications, affecting the eyes, kidneys, brain, and heart, can be deadly. Persons with HT double their risk of heart attack and have four times more risk of stroke than

TABLE 28.2. Normotensive populations.

Hunter-gatherers
Kalahari San (Bushmen)
Alaskan Aleuts
Greenland Eskimos
Australian Aborigines
Congo Pygmies
Tanzanian Hadza

Agriculturalists
Mexican Tarahumara Indians
Panamanian Cuna Indians
Kenyan Suba Tribesmen
Ugandan Tesot Tribesmen
Guatamalan Maya Indians
Gambian Tribesmen

Horticulturalists
Cook Island Polynesians
Highland New Guinea Tukisenta
Venezuelan Yanamamo Indians
Pukapukan Polynesians
Solomon Islanders
Malaysian Aborigines
New Guinea Chimbu
Brazilian Caraja Indians
Chinese Noso (Lolo) Aborigines
Surinam Indians
New Guinea Bomai
Brazilian Kren-Akorore Indians

Pastoralists
Kenyan Samburu
Kenyan Masai

do individuals with BPs of < 120/< 80. Untreated HT is a common cause of congestive heart failure; about half the cases of end-stage renal disease (kidney failure) are attributable to HT; and hypertensive retinopathy, very common in persons with untreated HT, can lead to blindness.

In 2002, the NIH published a list of lifestyle modifications which, together, can greatly reduce a person's risk of developing HT (Whelton et al., 2002). Comparing these recommendations with the corresponding features of ancestral life illustrates how deviations from the life ways for which our genes were selected can lead to chronic disease:

- **Maintain normal weight**. Body mass indices (weight/height) between 18.5 and 25 are considered normal. Studies of multiple hunter-gatherer groups reveal an average BMI of 21.5; for Americans, the 2002 average was 28!

- **Increase potassium and decrease sodium intake**. Americans consume more sodium (~4000 mg) than

potassium (~2500 mg) each day. In contrast, for Stone Agers, like all other terrestrial mammals, daily sodium intake (≤1000 mg/day) was far less than that of potassium (≥7500 mg/day).

- **Exercise 30 minutes or more on most days**. Paleolithic humans, who had no motorized equipment or draft animals, are estimated to have expended about 1240 kcal as physical activity each day. (The 1240 kcal figure is the averaged value for adult males and females). The corresponding estimate for Americans is 555 kcal per day.

- **Limit alcohol intake**. Hunter-gatherers studied in the last century were unable to make alcoholic beverages. Regular manufacture of such drinks almost certainly postdates the agricultural revolution.

- **Increase fruit and vegetable consumption**. In East Africa, the likely human homeland, fruit and vegetables provided about 50% of our ancestors' food energy intake during the period around 50 000 years ago. In the United States, currently, fruits and vegetables provide only about 16% of our daily food energy and the contribution in Europe is even less.

The take home message should be inescapable. The life patterns of Stone Age humans protected against, while our own lead to, development of high BP. Only a few of the normotensive populations in Table 28.2 were hunter-gatherers, but, except for access to alcoholic beverages, they all shared those characteristics which typified ancestral life and which recent epidemiological investigations have shown to reduce risk for HT.

Dental caries

We cannot be sure about the frequency of HT for the builders of Stonehenge nor for the Tang Dynasty Chinese, but teeth are the best preserved of all human remains at archaeological sites and, because of this, the prevalence of dental caries (cavities) can be very reliably established at different time periods throughout human experience. While HT causation is almost certainly a multifactorial process, the historical correlation of dental caries with sugar consumption indicates a less complicated genes × risk factor causal relationship.

Researchers have found that, for late Paleolithic humans before the emergence of agriculture, about 1–2% of adult teeth showed caries (Brothwell, 1963). At that time there were no refined sugars. Studies of recent hunter-gatherers revealed that they were very fond of honey, but because it was available seasonally and was both difficult and painful to obtain, the average individual in such cultures consumed only about 2 kg/year. It is likely that intake during the Stone Age was similar. During the Middle Ages bees had been domesticated, so honey was more available and about 10% of teeth from medieval times have cavities. Cane sugar appears to have been developed around 500 BCE in India, but it remained expensive – available almost exclusively to the upper classes, until sugar cane cultivation began in the West Indies and large-scale production was initiated. Even in 1815, average consumption in Britain, of honey *and* sugar, was only about 7 kg/year.

Industrialization made sugar cheap. By 1900 it had become a staple and, at that period, 60–70% of teeth in Britain were carious. In 2000, American sugar consumption was nearly 70 kg/year, but because of fluoridation, fluoride-containing toothpaste, and generally improved dental care, the dental cavity rate has fallen to 15–20% – only about 10 times what it was for our remote ancestors.

Lung cancer

The relationship of lung cancer to cigarette smoking is almost as straightforward as that of dental caries to sugar. Tobacco is naturally indigenous to the Americas, Australia, and a few Pacific Islands. Australian Aborigines have chewed tobacco for many millennia, but for most humans, tobacco exposure postdated the sixteenth-century voyages of exploration. Thereafter tobacco availability spread with amazing rapidity, but it was mainly utilized as snuff, for pipes and cigars, and as chewing tobacco. These forms of tobacco usage fostered cancers of the throat, larynx, and mouth: Presidents Ulysses Grant and Grover Cleveland, and also Kaiser Frederick III of Germany, were victims illustrating this effect.

Cigarettes apparently originated in Turkey and spread to Western Europe only after the Crimean War (1854–1856) in which Turkey, Britain, and France were allied against Russia. Turkish officers introduced cigarette smoking to their English and French counterparts from whom it spread to upper-class Westerners. Still, cigarettes remained a minimal component of overall tobacco usage (3% of all tobacco consumption in 1900) until World War I when cigarettes were issued, free, to soldiers as morale boosters. Thereafter cigarette usage skyrocketed to a peak in the mid twentieth century. Subsequent to the US Surgeon General's report in 1964, cigarette smoking in the United States gradually declined. About 23% of US adults smoked in 2006 versus about 46% in 1949, the peak year of consumption.

Cigars, pipes, chewing tobacco, and snuff expose the nose, oral cavity, throat, and larynx to tobacco's effects, but generally spare the lower portions of the respiratory tract. Conversely, cigarette smoke is usually inhaled, so that it affects the lungs. The incidence curves for lung cancer, which in the early 1900s was considered one of the rarest forms of primary neoplasm, parallel those for cigarette consumption with astonishing exactitude. There is a lag period of from 20 to 30 years (which indicates the time necessary for tobacco's carcinogens to induce malignancy), but both the rising and falling rates of pulmonary carcinoma follow the cigarette usage rate with near perfect alignment (Witschi, 2001).

The second most important risk factor for lung cancer is indoor exposure to radon, a radioactive gas, which may account for about 10% of American lung cancer deaths (Franklin and Samet, 2001). Radon exposure is another example of discordance: despite being called "cavemen," Stone Agers spent far less of their lives indoors than do affluent Westerners. Recently studied hunger-gatherers, even those in colder, upper-latitude climates, spent very little time in caves, rock shelters, or other places where radon might have accumulated. Nearly all their lives, especially during daytime, were spent outdoors. Living, studying, and working indoors allows exposure to radon gas, which originates from the ground and from building materials (stone, concrete, etc.) and accumulates in houses, stores, schools, office buildings, and the like.

Insulin resistance and diabetes

Most everyone knows that the prevalence of diabetes is climbing rapidly, up nearly five-fold, 500%, since 1960, but the relationship between insulin resistance and diabetes is less widely recognized. People who are insulin resistant must secrete more than the usual amount of insulin (from the pancreas) to metabolize a given amount of ingested carbohydrate. Insulin resistant individuals have a high risk of developing type II (adult onset) diabetes and they are also more likely to manifest HT, obesity, coronary heart disease, peripheral vascular disease, and the polycystic ovary syndrome.

Like type II diabetes, insulin resistance frequency is soaring and, also like diabetes, it is an affliction of affluence. Australian Aborigines, Japanese Ainu, Brazilian Amerindians, Efe pygmies from the Congo, and !Kung San Bushmen from Africa's Kalahari Desert were all hunter-gatherers who exhibited exceptional insulin sensitivity – the converse of insulin resistance. These groups inhabited widely separated geographic regions, but they shared the characteristics of being lean, exercising a great deal, and consuming few or no foods that have excessive insulin-raising (insulinotropic) effects.

As obesity is so strongly linked to type II diabetes, weight, and especially weight for height, BMI, is commonly thought of as particularly important. Actually neither weight nor BMI is the critical factor. Rather, body composition seems most closely related to both diabetes and insulin resistance. Individuals with a greater proportion of fat are more likely, and those with a greater proportion of muscle, less likely to develop insulin resistance. A descriptive expression is:

$$\text{Insulin Resistance} \quad \sim \quad \frac{\%\ \text{body fat}}{\%\ \text{muscle mass}}.$$

This relationship is easily explained (Eaton et al., 2002a). Fat cells and muscle cells have identical insulin receptors which "compete" for circulating insulin molecules. However, a given amount of fat, say 1 g, can extract much less glucose from the blood than can a similar gram of muscle despite equivalent insulin stimulation. In effect, insulin molecules interacting with adipocyte receptors are being underutilized in comparison with the blood sugar lowering impact they could have achieved had they interacted with myocyte insulin receptors.

Physical activity is beneficial because it tends to increase muscle tissue and decrease fat as proportions of body weight. Also, exercise increases muscle fitness, a correlate of muscle metabolic rate. A gram of fit muscle extracts more glucose than a gram of unfit muscle, given equal exposure to insulin. When muscle fitness is taken into account, a more complete descriptive expression is:

$$\text{Insulin Resistance} \sim \frac{\%\ \text{body fat mass}}{\%\ \text{body muscle mass} \times \text{muscle fitness}}.$$

Studies of recent hunter-gatherers showed them to be much fitter and far leaner than age-matched Westerners. Also, the skeletal remains of Stone Age humans reveal evidence of impressive muscularity, comparable to that of today's superior athletes (such as Olympians). These key factors affecting insulin resistance nicely illustrate the discordance hypothesis: our genetic makeup is designed for ancestral circumstances and deviation from the Stone Age pattern promotes dysfunction.

Figure 28.1 diagrams carbohydrate metabolism for Stone Age populations, whose end result was appropriate blood glucose reduction, and for modern couch potatoes who require extra insulin secretion to achieve adequate blood glucose regulation.

Most health-conscious individuals have heard that a pear-shaped body configuration (big hips and upper thighs) is preferable to an apple-shape (big midsection). This difference relates largely to regional blood flow patterns. Intra-abdominal (visceral) blood flow is 21% of total cardiac output at rest and much more following a meal. In contrast, cutaneous and subcutaneous blood flow is only 6% of total cardiac output at rest and less following a meal. It follows that the

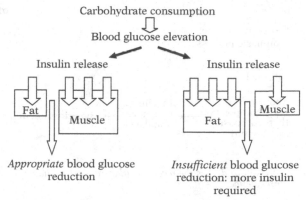

28.1. Carbohydrate metabolism.

insulin receptors of visceral fat cells are much better positioned than are those of subcutaneous fat cells when it comes to competing with muscle cell insulin receptors for circulating insulin molecules. The apple-shaped person has more visceral and the pear-shaped person fewer visceral fat cells as a proportion of total body fat content.

Insulin resistance appears to be a biphasic process. Phase I (Figure 28.2a) is a classic example of discordance at work. Body composition differing from the ancestral pattern (too much fat, too little muscle), exacerbated by low-level muscle fitness produces insulin receptor imbalance (too many fat cell receptors; too few and insufficiently active muscle cell receptors). These factors interact with our recently adopted high glycemic load diet to induce repetitive hyperinsulinemia. These phase I factors are responsible for the rising insulin resistance and type II diabetes prevalence during the past few decades.

It is phase II (Figure 28.2b) that fascinates geneticists and molecular biologists. Repetitive hyperinsulinemia activates genetically determined intracellular biochemical mechanisms (such as GLUT 4 trafficking) whose cumulative effect is to create intrinsic cellular resistance to the actions of insulin. Muscle, fat, and liver cells from phase I individuals react normally to insulin (at least early in the condition); the problem at this stage is tissue imbalance and muscle unfitness. However, in phase II these tissues become less sensitive to insulin stimulation. The adaptive mechanisms induced by phase I's repetitive hyperinsulinemia somehow make body tissues subnormally responsive. The genetic and biomolecular phenomena involved in this segue are complex and the subject of intense investigative scrutiny, but cannot, in themselves, account for rising rates of insulin resistance and type II diabetes.

It remains to discuss insulinotropic foods (Lindeberg et al., 2007). These share the characteristics of being newly adopted dietary constituents, in the evolutionary time frame, and of having special potency in raising serum insulin levels. The carbohydrate

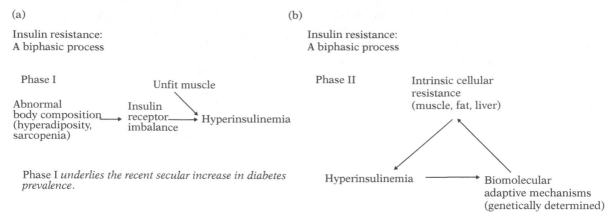

(a)

Insulin resistance:
A biphasic process

Phase I

Abnormal   Insulin   Unfit muscle
body composition   →   receptor   →   Hyperinsulinemia
(hyperadiposity,   imbalance
sarcopenia)

Phase I *underlies the recent secular increase in diabetes prevalence.*

(b)

Insulin resistance:
A biphasic process

Phase II   Intrinsic cellular
                            resistance
                            (muscle, fat, liver)

Hyperinsulinemia   →   Biomolecular
                         adaptive mechanisms
                         (genetically determined)

Phase II *cannot explain recent secular increase in diabetes prevalence.*

28.2(a and b). The development of insulin resistance.

available from sugars and refined cereal grains comes in a form which is rapidly absorbed from the intestinal tract, that is, these foods have a high "glycemic index." Also, both sugars and refined grains contain a high proportion of carbohydrate as a component of their total weight so they are high "glycemic load" foods. (Glycemic load = glycemic index × carbohydrate content per serving.) Rapid entry of glucose (from digested carbohydrate) – in large amounts – into the bloodstream causes excessive insulin release relative to the ancestral condition, so both refined cereals and sugars are insulinotropic. Dairy products are different: they have relatively low glycemic indices, yet produce disproportionately high insulin release. The mechanism involved is currently unknown, but they are also considered insulinotropic.

From our genes' standpoint all three food categories are Johnny-come-lately's. Refined cereal grains were unavailable to Stone Agers who lacked the necessary technology and who, in any case, consumed cereals only during times of food shortage. Whole grains have been dietary staples since the advent of agriculture, but truly efficient milling appeared no earlier than the late nineteenth century. Now 85% of the cereals consumed in the United States are refined. We estimate that Paleolithic humans consumed about 2 kg of sugar (as honey) a year. In 2000 average Americans consumed 70 kg/year. And Stone Agers had no domesticated animals (except, perhaps, dogs in the latest phases). Consequently, adults during the entire extent of human and prehuman evolutionary experience had no dairy foods whatever.

CONCLUSIONS

Evolution is increasingly accepted as the core around which our understanding of biology must be constructed,

and, as a discipline within the biological sciences, medicine should also be informed by evolutionary awareness. The discordance hypothesis is an attempt to connect the areas of disease causation and prevention with our growing knowledge of human evolution; in Kuhnian terms it represents a candidate paradigm (Kuhn, 1996). As we learn more about the specifics of Paleolithic experience and as our understanding of human pathophysiology becomes better refined, some of the material presented in this chapter will undoubtedly require modification, but the basic points will almost certainly be confirmed:

- Our genetic makeup is ancient.
- "Rapid" cultural change has created discordance between our genes and our lives.
- This discordance, or mismatch, fosters development of most complex chronic degenerative diseases.
- Preventing these diseases, and health promotion in general, involves reversion towards the basic essentials of ancestral existence (while, ideally, preserving the positive health effects of cultural evolution).

DISCUSSION POINTS

1. How similar are we genetically to preagricultural humans?
2. Why is the rate of type II diabetes rising around the world?
3. Is there a "natural" diet for humans?
4. How do NIH recommendations for preventing high blood pressure compare with our understanding of the ancestral human lifestyle?
5. What is the "Discordance Hypothesis?"

REFERENCES

Brothwell, D.R. (1963). The macroscopic dental pathology of some earlier human populations. In *Dental Anthropology*, Brothwell, D.R. (ed.). New York: Pergamon Press, pp. 271–318.

Cordain, L., Brand-Miller, J., Eaton, S.B., et al. (2002). Plant-animal subsistence ratios and macro-nutrient energy estimations in worldwide hunter-gatherer diets. *American Journal of Clinical Nutrition*, **56**, 181–191.

Cordain, L., Eaton, S.B., Sebastian, A., et al. (2005). Origins and evolution of the Western diet: health implications for the twenty-first century. *American Journal of Clinical Nutrition*, **81**, 341–354.

Eaton, S.B. (2006). The ancestral human diet: what was it and should it be a paradigm for contemporary nutrition? *Proceedings of the Nutrition Society*, **65**, 1–7.

Eaton, S.B. and Eaton, S.B.III (1999a). The evolutionary context of chronic degenerative diseases. In *Evolution in Health and Disease*, Stearns S.C. (ed.). Oxford: Oxford University Press, pp. 251–259.

Eaton, S.B. and Eaton, S.B.III (1999b). Breast cancer in evolutionary context. In *Evolutionary Medicine*, Trevathan W.R., Smith E.O. and McKenna J.J. (eds). Oxford: Oxford University Press, pp. 429–442.

Eaton, S.B. and Eaton, S.B.III (2003). An evolutionary perspective on human physical activity: implications for health. *Comparative Biochemistry and Physiology. Part A*, **136**, 153–159.

Eaton, S.B., Strassman B.I. and Nesse R.M. (2002a). Evolutionary health promotion. *Preventive Medicine*, **34**, 109–118.

Eaton, S.B., Cordain, L. and Lindeberg, S. (2002b). Evolutionary health promotion: a consideration of common counterarguments. *Preventive Medicine*, **34**, 119–123.

Ellison P.T. (2001). *On Fertile Ground. A Natural History of Human Reproduction*. Cambridge, MA: Harvard University Press.

Franklin, H. and Samet, J.M. (2001). Radon. *CA: A Cancer Journal for Clinicians*, **51**, 337–344.

Joint National Committee (2003). The seventh report of the joint national committee on prevention, detection, evaluation, and treatment of high blood pressure. *Journal of the American Medical Association*, **289**, 2560–2572.

Kuhn, T.S. (1996). *The Structure of Scientific Revolutions*, 3rd edn. Chicago: University of Chicago Press, p. 148.

Sebastian, A. (2005). Dietary protein content and the diet's net acid load: opposing effects on bone health. *American Journal of Clinical Nutrition*, **82**, 921–922.

Whelton, P.K., et al. (2002). Primary prevention of hypertension: clinical and public health advisory from the National High Blood Pressure Education Program. *Journal of the American Medical Association*, **288**, 1882–1888.

Witschi, H. (2001). A short history of lung cancer. *Toxological Sciences*, **64**, 4–6.

Yazdanbaksh, M., Kremser, P.G. and van Ree, R. (2002). Allergy, parasites, and the hygiene hypothesis. *Science*, **296**, 490–494.

29 Evolutionary Medicine and the Causes of Chronic Disease

Paul W. Ewald

The current integration of evolution with medicine is artificially narrow because it reflects the biases of medicine as a whole and the specializations of particular investigators. If evolutionary principles are to offer a fundamental framework for understanding medical issues they should help identify these biases and the areas over which integration needs to be broadened. The chapter discusses this problem as it relates to evolutionary interpretations of chronic diseases. It also illustrates how an evolutionary perspective can provide a broader framework for resolving the causes and control of chronic diseases, focusing on the greatest killers in prosperous societies – atherosclerosis and cancer.

OVERVIEW OF CHRONIC DISEASES

Chronic diseases account for about 70% of the mortality in the United States and other wealthy countries. Most of this disease-induced mortality is attributable to heart attacks, strokes, and cancer. In spite of this importance, the causes of chronic diseases remain largely unresolved. This situation represents a major short-coming of modern medicine because understanding the causes of disease enables prevention, and prevention is the most effective way of eliminating the damage inflicted by disease. We can therefore expect that a better understanding of the causes of chronic diseases will yield some of the most valuable contributions that any discipline can make to improvements in health.

Chronic diseases can be defined broadly as diseases that persist within individuals for a long time. The US National Center for Health Statistics considers "long" to mean persistence for at least three months. Although this dividing line between acute and chronic is somewhat arbitrary, a biological basis for distinguishing acute and chronic diseases can be made for infectious diseases. If an immune response can readily eliminate an infection, it generally does so within several weeks. Infectious diseases that are controlled (or kill the host) during this period are generally considered acute.

If pathogens have adaptations that allow them to evade the immune system, they can persist indefinitely. Illnesses caused by such persistent pathogens tend to be chronic. The three-month dividing line between acute and chronic diseases therefore corresponds roughly with an immunological basis for rapid resolution of infectious challenges versus protracted conflicts.

The causes of chronic diseases are not well understood largely because the processes of causation tend to be inconspicuous and difficult to evaluate (Cochran et al., 2000). For the past century the prevailing view of modern medicine has been that chronic diseases result largely from interactions between a person's genetic vulnerabilities and environmental influences. These interactions determine the rate at which the body succumbs to chronic disease.

Evolutionary medicine has recast this problem by asking why the process of natural selection has resulted in the existing collection of vulnerabilities to chronic disease (Nesse and Stearns, 2008). The investigative framework of evolutionary medicine integrates the mechanistic (proximate) explanations, which address how biological processes work, with evolutionary (ultimate) explanations, which address why biological evolution has resulted in these processes. Proximate causes of human diseases include human alleles and environmental factors that contribute to pathogenesis. Ultimate causes explain why humans have evolved to be vulnerable to such influences, and if an environmental cause is a parasite why the parasite has evolved to cause this damage.

In evolutionary medicine investigations of the causes of chronic disease generally focus on the differences between ancestral and modern environments. The prevailing emphasis is on the mismatch between our modern environment and our evolved biology. Aspects of modern diets that differ from ancestral diets, for example, are raised as explanations for diseases such as atherosclerosis and diabetes (Eaton et al., 1988; Nesse and Williams, 1995; Gerber and Crews, 1999; Leonard, 2008). Modern changes in nondietary exposures to cigarette smoke, ultraviolet

Human Evolutionary Biology, ed. Michael P. Muehlenbein. Published by Cambridge University Press. © Cambridge University Press 2010.

radiation in light skinned people, or birth control, are suggested as reasons for increased rates of cancers (Williams and Nesse, 1991; Eaton et al., 1994; Nesse and Williams, 1994; Diamond, 2005). Increased life span in modern environments is suggested as a reason for chronic diseases associated with old age (Nesse and Williams, 1995; Austad 1997; Finch, 2007; Austad and Finch, 2008). Human bodies are presumed to be vulnerable to such proximate causes of chronic diseases because exposure to these proximate causes has been too recent to purge these vulnerabilities by natural selection. Evolutionary medicine also considers the inability of natural selection to perfect adaptations, due, for example, to constraints on developmental processes or the limited power of natural selection to perfect adaptations (Nesse, 2008).

These arguments of evolutionary medicine generally accept the consensus of nonevolutionary medicine on the proximate causes of disease and develop evolutionary explanations for these proximate causes. This consensus of nonevolutionary medicine is generally that the proximate causes of chronic diseases result from an interaction between human genetics and noninfectious environmental factors, the familiar gene by environment explanations. But the proximate causes of chronic illness are still largely uncertain. If evolutionary medicine generates ultimate explanations for the wrong proximate explanations, the ultimate explanations will be wrong as well. It is therefore important for research in evolutionary medicine to consider the full range of feasible proximate explanations of disease to safeguard against offering incorrect ultimate explanations. Moreover, evolutionary insights may facilitate evaluation of alternative proximate mechanisms of pathogenesis by distinguishing those mechanisms that are evolutionarily feasible from those that are not.

These arguments apply in principle to acute diseases, but the causes of acute diseases are understood better than are the causes of chronic diseases; there is therefore less danger of proposing an evolutionary explanation for an incorrect proximate mechanism of acute diseases, and little need for new evolutionary insights to identify the causal agents of acute diseases. The overwhelming majority of acute diseases are caused by parasitism, broadly defined to include multicellular, cellular, and subcellular parasites (infectious diseases being defined as parasites that live in intimate association with host cells and tissues). The evolutionary study of acute disease has therefore focused on the degree of harmfulness (virulence) to which host–parasite associations evolve and the evolutionary function of disease manifestations – do the manifestations increase the evolutionary fitness of the host, the parasite, both, or neither (Ewald, 1980, 1994).

A discordance between the evolutionary study of acute and chronic disease arises because studies of acute diseases emphasize parasitic causes, whereas studies of chronic diseases emphasize the evolved characteristics of humans and environmental factors other than parasites. A central challenge for evolutionary medicine is to determine whether this discordance reflects a real difference between acute and chronic diseases, biases against considering infectious causes of chronic diseases, or both.

Although infectious causes of chronic diseases have been accepted for over a century, the clarity of hindsight shows that this category of causation has been underappreciated in evolutionary medicine. Illnesses such as peptic ulcers and cancers, for example, were explained by the mismatches between modern and ancient environments without consideration of infectious causation, even when evidence implicating infectious causation was well developed (e.g., Nesse and Williams, 1995). It is now broadly accepted that peptic ulcers, cancers of the stomach, liver, cervix, and oropharyngeal, and nasopharyngeal regions are caused largely by infectious processes (Ewald, 2000; Cochran et al., 2000; Greaves, 2000). The overlooking of infectious causation may be much more significant than these examples indicate because, as will be discussed in this chapter, evidence indicates that many other important chronic diseases, such as atherosclerosis, Alzheimer's disease, breast cancer, and schizophrenia, may be caused largely by infectious processes.

To avoid a bias against identifying infectious causes, disease causation can be considered schematically as a causal triangle with each apex of the triangle corresponding to one of the three categories of causation (Figure 29.1). The placement of any disease within the triangle emphasizes the relative importance of the three categories of disease causation. Diseases are placed closest to the vertex that corresponds to the primary cause of the disease, but the location within the triangle (instead of at the vertex itself), emphasizes that other categories contribute to pathogenesis. A primary cause is defined as one that is necessary for the disease to occur. Prevention of the primary cause(s) of a disease prevents the disease. Prevention of a secondary cause will reduce the frequency or severity of disease but will not prevent the disease itself. *Mycobacterium tuberculosis*, for example, is a primary cause of tuberculosis, because prevention of *Mycobacterium tuberculosis* infection will prevent tuberculosis. Poor nutrition and genetic vulnerabilities to *M. tuberculosis* are secondary (or exacerbating) causes of tuberculosis, because tuberculosis can still occur in individuals who have good nutrition or are genetically less vulnerable if other conditions such as immune suppression or intense exposure to *M. tuberculosis* occur. Sometimes more than one category of causation must be present for a disease to occur. Cancers for which pathogens are accepted primary

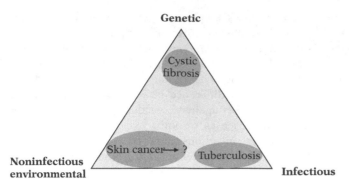

29.1. The triad of disease causation. The diagram emphasizes joint influences of the three categories of disease causation. The placement of a disease corresponds to the relative importance of the three categories with the closest apex indicating the primary cause as defined in the text. The vertices that are more distant represent exacerbating (i.e., secondary) causes or coprimary causes of secondary importance. Cystic fibrosis, for example, is a genetic disease that is exacerbated by a variety of different pulmonary infections and low-salt diet. Skin cancer refers to squamous cell skin cancers. The placement of tuberculosis and skin cancer is discussed in the text.

causes, for example, generally also require additional mutations. In this case environmental mutagens and the infectious agents are coprimary causes, and the cancer would be located roughly equidistant from the parasitic and environmental vertices. By considering this triangle of causation researchers can avoid falling into the trap of failing to consider the validity of one category of primary or secondary causation, just because the validity of another category has been identified.

The rest of this chapter will apply this perspective to specific chronic illnesses that have been discussed in the literature of evolutionary medicine. I emphasize feasible hypotheses of disease causation that have not been considered, especially when supportive evidence exists in the literature. As a result of the bias against infectious causation of chronic disease mentioned above, the overlooked hypotheses generally involve a failure to consider hypotheses of infectious causation and the interplay between infectious causation and the other categories of causation.

Atherosclerosis

By effecting heart attacks and strokes, atherosclerosis is the leading cause of mortality in the United States and other wealthy countries (http://www.cdc.gov/nccdphp/overview.htm). In accordance with its importance, vast economic and intellectual resources have been spent to identify risk factors for atherosclerosis. Dietary constituents have been implicated as exacerbating influences (cholesterol, saturated fats, dietary iron) and ameliorating influences (unsaturated fats, omega 3 fatty acids, garlic). The most important genetic risk factor for atherosclerosis that has been identified is the epsilon 4 (ϵ4) allele of the apolipoprotein E (*APOE*) gene (Ilveskoski et al., 1999; Mahley and Huang, 1999).

Almost all of the literature on the causes of atherosclerosis in evolutionary medicine attempts to explain human vulnerability to atherosclerosis in the context of such vulnerability to genetic and noninfectious environmental risk factors. These analyses, however,

lead to several paradoxes that suggest major shortcomings of these explanations. Each of these paradoxes is resolved when hypotheses of infectious causation are integrated into the analysis (Ewald, 2008).

One paradox arises from variation in the frequency of the ϵ4 allele of the *APOE* gene in humans and other primates. Its frequency ranges from about 5–40% in different human populations (Corbo and Scacchi, 1999; Fullerton et al., 2000). Comparisons of nucleotide sequences show that it is the ancestral allele of *Homo sapiens* and of primates generally (Fullerton et al., 2000) and that its relative frequency has declined over the past 200 000 years of human evolution relative to the other common alleles, ϵ2 and especially ϵ3 (Fullerton et al., 2000). The ϵ4 allele therefore is not an aberrant mutant allele. Some selection pressure in humans must have disfavored ϵ4 for a long time but differently in different human populations.

A diet-based hypothesis proposes that rich agricultural diets are responsible (Corbo and Scacchi, 1999). Accordingly, the ϵ4 allele frequency is about 5–15% in populations with several thousand years of agricultural experience and about 20–40% in populations that have been living largely as hunter-gatherers until the twentieth century (Corbo and Scacchi, 1999). These figures indicate that the allele has been declining more rapidly in agricultural settings. But ϵ4 is a minority allele even among recent hunter-gatherers. The most significant decline in ϵ4 frequency therefore occurred prior to the onset of agriculture.

Finch and Stanford (2004) suggest that this shift may have resulted from a shift toward increased meat eating associated with the early evolution of *H. sapiens*. Although this timing is consistent with the reduction in ϵ4 frequencies, it remains an untested hypothesis because the change in the makeup of the human diet during this time is unknown, as is the way in which any such change might alter the role of ϵ4 in lipid deposition.

The APOE proteins transport fat and cholesterol. Proximate explanations have therefore presumed that the association between ϵ4 and risk of atherosclerosis results from some difference between ϵ4 and the other

APOE proteins in transport of lipids. But a decade ago, Gérard et al. (1999) showed that people harboring the ε4 allele were more likely to be infected with the major candidate infectious cause of atherosclerosis: *Chlamydophila (Chlamydia) pneumoniae*. Recently the same research group has shown that the ε4 protein facilitates attachment of *C. pneumoniae* to human cells (Gérard et al., 2008). The evidence therefore supports the hypothesis that a fitness cost of the ε4 allele is increased vulnerability to *C. pneumoniae*. Unlike the agricultural diet hypothesis, the hypothesis that ε4 exacerbates atherosclerosis by increasing vulnerability to *C. pneumoniae* is consistent with the timing of the evolutionary decline in ε4 frequency, so long as *C. pneumoniae* has been a pathogen of humans throughout the evolution of *H. sapiens* (Ewald and Cochran, 2000; Ewald, 2008). The extent to which such a long-standing association has occurred should be testable using molecular phylogenetic reconstructions of *Chlamydophila*. The hypothesis that a shift toward an atherosclerosis-inducing diet occurred early during the evolution of *H. sapiens* is also consistent with this early shift away from ε4 (Finch and Stanford, 2004), but this hypothesis will be more difficult to test. Unlike the infectious causation hypothesis, this early *sapiens* diet hypothesis does not have empirical support from a proximate pathogenic mechanism, because we do not know whether the diet of early *H. sapiens* involved a shift toward constituents that fostered atherosclerosis. This argument illustrates the need for a broader consideration of alternative hypotheses in evolutionary medicine, because the proponents of the early *sapiens* diet hypothesis made no reference to the possibility that *C. pneumoniae* could be responsible for the reduction in the ε4 frequency (Finch and Stanford, 2004; Finch, 2007; Austad and Finch, 2008), even though the vulnerability of ε4 individuals to *C. pneumoniae* and the relevance of this vulnerability to atherosclerosis has been in the literature for about a decade (Gérard et al., 1999; Ewald, 2000; Ewald and Cochran, 2000).

Another paradox concerns the widely accepted hypothesis that cholesterol is a cause of atherosclerosis. Cholesterol accumulates within arterial walls early during the pathogenesis of atherosclerosis (Crowther, 2005). When it was thought that this accumulation was on the inside surface of the arteries it was reasonable to hypothesize that high cholesterol diets would lead to high levels of cholesterol in the blood, which in turn would lead to higher deposition of cholesterol. The link between elevated serum cholesterol and deposition of cholesterol within the arterial walls is less convincing, but the elevated cholesterol hypothesis can still be rescued if the cholesterol is somehow selectively transported to sites within arterial wall without being deposited on the lining of arteries.

At first glance, the beneficial effects of statins seems to offer strong support for the cholesterol hypothesis, because statins lower serum cholesterol levels and reduce rates of heart attacks. This reduction in heart attacks, however, was more powerful than could be accounted for by the cholesterol-lowering effects of statins (Mays and Dujovne, 2008). Importantly, statins are associated with reductions in inflammation and infection-induced mortality in atherosclerosis patients (Almog et al., 2007). These findings suggest that the beneficial effects of statins may arise not from their cholesterol-lowering effects, but from anti-inflammatory or antimicrobial effects (Jerwood and Cohen, 2008; Mays and Dujovne, 2008).

Taubes (2008) has pointed out another paradox of the cholesterol hypothesis. The drug Vytorin® was approved for lowering the risk of heart attacks based on its cholesterol-lowering effects. These effects arise from the two active components of Vytorin®. The first component is a statin called Zocor®. The second is a drug, called Zetia®, which lowers cholesterol by a different mechanism. The addition of Zetia® to Zocor® was approved because it lowered cholesterol more than Zocor® alone. The addition of Zetia®, however, did not lower the risk of heart attacks more than Zocor® alone. Taubes (2008) rightly points out that this finding serves as a critical test of the cholesterol hypothesis and that the cholesterol hypothesis failed the test.

These considerations suggest that the paradoxes over the role of cholesterol resulted from premature acceptance of a reasonable hypothesis rather than a more comprehensive consideration of the evidence in the context of the full range of feasible hypotheses. Integrating hypotheses of infectious causation resolves these paradoxes, by revealing that the incongruous findings are consistent with a more cohesive theory of pathogenesis. The accumulation of cholesterol within the walls of arteries, for example, is preceded by the accumulation of cholesterol-laden macrophages, called foam cells, within the arterial walls (Crowther, 2005). *Chlamydophila (Chlamydia) pneumoniae* can infect macrophages and transform them into foam cells by fostering the uptake of fat and cholesterol. Zetia® may fail to decrease cardiovascular events because elevated serum cholesterol is a correlate and perhaps a minor exacerbating cause of these events rather than a primary cause.

A similar resolution applies to dietary components that appear to reduce the risk of heart attacks. Consider garlic. Over the past two decades, investigations of apparently beneficial effects of garlic on atherosclerosis have focused almost entirely on alterations of lipids by a component of garlic: allicin. But published studies have not documented consistent associations. Recently the controversy was largely put to rest by the most thorough study yet on the relationship between garlic and lipid composition, which showed no significant effects (Gardner et al., 2007). The results of this study may be invoked by some to dismiss the hypothesized beneficial effects of garlic against atherosclerosis. Yet the results of that study were predicted by a hypothesis that invokes the antibiotic effects of garlic instead of a lipid-altering effect (Ewald, 2008). This antibiotic hypothesis has not yet been specifically tested because researchers have been so narrowly focused on the lipid hypothesis of atherosclerosis pathogenesis. Studies of the role of garlic on atherosclerosis need to take a more balanced approach by investigating the possibility that garlic may ameliorate atherosclerosis by inhibiting pathogens that are implicated as primary causes of atherosclerosis.

Unlike the other hypotheses that have been proposed to explain the atherosclerosis, the infectious hypotheses of atherosclerosis do not generate such interpretive paradoxes. Moreover, predictions of the infectious causation hypothesis have been confirmed by the most direct tests: the presence of C. pneumoniae and other pathogens in the foci of disease, elevated positivity in afflicted patients, and mechanisms that link infection with the other risk factors in a causal network (e.g., C. pneumoniae fostering foam cell formation in early stages of atherosclerosis, and ε4 increasing vulnerability to C. pneumoniae; see Ewald and Cochran, 2000 and Ewald, 2008 for more details).

These considerations of atherosclerosis illustrate the value of using the triangle of causation as a basis for analyzing disease causation. The evidence pertaining to ε4 illustrates how paradoxes associated with the most important genetic vulnerability to atherosclerosis are resolved when the analysis is broadened to include the infection vertex. The evidence pertaining to cholesterol shows how paradoxes associated with one of the most widely incriminated dietary agents of atherosclerosis are resolved when the analysis includes infection. The evidence pertaining to garlic shows how an ameliorating environmental influence could be inappropriately rejected because the spectrum of hypotheses under consideration was inappropriately narrow. In each case evolutionary considerations are useful in generating and evaluating alternative causal explanations.

Cancer

Cancer is characterized by the pathological proliferation and systemic spread of cells. It is widely accepted that cancers result primarily from mutations that promote cell proliferation and inhibit cell adhesion (leading to metastasis). Consequently, researchers writing from an evolutionary perspective have attempted to explain why the human body is vulnerable to such mutation-induced disregulation of proliferation and adhesion. As is the case with atherosclerosis,

arguments tend to implicate the mismatch between ancestral and modern environments. It is argued, for example, that the increased life spans of modern humans increases the prevalence of cancer because cancers tend to occur during the later decades of life. The proximate part of this argument focuses on the greater chance for the accumulation of oncogenic mutations and disregulation of cellular processes as life span increases; the evolutionary part focuses on the weakness of natural selection to counteract cancers that occur late in life (Greaves, 2000). To explain why genetic predispositions to cancer might be spreading in human populations, evolutionary hypotheses raise the possibility that the genetic predispositions are alleles that have been favored recently in human evolution and that an increased cancer risk has tagged along as a pleiotropic cost (Crespi and Summers, 2006).

Analyses of oncogenesis in evolutionary medicine focus on environmental mutagens and inherited predispositions (Crespi and Summers, 2005; Merlo et al., 2006; Frank, 2007; Greaves, 2008; Komarova and Wodarz, 2008). This overly narrow scope is found even in the most comprehensive text of biological topics in evolutionary medicine (Stearns and Koella, 2008), where cancer is included in the section on "Noninfectious and degenerative diseases," even though infectious causation can be definitively ruled out for only a miniscule proportion of human cancer. When infections are considered (Greaves, 2000, 2006), infection has not been integrated with mutation and inheritance into a unified theory of oncogenesis.

As was the case with peptic ulcers and atherosclerosis, considerations of cancer within evolutionary medicine has been lagging behind rather than advancing demonstrations of infectious causation. Over the past 30 years a major trend in the understanding of oncogenesis has been recognition of an increasing role for infectious causes of cancer, mirroring the trend for chronic diseases in general (Ewald, 2009). In the mid 1970s unicellular or subcelluar parasites were generally accepted as a cause of only one human cancer: Burkitt's lymphoma, which is caused by simultaneous infection of Epstein–Barr virus and Plasmodium falciparum. With regard to multicellular parasites, some trematodes were recognized as contributing to liver and bladder cancer. The total amount of cancer accepted as being caused by parasitism was therefore less than 1% of all human cancer. Over the past three decades, this percentage has increased steadily. Now about 20% of all human cancer is accepted by the World Health Organization as being caused by parasitism. Although overviews of cancer often presume that the remaining 80% are caused by something other than parasitism, this conclusion is not justified by the evidence. Because all three categories of disease causation may act in concert, identification of a cause in one category cannot be used as evidence that any other category is invalid. The documented importance of inherited predispositions, environmental mutagens, and the genes they mutate therefore does not imply that infectious agents are playing no causal role.

This idea is well illustrated by skin cancer. There is general agreement that ultraviolet rays contribute to skin cancer (Greaves, 2000; Kleinsmith, 2005; Karagas et al., 2007). One inherited genetic vulnerability for skin cancer is low amounts of protective skin pigment (Greaves, 2000, Kleinsmith, 2005). This vulnerability can be attributed largely to a mismatch between ancestral and modern environments – the ancestry of people with light skin color can be traced to northern latitudes, where exposure to ultraviolet radiation has tended to be low. Light skinned people tend to have high cancer rates when they live in more southerly environments with higher exposure to ultraviolet radiation. Evolutionary analyses of skin cancer have been largely restricted to such inherited and noninfectious environmental causes with little consideration of infectious causation. Yet evidence implicating nongenital serotypes of human papillomavirus (HPV) as a cause of squamous cell skin cancer has been published over the past decade. The evidence is particularly strong for people with suppressed immunity and the rare hereditary skin disease epidermodysplasia verruciformis, and probably involves disregulation of control over cellular division (Jenson et al., 2001; Karagas et al., 2006; Andersson et al., 2008). These considerations suggest that HPVs may be coprimary causes or exacerbating causes of squamous cell skin cancer (hence the arrow with the question mark in Figure 29.1). Considering only epidermodysplasia verruciformis and HPV the evidence supports a central location in the diagram.

Cells have four critical barriers to cancer. *Cell cycle arrest* keeps the cell from dividing. *Apoptosis* (cell suicide) can destroy proliferating cells before they progress to metastatic cancer. *Restriction of telomerase* can block oncogenesis by placing an upper limit on the total number of divisions that a cell lineage undergoes. *Cell adhesion* prevents metastasis.

Some viruses have evolved intricate mechanisms to interfere with these barriers, apparently to foster persistence of the viruses within their hosts. By causing the cells that they infect to divide, the genetic material of a virus can replicate in concert, while incurring little exposure to the immune system. By interfering with apoptosis the virus can keep the cell from destroying the virus via cell suicide. By increasing telomerase activity, the virus can push the infected cell toward immortality, perpetuating this profitable exploitation of the host cell for both resources and protection. By interfering with cell-to-cell adhesion, infected cells can spread to other parts of the body to facilitate further viral proliferation and transmission. This argument does not suggest that pathogens benefit from lethal cancer. Rather, the breakdown of the barriers to cancer favor persistence within the host but nudge infected cells toward cancer. An infection with any of the known oncogenic viruses is by itself insufficient to cause human cancer, because only a small proportion of the people who are infected with any one of these viruses will develop cancer. Other causes must therefore contribute.

Consideration of breast cancer offers an illustration of the need for an integrated approach to oncogenesis. The most widely accepted evolutionary explanation for breast cancer has been advanced by Boyd Eaton and his colleagues (Eaton et al., 1994; Eaton and Eaton, 1999; Greaves, 2008; Stearns et al., 2008; Trevathan et al., 2008). This explanation, which can be labeled the hormonal proliferation hypothesis, focuses on oncogenic effects of estrogen and, to a lesser extent, progesterone. It proposes that elevated rates of breast cancer in economically wealthy societies may result from enhanced cyclic exposure to these hormones.

The hormonal proliferation hypothesis points out that estrogen and perhaps progesterone contribute to the proliferation of cells in breast tissue (Potten et al., 1988) and attributes elevated rates of breast cancer in modern societies to a mismatch between modern and ancestral environments. It proposes that the mismatch results from modern birth control and modern diets. Birth control increases the number of menstrual cycles during a lifetime because women do not cycle during pregnancy and much of lactation. Rich diets apparently increase the number of cycles by advancing menarche and delaying menopause. Studies within post-industrial societies show that women who begin menstruation at younger ages and end menstruation at older ages have higher risks of breast cancer than women whose menstrual cycles occur over a more restricted range of years (Key and Pike, 1988; Trichopoulos et al., 1996). As a result of these differences, Eaton et al. (1994) estimate that women in the United States have nearly three times as many menstrual cycles during their lifetimes as women in hunter-gather societies. They estimate that US women have about 100 times more breast cancer.

As Eaton and his colleagues note, the risk of breast cancer is greater during pregnancy (Bruzzi et al., 1988; Williams et al., 1990). The increased risk persists during the first year after childbirth and declines gradually thereafter. Prior to menopause the declining risk of breast cancer with increasing time since last birth approaches but does not drop below that of nulliparous women (Bruzzi et al., 1988; Williams et al., 1990; Lambe et al., 1994). Protection associated with a greater number of births was significant only for women in their 40s (Bruzzi et al., 1988; Williams et al., 1990). These associations raise a paradox if one assumes that pregnancy protects against breast cancer: no protective effect of pregnancy is apparent after birth among premenopausal women. To the contrary, parous women have *greater* risks of premenopausal breast cancer than nulliparous women of the same age, but lower risks of cancer during the decade of life in which menopause generally occurs. Eaton and his colleagues accommodate this paradox by restricting their hypothesis to postmenopausal breast cancer. At best this restriction leaves much of the elevated breast cancer in post-industrial societies – that occurring premenopausally – without an adequate evolutionary explanation.

The restriction of the hormonal proliferation hypothesis to postmenopausal breast cancer raises another paradox. How can the proliferative effects of estrogen cause increased cancer rates only after estrogen levels have declined? Cognizant of this paradox, Eaton and Eaton (1999) conclude that the main oncogenic effects of estrogen result from effects of proliferation on mutation rather than the nudging of proliferation higher. This resolution represents a major restriction of the estrogen proliferation hypothesis to the proliferative effects on mutation, thus abandoning any direct contribution of hormone-induced proliferation on the runaway proliferation that characterizes cancer.

Another paradox arises because protection against peri- and postmenopausal breast cancer appears to be associated with pregnancies during the first decade or so after menarche. Eaton and Eaton (1999) resolve this paradox by concluding that the cancer-inducing proliferation is occurring in this time period. As a result of earlier menarche, women in post-industrial societies have more menstrual cycles prior to first pregnancies than women in hunter-gatherer societies, and thus may have more breast cell proliferation that could predispose them to a greater incidence of breast cancer. However, because the elevated rates of breast cancer in post-industrial societies generally occur postmenopausally, decades after this proliferation, the proliferation itself does not seem to be causing breast cancer directly by nudging up proliferation rates. Eaton and his colleagues therefore suggest that the mechanism by which proliferation increases cancer is mainly through increased mutation prior to the first birth.

After menopause proliferation of breast cells is twice as great for nulliparous women than for parous women. But this difference does not accord well with the hormonal proliferation hypothesis, because women are not experiencing the elevated levels of estrogen after menopause. Nor does this difference accord with the histological changes that are associated with parity, because at menopause breast cell types revert to the composition found in nulliparous woman (Russo, 1992). One can save the hormonal proliferation hypothesis by arguing that mutations that occur soon after puberty are particularly associated with reduced control of proliferation. But this reduced control of proliferation would have to occur after menopause but not detectably before menopause.

The protective association with parity for postmenopausal breast cancer and the exacerbating association of parity prior to menopause is a paradox that needs to be resolved by a general theory of breast cancer. If parity protects against breast cancer, why does it protect only after menopause? Higher rates of estrogen during pregnancy may play a role, but the paradox is not resolved by direct effects of estrogen on proliferation rates, because a large proportion of these cancers do not express the receptors for estrogen and progesterone. Also, the persistence of elevated breast cancer risk during the first year after parturition is not explainable as a direct effect of estrogen, because estrogen levels decline within a week or so after pregnancy (Doyle et al., 2007).

Consideration of infectious causation resolves these paradoxes. An immune suppression/infection hypothesis proposes that infectious agents may act in concert with mutations and inherited predispositions to cause breast cancer. It does not deny the potential applicability of estrogen and progesterone. But rather than focusing on the proliferative effects of these hormones, it emphasizes their suppressive effects on cell mediated immunity (Doyle et al., 2007, 2008), which may increase vulnerability to persistent viruses.

Three persistent viruses have been strongly associated with breast cancer: Epstein–Barr virus (EBV), HPV (particularly serotypes 16 and 18), and mouse mammary tumor virus (MMTV; human isolates are generally referred to as MMTV-like). Each of these viruses is found in about 25–50% of breast cancers but in less than about 10% of the normal breast tissue from the same patients (Wang et al., 1995; Bonnet et al., 1999; Fina et al., 2001; Lawson et al., 2001; Kleer et al., 2002; Damin et al., 2004; Kan et al., 2005; de Villiers et al., 2005). Each virus stimulates proliferation of infected cells and inhibits apoptosis (Subramanian and Robertson, 2002; Katz et al., 2005; Knight et al., 2005; Mileo et al., 2006; Guasparri et al., 2008; Hino et al., 2008). Both EBV and HPV are also known to promote cellular immortalization by enhancing telomerase activity (Ding et al., 2007; Liu et al., 2008; Tungteakkhun and Duerksen-Hughes, 2008; investigations of effects of MMTV on telomerase activity have not yet, to my knowledge, been conducted).

Each of these viruses interferes with cell-to-cell adhesion, thereby promoting metastasis. Expression of the cellular adhesion molecule, nm23H1, declines as breast and other cancers progress to lethal metastatic outcomes (Branca et al., 2006; Mileo et al., 2006; Sgouros et al., 2007). The MMTV inhibits expression of the nm23H1 (Ouatas et al., 2002) and EBV inhibits the ability of nm23H1 to suppress cell migration (Subramanian and Robertson, 2002; Murakami et al., 2005). Human papillomavirus downregulated and inactivated nm23H1 in keratinocytes (Mileo et al., 2006). In a study using esophageal cells, HPV16 did not inhibit nm23H1 but did inhibit a different cellular adhesion molecule, CD44v6 (Liu et al., 2005), decreased levels of which are associated with progression to metastatic cancer (Maula et al., 2001).

Whether any one of these viruses must act with any other to generate cancer is unknown. These associations could therefore explain anywhere from about half to nearly 100% of human breast cancer. Regardless of this number, evidence indicates that any oncogenic action of these viruses occurs together with oncogenic mutations, because oncogenic mutations are found in a substantial proportion of breast cancers. But consideration of the interdependence of the different steps of oncogenesis suggests that viral sabotaging of barriers to cancer generally occurs before oncogenic mutations. Activation of cell replication without inhibition of cellular senescence, for example, would have low oncogenic potential because it would generate only a limited number of cellular divisions. Similarly, inhibition of cellular senescence would be of limited value without inhibition of apoptosis, which would have the potential of destroying the infected cell. Without infectious causation, a specific sequence of very specific mutations would have to compromise these barriers without destroying the cell's viability. In contrast, infection by the three viruses mentioned above typically inhibit three or four of the key barriers to metastatic cancer simultaneously. This simultaneous inhibition of the three barriers to proliferation would allow for large numbers of infected cells to be generated. Once large numbers of precancerous cells are generated, the standard arguments about cancer evolution would apply: additional oncogenic mutations that further inhibit these barriers or other barriers to cancer would favor evolution of subsets of the precancerous cells toward cancer. Even if large numbers of precancerous cells in this population of precancerous cells lose their viability because of damaging mutations, oncogenesis can continue because many other infected cells remain to generate additional oncogenic mutations.

The immune suppression/infection hypothesis offers resolutions of the paradoxes raised by the hormonal proliferation hypothesis. The elevated risk of breast cancer during pregnancy may result from effects of the elevated reproductive hormones on the part of the immune system that controls virally infected cells. This effect might occur synergistically with any hormonal enhancement of the proliferation of infected cells. The associations with parity can be explained by effects of mate fidelity and duration of sexual activity on exposure to infection. Women who begin menses earlier will tend to have a longer exposure to sexually transmitted pathogens such as HPV and kissing transmitted pathogens such as EBV. Having children may reduce exposure to such pathogens because women who are raising families tend to have fewer sexual partners than single women. Moreover, nulliparous women may be nulliparous because they have had a relatively high exposure to sexually transmitted pathogens which can reduce fertility (as is the case with *Chlamydia trachomatis* and *Neisseria gonorrheae*).

The immune suppression/infectious hypothesis predicts a temporal pattern of mutations that is not predicted by the hormonal proliferation hypothesis. Specifically, if simultaneous inhibition of the cancer barriers by viruses is needed to initiate oncogenesis, the additional mutations of barriers to cancer should tend to occur later during oncogenesis. In contrast the hormonal proliferation hypothesis proposes that such mutations are initiating events and should therefore occur often at the onset of oncogenesis.

This prediction is testable, as illustrated by a recent study (Saal et al., 2008) of a gene referred to as *PTEN* (for **P**hosphatase and **TEN**sin homolog) in women who have inherited a mutation in a breast cancer susceptibility gene called *BRCA1* (for **BR**east **CA**ncer susceptibility gene type 1). The *PTEN* gene fosters cellular replication and inhibits apoptosis. The *BRCA1* gene encodes a protein that repairs mutations. Mutations in the *BRCA1* gene therefore contribute to cancer by reducing the repair of mutations in other genes that encode barriers to cancer, and women who inherit a mutant *BRCA1* allele have a higher net rate of mutation in their somatic cells. Understanding the function of these two genes led researchers to expect that women with the *BRCA1* mutation would be susceptible to *PTEN* mutations, which could then lead the mutated cells down the path to cancer. *PTEN* mutations do tend to occur in the breast cancers. But in contrast with the expectations of researchers studying this association, *PTEN* mutations occur late during oncogenesis, as expected if viral infection initiates oncogenesis and mutations complete the process after a sufficiently large population of precancerous cells has become established.

This example illustrates how hypotheses of infectious causation of cancer can be broadly distinguished from hypotheses that rely solely on mutation. Infections will be found early during oncogenesis, but oncogenic mutations will not be found in the earliest stages. Because infection initiated cancers still depend on mutations for cancer progression evidence that broadly implicates oncogenic mutation (e.g., age dependent risks of cancer; Frank, 2007) are consistent with infection-initiated oncogenesis. But despite this consistency the distinction between mutation-only hypotheses and infection-initiated hypotheses is critical because the infection-initiated hypotheses suggest that cancers can be prevented by preventing infection. The great practical benefit from cancer prevention relative to cancer treatment emphasizes the importance of testing the central prediction from the infection-initiated hypothesis, namely that infectious agents tend to be present and oncogenic mutations absent during the earliest phases of oncogenesis.

Pregnancy sickness

In some chronic diseases the failure to consider infectious causation has led to study of hypotheses that may explain the baseline occurrence of a phenomona, but not the most damaging occurrences. Pregnancy sickness offers an example. Pregnancy sickness, also called "morning sickness," is a sensation of nausea that is particularly common during the first trimester of pregnancy. Evolutionary considerations suggest that debilitating nausea would be selected against if it did not provide a compensating fitness benefit.

Hook (1976), Profet (1992), and Flaxman and Sherman (2000) developed the hypothesis that pregnancy sickness is an adaptation that reduces ingestion of noxious agents (toxins and pathogens) and thereby protects the developing embryo and the expectant mother. Such noxious agents tend to be dangerous or indicators of dangerous substances within food. Nausea in response to the smell or taste of secondary compounds may therefore cause the avoidance or regurgitation of toxin-laden and microbially contaminated foods. Fetal damage during the first trimester of pregnancy could have marked consequences because damage to small numbers of cells early during development may cause substantial damage to tissues or organs that develop from those cells, leading to birth defects or miscarriages. Heightened nausea in response to secondary chemicals in food may protect developing embryo and fetus from such damage. This hypothesis is supported by the timing of symptoms during pregnancy and reduced risks of miscarriage among women experiencing pregnancy sickness (Profet, 1992; Flaxman and Sherman, 2000).

This fetal protection hypothesis, however, raises a paradox. Sometimes the vomiting associated with pregnancy sickness is so severe that women will die of dehydration. One would not expect natural selection to favor lethal nausea as a defense against damage that is much less costly to fitness. When such paradoxes are addressed in the literature of evolutionary medicine, they are often explained away by the weakness of natural selection. Specifically, it is argued that natural selection is too weak to eliminate the damaging effects of an overzealous response, which might be maintained in the population as a result of the positive effects of a much more common, controlled response. This overzealous response hypothesis is not compelling because maternal death seems like too great a price to reduce exposure of babies to the secondary compounds in food and because control of nausea below life-threatening levels does not

seem to be a particularly difficult evolutionary hurdle. Natural selection would favor regulation of the response so that women gain the protective benefits without risking death. Although it is appropriate to consider the overzealous response hypothesis, it certainly should not be accepted without testing, especially when other more compelling hypotheses have not been considered or tested.

Consideration of infectious causation offers such an alternative hypothesis. Among infectious diseases one frequently finds manifestations that sometimes function as a defense but can be disregulated by infectious agents to such an extent that host death ensues. Diarrhea, for example, is known to facilitate the recovery by expelling the causal pathogens – suppression of diarrhea in patients experimentally infected with the diarrhea-causing pathogen, *Shigella sonnei*, prolonged recovery from infection (Dupont and Hornick, 1973). When humans are infected with *Vibrio cholerae*, however, the diarrhea-induced dehydration is as a rule the cause of death. Unlike *S. sonnei*, *V. cholerae* does not invade the intestinal lining. Moreover, because *V. cholerae* is a strong swimmer and can adhere to the lining of the small intestine, it can persist in the intestine in the face of the rapid fluid movements associated with diarrhea. In fact, *V. cholerae* induces the diarrhea by releasing a toxin that attaches to the cells that line the small intestine. This attachment results in the entrance of part of the toxin molecule into the cell, which eventually leads to the cell's release of fluid into the lumen of the intestine. The resultant diarrhea washes out competing bacteria from the entire intestinal tract. Strains of *V. cholerae* that produce large amounts of toxin thus produce a watery fecal material with large numbers of *V. cholerae* and few other micro-organisms. This watery *Vibrio*-laden stool can broadly contaminate the external environment, fostering transmission to other humans. For *V. cholerae* the generation of diarrhea is thus a sometimes lethal manipulation (*sensu* Ewald, 1980) of what is often a defensive response for its own benefit (summarized from Ewald, 1994).

Consideration of infectious causation therefore raises the possibility that a similar hypothesis might resolve the paradox of lethal pregnancy sickness. Specifically, it suggests the possibility that an infectious organism could manipulate the nausea defense, exacerbating it to the point that it may cause life-threatening dehydration. *Helicobacter pylori* is a logical candidate for such an organism because it infects the lining of the stomach and regurgitation would provide one of the most direct transmission routes. Epidemiological evidence implicates this route (Perry et al., 2006), as well as transmission from mother to offspring (Sinha et al., 2004; Yang et al., 2005, Delport and van der Merwe, 2007) and wife to husband (Fujimoto et al., 2007).

TABLE 29.1. Illnesses of uncertain cause, their associations with infection, and publications within evolutionary medicine that did not consider infectious causation of the specified illness.

Illness	Causal hypothesis presented in evolutionary medicine literature	Infectious agent or infectious disease	Publications
Skin cancer	Ultraviolet radiation, light skin color	HPV	Greaves (2000*, 2008*)
Breast cancer	Increased menstrual cycling	MMTV, EBV, HPV	Eaton et al. (1994) Eaton and Eaton (1999*)
Pre-eclampsia	Parent–offspring genetic conflict	Periodontitis, urinary tract infection	Haig (1993, 2008*)
Pregnancy sickness	Protection from dietary mutagens	*Helicobacter pylori*	Profet (1992); Sherman and Flaxman (2002*)
Schizophrenia	Mutations and environmental stress	*Toxoplasma gondii*	Nesse and Stearns (2008*) Trevathan et al. (2008*)
Atherosclerosis	Rich diet and lipid transport of APOE4	*Chlamydophila (Chlamydia) pneumoniae*	Nesse and Williams (1995*) Gerber and Crews (1999*) Finch and Stanford (2004*)
Alzheimer's disease (nonfamilial)	Immunological correlates and compensating benefits of APOE4	*Chlamydia pneumoniae*	Nesse and Williams (1995) Austad (1997), Finch (2007*) Austad and Finch (2008*)

Note: *Signifies works that were published after the infectious link to the illness was published. EBV, Epstein–Barr virus; HPV, human papillomavirus; MMTV, mouse mammary tumor virus.

Investigations of associations between *H. pylori* and severe pregnancy sickness have generated mixed results. About half of the studies document significant associations (Koçak et al., 1999; Reymunde et al., 2001; Kazerooni et al., 2002; Cevrioglu et al., 2004; Shirin et al., 2004; Karaer et al., 2008) and about half do not (Wu et al., 2000; Berker et al., 2003; Jacobson et al., 2003; McKenna et al., 2003; Weyermann et al., 2003; Lee et al., 2005). This discrepancy may result from differences in the strains in different populations of study subjects. In a study of 143 pregnant Turkish women (Noyan et al., 2004) the elevated positivity for *H. pylori* among dyspepsic women (75% vs. 64%) was not statistically significant. But among those women who tested positive for *H. pylori*, those who complained of dyspepsia were significantly more often infected with strains possessing a major virulence determinant – the cytotoxin-associated gene *A* (66% vs. 34%).

Taken together these results suggest that some strains of *H. pylori* may facilitate their own transmission by exaggerating a nausea response that evolved to protect the fetuses of uninfected women. As is the case with *V. cholerae*, the conflict of interest between *H. pylori* and its human host may sometimes lead to death through dehydration.

IMPLICATIONS FOR EVOLUTIONARY MEDICINE

Table 29.1 lists several other chronic illnesses of uncertain cause that have been discussed in the literature of evolutionary medicine. The list is not exhaustive,

but rather emphasizes conditions that have been considered often in the literature on evolutionary medicine or pose major unsolved health problems. The pathogens implicated are not yet broadly accepted as causes of the listed conditions, even though strong evidence of infectious causation exists.

Discussion of schizophrenia in evolutionary medicine-literature (e.g., Nesse and Stearns, 2008), generally has not incorporated the increasingly strong evidence of infectious causation (Ledgerwoood et al., 2003; Brown et al., 2004; Crespi et al., 2007; Hinze-Selch et al., 2007; Mortensen et al., 2007; Niebuhr et al., 2008). Longstanding claims of genetic causation based on familial studies have been accepted even though these associations can be explained by in utero infectious causation (Ledgerwoood et al., 2003). Specific genetic associations generally explain only vanishingly small proportions of schizophrenia. If primary causes of schizophrenia are largely infectious we can expect that the most common genetic predispositions to schizophrenia will tend to be vulnerabilities to such infectious causes (Ledgerwood et al., 2003). Accordingly one of the strongest genetic associations identified to date was recently found to encode an interleukin receptor, even though the researchers were not searching for a gene associated with the immune system and considered this association surprising (Lencz et al., 2007).

Pre-eclampsia is a complication in about 3% of pregnancies (Conde-Agudelo et al., 2008). In the evolutionary medical literature explanations of pre-eclampsia have been almost entirely restricted to genetic conflicts between mother and fetus (Haig, 1993, 2008). Yet a large literature on associations between infection and pre-eclampsia has accumulated. Particularly reliable are the associations between pre-eclampsia and both periodontal disease, which is caused largely by *Porphyromonas gingivalis*, and urinary tract infection (Conde-Agudelo and Belizan, 2000; Conde-Agudelo et al., 2008; Siqueira et al., 2008).

A thorough evolutionary examination of pre-eclampsia needs to resolve the maternal–fetal hypothesis with these infectious associations. Analyses need to consider whether there is any evidence implicating infectious causation that cannot be explained by maternal–fetus conflict and vice versa. Similarly, analyses need to consider whether the two categories of causation are both acting and perhaps interacting. Both categories of hypotheses need to be considered in light of other risk factors (Conde-Agudelo and Belizan, 2000) to address key characteristics of pre-eclampsia, such as its relatively low prevalence, greater damage in poor countries, and associations with parity.

The understanding of nonfamilial Alzheimer's disease parallels that of atherosclerosis. *Chlamydophila* (*Chlamydia*) *pneumoniae* has been identified in a large portion of Alzheimer's disease patients (Balin et al., 1998). The association of ε4 with Alzheimer's disease and the vulnerability to *C. pneumoniae* infection conferred by ε4 indicates that *C. pneumoniae* is probably a cause of sporadic Alzheimer's disease. Yet evolutionary discussions of Alzheimer's disease generally do not integrate the evidence implicating *C. pneumoniae* (Finch, 2007; Austad and Finch, 2008).

Without consideration of infectious causation the relative roles of infectious causation, inherited genes, and noninfectious environmental risk factors cannot be ascertained. Because the value of the evolutionary explanations depends on the validity of the proximate causes that they are attempting to explain, a failure to consider the full spectrum of the proximate causes raises concern about the validity of the evolutionary explanations. If such diseases are caused largely by infection, evolutionary explanations that do not incorporate infectious causation will be largely incorrect. This problem is particularly important if infectious causation is intimately related to genetic causation and noninfectious environmental causation, as appears to be the case with breast cancer and atherosclerosis. If, for example, the primary role of ε4 in atherosclerosis is to increase the infection rate of *C. pneumoniae* then the attempts to explain the association between ε4 allele and atherosclerosis or Alzheimer's disease strictly in terms of transport of lipids are futile.

Much is at stake. Will evolutionary medicine lead the health sciences to better understandings of medical issues and new discoveries? Or will it simply follow circuitous routes dictated by the biases of nonevolutionary medicine at any particular time,

drawing evolutionary interpretations of the current often incorrect consensuses? If evolutionary medicine follows the latter route, it is destined to present interpretations that are faulty because they are based on the faulty consensuses.

Taking the lead on these issues also offers the promise of practical benefits. If infectious causation is more pervasive than is currently accepted by modern medicine and evolutionary biology expedites the recognition of this pervasiveness then it will also expedite the recognition of ways to prevent these diseases. In many cases the practical benefits are obvious. Prevention of chronic diseases such as cancers, heart attacks, and strokes, would provide vast health benefits because such chronic diseases are the major sources of mortality in economically prosperous countries. It is reasonable to expect that such diseases could be prevented, for example, by vaccination because the track record of the health sciences at controlling or preventing infectious diseases has been very strong.

A broadened perspective within evolutionary medicine also provides more subtle insights into how diseases should be treated. Recognition that pregnancy sickness may often protect fetal development suggests that the symptoms of pregnancy sickness may often be best left untreated, but this suggestion does not apply to symptoms that could cause life-threatening dehydration (Flaxman and Sherman, 2000). Without consideration of infectious causation, evolutionary medicine may not be able to specify the boundary between appropriate and inappropriate treatment. If, however, *H. pylori* is responsible for the life-threatening symptoms of pregnancy sickness, this difficulty may be resolvable. The evolutionarily informed advice would be for women to be tested for *H. pylori* and treated with antibiotics if the test comes back positive. Ideally this testing should be done prior to pregnancy to avoid any negative effects of the antibiotics on fetal development. Once *H. pylori* is eliminated, dangerous symptoms of pregnancy sickness should be ameliorated.

Medicine has developed largely without a fundamental theoretical framework. This situation has made medicine vulnerable to the predispositions that are present among the majority of experts at any given time. From the early nineteenth century onward these predispositions have involved the unwarranted rejection of infectious diseases. Decade by decade new categories of diseases have been gradually added to the spectrum of infectious disease. Over the past half century this trend has continued with chronic diseases such as peptic ulcers, cervical cancer, and liver cancer being recognized as infectious diseases. Infectious causation of diseases such as breast cancer, atherosclerosis, and schizophrenia are on the threshold of acceptance. Evolutionary medicine is uniquely positioned to advance these trends because, unlike other areas of medicine, it is inclusive of all areas of medicine. It seeks to integrate conceptually all proximate and ultimate explanations of health and disease. Molecular, biochemical, cellular, developmental, morphological, and physiological processes provide the evidence on which this integration is based. But to accomplish this integrative goal, researchers in evolutionary medicine must broaden their perspectives not only to include these areas but to recognize what is being overlooked and unexplored. Evolutionary medicine has not yet taken on this challenge. Instead it has been characterized by narrowly circumscribed analyses that have failed to incorporate significant bodies of relevant evidence. In this chapter I have attempted to illustrate this problem and show how a more comprehensive and integrated consideration of the feasible categories of causation helps to resolve paradoxes, provide a more cohesive understanding of chronic illnesses, and suggest critical areas for future study.

DISCUSSION POINTS

1. How does the triangle of disease causation provide a conceptual basis for an integrated understanding of the role of various risk factors in chronic disease?
2. What is the distinction between primary and secondary causes of diseases and why is this distinction important to efforts to control disease?
3. What is the main difficulty with interpreting oncogenesis solely as a mutation-driven process, and how does integration of infectious causation ameliorate this problem?
4. How does the action of natural selection on infectious agents favor adaptations that nudge infected cells toward cancer?
5. Why do evolutionary considerations and mechanism of cellular invasion by *Chlamydophila pneumoniae* invoke infectious causation as a cause of the epsilon 4 diseases: atherosclerosis, sporadic Alzheimer's disease, and multiple sclerosis?
6. What evolutionary paradox arises from interpreting pregnancy sickness solely as an adaptation to reduce exposure of the fetus to damaging chemicals, and how does consideration of *Helicobacter pylori* resolve this paradox?

ACKNOWLEDGMENTS

The Rena Shulsky Foundation has generously supported the work on cancer as part of a project to develop a unified theory of oncogenesis. Discussions with Gregory Cochran over the past decade have profoundly influenced my thinking about the causes of chronic diseases. Carolyn Doyle provided references on breast cancer and valuable discussions on the topic. Holly A. Swain Ewald first drew my attention to viral disruption of cellular adhesion molecules, and has provided broadly insightful suggestions throughout the development of this work.

REFERENCES

Almog, Y., Novack, V., Eisinger, M., et al. (2007). The effect of statin therapy on infection-related mortality in patients with atherosclerotic diseases. *Critical Care Medicine*, **35**, 1635–1636.

Andersson, K., Waterboer, T., Kirnbauer, R., et al. (2008). Seroreactivity to cutaneous human papillomaviruses among patients with nonmelanoma skin cancer or benign skin lesions. *Cancer Epidemiology, Biomarkers and Prevention*, **17**, 189–195.

Austad, S. N. (1997). *Why We Age. What Science is Discovering About the Body's Journey Through Life*. New York: Wiley.

Austad, S. N. and Finch, C. E. (2008). The evolutionary context of human aging and degenerative disease. In *Evolution in Health and Disease*, S. S. Stearns and J. C. Koella (eds). New York: Oxford University Press, pp. 301–311.

Balin, B. J., Gerard, H. C., Arking, E. J., et al. (1998). Identification and localization of *Chlamydia pneumoniae* in the Alzheimer's brain. *Medical Microbiology and Immunology*, **187**, 23–42.

Berker, B., Soylemez, F., Cengiz, S. D., et al. (2003). Serologic assay of *Helicobacter pylori* infection. Is it useful in hyperemesis gravidarum? *Journal of Reproductive Medicine*, **48**, 809–812.

Bonnet, M., Guinebretiere, J.-M., Kremmer, E., et al. (1999). Detection of Epstein–Barr virus in invasive breast cancers. *Journal of the National Cancer Institute*, **91**, 1376–1381.

Branca, M., Giorgi, C., Ciotti, M., et al. (2006). Down-regulated nucleoside diphosphate (NDP) kinase nm23-H1 expression is unrelated to high-risk human papillomavirus (HPV) but associated with progression of cin and unfavourable prognosis in cervical cancer. *Journal of Clinical Pathology*, **59**, 1044–1051.

Brown, A. S., Begg, M. D., Gravenstein, S., et al. (2004). Serologic evidence of prenatal influenza in the etiology of schizophrenia. *Archives of General Psychiatry*, **61**, 774–780.

Bruzzi, P., Negri, E., La Vecchia, C., et al. (1988). Short term increase in risk of breast cancer after full term pregnancy. *British Medical Journal*, **297**, 1096–1098.

Cevrioglu, A. S., Altindis, M., Yilmazer, M., et al. (2004). Efficient and non-invasive method for investigating *Helicobacter pylori* in gravida with hyperemesis gravidarum: *Helicobacter pylori* stool antigen test. *Journal of Obstetrics and Gynaecology Research*, **30**, 136–141.

Cochran, G. M., Ewald, P. W. and Cochran, K. D. (2000). Infectious causation of disease: an evolutionary perspective. *Perspectives in Biology and Medicine*, **43**, 406–448.

Conde-Agudelo, A. and Belizan, J. M. (2000). Risk factors for preeclampsia in a large cohort of Latin American and Caribbean women. *British Journal of Obstetrics and Gynaecology*, **107**, 75–83.

Conde-Agudelo, A., Villar, J. and Lindheimer, M. (2008). Maternal infection and risk of preeclampsia: systematic review and metaanalysis. *American Journal of Obstetrics and Gynecology*, **198**, 7–22.

Corbo, R. M. and Scacchi, R. (1999). Apolipoprotein E (APOE) allele distribution in the world. Is APOE4 a "thrifty" allele? *Annals of Human Genetics*, **63**, 301–310.

Crespi, B. J. and Summers, K. (2005). Evolutionary biology of cancer. *Trends in Ecology and Evolution*, **20**, 545–552.

Crespi, B. J. and Summers, K. (2006). Positive selection in the evolution of cancer. *Biological Reviews of the Cambridge Philosophical Society*, **81**, 407–424.

Crespi, B. J., Summers, K. and Dorus, S. (2007). Adaptive evolution of genes underlying schizophrenia. *Proceedings of the Royal Society of London. Series B*, **274**, 2801–2810.

Crowther, M. A. (2005). Pathogenesis of atherosclerosis. *Hematology. Education Program of the American Society of Hematology*, **1**, 436–441.

Damin, A., Karam, R., Zettler, C., et al. (2004). Evidence for an association of human papillomavirus and breast carcinomas. *Breast Cancer Research and Treatment*, **84**, 131–137.

De Villiers, E.-M., Sandstrom, R., Zur Hausen, H., et al. (2005). Presence of papillomavirus sequences in condylomatous lesions of the mammillae and in invasive carcinoma of the breast. *Breast Cancer Research*, **7**, R1–R11.

Delport, W. and Van Der Merwe, S. W. (2007). The transmission of *Helicobacter pylori*: the effects of analysis method and study population on inference. *Best Practice and Research: Clinical Gastroenterology*, **21**, 215–236.

Diamond, J. (2005). Evolutionary biology: Geography and skin colour. *Nature*, **435**, 283–284.

Ding, L., Li, L., Yang, J., et al. (2007). Latent membrane protein 1 encoded by Epstein–Barr virus induces telomerase activity via p16INK4A/Rb/E2F1 and JNK signaling pathways. *Journal of Medical Virology*, **79**, 1153–1163.

Doyle, C., Swain Ewald, H. A. and Ewald, P. W. (2007). Premenstrual syndrome: an evolutionary perspective on its causes and treatment. *Perspectives in Biology and Medicine*, **50**, 181–202.

Doyle, C., Swain Ewald, H. A. and Ewald, P. W. (2008). An evolutionary perspective on premenstrual syndrome and its implications for investigating infectious causes of chronic disease. In *Evolutionary Medicine and Health. New Perspectives*, W. R. Trevathan, E. O. Smith and J. J. Mckenna (eds). New York: Oxford University Press, pp. 196–215.

Dupont, H. L. and Hornick, R. B. (1973). Adverse effect of Lomotil therapy in shigellosis. *Journal of the American Medical Association*, **226**, 1525–1528.

Eaton, S. B. and Eaton, S. B. III (1999). Breast cancer in evolutionary context. In *Evolutionary Medicine*, W. R. Trevathan, E. O. Smith and J. J. Mckenna (eds). New York: Oxford University Press, pp. 429–442.

Eaton, S. B., Shostak, M. and Konner, M. (1988). *The Paleolithic Prescription*. New York: Harper and Row.

Eaton, S. B., Pike, M. C., Short, R. V., et al. (1994). Women's reproductive cancers in evolutionary context. *Quarterly Review of Biology*, **69**, 353–367.

Ewald, P. W. (1980). Evolutionary biology and treatment of signs and symptoms of infectious disease. *Journal of Theoretical Biology*, **86**, 169–176.

Ewald, P. W. (1994). *Evolution of Infectious Disease*. New York: Oxford University Press.

Ewald, P. W. (2000). *Plague Time. How Stealth Infections Cause Cancer, Heart Disease, and Other Deadly Ailments*. New York: Free Press.

Ewald, P. W. (2008). An evolutionary perspective on the causes of chronic diseases: atherosclerosis as an illustration. In *Evolutionary Medicine and Health. New Perspectives*, W. R. Trevathan, E. O. Smith and J. J. Mckenna (eds). New York: Oxford University Press, pp. 350–367.

Ewald, P. W. (2009). An evolutionary perspective on parasitism as a cause of cancer. *Advances in Parasitology*, **61**, 28–43.

Ewald, P. W. and Cochran, G. M. (2000). *Chlamydia pneumoniae* and cardiovascular disease: an evolutionary perspective on infectious causation and antibiotic treatment. *Journal of Infectious Diseases*, **181**, S394–S401.

Fina, F., Romain, S., Ouafik, L. H., et al. (2001). Frequency and genome load of Epstein–Barr virus in 509 breast cancers from different geographical areas. *British Journal of Cancer*, **84**, 783–790.

Finch, C. E. (2007). *The Biology of Human Longevity: Inflammation, Nutrition, and Aging in the Evolution of Lifespans*. Boston: Academic.

Finch, C. E. and Stanford, C. B. (2004). Meat-adaptive genes and the evolution of slower aging in humans. *Quarterly Review of Biology*, **79**, 3–50.

Flaxman, S. M. and Sherman, P. W. (2000). Morning sickness: a mechanism for protecting mother and embryo. *Quarterly Review of Biology*, **75**, 113–148.

Frank, S. A. (2007). *Dynamics of Cancer. Incidence, Inheritance, and Evolution*. Princeton: Princeton University Press.

Fujimoto, Y., Furusyo, N., Toyoda, K., et al. (2007). Intrafamilial transmission of *Helicobacter pylori* among the population of endemic areas in Japan. *Helicobacter*, **12**, 170–176.

Fullerton, S. M., Clark, A. G., Weiss, K. M., et al. (2000). Apolipoprotein E variation at the sequence haplotype level: implications for the origin and maintenance of a major human polymorphism. *American Journal of Human Genetics*, **67**, 881–900.

Gardner, C. D., Lawson, L. D., Block, E., et al. (2007). Effect of raw garlic vs. commercial garlic supplements on plasma lipid concentrations in adults with moderate hypercholesterolemia: a randomized clinical trial. *Archives of Internal Medicine*, **167**, 346–353.

Gérard, H. C., Wang, G. F., Balin, B. J., et al. (1999). Frequency of apolipoprotein E (APOE) allele types in patients with *Chlamydia*-associated arthritis and other arthritides. *Microbial Pathogenesis*, **26**, 35–43.

Gérard, H. C., Fomicheva, E., Whittum-Hudson, J. A., et al. (2008). Apolipoprotein E4 enhances attachment of *Chlamydophila* (*Chlamydia*) *pneumoniae* elementary bodies to host cells. *Microbial Pathogenesis*, **44**, 279–285.

Gerber, L. M. and Crews, D. E. (1999). Evolutionary perspectives on chronic degenerative diseases. In *Evolutionary Medicine*, W. R. Trevathan, E. O. Smith and J. J. McKenna (eds). New York: Oxford University Press, pp. 443–469.

Greaves, M. (2000). *Cancer. The Evolutionary Legacy*. Oxford: Oxford University Press.

Greaves, M. (2006). Infection, immune responses and the aetiology of childhood leukaemia. *Nature Reviews Cancer*, **3**, 193–203.

Greaves, M. (2008). Cancer: evolutionary origins of vulnerability. In *Evolution in Health and Disease*, S.S. Stearns and J.C. Koella (eds). New York: Oxford University Press, pp. 277–287.

Guasparri, I., Bubman, D. and Cesarman, E. (2008). EBV LMP2A affects LMP1-mediated NF-κB signaling and survival of lymphoma cells by regulating TRAF2 expression. *Blood*, **111**, 3813–3820.

Haig, D. (1993). Genetic conflicts in human pregnancy. *Quarterly Review of Biology*, **68**, 495–532.

Haig, D. (2008). Intimate relations: evolutionary conflicts of pregnancy and childhood. In *Evolution in Health and Disease*, S.S. Stearns and J.C. Koella (eds). New York: Oxford University Press, pp. 65–76.

Hino, R., Uozaki, H., Inoue, Y., et al. (2008). Survival advantage of EBV-associated gastric carcinoma: surviving up-regulation by viral latent membrane protein 2A. *Cancer Research*, **68**, 1427–1435.

Hinze-Selch, D., Däubener, W., Eggert, L., et al. (2007). A controlled prospective study of *Toxoplasma gondii* infection in individuals with schizophrenia: beyond seroprevalence. *Schizophrenia Bulletin*, **33**, 782–788.

Hook, E.B. (1976). Changes in tobacco smoking and ingestion of alcohol and caffeinated beverages during early pregnancy: are these consequences in part of feto-protective mechanisms diminishing maternal exposure to embryo toxins. In *Birth Defects, Risks and Consequences*, S. Kelly (ed.). New York: Academic, pp. 173–183.

Ilveskoski, E., Perola, M., Lehtimaki, T., et al. (1999). Age-dependent association of apolipoprotein E genotype with coronary and aortic atherosclerosis in middle-aged men. An autopsy study. *Circulation*, **100**, 608–613.

Jacobson, G.F., Autry, A.M., Somer-Shely, T.L., et al. (2003). *Helicobacter pylori* seropositivity and hyperemesis gravidarum. *Journal of Reproductive Medicine*, **48**, 578–582.

Jenson, A.B., Geyer, S., Sundberg, J.P., et al. (2001). Human papillomavirus and skin cancer. *Journal of Investigative Dermatology Symposium Proceedings*, **6**, 203–206.

Jerwood, S. and Cohen, J. (2008). Unexpected antimicrobial effect of statins. *Journal of Antimicrobial Chemotherapy*, **61**, 362–364.

Kan, C.-Y., Iacopetta, B., Lawson, J., et al. (2005). Identification of human papillomavirus DNA gene sequences in human breast cancer. *British Journal of Cancer*, **93**, 946–948.

Karaer, A., Ozkan, O., Ozer, S., et al. (2008). Gastrointestinal symptoms and *Helicobacter pylori* infection in early pregnancy. A seroepidemiologic study. *Gynecologic and Obstetric Investigation*, **66**, 44–46.

Karagas, M.R., Nelson, H.H., Sehr, P., et al. (2006). Human papillomavirus infection and incidence of squamous cell and basal cell carcinomas of the skin. *Journal of the National Cancer Institute*, **98**, 1425–1426.

Karagas, M.R., Nelson, H.H., Zens, M.S., et al. (2007). Squamous cell and basal cell carcinoma of the skin in relation to radiation therapy and potential modification of risk by sun exposure. *Epidemiology*, **18**, 776–784.

Katz, E., Lareef, M.H., Rassa, J.C., et al. (2005). MMTV Env encodes an ITAM responsible for transformation of mammary epithelial cells in three-dimensional culture. *Journal of Experimental Medicine*, **201**, 431–439.

Kazerooni, T., Taallom, M. and Ghaderi, A.A. (2002). *Helicobacter pylori* seropositivity in patients with hyperemesis gravidarum. *International Journal of Gynaecology and Obstetrics*, **79**, 217–220.

Key, T.J.A. and Pike, M.C. (1988). The role of oestrogens and progestagens in the epidemiology and prevention of breast cancer. *European Journal of Cancer and Clinical Oncology*, **24**, 29–43.

Kleer, C., Tseng, M., Gutsch, D., et al. (2002). Detection of Epstein–Barr virus in rapidly growing fibroadenomas of the breast in immunosuppressed hosts. *Modern Pathology*, **15**, 759–764.

Kleinsmith, L.J. (2005). *Principles of Cancer Biology*. San Francisco: Pearson/Benjamin Cummings.

Knight, J., Sharma, N. and Robertson, E. (2005). Epstein–Barr virus latent antigen 3C can mediate the degradation of the retinoblastoma protein through an SCF cellular ubiqitin ligase. *Proceedings of the National Academy of Sciences of the United States of America*, **102**, 18562–18566.

Koçak, I., Akcan, Y., Ustün, C., et al. (1999). *Helicobacter pylori* seropositivity in patients with hyperemesis gravidarum. *International Journal of Gynaecology and Obstetrics*, **66**, 251–254.

Komarova, N.L. and Wodarz, D. (2008). Cancer as a micro-evolutionary process. In *Evolution in Health and Disease*, S.S. Stearns and J.C. Koella (eds). New York: Oxford University Press, pp. 289–299.

Lambe, M., Hsieh, C., Trichopoulos, D., et al. (1994). Transient increase in risk of breast cancer after giving birth. *New England Journal of Medicine*, **331**, 5–9.

Lawson, J.S., Tran, D. and Rawlinson, W.D. (2001). From Bittner to Barr: a viral, diet and hormone breast cancer aetiology hypothesis. *Breast Cancer Research*, **3**, 81–85.

Ledgerwood, L.G., Ewald, P.W. and Cochran, G.M. (2003). Genes, germs, and schizophrenia: an evolutionary perspective. *Perspectives in Biology and Medicine*, **46**, 317–348.

Lee, R.H., Pan, V.L. and Wing, D.A. (2005). The prevalence of *Helicobacter pylori* in the Hispanic population affected by hyperemesis gravidarum. *American Journal of Obstetrics Gynecology*, **193**, 1024–1027.

Lencz, T., Morgan, T.V., Athanasiou, M., et al. (2007). Converging evidence for a pseudoautosomal cytokine receptor gene locus in schizophrenia. *Molecular Psychiatry*, **12**, 572–580.

Leonard, W.R. (2008). Lifestyle, diet, and disease: comparative perspectives on the determinants of chronic health risks. In *Evolution in Health and Disease*, S.S. Stearns and J.C. Koella (eds). New York: Oxford University Press, pp. 265–276.

Liu, W.K., Chu, Y.L., Zhang, F., et al. (2005). The relationship between HPV16 and expression of CD44v6, nm23H1 in esophageal squamous cell carcinoma. *Archives of Virology*, **150**, 991–1001.

Liu, X., Roberts, J., Dakic, A., et al. (2008). HPV E7 contributes to the telomerase activity of immortalized and tumorigenic cells and augments E6-induced hTERT promoter function. *Virology*, **375**, 611–623.

Mahley, R.W. and Huang, Y.D. (1999). Apolipoprotein E: from atherosclerosis to Alzheimer's disease and beyond. *Current Opinion in Lipidology*, **10**, 207–217.

Maula, S., Huuhtanen, R. L., Blomqvist, C. P., et al. (2001). The adhesion molecule CD44v6 is associated with a high risk for local recurrence in adult soft tissue sarcomas. *British Journal of Cancer*, **84**, 244–252.

Mays, M. E. and Dujovne, C. A. (2008). Pleiotropic effects: should statins be considered an essential component in the treatment of dyslipidemia? *Current Atherosclerosis Reports*, **10**, 45–52.

McKenna, D., Watson, P. and Dornan, J. (2003). *Helicobacter pylori* infection and dyspepsia in pregnancy. *Obstetrics and Gynecology*, **102**, 845–849.

Merlo, L. M., Pepper, J. W., Reid, B. J., et al. (2006). Cancer as an evolutionary and ecological process. *Nature Reviews Cancer*, **6**, 924–935.

Mileo, A. M., Piombino, E., Severino, A., et al. (2006). Multiple interference of the human papillomavirus-16 E7 oncoprotein with the functional role of the metastasis suppressor Nm23-H1 protein. *Journal of Bioenergetics and Biomembranes*, **38**, 215–225.

Mortensen, P. B., Nørgaard-Pedersen, B., Waltoft, B. L., et al. (2007). Early infections of *Toxoplasma gondii* and the later development of schizophrenia. *Schizophrenia Bulletin*, **33**, 741–744.

Murakami, M., Lan, K., Subramanian, C., et al. (2005). Epstein–Barr virus nuclear antigen 1 interacts with Nm23-H1 in lymphoblastoid cell lines and inhibits its ability to suppress cell migration. *Journal of Virology*, **79**, 1559–1568.

Nesse, R. M. (2008). The importance of evolution for medicine. In *Evolutionary Medicine and Health. New Perspectives*, W. R. Trevathan, E. O. Smith and J. J. Mckenna (eds). New York: Oxford University Press, pp. 416–433.

Nesse, R. M. and Stearns, S. C. (2008). The great opportunity: evolutionary applications to medicine and public health. *Evolutionary Applications*, **1**, 28–48.

Nesse, R. M. and Williams, G. C. (1995). *Why We Get Sick. The New Science of Darwinian Medicine*. New York: Times Books.

Niebuhr, D. W., Millikan, A. M., Cowan, D. N., et al. (2008). Selected infectious agents and risk of schizophrenia among US military personnel. *American Journal of Psychiatry*, **165**, 99–106.

Noyan, V., Apan, T. Z., Yucel, A., et al. (2004). Cytotoxin associated gene *A*-positive *Helicobacter pylori* strains in dyspeptic pregnant women. *European Journal of Obstetrics, Gynecology and Reproductive Biology*, **116**, 186–189.

Ouatas, T., Clare, S. E., Hartsough, M. T., et al. (2002). MMTV-associated transcription factor binding sites increase nm23-H1 metastasis suppressor gene expression in human breast carcinoma cell lines. *Clinical and Experimental Metastasis*, **19**, 35–42.

Perry, S., De La Luz Sanchez, M., Yang, S., et al. (2006). Gastroenteritis and transmission of *Helicobacter pylori* infection in households. *Emerging Infectious Diseases*, **12**, 1701–1708.

Potten, C. S., Watson, R. J., Williams, G. T., et al. (1988). The effect of age and menstrual cycle upon proliferative activity in the normal human breast. *British Journal of Cancer*, **58**, 163–170.

Profet, M. (1992). Pregnancy sickness as adaptation: a deterrent to maternal ingestion of teratogens. In *The Adapted Mind*, J. E. Barkow, L. Cosmides, et al. (eds). New York: Oxford University Press, pp. 327–365.

Reeth, K. V., Brown, I., Essen, S., et al. (2004). Genetic relationships, serological cross-reaction and cross-protection between H1N2 and other influenza A virus subtypes endemic in European pigs. *Virus Research*, **103**, 115–124.

Reymunde, A., Santiago, N. and Perez, L. (2001). *Helicobacter pylori* and severe morning sickness. *American Journal of Gastroenterology*, **96**, 2279–2280.

Russo, J. (1992). Influence of age and parity on the development of the human breast. *Breast Cancer Research and Treatment*, **23**, 211–218.

Saal, L. H., Gruvberger-Saal, S. K., Persson, C., et al. (2008). Recurrent gross mutations of the PTEN tumor suppressor gene in breast cancers with deficient DSB repair. *Nature Genetics*, **40**, 102–107.

Sgouros, J., Galani, E., Gonos, E., et al. (2007). Correlation of nm23-H1 gene expression with clinical outcome in patients with advanced breast cancer. *In Vivo*, **21**, 519–522.

Sherman, P. W. and Flaxman, S. M. (2002). Nausea and vomiting of pregnancy in an evolutionary perspective. *American Journal of Obstetrics and Gynecology*, **186**, S190–S197.

Shirin, H., Sadan, O., Shevah, O., et al. (2004). Positive serology for *Helicobacter pylori* and vomiting in the pregnancy. *Archives of Gynecology and Obstetrics*, **270**, 10–14.

Sinha, S. K., Martin, B., Gold, B. D., et al. (2004). The incidence of *Helicobacter pylori* acquisition in children of a Canadian First Nations community and the potential for parent-to-child transmission. *Helicobacter*, **9**, 59–68.

Siqueira, F. M., Cota, L. O., Costa, J. E., et al. (2008). Maternal periodontitis as a potential risk variable for pre-eclampsia: a case-control study. *Journal of Periodontology*, **79**, 207–215.

Stearns, S. S. and Koella, J. C. (eds) (2008). *Evolution in Health and Disease*, 2nd edn. New York: Oxford University Press.

Stearns, S. C., Nesse, R. M. and Haig, D. (2008). Introducing evolutionary thinking for medicine. In *Evolution in Health and Disease*, S. S. Stearns and J. C. Koella (eds). New York: Oxford University Press, pp. 1–15.

Subramanian, C. and Robertson, E. S. (2002). The metastatic suppressor Nm23-H1 interacts with EBNA3C at sequences located between the glutamine- and proline-rich domains and can cooperate in activation of transcription. *Journal of Virology*, **76**, 8702–8709.

Taubes, G. (2008). What's cholesterol got to do with it? *New York Times*. January 27, p. 18.

Trevathan, W. R., Smith, E. O. and Mckenna, J. J. (2008). Background. In *Evolutionary Medicine and Health. New Perspectives*, W. R. Trevathan, E. O. Smith and J. J. Mckenna (eds). New York: Oxford University Press, pp. 1–54.

Trichopoulos, D., Li, F. and Hunter, D. (1996). What causes cancer? *Scientific American*, **275**, 80–87.

Tungteakkhun, S. S. and Duerksen-Hughes, P. J. (2008). Cellular binding partners of the human papillomavirus E6 protein. *Archives of Virology*, **153**, 397–408.

Wang, Y., Holland, J., Bleiweiss, I., et al. (1995). Detection of mammary tumor virus *ENV* gene-like sequences in human breast cancer. *Cancer Research*, **55**, 5173–5179.

Weyermann, M., Brenner, H., Adler, G., et al. (2003). *Helicobacter pylori* infection and the occurrence and severity of gastrointestinal symptoms during pregnancy. *American Journal of Obstetrics and Gynecology*, **189**, 526–531.

Williams, E., Jones, L., Vessey, M., et al. (1990). Short term increase in risk of breast cancer associated with full term pregnancy. *British Medical Journal*, **300**, 578–579.

Williams, G. C. and Nesse, R. M. (1991). The dawn of Darwinian medicine. *Quarterly Review of Biology*, **66**, 1–22.

Wu, C. Y., Tseng, J. J., Chou, M. M., et al. (2000). Correlation between *Helicobacter pylori* infection and gastrointestinal symptoms in pregnancy. *Advances in Therapy*, **17**, 152–158.

Yang, Y. J., Sheu, B. S., Lee, S. C., et al. (2005). Children of *Helicobacter pylori*-infected dyspeptic mothers are predisposed to *H. pylori* acquisition with subsequent iron deficiency and growth retardation. *Helicobacter*, **10**, 249–255.

30 Beyond Feast–Famine: Brain Evolution, Human Life History, and the Metabolic Syndrome

Christopher W. Kuzawa

EXPLAINING THE MODERN METABOLIC DISEASE EPIDEMIC: THE THRIFTY GENOTYPE HYPOTHESIS AND ITS LIMITATIONS

Today, more than 1 billion people are overweight or obese, and the related condition of cardiovascular disease (CVD) accounts for more deaths than any other cause (Mackay et al., 2004). Why this epidemic of metabolic disease has emerged so rapidly in recent history is a classic problem for anthropologists concerned with the role of culture change in disease transition (Ulijaszek and Lofink, 2006). In 1962, the geneticist James Neel proposed an explanation for this phenomenon that looked for clues in the "feast–famine" conditions that he believed our nomadic, foraging ancestors faced in the past. Given the unpredictability of food resources in natural ecologies, Neel suggested that a "thrifty" metabolism capable of efficiently storing excess dietary energy as body fat when food was abundant would have provided a survival advantage during later periods of shortage. In the wake of the rapid dietary and lifestyle change in recent generations, and the comparatively slow pace of genetic change, these foraging-adapted genes would now be "rendered detrimental by progress" (Neel, 1962), leading to obesity and diabetes. While a useful way to think about obesity generally, variations on this idea have been proposed to help explain the high rates of metabolic disease among populations believed to have experienced especially severe nutritional conditions in the past, including the diabetes-prone Pima Indians of the American Southwest (Knowler et al., 1982), and several groups of South Pacific Islanders who have among the highest rates of obesity in the world (Dowse and Zimmet, 1993; McGarvey, 1994).

The thrifty gene hypothesis was one of the first uses of evolutionary reasoning to shed light on a human disease, and is thus a classic example of evolutionary or "Darwinian" medicine (see Nesse and Williams, 1994; Stearns and Koella, 2008; Trevathan et al.,

2008). The idea that obesity and related diseases result from a "discordance" or "mismatch" between our ancient genes and our rapidly changing lifestyle and diet is now widely accepted, and it seems clear that obesity must be more common today in large part because we are eating more and expending less than our ancestors traditionally did (Eaton and Konner, 1985; see Chapter 28 of this volume). Despite the intuitive appeal of these ideas, the hypothesis is not without limitation. For one, it helps explain why we gain weight in a modern environment of nutritional abundance, but says very little about the syndrome of metabolic changes that account for the diseases that accompany obesity (Vague, 1955; Kissebah et al., 1982). While excess weight gain in general is unhealthy, it is specifically fat deposition in the visceral or abdominal region that accounts for the bulk of its adverse effects on health. Visceral fat has unique metabolic properties that contribute to prediabetes states like insulin resistance when deposition in this depot is excessive (Reaven, 1988; Wellen and Hotamisligil, 2003). The simple idea that our bodies are famine adapted may help explain why we are prone to gaining weight when we eat too much, but it says little about the metabolic symptoms that make weight gain unhealthy.

The sole focus of the hypothesis on genes is also outdated in light of newer evidence that biological susceptibility of developing these adult diseases is also elevated among individuals who experienced poor nutrition during early life (Barker et al., 1989; Gluckman and Hanson, 2006). A large research literature has demonstrated that the body's responses to early life nutrition subsequently influences that individual's risk for developing adult diseases like diabetes and CVD, a process described as nutritional "programming" (Lucas, 1991) or "induction" (Bateson, 2001). For instance, CVD and the rate of CVD mortality in adulthood are inversely related to size at birth – a measure of fetal nutrition – in places like the UK, Sweden, India, and the Philippines (Kuzawa and Adair, 2003;

Human Evolutionary Biology, ed. Michael P. Muehlenbein. Published by Cambridge University Press. © Cambridge University Press 2010.

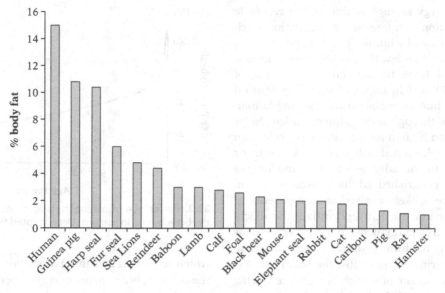

30.1. Percentage of body fat at birth in mammals. Adapted from Kuzawa (1998).

Yajnik, 2004; Lawlor et al., 2005). Individuals who were born small also experience durable changes in biological systems that contribute to CVD, as reflected in their higher risk of depositing fat in the visceral region, or of developing high blood pressure, impaired glucose tolerance or diabetes, or high cholesterol (Adair et al., 2001; Kuzawa and Adair, 2003). When the dietary intake of pregnant rats and sheep is restricted, their offspring have much the same set of outcomes as we see in humans born small (McMillen and Robinson, 2005; Fernandez-Twinn and Ozanne, 2006). This finding suggests that the body has an ability to induce a "thrifty phenotype" that is better suited to survival in nutrition-ally challenging environments. It also illustrates why a purely gene-based model of metabolic disease evolution is incomplete (Hales and Barker, 1992).

A third limitation of the thrifty genotype hypothesis is that it merely *assumes* that famine was the key source of selection on human metabolism without con-sidering the actual causes of nutritional stress and the ages at which it is most severe (Kuzawa, 1997). There are reasons to question whether our distant foraging ancestors experienced a major burden of famine, which archaeological and contemporary ethnographic data suggest only emerged as an important problem after the evolutionarily recent development of agri-cultural subsistence (Cohen and Armelagos, 1984; Benyshek and Watson, 2006). Given this, famine may have been a relatively weak force of selection on the human gene pool (Speakman, 2006).

Additional support for this interpretation comes from the growth pattern of body fat in humans, shown in Figure 30.1. In humans, body fat makes up a larger percentage of weight at birth than in any other

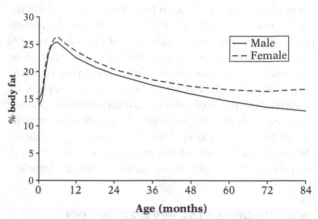

30.2. Age changes in percentage body fat in humans. Data from Davies and Preece (1989), pp. 95–97.

mammal studied thus far (Kuzawa, 1998). This is followed by a continued period of rapid fat deposition during the early postnatal months. In well-nourished populations, adiposity reaches peak levels during the first year of life before gradually declining to a nadir in childhood, when humans reach their lowest level of body fat in the life cycle (Figure 30.2). If the threat of famine is what drove the human tendency to build up fat reserves, it is not obvious why children's bodies should do so little to prepare for these difficult periods. The lower priority placed upon maintaining an energy reserve by middle childhood suggests that the back-ground risk of starvation faced by our ancestors – "famine" – was small in comparison to the nutritional stress experienced during the preceding developmental period. A prereproductive life history stage that does not prioritize energy storage is difficult to reconcile with the assumption that famine was the major nutritional challenge faced by human populations.

In this chapter, I review the causes, consequences, and adaptive solutions to the nutritional stress of infancy and childhood in hopes of shedding light on the evolution of human metabolism. Viewing human nutritional stress through a developmental lens helps identify ages when human metabolism has likely been under strongest selection. I will show how two traits that are central to the adverse health consequences of obesity in overnourished adults – visceral fat and insulin resistance – likely evolved as components of an energy backup system that reprioritizes the allocation of prized energy and glucose during a crisis, thus allowing the body to shunt resources from nonessential functions to critical organs like the brain (for discussion of the function of insulin resistance during pregnancy, see Haig, 1993). Demands placed on this system are greater during infancy and early childhood than at any other age of the life cycle, suggesting that some of the evolutionary seeds of adult metabolic disease may trace to selection pressures operative during early life. In addition, evidence for developmental plasticity in this backup system and the adult diseases that it contributes to suggests that its priorities may be adjusted in response to nutritional and other stressors experienced early in the life cycle. A developmental approach to the evolution of human metabolic disease underscores the importance of the body's internal strategies of allocating finite resources – in addition to the external stress of famine – as key to understanding the contemporary epidemic of metabolic disease.

A DEVELOPMENTAL PERSPECTIVE ON ENERGY STRESS IN HUMAN EVOLUTION: THE DEVELOPMENTAL BOTTLENECK

Humans have been described as naturally obese (Pond, 1997), a characterization which is particularly fitting at birth as human newborns are born with more body fat than any other species (Kuzawa, 1998). Attempts to explain our unusual "baby fat" have traditionally looked to our hairlessness for clues, and it is widely assumed that natural selection compensated for our loss of fur with a layer of insulative body fat (e.g., Hardy, 1960; Morris, 1967). A competing perspective notes that this excess adipose tissue is well-suited to serve as a backup energy supply for another distinctive human trait: our large brains (Kuzawa, 1998). Brains have among the highest metabolic rates of any tissue or organ in the body, and they are quickly damaged in the event of even temporary disruption in energy supply. Neuronal tissues are costly because they must be maintained in a state far from thermodynamic equilibrium,

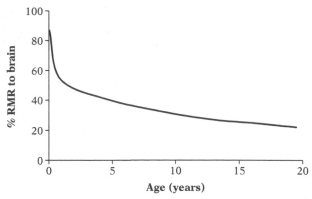

30.3. Percentage of resting metabolic rate (RMR) devoted to the brain by age in humans. Data adapted from Holliday (1986).

which requires a constant redistribution of ions across the cell membrane using energy-dependent ion pumps. Humans are exceptional in the quantity of this costly tissue that they must sustain, and it has been estimated that greater than 80% of the body's metabolism is devoted to the brain in the newborn (Figure 30.3).

It is a notable feature of human metabolism that total metabolic rate measured in adulthood (for instance, calories expended per day) is not increased above that of other mammals of similar body size despite our highly encephalized state (Armstrong, 1983). Thus, brain expansion must have been matched by a reduction in other energetically expensive tissues (Armstrong, 1983). Human evolution was marked by a movement to more energy-dense and easy to digest foods (Leonard and Robertson, 1992, 1994), which may have allowed a reduction in the size and energy requirements of the metabolically costly gut (Aiello and Wheeler, 1995). This reduction in gut expenditure may have helped offset our increased brain needs in adult life (Aiello and Wheeler, 1995), but this seems less likely during infancy and childhood. The size of the brain relative to the body is far larger in the human neonate and infant than in the adult, and as a result, the body's total energy requirements are likely elevated at this age relative to other similarly sized mammalian and primate newborns (Foley and Lee, 1991). Moreover, unlike energy expended on other tissues or systems, brain metabolism may not be reduced to conserve energy during a period of shortage, but instead must be maintained within narrow limits to avoid permanent damage (Owen et al., 1967). Thus, our large brains impose a double burden on metabolism during infancy: they increase demand for energy while restricting flexibility in metabolic expenditure when nutritional supply is disrupted.

Other factors that are commonly experienced during infancy can impede the supply of nutrients, ensuring that negative energy balance is a regular occurrence at this age in most populations. We are born with a naïve adaptive immune system, and must therefore come into contact with (and be infected by) specific pathogens to acquire the repertoire of antibodies necessary to protect us from future infection. Initially, newborns enjoy immune protection from several maternal sources. The first is in the form of maternal immunoglobulin G (IgG) antibodies acquired across the placenta. In addition, exclusively breast-fed infants are shielded from exposure to pathogens, and also enjoy passive immune protection from maternal secretory antibodies (sIgA) in breast milk. As a result, newborns are often quite healthy in the early postnatal months. However, both sources of passive immune protection eventually wane. As energy requirements outstrip the supply capacity of breast milk by roughly 6 months of age, less sterile complementary foods must be introduced, and infectious disease becomes unavoidable in all but the most sanitary environments. These childhood infections, in turn, are a source of nutritional stress, and indeed, it is primarily through their effects on nutritional status that they compromise health and contribute to mortality during infancy and childhood (see Scrimshaw, 1989, for review). Once sick, a child loses appetite and this may be compounded in some cultures by the withholding of food by caretakers (Scrimshaw, 1989). The common diarrheal diseases reduce nutrient absorption and digestion, while the fevers associated with many viral infections can increase metabolic rate and thus energy expenditure. The specific symptoms may vary by illness, but the pattern of nutritional depletion that accompanies infections has the effect of suppressing immune function, leaving the infant more prone to future infection and a compounding cycle of nutritional stress (Pelletier et al., 1995).

The human infant thus faces a profound energetic dilemma: at precisely the age when they are most dependent upon provisioning by caretakers to maintain the high and obligatory energy needs of their large brains, they are likely to be cut off from that supply chain as a result of illness and the nutritional stresses of weaning. It is this confluence of factors, and the synergy between nutritional stress and compromised immunity, that accounts for much of the high infant mortality in many societies (Pelletier et al., 1995). Natural selection likely favored neonatal adiposity as a strategy to prepare for these difficult periods. It is easy to imagine how infants with a genetic predisposition to deposit copious quantities of energy as fat prior to weaning would be better represented among the subset who survive to adulthood to reproduce and pass on their genes (Kuzawa, 1998).

It is important to note that these sources of energy stress have largely receded by mid childhood: children have already acquired antibodies against the major pathogens that they are likely to confront. As a result,

infections and periods of negative energy balance decline to a small fraction of their high prevalence during the postweaning period. Because older children are far less likely to have to rely upon energy reserves for survival, it is easy to see why the human body places lower priority on maintaining sizeable body fat stores by this age.

Thus, the unique energetic stressors of early life have left an imprint on the developmental pattern of fat deposition in humans, represented by intensive investment in the tissue in the run-up to the stresses of weaning, and a gradual decline in the priority given to maintaining this reserve at older and more resilient ages. But this merely shows that the body is prioritizing the allocation of energy to an energetic buffer. This energy backup must have also been accompanied by the evolution of an efficient distribution and delivery system. Because the brain accounts for 50–80% of the body's energy usage during the first few years of life, we should expect that this delivery system – human metabolism – will be organized around the goal of ensuring a constant supply of energy to this fragile and costly organ. The availability of dietary nutrients is variable and often unpredictable, and as we will see next, human metabolism manages this risk by rapidly modifying energy allocation in response to changes in intake and expenditure. Multiple cues signaling nutritional stress induce a similar response that achieves a common end: stored fats are mobilized for use, while glucose is spared for the brain at the expense of less critical functions like muscle.

TISSUE-SPECIFIC INSULIN RESISTANCE AS A LIFE HISTORY STRATEGY

Although long-term changes in fat stores are monitored by the brain through the effects of fat-derived hormones like leptin, short-term changes in energy metabolism are regulated in a more distributed fashion by organs like the pancreas, and also by the tissues themselves, which alter their own uptake and use of glucose in response to changes in circulating hormones and the supply of nutrients. A sudden glut of glucose or other substrate in the blood stream is sensed by the beta cells of the pancreas, which secrete insulin into the circulation. Insulin stimulates glucose uptake by tissues throughout the body, initiating a negative feedback process that helps re-establish normal basal glucose concentrations. The tissues that are not responsive to insulin are few, but prominently include the organ most vulnerable to energy shortage – the brain (Seaquist et al., 2001). While this may at first seem paradoxical, this tissue-specific pattern of insulin sensitivity provides the body with the ability to modify partitioning of glucose between the brain and other tissues by increasing

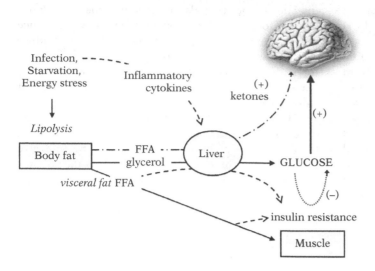

30.4. How body fat and insulin resistance protect the brain. During a crisis, stored triglycerides are broken down into free fatty acids (FFA) and glycerol. Glycerol enters the liver as a gluconeogenic substrate, thus increasing the pool of available glucose. Free fatty acids enter the liver where they are converted to ketone bodies, which the brain uses as a substitute energy substrate. Free fatty acids also induce insulin resistance in the liver and muscle, which spares glucose for the brain. When the nutritional stress is caused by infection, inflammatory cytokines reach the liver and also spare glucose by inducing insulin resistance in liver and muscle.

or decreasing insulin-mediated glucose uptake in the periphery (e.g., muscle, liver, body fat).

How this system of resource partitioning works is well illustrated by cases of accidental insulin overdose. When insulin-dependent diabetics inject themselves with too much insulin, this can be fatal, because it "overfeeds" the insulin-sensitive periphery, leaving nothing to fuel the insulin-independent brain (Waring and Alexander, 2004). Reducing insulin-mediated uptake in the periphery has the opposite effect, as it reduces glucose use in tissues like muscle, leaving more to be delivered to the brain. The body has two means of decreasing peripheral glucose use to spare the brain when confronted with nutritional stress. The first is to reduce insulin production, which reduces glucose uptake in the periphery. Secondly, the body can influence where glucose is used with greater precision by increasing or decreasing insulin sensitivity on a tissue-by-tissue basis. Skeletal muscle is a key player in this flexible system. Because it is the largest consumer of glucose, the body's response to changes in energy status prominently involves modulating muscle glucose uptake by changing insulin sensitivity in this tissue. The brain is an important arbiter of this response. The brain may not be insulin sensitive, but it does express insulin receptors and can also sense the adequacy of the supply of glucose that it receives. In the event that cerebral glucose flux is attenuated, the brain helps ensure its own glucose supply by inducing peripheral insulin resistance via effects on the sympathetic nervous system and other pathways. In addition to these direct effects of the brain, some of the change in glucose partitioning is co-ordinated directly by the affected peripheral tissues, as discussed in greater detail below.

From the perspective of adult metabolic disease, this reduced insulin-stimulated uptake of glucose by skeletal muscle forces the body to increase – and

eventually deplete – its production of insulin in type II diabetes (formally noninsulin dependent diabetes, or adult-onset diabetes). However, adopting instead the perspective of a young organism managing a finite energy supply, insulin resistance in a tissue like skeletal muscle allows the body to shift its priorities of glucose allocation – away from peripheral tissues and toward the fragile and energy demanding brain (Figure 30.4).

HOW VISCERAL FAT AND INSULIN RESISTANCE HELP PROTECT THE BRAIN DURING A CRISIS

Given these design features, it is not surprising that insulin resistance is triggered during states associated with negative energy balance, such as starvation, infection, or trauma (Jensen et al., 1987; Childs et al., 1990). During a fast, free fatty acids (FFA) are first mobilized from the metabolically active visceral adipose depot, which is secreted into the portal circulation that drains into the liver. Fat mobilization from this depot is achieved by innervation of the tissue by sympathetic nerve fibers and, unlike other fat depots, these effects are relatively insensitive to the anti-fat mobilizing effects of insulin (van Harmelen et al., 2002; Hucking et al., 2003). The liver interprets the sudden appearance of FFA in the portal circulation as a signal that the body is being forced to mobilize fat from this depot, and reprioritizes metabolism to protect the supply of glucose to the brain. This is achieved by increasing hepatic glucose production while also inducing insulin resistance in the liver and in skeletal muscle, thus reducing glucose uptake (Kabir et al., 2005). In addition, FFA in the circulation are taken up to be used as energy by tissues like muscle, which directly induces insulin resistance (Roden et al., 1996).

Similar changes in how the body uses glucose are seen in response to the stress of infection or trauma. The body's first response to an infection includes inflammation, which helps mobilize immune and non-immune resources as a first line of defense against tissue invasion and injury (Fernandez-Real and Ricart, 1999, 2003). Inflammation is initiated in the liver in response to signals indicating tissue injury, such as the appearance of proteins that signal the presence of bacteria, or by molecules called cytokines produced by immune cells like macrophages. Interestingly, in obese individuals visceral adipose tissue is an important source of these cytokines, which, like FFA, are secreted from fat cells into the portal circulation where they induce a similar constellation of changes in energy metabolism as they reach the liver (Tsigos et al., 1999). Why these overfilled fat cells secrete proinflammatory cytokines is currently a focus of intensive research, and there are interesting leads. One is that adipocytes and the immune cells that secrete inflammatory cytokines (macrophages) are closely related cells that share similar patterns of gene expression (Wellen and Hotamisligil, 2003). It is also of interest that fat cells secrete some cytokines as they swell in size, perhaps helping the cell avoid overfilling or rupturing (McCarty, 1999). Regardless of the mechanism underlying the inflammatory effects of excessive body fat, it seems very likely that these cells are now sending signals that, in the past, would have been reliable indicators of infection or trauma rather than obesity. It is ironic to think that the metabolism of the overfed individual today, whose swollen fat cells produce high levels of cytokines, may be tricked by their obesity into sensing an inflammatory challenge or threat, inappropriately initiating the same constellation of metabolic adjustments designed to cope with starvation or infection.

What is clear is that multiple signals of nutritional stress or trauma – whether FFA or cytokines – help link the highly labile fat depot of the abdomen with metabolic adjustments co-ordinated by the liver (Figure 30.4). Not unlike the effects of FFA in the portal circulation, these cytokines induce a state of peripheral insulin resistance, while also mobilizing FFA stored in visceral fat for use as energy. These FFA also bathe the liver, and thus further contribute to insulin resistance through the pathways described above (Pickup and Crook, 1998; Orban et al., 1999).

The evolutionary perspective adopted here, emphasizing the importance of brain metabolism and the pattern and sources of nutritional stress, helps clarify the logic underlying metabolic adjustments that lie at the heart of the metabolic syndrome: when the body is confronted with a challenge to energy balance, such as starvation or infection, the less critical, peripheral tissues shift to using fats mobilized from adipose tissue, which saves the more easily metabolized and desirable energy substrate of glucose for use by the brain. As peripheral insulin resistance helps the body control where glucose is used, it seems likely that these general features of human metabolism were forged in the evolutionary crucible of early life nutritional stress. As emphasized above, the challenge of fueling the brain, and thus the need to reduce insulin-mediated uptake in the periphery during undernutrition or infection, is most acute, common, and life-threatening at this age. The need to buffer brain glucose must place an important constraint on any response to nutritional stress during fetal life, infancy, and early childhood, when the brain demands the equivalent of 50% or more of the body's total available energy (e.g., Childs et al., 1990).

HOW PRENATAL NUTRITION HELPS FINE TUNE THE SETTINGS OF THE BODY'S ENERGY BACKUP SYSTEM

As previously alluded to, there is now a great deal of evidence that individuals born small have higher risk of developing conditions like diabetes and CVD. The present discussion provides an opportunity to speculate on the function of these biological changes, for they prominently involve resetting how the body prioritizes traits like visceral adiposity and insulin resistance (for a review see Kuzawa et al., 2008). Although most human research in this field has documented relationships between early life nutrition and metabolic disease risk factors in later childhood, adolescence, or adulthood, a handful of studies have now investigated these same outcomes in infants and young children – the age when the allocation of scarce glucose may be most critical for survival, and thus, when these metabolic pathways may have been under particularly strong evolutionary pressure. What these studies find is that, compared to their age mates who were better-nourished prior to birth, individuals born small tend to put on more visceral fat and become more insulin resistant as they gain weight (Soto et al., 2003; Ibanez et al., 2006). Metabolic changes in glucose use and visceral fat are already detectable in the first year of life, showing that a body with a prior history of prenatal nutritional stress is not only prone to adult disease, but also handles its energy and substrate differently during late infancy and early childhood.

Why the body adopts this strategy when prenatal nutrition is scarce is uncertain, but two possibilities seem plausible. Firstly, glucose-sparing adjustments could be important for protecting the brain during prenatal life when intrauterine nutrition is compromised (Hales and Barker, 1992). The challenge of protecting the brain, as outlined above, does not begin at birth, and nutritional shortfall in utero can result when fetal demand for energy outstrips maternal supply across the placenta (Harding, 2001). By this reasoning, an adjustment made in utero to boost immediate survival may have unintended side-effects after birth (Hales and Barker, 1992).

The second and not mutually exclusive possibility is that some of the adjustments have been shaped for benefits accrued after birth (Gluckman et al., 2005). To the extent that poor prenatal nutrition or maternal stress are correlated with the world that the fetus will be born into, these signals could allow it to modify the priorities of energy use, including the settings of this backup system, to match the severity of postnatal nutritional stress or challenge that it is likely to face (for more on fetal nutrition as an anticipatory cue see Bateson, 2001; Kuzawa, 2005; Wells, 2007; Kuzawa, 2008). These adjustments could help the infant manage its precarious metabolic state and survive this developmental bottleneck of early life nutritional stress when conditions are difficult. But this shift in metabolic priorities would also have the effect of accentuating the metabolic derangements that accompany weight gain when nutrition is chronically abundant. When a previously undernourished body is confronted with excess nutrition, a strategy of prioritizing visceral fat deposition and sparing glucose becomes a liability, increasing risk of developing the metabolic syndrome, diabetes, and CVD as an adult. In other words, if there are "thrifty genes" they code not for static properties but for flexible strategies, or reaction norms. This ability to flex may help match the body's metabolism to nutritional stress early in life, but it also helps plant the seeds for metabolic diseases caused when nutrition is chronically abundant later in life.

BEYOND FEAST–FAMINE: FRAGILE BRAINS AND METABOLIC PLASTICITY

In proposing the thrifty gene hypothesis, Neel made the elegant point that we gain insights into modern human diseases by reconstructing the environments and stressors that our ancestors faced, as this provides a sense for what our bodies and biology are designed to expect. I have tried to show how we can approach human metabolism from a developmental perspective to hone this exercise in reverse-engineering. Rather than inferring past selection from the default clinical perspective of adult obesity or diabetes, the developmental approach developed in this chapter began by isolating ages of high mortality to identify when metabolism is most critical for survival. Identifying the sources of nutritional stress at the ages of peak nutritional mortality provides clues into the types of biological responses or strategies that would have been available to natural selection to work around this

stress, thereby increasing survival and thus genetic fitness. Because the nutritional stress of weaning occurs prior to reproductive maturity, it represents a "developmental bottleneck" through which any metabolic genes must first pass before being passed on to offspring (Kuzawa, 1998). This selection would help ensure that traits that increase early life survival would stabilize into the pattern of human ontogeny across many generations of differential survival. I have argued that body fat, especially the rapidly mobilized visceral fat depot, and the ability to reprioritize glucose allocation by inducing insulin resistance, may be two examples of metabolic traits that could be beneficial to an infant faced with the challenge of buffering its most critical functions like the brain during the common nutritional stressors confronted at this age.

The developmental influences on these metabolic traits show that the genes favored by natural selection do not determine the organism's metabolic state in a simple one-to-one fashion. Not only does adult health depend upon the interaction between one's genome and lifestyle, as long appreciated, but the body also appears to have a capacity to adjust the settings and priorities of its metabolism in response to early nutritional experiences and to cues conveyed by the mother across the placenta. These biological responses may be viewed as examples of the well-known importance of developmental plasticity as a mode of human adaptability (Lasker, 1969; Frisancho, 1993; also see Chapter 2 of this volume). Organisms must cope with ecological and social change on a variety of time scales, spanning rapid and reversible changes (e.g., breakfast followed by a fast until lunch) to much longer trends that take years, decades, generations, or longer to unfold (e.g., extended droughts, migrating to a new ecology, an ice age) (Potts, 1998). Genes are effective at tracking the slowest, most gradual changes in ecology and diet, while homeostasis buffers rapid and reversible changes. Developmental adjustments made in response to early life nutrition operate on an intermediate time scale, allowing the organism to change its settings more rapidly than could be achieved by natural selection, but in a fashion that is more durable and stable than what the body achieves via homeostasis (Kuzawa, 2005).

In this sense, our metabolisms may not be all that different from other systems that have a capacity to adjust long-term settings to local conditions via developmental responses, illustrated classically by the important effects of developmental experience on systems like the brain, the immune system, and the skeleton. Many of the body's systems are built from a genetic architecture that evolved through natural selection, but that – by design – rely upon information about local ecology acquired during early development to complete their construction. There is little doubt that the size, configuration, and attachments of the body's skeletal elements evolved through natural selection operating on gene frequencies. However, because individuals are idiosyncratic in their behavior and thus the mechanical loadings that they will impose on their skeletons, natural selection constructed a system that has an exquisite capacity to compensate for biomechanical strain, and to organize developmentally around the pattern of use and disuse in the individual. Perhaps it should come as no surprise that the body's priorities of energy allocation, which are so critical for survival, should have a similar capacity. After all, humans do not merely vary in their culture, pathogen ecology, and pattern of physical activity, but also in their diet, nutritional sufficiency, and exposure to metabolically demanding stressors. These physiological and metabolic "loadings" may be hidden from view, but they are just as critical as components of the individual's adaptive strategy. The ability to adjust metabolic settings to local conditions might help humans prepare for important adaptive challenges, like the severity of infant and early childhood nutritional stress outlined here. However, this same flexibility may also carry longer-term costs to health in modern environments if scarcity-adapted metabolic priorities adopted early in life are later confronted with chronic overnutrition and weight gain (Gluckman et al., 2005).

It remains to be determined which of these ideas about the function of these traits are correct in detail. What does seem clear is that our metabolisms are designed to do more than survive the crisis of famine. Identifying the ages and causes of nutritional stress and nutritional mortality – both external and internal to the body itself – will be critical if we hope to understand how natural selection responded to this stress, and the legacy of these adaptations for the current global plague of metabolic disease.

DISCUSSION POINTS

1. Human babies are born with more body fat than any other mammal – including seals. What might explain the evolution of this unusual trait?
2. The human brain is energetically very costly, especially early in the life cycle. What are some of the ways that evolution may have changed human biology to allow the evolution of a large brain size in humans?
3. It is well known that diseases like diabetes and cardiovascular disease "run in families," reflecting a genetic contribution. Yet, these diseases have emerged as major public health problems in a few short generations – which is not enough time for gene frequencies to change substantially. How do we reconcile these observations?
4. This chapter argues that developmental adaptations made by the fetus to nutritional stress could contribute to a higher risk for diabetes later in life. What are these adaptations and how might they be beneficial early in life?
5. Chronic diseases often negatively impact health and survival late in the life cycle – when one's genome has already been passed on via offspring to the next generation. Discuss how selection pressures operating early in the life cycle might influence the evolution of diseases with late life negative impacts.
6. If we ask "What are the causes of the diabetes epidemic?", a public-health or medicine-inspired answer to this question might note that many people are eating too much and gaining weight, among other factors, which leads to insulin resistance. What types of answers does an evolutionary approach to this problem inspire, and how are they different from a public health or medical approach?

ACKNOWLEDGMENTS

Dan Benyshek, Dan Eisenberg, Peter Ellison, Lee Gettler, David Haig, Elizabeth Quinn, Melanie Vento, and two anonymous reviewers provided helpful comments on various drafts of this paper.

REFERENCES

Adair, L. S., Kuzawa, C. W. and Borja, J. (2001). Maternal energy stores and diet composition during pregnancy program adolescent blood pressure. *Circulation*, **104**, 1034–1039.

Aiello, L. and Wheeler, P. (1995). The expensive-tissue hypothesis; the brain and the digestive system in human and primate evolution. *Current Anthropology*, **36**, 199–221.

Armstrong, E. (1983). Relative brain size and metabolism in mammals. *Science*, **220**, 1302–1304.

Barker, D. J., Osmond, C., Golding, J., et al. (1989). Growth in utero, blood pressure in childhood and adult life, and mortality from cardiovascular disease. *British Medical Journal*, **298**, 564–567.

Bateson, P. (2001). Fetal experience and good adult design. *International Journal of Epidemiology*, **30**, 928–934.

Benyshek, D. C. and Watson, J. T. (2006). Exploring the thrifty genotype's food-shortage assumptions: a cross-cultural comparison of ethnographic accounts of food security among foraging and agricultural societies. *American Journal of Physical Anthropology*, **131**, 121–126.

Childs, C., Heath, D. F., Little, R. A., et al. (1990). Glucose metabolism in children during the first day after burn injury. *Archives of Emergency Medicine*, **7**, 135–147.

Cohen, M. N. and Armelagos, G. J. (1984). *Paleopathology at the Origins of Agriculture*. New York: Academic Press.

Davies, P. S. W. and Preece, M. A. (1989). Body composition methods in children: methods of assessment. In *The Physiology of Human Growth*, J. M. Tanner and M. A. Preece (eds). Cambridge: Cambridge University Press, pp. 95–107.

Dowse, G. and Zimmet, P. (1993). The thrifty genotype in non-insulin dependent diabetes. *British Medical Journal*, **306**, 532–533.

Eaton, S. B. and Konner, M. (1985). Paleolithic nutrition. A consideration of its nature and current implications. *New England Journal of Medicine*, **312**, 283–289.

Fernandez-Real, J. M. and Ricart, W. (1999). Insulin resistance and inflammation in an evolutionary perspective: the contribution of cytokine genotype/phenotype to thriftiness. *Diabetologia*, **42**, 1367–1374.

Fernandez-Real, J. M. and Ricart, W. (2003). Insulin resistance and chronic cardiovascular inflammatory syndrome. *Endocrine Reviews*, **24**, 278–301.

Fernandez-Twinn, D. S. and Ozanne, S. E. (2006). Mechanisms by which poor early growth programs type-2 diabetes, obesity and the metabolic syndrome. *Physiology and Behavior*, **88**, 234–243.

Foley, R. and Lee, P. (1991). Ecology and energetics of encephalization in hominid evolution. *Proceedings of the Royal Society of London. Series B*, **334**, 223–232.

Frisancho, A. (1993). *Human Adaptation and Accommodation*. Ann Arbor, MI: University of Michigan Press.

Gluckman, P. D. and Hanson, M. A. (2006). *Developmental Origins of Health and Disease*. Cambridge: Cambridge University Press.

Gluckman, P. D., Hanson, M. A., Morton, S. M., et al. (2005). Life-long echoes – a critical analysis of the developmental origins of adult disease model. *Biology of the Neonate*, **87**, 127–139.

Haig, D. (1993). Genetic conflicts in human pregnancy. *Quarterly Review of Biology*, **68**, 495–532.

Hales, C. N. and Barker, D. J. (1992). Type 2 (non-insulin-dependent) diabetes mellitus: the thrifty phenotype hypothesis. *Diabetologia*, **35**, 595–601.

Harding, J. E. (2001). The nutritional basis of the fetal origins of adult disease. *International Journal of Epidemiology*, **30**, 15–23.

Hardy, A. (1960). Was man more aquatic in the past? *New Scientist*, **7**, 642.

Holliday, M. (1986). Body composition and energy needs during growth. In *Human Growth: a Comprehensive Treatise*, F. Falkner and J. M. Tanner (eds). New York: Plenum Press, pp. 117–139.

Hucking, K., Hamilton-Wessler, M., Ellmerer, M., et al. (2003). Burst-like control of lipolysis by the sympathetic nervous system in vivo. *Journal of Clinical Investigation*, **111**, 257–264.

Ibanez, L., Ong, K., Dunger, D. B., et al. (2006). Early development of adiposity and insulin resistance following catch-up weight gain in small-for-gestational-age children. *Journal of Clinical Endocrinology and Metabolism*, **91**, 2153–2158.

Jensen, M. D., Haymond, M. W., Gerich, J. E., et al. (1987). Lipolysis during fasting. Decreased suppression by insulin and increased stimulation by epinephrine. *Journal of Clinical Investigation*, **79**, 207–213.

Kabir, M., Catalano, K. J., Ananthnarayan, S., et al. (2005). Molecular evidence supporting the portal theory: a causative link between visceral adiposity and hepatic insulin resistance. *American Journal of Physiology. Endocrinology and Metabolism*, **288**, E454–E461.

Kissebah, A. H., Vydelingum, N., Murray, R., et al. (1982). Relation of body fat distribution to metabolic complications of obesity. *Journal of Clinical Endocrinology and Metabolism*, **54**, 254–260.

Knowler, W., Savage, P., Nagulesparan, M., et al. (1982). Obesity, insulin resistance and diabetes mellitus in the Pima Indians. In *The Genetics of Diabetes Mellitus*, J. Kobberling and R. Tattersall (eds). London: Academic Press, pp. 243–250.

Kuzawa, C. W. (1997). Body fat as system: an evolutionary and developmental consideration of the growth and function of body fat (abstract). *American Journal of Physical Anthropology*, **24**(suppl.), 148.

Kuzawa, C. W. (1998). Adipose tissue in human infancy and childhood: an evolutionary perspective. *American Journal of Physical Anthropology*, **27**(suppl.), 177–209.

Kuzawa, C. W. (2005). Fetal origins of developmental plasticity: are fetal cues reliable predictors of future nutritional environments? *American Journal of Human Biology*, **17**, 5–21.

Kuzawa, C. W. (2008). The developmental origins of adult health: Intergenerational inertia in adaptation and disease. In *Evolutionary Medicine and Health: New Perspectives*, W. Trevathan, E. Smith and J. McKenna (eds). New York: Oxford University Press, pp. 325–349.

Kuzawa, C. W. and Adair, L. S. (2003). Lipid profiles in adolescent Filipinos: relation to birth weight and maternal energy status during pregnancy. *American Journal of Clinical Nutrition*, **77**, 960–966.

Kuzawa, C. W., Gluckman, P. D., Hanson, M. A., et al. (2008). Evolution, developmental plasticity and metabolic disease. In *Evolution in Health and Disease*, S. C. Stearns and J. C. Koella (eds), 2nd edn. Oxford: Oxford University Press, pp. 253–264.

Lasker, G. W. (1969). Human biological adaptability. The ecological approach in physical anthropology. *Science*, **166**, 1480–1486.

Lawlor, D. A., Ronalds, G., Clark, H., et al. (2005). Birth weight is inversely associated with incident coronary heart disease and stroke among individuals born in the 1950s: findings from the Aberdeen Children of the 1950s Prospective Cohort Study. *Circulation*, **112**, 1414–1418.

Leonard, W. and Robertson, M. (1992). Nutritional requirements and human evolution: a bioenergetics model. *American Journal of Human Biology*, **4**, 179–195.

Leonard, W. and Robertson, M. (1994). Evolutionary perspectives on human nutrition: the influence of brain and body size on diet and metabolism. *American Journal of Human Biology*, **6**, 77–88.

Lucas, A. (1991). Programming by early nutrition in man. *Ciba Foundation Symposium*, **156**, 38–50, discussion 50–55.

Mackay, J., Mensah, G. A., Mendis, S., et al. (2004). *The Atlas of Heart Disease and Stroke*. Geneva: World Health Organization.

McCarty, M. F. (1999). Interleukin-6 as a central mediator of cardiovascular risk associated with chronic inflammation, smoking, diabetes, and visceral obesity: down-regulation with essential fatty acids, ethanol and pentoxifylline. *Medical Hypotheses*, **52**, 465–477.

McGarvey, S. (1994). The thrifty gene concept and adiposity studies in biological anthropology. *Journal of Polynesian Society*, **103**, 29–42.

McMillen, I. C. and Robinson, J. S. (2005). Developmental origins of the metabolic syndrome: prediction, plasticity, and programming. *Physiological Reviews*, **85**, 571–633.

Morris, D. (1967). *The Naked Ape*. New York: McGraw-Hill.

Neel, J. (1962). Diabetes mellitus: A "thrifty" genotype rendered detrimental by "progress"? *American Journal of Human Genetics*, **14**, 353–362.

Nesse, R. M. and Williams, G. C. (1994). *Why We Get Sick: the New Science of Darwinian Medicine*. New York: Times Books.

Orban, Z., Remaley, A. T., Sampson, M., et al. (1999). The differential effect of food intake and β-adrenergic stimulation on adipose-derived hormones and cytokines in man. *Journal of Clinical Endocrinology and Metabolism*, **84**, 2126–2133.

Owen, O. E., Morgan, A. P., Kemp, H. G., et al. (1967). Brain metabolism during fasting. *Journal of Clinical Investigation*, **46**, 1589–1595.

Pelletier, D., Frongillo, E., Shroeder, D., et al. (1995). The effects of malnutrition on child mortality in developing countries. *Bulletin of the World Health Organization*, **73**, 443–448.

Pickup, J. C. and Crook, M. A. (1998). Is type II diabetes mellitus a disease of the innate immune system? *Diabetologia*, **41**, 1241–1248.

Pond, C. (1997). The biological origins of adipose tissue in humans. In *The Evolving Female. A Life History Perspective*, M. Morbeck, A. Galloway and A. Zihlman (eds). Princeton: Princeton University Press, pp. 147–162.

Potts, R. (1998). Environmental hypotheses of hominin evolution. *American Journal of Physical Anthropology*, **27** (suppl.), 93–136.

Reaven, G. (1988). Role of insulin resistance in human disease. *Diabetes*, **37**, 1595–1607.

Roden, M., Price, T. B., Perseghin, G., et al. (1996). Mechanism of free fatty acid-induced insulin resistance in humans. *Journal of Clinical Investigation*, **97**, 2859–2865.

Scrimshaw, N. (1989). Energy cost of communicable diseases in infancy and childhood. In *Activity, Energy Expenditure, and Energy Requirements of Infants and Children*, B. Schurch and N. Scrimshaw (eds). Switzerland: Nestlé Foundation, pp. 215–238.

Seaquist, E. R., Damberg, G. S., Tkac, I., et al. (2001). The effect of insulin on in vivo cerebral glucose concentrations and rates of glucose transport/metabolism in humans. *Diabetes*, **50**, 2203–2209.

Soto, N., Bazaes, R. A., Pena, V., et al. (2003). Insulin sensitivity and secretion are related to catch-up growth in small-for-gestational-age infants at age one year: results from a prospective cohort. *Journal of Clinical Endocrinology and Metabolism*, **88**, 3645–3650.

Speakman, J. R. (2006). Thrifty genes for obesity and the metabolic syndrome – time to call off the search? *Diabetes and Vascular Disease Research*, **3**, 7–11.

Stearns, S. C. and Koella, J. C. (2008). *Evolution in Health and Disease*. Oxford: Oxford University Press.

Trevathan, W., Smith, E. O. and McKenna, J. J. (2008). *Evolutionary Medicine and Health: New Perspectives*. New York: Oxford University Press.

Tsigos, C., Kyrou, I., Chala, E., et al. (1999). Circulating tumor necrosis factor alpha concentrations are higher in abdominal versus peripheral obesity. *Metabolism*, **48**, 1332–1335.

Ulijaszek, S. and Lofink, H. (2006). Obesity in biocultural perspective. *Annual Review of Anthropology*, **35**, 337–360.

Vague, J. (1955). [The distinction of android and gynoid obesities as the base for their prognosis.] *La semaine des hôpitaux*, **31**, 3503–3509.

van Harmelen, V., Dicker, A., Ryden, M., et al. (2002). Increased lipolysis and decreased leptin production by human omental as compared with subcutaneous preadipocytes. *Diabetes*, **51**, 2029–2036.

Waring, W. S. and Alexander, W. D. (2004). Emergency presentation of an elderly female patient with profound hypoglycaemia. *Scottish Medical Journal*, **49**, 105–107.

Wellen, K. E. and Hotamisligil, G. S. (2003). Obesity-induced inflammatory changes in adipose tissue. *Journal of Clinical Investigation*, **112**, 1785–1788.

Wells, J. (2007). Environmental quality, developmental plasticity and the thrifty phenotype: a review of evolutionary models. *Evolutionary Bioinformatics*, **3**, 109–120.

Yajnik, C. S. (2004). Early life origins of insulin resistance and type 2 diabetes in India and other Asian countries. *Journal of Nutrition*, **134**, 205–210.

31 Human Longevity and Senescence

Douglas E. Crews and James A. Stewart

INTRODUCTION

From a life history perspective, growth and development are processes that enhance the functional capacities of an organism through attainments of competencies and physiological signaling across somatic systems, the "growth span" (Arking, 2006). The period following growth and development is the mature state. This may usefully be described as the "health span" during which the soma is most able to maintain itself, while also reproducing and investing in reproductive effort. Senescence follows the health span. It is a later occurring age-independent, event-driven process measured in biological time as cascades of physical and chemical events that increase risks of death in the next time interval, the "senescent span" (Crews, 2003; Arking, 2006).

Senescence is a biological process that only the living experience; it is the final period of life history (LH)[1] for most organisms whether unicellular or possessing both a soma and a germline.

Senescent changes tend to be cumulative (increasing vulnerability to challenge over time), progressive (losses are gradual), intrinsic (not resulting from modifiable environmental factors), and deleterious (reduced function increases mortality risk in the next time interval) (Arking, 2006). Senescence may be measured in

biological time using physiological biomarkers, cascades of inter-related metabolic events (that contrary to developmental integration, lead toward disintegration), cell and transcription cycles (the fundamental units of biological time), gene expression patterns, changes in the proteome, epigenetic effects, and other factors (Arking, 2006). Measures of senescence do not depend upon the passage of absolute time. Rather they are secondary to current competencies, organism-wide integration via messenger molecules, and the ability of receptors to receive various signals and modulate cellular functions (Finch and Rose, 1995). Here we examine human senescence within the context of overall organismic, large-bodied mammalian, and primate evolution, and humankind's unique biocultural adaptations. Biocultural interactions extensively altered how today's humans and earlier hominins interacted with their environments by modifying multiple ecological relationships. We start with a brief background on the development of evolutionary theories of life span and how human longevity intrigued early evolutionary thinkers and continues to intrigue modern biologists. Following sections explore general evolutionary theories, the biology of senescence across organisms, some senescent mechanisms, aspects of comparative biology, and briefly review similarities in primate LH, before reviewing trends in later *Homo* and human LH and life span, biocultural evolution in constructed environments, and our species' path to modern life styles and longevity.

EVOLUTIONARY MODELS

Historical evolutionists

Darwin (Charles Robert, 1809–1882) noted that senescence presents multiple problems for evolutionary theory (Darwin, 1859). However, it was in papers presented between 1881 and 1889 that Weismann (Friedrich Leopold August, 1834–1914) first proposed an integrated evolutionary theory of senescence and longevity. Weismann (1881–1889) suggested that death

---

[1] *Life history (LH)*: In its simplest form, LH has been defined simply as things to which *r*, the rate of population growth, is most sensitive (Stearns, 1992, p. 32). It includes all aspects of a species growth, reproduction, and survivorship. Individual and population traits, including length of gestation, litter/clutch size, maturity at birth/hatching, ages at first menses (menarche), first reproduction, weaning, last reproduction, and menopause, birth-spacing, interbirth intervals, age at puberty, length of growth and development period, age-specific fecundity/fertility, and age at death are the data for studying LH evolution and examined as LH covariates. Understanding the coevolution of these traits, along with age-specific mortality, survivorship, and reproduction is the goal of LH research. Major components of LH research include understanding trade-offs between somatic maintenance and reproductive efforts, variable selective pressures on LH traits, and biocultural and behavioral influences on human LH. Life history theory is the study of the evolution and function of life states and behaviors related to these stages (Stearns, 1992).

Human Evolutionary Biology, ed. Michael P. Muehlenbein. Published by Cambridge University Press. © Cambridge University Press 2010.

was adaptive in environments with limited resources even if organisms began with intrinsically long life spans of indeterminate length. In such a setting, natural selection (NS) would favor organisms that passed on their germlines before others and before environmental insults and trauma could kill them, while investing less in their somas than longer-lived, slow reproducing individuals. According to Weismann, those with bodies better able to procure scarce resources for investing in reproduction early in life would create an array of living organisms whose life spans were inversely proportional to their rates of maturation and their prereproductive and reproductive-age mortality rates. He saw that the quicker a species achieved LH landmarks, the shorter its average and maximum life spans need be.

Many extant species show a semelparous LH, death occurs after a single lifetime reproductive event. This is particularly so for insects, annual plants, and some fish species (salmon). Unless an organism must contribute to raising its young, there is little push for somatic maintenance after reproduction (Weismann, 1881–1889; Crews, 2003, 2005a, 2007; Larke and Crews, 2006). As larger organisms commonly develop more complex somas, provide parental investment[2] (PI) to their offspring, who develop slowly, and have more than a single reproductive cycle (iteroparous), their deaths are postponed to later ages by NS. Weismann suggested that for most organisms, there is no advantage to keeping ancient, less competitive individuals around to deplete resources needed for younger generations. As with many of Weismann's conjectures, such group-selection models are not accepted today. However, he did presage Dawkin's (1989) selfish genes and Wilson's (1975) group selection approaches and conceptualized a disposable soma, as was later developed more fully by Kirkwood (1977). Additionally, many later theoretical perspectives on senescence, including those of P. Medawar (1952), G. C. Williams (1957), T. B. L. Kirkwood (1977), R. Holliday (1997), and W. D. Hamilton (1966), were presaged by Weismann's (1881–1889) insights into evolutionary processes and senescence.

Modern evolutionists

Age-specific gene action and pleiotropy

Medawar (1952) first suggested age-specific gene action and mutation accumulation contributed to senescence. As NS diminishes at ages beyond maximum reproductive potential (MRP[3]: see Crews, 2003, 2007; Larke and Crews, 2006) and ages when the majority of reproductive effort occurs, alleles with deleterious effects at older ages, but with little or no effects at younger ones, may remain in the gene pool (Williams, 1957; Kirkwood, 2002). Alleles with late-acting detrimental effects may then accumulate beyond ages at which NS is capable of eliminating them. Consequently, broad senescent changes may result from these deleterious effects at ages when reproductive potential declines toward zero. Some may doubt that such late-acting mutations are responsible for population-level senescent processes. However, as pointed out by Lasker and Crews (1996), variable mutations and mutation rates across populations and environments, coupled with linkage disequilibrium, will lead to different alleles accumulating in various populations. As long as these are not linked to other alleles that reduce, or they reduce fitness themselves, gene flow will spread these alleles and, importantly, their linked haplotypes, across a species. [For example multiple viral DNA segments have hitchhiked within the human genome for millions of years, up to 8% of the total (Feschotte, 2010).] When linked haplotypes show benefits to reproductive success (RS) in many environments, NS may maintain or increase their frequency in local settings. Population bottlenecks, founder effects, and various forms of genetic drift also may increase frequencies of late-acting detrimental alleles within local populations. Thus late-acting detrimental alleles may "hitchhike" along with beneficial alleles within linked haplotypes, hidden till life spans increase to ages at which they are expressed.

In addition to addressing mutation accumulation following Medawar (1952), Williams (1957) recognized that pleiotropy also might lead to senescence. Alleles beneficial during the growth span or that enhance reproduction and MRP during the health span may be incompatible with long life. Alleles enhancing senescence in mid- to late-life, but improving survival during infancy, childhood, and attainment of MRP exhibit antagonistic pleiotropy (early benefits with late-acting deleterious effects). Due to trade-offs between reproduction and somatic maintenance, once an organism begins reproducing and fledging offspring, NS gradually loses its ability to eliminate alleles that are deleterious to the organism's extended somatic maintenance. For example, apoptosis is crucial during embryonic development for sculpting the developing organism as stem cells create more new cells than are

---

[2] *Parental investment (PI)*: All expenditures of parents' time, energy, and future reproductive effort on one offspring at a cost to parents' abilities to invest in their own somas or other possible offspring and enhance their own fitness (Hamilton, 1964, 1966).

[3] *Maximum reproductive potential (MRP)*: Is the point in life at which an organism is sufficiently mature to not only bear/sire offspring, but also best able to rear and fledge any offspring produced with maximum efficiency (Crews and Gerber, 1994; Gerber and Crews, 1999; Harper and Crews, 2000; Crews, 2003; Larke and Crews, 2006).

needed. While the soma is sculpted, external markers (ligands) regulate apoptosis of excess cells. Once adulthood is achieved, animals are no longer growing or developing and rates of apoptosis are adjusted accordingly as morphogenesis is completed. During adulthood most cells that are damaged, malfunctioning or show damaged DNA are marked for apoptosis by internal cues (e.g., mitochondrial factors, p53). However, proliferation of new cells to replace those that are worn out and/or lost through apoptosis may not be balanced and tissue deterioration may occur. Loss of cells from apoptosis and accumulation of senescent cells may jointly contribute to senescence and tissue deterioration (Campisi, 2003). Williams (1957) also suggested that as life span increases NS bumps up against detrimental alleles that have been retained through mutation accumulation, at what was previously the upper limit of reproductive effort. When this occurs, alleles at other loci that modify the late-acting detrimental effects of accumulated alleles, or alternate alleles at the same loci, may increase in frequency if they promote longer life along with greater RS. However, any alleles that improve prereproductive survival or improve RS in early life will not be eliminated via NS, regardless of their late life effects.

Disposable somas and immortal genomes

Accumulated evidence now suggests that senescence may occur in most organisms. For many decades fissile organisms were thought to not show senescence and therefore be in a sense immortal. However, recent research on *Escherichia coli* and fissile single celled prokaryotes suggest they do senesce (Stewart et al., 2005). Not only do eukaryotic, sexually reproducing species with telomeres to shorten with each round of mitosis, LH stages to grow and develop through, antagonistic pleiotropy, vital macromolecules to oxidize, and proteomes that are degraded over time show senescence, it now seems that all living organisms may show senescence.

Kirkwood's (1977, 1990, 2002, 2005) and Kirkwood and Holliday's (1979) elaborations of Weismann's disposable soma theory of senescence proposes a distinction between evolutionary theories and physiological descriptions of somatic senescence (see also Crews, 2003). An organism's resources, such as energy, are limited and must be divided between self-preservation (the soma) and continuance of the immortal germline across generations (reproduction). During the growth span resources are invested into the soma. Throughout the growth span, the soma and germline share the same goal, producing a mature soma to pass on the germline. Because the soma is the organ by which the immortal germline ultimately must reproduce itself, over the growth span trade-offs between somatic growth/development and transmitting the germline

are minimized. However, once sexual maturation and MRP are attained, a trade-off or competition between investing in maintaining the "disposable soma" or reproducing the "immortal germline" through offspring ensues (Kirkwood, 1990, 2002, 2005). Models proposing LH trade-offs between the soma and reproductive effort mainly focus upon the postgrowth span of LH. The soma is not disposable prior to achieving reproductive maturation and RS. Trade-offs expressed during the growth phase are associated with early LH strategies of high versus low early life reproduction and rapid versus slow growth and development. In some species LH patterns have deep evolutionary roots reflecting phylogenetic inertia. For example, mayflies, fruit flies, annual plants, many insects, amphibians, and fish show rapid life histories, little to no PI, rapid growth, and high numbers of eggs, seeds, and sperm upon reproductive effort; along with rapid senescence and high intrinsic mortality. Conversely, other species of insects, plants, amphibians, reptiles, fish, and mammals, notably cicadas, wasps and other hymenoptera species, redwood and bristlecone pine trees, turtles, sharks, whales, and elephants show a different LH pattern. There are no phylogenetic continuums of LH strategies nor does phylogeny necessarily predict LH, different species within even a single genus may show different LH patterns and the same species may show markedly different LH patterns under different environmental constraints.

Early in life and during the period of maximum reproductive effort, NS favors somatic maintenance in species that reproduce multiple times (iteroparous) or that show PI in their offspring after birth/hatching (Hamilton, 1964, 1966; Crews, 2003, 2005a, 2008; Larke and Crews, 2006). As life continues, investment in reproductive effort eventually takes priority over somatic maintenance and senescent processes may not be eliminated, because additional somatic investments yield trivial reproductive benefits. Eventually, the net effect of NS declines to zero. Metazoans have complex, inter-related and interdependent cells, organs, and systems within their somas. Senescence of cells begins early in the life span as strategic resources are directed toward replicating the immortal germline at the expense of the soma (Austad, 2005a; Kirkwood, 1977; Hayflick, 1979). Among species with high PI and multiple reproductive cycles, NS works to maintain the parental soma because it is the sole reproductive organ and source of continued investment in reproductive effort and PI. Simple mathematical formulas, such as the Gompertz or Weibull equations, model the increase in somatic susceptibility over time from multiple, inter-related, and unrelated intrinsic and extrinsic factors that combine to eventually kill all organisms. All metazoans senesce after they attain MRP. Although some processes of senescence even may be initiated as early as fetal life, most occur after MRP. For some species, a previously unidentified LH stage may lie just beyond the senescent phase. During this "late-life stage," the rate of increase in senescence and mortality hazards may slow among some experimental and free-living populations (Carey et al., 1992; Carnes and Olshansky, 2001; Rose et al., 2005b; Perls, 2006; Rauser et al., 2006).

Rauser et al. (2006) suggest that such a late-life phase may begin when an organism's survivorship and fecundity deteriorations cease and mortality rate increases approach zero. The Gompertz equation ($\mu(x) = Ae^{\alpha x}$) predicts that as organisms age, their mortality rates increase exponentially (Hamilton, 1966, reviewed in Box 13.1 in Crews and Ice, 2011). However, the slope of the Gompertz equation decreases slightly after about age 85, suggesting decreasing increases in mortality may happen in late life. Late-life survival is better modeled by Hamilton (1966):

$$s(x) = \sum_{y=x+1} e^{-ry} l(y) m(y) \text{ and } \bar{\mu}(x) = \frac{Ae^{\alpha x}}{1 + [\sigma^2 A(e^{\alpha x} - 1)]\alpha^{-1}},$$

where $s(x)$ is the variance in survival and $\bar{\mu}(x)$ is the average age at death. Other variables in the equations are, x representing age, r the Malthusian growth rate, $l(y)$ the specified survivorship to age x, and $m(y)$ age-specific fecundity, where y is the net expected reproduction at all ages x and higher. A is an age-independent parameter representing baseline mortality within the population, and α is an age-independent parameter representing senescence (rate of aging in Hamilton's terms). Data from some laboratories and experimental species suggest mortality rates at later ages may follow patterns represented by Hamilton's (1966) equations. This leads to suggestions that a late-life phase with nonincreasing mortality rates may occur late in the LH of some multicellular organisms (Hamilton, 1964, 1966; Rose et al., 2005b; Rauser et al., 2006). Hamilton (1966) suggested such a phenomenon might reflect declining NS over time. Humans may show some late-life stabilization of age-specific mortality rates as age-related deterioration slows or ceases after 85 or 100 years (Perls, 2001; Doblhammer and Kytir, 2001; Rose et al., 2005b; Crimmins and Finch, 2006; Rauser et al., 2006).

THE MODERN SYNTHESIS AND SENESCENT BIOLOGY

The synthesis of evolutionary theory with genetics combined with developments in multivariate statistical analyses of genes and population genetics during the mid-twentieth century allowed some analyses of the complex gene–environment interactions producing senescence. During the late twentieth century,

evolutionary theories of senescence advanced in parallel with molecular biology. Cellular and comparative biology, breeding experiments with animal models and, more recently, growth of cells in laboratories, analyses of DNA, and transfers of DNA into and out of animal models have improved understandings of cellular senescent biology. It is now clear that across organisms, multiple alleles at many loci and differential processing of proteins produce complex molecular pathways modulating physiological functioning. Some of these are incompatible with longevity; others are associated with high somatic maintenance and extended life spans. Among humans both culture and environment influence and alter these multiple genetic/protein-based predispositions. Across species and populations average age at first reproduction and average life span are closely associated. Variation in life spans and reproductive potential within and between populations and species shows that environments modulate and extensive genetic variation underlies LH (Carey and Judge, 2001; Perls, 2001, 2006; Perls et al., 2002; Crews, 2003; Pak et al., 2003; Perls and Terry, 2003; Bredesen, 2004; Holliday, 2004; Austad, 2005a, 2005b; Kirkwood et al., 2005; Rose et al., 2005a, 2005b; Christensen, et al., 2006; Harper et al., 2006a).

Cellular biology

Cell replicative ability

Hayflick (1965) reported that mitosis by dividing cells (fibroblasts) cultured in vitro was limited. Contrary to the then prevailing opinion, that dividing cells possessed an unlimited capacity for self-reproduction, Hayflick observed that fibroblasts from newborns doubled in vitro only about 50–60 times. How, and if, Hayflick's eponymous limit is associated with human senescence was then, and remains today, largely unknown. Some researchers suggest that dividing cells may come close to their in vitro doubling limit when they produce and maintain a mature adult reproductive human. However, current estimates are that only about 42 cell doublings by the zygote are needed to produce a mature infant and only another 4–6 to grow an adult soma and maintain it thereafter (Dennison et al., 1997). This would leave the average person with more than sufficient cell doublings to maintain their soma over 10 decades. Limited cell doublings as defined by Hayflick (1965) may still be related to cell culture methods. By dividing the cell population, the number of primary fibroblasts left to produce new cells is halved also (Cristofalo et al., 1998, 1999, 2003; Clark, 1999). Among the criteria for an eponymous limit is a specified time during which the culture must cover its environment; without a time limit, cultures may continue dividing slowly to confluence.

Animals do not appear to "run out" of cells over their life spans due to limited cell proliferations in vivo, nor do doublings in cell culture correlate as strongly with age as was originally suggested (Hart and Tuturro, 1983). For example, total doublings for cells cultured from the same individual may show an average of about 50 doublings, but a range from 10 to 90 doublings (Hart and Tuturro, 1983). Cells from different individuals of the same age also show wide variation in their total numbers of doublings; cells from infants average about 50–60, but range from 25–120, while those from persons aged 65 years average about 20 doublings, with a range from 5–50. Across individuals of all ages the average is about 35 doublings, but again the range is large, from 0–120 (Hart and Tuturro, 1983). Additionally, no significant correlation of fibroblast doublings and age of cell donors, was observed in a small sample of healthy Baltimore Longitudinal Study of Aging participants ($n = 42$) (Cristofalo et al., 1998, 1999). Also problematic is the finding that fibroblasts from some very old individuals show no doubling potential at all, yet these individuals continue to survive. Given these accumulated data, relationships between in vitro doublings and in vivo senescence are unclear and may be less important than other processes (Cristofalo et al., 2003).

During terminal passages in vitro cells do tend to show multiple alterations that are similar to those of senescent cells in vivo (Clark, 1999; Aviv, 2002; Campisi, 2003). Data on telomere shortening suggest that in vivo skin, thyroid, immune, and digestive cells may fail to replace themselves adequately with increasing age (Clark, 1999). In general, loss of proliferate capacity in vitro is not thought to produce or reflect processes of senescence. Loss of replicative capacity shows how a complex molecular system fails to maintain itself infinitely due to intrinsic fragility and random losses of function. Average numbers of doublings in vitro are correlated loosely with age of cell donors but are not highly predictive of individual life spans. Loss of replicative capacity does not appear to underlie senescence in humans or mammals. However, senescent cells with traits similar to those of cells experiencing replicative senescence do accumulate with time in human organs.

Telomeres

Cells of patients with syndromes mimicking some aspects of somatic senescence, e.g., ataxia telangiectasia, Werner's syndrome, Bloom's syndrome, trisomy 21, dyskeratosis congetialia, aplastic anemia (Blasco, 2005), show progressive telomere shortening. Telomere shortening also associates with heart failure, immunosenescence, digestive tract atrophies, infertility, reduced viability of stem cells, reduced angiogenic potential, reduced wound-healing, and loss of body mass (Aviv, 2002; Blasco, 2005). However, other biomarkers of senescence are not highly correlated with telomere shortening (Blasco, 2005).

It is now clear that shortening of telomeres limits cellular replicative life spans and leads to the Hayflick limit of human fibroblasts in vitro (Olovnikov et al., 1996; Shay, 1997; Aviv, 2002; deLange, 2002). Dividing cells that reach their replicative limit in vivo become senescent and tend to accumulate in somatic tissues. Such data suggest that chromosome shortening measures or represents intrinsic cellular senescence, an internal biological clock in dividing cell clones. Still, cell replicative capacity is not closely associated with individual life spans. An unknown number of variables – including cell type, donor age, oxidative stress, and heredity – influence correlations between telomere shortening and age, both across and within individuals. Telomere length varies across peripheral lymphocytes from the same individuals, between similar cells from different individuals of the same age, and across cell cultures of fibroblasts from the same donor (Harley et al., 1990; Hastie et al., 1990; Slagboom et al., 1994). Studies of human peripheral lymphocytes suggest an average loss of telomeric DNA of about 31–33 base pairs (bp)/year (Hastie et al., 1990; Slagboon et al., 1994). The average, across various types of human cells, ranges from 40 to 200 bp/division (Allsopp, 1996), compared with an average of 50 bp/doubling in cultured cells. Telomere length is sexually dimorphic in humans at birth, women's telomeres are longer then men's (Aviv, 2002). Men also appear to senesce faster than women until late life (Foley, 1986; Perls, 1995; Anderson-Ranberg et al., 1999), although this may not be so (Graves et al., 2006). Men may require more cell doublings to accommodate their 15–20% larger body sizes, exhausting their shorter telomeres earlier in life. Although attrition of telomeric DNA tandem repeats (TTAGGG) paces in vitro replicative history of primate, including human, cells, a similar association is not observed in rodent cells (Aviv, 2002). Available evidence does not suggest a causal relationship between telomere attrition and biological senescence in humans (Aviv, 2002). One cause, oxidative stress, is a promoter of cellular senescence, thus telomere length also may indicate exposures to reactive oxygen species.

Apoptosis

Senescence occurs at the molecular level (Arking, 2006), but is expressed as somatic aging, cellular dysfunction and death (Campisi, 2003). Senescent cells accumulate in most tissues with increasing age. This may partly reflect sluggish apoptosis (programmed cell death). Tissues composed of dividing and nondividing cells both accumulate senescent cells with age: the former partly due to the Hayflick limit and loss of telomeric DNA, the latter secondary to intrinsic molecular stress, wear-and-tear, oxidative stress, and other processes. Apoptosis is associated with the appearance of distinctive molecular markers. Among most sexually reproducing organisms the force of NS continually decreases over the life span after maturation. It seems likely that over evolutionary history, alleles beneficially affecting cellular processes and leading to greater cell stability would have evolved via NS. However, increasing cell and organ dysfunction is observed with increasing age in most modern species. This appears to be due to intrinsic molecular alterations that disrupt cellular processes rather than to processes at the organ level that lead to systemic organ failure. In complex organisms, it seems that both apoptosis and cellular senescence may have evolved to suppress the transformation of functional cells to cancerous cells or cells that might harm the entire organ in other ways (e.g., autoimmunity, reduced OXPHOS, mutant mtDNA). Multiple alleles influence apoptosis (Bredesen, 2004) and create conditions that mimic many of the effects of senescence seen in humans, e.g., Werner and Bloom syndromes, and other pathologies associated with shortened telomeres (Blasco, 2005). Continued apoptosis is one method for maintaining the soma in response to continued cellular damage and deterioration. However, inappropriate apoptosis may limit life spans via antagonistic pleiotropy and loss of nonsenescent cells. Lack of apoptosis may allow senescent cells to survive and deteriorate or compromise tissue and organ function or become cancerous. Apoptosis may be delicately balanced to eliminate most senescent and damaged cells, but over the life span allows some to survive and damage the organism.

ANIMAL MODELS AND COMPARATIVE BIOLOGY

Humans are complex organisms. They show several LH attributes that seem to differ significantly from those of most other organisms (Austad, 1991; Hawkes et al., 1997, 1998; Alvarez, 2000; Smith and Bogin, 2000; Carey and Judge, 2001; Bird and Bird, 2002; Crews, 2003; Crews and Gerber, 2003; Howell and Pfeiffer, 2006; Larke and Crews, 2006; Christensen et al., 2006). One problem when applying results from experimental animal models and molecular experiments to humans is that, thus far, only a limited number of organisms have been studied (Austad, 1994, 1997). Yeasts, nematodes, fruit flies, and mice are the most frequently used animal models for studying longevity enhancement and senescent process. All are "demonstrably inept at aging successfully" (Austad, 2003, p. 1327). As mammals, mice share many homologous gene loci with humans. However, as a clade

composed of short-lived, litter-baring, low PI species, they do not show a similar proportional timing or pattern of LH to even the smallest of primates. Multiple LH features observed among insects and rodents are not found in primates and humans (e.g., dauer stages, litters). Such laboratory species tend to reveal basic processes of cellular senescence. What they do not illustrate is how senescence is expressed as aging and longevity in humans. Humankind uses biocultural and genetic adaptations, physiological and biological systems, behaviors and constructed environments in their responses to the age-independent, event-driven processes of senescence that occur over biological time. The same cascades of senescent physical and chemical events likely occur in all iteroparous sexually reproducing species (Arking, 2006). How evolution and phenotypic adaptability have slowed these processes and risks of death from altered gene expression, proteomic disintegration, and disruptive metabolic events is what differentiate species-specific and population-specific LH.

In general, phylogeny is poorly associated with longevity (Rose, 1991). However, comparative biology has produced multiple insights on human senescence, and likely will continue to do so (Austad, 1997, 2003; Crews, 2003). However, Austad (2003, p. 1327) points out that "All of the best known and most widely employed models for aging are short-lived taxa ..." Although commonly studied animals provide limited information on how human LH, aging, and longevity are structured (Austad, 2003), they do show that high "lifelong demographic heterogeneity" exists between and within species (Vaupel, 2001; Rose et al., 2005a, 2005b). Lifelong demographic heterogeneity is measured by the variability in mortality/survival patterns observed between cohorts and groups within populations; each requires a different Gompertz equation to fit its variable LH experiences (Carnes and Olshansky, 2001). Experimental data from insects, rodents, and humans indicate broad LH variation within populations large and small, and, by extension, different cohorts within populations (Doblhammer, 2000; Stearns et al., 2000; Charlesworth, 2001; Arking et al., 2002; Kirkwood, 2005; Kirkwood et al., 2005; Rose et al., 2005a, 2005b; Perls, 2006). This illustrates the multiple influences of environments and genes on LH. They also illustrate one pitfall of comparing demography, genes, or biology across, within, and between species. Populations within species inhabit a variety of local and regional environments (Rose et al., 2005b), producing multiple adaptive landscapes and variable adaptations.

Drosophila
Lacking a close evolutionarily relationship, fruit flies do not recapitulate age-related declines and diseases observed among humans. However, these organisms

show commonalities in their patterns of molecular and cellular senescence, illustrating genetic and environmental influences on diminishing cellular functions with age (Arking et al., 2002; Grotewiel et al., 2005). Artificial selection produces populations showing variable longevity in multiple *Drosophila* species (Luckinbill et al., 1984; Rose, 1991; Pletcher and Curtsinger, 1998; Service, 2000; Charlesworth, 2001; Arking et al., 2002; Rose et al., 2005a). Selecting long-lived or late-reproducing fruit flies produces significant increases in mean and maximum life spans after a few generations. Given the controlled laboratory environment, allelic variation must contribute to improved longevity. As yet, directional artificial selection on longevity has not shown an upper limit to life span (Curtsinger et al., 1992; Arking et al., 2002). Among a single strain of *Drosophila*, a variety of stressful external stimuli may induce a variety of longevity phenotypes, suggesting that multiple inter-related mechanisms modulate their LH and life span (Arking et al., 2002). Combinations of longevity-enhancing alleles also may recombine, predisposing latter generations to increasing life spans. These results show that simple mechanisms influence longevity, and that life span is not intrinsically limited in most species. Current environments, evolutionary history, and multiple evolutionarily stable strategies across populations within species may produce inertial barriers that limit life span. Genotypic variation for survival, initial, and age-specific mortality rates is high in fruit flies. However, specific DNA segments do not correlate highly with demographic and functional measures of senescence (Burger and Promislow, 2006). Responses of young flies to early life stressors may alter their homeobox genes rather than other DNA segments (Burger and Promislow, 2006), possibly modulating precursors of senescence during early life.

Drosophila species illustrate the relative simplicity of modulating longevity in a laboratory setting (Arking et al., 2002). They also reveal a reservoir of genetic variation for senescence-slowing alleles not expressed in wild populations. Senescence-slowing alleles are not advantaged in wild populations because longevity enhancement often is associated with reduced RS. In natural settings NS limits frequencies of "longevity-enhancing" alleles because they reduce RS. Unlike *Drosophila*, humans, particularly men, may increase their RS by living longer; women also may, but not as much as men (Marlowe, 2000; Crews, 2003, 2008). The reservoir of longevity-enhancing alleles seen in many wild-living populations already may be expressed as extended longevity in humans (Crews, 2003).

Rodentia

Harper et al. (2006b) question the applicability of research on laboratory-bred animals to human longevity (see also Austad, 2003). On a calorie-restricted (CR) diet, mice with wild grandsires show changes in corticosterone and testosterone that resemble those of standard laboratory animals, however their mean longevity does not differ significantly from mice-eating standard laboratory chow (Harper et al., 2006b). Calorie-restricted laboratory mice show higher mortality during early life, lower mortality in later life, and have fewer cancers than *ad libitum* fed mice. Understanding how CR changes physiology and longevity of laboratory-bred mice may aid in understanding processes of human senescence. Laboratory selection for either long-life or late reproduction in rodents extends average and maximum longevity. As with fruit flies, a reservoir of longevity-enhancing alleles is exposed in laboratory rodents. Wild rodents show a range of LH, body sizes, and longevities with the largest and longest-lived inhabiting isolated islands and tropical settings. Without major predators and when isolated from competition, even rodents show longer lives and larger body sizes. This suggests that as populations develop mechanisms that reduce predation or competition for resources, isolation for rodents and culture for humans, longer lives and increased body sizes ensue.

Nonhuman primates

Most primates show average life spans for their body sizes above those predicted by regression models based on extant mammals (Finch, 1990; Carey and Judge, 2001). Based upon longevity of 587 captive mammalian species, primates are not necessarily the longest-lived for their body sizes (Austad and Fischer, 2005). Even small- and medium-build primates tend toward longer life spans than predicted for mammals of their size. They also tend to have larger brains for their body sizes than do most other mammals. Primates generally birth single offspring, show high PI, arboreal adaptations, and live in social groups, traits reducing susceptibility to environmentally imposed mortality and predation (Shea, 1998; Carey and Judge, 2001; Austad and Fischer, 2005). Among prosimians, monkeys, and apes multiple LH strategies are observed. However, being long-lived and slow-reproducing mammals for their body sizes with large brains are common traits, suggesting that these reflect some similar evolutionary pressures that have influenced LH.

In general, primates progress through their LH phases and transitions more slowly then similar-sized nonprimate mammals (Austad, 1997; Shea, 1998; Carey and Judge, 2001; Leigh, 2004; Austad and Fischer, 2005; Crews and Bogin, 2010). Most primates lie above regressions lines predicting average life spans of mammals from average birthweight, adult encephalization index (ratio of brain size to body size, see Austad and Fischer, 2005), adult body size, or cranial capacity (Finch, 1990; Carey and Judge, 2001; Leigh, 2004). This is most obvious among large-bodied apes. They may fall above such regression lines by a standard deviation or more. Encephalization appears to characterize primates, along with extended offspring dependency, high PI, and extended longevity. This suite of LH characteristics has been identified as a possible base for the modern human LH strategy (Shea, 1998; Bogin, 1999; Smith and Bogin, 2000; Crews, 2003; Crews and Bogin, 2010).

A high encephalization index is associated with greater behavioral adaptability. This allows organisms to use more self-motivated and opportunistic behaviors when responding to their environments. Even early hominins had and modern apes show high encephalization. Encephalization among *Homo* provided a basis for developing cultural activities leading to biocultural evolution. Primates generally birth a single offspring. Although twinning is the norm in some New World monkeys (marmosets and tamarins), it is rare in others (Cebids). Twinning among humans (about 3–5% of births) is relatively common. Providing high PI to one offspring over an extended period of time is energetically more efficient for parents than providing for two or more over a shorter period. All primates share a suite of LH traits found across long-lived species (Shea, 1998). Today humans express a high PI evolutionarily stable strategy (ESS) (Crews, 2003, 2008; Larke and Crews, 2006). Human infants are among the most physically altricial of primate and mammalian offspring. Even so, humans show the shortest birth intervals, highest RS, and youngest weaning ages of extant large-bodied apes. These are LH traits correlated with fitness and evolutionary success. Since the Miocene, monkey genera have been more successful than have apes at expanding their ranges and competing for survival (Leigh, 2004). Monkeys, like humans, have shorter birth intervals, younger ages at weaning, and higher RS than do extant large-bodied apes. They also are smaller and show less postweaning PI and provisioning. Among the *Hominidae*, humans show more monkey-like LH adaptations than do other large-bodied apes.

HUMAN LONGEVITY AND SENESCENCE

Foraging, scavenging, gathering, hunting, horticulture, and eventually agriculture altered patterns of LH, particularly survival, reproduction, and PI among hominins. Changing subsistence strategies and technologies altered energy availability and the human niche. Cultural developments and niche construction modified human environmental exposures (Odling-Smee et al., 2003; Armelagos et al., 2005) producing differential longevity across populations and groups (Crews, 2003, 2005a, 2008). Sociocultural adaptations by humans promoted "biocultural

evolution"[4] (Goodman and Leatherman, 1999; Carey and Judge, 2001; Aiello and Wells, 2002; Crews, 2003, 2005a, 2008; Crews and Gerber, 2003; Larke and Crews, 2006; Arking, 2007). Physical development in utero, physical maturity at birth, length and timing of postnatal growth and development, ages at menarche, first and last reproduction, and menopause, reproductive effort, fertility, RS, rates of senescence, and longevity continue to vary across the four remaining large-bodied primate species and local populations (Finch, 1990; Carey and Judge, 2001; Crews, 2003; Leigh, 2004; Larke and Crews, 2006). Alterations in early LH parameters may alter all later aspects of LH, eventually leading to changes in late-life survival and reproductive patterns (Crews, 2003; Finch and Crimmins, 2004; Crimmins and Finch, 2005, 2006; Finch, 2005; Larke and Crews, 2006). For example, within cohorts of modern populations, low childhood mortality and infectious disease morbidity tend to produce lower adult mortality and correlate with extended longevity (Finch and Crimmins, 2004; Crimmins and Finch, 2005, 2006; Finch, 2005). This was predicted by Weismann (1881–1889) who surmised that life spans are inversely proportional to rates of maturation and prereproductive mortality. These and other results suggest that models based upon retention of reserve capacity (RC) into older ages and life in cushier environments partly explain increases in later *Homo* and particularly modern human longevity (Crews, 2003, 2005a; Caspari and Lee, 2004; Finch and Crimmins, 2004; Crimmins and Finch, 2006; Larke and Crews, 2006; Bogin, 2009; Crews and Bogin, 2010). Such results do not support models suggesting that adult or postreproductive mortality rates influence patterns of growth and development. Rather the opposite, lower mortality and infectious disease rates during the growth phase potentiate lower adult and late-life mortality and improved survival.

Genes, evolution, and cell biology

Biological, observational, and laboratory research have advanced understandings of mechanisms by which many species (e.g., elephants, whales, sharks, tortoises, humans) have extended their life spans past 50 years, while other species have not. However, specific genetic modulators and biomarkers have not shown strong associations with longevity. Genetic variability between populations of individuals within species and

---

[4] *Biocultural evolution*: The interplay of culture and biology on human evolutionary processes. The scientific study of relationships between culture and human biology that have shaped modern human variation and the resources available to humans across cultures. Only among humans are biocultural pressures on evolution observed.

across species in the organization of their DNA and genes hinders comparisons across humans and other organisms (Arking et al., 2002; Hahn et al., 2007). For example, chimpanzees and humans show relatively high population-level differences in their genetic structures (Fischer et al., 2006). Additionally, among humans, genetic variation is structured by cultural behaviors. For example, there is little Y-chromosome variation among members of the Cohen priesthood (Hammer et al., 1997). Alleles at the BRCA 1 and 2 and the apolipoprotein (APO) E loci, along with human leukocyte antigen (HLA) haplotypes show variable frequencies across populations. Some genes are associated with early mortality, while others are found at high frequencies among long-lived individuals of most populations. For example, across samples from different populations, different HLA haplotypes are associated with longevity. Genetic variation within and across human samples limits any utility of comparing alleles associated with longevity to nonhuman primates or other animals (Fischer et al., 2006). There are multiple human longevity genotypes and phenotypes. Local selection differentials, population size, genetic/phenotypic heterogeneity, and multiple genetic and population processes structure correlations of LH characteristics with specific alleles (Rose et al., 2005b). Population sizes, mating patterns, migration and sex distributions influence genetic variation in all species. Along with demographic, evolutionary, and environmental factors, sociocultural behaviors have structured human genetic variability for LH.

Humans appear to show some unique and many similar LH traits compared to nonhuman primates and other long-lived animals. Humans are not the most long-lived mammals, whales probably are. Verifying that a LH trait is unique to a species requires broad comparative surveys, quantitative assessments, and measurements of variation within many different species. Uniqueness can not be assumed or dictated by arbitrary definitions. Sometimes researchers define a trait in a manner that precludes it from applying to more than one species. Such definitional fiats hinder comparative research in LH evolution. In addition, some LH associations observed in modern populations may not have been crafted by NS. Rather they may represent only current bioculturally constructed correlations. That is, evolutionary origins of some LH traits may be unrelated to current patterns and distributions, representing instead selective pressures on other traits or at previous points in human evolution.

The human genome project shows that only about 23 000 segments of DNA code for proteins in humans (http://www.ornl.gov/sci/techresources/Human-Genome/project/info.shtml). Subsequent research indicates that variable reading frames, pre- and post-translation editing, and different lengths of final polypeptides

lead to about 300 000–400 000 different protein forms in humans (Jaenisch and Bird, 2003). As understandings of cellular senescence expanded, mitochondrial (mtDNA) came under closer scrutiny. Some senescent processes, e.g., loss of muscle mass (sarcopenia), are associated directly with mtDNA mutations. However, no single mutation is associated with a majority of senescent processes (Pak et al., 2003). A number of polymorphisms (e.g., mtDNA, APOE, HLA alleles/haplotypes, presenilins, BRCA 1 and 2 alleles) are associated with senescent-related diseases (e.g., sarcopenia, autoimmunity, dementias, cancer) that shorten life span. Others, such as low levels of APOC3 are associated with longer life spans (Atzmon et al., 2006). A monotonic over-representation of the APOC3–641CC genotype is observed at older ages in humans. Relatively infrequent (10%) at age 60 years, this allele is found at 25% among survivors to 100 years (Atzmon et al., 2006). APOC3–641CC predisposes to a lipid phenotype compatible with long-life in modern settings. This association may not have occurred in other populations at other times during hominin evolution. Among some populations with low dietary fat and cholesterol intakes, such as the Yanomami and other American Indian populations, a high frequency of the APOE*4 allele, apparently with no APOE*2 alleles prior to European contact, suggest the APOE*4 allele may have been advantaged because it maintains cholesterol levels (Crews et al., 1993). In such low cholesterol settings, APOE*4 alleles might have helped maintain cholesterol levels for generating steroid hormones during gestation, growth, development, and reproductive effort. The HLA haplotypes and APOE alleles show age-related frequency differences across populations. Different HLA haplotypes are associated with late-life survival in each population examined. In modern Westernized populations, the APOE*4 allele is associated with high lipid levels, predisposes to cardiovascular diseases in middle-age, and Alzheimer's dementia in old age. The APOE*4 allele shows a decreasing frequency with increasing age through at least the eighth decade of life. Although some associations are similar, multiple different alleles at single loci and variable haplotypes are associated with extended longevity in different human populations in variable environments and cultures. Such a high degree of variation defies attempts to identify species-wide genetic modulators of senescence, although some must exist.

Associations with longevity suggest that many alleles may similarly predispose to long-term survival as does APOC3–641CC. Besides specific alleles, epigenetic regulation of gene expression affects multiple loci (Jaenisch and Bird, 2003). Epigenetic alterations affecting gene regulation occur during embryonic and fetal development, but may continue throughout life as cellular mitosis replicates imprinted loci (Barker, 1998; Cameron and Demerath, 2002; Kuzawa, 2005). Multiple epigenetic factors influence health and life span. Associations of diet with cancers and the metabolic syndrome and of smoking with lung cancer are examples. All cells have multiple, often overlapping, and highly conserved housekeeping gene systems (e.g., albumin, ubiquitin, SOD, uric acid, glutathione, and other antioxidants) (Eisenberg and Levanon, 2003). Proteolytic systems, for example the proteasome, are important for cellular housekeeping. The proteasome degrades normal, misfolded, and oxidized proteins recycling their amino acids. It also degrades old proteins that accumulate in cells with age. Both in vitro and in vivo, senescent alterations are correlated with impaired proteasome function (Chondrogianni and Gonos, 2005). Proteasome function regulates cellular homeostasis, helping to determine cellular senescence along with hundreds of other loci.

Uniquely human life history traits

Humans, like a number of other mammalian and nonmammalian species show long lives. Comparisons with other long-lived species are one avenue for determining unique and conserved aspects of human LH (Austad, 1997; Carey and Judge, 2001; Crews, 2003; Austad and Fischer, 2005). Concepts such as a minimum necessary life span (MNLS: the amount of time members of a species must survive to complete the "necessary tasks of life" as described by Weismann [1881–1889]: gestation, growth, development, maturation, and reproductive effort), PI, and the costs of reproduction help focus this research. Elsewhere, we estimated a MNLS of about 45 years for humans and about 35–42 years for chimps and gorillas (Crews, 2003; Larke and Crews, 2006). Based upon Weismann's (1881–1889) concepts, a MNLS is similar to a Darwinian necessary life span.

One woman already has lived three times the estimated human MNLS (about 122.5 years). As yet, no chimp or gorilla has doubled their estimated MNLS. However, the chimp who played Cheeta in the 1950s Tarzan films celebrated his 75th birthday on April 9, 2007. Among other large-bodied apes, on December 22, 2006, the world's first captive-born gorilla (Colo) celebrated her 50th birthday at the Columbus Zoo (*Columbus Dispatch*, Ohio); on January 17, 2009 the oldest male gorilla in a North American zoo (Timmy) celebrated his 50th birthday at the Louisville Zoo (*Columbus Dispatch*, Ohio); and the world's oldest captive orangutan (Nonja) died at age 55 in December 2007 at the Miami Metropolitan Zoo. Based upon estimated ages in a wild population observed over 32 years, Wich et al. (2004) suggest orangutan males may survive

58 and females 53 years (without observable menopause and an interbirth interval of 9.3 years). Survival through 5–6 decades, about a 30–35 year female reproductive span, and multiple birth cycles is the basic LH pattern among not only living large-bodied apes, but many large-bodied mammals. Among men a 60-year-plus reproductive span, highly variable RS, high paternal PI, extended life spans, and survival past 7 decades of life are not shared with other large-bodied primates, nor are life spans of 100+ years and female survival to twice their age of last menses. Along with altricial offspring and high PI, these traits differentiate humans from other hominins and hominoids.

Others have identified suites of unique human LH characteristics (Shea, 1998; Smith and Bogin, 2000; Carey and Judge, 2001; Crews, 2003). Humans are generalists with some LH traits linking them to other mammals, nonhuman primates, and extant and extinct hominins and hominoids. Evolutionary and allometric constraints on the size of ovaries and number of available primary oocytes may lead to reproductive decline with increasing age among all large-bodied mammals (Jones, 1975; Graham et al., 1979; Austad, 1994; Leidy, 1994; Packer et al., 1998). Although some have attempted to make a special case for human reproductive decline and menopause by suggesting it is uniquely human (Pavelka et al., 1991; Turke, 1997; Hawkes et al., 1997, 1998), oocyte depletion over the life course appears universal among female mammals (Graham et al., 1979; Leidy, 1994; Austad, 1997; Packer et al., 1998). Austad (1994, 1997) describes reproductive decline with increasing age among mammalian females in such a fashion that one must conclude it is a pleisomorphic LH trait (see also Packer et al., 1998). The main reason few female mammals show menopause as currently defined is that few commonly survive past their fifth decade of life, at which age almost all women are postmenopausal (see Figure 3.5 in Crews, 2003, p. 114).

Similar to oocyte depletion, gestation lengths of about 9 months and completion of 95% brain growth by about 5–6 years are shared among living large-bodied apes. These commonalities suggest similar trait states may have characterized earlier hominins and hominoids, perhaps representing basic large-bodied ape characteristics. Gestation length may be a symplesiomorphic or synapomorphic LH trait within the large-bodied ape clade. Elephants show longer gestation lengths, while marsupials have utilized shorter gestation lengths and pouches for extended infant care. Such examples show that gestation length is an evolutionarily malleable trait. Both human and marsupial newborns are altricial. Evolving hominins and marsupials may have faced a similar problem – gestating neonates in a limited amount of time – and hit upon a similar solution, altricial newborns. Although their developmental processes are distinctive, their solution, high PI postbirthing, is the same. Sociality and food sharing characterize many animal species. However, they are not associated with extended life spans in most. Extended life spans and multigeneration social groups are shared by humans, whales, elephants, chimps, gorillas, lions, and other social animals. Neither is uniquely human, both are pleisomorphic and synapomorphic mammalian LH traits.

Humans invest early in life in their neurological development by slowing their somatic development in utero and producing secondarily altricial offspring (Smith, 1989). Born less physically developed, compared to other large-bodied apes, humans require an extended growth period. Investments in neurological tissues in utero and the first five years of life delay physical development. Altricial newborns, early neurological development, and slow physical development require late attainment of reproductive maturity compared to more rapidly developing large-bodied apes (Crews, 2003; Crews and Bogin, 2010). They also necessitate extensive postweaning PI, thereby extending the human MNLS. Slow developing secondarily altricial offspring with early neurological development are uniquely human, an autapomorphic trait (see Smith, 1989; Peccei, 1995; Crews, 2003; Crews and Bogin, 2010). Secondarily altricial offspring are not seen in any other primate genus or species. These secondarily altricial offspring are the basis of human LH evolution and likely evolved with bipedalism and culture. Thus, they require their own evolutionary explanation perhaps as a response to early bipeds' limited pelvic width. Freeloading altricial offspring are dependent on their parents' abilities to provide continual PI over 15–20 years to offspring. Although PI occurs in all primates, its expansion to 15–20+ years is uniquely human.

Modern humans also show adaptations in their gut morphology that differ from other large-bodied apes. Other apes have their greatest gut volume in their colons, 45%, with only 14–29% small intestine (Milton, 1999). In humans, 56% of the total gut is small intestine, while the colon accounts for only 17–23% (Milton, 1999). Based upon gut proportions, large-brained hominins adapted to high calorie diets digested mainly in the small intestine. Most other large-bodied apes favor low-energy foods digested mainly in the colon. Humans show greater plasticity in diet than other large-bodied apes. Their morphology supports the high-energy diet model of hominin evolution (Leonard, 2002; Aiello and Wells, 2002), as do isotopic analyses of human remains (Sponheimer and Lee-Thorp, 1999). Early *Homo* and later humans likely relied upon a variety of protein-rich foods as our modern LH pattern developed; likely including easily obtained marine resources (Kingdon, 1993). Only four great ape species survived the Miocene ape extinction. Different subsistence strategies practiced by each today likely reflect the unique and marginal subsistence strategies adopted by each in the late Miocene (Leigh, 2004).

Rounding out the suite of coadapted human LH traits are provisioning and care of nonkin, constant food-sharing, and shared childcare (Crews, 2003, 2005a; Crews and Gerber, 2003; Bogin 2009; Crews and Bogin, 2010). Modern-day humans are not exact copies of earlier hominins and simple linear models based upon earlier forms poorly fit LH traits of extant primates, as do models based upon extant human foragers poorly fit our ancient ancestors (Crews 2003, 2005a; Crews and Gerber, 2003). Earlier hominins did not possess the current suite of human biocultural adaptations or our highly constructed niches and built environments. Additionally, the four extant large-bodied ape species are not an ideal comparative sample of all large-bodied primates that ever existed for extrapolating Hominin evolution. They are just the ones who managed to survive into the present due to their adoption of unique subsistence strategies and LH patterns (Leigh, 2004 presents some possible reasons). All four have followed separate evolutionary pathways since bipedal apes formed their own lineage some 7 mya.

Human life span

Based upon recovered burials and bioarchaeological reconstructions, Neolithic agriculturalists averaged about 25 years of life. However, nineteenth-century industrialists could expect to survive about 45 years. In some modern populations, women average 85 years of life, while men approach 80 years (Crews, 2005b). Based upon estimated ages and survival patterns observed in living foraging/hunting populations, some suggest that once people survive childhood and their second decade of life (to age 20 years) in such settings, they are likely to live into their 5–6th decade of life (40–50 years) (Hawkes et al., 1997, 1998; Carey and Judge, 2001). Since the late Paleolithic, it is likely that in most human groups some individuals may have survived to complete 50 years of life or more. Nevertheless, it is unlikely that many survived to 70 or 80 years; ages that few survive to today in many populations. Among preagricultural groups of foragers, gathers, and hunters of the Paleolithic with small local populations ($N \sim 500$), it is not probable that many survived even to their 50s. Estimates of survival probabilities based upon estimated birth and death rates among today's foraging populations are not reliable. Actual birthdates seldom are known and data for constructing stable life tables are not available for recent generations. Multiple external factors influence such estimates. In contemporary cosmopolitan settings, life expectancy exceeds 85 years among women in only a couple of settings (Sweden and Japan) (Crews, 2005b; Larke and Crews, 2006). Some expect that maximum life span will achieve 130–150 years by the mid twenty-second century. A total wager of $300 between two well-known gerontological researchers, Steven Austad and S. Jay Olshansky, in 2000, was widely reported. Austad expects someone alive today to survive over 14 decades of life. If someone does, in 2150 his heirs will receive the original wager and accumulated interest; if not, Olshansky's descendents win. This bet is significant not only for what it says about predicted life expectancy in the future, but also for what it implies about PIs in future descendents by provisioning them via social structures current in modern societies. In a recent report, Vaupel and colleagues (Christensen et al., 2009) suggest that if current trends continue, those born since 2000 in cosmopolitan settings may achieve an expectation of life of 100 years.

Reproduction and life span

Across human populations, the number of offspring successfully birthed and reared to reproductive age varies positively, negatively, and not at all with longevity of women (Larke and Crews, 2006). Confounding arises partly because both early and late fecundity are associated with longevity, while total parity correlates poorly with longevity (Larke and Crews, 2006). Variable associations of RS with longevity in women indicate multiple confounding factors (Larke and Crews, 2006). For one, RS is determined less by fecundity, and more by the quality/quantity of PI a women has to invest in offspring. Human PI is more complex, extensive, and longer lasting (both relatively and absolutely) than for any other mammal. This supports a model that human LH coevolved with altricial offspring and their need for extensive long-term PI. Shaped by biocultural factors, this need influenced all of human physiology, LH variation, reproduction, senescence, and patterns of longevity (see also Crews and Bogin, 2010).

Analyzing wealth, longevity, reproductive, and other LH variables, Larke and Crews (2006) showed that resources and the timing of reproductive events significantly influence life expectancy. Wealth and age at menopause correlated, while mean age at childbearing did not significantly correlate, with life span. When entered jointly in a regression model with age at first birth and wealth, mean age at childbearing significantly positively predicts life expectancy. Colinearity with mean age at first birth partly explains reports that mean age at childbearing is associated with longevity. Across national populations, the strongest correlates with life expectancy are wealth and age at first birth (see Figures 2 and 4 in Larke and Crews, 2006). When entered into multiple regression models with age at first birth and age at mean childbearing, wealth's effect on life expectancy reduces below statistical significance. These results support suggestions by Crews (2003, 2008; see also Larke and Crews, 2006; Bogin, 2009; Crews and Bogin, 2010) that later reproduction leads to increased RC and, ultimately, greater longevity. Across nations, delayed reproduction is associated with longer life spans, suggesting that accumulating resources before reproducing produces greater RC. High correlations of per capita income and mean age at first birth with life expectancy suggest that cultural mechanisms provide opportunities to accumulate RC and increase women's reproductive opportunities at later ages. More benign, culturally constructed environments lead to greater longevity by providing women opportunities to increase their RC and their RS at later ages.

Benign/cushy environments

Human niche construction has created built environments for humans to reside and reproduce in for over 200 000 years (Odling-Smee et al., 2003; Crews, 2007, 2008; Crews and Bogin, 2010). Built environments allow people to live longer than ever before. Today's constructed environments allow humans to retain their high functioning stress responses, immune, and housekeeping systems well into their later years. A less positive result of constructed environments is that children in modern environments have high asthma and allergy rates. Exposures to pathogens and allergens among individuals growing and developing in less-constructed environments may protect children from such aliments. Modern urban slums may be "dirtier" and produce more infections and parasitic exposures than traditional farms and tropical settings. In cosmopolitan settings, less exposure to childhood diseases produces fewer immune responses and less inflammation, contributing to retention of RC and extended life spans (Crews, 2003, 2007, 2008; Drenos and Kirkwood, 2004; Finch and Crimmins, 2004; Crimmins and Finch, 2005, 2006; Finch, 2005; Arking, 2006; Drenos et al., 2006).

Well-nourished humans living in built environments experience more benign living conditions than did their evolutionary forbearers. Laboratory, domesticated, some island, and provisioned animal populations live better than their wild-living counterparts and less isolated populations. Benign settings provided opportunities for slowing senescence and enhancing longevity not available in more stressful environments. Large increases in life span should not occur in high stress settings (Crews and Bogin, 2010). In such circumstances, organisms are frequently at or close to their limits of survival, metabolic efficiency is at a premium, and calories are best invested in reproductive effort (Parsons, 2002, 2003). Multiple genes for cellular stress resistance, some predisposing to high metabolic efficiency, enabling adaptation to energy-poor environments, were adapted to harsher environments with different selective pressures than prevail today (Parsons, 1996, 2002, 2003; Crews, 2003, 2008; Arking, 2006; Drenos et al., 2006). In natural-living populations, high vitality, stress resistance, and resilience enhance RC and survival, and continued survival depends upon substantial investments in allostatic responses. Genes for stress resistance, immune competence, and energetic and metabolic efficiency selectively advantaged in environments of evolutionary adaptation underlie modern human allostatic processes. Today LH and longevity vary with ecological circumstances. As culturally based niche construction created cushy built environments, extended life spans evolved in hominins. Similarly, only after defenses against predation (i.e., large body size, body armor) were established did extended life spans appear in whales, tortoises/turtles, sharks, and elephants.

Older humans express their innate inclinations toward slower senescence and later mortality when living in modern-day cushy environments (Perls 2001, 2006; Drenos and Kirkwood, 2004; Crews, 2005a, 2007, 2008; Drenos et al., 2006). A "late-life" phase of LH following the senescent stage may occur among some species, laboratory organisms, and very old humans, in cushy/benign environments (Rose et al., 2005a; Drenos et al., 2006; Rauser et al., 2006). Social inequality likely increased as the Neolithic revolution led to settled agriculture and large-scale niche construction. Not all participate equally in benefits from cushy environments. Inequalities within cosmopolitan societies and between them and traditional ones contributed to the rapid spread of the third epidemiological transition worldwide (Armelagos et al., 2005). Variable disease patterns and life spans even among individuals living in the same environment reflect underlying demographic and physiological heterogeneity.

Frailty and inflammation

As cushier environments developed over recent human history, reduced disease exposures were a cohort phenomenon (Finch and Crimmins, 2004; Crews, 2005b). Swedish demographic data from 1751 to the present show that low early life mortality rates are associated with low adult mortality and greater longevity of the same cohort (Finch and Crimmins, 2004). This may be a general population phenomenon (Crimmins and Finch, 2006). Finch and Crimmins (2004; Crimmins and Finch, 2005, 2006) suggest reduced lifetime exposures to infectious diseases and inflammation contributed to large-scale historical declines in old-age

mortality and increased longevity. Inflammation is described as the "fire of frailty" (Walston and Fried, 1999; Fried et al., 2004; Walston, 2005). Reducing lifelong inflammatory responses reduces secondary somatic damage, thereby improving RC and stamina. Caruso et al. (2005) suggest individuals genetically predisposed to weak inflammatory activity have a better chance of living longer, provided infectious diseases do not kill them. Variable alleles at multiple cytokine loci influence the type and intensity of immune–inflammatory responses, thereby affecting individual life expectancy. At least one proinflammatory protein (APOE4) is associated with earlier mortality among middle-aged and older adults. It appears that cushier environments with fewer pathogens and genetic predispositions to weaker inflammatory activity jointly predispose to longer life.

Cohort morbidity phenotypes also may explain seasonal variation in mortality (Finch and Crimmins, 2004) in agricultural populations. Agricultural diets have ratios of omega-6 to omega-3 fatty acids (2:1) that differ from the diet proposed for ancient hominins (10:1) (FAOUN/WHO, 1980). An altered omega-6 to omega-3 fatty acid ratio from agricultural diets may produce nutritional inflammation. Conversely, by increasing access to fresh fruits and vegetables throughout the year, agriculture may have contributed to reduced stress from ROS and deficiency diseases and slowed some senescent processes.

DISCUSSION

Humans, like all living beings, resulted from evolutionary processes by which radiations of earlier taxonomic groups created new ones. As the kingdom *Animalea* arose from earlier precursors, basic aspects of cellular function were highly conserved through successive radiations. The same evolutionary forces shaped rock-eating lichens, rock lobsters, rock-tool makers, and rocket scientists. They also led to social organisms, nest builders, niche constructors, food sharers, nutcrackers, fire starters, and both microbial and human wine and cheese makers. Evolution produced multiple species dependent upon complex patterns of food acquisition – ants, bees, chimpanzees, humans, lions, wild dogs – using the same basic DNA and cellular structures. This is why animal models and their rates of senescence reveal biomolecular strata occurring within human cells and somas that lead to biological senescence.

Short-lived laboratory organisms are incompetent at successfully aging (Austad, 2003). Still, their DNA, proteins, macromolecular structures, and processes of cellular degradation are similar to those of most organisms. *Drosophila* show broad additive and nonadditive genetic variance for LH traits, ranging from ages at first reproduction and intrinsic dysfunction to fecundity and life span (Arking et al., 2002). Research on CR rodents shows that a nutritionally adequate but CR diet slows LH, reduces age-specific fertility, and increases somatic maintenance. Conversely, laboratory mice eating energy-rich *ad libitum* diets show more rapid LH, early reproduction, and decreased life spans. Even among nonhuman primates, CR improves immune function and reduces inflammation (Ingram et al., 2004, 2005; Nikolich-Zugich and Messaoudi, 2005). The laboratory environment differs from that of wild animals. Laboratory populations have fewer stressors than nonlaboratory. Calorie restriction generally extends mean and maximum life spans of laboratory rodents, but it does not do so for the wild-born. Laboratory experiments show that multiple biomolecular processes, for example telomere shortening and replicative capacity in vitro and antioxidant levels, are related to but may not be directly responsible for in vivo senescence. We expect multiple cellular processes to correlate with whole organism senescence because most are secondary processes, having evolved within already senescing organisms. Others likely result from antagonistic pleiotropy, mutation accumulation, and evolutionary inertia, representing ubiquitous senescence in living cells. At the cellular level, senescence reflects the loss of cellular integrity that results from ROS, mutation, DNA and protein errors, losses of regulatory, repair and housekeeping functions, immunity and self-recognition systems, telomere shortening and lack of replacement of cells, and waste accumulation, changes that characterize all aging cells. Human senescence begins with these same biological processes. Over evolutionary time humans developed additional protections against such senescent alterations, allowing them to show a unique pattern of senescence and LH. Nevertheless, their underlying biology remains that of a large-bodied mammal and ape. The bioculturally constructed niche modern humans have created protects them from many extrinsic stressors experienced by other species and previous hominins. This allows those with intrinsic stamina and RC to express predispositions to longevity. Phenotypes less prone to stress-related, environmental, and intrinsic damage, biological wear and tear, and with high immune competence and mechanical reliability may now express their predispositions and live long lives.

The human constructed niche and longevity

Humankind currently lives in constructed niches and built environments different from those of their ancestors, nonlaboratory organisms, most nonhuman primates, and nondomestic animals. Exceptions include some hominoids raised by humans during language studies, e.g., Washoe, and those like Cheeta, Timmy, and Colo. Learned, shared, adaptive, and accumulated behaviors and beliefs (culture) drive humans to create microenvironmental niches suiting their current perceived needs and tastes. Most organisms must respond to more immediate evolutionary pressures driving their behavior, form, and function. Humankind uses extrasomatic mechanisms to maintain intrinsic integrity. Animal skins and fire reduced dependence on physiological resources. Tools and fire increased the range of obtainable foods and relaxed selection for large teeth and jaws. Each generation, NS favors individuals with phenotypes that increase their likelihood of producing and rearing fertile offspring in their specific environment. Modern humans constantly adapt to a variety of culturally created microenvironments. Genetic adaptations of LH to these new environments may be slow. Easier to detect may be developmental changes that occur *without* genetic alterations through changes in hormonal release, timing, or response (Finch and Rose, 1995; Crews and Bogin, 2010) and patterns of phenotypic plasticity exposed as response to changing environments (Weiss and Buchannan, 2003; Hallgrimsson and Hall, 2005).

Human life history plasticity

Human developmental milestones and the timing and pattern of LH events are plastic (Finch and Rose, 1995; Barker, 1998; Bogin, 1999; Cameron and Demerath, 2002; Crews, 2003; Finch and Crimmins, 2004; Weiss and Buchannan, 2003; Hallgrimsson and Hall, 2005; Kirkwood et al., 2005; Crews and Bogin, 2010). To some degree, the mid-childhood and adolescent growth spurts are genetically programmed. However, they are still sufficiently flexible that their timing, duration, and rapidity vary across populations (Bogin, 1999; Crews and Bogin, 2010). Flexibility is the rule rather than exception in human LH milestones, first and last menses, age-at-first birth, age at death, and the point of MRP (Larke and Crews, 2006). Using culturally derived adaptive strategies humans change their environments and allow their somas to respond to new and old stressors in different ways (Baker, 1984).

Cultural processes led to animal domestication, increasing the general availability of protein and fats (Larsen, 2003). This may have led to increased height, weight, and improved nutrition. Animal domestication also brought novel infectious diseases (zoonoses) to humans, increasing frequencies of alleles promoting disease resistance. Animal and plant domestication and, subsequently, agriculture exposed humans to more than just new nutritional stresses. Population density increased and permanent settlements followed increased availability of calories from agriculture. This created an ecological setting wherein infectious diseases and nutritional deficiencies dramatically increased (Larsen, 2000). As population densities and trading networks expanded, infectious diseases flourished and spread in villages, towns, and cities that developed (Fenner, 1970). Skeletal remains suggest measures of health declined following the innovation of agriculture until recent epidemiological transitions (Larsen, 2000; Howell and Pfeiffer, 2006).

Human LH patterns respond to culturally derived adaptive strategies. As humans created new micro-environments within which NS and developmental plasticity occurred, they exposed a reservoir of genetic and phenotypic plasticity for LH. In the classic Livingstonian model, culturally derived behaviors allow humans to change their environment, eliminating, creating, and altering selection pressures on themselves (Livingstone, 1958; Baker, 1984). As cultural forces created cushy living environments, selection relaxed for some traits, created new stressors, and allowed a fuller expression of human developmental and LH plasticity (Crews, 2007).

Culture, diet, and life history

Large-brained neonates and toddlers require calories and nutrients beyond what is provided by diets of other large-bodied apes (Aiello and Wells, 2002; Leonard, 2002). One cultural development, using fire, allowed later *Homo* to consume toxic and difficult to process foods (Wrangham et al., 1999; Wrangham and Conklin-Brittain, 2003). Fire allowed intensive exploitation of old and new food sources and influenced human LH, skeletal biology, and digestive anatomy and physiology. Humankind's shorter colon than other apes reflects bioculturally mediated selective pressures and adaptation. Compared to other mammals and large-bodied apes, modern human gut anatomy suggests modifications toward high throughputs of calorie-dense foods (Milton, 1999). Dietary changes toward meat eating and related technological innovations coincided with brain expansion of the *Homo* species and our modern LH. Use of fire and meat consumption likely predisposed evolving hominins to small intestines of greater length as they digested more cooked and uncooked protein. Conversely, shorter colons likely followed diets with less fiber content. Cultural innovations improved physiological extractive efficiency compared to other large-bodied apes, and also allowed *Homo* species, including eventually humans, to consume more calorie-dense and fewer low-calorie foods (Larsen, 2003).

New technologies and richer diets, freed somatic resources previously invested in food processing for investments elsewhere. They were available for promoting early neurological development, slow steady growth over an extended growth phase, somatic

maintenance, development of RC, and additional reproductive effort (Crews, 2003; Bogin, 2009; Crews and Bogin, 2010). As RS and fitness became dependent upon learned activities and social relations (culture), biocultural selection pressures for neurological development likely increased predispositions toward processes that enhanced mental abilities. Based upon available data, *Homo ergaster* may have birthed less precocial more altricial young than did earlier hominins. Altricial human infants require high postgestational PI to support their brain and somatic growth and development. Culturally constructed environments provide a safe haven for altricial offspring to not only grow and develop, but to be a viable ESS (Crews, 2003; Odling-Smee et al., 2003). Without a culturally constructed environmental niche in which to survive, humankind's altricial offspring may not have been an ESS. Human encephalization seems to have followed cultural developments toward more stimulating environments with fire, tools, social activities, and complex social relationships. Laboratory experiments with rodents demonstrate that more complex (enriched) environments lead to more highly networked brains. More highly networked brains seem to work in ways considered more intelligent by humans. Animals with more networked brains solve experimental tasks considered relevant to humans more rapidly. Cultural developments, technologies, and processes enriched and complicated the lives of evolving *Homo* species, likely leading to a more intelligent longer-lived species.

Many of today's "modern" human populations tend to consume only a small portion of edibles in their environments. In today's "more marginal" environments, among predecessors of recently "modernized" groups, and in more densely populated settings, a broader array of edibles – larvae, insects, spiders, amphibians, reptiles, bats, and other less desirable animals and plants – are and were consumed. Archaeologically and historically, the amount of time and human energy devoted to food preparation in cosmopolitan societies *decreased* as production increased. At the same time, rates of acute, infectious diseases and childhood death rates declined. Among humans, time of weaning is highly variable and related to multiple cultural and biological, and physiological factors. This LH event is relatively easy to determine bioarchaeologically (Larsen, 2000). Selective pressures on age at weaning apparently decreased as humans developed adequate and effective weaning foods. Availability of quality weaning foods relaxed selective pressure on humans, releasing a mortality bottleneck limiting the proportion of children attaining adulthood. During much of early hominin evolution, those who survived infancy likely faced high selective pressures after weaning through puberty, as is seen in marginal environments yet today. Those who survived likely possessed genetic predispositions promoting metabolic efficiency, efficient cellular housekeeping, less inflammation, better DNA repair, and more efficient immune- and stress-responses. In such settings, those attaining adulthood likely were hardier and better able to combat prevailing environmental stressors. As childhood environmental assaults were reduced and less energy was expended on daily food gathering and preparation, resources were freed to be retained as RC through adulthood and into late life. As life styles became cushier, humans produced and retained more total energetic capacity over their early lives (Crews, 2003, 2007; Drenos et al., 2006; Bogin, 2009; Crews and Bogin, 2010). Eventually, humans capable of retaining RC into later decades of life showed increased life spans (Crews, 2003; Bogin, 2009; Crews and Bogin, 2010).

Evolutionarily the majority of life span extension occurred over a relatively short time. It is not likely that NS, as described in traditional or neo-Darwinian models, greatly reshaped the distribution of senescence-slowing or longevity-enhancing alleles in this period. Prior to historical times, evolutionary alterations in human LH occurred as cultural became humankind's primary adaptive response, biocultural feedback loops began determining the pace and pattern of early LH, and altricial offspring became an ESS. Altricial offspring require more time to grow and develop. The need to nourish altricial offspring resonated throughout human LH, slowing the pace of infancy, inserting childhood and adolescent periods and a puberty growth spurt, extending the ages of reproductive effort, and ultimately increasing selective pressures on somatic maintenance (Bogin, 2009; Crews and Bogin, 2010). It is not likely that selection acted directly to increase human life span, because "selection to lengthen life is generally weak" (Stearns, 1992, p. 20; see also Carey and Judge, 2001; Crews and Ice, 2010). Rather, NS molds somatic responses to slow and halt already existing processes of senescence. Secondly, NS acts most strongly in early life and the period of maximum reproductive effort. Its effects on late-life are neither positive nor negative because late-life is beyond the purview of NS.

Models for the evolution of human life history

Models proposing "man the hunter" and "women the gatherer" (e.g., Lee and DeVore, 1969; Lovejoy, 1981) were replaced by models suggesting that grandmothers or future reproduction of relatives (Hill and Hurtado, 1991; Hawkes et al., 1997, 1998; Alvarez, 2000), adult mortality (Charnov, 1993; Stearns et al., 2000), or learning and embodied capital (Kaplan, 1996; Kaplan et al., 2000, 2003; Bird and Bird, 2002; Kaplan and Robson, 2002) altered evolutionary pressures on

humans and promoted modern human patterns of LH, senescence, and longevity. Their general focus has been to obtain data from modern human populations and extrapolate it into the past, with little reference to comparative mammalian biology. These extrapolations often are based upon selection extending longevity; an unlikely process if selection to lengthen life is generally weak (Stearns, 1992; Carey and Judge, 2001; Crews and Ice, 2010).

Models based upon ecology observations of multiple extant species and populations reveal a commonality of senescent processes across sexually reproducing animals (Jones, 1975; Garvilov and Garvilova, 1991; Charnov, 1993; Finch and Rose, 1995; Austad, 1997; Arking et al., 2002; Crews, 2003; Holliday, 2004; Austad and Fischer, 2005). Human longevity is not unique. Whales, tortoises, sharks, and elephants are equally, or *more*, long-lived. Among chimpanzees, bonobos, gorillas, modern hunter-gatherers, foragers, and agriculturalists females migrate from their natal groups upon maturity. A behavior that likely characterized earlier hominins and most modern humans. In such social settings, grandchildren seldom see their maternal grandmother. Thus, in most species, maternal grandmothers live in different locations and do not have opportunities to invest in their daughter's grandchildren as proposed elsewhere (Hawkes et al., 1997, 1998; Alvarez, 2000).

As parts of a species' LH strategy, correlations among LH traits and their timing are expected, but correlation is not causation. Thus, ages at menarche, menopause, first birth, MRP, and the lengths of life span, growth span, infancy and childhood, and prereproductive and adult mortality rates should be correlated. Longer periods of growth and development influence how investments in somatic maintenance, and the length of the reproductive phase are apportioned. Events occurring in early life determine the MNLS, the pace of growth and development, and the age at MRP, determining in part the reproductive span and how long intrinsic adult mortality must be delayed to allow sufficient reproductive effort to accomplish the necessary tasks of life. Mammals require growth and reproductive phases for a successful ESS. Over time, somatic stabilizing factors evolved to protect the developing human reproductive unit during its maturation and reproduction phases in relatively harsh environments. By slowing intrinsic senescent processes they delayed somatic dysfunction and altered intrinsic and extrinsic mortality rates, producing correlations with other coadapted LH traits.

As a population's gene pool evolves toward a better fit to local environmental pressures, any alleles improving survival during the growth and reproductive phases of life should be advantaged. Alleles with late-acting effects on survival, beyond the ages of reproductive effort, whether positive or negative, are neither disadvantaged nor advantaged. Because their frequencies are unaffected by NS, multiple alleles predisposing to better or worse late-life survival, may be active during the senescent phase of LH. As biocultural processes continue to provide a means for older persons to increase their own RS and IF, additional advantages accrue to alleles predisposing to improved late-life survival. Considering that having a significant number of elders in any one local population is a relatively recent phenomenon (Caspari and Lee, 2004), it is not likely that such alleles were greatly advantaged until the advent of modern humans.

Across human populations, older surviving men have more offspring than men who die at younger ages. Greater variation in male than female RS allows men to benefit more from long life than women. There are intrinsic constraints on completed fertility of women, but not men. This may be more relevant than menopause and grandmothers to life span and LH evolution in humans (Marlowe, 2000; Crews, 2003, 2008). Genghis Kahn and his male relatives are hypothesized to have contributed their Y-chromosome and to be the direct ancestors of about 8% of contemporary Chinese men living in central Asia (about 16.7 million or 0.5% of all living men) (Zerjel et al., 2003). Conversely, a significant number of men fail to ever sire children. This may be related to why men may show lower rates of senescence at extreme old age than do women (Gavrilov and Gavrilova, 1991; Carey, 1995; Graves et al., 2006). Alleles predisposing to somatically competent men likely were advantaged in environments of evolutionary adaptation. In today's cushy environments these competencies may be expressed as greater RC at older ages and longevity.

CONCLUSIONS

Natural selection for longevity may not be the reason for modern humankind's seemingly slow senescence. It is unlikely that modern humans would senescence as slowly in any of our ancestral environments as they do in today's constructed niches. That is, Cheeta would never have attained 75 years in the wild. Humans continue to face high selective pressures, but today's individuals are descended from ancestors who were reproductively successful in harsher environments. Today, these pressures are avoided by living in built environments created as sociocultural adaptations to external stressors. Most current scenarios for human senescence and longevity fail to consider these factors. Contrary to many evolutionary models based on selection for longevity, a biocultural model suggests that cultural developments, niche construction, and culturally derived adaptive strategies altered the ecological

settings of early and late *Homo* species reducing intrinsic dysfunction and extrinsic mortality rates. These changes allowed individuals to avoid extrinsic stressors and retain RC into later life. They also provided a niche wherein secondarily altricial offspring could survive and prosper as an ESS. Secondarily altricial human newborns continue growth and development well into their second decade of life. The second decade of life is the first decade of reproductive effort among other large-bodied apes. Biocultural evolution during the earliest phases of life altered all later aspects of LH, modulating growth and development, reproduction, and length of life (Bogin, 2009; Crews and Bogin, 2010).

Human LH, senescence, and longevity are based upon a suite of characteristics – encephalization, non-genetically programmed behaviors as survival strategies, sociality, and group living – shared with many primate species (Shea, 1998; Crews, 2003). Parental investment, social groupings, incipient culture, group hunting, group protection, and sharing of meat are all seen among other large-bodied apes. As the hominin large-bodied ape clade evolved, several uniquely hominin trends, beyond bipedalism and culture, emerged. Culture and hands freed from locomotion led to additional uniquely human traits, extensive use of tools and fire, niche construction, secondarily altricial offspring, slow development to reproductive maturity, mutually supporting social networks of kin, and lifelong PI.

In modern cosmopolitan settings, most women live to experience menopause; however, in the late nineteenth and early twentieth centuries only 20% of Native American, 30% of Black, and 60% of White US women survived to age 55, respectively (see Figure 3.5 in Crews, 2003, p. 114). Even in the twenty-first century many women in Third-world settings do not survive past menopause. It seems unlikely that many females did so among earlier *Homo* species. Nor are life spans of over 50 years often observed among wild living non-human large-bodied female apes or mammals of known age. Ovarian depletion, reduced fertility in the late fourth and early fifth decades, and menopause occur in all female mammals who survive sufficiently long, past their fourth decade of life (Austad, 1994; Leidy, 1994; Rogers, 1993; Packer et al., 1998). Contrary to several selective models of human LH, lifeways of modern-day traditional living populations are not a window on earlier *Homo* and human lifeways. Nor do they represent the lifeways of hominins when our current LH developed. Every extant population is exposed to modern cultures, lifeways, and multiple outside influences (Crews and Gerber, 2003). Societies not so influenced prior to the arrival of missionaries and anthropologists included residents of tropical South America and Africa and highland New Guinea at first contact. Although isolated, none was struggling to feed itself, suggesting little energetic trade-off between reproduction and self-maintenance. They lived in highly integrated societies. They had sufficient free time to elaborate extensive social activities. Their traditional lifeways do not conform to evolutionary scenarios based upon scarcity of resources or difficult processing to extract limited resources as are reported among some modern-day foragers (Kaplan, 1996; Kaplan et al., 2000, 2003; Bird and Bird, 2002).

It is unclear if late-life stages when both senescence and mortality rates slow their increases exist among multiple species. It is certain that no currently popular model of human LH evolution explains a mortality plateau in the last decades of human life. Conversely, biocultural evolution leading to the retention of somatic resources over a lifetime of living in cushy environments is compatible with a late-life mortality plateau. Understanding human senescence and longevity depends as much on comparisons with short-lived and nonmammalian taxa, as it does on comparisons with other long-lived and mammalian organisms. Increased focus on how biocultural influences on humans changed survival probabilities and longevity will be a positive development toward understanding LH evolution across all species.

Understanding how modern humans became relatively long-lived compared to other apes requires comparisons with other primates, mammals, and life forms and examinations of across population human LH variation. Among humans RS, reproductive effort, RC, stress resistance, life span, longevity, and senescence are highly variable individual traits. Not even identical twins show the same ages at death and familial correlations for longevity often are low. Mechanistic models for human LH evolution often are retrospective and do not account for the influences of biocultural evolution. Biological processes contribute to human senescence via mostly unknown genes and unidentified physiological pathways. Pushed by the grand theory of evolution by NS, existing models of human LH evolution often lack mid-level theory, failing to identify or hypothesize genetic or physiological mechanisms, or physiological pathways linking genes and LH. No system of accounting by which NS may tally the influences of epigenetic factors and learning or the future RS of offspring and calculate the probability of grand-offspring has been proposed. Clearly intrinsic factors acting throughout the growth phase influence adult mortality and human life spans. Natural selection likely acted on early LH factors linked to RS that also improved resistance to extrinsic stressors, retention of RC, and ultimately extended life spans in culturally created settings (Crews and Bogin, 2010). However, NS may never have acted directly to extend human life span or postreproductive survival. Rather, life spans extending well beyond those of other large-bodied apes appear to be an epiphenomenon of NS acting on aspects of early LH, altricial offspring, slow growth, and an extended growth phase that stretched out the period for attaining MRP (Bogin, 2009; Crews and Bogin, 2010). This early LH is now played out in culturally constructed environments that allow retention of RC and lead to somatic maintenance into the ninth, tenth, and later decades of human life.

DISCUSSION POINTS

1. The disposable soma theory for the evolution of senescence is based upon what aspects of reproductive biology? How have recent developments in understanding the living histories of single-celled organisms altered this view? How would you reframe this model in light of these new results? What are two mechanisms by which senescence-promoting alleles may be retained in the genome over evolutionary time?

2. What methods might one use to measure phenotypic aging and why should it be measured? What methods might one use to measure biological senescence at the cellular level and why should it be measured? What are some differences between senescence and aging?

3. Given what you have read and know of modern human life history and variation, describe how because humans are cultural animals their life history may evolve differently from nonculture-using animals.

4. What are some biological factors that pattern life history (LH) of animals and mammals? Does modern human LH show factors other than biology that influence their LH? Are humans so unique in their LH that we need special models to explain their LH evolution?

5. Austad critiqued earlier definitions of senescence and aging for not including any mention of reproduction and fertility. Why are reproduction and fertility important for defining senescence and aging? What are the meanings of "age of maximum reproductive potential" (MRP) and "minimum necessary life span" (MNLS)? How does the age of MRP influence the MNLS?

ACKNOWLEDGMENTS

Sincere thanks to the editor Michael P. Muehlenbein for his efforts in bringing this comprehensive volume together and inviting us to include this work. We also acknowledge the insightful comments and suggestions of two reviewers that improved this contribution. Finally, we wish to thank Courtney McFadden and Rhea Alleyne for typing multiple versions of this manuscript.

REFERENCES

Aiello, L. C. and Wells, J. C. K. (2002). Energetics and the evolution of the genus *Homo*. *Annual Review of Anthropology*, **31**, 323–338.

Allsopp, R. C. (1996). Models of initiation of replicative senescence by loss of telomeric DNA. *Experimental Gerontology*, **31**, 235–243.

Alvarez, H. P. (2000). Grandmother hypothesis and primate life histories. *American Journal of Physical Anthropology*, **113**, 435–450.

Anderson-Ranberg, K., Christensen, K., Jeune, B., et al. (1999). Declining physical abilities with age: a cross-sectional study of older twins and centenarians in Denmark. *Age and Ageing*, **28**, 373–377.

Armelagos, G. J., Brown, P. J. and Turner, B. (2005). Evolutionary, historical and political economic perspectives on health and disease. *Social Science and Medicine*, **61**, 755–765.

Arking, R. (2006). *The Biology of Aging: Observations and Principles*, 3rd edn. New York: Oxford University Press.

Arking, R. (2007). Human energetics and the predicted response to dietary restriction. *Rejuvenation Research*, **10**, 261–279.

Arking, R., Novoseltseva, J., Hwangbo, D. S., et al. (2002). Different age-specific demographic profiles are generated in the same normal-lived *Drosophila* strain by different longevity stimuli. *Journals of Gerontology: Biological Sciences*, **57**, B390–B398.

Atzmon, G., Rincon, M., Schechter, C. B., et al. (2006). Lipoprotein genotype and conserved pathway for exceptional longevity in humans. *PLoS Biology*, **4**, e113.

Austad, S. N. (1991). Mammalian aging metabolism and ecology: evidence from the bats and marsupials. *Journal of Gerontology*, **46**, B46–B53.

Austad, S. N. (1994). Menopause – an evolutionary perspective. *Experimental Gerontology*, **29**, 255–263.

Austad, S. N. (1997). Comparative aging and life histories in mammals. *Experimental Gerontology*, **32**, 23–38.

Austad, S. N. (2003). Introduction to animal models. *Experimental Gerontology*, **38**, 1327.

Austad, S. N. (2005a). Cell senescence in the dock. *Science*, **308**, 54.

Austad, S. N. (2005b). Diverse aging rates in metazoans: targets for functional genomics. *Mechanisms of Ageing and Development*, **126**, 43–49.

Austad, S. N. and Fischer, K. E. (2005). Primate longevity: its place in the mammalian scheme. *American Journal of Primatology*, **28**, 251–261.

Aviv, A. (2002). Telomeres, sex, reactive oxygen species, and human cardiovascular aging. *Journal of Molecular Medicine*, **80**, 689–695.

Baker, P. T. (1984). The adaptive limits of human populations. *Man*, **19**, 1–14.

Barker, D. J. P. (1998). *Mothers, Babies, and Health in Later Life*, 2nd edn. Edinburgh: Churchill Livingstone.

Bird, R. B. and Bird, D. W. (2002). Constraints of knowing or constraints of growing? Fishing and collecting by the children of Mer. *Human Nature*, **13**, 239–267.

Blasco, M. A. (2005). Telomeres and human disease: ageing, cancer and beyond. *Nature Reviews Genetics*, **6**, 611–622.

Bogin, B. (1999). *Patterns of Human Growth*. New York: Cambridge University Press.

Bogin, B. (2009). Childhood, adolescence, and longevity: a model of reserve capacity in human life history. *American Journal of Human Biology*, **21**, 567–577.

Bredesen, D. E. (2004). The non-existent aging program: how does it work? *Aging Cell*, **3**, 255–259.

Burger, J. M. S. and Promislow D. E. L. (2006). Are functional and demographic senescence genetically independent? *Experimental Gerontology*, **41**, 1108–1016.

Cameron, N. and Demerath, E. (2002). Critical periods in human growth and their relationship to disease of aging. *Yearbook of Physical Anthropology*, **45**, 159–184.

Campisi, J. (2003). Cellular senescence and apoptosis: how cellular responses might influence aging phenotypes. *Experimental Gerontology*, **38**, 5.

Carey, J. R. (1995). A male–female longevity paradox in Medfly cohorts. *Journal of Animal Ecology*, **64**, 107–116.

Carey, J. R. and Judge, D. S. (2001). Life span extension in humans is self-reinforcing: a general theory of longevity. *Population and Development Review*, **27**, 411–436.

Carey, J. R., Liedo, P., Orozco, D., et al. (1992). Slowing of mortality-rates at older ages in large Medfly cohorts. *Science*, **258**, 457–461.

Carnes, B. A. and Olshansky, S. J. (2001). Heterogeneity and its biodemographic implications for longevity and mortality. *Experimental Gerontology*, **36**, 419–430.

Caruso, C., Candore, G., Colonna-Romano, G., et al. (2005). Inflammation and life-span. *Science*, **307**, 208.

Caspari, R. and Lee, S. H. (2004). Older age becomes common late in human evolution. *Proceedings of the National Academy of Sciences of the United States of America*, **101**, 10895–10900.

Charlesworth, B. (2001). Patterns of age-specific means and genetic variances of mortality rates predicted by the mutation-accumulation theory of aging. *Journal of Theoretical Biology*, **210**, 47–65.

Charnov, E. L. (1993). *Life History Invariants: Some Explanations of Symmetry in Evolutionary Ecology*. Oxford: Oxford University Press.

Chondrogianni, N. and Gonos, E. S. (2005). Proteasome dysfunction in mammalian aging: steps and factors involved. *Experimental Gerontology*, **40**, 931–938.

Christensen, K., Johnson, T. E. and Vaupel, J. W. (2006). The quest for genetic determinants of human longevity: challenges and insights. *Nature Reviews Genetics*, **7**, 436–448.

Christensen, K., Doblhammer, G., Rau, R., et al. (2009) Ageing populations: the challenges ahead. *Lancet*, **374**, 1196–1208.

Clark, W. R. (1999). *A Means to an End: the Biological Basis of Aging and Death*. New York: Oxford University Press.

Crews, D. E. (2003). *Human Senescence: Evolutionary and Biocultural Perspectives*. New York: Cambridge University Press.

Crews, D. E. (2005a). Evolutionary perspectives on human longevity and frailty. In *Longevity and Frailty*, J. R. Carey, J. M. Robine and Y. Christen (eds). New York: Springer-Verlag, pp. 57–65.

Crews, D. E. (2005b). Artificial environments and an aging population: designing for age-related functional loss.

Journal of Physiological Anthropology and Applied Human Sciences, **24**, 103–109.

Crews, D. E. (2007). Senescence, aging, and disease. *Journal of Physiological Anthropology*, **26**, 365–372.

Crews, D. E. (2008). Co-evolution of human culture, mating strategies and longevity. In *Aging Related Problems in Past and Present Populations*, C. Susanne and E. Bodzsar (eds). Budapest: Eötvös University Press, pp. 9–29.

Crews, D. E. and Bogin, B. (2010). Growth, development, senescence, and aging: a life history perspective. In *A Companion to Biological Anthropology*, C. S. Larsen (ed.). New York: Blackwell Publishing, in press.

Crews, D. E. and Gerber, L. M. (1994). Chronic degenerative diseases, aging. In *Biological Anthropology and Aging: Perspectives on Human Variation over the Life Span*, D. E. Crews and R. M. Garutto (eds). Oxford: Oxford University Press, pp. 154–181.

Crews, D. E. and Gerber, L. M. (2003). Reconstructing life histories of hominids and humans. *Collegium Antropologicum*, **27**, 7–22.

Crews, D. E. and Ice, G. J. (2011). Aging, senescence, and human variation. In *Textbook of Human Biology: an Evolutionary and Biocultural Perspective*, vol. 2, S. Stinson, B. Bogin and D. O'Rourke (eds). New York: Wiley-Liss, in press.

Crimmins, E. M. and Finch, C. E. (2005). Early life conditions affect historical change in old-age mortality. In *Longevity and Frailty*, J. R. Carey, J. M. Robine and Y. Christen (eds). New York: Springer-Verlag, pp. 99–106.

Crimmins, E. M. and Finch, C. E. (2006). Infection, inflammation, height and longevity. *Proceedings of the National Academy of Sciences of the United States of America*, **103**, 498–503.

Cristofalo, V. J., Allen, R. G., Pignolo, R. J., et al. (1998). Relationship between donor age and the replicative life-span of human cells in culture: a reevaluation. *Proceedings of the National Academy of Sciences of the United States of America*, **95**, 10614–10619.

Cristofalo, V. G., Tresini, M., Francis, M. K., et al. (1999). Biological theories of senescence. In *Handbook of Theories of Aging*, V. L. Bengston and K. W. Shaie (eds). New York: Springer Publishing, pp. 98–112.

Cristofalo, V., Beck, J. and Allen, R. G. (2003). Cell senescence: an evaluation of replicative senescence in culture as a model for cell aging in situ. *Journals of Gerontology Series A*, **58A**, 776.

Curtsinger, J. W., Fukui, H. H., Townsend, D. R., et al. (1992). Demography of genotypes: failure of the limited life-span paradigm in *Drosophila melanogaster*. *Science*, **258**, 461–463.

Darwin, C. R. (1859). *On the Origin of Species by Means of Natural Selection, or the Preservation of Favoured Races in the Struggle for Life*. London: John Murray.

Dawkins, R. (1989). *The Selfish Gene*. New York: Oxford University Press.

deLange, T. (2002). Protection of mammalian telomeres. *Oncogene*, **21**, 532–540.

Dennison, E., Fall, C., Cooper, C., et al. (1997). Prenatal factors influencing long-term outcome. *Hormone Research*, **48**(suppl.), 25–26.

Doblhammer, G. (2000). Reproductive history and mortality later in life: a comparative study of England and Wales and Austria. *Population Studies*, **54**, 169–176.

Doblhammer, G. and Kytir, J. (2001). Compression or expansion of morbidity? Trend in healthy-life expectancy in the elderly Austrian population between 1978 and 1998. *Social Science and Medicine*, **52**, 385–391.

Drenos, F. and Kirkwood, T. B. L. (2004). Modeling the disposable soma theory of ageing. *Mechanisms of Aging and Development*, **126**, 99–103.

Drenos, F., Westendorp, R. and Kirkwood, T. B. L. (2006). Trade-off mediated effects on the genetics of human survival caused by increasingly benign living conditions. *Biogerontology*, **7**, 287–295.

Eisenberg, E. and Levanon, E. Y. (2003). Human housekeeping genes are compact. *Trends in Genetics*, **19**, 362–365.

Fenner, F. (1970). The effects of changing social organization on the infectious diseases of man. In *The Impact of Civilization on the Biology of Man*, S. V. Boyden (ed.). Canberra: Australian National University Press.

Feschotte, C. (2010). Virology: Bornavirus enters the genome. *Nature*, **463**, 39–40.

Finch, C. E. (1990). *Longevity, Senescence, and the Genome*. Chicago: University of Chicago Press.

Finch, C. E. (2005). Developmental origins of aging in the brain and blood vessels: an overview. *Neurobiology of Aging*, **26**, 281–291.

Finch, C. E. and Crimmins, E. M. (2004). Inflammatory exposure and historical changes in human life-spans. *Science*, **305**, 1736–1739.

Finch, C. E. and Rose, M. R. (1995). Hormones and physiological architecture of life history evolution. *Quarterly Review of Biology*, **70**, 1–52.

Fischer, A., Pollack, J., Thalmann, O., et al. (2006). Demographic history and genetic differentiation in apes. *Current Biology*, **16**, 1133–1138.

Foley, C. J. (1986). Aging – a coming of age. *New York State Journal of Medicine*, **86**, 617–618.

Food and Agriculture Organization of the United Nations and the World Health Organization (FAOUN/WHO) (1980). *Dietary Fats and Oils in Human Nutrition*. Rome: Food and Agriculture Organization of the United Nations.

Fried, L. P., Ferrucci, L., Darer, J., et al. (2004). Untangling the concepts of disability, frailty, and comorbidity: implications for improved targeting and care. *Journals of Gerontology Series A*, **59**, 255–263.

Gavrilov, L. A. and Gavrilova, N. S. (1991). *The Biology of Life Span: a Quantitative Approach*, V. P. Skulachev (ed.), J. Payne and L. Payne (trans). New York: Harwood Academic.

Gerber, L. M. and Crews, D. E. (1999). Evolutionary perspectives on chronic diseases: changing environments, life styles, and life expectancy. In *Evolutionary Medicine*. W. R. Trevathan, J. J. McKenna and N. Smith (eds). New York: Oxford University Press, pp. 443–469.

Goodman, A. and Leatherman, T. (1999). *Building a Biocultural Synthesis*. Ann Arbor, MI: University of Michigan Press.

Graham, C. E., Kling, O. R. and Steiner, R. A., et al. (1979). Reproductive senescence in female non-human primates.

In *Aging in Non-human Primates*, D.J. Bowden (ed.). New York: Van Nostrand Reinhold, pp. 183–209.

Graves, B.M., Strand, M. and Lindsay, A.R. (2006). A reassessment of sexual dimorphism in human senescence: theory, evidence, and causation. *American Journal of Human Biology*, **18**, 161–168.

Grotewiel, M.S., Martin, I., Bhandari, P., et al. (2005). Functional senescence in *Drosophila melanogaster*. *Ageing Research Reviews*, **4**, 332–337.

Hahn, M.W., Demuth, J.P. and Han, S.-G. (2007). Accelerated rate of gene gain and loss in primates. *Genetics*, **177**, 1941–1949.

Hallgrimsson, B. and Hall, B.K. (eds) (2005). *Variation: a Central Concept in Biology*. New York: Elsevier.

Hamilton, W.D. (1964). The genetical evolution of social behavior. *Journal of Theoretical Biology*, **7**, 1–16.

Hamilton, W.D. (1966). The moulding of senescene by natural selection. *Journal of Theoretical Biology*, **12**, 12–45.

Hammer, M.F., Skorecki, K., Selig, S., et al. (1997). Y chromosomes of Jewish priests. *Nature*, **385**, 32–35.

Harley, C.B., Futcher, A.B. and Greider, C.W (1990). Telomeres shorten during aging of human fibroblasts. *Nature*, **345**, 458–460.

Harper, G.J. and Crews, D.E. (2000). Aging, senescence, and human variation. In *Human Biology*, S. Stinson, R. Huss-Ashmore and D. O'Rourke (eds). New York: Wiley-Liss, pp. 465–505.

Harper, J.M., Salmon, A.B., Chang, Y., et al. (2006a). Stress resistance and aging: influence of genes and nutrition. *Mechanisms of Ageing and Development*, **127**, 687–694.

Harper, J.M., Leathers, C.W. and Austad, S.N. (2006b). Does caloric restriction extend life in wild mice? *Aging Cell*, **5**, 441–449.

Hart, R.W. and Tuturro, A. (1983). Theories of aging. In *Review of Biological Research in Aging*, vol. 1. New York: Alan R. Liss, pp. 5–18.

Hastie, N.D., Dempster, M., Dunlop, M.G., et al. (1990). Telomere reduction human colorectal carcinoma and with ageing. *Nature*, **346**, 866–868.

Hawkes, K., O'Connell, J.F. and Blurton Jones, N.G., et al. (1997). Hazda women's time allocation, offspring provisioning, and the evolution of long postmenopausal life span. *Current Anthropology*, **48**, 551–577.

Hawkes, K., O'Connell, J.F., Blurton Jones, N.G., et al. (1998). Grandmothering, menopause, and the evolution of human life histories. *Proceedings of the National Academy of Sciences of the United States of America*, **95**, 1336–1339.

Hayflick, L. (1965). The limited in vitro lifetime of human diploid cell strains. *Experimental Cell Research*, **37**, 614–636.

Hayflick, L. (1979). The cell biology of aging. *Journal of Investigative Dermatology*, **73**, 8–14.

Hill, K. and Hurtado, A.M. (1991). The evolution of reproductive senescence and menopause in human females: an evaluation of the grandmother hypothesis. *Human Nature*, **2**, 313–350.

Holliday, R. (1997). Understanding ageing. *Philosophical Transactions of the Royal Society of London. Series B*, **352**, 1793–1797.

Holliday, R. (2004). The multiple and irreversible causes of aging. *Journals of Gerontology Series A*, **59A**, 568–572.

Howell, N. and Pfeiffer, S. (2006). Sexual revolutions: gender and labor at the dawn of agriculture. *Journal of Comparative Family Studies*, **37**, 492–493.

Ingram, D.K., Anson, R.M., De Cabo, R., et al. (2004). Development of calorie restriction mimetics as a prolongevity strategy. *Annals of the New York Academy of Sciences*, **1019**, 412–423.

Ingram, D.K., de Cabo, R., Anson, R.M., et al. (2005). Calorie restriction in nonhuman primates: impacts on aging, disease, and frailty. In *Longevity and Frailty*, J.R. Carey, J.M. Robine, J.P. Michel, et al. (eds). New York: Springer-Verlag, pp. 39–56.

Jaenisch, R. and Bird, A. (2003). Epigenetic regulation of gene expression: how the genome integrates intrinsic and environmental signals. *Nature Genetics*, **33**, 245.

Jones, E.C. (1975). The post-fertile life of non-human primates and other mammals. In *Psychosomatics in Perimenopause*, A. Haspels and H. Musaph (eds). Baltimore: University Park Press, pp. 13–39.

Kaplan, H. (1996). A theory of fertility and parental investment in traditional and modern human societies. *Yearbook of Physical Anthropology*, **39**, 91–135.

Kaplan, H.S. and Robson A.J. (2002). The emergence of humans: the coevolution of intelligence and longevity with intergenerational transfers. *Proceedings of the National Academy of Sciences of the United States of America*, **99**, 10221–10226.

Kaplan, H., Hill, K., Lancaster, J., et al. (2000). A theory of human life history evolution: diet, intelligence, and longevity. *Evolutionary Anthropology*, **9**, 158–185.

Kaplan, H., Lancaster, J. and Robson, A. (2003). Embodied capital and the evolutionary economics of the human life span. *Population and Development Review*, **29**, 152–183.

Kingdon, J. (1993). *Self-Made Man: Human Evolution from Eden to Extinction*. New York: John Wiley and Sons.

Kirkwood, T.B.L. (1977). Evolution of ageing. *Nature*, **270**, 301–304.

Kirkwood, T.B.L. (1990). The disposable soma theory of aging. In *Genetic Effects on Aging II*, D.E. Harrison (ed.). Caldwell, NJ: Teaford Press, pp. 9–19.

Kirkwood, T.B.L. (2002). Evolution of ageing. *Mechanisms of Ageing and Development*, **123**, 737–745.

Kirkwood, T.B.L. (2005). Asymmetry and the origins of ageing. *Mechanisms of Ageing and Development*, **126**, 533–534.

Kirkwood, T.B. and Holliday, R. (1979). The evolution of ageing and longevity. *Proceedings of the Royal Society of London. Series B*, **205**, 531–546.

Kirkwood, T.B.L., Feder, M., Finch, C.E., et al. (2005). What accounts for the wide variation in life span of genetically identical organisms reared in a constant environment? *Mechanisms of Ageing and Development*, **126**, 439–443.

Kuzawa, C.W. (2005). Fetal origins of developmental plasticity: are fetal cues reliable predictors of future nutritional environments? *American Journal of Human Biology*, **17**, 5–21.

Larke, A. and Crews, D.E. (2006). Parental investment, late reproduction, increased reserve capacity are associated with longevity in humans. *Journal of Physiological Anthropology*, **25**, 119–131.

Larsen, C. S. (2000). Reading the bones of La Flordia, *Scientific American*, **282**, 80–85.

Larsen, C. S. (2003). Animal source foods and human health during evolution. *Journal of Nutrition*, **133**, 3893S–3897S.

Lasker, G. W. and Crews, D. E. (1996). Behavioral influences on the evolution of human genetic diversity. *Molecular Phylogenetics and Evolution*, **5**, 232–240.

Lee, R. B. and DeVore, I. (1968). *Man the Hunter*. Chicago: Aldine de Gruyter.

Leidy, L. E. (1994). Biological aspects of menopause: across the lifespan. *Annual Review of Anthropology*, **23**, 231–253.

Leigh, S. R. (2004). Brain growth, life history, and cognition in primate and human evolution. *American Journal of Primatology*, **62**, 139–164.

Leonard, W. R. (2002). Food for thought: dietary change was a driving force in human evolution. *Scientific American*, **287**, 1068–1115.

Livingstone, F. B. (1958). Anthropological implications of sickle cell gene distribution in Africa. *American Anthropologist*, **60**, 533–557.

Lovejoy, C. O. (1981). The origin of man. *Science*, **211**, 341–350.

Luckinbill, L. S., Arking, R., Clare, M. J., et al. (1984). Selection for delayed senescence in *Drosophila melanogaster*. *Evolution*, **38**, 996–1003.

Marlowe, F. (2000). The patriarch hypothesis: an alternative explanation of menopause. *Human Nature*, **11**, 27–42.

Medawar, P. B. (1952). *An Unsolved Problem of Biology*. London: Lewis.

Milton, K. (1999). A hypothesis to explain the role of meat-eating in human evolution. *Evolutionary Anthropology*, **8**, 11–21.

Nikolich-Zugich, J. and Messaoudi, I. (2005). Mice and flies and monkeys too: caloric restriction rejuvenates the aging immune system of non-human primates, *Experimental Gerontology*, **40**, 884–893.

Odling-Smee, F. J., Laland, K. N. and Feldman, M. W. (eds) (2003). *Niche Construction*. Princeton: Princeton University Press.

Olovnikov, A. M. (1996). Telomeres, telomerase and aging: origin of the theory. *Experimental Gerontology*, **31**, 443–448.

Packer, C., Tatar, M. and Collin, A. (1998). Reproductive cessation in female mammals. *Nature*, **392**, 807–810.

Pak, J. W., Herbst, A., Bua, E., et al. (2003). Mitochondrial DNA mutations as a fundamental mechanism in physiological declines associated with aging. *Aging Cell*, **2**, 1.

Parsons, P. A. (1996). The limit to human longevity: an approach through a stress theory of ageing. *Mechanisms of Ageing and Development*, **87**, 211–218.

Parsons, P. A. (2002). Life span: does the limit to survival depend upon metabolic efficiency under stress? *Biogerontology*, **3**, 233–241.

Parsons, P. A. (2003). From the stress theory of aging to energetic and evolutionary expectations for longevity. *Biogerontology*, **4**, 63–73.

Pavelka, M., McDonald, S. and Fedigan, L. (1991). Menopause: a comparative life history perspective. *Yearbook of Physical Anthropology*, **34**, 13–38.

Peccei, J. S. (1995). The origin and evolution of menopause: the altriciality-life span hypothesis. *Ethology and Sociobiology*, **16**, 425–449.

Perls, T. T. (1995). The oldest old. *Scientific American*, **272**, 70–75.

Perls, T. T. (2001). Genetic and phenotypic markers among centenarians. *Journals of Gerontology Series A*, **56**, M67–M70.

Perls, T. T. (2006). The different paths to 100. *American Journal of Clinical Nutrition*, **83**, 84–87.

Perls, T. and Terry, D. (2003). Genetics of exceptional longevity. *Experimental Gerontology*, **38**, 725–730.

Perls, T., Kunkel, L. M. and Puca, A. A. (2002). The genetics of exceptional human longevity. *Journal of the American Geriatrics Society*, **50**, 59–68.

Pletcher, S. D. and Curtsinger, J. W. (1998). Age-specific properties of spontaneous mutations affecting mortality in *Drosophila melanogaster*. *Genetics*, **148**, 287–303.

Rauser, C. L., Mueller, L. D. and Rose, M. R. (2006). The evolution of late life. *Ageing Research Reviews*, **5**, 14–32.

Rogers, A. R. (1993). Why menopause? *Evolutionary Ecology*, **7**, 406–420.

Rose, M. R. (1991). *Evolutionary Biology of Aging*. New York: Oxford University Press.

Rose, M. R., Passananti, H. B., Chippindale, A. K., et al. (2005a). The effects of evolution are local: evidence from experimental evolution in *Drosophila*. *Integrative and Comparative Biology*, **45**, 486–491.

Rose, M. R., Rauser, C. L. and Mueller, L. D. (2005b). Late life: a new frontier for physiology. *Physiological and Biochemical Zoology: PBZ*, **78**, 869–878.

Service, P. M. (2000). Heterogeneity in individual mortality risk and its importance for evolutionary studies of senescence. *American Naturalist*, **156**, 1–13.

Shay, J. W. (1997). Molecular pathogenesis of aging and cancer: are telomeres and telomerase the connection? *Journal of Clinical Pathology*, **50**, 799–800.

Shea, B. (1998). Growth in non-human primates. In *The Cambridge Encyclopedia of Human Growth and Development*, S. J. Ulijaszek, F. E. Johnston and M. A. Preece (eds). New York: Cambridge University Press, pp. 100–103.

Slagboom, P. E., Droog, S. and Broomsma, D. I. (1994). Genetic determination of telomere size in humans: a twin study of three age groups. *American Journal of Human Genetics*, **55**, 876–882.

Smith, B. H. (1989). Growth and development and its significance for early hominid behaviour. *OSSA: International Journal of Skeletal Research*, **14**, 63–96.

Smith, B. H. (1991). Dental development and the evaluation of life history in *Hominidae*. *American Journal of Physical Anthropology*, **86**, 157–174.

Smith, B. H. and Bogin, B. (2000). Evolution of the human life cycle. In *Human Biology: an Evolutionary and Biocultural Perspective*, S. Stinson, B. Bogin, R. Huss-Ashmore, et al. (eds). New York: Wiley-Liss, pp. 377–424.

Sponheimer, M. and Lee-Thorp, J. (1999). Isotopic evidence for the diet of an early hominid, *Australopithecus africanus*. *Science*, **283**, 368–370.

Stearns, S. (1992). *The Evolution of Life Histories*. New York: Oxford University Press.

Stearns, S. C., Ackermann, M., Doebeli, M., et al. (2000). Experimental evolution of aging, growth, and reproduction in fruitflies. *Proceedings of the National Academy of Sciences of the United States of America*, **97**, 3309–3313.

Stewart, E J., Madden, R., Paul G., et al. (2005). Aging and death in an organism that reproduces by morphologically symmetric division. *PLoS Biology*, **3**(2), 295–300.

Turke, P. W. (1997). Hypothesis: menopause discourages infanticide and encourages continued investment by agnates. *Evolution and Human Behavior*, **18**, 3–13.

Vaupel, J. W. (2001). Demographic insights into longevity. *Population: an English Selection*, **13**, 245–259.

Walston, J. D. (2005). Biological markers and the molecular biology of frailty. In *Longevity and Frailty*, J. R. Carey, J. M. Robine, J. P. Michel, et al. (eds). New York: Springer-Verlag, pp. 39–56.

Waltson, J. and Fried, L. P. (1999). Frailty and the older man. *Medical Clinics of North America*, **83**, 1173–1194.

Weiss, K. M. and Buchanan, A. (2003). Evolution by phenotype: a biomedical perspective. *Perspectives in Biology and Medicine*, **46**, 159–182.

Weismann, A. (1881–1889). *Essays upon Heredity and Kindred Biological Problems*, vol. 2, E. B. Poulton and A. E. Shipley (eds), 1892 translation. Oxford: Clarendon Press.

Wich, S. A., Utami-Atmoko, S. S., Setia, T. M., et al. (2004). Life history of wild Sumatran orangutans (*Pongo abelii*). *Journal of Human Evolution*, **47**, 385–398.

Williams, G. C. (1957). Pleiotropy, natural selection, and the evolution of senescence. *Evolution*, **11**, 398–411.

Wilson, E. O. (1975). *Sociobiology: the New Synthesis*. Cambridge, MA: Harvard University Press.

Wrangham, R. W. and Conklin-Brittain, N. (2003). Cooking as a biological trait. *Comparative Biochemistry and Physiology: a Molecular and Integrative Physiology*, **136**, 35–46.

Wrangham R. W., Jones, J. H., Laden, G., et al. (1999). The raw and the stolen: cooking and the ecology of human origins. *Current Anthropology*, **40**, 567–594.

Zerjel, T., Xue, Y. and Bertorelle, G. (2003). The genetic legacy of the Mongols. *American Journal of Human Genetics*, **72**, 717–721.

32 Evolutionary Psychiatry: Mental Disorders and Behavioral Evolution

Brant Wenegrat

INTRODUCTION

Since W. D. Hamilton's seminal work on kin-selection (Hamilton, 1964a, 1964b) advances in the field of behavioral evolution have changed our view of animal and human social behavior. A wide range of social behaviors – altruistic, co-operative, aggressive, parental, and sexual – are now seen as products of past selective forces, many of which were generated by social interactions. Fifty years ago, it was possible for Behaviorists to argue that humans are blank slates whose nature – if such exists – is wholly determined by cultural/environment forces. The most basic emotional attitudes – for example, sexual preferences and parental concern for children – were said to result from training and cultural expectation. That position is no longer tenable, in large part due to concepts from behavioral evolution.

At the same time, important advances have been made in psychiatry. Over the last 40 years, psychiatric researchers have identified mental disorders that can be reliably diagnosed, may affect Darwinian fitness, and are partly caused by genes. A large proportion of people in most or all societies suffer from one or another of these mental disorders. The research on diagnosis culminated in the 1994 publication of the *Diagnostic and Statistical Manual of the American Psychiatric Association, Fourth Edition*, known as the DSM-IV, which lists objective criteria for diagnosing disorders (American Psychiatric Association, 1994). The DSM-IV criteria for specific disorders are described later in this chapter, along with pertinent prevalence, genetic, and fitness data. To give readers a better sense of some disorders, case examples are given.

Insofar as concepts from behavioral evolution can explain human behavior, they might shed light on disorders like those in the DSM-IV (Troisi, 2005). These concepts deal with social life, and social events and experiences are known to trigger disorders, or cause them in some cases. Concomitant social deficits account for the disabilities resulting from these disorders, and psychotherapy, which is a social process, can play a role in treatment. Concepts explaining social life seem especially suited to shed light on disorders with social precipitants, causes, consequences, and remedies.

In this chapter, I review some of the efforts to understand mental disorders and possible treatments for them in terms of concepts drawn from behavioral evolution, a field known as evolutionary psychiatry. Authors and researchers working in this field employ five disparate paradigms (see Nettle, 2004). Although the differences between them are sometimes ignored, these paradigms have differing aims and relative strengths and weaknesses. Many debates in the field result from application of disparate paradigms to the same disorder. This chapter will cite examples from each of the five major paradigms to illustrate their differences, logic, goals, and weaknesses.

THE ADAPTATIONIST PARADIGM

The *adaptationist paradigm* aims to view mental disorders, their milder variations, or illness-related traits as behavioral adaptations that confer or have conferred inclusive fitness benefits. Widely cited models of schizophrenia and schizotypal disorder, depression and mania, and anorexia nervosa illustrate the goals and pitfalls of this paradigm.

Schizophrenia and schizotypal disorder

According to the DSM-IV, schizophrenics suffer from social–emotional deficits (emotional flatness, apathy, loss of social interests), odd or magical thinking, and psychotic symptoms, such as hallucinations and delusions (American Psychiatric Association, 1994). Their speech and behavior may be eccentric, disorganized, or even unintelligible. The symptoms must cause social or occupational impairment, have lasted for six months or longer, and not be attributable to another condition, such as drug abuse. A typical case of schizophrenia is described in Box 32.1.

Human Evolutionary Biology, ed. Michael P. Muehlenbein. Published by Cambridge University Press. © Cambridge University Press 2010.

BOX 32.1. A case of schizophrenia.

LN, a 41-year-old woman, had been normal until adolescence, when she began to withdraw from friends and became increasing "odd" and emotionally inexpressive. She completed high school and was admitted to college, but during her first semester she became convinced her professors were trying to kill her. She thought students in her dormitory were spying on her, and that there were coded messages about her in the campus paper. She thought that campus football scores somehow foretold her future. She was admitted to the student infirmary and diagnosed with schizophrenia. LN withdrew from college and returned to her parents' home. Over the next 20 years, her delusions were generally controlled by medications, but she had no interest in meeting other people and didn't like leaving the house. She never worked or dated. When her elderly parents needed to sell their house, LN went to live with her sister, in another town. Her sister, who had two children, could not manage LN's disruptive behavior and had to place her in a group home for the mentally ill. LN's maternal grandmother and a maternal aunt were also schizophrenic and a maternal uncle was thought to be schizotypal.

Schizophrenia runs in families (Gottesman and Shields, 1982). For example, first-degree relatives of a schizophrenic (children or full siblings) are 10 times more likely to develop this disorder than individuals drawn from the general population. If one identical twin is schizophrenic, the chances are 50% that the other will also be ill. Relatives of schizophrenics are also at increased risk for personality disorders that are thought to represent milder forms of the same trait or illness. Schizoid personalities, for example, are emotionally flat and socially disengaged, but their thinking and speech are normal. Schizotypal personalities are socially disengaged, and in addition their thinking is odd or magical and they have strange beliefs (American Psychiatric Association, 1994). They resemble schizophrenics, but lack the psychotic symptoms. A typical case of schizotypal personality disorder is described in Box 32.2. Schizophrenia and related personality disorders are referred to as the "schizophrenia spectrum disorders" and are generally believed to be caused by multiple illness-promoting alleles, each of small effect, acting at multiple loci (see Crow, 2007). Environmental insults – such as in utero viral exposures – interact with genetic risks. The higher the dose of illness-promoting genes and environment insults, the more likely an individual will develop schizophrenia or schizotypal personality disorder; the lower the dose, the more likely the person will be relatively normal.

As one might guess from the case described in Box 32.1, schizophrenics – especially males – living in modern societies have fewer children than average (Bassett et al., 1996; Avila et al., 2001). However, if schizophrenia impairs reproduction, how do alleles that cause it maintain their place in the gene pool? Why don't they disappear?

One potential answer is that relatives of schizophrenics with mild to moderate traits related to schizophrenia might have selective advantages that offset the handicaps resulting from full-blown illness. Selection against schizophrenics might be balanced by kin advantage. Kellett (1973), for example, speculated that schizoid personalities enjoy a fitness advantage in some societies, in which weaker emotional ties to larger-than-family groups might be beneficial. Along a similar line, Sullivan and Allen (1999, 2004) argued that individuals with schizoid personalities are sometimes better able to balance their own needs and interests against those of their social group, to which they are less committed.

An advantage might also accrue from creative eccentricity. The notion that there is a close relationship between creative genius and madness goes far back in Western culture (Burns, 2006). In fact, while schizophrenics themselves are not especially creative, some of their healthy relatives do have unusual gifts, which might lead to increased fitness in some societies (Karlsson, 1966; Anthony, 1968; Heston and Denny, 1968; Kauffman et al., 1979; Rabin et al., 1979; Rothenberg, 1983; Ludwig, 1995; Nettle, 2001). In a sample of British adults, Nettle and Clegg (2006) showed that a measure of schizotypal eccentricity correlates with artistic and creative activities, which correlate in turn with numbers of sexual partners.

Studies of the fertility of relatives of schizophrenics have had mixed results. Some show increased reproductive fitness, while others do not (Kendler et al., 1998; Avila et al., 2001; Haukka et al., 2003). Proving increased fitness in families of schizophrenics is an obvious first step in validating advantages of illness-related traits.

Stevens and Price (2000a, 2000b) hypothesized that schizotypal oddity and eccentricity served an adaptive function in early human groups, even if it is a handicap today. According to these authors, ancestral groups repeatedly grew to the point that they outstripped their resources, leading to in-group conflict. At such times, schizotypal persons could catalyze group splitting. Because of their odd, other-worldly manner and strange beliefs and convictions, schizotypal persons may take on cultic status in times of social stress. Such a person could form the nidus for an incipient new group emerging from an old one.

Although Stevens and Price (2000a, 2000b) cite studies showing that schizotypal traits promote formation of cult-like groups, other research undermines their hypothesis. Modern-day studies of hunter-gatherer tribes

> **BOX 32.2.** A case of schizotypal personality disorder.
>
> AH came to psychiatric attention for the first time in his late 50s. He had been living with his elderly mother in her apartment until her sudden death from a stroke. He had completed high school and two years of junior college and worked as a cab driver for the previous 25 years. He had never married or had intercourse, and he had never had friends or socialized, other than with his mother. He didn't experience this as a problem. He felt uncomfortable having passengers in his cab and feared they would try to cheat or influence him, but he was able to keep a professional – if not friendly – demeanor. After dropping off a passenger, he would "review" his thoughts to see that they hadn't been altered by contact with other people. After AH's mother died, the apartment manager – who had tolerated AH's glaring, unpleasant manner for his mother's sake – asked him to leave. AH created a disturbance and the police were called. As part of the ensuing court process, AH was referred for a psychiatric evaluation.

show that although group splitting can occur, it is an infrequent occurrence (Chagnon, 1979). When groups split, moreover, kinship – not cult beliefs – determine who joins which daughter group. Studies in Western societies show that people join cults because they are socially isolated, not because they are captivated by the cult ideology or entranced by the leader (Galanter, 1978, 1989; Wenegrat, 1989). Accepting cult ideology and the leader's authority is the price paid for group membership, not the motivation. Finally, in his comprehensive cross-cultural study of shamanism, I. M. Lewis found that shamans in traditional societies are distinguished from others by virtue of their excellent social skills, not by their eccentricity (Lewis, 1971). Social skills, not eccentricity, are needed to establish a credible claim to cultic or magical powers.

Depression and mania

According to the DSM-IV, major depression is characterized by depressed mood most of the day, diminished interest and pleasure in usually enjoyable activities, insomnia or excessive sleeping, agitation or mental and physical slowing, persistent feelings of worthlessness or guilt, impaired thinking or concentration, and frequent thoughts of death (American Psychiatric Association, 1994). Appetite may be increased or decreased, with consequent changes in weight, and sexual interest is diminished or lost. Depressed individuals may harm or kill themselves. As is the case with schizophrenia, these symptoms impair social or occupational functioning and they cannot be accounted for by other disorders or illnesses. Severely depressed patients may experience hallucinations or delusions with depressive themes; for example, they may hear voices condemning them and believe they are already dead. Depressive episodes may come and go, even without treatment, but generally recur. A case of severe major depression without psychotic features is described in Box 32.3.

Since major depression is partly caused by genes (Sullivan et al., 2000) and leads to decreased fitness (Cuijpers and Smit, 2002), it presents the same conundrum as schizophrenia: how are alleles promoting it maintained in the population?

Pointing out that some depressive episodes are triggered by real or threatened loss of social standing, Stevens and Price (2000b) argue that depression is an adaptive strategy in that situation. By withdrawing from competition, the depressed individual avoids further conflict and losses. He or she submits to the fact he or she has failed, and ceases to struggle further. Depression, by this account, communicates submission. In a related vein, Gardner hypothesized that depression manifests an evolutionarily ancient "PSALIC," or "Propensity State Antedating Language in Communication;" a primitive signaling system, in other words (Gardner, 1988).

Watson and Andrews (2002) suggest that depression evolved as an adaptive response to intractable social problems. The mental effect of depression is to focus the individual's attention on his or her problems. It also sends a signal to potential helpers that more assistance is needed. Since uninterrupted depression may even end in death, those who have an interest in the depressed person's survival are virtually extorted to offer increased aid. Hagen (1999) made a similar argument regarding postpartum depression. According to Hagen, postpartum depression is a way that mothers signal their partners and other persons that they need assistance caring for their newborns.

Individuals who have episodes of mania or hypomania – which usually alternate with periods of depression – are given the DSM-IV diagnosis bipolar disorder (American Psychiatric Association, 1994). Mania is characterized by an increase in energy, euphoria, high self-esteem and even delusions of grandeur, disregard for risks and consequent reckless behavior, and initiation of projects that are patently unrealistic. A manic patient, for instance, may believe he or she alone has the answer to the world's problems. He or she may plan to run for President, spend all his money on campaign posters, attack the banker who refuses to finance his television adverts, and fight with policemen who are called to quell the disturbance. Hypomania is a milder form of mania, which doesn't cause marked social or occupational impairment or require hospitalization. Bipolar disorder is partly genetically caused (Craddock and Forty, 2006). Adaptationist models of low-level forms of mania mirror those of

BOX 32.3. A case of severe depression.

ME experienced his first severe depressive episode 2 years after graduating from college, at age 26. This episode started after he failed to get a work promotion he expected. He was pervasively sad, he lost 20 pounds, he lost interest in sports, sex, and friends, and he slept only 4 hours nightly, although he was exhausted during the day. He was consumed by feelings of shame and he saw himself as a failure and disappointment to everyone. After two months of illness, he stopped going to work. His girlfriend eventually got him to seek medical care. He recovered with antidepressants and psychotherapy. Three years later, when he was off medications and out of therapy – the

depressive symptoms returned. His ex-girlfriend – they had broken up six months before – saw he was depressed again and tried to get him back into treatment, but ME refused. He again stopped going to work, neglected his hygiene, and started losing weight. He cried frequently and thought that he was worthless. His ex-girlfriend called ME's landlord when he didn't answer his phone for a couple of days. Using his passkey, the landlord found ME lying on his bed, malodorous, disheveled, and emaciated. ME improved dramatically after a month in the hospital. ME's father and sister had also been treated for depressive episodes, which were not as severe as ME's.

depression. According to such models, mania might be adaptive when social standing is rising (Gardner, 1988; Stevens and Price, 2000a), and milder manic traits may help individuals achieve positions of leadership (Gardner, 1982; Akiskal and Akiskal, 2005).

These models of mood disorders run counter to some of the data. Contrary to the notion that depression is a way of avoiding hostility, many depressed individuals are markedly irritable (Painuly et al., 2005). Contrary to the notion that depression leads to assistance, depressed individuals are seen in a negative light (Coyne, 1976) and lose important relationships (Reich, 2003). Moreover, Nettle (2004) points out that the high heritability of depressive disorders, the tendency of depression to occur out of the blue – not in response to stressors, the chronicity of the illness once it is established, the absence of concrete evidence that it serves a useful function, and the plethora of evidence for adverse fitness effects are inconsistent with depression being a true biological adaptation.

Anorexia nervosa

The adaptationist paradigm has also been applied to anorexia nervosa, a disorder that occurs mostly in young women and is characterized by a preoccupation with thinness and a terror of becoming fat (American Psychiatric Association, 1994). By fasting and exercising, anorexics reduce their weight to the point that they stop menstruating and become temporarily infertile. Untreated anorexia may end in death from starvation. A typical case of anorexia nervosa is described in Box 32.4. Like schizophrenia, major depression, and bipolar disorder, anorexia nervosa is partly caused by genes (Bulik et al., 2006). Several authors argued that anorexia nervosa is an adaptive strategy for suppressing reproduction, when circumstances are such that reproduction would be disadvantageous (Wasser and Barash, 1983; Surbey, 1987; Voland and Voland, 1989). As the case described in Box 32.4 illustrates, this model has to explain how an illness so devastating can possibly increase fitness. Furthermore, contrary to the widely

held belief that it has a long history, anorexia nervosa as it is now known first appeared in Western Europe and the United States during the late nineteenth century, and its subsequent prevalence has largely been determined by cultural dictates regarding ideal weights (reviewed in Wenegrat, 1996).

Beyond the obvious weak points of adaptationist models, some of which were cited above, the adaptationist paradigm faces some general challenges. Firstly, adaptationist models may not be needed or called for. Since psychiatric disorders are promoted by many alleles, each with a small effect size, persistence of alleles leading to mental disorders can be explained without recourse to putative positive benefits (Keller and Miller, 2006). Occasional germline mutations can replenish any alleles lost due to mental illness.

Secondly, even if risk alleles for psychiatric disorders do confer positive benefits, the adaptationist paradigm might focus too much on illness and illness-related traits: Genes leading to mental disorders might promote advantageous behaviors unrelated to illness per se or confer advantages in other realms entirely. For example, a short allelic variant of the promoter region of the serotonin transporter gene increases the risk of depression, especially following stressors (Jacobs et al., 2006; Wilhelm et al., 2006), but serotonergic dysfunction may lead to greater impulsiveness and earlier reproduction (Stein, 2006). A considerable body of evidence suggests that immune system genes may contribute to schizophrenia (Strous and Shoenfeld, 2006). If so, some genes that promote schizophrenia may protect their carriers against viruses, cancer, or autoimmune disorders. Neither of these possibilities are captured by adaptationist approaches to mental disorders or illness-related traits.

THE ETHOLOGICAL PARADIGM

Ethology is the study of species-typical animal behavior. Human ethology refers to the study of behaviors typical of *Homo sapiens* (Eibl-Eisbesfeldt, 1989).

Being species-typical, such behaviors must result from the action of human genotypes in "normal" human environments. The *ethological paradigm* asks if signs and symptoms of mental disorders can be described efficiently by reference to human ethology.

For example, one typical human behavior evident in all cultures is to form close-knit social groups, members of which are considered superior to outsiders (Tinbergen, 1976). In traditional tribes, for instance, outsiders may be killed, especially if they are male. The name of the tribe is frequently the term for human beings (Berger and Luckmann, 1966). In such societies, small groups are composed of kin. Early studies of hunter-gatherer tribes, for instance, showed that members were only slightly less related on average than first cousins (Chagnon, 1979). In such conditions, positive attitudes toward group members and invidious attitudes toward outsiders are consistent with kin altruism.

Persons in modern societies may not kill outsiders (though, sadly, they sometimes do), but they nonetheless cohere in small, invidious social groups. Spontaneous group formation and resultant outward hostility has been shown in psychology laboratories, occurring among total strangers (Tajfel and Billig, 1974; Tajfel, 1981, 1982). Formation of social groups is thought to have protected early hominids from predation and early humans from competing groups, and to reflect innate psychological and behavioral factors (Alexander, 1979).

Human social groups, even those created in the laboratory, develop consensual worldviews (reviewed in Wenegrat, 2001). Reality, in other words, is socially constructed within the human group. Lumsden and Wilson (1981) argued that group enculturation was probably highly adaptive and led to the selection of conformity-promoting genotypes. The neurological basis of social imitation is only now beginning to be understood (see Meltzoff and Prinz, 2002). In real groups and in groups created in laboratories, individuals who fail to conform may be seen as outgroup members and become the object of aggression.

Many signs and symptoms of schizophrenia spectrum disorders manifest failure of normal group-orientation (Wenegrat, 1990a). That is, they represent failure of a particular piece of the species-typical human behavioral repertoire. Individuals with these disorders fail to fully engage in larger-than-family social groups or to adopt their worldviews. They are generally normal children, but when they reach adolescence – when normal individuals orient themselves in larger-than-family social groups – their social disengagement and their inability to learn subtle social rules and the consensual worldviews of these larger groups first become apparent. They may seem odd to others. They start to feel uncomfortable outside their families of origin and fearful of others within the groups they might join.

Like some adaptationist models described in the previous section, this ethological model locates the pathology of the schizophrenia spectrum in larger-than-family groups, but it does not presume that the deficit is or has been adaptive. The emphasis is on placing deficits seen in illness in the context of the normal human behavioral repertoire.

Attachment theory

John Bowlby, whose "Attachment Theory" has had a major impact in psychiatry, studied relationships between caretakers and infants, and infants' responses to being separated (Bowlby, 1969, 1973, 1980). He discovered that infants separated from their principal attachment figure, who don't have other familiar attachment figures available, engage in stereotyped emotional displays. In the first, or "Protest" phase, the infant cries or screams and attempts to follow or find the missing attachment figure. This phase may last several days. If the caretaker doesn't return, the "Despair" phase will set in. Active efforts to signal or find the caretaker cease, and the infant becomes apathetic and socially withdrawn. The third, or "Detachment," phase starts after separations lasting two weeks or more. The child in this phase is docile, apathetic, and detached from his or her social environment. This phase may last a long time, even if the attachment figure reappears.

Bowlby noted the similarities between children's attachment behaviors and separation responses and those shown by immature primates. He considered attachment behaviors and separation responses as evolved behavioral patterns common to humans and primates.

The resemblance between the phases of the separation response and certain types of depression were evident to Bowlby. The Protest phase, for instance, resembles agitated depressions in adults. The Despair and Detachment phases resemble "retarded" depressions, which are depressions characterized by psychomotor slowing, apathy, and withdrawal. This led to the hypothesis that adult depression is a manifestation of early attachment behaviors somehow reactivated, and to the hypothesis that early attachment disruptions predispose to later depression. The latter hypothesis has found some empirical support, though depressive illness also occurs in people who had no attachment disruptions (Gilmer and McKinney, 2003).

This model of depression resembles an adaptationist model that treats depression as a response to loss of attention and interest from others, but there is no supposition that depression as such is adaptive.

Borderline personality disorder, pathological mourning, and some factitious illnesses can also be understood as miscarried attachment behaviors (Wenegrat, 1990a). For example, borderline personality disorder is characterized by clinging, dependent behavior, intolerance of rejection, anxiety, depression, poorly controlled aggression, and self-mutilation or injury – such as cutting or burning oneself – in response to stress. Nonhuman primates deprived of maternal care show behavioral signs of depression and increased aggression and fearfulness. When anxious, they bite and hit themselves. Studies of borderline personality disorder patients reveal a very high incidence of early abuse or neglect (Paris et al., 1994).

Ethological models may cast light on salient aspects of serious mental disorders, but they do not account for all of their signs and symptoms. Schizophrenics, for instance, hear voices of people arguing; a model of schizophrenia as a failure of group social strategies cannot account for this or many other such symptoms.

THE INNATE-MODULE PARADIGM

Evolved behavior patterns depend on innate perceptual, cognitive, emotional, and motivational processes. For example, human males seek exclusive sexual access to fertile women (Trivers, 1985). To do so, they must recognize signs of female fertility; experience sexual attraction to women in general and to fertile women in particular; be able to understand their immediate social milieu – including social alliances, sexual rules, and mores, and styles and manners of courtship; and feel emotions promoting stable, exclusive relationships, such as affection and jealousy. Insofar as the human brain evolved to solve social problems (Dunbar, 1998), capacities like those listed here may depend on evolved neural circuitry specialized for each purpose. According to researchers broadly identified as Evolutionary Psychologists, specialized mental capacities that evolved for particular functions – or innate psychological modules – comprise the basic structures of human mental life (Barkow et al., 1992; Hirschfeld and Gelman, 1994; Pinker, 1997; Cosmides and Tooby, 1997). The *innate-module paradigm* asks if mental disorders can be understood as dysfunctions of discrete evolved mental modules.

For example, studies have shown that humans more readily learn fear responses to objects and situations that posed threats to our ancestors – such as snakes, spiders, threatening animals, darkness, and high places – than to far more dangerous modern-day objects and situations, such as guns, knives, and high-speed driving (Marks, 1987; Marks and Tobena, 1990). Also, fears of ancestral threats don't diminish as rapidly in the absence of reinforcement as fears of modern-day threats. Evidence like this suggests that common troubling phobias – such as phobias of snakes, spiders, animals, darkness, and high places (American Psychiatric Association, 1994) – reflect evolved mental modules that promote acquisition of fears advantageous to human ancestors.

Nesse (2001) compared evolved anxiety modules to human-made smoke detectors. As an occasional false alarm is better than failure to sound an alarm in the event of a fire, smoke detectors are set to go off at a very low threshold. An evolved anxiety module that served to protect from a dire threat might likewise be tuned to sound an occasional false alarm. Individuals whose experience or genes equipped them with slightly more sensitive innate anxiety modules would be plagued by repeated false alarms, to their detriment.

The cerebral modularity of linguistic functions has been known since the nineteenth century. Psycholinguistic studies support the view that humans have innate syntactical language processors. According to Timothy Crow, human vulnerability to psychotic disorders is a by-product of the modular organization of language (Crow, 1995, 1997, 2002, 2007). Green and Phillips (2004) cite evidence that specialized neural networks – comprising an evolved social threat detector – rapidly process signs of potential aggression from others. Clinical paranoia might result from excessive activity of this evolved mental module, which causes ambiguous stimuli to be perceived as threatening. Schizophrenics with persecutory delusions have structural and functional abnormalities in some of the neural systems described by Green and Phillips.

High and low moods reflect changes in activity in specialized neural circuits (Morgane et al., 2005), which can be conceptualized as mood modules. Several authors have analyzed potential adaptive functions served by normal moods (or normally functioning modules) to shed light on mood disorders (Nesse, 2000; Sloman and Gilbert, 2000; Nettle, 2004; Dickinson and Eva, 2006; Freed and Mann, 2007). To give just one example, Nesse (2000) argued that normal moods serve to insure productive effort allocations. In unpropitious circumstances, when goals and desires cannot be fulfilled, mood is normally lower. At such times, pessimism and low motivation serve to conserve resources that would otherwise be wasted. In propitious circumstances, mood is normally higher (see Tiger, 1979). Optimism and increased energy ensure that strong efforts are made to utilize opportunities. Nesse's view explains the key role of low expectations in depressive disorders (Overmier, 2002).

The innate-module paradigm has been applied to autism, a severe disorder appearing in early childhood or even infancy (American Psychiatric Association, 1994; Tidmarsh and Volkmar, 2003). Autism – like schizophrenia – occurs in milder forms; hence psychiatrists now refer to autistic spectrum disorders. Severely disturbed children avoid physical closeness, fail to make eye contact or respond to emotional gestures, and ignore their parents' voices. They lack interest in other children or in normal childhood games, such as hide-and-seek. They show stereotyped bodily movements, such as clapping, rocking, or swaying, and are fascinated by repetitive motions, such as of fans or pendulums. Older children may become obsessed with collections or interests, such as bottle-caps or baseball statistics, to the exclusion of other activities and interests. Autistic children resist change and prefer unvarying routines. New restaurants, stores, or routes may meet with strong objections. Language skills are limited. Higher functioning patients, with the mildest forms of autism, generally become independent adults with deficits in social skills, but most affected children have a poor prognosis. When it was first described, autism was thought to be extremely rare, but more recent estimates show that its incidence is about 1 in 1000 children. Another 5 in 1000 may have spectrum disorders closely related to autism. While there is a known genetic component to the disorder, environmental factors may also play causal roles.

A critical step in anticipating and altering other people's behavior is to correctly infer their mental state – that is, their beliefs, moods, and desires – from their current and past behavior (Dennett, 1987). Since modern concepts of behavioral evolution suggest that the human brain was shaped by social necessities, such inferences could depend on evolved neural structures specialized for that purpose. There might be mental modules adapted to forming accurate theories of others' minds.

Baron-Cohen (1995) and others suggested that a deficit in a "theory of mind module" can account for the social deficits seen in autistic disorders. Among the findings cited by Baron-Cohen, autistic children fail to demonstrate joint-attention behaviors: they do not monitor or follow an adult's gaze, nor do they point to objects or show objects to adults as a way of directing attention. To engage in such behaviors, children must surmise an adult's attentional focus. They fail "false belief" tests. A typical false belief test involves seeing one person, say Sally, put a marble in one place, and later, while Sally is away, seeing Anne move the object elsewhere. When asked where Sally thinks the object is located, very young normal children point to where Anne placed it. Around the age of three or four years, however, they correctly point to where Sally had put it before leaving. They understand Sally's viewpoint, and that Sally has a false belief. Unlike normal children, or mentally handicapped children without autistic features, autistic children fail such tests at much later ages.

In the early 1990s, a research group in Genoa studying macaques discovered so-called mirror neurons in a part of the frontal lobe corresponding to Broca's area in human beings (di Pellegrino et al., 1992). Mirror neurons fire just before animals perform specific motor acts – grasping an object, for example – and also when they see the same acts performed by others. Such neurons seem to encode the concept of the act, whether in self or other. Related neurons in the temporal sulcus even encode for intentions (Jellema et al., 2002). For example, such a neuron might fire when another animal is observed trying to open a cylinder, regardless of the method – twisting the top or pulling it, or whether the act is completed or not.

Although the evidence is indirect, a system of mirror neurons appears to exist in humans (Keysers and Gazzola, 2006). For example, functional brain scans show that human brain regions corresponding to those where mirror neurons have been found in monkeys show similar activations in preparation for motor acts and while observing the same acts. Transcranial magnetic stimulation (TMS) is a procedure for stimulating regions of cerebral cortex with changing magnetic fields administered through the scalp. During TMS, observing a motor act triggers corresponding peripheral nerve activity: A subject who sees someone grasping may show increased activity in nerves to his own hand.

Mirror neurons – and neurons encoding intentions, rather than just actions – encode simple theories of mind. In an animal study cited above, for instance, neuronal firing signified an inference that an act was intended to open a cylinder – regardless of the method used (twisting or pulling) or the actual outcome. Hence, mirror neuron systems may play key roles in social cognition, including theory of mind (Keysers and Gazzola, 2006).

The same techniques that reveal mirror neuron systems in healthy humans reveal deficits in autistic children (Fecteau et al., 2006). For example, with TMS, they fail to show peripheral nervous activation while viewing others' actions (Théoret et al., 2005). Mirror neuron deficits might disable a theory of mind module in autistics, or the deficit might be elsewhere and reflected in mirror neurons. In any event, if these findings hold up in future research, they will bolster the theory-of-mind hypothesis of autism.

Like ethological models, innate-module models may fail to account for critical signs and symptoms of complex mental disorders. Buller (2005), for instance, argued that Baron-Cohen's treatment of autism ignores serious deficits unrelated to theory-of-mind. Also, some of these models may be criticized on the ground that essential mental capacities – even those dependent on localized brain circuits – may not be adaptations in a technical sense (see Buller, 2005). For example, specific brain regions are needed to read and write. Furthermore, we can identify children with highly specific deficits in reading or writing abilities. Yet, reading or writing "modules" cannot be adaptations, in the sense that they evolved because of advantages accruing from written language. Instead, reading and writing are epiphenomenal skills enabled by brain regions serving more primitive language and visual functions. In the absence of definite evidence that capacities are adaptations, modular models of illness – however instructive they may be – are social-deficit models dressed up in Darwinian garb.

THE MISMATCH PARADIGM

The *mismatch paradigm* starts with the fact that human behavior evolved before the dawn of agriculture, some 8000–10 000 years ago, and that too little time has since elapsed for large-scale evolutionary changes to have occurred. If this is true, human behavior evolved in the setting of small hunter-gatherer groups composed of close kin moving about in small ranges. Such groups may have resembled extant hunter-gatherer tribes which have been studied intensively by modern-day anthropologists.

Modern society is a far cry from the small hunter-gatherer band. Unlike hunter-gatherers, people in modern cities are exposed every day to strangers with whom they have little in common, may live far from kin, play ambiguous or conflicting roles in multiple social groups (such as at home, at work, or in their place of worship), and lack clear authority figures or a sense of their social placement. If human behavior is adapted to life in small groups, then some mental disorders in the DSM-IV might result from our living in settings to which we are ill adapted: a mismatch, in other words, between our current setting and our evolved psychology.

The mismatch paradigm has been applied to drug abuse. Neural pathways in the anterior cingulate gyrus, the frontal cortex, the nucleus accumbens, and related regions are known to mediate reward and pleasure sensations (Baler and Volkow, 2006). In a drug-free environment, these ancient pathways promote adaptive reward-seeking behaviors. However, modern-day societies expose people to drugs that affect neurotransmitters – especially dopamine, glutamate, and endogenous opioids – or neurotransmitter receptors employed in these pleasure pathways, in effect hijacking the pathways to promote drug-seeking behavior (Lüscher and Ungless, 2006). For example, cocaine and amphetamine congeners (e.g., amphetamine, methamphetamine, Ecstasy) act to release dopamine from nerve terminals, increasing the amount of dopamine in synapses, and opiates (e.g., morphine and heroin) stimulate receptors that are otherwise only responsive to endogenous opioids (endorphins and enkephalins). Sedative hypnotics (alcohol, barbiturates, benzodiazepines) subject to abuse alter the balance between excitatory (mostly glutamatergic) and inhibitory (mostly gabaergic) neurotransmission and also stimulate dopamine and opioid release. In drug addicts, neural pathways that might have promoted fitness promote self-destructive behaviors that lower inclusive fitness. Heroin addicts, for example, neglect or abuse their children while ruining their own health. Mothers addicted to cocaine – which has especially powerful effects on pleasure pathways – may let their children die in order to stay high.

Unlike other drugs of abuse, alcohol was present in the ancestral environment. Ripe, fermenting fruit is especially rich in calories and nutrients and can be localized by the odor of alcohol it emits. In frugivores (fruit eaters) – including nonhuman primates and early hominids – preference for the smell and taste of ethanol may be adaptive, since ethanol is linked with nutritional rewards (Dudley, 2000, 2002, 2004). Opportunistic, fruit-induced drunkenness has been observed in several frugivorous species; most famously, elephants.

Human fermentation of fruits and grains appeared in Mesopotamia 6000 years ago. Alcohol distillation, to produce more potent beverages, was discovered in seventh-century China and spread to Central Asia, the Middle East, and later Europe. For the first time, a preference for the smell and taste of ethanol could lead to overconsumption. With alcohol freely available, genetic differences that were previously fitness-neutral began to moderate the risk of becoming addicted. Research, for instance, shows that genetic differences in enzymes involved in degrading alcohol correlate with ethnic and racial differences in alcoholism risk, though not with other health indices (Mulligan et al., 2003). According to this model, alcoholism belongs with obesity and diabetes: disorders fueled by preferences that evolved in conditions of scarcity, operating now in conditions of oversupply.

The mismatch paradigm may explain societal differences in the outcome of schizophrenia. Schizophrenics living in traditional rural societies are more likely to marry and reproduce, to contribute to child-rearing, and to avoid isolation than schizophrenics living in modern urban societies (Sartorius et al., 1996). As discussed above, individuals with schizophrenia spectrum disorders don't feel part of larger-than-family social groups as healthy people do, nor do they conform to the consensual worldviews associated with such groups. They are loners with odd ideas. Large complex societies with competing and often incoherent worldviews may render schizophrenia-promoting genes much more dysfunctional than they were in the environment in which human beings evolved. In a related vein, complex urban environments – in contrast to rural ones – may exacerbate or even cause psychotic disorders, especially in individuals with strong genetic risks (Peen and Dekker, 2004).

Two mismatch models attribute anorexia nervosa to the novel power of media acting on evolved psychological processes. In a well-nourished society, slenderness is a sign of female youthfulness and hence fertility (Singh, 1993). In such a society, same-sex competitive urges cause women to value slimness. Mass media present them with images of extremely thin high-status women, such as models and actresses, which they try to emulate. The result is self-starvation (Abed, 1998). Alternatively, some of the psychological and behavioral traits of patients with anorexia nervosa – such as their restlessness and subjectively increased energy and their denial of thinness – may be evolved responses to weight loss that helped ancestral humans flee from local famines. In the modern world, mass media disseminates unrealistic standards of feminine thinness, girls diet to meet them, and adaptations to famine are set into motion (Guisinger, 2003).

Humans evolved in social groups composed of long-term familiars. Modern humans, by contrast, may spend days surrounded by strangers with whom they can't communicate. Emigrants or people living or working abroad are more likely than other people to develop paranoid delusions (Allodi, 1982; Kendler, 1982). In such cases, returning home may be curative. Exposure to strangers may also contribute to social anxiety disorder. Many patients with social anxiety, for example, are limited by anxiety when speaking in front of groups (American Psychiatric Association, 1994). If students, they cannot take courses which involve giving presentations. If working, they cannot take jobs or accept promotions requiring public speaking. Humans in ancestral groups would seldom if ever have needed to speak to groups of people who weren't long-term familiars.

In testing whether environmental factors cause disease, the first step is usually to see whether individuals exposed to such factors have higher rates of illness than those who have not been exposed. This is how epidemiologists established that cigarette smoking causes lung cancer and heart disease, and that asbestos exposure produces certain malignant tumors. Currently, mismatch models suffer from lack of definite evidence that people in modern societies are actually mentally worse off than people who live in societies more akin to ancestral groups. As noted above, a tentative case can be made for increased risk of psychosis in urban societies, but much more data is needed before even tentative statements can be made regarding mood and anxiety disorders.

THE PSYCHOTHERAPY PARADIGM

Even with modern medicines, psychotherapy is an essential component in the treatment of most mental disorders (see, for example, Fava et al., 2005; Rosenbaum et al., 2006). There are many different types and schools of psychotherapy, but all of them aim to change feelings, beliefs, and relationship styles in a healthy direction.

A therapist has to decide which of his or her patient's feelings, beliefs, and relationship styles lead to distress in life, how best to talk about them, and how they can be changed. Explicitly or not, such decisions are theory-driven. For example, if a patient whose mother just died seems upset but denies it, a therapist might suggest that he's not acknowledging how his mother's death has affected him. The suggestion is based on theories of parent–child relationships, loss and normal grief, emotional control, and self-deception. Although different schools of therapy produce equivalent outcomes, therapists who are most aware of the theories that guide their decisions appear to be most effective in helping patients change (Wampold, 2001). Therapists who cannot articulate their reasons for interventions are less likely to help most patients.

Although current therapies are certainly beneficial, the conceptual frameworks used by modern-day psychotherapists share limitations or weaknesses (see, for example, Fancher, 1995). If therapists were armed with a more robust conceptual framework, could they offer more effective help to patients? The *psychotherapy paradigm* asks if modern concepts of behavioral evolution can help therapists better identify feelings, beliefs, events, and relationship styles relevant to illness and to mental health and to better understand these in particular patients. It asks if these concepts can offer a robust theoretical framework for informing the kinds of decisions psychotherapists make.

On the face of it, the answer is likely to be positive. Patients in psychotherapy present with feelings and thoughts, and with relationship problems, directly related to issues addressed in the overall framework of behavioral evolution: among many others, attachment and separation, parent–child conflict and manipulation, conflict with siblings and children, conflicts with other kin, regulation of sexual appetites, same-sex and other-sex conflicts, resource competition, status competition, membership in social groups or ostracism from them, social reputation, reciprocity obligations and moralistic aggression, control of aggression and fear of other peoples' behavior, and self- and other-deception. All of these have been treated from a Darwinian viewpoint. In fact, the Darwinian framework is singular in explicitly tackling the range of issues encountered by psychotherapists.

Adaptationist, mismatch, and ethological models cited in previous sections may guide therapy of specific disorders (Glantz and Pearce, 1989; Sloman and Gilbert, 2000; Stevens and Price, 2000a). However, evolutionary models of normal behavior may prove to be more valuable in relation to psychotherapy than models of illness per se. For example, evolutionary concepts of normal behavior have been applied in relation to classical Freudian theory (see, for example, Slavin, 1985; Badcock, 1986; Slavin and Kriegman, 1988, 1992), to the Jungian notion of archetypes (see, for example, Wenegrat, 1990b; Stevens, 2002), and to some aspects of object-relations theory, a branch of psychoanalysis (see Wenegrat, 1990a). Evolutionary treatments of normal behavior can also shed light on aspects of psychotherapy generally, such as mental suggestion (Wenegrat, 2001) and unconsciousness, self-deception and mental inconsistency (see, for example, Trivers, 1985; Kurzban and Aktipis, 2007).

Findings in social psychology and social cognition will prove essential to the psychotherapy paradigm. In many cases, these findings flesh out more abstract concepts from Behavioral Evolution. For example, social cognitive research has uncovered mental processes that produce co-ordinated behavior in larger-than-family groups (reviewed in Wenegrat, 2001). This research and the related evolutionary ideas turn out to be highly relevant to the process of psychotherapy.

Of course, evolutionary concepts cannot be used in therapy except in relation to cultural factors and differences. Evolved behavioral patterns take on particular forms, depending on the culture in which they are expressed. Some cultures, for example, favor paternal investment; other cultures don't (Kurland, 1979). The same paternal behavior – care or neglect – has a different meaning depending on whether the culture is of one type or the other. However, the same is true for other theoretical frameworks psychotherapists now use: none can be applied in the absence of cultural knowledge.

A more serious challenge for psychotherapy models is avoiding triviality or even circular reasoning, especially when relying on models of mental disorders.

To give a simple example, the ethological attachment model of depression implies that therapists should pay special attention to patients' concerns about lost relationships. Adaptationist and ethological social rank models direct attention to patients' concerns about social status. But the therapeutic enterprise is over 100 years old now, and much has already been learned. Therapists already know that depressed patients ruminate about lost or threatened relationships and/or their social status, and that talking about these issues is vital to their recovery. In fact, the plausibility of the attachment models and the social rank models rests in large part on clinical knowledge gleaned by psychotherapists.

SUMMARY

Psychiatric disorders are common, occur in all societies, and have fitness consequences. Concepts from behavioral evolution – so successful in accounting for human behavior generally – may shed light on their symptoms, causes, and possible treatments.

Most work in the field of evolutionary psychiatry has relied on one of five paradigms. The adaptationist paradigm aims to view mental disorders, their milder variations, or illness-related traits as behavioral adaptations that confer or have conferred inclusive fitness benefits. The best-known adaptationist models are of schizotypal oddity and depressive episodes, but there are also adaptationist models of several other disorders, such as anorexia nervosa. As we have seen, the evidence is weak that any mental disorder or a milder variant is an adaptation, in the technical sense, but research on reproductive fitness of patients and relatives and its correlation with illness-related traits – exemplified by a study conducted by Nettle and Clegg (2006) – may lead to future progress. Future research could be strengthened by using life-history data to estimate inclusive fitness, which is a broader measure of Darwinian fitness than reproductive success (see, for example, Madan, 2002). Whether adaptationist models of mental illness are really needed at all – that is, whether persistence of genes promoting mental illness really implies some benefits – and whether such models are focused on the right kinds of benefits are two major open questions.

The ethological paradigm asks if a mental disorder can be described economically in relation to elements of the normal behavioral repertoire observed in all societies. As we have seen, mental disorders can be characterized as miscarriages of attachment behaviors and group-related behaviors. However, ethological models generally don't encompass the full complexity of DSM-IV disorders.

The innate-module paradigm asks if mental disorders can be explained in terms of evolved mental modules, as these are hypothesized by Evolutionary

Psychologists. Anxiety, mood, and autistic spectrum disorders have been analyzed as modular dysfunctions. Like ethological models, innate-module models often fall short of encompassing the full range of illness phenomena, and they are also subject to the conceptual problem vexing evolutionary versions of modular theory: the difficulty of proving that given mental capacities are technically adaptations.

The mismatch paradigm tries to explain disorders in terms of potential mismatches between our current environment and our evolved psychology. This viewpoint is most convincingly applied to drug addiction and to alcoholism. It may also shed light on some little-noticed epidemiological features of mental disorders, such as cultural differences in the outcome of schizophrenia and the high incidence of paranoia in immigrants. As discussed above, cross-cultural incidence data will be essential in proving that features of modern society cause any mental disorder.

Finally, the psychotherapy paradigm asks if modern concepts of behavioral evolution can help therapists understand feelings, beliefs, and relationship styles relevant to illness and to mental health. It asks if these concepts can offer an overarching framework for the types of decisions psychotherapists make. The psychotherapy enterprise is over 100 years old now, and much is already discovered. To be successful, the psychotherapy paradigm will have to do more than tell therapists what therapists already know, while avoiding circular reasoning.

It seems likely that all five paradigms will figure in the future of evolutionary psychiatry. Adaptationist models may explain some disorders, while ethological, innate-module, or mismatch models explain others. Several paradigms may apply to a single disorder: For example, creativity may explain the persistence of certain risk genes in the families of schizophrenics, while dysfunction of social cognition accounts for who actually falls ill. Cultural variation in the outcome of schizophrenia may be due to factors explained by mismatch models. Adaptationist models might account for mild depression, while ethological or innate-module models better describe severe depression. The triggers for mild depression and the causes of ethological or modular dysfunctions that occur in severe depression might both be analyzed in terms of mismatch models.

Above all the others, the psychotherapy paradigm probably holds the key to the future of evolutionary psychiatry. Although it can utilize work from each of the other paradigms, the psychotherapy paradigm doesn't require robust models of mental disorders. As we have seen, for the most part such models are lacking. Existing evolutionary accounts of normal behavior are equally – if not more – relevant to psychotherapy than models of mental disorders narrowly conceived. Furthermore, many practicing therapists are keenly aware of flaws in their current conceptions of health, behaviour, and therapy processes. If strong concepts and research findings from behavioral evolution – along with related ideas from social and cognitive psychology – can be broadly disseminated while avoiding the obvious pitfalls, they should be warmly received. To paraphrase Sigmund Freud, who was writing of dream analysis, the psychotherapy paradigm provides a royal road through which Darwinian concepts can enter the minds of clinicians.

DISCUSSION POINTS

1. Compare and contrast adaptationist and mismatch models of anorexia nervosa (AN). Identify some life events for which these models predict different effects on the course of patients with AN. For example, what would these models predict for a mother with AN as she approaches menopause? Identify social conditions which these models predict would affect prevalence rates. Do online research to see if some of these predictions have been studied. (Suggestion: start by searching Medline at www.ncbi.nlm.nih.gov/pubmed.)

2. Argue for and against the following proposition: "Evolutionary psychiatry is bedeviled by confusion between extreme mental states and their normal counterparts." In arguing this point, consider each of the major disorders described in this chapter and each of the separate paradigms. Which paradigm(s) are best equipped to deal with normal states? Which paradigms are best equipped to deal with extreme states?

3. The risk of schizophrenia is increased in individuals whose mothers had influenza in the early second trimester. Likewise, the risk of schizophrenia in offspring of schizophrenics appears to be increased by complications at birth. Which of the evolutionary models of schizophrenia described in this chapter can most readily accommodate these facts? How could such findings be reconciled with other models?

4. Schizophrenics and major depressives have high lifetime risks of suicide. How might you reconcile adaptationist models of these disorders with these high suicide risks? Under what conditions might suicide have positive effects on inclusive fitness? With these conditions in mind, why should schizophrenics and major depressives be more likely to kill themselves than anyone else?

5. Describe several novel predictions derived from adaptationist models of mental disorders. Do online research using databases like Medline to find existing data pertaining to your predictions. If you can't find such data, briefly outline studies to test your predictions.

6. Describe several novel predictions derived from mismatch models of mental disorders. Do online research using databases like Medline to find existing data pertaining to your predictions. If you can't find such data, briefly outline studies to test your predictions that could be performed in a single country.

7. Psychotherapy often focuses on patients' responses to disturbing events. From the point of view of behavioral evolution, describe some common events that some or all individuals might find especially disturbing. Patients sometimes misconstrue minor events as disturbing ones. Keeping in mind Nesse's "false alarm" hypothesis of anxiety disorders, which individuals might be especially likely to perceive the events you've described when they've not really occurred?

REFERENCES

Abed, R. T. (1998). The sexual competition hypothesis for eating disorders. *British Journal of Medical Psychology*, **71**(4), 525–547.

Akiskal, K. K. and Akiskal, H. S. (2005). The theoretical underpinnings of affective temperaments: implications for evolutionary foundations of bipolar disorder and human nature. *Journal of Affective Disorders*, **85**(1–2), 231–239.

Alexander, R. D. (1979). *Darwinism and Human Affairs*. Seattle: University of Washington Press.

Allodi, F. (1982). Acute paranoid reaction (*Boufee Delirante*) in Canada. *Canadian Journal of Psychiatry*, **27**(5), 366–373.

American Psychiatric Association (1994). *Diagnostic and Statistical Manual of the American Psychiatric Association, Fourth Edition*. Washington, DC: American Psychiatric Association.

Anthony, E. J. (1968). The developmental precursors of adult schizophrenia. *Journal of Psychiatric Research*, **6**(suppl. 1), 293–316.

Avila, M., Thaker, G. and Adami, H. (2001). Genetic epidemiology and schizophrenia: a study of reproductive fitness. *Schizophrenia Research*, **47**(2–3), 233–241.

Badcock, C. R. (1986). *The Problem of Altruism: Freudian–Darwinian Solutions*. New York: Basil Blackwell.

Baler, R. D. and Volkow, N. D. (2006). Drug addiction: the neurobiology of disrupted self-control. *Trends in Molecular Medicine*, **12**(12), 559–566.

Barkow, J. H., Cosmides, L. and Tooby, J. (eds) (1992). *The Adapted Mind: Evolutionary Psychology and the Generation of Culture*. New York, Oxford University Press.

Baron-Cohen, S. (1995). *Mindblindness: an Essay on Autism and Theory of Mind*. Cambridge, MA: MIT Press.

Bassett, A. S., Bury, A., Hodgkinson, K. A., et al. (1996). Reproductive fitness in familial schizophrenia. *Schizophrenia Research*, **21**(3), 151–160.

Berger, P. L. and Luckmann, T. (1966). *The Social Construction of Reality: a Treatise in the Sociology of Knowledge*. Garden City, NJ: Doubleday.

Bowlby, J. (1969). *Attachment and Loss*, vol. 1. New York: Basic Books.

Bowlby, J. (1973). *Attachment and Loss*, vol. 2. New York: Basic Books.

Bowlby, J. (1980). *Attachment and Loss*, vol. 3. New York: Basic Books.

Bulik, C. M., Sullivan, P. F., Tozzi, F., et al. (2006). Prevalence, heritability, and prospective risk factors for anorexia nervosa. *Archives of General Psychiatry*, **63**(3), 305–312.

Buller, D. J. (2005). *Adapting Minds: Evolutionary Psychology and the Persistent Quest for Human Nature*. Cambridge, MA: MIT Press.

Burns, J. K. (2006). Psychosis: a costly by-product of social brain evolution in *Homo Sapiens*. *Progress in Neuro-Psychopharmacology and Biological Psychiatry*, **30**(5), 797–814.

Chagnon, N. A. (1979). Mate competition, favoring close kin, and village fissioning among the Yänomämo indians. *Evolutionary Biology and Human Social Behavior: an Anthropological Perspective*, N. A. Chagnon and W. Irons (eds). North Scituate, MA: Duxbury Press, pp. 86–131.

Cosmides, L. and Tooby, J. (1997). The modular nature of human intelligence. *The Origin and Evolution of Intelligence*. A. B. Scheibel and J. W. Schopf (eds). Sudbury, MA: Jones and Bartlett, pp. 71–102.

Coyne, J. C. (1976). Depression and the response of others. *Journal of Abnormal Psychology*, **85**(2), 186–193.

Craddock, N. and Forty, L. (2006). Genetics of affective (mood) disorders. *European Journal of Human Genetics*, **14**(6), 660–668.

Crow, T. J. (1995). A Darwinian approach to the origins of psychosis. *British Journal of Psychiatry*, **167**(1), 12–25.

Crow, T. J. (1997). Is schizophrenia the price that *Homo sapiens* pays for language? *Schizophrenia Research*, **28** (2–3), 127–141.

Crow, T. J. (ed.) (2002). *The Speciation of Modern Homo Sapiens*. Oxford: Oxford University Press.

Crow, T. J. (2007). Why genetic linkage has not solved the problem of psychosis: review and hypothesis. *American Journal of Psychiatry*, **164**(1), 13–21.

Cuijpers, P. and Smit, H. (2002). Excess mortality in depression: a meta-analysis of community studies. *Journal of Affective Disorders*, **72**(3), 227–236.

Dennett, D. (1987). *The Intentional Stance*. Cambridge, MA: MIT Press.

di Pellegrino, G., Fadiga, L., Fogassi, L., et al. (1992). Understanding motor events: a neurophysiological study. *Experimental Brain Research*, **91**(1), 176–180.

Dickinson, M. J. and Eva, F. J. (2006). Anxiety and depression may have an evolutionary role as negative reinforcers, encouraging socialization. *Medical Hypotheses*, **66** (4), 796–800.

Dudley, R. (2000). Evolutionary origins of human alcoholism in primate frugivory. *Quarterly Review of Biology*, **75** (1), 3–15.

Dudley, R. (2002). Fermenting fruit and the historical ecology of ethanol ingestion: is alcoholism in modern humans an evolutionary hangover? *Addiction*, **97**(4), 381–388.

Dudley, R. (2004). Ethanol, fruit ripening, and the historical origins of human alcoholism in primate frugivory. *Integrative Comprehensive Biology*, **44**(4), 315–323.

Dunbar, R. I. M. (1998). The social brain hypothesis. *Evolutionary Anthropology: Issues, New, and Reviews*, **6**(5), 178–190.

Eibl-Eisbesfeldt, I. (1989). *Human Ethology*. Hawthorne, NY: Aldine de Gruyter.

Fancher, R. T. (1995). *Cultures of Healing: Correcting the Image of American Mental Health Care*. New York: W. H. Freeman.

Fava, G. A., Ruini, C. and Rafanelli, C. (2005). Sequential treatment of mood and anxiety disorders. *Journal of Clinical Psychiatry*, **66**(11), 1392–1400.

Fecteau, S., Lepage, J.-F. and Théoret, H. (2006). Autism spectrum disorder: seeing is not understanding. *Current Biology*, **16**(4), R131–R133.

Freed, P. J. and Mann, J. J. (2007). Sadness and loss: toward a neurobiopsychosocial model. *American Journal of Psychiatry*, **164**(1), 28–34.

Galanter, M. (1978). The "Relief Effect": a sociobiological model for neurotic distress and large-group theory. *American Journal of Psychiatry*, **135**(5), 588–591.

Galanter, M. (1989). *Cults: Faith, Healing, and Coercion*. New York: Oxford University Press.

Gardner, R. (1982). Mechanisms in manic depressive disorder: an evolutionary model. *Archives of General Psychiatry*, **39**(12), 1436–1441.

Gardner, R. (1988). Psychiatric syndromes as infrastructure for intra-specific communication. In *Social Fabrics of the Mind*, M. R. A. Chance (ed.). Hillsdale, NJ: Lawrence Erlbaum Associates, pp. 197–226.

Gilmer, W. S. and McKinney, W. T. (2003). Early experience and depressive disorders: human and non-human primate studies. *Journal of Affective Disorders*, **75**(2), 97–113.

Glantz, K. and Pearce, J. (1989). *Exiles from Eden*. New York: W. W. Norton.

Gottesman, I. I. and Shields, J. (1982). *Schizophrenia: the Epigenetic Puzzle*. Cambridge: Cambridge University Press.

Green, M. J. and Phillips, M. L. (2004). Social threat perception and the evolution of paranoia. *Neuroscience and Biobehavioral Reviews*, **28**(3), 333–342.

Guisinger, S. (2003). Adapted to flee famine: adding an evolutionary perspective on anorexia nervosa. *Psychological Review*, **110**(4), 745–761.

Hagen, E. (1999). The functions of post-partum depression. *Evolution and Human Behavior*, **20**(5), 325–359.

Hamilton, W. D. (1964a). The genetical evolution of social behavior, I. *Journal of Theoretical Biology*, **7**(1), 1–16.

Hamilton, W. D. (1964b). The genetical evolution of social behaviour, II. *Journal of Theoretical Biology*, **7**(1), 17–52.

Haukka, J., Suvisaari, J. and Lönnqvist, J. (2003). Fertility of patients with schizophrenia, their siblings, and the general population: a cohort study from 1950–1959 in Finland. *American Journal of Psychiatry*, **160**(3), 460–463.

Heston, L. and Denny, D. (1968). Interactions between early life experiences and biological features in schizophrenia. In *The Transmission of Schizophrenia*, D. Rosenthal and S. S. Kety (eds). Oxford: Pergamon Press, pp. 363–376.

Hirschfeld, L. A. and Gelman, S. A. (eds) (1994). *Mapping the Mind: Domain Specificity in Cognition and Culture*. New York: Cambridge University Press.

Jacobs, N., Kenis, G., Peeters, F., et al. (2006). Stress-related negative affectivity and genetically altered serotonin transporter function: evidence of synergism in shaping risk of depression. *Archives of General Psychiatry*, **63**(9), 989–996.

Jellema, T., Baker, C. I., Oram, M. W., et al. (2002). Cell populations in the banks of the superior temporal sulcus of the macaque and imitation. In *The Imitative Mind: Development, Evolution, and Brain Bases*, A. N. Meltzoff and W. Prinz (eds). Cambridge: Cambridge University Press, pp. 267–290.

Karlsson, J. (1966). *The Biological Basis for Schizophrenia*. Springfield, IL: Charles C. Thomas.

Kauffman, C., Grunebaum, H., Cohler, B., et al. (1979). Superkids: competent children of psychotic mothers. *American Journal of Psychiatry*, **136**(11), 1398–1402.

Keller, M. C. and Miller, G. (2006). Resolving the paradox of common, harmful, heritable mental disorders: which evolutionary genetic models work best? *Behavioral and Brain Sciences*, **29**, 385–452.

Kellett, J. M. (1973). Evolutionary theory for the dichotomy of the functional psychoses. *Lancet*, **1**(7808), 860–863.

Kendler, K., Karkowski, L. and Walsh, D. (1998). The structure of psychosis: latent class analysis of probands from the Roscommon family study. *Archives of General Psychiatry*, **55**(6), 492–499.

Kendler, K. S. (1982). Demography of paranoid psychosis (delusional disorder): a review and comparison with schizophrenia and affective illness. *Archives of General Psychiatry*, **39**(8), 890–902.

Keysers, C. and Gazzola, V. (2006). Towards a unifying neural theory of social cognition. *Progress in Brain Research*, **156**, 379–401.

Kurland, J. A. (1979). Paternity, mother's brother, and human sociality. In *Evolutionary Biology and Human Social Behavior: an Anthropological Perspective*, N. A. Chagnon and W. Irons (eds). North Scituate, MA: Duxbury Press, pp. 145–177.

Kurzban, R. and Aktipis, C. A. (2007). Modularity and the social mind: are psychologists too Self-ish? *Personality and Social Psychology Review*, **11**, 131–149.

Lewis, I. M. (1971). *Ecstatic Religion: a Study of Shamanism and Spirit Possession*. London: Routledge.

Ludwig, A. M. (1995). *The Price of Greatness: Resolving the Creativity and Madness Controversy*. New York: Guilford Press.

Lumsden, C. J. and Wilson, E. O. (1981). *Genes, Mind, and Culture: the Coevolutionary Process*. Cambridge, MA: Harvard University Press.

Lüscher, C. and Ungless, M. (2006). The mechanistic classification of addictive drugs. *PLoS Medicine*, **3**(11), e437.

Madan, K. O. (2002). Hamilton goes empirical: estimation of inclusive fitness from life-history data. *Proceedings of the Royal Society of London. Series B*, **270**, 307–311.

Marks, I. M. (1987). *Fears, Phobias, and Rituals*. New York: Oxford University Press.

Marks, I. M. and Tobena, A. (1990). Learning and unlearning fear: a clinical and evolutionary perspective. *Neuroscience and Biobehavioral Reviews*, **14**(4), 365–384.

Meltzoff, A. N. and Prinz, W. (eds) (2002). *The Imitative Mind: Development, Evolution, and Brain Bases*. New York: Cambridge University Press.

Morgane, P. J., Galler, J. R. and Mokler, D. J. (2005). A review of systems and networks of the limbic forebrain/limbic midbrain. *Progress in Neurobiology*, **75**(2), 143–160.

Mulligan, C. J., Robin, R. W., Osier, M. V., et al. (2003). Allelic variation at alcohol metabolism genes (*ADH1B*, *ADH1C*, *ALDH2*) and alcohol dependence in an American Indian population. *Human Genetics*, **113**(4), 325–336.

Nesse, R. M. (2000). Is depression an adaptation? *Archives of General Psychiatry*, **57**(1), 14–20.

Nesse, R. M. (2001). The smoke detector principle: natural selection and the regulation of defensive responses. *New York Academy of Sciences*, **935**, 75–85.

Nettle, D. (2001). *Strong Imagination: Madness, Creativity and Human Nature*. New York: Oxford University Press.

Nettle, D. (2004). Evolutionary origins of depression: a review and reformulation. *Journal of Affective Disorders*, **81**(2), 91–102.

Nettle, D. and Clegg, H. (2006). Schizotypy, creativity and mating success in humans. *Proceedings of the Royal Society of London. Series B*, **273**(1586), 611–615.

Overmier, J. B. (2002). On learned helplessness. *Integrative Physiological and Behavioral Science*, **37**(1), 4–8.

Painuly, N., Sharan, P. and Mattoo, S. K. (2005). Relationship of anger and anger attacks with depression: a brief review. *European Archives Psychiatry and Clinical Neuroscience*, **255**(4), 215–222.

Paris, J., Zweig-Frank, H. and Guzder, J. (1994). Psychological risk factors for borderline personality disorder in female patients. *Comprehensive Psychiatry*, **35**(4), 301–305.

Peen, J. and Dekker, J. (2004). Is urbanicity an environmental risk-factor for psychiatric disorders? *Lancet*, **363** (9426), 2012–2013.

Pinker, S. (1997). *How the Mind Works*. New York: Norton.

Rabin, A. I., Doneson, S. L. and Jentons, R. L. (1979). Studies of psychological functions in schizophrenia. In *Disorders of the Schizophrenic Spectrum*, L. Bellak (ed.). New York: Basic Books, pp. 181–231.

Reich, G. (2003). Depression and couples relationship. *Psychotherapeutics*, **48**(1), 2–14.

Rosenbaum, B., Valbak, K., Harder, S., et al. (2006). Treatment of patients with first-episode psychosis: two-year outcome data from the Danish National Schizophrenia Project. *World Psychiatry*, **5**(2), 100–103.

Rothenberg, A. (1983). Psychopathology and creative cognition: a comparison of hospitalized patients, Nobel laureates, and controls. *Archives of General Psychiatry*, **40** (9), 937–942.

Sartorius, N., Gulbinat, W., Harrison, G., et al. (1996). Long-term follow-up of schizophrenia in 16 countries. A description of the International Study of Schizophrenia conducted by the World Health Organization. *Social Psychiatry and Psychiatric Epidemiology*, **31**(5), 249–258.

Singh, D. (1993). Adaptive significance of female physical attractiveness: role of waist-to-hip ratio. *Journal of Personality and Social Psychology*, **65**(2), 293–307.

Slavin, M. O. (1985). The origins of psychic conflict and adaptive function of repression: an evolutionary biological view. *Psychoanalysis and Contemporary Thought*, **8**(3), 407–440.

Slavin, M. O. and Kriegman, D. (1988). Freud, biology, and sociobiology. *American Psychologist*, **43**(8), 658–661.

Slavin, M. O. and Kriegman, D. (1992). *The Adaptive Design of the Human Psyche: Psychoanalysis, Evolutionary Biology, and the Therapeutic Process*. New York: Guilford Press.

Sloman, L. and Gilbert, P. (eds) (2000). *Subordination and Defeat: an Evolutionary Approach to Mood Disorders and their Therapy*. Mahway, NJ: Lawrence Erlbaum Associates.

Stein, D. J. (2006). Evolutionary theory, psychiatry and psychopharmacology. *Progress in Neuro-Psychopharmacology and Biological Psychiatry*, **30**(5), 766–773.

Stevens, A. (2002). *Archetype Revisited: an Updated Natural History of the Self*. London: Brunner-Routledge.

Stevens, A. and Price, J. (2000a). *Evolutionary Psychiatry: a New Beginning*. London: Routledge.

Stevens, A. and Price, J. (2000b). *Prophets, Cults and Madness*. London: Gerald Duckworth.

Strous, R. D. and Shoenfeld, Y. (2006). Schizophrenia, autoimmunity and immune system dysregulation: a comprehensive model updated and revisited. *Journal of Autoimmunity*, **27**, 71–80.

Sullivan, P. F., Neale, M. C. and Kendler, K. S. (2000). Genetic epidemiology of major depression: review and meta-analysis. *American Journal of Psychiatry*, **157**(10), 1552–1562.

Sullivan, R. J. and Allen, J. S. (1999). Social deficits associated with schizophrenia defined in terms of interpersonal Machiavellianism. *Acta Psychiatrica Scandinavica*, **99**(2), 148–154.

Sullivan, R. J. and Allen, J. S. (2004). Natural selection and schizophrenia. *Behavioral and Brain Sciences*, **27**, 865–866.

Surbey, M. K. (1987). Anorexia nervosa, amenorrhea and adaptation. *Ethology and Sociobiology*, **8**(suppl.), 47S–61S.

Tajfel, H. (1981). *Human Groups and Social Categories: Studies in Social Psychology*. Cambridge: Cambridge University Press.

Tajfel, H. (1982). Social psychology of intergroup relations. *Annual Review of Psychology*, **33**, 1–39.

Tajfel, H. and Billig, M. (1974). Familiarity and categorization in intergroup behavior. *Journal of Experimental Social Psychology*, **10**(2), 159–170.

Théoret, H., Halligan, E., Kobayashi, M., et al. (2005). Impaired motor facilitation during action observation in individuals with autism spectrum disorders. *Current Biology*, **15**(3), R84–R85.

Tidmarsh, L. and Volkmar, F. R. (2003). Diagnosis and epidemiology of autism spectrum disorders. *Canadian Journal of Psychiatry*, **48**(8), 517–525.

Tiger, L. (1979). *Optimism: the Biology of Hope*. New York: Simon and Schuster.

Tinbergen, N. (1976). Ethology in a changing world. In *Growing Points in Ethology*, P. P. G. Bateson and R. A. Hinde (eds). New York: Cambridge University Press, pp. 507–528.

Trivers, R. L. (1985). *Social Evolution*. Menlo Park, CA: Benjamin Cummings.

Troisi, A. (2005). The concept of alternative strategies and its relevance to psychiatry and clinical psychology. *Neuroscience and Biobehavioral Reviews*, **29**(1), 159–168.

Voland, E. and Voland, R. (1989). Evolutionary biology and psychiatry: the case for anorexia nervosa. *Ethology and Sociobiology*, **10**(4), 223–240.

Wampold, B. E. (2001). *The Great Psychotherapy Debate: Models, Methods, and Findings*. Mahwah, NJ: Lawrence Erlbaum Associates.

Wasser, S. K. and Barash, D. P. (1983). Reproductive suppression among female mammals: implications for biomedicine and sexual selection theory. *Quarterly Review of Biology*, **58**(4), 513–538.

Watson, P. J. and Andrews, P. W. (2002). Towards a revised evolutionary adaptationist analysis of depression: the social navigation hypothesis. *Journal of Affective Disorders*, **72**(1), 1–14.

Wenegrat, B. (1989). Religious cult membership: a sociobiologic model. In *Cults and New Religious Movements: a Report of the American Psychiatric Association*, M. Galanter (ed.). Washington, DC: American Psychiatric Press, pp. 197–210.

Wenegrat, B. (1990a). *Sociobiological Psychiatry: Normal Behavior and Psychopathology*. Lexington, MA: D. C. Heath.

Wenegrat, B. (1990b). *The Divine Archetype: the Sociobiology and Psychology of Religion*. Lexington, MA: D. C. Heath.

Wenegrat, B. (1996). *Illness and Power: Women's Mental Disorders and the Battle Between the Sexes*. New York: New York University Press.

Wenegrat, B. (2001). *Theater of Disorder: Patients, Doctors, and the Construction of Illness*. New York: Oxford University Press.

Wilhelm, K., Mitchell, P. B., Niven, H., et al. (2006). Life events, first depression onset and the serotonin transporter gene. *British Journal of Psychiatry*, **188**, 210–215.

33 Industrial Pollutants and Human Evolution

Lawrence M. Schell

INTRODUCTION: THE ANTHROPOLOGICAL STUDY OF INDUSTRIAL POLLUTANTS

Human experience with industrial pollutants is very recent and very brief relative to the span of our species' evolution. Marked increases in pollution exposure began in the mid 1700s with the industrial revolution. It seems paradoxical that *Homo sapiens* evolved in response to features of decidedly nonindustrialized environments but is now largely an urban species living in industrialized societies. Are we somehow prepared by our evolutionary history to deal with industrialization, or is this an adaptive challenge confronting modern *Homo sapiens*?

This large question is dissected into more scientifically manageable ones that evolutionary human biologists can use to structure research. For example, using demographic measures of species success (Gage, 2005), *Homo sapiens* appears to be flourishing. This conclusion is based on comparisons of populations' mortality profiles, over space and in different historical periods. Using other measures of adaptation however, such as patterns of growth and development or morbidity patterns, the adjustment may appear far less than complete. Although industrialization produces economic benefits and related health benefits, it also produces pollutants that are detrimental to biological systems particularly among the socioeconomically disadvantaged who experience greater exposure. Should we consider all the impacts together in a summary measure of success such as life span, or should we "unpack" the urban environment by determining the specific constituents of urban environments, and evaluate the effects of each constituent on specific outcomes (Schell and Ulijaszek, 1999)? This latter approach is taken here. By analyzing the question in terms of effects of specific industrial pollutants, we may be able to see the direct effects of these elements, remembering that the relationship between industrialization and health is dynamic and could change (Schell and Ulijaszek, 1999).

The questions are these: Are there features of industrialization that pose biological challenges to

humans? And, given that the study of pollutants is not traditional in anthropology, what is the right way to determine if industrial pollutants have any effects on human biology, evolution, and health? An important and related question is, how do anthropologists determine successful adaptation, what are the appropriate measures?

This chapter will address these questions but the conclusions are provisional. The research necessary to answer these questions is still developing across a broad range of disciplines. Typically, reviews of research on pollutants are organized to describe the effects of one pollutant at a time. In contrast, this review is organized by endpoints that relate to challenges to reproduction: human growth and aging, morbidity and mortality which are areas anthropologists have studied to understand human evolution and biologic variation (Stinson et al., 2000). Due to space limitations this is not a comprehensive review of all pollutants, or of all effects of any one pollutant. An effort is made to include contrary studies in each case to emphasize methodological issues that are raised by the diversity in results. The review is intended to introduce the subject including the inherent methodological difficulties that may produce variety in the results. Unfortunately, the review cannot include a description of global warming and its impact on human evolution although global warming is clearly related to industrial pollution and may have a tremendous impact on our species.

An important consideration when evaluating pollution's effects on human evolution is how much of an impact do pollutants make? This is a difficult question to answer for several reasons. Most research on pollutants concerns disease outcomes rather than fertility or other effects of import to evolution (changes in gene frequencies in a population). In large populations such as those in industrialized societies where pollution is generally greatest, fertility is low and influenced greatly by social factors. Natural selection in response to pollution and modern stressors in such populations has not been studied and probably cannot be because

Human Evolutionary Biology, ed. Michael P. Muehlenbein. Published by CAMBRIDGE UNIVERSITY PRESS. © Cambridge University Press 2010.

of the limitations of modern methods. However, evolution may be affected without natural selection operating as the two are not synonymous, and evidence for nonadaptive genetic changes, such as mutations, in response to pollutants are evidence of evolutionary change too.

HOW WILL WE EVER KNOW? METHODS, METHODS, METHODS

Knowing the answer to the questions of effects of many environmental features, including pollution, depends on the suitability of methods. Generally we observe only statistical relationships between pollutants and health effects because we would not knowingly create an experiment by exposing persons to possibly injurious pollutants. With observational studies there are five elements that should be present (usually requiring more than one study) to determine if the pollutant actually causes the biological effect, and these are known as Hill's Criteria of causation. The criteria most often emphasized are: (1) a dose–response relationship between putative cause and effect; (2) a biologically plausible relationship; (3) replication of results; (4) a temporal sequence (cause precedes effect); (5) a strong relationship; and (6) consistency across tests of the relationship. When all six conditions are met we are much more certain that a relationship observed between a pollutant and a health effect has a causal basis.

Sometimes there is a reason to relax at least one of these criteria. For example, some biologically implausible relationships between a pollutant and a health effect have turned out to be true, and in pursuing the implausible we learn about new biological processes. In this vein, sometimes a straight dose–response relationship does not fit the data and a U- or J-shaped relationship is evident. Today the matter of hormetic effects, that is a reversal in the direction or strength of a dose–response relationship that occurs at very low levels of exposure, is under close investigation (Calabrese, 2005).

The old standby approach of comparing industrialized and nonindustrialized populations is no longer used to discern effects of specific pollutants for several good reasons (Schell and Denham, 2003). Firstly, industrialized places vary amongst themselves as do nonindustrialized ones. Further, they do so across several dimensions and continuously. To classify settlements into just two categories is a form of typology, and typology steamrollers over variation. A typological approach to understand the impact of industrialization ignores the diversity inherent in such populations, and thus it cannot use that diversity to understand the different impacts of industrialization. Human

biologists have left biological typology behind and may now leave environmental typology behind as well. A second reason is the effect of offsetting impacts. Industrial pollutants may have detrimental effects that pose challenges to adaptation yet at the same time the delivery of health care may be more intense in industrialized areas. The two opposing forces can balance one another producing a net effect of zero in a general health measure. However, this does not mean that the pollutants are innocuous. The third and final reason is that the aggregate approach does not recognize the effect of social arrangements, by which I mean all the terms that are used today to indicate that power relations in societies structure resources, exposures and risks that produce biological differences along lines of social difference (Schell, 1998). Health disparities (minority, "racial" [sic], or ethnic disparity) is the term used in public health to describe this. Thus, large aggregates may seem healthy but the analysis of health by social gradients within these larger aggregates may reveal the true biological challenges.

A better approach is to measure the amount of exposure to a particular pollutant and compare that to specific effect of interest to human biologists (fertility, menstrual cycle characteristics, sperm counts, prereproductive mortality, growth, development, etc.). Human biologists have excellent methods for measuring effects in physiological systems, but measuring the environmental factor is often not as rigorous. This is an example of unbalanced precision. Comparisons of urban and rural populations, the environmental typology referred to just above, involves not measuring environmental features as they are assumed to be the same within each environmental category. This approach prevents observation of effects of the specific features of each. Measuring those features should produce more accurate descriptions of their effects, and yield more accurate and reliable information.

WHAT ARE INDUSTRIAL POLLUTANTS?

Pollution is usually defined as an unwanted material or energy that is considered a threat to well-being. Pollutants are produced by industrial processes, but many pollutants, carbon dioxide (CO_2), methane, dust, etc., are produced by natural processes as well (Waldbott, 1978). For this reason alone humans may have physiological mechanisms to deal with exposure to some pollutants. Some of the best information on pollutants is available through US government websites (for example http://www.atsdr.cdc.gov/toxpro2.html).

Persistent organic pollutants (POPs) are a group of compounds that are toxic, are resistant to breakdown in the environment (indeed many were created to have this property for industrial or agricultural

TABLE 33.1. Persistent organic pollutants. The US Environmental Protection Agency's "Dirty Dozen."

Pollutant	Pesticide	Industrial chemical	By-product
Aldrin	X		
Chlordane	X		
Dichlorodiphenyl-trichloroethane (DDT)	X		
Dieldrin			
Endrin			
Heltachlor			
Hexachlorobenzene	X	X	X
Mirex	X		
Toxaphene	X		
Polychlorinated biphenyls (PCBs)		X	X
Polychlorinated dibenzo-p-dioxin			X
Polychlorinated dibenzo-p-furans			X

applications). Generally POPs are lipophilic and therefore are present in dietary fats and are stored in adipose tissue. As a result of this biologic process, POPs typically increase in tissue concentration up the food chain in a process termed biomagnification (Table 33.1; http://www.epa.gov/oppfead1/international/pops.htm). Thus, animals near the top of the food chain, piscivores and carnivores, can have substantial concentrations of POPs because in a lifetime they consume many animals that have consumed other animals that have consumed small quantities of POPs. Notable POPs are polychlorinated biphenyls (PCBs) and dioxins which have similar structures (Figure 33.1).

Phthalates are compounds used widely in consumer products primarily to soften plastic items such as shower curtains, garden hoses, food containers, but also can be present in floor tiles, furniture upholstery, shoes, hairspray, and other products used in daily life (Yang et al., 2006). They are found in polyvinylchloride, which is widely used and familiar to us as plastic plumbing and vinyl siding. Phthalates are also found in plastics used in medical tubing, and are found in air, water, and food (Duty et al., 2003).

Lead is a base metal that is completely unnecessary for human health. Lead dust is more concentrated alongside roadways as a result of deposits from vehicle exhaust, and in, on, and around older homes painted with lead-based paint. Depressed, older, inner-city neighborhoods with both features have populations with higher lead levels relative to other populations despite widespread reductions in environmental lead achieved over the past 25 years.

Humans are exposed to lead and POPs chiefly through the diet, although airborne and dermal routes of exposure may contribute to some extent also. Humans consume POPs in commercially available foods, fish caught locally in contaminated waterways, as well as through packaging. Nonoccupational lead exposure occurs largely through ingestion especially by placing nonfood items in the mouth, which is very common among toddlers. Respiration of lead particles makes a far smaller contribution to the intake. Exposure to industrial pollutants is widespread. The US Centers for Disease Control and Prevention (2005) reported that evidence of tobacco smoke, lead, mercury, and phthalates is detected in nearly the entire US population and nearly 150 chemicals were detected in some portion of the US population.

Air pollution is a heterogeneous mixture including oxides of nitrogen (NO_x), oxides of sulfur (SO_x), carbon monoxide (CO), ozone (O_3), and particulate matter (PM). Particulate matter is usually classified by particle size. Particulate matter is not only a pulmonary irritant but also is a vehicle to bring specific compounds into the lung and then the bloodstream where it can produce systemic effects.

Noise is the only pollutant included in this review that is not a material but a form of energy (radiation and light are others). Noise is defined as unwanted sound, and is measured in terms of duration, frequency, and volume (decibels). Noise produces both auditory and nonauditory effects, the former related more to the nerve damage produced by the energy content, and the latter by its ability to create annoyance, stimulate the autonomic nervous system and to produce stress (Kryter, 1985). Thus, although noise is measured as energy, its nonauditory effects are produced by stress.

It is impossible to include here detailed descriptions of the routes of exposure to these pollutants, although such maps provide ample interesting evidence of the social input to pollutant exposure of interest to anthropologists (Schell and Czerwinski, 1998). Likewise, details of pollutant measurement methods would require substantial additional space, although they pertain to the quality of research discussed. Readers are referred to the many excellent reviews and texts on this subject (National Research Council, 1999).

POLLUTANTS AND PRENATAL GROWTH: OPPORTUNITY FOR FETAL PROGRAMMING AND EFFECTS

Alterations in the pattern of prenatal growth are most commonly discerned as changes in birthweight, length

33.1. Polychlorinated biphenyl (a), dioxin (b), thyroxine (c), and estradiol (d), showing similarities in structure that are thought to underlie some of the hormonal effects of the toxicants. The numbered positions on each ring of the polychlorinated biphenyl may be substituted with a chlorine. Depending on the location and number of substitutions, the molecule will vary in persistence and toxicological properties. Molecules unchlorinated at ortho positions (2,2′,6,6′) can resemble dioxin and dibenzofurans which are considered highly toxic.

of gestation, and the occurrence of congenital malformations. Alterations in prenatal growth signify adverse circumstances for growth and development and possibly health generally (Schell, 1997). Alterations in growth patterns also suggest the action of fetal programming with significant effects occurring far later in life (Kuzawa and Pike, 2005; Newbold et al., 2006). Fetal programming refers to alterations in fetal physiology through epigenetic processes, wrought by information transferred from mother to fetus. Programming can result in altered patterns of postnatal growth and functioning, and much later affect adult health. Programming usually refers to changes that are adaptive in the short term in that they are thought to allow fetal survival or anticipate conditions in postnatal life, but may involve changes with a cost, and could be detrimental to health in later life. As such they involve a classic trade-off between resources devoted to promote growth versus those devoted to promote reproduction. Generally most writers on programming do not include

unsuccessful changes as programming but consider these changes as the result of constraint. These would be changes in fetal growth induced by the environment that produce malformations or other clearly detrimental effects. Distinguishing between altered fetal growth that is a part of fetal programming from altered fetal growth that is only detrimental, is difficult on theoretical grounds (Ellison and Jasienska, 2007).

The classic work on programming concerns the influence of nutrition on adult health and disease. However, a recent review of research on pollutants and reproduction emphasized the role of pollutants in modifying the epigenome (the biochemical reactions that influence and constitute gene expression) (Woodruff et al., 2008). As more studies on pollutants and gene expression find links of exposure to later adult health, pollutants should be considered another influence on fetal programming.

Pollutants can alter prenatal growth and or produce effects pertinent to reproduction later in life.

Although such effects have not been considered programming as just described, they have all the outward signs of it except the production of benefit at some point in the life span. The classic case of chemical modification of long-term development is diethylstilbestrol (DES) (Newbold, 2004). This is a synthetic sex hormone with estrogenic effects prescribed to many pregnant women in the mid twentieth century to prevent miscarriage but was ineffective. However, it did promote the development of vaginal cancer in young adult women who were in utero while their mothers were taking the drug. Thus, we know that exogenous hormones can influence the developing reproductive system and produces significant late-occurring effects including changes in behavior and adult-onset disease. Hormonally active agents in the environment are of great concern (National Research Council, 1999).

There have been several reviews of studies on air pollution and prenatal growth (Schell and Hills, 2004; Lacasana et al., 2005; Sram et al., 2005) and in addition to the main finding of reduced prenatal growth, it is apparent that the strength of biological effects varies by pollutant. There are mixed findings for effects of nitrogen dioxide (NO_2), carbon monoxide (CO), and sulfur dioxide (SO_2). At this time the strongest evidence is for an effect of particulate matter including increased frequency of low birthweight and prematurity (Xu et al., 1995; Bobak, 2000; Lee et al., 2003; Leem et al., 2006), small for gestation age (Parker et al., 2005), and intrauterine growth retardation (Dejmek et al., 1999) as well as to reductions in mean birthweight (Chen et al., 2002; Yang et al., 2003; Gouveia et al., 2004; Parker et al., 2005). The validity of air pollution's role is supported by the worldwide distribution of studies in this literature: e.g., Canada, China, Czech Republic, Korea, Taiwan (Xu et al., 1995; Bobak and Leon, 1999; Bobak, 2000; Lee et al., 2003; Liu et al., 2003; Yang et al., 2003). When results are replicated in different societies it suggests that the statistical associations are not produced by bias from uncontrolled variables especially socially mediated ones. Some uncontrolled variables exist in most observational studies, but the same configuration of socially mediated variables is less likely to occur when many different societies are studied.

The effect of lead on prenatal growth is now well established and the problem of impaired cognitive development has been included in US Government publications on the risks of lead exposure (Centers for Disease Control and Prevention, 1991). Studies conducted in a variety of settings, including those where lead exposure does not covary closely with poverty as it has in the United States (Bornschein et al., 1987; Bellinger et al., 1991a, 1991b; Gonzalez-Cossio et al., 1997; Schell and Stark, 1999) show that lead impairs cognitive development even at rather low levels of blood lead, although contrary studies can be found also (for review see Andrews et al., 1994). The case of lead is instructive because the size of the reduction in birthweight or head circumference may be small, but the size of the affected population is enormous.

Evidence for an effect of POPs on prenatal growth is extensive but not consistent (see Schell 1991, 1999 for reviews, and for some evidence of the complexity of the analysis also see Berkowitz et al., 1996; Longnecker et al., 2001; Eskenazi et al., 2003). The strongest evidence for an effect comes from studies of the offspring of Japanese and Taiwanese mothers who consumed rice oil contaminated with a mixture of POPs including dibenzofurans and PCBs. The contamination caused a disease called Yusho in Japan and Yu-cheng in Taiwan. Offspring in utero at the time the rice oil was consumed were smaller at birth in both cases (Yamaguchi et al., 1971), and offspring exposed in utero after consumption ended were smaller also probably from POPs stored in mother's adipose tissue from the earlier exposure (Yen et al., 1989). Studies of birth size in relation to fish consumption (Dar et al., 1992), a common route of POP exposure, are complicated by the nutrients in fish that promote prenatal growth (Jacobson, 2004).

Noise is a consistent by-product of many industrial processes and transportation in industrialized societies. Changes in mean birthweight and the frequency of low birthweight may be influenced by noise pollution because noise is capable of producing a stress response during pregnancy (Welch and Welch, 1970; Committee on Environmental Health of the American Academy of Pediatrics, 1997). Many studies of populations living near airports find that noise stress from airplane takeoffs and landings is associated with reduced prenatal growth (for review, see Schell and Denham 2003). Studies conducted in Japan have analyzed large numbers of births and found a clear dose–response relationship between noise levels and the frequency of small newborns (Ando and Hattori, 1973) as well as temporal associations between the introduction of loud jet aircraft and the frequency of small newborns (Ando, 1988). Repeated associations of the same direction and magnitude, a temporal association and a dose–response curve all support the theory that noise stress affects prenatal growth. No single study is definitive by itself, but when the entire body of work is considered in conjunction with laboratory and field studies of short-term physiological responses to airport noise (Evans et al., 2001) most of the criteria for a causal inference are met. There are, however, studies that have not found effects. Some have studied women exposed to lower levels of noise than is found around airports or noise that is expected and therefore not necessarily very stressful (Wu et al., 1996).

What is true for all studies of environmental influences is that findings pertain to a specific range of exposures, and the effects reported in any single study depend on the range of exposure present. Some variation in results across studies of different exposures and social circumstances is therefore expected and evident.

ENDOCRINE SYSTEM EFFECTS: GROWTH, MATURATION, AND MENARCHE

Many POPs are known to alter the development of some parameter of the endocrine system that suggests effects may develop later in life (Brouwer et al., 1999; Stein et al., 2002). Growth, sexual maturation, menstrual cycle characteristics, sperm quality by numerous measures, fertility, and fecundability have been studied in one or more populations, usually retrospectively, following known exposure. Despite the weakness inherent in retrospective studies, there is ample evidence for effects on these parameters.

Effects of some pollutants on postnatal physical growth have been reviewed (Schell and Knutson, 2002). Blood lead clearly influences postnatal growth and development even at moderate to low levels (Ignasiak et al., 2006). Also, POPs can affect growth, the strongest evidence coming from studies of children exposed to rice oil contaminated with a variety of POPs including dibenzofurans and PCBs (Rogan et al., 1988). Studies of other samples are supportive of 1,1-dichloro-2,2-bis(p-chlorophenyl)ethylene (DDE)-related effects (Gladen et al., 2000; Longnecker et al., 2000) and PCBs may play a role as well (Blanck et al., 2002; Gallo et al., 2002). Air pollution is related to poorer physical growth in many studies (reviewed in Schell and Hills, 2004); for particularly good studies see Jedrychowski et al. (1999) and Jedrychowski (2000).

Sexual maturation holds special interest as alterations in age at menarche relate to the length of the reproductive span and may affect fertility. Alterations in age at menarche also signal significant effects on the system of endocrine and central nervous system (CNS) control over maturation and other related endpoints such as ovulation, fertilization, implantation, and reproduction generally. Age at menarche is also related to adult-onset diseases.

In human populations, lead and persistent organic pollutants have been associated with changes in age at menarche. Analyses of data representing the US population showed that the level of lead in the blood at the time of interview was associated with delayed menarche among girls 8–18 years of age (Selevan et al., 2003; Wu et al., 2003). The delays varied from one to six months. Also attainment of Tanner breast and pubic hair stages was associated with very small increases of blood lead levels from 1 to 3 g/dL. The investigators controlled for socioeconomic factors and body size statistically.

Delay in reaching menarche in relation to blood lead level was observed also among Akwesasne Mohawk adolescents. Also, PCBs were associated with a significantly earlier age at menarche (Denham et al., 2005). In this study, not all PCBs were associated with changes in age at menarche, and the effect was found with a group of estrogenic congeners identified as such through laboratory studies. Polybrominated biphenyls (PBBs) have been associated with earlier age at menarche and attainment of pubic hair stages in breast-fed girls (Blanck et al., 2002). However, Flemish adolescents experienced delayed sexual maturation in relation to polychlorinated aromatic hydrocarbons (Staessen et al., 2001; Den Hond et al., 2002). Furthermore, age at menarche among girls in Seveso, Italy who had been exposed to dioxin from an industrial explosion, was not altered (Warner et al., 2004). Interestingly, the authors noted that most effects of dioxin seen in laboratory studies were from prenatal exposure, whereas the sample of Seveso girls had had postnatal exposure only.

These results illustrate the variation in effects that can be found in studies of many groups of pollutants. Here there is a variety of chemical substances involved in uncontrolled exposures at possibly different ages and at different doses. As these results are based on observational studies of humans, the cause and effect relationship cannot be concluded. Support for a true causal relationship is derived from experimental studies in which laboratory animals were dosed with a toxicant and experienced delayed maturation (for example, with lead, McGivern et al., 1991; Corpas et al., 2002; Dearth et al., 2002), suggesting that the associations in humans do not occur by chance and are not caused by some other unmeasured factor (Goldman et al., 2000).

ENDOCRINE SYSTEM EFFECTS: FEMALE REPRODUCTIVE FUNCTIONING

Some of the earliest studies of PCBs found substantially reduced fertility and increased newborn mortality among rhesus macaques (Allen et al., 1979). Animal models continue to be important in studying reproductive effects as PCBs have been related to alterations of the menstrual cycle in nonhuman primates (Willes et al., 1980), and studies in other model species (rats, mice, and mink) are confirmatory (http://www.atsdr.cdc.gov/DT/pcb007.html).

Strong evidence of effects among humans comes from studies following the Yu-cheng accident. Women who had consumed rice oil that had been contaminated with a combination of POPs (dioxin, dibenzofurans and PCBs) had significantly more abnormal menstrual bleeding (Yu et al., 2000). Girls who had been exposed in utero had higher rates of irregular menstrual cycles, and serum estradiol and follicle-stimulating hormone (FSH) were higher too (Yang et al., 2005).

Premenarchael girls in Seveso, Italy exposed to dioxin had longer menstrual cycles after they reached menarche compared to women who were postmenarcheal when exposed (Eskenazi et al., 2002). However, women exposed to PCBs through consumption of contaminated fish from the US Great Lakes and women exposed through the consumption of Baltic fish, had cycles of reduced length (Mendola et al., 1997; Axmon et al., 2004).

Carefully constructed studies have shown that PCBs are associated with reduced female fecundability (Buck et al., 2000) and longer time to conception (Courval et al., 1999). Higher miscarriage rates have been reported also (Gerhard et al., 1998), and in the Yu-cheng cohort, more stillbirths and preadolescent deaths of offspring (Yu et al., 2000). Though PCBs were not related to menstrual cycle length or other cycle parameters, dichlorodiphenyltrichloroethane (DDT) level (measured as DDE) was significantly associated with shorter cycles, including the luteal phase length alone (Windham et al., 2005).

Although POPs are involved in all these studies, they can differ considerably in their effects, and in the laboratory have been shown to produce estrogenic or anti-estrogenic, androgenic or anti-androgenic effects. Some are known to act through different receptors. The timing of exposure is likely to produce different effects as well. Seemingly contradictory results (delayed or accelerated sexual maturation) may be consistent with exposure to different types or combinations of POPs, and especially to the timing of exposure in the context of sexual differentiation and development prenatal and possibly during other critical periods.

EFFECTS ON SPERM AND THE MALE REPRODUCTIVE SYSTEM DEVELOPMENT

This is a controversial topic (Toppari et al., 1996; Giwercman and Bonde, 1998; Chia, 2000). Numerous studies have related an apparent decline in some measures of sperm quality (number, motility, frequency of sperm deformations) to an increase in exposure to pollutants while others have not (Carlsen et al., 1992; Fisch et al., 1996; also see Chapter 21 of this volume). Some of the controversial elements of this body of work include whether there has in fact been a decline in sperm quality or a change in sampling the population (for example, including more older men) that mimics such an effect, and whether there is any relation between concurrent pollutant exposure and sperm quality (Comhaire et al., 2007). The question is not

answered well by studies of human populations because of the normal variation in sperm quality, the large variety of pollutants that might be tested for influence, and the presence of other influences that may go unmeasured. Many studies focus on one or two pollutants leaving questions concerning the roles of other pollutants unanswered, and there are a number of measures of sperm quality being used. The result is a large body of literature on humans that is uneven, containing many reports of relationships and other reports of their absence. Studies of laboratory animals and in vitro studies provide the most control and weigh heavily in favor of the ability of many pollutants to detrimentally influence sperm or testicular development in some way (Andric et al., 2000a, 2000b; Goldman et al., 2000; Faroon et al., 2001; Gray et al., 2001; Corpas et al., 2002; Veeramachaneni, 2008; Woodruff et al., 2008).

A fairly consistent finding is that some measures of sperm quality are affected by air pollution levels. In a careful study of young men from Teplice, Czech Republic, which is highly industrialized, clinical measures of sperm quality (number, motility, frequency of malformations) did not vary systematically in relation to episodes of air pollution. However, the sperm chromatin structure assay showed that the percentage of sperm with DNA fragmentation was significantly elevated in relation to such episodes (Rubes et al., 2005). Such fragmentation may cause increased rates of male mediated reproductive effects including infertility and miscarriage. Comparing Teplice men with men from an unindustrialized area, however, showed more abnormal chromatin, a lower proportion of motile sperm, sperm with normal morphology, and normal sperm head shape among the Teplice men, although mean and median sperm concentration and count did not differ (Selevan et al., 2000).

Evidence for effects of different PCB congeners on measures of sperm quality is mixed. Follow-up of Yu-cheng boys exposed in utero show that they had increased abnormal sperm morphology, reduced motility, and reduced hamster oocyte penetration capacity compared to controls (Guo et al., 2000) as well as reduced penile length at 11–14 years of age (Guo et al., 1993). Studies of males chronically exposed to lower levels of POPs have produced less clear results. Sperm chromatin integrity was related to levels of PCB (CB-153) in a consistent dose–response relationship among European men but not in Inuit men from Greenland, and no relationships with DDE levels were detected (Spano et al., 2005). Significantly less sperm motility was found in men with relatively high levels of marker PCBs than in the less exposed (Rignell-Hydbom et al., 2004) and no differences were related to levels of DDE. Studies comparing fishermen on the east coast of Sweden, where PCB and other POP contamination levels are high, to those on the west coast, where they are lower, have found negative associations of this exposure with sperm motility, sperm chromatin integrity, and Y:X chromosome ratio but not with sperm concentration or semen volume; couple fertility did not differ (Axmon et al., 2008). Other studies have reported no relationship between PCB exposure and fertility and sperm counts (Emmett et al., 1988a, 1988b; Buck et al., 1997). It may be that metabolites are more potent than the parent compounds as sperm count and motility have been associated with PCB metabolite levels (Dallinga et al., 2002).

A set of recent studies was based on a sample of men using the andrology laboratory at Massachusetts General Hospital. Using the Comet assay that measures DNA integrity, they showed that sperm damage was associated with specific urinary phthalate metabolites after statistical adjustment for appropriate covariates (Duty et al., 2003; Hauser et al., 2007). An additional study examined the effect of a combination of certain PCBs and phthalates and found that the combination was strongly related to "below reference value sperm motility" (Hauser et al., 2005). The authors used standard World Health Organization measures of sperm motility, and the measured levels of phthalate metabolites in this sample were typical of levels in US men, generally suggesting that the findings may apply widely. Other studies have found elevated serum estradiol levels and decreased sperm motility in relation to exposure to pesticides and solvents (Oliva et al., 2001).

As a conclusion to the different results from studies of humans, consider the study by Kuriyama and Chahoud (2004) who found that low doses in utero of one PCB (#118) produced smaller testes, epididymides and larger seminal vesicles, decreased sperm and spermatid numbers, and impaired daily sperm production in adult male rats. Whereas other studies have shown that large doses produce increased sperm production and testis weight (Cooke et al., 1996; Kim, 2001). Researchers have postulated that effects at different doses of endocrine mimics may produce very contrary effects (nonlinear, U-shaped, etc.), (Calabrese and Baldwin, 2003; Kuriyama and Chahoud, 2004). The study of low dose effects from POPs has raised serious questions about the nature of dose–response generally, and the monotonic dose–response relationship, once considered a hallmark of causal relationship between exposure and an effect, may not be that hallmark in every instance.

Ano-genital distance, a measure usually much greater in males than females, is a frequently used measure of sexual differentiation in toxicological studies of rodent sexual development (Sharpe, 2001). Young boys' ano-genital distance, corrected for body weight, was inversely related to levels phthalate metabolites in the blood (Swan et al., 2005) suggesting that some phthalate are anti-androgenic. Boys with short distances for body weight also had higher frequency of undescended testicles and scrotal sacs characterized as small or indistinct from surrounding tissues.

The relationship of blood lead levels to measures of male fertility, particularly sperm quality or function, has been well established through many studies of humans and in experimental work with animals. Among workers at a lead–zinc smelter, sperm count and concentration were related to the concentration of blood lead especially among those men with a specific genotype of δ-aminolevulinic acid dehydratase (Alexander et al., 1998). Analysis of the fertility potential of sperm in conjunction with in vitro fertilization efforts found that seminal plasma lead levels were related to decreased human sperm function (Benoff et al., 2003), and decreased success with in vitro fertilization (IVF). These results on sperm function agree to a great extent with findings from studies of laboratory animals (as reviewed in Benoff et al., 2003) and with findings from studies of fertility and lead levels. Importantly, the findings from studies of male partners of IVF patients are relevant to the general population because the lead levels are similar to reference values and are below levels associated with decreased sperm function and fertility through occupational exposure (Lancranjan et al., 1975).

DIABETES AND OBESITY

Diabetes and obesity are associated with reduced fertility in both men and women. Factors that contribute to increased obesity and diabetes may indirectly influence fitness. In recent years independent studies from different parts of the world, the United States, Belgium, and Sweden, have demonstrated that levels of dioxins and PCBs are related to diabetes and obesity.

Using data from the US National Health Examination Survey, diabetes prevalence was strongly and positively associated with lipid-adjusted serum concentrations of the six POPs found in 80% or more of the sample. The analysis included adjustment for appropriate control variables including age, sex, race, body mass index (BMI), waist circumference, poverty level, etc. A strong dose–response relationship was evident in the data. The risk of diabetes increased steadily with increasing POP level and was several-fold greater in groups above the 75th percentile of toxicant level compared to those below the 25th percentile (Lee et al., 2006).

In a Belgian sample, levels of POPs were between 62% and 39% higher in diabetic patients compared to controls, a highly significant difference before and after statistical adjustment for appropriate covariates (Fierens et al., 2003). The risk of diabetes was significantly greater in the top tercile of toxicant level; risk ratios varied between 5.1 and 13.3 depending on the compound or group of compounds, In a Swedish sample there were positive though weaker relationships between diabetes risk and chlorobiphenyl and DDE levels (Rylander et al., 2005).

AIR POLLUTION: A SPECIAL CASE

Anthropological concern with the evolutionary significance of pollutants requires focusing on influences on fitness. Evidence of an association between air pollution and standard measures of fitness is not a subject of study in public health where most of the research is concentrated. Most studies of the effects of air pollution on mortality focus on adults and particularly the elderly where effects are thought to be greater and of greater significance to public health. Nevertheless, in the air pollution literature there is work on some indirect influences on fitness such as sperm quality and effects on the age of attainment of sexual maturation, both of which influence the length of the reproductive span. However, evidence is now accumulating that indicates that some types of air pollution have more direct effects that pertain to human evolution.

Prereproductive mortality

Studies of air pollution have a long history and there is arguably more information on its effects than any other pollutant. The link between air pollution and adult mortality is soundly established by population-based studies (Dockery et al., 1993; Pope, 1999). Anthropological concern with the evolutionary significance of pollutants requires greater attention to influences on fitness especially to prereproductive mortality.

The first well-documented episode of pollution related mortality was the London smog of 1950 formed by particulates and SO_2 from home heating fires. Adult mortality was increased and infant mortality was increased significantly as well, although this point was not fully appreciated against the higher death rates among the elderly. Recently studies of infant deaths in Mexico City have pointed to the role of small particulate matter ($PM_{2.5}$, e.g., particulates smaller than 2.5 μm in diameter) (Loomis et al., 1999). An increase of 10 units in the level of particulates less than 2.5 μm in diameter ($PM_{2.5}$) was linked to a 4–7% excess mortality in infants (excluding accidents and other sources of mortality clearly unrelated to air pollution).

Genotoxicity

The genotoxicity of air pollution was discovered through investigations of the relationship between air pollution and lung cancer (Pope et al., 2002). Several studies have demonstrated that air pollution, particularly exposure to polycyclic aromatic hydrocarbons (PAHs), is associated with increases in measures of genotoxicity, specifically changes in carcinogen-DNA adducts and carcinogen-protein adducts using in vitro methods. These are mutations in somatic cells but the ability of air pollution to affect genes in human germ cells is a matter pertaining to human evolution and therefore of importance to anthropologists.

Air pollution can contribute to chromosomal abnormalities. Chromosomal abnormality frequency was calculated for 60 African-American and Dominican mother–infant pairs living in low income areas of New York (Bocskay et al., 2005). Airborne PAH level measured by air monitoring was significantly and positively associated with stable chromosomal aberrations in cord blood though not with unstable ones which are less significant as a cancer risk.

The genotoxicity of PAHs is supported by comparisons of levels of airborne PAHs and genotoxicity of particles (greater than 10 μm) collected in Antwerp, from an industrial site, and a rural site (Brits et al., 2004). The most damaging particles came from the urban area and were highest in PAH concentration. Genetic damage could result from airborne particles or from compounds adhered to them because particles more easily enter the distal lung tissue where compounds are absorbed into the bloodstream and can be carried to the liver. There they may be metabolized into chemicals more damaging to DNA and then carried to the spermatogonial stem cells in the testes (Samet et al., 2004). Studies of human lung cell cultures also show dose-dependent increases in genotoxicity in terms of DNA breaks (Dybdahl et al., 2004; Karlsson et al., 2004).

Studies with animal models allow for manipulation of exposures and stronger inferences regarding causal connections. The most informative studies to date demonstrated that heritable mutations in mammals may be caused by particulate air pollution, and that the primary contributor is the PAH compounds. Laboratory mice 6–8 weeks old housed downwind from steel mills for 20 weeks had 1.5–2.0 times the number of heritable mutations compared to an unexposed group, and primarily inherited through the paternal line (Somers et al., 2002). In a follow-up study, mice caged in an industrial environment breathing unfiltered air had significantly elevated mutation rates compared to mice kept in rural areas with unfiltered or mice breathing HEPA-filtered air in either rural or industrial environments (Somers et al., 2004). As

HEPA filtration reduced the expanded simple tandem repeat (ESTR) mutation rates at the industrial site, the airborne agents that were filtered were responsible for the elevated rate, and these were most likely PAHs.

Taken together, these studies demonstrate that some component of air pollution, probably PAHs, is able to produce heritable mutations in mammals. The impact of these mutations in germ cells on human mortality, morbidity, and evolution has yet to be investigated.

GENETIC VARIATION IN SUSCEPTIBILITY TO INDUSTRIAL POLLUTANTS

In the simplest terms, for natural selection to occur genetic variation in a trait must be linked causally to variation in the contribution of offspring number to the following generation. This may be due to differential fertility, mortality, or by some more indirect path. If genetic susceptibility to the effects of industrial pollutants exists, and these effects may affect fertility or mortality before the end of the reproductive span, then selection may occur for variations conferring protection against such effects.

Much of the evidence for genetic variation in susceptibility to pollutants is drawn from studies of PAHs and aromatic amines. Both are generated by combustion, the first from fossil fuels and the second from cigarette smoke and other sources. The genes most often found related to variation in response to carcinogens are the metabolic genes (P450), glutathione S-transferase (GST), and the N-acetyltransferase (NAT) (Perera, 1997). Common genetic polymorphisms in P450s and NAT2 have an impact on cancer susceptibility. Increased lung cancer risk upon exposure to a carcinogen, especially at low dose, is associated with one or more of the CYP1A1 polymorphisms.

DNA repair varies tremendously. The activity of two DNA repair enzymes, O^6-alkyldeoxyguanine-DNA alkyl-transferase and uracil DNA glucosylase, differs by over 100-fold within humans (Perera, 1997), which is linked to increased cancer risk. Genetic variation in receptors involved in toxicokinetics of carcinogens also impact risk of cancer. An important receptor for effects of many aromatic hydrocarbons such as dioxin and some PCBs, as well as PAHs, is the arylhydrocarbon (Ah) receptor. Individuals with this high-affinity binding receptor upregulate CYP1A1, CYP1A2, and other genes.

SUMMARY AND CONCLUSION

Pollutants exist in a huge variety and their effects depend on their chemical structure, how they are metabolized, the dose or exposure, and when in development it occurs. They do not all produce the same constellation of effects. Some groups of pollutants do alter parameters that pertain to reproduction and fitness. Germ-cell mutations, reduced fertility in males and females, changes in menstrual cycle characteristics and measures of male potency, as well as in the pattern or timing of sexual reproduction have been linked to one or more pollutants in several different studies of humans. Research ethics do not permit randomized control trials of humans and pollutants, which is appropriate but makes it virtually impossible to formally establish cause and effect. Yet, there is plentiful supporting evidence from laboratory studies of the same pollutants among appropriate animal models that is substantial and cannot be dismissed. Certainly there are unknown factors that create variability in results. Further research will identify these and in so doing will increase knowledge of normal and disturbed reproduction.

Since Silent Spring was published (Carson, 1962), scientists have known something of the power of pesticides to alter reproduction. More recently, Our Stolen Future (Colborn et al., 1996) has reawakened concern about the long-term effects of pesticides and pollutants generally on human reproduction and health. The US's National Research Council (1999) report, Hormonally Active Agents in the Environment, has substantiated many concerns about chemicals and human reproduction, and even a cursory examination of government websites dealing with toxicants finds many reports of effects on reproduction. Differences in reproduction is not by itself evidence of evolution, but it is a required element. Understanding how genetic variation associates with reproductive effects is the next step in understanding how pollution may affect human evolutionary trajectories.

DISCUSSION POINTS

1. Which pollutant seems most likely to influence human evolution?
2. What evidence is there that pollutants really are bad for human populations, and what standard is applied to decide this?
3. Is it possible that humans are adapted to any of the pollutants discussed, and if so, how could this have happened?
4. What evidence is required to prove that a pollutant influences human evolution and is it possible to collect that information in less than a generation with methods currently available?
5. Why is it so difficult to know the effects of pollutants?

REFERENCES

Alexander, B. H., Checkoway, H., Costa-Mallen, P., et al. (1998). Interaction of blood lead and δ-aminolevulinic acid dehydratase genotype on markers of heme synthesis and sperm production in lead smelter workers. *Environmental Health Perspectives*, **106**, 213–216.

Allen, J. R., Barsotti, D. A., Lambrecht, L. K., et al. (1979). Reproductive effects of halogenated aromatic hydrocarbons on nonhuman primates. *Annals of the New York Academy of Sciences*, **320**, 419–425.

Ando, Y. (1988). Effects of daily noise on fetuses and cerebral hemisphere specialization in children. *Journal of Sound and Vibration*, **127**, 411–417.

Ando, Y. and Hattori, H. (1973). Statistical studies on the effects of intense noise during human fetal life. *Journal of Sound and Vibration*, **27**, 101–110.

Andrews, K. W., Savitz, D. A. and Hertz-Picciotto, I. (1994). Prenatal lead exposure in relation to gestational age and birth weight: a review of epidemiologic studies. *American Journal of Industrial Medicine*, **26**, 13–32.

Andric, S. A., Kostic, T. S., Dragisic, S. M., et al. (2000a). Acute effects of polychlorinated biphenyl-containing and -free transformer fluids on rat testicular steroidogenesis. *Environmental Health Perspectives*, **108**, 955–959.

Andric, S. A., Kostic, T. S., Stojilkovic, S. S., et al. (2000b). Inhibition of rat testicular androgenesis by a polychlorinated biphenyl mixture, Aroclor 1248. *Biology of Reproduction*, **62**, 1882–1888.

Axmon, A., Rylander, L., Stromberg, U., et al. (2004). Altered menstrual cycles in women with a high dietary intake of persistent organochlorine compounds. *Chemosphere*, **56**, 813–819.

Axmon, A., Rylander, L. and Rignell-Hydbom, A. (2008). Reproductive toxicity of seafood contaminants: prospective comparisons of Swedish east and west coast fishermen's families. *Environmental Health*, **7**, 20.

Bellinger, D. C., Leviton, A., Rabinowitz, M., et al. (1991a). Weight gain and maturity in fetuses exposed to low levels of lead. *Environmental Research*, **54**, 151–158.

Bellinger, D. C., Sloman, J., Leviton, A., et al. (1991b). Low-level lead exposure and children's cognitive function in the preschool years. *Pediatrics*, **87**, 219–227.

Benoff, S., Centola, G. M., Millan, C., et al. (2003). Increased seminal plasma lead levels adversely affect the fertility potential of sperm in IVF. *Human Reproduction*, **18**, 374–383.

Berkowitz, G. S., Lapinski, R. H. and Wolff, M. S. (1996). The role of DDE and polychlorinated biphenyl levels in preterm birth. *Archives of Environmental Contamination and Toxicology*, **30**, 139–141.

Blanck, H. M., Marcus, M., Rubin, C., et al. (2002). Growth in girls exposed in utero and postnatally to polybrominated biphenyls and polychlorinated biphenyls. *Epidemiology*, **13**, 205–210.

Bobak, M. (2000). Outdoor air pollution, low birth weight, and prematurity. *Environmental Health Perspectives*, **108**, 173–176.

Bobak, M. and Leon, D. A. (1999). Pregnancy outcomes and outdoor air pollution: an ecological study in districts of the Czech Republic 1986–8. *Occupational and Environmental Medicine*, **56**, 539–543.

Bocskay, K. A., Tang, D., Orjuela, M. A., et al. (2005). Chromosomal aberrations in cord blood are associated with prenatal exposure to carcinogenic polycyclic aromatic hydrocarbons. *Cancer Epidemiology, Biomarkers and Prevention*, **14**, 506–511.

Bornschein R. L., Succop P. A., Dietrich K. N., et al. (1987). Prenatal lead exposure and pregnancy outcomes in the Cincinnati Lead Study. In *Heavy Metals in the Environment*, S. E. Lindenburg and T. P. Hutchinson (eds). Edinburgh: CEP Consultants, pp. 156–158.

Brits, E., Schoeters, G. and Verschaeve, L. (2004). Genotoxicity of PM10 and extracted organics collected in an industrial, urban and rural area in Flanders, Belgium. *Environmental Research*, **96**, 109–118.

Brouwer, A., Longnecker, M. P., Birnbaum, L. S., et al. (1999). Characterization of potential endocrine-related health effects at low-dose levels of exposure to PCBs. *Environmental Health Perspectives*, **107**, 639–649.

Buck, G. M., Sever, L. E., Mendola, P., et al. (1997). Consumption of contaminated sport fish from Lake Ontario and time-to-pregnancy. New York State Angler Cohort. *American Journal of Epidemiology*, **146**, 949–954.

Buck, G. M., Vena, J. E., Schisterman, E. F., et al. (2000). Parental consumption of contaminated sport fish from Lake Ontario and predicted fecundability. *Epidemiology*, **11**, 388–393.

Calabrese, E. J. (2005). Historical blunders: how toxicology got the dose–response relationship half right. *Cellular and Molecular Biology (Noisy-le-Grand, France)*, **51**, 643–654.

Calabrese, E. J. and Baldwin, L. A. (2003). Hormesis: the dose–response revolution. *Annual Review of Pharmacology and Toxicology*, **43**, 175–197.

Carlsen, E., Giwercman, A., Keiding, N., et al. (1992). Evidence for decreasing quality of semen during past 50 years. *British Medical Journal*, **305**, 609–613.

Carson, R. (1962). *Silent Spring*. Boston: Houghton Mifflin.

Centers for Disease Control and Prevention (1991). *Preventing Lead Poisoning in Young Children*. Atlanta, GA: US Department of Health and Human Services, Public Health Service, pp. 1–26.

Centers for Disease Control and Prevention (2005). *Third National Report on Human Exposure to Environmental Chemicals*. Atlanta, GA: Centers for Disease Control and Prevention.

Chen, L., Yang, W., Jennison, B. L., et al. (2002). Air pollution and birth weight in northern Nevada, 1991–1999. *Inhalation Toxicology*, **14**, 141–157.

Chia, S.-E. (2000). Endocrine disruptors and male reproductive function – a short review. *International Journal of Andrology*, **23**, 45–46.

Colborn, T., Dumanoski, D. and Peterson Myers, J. (1996). *Our Stolen Future: are we Threatening our Fertility, Intelligence, and Survival?* New York: Dutton.

Comhaire, F., Mahmoud, A. M. A. and Schoonjans, F. (2007). Sperm quality, birth rates and the environment in Flanders (Belgium). *Reproductive Toxicology (Elmsford, N.Y.)*, **23**, 133–137.

Committee on Environmental Health of the American Academy of Pediatrics (1997). Noise: a hazard for the fetus and newborn. *Pediatrics*, **100**, 1–4.

Cooke, P.S., Zhao, Y.D. and Hansen, L. (1996). Neonatal polychlorinated biphenyl treatment increases adult testis size and sperm production in the rat. *Toxicology and Applied Pharmacology*, **136**, 112–117.

Corpas, I., Castillo, M., Marquina, D., et al. (2002). Lead intoxication in gestational and lactation periods alters the development of male reproductive organs. *Ecotoxicology and Environmental Safety*, **53**, 259–266.

Courval, J.M., DeHoog, J.V., Stein, A.D., et al. (1999). Sport-caught fish consumption and conception delay in licensed Michigan anglers. *Environmental Research*, **80**, S183–S188.

Dallinga, J.W., Moonen, E.J.C., Dumoulin, J.C.M., et al. (2002). Decreased human semen quality and organochlorine compounds in blood. *Human Reproduction*, **17**, 1973–1979.

Dar, E., Kanarek, M.S., Anderson, H.A., et al. (1992). Fish consumption and reproductive outcomes in Green Bay, Wisconsin. *Environmental Research*, **59**, 189–201.

Dearth, R.K., Hiney, J.K., Srivastava, V., et al. (2002). Effects of lead (Pb) exposure during gestation and lactation on female pubertal development in the rat. *Reproductive Toxicology*, **16**, 343–352.

Dejmek, J., Selevan, S.G., Benes, I., et al. (1999). Fetal growth and maternal exposure to particulate matter during pregnancy. *Environmental Health Perspectives*, **107**, 475–480.

Den Hond, E., Roels, H.A., Hoppenbrouwers, K., et al. (2002). Sexual maturation in relation to polychlorinated aromatic hydrocarbons: Sharpe and Skakkebaek's hypothesis revisited. *Environmental Health Perspectives*, **110**, 771–776.

Denham, M., Schell, L.M., Deane, G., et al. (2005). Relationship of lead, mercury, mirex, dichlorodiphenyldichloroethylene, hexachlorobenzene, and polychlorinated biphenyls to timing of menarche among Akwesasne Mohawk Girls. *Pediatrics*, **115**, e127–e134.

Dockery, D.W., Pope, C.A. III, Xu, X., et al. (1993). An association between air pollution and mortality in six US cities. *New England Journal of Medicine*, **329**, 1753–1759.

Duty, S.M., Singh, N.P., Silva, M.J., et al. (2003). The relationship between environmental exposures to phthalates and DNA damage in human sperm using the neutral comet assay. *Environmental Health Perspectives*, **111**, 1164–1169.

Dybdahl, M., Risom, L., Bornholdt, J., et al. (2004). Inflammatory and genotoxic effects of diesel particles in vitro and in vivo. *Mutation Research*, **562**, 119–131.

Ellison, P.T. and Jasienska, G. (2007). Constraint, pathology, and adaptation: how can we tell them apart? *American Journal of Human Biology*, **19**, 622–630.

Emmett, E.A., Maroni, M., Jefferys, J., et al. (1988a). Studies of transformer repair workers exposed to PCBs: II. Results of clinical laboratory investigations. *American Journal of Industrial Medicine*, **14**, 47–62.

Emmett, E.A., Maroni, M., Schmith, J.M., et al. (1988b). Studies of transformer repair workers exposed to PCBs: I.

Study design, PCB concentrations, questionnaire, and clinical examination results. *American Journal of Industrial Medicine*, **13**, 415–427.

Eskenazi, B., Warner, M., Mocarelli, P., et al. (2002). Serum dioxin concentrations and menstrual cycle characteristics. *American Journal of Epidemiology*, **156**, 383–392.

Eskenazi, B., Mocarelli, P., Warner, M., et al. (2003). Maternal serum dioxin levels and birth outcomes in women of Seveso, Italy. *Environmental Health Perspectives*, **111**, 947–953.

Evans, G.W., Lercher, P., Meis, M., et al. (2001). Community noise exposure and stress in children. *Journal of the Acoustical Society of America*, **109**, 1023–1027.

Faroon, O.M., Keith, S., Jones, D., et al. (2001). Effects of polychlorinated biphenyls on development and reproduction. *Toxicology and Industrial Health*, **17**, 63–93.

Fierens, S., Mairesse, H., Heilier, J.F., et al. (2003). Dioxin/polychlorinated biphenyl body burden, diabetes and endometriosis: findings in a population-based study in Belgium. *Biomarkers*, **8**, 529–534.

Fisch, H., Goluboff, E.T., Olson, J.H., et al. (1996). Semen analyses in 1283 men from the United States over a 25-year period: no decline in quality [see comments]. *Fertility and Sterility*, **65**, 1009–1014.

Gage, T.B. (2005). Are modern environments really bad for us? Revisiting the demographic and epidemiologic transitions. *American Journal of Physical Anthropology*, **41** (suppl.), 96–117.

Gallo, M.V., Ravenscroft, J., Denham, M., et al. (2002). Environmental contaminants and growth of Mohawk adolescents at Akwesasne. In *Human Growth from Conception to Maturity*, G. Gilli, L.M. Schell and L. Benso (eds). London: Smith-Gordon, pp. 279–287.

Gerhard, I., Daniel, V., Link, S., et al. (1998). Chlorinated hydrocarbons in women with repeated miscarriages. *Environmental Health Perspectives*, **106**, 675–681.

Giwercman, A. and Bonde, J.P. (1998). Declining male fertility and environmental factors. *Endocrinology and Metabolism Clinics of North America*, **27**, 807–830.

Gladen, B.C., Ragan, N.B. and Rogan, W.J. (2000). Pubertal growth and development and prenatal and lactational exposure to polychlorinated biphenyls and dichlorodiphenyl dichloroethene. *Journal of Pediatrics*, **136**, 490–496.

Goldman, J.M., Laws, S.C., Balchak, S.K., et al. (2000). Endocrine-disrupting chemicals: prepubertal exposures and effects on sexual maturation and thyroid activity in the female rat. A focus on the EDSTAC recommendations. *Critical Reviews in Toxicology*, **30**, 135–196.

Gonzalez-Cossio, T., Peterson, K.E., Sanin, L.-H., et al. (1997). Decrease in birth weight in relation to maternal bone-lead burden. *Pediatrics*, **100**, 856–862.

Gouveia, N., Bremner, S.A. and Novaes, H.M. (2004). Association between ambient air pollution and birth weight in São Paulo, Brazil. *Journal of Epidemiology and Community Health*, **58**, 11–17.

Gray, L.E., Ostby, J., Furr, J., et al. (2001). Effects of environmental antiandrogens on reproductive development in experimental animals. *Human Reproduction Update*, **7**, 248–264.

Guo, Y.-L. L., Lai, T. J., Ju, S. H., et al. (1993). Sexual developments and biological findings in Yu-cheng children. *Organohalogen Compounds*, **14**, 235–238.

Guo, Y. L., Ping-Chi, H., Chao-Chin, H., et al. (2000). Semen quality after prenatal exposure to polychlorinated biphenyls and dibenzofurans. *Lancet*, **356**, 1240–1241.

Hauser, R., Williams, P., Altshul, L., et al. (2005). Evidence of interaction between polychlorinated biphenyls and phthalates in relation to human sperm motility. *Environmental Health Perspectives*, **113**, 425–430.

Hauser, R., Meeker, J. D., Singh, N. P., et al. (2007). DNA damage in human sperm is related to urinary levels of phthalate monoester and oxidative metabolites. *Human Reproduction*, **22**, 688–695.

Ignasiak, Z., Slawinska, T., Rozek, K., et al. (2006). Lead and growth status of school children living in the copper basin of south-western Poland: differential effects on bone growth. *Annals of Human Biology*, **33**, 401–414.

Jacobson, S. (2004). Specificity of the neuropsychological effects of prenatal exposure to PCBs, methylmercury, and lead on infant cognitive development: effects of PBTs on neuropsychological function in children in circumpolar regions. *Neurotoxicology*, **25**, 672–673.

Jedrychowski, W. (2000). Environmental respiratory health in central and eastern Europe. *Central European Journal of Public Health*, **8**, 33–39.

Jedrychowski, W., Flak, E. and Mroz, E. (1999). The adverse effect of low levels of ambient air pollutants on lung function growth in preadolescent children. *Environmental Health Perspectives*, **107**, 669–674.

Karlsson, H. L., Nygren, J. and Moller, L. (2004). Genotoxicity of airborne particulate matter: the role of cell-particle interaction and of substances with adduct-forming and oxidizing capacity. *Mutation Research*, **565**, 1–10.

Kim, I. S. (2001). Effects of exposure of lactating female rats to polychlorinated biphenyls (PCBs) on testis weight, sperm production and Sertoli cell numbers in the adult male offspring. *Journal of Veterinary Medical Science*, **63**, 5–9.

Kryter, K. D. (1985). *The Effects of Noise on Man*, 2nd edn. New York: Academic Press.

Kuriyama, S. N. and Chahoud, I. (2004). In utero exposure to low-dose 2,3′,4,4′,5-pentachlorobiphenyl (PCB 118) impairs male fertility and alters neurobehavior in rat offspring. *Toxicology*, **202**, 185–197.

Kuzawa, C. W. and Pike, I. L. (2005). Introduction: fetal origins of developmental plasticity. *American Journal of Human Biology*, **17**, 1–4.

Lacasana, M., Esplugues, A. and Ballester, F. (2005). Exposure to ambient air pollution and prenatal and early childhood health effects. *European Journal of Epidemiology*, **20**, 183–199.

Lancranjan, I., Popescu, H. I., Gavanescu, O., et al. (1975). Reproductive ability of workmen occupationally exposed to lead. *Archives of Environmental Health*, **30**, 396–401.

Lee, B. E., Ha, E. H., Park, H. S., et al. (2003). Exposure to air pollution during different gestational phases contributes to risks of low birth weight. *Human Reproduction*, **18**, 638–643.

Lee, D. H., Lee, I. K., Song, K., et al. (2006). A strong dose–response relation between serum concentrations of persistent organic pollutants and diabetes: results from the National Health and Examination Survey 1999–2002. *Diabetes Care*, **29**, 1638–1644.

Leem, J. H., Kaplan, B. M., Shim, Y. K., et al. (2006). Exposures to air pollutants during pregnancy and preterm delivery. *Environmental Health Perspectives*, **114**, 905–910.

Liu, S., Krewski, D., Shi, Y., et al. (2003). Association between gaseous ambient air pollutants and adverse pregnancy outcomes in Vancouver, Canada. *Environmental Health Perspectives*, **111**, 1773–1778.

Longnecker, M. P., Klebanoff, M. A., Brock, J., et al. (2000). Background-level in utero exposure to the ubiquitous DDT metabolite DDE is associated with reduced height at age 7 years. *Acta Medica Auxologica*, **32**, 73.

Longnecker, M. P., Klebanoff, M. A., Zhou, H., et al. (2001). Association between maternal serum concentration of the DDT metabolite DDE and preterm and small-for-gestational-age babies at birth. *Lancet*, **358**, 110–114.

Loomis, D., Castillejos, M., Gold, D. R., et al. (1999). Air pollution and infant mortality in Mexico City. *Epidemiology*, **10**, 118–123.

McGivern, R. F., Sokol, R. Z. and Berman, N. G. (1991). Prenatal lead exposure in the rat during the third week of gestation: long-term behavioral, physiological, and anatomical effects associated with reproduction. *Toxicology and Applied Pharmacology*, **110**, 206–215.

Mendola, P., Buck, G. M., Sever, L. E., et al. (1997). Consumption of PCB-contaminated freshwater fish and shortened menstrual cycle length. *American Journal of Epidemiology*, **146**, 955–960.

National Research Council (1999). *Hormonally Active Agents in the Environment*. Washington, DC: National Academy Press.

Newbold, R. R. (2004). Lessons learned from perinatal exposure to diethylstilbestrol. *Toxicology and Applied Pharmacology*, **199**, 142–150.

Newbold, R. R., Padilla-Banks, E. and Jefferson, W. N. (2006). Adverse effects of the model environmental estrogen diethylstilbestrol are transmitted to subsequent generations. *Endocrinology*, **147**, S11–S17.

Oliva, A., Spira, A. and Multigner, L. (2001). Contribution of environmental factors to the risk of male infertility. *Human Reproduction*, **16**, 1768–1776.

Parker, J. D., Woodruff, T. J., Basu, R., et al. (2005). Air pollution and birth weight among term infants in California. *Pediatrics*, **115**, 121–128.

Perera, F. P. (1997). Environment and cancer: who are susceptible? *Science*, **278**, 1068–1073.

Pope, C. A. III. (1999). Mortality and air pollution: associations persist with continued advances in research methodology. *Environmental Health Perspectives*, **107**, 613–614.

Pope, C. A. III, Burnett, R. T., Thun, M. J., et al. (2002). Lung cancer, cardiopulmonary mortality, and long-term exposure to fine particulate air pollution. *Journal of the American Medical Association*, **287**, 1132–1141.

Rignell-Hydbom, A., Rylander, L., Giwercman, A., et al. (2004). Exposure to CB-153 and *p,p′*-DDE and male reproductive function. *Human Reproduction*, **19**, 2066–2075.

Rogan, W. J., Gladen, B. C., Hung, K.-L., et al. (1988). Congenital poisoning by polychlorinated biphenyls and their contaminants in Taiwan. *Science*, **241**, 334–336.

Rubes, J., Selevan, S. G., Evenson, D. P., et al. (2005). Episodic air pollution is associated with increased DNA fragmentation in human sperm without other changes in semen quality. *Human Reproduction*, **20**, 2776–2783.

Rylander, L., Rignell-Hydbom, A. and Hagmar, L. (2005). A cross-sectional study of the association between persistent organochlorine pollutants and diabetes. *Environmental Health*, **4**, 28.

Samet, J. M., DeMarini, D. M. and Malling, H. V. (2004). Biomedicine. Do airborne particles induce heritable mutations? *Science*, **304**, 971–972.

Schell, L. M. (1991). Effects of pollutants on human prenatal and postnatal growth: noise, lead, polychlorinated compounds and toxic wastes. *Yearbook of Physical Anthropology*, **34**, 157–188.

Schell, L. M. (1997). Using patterns of child growth and development to assess communitywide effects of low-level exposure to toxic materials. *Toxicology and Industrial Health*, **13**, 373–378.

Schell, L. M. (1998). Culture as a stressor: a revised model of biocultural interaction. *American Journal of Physical Anthropology*, **102**, 67–77.

Schell, L. M. (1999). Human physical growth and exposure to toxicants: lead and polychlorinated biphenyls. In *Human Growth in Context*, F. E. Johnston, P. B. Eveleth and B. S. Zemel (eds). London: Smith-Gordon, pp. 221–238.

Schell, L. M. and Czerwinski, S. A. (1998). Environmental health, social inequality and biological differences. In *Human Biology and Social Inequality*, S. Strickland and P. Shetty (eds). Cambridge: Cambridge University Press, pp. 114–131.

Schell, L. M. and Denham, M. (2003). Environmental pollution in urban environments and human biology. *Annual Review of Anthropology*, **32**, 111–134.

Schell, L. M. and Hills, E. A. (2004). Urban pollution, disease and the health of children. In *The Changing Face of Disease: Implications for Society*, N. Mascie-Taylor, J. Peters and S. T. McGarvey (eds). Boca Raton, FL: CRC Press, pp. 85–103.

Schell, L. M. and Knutson, K. L. (2002). Environmental effects on growth. In *Human Growth and Development*, N. Cameron (ed.). New York: Academic Press, pp. 165–195.

Schell, L. M. and Stark, A. D. (1999). Pollution and child health. In *Urbanism, Health and Human Biology in Industrialised Countries*, L. M. Schell and S. J. Ulijaszek. Cambridge: Cambridge University Press, pp. 136–157.

Schell, L. M. and Ulijaszek, S. J. (1999). Urbanism, urbanisation, health and human biology: an introduction. In *Urbanism, Health and Human Biology in Industrialised Countries*, L. M. Schell and S. J. Ulijaszek (eds). Cambridge: Cambridge University Press, pp. 3–20.

Selevan, S. G., Borkovec, L., Slott, V. L., et al. (2000). Semen quality and reproductive health of young Czech men exposed to seasonal air pollution. *Environmental Health Perspectives*, **108**, 887–894.

Selevan, S. G., Rice, D. C., Hogan, K. A., et al. (2003). Blood lead concentration and delayed puberty in girls. *New England Journal of Medicine*, **348**, 1527–1536.

Sharpe, R. M. (2001). Hormones and testis development and the possible adverse effects of environmental chemicals. *Toxicology Letters*, **120**, 221–232.

Somers, C. M., Yauk, C. L., White, P. A., et al. (2002). Air pollution induces heritable DNA mutations. *Proceedings of the National Academy of Sciences of the United States of America*, **99**, 15904–15907.

Somers, C. M., McCarry, B. E., Malek, F., et al. (2004). Reduction of particulate air pollution lowers the risk of heritable mutations in mice. *Science*, **304**, 1008–1010.

Spano, M., Toft, G., Hagmar, L., et al. (2005). Exposure to PCB and *p,p*′-DDE in European and Inuit populations: impact on human sperm chromatin integrity. *Human Reproduction*, **20**, 3488–3499.

Sram, R. J., Binkova, B., Dejmek, J., et al. (2005). Ambient air pollution and pregnancy outcomes: a review of the literature. *Environmental Health Perspectives*, **113**, 375–382.

Staessen, J. A., Nawrot, T., Hond, E. D., et al. (2001). Renal function, cytogenetic measurements, and sexual development in adolescents in relation to environmental pollutants: a feasibility study of biomarkers. *Lancet*, **357**, 1660–1669.

Stein, J., Schettler, T., Wallinga, D., et al. (2002). In harm's way: toxic threats to child development. *Journal of Developmental and Behavioral Pediatrics*, **23**, S13–S22.

Stinson, S., Bogin, B., Huss-Ashmore, R., et al. (2000). *Human Biology: an Evolutionary and Biocultural Perspective*. New York: John Wiley and Sons.

Swan, S. H., Main, K. M., Liu, F., et al. (2005). Decrease in anogenital distance among male infants with prenatal phthalate exposure. *Environmental Health Perspectives*, **113**, 1056–1061.

Toppari, J., Larsen, J. C., Christiansen, P., et al. (1996). Male reproductive health and environmental xenoestrogens. *Environmental Health Perspectives*, **104**, 741–803.

Veeramachaneni, D. N. (2008). Impact of environmental pollutants on the male: effects on germ cell differentiation. *Animal Reproduction Science*, **105**, 144–157.

Waldbott, G. L. (1978). *Health Effects of Environmental Pollutants*, 2nd edn. St. Louis: Mosby.

Warner, M., Samuels, S., Mocarelli, P., et al. (2004). Serum dioxin concentrations and age at menarche. *Environmental Health Perspectives*, **112**, 1289–1292.

Welch, B. L. and Welch, A. M. (1970). *Physiological Effects of Noise*. New York: Plenum Press.

Willes, R. F., Rice, D. C. and Truelove, J. F. (1980). Chronic effects of lead in nonhuman primates. In *Lead Toxicity*, R. L. Singhal and J. A. Thomas (eds). Baltimore: Urban and Schwarzenberg, pp. 213–240.

Windham, G. C., Lee, D., Mitchell, P., et al. (2005). Exposure to organochlorine compounds and effects on ovarian function. *Epidemiology*, **16**, 182–190.

Woodruff, T. J., Carlson, A., Schwartz, J. M., et al. (2008). Proceedings of the Summit on Environmental Challenges to Reproductive Health and Fertility: executive summary. *Fertility and Sterility*, **89**, 281–300.

Wu, T.-N., Chen, L.-J., Lai, J.-S., et al. (1996). Prospective study of noise exposure during pregnancy on birth weight. *American Journal of Epidemiology*, **143**, 792–796.

Wu, T., Buck, G. M. and Mendola, P. (2003). Blood lead levels and sexual maturation in US girls: the third National Health and Nutrition Examination Survey, 1988–1994. *Environmental Health Perspectives*, **111**, 737–741.

Xu, X., Ding, H. and Wang, X. (1995). Acute effects of total suspended particles and sulfur dioxides on preterm delivery: a community-based cohort study. *Archives of Environmental Health*, **50**, 407–415.

Yamaguchi, A., Yoshimura, T. and Kuratsune, M. (1971). A survey on pregnant women having consumed rice oil contaminated with chlorobiphenyls and their babies. *Fukuoka Igaku Zasshi*, **62**, 117–122.

Yang, C.-Y., Tseng, Y.-T. and Chang, C.-C. (2003). Effects of air pollution on birth weight among children born between 1995 and 1997 in Kaohsiung, Taiwan. *Journal of Toxicology and Environmental Health, Part A*, **66**, 807–816.

Yang, C.-Y., Yu, M.-L., Guo, H.-R., et al. (2005). The endocrine and reproductive function of the female Yucheng adolescents prenatally exposed to PCBs/PCDFs. *Chemosphere*, **61**, 355–360.

Yang, M., Park, M. S. and Lee, H. S. (2006). Endocrine disrupting chemicals: human exposure and health risks. *Journal of Environmental Science and Health. Part C*, **24**, 183–224.

Yen, Y. Y., Lan, S. J., Ko, Y. C., et al. (1989). Follow-up study of reproductive hazards of multiparous women consuming PCBs-contaminated rice oil in Taiwan. *Bulletin of Environmental Contamination and Toxicology*, **43**, 647–655.

Yu, M.-L., Guo, Y.-L. L., Hsu, C.-C., et al. (2000). Menstruation and reproduction in women with polychlorinated biphenyl (PCB) poisoning: long-term follow-up interviews of the women from Taiwan Yucheng cohort. *International Journal of Epidemiology*, **29**, 672–677.

34 Acculturation and Health

Thomas W. McDade and Colleen H. Nyberg

INTRODUCTION

We currently live in an era of rapid globalization, with few societies beyond its reach. In 1950, 30% of the world's population lived in urban areas; today the proportion of urban residents is nearly 50%, and will exceed 60% by 2030 (United Nations, 2001). Expansions in international trade, market economies, and formal systems of education provide opportunities for some and social inequality for many, while a global mass media fuels new consumer desires and expectations (Ger and Belk, 1996; Navarro, 1999). Populations that were once relatively isolated – geographically, linguistically, culturally – are becoming increasingly exposed to, or interfacing with, novel environments and lifestyles that may differ considerably from their own. What implications do these processes have for human biology and health? For our understanding of processes related to human adaptation?

These questions have been hotly debated by anthropologists, epidemiologists, and economists for over half a century, and simple answers are not forthcoming. Most research focuses on the health impact of recent transitions related to globalization, but it is important to recognize that human biology has been shaped by dynamic relationships with cultural, economic, and broader ecological factors since the Neolithic Revolution. For example, rises in infectious disease associated with shifts from hunting and gathering to sedentization and agricultural intensification beginning about 10 000 years ago can be considered the first epidemiologic transition (Barrett et al., 1998). Links between cultural/economic transitions and health are fundamental to human evolutionary biology since morbidity and mortality are important correlates of reproductive fitness, and because cultural/economic factors define, in large part, the environments to which humans adapt.

Field-based research on cultural/economic changes and health has provided an opportunity to explore mechanisms of human adaptation, and to consider the local consequences of broader cultural and economic processes. For some scholars, isolated indigenous populations represent a Rousseauian vision of humans living in harmony with their environment, with incursions leading to losses of autonomy and disruptions in a delicate homeostasis that impair health (Wirsing, 1985). For others, health improves when cultural and economic trends provide reliable access to nutritional and health care resources (Dennett and Connell, 1988; Gage, 2005). Still others recognize both the challenges and opportunities of change, and explicitly consider the multiple, and potentially conflicting, paths through which cultural and economic factors may shape health (Berry et al., 1986; McElroy, 1990; Godoy et al., 2005c; Steffen et al., 2006).

Many terms have been used to describe the processes of change that affect health, including "acculturation," "Westernization," "modernization," and "market integration." While these terms reflect significant differences in emphasis, they all share a common interest in linking individual well-being to surrounding social, cultural, economic, and political dynamics. "Acculturation" foregrounds changes in local culture resulting from regular interaction between two or more autonomous cultural systems (Social Science Research Council Summer Seminar, 1954). "Westernization" and "modernization" typically focus on the development of a cash economy, secularized government, systems of formal education, and urban residential units (Levy, 1966; Spindler, 1984). "Market integration" focuses more specifically on the emergence of market-based systems of exchange, and the degree of participation in, and dependence on, markets to meet subsistence needs (Godoy, 2001).

Obviously, these are overlapping, closely related processes of change, and in this review we use the more general term "cultural and economic transitions" to encapsulate all of them. After introducing key concepts that serve as a foundation for research in this area, we review the state of current knowledge regarding the health impact of cultural and economic transitions in populations around the world. We discuss the mechanisms through which these transitions affect human biology and health, and highlight the models that have been used to reveal these associations.

Human Evolutionary Biology, ed. Michael P. Muehlenbein. Published by Cambridge University Press. © Cambridge University Press 2010.

DEVELOPMENT OF KEY CONCEPTS

History

Despite elegant work by Boas (1911) with immigrant populations demonstrating biological plasticity in response to environmental change, prevailing conceptualizations of human populations as genetically fixed and bounded entities dominated into the 1940s (Little, 1982; Johnston and Little, 2000). The predominance of such typological thinking left little room for serious consideration of cultural factors as determinants of human biological variation. However, in the 1950s and 1960s, evidence for adaptive biological responsiveness to climatic stressors – cold, heat, altitude – began to accumulate in a wide range of human populations, thereby documenting considerable phenotypic plasticity and setting the stage for an ecological approach to human biology.

With a renewed emphasis on theory and hypothesis testing, the "new physical anthropology" (Washburn, 1951) encouraged a series of studies – many as part of the International Biological Programme – on biobehavioral mechanisms of adaptation to diverse environments around the world (Baker and Weiner, 1966). Climate, disease, nutrition, and other aspects of the physical ecology were investigated as primary stressors with implications for human development and health. From this perspective it was but a small step to acknowledge cultural factors as key organizers of exposure to, and buffers against, environmental stressors (Baker et al., 1986; Huss-Ashmore, 2000).

These historical developments set the stage for culture change and health as a major area of ongoing research in human biology, with the field currently confronting two significant challenges. The first concerns the recognition that the majority of research on culture and health relies on relatively simplistic understandings of culture, and that more sophisticated operational definitions of cultural factors are needed in future research (McDade, 2002; Dressler, 2005). The second concerns the importance of political–economic factors in structuring exposure to stressors associated with change, and in constraining ability to respond to those stressors in ways that attenuate their adverse impact on health (Singer, 1996). An emphasis on plasticity and the mechanisms through which humans are able survive in a wide range of environments can divert attention from the quality of these environments – many of which are marginal, at best – and runs the risk of naturalizing or justifying social inequality. Human biologists have recognized this risk, and a number of studies have attempted to consider the local effects of global processes, the biological costs of adaptation, and the social and economic relations underlying adaptive capacity and stress exposure (Leatherman and Goodman, 1997; Goodman and Leatherman, 1998).

Culture, adaptation, and health

Although frequently invoked as a causal factor in shaping human behavior, biology, and health, culture is rarely operationalized and a consensus definition remains elusive. For most purposes, culture can be defined as socially transmitted systems of shared knowledge, beliefs, values, and/or behaviors that vary systematically across recognizable social groups (Tylor, 1924). Culture is a primary component of the human adaptive strategy. For the past 20000 years humans have inhabited a wider ecological and geographic range than any other vertebrate, and – for better and for worse – over the past 10000 years humans have become Earth's dominant species due to the emergence of sophisticated technological and social systems (Alexander, 1990; Boyd and Richerson, 2005). Culture reduces the costs of learning and increases the breadth and accuracy of knowledge accumulated across generations, and it is a necessary input for normal cognitive and neurological development (Changeux, 1997; Tomasello, 1999). Culture is thus often seen as providing solutions to challenges posed to reproduction and survival, but it simultaneously creates new problems in these same domains (Schell, 1997). For example, cultural factors may allocate risks and resources along lines of residence, occupation, and/or social class. In some situations, inequality in the distribution of resources may be so pronounced that reproduction and survival are compromised for significant numbers of individuals. Thus, culture is both a buffer to adversity and an unequal allocator of risk to health, providing adaptive opportunities as well as challenges (Thomas, 1992; Schell, 1997). Lasker (1969) outlined three levels at which organisms adapt to shifting ecological circumstances: (1) genetic modifications resulting from natural selection across generations; (2) ontogenetic modifications that emerge over an individual life course; and (3) short-term physiological and behavioral responses. Recently, Kuzawa (2005) has proposed "phenotypic inertia" as a fourth level of adaptability, intermediate in time course between genetic and ontogenetic changes, which allows for intergenerational transmission of information through nongenetic mechanisms. Humans are unique among animals in the degree to which we rely on behavioral responses and extrasomatic adaptations to facilitate survival and reproductive success, an attribute that provides considerable flexibility in response to environmental change.

Culture is thus a key mechanism of human adaptability, but it also constructs in large part the settings to which humans must adapt. For humans, culture *is* the environment. Shifts in the cultural landscape can therefore be expected to result in phenotypic changes at multiple levels by acting through genetic, intergenerational, ontogenetic, and/or physiological/behavioral mechanisms.

There are a number of cases demonstrating the impact of cultural factors on human genetic diversity (Durham, 1991). For example, classic research in West Africa suggests that the intensification of agriculture led to changes in the population frequency of alleles protective against malaria (Livingstone, 1958). Clearing wide patches of tropical rainforest provided abundant breeding grounds for the mosquito vector, which encouraged the transmission of malaria and conferred a selective advantage upon resistant individuals.

Although cultural transitions have left their mark on the human genome – particularly in a deep evolutionary time frame – it is important to emphasize that contemporary human populations adjust to environmental changes primarily through nongenetic mechanisms. This is due to the slow pace of genetic change, as well as the exceptional degree of phenotypic plasticity that is a defining trait of the human species. Behavioral and developmental mechanisms provide more rapid, flexible, and mutable options for adaptive responsiveness such that genetic changes come into play only as a "last resort" (Slobodkin, 1968; Huss-Ashmore, 2000).

Recent interest in the early life origins of adult disease (see Chapters 2 and 30 of this volume) has provided evidence that "intergenerational inertia" may be an important mechanism linking cultural/economic transitions and health. For example, in the Philippines, adolescents have high levels of cholesterol despite low rates of obesity and relatively low intakes of dietary fat, and unexpectedly high rates of hypertension at any given level of body mass (Colin Bell et al., 2002; Kuzawa et al., 2003). These results – and similar findings from other populations as well as research with animal models – suggest that prenatal undernutrition leads to permanent changes in organ structure and physiological function that lower the energy needs of the individual as an adult. In teleological terms, the developing organism uses the prenatal environment to predict future resource availability such that a "thrifty phenotype" may be an adaptive response to an impoverished nutritional ecology.

But when calorie-dense foods are readily available later in life, prenatal undernutrition may make individuals particularly vulnerable to diseases related to energy surplus, including obesity, cardiovascular disease, and diabetes. Globally, industrial agriculture and commercial food production are increasing the availability of concentrated sources of calories (e.g., refined sugar, cooking oils), and levels of physical activity are dropping (Popkin, 1994, 2003). Populations experiencing rapid nutritional and lifestyle changes may be particularly vulnerable to the contribution of "developmental mismatch" to chronic degenerative diseases. Birthweights in India, for example, are among the lowest in the world, yet the country is currently experiencing an epidemic of type II diabetes despite low levels of obesity (Yajnik, 2004).

Developmental inertia is a relatively unexplored mechanism linking cultural and economic transitions to health, but it is likely to be particularly important for understanding secular trends in growth and chronic disease.

Similar developmental mismatches – albeit within a shorter time frame – have been invoked as mechanisms to explain a wider range of associations between culture change and health. Early work on the social origins of hypertension, for example, suggested that incongruity between the cultural environment an individual is socialized into as a child and the one he or she functions in as an adult, may lead to upregulation of stress pathways and poor health (Cassel, 1974). Similarly, individuals who grow up at high altitude acclimatize by developing larger lung volumes, whereas individuals who migrate to high altitude as adults do not show the same level of functional capacity in hypoxic environments (Frisancho, 1978).

The vast majority of research on how environments affect health focuses on the final mechanism of adaptation: short-term (i.e., nondevelopmental) physiological and behavioral responses to challenge. This reflects the importance of phenotypic plasticity to the human adaptive strategy, and is particularly the case for research on culture and health, since most investigators are interested in the impact of relatively rapid cultural and economic transitions, typically occurring within a single generation. This body of research is the primary focus of this chapter, with numerous examples discussed below.

It is important to acknowledge that the relationships among environmental stressors, processes of adaptation, and health are not always straightforward. For example, the development of a thrifty phenotype in a nutritionally marginal environment may well represent an adaptive process, but the consequences for health in a nutrient-rich environment (obesity, chronic disease) are hardly positive. Similarly, even though a thrifty phenotype may be beneficial in a nutritionally marginal environment compared to a more spendthrift phenotype, undernutrition early in life compromises work capacity, reproductive performance, cognitive function, and longevity (Martorell, 1995; Pelletier et al., 1995) – again, outcomes that are clearly suboptimal. For these reasons, some scholars eschew the term adaptation altogether when making reference to human biological responses to social, economic, and cultural contexts, particularly when these contexts are impoverished (Pelto and Pelto, 1989; Frisancho, 1993; Singer, 1996).

Several general hypotheses follow from this discussion of the mechanisms of adaptation that can be used as a guide for research into the impact of cultural and economic transitions on human biology and health. Firstly, since changes in gene frequency in response to shifting selection pressures represent an exceedingly slow mechanism of "last resort," few studies are likely to find genetic signatures of natural selection in response to contemporary changes in the cultural and economic environment. This does not deny the importance of genetic mechanisms of adaptation in an evolutionary time frame, nor does it discount the possibility that the unique evolutionary history of a particular population may shape its response to contemporary environmental changes in significant ways (Snodgrass et al., 2007).

Secondly, intergenerational and developmental processes provide a degree of phenotypic plasticity for infants and juveniles that is not available to adults. With fewer mechanisms of adaptation online, adults can be expected to pay a higher biological toll in response to rapid environmental change than they would have paid if they confronted the same changes earlier in life. Similarly, conditions experienced early in life – prenatally and in infancy – can be expected to have long-term as well as intergenerational implications for health in the context of changing environments.

Thirdly, the pace of change should be associated with its impact on health. Rapid cultural and economic transitions pose more immediate adaptive challenges by opening up larger gaps between old and new environments, and by constraining options for phenotypic adjustments that strive to optimize fitness. Genetic, intergenerational, and developmental mechanisms of adaptation, for example, do not apply in situations of rapid change, and behavioral responses – particularly those embedded in cultural systems handed down across generations – may have to overcome considerable inertia before they can serve as effective buffers. Related to this point, the adverse health impact of environmental change should be greatest in the short term, and should attenuate over time through phenotypic responsiveness on multiple levels.

CULTURAL AND ECONOMIC TRANSITIONS AND HEALTH: WHAT DO WE KNOW?

In the past 25 years, over 200 peer-reviewed studies have been published investigating the impact of cultural and economic transitions on health in diverse populations around the world[1]. Few consistent patterns of association emerge from this literature, cautioning against any attempts to draw sweeping conclusions regarding the impact of cultural and economic change on health in contemporary populations. Here we summarize key

features of this literature prior to discussing the pathways linking change and health.

Topical foci

The major emphases of recent studies have been on outcomes related to child growth, chronic degenerative diseases such as obesity, cardiovascular disease, and diabetes, and mental health. Changing patterns of infectious disease have also received considerable attention, while other health outcomes addressed in recent studies include violent deaths, and risk behaviors such as substance abuse.

Child growth

Measures of growth and nutritional status are sensitive indicators of past, present, and future health, and have therefore served as key outcomes for research on the impact of cultural and economic transitions. In part due to the relative ease of anthropometric data collection, studies of child growth have been conducted in a wide range of populations, with mixed results (Panter-Brick et al., 1996). In some settings, cash earnings from market activities are used to purchase food and to supplement traditional diets, which may improve nutritional status and reduce growth stunting (Santos and Coimbra, 1991; Stinson, 1983; Norgan, 1995; Reyes et al., 2003; Godoy et al., 2005c; Foster et al., 2005). In others, market participation leads to an increase in growth stunting through nutritional stress or an increase in infectious disease that acts synergistically with undernutrition to impair growth (Fitton, 2000). In the Andes, the impaired statural growth of many high-altitude populations was revealed to be the result of poverty and disease, rather than primarily high altitude stress, underscoring the role that inequalities in access to the benefits of economic development play in shaping child growth (Leonard, 1995; Leatherman, 1998). While one in four children globally suffer from undernutrition (de Onis et al., 2004), in many developing nations the prevalence of childhood obesity and associated metabolic disorders is growing rapidly (Gracey, 2003).

Chronic degenerative diseases

One of the most consistent findings of recent research is the positive association between market integration and the emergence of chronic degenerative diseases. Increases in adult body weight and the prevalence of obesity have been demonstrated in several populations, and at differing levels of articulation with market economies (Zimmet et al., 1980; Baker et al., 1986; Norgan, 1994; Shephard and Rode, 1996). Similarly, blood pressure is higher for adults in affluent countries and in populations that have moved away from subsistence-based economies. Mortality rates from cardiovascular

---

[1] This estimate is based on searches in electronic databases (PubMed, PsychInfo, Google Scholar), as well as anthropological journals and bibliographies of past publications on the topic. A variety of search terms were used to capture the most inclusive sample of articles, including *acculturation*, *Westernization*, *modernization*, *market integration*, *urbanization*, *industrialization*, and *culture change*. Studies were included if they contained individual-level health outcomes (i.e., no regional or national level data sets) and were studies of in situ change (i.e., not studies of migrants).

disease are also elevated (Scotch, 1963; Waldron et al., 1982; Poulter et al., 1990; Dressler and Bindon, 1997; Egusa et al., 2002; Steffan et al., 2006). Often, the "modernization" gradient in high blood pressure is more pronounced in men than in women, although the precise mechanisms underlying this difference remain unclear (Schall, 1995). Accompanying changes in body composition and blood pressure is the emergence of type II diabetes in a range of populations, including Pacific Islanders (Prior and Rose, 1966), Australian Aborigines (Norgan, 1994), Pima Indians (Ravussin et al., 1994), and among circumpolar peoples (Shephard and Rode, 1996).

Although changes in diet and reductions in physical activity levels are obvious contributors to cardiovascular and metabolic diseases, psychosocial distress associated with culture change may also be a powerful factor (McGarvey and Schendel, 1986; Steffan et al., 2006). In addition, mismatch between prenatal nutritional environments and those encountered later in life may be an important, though relatively unexplored, mechanism leading to the development of chronic disease (see Chapter 30 of this volume).

While broad-scale epidemiological surveys indicate that rates of cancer are increasing dramatically throughout the developing world, scant data are available directly linking this increase to cultural and economic transitions. As with cardiovascular and metabolic diseases, changes in diet and activity levels are thought to be major contributors: excessive bodyweight has been recognized by the World Cancer Research Fund as an important risk factor for many malignancies (World Cancer Research Fund and American Institute for Cancer Research, 2007; Renehan et al., 2008). However, the link between obesity and certain cancers may vary dramatically between sexes and populations of different geographic origin.

In addition, the adoption of risky behaviors such as smoking and alcohol abuse have also been hypothesized to contribute. For example, high rates of lung cancer among indigenous Siberian populations have been linked to smoking, while increased rates of reproductive carcinomas have been attributed to high rates of sexually transmitted diseases and high-risk sexual behavior in Arctic communities (Freitag et al., 1991). China, which has the largest global production and consumption of tobacco, experiences over 1 million smoking-related deaths each year, most the result of lung cancer (Zhang and Cai, 2003; Jiang et al., 2008).

Mental health and substance abuse

Several studies have documented increased rates of depression and suicide, as well as increased levels of physiological indicators of stress, in populations in transition (Graves and Graves, 1979; Brown, 1981; Dressler, 1991; Shephard and Rode, 1996). Epidemics of adolescent suicide among circumpolar populations, in Samoa, and in Micronesia have fueled

speculation that the encroachment of Western institutions and values, the decline of local communities, the disintegration of extended family networks, and a lack of economic opportunities have contributed to adolescent stress (Kraus and Buffler, 1976; Macpherson and Macpherson, 1987; Rubinstein, 1992; McDade, 2002). A model of acculturative stress proposed by Berry and Kim (1998) suggests that degree of distress depends largely on whether those acculturating wish to maintain their traditional identity, and the degree to which they want to integrate and interact with those in the new culture (Berry and Kim, 1998). Research on response to colonial stressors among Inuit youth highlights the role of coping behaviors in buffering stress-related health outcomes (O'Neil, 1986), while several studies in the Samoa Project have documented physiological responses to the psychological and behavioral changes accompanying rapid social and economic transition (McGarvey and Schendel, 1986; Baker et al., 1986; Pearson et al., 1993). Recent research with an Amazonian horticulturalist population in the early stages of transition suggests that greater participation in the market economy is significantly linked to anger, fear, and sadness (Godoy et al., 2005c).

The effects of psychological distress may also manifest through the emergence of excessive alcohol consumption and substance abuse. While these behaviors may initially serve as coping mechanisms in response to the experience of acculturative stress, the consequences for individual health and community well-being can be devastating (Berry, 1970; Neff, 1993; Dressler et al., 2004). Sporadic episodes of binge drinking are common among communities undergoing rapid social and economic transitions, and may be most problematic (Graves, 1967; Singer et al., 1992). Cigarette smoking has been documented in many transitioning populations, but precise estimates of duration and frequency of smoking are not known. The abuse – rather than controlled social or ritual use – of marijuana, stimulants, cocaine, depressants, tranquilizers, and hallucinogens has also become more common among many indigenous communities (Shephard and Rode, 1996; Curtis et al., 2005). Alcohol and substance abuse have also been implicated in escalating rates of accidents and violent crimes among modernizing populations (Curtis et al., 2005).

Infectious and parasitic diseases

Infectious disease persists as a major contributor to morbidity and mortality around the world. While historical improvements in standard of living have reduced rates of infectious disease (McKeowen, 1976), contemporary associations with cultural and economic transitions are mixed. Economic development may facilitate prevention and treatment by providing access to health-care resources, yet it may also contribute to disease transmission and severity through degradation of the local ecology (Walsh et al., 1993; Brinkman, 1994; Santos et al. 1997). Rapid urbanization may contribute to increases in population density, erosion of basic infrastructure, deforestation, and changing subsistence and agricultural patterns with direct implications for pathogen exposure (Armelagos et al., 1996; Garruto et al., 1999). For example, in Guyana the conversion of cattle-grazing fields to a more economically viable cash crop, rice, provided an ideal breeding ground for mosquitoes. Consequently, malaria rates soared (Desowitz, 1976; Brown and Whitaker, 1994). The case of the Aché of Paraguay underscores the profound impact of infectious disease on immunologically naïve Amerindian populations: two-thirds of the adult population developed tuberculosis within just 20 years of sustained contact with outsiders (Hurtado et al., 2003).

Although macroparasites have received less attention than other infectious diseases, they may play a major role in malnutrition, growth stunting, and comorbidity with other infections (Lawrence et al., 1980). In cases where urbanization is associated with higher parasite loads, increased infection is often attributed to crowding, poor sanitation, and proximity to animals and fecal material (Confalonieri et al., 1991; Fitton, 2000). Type of water source may also be a significant predictor of parasite exposure: in the Philippines, neighborhoods which relied on rain water rather than irrigation canals had a higher prevalence of the water-borne parasite schistosomiasis (Payne et al., 2006). In other populations, increased market integration and education have been linked to a reduction of intensity in intestinal parasites, a pattern observed among the Tsimane' of the Bolivian Amazon (Tanner, 2005).

Regional foci

While research into the health impact of cultural and economic transitions has been conducted globally, the Pacific Islands, Latin America, and the Arctic have served as particularly important testing grounds for exploring the physiological and psychological correlates of lifestyle change. Many of the projects in these areas represent the legacy of human adaptability research initiated by the International Biological Programme and the Man and the Biosphere Programme (Little et al., 1997).

Pacific Islands

Rapid economic development in the Samoan islands occurring post-World War II provided a compelling backdrop for a comprehensive assessment of the impact of lifestyle changes on health. The Samoan Studies Project – initiated in 1975 – is distinguished by the breadth of economic, demographic, cultural,

biological, and health data it compiled (Baker et al., 1986). The project took advantage of the recent emergence of a "modernization gradient" across the islands to reveal a general pattern of increasing obesity, hypertension, adverse lipid profiles, as well as adverse stress hormone profiles in individuals residing in more Westernized areas. Several studies have continued to track these trends over time (McGarvey et al., 1993; Keighley et al., 2007), and to explore culture change as a source of adolescent stress (McDade et al., 2000; McDade, 2002).

Similar findings emerged from the Tokelau Island Study, in which a cohort of central Pacific islanders were surveyed before migration to New Zealand in 1967 and then followed post-migration for over two decades. The health consequences of migration included a shift toward a more carbohydrate- and sugar-laden diet, a greater likelihood of developing gout, adverse lipid profiles, an increase in the prevalence of type II diabetes, and a modest increase in blood pressure (Stanhope and Prior, 1980; Prior et al., 1987; Salmond et al., 1989). A similar study in the Solomon Islands revealed more modest post-migration changes in blood pressure and body composition (Page et al., 1974; Friedlander, 1987).

In Papua New Guinea, the relative isolation of the highlands has provided a baseline for comparison with the more rapidly transitioning settlements in coastal areas, and numerous studies have demonstrated links between economic integration and elevated blood pressure, body mass index (BMI), and reduced activity levels (Jenkins, 1989; Schall, 1995; Yamauchi et al., 2001; Ulijaszek, 2007). Other studies based in Papua New Guinea have cautioned against the assumption that isolated highland populations represent idealized states of good health and harmony with the natural environment: many of these populations have long histories of contact with outsiders, and rates of undernutrition and infectious disease are high (Dennett and Connell, 1988).

Arctic

Circumpolar populations provide case studies in mechanisms of adaptation to extreme cold, but arctic research has also served as a major foundation for exploring the ramifications of acculturation on psychological and physical health. The Inuit of northern Canada, for example, are frequently cited as a classic example of how acculturation can provoke mental distress and undermine health through risk behaviors such as smoking and alcoholism (Berry, 1970; Lynge, 1976; O'Neil, 1986; Shephard and Rode, 1996). In addition to increased rates of infectious diseases introduced through contact with outsiders, circumpolar peoples have experienced soaring rates of chronic diseases such as hypertension, diabetes, and cancer due to changes in diet and physical activity (Shephard and Rode, 1996). More recent research has integrated a political–economic perspective to explore the influence of the collapse of the former Soviet Union on the health of indigenous populations (Leonard et al., 2002; Sorensen et al., 2005; Snodgrass et al., 2006).

Latin America

Early research in Latin America focused largely on adaptation to high altitude, but attention soon shifted to patterns of social relations and economic stratification that shaped resource access, with implications for adaptive processes and health (Leatherman et al., 1986; Leatherman, 1998). For example, these investigations revealed that factors linked to poverty – undernutrition, infectious disease – played a stronger role than high altitude hypoxia in contributing to growth stunting (Leatherman et al., 1995; Leonard, 1989; Stinson, 1996). Research in Amazonia has considered a wide breadth of topics, including changing patterns of infectious and parasitic diseases, blood pressure, diversity of dietary strategies within shifting social and ecological contexts, and socioeconomic factors affecting adult nutritional status and work capacity (Reed et al., 1970; Dufour, 1983; Fleming-Moran et al., 1991; Fitton and Crews, 1995; Dufour et al., 1997). Ongoing research in lowland Bolivia is collecting repeat observations from a population at the early stages of market integration in an attempt to explicitly distinguish the effects of market, cultural, and ecological factors on health (Foster et al., 2005; Godoy et al., 2005c; McDade et al., 2005).

Africa

Despite a high level of anthropological and public health interest in Africa, relatively few studies have explicitly investigated the health effects of cultural and economic transitions. Exceptions include important comparative research on blood pressure, investigations into changing patterns of fertility, and the development of demographic models of shifting population dynamics (Scotch, 1963; Howell, 1979; Caldwell and Caldwell, 1987; Hyatt and Milne, 1993). The Turkana project has explored changing subsistence and settlement patterns among Turkana pastoralists in northern Kenya, and has revealed better health among nomads with reference to child growth and maternal nutritional status (Leslie et al., 1993; Little et al., 1993; Shell-Duncan and Obiero, 2000). The recent and rapid emergence of maternal obesity represents a new public health problem in urbanizing Africa, where child stunting and wasting remain prevalent (Garrett and Ruel, 2005; Villamor et al., 2006). Despite the enormous importance of these studies in underscoring the complexity of the "nutrition transition" in Africa, most data have been collected at the population level and therefore cannot explore the within-household dynamics that may contribute to the simultaneous existence of malnutrition and overweight.

Defining "change"

Deriving a meaningful operational definition of "change" is a major challenge associated with research into the health consequences of cultural and economic transitions. The vast majority of studies to date (well over three-quarters of those reviewed for this chapter) employ group-level, ecological comparisons of individuals sharing a common biological and cultural history, but currently living in divergent cultural and/or economic conditions. Comparisons across urban and rural populations are particularly common.

For obvious logistical reasons, cross-sectional designs are frequent, and were used in almost 90% of the studies reviewed for this chapter. In these studies, "change" is captured by making cross-sectional comparisons across two or more environments, where it is assumed that the baseline environment will in the future become similar to the environment used for comparison. Change is thus anticipated, but not directly measured, within individuals. Only prospective studies can measure changes in cultural and economic environments within individuals across time. This approach has been applied most productively to studies of migrants, where dramatic environmental changes can be captured in relatively short lengths of time. Prospective studies of in situ change are less common, but are important resources for assessing causal linkages between cultural and economic transitions and health (Little and Johnson, 1987; Adair and Popkin, 1988; Godoy et al., 2005a, 2005b).

CULTURAL AND ECONOMIC TRANSITIONS, ADAPTATION, AND HEALTH: PATHWAYS

There are several pathways through which cultural and economic transitions impact health (Figure 34.1). These processes are not mutually exclusive, and they interrelate in complex and variable ways across different

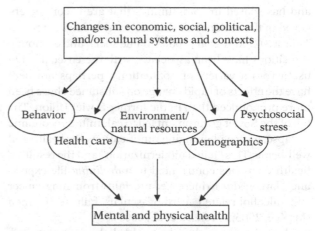

34.1. Primary pathways through which cultural and economic transitions affect human health.

geographic, ecological, and historical contexts to shape health through the mechanisms of adaptation outlined above.

Behavior

Cultural and economic transitions initiate behavioral changes that may have dramatic implications for patterns of diet and physical activity. A number of studies have documented shifts in energy balance resulting from reductions in physical activity and increased consumption of calorie-dense foods as populations move from a labor-intensive, subsistence-based diet to a partial or complete reliance on commercially prepared foods. In places like the Samoan Islands, these behavioral changes have fueled the rapid emergence of epidemics of obesity and chronic degenerative diseases (McGarvey et al., 1993; Keighley et al., 2007). On the other hand, for marginally nourished populations, the availability of commercial foods – and the means to acquire them – has been associated with better nutritional status by improving dietary diversity and the quantity of macronutrients (DeWalt, 1983).

Breast-feeding is a critical determinant of infant health in low-income settings, and contextual factors that discourage breast-feeding may therefore have dramatic implications for infant morbidity and mortality (McDade and Worthman, 1998). The availability of infant formula may discourage breast-feeding, particularly when there are competing demands for a mother's time and energy (Nerlove, 1974), and when formula is perceived to represent a more advanced, technological, or "modern" option (Manderson, 1982).

In addition to changes in patterns of diet and physical activity, health may be affected by the emergence of risk behaviors such as smoking or excessive alcohol consumption. Alcoholism was pronounced among Native Americans undergoing forced assimilation in the United States (Graves, 1967), emerged as an adolescent response to colonial stress in the Arctic (O'Neil, 1986), and has soared in communities that are becoming engaged in the regional market economy in lowland South America (Godoy et al., 2006). While social and economic transitions have been associated with increased alcohol use across a variety of populations, perhaps nowhere have the effects of rapid change on substance abuse been more pronounced than in the former Soviet Union. The fall of the Soviet government led to calamitous economic collapse, with wide-ranging ramifications for health and well-being. This "failed modernization" and the resulting health crisis has contributed to a *declining* life expectancy, largely due to increased mortality from lung cancer and alcohol-related cardiovascular failure in men (McKee, 2005; Sorensen et al., 2005).

Associations with risky health behaviors may also be more indirect. For example, an increase in the prevalence of disordered eating behaviors, including self-induced vomiting to lose weight, followed prolonged television viewing among adolescent girls in Fiji (Becker, 2004). Becker notes that "the impact of televisions appears especially profound, given the longstanding cultural traditions that previously had appeared protective against dieting, purging, and body dissatisfaction" (Becker et al., 2002, p. 511).

Health care

For many populations, early stages of acculturation and market integration provide for the first time reliable access to primary health care, vaccines and antibiotics effective against a wide range of infectious diseases, and new explanatory models for the prevention of disease grounded in germ theory (Akin et al., 1985; Gage, 2005). The implications for child health and survival may be particularly dramatic. For example, among Venezuelan lowlanders, economic changes provided households tied to the market with greater access to medical services, and cash from wage labor allowed them to purchase health care (Holmes, 1985). A by-product of emerging markets is the credit system, which can smooth shocks by providing the means to borrow money to pay for medicines and health services during times of need (Amin, 1997; Godoy, 2001). Access to a well-functioning credit system may also buffer vulnerable individuals from nutritional shortfalls, and prevent the further deterioration of health from fluctuating food consumption that impairs ability to work, whether at a subsistence level or for cash wages (Rose, 1999; Morduch, 1995).

In contrast, access to Western biomedical models and treatments may actually impair health, as in the case of the global distribution and promotion of infant formula as a substitute for breast milk (Post and Baer, 1978). Women in rural Kenya who deliver their babies in hospitals are twice as likely to feed their infants formula (Cosminsky, 1985). In rural Mexico, women who seek care in a local hospital introduce supplemental foods to their infants earlier than women not near the hospital since medicines used in hospitals are believed to enter breast milk and make breast-feeding dangerous (Millard and Graham, 1985). In other settings, formula may be given to mothers in hospitals or prescribed by physicians, thereby associating formula with modern medicine (Manderson, 1982; Simpson, 1985). The rapid spread of formula can be attributed to an aggressive global marketing campaign mounted by formula manufacturers, and it was not until growing concerns regarding the contribution of formula use to rising rates of infant mortality that the campaign was stopped.

Another level of complexity is added when tensions arise between mothers and public health policy makers regarding recommendations for breast-feeding practices in the context of epidemic HIV, especially in Africa. For HIV-positive mothers, breast-feeding confers a 15–20% risk of mother-to-child-transmission (Atashili et al., 2008). However, this risk must be balanced against the concerns of malnutrition and mortality attributed to an increased susceptibility to infectious disease contracted from water-based formulas. Careful evaluation of the pros and cons led the World Health Organization to recommend exclusive breast-feeding for 6 months of life even for HIV-positive mothers, as breast-feeding remains a major protector of child survival in sub-Saharan Africa (Coutsoudis et al., 2008; David et al., 2008).

Environment and natural resources

Rapid alterations in the local physical environment can have significant implications for health. Although few populations rely exclusively on local natural resources for subsistence, rapid population growth and intensification of agriculture put pressure on natural resources. Among slash and burn horticulturalists, market integration is associated with more rapid deforestation and intensification of farm production (Vadez et al., 2004), while constraints on land use encourage nomads to shift from pastoralism to farming (Little and Leslie, 1999; McCabe, 2003). These changes may have implications for the sustainability of natural resource use, as well as for the quality and quantity of nutrients extracted from the local environment.

In addition, increases in population density resulting from and contributing to agricultural intensification facilitate the spread of infectious disease, as does more frequent contact with outsiders, particularly in the absence of improvements in sanitation and hygiene (Jenkins, 1989). Initial exposure to novel infectious diseases has in some cases had devastating consequences for indigenous populations, leading to dramatic rates of mortality, as well as social and economic upheaval (McNeill, 1977). Recent research in lowland Bolivia has documented higher concentrations of C-reactive protein (an indicator of inflammation) in children regularly attending school compared to children not in school, suggesting that the relatively recent introduction of formal schooling may promote pathogen transmission in this population (McDade et al., 2005).

In urban (and urbanizing) environments, industrial development and rapid increases in population density may outpace the management of organic waste and other forms of pollution, such as noise, air, mercury, and lead (Schell and Denham, 2003). Exposure to environmental pollutants may be particularly severe for children, who may experience impairments in endocrine function, physical growth, and cognitive development as a result. In a community-based study of Akwesasne Mohawk youth, increased blood lead levels predicted a substantial delay in the timing of menarche (Denham et al., 2005). Studies of the organic pollutants polychlorinated biphenyls (PCBs), have revealed reduced birthweight in infants born to PCB-exposed mothers (Schell et al., 2006), and a negative association between adolescent PCB levels and measures of long-term memory (Newman et al., 2006). The potential for intergenerational transmission of these pollutants may be especially profound, as these lipophilic toxins are concentrated in human fat cells where they can be transferred in utero to the developing fetus and to infants via breast milk (Schell and Knutson, 2002; also see Chapter 33 of this volume).

Closely related to changes in the physical environment is the possibility that cultural knowledge regarding the use and management of local natural resources will be lost as a result of acculturation. Local ecological knowledge serves as a guide for activities central to survival and well-being, including effective habitat management, subsistence and food procurement, and attempts to prevent and cure disease (Brookfield and Padoch, 1994; Atran and Medin, 1997; Etkin, 2000; Pandey, 2001). Cultural and economic transitions may threaten this knowledge if formal schooling and integration into emerging market economies prioritize alternative sources of information and provide access to substitute products not made from local resources (Godoy et al., 1998; Reyes-Garcia et al., 2005). Recent research in lowland Bolivia has demonstrated the potential cost to health if this knowledge is lost: adults with better knowledge regarding the availability and potential uses of local ethnobotanical resources have healthier children, as indicated by higher measures of nutritional status and lower levels of infection (McDade et al., 2007).

Demographic processes

Classic demographic transition theory states that as living conditions improve in preindustrial populations mortality rates will drop – particularly for infants and children – and life expectancy will increase (Caldwell, 1976). Declining fertility rates will follow, thus completing the transition from high rates of birth and death, to low rates of birth and death in industrialized settings. Over the last 200 years, global life expectancy in industrialized populations has more than doubled, from about 25 to 65 years for men and 70 years for women (Oeppen and Vaupel, 2002), and rates of total fertility have dropped to about 2 children in industrialized nations since World War II. In comparison, total fertility rates in the developing world are far higher, likely due to the lack of contraceptives and less access to education in these populations (Caldwell and Caldwell, 1987). This represents an idealized model of the complex relationships among socioeconomic and demographic changes that may apply to many populations in transition, with important implications for health. Local processes of demographic change may also deviate substantially from this model.

As standard of living improves adolescent girls and boys reach puberty earlier, lowering the age at which successful reproduction is possible (although social, cultural, and/or economic factors may encourage delayed childbearing). Lactational amenorrhea – induced by a neuroendocrine cascade initiated by frequent nipple stimulation – increases interbirth intervals in noncontracepting populations (Konner and Worthman, 1980; Vitzthum, 1994). The availability and use of supplemental infant foods may interrupt this process and lead to rapid population growth, as well as increases in rates of infant morbidity (McDade and Worthman, 1998). Sexually transmitted diseases and patterns of secondary sterility may also reduce fecundity (Caldwell and Caldwell, 1983; Ellison et al., 1986; Hill and Hurtado, 1996).

Cultural transitions may also encourage shifts in the number of children desired by adults, and in patterns of parental investment. Greater parental investment in offspring education, a smaller desired family size, and a delay in the onset of reproduction, as well as a reduction in total number of offspring can result in reduced- and below-replacement fertility (Levine, 1994; Leslie et al., 1994; Kaplan et al., 2002). In Kenya, for example, increased female education was associated with a substantial decrease in the number of births (Hyatt and Milne, 1993). In such situations, parents may perceive increased "costs" associated with raising children that motivate shifts in patterns of fertility and child rearing (Caldwell and Caldwell, 1987).

Rapid population growth following accelerations in reproductive schedules and/or declines in infant mortality may increase population density, thereby facilitating the transmission of pathogens and increasing rates of infectious diseases (Barrett et al., 1998). Population growth may also put strain on local natural resources, and serve as an impetus for outmigration and urbanization (Armelagos et al., 1991, 2006).

Demographic data from the Aché – a small indigenous population of eastern Paraguay – underscore many of these points. Until relatively recently, the Aché lived in small, mobile bands and subsisted by hunting animals and gathering plants and insects in the forest (Hill and Hurtado, 1996). During this period, age-specific mortality rates were high compared to contemporary industrial populations, but similar to other foraging populations. Even though rates of infant and child mortality were particularly elevated, approximately one third of all Aché born lived to the age of 60. First contact with outsiders in the 1970s brought epidemics of infectious diseases that killed nearly 40% of the population, and since that time the Aché have settled into life on reservations near the forest. State and missionary interventions, changes in diet, and access to Western medicines have resulted in reduced mortality rates compared to life in the forest at all ages except for the first year of life, when rates of infant mortality are actually higher on the reservation.

During life in the forest, the total fertility rate was 8 live births per woman, with an interbirth interval of 37.6 months, and an average age of first birth of 19.5 years (Hill and Hurtado, 1996). On the reservation, women start reproducing nearly 2 years earlier at 17.7 years of age, and have more live births (8.5) at interbirth intervals that are 6 months shorter (31.5 months). This accelerated reproductive schedule may have implications for women's health, infant, and child well-being, as well as population growth for the Aché.

Psychosocial stress

Changes in neuroendocrine activity are an essential part of the "fight-or-flight" response in all vertebrates. For most species these are critical adaptive responses to social and ecological stressors, but the metabolic consequences may be more profound for humans who experience the same hormonal activation in response to purely psychosocial stressors (Sapolsky, 2004). Stress has been shown to be an important determinant of mental and physical health for humans (McEwen and Lasley, 2002; Cohen et al., 2007), and may therefore be a significant mediator of the health impact of cultural and economic transitions. Psychosocial stressors initiate the activation of multiple neuroendocrine pathways, including the sympathetic adrenal medullary system (SAM), and the hypothalamic-pituitary-adrenocortical (HPA) axis, which in turn cause changes in cardiovascular, metabolic, and immune system activity (Johnson et al., 1992; McEwen, 1998). Short-term responses facilitate adaptation to a wide range of stressors, but frequent or excessive demands contribute to poorer health outcomes. The operationalization of stress poses significant challenges to cross-cultural research, since self-report measures of perceived stress or symptoms of emotional distress (e.g., depression, anxiety) commonly used in Western countries may not be valid in different linguistic and cultural contexts. Physiological measures provide relatively objective indicators of stress that are comparable and interpretable across different populations (despite cultural and linguistic differences), and that provide insight into the biological pathways that may lead to impaired health (Ice and James, 2007). For this reason a number of studies have investigated the impact of cultural and economic transitions on biomarkers of stress.

The potential impact of rapid social change on stress and health has long been recognized, with early work emphasizing ambiguity and conflict resulting from discordance between traditions and expectations set up during childhood socialization, and a rapidly changing cultural environment encountered in adulthood (Scotch and Geiger, 1969; Henry and Cassel, 1969; Cassel, 1974). Several studies demonstrate significant effects of acculturation on the frequency of self-reported symptoms of physical and emotional distress (Graves and Graves, 1979, 1985; Hanna and Fitzgerald, 1993). Extensive work with Samoan populations has associated the "modernization" gradient with increases in blood pressure and catecholamine levels (McGarvey and Baker, 1979; Baker et al., 1986; Pearson et al., 1993; McGarvey, 2001). Similarly, individual-level measures of stressors associated with culture change predict increases in blood pressure, catecholamine concentrations, and reductions in cell-mediated immunocompetence (Dressler et al., 1987a,b; James et al., 1987; McDade, 2002). A recent meta-analysis suggests that acculturative distress is a stronger predictor of increases in blood pressure than other obvious correlates of lifestyle such as dietary change and physical activity levels (Steffen et al., 2006).

The vast majority of research investigating associations between acculturation and stress has been conducted with adults, with a few noteworthy exceptions. In rural Samoa, school attendance has been associated with increases in catecholamine production in young children (Sutter, 1980). Recent work has shown that status inconsistency is associated with psychosocial stress (as indicated by a biomarker of cell-mediated immune function) in Samoan adolescents (McDade, 2002). In a study of exposure to tourists in Nepal, children living along a trekker's trail were taller and heavier with greater BMI than rural children. However, these children also had significantly higher blood pressure, with girls exhibiting greater increases associated with living along the trail than boys (Pollard et al., 2000).

In addition to affecting stressor exposure, cultural and economic changes may alter sources of social support that individuals can draw on to buffer the adverse health effects of stress. For example, in the Andes, a strong involvement in wage labor, particularly among migrants, impeded access to familial or extra-familial sources of support (Carey and Thomas, 1987). Similarly, Polynesian men who retained a more traditional group orientation reaped the benefits of better health while "modernizing," as opposed to those men who did not retain social contacts (Graves and Graves, 1979).

Recent research in industrialized countries has underscored the importance of social capital as an attribute of communities that may be protective for health (Kawachi, 1999; Subramanian et al., 2005). Social capital refers to institutions, norms of trust and reciprocity, and social networks that facilitate collective action for the common good and that can be drawn on by individuals in times of need (Kawachi, 2000). In preindustrialized settings social capital may be a particularly important resource for protecting against shocks such as illness or poor foraging/agricultural returns (Godoy et al., 2005a). This has led to the hypothesis that as markets develop in preindustrialized settings and socioeconomic stratification emerges, health may suffer due to an erosion of social capital. Associations are mixed, with market integration leading to distrust, conflict, and attenuation of norms of sharing in some populations (Putsche, 2000; Bury, 2004). In other settings, new market activities work within existing structures of social capital, and norms of sharing and reciprocity help distribute the benefits of market participation and attenuate the negative effects of socioeconomic stratification (Le Ferrara, 2003).

ACCULTURATION AND HEALTH: MODELS

Investigators have employed a number of conceptual and analytical models in their efforts to evaluate the impact of cultural and economic transitions on health. These models serve as guides for hypothesis testing and the interpretation of results, and reveal assumptions regarding how contextual factors relate to health. Four general classes of models have been used to guide prior research (Figure 34.2).

Additive models: more is better (or worse)

Of fundamental importance is a meaningful representation of aspects of the surrounding environment that may have implications for individual well-being. Additive models (Figure 34.2a) assume that exposure to, or

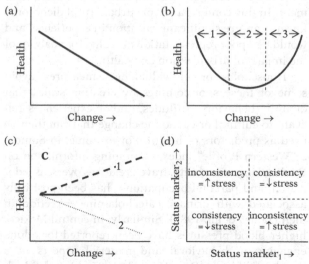

34.2. Four general types of conceptual models **(a–d)** used to evaluate the association between measures of cultural and economic change and measures of health.

engagement with, nontraditional ways of living is associated with health in a roughly linear fashion in a given population, such that a dose–response relationship can be detected between environmental changes and changes in health. The vast majority of these studies employ ecological comparisons, although a significant number include individual-level measures of experience as well.

Ecological models employ group-level comparisons of individuals sharing a common history, but currently living in divergent geographic locations that differ in some meaningful way. In these studies, geographic location provides a proxy for exposure to, and engagement in, novel social, cultural, and/or economic systems. This approach has been critical in underscoring the importance of acculturation to health in diverse populations around the world, and has provided a solid foundation for future research (Patrick et al., 1983; Jenner et al., 1987; Wessen et al., 1992; Ulijaszek et al., 2005).

Ecological studies are easy to implement, but they are limited in that causation is impossible to ascertain with group-level comparisons, and confounding is possible as group differences in health may be due to factors other than acculturation. In addition, assumptions about the nature of the difference in experience between groups need to be evaluated, and substantial heterogeneity within the comparison groups may be obscured.

For example, a number of studies use urban residence as an indicator of acculturation, but in the Philippines, type of neighborhood within an urban area is a major determinant of access to basic infrastructure, health care services, educational opportunities, and socioeconomic resources (McDade and Adair, 2001; Dahly and Adair, 2007). Rural settlement types are comparably diverse, with some rural areas looking more like urban areas on multiple measures of environment. In this context, a simple urban–rural dichotomy fails to capture significant variation in experience, and would be a poor representation of reality for analyses of the impact of urbanization on health.

Household- or individual-level measures avoid some of these shortcomings by creating scales that quantify behaviors, attitudes, and/or experiences relevant to cultural or economic change that can then be used as predictors of health. For example, in Samoa, a "Western Profile" index – summing information on English language ability, travel, relatives overseas, education, and parental occupation – has been positively associated with urinary catecholamine excretion in men (James et al., 1987). Similarly, in central Mexico, higher blood pressures have been reported for adults engaging in behavioral and material aspects of a "modern" lifestyle (Dressler et al., 1987a,b). Individual-level assessments recognize the diversity of experience within communities, and provide measures of experience that are more proximate to health outcomes than possible with ecological comparisons.

Curvilinear models

In general, additive models attempt to represent the association between change and health in a linear fashion: cultural and economic transitions lead to rapid changes in the local sociocultural landscape, posing adaptive challenges that may compromise health. A variant of the additive model is an explicit consideration of curvilinear or threshold associations (Figure 34.2b). These models are similar in that they investigate the impact of a single dimension of culture change on health, but they differ in that they do not assume a linear, dose–response relationship. Within a single population, early stages of acculturation may have negative effects on health, with later stages having positive effects (or vice versa). Alternatively, the association between acculturation and health may reach the point of diminishing returns, resulting in a J-shaped or threshold pattern.

For example, among Filipino immigrants to Hawai'i, intermediate levels of "culture contact" were associated with elevated stress hormone levels, while more favorable stress hormone profiles were found in immigrants with low and high levels of contact (Brown, 1982). Recent research in Brazil reveals a curvilinear relationship between body composition and socioeconomic status that differs between men and women: People living in the *favela* – an impoverished squatter neighborhood – were the smallest and leanest, with adiposity increasing and then leveling off for men with higher levels of socioeconomic resources. Adiposity also increased with socioeconomic status for women in the middle class, but then dropped for upper-class women, who were lean compared to women in the middle class (Dos Santos et al., 2001).

Findings like these hint at the complexity of cultural and economic change, and curvilinear associations with health may be more common than often assumed based on research with additive models. Synchronically, a linear relationship between acculturation and health may be evident if the range of cultural variation is truncated. If Figure 34.2b represents the association between acculturation and health in a given population at a given point in time, this relationship would look like a negative linear relationship if analyses were limited to the lowest third of the acculturation distribution. Recent research in the islands of Samoa underscores this point (Ezeamama et al., 2006). In the more economically developed area of American Samoa, individual-level measures of socioeconomic status were inversely related to risk factors for cardiovascular disease, a pattern similar to that found in most Western nations. But in nearby

Samoa, where economic development has progressed more slowly, socioeconomic resources were positively associated with cardiovascular risk.

Diachronically, a different pattern of relationship between change and health may emerge. If the x axis in Figure 34.2b represents cultural change over time, an analysis of the association between acculturation and health in the earliest stages of acculturation (indicated by #1) would reveal a negative association. The same analysis conducted at a later point in time (indicated by #3) would reveal a positive association. Three decades of data collection in the Samoan islands has been able to document a shift in the association between socioeconomic status and cardiovascular disease risk, underscoring the value of longitudinal research. Unfortunately, few studies have the breadth or time depth to consider the full range of variation in cultural and economic factors that predict health in a population, and this fact may be a major reason for inconsistency in results drawn from studies conducted in different regions, and at different points in time.

Heterogeneity models

In their simplest form, additive and curvilinear models assume that cultural and economic transitions have uniform affects on the health of all individuals in a population. Clearly, this need not be the case, and analyses vary in the degree to which they explicitly evaluate the degree of heterogeneity in responses to change.

Historically, the emergence of agriculture has been a fundamental transition for human populations with dramatic implications for health (Cohen, 1989). The intensification of agriculture, and related trends toward population growth, central organization, and labor specialization promote the unequal distribution of resources, including resources related to health. For example, paleopathological evidence from Chile circa 1000 BP indicates that the intensification of agriculture was associated with a general decline in health (#2 in Figure 34.2c), but this decline was not equally distributed. Shaman men – a privileged class – showed little evidence of poor health (#1 in Figure 34.2c) and were much better off than men in general, who in turn were in better health than women in this population (Allison, 1984). This stratification in health is not evident in analyses from earlier historical periods when the society was more egalitarian, and emerges only with the social and economic transitions associated with agricultural intensification. It is important to inspect for within-population heterogeneity in associations between change and health since averaging across subgroups may lead to the mistaken conclusion that change is not associated with health.

Acculturation may lead to stratification in health across social groups within a population, but it may also have divergent effects on different health outcomes. For example, as noted above, Westernization in the islands of Samoa has been associated with a general pattern of increased burdens of morbidity in adults as indicated by multiple measures of cardiovascular and metabolic disease risk (Baker et al., 1986). This relationship can be represented by line #2 in Figure 34.2c. However, these transitions are also associated with declines in mortality – infant mortality in particular – which are in turn associated with increases in life expectancy, an undisputable indicator of positive health. These improvements in health can be represented by line #1 in Figure 34.2c, and underscore the point that it cannot be assumed that the implications of cultural and economic transitions for health will be uniform across health indicators.

Inconsistency models

Inconsistency models diverge from other approaches in that they simultaneously explore multiple dimensions of change in an attempt to capture the tensions, conflicts, and ambiguities that arise within individuals living in the context of cultural and economic transitions. These models do not assume that acculturation is related to health in an additive fashion; instead, it is the social and/or psychological dissonance associated with rapid change that may be causal. Inconsistency models rely primarily on psychosocial pathways linking experience and health, and have therefore been applied most frequently to physiological outcomes related to stress.

Models of status inconsistency recognize that an individual's social status is derived from multiple sources, and that these status markers are used to advertise a certain style of life or social worth (Dressler, 1988). Changes in the sociocultural landscape foster the emergence of new, locally meaningful markers of social status. To the extent that these markers are in agreement within individuals, a coherent social identity is projected. However, disagreement – or status inconsistency – may invalidate an individual's claim to a certain social standing, thereby resulting in stress (Figure 34.2d). Models of status inconsistency represent more dynamic measures of individual experience vis-à-vis culture change than additive models that may be particularly relevant to the experience of stress.

In the context of globalization, Western material goods and lifestyles tend to be associated with high status around the world, while developing economies cannot provide enough high-paying jobs to support the demand for increased levels of consumption. This sets the stage for "lifestyle incongruity," or inconsistency between emerging symbolic signs of prestige (e.g., material style of life, travel, etc.) and occupational/educational class. Independent of potentially confounding variables, adults who are incongruent on these two dimensions of social status have been found to have elevated blood pressure in Samoa, St. Lucia, Mexico, and Brazil (Dressler, 1982, 1990; Dressler and Bindon, 1997; Dressler et al., 1987a, 1987b).

Other formulations of status inconsistency have been developed for specific ethnographic settings. For men in American Samoa, having a high status job but relatively low educational status has been associated with elevated blood pressure (McGarvey and Schendel, 1986). Increased blood pressure among young men has also been attributed to increased financial demands from extended family (Bitton et al., 2006), while for Samoan women, "living beyond one's means" has been associated with elevated blood pressure (Chin-Hong and McGarvey, 1996). For Samoan migrants in California, economic status and leadership in the local community have been identified as two important dimensions of social status, with inconsistency across these dimensions associated with increased blood pressure (Janes, 1990). For Samoan adolescents, discordance between a traditional marker of social status (having a *matai* chief in the household) and a new marker of social status (familiarity with Western lifestyles) has been associated with increased Epstein–Barr virus (EBV) antibody titers, indicating suppressed cell-mediated immune function (McDade, 2002). Inconsistency models are significant in that they attempt to capture complex sociocultural processes in such a way that they can be analyzed in relation to stress and health.

CONCLUSIONS

Culture is a key part of the human adaptive strategy, and a driving force in constructing the environments to which humans adapt. By studying biology in relation to dynamic social, cultural, and political–economic contexts we illuminate the pathways through which globalization affects individual health. We also gain insight into the mechanisms of adaptation, their limits, and their role in shaping the health consequences of change.

Is acculturation good or bad for health? The answer to this question depends on who you are, where you are, and how you define health. Transitions in cultural and economic contexts are not uniform or monolithic. While aggregate, population-level analyses identify general patterns of association between standard of living and key demographic and health indicators, research from around the globe demonstrates significant inter- and intra-population variation in processes of change and their links to health.

Cross-sectional, ecological comparisons have dominated research to this point, and have provided a firm foundation on which to build more sophisticated models that attempt to capture this variation. There is tremendous potential to develop new conceptual and analytic models that link context, behavior, and health at the individual level, across time. By thinking critically and creatively about how to define and measure shifting social, cultural, and economic contexts, and by linking these measures to physiology and health, we greatly enrich our understanding of human biology and processes of adaptation.

DISCUSSION POINTS

1. Is globalization good or bad for health? Why is a definitive answer to this question elusive?
2. What are the primary mechanisms of human adaptation, and how do they provide insight into the impact of cultural/economic transitions on human health?
3. What approaches have been applied to model the impact of cultural/economic transitions on human biology and health? Which models best capture which types of issues?
4. What are the gaps in current knowledge? What future research is needed to address these gaps?

REFERENCES

Adair, L. S. and Popkin, B. (1988). *A Longitudinal Analysis of Patterns and Determinants of Women's Nutrition in the Philippines*. Maternal Nutrition and Health Care Program Research Reports Series, no. 1. Washington, DC: International Center for Research on Women.

Akin, J. S., Griffin, C. C., Guilkey, D. K., et al. (1985). *The Demand for Primary Health Services in the Third World*. Totowa, NJ: Rowman and Allenheld.

Alexander, R. D. (1990). How did humans evolve? Reflections on the uniquely unique species. *University of Michigan Musuem of Zoology Special Publications*, **1**, 1–38.

Allison, M. J. (1984). Paleopathology in Peruvian and Chilean populations. In *Paleopathology at the Origins of Agriculture*, M. Cohen (ed.). New York: Academic Press, pp. 512–529.

Amin, R. (1997). NGO-promote women's credit program. *Women and Health*, **25**(1), 71–88.

Armelagos, G. J., Goodman, A. H. and Jacobs, K. (1991). The origins of agriculture: population growth during a period of declining health. *Population and Environment*, **13**(1), 9–22.

Armelagos, G. J., Barnes, K. C. and Lin, J. (1996). Disease in human evolution: the re-emergence of infectious disease in the third epidemiological transition. *AnthroNotes*, **18**(3), 1–7.

Armelagos, G. J., Brown, P. J. and Turner, B. (2006). Evolutionary, historical and political economic perspectives on health and disease. *Social Science and Medicine*, **61**(4), 755–765.

Atashili, J., Kalilani, L., Seksaria, V., et al. (2008). Potential impact of infant feeding recommendations on mortality and HIV-infection in children born to HIV-infected mothers in Africa: a simulation. *BMC Infectious Diseases*, 8, 66.

Atran, S. and Medin, D. (1997). Knowledge and action: cultural models of nature and resource management in Mesoamerica. In *Environment, Ethics, and Behavior*, M. Bazerman, D. Messick, A. Tinbrunsel, et al. (eds). San Francisco: New Lexington Press, pp. 171–120.

Baker, P. T. and Weiner, J. S. (eds) (1966). *The Biology of Human Adaptability*. Oxford: Clarendon Press.

Baker, P. T., Hanna, J. M. and Baker, T. S. (eds) (1986). *The Changing Samoans: Behavior and Health in Transition*. New York: Oxford University Press.

Barrett, R. L., Kuzawa, C. W., McDade, T. W., et al. (1998). Emerging and re-emerging infectious diseases: the third epidemiological transition. *Annual Review of Anthropology*, 27, 247–271.

Becker, A. E. (2004). Television, disordered eating, and young women in Fiji: negotiating body image and identity during rapid social change. *Culture, Medicine, and Psychiatry*, 28(4), 533–559.

Becker, A. E., Burwell, R. A., Gilman, S. E., et al. (2002). Eating behaviours and attitudes following prolonged exposure to television among ethnic Fijian adolescent girls. *British Journal of Psychiatry*, 180, 509–514.

Berry, J. W. (1970). Acculturative stress in northern Canada: ecological, cultural, and psychological factors. In *Circumpolar Health*, R. J. Shepard and S. Itoh (eds). Toronto: University of Toronto Press, pp. 490–497.

Berry, J. W. and Kim, U. (1998). Acculturation and mental health. In *Health and Cross-Cultural Psychological: Towards Applications*, P. Dasen, J. W. Berry and N. Sartorious (eds). Thousand Oaks, CA: Sage Publications, pp. 207–236.

Berry, J. W., Trimble, J. E. and Olmedo, E. L. (1986). Assessment of acculturation. In *Field Methods in Cross-Cultural Research*, W. Lonner and J. W. Berry (eds). Thousand Oaks, CA: Sage Publications, pp. 219–342.

Bitton, A., McGarvey, S. T. and Viali, S. (2006). Anger expression and lifestyle incongruity interactions on blood pressure in Samoan adults. *American Journal of Human Biology*, 18, 369–376.

Boas, F. (1911). *Changes in the Bodily Form of Descendants of Immigrants (Senate Document 208)*. Washington, DC: Government Printing Office.

Boyd, R. and Richerson, P. J. (2005). *The Origin and Evolution of Cultures*. New York: Oxford University Press.

Brinkman, U. K. (1994). Economic development and tropical disease. *Annals of the New York Academy of Sciences*, 740, 303–311.

Brookfield, H. and Padoch, C. (1994). Appreciating agrodiversity: a look at the dynamism and diversity of indigenous farming practices. *Environment*, 36, 6–43.

Brown, D. E. (1981). General stress in anthropological fieldwork. *American Anthropologist*, 83, 74–92.

Brown, D. E. (1982). Physiological stress and culture change in a group of Filipino-Americans: a preliminary investigation. *Annals of Human Biology*, 9(6), 553–563.

Brown, P. J. and Whitaker, E. D. (1994). Health implications of modern agricultural transformations: malaria and pellagra in Italy. *Human Organization*, 53(4), 346–351.

Bury, J. (2004). Livelihoods in transition: transnational gold mining operations and local change in Cajamarca, Peru. *Geographical Journal*, 170(1), 78–91.

Caldwell, J. C. (1976). Toward a restatement of demographic transition theory. *Population and Development Review*, 2(3/4), 321–366.

Caldwell, J. C. and Caldwell, P. (1983). The demographic evidence for the incidence and cause of abnormally low fertility in tropical Africa. *World Health Statistics Quarterly*, 36, 2–34.

Caldwell, J. C. and Caldwell, P. (1987). The cultural context of high fertility in sub-Saharan Africa. *Population and Development Review*, 13(3), 409–437.

Carey, J. W. and Thomas, R. B. (1987). Social influences on morbidity and mortality patterns: a case from Peru. *American Journal of Physical Anthropology*, 72, 186–187.

Cassel, J. C. (1974). Psychiatric epidemiology. In *American Handbook of Psychiatry*, S. Arieti and E. Caplin (eds). New York: Basic Books, pp. 401–410.

Changeux, J. P. (1997). *Neuronal Man: The Biology of Mind*. Princeton: Princeton University Press.

Chin-Hong, P. V. and McGarvey, S. T. (1996). Lifestyle incongruity and adult blood pressure in Western Samoa. *Psychosomatic Medicine*, 58, 130–137.

Cohen, M. N. (1989). *Health and the Rise of Civilization*. New Haven: Yale University Press.

Cohen, S. M., Janicki-Deverts, D. and Miller, G. E. (2007). Psychological stress and disease. *Journal of the American Medical Association*, 298(14), 1685–1687.

Colin Bell, A., Adair, L. S. and Popkin, B. M. (2002). Ethnic differences in the association between body mass index and hypertension. *American Journal of Epidemiology*, 155(4), 346–353.

Confalonieri, U., Ferreira, L. F. and Araujo, A. (1991). Intestinal helminths in lowland South American Indians: some evolutionary interpretations. *Human Biology*, 63, 863–873.

Cosminsky, S. (1985). Infant feeding practices in rural Kenya. In *Breastfeeding, Child Health, and Child Spacing: Cross-cultural Perspectives*, V. Hull and M. Simpson (eds). London: Croom Helm, pp. 35–54.

Coutsoudis, A., Coovadia, H. M. and Wilfert, C. M. (2008). HIV, infant feeding, and more perils for poor people: new WHO guidelines encourage review of formula milk policies. *Bulletin of the World Health Organization*, 86(3), 210–214.

Curtis, T., Kvernmo, S. and Bjerregaard, P. (2005). Changing living conditions, life style and health. *International Journal of Circumpolar Health*, 64(5), 442–450.

Dahly, D. L. and Adair, L. S. (2007). Quantifying the urban environment: a scale measure of urbanicity outperforms the urban-rural dichotomy. *Social Science and Medicine*, 64, 1407–1419.

David, S., Abbas-Chorfa, F., Vanhems, P., et al. (2008). Promotion of WHO feeding recommendations: a model evaluating the effects on HIV-free survival in African children. *Journal of Human Lactation*, 24(2), 140–149.

Denham, M., Schell, L. M., Deane, G., et al. (2005). Relationship of lead, mercury, mirex, dichlorophenyldichloroethylene, hexachlorobenzene, and polychlorinated biphenyls to timing of menarche among Akwesasne Mohawk girls. *Pediatrics*, **115**(2), 127–134.

Dennett, G. and Connell, J. (1988). Acculturation and health in the highlands of Papua New Guinea. *Current Anthropology*, **29**, 273–299.

de Onis, M., Blossner, M., Borghi, E., et al. (2004). Estimates of global prevalence of childhood underweight in 1990 and 2015. *Journal of the American Medical Association*, **291**, 2600–2606.

Desowitz, R. S. (1976). How the wise man brought malaria to Africa. *Natural History*, **85**(8), 36–44.

DeWalt, K. W. (1983). *Nutritional Strategies and Agricultural Change in a Mexican Community*. Ann Arbor, MI: UMI Research Press.

Dos Santos, J. E., Oths, K. S. and Dressler, W. W. (2001). Socioeconomic factors and adult body composition in a developing society. *Revista Brasileira de Hipertensao*, **8**, 173–178.

Dressler, W. W. (1982). *Hypertension and Culture Change: Acculturation and Disease in the West Indies*. South Salem, NY: Redgrave.

Dressler, W. W. (1988). Social consistency and psychological distress. *Journal of Health and Social Behavior*, **29**, 79–91.

Dressler, W. W. (1990). Lifestyle, stress, and blood pressure in a southern black community. *Psychosomatic Medicine*, **52**, 182–198.

Dressler, W. W. (1991). *Stress and Adaptation in the Context of Culture*. Albany, NY: State University of New York Press.

Dressler, W. W. (2005). What's cultural about biocultural research? *Ethos*, **33**, 20–45.

Dressler, W. W. and Bindon, J. R. (1997). Social status, social context, and arterial blood pressure. *American Journal of Physical Anthropology*, **102**, 55–66.

Dressler, W. W., Dos Santo, J. E., Gallagher P. N., et al. (1987a). Arterial blood pressure and modernization in Brazil. *American Anthropologist*, **89**, 389–409.

Dressler, W. W., Mata, A., Chavez, A., et al. (1987b). Arterial blood pressure and individual modernization in a Mexican community. *Social Science and Medicine*, **24**, 679–687.

Dressler, W. W., Ribiero, R. P., Balieiro, M. C., et al. (2004). Eating, drinking and being depressed: the social, cultural and psychological context of alcohol consumption and nutrition in a Brazilian community. *Social Science and Medicine*, **59**(4), 709–720.

Dufour, D. L. (1983). Nutrition in the northwestern Amazon: household dietary intake and time-energy expenditure. In *Adaptive Responses of Native Amazonians*, R. B. Hames and W. T. Vickers (eds). New York: Academic Press, pp. 329–356.

Dufour, D. L., Staten, L. K., Reina, J. C., et al. (1997). Living on the edge: dietary strategies of economically impoverished women in Cali, Colombia. *American Journal of Physical Anthropology*, **102**, 5–15.

Durham, W. (1991). *Coevolution: Genes, Culture, and Human Diversity*. Stanford: Stanford University Press.

Egusa, G., Watanabe, H., Ohshita, K., et al. (2002). Influence of the extent of westernization of lifestyle on the progression of preclinical artherosclerosis in Japanese subjects. *Journal of Artherosclerosis and Thrombosis*, **9**(6), 299–304.

Ellison, P. T., Peacock, N. R. and Lager, C. (1986). Salivary progesterone and luteal function in two low-fertility populations in Eastern Zaire. *Human Biology*, **58**(4), 473–483.

Etkin, N. L. (2000). *Eating on the Wild Side: the Pharmacologic, Ecologic, and Social Implications of Using Noncultigens*. Tucson: University of Arizona Press.

Ezeamama, A. E., Viali, S., Tuitele, J., et al. (2006). The influence of socioeconomic factors on cardiovascular disease risk factors in the context of economic development in the Samoan archipelago. *Social Science and Medicine*, **63**(10), 2533–2545.

Fitton, L. J. (2000). Parasites and culture change in Ecuador. *American Journal of Human Biology*, **12**, 465–477.

Fitton, L. J. (2005). Aging in Amazonia: blood pressure and culture change among the Cofan of Ecuador. *Journal of Cross Cultural Gerontology*, **20**(2), 159–179.

Fitton, L. J. and Crews, D. E. (1995). Blood pressure in the Cofan Indians of Ecuador: another low blood pressure population. *American Journal of Human Biology*, **7**, 122.

Fleming-Moran, M., Sanots, R. V. and Coimbra, C. E. A. (1991). Blood pressure levels of the Surui Indians of the Brazilian Amazon: group- and sex-specific effects resulting from body composition, health status, and age. *Human Biology*, **63**(6), 835–861.

Foster, Z., Byron, E., Reyes-Garcia, V., et al. (2005). Physical growth and nutritional status of Tsimane' Amerindian children of lowland Bolivia. *American Journal of Physical Anthropology*, **126**, 343–351.

Friedlander, J. S. (1987). *The Solomon Islands Project: a Long-Term Study of Health, Human Biology, and Culture Change*. Oxford: Clarendon Press, pp. 352–362.

Frisancho, A. R. (1978). Human growth and development among high-altitude populations. In *The Biology of High-Altitude Peoples*, P. T. Baker (ed.). New York: Cambridge University Press.

Frisancho, A. R. (1993). *Adaptation and Accommodation*. Ann Arbor, MI: University of Michigan Press.

Gage, T. B. (2005). Are modern environments really bad for us? Revisiting the demographic and epidemiologic transitions. *Yearbook of Physical Anthropology*, **128**(S41), 96–117.

Garrett, J. and Ruel, M. T. (2005). The coexistence of child undernutrition and maternal overweight: prevalence, hypotheses, and programme and policy implications. *Maternal and Child Nutrition*, **1**(3), 185–196.

Garruto, R. M., Little, M. A., James, G. D., et al. (1999). Natural experimental models: the global search for biomedical paradigms among traditional, modernizing, and modern populations. *Proceedings of the National Academy of Sciences of the United States of America*, **96**(18), 10536–10543.

Ger, G. and Belk, R. W. (1996). I'd like to buy the world a Coke: consumptionscapes of the less affluent world. *Journal of Consumer Policy*, **19**(3), 271.

Godoy, R. A. (2001). *Indians, Markets, and Rain Forests: Theoretical, Comparative, and Quantitative Explorations in the Neotropics*. New York: Columbia University Press.

Godoy, R. A., Brokaw, N., Wilkie, D., et al. (1998). Of trade and cognition: markets and the loss of folk knowledge among the Tawahka Indians of the Honduran rain forest. *Journal of Anthropological Research*, **54**, 219–233.

Godoy, R. A., Byron, E., Reyes-Garcia, V., et al. (2005a). Income inequality and adult nutritional status: anthropometric evidence from a pre-industrial society in the Bolivian Amazon. *Social Science and Medicine*, **61**(5), 907–919.

Godoy, R. A., Reyes-Garcia, V., Vadez, V., et al. (2005b). Human capital, wealth, and adult nutritional status among a foraging-horticultural society of the Bolivian Amazon. *Economics and Human Biology*, **3**(1), 139–163.

Godoy, R. A., Reyes-Garcia, V., Byron, E., et al. (2005c). The effect of market economies on the well-being of indigenous peoples and their use of renewable natural resources. *Annual Review of Anthropology*, **34**, 121–138.

Godoy, R. A., Reyes-García, V., Leonard, W. R., et al. (2006). Does village inequality in modern income harm the psyche? Anger, fear, sadness, and alcohol consumption in a pre-industrial society. *Social Science and Medicine*, **63**, 359–372.

Goodman, A. H. and Leatherman, T. L. (eds) (1998). *Building a New Biocultural Synthesis: Political–Economic Perspectives on Human Biology*. Ann Arbor, MI: University of Michigan Press.

Gracey, M. (2003). Child health implications of worldwide urbanization. *Review of Environmental Health*, **18**(1), 51–63.

Graves, T. D. (1967). Acculturation, access, and alcohol in a tri-ethnic community. *American Anthropologist*, **69**(3/4), 306–321.

Graves, T. D. and Graves, N. B. (1979). Stress and health: modernization in a traditional Polynesian society. *Medical Anthropology*, **3**, 23–60.

Graves, T. D. and Graves, N. B. (1985). Stress and health among Polynesian migrants to New Zealand. *Journal of Behavioral Medicine*, **8**(1), 1–19.

Hanna, J. M. and Fitzgerald, M. H. (1993). Acculturation and symptoms: a comparative study of reported health symptoms in three Samoan communities. *Social Science and Medicine*, **36**(9), 1169–1180.

Henry, J. P. and Cassel, J. C. (1969). Psychosocial factors in essential hypertension recent epidemiologic and animal experimental evidence. *American Journal of Epidemiology*, **90**, 171–200.

Hill, K. and Hurtado, A. M. (1996). *Aché Life History*. New York: Aldine de Gruyter.

Holmes, R. (1985). Nutritional status and cultural change in Venezuela's Amazon territory. In *Change in the Amazon Basin*, vol. 2, J. Hemming (ed.). Manchester: Manchester University Press, pp. 237–255.

Howell, N. (1979). *Demography of the Dobe !Kung*. New York: Aldine de Gruyter.

Hurtado, A. M., Hill, K. R., Rosenblatt, W., et al. (2003). Longitudinal study of tuberculosis outcomes among immunologically naïve Aché natives of Paraguay. *American Journal of Physical Anthropology*, **121**(2), 134–150.

Huss-Ashmore, R. (2000). Theory in human biology: evolution, ecology, adaptability, and variation. In *Human Biology: an Evolutionary and Biocultural Perspective*,

S. Stinson, B. Bogin, R. Huss-Ashmore, et al. (eds). New York: Wiley-Liss, pp. 1–26.

Hyatt, D. E. and Milne, W. J. (1993). Determinants of fertility in urban and rural Kenya: estimates and a simulation of the impact of education policy. *Environment and Planning*, **25**(3), 371–382.

Ice, G. H. and James, G. D. (eds) (2007). *Measuring Stress in Humans: a Practical Guide for the Field*. Cambridge: Cambridge University Press.

James, G. D., Baker, P. T., Jenner, D. A., et al. (1987). Variation in lifestyle characteristics and catecholamine excretion rates among young Western Samoan men. *Social Science and Medicine*, **25**(9), 981–986.

Janes, C. R. (1990). *Migration, Social Change, and Health: a Samoan Community in Urban California*. Stanford: Stanford University Press.

Jenkins, C. (1989). Culture change and epidemiological patterns among the Hagahai, Papua New Guinea. *Human Ecology*, **17**(1), 27–57.

Jenner, D. A., Harrison, G. A., Prior, I. A. M., et al. (1987). Inter-population comparisons of catecholamine excretion. *Annals of Human Biology*, **14**(1), 1–9.

Jiang, J., Liu, B., Nasca, P. C., et al. (2008). Age-related effects of smoking on lung-cancer mortality: a nation-wide case control comparison in 103 population centers in China. *Annals of Epidemiology*, **18**(6), 484–491.

Johnson, E. O., Karmilaris, T. C., Chrousos, G. P., et al. (1992). Mechanisms of stress: a dynamic overview. *Neuroscience and Biobehavioral Reviews*, **16**, 115–130.

Johnston, F. E. and Little, M. A. (2000). History of human biology in the United States of America. In *Human Biology: an Evolutionary and Biocultural Perspective*, S. Stinson, B. Bogin, R. Huss-Ashmore, et al. (eds). New York: Wiley-Liss, pp. 27–46.

Kaplan, H., Lancaster, J. B., Tucker, W. T., et al. (2002). Evolutionary approach to below replacement fertility. *American Journal of Human Biology*, **14**, 233–256.

Kawachi, I. (1999). Social capital and community effects on population and individual health. *Annals of the New York Academy of Sciences*, **896**, 120–130.

Kawachi, I. (2000). Income inequality and health. In *Social Epidemiology*, L. F. Berkman and I. Kawachi (eds). New York: Oxford University Press.

Keighley, E. D., McGarvey, S. T., Quested, C., et al. (2007). Nutrition and health in modern Samoans: temporal trends and adaptive perspectives. In *Health Change in the Asian-Pacific Region: Biocultural and Epidemiological Approaches*, R. Ohtsuka and S. J. Ulijaszek (eds). Cambridge: Cambridge University Press, pp. 147–191.

Konner, M. and Worthman, C. (1980). Nursing frequency, gonadal function, and birth spacing among !Kung hunter-gatherers. *Science*, **207**, 788–791.

Kraus, R. and Buffler, P. (1976). Suicide in Alaskan Natives: a preliminary report. In *Circumpolar Health*, R. J. Shepard and S. Itoh (eds). Toronto: University of Toronto Press, pp. 556–557.

Kuzawa, C. W. (2005). Fetal origins of developmental plasticity: are fetal cues reliable predictors of future nutritional environments? *American Journal of Human Biology*, **17**(1), 5–21.

Kuzawa, C. W., Adair, L. S., Avila, J. L., et al. (2003). Athero-genic lipid profiles in Filipino adolescents with low body mass body mass index and low index and low dietary fat intake. *American Journal of Biology*, **15**(5), 688–696.

La Ferrara, E. (2003). The return of the native. *Current Anthropology*, **44**, 389–402.

Lasker, G. W. (1969). Human biological adaptability: the ecological approach in physical anthropology. *Science*, **166**, 1480–1486.

Lawrence, D. N., Neel, J. V., Stanley, H. A., et al. (1980). Epidemiological studies among Amerindian populations of Amazonia: parasitoses in newly contacted and acculturating villages. *American Journal of Tropical Medicine*, **29**, 530–537.

Leatherman, T. L. (1994). Health implications of changing agrarian economies in the southern Andes. *Human Organization*, **53**(4), 371–380.

Leatherman, T. L. (1998). Changing biocultural perspectives on health in the Andes. *Social Science and Medicine*, **47**(9), 1397.

Leatherman, T. L., Luerssen, J. S., Markowitz, L., et al. (1986). Illness and political economy: an Andean dialectic. *Cultural Survival Quarterly*, **10**(3), 19–21.

Leatherman, T. L., Carey, J. W. and Thomas, R. B. (1995). Socioeconomic change and patterns of growth in the Andes. *American Journal of Physical Anthropology*, **97**, 307–321.

Leatherman, T. L. and Goodman, A. H. (1997). Expanding the biocultural synthesis toward a biology of poverty. *American Journal of Physical Anthropology*, **102**, 1–3.

Leonard, W. R. (1989). Nutritional determinants of high-altitude growth in Nunoa, Peru. *American Journal of Physical Anthropology*, **80**(3), 341–352.

Leonard, W. R. (1995). Growth differences between children of highland and coastal Ecuador. *American Journal of Physical Anthropology*, **98**(1), 47–57.

Leonard, W. L., Spencer, G. J., Galloway, V. A., et al. (2002). Declining growth status of indigenous Siberian children in post-Soviet Russia. *Human Biology*, **74**(2), 197–209.

Leslie, P. W., Campbell, K. L. and Little, M. A. (1993). Pregnancy loss in nomadic and settled women in Turkana, Kenya: a prospective study. *Human Biology*, **65**(2), 237–254.

Leslie, P. W., Campbell, K. L. and Little, M. A. (1994). Reproductive function in nomadic and settled women of Turkana, Kenya. *Annals of the New York Academy of Sciences*, **709**, 218–220.

Levine, R. A. (1994). Gusii fertility, marriage, and family. In *Child Care and Culture: Lessons from Africa*, R. A. Levine, S. Dixon, S. LeVine et al. (eds). Cambridge: Cambridge University Press, pp. 92–120.

Levy, M. J. (1966). *Modernization and the Structure of Societies: a Setting for the Study of International Affairs*. Princeton: Princeton University Press.

Little, M. A. (1982). Development of ideas on human ecology and adaptation. In *History of American Physical Anthropology, 1930–1980*, F. J. Spencer (ed.). New York: Academic Press, pp. 405–433.

Little, M. A. and Johnson, F. E. (1987). Mixed longitudinal growth of nomadic Turkana pastoralists. *Human Biology*, **59**(4), 695–707.

Little, M. A. and Leslie, P. W. (1999). *Turkana Herders of the Dry Savanna. Ecology and Biobehavioral Response of Nomads to an Uncertain Environment*. New York: Oxford University Press.

Little, M. A., Gray, S. J. and Leslie, P. W. (1993). Growth of nomadic and settled Turkana infants of northwest Kenya. *American Journal of Physical Anthropology*, **92**(3), 273–289.

Little, M. A., Leslie, P. W. and Baker, P. T. (1997). Multidisciplinary research of human biology and behavior. In *History of Physical Anthropology: an Encyclopedia*, vol. 2, F. J. Spencer (ed.). New York: Garland Publishing, p. 196.

Livingstone, F. B. (1958). Anthropological implications of sickle cell gene distribution in West Africa. *American Anthropologist*, **60**, 533–562.

Lynge, I. (1976). Alcohol problems in western Greenland. In *Circumpolar Health*, R. Shepard and S. Itoh (eds). Toronto: University of Toronto Press, pp. 543–547.

Macpherson, C. and Macpherson, L. (1987). Towards an explanation of recent trends in suicide in Western Samoa. *Man (New Series)*, **22**, 305–330.

Manderson, L. (1982). Bottle feeding and ideology in colonial Malaya: the production of change. *International Journal of Health Services*, **12**, 597–616.

Martorell, R. (1995). Results and implications of the INCAP follow-up study. *Journal of Nutrition*, **125**(4 suppl.), 1127S–1138S.

McCabe, T. J. (2003). Sustainability and livelihood diversification among the Maasai of Northern Tanzania. *Human Organization*, **62**(3), 100–111.

McDade, T. W. (2001). Lifestyle incongruity, social integration, and immune function in Samoan adolescents. *Social Science and Medicine*, **53**, 1351–1362.

McDade, T. W. (2002). Status incongruity in Samoan youth: a biocultural analysis of culture change, stress, and immune function. *Medical Anthropology Quarterly*, **16**, 123–150.

McDade, T. W. and Adair, L. S. (2001). Defining the "urban" in urbanization and health: a factor analysis approach. *Social Science and Medicine*, **53**, 55–70.

McDade, T. W. and Worthman, C. W. (1998). The weanling's dilemma reconsidered: a biocultural analysis of breast-feeding ecology. *Journal of Developmental and Behavioral Pediatrics*, **19**(4), 286–299.

McDade, T. W., Stallings, J. F. and Worthman, C. W. (2000). Culture change and stress in Western Samoan youth: methodological issues in the cross-cultural study of stress and immune function. *American Journal of Human Biology*, **12**, 792–802.

McDade, T. W., Leonard, W. R., Burhop, J., et al. (2005). Predictors of C-reactive protein in Tsimane' 2–15 year-olds in lowland Bolivia. *American Journal of Physical Anthropology*, **128**, 906–913.

McDade, T. W., Reyes-Garcia, V., Blackinton, P., et al. (2007). Ethnobotanical knowledge is associated with indices of child health in the Bolivian Amazon. *Proceedings of the National Academy of Sciences of the United States of America*, **104**, 6134–6139.

McElroy, A. (1990). Biocultural models in studies of human health and adaptation. *Medical Anthropology Quarterly*, **4**(3), 243–265.

McEwen, B. (1998). Protective and damaging effects of stress mediators. *Seminars in Medicine of the Beth Israel Deaconess Medical Center*, **338**(3), 171–179.

McEwen, B. and Lasley, E. (2002). *The End of Stress as We Know It*. Washington, DC: The Joseph Henry Press.

McGarvey, S.T. (2001). Cardiovascular disease (CVD) risk factors in Samoa and American Samoa, 1990–95. *Pacific Health Dialog*, **8**(1), 157–162.

McGarvey, S.T. and Baker, P.T. (1979). The effects of modernization and migration on Samoan blood pressure. *Human Biology*, **51**(4), 461–479.

McGarvey, S.T. and Schendel, D.E. (1986). Blood pressure of Samoans. In *The Changing Samoans: Behavior and Health in Transition*, P.T. Baker, J.M. Hanna and T.S. Baker (eds). New York: Oxford University Press, pp. 351–393.

McGarvey, S.T., Levinson, P.D., Bausserman, L., et al. (1993). Population change in adult obesity and blood lipids in American Samoa from 1976–1978 to 1990. *American Journal of Human Biology*, **5**(11), 17–30.

McKee, M. (2005). Understanding population health: lessons from the former Soviet Union. *Clinical Medicine*, **5**(4), 374–378.

McKeowen, T. (1976). *The Modern Rise of Population*. New York: Academic Press.

McNeill, W.H. (1977). *Plagues and People*. London: Anchor Books.

Millard, A. and Graham, M. (1985). Breastfeeding in two Mexican villages: social and demographic perspectives. In *Breastfeeding, Child Health, and Child Spacing: Cross-Cultural Perspectives*, V. Hull and M. Simpson (eds). London: Croom Helm, pp. 55–77.

Morduch, J. (1995). Income smoothing and consumption smoothing. *Journal of Economic Perspectives*, **9**(3), 103–114.

Navarro, V. (1999). Health and inequity in the world in the era of "globalization." *International Journal of Health Services*, **29**(2), 215–226.

Neff, J. (1993). Life stressors, drinking patterns, and depressive symptomology. *Addictive Behaviors*, **18**, 373–387.

Nerlove, S.B. (1974). Women's workload and infant feeding practices: a relationship with demographic implications. *Ethnology*, **13**, 207–214.

Newman, J., Aucompaugh, A.G., Schell, L.M., et al. (2006). PCBs and cognitive functioning of Mohawk adolescents. *Neurotoxicology and Teratology*, **28**(4), 439–445.

Norgan, N.G. (1994). Interpretation of low body-mass indices: Australian Aborigines. *American Journal of Physical Anthropology*, **94**, 229–237.

Norgan, N.G. (1995). Changes in patterns of growth and nutritional anthropometry in two rural modernizing Papua New Guinea communities. *Annals of Human Biology*, **22**(6), 491–513.

O'Neil, J.D. (1986). Colonial stress in the Canadian Arctic: an ethnography of young adults changing. In *Anthropology and Epidemiology: Interdisciplinary Approaches to the Study of Health and Diseases*, C.R. Janes, R. Stall and S.M. Gifford (eds). Boston: D. Reidel, pp. 249–274.

Oeppen, J. and Vaupel, J.W. (2002). Demography. Broken limits to life expectancy. *Science*, **296**(5570), 1029–1031.

Page, L., Damon, A. and Moellering, R.C. (1974). Antecedents of cardiovascular disease in six Solomon Islands societies. *Circulation*, **49**(6), 1132–1146.

Pandey, D.N. (2001). A bountiful harvest of rainwater. *Science*, **293**, 1763.

Panter-Brick, C., Todd, A. and Baker, R. (1996). Growth status of homeless Nepali boys: do they differ from rural and urban controls? *Social Science and Medicine*, **43**(4), 441–451.

Patrick, R.C., Prior, I.A.M., Smith, J.C., et al. (1983). Relationship between blood pressure and modernity among Ponapeans. *International Journal of Epidemiology*, **12**(1), 36–44.

Payne, G., Carabin, H., Tallo, V., et al. (2006). Concurrent comparison of three water contact measurement tools in four endemic villages of the Philippines. The Schistosomiasis Transmission Ecology in the Philippines Project (STEP). *Tropical Medicine and International Health*, **11**(6), 834–842.

Pearson, J.D., James, G.D. and Brown, D.E. (1993). Stress and changing lifestyles in the Pacific: physiological stress responses of Samoans in rural and urban settings. *American Journal of Human Biology*, **5**, 49–60.

Pelletier, D.L., Frongillo, E.A., Schroeder, D.G., et al. (1995). The effects of malnutrition on child mortality in developing countries. *Bulletin of the World Health Organization*, **73**(4), 443–448.

Pelto, G.H. and Pelto, P.J. (1989). Small but healthy? An anthropological perspective. *Human Organization*, **48**, 11–15.

Pollard, T.M., Ward, G.A., Thornley, J., et al. (2000). Modernisation and children's blood pressure: on and off the tourist trail in Nepal. *American Journal of Human Biology*, **4**, 478–486.

Popkin, B.M. (1994). The nutrition transition in low income countries: an emerging crisis. *Nutritional Review*, **52**, 285–298.

Popkin, B.M. (2003). Dynamics of the nutrition and its implications for the developing world. *Forum Nutrition*, **56**, 262–264.

Post, J.E. and Baer, E. (1978). Demarketing infant formula: consumer products in the developing world. *Journal of Contemporary Business*, **7**, 17–35.

Poulter, N.R., Khaw, K.T., Hopwood, B.E., et al. (1990). The Kenyan Luo migration study: observations on the initiation of a rise in blood pressure. *British Journal of Medicine*, **300**(6730), 967–972.

Prior, I.A.M. and Rose, B.S. (1966). Hyperuricemia, gout, and diabetic abnormality in Polynesian people. *Lancet*, **1**, 333.

Prior, I.A., Welby, T.J., Ostbye, T., et al. (1987). Migration and gout: the Tokelau Island migrant study. *British Medical Journal*, **295**(596), 457–461.

Putsche, L. (2000). A reassessment of resource depletion, market dependency, and culture change on a Shipibo Reserve in the Peruvian Amazon. *Human Ecology*, **28**(1), 131–140.

Ravussin, E., Valencia, M.E., Esparza, J., et al. (1994). Effects of a traditional lifestyle on obesity in Pima Indians. *Diabetes Care*, **17**(9), 1067–1074.

Reed, D., Labarthe, D. and Stallone, R. (1970). Health effects of westernization and migration among Chamorros. *American Journal of Epidemiology*, **92**, 94–112.

Renehan, A. G., Tyson, M., Egger, M., et al. (2008). Body-mass index and incidence of cancer: a systematic review and meta-analysis of prospective observational studies. *Lancet*, **371**, 569–578.

Reyes, M. E., Tan, S. K. and Malina, R. M. (2003). Urban-rural contrasts in the growth status of school children in Oaxaca, Mexico. *Annals of Human Biology*, **30**(6), 693–713.

Reyes-Garcia, V., Vadez, V., Byron, E., et al. (2005). Market economy and the loss of folk knowledge of plant uses: estimates from the Tsimane' of the Bolivian Amazon. *Current Anthropology*, **46**, 651–656.

Rose, E. (1999). Consumption smoothing and excess female mortality in rural India. *Review of Economics and Statistics*, **81**(1), 41–49.

Rubinstein, D. H. (1992). Suicide in Micronesia and Samoa: a critique of explanations. *Pacific Studies*, **15**(1), 51–75.

Salmond, C. E., Prior, I. A. and Wessen, A. F. (1989). Blood pressure patterns and migration: a 14-year cohort study of adult Tokelauans. *American Journal of Epidemiology*, **130** (1), 37–51.

Santos, R. V. and Coimbra, C. E. A. (1991). Socioeconomic transition and physical growth of Tupi-Monde American children in the Aipuna Park, Brazilian Amazon. *Human Biology*, **63**(6), 795–819.

Santos, R. V., Flowers, N. M., Coimbra, C. E. A. et al. (1997). Tapirs, tractors, and tapes: the changing economy and ecology of the Xavante Indians of Central Brazil. *Human Ecology*, **25**(4), 545–566.

Sapolsky, R. (2004). Social status and health in humans and other primates. *Annual Review of Anthropology*, **33**, 393–418.

Schall, J. I. (1995). Sex differences in the response of blood pressure to modernization. *American Journal of Human Biology*, **7**, 159–172.

Schell, L. M. (1997). Culture as a stressor: a revised model of biocultural interaction. *American Journal of Physical Anthropology*, **102**, 67–77.

Schell, L. M. and Denham, M. (2003). Environmental pollution in urban environments and human biology. *Annual Review of Anthropology*, **32**, 111–134.

Schell, L. M. and Knutson, K. L. (2002). Environmental effects on growth. In *Human Growth and Development*, N. Cameron (ed.). London: Academic Press, pp. 165–196.

Schell, L. M., Gallo, M. V., Denham, M., et al. (2006). Effects of pollution on human growth and development: an introduction. *Journal of Physiological Anthropology*, **25**(1), 103–112.

Schwaner, T. D. and Dixon, C. F. (1974). Helminthiasis as a measure of cultural change in the Amazon basin. *Biotropica*, **6**, 32–37.

Scotch, N. A. (1963). Sociocultural factors in the epidemiology of Zulu hypertension. *American Journal of Public Health*, **53**(8), 1205–1213.

Scotch, N. A. and Geiger, H. J. (1969). The epidemiology of essential hypertension: a review with special attention to psychologic and cultural factors in etiology. *Journal of Chronic Diseases*, **16**, 1183–1213.

Shell-Duncan, B. and Obiero, W. O. (2000). Child nutrition in the transition from nomadic pastoralism to settled lifestyles: individual, household, and community-level factors. *American Journal of Physical Anthropology*, **113** (2), 183–200.

Shephard, R. J. and Rode, A. (1996). *The Health Consequences of "Modernization": Evidence from Circumpolar Peoples*. Cambridge: Cambridge University Press.

Simpson, M. (1985). Breastfeeding, infant growth, and return to fertility in an Iranian city. In *Breastfeeding, Child Health, and Child Spacing: Cross-Cultural Perspectives*, V. Hull and M. Simpson (eds). London: Croom Helm, pp. 109–138.

Singer, M. (1996). Farewell to adaptationism: unnatural selection and the politics of biology. *Medical Anthropology Quarterly*, **10**, 496–515.

Singer, M., Valentin, F., Baer, H., et al. (1992). Why does Juan Garcia have a drinking problem? *Medical Anthropology*, **14**, 17–108.

Slobodkin, L. B. (1968). Toward a predictive theory of evolution. In *Population Biology and Evolution*, R. C. Lewontin (ed.). Syracuse, NY: Syracuse University Press, pp. 185–205.

Snodgrass, J. J., Leonard, W. R., Sorensen, M. V., et al. (2006). The emergence of obesity among indigenous Siberians. *Journal of Physiological Anthropology*, **25**(1), 75–84.

Snodgrass, J. J., Sorensen, M. V., Tarskaia, L. A., et al. (2007). Adaptive dimensions of health research among indigenous Siberians. *American Journal of Human Biology*, **19**, 165–180.

Social Science Research Council Summer Seminar (1954). Acculturation: an exploratory formulation. *American Anthropologist*, **56**, 973–1002.

Sorensen, M. V., Snodgrass, J. J., Leonard, W. R., et al. (2005). Health consequences of postsocial transition: dietary and lifestyle determinants of plasma lipids in Yakutia. *American Journal of Human Biology*, **17**(5), 576–592.

Spindler, L. S. (1984). *Culture Change and Modernization*. Prospect Heights, IL: Waveland Press.

Stanhope, J. M. and Prior, I. A. (1980). The Tokelau island migrant study: prevalence and incidence of diabetes mellitus. *New Zealand Medical Journal*, **92**(673), 417–421.

Steffan, P. R., Smith, T. B., Larson, M., et al. (2006). Acculturation to Western society as a risk factor for high blood pressure: a meta-analytic review. *Psychosomatic Medicine*, **68**, 386–397.

Stinson, S. (1983). Socioeconomic status and child growth in rural Bolivia. *Ecology of Food and Nutrition*, **13**, 179–187.

Stinson, S. (1996). Early childhood growth of Chachi Amerindians and Afro-Ecuadorians in northwest Ecuador. *American Journal of Physical Anthropology*, **8**, 43–53.

Subramanian, S. V., Kim, D. and Kawachi, I. (2005). Covariation in the socioeconomic determinants of self rated health and happiness: a multilevel analysis of individuals and communities in the USA. *Journal of Epidemiology and Community Health*, **59**(9), 664–669.

Sutter, F. K. (1980). Communal versus individual socialization at home and in school in rural and urban Western Samoa, Thesis. University of Hawai'i, Honolulu.

Tanner, S. (2005). A population in transition: health, culture change, and intestinal parasitism among the Tsimane' of lowland Bolivia. PhD dissertation. University of Michigan, Ann Arbor.

Thomas, R. B. (1992). The evolution of human adaptability paradigms: towards a biology of poverty. In *Building a New Biocultural Synthesis: Political–Economic Perspectives on Human Biology*, A. H. Goodman and T. L. Leatherman (eds). Ann Arbor, MI: University of Michigan Press.

Tomasello, M. (1999). *The Cultural Origins of Human Cognition*. Cambridge, MA: Harvard University Press.

Tylor, E. B. (1924). (original 1871) *Primitive Culture*, 2 vols, 7th edn. New York: Bretano's.

Ulijaszek, S. J. (2007). Frameworks of population obesity and the use of cultural consensus modeling in the study of environments contributing to obesity. *Economics and Human Biology*, **5**, 443–457.

Ulijaszek, S. J., Koziel, S. and Hermanussen, M. (2005). Village distance from urban centre as the prime modernization variable in differences in blood pressure and body mass index of adults of the Purari delta of the Gulf Province, Papua New Guinea. *Annals of Human Biology*, **32**(3), 326–338.

United Nations (2001). *World Urbanization Prospects: the 1999 Revision*. New York: United Nations.

Vadez, V., Reyes-Garcia, V., Godoy, R. A., et al. (2004). Does integration to the market threaten agricultural diversity? Panel and cross-sectional data from a horticultural-foraging society in the Bolivian Amazon. *Human Ecology*, **32**(5), 635–646.

Villamor, E., Msamanga, G., Urassa, W., et al. (2006). Trends in obesity, underweight, and wasting among women attending prenatal clinics in urban Tanzania, 1995–2004. *American Journal of Clinical Nutrition*, **83**(6), 1387–1394.

Vitzthum, V. (1994). Comparative study of breastfeeding structure and its relation to human reproductive ecology. *Yearbook of Physical Anthropology*, **37**, 307–349.

Waldron, I., Nowotarski, M., Friemer, M., et al. (1982). Cross-cultural variation in blood pressure: a quantitative analysis of the relationships of blood pressure to culture characteristics, salt consumption, and body weight. *Social Science and Medicine*, **16**(4), 419–430.

Walsh, J. F., Molyneux, D. H. and Birley, M. H. (1993). Deforestation: effects on vector-borne disease. *Parasitology*, **106**, S55–S75.

Washburn, S. L. (1951). The new physical anthropology. *New York Academy of Sciences, Series II*, **13**, 298–304.

Wessen, A. F., Hooper, A., Huntsman, J., et al. (eds) (1992). *Migration and Health in a Small Society: the Case of Tokelau*. New York: Oxford University Press.

Wirsing, R. (1985). The health of traditional societies and the effects of acculturation. *Current Anthropology*, **26**(3), 303–322.

World Cancer Research Fund and American Institute for Cancer Research (2007). *Food, Nutrition, Physical Activity, and the Prevention of Cancer: A Global Perspective*. Washington, DC: American Institute for Cancer Research.

Yajnik, C. S. (2004). Early life origins of insulin resistance and type 2 diabetes in India and other Asian countries. *Journal of Nutrition*, **134**(1), 205–210.

Yamauchi, T., Umezaki, M. and Ohsuka, R. (2001). Physical activity and subsistence pattern of the Huli, a Papua New Guinea population. *American Journal of Physical Anthropology*, **114**, 258–268.

Zhang, H. and Cai, B. (2003). The impact of tobacco on lung health in China. *Respirology*, **8**(1), 17–21.

Zimmet, P. Z., Taylor, R., Jackson, L., et al. (1980). Blood pressure studies in rural and urban Samoa. *Medical Journal of Australia*, **2**(4), 202–205.

Index

Locators in **bold** refer to major content
Locators in *italic* refer to figures/tables
Locators for headings which also have subheadings refer to general aspects of that topic only.

Printed in the United States
By Bookmasters